Student Solutions Manual

Precalculus

TENTH EDITION

Ron Larson
The Pennsylvania State University, The Behrend College

Prepared by

Ron Larson
The Pennsylvania State University, The Behrend College

CENGAGE Learning

Australia • Brazil • Mexico • Singapore • United Kingdom • United States

CENGAGE
Learning

WCN: 01-100-101

ISBN: 978-1-337-28078-5

Cengage Learning
20 Channel Center Street
Boston, MA 02210
USA

Cengage Learning is a leading provider of customized learning solutions with office locations around the globe, including Singapore, the United Kingdom, Australia, Mexico, Brazil, and Japan. Locate your local office at: **www.cengage.com/global**.

Cengage Learning products are represented in Canada by Nelson Education, Ltd.

To learn more about Cengage Learning Solutions, visit **www.cengage.com**.

Purchase any of our products at your local college store or at our preferred online store **www.cengagebrain.com**.

Printed at CLDPC, USA, 05-22

CONTENTS

CONTENTS

CHAPTER 1
Functions and Their Graphs

CHAPTER 1
Functions and Their Graphs

Section 1.1 Rectangular Coordinates

1. Cartesian 

3. Distance Formula

5.

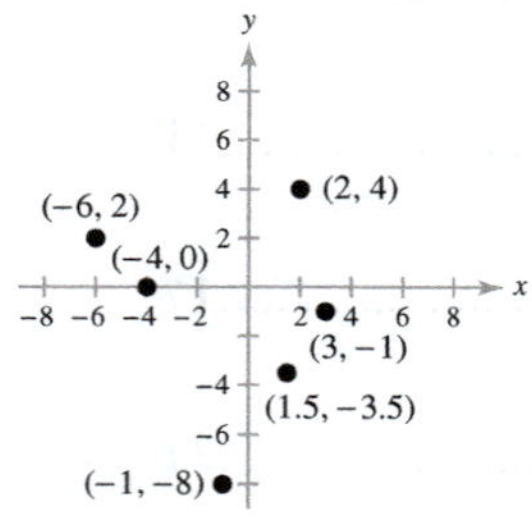

7. $(-3, 4)$

9. $x > 0$ and $y < 0$ in Quadrant IV.

11. $x = -4$ and $y > 0$ in Quadrant II.

13. $x + y = 0, x \neq 0, y \neq 0$ means $x = -y$ or $y = -x$.
This occurs in Quadrant II or IV.

15.

Year, x	Number of Stores, y
2008	7720
2009	8416
2010	8970
2011	10,130
2012	10,773
2013	10,942
2014	11,453

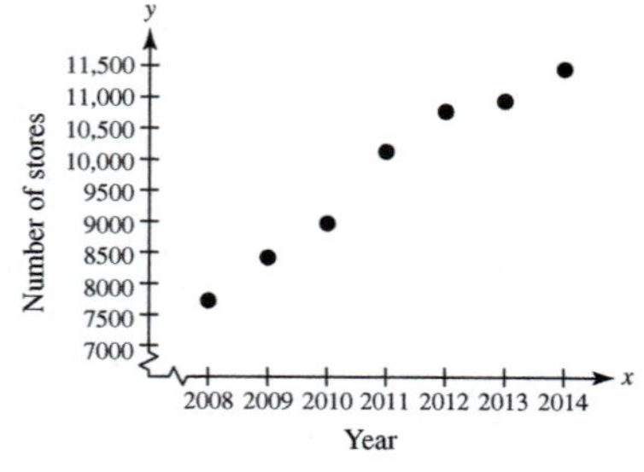

17. $$\begin{aligned} d &= \sqrt{(x_2 - x_1)^2 + (y_2 - y_1)^2} \\ &= \sqrt{(3 - (-2))^2 + (-6 - 6)^2} \\ &= \sqrt{(5)^2 + (-12)^2} \\ &= \sqrt{25 + 144} \\ &= 13 \text{ units} \end{aligned}$$

19. $$\begin{aligned} d &= \sqrt{(x_2 - x_1)^2 + (y_2 - y_1)^2} \\ &= \sqrt{(-5 - 1)^2 + (-1 - 4)^2} \\ &= \sqrt{(-6)^2 + (-5)^2} \\ &= \sqrt{36 + 25} \\ &= \sqrt{61} \text{ units} \end{aligned}$$

21. $$\begin{aligned} d &= \sqrt{(x_2 - x_1)^2 + (y_2 - y_1)^2} \\ &= \sqrt{\left(2 - \frac{1}{2}\right)^2 + \left(-1 - \frac{4}{3}\right)^2} \\ &= \sqrt{\left(\frac{3}{2}\right)^2 + \left(-\frac{7}{3}\right)^2} \\ &= \sqrt{\frac{9}{4} + \frac{49}{9}} \\ &= \sqrt{\frac{277}{36}} \\ &= \frac{\sqrt{277}}{6} \text{ units} \end{aligned}$$

23. (a) $(1, 0), (13, 5)$

$$\begin{aligned} \text{Distance} &= \sqrt{(13 - 1)^2 + (5 - 0)^2} \\ &= \sqrt{12^2 + 5^2} = \sqrt{169} = 13 \end{aligned}$$

$(13, 5), (13, 0)$

$$\text{Distance} = |5 - 0| = |5| = 5$$

$(1, 0), (13, 0)$

$$\text{Distance} = |1 - 13| = |-12| = 12$$

(b) $5^2 + 12^2 = 25 + 144 = 169 = 13^2$

25. $d_1 = \sqrt{(4-2)^2 + (0-1)^2} = \sqrt{4+1} = \sqrt{5}$

$d_2 = \sqrt{(4+1)^2 + (0+5)^2} = \sqrt{25+25} = \sqrt{50}$

$d_3 = \sqrt{(2+1)^2 + (1+5)^2} = \sqrt{9+36} = \sqrt{45}$

$(\sqrt{5})^2 + (\sqrt{45})^2 = (\sqrt{50})^2$

27. $d_1 = \sqrt{(1-3)^2 + (-3-2)^2} = \sqrt{4+25} = \sqrt{29}$

$d_2 = \sqrt{(3+2)^2 + (2-4)^2} = \sqrt{25+4} = \sqrt{29}$

$d_3 = \sqrt{(1+2)^2 + (-3-4)^2} = \sqrt{9+49} = \sqrt{58}$

$d_1 = d_2$

29. (a)

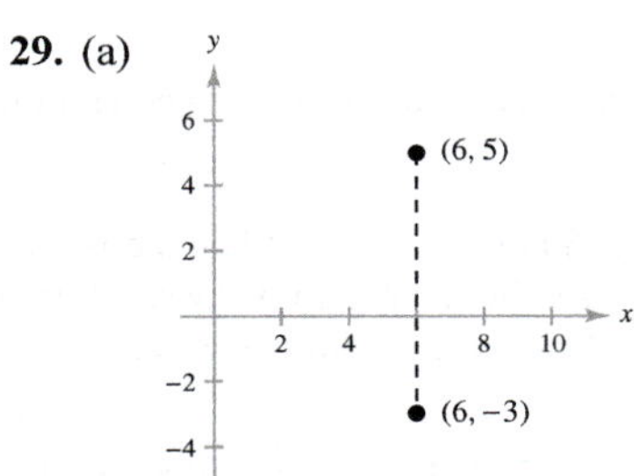

(b) $d = \sqrt{(5-(-3))^2 + (6-6)^2} = \sqrt{64} = 8$

(c) $\left(\frac{6+6}{2}, \frac{5+(-3)}{2}\right) = (6, 1)$

31. (a)

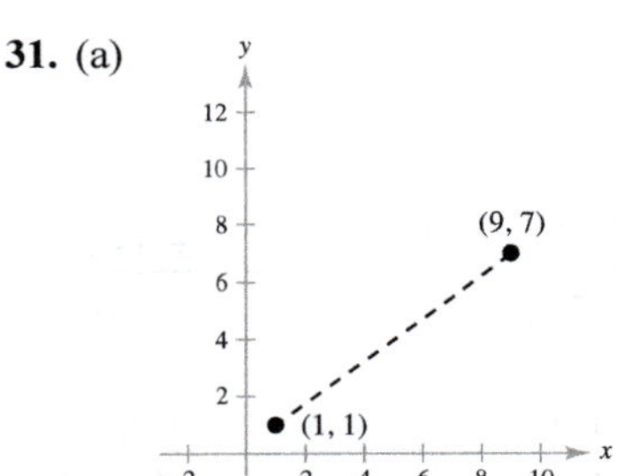

(b) $d = \sqrt{(9-1)^2 + (7-1)^2} = \sqrt{64+36} = 10$

(c) $\left(\frac{9+1}{2}, \frac{7+1}{2}\right) = (5, 4)$

33. (a)

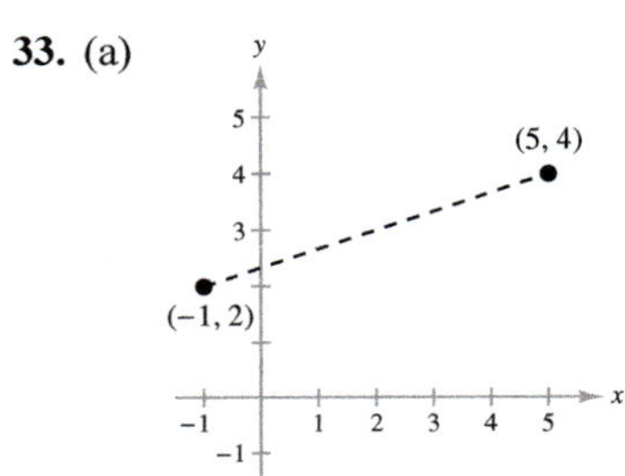

(b) $d = \sqrt{(5+1)^2 + (4-2)^2}$

$= \sqrt{36+4} = 2\sqrt{10}$

(c) $\left(\frac{-1+5}{2}, \frac{2+4}{2}\right) = (2, 3)$

35. (a)

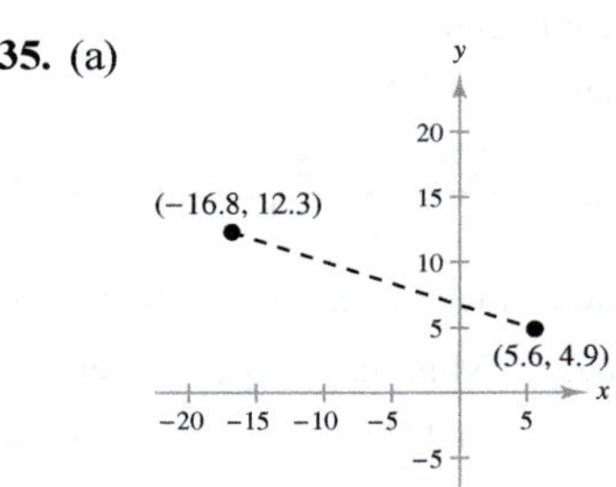

(b) $d = \sqrt{(-16.8-5.6)^2 + (12.3-4.9)^2}$

$= \sqrt{501.76 + 54.76} = \sqrt{556.52}$

(c) $\left(\frac{-16.8+5.6}{2}, \frac{12.3+4.9}{2}\right) = (-5.6, 8.6)$

37. $d = \sqrt{120^2 + 150^2}$

$= \sqrt{36{,}900}$

$= 30\sqrt{41}$

≈ 192.09

The plane flies about 192 kilometers.

39. midpoint $= \left(\frac{x_1+x_2}{2}, \frac{y_1+y_2}{2}\right)$

$= \left(\frac{2010+2014}{2}, \frac{35{,}123+45{,}998}{2}\right)$

$= (2012, 40{,}560.5)$

In 2012, the sales for the Coca-Cola Company were about $40,560.5 million.

41. $(-2+2,\ -4+5) = (0, 1)$

$(2+2,\ -3+5) = (4, 2)$

$(-1+2,\ -1+5) = (1, 4)$

43. $(-7+4, -2+8) = (-3, 6)$

$(-2+4, 2+8) = (2, 10)$

$(-2+4, -4+8) = (2, 4)$

$(-7+4, -4+8) = (-3, 4)$

45. (a) The minimum wage had the greatest increase from 2000 to 2010.

(b) Minimum wage in 1985: \$3.35

Minimum wage in 2000: \$5.15

Percent increase: $\left(\dfrac{5.15 - 3.35}{3.35}\right) \times 100 \approx 53.7\%$

Minimum wage in 2000: \$4.25

Minimum wage in 2015: \$7.25

Percent increase: $\left(\dfrac{7.25 - 5.15}{5.15}\right) \times 100 \approx 40.8\%$

So, the minimum wage increased 53.7% from 1985 to 2000 and 40.8% from 2000 to 2015.

(c) $\begin{matrix}\text{Minimum wage}\\ \text{in 2030}\end{matrix} = \begin{matrix}\text{Minimum wage}\\ \text{in 2015}\end{matrix} + \left(\begin{matrix}\text{Percent}\\ \text{increase}\end{matrix}\right)\left(\begin{matrix}\text{Minimum wage}\\ \text{in 2015}\end{matrix}\right) \approx \$7.25 + 0.408(\$7.25) \approx \$10.21$

So, the minimum wage will be about \$10.21 in the year 2030.

(d) Answers will vary. *Sample answer:* Yes, the prediction is reasonable because the percent increase is over an equal time period of 15 years.

47. True. Because $x < 0$ and $y > 0$, $2x < 0$ and $-3y < 0$, which is located in Quadrant III.

49. True. Two sides of the triangle have lengths $\sqrt{149}$ and the third side has a length of $\sqrt{18}$.

51. Answers will vary. *Sample answer:* When the x-values are much larger or smaller than the y-values, different scales for the coordinate axes should be used.

53. Because $x_m = \dfrac{x_1 + x_2}{2}$ and $y_m = \dfrac{y_1 + y_2}{2}$ we have:

$$2x_m = x_1 + x_2 \qquad 2y_m = y_1 + y_2$$

$$2x_m - x_1 = x_2 \qquad 2y_m - y_1 = y_2$$

So, $(x_2, y_2) = (2x_m - x_1, 2y_m - y_1)$.

55. The midpoint of the given line segment is $\left(\dfrac{x_1 + x_2}{2}, \dfrac{y_1 + y_2}{2}\right)$.

The midpoint between (x_1, y_1) and $\left(\dfrac{x_1 + x_2}{2}, \dfrac{y_1 + y_2}{2}\right)$ is $\left(\dfrac{x_1 + \dfrac{x_1 + x_2}{2}}{2}, \dfrac{y_1 + \dfrac{y_1 + y_2}{2}}{2}\right) = \left(\dfrac{3x_1 + x_2}{4}, \dfrac{3y_1 + y_2}{4}\right)$.

The midpoint between $\left(\dfrac{x_1 + x_2}{2}, \dfrac{y_1 + y_2}{2}\right)$ and (x_2, y_2) is $\left(\dfrac{\dfrac{x_1 + x_2}{2} + x_2}{2}, \dfrac{\dfrac{y_1 + y_2}{2} + y_2}{2}\right) = \left(\dfrac{x_1 + 3x_2}{4}, \dfrac{y_1 + 3y_2}{4}\right)$.

So, the three points are $\left(\dfrac{3x_1 + x_2}{4}, \dfrac{3y_1 + y_2}{4}\right)$, $\left(\dfrac{x_1 + x_2}{2}, \dfrac{y_1 + y_2}{2}\right)$, and $\left(\dfrac{x_1 + 3x_2}{4}, \dfrac{y_1 + 3y_2}{4}\right)$.

57. Use the Midpoint Formula to prove the diagonals of the parallelogram bisect each other.

$$\left(\frac{b + a}{2}, \frac{c + 0}{2}\right) = \left(\frac{a + b}{2}, \frac{c}{2}\right)$$

$$\left(\frac{a + b + 0}{2}, \frac{c + 0}{2}\right) = \left(\frac{a + b}{2}, \frac{c}{2}\right)$$

59. (a) **First Set**

$$d(A, B) = \sqrt{(2-2)^2 + (3-6)^2} = \sqrt{9} = 3$$

$$d(B, C) = \sqrt{(2-6)^2 + (6-3)^2} = \sqrt{16+9} = 5$$

$$d(A, C) = \sqrt{(2-6)^2 + (3-3)^2} = \sqrt{16} = 4$$

Because $3^2 + 4^2 = 5^2$, A, B, and C are the vertices of a right triangle.

Second Set

$$d(A, B) = \sqrt{(8-5)^2 + (3-2)^2} = \sqrt{10}$$

$$d(B, C) = \sqrt{(5-2)^2 + (2-1)^2} = \sqrt{10}$$

$$d(A, C) = \sqrt{(8-2)^2 + (3-1)^2} = \sqrt{40}$$

A, B, and C are the vertices of an isosceles triangle or are collinear: $\sqrt{10} + \sqrt{10} = 2\sqrt{10} = \sqrt{40}$.

(b)

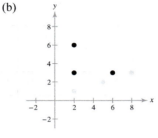

First set: Not collinear

Second set: Collinear.

(c) A set of three points is collinear when the sum of two distances among the points is exactly equal to the third distance.

Section 1.2 Graphs of Equations

1. solution or solution point

3. intercepts

5. circle; (h, k); r

7. (a) $(0, 2)$: $2 \stackrel{?}{=} \sqrt{0+4}$

$$2 = 2$$

Yes, the point *is* on the graph.

(b) $(5, 3)$: $3 \stackrel{?}{=} \sqrt{5+4}$

$$3 \stackrel{?}{=} \sqrt{9}$$

$$3 = 3$$

Yes, the point *is* on the graph.

9. (a) $(2, 0)$: $(2)^2 - 3(2) + 2 \stackrel{?}{=} 0$

$$4 - 6 + 2 \stackrel{?}{=} 0$$

$$0 = 0$$

Yes, the point *is* on the graph.

(b) $(-2, 8)$: $(-2)^2 - 3(-2) + 2 \stackrel{?}{=} 8$

$$4 + 6 + 2 \stackrel{?}{=} 8$$

$$12 \neq 8$$

No, the point *is not* on the graph.

11. (a) $(1, 5)$: $5 \stackrel{?}{=} 4 - |1 - 2|$

$$5 \stackrel{?}{=} 4 - 1$$

$$5 \neq 3$$

No, the point *is not* on the graph.

(b) $(6, 0)$: $0 \stackrel{?}{=} 4 - |6 - 2|$

$$0 \stackrel{?}{=} 4 - 4$$

$$0 = 0$$

Yes, the point *is* on the graph.

13. (a) $(3, -2)$: $(3)^2 + (-2)^2 \stackrel{?}{=} 20$

$$9 + 4 \stackrel{?}{=} 20$$

$$13 \neq 20$$

No, the point *is not* on the graph.

(b) $(-4, 2)$: $(-4)^2 + (2)^2 \stackrel{?}{=} 20$

$$16 + 4 \stackrel{?}{=} 20$$

$$20 = 20$$

Yes, the point *is* on the graph.

15. $y = -2x + 5$

x	-1	0	1	2	$\frac{5}{2}$
y	7	5	3	1	0
(x, y)	$(-1, 7)$	$(0, 5)$	$(1, 3)$	$(2, 1)$	$\left(\frac{5}{2}, 0\right)$

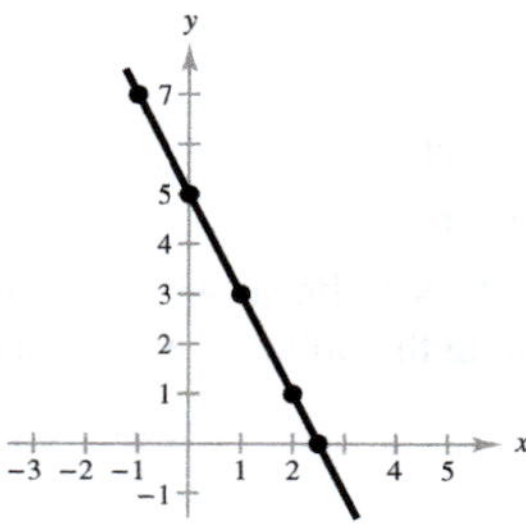

17. $y = x^2 - 3x$

x	-1	0	1	2	3
y	4	0	-2	-2	0
(x, y)	$(-1, 4)$	$(0, 0)$	$(1, -2)$	$(2, -2)$	$(3, 0)$

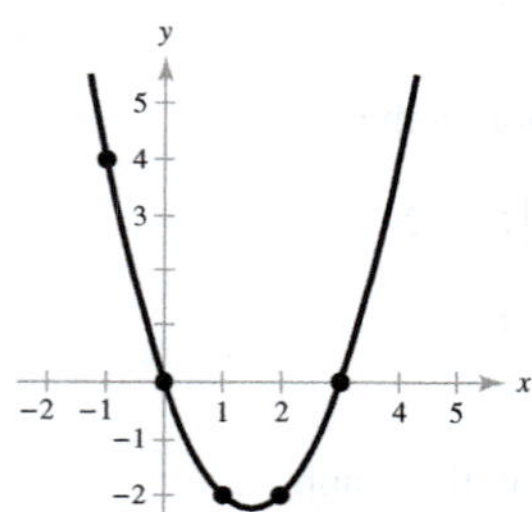

19. x-intercept: $(3, 0)$

y-intercept: $(0, 9)$

21. x-intercept: $(-2, 0)$

y-intercept: $(0, 2)$

23. $y = 5x - 6$

Let $y = 0$.

$0 = 5x - 6$

$6 = 5x$

$\frac{6}{5} = x$.

x-intercept: $\left(\frac{6}{5}, 0\right)$

Let $x = 0$.

$y = 5(0) - 6$

$y = 0 - 6$

$y = -6$

y-intercept: $(0, -6)$

25. $y = \sqrt{x + 4}$

Let $y = 0$.

$0 = \sqrt{x + 4}$

$0 = x + 4$

$-4 = x$

x-intercept: $(-4, 0)$

Let $x = 0$.

$y = \sqrt{0 + 4}$

$y = \sqrt{4}$

$y = 2$

y-intercept: $(0, 2)$

27. $y = |3x - 7|$

Let $y = 0$.

$0 = |3x - 7|$

$0 = 3x - 7$

$7 = 3x$

$\frac{7}{3} = x$

x-intercept: $\left(\frac{7}{3}, 0\right)$

Let $x = 0$.

$y = |3(0) - 7|$

$y = |-7|$

$y = 7$

y-intercept: $(0, 7)$

29. $y = 2x^3 - 4x^2$

Let $y = 0$.

$0 = 2x^3 - 4x^2$

$0 = 2x^2(x - 2)$

$x = 0$ or $x = 2$

x-intercepts: $(0, 0), (2, 0)$

Let $x = 0$.

$y = 2(0)^3 - 4(0)^2$

$y = 0 - 0$

$y = 0$

y-intercept: $(0, 0)$

31. $y^2 = 6 - x$

Let $y = 0$.

$0^2 = 6 - x$

$x = 6$

x-intercept: $(6, 0)$

Let $x = 0$.

$y^2 = 6 - (0)$

$y^2 = 6$

$y = \pm\sqrt{6}$

y-intercepts: $(0, \pm\sqrt{6})$

33. $x^2 - y = 0$

$(-x)^2 - y = 0 \Rightarrow x^2 - y = 0 \Rightarrow$ y-axis symmetry

$x^2 - (-y) = 0 \Rightarrow x^2 + y = 0 \Rightarrow$ No x-axis symmetry

$(-x)^2 - (-y) = 0 \Rightarrow x^2 + y = 0 \Rightarrow$ No origin symmetry

35. $y = x^3$

$y = (-x)^3 \Rightarrow y = -x^3 \Rightarrow$ No y-axis symmetry

$-y = x^3 \Rightarrow y = -x^3 \Rightarrow$ No x-axis symmetry

$-y = (-x)^3 \Rightarrow -y = -x^3 \Rightarrow y = x^3 \Rightarrow$ Origin symmetry

37. $y = \dfrac{x}{x^2 + 1}$

$y = \dfrac{-x}{(-x)^2 + 1} \Rightarrow y = \dfrac{-x}{x^2 + 1} \Rightarrow$ No y-axis symmetry

$-y = \dfrac{x}{x^2 + 1} \Rightarrow y = \dfrac{-x}{x^2 + 1} \Rightarrow$ No x-axis symmetry

$-y = \dfrac{-x}{(-x)^2 + 1} \Rightarrow -y = \dfrac{-x}{x^2 + 1} \Rightarrow y = \dfrac{x}{x^2 + 1} \Rightarrow$ Origin symmetry

39. $xy^2 + 10 = 0$

$(-x)y^2 + 10 = 0 \Rightarrow -xy^2 + 10 = 0 \Rightarrow$ No y-axis symmetry

$x(-y)^2 + 10 = 0 \Rightarrow xy^2 + 10 = 0 \Rightarrow$ x-axis symmetry

$(-x)(-y)^2 + 10 = 0 \Rightarrow -xy^2 + 10 = 0 \Rightarrow$ No origin symmetry

41.

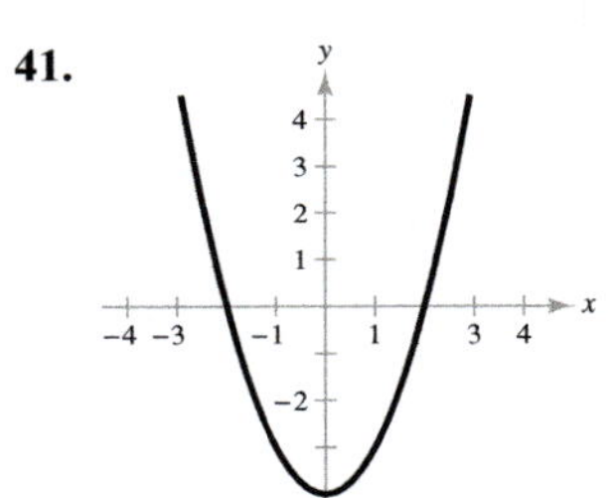

43.

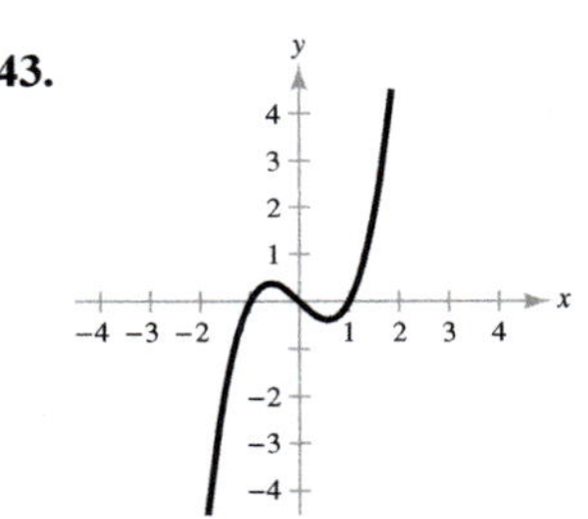

45. $y = -3x + 1$

x-intercept: $\left(\frac{1}{3}, 0\right)$

y-intercept: $(0, 1)$

No symmetry

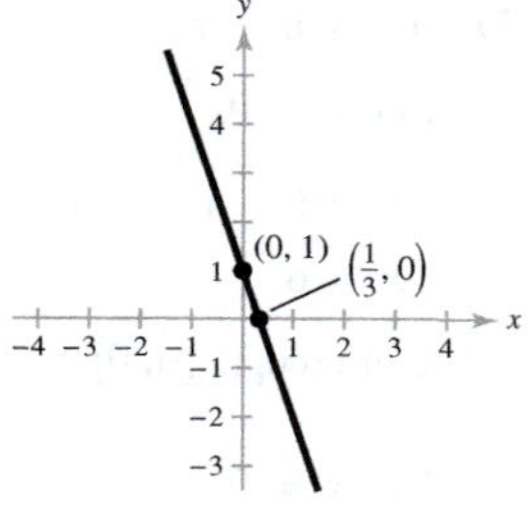

47. $y = x^2 - 2x$

x-intercepts: $(0, 0), (2, 0)$

y-intercept: $(0, 0)$

No symmetry

x	-1	0	1	2	3
y	3	0	-1	0	3

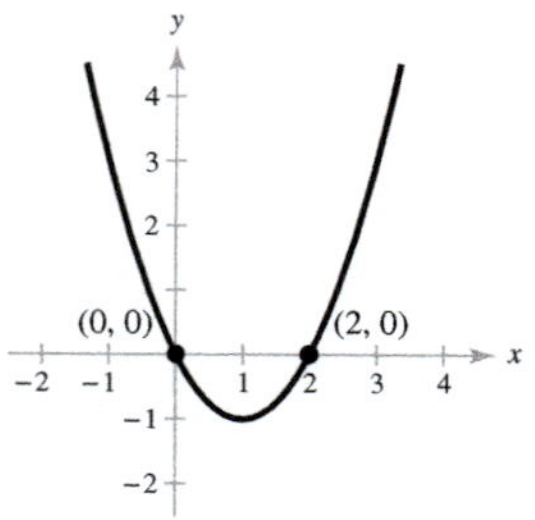

49. $y = x^3 + 3$

x-intercept: $\left(\sqrt[3]{-3}, 0\right)$

y-intercept: $(0, 3)$

No symmetry

x	-2	-1	0	1	2
y	-5	2	3	4	11

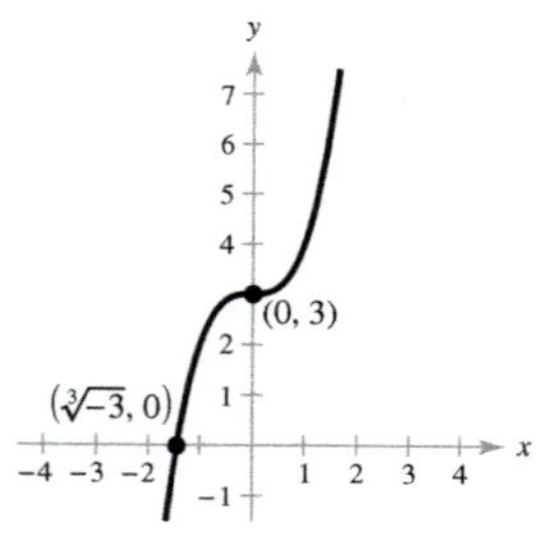

51. $y = \sqrt{x - 3}$

x-intercept: $(3, 0)$

y-intercept: none

No symmetry

x	3	4	7	12
y	0	1	2	3

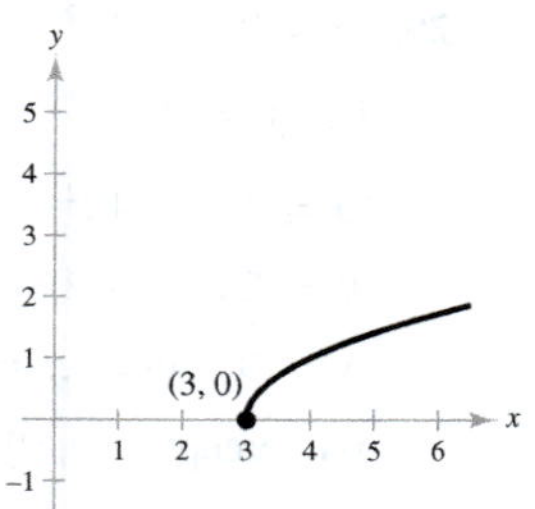

53. $y = |x - 6|$

x-intercept: $(6, 0)$

y-intercept: $(0, 6)$

No symmetry

x	-2	0	2	4	6	8	10
y	8	6	4	2	0	2	4

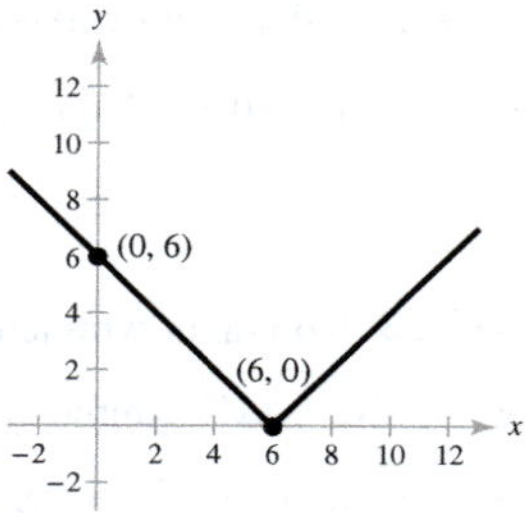

55. $x = y^2 - 1$

x-intercept: $(-1, 0)$

y-intercepts: $(0, -1), (0, 1)$

x-axis symmetry

x	-1	0	3
y	0	±1	±2

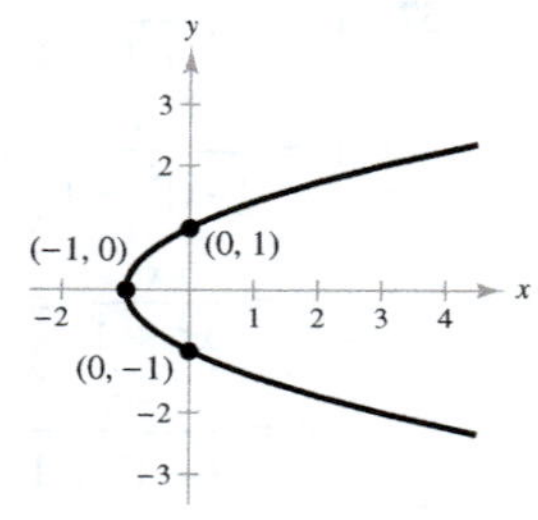

57. $y = 3 - \frac{1}{2}x$

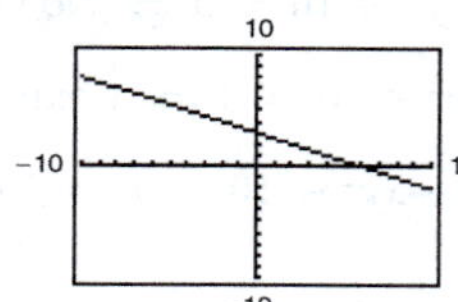

Intercepts: $(6, 0), (0, 3)$

59. $y = x^2 - 4x + 3$

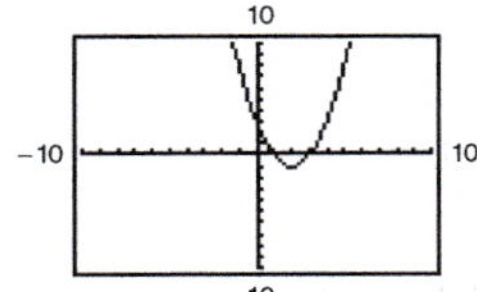

Intercepts: $(3, 0), (1, 0), (0, 3)$

61. $y = \dfrac{2x}{x - 1}$

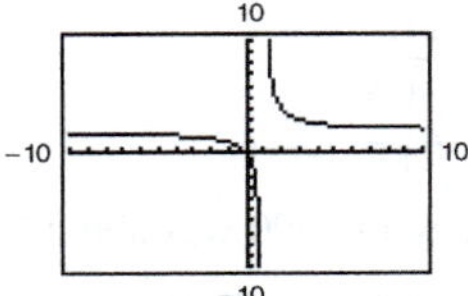

Intercept: $(0, 0)$

63. $y = \sqrt[3]{x + 1}$

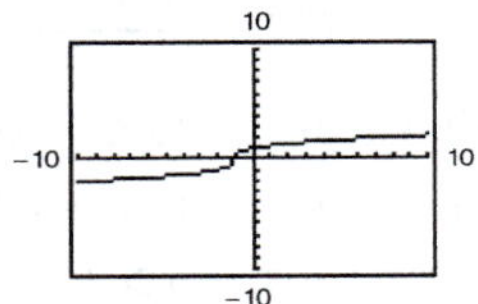

Intercepts: $(-1, 0), (0, 1)$

65. $y = |x + 3|$

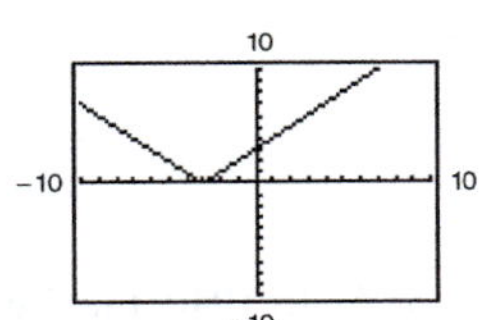

Intercepts: $(-3, 0), (0, 3)$

67. Center: $(0, 0)$; Radius: 3

$$(x - 0)^2 + (y - 0)^2 = 3^2$$
$$x^2 + y^2 = 9$$

69. Center: $(-4, 5)$; Radius: 2

$$(x - h)^2 + (y - k)^2 = r^2$$
$$[x - (-4)]^2 + [y - 5]^2 = 2^2$$
$$(x + 4)^2 + (y - 5)^2 = 4$$

71. Center: $(3, 8)$; Solution point: $(-9, 13)$

$$\begin{aligned} r &= \sqrt{(x - h)^2 + (y - k)^2} \\ &= \sqrt{(-9 - 3)^2 + (13 - 8)^2} \\ &= \sqrt{(-12)^2 + (5)^2} \\ &= \sqrt{144 + 25} \\ &= \sqrt{169} \\ &= 13 \end{aligned}$$

$$(x - h)^2 + (y - k)^2 = r^2$$
$$(x - 3)^2 + (y - 8)^2 = 13^2$$
$$(x - 3)^2 + (y - 8)^2 = 169$$

73. Endpoints of a diameter: $(3, 2), (-9, -8)$

$$\begin{aligned} r &= \frac{1}{2}\sqrt{(-9 - 3)^2 + (-8 - 2)^2} \\ &= \frac{1}{2}\sqrt{(-12)^2 + (-10)^2} \\ &= \frac{1}{2}\sqrt{144 + 100} \\ &= \frac{1}{2}\sqrt{244} = \frac{1}{2}(2)\sqrt{61} = \sqrt{61} \end{aligned}$$

$$(h, k): \left(\frac{3 + (-9)}{2}, \frac{2 + (-8)}{2}\right) = \left(\frac{-6}{2}, \frac{-6}{2}\right) = (-3, -3)$$

$$(x - h)^2 + (y - k)^2 = r^2$$
$$[x - (-3)]^2 + [y - (-3)]^2 = (\sqrt{61})^2$$
$$(x + 3)^2 + (y + 3)^2 = 61$$

75. $x^2 + y^2 = 25$

Center: $(0, 0)$, Radius: 5

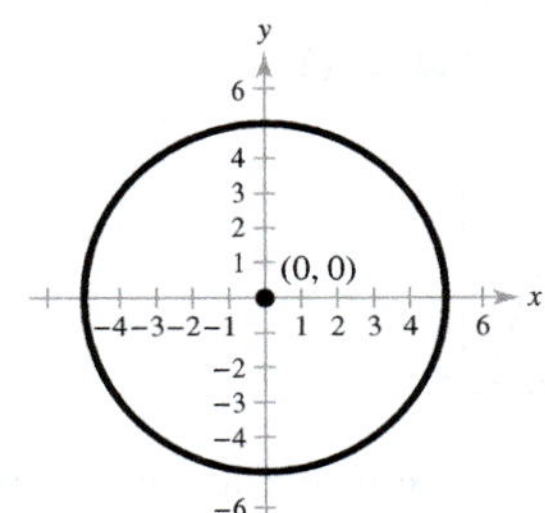

77. $(x - 1)^2 + (y + 3)^2 = 9$

Center: $(1, -3)$, Radius: 3

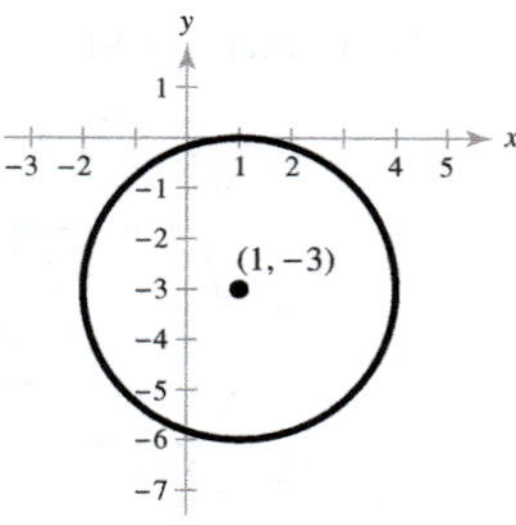

79. $\left(x - \frac{1}{2}\right)^2 + \left(y - \frac{1}{2}\right)^2 = \frac{9}{4}$

Center: $\left(\frac{1}{2}, \frac{1}{2}\right)$, Radius: $\frac{3}{2}$

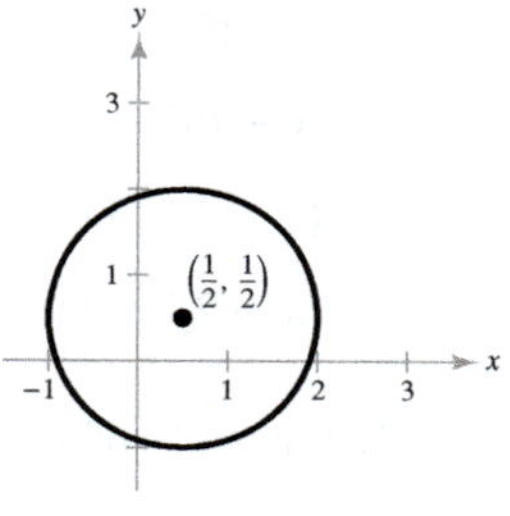

81. $y = 1{,}200{,}000 - 80{,}000t,\ 0 \le t \le 10$

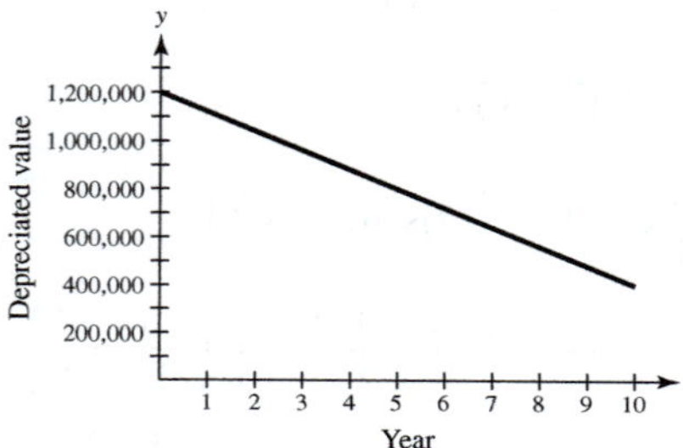

83. (a)

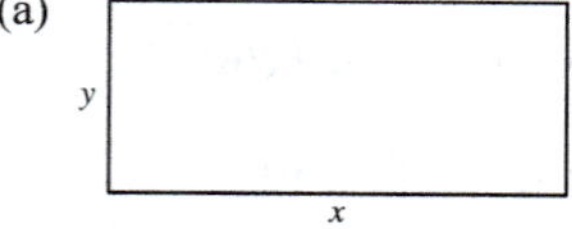

(b) $2x + 2y = \frac{1040}{3}$

$2y = \frac{1040}{3} - 2x$

$y = \frac{520}{3} - x$

$A = xy = x\left(\frac{520}{3} - x\right)$

(c)

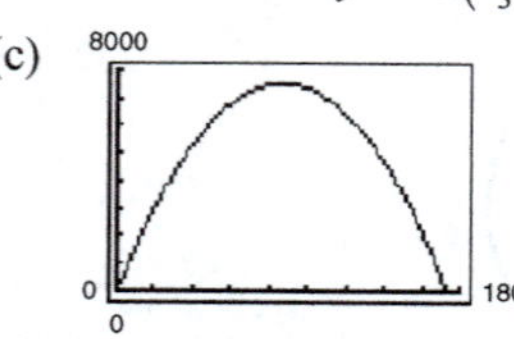

(d) When $x = y = 86\frac{2}{3}$ yards, the area is a maximum of $7511\frac{1}{9}$ square yards.

(e) A regulation NFL playing field is 120 yards long and $53\frac{1}{3}$ yards wide. The actual area is 6400 square yards.

85. (a)

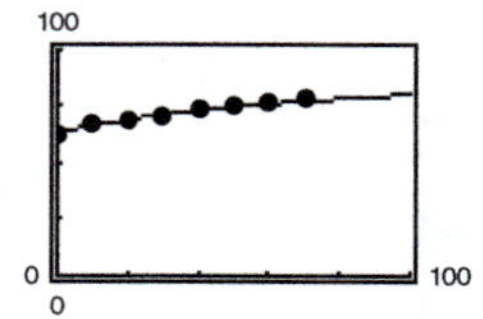

The model fits the data well.

(b) Graphically: The point $(50, 74.7)$ represents a life expectancy of 74.7 years in 1990.

Algebraically: $y = \dfrac{63.6 + 0.97(50)}{1 + 0.01(50)}$

$= \dfrac{112.1}{1.5}$

$= 74.7$

So, the life expectancy in 1990 was about 74.7 years.

(c) Graphically: The point $(24.2, 70.1)$ represents a life expectancy of 70.1 years during the year 1964.

Algebraically: $y = \dfrac{63.6 + 0.97t}{1 + 0.01t}$

$70.1 = \dfrac{63.6 + 0.97t}{1 + 0.01t}$

$70.1(1 + 0.01t) = 63.6 + 0.97t$

$70.1 + 0.701t = 63.6 + 0.97t$

$6.5 = 0.269t$

$t = 24.2$

When $y = 70.1$, $t = 24.2$ which represents the year 1964.

(d) $y = \dfrac{63.6 + 0.97(0)}{1 + 0.01(0)}$

$= \dfrac{63.6}{1} = 63.6$

The y-intercept is $(0, 63.6)$. In 1940, the life expectancy of a child (at birth) was 63.6 years.

(e) Answers will vary.

87. False. The line $y = x$ is symmetric with respect to the origin.

89. True. Depending upon the center and radius, the graph of a circle could intersect one, both, or neither axis.

91. $y = ax^2 + bx^3$

(a) $y = a(-x)^2 + b(-x)^3$

$= ax^2 - bx^3$

To be symmetric with respect to the y-axis; a can be any non-zero real number, b must be zero.

Sample answer: $a = 1, b = 0$

(b) $-y = a(-x)^2 + b(-x)^3$

$-y = ax^2 - bx^3$

$y = -ax^2 + bx^3$

To be symmetric with respect to the origin; a must be zero, b can be any non-zero real number.

Sample answer: $a = 0, b = 1$

Section 1.3 Linear Equations in Two Variables

1. linear

3. point-slope

5. perpendicular

7. linear extrapolation

9. (a) $m = \frac{2}{3}$. Because the slope is positive, the line rises. Matches L_2.

(b) m is undefined. The line is vertical. Matches L_3.

(c) $m = -2$. The line falls. Matches L_1.

11.

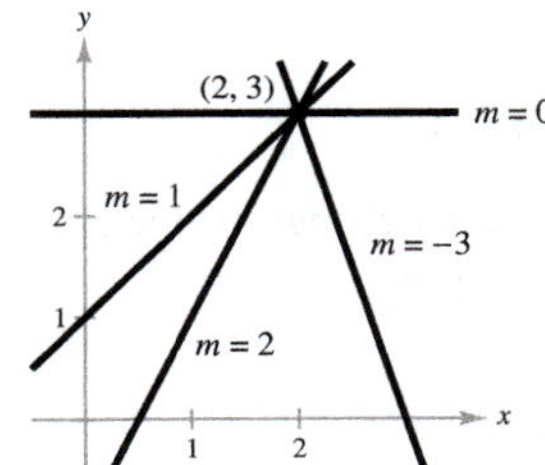

13. Two points on the line: $(0, 0)$ and $(4, 6)$

$$\text{Slope} = \frac{y_2 - y_1}{x_2 - x_1} = \frac{6}{4} = \frac{3}{2}$$

15. $y = 5x + 3$

Slope: $m = 5$

y-intercept: $(0, 3)$

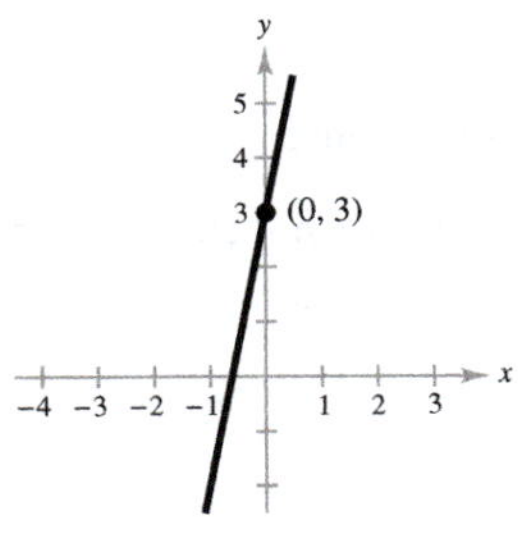

17. $y = -\frac{3}{4}x - 1$

Slope: $m = -\frac{3}{4}$

y-intercept: $(0, -1)$

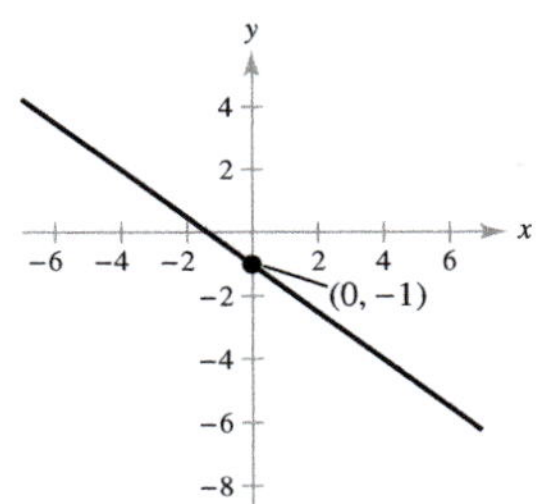

19. $y - 5 = 0$

$y = 5$

Slope: $m = 0$

y-intercept: $(0, 5)$

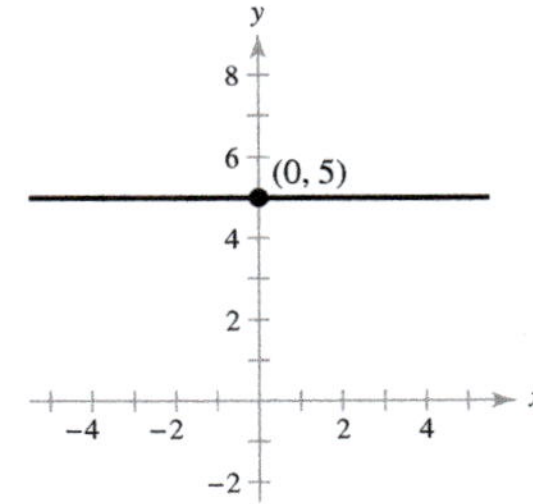

21. $5x - 2 = 0$

$x = \frac{2}{5}$, vertical line

Slope: undefined

y-intercept: none

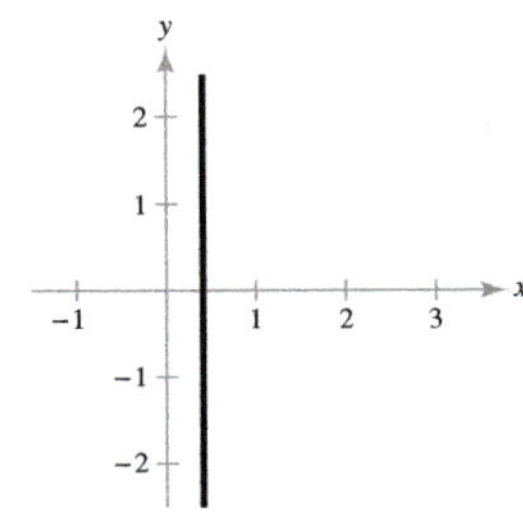

23. $7x - 6y = 30$

$$-6y = -7x + 30$$

$$y = \frac{7}{6}x - 5$$

Slope: $m = \frac{7}{6}$

y-intercept: $(0, -5)$

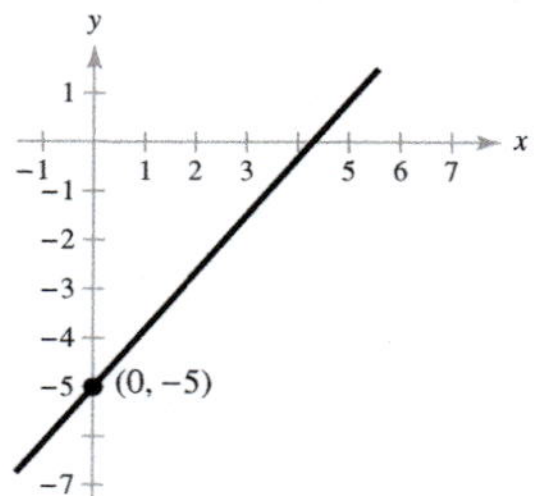

25. $m = \dfrac{0 - 9}{6 - 0} = \dfrac{-9}{6} = -\dfrac{3}{2}$

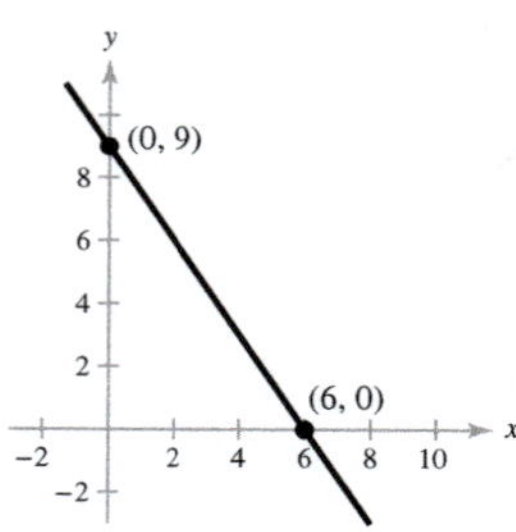

27. $m = \dfrac{6 - (-2)}{1 - (-3)} = \dfrac{8}{4} = 2$

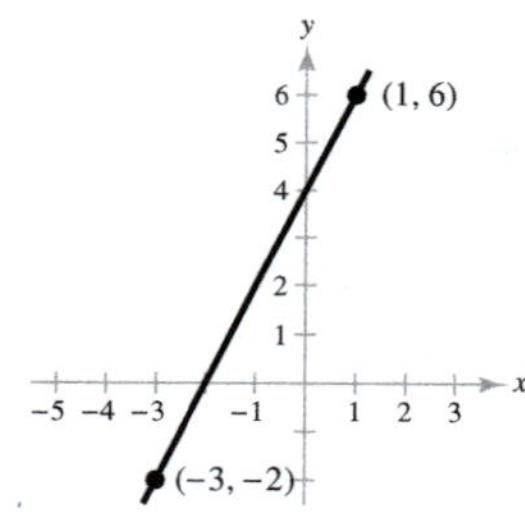

29. $m = \dfrac{-7 - (-7)}{8 - 5} = \dfrac{0}{3} = 0$

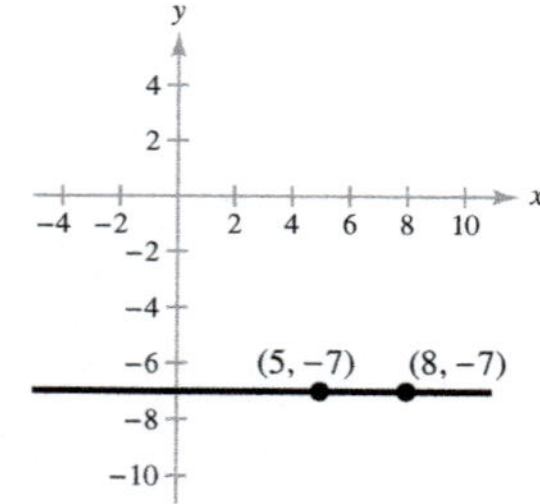

31. $m = \dfrac{4 - (-1)}{-6 - (-6)} = \dfrac{5}{0}$

m is undefined.

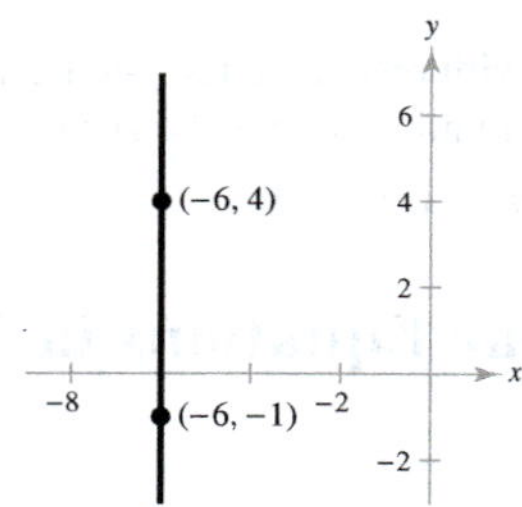

33. $m = \dfrac{1.6 - 3.1}{-5.2 - 4.8} = \dfrac{-1.5}{-10} = 0.15$

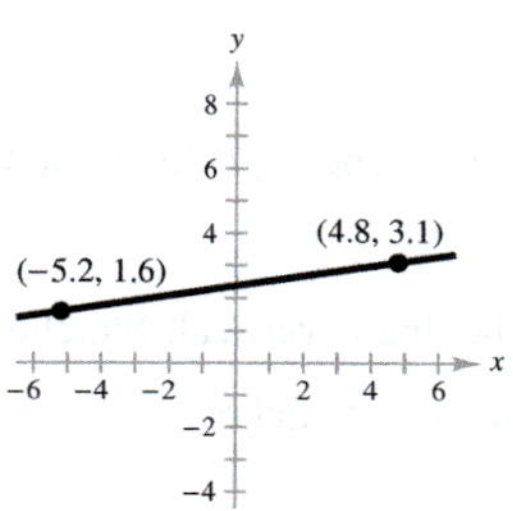

35. Point: $(5, 7)$, Slope: $m = 0$

Because $m = 0$, y does not change. Three other points are $(-1, 7)$, $(0, 7)$, and $(4, 7)$.

37. Point: $(-5, 4)$, Slope: $m = 2$

Because $m = 2 = \frac{2}{1}$, y increases by 2 for every one unit increase in x. Three additional points are $(-4, 6)$, $(-3, 8)$, and $(-2, 10)$.

39. Point: $(4, 5)$, Slope: $m = -\frac{1}{3}$

Because $m = -\frac{1}{3}$, y decreases by 1 unit for every three unit increase in x. Three additional points are $(-2, 7)$, $\left(0, -\frac{19}{4}\right)$, and $(1, 6)$.

41. Point: $(-4, 3)$, Slope is undefined.

Because m is undefined, x does not change. Three points are $(-4, 0)$, $(-4, 5)$, and $(-4, 2)$.

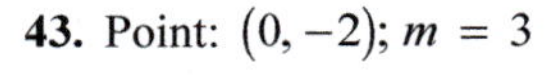

43. Point: $(0, -2)$; $m = 3$

$$y + 2 = 3(x - 0)$$
$$y = 3x - 2$$

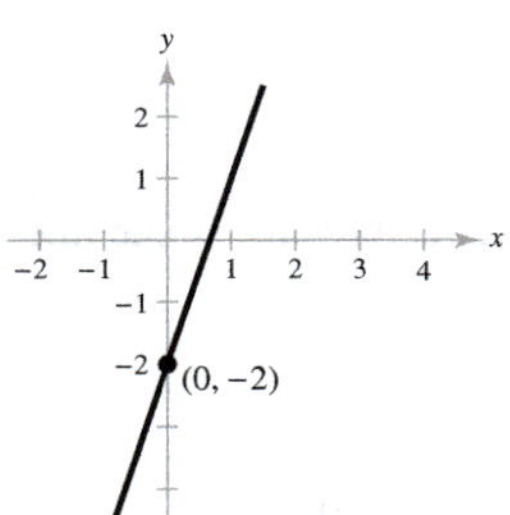

45. Point: $(-3, 6)$; $m = -2$

$$y - 6 = -2(x + 3)$$
$$y = -2x$$

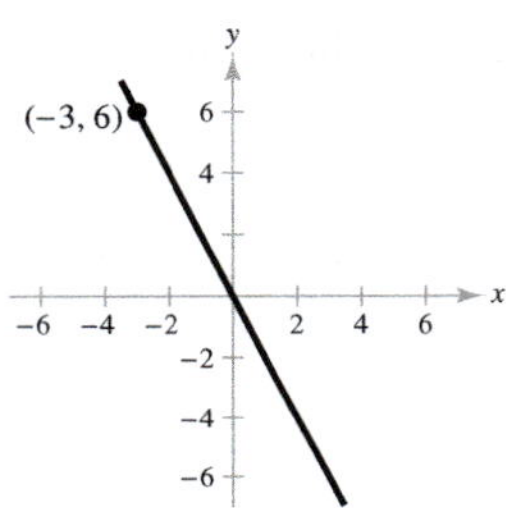

47. Point: $(4, 0)$; $m = -\frac{1}{3}$

$$y - 0 = -\tfrac{1}{3}(x - 4)$$
$$y = -\tfrac{1}{3}x + \tfrac{4}{3}$$

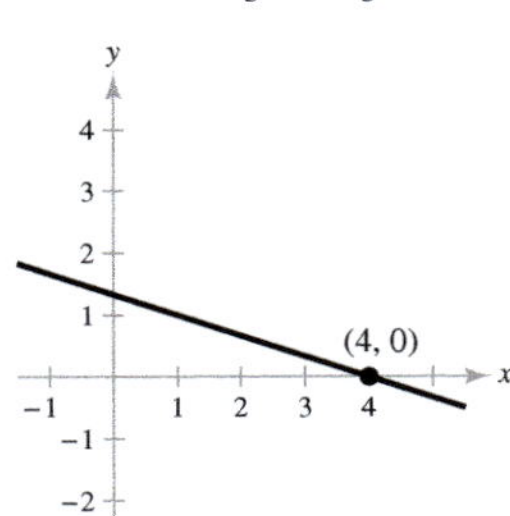

49. Point: $(2, -3)$; $m = -\dfrac{1}{2}$

$$y - (-3) = -\frac{1}{2}(x - 2)$$
$$y + 3 = -\frac{1}{2}x + 1$$
$$y = -\frac{1}{2}x - 2$$

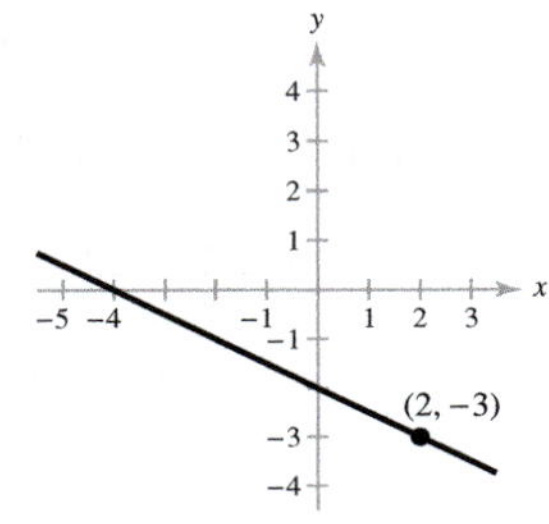

51. Point: $\left(4, \frac{5}{2}\right)$; $m = 0$

$$y - \tfrac{5}{2} = 0(x - 4)$$
$$y - \tfrac{5}{2} = 0$$
$$y = \tfrac{5}{2}$$

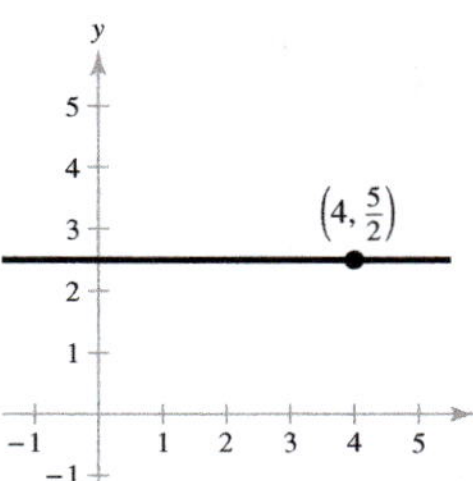

53. Point: $(-5.1, 1.8)$; $m = 5$

$$y - 1.8 = 5(x - (5.1))$$
$$y = 5x + 27.3$$

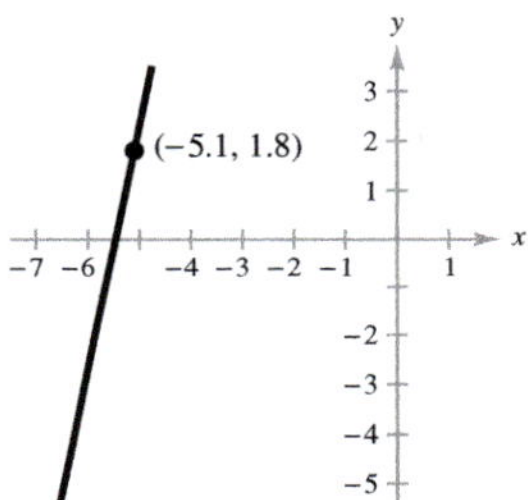

55. $(5, -1)$, $(-5, 5)$

$$y + 1 = \frac{5 + 1}{-5 - 5}(x - 5)$$
$$y = -\frac{3}{5}(x - 5) - 1$$
$$y = -\frac{3}{5}x + 2$$

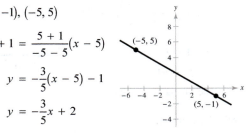

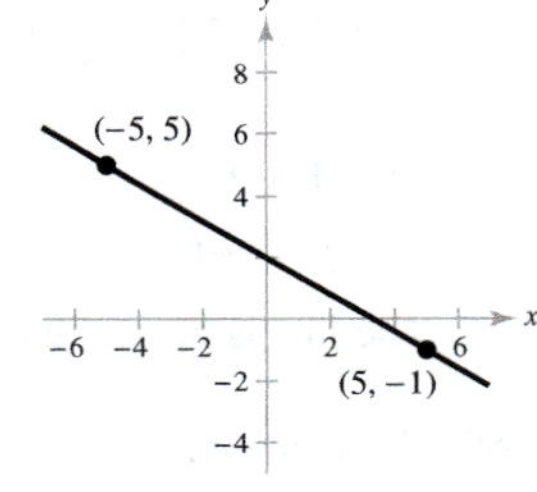

57. $(-7, 2)$, $(-7, 5)$

$$m = \frac{5 - 2}{-7 - (-7)} = \frac{3}{0}$$

m is undefined.

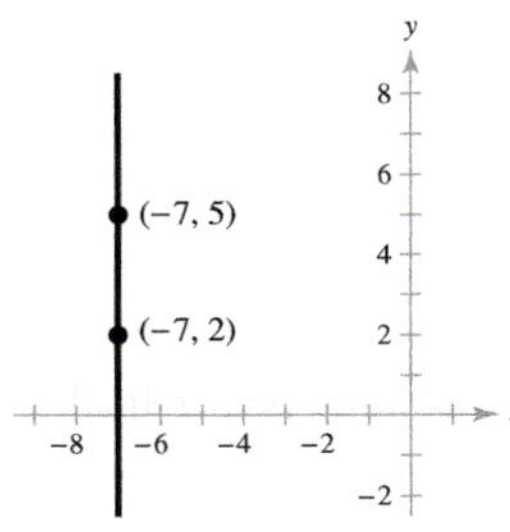

59. $\left(2, \frac{1}{2}\right), \left(\frac{1}{2}, \frac{5}{4}\right)$

$$y - \frac{1}{2} = \frac{\frac{5}{4} - \frac{1}{2}}{\frac{1}{2} - 2}(x - 2)$$

$$y = -\frac{1}{2}(x - 2) + \frac{1}{2}$$

$$y = -\frac{1}{2}x + \frac{3}{2}$$

61. $(1, 0.6), (-2, -0.6)$

$$y - 0.6 = \frac{-0.6 - 0.6}{-2 - 1}(x - 1)$$

$$y = 0.4(x - 1) + 0.6$$

$$y = 0.4x + 0.2$$

63. $(2, -1), \left(\frac{1}{3}, -1\right)$

$$y + 1 = \frac{-1 - (-1)}{\frac{1}{3} - 2}(x - 2)$$

$$y + 1 = 0$$

$$y = -1$$

The line is horizontal.

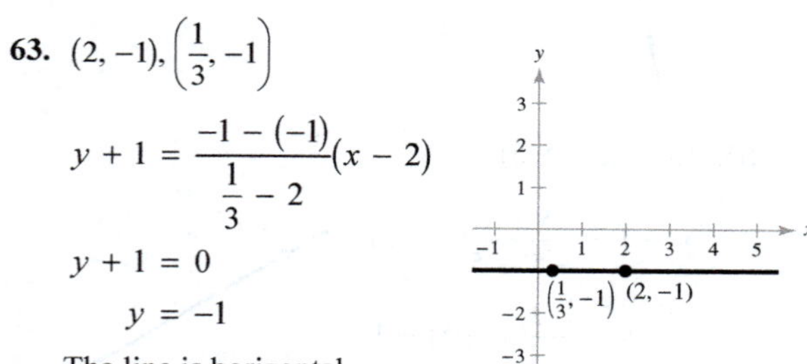

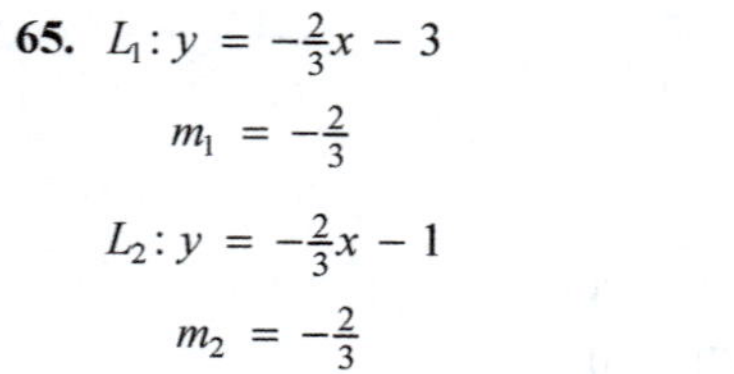

65. $L_1: y = -\frac{2}{3}x - 3$

$$m_1 = -\frac{2}{3}$$

$L_2: y = -\frac{2}{3}x - 1$

$$m_2 = -\frac{2}{3}$$

The slopes are equal, so the lines are parallel.

67. $L_1: y = \frac{1}{2}x - 3$

$$m_1 = \frac{1}{2}$$

$L_2: y = -\frac{1}{2}x + 1$

$$m_2 = -\frac{1}{2}$$

The lines are neither parallel nor perpendicular.

69. $L_1: (0, -1), (5, 9)$

$$m_1 = \frac{9 + 1}{5 - 0} = 2$$

$L_2: (0, 3), (4, 1)$

$$m_2 = \frac{1 - 3}{4 - 0} = -\frac{1}{2}$$

The slopes are negative reciprocals, so the lines are perpendicular.

71. $L_1: (-6, -3), (2, -3)$

$$m_1 = \frac{-3 - (-3)}{2 - (-6)} = \frac{0}{8} = 0$$

$L_2: \left(3, -\frac{1}{2}\right), \left(6, -\frac{1}{2}\right)$

$$m_2 = \frac{-\frac{1}{2} - \left(-\frac{1}{2}\right)}{6 - 3} = \frac{0}{3} = 0$$

L_1 and L_2 are both horizontal lines, so they are parallel.

73. $4x - 2y = 3$

$$y = 2x - \frac{3}{2}$$

Slope: $m = 2$

(a) $(2, 1), m = 2$

$$y - 1 = 2(x - 2)$$

$$y = 2x - 3$$

(b) $(2, 1), m = -\frac{1}{2}$

$$y - 1 = -\frac{1}{2}(x - 2)$$

$$y = -\frac{1}{2}x + 2$$

75. $3x + 4y = 7$

$$y = -\frac{3}{4}x + \frac{7}{4}$$

Slope: $m = -\frac{3}{4}$

(a) $\left(-\frac{2}{3}, \frac{7}{8}\right), m = -\frac{3}{4}$

$$y - \frac{7}{8} = -\frac{3}{4}\left(x - \left(-\frac{2}{3}\right)\right)$$

$$y = -\frac{3}{4}x + \frac{3}{8}$$

(b) $\left(-\frac{2}{3}, \frac{7}{8}\right), m = \frac{4}{3}$

$$y - \frac{7}{8} = \frac{4}{3}\left(x - \left(-\frac{2}{3}\right)\right)$$

$$y = \frac{4}{3}x + \frac{127}{72}$$

77. $y + 5 = 0$

$y = -5$

Slope: $m = 0$

(a) $(-2, 4), m = 0$

$y = 4$

(b) $(-2, 4)$, m is undefined.

$x = -2$

79. $x - y = 4$

$y = x - 4$

Slope: $m = 1$

(a) $(2.5, 6.8), m = 1$

$y - 6.8 = 1(x - 2.5)$

$y = x + 4.3$

(b) $(2.5, 6.8), m = -1$

$y - 6.8 = (-1)(x - 2.5)$

$y = -x + 9.3$

81. $\frac{x}{3} + \frac{y}{5} = 1$

$(15)\left(\frac{x}{3} + \frac{y}{5}\right) = 1(15)$

$5x + 3y - 15 = 0$

83. $\frac{x}{-1/6} + \frac{y}{-2/3} = 1$

$6x + \frac{3}{2}y = -1$

$12x + 3y + 2 = 0$

85. $\frac{x}{c} + \frac{y}{c} = 1, c \neq 0$

$x + y = c$

$1 + 2 = c$

$3 = c$

$x + y = 3$

$x + y - 3 = 0$

87. (a) $m = 135$. The sales are increasing 135 units per year.

(b) $m = 0$. There is no change in sales during the year.

(c) $m = -40$. The sales are decreasing 40 units per year.

89. $y = \frac{6}{100}x$

$y = \frac{6}{100}(200) = 12$ feet

91. $(16, 3000), m = -150$

$V - 3000 = -150(t - 16)$

$V - 3000 = -150t + 2400$

$V = -150t + 5400, 16 \leq t \leq 21$

93. The C-intercept measures the fixed costs of manufacturing when zero bags are produced.

The slope measures the cost to produce one laptop bag.

95. Using the points $(0, 875)$ and $(5, 0)$, where the first coordinate represents the year t and the second coordinate represents the value V, you have

$$m = \frac{0 - 875}{5 - 0} = -175$$

$V = -175t + 875, 0 \leq t \leq 5.$

97. Using the points $(0, 32)$ and $(100, 212)$, where the first coordinate represents a temperature in degrees Celsius and the second coordinate represents a temperature in degrees Fahrenheit, you have

$$m = \frac{212 - 32}{100 - 0} = \frac{180}{100} = \frac{9}{5}.$$

Since the point $(0, 32)$ is the F- intercept, $b = 32$, the equation is $F = 1.8C + 32$ or $C = \frac{5}{9}F - \frac{160}{9}$.

99. (a) Total Cost = cost for fuel and maintainance + cost for operator + purchase cost

$C = 9.5t + 11.5t + 42{,}000$

$C = 21t + 42{,}000$

(b) Revenue = Rate per hour · Hours

$R = 45t$

(c) $P = R - C$

$P = 45t - (21t + 42{,}000)$

$P = 24t - 42{,}000$

(d) Let $P = 0$, and solve for t.

$0 = 24t - 42{,}000$

$42{,}000 = 24t$

$1750 = t$

The equipment must be used 1750 hours to yield a profit of 0 dollars.

101. False. The slope with the greatest magnitude corresponds to the steepest line.

103. Find the slope of the line segments between the points A and B, and B and C.

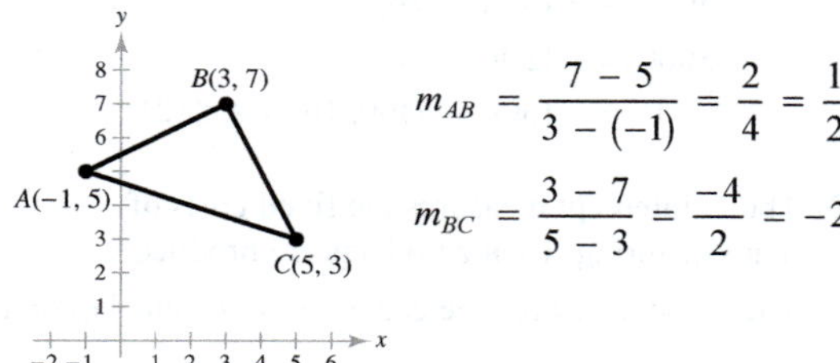

$$m_{AB} = \frac{7-5}{3-(-1)} = \frac{2}{4} = \frac{1}{2}$$

$$m_{BC} = \frac{3-7}{5-3} = \frac{-4}{2} = -2$$

Since the slopes are negative reciprocals, the line segments are perpendicular and therefore intersect to form a right angle. So, the triangle is a right triangle.

105. Since the scales for the y-axis on each graph is unknown, the slopes of the lines cannot be determined.

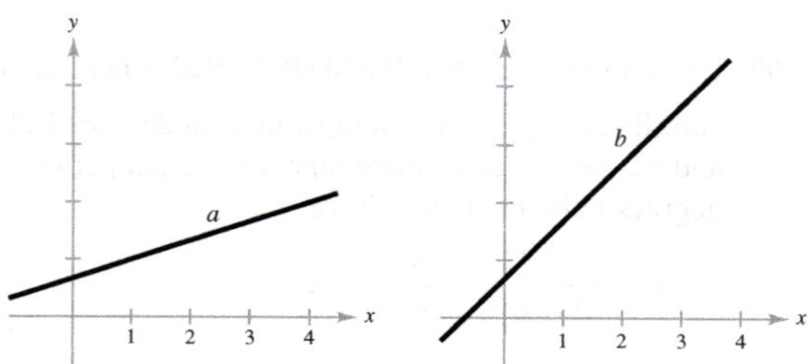

107. No, the slopes of two perpendicular lines have opposite signs. (Assume that neither line is vertical or horizontal.)

109. The line $y = 4x$ rises most quickly.

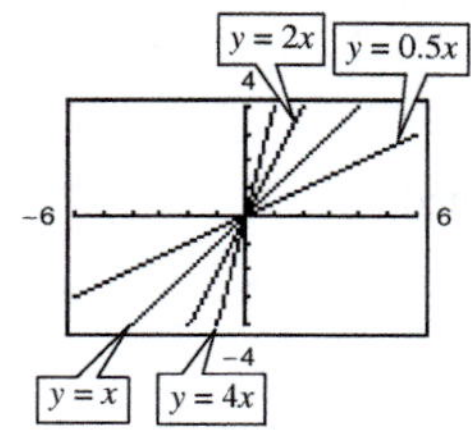

The line $y = -4x$ falls most quickly.

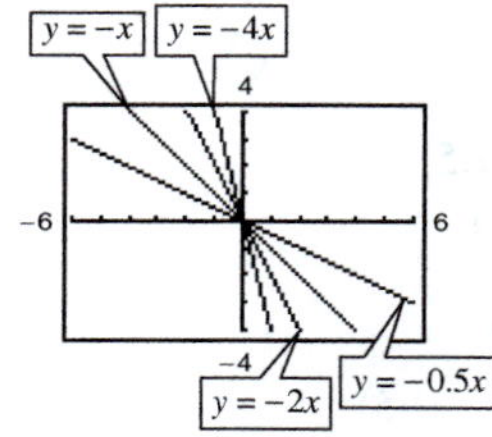

The greater the magnitude of the slope (the absolute value of the slope), the faster the line rises or falls.

111. Set the distance between $(4, -1)$ and (x, y) equal to the distance between $(-2, 3)$ and (x, y).

$$\sqrt{(x-4)^2 + [y-(-1)]^2} = \sqrt{[x-(-2)]^2 + (y-3)^2}$$

$$(x-4)^2 + (y+1)^2 = (x+2)^2 + (y-3)^2$$

$$x^2 - 8x + 16 + y^2 + 2y + 1 = x^2 + 4x + 4 + y^2 - 6y + 9$$

$$-8x + 2y + 17 = 4x - 6y + 13$$

$$-12x + 8y + 4 = 0$$

$$-4(3x - 2y - 1) = 0$$

$$3x - 2y - 1 = 0$$

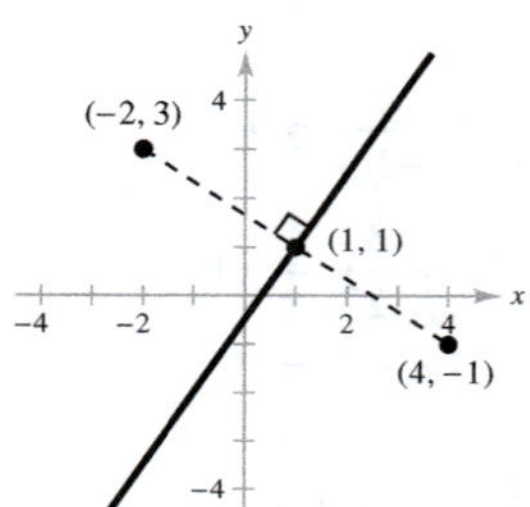

This line is the perpendicular bisector of the line segment connecting $(4, -1)$ and $(-2, 3)$.

113. Set the distance between $\left(3, \frac{5}{2}\right)$ and (x, y) equal to the distance between $(-7, 1)$ and (x, y).

$$\sqrt{(x-3)^2 + \left(y-\frac{5}{2}\right)^2} = \sqrt{[x-(-7)]^2 + (y-1)^2}$$

$$(x-3)^2 + \left(y-\frac{5}{2}\right)^2 = (x+7)^2 + (y-1)^2$$

$$x^2 - 6x + 9 + y^2 - 5y + \frac{25}{4} = x^2 + 14x + 49 + y^2 - 2y + 1$$

$$-6x - 5y + \frac{61}{4} = 14x - 2y + 50$$

$$-24x - 20y + 61 = 56x - 8y + 200$$

$$80x + 12y + 139 = 0$$

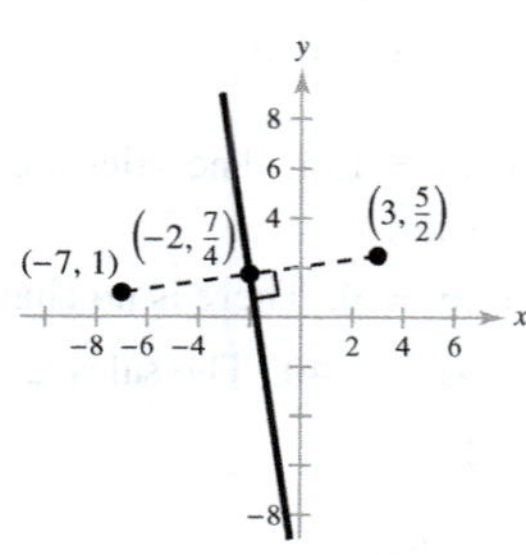

This line is the perpendicular bisector of the line segment connecting $\left(3, \frac{5}{2}\right)$ and $(-7, 1)$.

Section 1.4 Functions

1. domain; range; function

3. implied domain

5. Yes, the relationship is a function. Each domain value is matched with exactly one range value.

7. No, it does not represent a function. The input values of 10 and 7 are each matched with two output values.

9. (a) Each element of A is matched with exactly one element of B, so it does represent a function.

(b) The element 1 in A is matched with two elements, -2 and 1 of B, so it does not represent a function.

(c) Each element of A is matched with exactly one element of B, so it does represent a function.

(d) The element 2 in A is not matched with an element of B, so the relation does not represent a function.

11. $x^2 + y^2 = 4 \Rightarrow y = \pm\sqrt{4 - x^2}$

No, y *is not* a function of x.

13. $y = \sqrt{16 - x^2}$

Yes, y *is* a function of x.

15. $y = 4 - |x|$

Yes, y *is* a function of x.

17. $y = -75$ or $y = -75 + 0x$

Yes, y *is* a function of x.

19. $f(x) = 3x - 5$

(a) $f(1) = 3(1) - 5 = -2$

(b) $f(-3) = 3(-3) - 5 = -14$

(c) $f(x + 2) = 3(x + 2) - 5$
$= 3x + 6 - 5$
$= 3x + 1$

21. $g(t) = 4t^2 - 3t + 5$

(a) $g(2) = 4(2)^2 - 3(2) + 5$
$= 15$

(b) $g(t - 2) = 4(t - 2)^2 - 3(t - 2) + 5$
$= 4t^2 - 19t + 27$

(c) $g(t) - g(2) = 4t^2 - 3t + 5 - 15$
$= 4t^2 - 3t - 10$

23. $f(y) = 3 - \sqrt{y}$

(a) $f(4) = 3 - \sqrt{4} = 1$

(b) $f(0.25) = 3 - \sqrt{0.25} = 2.5$

(c) $f(4x^2) = 3 - \sqrt{4x^2} = 3 - 2|x|$

25. $q(x) = \dfrac{1}{x^2 - 9}$

(a) $q(0) = \dfrac{1}{0^2 - 9} = -\dfrac{1}{9}$

(b) $q(3) = \dfrac{1}{3^2 - 9}$ is undefined.

(c) $q(y + 3) = \dfrac{1}{(y + 3)^2 - 9} = \dfrac{1}{y^2 + 6y}$

27. $f(x) = \dfrac{|x|}{x}$

(a) $f(2) = \dfrac{|2|}{2} = 1$

(b) $f(-2) = \dfrac{|-2|}{-2} = -1$

(c) $f(x - 1) = \dfrac{|x - 1|}{x - 1} = \begin{cases} -1, & \text{if } x < 1 \\ 1, & \text{if } x > 1 \end{cases}$

29. $f(x) = \begin{cases} 2x + 1, & x < 0 \\ 2x + 2, & x \geq 0 \end{cases}$

(a) $f(-1) = 2(-1) + 1 = -1$

(b) $f(0) = 2(0) + 2 = 2$

(c) $f(2) = 2(2) + 2 = 6$

31. $f(x) = -x^2 + 5$

$f(-2) = -(-2)^2 + 5 = 1$

$f(-1) = -(-1)^2 + 5 = 4$

$f(0) = -(0)^2 + 5 = 5$

$f(1) = -(1)^2 + 5 = 4$

$f(2) = -(2)^2 + 5 = 1$

x	-2	-1	0	1	2
$f(x)$	1	4	5	-4	1

33. $f(x) = \begin{cases} -\frac{1}{2}x + 4, & x \le 0 \\ (x - 2)^2, & x > 0 \end{cases}$

$f(-2) = -\frac{1}{2}(-2) + 4 = 5$

$f(-1) = -\frac{1}{2}(-1) + 4 = 4\frac{1}{2} = \frac{9}{2}$

$f(0) = -\frac{1}{2}(0) + 4 = 4$

$f(1) = (1 - 2)^2 = 1$

$f(2) = (2 - 2)^2 = 0$

x	-2	-1	0	1	2
$f(x)$	5	$\frac{9}{2}$	4	1	0

35. $15 - 3x = 0$

$3x = 15$

$x = 5$

37. $\dfrac{3x - 4}{5} = 0$

$3x - 4 = 0$

$x = \dfrac{4}{3}$

39. $f(x) = x^2 - 81$

$x^2 - 81 = 0$

$x^2 = 81$

$x = \pm 9$

41. $x^3 - x = 0$

$x(x^2 - 1) = 0$

$x(x + 1)(x - 1) = 0$

$x = 0, x = -1, \text{ or } x = 1$

43. $f(x) = g(x)$

$x^2 = x + 2$

$x^2 - x - 2 = 0$

$(x - 2)(x + 1) = 0$

$x - 2 = 0 \quad x + 1 = 0$

$x = 2 \qquad x = -1$

45. $f(x) = g(x)$

$x^4 - 2x^2 = 2x^2$

$x^4 - 4x^2 = 0$

$x^2(x^2 - 4) = 0$

$x^2(x + 2)(x - 2) = 0$

$x^2 = 0 \Rightarrow x = 0$

$x + 2 = 0 \Rightarrow x = -2$

$x - 2 = 0 \Rightarrow x = 2$

47. $f(x) = 5x^2 + 2x - 1$

Because $f(x)$ is a polynomial, the domain is all real numbers x.

49. $g(y) = \sqrt{y + 6}$

Domain: $y + 6 \ge 0$

$y \ge -6$

The domain is all real numbers y such that $y \ge -6$.

51. $g(x) = \dfrac{1}{x} - \dfrac{3}{x + 2}$

The domain is all real numbers x except

53. $f(s) = \dfrac{\sqrt{s - 1}}{s - 4}$

Domain: $s - 1 \ge 0 \Rightarrow s \ge 1$ and $s \ne 4$

The domain consists of all real numbers s, such that $s \ge 1$ and $s \ne 4$.

55. $f(x) = \dfrac{x - 4}{\sqrt{x}}$

The domain is all real numbers x such that $x > 0$ or $(0, \infty)$.

57. (a)

Height, x	Volume, V
1	484
2	800
3	972
4	1024
5	980
6	864

The volume is maximum when $x = 4$ and $V = 1024$ cubic centimeters.

(b)

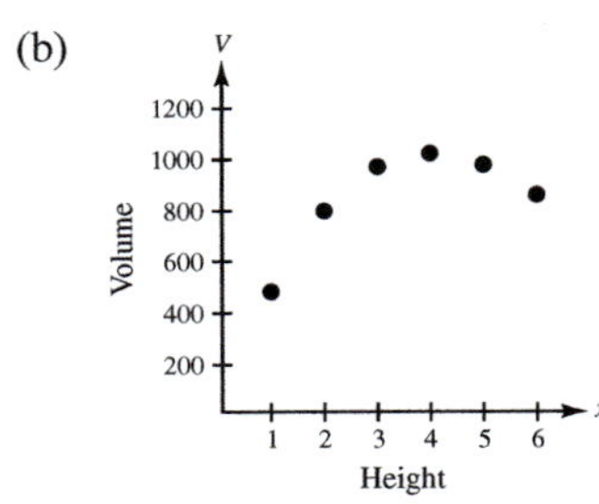

V is a function of x.

(c) $V = x(24 - 2x)^2$

Domain: $0 < x < 12$

59. $A = s^2$ and $P = 4s \Rightarrow \dfrac{P}{4} = s$

$$A = \left(\frac{P}{4}\right)^2 = \frac{P^2}{16}$$

61. $y = -\frac{1}{10}x^2 + 3x + 6$

$y(25) = -\frac{1}{10}(25)^2 + 3(25) + 6 = 18.5$ feet

If the child holds a glove at a height of 5 feet, then the ball *will* be over the child's head because it will be at a height of 18.5 feet.

63. $A = \dfrac{1}{2}bh = \dfrac{1}{2}xy$

Because $(0, y)$, $(2, 1)$, and $(x, 0)$ all lie on the same line, the slopes between any pair are equal.

$$\frac{1 - y}{2 - 0} = \frac{0 - 1}{x - 2}$$
$$\frac{1 - y}{2} = \frac{-1}{x - 2}$$
$$y = \frac{2}{x - 2} + 1$$
$$y = \frac{x}{x - 2}$$

So, $A = \dfrac{1}{2}x\left(\dfrac{x}{x - 2}\right) = \dfrac{x^2}{2(x - 2)}$.

The domain of A includes x-values such that $x^2/[2(x - 2)] > 0$. By solving this inequality, the domain is $x > 2$.

65. For 2008 through 2011, use

$$p(t) = 2.77t + 45.2.$$

2008: $p(8) = 2.77(8) + 45.2 = 67.36\%$

2009: $p(9) = 2.77(9) + 45.2 = 70.13\%$

2010: $p(10) = 2.77(10) + 45.2 = 72.90\%$

2011: $p(11) = 2.77(11) + 45.2 = 75.67\%$

For 2011 through 2014, use

$$p(t) = 1.95t + 55.9.$$

2012: $p(12) = 1.95(12) + 55.9 = 79.30\%$

2013: $p(13) = 1.95(13) + 55.9 = 81.25\%$

2014: $p(14) = 1.95(14) + 55.9 = 83.20\%$

67. (a) Cost = variable costs + fixed costs

$C = 12.30x + 98{,}000$

(b) Revenue = price per unit $\times$ number of units

$R = 17.98x$

(c) Profit = Revenue − Cost

$P = 17.98x - (12.30x + 98{,}000)$

$P = 5.68x - 98{,}000$

69. (a)

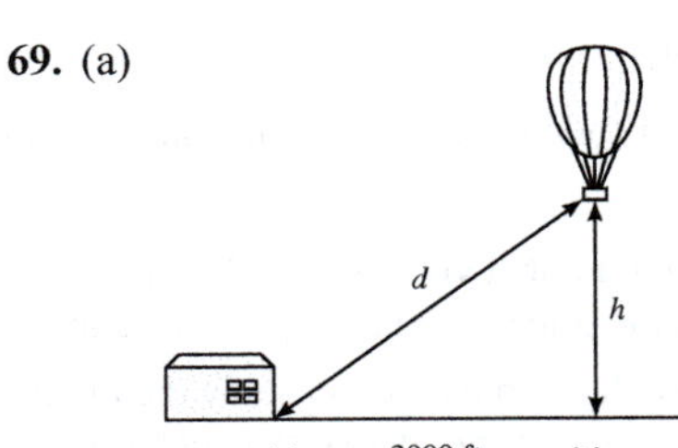

(b) $(3000)^2 + h^2 = d^2$

$$h = \sqrt{d^2 - (3000)^2}$$

Domain: $d \geq 3000$ (because both $d \geq 0$ and $d^2 - (3000)^2 \geq 0$)

71. (a) $R = n(\text{rate}) = n[8.00 - 0.05(n - 80)],\ n \geq 80$

$$R = 12.00n - 0.05n^2 = 12n - \frac{n^2}{20} = \frac{240n - n^2}{20},\ n \geq 80$$

(b)

n	90	100	110	120	130	140	150
$R(n)$	\$675	\$700	\$715	\$720	\$715	\$700	\$675

The revenue is maximum when 120 people take the trip.

73.

$$f(x) = x^2 - 2x + 4$$
$$f(2 + h) = (2 + h)^2 - 2(2 + h) + 4 = 4 + 4h + h^2 - 4 - 2h + 4 = h^2 + 2h + 4$$
$$f(2) = (2)^2 - 2(2) + 4 = 4$$
$$f(2 + h) - f(2) = h^2 + 2h$$
$$\frac{f(2 + h) - f(2)}{h} = \frac{h^2 + 2h}{h} = h + 2,\ h \neq 0$$

75.

$$f(x) = x^3 + 3x$$
$$f(x + h) = (x + h)^3 + 3(x + h) = x^3 + 3x^2h + 3xh^2 + h^3 + 3x + 3h$$
$$\frac{f(x + h) - f(x)}{h} = \frac{(x^3 + 3x^2h + 3xh^2 + h^3 + 3x + 3h) - (x^3 + 3x)}{h}$$
$$= \frac{h(3x^2 + 3xh + h^2 + 3)}{h}$$
$$= 3x^2 + 3xh + h^2 + 3,\ h \neq 0$$

77.

$$g(x) = \frac{1}{x^2}$$
$$\frac{g(x) - g(3)}{x - 3} = \frac{\frac{1}{x^2} - \frac{1}{9}}{x - 3}$$
$$= \frac{9 - x^2}{9x^2(x - 3)}$$
$$= \frac{-(x + 3)(x - 3)}{9x^2(x - 3)}$$
$$= -\frac{x + 3}{9x^2},\ x \neq 3$$

79. $f(x) = \sqrt{5x}$

$$\frac{f(x) - f(5)}{x - 5} = \frac{\sqrt{5x} - 5}{x - 5},\ x \neq 5$$

81. By plotting the points, we have a parabola, so $g(x) = cx^2$. Because $(-4, -32)$ is on the graph, you have $-32 = c(-4)^2 \Rightarrow c = -2$. So, $g(x) = -2x^2$.

83. Because the function is undefined at 0, we have $r(x) = c/x$. Because $(-4, -8)$ is on the graph, you have $-8 = c/-4 \Rightarrow c = 32$. So, $r(x) = 32/x$.

85. False. The equation $y^2 = x^2 + 4$ is a relation between x and y. However, $y = \pm\sqrt{x^2 + 4}$ does not represent a function.

87. False. The range is $[-1, \infty)$.

89. The domain of $f(x) = \sqrt{x - 1}$ includes $x = 1$, $x \geq 1$ and the domain of $g(x) = \dfrac{1}{\sqrt{x - 1}}$ does not include $x = 1$ because you cannot divide by 0. The domain of $g(x) = \dfrac{1}{\sqrt{x - 1}}$ is $x > 1$. So, the functions do not have the same domain.

91. No; x is the independent variable, f is the name of the function.

93. (a) Yes. The amount that you pay in sales tax will increase as the price of the item purchased increases.

(b) No. The length of time that you study the night before an exam does not necessarily determine your score on the exam.

Section 1.5 Analyzing Graphs of Functions

1. Vertical Line Test

3. decreasing

5. average rate of change; secant

7. Domain: $(-2, 2]$; Range: $[-1, 8]$

(a) $f(-1) = -1$

(b) $f(0) = 0$

(c) $f(1) = -1$

(d) $f(2) = 8$

9. Domain: $(-\infty, \infty)$; Range: $(-2, \infty)$

(a) $f(2) = 0$

(b) $f(1) = 1$

(c) $f(3) = 2$

(d) $f(-1) = 3$

11. A vertical line intersects the graph at most once, so *y is* a function of *x*.

13. A vertical line intersects the graph more than once, so *y is not* a function of *x*.

15.
$$\begin{aligned} f(x) &= 3x + 18 \\ 3x + 18 &= 0 \\ 3x &= -18 \\ x &= -6 \end{aligned}$$

17.
$$\begin{aligned} f(x) &= 2x^2 - 7x - 30 \\ 2x^2 - 7x - 30 &= 0 \\ (2x + 5)(x - 6) &= 0 \end{aligned}$$
$$2x + 5 = 0 \quad \text{or} \quad x - 6 = 0$$
$$x = -\tfrac{5}{2} \qquad\qquad x = 6$$

19.
$$f(x) = \frac{x + 3}{2x^2 - 6}$$
$$\begin{aligned} \frac{x + 3}{2x^2 - 6} &= 0 \\ x + 3 &= 0 \\ x &= -3 \end{aligned}$$

21.
$$\begin{aligned} f(x) &= \tfrac{1}{3}x^3 - 2x \\ \tfrac{1}{3}x^3 - 2x &= 0 \\ (3)\left(\tfrac{1}{3}x^3 - 2x\right) &= 0(3) \\ x^3 - 6x &= 0 \\ x\left(x^2 - 6\right) &= 0 \end{aligned}$$
$$x = 0 \quad \text{or} \quad x^2 - 6 = 0$$
$$x^2 = 6$$
$$x = \pm\sqrt{6}$$

23.
$$\begin{aligned} f(x) &= x^3 - 4x^2 - 9x + 36 \\ x^3 - 4x^2 - 9x + 36 &= 0 \\ x^2(x - 4) - 9(x - 4) &= 0 \\ (x - 4)\left(x^2 - 9\right) &= 0 \end{aligned}$$
$$x - 4 = 0 \Rightarrow x = 4$$
$$x^2 - 9 = 0 \Rightarrow x = \pm 3$$

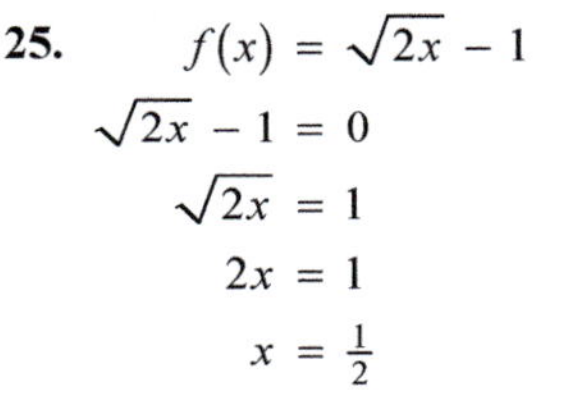

25.
$$\begin{aligned} f(x) &= \sqrt{2x} - 1 \\ \sqrt{2x} - 1 &= 0 \\ \sqrt{2x} &= 1 \\ 2x &= 1 \\ x &= \tfrac{1}{2} \end{aligned}$$

27. (a)

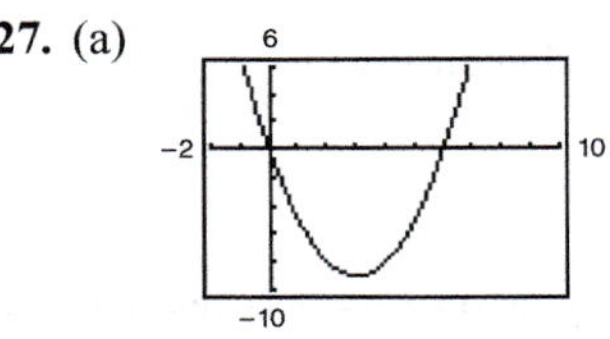

Zeros: $x = 0, 6$

(b)
$$\begin{aligned} f(x) &= x^2 - 6x \\ x^2 - 6x &= 0 \\ x(x - 6) &= 0 \end{aligned}$$
$$x = 0 \Rightarrow x = 0$$
$$x - 6 = 0 \Rightarrow x = 6$$

29. (a)

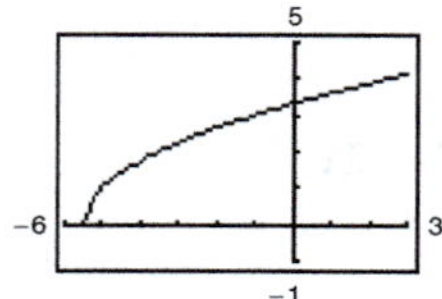

Zero: $x = -5.5$

(b) $$f(x) = \sqrt{2x + 11}$$
$$\sqrt{2x + 11} = 0$$
$$2x + 11 = 0$$
$$x = -\tfrac{11}{2}$$

31. (a)

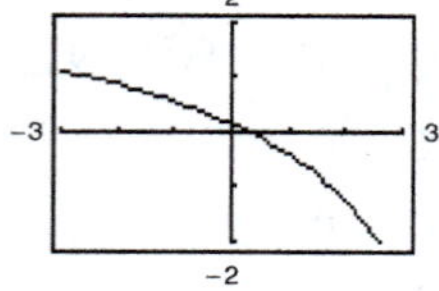

Zero: $x = 0.3333$

(b) $$f(x) = \frac{3x - 1}{x - 6}$$
$$\frac{3x - 1}{x - 6} = 0$$
$$3x - 1 = 0$$
$$x = \frac{1}{3}$$

33. $f(x) = -\frac{1}{2}x^3$

The function is decreasing on $(-\infty, \infty)$.

35. $f(x) = \sqrt{x^2 - 1}$

The function is decreasing on $(-\infty, -1)$ and increasing on $(1, \infty)$.

37. $f(x) = |x + 1| + |x - 1|$

The function is increasing on $(1, \infty)$.

The function is constant on $(-1, 1)$.

The function is decreasing on $(-\infty, -1)$.

39. $f(x) = \begin{cases} 2x + 1, & x \le -1 \\ x^2 - 2, & x > -1 \end{cases}$

The function is decreasing on $(-1, 0)$ and increasing on $(-\infty, -1)$ and $(0, \infty)$.

41. $f(x) = 3$

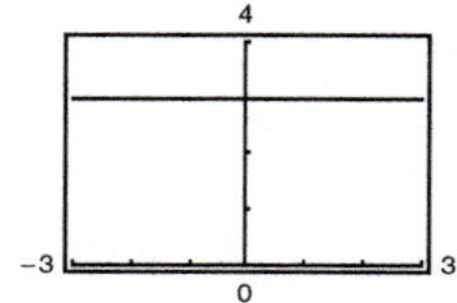

Constant on $(-\infty, \infty)$

x	−2	−1	0	1	2
$f(x)$	3	3	3	3	3

43. $g(x) = \frac{1}{2}x^2 - 3$

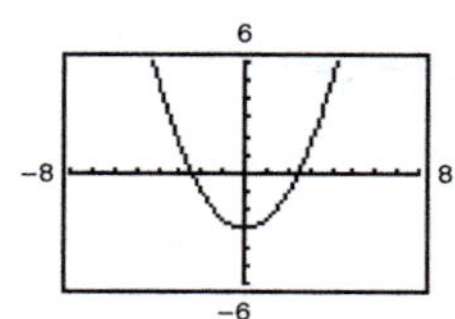

Decreasing on $(-\infty, 0)$.

Increasing on $(0, \infty)$.

x	−2	−1	0	1	2
$g(x)$	−1	$-\frac{5}{2}$	−3	$-\frac{5}{2}$	−1

45. $f(x) = \sqrt{1 - x}$

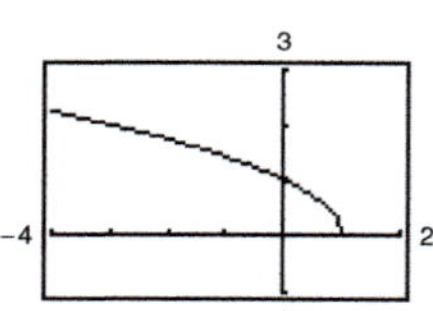

Decreasing on $(-\infty, 1)$

x	−3	−2	−1	0	1
$f(x)$	2	$\sqrt{3}$	$\sqrt{2}$	1	0

47. $f(x) = x^{3/2}$

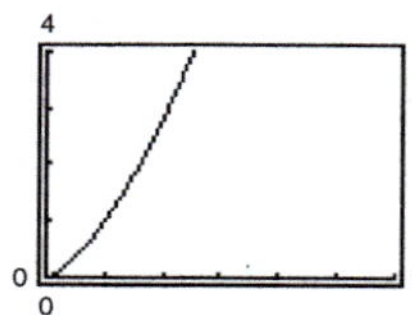

Increasing on $(0, \infty)$

x	0	1	2	3	4
$f(x)$	0	1	2.8	5.2	8

49. $f(x) = x(x + 3)$

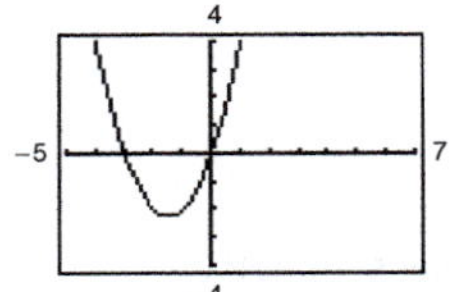

Relative minimum: $(-1.5, -2.25)$

51. $h(x) = x^3 - 6x^2 + 15$

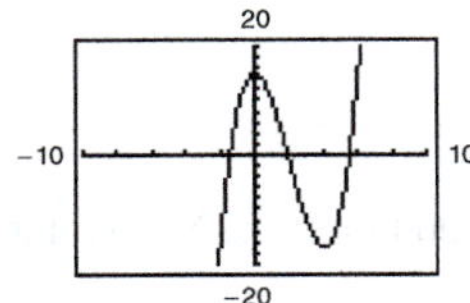

Relative minimum: $(4, -17)$

Relative maximum: $(0, 15)$

53. $h(x) = (x - 1)\sqrt{x}$

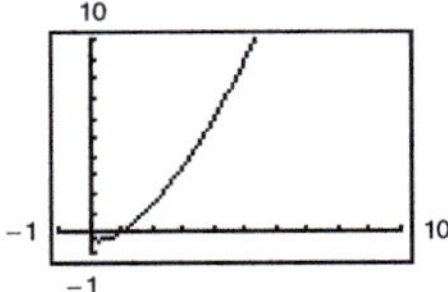

Relative minimum: $(0.33, -0.38)$

55. $f(x) = 4 - x$

$f(x) \geq 0$ on $(-\infty, 4]$

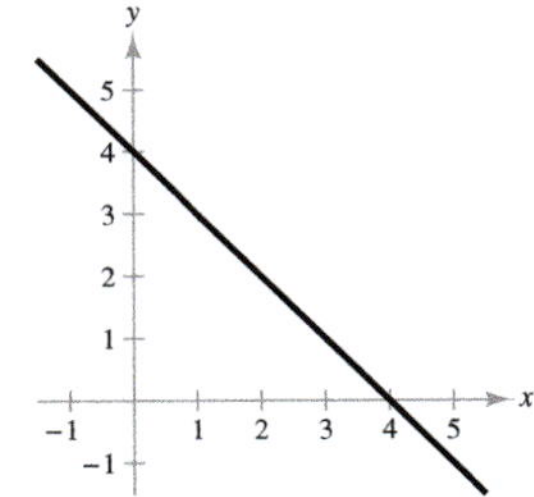

57. $f(x) = 9 - x^2$

$f(x) \geq 0$ on $[-3, 3]$

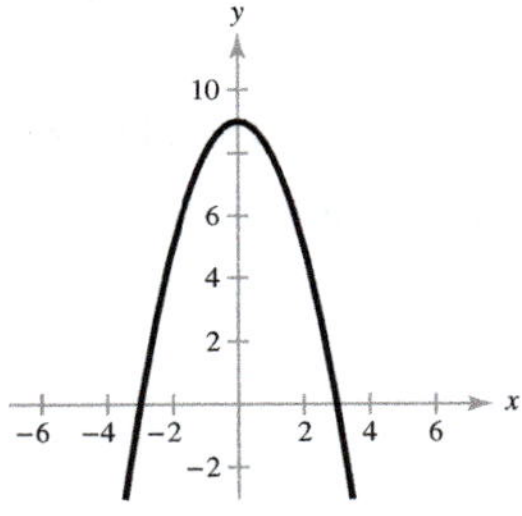

59. $f(x) = \sqrt{x - 1}$

$f(x) \geq 0$ on $[1, \infty)$

$$\sqrt{x - 1} \geq 0$$
$$x - 1 \geq 0$$
$$x \geq 1$$

$[1, \infty)$

61. $f(x) = -2x + 15$

$$\frac{f(3) - f(0)}{3 - 0} = \frac{9 - 15}{3} = -2$$

The average rate of change from $x_1 = 0$ to $x_2 = 3$ is -2.

63. $f(x) = x^3 - 3x^2 - x$

$$\frac{f(2) - f(-1)}{2 - (-1)} = \frac{-6 - (-3)}{3} = \frac{-3}{3} = -1$$

The average rate of change from $x_1 = -1$ to $x_2 = 2$ is -1.

65. (a) 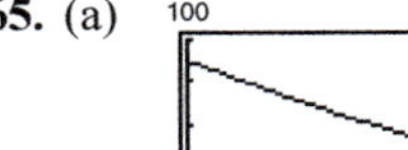

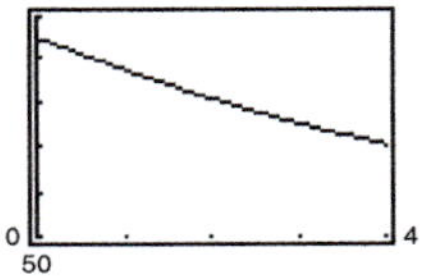

(b) To find the average rate of change of the amount the U.S. Department of Energy spent for research and development from 2010 to 2014, find the average rate of change from $(0, f(0))$ to $(4, f(4))$.

$$\frac{f(4) - f(0)}{4 - 0} = \frac{70.5344 - 95.08}{4}$$
$$= \frac{-24.5456}{4}$$
$$= -6.1364$$

The amount the U.S. Department of Energy spent on research and development for defense decreased by about \$6.14 billion each year from 2010 to 2014.

67. $s_0 = 6, v_0 = 64$

(a) $s = -16t^2 + 64t + 6$

(b)

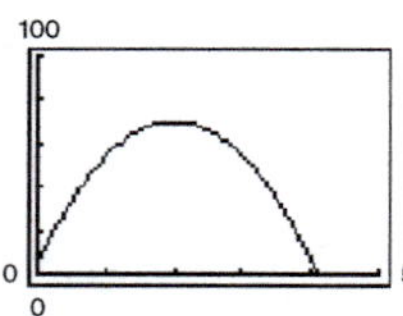

(c) $\dfrac{s(3) - s(0)}{3 - 0} = \dfrac{54 - 6}{3} = 16$

(d) The slope of the secant line is positive.

(e) $s(0) = 6, m = 16$

Secant line: $y - 6 = 16(t - 0)$

$y = 16t + 6$

(f)

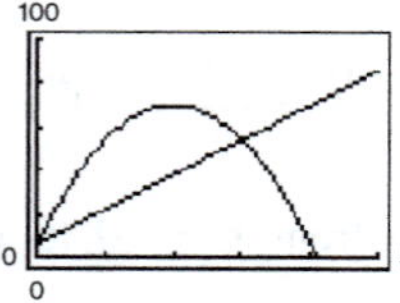

69. $v_0 = 120, s_0 = 0$

(a) $s = -16t^2 + 120t$

(b)

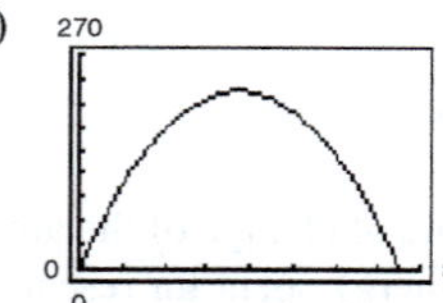

(c) The average rate of change from $t = 3$ to $t = 5$:

$\dfrac{s(5) - s(3)}{5 - 3} = \dfrac{200 - 216}{2} = -\dfrac{16}{2} = -8$ feet per second

(d) The slope of the secant line through $(3, s(3))$ and $(5, s(5))$ is negative.

(e) The equation of the secant line: $m = -8$

Using $(5, s(5)) = (5, 200)$ we have

$y - 200 = -8(t - 5)$

$y = -8t + 240.$

(f)

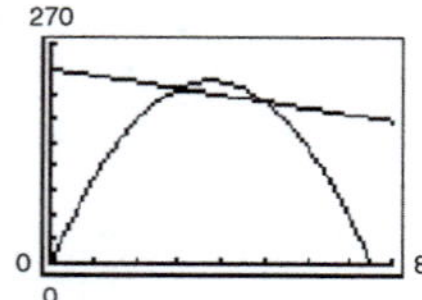

71. $f(x) = x^6 - 2x^2 + 3$

$f(-x) = (-x)^6 - 2(-x)^2 + 3$

$= x^6 - 2x^2 + 3$

$= f(x)$

The function is even. y-axis symmetry.

73. $h(x) = x\sqrt{x + 5}$

$h(-x) = (-x)\sqrt{-x + 5}$

$= -x\sqrt{5 - x}$

$\neq h(x)$

$\neq -h(x)$

The function is neither odd nor even. No symmetry.

75. $f(s) = 4s^{3/2} = 4(-s)^{3/2} \neq f(s) \neq -f(s)$

The function is neither odd nor even. No symmetry.

77.

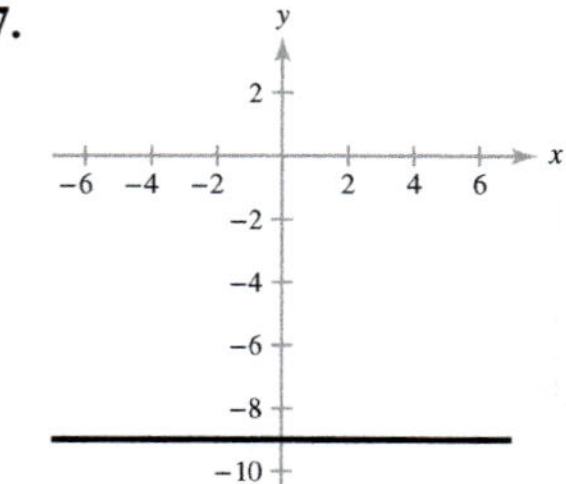

The graph of $f(x) = -9$ is symmetric to the y-axis, which implies $f(x)$ is even.

$f(-x) = -9 = f(x)$

The function is even.

79. $f(x) = -|x - 5|$

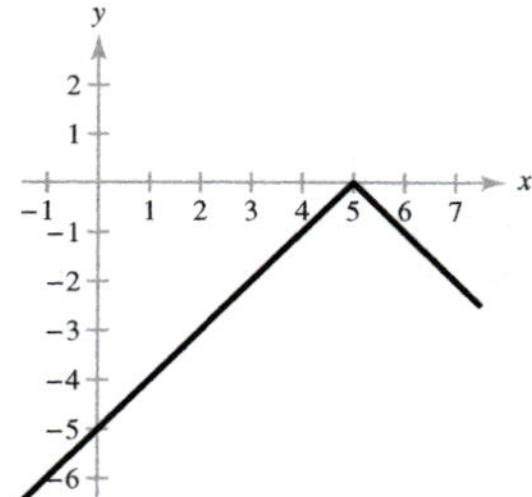

The graph displays no symmetry, which implies $f(x)$ is neither odd nor even.

$f(x) = -|(-x) - 5| = -|-x - 5| \neq f(x) \neq -f(x)$

The function is neither even nor odd.

81. $f(x) = \sqrt[3]{4x}$

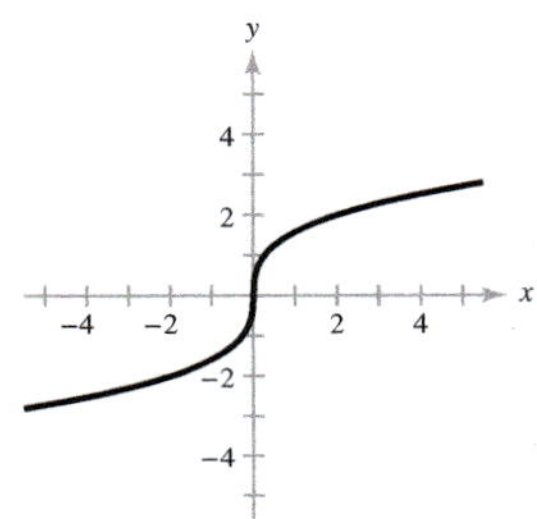

The graph displays origin symmetry, which implies $f(x)$ is odd.

$$\begin{aligned} f(-x) &= \sqrt[3]{4(-x)} \\ &= \sqrt[3]{-4x} \\ &= -\sqrt[3]{4x} \\ &= -f(x) \end{aligned}$$

The function is odd.

83. $h = \text{top} - \text{bottom}$

$$\begin{aligned} &= 3 - (4x - x^2) \\ &= 3 - 4x + x^2 \end{aligned}$$

85. $L = \text{right} - \text{left}$

$$= 2 - \sqrt[3]{2y}$$

87. The error is that $-2x^3 - 5 \neq -(2x^3 - 5)$.

The correct process is as follows.

$f(x) = 2x^3 - 5$

$f(-x) = 2(-x)^3 - 5 = -2x^3 - 5 = -(2x^3 + 5)$

$f(-x) \neq -f(x)$ and $f(-x) \neq f(x)$, so the function $f(x) = 2x^3 - 5$ is neither odd nor even.

89. (a) For the average salary of college professors, a scale of \$10,000 would be appropriate.

(b) For the population of the United States, use a scale of 10,000,000.

(c) For the percent of the civilian workforce that is unemployed, use a scale of 10%.

(d) For the number of games a college football team wins in a single season, single digits would be appropriate.

For each of the graphs, using the suggested scale would show yearly changes in the data clearly.

91. False. The function $f(x) = \sqrt{x^2 + 1}$ has a domain of all real numbers.

93. True. A graph that is symmetric with respect to the y-axis cannot be increasing on its entire domain.

95. $\left(-\frac{5}{3}, -7\right)$

(a) If f is even, another point is $\left(\frac{5}{3}, -7\right)$.

(b) If f is odd, another point is $\left(\frac{5}{3}, 7\right)$.

97. (a) $y = x$

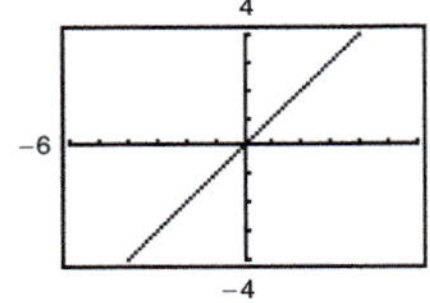

(b) $y = x^2$

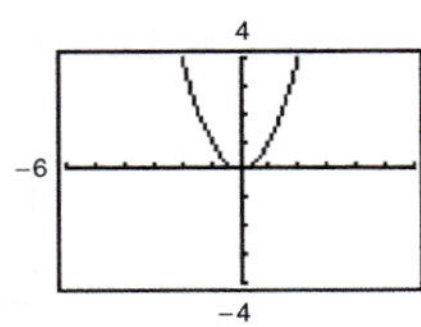

(c) $y = x^3$

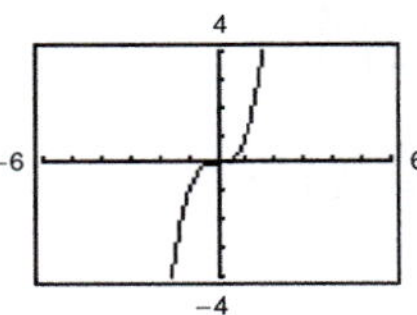

(d) $y = x^4$

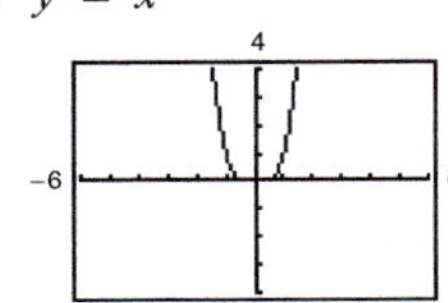

(e) $y = x^5$

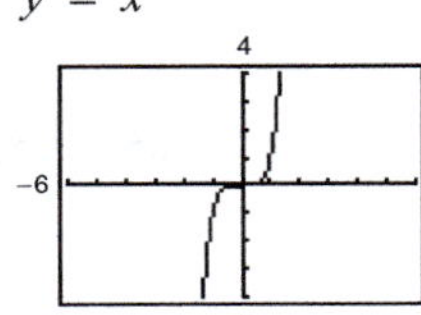

(f) $y = x^6$

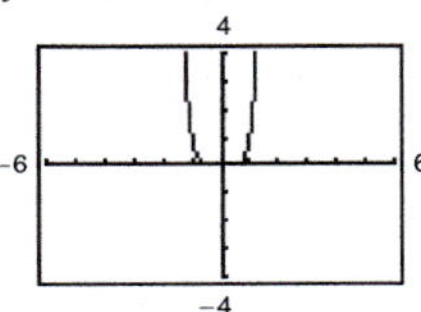

All the graphs pass through the origin. The graphs of the odd powers of x are symmetric with respect to the origin and the graphs of the even powers are symmetric with respect to the y-axis. As the powers increase, the graphs become flatter in the interval $-1 < x < 1$.

99. (a) Even. The graph is a reflection in the x-axis.

(b) Even. The graph is a reflection in the y-axis.

(c) Even. The graph is a vertical translation of f.

(d) Neither. The graph is a horizontal translation of f.

Section 1.6 A Library of Parent Functions

1. Greatest integer function

3. Reciprocal function

5. Square root function

7. Absolute value function

9. Linear function

11. (a) $f(1) = 4, f(0) = 6$

$(1, 4), (0, 6)$

$$m = \frac{6 - 4}{0 - 1} = -2$$

$$y - 6 = -2(x - 0)$$

$$y = -2x + 6$$

$$f(x) = -2x + 6$$

(b)

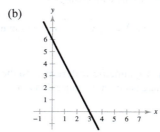

13. (a) $f\left(\frac{1}{2}\right) = -\frac{5}{3}, f(6) = 2$

$\left(\frac{1}{2}, -\frac{5}{3}\right), (6, 2)$

$$m = \frac{2 - \left(-\frac{5}{3}\right)}{6 - \left(\frac{1}{2}\right)}$$

$$= \frac{\frac{11}{3}}{\frac{11}{2}} = \left(\frac{11}{3}\right) \cdot \left(\frac{2}{11}\right) = \frac{2}{3}$$

$$f(x) - 2 = \frac{2}{3}(x - 6)$$

$$f(x) - 2 = \frac{2}{3}x - 4$$

$$f(x) = \frac{2}{3}x - 2$$

(b)

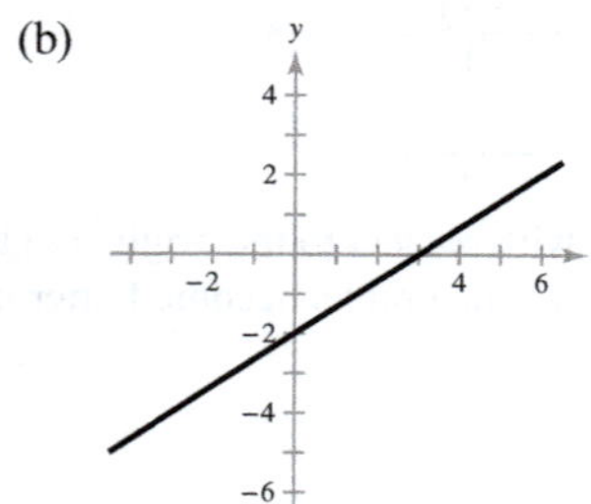

15. $f(x) = 2.5x - 4.25$

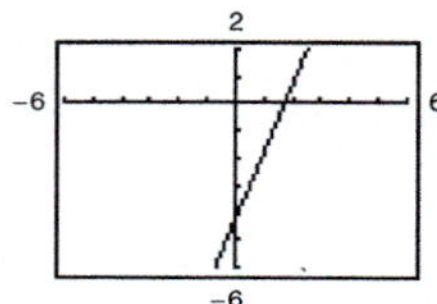

17. $g(x) = x^2 + 3$

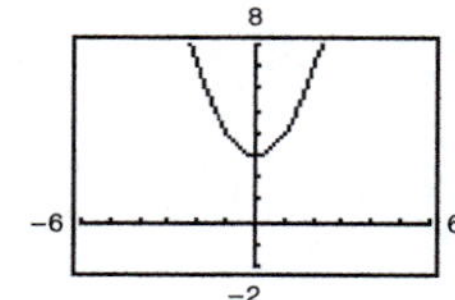

19. $f(x) = x^3 - 1$

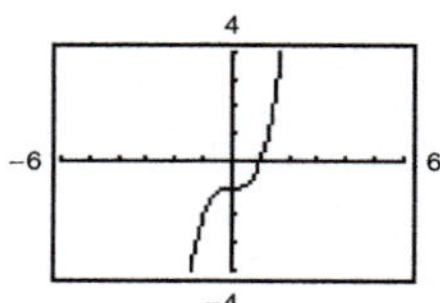

21. $f(x) = \sqrt{x} + 4$

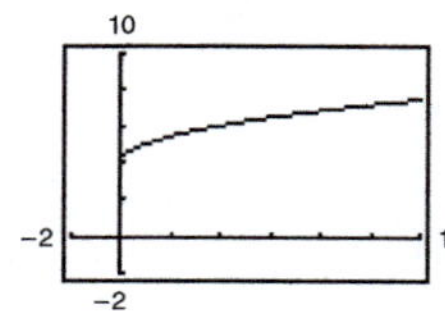

23. $f(x) = \dfrac{1}{x - 2}$

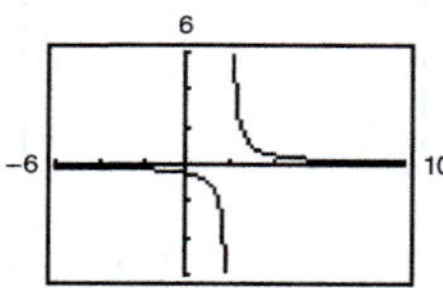

25. $g(x) = |x| - 5$

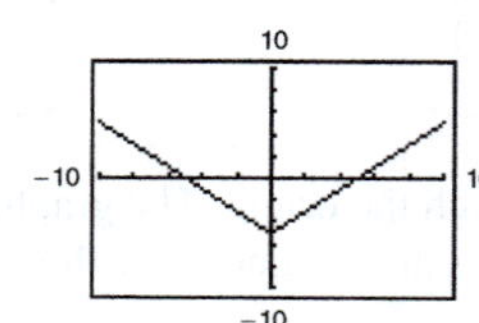

27. $f(x) = [\![x]\!]$

(a) $f(2.1) = 2$

(b) $f(2.9) = 2$

(c) $f(-3.1) = -4$

(d) $f\left(\frac{7}{2}\right) = 3$

29. $k(x) = [\![2x + 1]\!]$

(a) $k\left(\frac{1}{3}\right) = [\![2\left(\frac{1}{3}\right) + 1]\!] = [\![\frac{5}{3}]\!] = 1$

(b) $k(-2.1) = [\![2(-2.1) + 1]\!] = [\![-3.1]\!] = -4$

(c) $k(1.1) = [\![2(1.1) + 1]\!] = [\![3.2]\!] = 3$

(d) $k\left(\frac{2}{3}\right) = [\![2\left(\frac{2}{3}\right) + 1]\!] = [\![\frac{7}{3}]\!] = 2$

31. $g(x) = -[\![x]\!]$

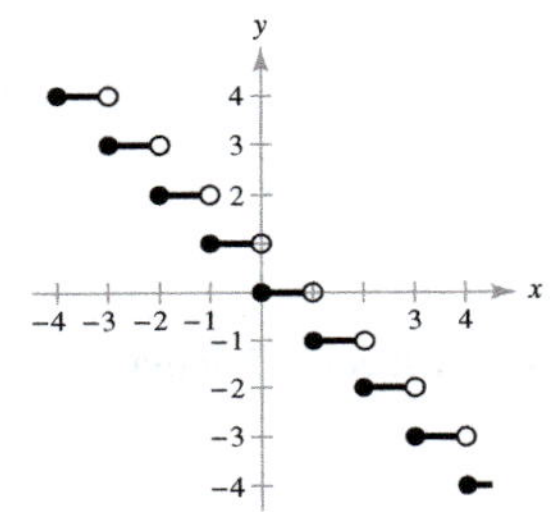

33. $g(x) = [\![x]\!] - 1$

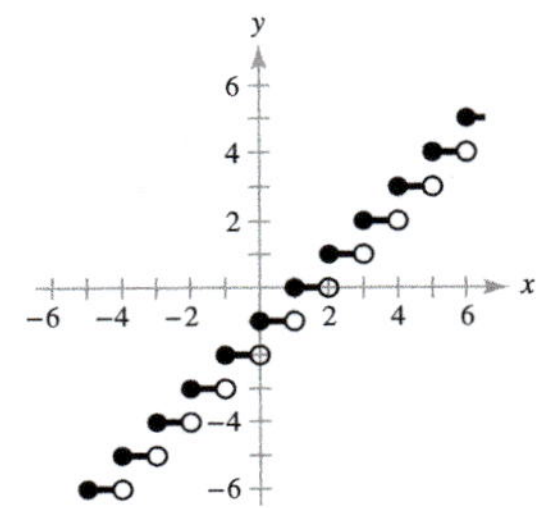

35. $g(x) = \begin{cases} x + 6, & x \le -4 \\ \frac{1}{2}x - 4, & x > -4 \end{cases}$

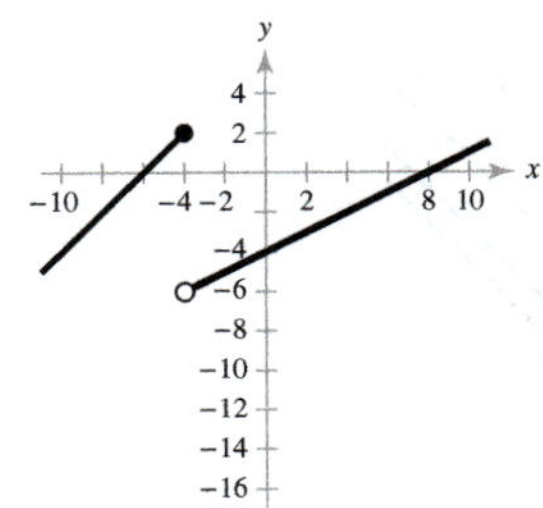

37. $f(x) = \begin{cases} 1 - (x - 1)^2, & x \le 2 \\ \sqrt{x - 2}, & x > 2 \end{cases}$

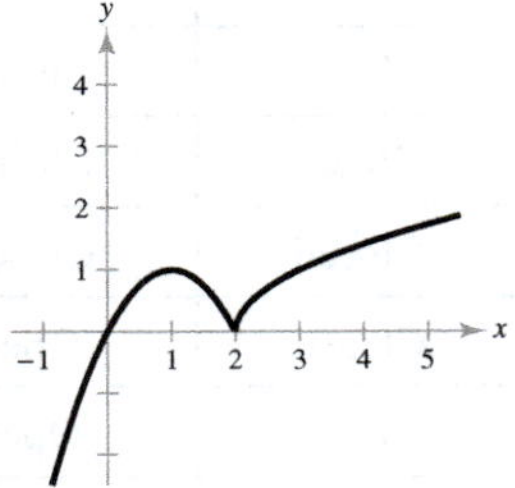

39. $h(x) = \begin{cases} 4 - x^2, & x < -2 \\ 3 + x, & -2 \le x < 0 \\ x^2 + 1, & x \ge 0 \end{cases}$

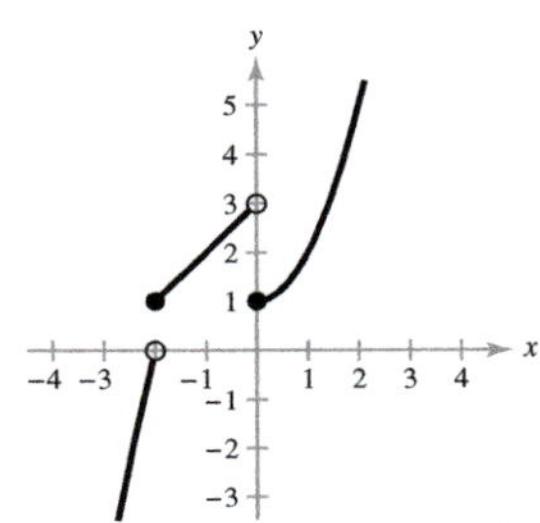

41. $s(x) = 2\left(\frac{1}{4}x - [\![\frac{1}{4}x]\!]\right)$

(a)

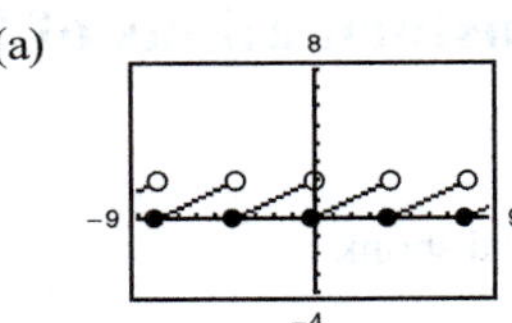

(b) Domain: $(-\infty, \infty)$; Range: $[0, 2)$

43. (a) $W(30) = 14(30) = 420$

$W(40) = 14(40) = 560$

$W(45) = 21(45 - 40) + 560 = 665$

$W(50) = 21(50 - 40) + 560 = 770$

(b) $W(h) = \begin{cases} 14h, & 0 < h \le 36 \\ 21(h - 36) + 504, & h > 36 \end{cases}$

(c) $W(h) = \begin{cases} 16h, & 0 < h \le 40 \\ 24(h - 40) + 640, & h > 40 \end{cases}$

45. Answers will vary. *Sample answer:*

Interval	*Input Pipe*	*Drain Pipe 1*	*Drain Pipe 2*
[0, 5]	Open	Closed	Closed
[5, 10]	Open	Open	Closed
[10, 20]	Closed	Closed	Closed
[20, 30]	Closed	Closed	Open
[30, 40]	Open	Open	Open
[40, 45]	Open	Closed	Open
[45, 50]	Open	Open	Open
[50, 60]	Open	Open	Closed

47. For the first two hours, the slope is 1. For the next six hours, the slope is 2. For the final hour, the slope is $\frac{1}{2}$.

$$f(t) = \begin{cases} t, & 0 \le t \le 2 \\ 2t - 2, & 2 < t \le 8 \\ \frac{1}{2}t + 10, & 8 < t \le 9 \end{cases}$$

To find $f(t) = 2t - 2$, use $m = 2$ and $(2, 2)$.

$y - 2 = 2(t - 2) \Rightarrow y = 2t - 2$

To find $f(t) = \frac{1}{2}t + 10$, use $m = \frac{1}{2}$ and $(8, 14)$.

$y - 14 = \frac{1}{2}(t - 8) \Rightarrow y = \frac{1}{2}t + 10$

Total accumulation $=$ 14.5 inches

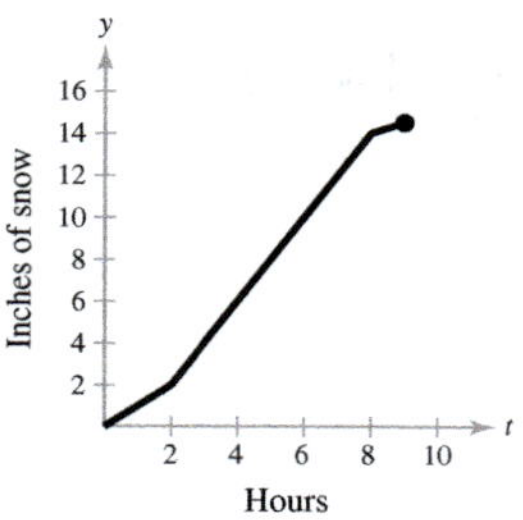

49. False. A piecewise-defined function is a function that is defined by two or more equations over a specified domain. That domain may or may not include x- and y-intercepts.

Section 1.7 Transformations of Functions

1. rigid

3. vertical stretch; vertical shrink

5. (a) $f(x) = |x| + c$ Vertical shifts

$c = -2$: $f(x) = |x| - 2$ 2 units down

$c = -1$: $f(x) = |x| - 1$ 1 unit down

$c = 1$: $f(x) = |x| + 1$ 1 unit up

$c = 2$: $f(x) = |x| + 2$ 2 units up

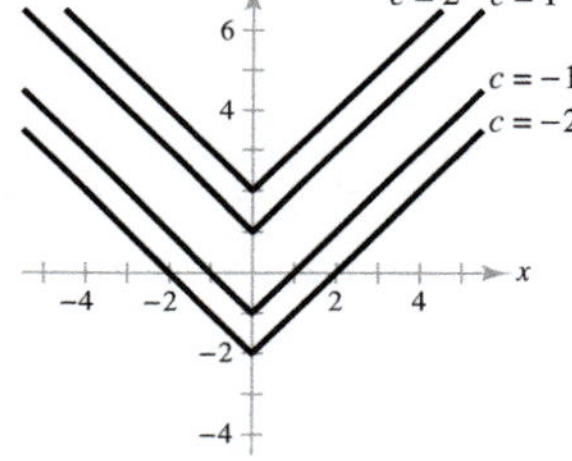

(b) $f(x) = |x - c|$ Horizontal shifts

$c = -2$: $f(x) = |x - (-2)| = |x + 2|$ 2 units left

$c = -1$: $f(x) = |x - (-1)| = |x + 1|$ 1 unit left

$c = 1$: $f(x) = |x - (1)| = |x - 1|$ 1 unit right

$c = 2$: $f(x) = |x - (2)| = |x - 2|$ 2 units right

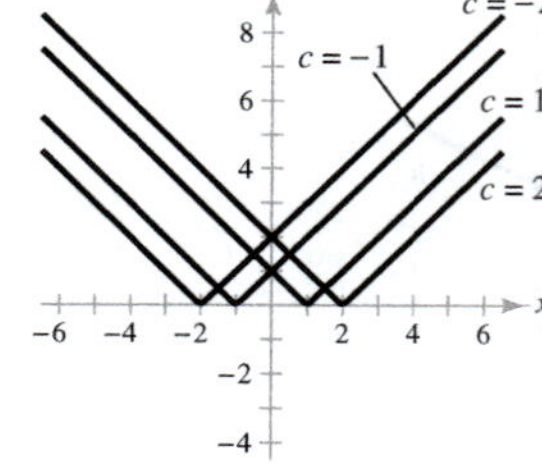

7. (a) $f(x) = [\![x]\!] + c$ Vertical shifts

$c = -4$: $f(x) = [\![x]\!] - 4$ 4 units down

$c = -1$: $f(x) = [\![x]\!] - 1$ 1 unit down

$c = 2$: $f(x) = [\![x]\!] + 2$ 2 units up

$c = 5$: $f(x) = [\![x]\!] + 5$ 5 units up

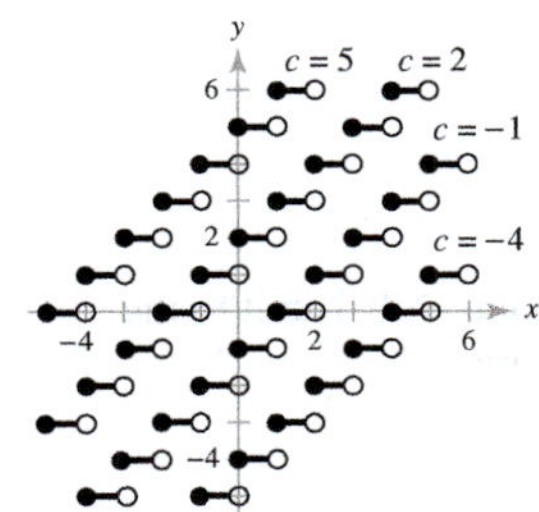

(b) $f(x) = [\![x + c]\!]$ Horizontal shifts

$c = -4$: $f(x) = [\![x - (-4)]\!] = [\![x + 4]\!]$ 4 units left

$c = -1$: $f(x) = [\![x - (-1)]\!] = [\![x + 1]\!]$ 1 unit left

$c = 2$: $f(x) = [\![x - (2)]\!] = [\![x - 2]\!]$ 2 units right

$c = 5$: $f(x) = [\![x - (5)]\!] = [\![x - 5]\!]$ 5 units right

9. (a) $y = f(-x)$

Reflection in the y-axis

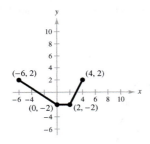

(b) $y = f(x) + 4$

Vertical shift 4 units upward

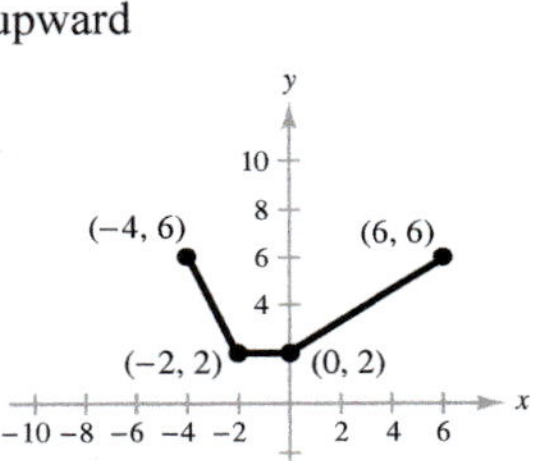

(c) $y = 2f(x)$

Vertical stretch (each y-value is multiplied by 2)

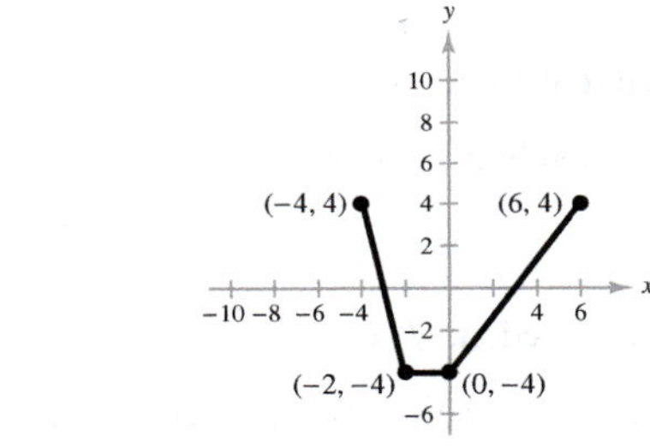

(d) $y = -f(x - 4)$

Reflection in the x-axis and a horizontal shift 4 units to the right

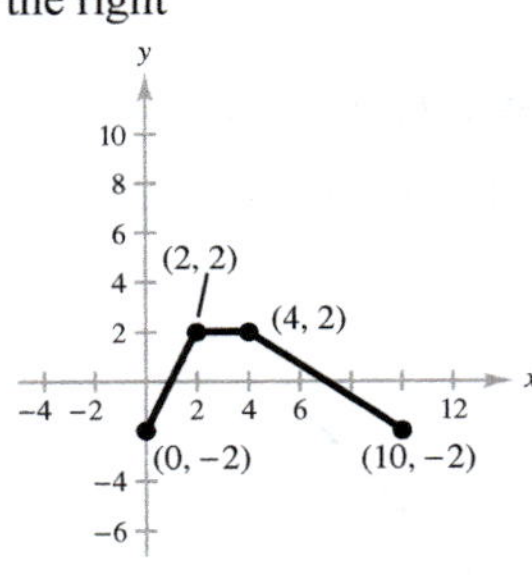

(e) $y = f(x) - 3$

Vertical shift 3 units downward

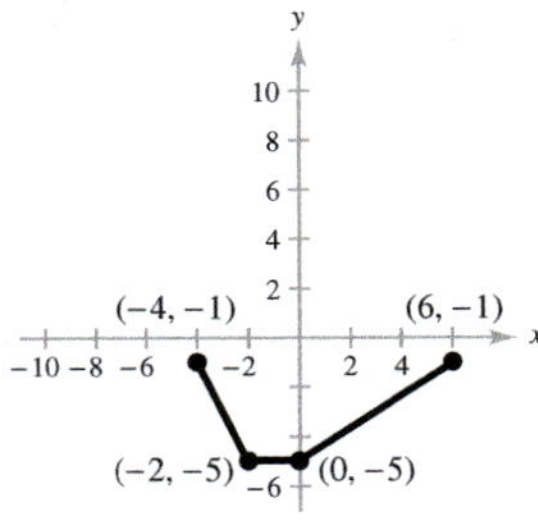

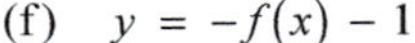

(f) $y = -f(x) - 1$

Reflection in the x-axis and a vertical shift 1 unit downward

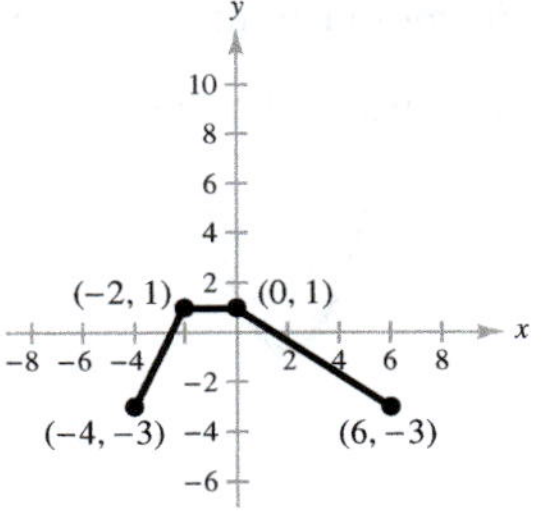

(g) $y = f(2x)$

Horizontal shrink (each x-value is divided by 2)

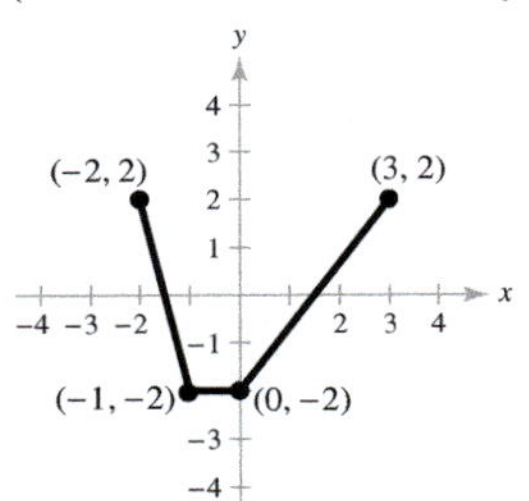

11. Parent function: $f(x) = x^2$

(a) Vertical shift 1 unit downward

$g(x) = x^2 - 1$

(b) Reflection in the x-axis, horizontal shift 1 unit to the left, and a vertical shift 1 unit upward

$g(x) = -(x + 1)^2 + 1$

13. Parent function: $f(x) = |x|$

(a) Reflection in the x-axis and a horizontal shift 3 units to the left

$g(x) = -|x + 3|$

(b) Horizontal shift 2 units to the right and a vertical shift 4 units downward

$g(x) = |x - 2| - 4$

15. Parent function: $f(x) = x^3$

Horizontal shift 2 units to the right

$y = (x - 2)^3$

17. Parent function: $f(x) = x^2$

Reflection in the x-axis

$y = -x^2$

19. Parent function: $f(x) = \sqrt{x}$

Reflection in the x-axis and a vertical shift 1 unit upward

$y = -\sqrt{x} + 1$

21. $g(x) = x^2 + 6$

(a) Parent function: $f(x) = x^2$

(b) A vertical shift 6 units upward

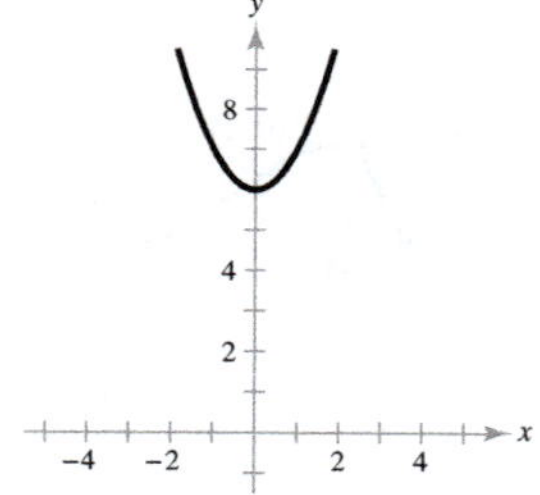

(d) $g(x) = f(x) + 6$

23. $g(x) = -(x - 2)^3$

(a) Parent function: $f(x) = x^3$

(b) Horizontal shift of 2 units to the right and a reflection in the x-axis

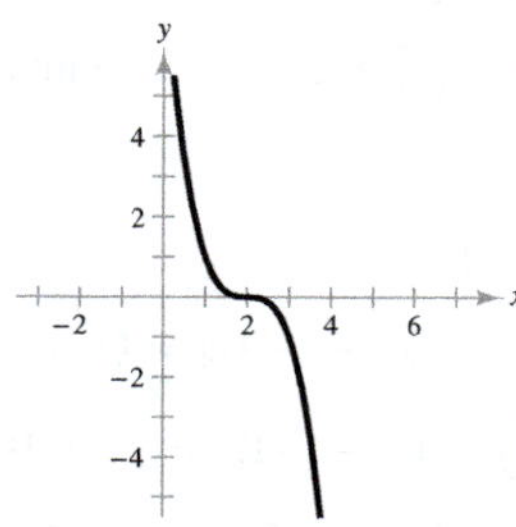

(d) $g(x) = -f(x - 2)$

25. $g(x) = -3 - (x + 1)^2$

(a) Parent function: $f(x) = x^2$

(b) Reflection in the x-axis, a vertical shift 3 units downward and a horizontal shift 1 unit left

(c)

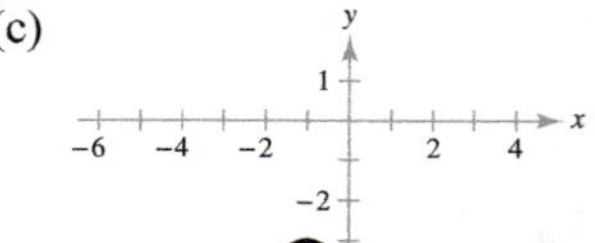

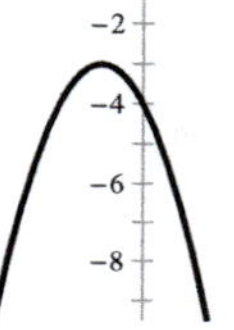

(d) $g(x) = -f(x + 1) - 3$

27. $g(x) = |x - 1| + 2$

(a) Parent function: $f(x) = |x|$

(b) A horizontal shift 1 unit right and a vertical shift 2 units upward

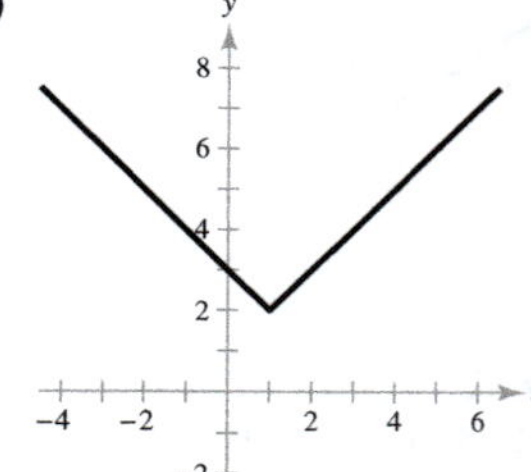

(d) $g(x) = f(x - 1) + 2$

29. $g(x) = 2\sqrt{x}$

(a) Parent function: $f(x) = \sqrt{x}$

(b) A vertical stretch (each y value is multiplied by 2)

(c)

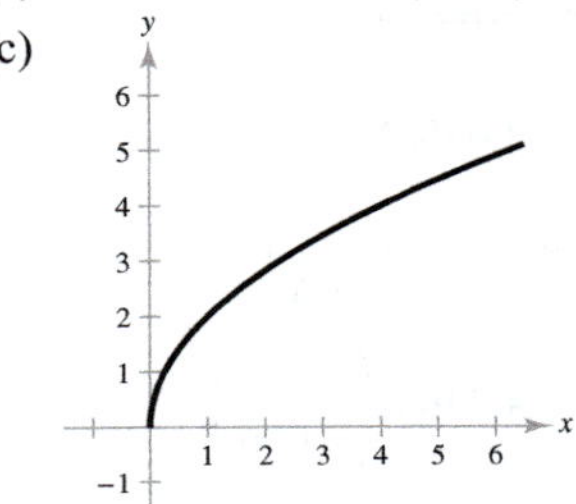

(d) $g(x) = 2f(x)$

31. $g(x) = 2[\![x]\!] - 1$

(a) Parent function: $f(x) = [\![x]\!]$

(b) A vertical shift of 1 unit downward and a vertical stretch (each y value is multiplied by 2)

(c)

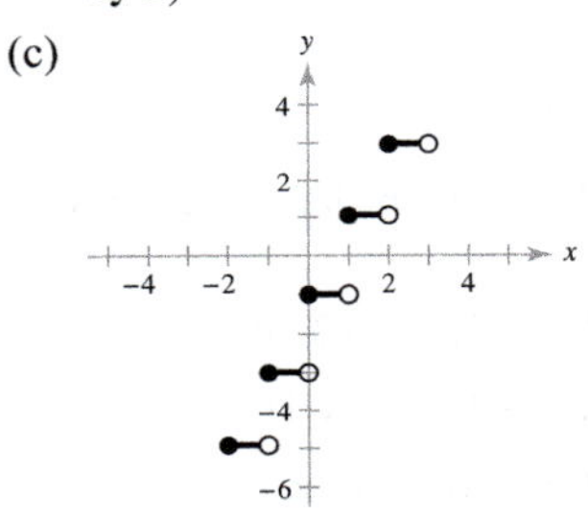

(d) $g(x) = 2f(x) - 1$

33. $g(x) = |2x|$

(a) Parent function: $f(x) = |x|$

(b) A horizontal shrink

(c)

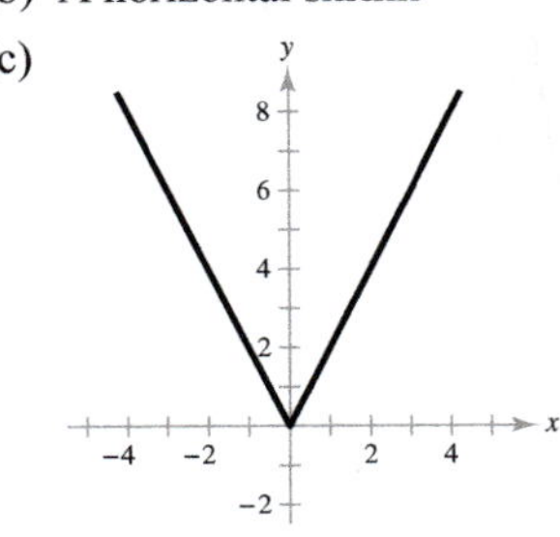

(d) $g(x) = f(2x)$

35. $g(x) = -2x^2 + 1$

(a) Parent function: $f(x) = x^2$

(b) A vertical stretch, reflection in the x-axis and a vertical shift 1 unit upward

(c)

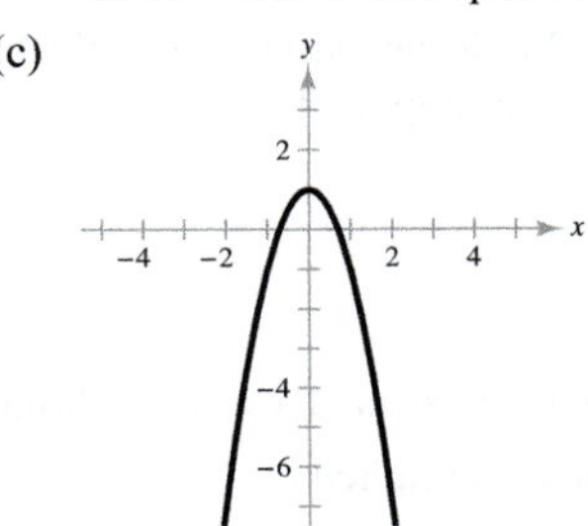

(d) $g(x) = -2f(x) + 1$

37. $g(x) = 3|x - 1| + 2$

(a) Parent function: $f(x) = |x|$

(b) A horizontal shift of 1 unit to the right, a vertical stretch, and a vertical shift 2 units upward

(c)

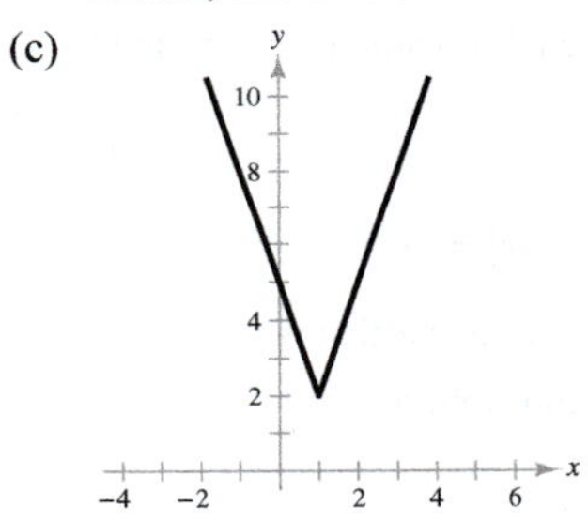

(d) $g(x) = 3f(x - 1) + 2$

39. $g(x) = (x - 3)^2 - 7$

41. $f(x) = x^3$ moved 13 units to the right

$g(x) = (x - 13)^3$

43. $g(x) = -|x| + 12$

45. $f(x) = \sqrt{x}$ moved 6 units to the left and reflected in both the x- and y-axes

$g(x) = -\sqrt{-x + 6}$

47. $f(x) = x^2$

(a) Reflection in the x-axis and a vertical stretch (each y-value is multiplied by 3)

$g(x) = -3x^2$

(b) Vertical shift 3 units upward and a vertical stretch (each y-value is multiplied by 4)

$g(x) = 4x^2 + 3$

49. $f(x) = |x|$

(a) Reflection in the x-axis and a vertical shrink $\left(\text{each } y\text{-value is multiplied by } \frac{1}{2}\right)$

$g(x) = -\frac{1}{2}|x|$

(b) Vertical stretch (each y-value is multiplied by 3) and a vertical shift 3 units downward

$g(x) = 3|x| - 3$

51. Parent function: $f(x) = x^3$

Vertical stretch (each y-value is multiplied by 2)

$g(x) = 2x^3$

53. Parent function: $f(x) = x^2$

Reflection in the x-axis, vertical shrink $\left(\text{each } y\text{-value is multiplied by } \frac{1}{2}\right)$

$g(x) = -\frac{1}{2}x^2$

55. Parent function: $f(x) = \sqrt{x}$

Reflection in the y-axis, vertical shrink $\left(\text{each } y\text{-value is multiplied by } \frac{1}{2}\right)$

$g(x) = \frac{1}{2}\sqrt{-x}$

57. Parent function: $f(x) = x^3$

Reflection in the x-axis, horizontal shift 2 units to the right and a vertical shift 2 units upward

$g(x) = -(x - 2)^3 + 2$

59. Parent function: $f(x) = \sqrt{x}$

Reflection in the x-axis and a vertical shift 3 units downward

$g(x) = -\sqrt{x} - 3$

61. (a)

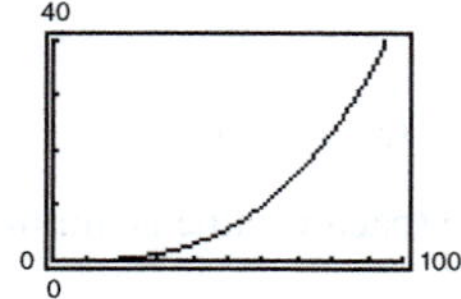

(b) $H(x) = 0.00004636x^3$

$$H\left(\frac{x}{1.6}\right) = 0.00004636\left(\frac{x}{1.6}\right)^3$$
$$= 0.00004636\left(\frac{x^3}{4.096}\right)$$
$$= 0.0000113184x^3 = 0.00001132x^3$$

The graph of $H\left(\frac{x}{1.6}\right)$ is a horizontal stretch of the graph of $H(x)$.

63. False. $y = f(-x)$ is a reflection in the y-axis.

65. True. Because $|x| = |-x|$, the graphs of $f(x) = |x| + 6$ and $f(x) = |-x| + 6$ are identical.

67. $y = f(x + 2) - 1$

Horizontal shift 2 units to the left and a vertical shift 1 unit downward

$(0, 1) \to (0 - 2, 1 - 1) = (-2, 0)$

$(1, 2) \to (1 - 2, 2 - 1) = (-1, 1)$

$(2, 3) \to (2 - 2, 3 - 1) = (0, 2)$

69. Since the graph of $g(x)$ is a horizontal shift one unit to the right of $f(x) = x^3$, the equation should be $g(x) = (x - 1)^3$ and not $g(x) = (x + 1)^3$.

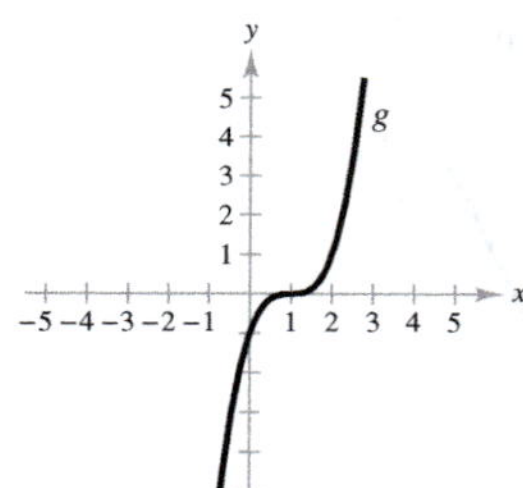

71. (a) The profits were only $\frac{3}{4}$ as large as expected:

$g(t) = \frac{3}{4}f(t)$

(b) The profits were \$10,000 greater than predicted:

$g(t) = f(t) + 10{,}000$

(c) There was a two-year delay: $g(t) = f(t - 2)$

Section 1.8 Combinations of Functions: Composite Functions

1. addition; subtraction; multiplication; division

3.

x	0	1	2	3
f	2	3	1	2
g	-1	0	$\frac{1}{2}$	0
$f + g$	1	3	$\frac{3}{2}$	2

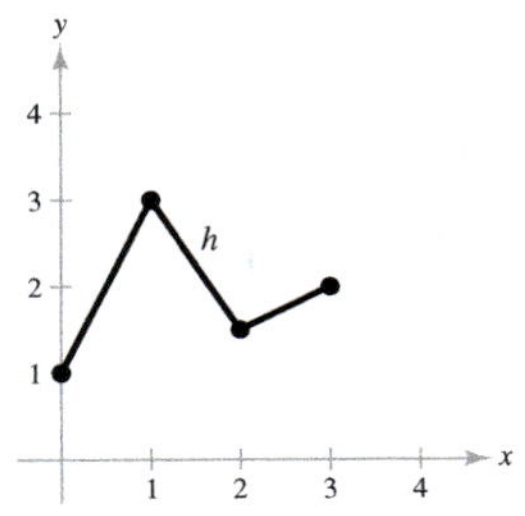

5. $f(x) = x + 2,\ g(x) = x - 2$

(a) $(f + g)(x) = f(x) + g(x)$
$= (x + 2) + (x - 2)$
$= 2x$

(b) $(f - g)(x) = f(x) - g(x)$
$= (x + 2) - (x - 2)$
$= 4$

(c) $(fg)(x) = f(x) \cdot g(x)$
$= (x + 2)(x - 2)$
$= x^2 - 4$

(d) $\left(\frac{f}{x}\right)(x) = \frac{f(x)}{g(x)} = \frac{x + 2}{x - 2}$

Domain: all real numbers x except $x = 2$

7. $f(x) = x^2,\ g(x) = 4x - 5$

(a) $(f + g)(x) = f(x) + g(x)$
$= x^2 + (4x - 5)$
$= x^2 + 4x - 5$

(b) $(f - g)(x) = f(x) - g(x)$
$= x^2 - (4x - 5)$
$= x^2 - 4x + 5$

(c) $(fg)(x) = f(x) \cdot g(x)$
$= x^2(4x - 5)$
$= 4x^3 - 5x^2$

(d) $\left(\frac{f}{g}\right)(x) = \frac{f(x)}{g(x)}$
$= \frac{x^2}{4x - 5}$

Domain: all real numbers x except $x = \frac{5}{4}$

9. $f(x) = x^2 + 6,\ g(x) = \sqrt{1 - x}$

(a) $(f + g)(x) = f(x) + g(x) = x^2 + 6 + \sqrt{1 - x}$

(b) $(f - g)(x) = f(x) - g(x) = x^2 + 6 - \sqrt{1 - x}$

(c) $(fg)(x) = f(x) \cdot g(x) = (x^2 + 6)\sqrt{1 - x}$

(d) $\left(\frac{f}{g}\right)(x) = \frac{f(x)}{g(x)} = \frac{x^2 + 6}{\sqrt{1 - x}} = \frac{(x^2 + 6)\sqrt{1 - x}}{1 - x}$

Domain: $x < 1$

11. $f(x) = \frac{x}{x + 1},\ g(x) = x^3$

(a) $(f + g)(x) = \frac{x}{x + 1} + x^3 = \frac{x + x^4 + x^3}{x + 1}$

(b) $(f - g)(x) = \frac{x}{x + 1} - x^3 = \frac{x - x^4 - x^3}{x + 1}$

(c) $(fg)(x) = \frac{x}{x + 1} \cdot x^3 = \frac{x^4}{x + 1}$

(d) $\left(\frac{f}{g}\right)(x) = \frac{x}{x + 1} \div x^3 = \frac{x}{x + 1} \cdot \frac{1}{x^3} = \frac{1}{x^2(x + 1)}$

Domain: all real numbers x except $x = 0$ and $x = -1$

For Exercises 13–23, $f(x) = x + 3$ and $g(x) = x^2 - 2$.

13. $(f + g)(2) = f(2) + g(2)$
$= (2 + 3) + (2^2 - 2)$
$= 7$

15. $(f - g)(0) = f(0) - g(0)$
$= (0 + 3) - ((0)^2 - 2)$
$= 5$

17. $(f - g)(3t) = f(3t) - g(3t)$
$= ((3t) + 3) - ((3t)^2 - 2)$
$= 3t + 3 - (9t^2 - 2)$
$= -9t^2 + 3t + 5$

19. $(fg)(6) = f(6)g(6)$
$= ((6) + 3)((6)^2 - 2)$
$= (9)(34)$
$= 306$

21. $(f/g)(5) = f(5) / g(5)$
$= ((5) + 3) / ((5)^2 - 2)$
$= \dfrac{8}{23}$

23. $(f/g)(-1) - g(3) = f(-1) / g(-1) - g(3)$
$= ((-1) + 3) / ((-1)^2 - 2) - ((3)^2 - 2)$
$= (2/-1) - 7$
$= -2 - 7 = -9$

25. $f(x) = 3x,\ g(x) = -\dfrac{x^3}{10}$

$(f + g)(x) = 3x - \dfrac{x^3}{10}$

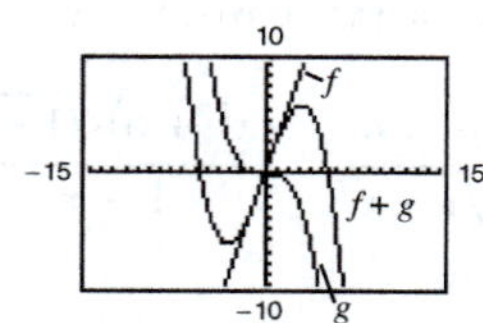

For $0 \le x \le 2$, $f(x)$ contributes most to the magnitude.

For $x > 6$, $g(x)$ contributes most to the magnitude.

27. $f(x) = 3x + 2,\ g(x) = -\sqrt{x + 5}$

$(f + g)x = 3x - \sqrt{x + 5} + 2$

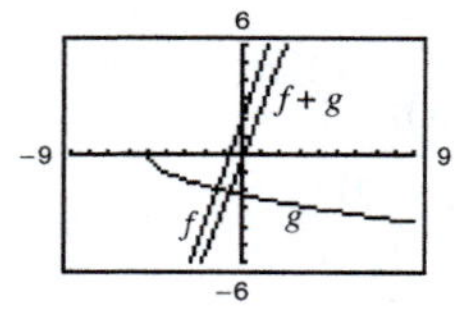

For $0 \le x \le 2$, $f(x)$ contributes most to the magnitude.

For $x > 6$, $f(x)$ contributes most to the magnitude.

29. $f(x) = x + 8,\ g(x) = x - 3$

(a) $(f \circ g)(x) = f(g(x)) = f(x - 3) = (x - 3) + 8 = x + 5$

(b) $(g \circ f)(x) = g(f(x)) = g(x + 8) = (x + 8) - 3 = x + 5$

(c) $(g \circ g)(x) = g(g(x)) = g(x - 3) = (x - 3) - 3 = x - 6$

31. $f(x) = x^2,\ g(x) = x - 1$

(a) $(f \circ g)(x) = f(g(x)) = f(x - 1) = (x - 1)^2$

(b) $(g \circ f)(x) = g(f(x)) = g(x^2) = x^2 - 1$

(c) $(g \circ g)(x) = g(g(x)) = g(x - 1) = x - 2$

33. $f(x) = \sqrt[3]{x-1},\ g(x) = x^3 + 1$

(a) $(f \circ g)(x) = f(g(x))$
$= f(x^3 + 1)$
$= \sqrt[3]{(x^3 + 1) - 1}$
$= \sqrt[3]{x^3} = x$

(b) $(g \circ f)(x) = g(f(x))$
$= g(\sqrt[3]{x-1})$
$= (\sqrt[3]{x-1})^3 + 1$
$= (x - 1) + 1 = x$

(c) $(g \circ g)(x) = g(g(x))$
$= g(x^3 + 1)$
$= (x^3 + 1)^3 + 1$
$= x^9 + 3x^6 + 3x^3 + 2$

35. $f(x) = \sqrt{x+4}$ Domain: $x \ge -4$
$g(x) = x^2$ Domain: all real numbers x

(a) $(f \circ g)(x) = f(g(x)) = f(x^2) = \sqrt{x^2 + 4}$
Domain: all real numbers x

(b) $(g \circ f)(x) = g(f(x)) = g(\sqrt{x+4}) = (\sqrt{x+4})^2 = x + 4$
Domain: $x \ge -4$

37. $f(x) = x^3$ Domain: all real numbers x
$g(x) = x^{2/3}$ Domain: all real numbers x

(a) $(f \circ g)(x) = f(g(x)) = f(x^{2/3}) = (x^{2/3})^3 = x^2$
Domain: all real numbers x.

(b) $(g \circ f)(x) = g(f(x)) = g(x^3) = (x^3)^{2/3} = x^2$
Domain: all real numbers x.

39. $f(x) = |x|$ Domain: all real numbers x
$g(x) = x + 6$ Domain: all real numbers x

(a) $(f \circ g)(x) = f(g(x)) = f(x + 6) = |x + 6|$
Domain: all real numbers x

(b) $(g \circ f)(x) = g(f(x)) = g(|x|) = |x| + 6$
Domain: all real numbers x

41. $f(x) = \dfrac{1}{x}$ Domain: all real numbers x except $x = 0$
$g(x) = x + 3$ Domain: all real numbers x

(a) $(f \circ g)(x) = f(g(x)) = f(x + 3) = \dfrac{1}{x+3}$
Domain: all real numbers x except $x = -3$

(b) $(g \circ f)(x) = g(f(x)) = g\left(\dfrac{1}{x}\right) = \dfrac{1}{x} + 3$
Domain: all real numbers x except $x = 0$

43. $f(x) = \frac{1}{2}x,\ g(x) = x - 4$

(a)

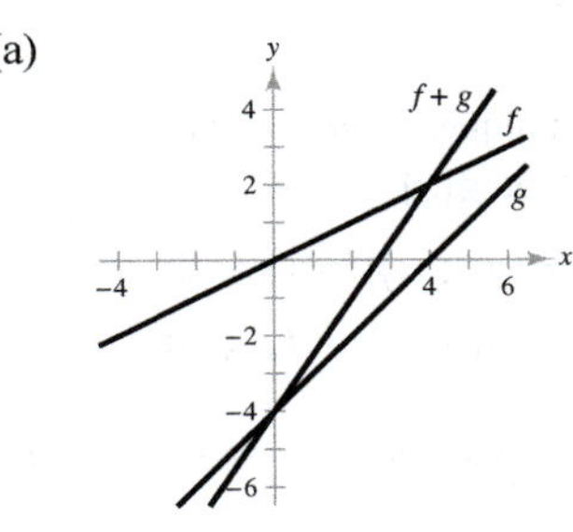

(b)

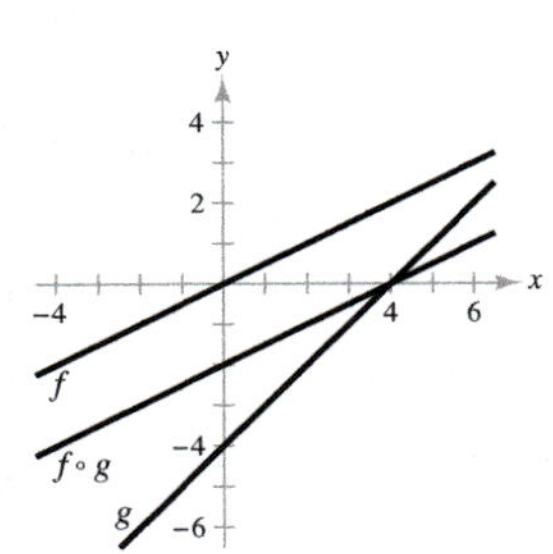

45. (a) $(f + g)(3) = f(3) + g(3) = 2 + 1 = 3$

(b) $\left(\dfrac{f}{g}\right)(2) = \dfrac{f(2)}{g(2)} = \dfrac{0}{2} = 0$

47. (a) $(f \circ g)(2) = f(g(2)) = f(2) = 0$

(b) $(g \circ f)(2) = g(f(2)) = g(0) = 4$

49. $h(x) = (2x^2 + 1)^2$

One possibility: Let $f(x) = x^2$ and $g(x) = 2x + 1$, then $(f \circ g)(x) = h(x)$.

51. $h(x) = \sqrt[3]{x^2 - 4}$

One possibility: Let $f(x) = \sqrt[3]{x}$ and $g(x) = x^2 - 4$, then $(f \circ g)(x) = h(x)$.

53. $h(x) = \dfrac{1}{x + 2}$

One possibility: Let $f(x) = 1/x$ and $g(x) = x + 2$, then $(f \circ g)(x) = h(x)$.

55. $h(x) = \dfrac{-x^2 + 3}{4 - x^2}$

One possibility: Let $f(x) = \dfrac{x + 3}{4 + x}$ and $g(x) = -x^2$, then $(f \circ g)(x) = h(x)$.

57. (a) $T(x) = R(x) + B(x) = \frac{3}{4}x + \frac{1}{15}x^2$

(b)

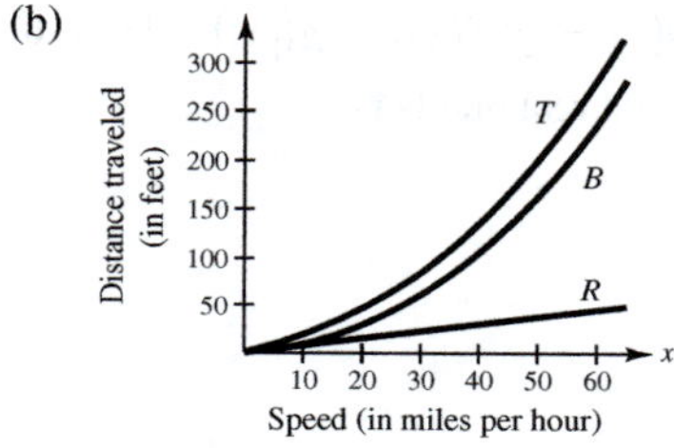

(c) $B(x)$; As x increases, $B(x)$ increases at a faster rate.

59. (a) $c(t) = \dfrac{b(t) - d(t)}{p(t)} \times 100$

(b) $c(16)$ represents the percent change in the population due to births and deaths in the year 2016.

61. (a) $r(x) = \dfrac{x}{2}$

(b) $A(r) = \pi r^2$

(c) $(A \circ r)(x) = A(r(x)) = A\left(\dfrac{x}{2}\right) = \pi\left(\dfrac{x}{2}\right)^2$

$(A \circ r)(x)$ represents the area of the circular base of the tank on the square foundation with side length x.

63. (a) $f(g(x)) = f(0.03x) = 0.03x - 500{,}000$

(b) $g(f(x)) = g(x - 500{,}000) = 0.03(x - 500{,}000)$

$g(f(x))$ represents your bonus of 3% of an amount over \$500,000.

65. False. $(f \circ g)(x) = 6x + 1$ and $(g \circ f)(x) = 6x + 6$

67. Let O = oldest sibling, M = middle sibling, Y = youngest sibling.

Then the ages of each sibling can be found using the equations:

$O = 2M$

$M = \frac{1}{2}Y + 6$

(a) $O(M(Y)) = 2\left(\frac{1}{2}(Y) + 6\right) = 12 + Y$; Answers will vary.

(b) Oldest sibling is 16: $O = 16$

Middle sibling: $O = 2M$

$16 = 2M$

$M = 8$ years old

Youngest sibling: $M = \frac{1}{2}Y + 6$

$8 = \frac{1}{2}Y + 6$

$2 = \frac{1}{2}Y$

$Y = 4$ years old

69. Let $f(x)$ and $g(x)$ be two odd functions and define $h(x) = f(x)g(x)$. Then

$$\begin{aligned} h(-x) &= f(-x)g(-x) \\ &= [-f(x)][-g(x)] \quad \text{because } f \text{ and } g \text{ are odd} \\ &= f(x)g(x) \\ &= h(x). \end{aligned}$$

So, $h(x)$ is even.

Let $f(x)$ and $g(x)$ be two even functions and define $h(x) = f(x)g(x)$. Then

$$\begin{aligned} h(-x) &= f(-x)g(-x) \\ &= f(x)g(x) \quad \text{because } f \text{ and } g \text{ are even} \\ &= h(x). \end{aligned}$$

So, $h(x)$ is even.

71. (a) Answer not unique. *Sample answer*:

$f(x) = x + 3,\ g(x) = x + 2$

$(f \circ g)(x) = f(g(x)) = (x + 2) + 3 = x + 5$

$(g \circ f)(x) = g(f(x)) = (x + 3) + 2 = x + 5$

(b) Answer not unique. *Sample answer*: $f(x) = x^2$, $g(x) = x^3$

$(f \circ g)(x) = f(g(x)) = (x^3)^2 = x^6$

$(g \circ f)(x) = g(f(x)) = (x^2)^3 = x^6$

73. (a) $g(x) = \frac{1}{2}[f(x) + f(-x)]$

To determine if $g(x)$ is even, show $g(-x) = g(x)$.

$g(-x) = \frac{1}{2}[f(-x) + f(-(-x))] = \frac{1}{2}[f(-x) + f(x)] = \frac{1}{2}[f(x) + f(-x)] = g(x)$ ✓

$h(x) = \frac{1}{2}[f(x) - f(-x)]$

To determine if $h(x)$ is odd show $h(-x) = -h(x)$.

$h(-x) = \frac{1}{2}[f(-x) - f(-(-x))] = \frac{1}{2}[f(-x) - f(x)] = -\frac{1}{2}[f(x) - f(-x)] = -h(x)$ ✓

(b) Let $f(x) = a$ function

$f(x) =$ even function + odd function.

Using the result from part (a) $g(x)$ is an even function and $h(x)$ is an odd function.

$f(x) = g(x) + h(x) = \frac{1}{2}[f(x) + f(-x)] + \frac{1}{2}[f(x) - f(-x)] = \frac{1}{2}f(x) + \frac{1}{2}f(-x) + \frac{1}{2}f(x) - \frac{1}{2}f(-x) = f(x)$ ✓

(c) $f(x) = x^2 - 2x + 1$

$f(x) = g(x) + h(x)$

$$g(x) = \frac{1}{2}[f(x) + f(-x)] = \frac{1}{2}\left[x^2 - 2x + 1 + (-x)^2 - 2(-x) + 1\right]$$
$$= \frac{1}{2}\left[x^2 - 2x + 1 + x^2 + 2x + 1\right] = \frac{1}{2}\left[2x^2 + 2\right] = x^2 + 1$$

$$h(x) = \frac{1}{2}[f(x) - f(-x)] = \frac{1}{2}\left[x^2 - 2x + 1 - \left((-x)^2 - 2(-x) + 1\right)\right]$$
$$= \frac{1}{2}\left[x^2 - 2x + 1 - x^2 - 2x - 1\right] = \frac{1}{2}[-4x] = -2x$$

$f(x) = (x^2 + 1) + (-2x)$

$k(x) = \dfrac{1}{x + 1}$

$k(x) = g(x) + h(x)$

$$g(x) = \frac{1}{2}[k(x) + k(-x)] = \frac{1}{2}\left[\frac{1}{x+1} + \frac{1}{-x+1}\right]$$
$$= \frac{1}{2}\left[\frac{1 - x + x + 1}{(x+1)(1-x)}\right] = \frac{1}{2}\left[\frac{2}{(x+1)(1-x)}\right]$$
$$= \frac{1}{(x+1)(1-x)} = \frac{-1}{(x+1)(x-1)}$$

$$h(x) = \frac{1}{2}[k(x) - k(-x)] = \frac{1}{2}\left[\frac{1}{x+1} - \frac{1}{1-x}\right]$$
$$= \frac{1}{2}\left[\frac{1 - x - (x+1)}{(x+1)(1-x)}\right] = \frac{1}{2}\left[\frac{-2x}{(x+1)(1-x)}\right]$$
$$= \frac{-x}{(x+1)(1-x)} = \frac{x}{(x+1)(x-1)}$$

$$k(x) = \left(\frac{-1}{(x+1)(x-1)}\right) + \left(\frac{x}{(x+1)(x-1)}\right)$$

Section 1.9 Inverse Functions

1. inverse

3. range; domain

5. one-to-one

7. $f(x) = 6x$

$f^{-1}(x) = \frac{x}{6} = \frac{1}{6}x$

$f(f^{-1}(x)) = f\left(\frac{x}{6}\right) = 6\left(\frac{x}{6}\right) = x$

$f^{-1}(f(x)) = f^{-1}(6x) = \frac{6x}{6} = x$

9. $f(x) = 3x + 1$

$f^{-1}(x) = \frac{x - 1}{3}$

$f(f^{-1}(x)) = f\left(\frac{x - 1}{3}\right) = 3\left(\frac{x - 1}{3}\right) + 1 = x$

$f^{-1}(f(x)) = f^{-1}(3x + 1) = \frac{(3x + 1) - 1}{3} = x$

11. $f(x) = x^2 - 4, x \geq 0$

$f^{-1}(x) = \sqrt{x + 4}$

$f(f^{-1}(x)) = f\left(\sqrt{x + 4}\right) = \left(\sqrt{x + 4}\right)^2 - 4 = (x + 4) - 4 = x$

$f^{-1}(f(x)) = f^{-1}(x^2 - 4) = \sqrt{(x^2 - 4) + 4} = \sqrt{x^2} = x$

13. $f(x) = x^3 + 1$

$f^{-1}(x) = \sqrt[3]{x - 1}$

$f(f^{-1}(x)) = f\left(\sqrt[3]{x - 1}\right) = \left(\sqrt[3]{x - 1}\right)^3 + 1 = (x - 1) + 1 = x$

$f^{-1}(f(x)) = f^{-1}(x^3 + 1) = \sqrt[3]{(x^3 + 1) - 1} = \sqrt[3]{x^3} = x$

15. $(f \circ g)(x) = f(g(x)) = f(4x + 9) = \frac{4x + 9 - 9}{4} = \frac{4x}{4} = x$

$(g \circ f)(x) = g(f(x)) = g\left(\frac{x - 9}{4}\right) = 4\left(\frac{x - 9}{4}\right) + 9 = x - 9 + 9 = x$

17. $f(g(x)) = f\left(\sqrt[3]{4x}\right) = \frac{\left(\sqrt[3]{4x}\right)^3}{4} = \frac{4x}{4} = x$

$g(f(x)) = g\left(\frac{x^3}{4}\right) = \sqrt[3]{4\left(\frac{x^3}{4}\right)} = \sqrt[3]{x^3} = x$

19.

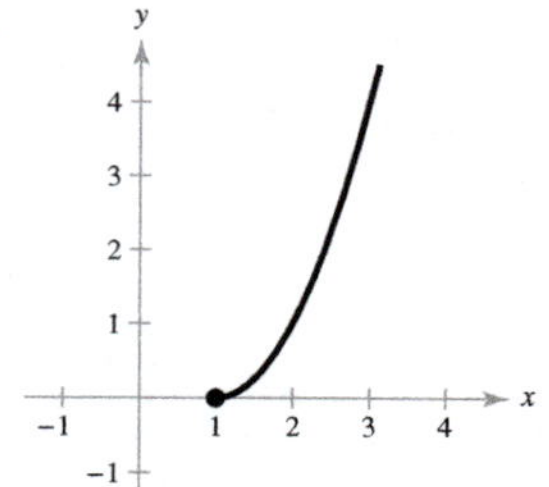

21. $f(x) = x - 5, g(x) = x + 5$

(a) $f(g(x)) = f(x + 5) = (x + 5) - 5 = x$

$g(f(x)) = g(x - 5) = (x - 5) + 5 = x$

(b)

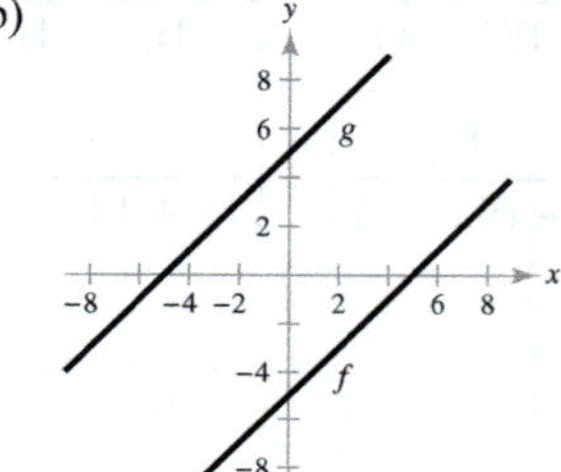

23. $f(x) = 7x + 1, g(x) = \dfrac{x - 1}{7}$

(a) $f(g(x)) = f\left(\dfrac{x - 1}{7}\right) = 7\left(\dfrac{x - 1}{7}\right) + 1 = x$

$g(f(x)) = g(7x + 1) = \dfrac{(7x + 1) - 1}{7} = x$

(b)

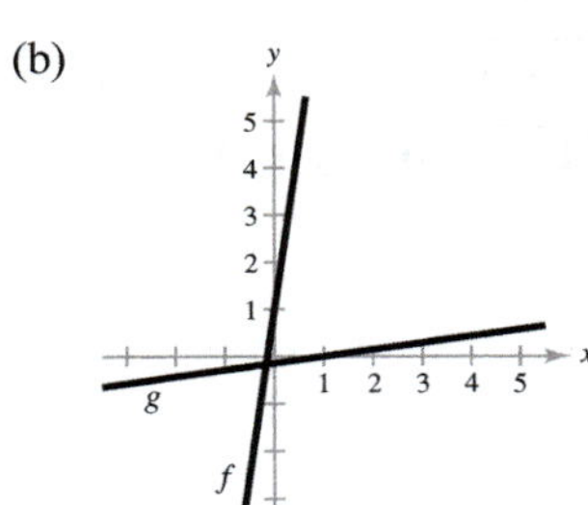

25. $f(x) = x^3, g(x) = \sqrt[3]{x}$

(a) $f(g(x)) = f(\sqrt[3]{x}) = (\sqrt[3]{x})^3 = x$

$g(f(x)) = g(x^3) = \sqrt[3]{(x^3)} = x$

(b)

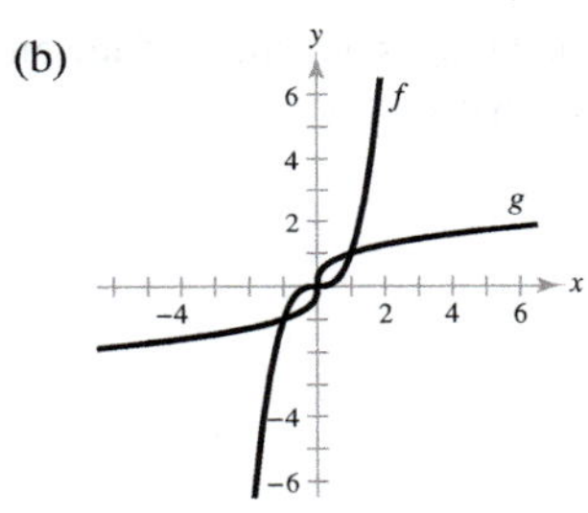

27. $f(x) = \sqrt{x + 5}, g(x) = x^2 - 5, x \geq 0$

(a) $f(g(x)) = f(x^2 - 5), x \geq 0$

$= \sqrt{(x^2 - 5) + 5} = x$

$g(f(x)) = g(\sqrt{x + 5})$

$= (\sqrt{x + 5})^2 - 5 = x$

(b)

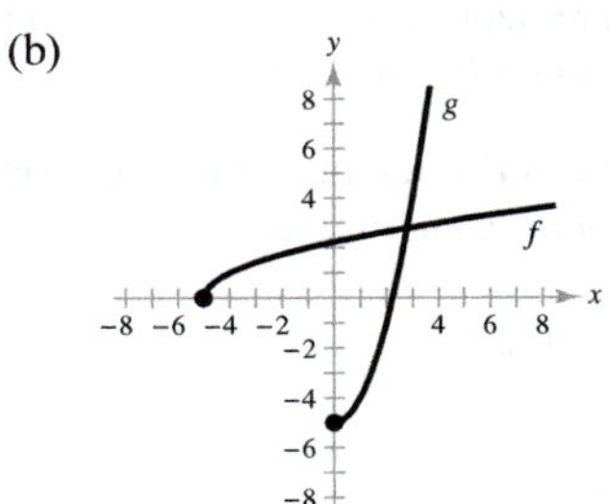

29. $f(x) = \dfrac{1}{x}, g(x) = \dfrac{1}{x}$

(a) $f(g(x)) = f\left(\dfrac{1}{x}\right) = \dfrac{1}{1/x} = 1 \div \dfrac{1}{x} = 1 \cdot \dfrac{x}{1} = x$

$g(f(x)) = g\left(\dfrac{1}{x}\right) = \dfrac{1}{1/x} = 1 \div \dfrac{1}{x} = 1 \cdot \dfrac{x}{1} = x$

(b)

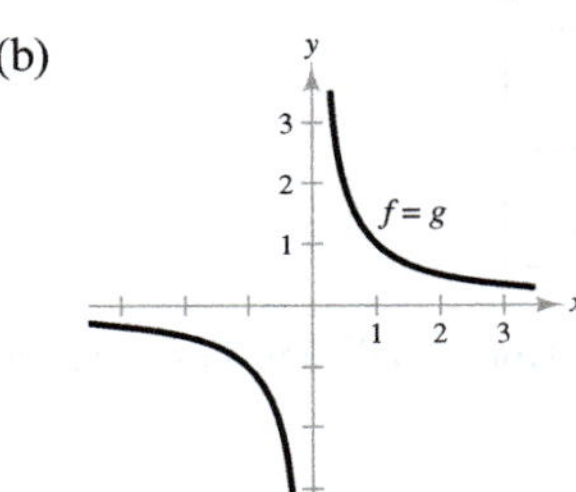

31. $f(x) = \dfrac{x - 1}{x + 5}, g(x) = -\dfrac{5x + 1}{x - 1}$

(a) $f(g(x)) = f\left(-\dfrac{5x + 1}{x - 1}\right) = \dfrac{\left(-\dfrac{5x + 1}{x - 1} - 1\right)}{\left(-\dfrac{5x + 1}{x - 1} + 5\right)} \cdot \dfrac{x - 1}{x - 1} = \dfrac{-(5x + 1) - (x - 1)}{-(5x + 1) + 5(x - 1)} = \dfrac{-6x}{-6} = x$

$g(f(x)) = g\left(\dfrac{x - 1}{x + 5}\right) = -\dfrac{\left[5\left(\dfrac{x - 1}{x + 5}\right) + 1\right]}{\left[\dfrac{x - 1}{x + 5} - 1\right]} \cdot \dfrac{x + 5}{x + 5} = -\dfrac{5(x - 1) + (x + 5)}{(x - 1) - (x + 5)} = -\dfrac{6x}{-6} = x$

(b)

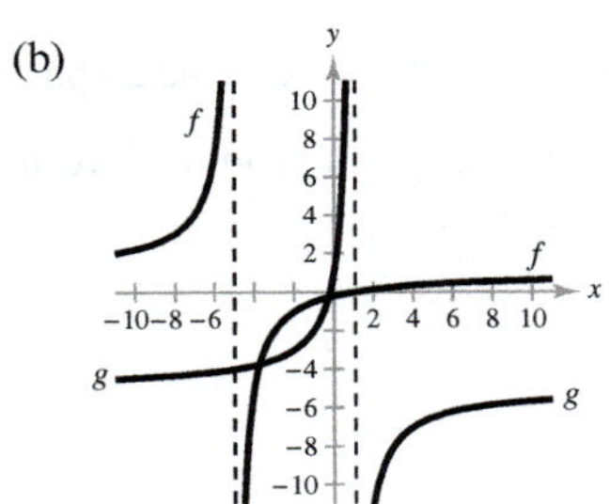

33. No, $\{(-2, -1), (1, 0), (2, 1), (1, 2), (-2, 3), (-6, 4)\}$ does not represent a function. -2 and 1 are paired with two different values.

35.

x	3	5	7	9	11	13
$f^{-1}(x)$	-1	0	1	2	3	4

37. Yes, because no horizontal line crosses the graph of *f* at more than one point, *f has* an inverse.

39. No, because some horizontal lines cross the graph of *f* twice, *f does not* have an inverse.

41. $g(x) = (x + 3)^2 + 2$

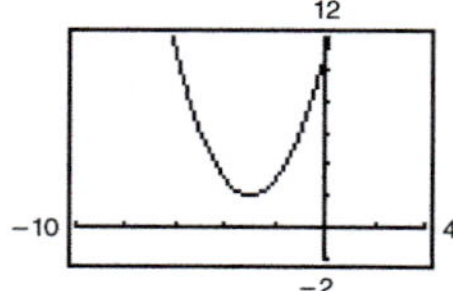

g does not pass the Horizontal Line Test, so *g does not* have an inverse.

43. $f(x) = x\sqrt{9 - x^2}$

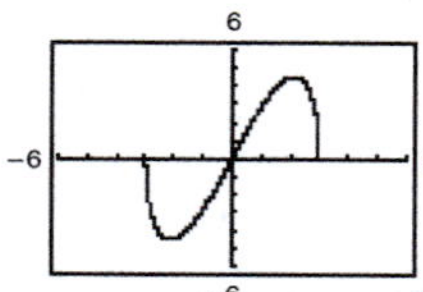

f does not pass the Horizontal Line Test, so *f does not* have an inverse.

45. (a)

$$\begin{aligned} f(x) &= x^5 - 2 \\ y &= x^5 - 2 \\ x &= y^5 - 2 \\ y &= \sqrt[5]{x + 2} \\ f^{-1}(x) &= \sqrt[5]{x + 2} \end{aligned}$$

(b)

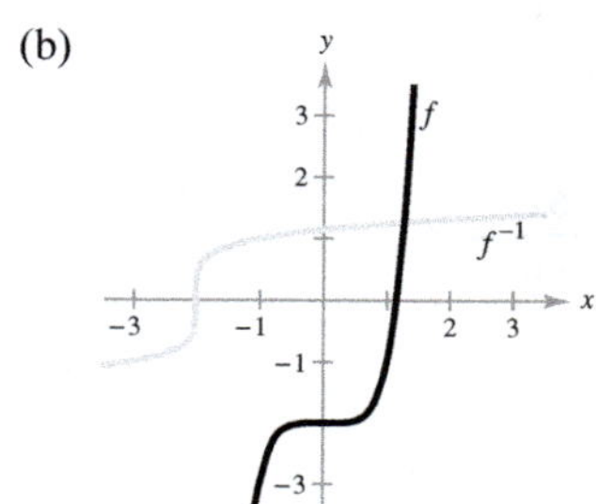

(c) The graph of f^{-1} is the reflection of the graph of *f* in the line $y = x$.

(d) The domains and ranges of *f* and f^{-1} are all real numbers.

47. (a)

$$\begin{aligned} f(x) &= \sqrt{4 - x^2}, 0 \le x \le 2 \\ y &= \sqrt{4 - x^2} \\ x &= \sqrt{4 - y^2} \\ x^2 &= 4 - y^2 \\ y^2 &= 4 - x^2 \\ y &= \sqrt{4 - x^2} \\ f^{-1}(x) &= \sqrt{4 - x^2}, 0 \le x \le 2 \end{aligned}$$

(b)

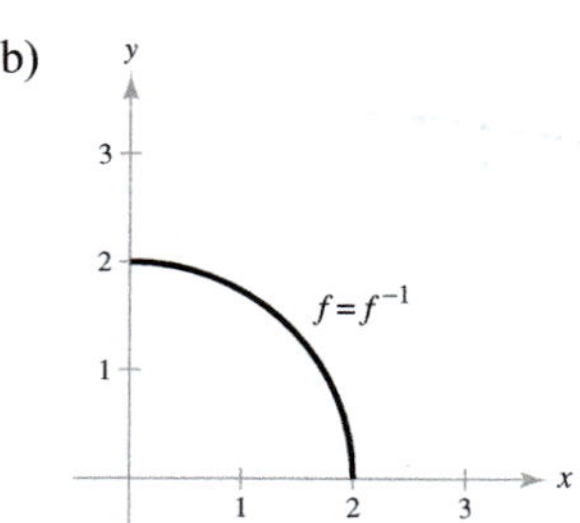

(c) The graph of f^{-1} is the same as the graph of *f*.

(d) The domains and ranges of *f* and f^{-1} are all real numbers *x* such that $0 \le x \le 2$.

49. (a)

$$\begin{aligned} f(x) &= \frac{4}{x} \\ y &= \frac{4}{x} \\ x &= \frac{4}{y} \\ xy &= 4 \\ y &= \frac{4}{x} \\ f^{-1}(x) &= \frac{4}{x} \end{aligned}$$

(b)

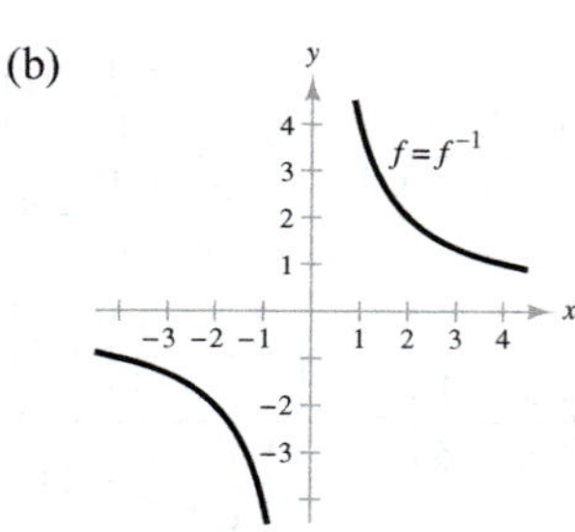

(c) The graph of f^{-1} is the same as the graph of *f*.

(d) The domains and ranges of *f* and f^{-1} are all real numbers except for 0.

51. (a)
$$f(x) = \frac{x+1}{x-2}$$
$$y = \frac{x+1}{x-2}$$
$$x = \frac{y+1}{y-2}$$
$$x(y-2) = y+1$$
$$xy - 2x = y+1$$
$$xy - y = 2x+1$$
$$y(x-1) = 2x+1$$
$$y = \frac{2x+1}{x-1}$$
$$f^{-1}(x) = \frac{2x+1}{x-1}$$

(b)

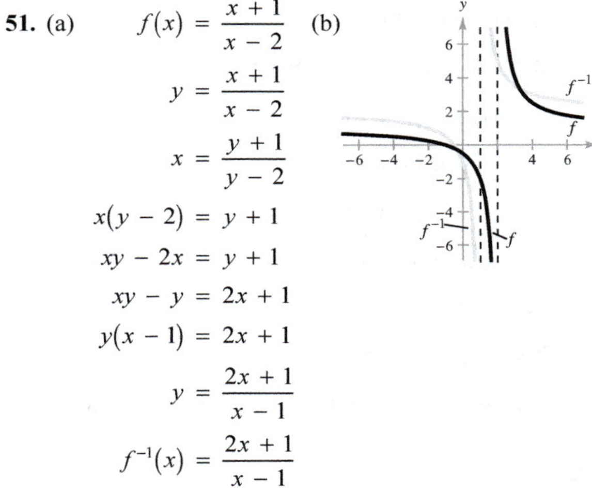

(c) The graph of f^{-1} is the reflection of graph of f in the line $y = x$.

(d) The domain of f and the range of f^{-1} is all real numbers except 2.

The range of f and the domain of f^{-1} is all real numbers except 1.

53. (a)
$$f(x) = \sqrt[3]{x-1}$$
$$y = \sqrt[3]{x-1}$$
$$x = \sqrt[3]{y-1}$$
$$x^3 = y-1$$
$$y = x^3+1$$
$$f^{-1}(x) = x^3+1$$

(b)

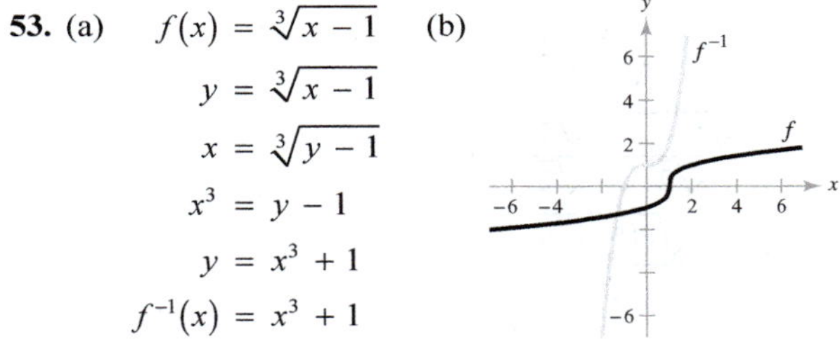

(c) The graph of f^{-1} is the reflection of the graph of f in the line $y = x$.

(d) The domains and ranges of f and f^{-1} are all real numbers.

55.
$$f(x) = x^4$$
$$y = x^4$$
$$x = y^4$$
$$y = \pm\sqrt[4]{x}$$

This does not represent y as a function of x. f does not have an inverse.

57.
$$g(x) = \frac{x+1}{6}$$
$$y = \frac{x+1}{6}$$
$$x = \frac{y+1}{6}$$
$$6x = y+1$$
$$y = 6x-1$$

This is a function of x, so g has an inverse.

$$g^{-1}(x) = 6x-1$$

59.
$$p(x) = -4$$
$$y = -4$$

Because $y = -4$ for all x, the graph is a horizontal line and fails the Horizontal Line Test. p does not have an inverse.

61.
$$f(x) = (x+3)^2, x \ge -3 \Rightarrow y \ge 0$$
$$y = (x+3)^2, x \ge -3, y \ge 0$$
$$x = (y+3)^2, y \ge -3, x \ge 0$$
$$\sqrt{x} = y+3, y \ge -3, x \ge 0$$
$$y = \sqrt{x} - 3, x \ge 0, y \ge -3$$

This is a function of x, so f has an inverse.

$$f^{-1}(x) = \sqrt{x} - 3, x \ge 0$$

63.
$$f(x) = \begin{cases} x+3, & x < 0 \\ 6-x, & x \ge 0 \end{cases}$$

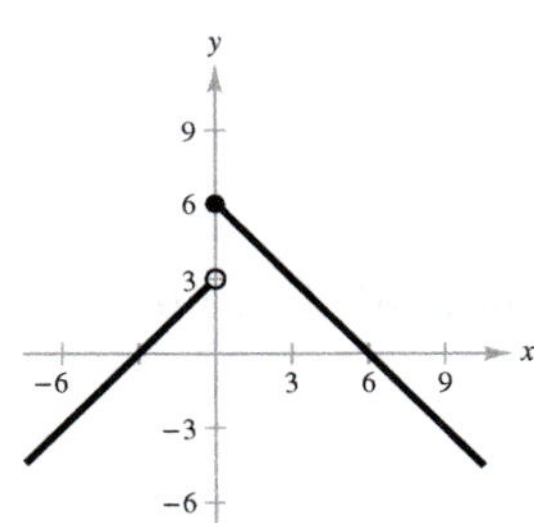

This graph fails the Horizontal Line Test, so f does not have an inverse.

65.
$$h(x) = |x+1| - 1$$

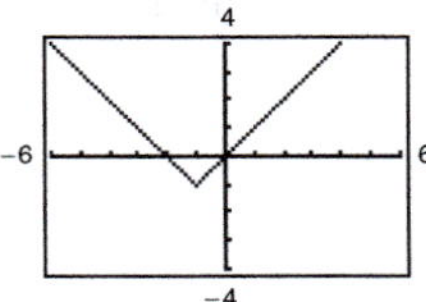

The graph fails the Horizontal Line Test, so h does not have an inverse.

67. $f(x) = \sqrt{2x + 3} \Rightarrow x \geq -\frac{3}{2}, y \geq 0$

$$y = \sqrt{2x + 3}, x \geq -\frac{3}{2}, y \geq 0$$

$$x = \sqrt{2y + 3}, y \geq -\frac{3}{2}, x \geq 0$$

$$x^2 = 2y + 3, x \geq 0, y \geq -\frac{3}{2}$$

$$y = \frac{x^2 - 3}{2}, x \geq 0, y \geq -\frac{3}{2}$$

This is a function of x, so f has an inverse.

$$f^{-1}(x) = \frac{x^2 - 3}{2}, x \geq 0$$

69.
$$f(x) = \frac{6x + 4}{4x + 5}$$
$$y = \frac{6x + 4}{4x + 5}$$
$$x = \frac{6y + 4}{4y + 5}$$
$$x(4y + 5) = 6y + 4$$
$$4xy + 5x = 6y + 4$$
$$4xy - 6y = -5x + 4$$
$$y(4x - 6) = -5x + 4$$
$$y = \frac{-5x + 4}{4x - 6}$$
$$= \frac{5x - 4}{6 - 4x}$$

This is a function of x, so f has an inverse.

$$f^{-1}(x) = \frac{5x - 4}{6 - 4x}$$

71. $f(x) = |x + 2|$

domain of $f: x \geq -2$, range of $f: y \geq 0$

$$f(x) = |x + 2|$$
$$y = |x + 2|$$
$$x = y + 2$$
$$x - 2 = y$$

So, $f^{-1}(x) = x - 2$.

domain of $f^{-1}: x \geq 0$, range of $f^{-1}: y \geq -2$

73. $f(x) = (x + 6)^2$

domain of $f: x \geq -6$, range of $f: y \geq 0$

$$f(x) = (x + 6)^2$$
$$y = (x + 6)^2$$
$$x = (y + 6)^2$$
$$\sqrt{x} = y + 6$$
$$\sqrt{x} - 6 = y$$

So, $f^{-1}(x) = \sqrt{x} - 6$.

domain of $f^{-1}: x \geq 0$, range of $f^{-1}: y \geq -6$

75. $f(x) = -2x^2 + 5$

domain of $f: x \geq 0$, range of $f: y \leq 5$

$$f(x) = -2x^2 + 5$$
$$y = -2x^2 + 5$$
$$x = -2y^2 + 5$$
$$x - 5 = -2y^2$$
$$5 - x = 2y^2$$
$$\sqrt{\frac{5 - x}{2}} = y$$
$$\frac{\sqrt{5 - x}}{\sqrt{2}} \cdot \frac{\sqrt{2}}{\sqrt{2}} = y$$
$$\frac{\sqrt{2(5 - x)}}{2} = y$$

So, $f^{-1}(x) = \frac{\sqrt{-2(x - 5)}}{2}$.

domain of $f^{-1}(x): x \leq 5$, range of $f^{-1}(x): y \geq 0$

77. $f(x) = |x - 4| + 1$

domain of $f: x \geq 4$, range of $f: y \geq 1$

$$f(x) = |x - 4| + 1$$
$$y = x - 3$$
$$x = y - 3$$
$$x + 3 = y$$

So, $f^{-1}(x) = x + 3$.

domain of $f^{-1}: x \geq 1$, range of $f^{-1}: y \geq 4$

In Exercises 79–83, $f(x) = \frac{1}{8}x - 3$, $f^{-1}(x) = 8(x + 3)$, $g(x) = x^3$, $g^{-1}(x) = \sqrt[3]{x}$.

79. $$\begin{aligned}(f^{-1} \circ g^{-1})(1) &= f^{-1}(g^{-1}(1)) \\ &= f^{-1}(\sqrt[3]{1}) \\ &= 8(\sqrt[3]{1} + 3) = 32\end{aligned}$$

81. $$\begin{aligned}(f^{-1} \circ f^{-1})(4) &= f^{-1}(f^{-1}(4)) \\ &= f^{-1}(8[4 + 3]) \\ &= 8[8(4 + 3) + 3] \\ &= 8[8(7) + 3] \\ &= 8(59) = 472\end{aligned}$$

83. $$\begin{aligned}(f \circ g)(x) &= f(g(x)) = f(x^3) = \tfrac{1}{8}x^3 - 3 \\ y &= \tfrac{1}{8}x^3 - 3 \\ x &= \tfrac{1}{8}y^3 - 3 \\ x + 3 &= \tfrac{1}{8}y^3 \\ 8(x + 3) &= y^3 \\ \sqrt[3]{8(x + 3)} &= y \\ (f \circ g)^{-1}(x) &= 2\sqrt[3]{x + 3}\end{aligned}$$

In Exercises 85–87, $f(x) = x + 4$, $f^{-1}(x) = x - 4$, $g(x) = 2x - 5$, $g^{-1}(x) = \dfrac{x + 5}{2}$.

85. $$\begin{aligned}(g^{-1} \circ f^{-1})(x) &= g^{-1}(f^{-1}(x)) \\ &= g^{-1}(x - 4) \\ &= \frac{(x - 4) + 5}{2} \\ &= \frac{x + 1}{2}\end{aligned}$$

87. $$\begin{aligned}(f \circ g)(x) &= f(g(x)) \\ &= f(2x - 5) \\ &= (2x - 5) + 4 \\ &= 2x - 1\end{aligned}$$

$$(f \circ g)^{-1}(x) = \frac{x + 1}{2}$$

Note: Comparing Exercises 85 and 87, $(f \circ g)^{-1}(x) = (g^{-1} \circ f^{-1})(x)$.

89. (a) $$\begin{aligned}y &= 10 + 0.75x \\ x &= 10 + 0.75y \\ x - 10 &= 0.75y \\ \frac{x - 10}{0.75} &= y\end{aligned}$$

So, $f^{-1}(x) = \dfrac{x - 10}{0.75}$.

x = hourly wage, y = number of units produced

(b) $y = \dfrac{24.25 - 10}{0.75} = 19$

So, 19 units are produced.

91. False. $f(x) = x^2$ is even and does not have an inverse.

93.

x	1	3	4	6
f	1	2	6	7

x	1	2	6	7
$f^{-1}(x)$	1	3	4	6

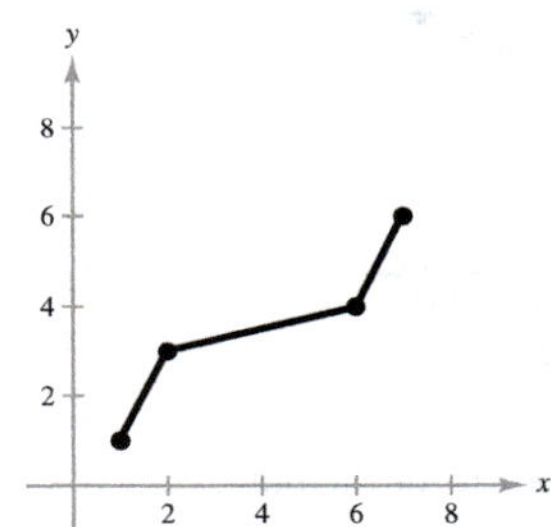

95. Let $(f \circ g)(x) = y$. Then $x = (f \circ g)^{-1}(y)$. Also,

$$\begin{aligned}(f \circ g)(x) = y \Rightarrow f(g(x)) &= y \\ g(x) &= f^{-1}(y) \\ x &= g^{-1}(f^{-1}(y)) \\ x &= (g^{-1} \circ f^{-1})(y).\end{aligned}$$

Because f and g are both one-to-one functions, $(f \circ g)^{-1} = g^{-1} \circ f^{-1}$.

97. If $f(x) = k(2 - x - x^3)$ has an inverse and $f^{-1}(3) = -2$, then $f(-2) = 3$. So,

$$\begin{aligned}f(-2) = k(2 - (-2) - (-2)^3) &= 3 \\ k(2 + 2 + 8) &= 3 \\ 12k &= 3 \\ k &= \tfrac{3}{12} = \tfrac{1}{4}.\end{aligned}$$

So, $k = \frac{1}{4}$.

99.

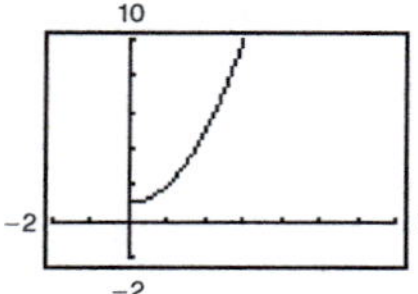

There is an inverse function $f^{-1}(x) = \sqrt{x - 1}$ because the domain of f is equal to the range of f^{-1} and the range of f is equal to the domain of f^{-1}.

101. This situation could be represented by a one-to-one function if the runner does not stop to rest. The inverse function would represent the time in hours for a given number of miles completed.

Section 1.10 Mathematical Modeling and Variation

1. variation; regression

3. least squares regression

5. directly proportional

7. combined

9. (a)

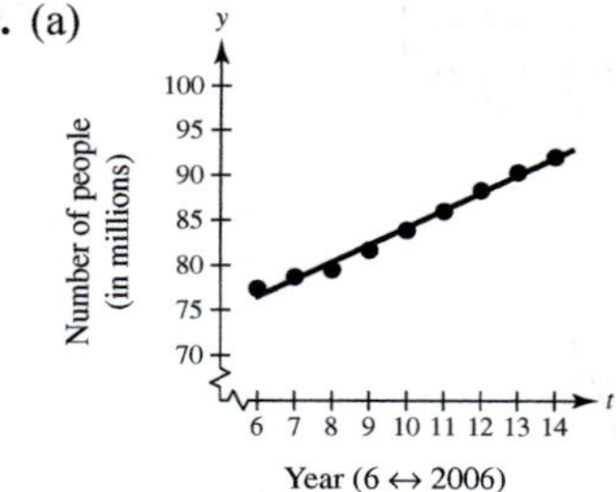

(b) The model is a good fit for the data.

11.

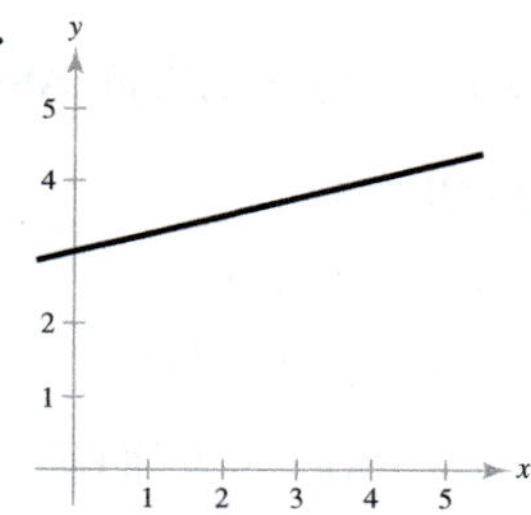

Using the point $(0, 3)$ and $(4, 4)$, $y = \frac{1}{4}x + 3$.

13.

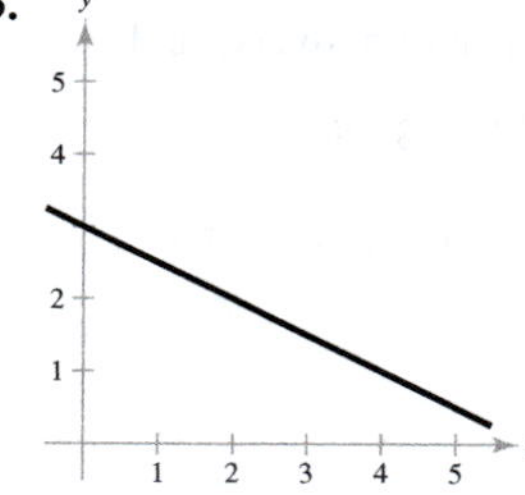

Using the points $(2, 2)$ and $(4, 1)$, $y = -\frac{1}{2}x + 3$.

15.

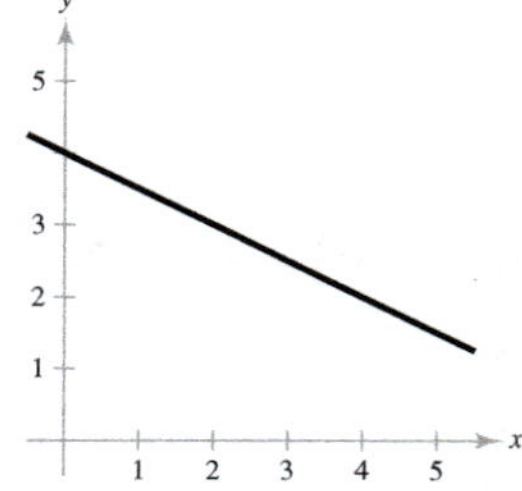

The line appears to pass through $(2, 3)$ and $(4, 2)$ so its equation is $y = -\frac{1}{2}x + 4$.

17. (a) and (b)

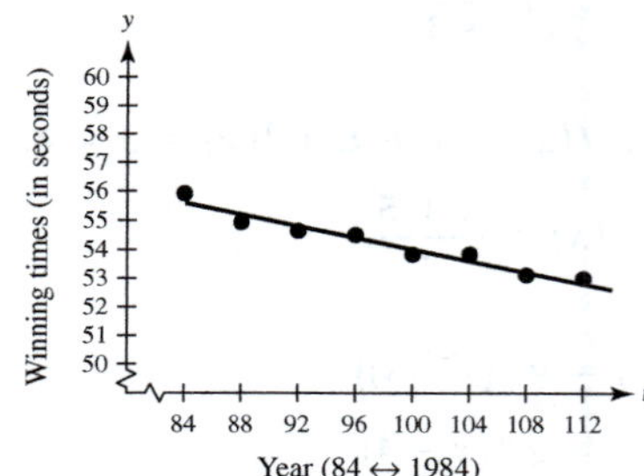

(b) Let 1984 correspond to $t = 84$, using the points $(84, 55.92)$ and $(112, 53.00)$:

$$\begin{aligned} m &= \frac{53.00 - 55.92}{112 - 84} \\ &= -0.1 \\ y - 53.00 &= -0.1(t - 112) \\ y - 53 &= -0.1t + 11.2 \\ y &= -0.1t + 64 \end{aligned}$$

(c) $y = -0.1t + 64.2$

(d) The models are similar.

19. $y = kx$

$14 = k(2)$

$7 = k$

$y = 7x$

21. $y = kx$

$1 = k(5)$

$\frac{1}{5} = k$

$y = \frac{1}{5}x$

23. $y = kx$

$8\pi = k(4)$

$2\pi = k$

$y = 2\pi x$

25. $k = 1$

x	2	4	6	8	10
$y = kx^2$	4	16	36	64	100

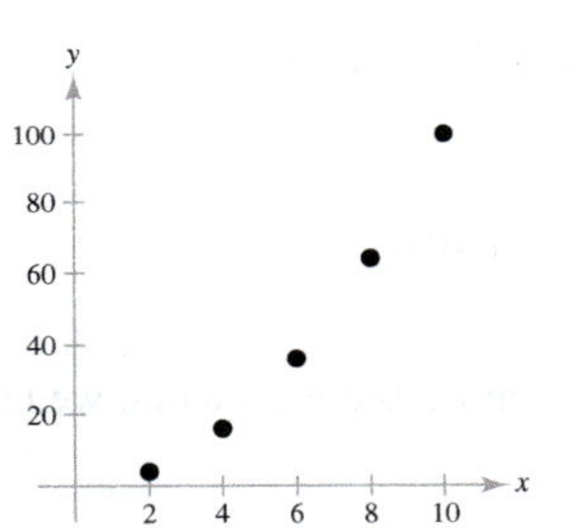

27. $k = \frac{1}{2}$

x	2	4	6	8	10
$y = \frac{1}{2}x^3$	4	32	108	256	500

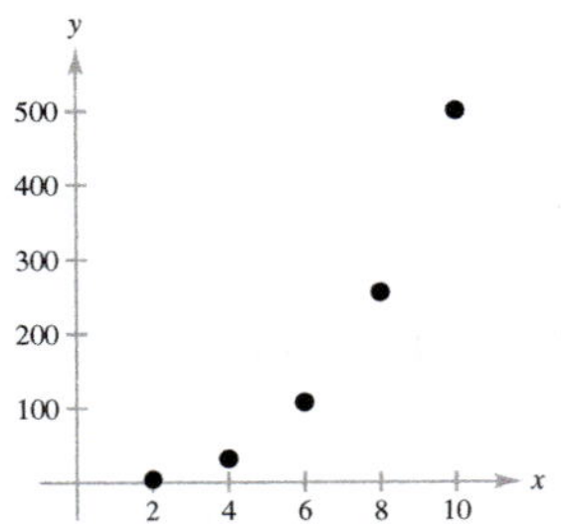

29. $k = 2, n = 1$

x	2	4	6	8	10
$y = \frac{2}{x}$	1	$\frac{1}{2}$	$\frac{1}{3}$	$\frac{1}{4}$	$\frac{1}{5}$

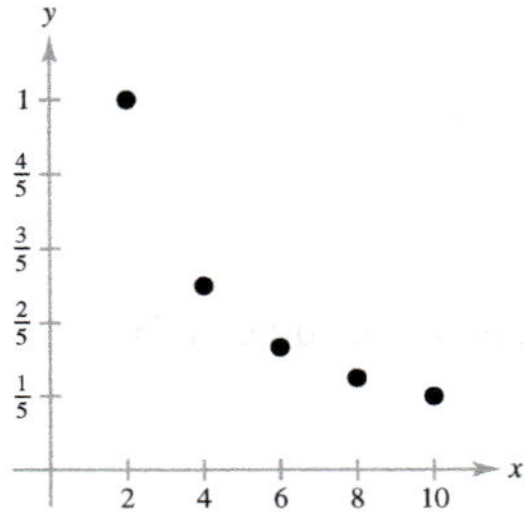

31. $k = 10$

x	2	4	6	8	10
$y = \frac{k}{x^2}$	$\frac{5}{2}$	$\frac{5}{8}$	$\frac{5}{18}$	$\frac{5}{32}$	$\frac{1}{10}$

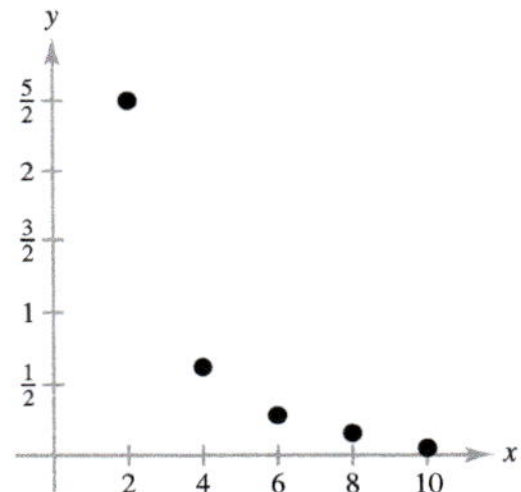

33. The graph appears to represent $y = 4/x$, so y varies inversely as x.

35. $y = \frac{k}{x}$

$1 = \frac{k}{5}$

$5 = k$

$y = \frac{5}{x}$

This equation checks with the other points given in the table.

37. $y = kx$

$-7 = k(10)$

$-\frac{7}{10} = k$

$y = -\frac{7}{10}x$

This equation checks with the other points given in the table.

39. $A = kr^2$

41. $y = \dfrac{k}{x^2}$

43. $F = \dfrac{kg}{r^2}$

45. $R = k(T - T_e)$

47. $P = kVI$

49. y is directly proportional to the square of x.

51. $A = \frac{1}{2}bh$

The area of a triangle is jointly proportional to its base and height.

53.
$$\begin{aligned} y &= kx \\ 54 &= k(3) \\ 18 &= k \\ y &= 18x \end{aligned}$$

55.
$$\begin{aligned} y &= \frac{k}{x} \\ 3 &= \frac{k}{25} \\ 75 &= k \\ y &= \frac{75}{x} \end{aligned}$$

57.
$$\begin{aligned} z &= kxy \\ 64 &= k(4)(8) \\ 2 &= k \\ z &= 2xy \end{aligned}$$

59.
$$\begin{aligned} P &= \frac{kx}{y^2} \\ \frac{28}{3} &= \frac{k(42)}{9^2} \\ \frac{28}{3} \cdot \frac{81}{42} &= k \\ \frac{2 \cdot 27}{3} &= k \\ 18 &= k \\ P &= \frac{18x}{y^2} \end{aligned}$$

61.
$$\begin{aligned} I &= kP \\ 113.75 &= k(3250) \\ 0.035 &= k \\ I &= 0.035P \end{aligned}$$

63.
$$\begin{aligned} y &= kx \\ 33 &= k(13) \\ \tfrac{33}{13} &= k \\ y &= \tfrac{33}{13}x \end{aligned}$$

When $x = 10$ inches, $y \approx 25.4$ centimeters.

When $x = 20$ inches, $y \approx 50.8$ centimeters.

65.
$$\begin{aligned} d &= kF \\ 0.12 &= k(220) \\ \tfrac{3}{5500} &= k \\ d &= \tfrac{3}{5500}F \\ 0.16 &= \tfrac{3}{5500}F \\ \tfrac{880}{3} &= F \end{aligned}$$

The required force is $293\frac{1}{3}$ newtons.

67.
$$\begin{aligned} d &= kF \\ 1.9 &= k(25) \Rightarrow k = 0.076 \\ d &= 0.076F \end{aligned}$$

When the distance compressed is 3 inches, we have

$3 = 0.076F$

$F \approx 39.47.$

No child over 39.47 pounds should use the toy.

69.
$$\begin{aligned} d &= kv^2 \\ 0.02 &= k\left(\frac{1}{4}\right)^2 \\ k &= 0.32 \\ d &= 0.32v^2 \\ 0.12 &= 0.32v^2 \\ v^2 &= \frac{0.12}{0.32} = \frac{3}{8} \\ v &= \frac{\sqrt{3}}{2\sqrt{2}} = \frac{\sqrt{6}}{4} \approx 0.61 \text{ mi/hr} \end{aligned}$$

71. (a)

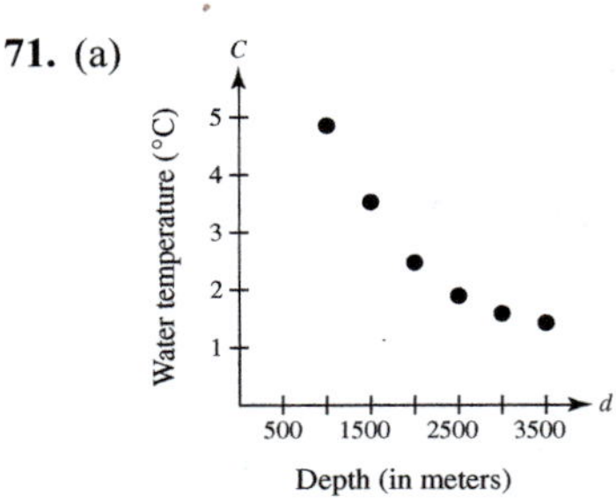

(b) The data shows an inverse variation model fits.

(c) $4.85 = \dfrac{k_1}{1000}$ $\quad 3.525 = \dfrac{k_2}{1500}$ $\quad 2.468 = \dfrac{k_3}{2000}$ $\quad 1.888 = \dfrac{k_4}{2500}$ $\quad 1.583 = \dfrac{k_5}{3000}$ $\quad 1.422 = \dfrac{k_6}{3500}$

$4850.0 = k_1$ $\quad 5287.5 = k_2$ $\quad 4936.0 = k_3$ $\quad 4720.0 = k_5$ $\quad 4749.0 = k_5$ $\quad 4977.0 = k_6$

Mean: $k = \dfrac{4850 + 5287.5 + 4936 + 4720 + 4977}{6} \approx 4919.92$, Model: $C = \dfrac{4919.92}{d}$

(d) $C = \dfrac{4919.92}{d}$

$3 = \dfrac{4919.92}{d}$

$d = \dfrac{4919.92}{3} \approx 1639.97$ meters

The temperature is $3°$ C at approximately 1640 meters.

73. $f_0 = k\dfrac{\sqrt{T_0}}{l_0\sqrt{p}}$ where f_0 = original frequency, T_0 = original tension, l_0 = original length of string, and p = mass density

$f_0 = 100$

(a) The frequency of a string with four times the tension would double the frequency in order to maintain the direct proportion.

Let $f_{\text{new}} = \dfrac{k \cdot \sqrt{T_{\text{new}}}}{l_{\text{new}} \cdot \sqrt{p}}$ and $T_{\text{new}} = 4T_0, l_{\text{new}} = l_0$

$$f_{\text{new}} = \frac{k \cdot \sqrt{4T_0}}{l_0 \cdot \sqrt{p}} = 2 \cdot \frac{k \cdot \sqrt{T_0}}{l_0 \cdot \sqrt{p}} = 2 \cdot f_0 = 2 \cdot 100 = 200Hz$$

(b) The frequency of a string with two times the length would half the frequency in order to maintain the inverse proportion.

Let $f_{\text{new}} = \dfrac{k \cdot \sqrt{T_{\text{new}}}}{l_{\text{new}} \cdot \sqrt{p}}$ and $l_{\text{new}} = 2l_0, \text{T}_{\text{new}} = T_0$

$$f_{\text{new}} = \frac{k \cdot \sqrt{T_0}}{2l_0 \cdot \sqrt{p}} = \frac{1}{2} \cdot \frac{k \cdot \sqrt{T_0}}{l_0 \cdot \sqrt{p}} = \frac{1}{2}f_0 = \frac{1}{2} \cdot 100 = 50Hz$$

(c) The frequency of a string with four times the tension would double the frequency, and two times the length would half the frequency in order to maintain the inverse proportion. Therefore the frequency would remain the same,

Let $f_{\text{new}} = \dfrac{k \cdot \sqrt{T_{\text{new}}}}{l_{\text{new}} \cdot \sqrt{p}}$ and $T_{\text{new}} = 4T_0, l_{\text{new}} = 2l_0$

$$f_{\text{new}} = \frac{k \cdot \sqrt{4T_0}}{2l_0 \cdot \sqrt{p}} = \frac{2 \cdot k\sqrt{T_0}}{2 \cdot l_0 \cdot \sqrt{p}} = \frac{2}{2}f_0 = 1 \cdot 100 = 100Hz$$

75. True. If $y = k_1x$ and $x = k_2z$, then $y = k_1(k_2z) = (k_1k_2)z$.

77. π is a constant and not a variable, so S does not vary with π. In the formula $S = 4\pi r^2$, the surface area, S is directly proportional to the square of the radius, r.

79. y is directly proportional to t since $y = 2x + 2$ and $t = x + 1$, then $y = 2(x + 1)$ becomes $y = 2t$.

Review Exercises for Chapter 1

1.

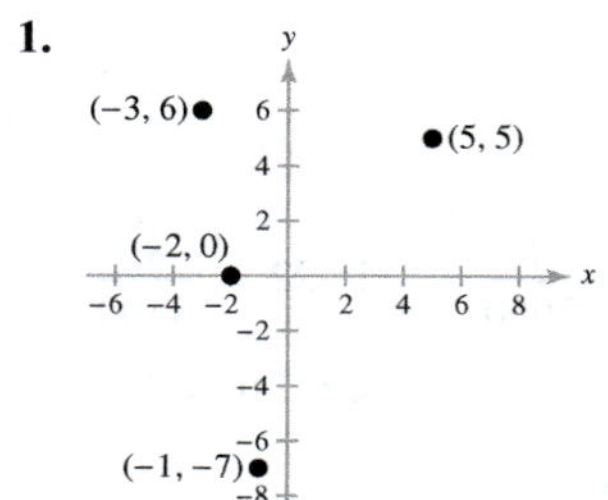

3. $x > 0$ and $y = -2$ in Quadrant IV.

5. (a)

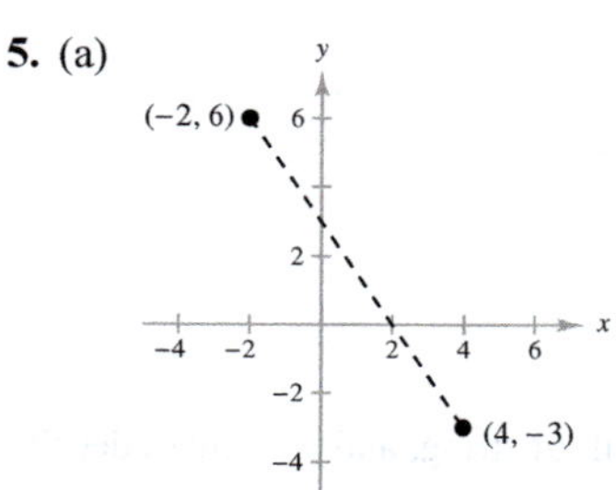

(b) $d = \sqrt{(-2 - 4)^2 + (6 + 3)^2}$

$= \sqrt{36 + 81} = \sqrt{117} = 3\sqrt{13}$

(c) Midpoint: $\left(\frac{-2 + 4}{2}, \frac{6 - 3}{2}\right) = \left(1, \frac{3}{2}\right)$

7. $y = 3x - 5$

x	−2	−1	0	1	2
y	−11	−8	−5	−2	1

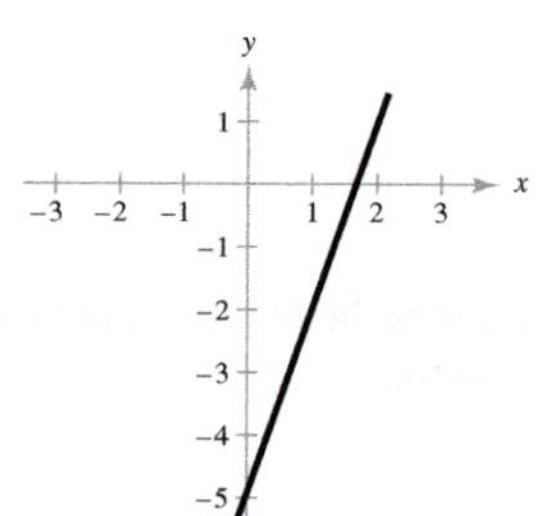

9. $y = x^2 - 3x$

x	−1	0	1	2	3	4
y	4	0	−2	−2	0	4

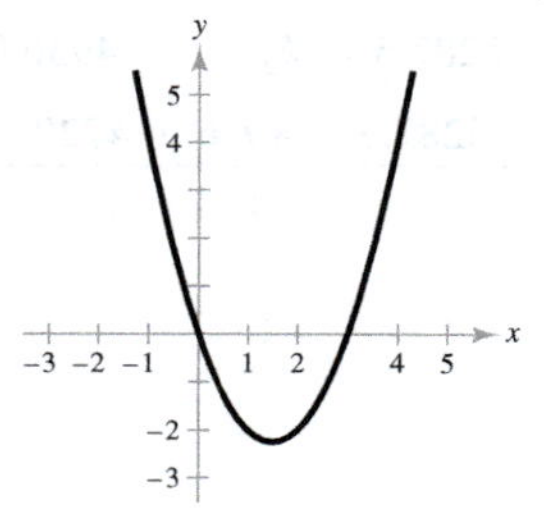

11. $y = 2x + 7$

x-intercept: Let $y = 0$.

$0 = 2x + 7$

$x = -\frac{7}{2}$

$\left(-\frac{7}{2}, 0\right)$

y-intercept: Let $x = 0$.

$y = 2(0) + 7$

$y = 7$

$(0, 7)$

13. $y = (x - 3)^2 - 4$

x-intercepts: $0 = (x - 3)^2 - 4 \Rightarrow (x - 3)^2 = 4$

$\Rightarrow x - 3 = \pm 2$

$\Rightarrow x = 3 \pm 2$

$\Rightarrow x = 5$ or $x = 1$

$(5, 0), (1, 0)$

y-intercept: $y = (0 - 3)^2 - 4$

$y = 9 - 4$

$y = 5$

$(0, 5)$

15. $y = -4x + 1$

Intercepts: $\left(\frac{1}{4}, 0\right), (0, 1)$

$y = -4(-x) + 1 \Rightarrow y = 4x + 1 \Rightarrow$ No y-axis symmetry

$-y = -4x + 1 \Rightarrow y = 4x - 1 \Rightarrow$ No x-axis symmetry

$-y = -4(-x) + 1 \Rightarrow y = -4x - 1 \Rightarrow$ No origin symmetry

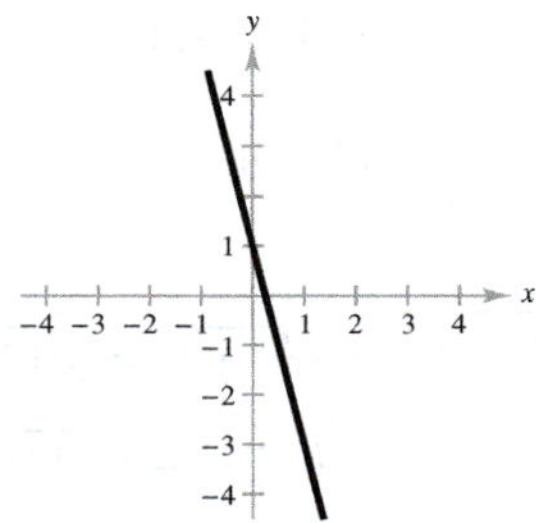

17. $y = 6 - x^2$

Intercepts: $\left(\pm\sqrt{6}, 0\right), (0, 6)$

$y = 6 - \left(-x^2\right) \Rightarrow y = 6 - x^2 \Rightarrow$ y-axis symmetry

$-y = 6 - x^2 \Rightarrow y = -6 + x^2 \Rightarrow$ No x-axis symmetry

$-y = 6 - (-x)^2 \Rightarrow y = -6 + x^2 \Rightarrow$ No origin symmetry

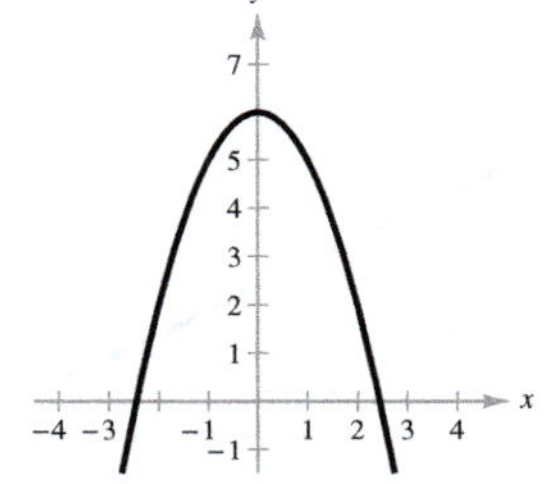

19. $y = x^3 + 5$

Intercepts: $\left(\sqrt[3]{-5}, 0\right), (0, 5)$

$y = (-x)^3 + 5 \Rightarrow y = -x^3 + 5 \Rightarrow$ No y-axis symmetry

$-y = x^3 + 5 \Rightarrow y = -x^3 - 5 \Rightarrow$ No x-axis symmetry

$-y = (-x)^3 + 5 \Rightarrow y = x^3 - 5 \Rightarrow$ No origin symmetry

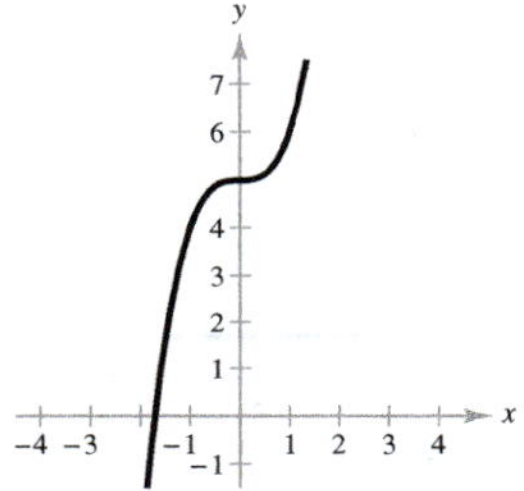

21. $y = \sqrt{x + 5}$

Domain: $[-5, \infty)$

Intercepts: $(-5, 0), \left(0, \sqrt{5}\right)$

$y = \sqrt{-x + 5} \Rightarrow$ No y-axis symmetry

$-y = \sqrt{x + 5} \Rightarrow y = -\sqrt{x + 5} \Rightarrow$ No x-axis symmetry

$-y = \sqrt{-x + 5} \Rightarrow y = -\sqrt{-x + 5} \Rightarrow$ No origin symmetry

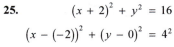

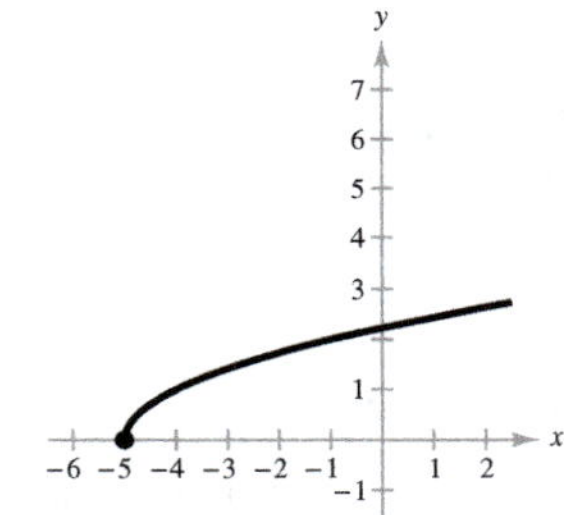

23. $x^2 + y^2 = 9$

Center: $(0, 0)$

Radius: 3

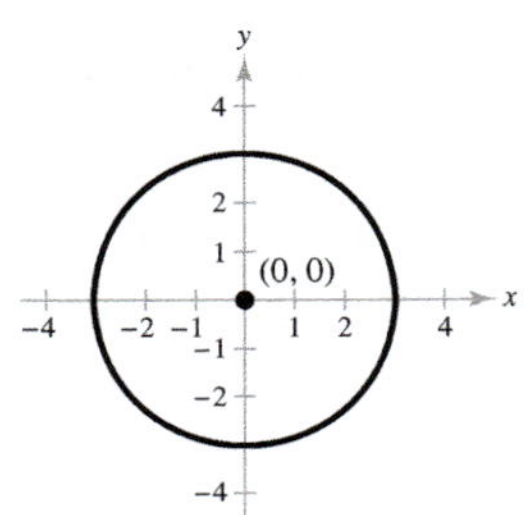

25. $(x + 2)^2 + y^2 = 16$

$\left(x - (-2)\right)^2 + (y - 0)^2 = 4^2$

Center: $(-2, 0)$

Radius: 4

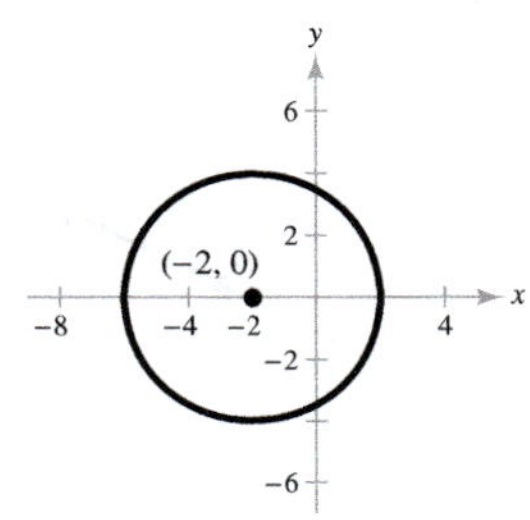

27. Endpoints of a diameter: $(0, 0)$ and $(4, -6)$

Center: $\left(\dfrac{0 + 4}{2}, \dfrac{0 + (-6)}{2}\right) = (2, -3)$

Radius:

$r = \sqrt{(2 - 0)^2 + (-3 - 0)^2} = \sqrt{4 + 9} = \sqrt{13}$

Standard form: $(x - 2)^2 + (y - (-3))^2 = \left(\sqrt{13}\right)^2$

$$(x - 2)^2 + (y + 3)^2 = 13$$

29. 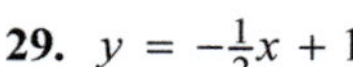

$y = -\frac{1}{2}x + 1$

Slope: $m = -\frac{1}{2}$

y-intercept: $(0, 1)$

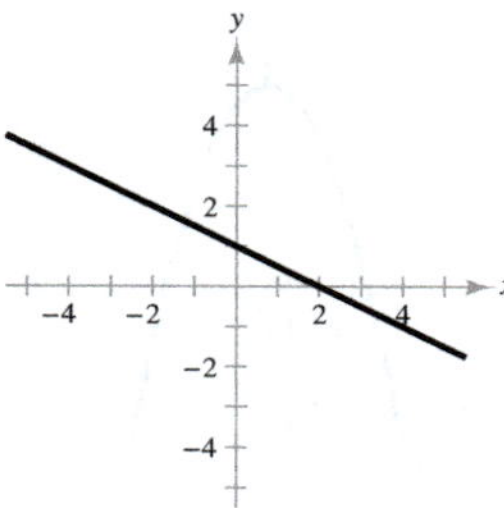

31. $y = 1$

Slope: $m = 0$

y-intercept: $(0, 1)$

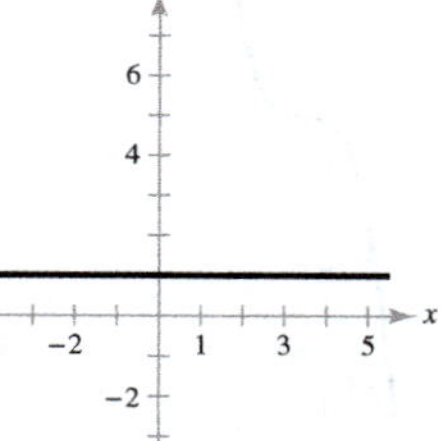

33. $(5, -2), (-1, 4)$

$$m = \frac{4 - (-2)}{-1 - 5} = \frac{6}{-6} = -1$$

35. $(6, -5)$, $m = \frac{1}{3}$

$$y - (-5) = \tfrac{1}{3}(x - 6)$$

$$y + 5 = \tfrac{1}{3}x - 2$$

$$y = \tfrac{1}{3}x - 7$$

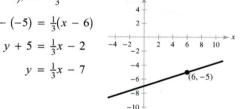

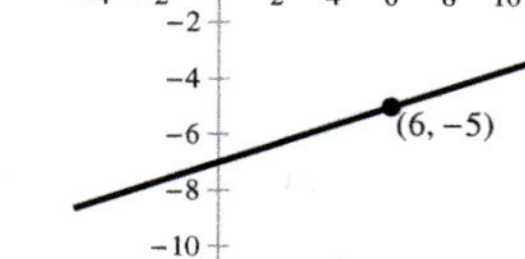

37. $(-6, 4), (4, 9)$

$$m = \frac{9 - 4}{4 - (-6)} = \frac{5}{10} = \frac{1}{2}$$

$$y - 4 = \frac{1}{2}(x - (-6))$$

$$y - 4 = \frac{1}{2}(x + 6)$$

$$y - 4 = \frac{1}{2}x + 3$$

$$y = \frac{1}{2}x + 7$$

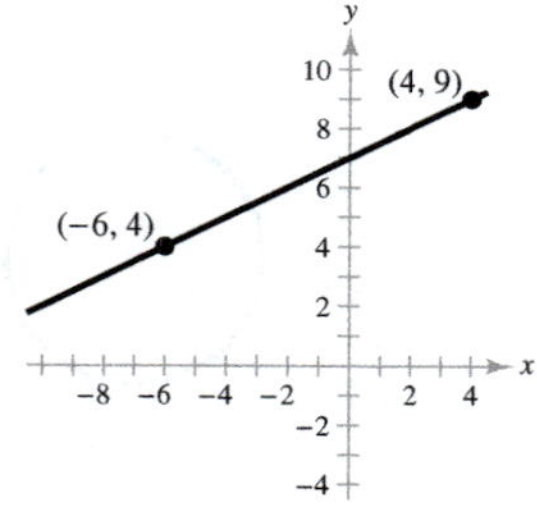

39. Point: $(3, -2)$

$$5x - 4y = 8$$

$$y = \tfrac{5}{4}x - 2$$

(a) Parallel slope: $m = \frac{5}{4}$

$$y - (-2) = \tfrac{5}{4}(x - 3)$$

$$y + 2 = \tfrac{5}{4}x - \tfrac{15}{4}$$

$$y = \tfrac{5}{4}x - \tfrac{23}{4}$$

(b) Perpendicular slope: $m = -\frac{4}{5}$

$$y - (-2) = -\tfrac{4}{5}(x - 3)$$

$$y + 2 = -\tfrac{4}{5}x + \tfrac{12}{5}$$

$$y = -\tfrac{4}{5}x + \tfrac{2}{5}$$

41. *Verbal Model:* Sale price = (List price) − (Discount)

Labels: Sale price $= S$

List price $= L$

Discount $= 20\%$ of $L = 0.2L$

Equation: $S = L - 0.2L$

$S = 0.8L$

43. $16x - y^4 = 0$

$$y^4 = 16x$$

$$y = \pm 2\sqrt[4]{x}$$

No, y is not a function of x. Some x-values correspond to two y-values.

45. $y = \sqrt{1 - x}$

Yes, the equation represents y as a function of x. Each x-value, $x \le 1$, corresponds to only one y-value.

47. $f(x) = x^2 + 1$

(a) $f(2) = (2)^2 + 1 = 5$

(b) $f(-4) = (-4)^2 + 1 = 17$

(c) $f(t^2) = (t^2)^2 + 1 = t^4 + 1$

(d) $f(t + 1) = (t + 1)^2 + 1 = t^2 + 2t + 2$

49. $f(x) = \sqrt{25 - x^2}$

Domain: $25 - x^2 \geq 0$

$(5 + x)(5 - x) \geq 0$

Critical numbers: $x = \pm 5$

Test intervals: $(-\infty, -5), (-5, 5), (5, \infty)$

Test: Is $25 - x^2 \geq 0$?

Solution set: $-5 \leq x \leq 5$

Domain: all real numbers x such that $-5 \leq x \leq 5$, or $[-5, 5]$

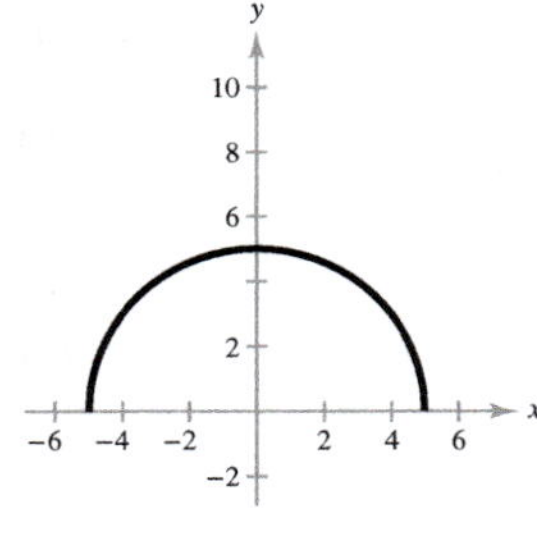

51. $v(t) = -32t + 48$

$v(1) = 16$ feet per second

53. $f(x) = 2x^2 + 3x - 1$

$$\frac{f(x + h) - f(x)}{h} = \frac{\left[2(x + h)^2 + 3(x + h) - 1\right] - (2x^2 + 3x - 1)}{h}$$

$$= \frac{2x^2 + 4xh + 2h^2 + 3x + 3h - 1 - 2x^2 - 3x + 1}{h}$$

$$= \frac{h(4x + 2h + 3)}{h}$$

$$= 4x + 2h + 3, \quad h \neq 0$$

55. $y = (x - 3)^2$

A vertical line intersects the graph no more than once, so y *is* a function of x.

57. $f(x) = 3x^2 - 16x + 21$

$3x^2 - 16x + 21 = 0$

$(3x - 7)(x - 3) = 0$

$3x - 7 = 0$ or $x - 3 = 0$

$x = \frac{7}{3}$ or $x = 3$

59. $f(x) = |x| + |x + 1|$

f is increasing on $(0, \infty)$.

f is decreasing on $(-\infty, -1)$.

f is constant on $(-1, 0)$.

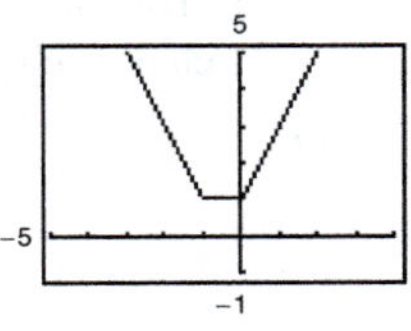

61. $f(x) = -x^2 + 2x + 1$

Relative maximum: $(1, 2)$

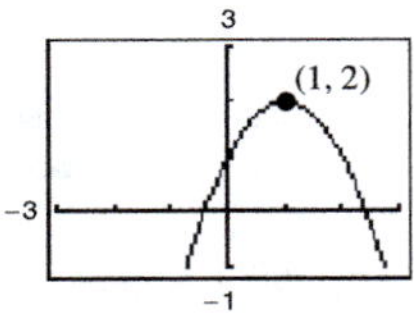

63. $f(x) = -x^2 + 8x - 4$

$$\frac{f(4) - f(0)}{4 - 0} = \frac{12 - (-4)}{4} = 4$$

The average rate of change of f from $x_1 = 0$ to $x_2 = 4$ is 4.

65. $f(x) = x^4 - 20x^2$

$f(-x) = (-x)^4 - 20(-x)^2 = x^4 - 20x^2 = f(x)$

The function is even, so the graph has y-axis symmetry.

67. (a) $f(2) = -6, f(-1) = 3$

Points: $(2, -6), (-1, 3)$

$$m = \frac{3 - (-6)}{-1 - 2} = \frac{9}{-3} = -3$$

$y - (-6) = -3(x - 2)$

$y + 6 = -3x + 6$

$y = -3x$

$f(x) = -3x$

(b)

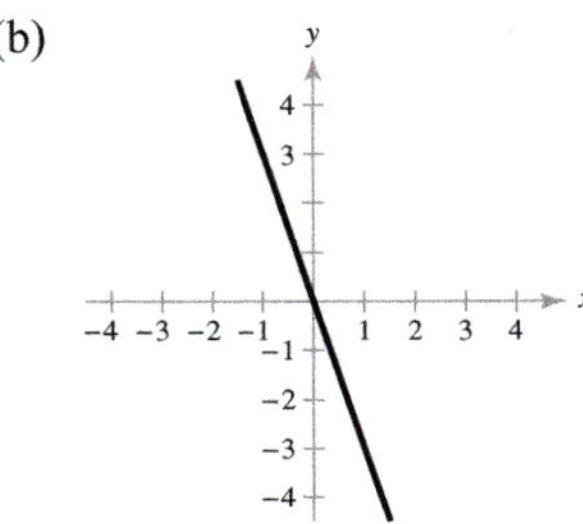

69. $g(x) = [\![x]\!] - 2$

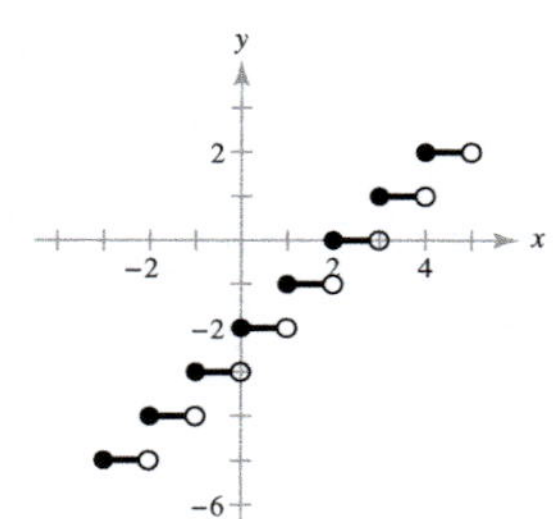

71. (a) $f(x) = x^2$

(b) $h(x) = x^2 - 9$

Vertical shift 9 units downward

(c)

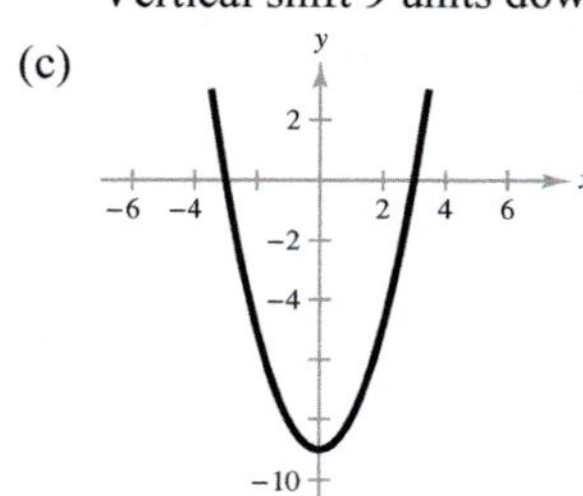

(d) $h(x) = f(x) - 9$

73. (a) $f(x) = \sqrt{x}$

(b) $h(x) = -\sqrt{x} + 4$

Reflection in the x-axis and a vertical shift 4 units upward

(c)

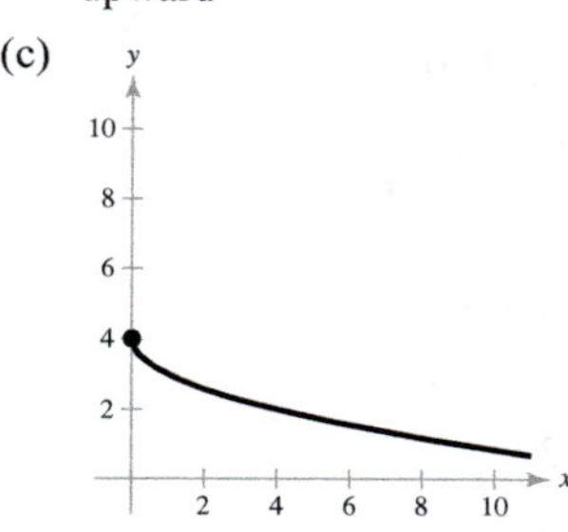

(d) $h(x) = -f(x) + 4$

75. (a) $f(x) = x^2$

(b) $h(x) = -(x + 2)^2 + 3$

Reflection in the x-axis, a horizontal shift 2 units to the left, and a vertical shift 3 units upward

(c)

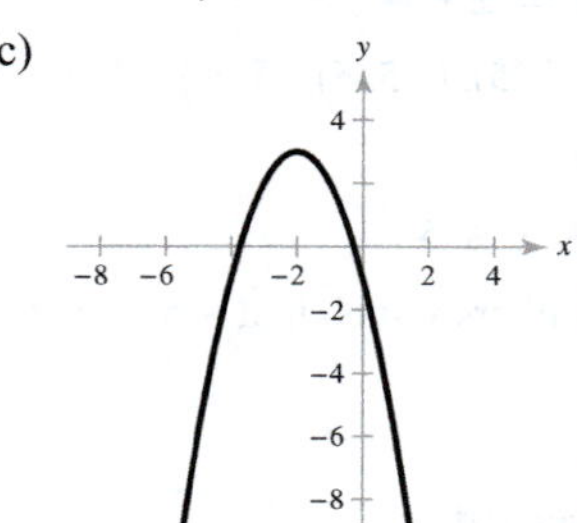

(d) $h(x) = -f(x + 2) + 3$

77. (a) $f(x) = [\![x]\!]$

(b) $h(x) = -[\![x]\!] + 6$

Reflection in the x-axis and a vertical shift 6 units upward

(c)

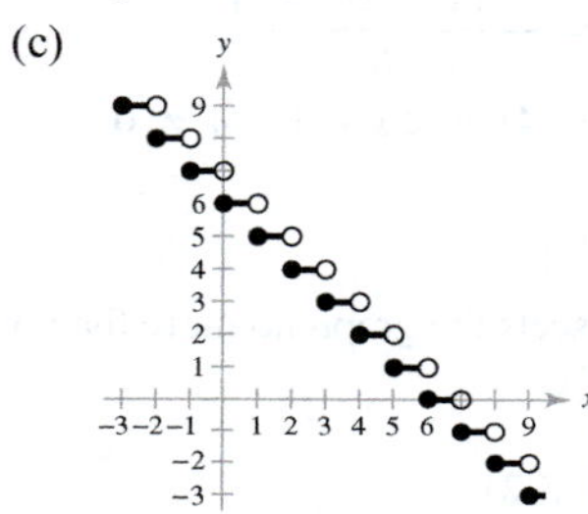

(d) $h(x) = -f(x) + 6$

79. (a) $f(x) = [\![x]\!]$

(b) $h(x) = 5[\![x - 9]\!]$

Horizontal shift 9 units to the right and a vertical stretch (each y-value is multiplied by 5)

(c)

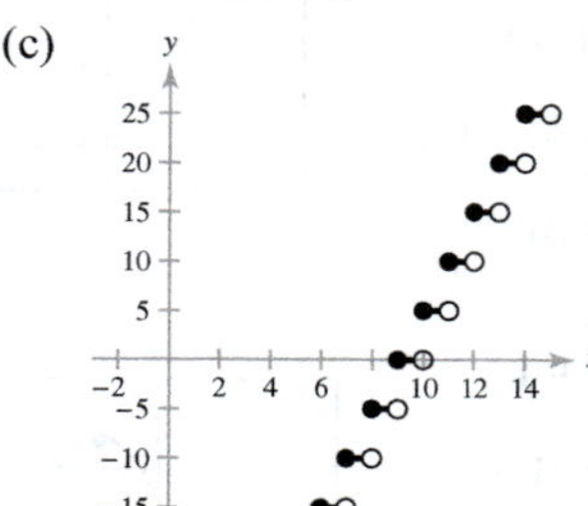

(d) $h(x) = 5f(x - 9)$

81. $f(x) = x^2 + 3,\ g(x) = 2x - 1$

(a) $(f + g)(x) = (x^2 + 3) + (2x - 1) = x^2 + 2x + 2$

(b) $(f - g)(x) = (x^2 + 3) - (2x - 1) = x^2 - 2x + 4$

(c) $(fg)(x) = (x^2 + 3)(2x - 1) = 2x^3 - x^2 + 6x - 3$

(d) $\left(\dfrac{f}{g}\right)(x) = \dfrac{x^2 + 3}{2x - 1}$, Domain: $x \neq \dfrac{1}{2}$

83. $f(x) = \frac{1}{3}x - 3,\ g(x) = 3x + 1$

The domains of f and g are all real numbers.

(a) $$\begin{aligned}(f \circ g)(x) &= f(g(x))\\ &= f(3x + 1)\\ &= \tfrac{1}{3}(3x + 1) - 3\\ &= x + \tfrac{1}{3} - 3\\ &= x - \tfrac{8}{3}\end{aligned}$$

Domain: all real numbers

(b) $$\begin{aligned}(g \circ f)(x) &= g(f(x))\\ &= g\left(\tfrac{1}{3}x - 3\right)\\ &= 3\left(\tfrac{1}{3}x - 3\right) + 1\\ &= x - 9 + 1\\ &= x - 8\end{aligned}$$

Domain: all real number

In Exercise 85, use the following functions. $f(x) = x - 100,\ g(x) = 0.95x$

85. $(f \circ g)(x) = f(0.95x) = 0.95x - 100$ represents the sale price if first the 5% discount is applied and then the \$100 rebate.

87. $$\begin{aligned}f(x) &= 3x + 8\\ y &= 3x + 8\\ x &= 3y + 8\\ x - 8 &= 3y\\ y &= \frac{x - 8}{3}\\ y &= \frac{1}{3}(x - 8)\end{aligned}$$

So, $f^{-1}(x) = \dfrac{1}{3}(x - 8) = \dfrac{x - 8}{3}$.

$$f(f^{-1}(x)) = f\left(\frac{1}{3}(x - 8)\right) = 3\left(\frac{1}{3}(x - 8)\right) + 8 = x - 8 + 8 = x$$

$$f^{-1}(f(x)) = f^{-1}(3x + 8) = \frac{1}{3}(3x + 8 - 8) = \frac{1}{3}(3x) = x$$

89. $f(x) = (x - 1)^2$

No, the function does not have an inverse because the horizontal line test fails.

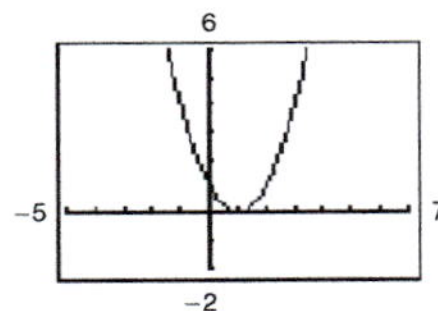

91. (a) $$\begin{aligned}f(x) &= \tfrac{1}{2}x - 3\\ y &= \tfrac{1}{2}x - 3\\ x &= \tfrac{1}{2}y - 3\\ x + 3 &= \tfrac{1}{2}y\\ 2(x + 3) &= y\\ f^{-1}(x) &= 2x + 6\end{aligned}$$

(b)

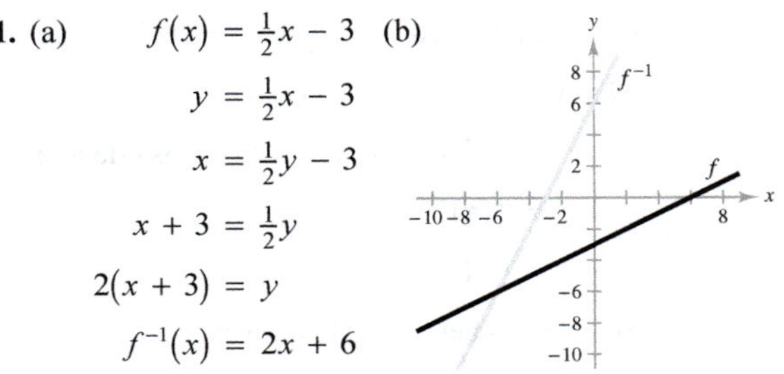

(c) The graph of f^{-1} is the reflection of the graph of f in the line $y = x$.

(d) The domains and ranges of f and f^{-1} are the set of all real numbers.

93. $f(x) = 2(x - 4)^2$ is increasing on $(4, \infty)$.

Let $f(x) = 2(x - 4)^2, x > 4$ and $y > 0$.

$$y = 2(x - 4)^2$$
$$x = 2(y - 4)^2, x > 0, y > 4$$
$$\frac{x}{2} = (y - 4)^2$$
$$\sqrt{\frac{x}{2}} = y - 4$$
$$\sqrt{\frac{x}{2}} + 4 = y$$
$$f^{-1}(x) = \sqrt{\frac{x}{2}} + 4, x > 0$$

95. (a) and (b)

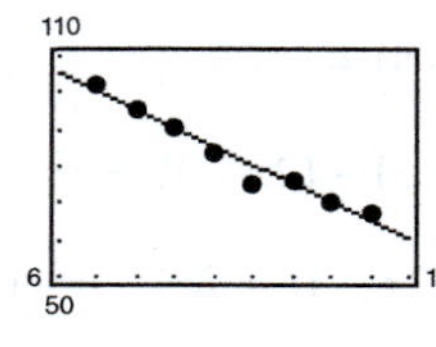

$N = -5.02t + 135.6$

97.
$$C = khw^2$$
$$28.80 = k(16)(6)^2$$
$$k = 0.05$$
$$C = (0.05)(14)(8)^2 = \$44.80$$

99. True. If $f(x) = x^3$ and $g(x) = \sqrt[3]{x}$, then the domain of g is all real numbers, which is equal to the range of f and vice versa.

Problem Solving for Chapter 1

1. (a) $W_1 = 0.07S + 2000$

(b) $W_2 = 0.05S + 2300$

(c)

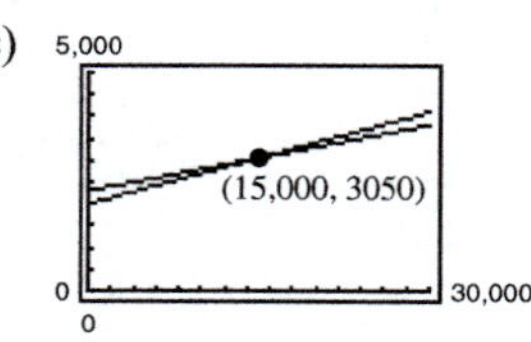

Point of intersection: $(15,000, 3050)$

Both jobs pay the same, \$3050, if you sell \$15,000 per month.

(d) No. If you think you can sell \$20,000 per month, keep your current job with the higher commission rate. For sales over \$15,000 it pays more than the other job.

3. (a) Let $f(x)$ and $g(x)$ be two even functions.

Then define $h(x) = f(x) \pm g(x)$.

$$h(-x) = f(-x) \pm g(-x)$$
$$= f(x) \pm g(x) \text{ because } f \text{ and } g \text{ are even}$$
$$= h(x)$$

So, $h(x)$ is also even.

(b) Let $f(x)$ and $g(x)$ be two odd functions.

Then define $h(x) = f(x) \pm g(x)$.

$$h(-x) = f(-x) \pm g(-x)$$
$$= -f(x) \pm g(x) \text{ because } f \text{ and } g \text{ are odd}$$
$$= -h(x)$$

So, $h(x)$ is also odd. $\left(\text{If } f(x) \neq g(x)\right)$

(c) Let $f(x)$ be odd and $g(x)$ be even. Then define $h(x) = f(x) \pm g(x)$.

$$h(-x) = f(-x) \pm g(-x)$$
$$= -f(x) \pm g(x) \text{ because } f \text{ is odd and } g \text{ is even}$$
$$\neq h(x)$$
$$\neq -h(x)$$

So, $h(x)$ is neither odd nor even.

5. $f(x) = a_{2n}x^{2n} + a_{2n-2}x^{2n-2} + \cdots + a_2x^2 + a_0$

$f(-x) = a_{2n}(-x)^{2n} + a_{2n-2}(-x)^{2n-2} + \cdots + a_2(-x)^2 + a_0 = a_{2n}x^{2n} + a_{2n-2}x^{2n-2} + \cdots + a_2x^2 + a_0 = f(x)$

So, $f(x)$ is even.

7. (a) April 11: 10 hours

April 12: 24 hours

April 13: 24 hours

April 14: $23\frac{2}{3}$ hours

Total: $81\frac{2}{3}$ hours

(b) $\text{Speed} = \dfrac{\text{distance}}{\text{time}} = \dfrac{2100}{81\frac{2}{3}} = \dfrac{180}{7} = 25\frac{5}{7}$ mph

(c) $D = -\dfrac{180}{7}t + 3400$

Domain: $0 \le t \le \dfrac{1190}{9}$

Range: $0 \le D \le 3400$

(d)

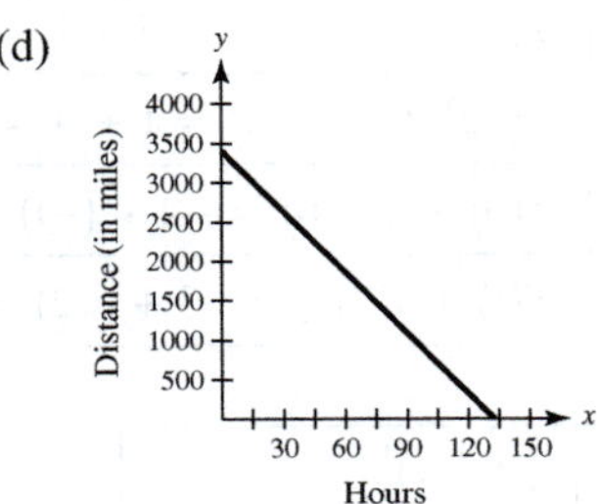

9. (a)–(d) Use $f(x) = 4x$ and $g(x) = x + 6$.

(a) $(f \circ g)(x) = f(x + 6) = 4(x + 6) = 4x + 24$

(b) $(f \circ g)^{-1}(x) = \dfrac{x - 24}{4} = \dfrac{1}{4}x - 6$

(c) $f^{-1}(x) = \dfrac{1}{4}x$

$g^{-1}(x) = x - 6$

(d) $(g^{-1} \circ f^{-1})(x) = g^{-1}\left(\dfrac{1}{4}x\right) = \dfrac{1}{4}x - 6$

(e) $f(x) = x^3 + 1$ and $g(x) = 2x$

$(f \circ g)(x) = f(2x) = (2x)^3 + 1 = 8x^3 + 1$

$(f \circ g)^{-1}(x) = \sqrt[3]{\dfrac{x - 1}{8}} = \dfrac{1}{2}\sqrt[3]{x - 1}$

$f^{-1}(x) = \sqrt[3]{x - 1}$

$g^{-1}(x) = \dfrac{1}{2}x$

$(g^{-1} \circ f^{-1})(x) = g^{-1}\left(\sqrt[3]{x - 1}\right) = \dfrac{1}{2}\sqrt[3]{x - 1}$

(f) Answers will vary.

(g) Conjecture: $(f \circ g)^{-1}(x) = (g^{-1} \circ f^{-1})(x)$

11. 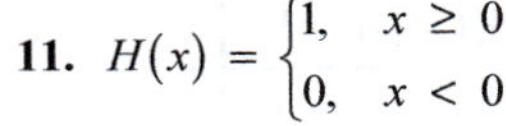
$H(x) = \begin{cases} 1, & x \ge 0 \\ 0, & x < 0 \end{cases}$

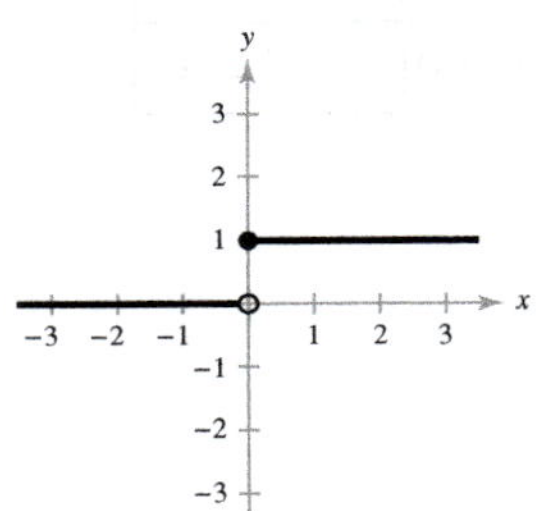

(a) $H(x) - 2$

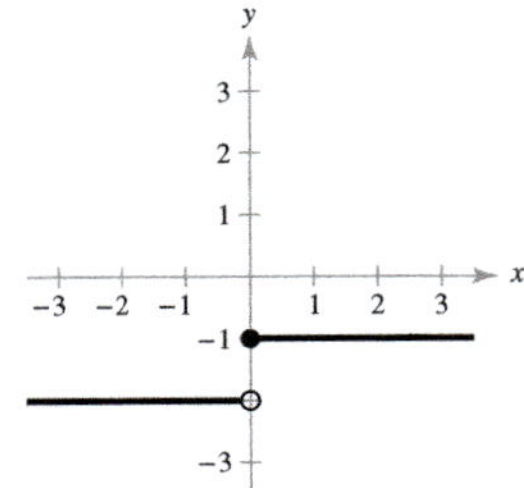

(b) $H(x - 2)$

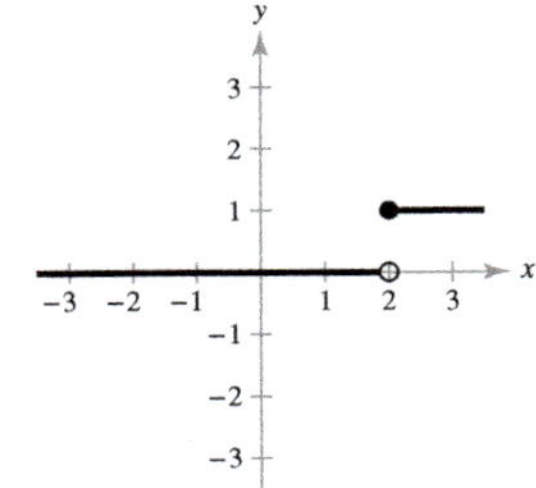

(c) $-H(x)$

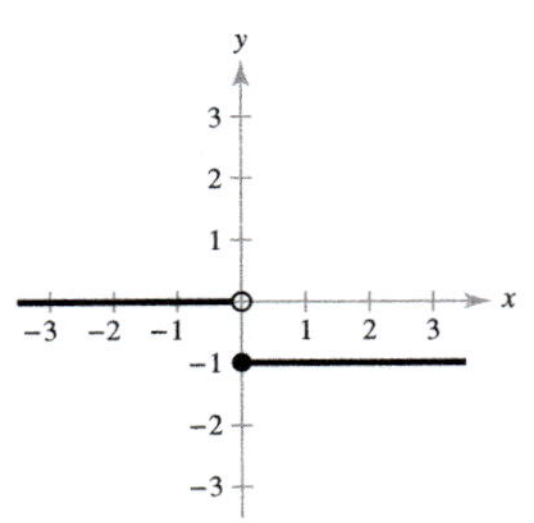

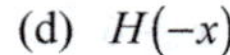
(d) $H(-x)$

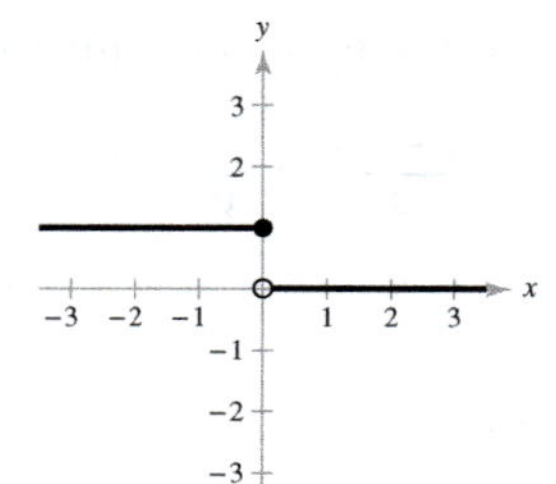

(e) $\frac{1}{2}H(x)$

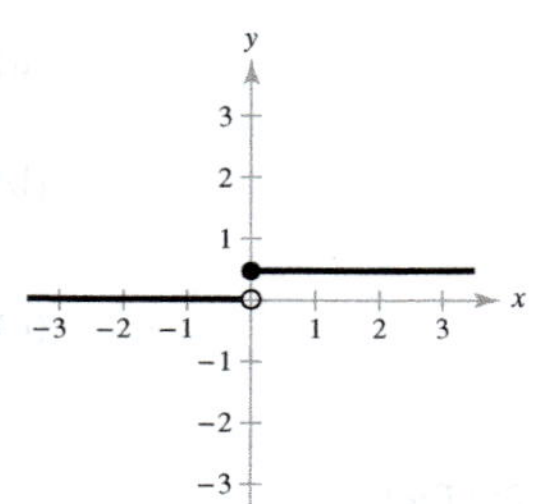

(f) $-H(x-2)+2$

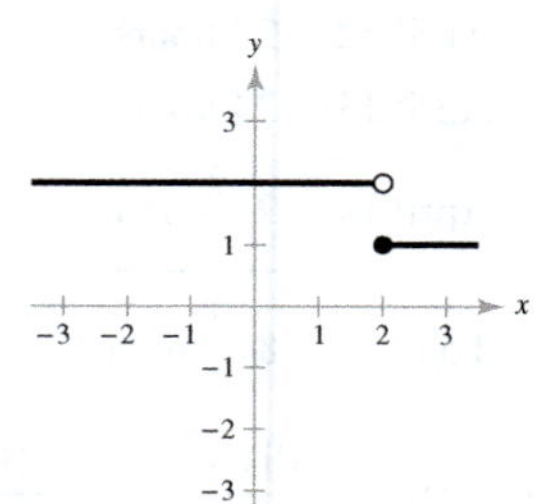

13. $(f \circ (g \circ h))(x) = f((g \circ h)(x)) = f(g(h(x))) = (f \circ g \circ h)(x)$

$((f \circ g) \circ h)(x) = (f \circ g)(h(x)) = f(g(h(x))) = (f \circ g \circ h)(x)$

15.

x	$f(x)$	$f^{-1}(x)$
−4	—	2
−3	4	1
−2	1	0
−1	0	—
0	−2	−1
1	−3	−2
2	−4	—
3	—	—
4	—	−3

(a)

x	$f(f^{-1}(x))$
−4	$f(f^{-1}(-4)) = f(2) = -4$
−2	$f(f^{-1}(-2)) = f(0) = -2$
0	$f(f^{-1}(0)) = f(-1) = 0$
4	$f(f^{-1}(4)) = f(-3) = 4$

(b)

x	$(f + f^{-1})(x)$
−3	$f(-3) + f^{-1}(-3) = 4 + 1 = 5$
−2	$f(-2) + f^{-1}(-2) = 1 + 0 = 1$
0	$f(0) + f^{-1}(0) = -2 + (-1) = -3$
1	$f(1) + f^{-1}(1) = -3 + (-2) = -5$

(c)

x	$(f \cdot f^{-1})(x)$
−3	$f(-3)f^{-1}(-3) = (4)(1) = 4$
−2	$f(-2)f^{-1}(-2) = (1)(0) = 0$
0	$f(0)f^{-1}(0) = (-2)(-1) = 2$
1	$f(1)f^{-1}(1) = (-3)(-2) = 6$

(d)

x	$\lvert f^{-1}(x)\rvert$
−4	$\lvert f^{-1}(-4)\rvert = \lvert 2\rvert = 2$
−3	$\lvert f^{-1}(-3)\rvert = \lvert 1\rvert = 1$
0	$\lvert f^{-1}(0)\rvert = \lvert -1\rvert = 1$
4	$\lvert f^{-1}(4)\rvert = \lvert -3\rvert = 3$

Practice Test for Chapter 1

1. Find the equation of the line through $(2, 4)$ and $(3, -1)$.

2. Find the equation of the line with slope $m = 4/3$ and y-intercept $b = -3$.

3. Find the equation of the line through $(4, 1)$ perpendicular to the line $2x + 3y = 0$.

4. If it costs a company \$32 to produce 5 units of a product and \$44 to produce 9 units, how much does it cost to produce 20 units? (Assume that the cost function is linear.)

5. Given $f(x) = x^2 - 2x + 1$, find $f(x - 3)$.

6. Given $f(x) = 4x - 11$, find $\dfrac{f(x) - f(3)}{x - 3}$.

7. Find the domain and range of $f(x) = \sqrt{36 - x^2}$.

8. Which equations determine y as a function of x?

(a) $6x - 5y + 4 = 0$

(b) $x^2 + y^2 = 9$

(c) $y^3 = x^2 + 6$

9. Sketch the graph of $f(x) = x^2 - 5$.

10. Sketch the graph of $f(x) = |x + 3|$.

11. Sketch the graph of $f(x) = \begin{cases} 2x + 1, & \text{if } x \geq 0, \\ x^2 - x, & \text{if } x < 0. \end{cases}$

12. Use the graph of $f(x) = |x|$ to graph the following:

(a) $f(x + 2)$

(b) $-f(x) + 2$

13. Given $f(x) = 3x + 7$ and $g(x) = 2x^2 - 5$, find the following:

(a) $(g - f)(x)$

(b) $(fg)(x)$

14. Given $f(x) = x^2 - 2x + 16$ and $g(x) = 2x + 3$, find $f(g(x))$.

15. Given $f(x) = x^3 + 7$, find $f^{-1}(x)$.

16. Which of the following functions have inverses?

(a) $f(x) = |x - 6|$

(b) $f(x) = ax + b, a \neq 0$

(c) $f(x) = x^3 - 19$

17. Given $f(x) = \sqrt{\dfrac{3 - x}{x}}$, $0 < x \le 3$, find $f^{-1}(x)$.

Exercises 18–20, true or false?

18. $y = 3x + 7$ and $y = \frac{1}{3}x - 4$ are perpendicular.

19. $(f \circ g)^{-1} = g^{-1} \circ f^{-1}$

20. If a function has an inverse, then it must pass both the Vertical Line Test and the Horizontal Line Test.

CHAPTER 2
Polynomial and Rational Functions

CHAPTER 2
Polynomial and Rational Functions

Section 2.1 Quadratic Functions and Models

1. polynomial

3. quadratic; parabola

5. $f(x) = x^2 - 2$ opens upward and has vertex $(0, -2)$. Matches graph (b).

6. $f(x) = (x + 1)^2 - 2$ opens upward and has vertex $(-1, -2)$. Matches graph (a).

7. $f(x) = -(x - 4)^2$ opens downward and has vertex $(4, 0)$. Matches graph (c).

8. $f(x) = 4 - (x - 2)^2 = -(x - 2)^2 + 4$ opens downward and has vertex $(2, 4)$. Matches graph (d).

9. (a) $y = \frac{1}{2}x^2$

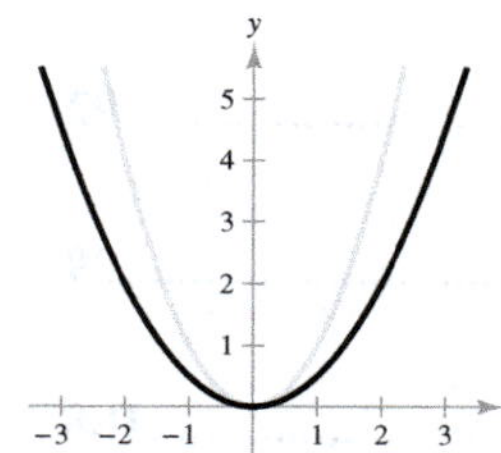

Vertical shrink

(b) $y = -\frac{1}{8}x^2$

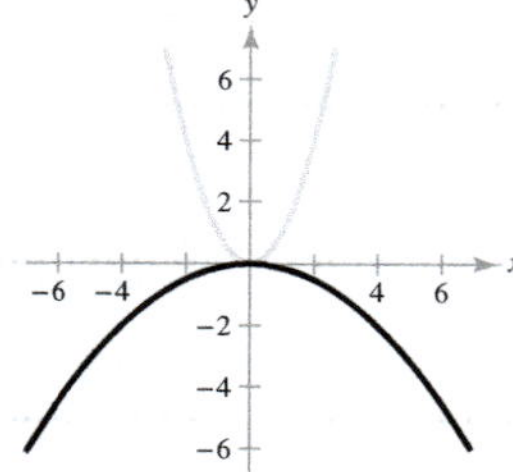

Vertical shrink and a reflection in the x-axis

(c) $y = \frac{3}{2}x^2$

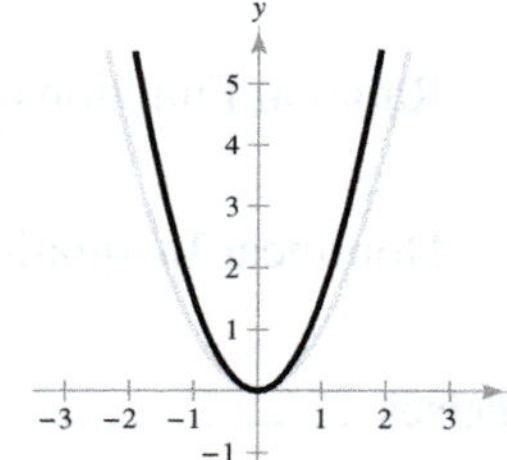

Vertical stretch

(d) $y = -3x^2$

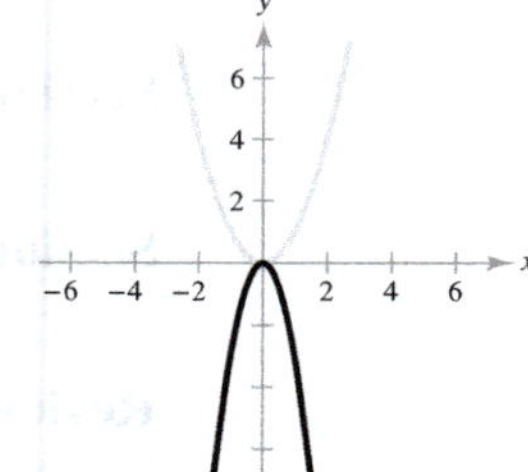

Vertical stretch and a reflection in the x-axis

11. (a) $y = (x - 1)^2$

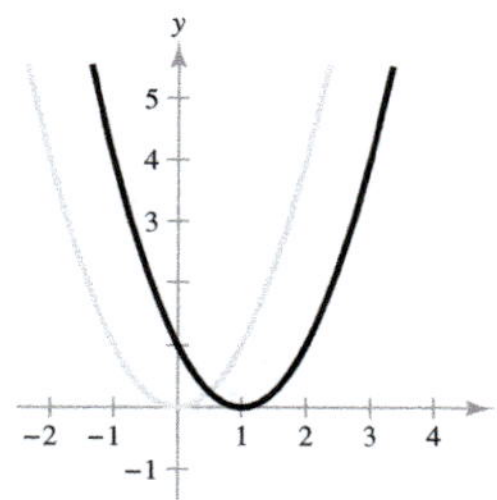

Horizontal shift one unit to the right

(b) $y = (3x)^2 + 1$

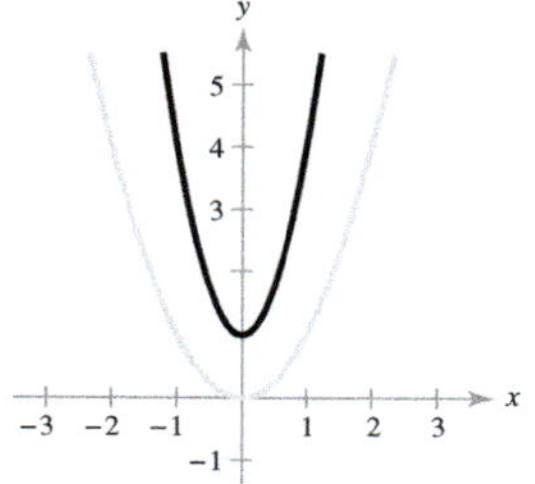

Horizontal shrink and a vertical shift one unit upward

(c) $y = \left(\frac{1}{3}x\right)^2 - 3$

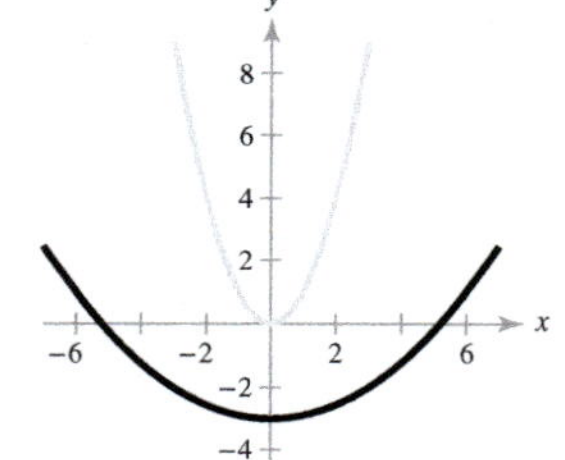

Horizontal stretch and a vertical shift three units downward

(d) $y = (x + 3)^2$

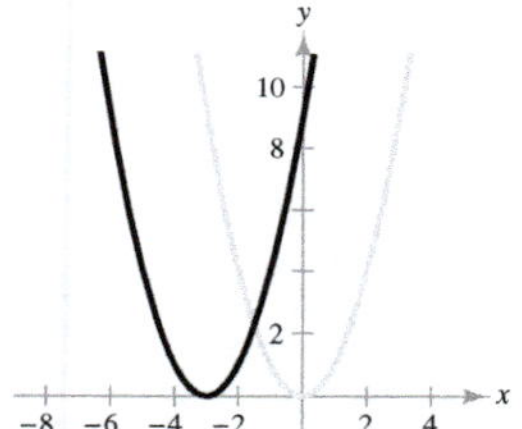

Horizontal shift three units to the left

13. $f(x) = x^2 - 6x$

$= (x^2 - 6x + 9) - 9$

$= (x - 3)^2 - 9$

Vertex: $(3, -9)$

Axis of symmetry: $x = 3$

Find x-intercepts:

$x^2 - 6x = 0$

$x(x - 6) = 0$

$x = 0$

$x - 6 = 0 \Rightarrow x = 6$

x-intercepts: $(0, 0), (6, 0)$

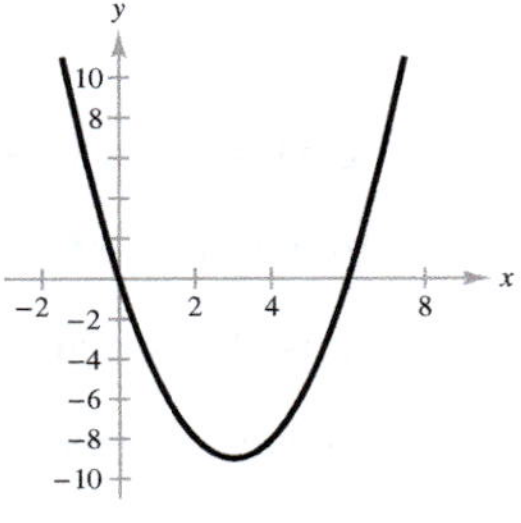

15. $h(x) = x^2 - 8x + 16 = (x - 4)^2$

Vertex: $(4, 0)$

Axis of symmetry: $x = 4$

Find x-intercepts:

$(x - 4)^2 = 0$

$x - 4 = 0$

$x = 4$

x-intercept: $(4, 0)$

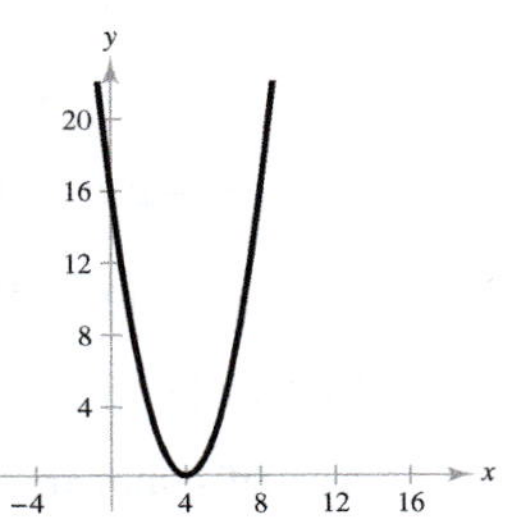

17. $f(x) = x^2 - 6x + 2$

$= (x^2 - 6x + 9) - 9 + 2$

$= (x^2 - 6x + 9) - 7$

$= (x - 3)^2 - 7$

Vertex: $(3, -7)$

Axis of symmetry: $x = 3$

Find x-intercepts:

$x^2 - 6x + 2 = 0$

$x^2 - 6x = -2$

$x^2 - 6x + 9 = -2 + 9$

$(x - 3)^2 = 7$

$x = 3 \pm \sqrt{7}$

x-intercepts: $(3 \pm 7, 0)$

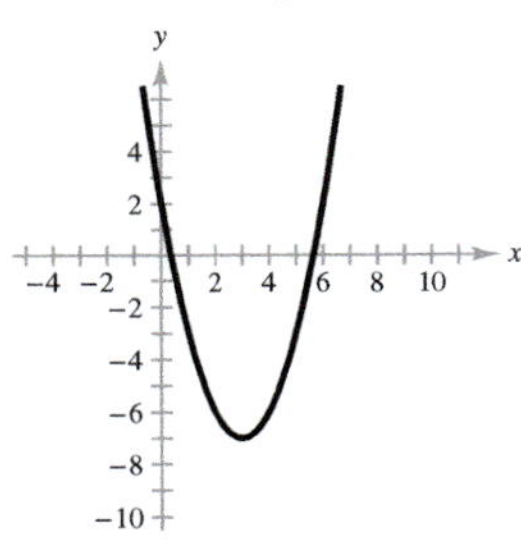

19. $f(x) = x^2 - 8x + 21$

$= (x^2 - 8x + 16) - 16 + 21$

$= (x - 4)^2 + 5$

Vertex: $(4, 5)$

Axis of symmetry: $x = 4$

Find x-intercepts:

$x^2 - 8x + 21 = 0$

$x^2 - 8x = -21$

$x^2 - 8x + 16 = -21 + 16$

$(x - 4)^2 = -5$

$x - 4 = \pm\sqrt{-5}$

$x = 4 \pm \sqrt{5}i$

Not a real number

No x-intercepts

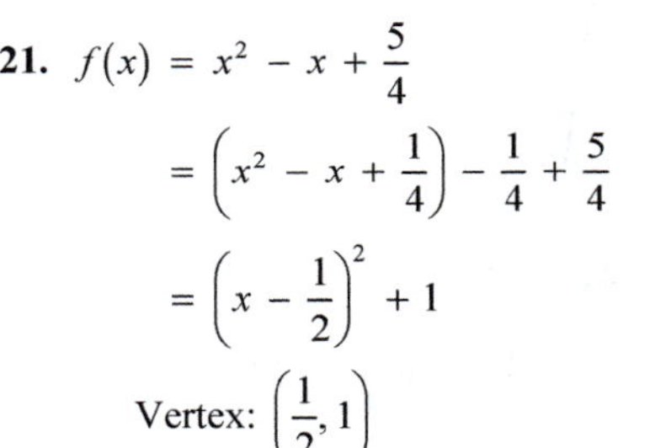

21. $f(x) = x^2 - x + \frac{5}{4}$

$= \left(x^2 - x + \frac{1}{4}\right) - \frac{1}{4} + \frac{5}{4}$

$= \left(x - \frac{1}{2}\right)^2 + 1$

Vertex: $\left(\frac{1}{2}, 1\right)$

Axis of symmetry: $x = \frac{1}{2}$

Find x-intercepts:

$x^2 - x + \frac{5}{4} = 0$

$x = \frac{1 \pm \sqrt{1 - 5}}{2}$

Not a real number

No x-intercepts

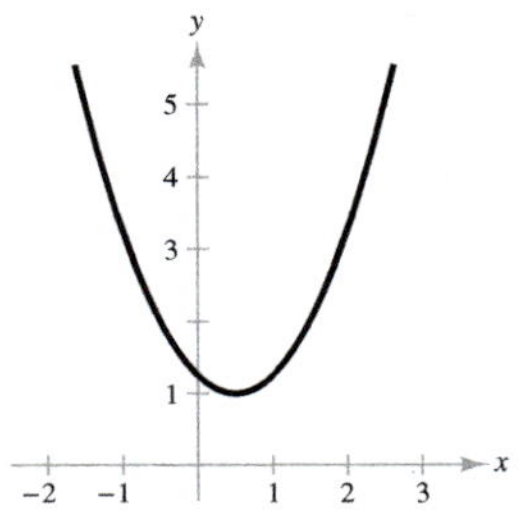

23. $f(x) = -x^2 + 2x + 5$

$= -(x^2 - 2x + 1) - (-1) + 5$

$= -(x - 1)^2 + 6$

Vertex: $(1, 6)$

Axis of symmetry: $x = 1$

Find x-intercepts:

$-x^2 + 2x + 5 = 0$

$x^2 - 2x - 5 = 0$

$$x = \frac{2 \pm \sqrt{4 + 20}}{2}$$

$$= 1 \pm \sqrt{6}$$

x-intercepts: $(1 - \sqrt{6}, 0), (1 + \sqrt{6}, 0)$

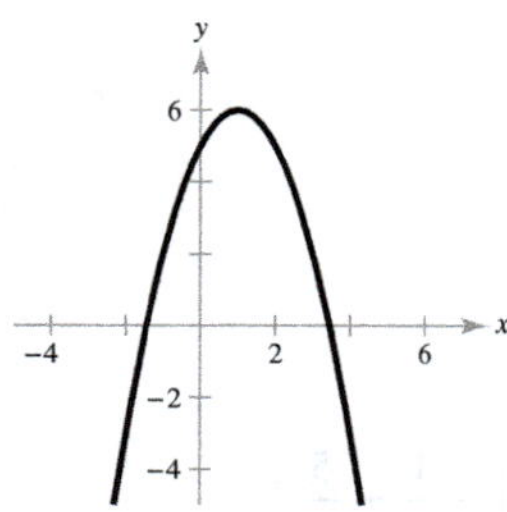

25. $h(x) = 4x^2 - 4x + 21$

$= 4\left(x^2 - x + \frac{1}{4}\right) - 4\left(\frac{1}{4}\right) + 21$

$= 4\left(x - \frac{1}{2}\right)^2 + 20$

Vertex: $\left(\frac{1}{2}, 20\right)$

Axis of symmetry: $x = \frac{1}{2}$

Find x-intercepts:

$4x^2 - 4x + 21 = 0$

$$x = \frac{4 \pm \sqrt{16 - 336}}{2(4)}$$

Not a real number

No x-intercepts

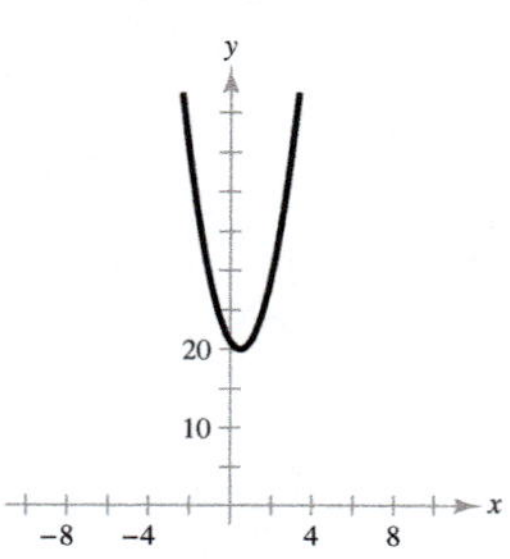

27. $f(x) = -(x^2 + 2x - 3) = -(x + 1)^2 + 4$

Vertex: $(-1, 4)$

Axis of symmetry: $x = -1$

x-intercepts: $(-3, 0), (1, 0)$

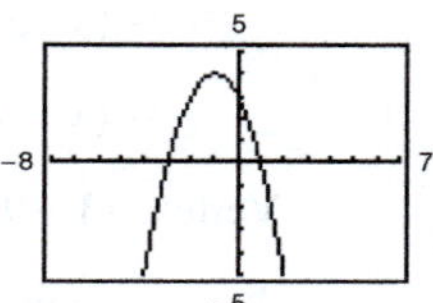

29. $g(x) = x^2 + 8x + 11 = (x + 4)^2 - 5$

Vertex: $(-4, -5)$

Axis of symmetry: $x = -4$

x-intercepts: $(-4 \pm \sqrt{5}, 0)$

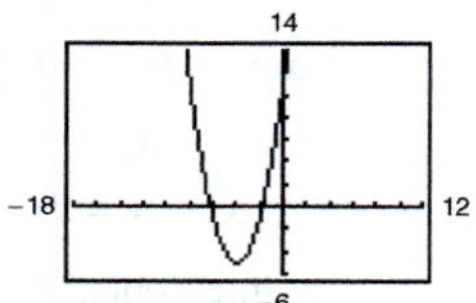

31. $f(x) = -2x^2 + 12x - 18$

$= -2(x^2 - 6x + 9 - 9) - 18$

$= -2(x^2 - 6x + 9) + 18 - 18$

$= -2(x^2 - 6x + 9)$

$= -2(x - 3)^2$

Vertex: $(3, 0)$

Axis of symmetry: $x = 3$

x-intercept: $(3, 0)$

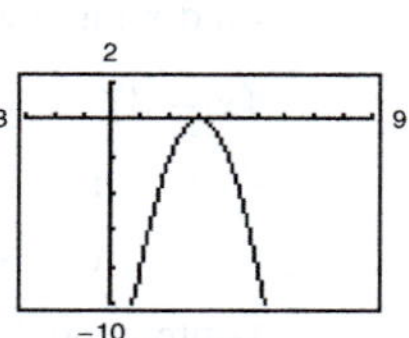

33. $g(x) = \frac{1}{2}(x^2 + 4x - 2) = \frac{1}{2}(x + 2)^2 - 3$

Vertex: $(-2, -3)$

Axis of symmetry: $x = -2$

x-intercepts: $(-2 \pm \sqrt{6}, 0)$

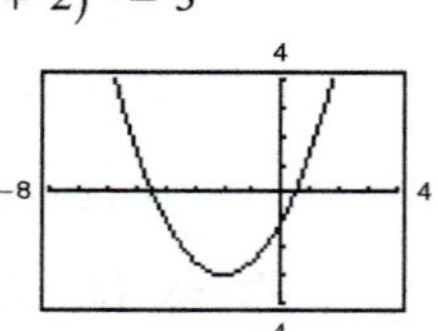

35. $(-2, -1)$ is the vertex.

$f(x) = a(x + 2)^2 - 1$

Because the graph passes through $(0, 3)$,

$3 = a(0 + 2)^2 - 1$

$3 = 4a - 1$

$4 = 4a$

$1 = a.$

So, $y = (x + 2)^2 - 1.$

37. $(-2, 5)$ is the vertex.

$$f(x) = a(x + 2)^2 + 5$$

Because the graph passes through $(0, 9)$,

$$9 = a(0 + 2)^2 + 5$$
$$4 = 4a$$
$$1 = a.$$

So, $f(x) = 1(x + 2)^2 + 5 = (x + 2)^2 + 5.$

39. $(1, -2)$ is the vertex.

$$f(x) = a(x - 1)^2 - 2$$

Because the graph passes through $(-1, 14)$,

$$14 = a(-1 - 1)^2 - 2$$
$$14 = 4a - 2$$
$$16 = 4a$$
$$4 = a.$$

So, $f(x) = 4(x - 1)^2 - 2.$

41. $(5, 12)$ is the vertex.

$$f(x) = a(x - 5)^2 + 12$$

Because the graph passes through $(7, 15)$,

$$15 = a(7 - 5)^2 + 12$$
$$3 = 4a \Rightarrow a = \tfrac{3}{4}.$$

So, $f(x) = \frac{3}{4}(x - 5)^2 + 12.$

43. $\left(-\frac{1}{4}, \frac{3}{2}\right)$ is the vertex.

$$f(x) = a\left(x + \tfrac{1}{4}\right)^2 + \tfrac{3}{2}$$

Because the graph passes through $(-2, 0)$,

$$0 = a\left(-2 + \tfrac{1}{4}\right)^2 + \tfrac{3}{2}$$
$$-\tfrac{3}{2} = \tfrac{49}{16}a \Rightarrow a = -\tfrac{24}{49}.$$

So, $f(x) = -\frac{24}{49}\left(x + \frac{1}{4}\right)^2 + \frac{3}{2}.$

45. $\left(-\frac{5}{2}, 0\right)$ is the vertex.

$$f(x) = a\left(x + \tfrac{5}{2}\right)^2$$

Because the graph passes through $\left(-\frac{7}{2}, -\frac{16}{3}\right)$,

$$-\tfrac{16}{3} = a\left(-\tfrac{7}{2} + \tfrac{5}{2}\right)^2$$
$$-\tfrac{16}{3} = a.$$

So, $f(x) = -\frac{16}{3}\left(x + \frac{5}{2}\right)^2.$

47. $y = x^2 - 2x - 3$

$$0 = x^2 - 2x - 3$$
$$0 = (x - 3)(x + 1)$$
$$x = 3 \text{ or } x = -1$$

x-intercepts: $(3, 0), (-1, 0)$

49. $y = 2x^2 + 5x - 3$

$$0 = 2x^2 + 5x - 3$$
$$0 = (2x - 1)(x + 3)$$
$$2x - 1 = 0 \Rightarrow x = \tfrac{1}{2}$$
$$x + 3 = 0 \Rightarrow x = -3$$

x-intercepts: $\left(\frac{1}{2}, 0\right), (-3, 0)$

51. $f(x) = x^2 - 4x$

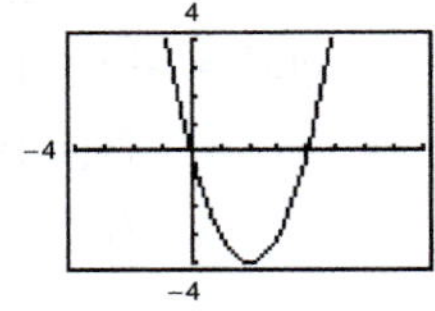

x-intercepts: $(0, 0), (4, 0)$

$$0 = x^2 - 4x$$
$$0 = x(x - 4)$$
$$x = 0 \quad \text{or} \quad x = 4$$

The x-intercepts and the solutions of $f(x) = 0$ are the same.

53. $f(x) = x^2 - 9x + 18$

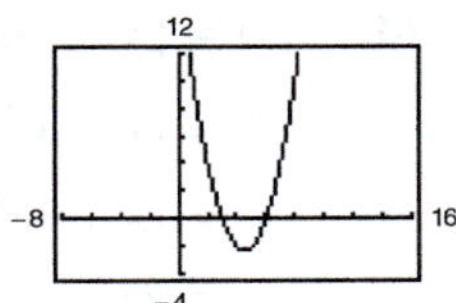

x-intercepts: $(3, 0), (6, 0)$

$$0 = x^2 - 9x + 18$$
$$0 = (x - 3)(x - 6)$$
$$x = 3 \quad \text{or} \quad x = 6$$

The x-intercepts and the solutions of $f(x) = 0$ are the same.

55. $f(x) = 2x^2 - 7x - 30$

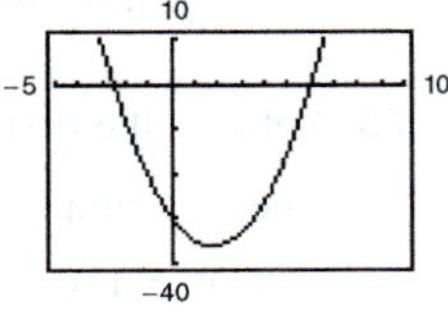

x-intercepts: $\left(-\frac{5}{2}, 0\right), (6, 0)$

$$0 = 2x^2 - 7x - 30$$
$$0 = (2x + 5)(x - 6)$$
$$x = -\tfrac{5}{2} \quad \text{or} \quad x = 6$$

The x-intercepts and the solutions of $f(x) = 0$ are the same.

57. $f(x) = [x - (-3)](x - 3)$ opens upward

$= (x + 3)(x - 3)$

$= x^2 - 9$

$g(x) = -[x - (-3)](x - 3)$

$= -(x + 3)(x - 3)$

$= -(x^2 - 9)$

$= -x^2 + 9$ opens downward

Note: $f(x) = a(x + 3)(x - 3)$ has x-intercepts $(-3, 0)$ and $(3, 0)$ for all real numbers $a \neq 0$.

59. $f(x) = [x - (-1)](x - 4)$ opens upward

$= (x + 1)(x - 4)$

$= x^2 - 3x - 4$

$g(x) = -[x - (-1)](x - 4)$ opens downward

$= -(x + 1)(x - 4)$

$= -(x^2 - 3x - 4)$

$= -x^2 + 3x + 4$

Note: $f(x) = a(x + 1)(x - 4)$ has x-intercepts $(-1, 0)$ and $(4, 0)$ for all real numbers $a \neq 0$.

61. $f(x) = [x - (-3)]\left[x - \left(-\frac{1}{2}\right)\right](2)$ opens upward

$= (x + 3)\left(x + \frac{1}{2}\right)(2)$

$= (x + 3)(2x + 1)$

$= 2x^2 + 7x + 3$

$g(x) = -(2x^2 + 7x + 3)$ opens downward

$= -2x^2 - 7x - 3$

Note: $f(x) = a(x + 3)(2x + 1)$ has x-intercepts $(-3, 0)$ and $\left(-\frac{1}{2}, 0\right)$ for all real numbers $a \neq 0$.

63. Let x = the first number and y = the second number.

Then the sum is

$x + y = 110 \Rightarrow y = 110 - x.$

The product is $P(x) = xy = x(110 - x) = 110x - x^2$.

$P(x) = -x^2 + 110x$

$= -(x^2 - 110x + 3025 - 3025)$

$= -\left[(x - 55)^2 - 3025\right]$

$= -(x - 55)^2 + 3025$

The maximum value of the product occurs at the vertex of $P(x)$ and is 3025. This happens when $x = y = 55$.

65. Let x = the first number and y = the second number.

Then the sum is

$x + 2y = 24 \Rightarrow y = \dfrac{24 - x}{2}.$

The product is $P(x) = xy = x\left(\dfrac{24 - x}{2}\right)$.

$P(x) = \frac{1}{2}(-x^2 + 24x)$

$= -\frac{1}{2}(x^2 - 24x + 144 - 144)$

$= -\frac{1}{2}\left[(x - 12)^2 - 144\right] = -\frac{1}{2}(x - 12)^2 + 72$

The maximum value of the product occurs at the vertex of $P(x)$ and is 72. This happens when $x = 12$ and $y = (24 - 12)/2 = 6$. So, the numbers are 12 and 6.

67. $y = -\frac{4}{9}x^2 + \frac{24}{9}x + 12$

The vertex occurs at $-\dfrac{b}{2a} = \dfrac{-24/9}{2(-4/9)} = 3.$

The maximum height is

$y(3) = -\frac{4}{9}(3)^2 + \frac{24}{9}(3) + 12 = 16$ feet.

69. $C = 800 - 10x + 0.25x^2 = 0.25x^2 - 10x + 800$

The vertex occurs at $x = -\dfrac{b}{2a} = -\dfrac{-10}{2(0.25)} = 20.$

The cost is minimum when $x = 20$ fixtures.

71. $R(p) = -25p^2 + 1200p$

(a) $R(20) = \$14{,}000$ thousand $= \$14{,}000{,}000$

$R(25) = \$14{,}375$ thousand $= \$14{,}375{,}000$

$R(30) = \$13{,}500$ thousand $= \$13{,}500{,}000$

(b) The revenue is a maximum at the vertex.

$-\dfrac{b}{2a} = \dfrac{-1200}{2(-25)} = 24$

$R(24) = 14{,}400$

The unit price that will yield a maximum revenue of \$14,400,000 is \$24.

73. (a) 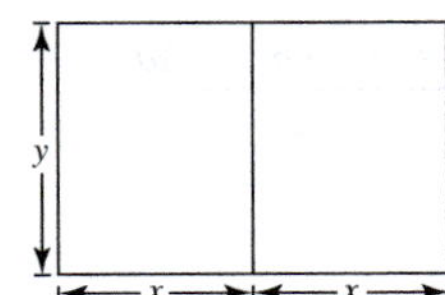

$$4x + 3y = 200 \Rightarrow y = \frac{1}{3}(200 - 4x) = \frac{4}{3}(50 - x)$$

$$A = 2xy = 2x\left[\frac{4}{3}(50 - x)\right] = \frac{8}{3}x(50 - x) = \frac{8x(50 - x)}{3}$$

(b) To find the dimensions that produce a maximum enclosed area, you can find the vertex.

To do this, either write the quadratic function in standard form or use $x = -\frac{b}{2a}$, so the coordinates of the vertex are $\left(-\frac{b}{2a}, f\left(-\frac{b}{2a}\right)\right)$.

$$A = \frac{8}{3}x(50 - x) = -\frac{8}{3}(x^2 - 50x) = -\frac{8}{3}(x^2 - 50x + 625 - 625) = -\frac{8}{3}\left[(x - 25)^2 - 625\right] = -\frac{8}{3}(x - 25)^2 + \frac{5000}{3}$$

So the vertex is $\left(25, \frac{5000}{3}\right)$ from the standard form, or is $x = -\frac{b}{2a} = -\frac{\frac{400}{3}}{2\left(-\frac{8}{3}\right)} = \frac{400}{16} = 25$ and $A(25) = \frac{5000}{3}$.

When $x = 25$ feet and $y = \frac{(200 - 4(25))}{3} = \frac{100}{3}$ feet.

The dimensions are $2x = 50$ feet by $33\frac{1}{3}$ feet, and the maximum enclosed area is $\frac{5000}{3} \approx 1666.67$ square feet.

75. True. The equation $-12x^2 - 1 = 0$ has no real solution, so the graph has no x-intercepts.

77. $f(x) = -x^2 + bx - 75$, maximum value: 25

The maximum value, 25, is the y-coordinate of the vertex.

Find the x-coordinate of the vertex:

$$x = -\frac{b}{2a} = -\frac{b}{2(-1)} = \frac{b}{2}$$

$$f(x) = -x^2 + bx - 75$$

$$f\left(\frac{b}{2}\right) = -\left(\frac{b}{2}\right)^2 + b\left(\frac{b}{2}\right) - 75$$

$$25 = -\frac{b^2}{4} + \frac{b^2}{2} - 75$$

$$100 = \frac{b^2}{4}$$

$$400 = b^2$$

$$\pm 20 = b$$

79. $f(x) = ax^2 + bx + c$

$$= a\left(x^2 + \frac{b}{a}x\right) + c$$

$$= a\left(x^2 + \frac{b}{a}x + \frac{b^2}{4a^2} - \frac{b^2}{4a^2}\right) + c$$

$$= a\left(x + \frac{b}{2a}\right)^2 - \frac{b^2}{4a} + c$$

$$= a\left(x + \frac{b}{2a}\right)^2 + \frac{4ac - b^2}{4a}$$

$$f\left(-\frac{b}{2a}\right) = a\left(\frac{b^2}{4a^2}\right) + b\left(-\frac{b}{2a}\right) + c$$

$$= \frac{b^2}{4a} - \frac{b^2}{2a} + c$$

$$= \frac{b^2 - 2b^2 + 4ac}{4a} = \frac{4ac - b^2}{4a}$$

So, the vertex occurs at

$$\left(-\frac{b}{2a}, \frac{4ac - b^2}{4a}\right) = \left(-\frac{b}{2a}, f\left(-\frac{b}{2a}\right)\right).$$

81. If $f(x) = ax^2 + bx + c$ has two real zeros, then by the Quadratic Formula they are $x = \dfrac{-b \pm \sqrt{b^2 - 4ac}}{2a}$.

The average of the zeros of f is

$$\frac{\dfrac{-b - \sqrt{b^2 - 4ac}}{2a} + \dfrac{-b + \sqrt{b^2 - 4ac}}{2a}}{2} = \frac{\dfrac{-2b}{2a}}{2} = -\frac{b}{2a}.$$

This is the x-coordinate of the vertex of the graph.

Section 2.2 Polynomial Functions of Higher Degree

1. continuous

3. n; $n - 1$

5. touches; crosses

7. standard

9. $f(x) = -2x^2 - 5x$ is a parabola with x-intercepts $(0, 0)$ and $\left(-\frac{5}{2}, 0\right)$ and opens downward. Matches graph (c).

10. $f(x) = 2x^3 - 3x + 1$ has intercepts $(0, 1), (1, 0), \left(-\frac{1}{2} - \frac{1}{2}\sqrt{3}, 0\right)$ and $\left(-\frac{1}{2} + \frac{1}{2}\sqrt{3}, 0\right)$. Matches graph (f).

11. $f(x) = -\frac{1}{4}x^4 + 3x^2$ has intercepts $(0, 0)$ and $\left(\pm 2\sqrt{3}, 0\right)$. Matches graph (a).

12. $f(x) = -\frac{1}{3}x^3 + x^2 - \frac{4}{3}$ has y-intercept $\left(0, -\frac{4}{3}\right)$. Matches graph (e).

13. $f(x) = x^4 + 2x^3$ has intercepts $(0, 0)$ and $(-2, 0)$. Matches graph (d).

14. $f(x) = \frac{1}{5}x^5 - 2x^3 + \frac{9}{5}x$ has intercepts $(0, 0), (1, 0), (-1, 0), (3, 0), (-3, 0)$. Matches graph (b).

15. $y = x^3$

(a) $f(x) = (x - 4)^3$

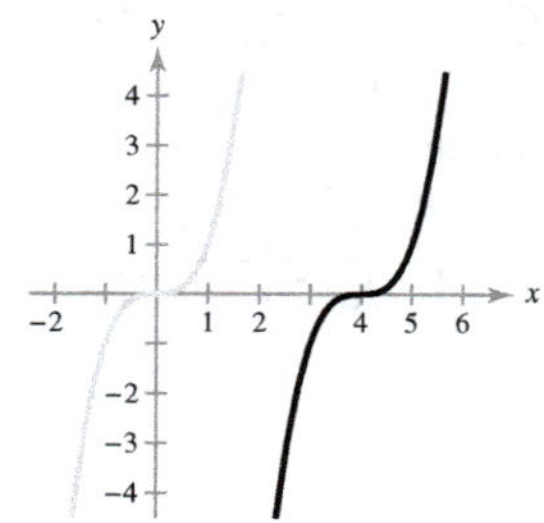

Horizontal shift four units to the right

(b) $f(x) = x^3 - 4$

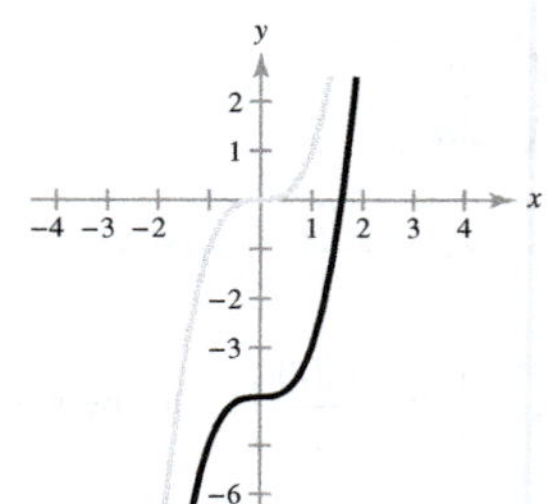

Vertical shift four units downward

(c) $f(x) = -\dfrac{1}{4}x^3$

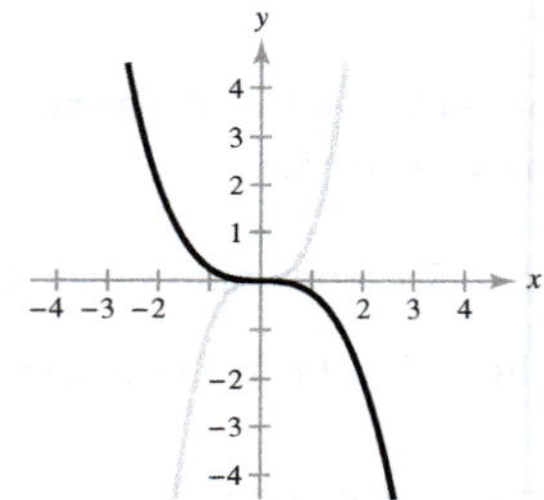

Reflection in the x-axis and a vertical shrink (each y-value is multiplied by $\frac{1}{4}$)

(d) $f(x) = (x - 4)^3 - 4$

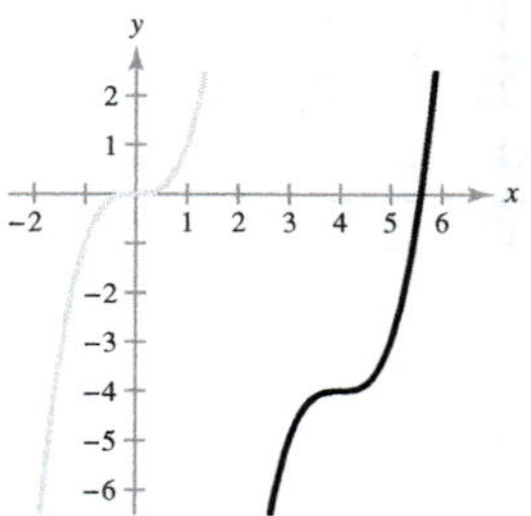

Horizontal shift four units to the right and vertical shift four units downward

17. $y = x^4$

(a) $f(x) = (x + 3)^4$

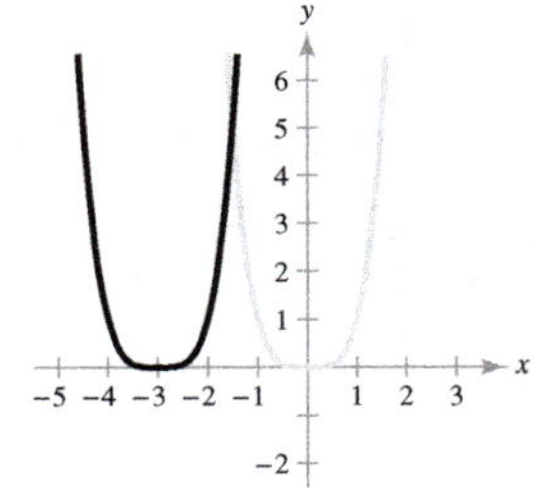

Horizontal shift three units to the left

(b) $f(x) = x^4 - 3$

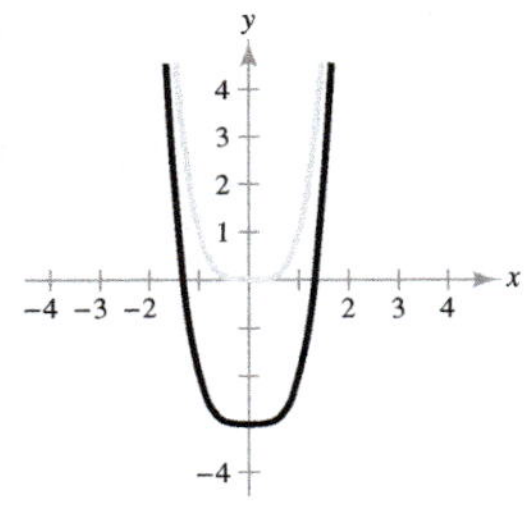

Vertical shift three units downward

(c) $f(x) = 4 - x^4$

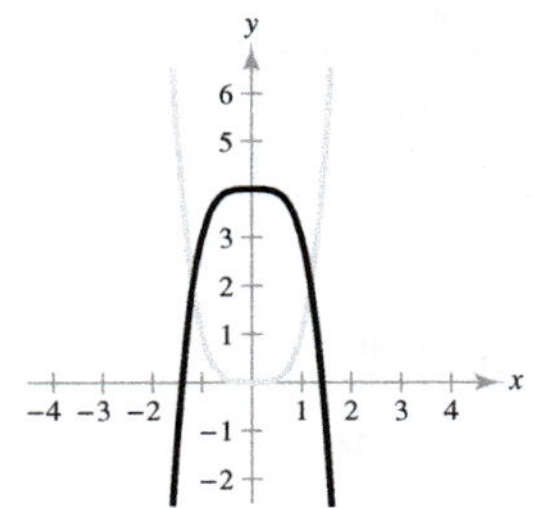

Reflection in the x-axis and then a vertical shift four units upward

(d) $f(x) = \frac{1}{2}(x - 1)^4$

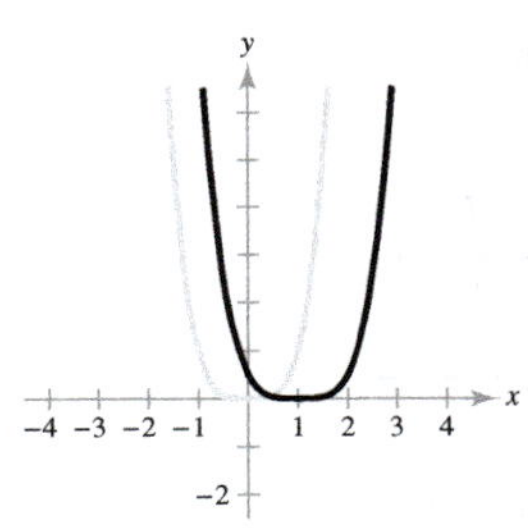

Horizontal shift one unit to the right and a vertical shrink (each y-value is multiplied by $\frac{1}{2}$)

(e) $f(x) = (2x)^4 + 1$

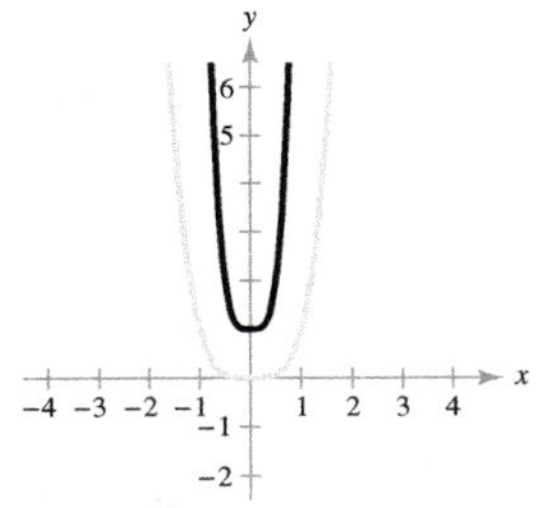

Vertical shift one unit upward and a horizontal shrink (each y-value is multiplied by 16)

(f) $f(x) = \left(\frac{1}{2}x\right)^4 - 2$

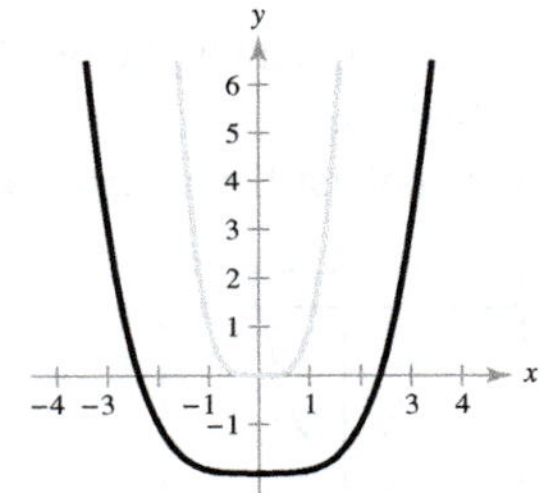

Vertical shift two units downward and a horizontal stretch (each y-value is multipied by $\frac{1}{16}$)

19. $f(x) = 12x^3 + 4x$

Degree: 3

Leading coefficient: 12

The degree is odd and the leading coefficient is positive. The graph falls to the left and rises to the right.

21. $g(x) = 5 - \frac{7}{2}x - 3x^2$

Degree: 2

Leading coefficient: -3

The degree is even and the leading coefficient is negative. The graph falls to the left and falls to the right.

23. $g(x) = 6x - x^3 + x^2$

Degree: 3

Leading coefficient: -1

The degree is odd and the leading coefficient is negative. The graph rises to the left and falls to the right.

25. $f(x) = 9.8x^6 - 1.2x^3$

Degree: 6

Leading coefficient: 9.8

The degree is even and the leading coefficient is positive. The graph rises to the left and rises to the right.

27. $f(s) = -\frac{7}{8}(s^3 + 5s^2 - 7s + 1)$

Degree: 3

Leading coefficient: $-\frac{7}{8}$

The degree is odd and the leading coefficient is negative. The graph rises to the left and falls to the right.

29. $f(x) = 3x^3 - 9x + 1$; $g(x) = 3x^3$

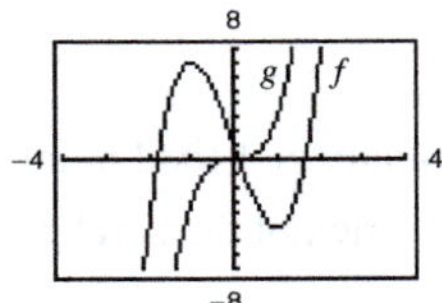

31. $f(x) = -(x^4 - 4x^3 + 16x)$; $g(x) = -x^4$

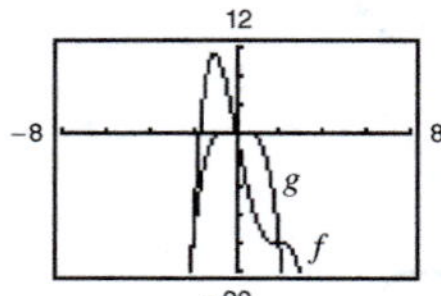

33. $f(x) = x^2 - 36$

(a) $0 = x^2 - 36$

$0 = (x + 6)(x - 6)$

$x + 6 = 0 \qquad x - 6 = 0$

$x = -6 \qquad x = 6$

Zeros: ± 6

(b) Each zero has a multiplicity of one (odd multiplicity).

(c) Turning points: 1 (the vertex of the parabola)

(d)

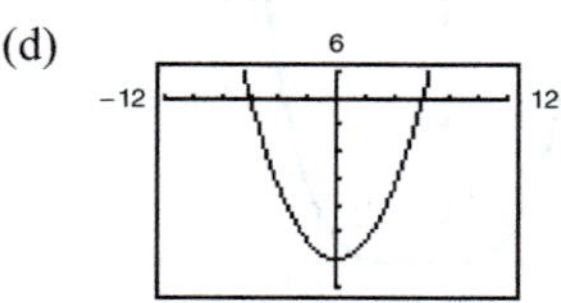

35. $h(t) = t^2 - 6t + 9$

(a) $0 = t^2 - 6t + 9 = (t - 3)^2$

Zero: $t = 3$

(b) $t = 3$ has a multiplicity of 2 (even multiplicity).

(c) Turning points: 1 (the vertex of the parabola)

(d)

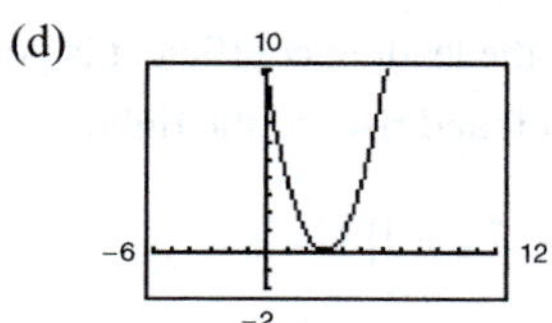

37. $f(x) = \frac{1}{3}x^2 + \frac{1}{3}x - \frac{2}{3}$

(a) $0 = \frac{1}{3}x^2 + \frac{1}{3}x - \frac{2}{3}$

$= \frac{1}{3}(x^2 + x - 2)$

$= \frac{1}{3}(x + 2)(x - 1)$

Zeros: $x = -2, x = 1$

(b) Each zero has a multiplicity of 1 (odd multiplicity).

(c) Turning points: 1 (the vertex of the parabola)

(d)

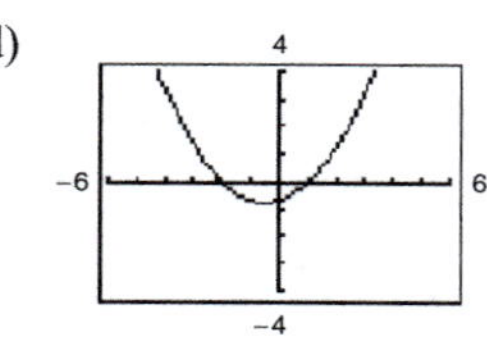

39. $g(x) = 5x(x^2 - 2x - 1)$

(a) $0 = 5x(x^2 - 2x - 1)$

$0 = x(x^2 - 2x - 1)$

For $x^2 - 2x - 1 = 0, a = 1, b = -2, c = -1$.

$$x = \frac{-(-2) \pm \sqrt{(-2)^2 - 4(1)(-1)}}{2(1)}$$

$$= \frac{2 \pm \sqrt{8}}{2}$$

$$= 1 \pm \sqrt{2}$$

Zeros: $x = 0, x = 1 \pm \sqrt{2}$

(b) Each zero has a multiplicity of 1 (odd multiplicity).

(c) Turning points: 2

(d)

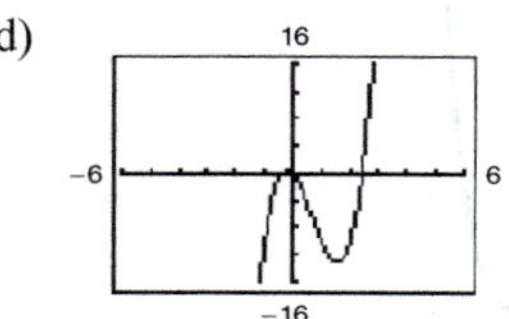

41. $f(x) = 3x^3 - 12x^2 + 3x$

(a) $0 = 3x^3 - 12x^2 + 3x = 3x(x^2 - 4x + 1)$

Zeros: $x = 0, x = 2 \pm \sqrt{3}$ (by the Quadratic Formula)

(b) Each zero has a multiplicity of 1 (odd multiplicity).

(c) Turning points: 2

(d)

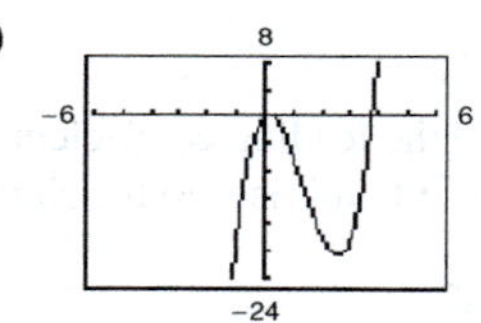

43. $g(t) = t^5 - 6t^3 + 9t$

(a) $0 = t^5 - 6t^3 + 9t = t(t^4 - 6t^2 + 9) = t(t^2 - 3)^2$

$= t(t + \sqrt{3})^2(t - \sqrt{3})^2$

Zeros: $t = 0, t = \pm\sqrt{3}$

(b) $t = 0$ has a multiplicity of 1 (odd multiplicity).

$t = \pm\sqrt{3}$ each have a multiplicity of 2 (even multiplicity).

(c) Turning points: 4

(d)

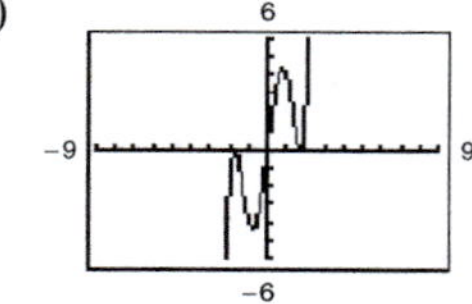

45. $f(x) = 3x^4 + 9x^2 + 6$

(a) $0 = 3x^4 + 9x^2 + 6$

$0 = 3(x^4 + 3x^2 + 2)$

$0 = 3(x^2 + 1)(x^2 + 2)$

No real zeros

(b) No multiplicity

(c) Turning points: 1

(d)

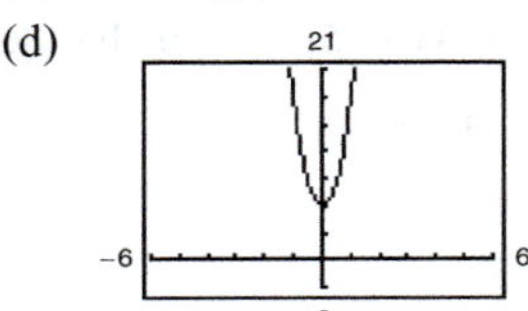

47. $g(x) = x^3 + 3x^2 - 4x - 12$

(a) $0 = x^3 + 3x^2 - 4x - 12 = x^2(x + 3) - 4(x + 3)$

$= (x^2 - 4)(x + 3) = (x - 2)(x + 2)(x + 3)$

Zeros: $x = \pm 2, x = -3$

(b) Each zero has a multiplicity of 1 (odd multiplicity).

(c) Turning points: 2

(d)

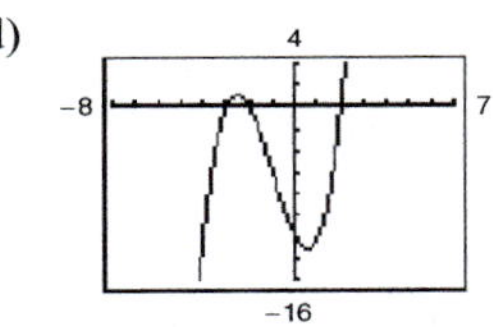

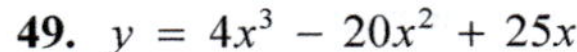

49. $y = 4x^3 - 20x^2 + 25x$

(a)

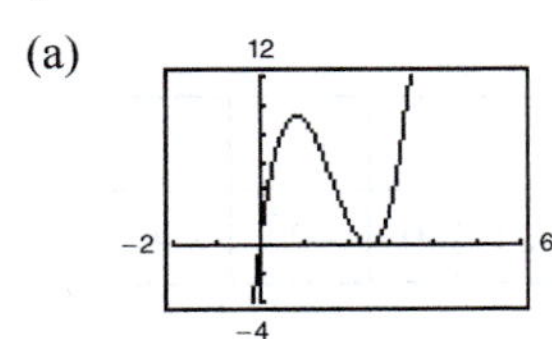

(b) x-intercepts: $(0, 0), \left(\frac{5}{2}, 0\right)$

(c) $0 = 4x^3 - 20x^2 + 25x$

$0 = x(4x^2 - 20x + 25)$

$0 = x(2x - 5)^2$

$x = 0, \frac{5}{2}$

(d) The solutions are the same as the x-coordinates of the x-intercepts.

51. $y = x^5 - 5x^3 + 4x$

(a)

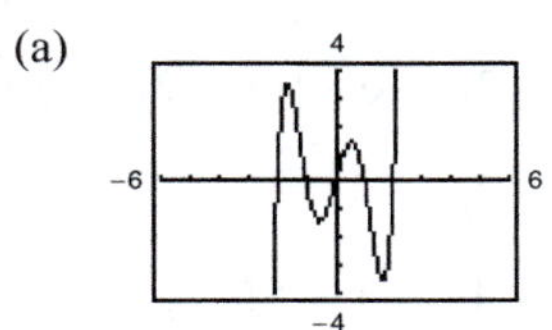

(b) x-intercepts: $(0, 0), (\pm 1, 0), (\pm 2, 0)$

(c) $0 = x^5 - 5x^3 + 4x$

$0 = x(x^2 - 1)(x^2 - 4)$

$0 = x(x + 1)(x - 1)(x + 2)(x - 2)$

$x = 0, \pm 1, \pm 2$

(d) The solutions are the same as the x-coordinates of the x-intercepts.

53. $f(x) = (x - 0)(x - 7)$

$= x^2 - 7x$

Note: $f(x) = ax(x - 7)$ has zeros 0 and 7 for all real numbers $a \neq 0$.

55. $f(x) = (x - 0)(x + 2)(x + 4)$

$= x(x^2 + 6x + 8)$

$= x^3 + 6x^2 + 8x$

Note: $f(x) = ax(x + 2)(x + 4)$ has zeros $0, -2,$ and -4 for all real numbers $a \neq 0$.

57. $f(x) = (x - 4)(x + 3)(x - 3)(x - 0)$

$= (x - 4)(x^2 - 9)x$

$= x^4 - 4x^3 - 9x^2 + 36x$

Note: $f(x) = a(x^4 - 4x^3 - 9x^2 + 36x)$ has zeros $4, -3, 3,$ and 0 for all real numbers $a \neq 0$.

59. $f(x) = \left[x - (1 + \sqrt{2})\right]\left[x - (1 - \sqrt{2})\right]$

$= \left[(x - 1) - \sqrt{2}\right]\left[(x - 1) + \sqrt{2}\right]$

$= (x - 1)^2 - (\sqrt{2})^2$

$= x^2 - 2x + 1 - 2$

$= x^2 - 2x - 1$

Note: $f(x) = a(x^2 - 2x - 1)$ has zeros $1 + \sqrt{2}$ and $1 - \sqrt{2}$ for all real numbers $a \neq 0$.

61. $f(x) = (x - 2)\left[x - \left(2 + \sqrt{5}\right)\right]\left[x - \left(2 - \sqrt{5}\right)\right]$

$= (x - 2)\left[(x - 2) - \sqrt{5}\right]\left[(x - 2) + \sqrt{5}\right]$

$= (x - 2)\left[(x - 2)^2 - 5\right]$

$= (x - 2)\left[x^2 - 4x + 4 - 5\right]$

$= (x - 2)\left(x^2 - 4x - 1\right)$

$= x^3 - 6x^2 + 7x + 2$

Note: $f(x) = a\left(x^3 - 6x^2 + 7x + 2\right)$ has zeros 2, $2 + \sqrt{5}$, and $2 - \sqrt{5}$ for all real numbers $a \neq 0$.

63. $f(x) = (x + 3)(x + 3) = x^2 + 6x + 9$

Note: $f(x) = a\left(x^2 + 6x + 9\right)$, $a \neq 0$, has degree 2 and zero $x = -3$.

65. $f(x) = (x - 0)(x + 5)(x - 1)$

$= x\left(x^2 + 4x - 5\right)$

$= x^3 + 4x^2 - 5x$

Note: $f(x) = ax\left(x^2 + 4x - 5\right)$, $a \neq 0$, has degree 3 and zeros $x = 0, -5,$ and 1.

67. $f(x) = (x - (-5))^2(x - 1)(x - 2) = x^4 + 7x^3 - 3x^2 - 55x + 50$

or $f(x) = (x - (-5))(x - 1)^2(x - 2) = x^4 + x^3 - 15x^2 + 23x - 10$

or $f(x) = (x - (-5))(x - 1)(x - 2)^2 = x^4 - 17x^2 + 36x - 20$

Note: Any nonzero scalar multiple of these functions would also have degree 4 and zeros $x = -5, 1,$ and 2.

69. $f(x) = (x - 0)(x - 0)(x - 0)\left(x - \sqrt{3}\right)\left(x - \left(-\sqrt{3}\right)\right)$

$= x^3\left(x - \sqrt{3}\right)\left(x + \sqrt{3}\right)$

$= x^3\left(x^2 - 3\right)$

$= x^5 - 3x^3$

Note: $f(x) = a\left(x^5 - 3x^3\right)$, $a \neq 0$, has degree 5 and zeros $x = 0, \sqrt{3},$ and $-\sqrt{3}$.

71. $f(t) = \frac{1}{4}\left(t^2 - 2t + 15\right) = \frac{1}{4}(t - 1)^2 + \frac{7}{2}$

(a) Rises to the left; rises to the right

(b) No real zeros (no x-intercepts)

(c)

t	-1	0	1	2	3
$f(t)$	4.5	3.75	3.5	3.75	4.5

(d) The graph is a parabola with vertex $\left(1, \frac{7}{2}\right)$.

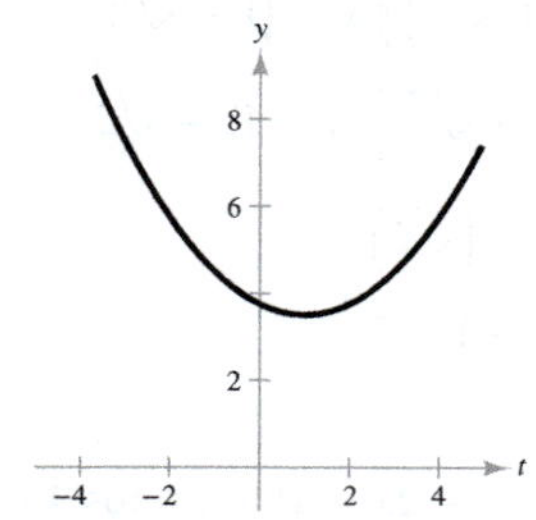

73. $f(x) = x^3 - 25x = x(x + 5)(x - 5)$

(a) Falls to the left; rises to the right

(b) Zeros: 0, -5, 5

(c)

x	-2	-1	0	1	2
$f(x)$	42	24	0	-24	-42

(d)

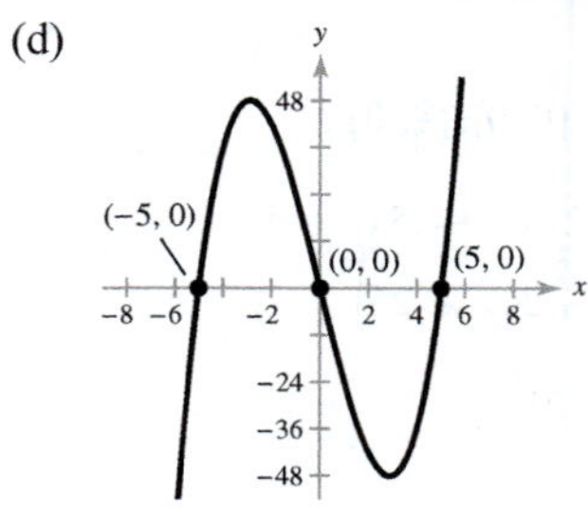

75. $f(x) = -8 + \frac{1}{2}x^4 = \frac{1}{2}(x^4 - 16)$
$= \frac{1}{2}(x^2 + 4)(x - 2)(x + 2)$

(a) Rises to the left; rises to the right

(b) Zeros $x = \pm 2$:

(c)

x	-2	-1	0	1	2
$f(x)$	0	$-\frac{15}{2}$	-8	$-\frac{15}{2}$	0

(d)

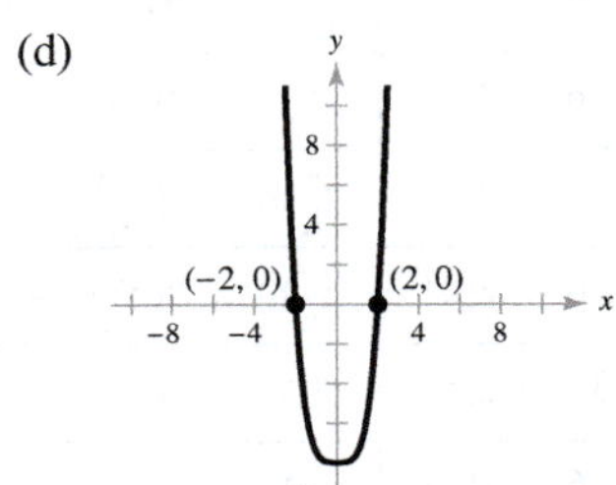

77. $f(x) = 3x^3 - 15x^2 + 18x = 3x(x - 2)(x - 3)$

(a) Falls to the left; rises to the right

(b) Zeros: 0, 2, 3

(c)

x	0	1	2	2.5	3	3.5
$f(x)$	0	6	0	-1.875	0	7.875

(d)

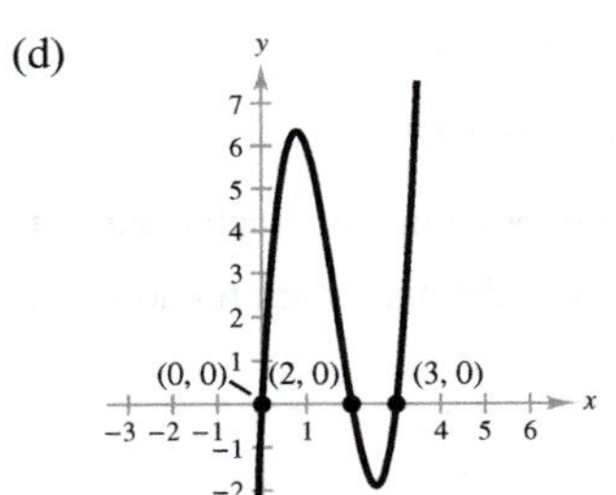

79. $f(x) = -5x^2 - x^3 = -x^2(5 + x)$

(a) Rises to the left; falls to the right

(b) Zeros: $0, -5$

(c)

x	-5	-4	-3	-2	-1	0	1
$f(x)$	0	-16	-18	-12	-4	0	-6

(d)

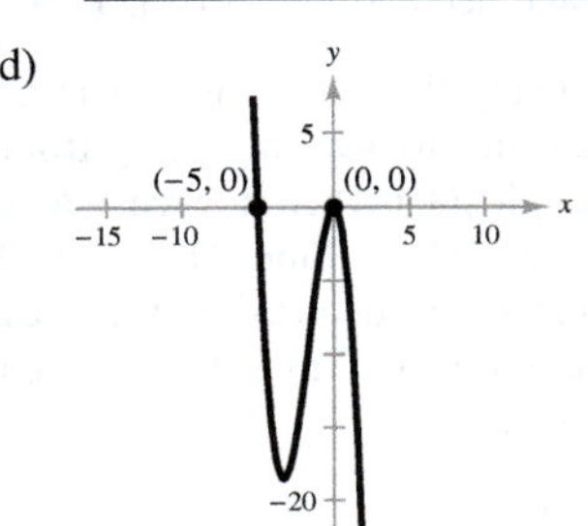

81. $f(x) = 9x^2(x + 2)^2$

(a) Falls to the left, rises to the right

(b) Zeros: $x = 0, \ -2$

(c)

x	-3	-2	-1	0	1
$f(x)$	81	0	9	0	81

(d)

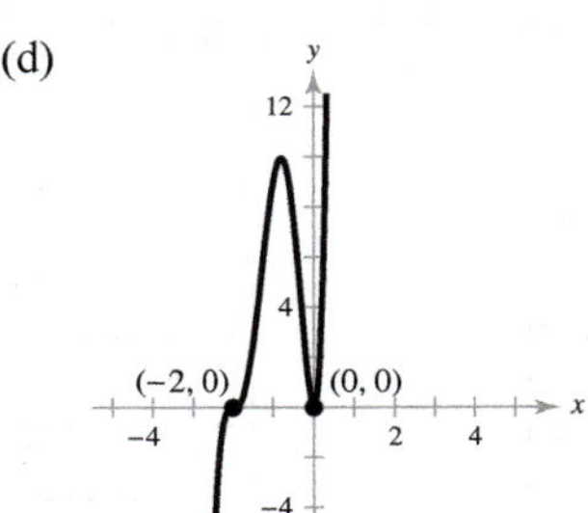

83. $g(t) = -\frac{1}{4}(t - 2)^2(t + 2)^2$

(a) Falls to the left; falls to the right

(b) Zeros: $2, -2$

(c)

t	-3	-2	-1	0	1	2	3
$g(t)$	$-\frac{25}{4}$	0	$-\frac{9}{4}$	-4	$-\frac{9}{4}$	0	$-\frac{25}{4}$

(d)

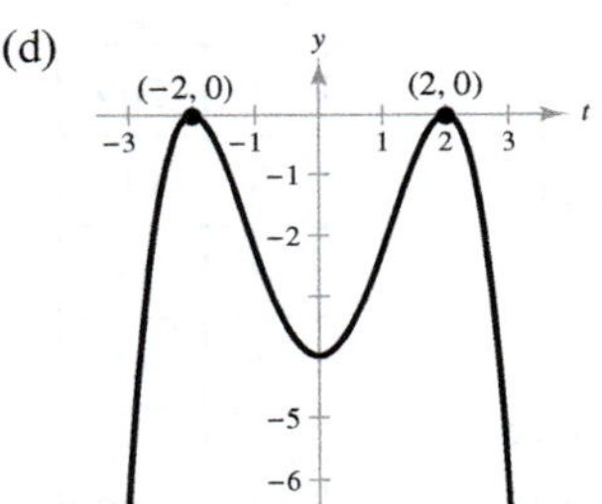

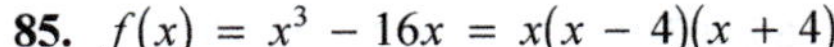
85. $f(x) = x^3 - 16x = x(x - 4)(x + 4)$

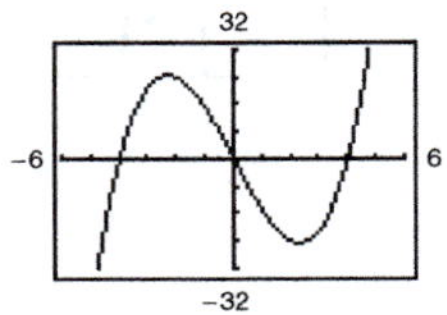

Zeros: 0 of multiplicity 1; 4 of multiplicity 1; and −4 of multiplicity 1

87. $g(x) = \frac{1}{5}(x + 1)^2(x - 3)(2x - 9)$

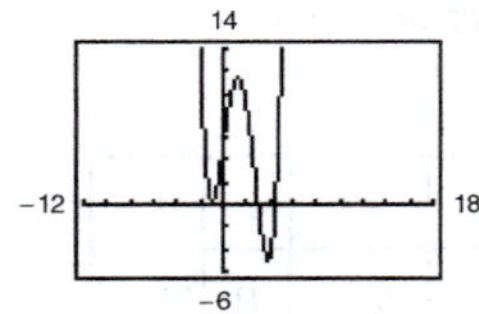

Zeros: -1 of multiplicity 2; 3 of multiplicity 1; $\frac{9}{2}$ of multiplicity 1

89. $f(x) = x^3 - 3x^2 + 3$

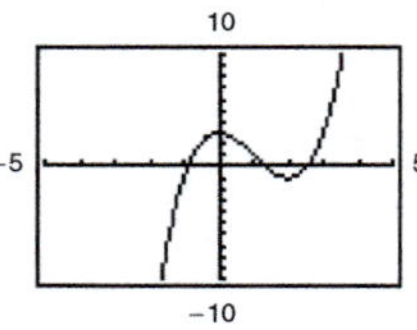

x	y
−3	−51
−2	−17
−1	−1
0	3
1	1
2	−1
2	−21
3	3
4	19

The function has three zeros. They are in the intervals $[-1, 0]$, $[1, 2]$, and $[2, 3]$. They are $x \approx -0.879, 1.347, 2.532$.

91. $g(x) = 3x^4 + 4x^3 - 3$

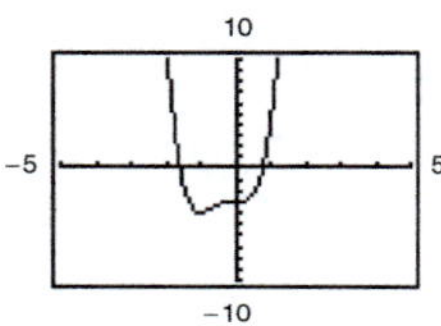

x	y
−4	509
−3	132
−2	13
−1	−4
0	−3
1	4
2	77
3	348

The function has two zeros. They are in the intervals $[-2, -1]$ and $[0, 1]$. They are $x \approx -1.585, 0.779$.

93. (a) Volume $= l \cdot w \cdot h$

height $= x$

length $=$ width $= 36 - 2x$

So, $V(x) = (36 - 2x)(36 - 2x)(x) = x(36 - 2x)^2$.

(b) Domain: $0 < x < 18$

The length and width must be positive.

(c)

Box Height	Box Width	Box Volume, V
1	$36 - 2(1)$	$1[36 - 2(1)]^2 = 1156$
2	$36 - 2(2)$	$2[36 - 2(2)]^2 = 2048$
3	$36 - 2(3)$	$3[36 - 2(3)]^2 = 2700$
4	$36 - 2(4)$	$4[36 - 2(4)]^2 = 3136$
5	$36 - 2(5)$	$5[36 - 2(5)]^2 = 3380$
6	$36 - 2(6)$	$6[36 - 2(6)]^2 = 3456$
7	$36 - 2(7)$	$7[36 - 2(7)]^2 = 3388$

The volume is a maximum of 3456 cubic inches when the height is 6 inches and the length and width are each 24 inches. So the dimensions are $6 \times 24 \times 24$ inches.

(d)

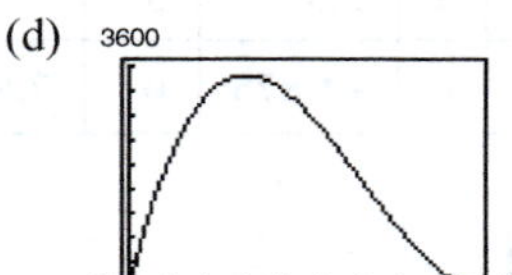

The maximum point on the graph occurs at $x = 6$. This agrees with the maximum found in part (c).

95. (a)

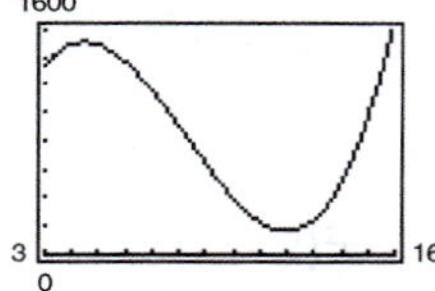

Using trace and zoom features, the relative maximum is approximately $(4.44, 1512.60)$ and the relative minimum is approximately $(11.97, 189.37)$.

(b) The revenue is increasing on $(3, 4.44)$ and $(11.97, 16)$ and decreasing on $(4.44, 11.97)$.

(c) Answers will vary. *Sample answer:* The revenue for the software company was increasing from 2003 to midway through 2004 when it reached a maximum of approximately \$1.5 trillion. Then from 2004 to 2012 the revenue was decreasing. It decreased to \$189 million. From 2012 to 2016 the revenue has been increasing.

97. $R = \frac{1}{100{,}000}\left(-x^3 + 600x^2\right)$

The point of diminishing returns (where the graph changes from curving upward to curving downward) occurs when $x = 200$. The point is $(200, 160)$ which corresponds to spending $\$2{,}000{,}000$ on advertising to obtain a revenue of $\$160$ million.

99. True. A polynomial function only falls to the right when the leading coefficient is negative.

101. False. The range of an even function cannot be $(-\infty, \infty)$. An even function's graph will fall to the left and right or rise to the left and right.

103. Answers will vary. *Sample answers:*

$a_4 < 0$ $\qquad$ $a_4 > 0$

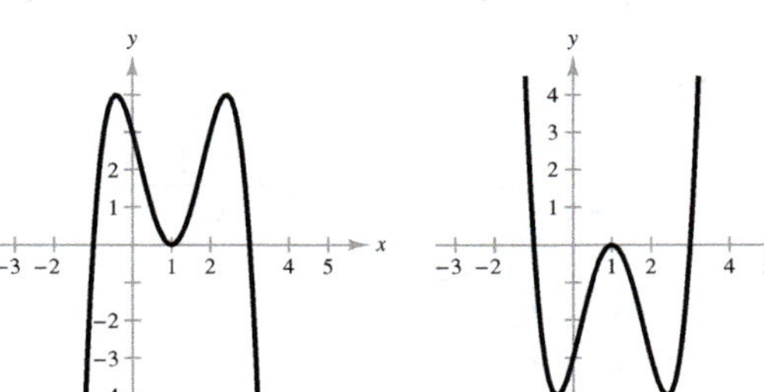

105. $f(x) = x^4$; $f(x)$ is even.

(a) $g(x) = f(x) + 2$

Vertical shift two units upward

$$\begin{aligned} g(-x) &= f(-x) + 2 \\ &= f(x) + 2 \\ &= g(x) \end{aligned}$$

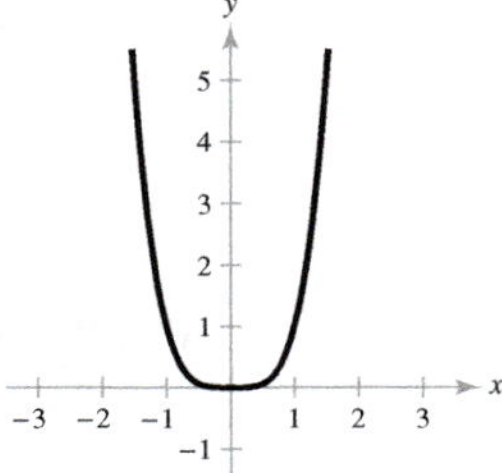

Even

(b) $g(x) = f(x + 2)$

Horizontal shift two units to the left

Neither odd nor even

(c) $g(x) = f(-x) = (-x)^4 = x^4$

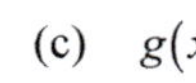

Reflection in the y-axis. The graph looks the same.

Even

(d) $g(x) = -f(x) = -x^4$

Reflection in the x-axis

Even

(e) $g(x) = f\left(\frac{1}{2}x\right) = \frac{1}{16}x^4$

Horizontal stretch

Even

(f) $g(x) = \frac{1}{2}f(x) = \frac{1}{2}x^4$

Vertical shrink

Even

(g) $g(x) = f\left(x^{3/4}\right) = \left(x^{3/4}\right)^4 = x^3, x \ge 0$

Neither odd nor even

(h) $g(x) = (f \circ f)(x) = f(f(x)) = f\left(x^4\right) = \left(x^4\right)^4 = x^{16}$

Even

107.

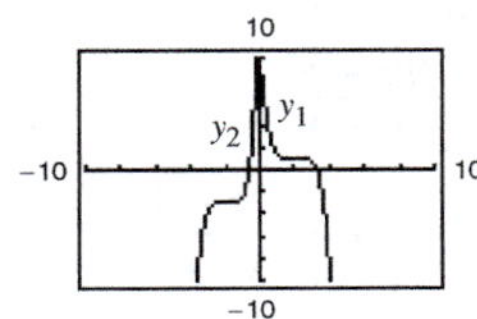

(a) $y_1 = -\frac{1}{3}(x - 2)^5 + 1$ is decreasing and

$y_2 = \frac{3}{5}(x + 2)^5 - 3$ is increasing.

(b) It is possible for $g(x) = a(x - h)^5 + k$ to be strictly increasing if $a > 0$ and strictly decreasing if $a < 0$.

(c) f cannot be written in the form $f(x) = a(x - h)^5 + k$ because f is not strictly increasing or strictly decreasing.

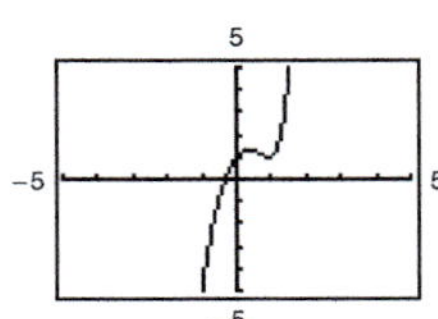

Section 2.3 Polynomial and Synthetic Division

1. $f(x)$ is the dividend; $d(x)$ is the divisor; $q(x)$ is the quotient; $r(x)$ is the remainder

3. improper

5. Factor

7. $y_1 = \dfrac{x^2}{x+2}$ and $y_2 = x - 2 + \dfrac{4}{x+2}$

$$
\begin{array}{r}
x - 2 \\
x + 2\overline{\big)\, x^2 + 0x + 0} \\
\underline{x^2 + 2x} \\
-2x + 0 \\
\underline{-2x - 4} \\
4
\end{array}
$$

So, $\dfrac{x^2}{x+2} = x - 2 + \dfrac{4}{x+2}$ and $y_1 = y_2$.

9. $y_1 = \dfrac{x^2 + 2x - 1}{x+3}$, $y_2 = x - 1 + \dfrac{2}{x+3}$

(a) and (b)

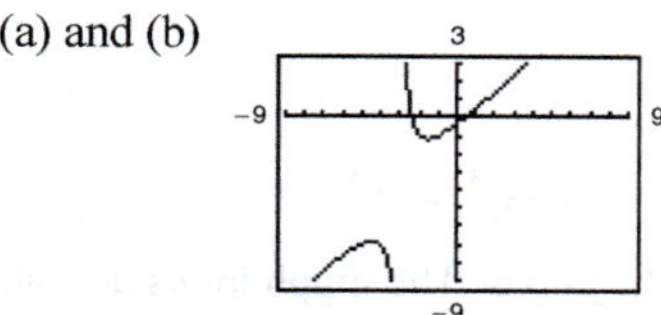

(c)
$$
\begin{array}{r}
x - 1 \\
x + 3\overline{\big)\, x^2 + 2x - 1} \\
\underline{x^2 + 3x} \\
-x - 1 \\
\underline{-x - 3} \\
2
\end{array}
$$

So, $\dfrac{x^2 + 2x - 1}{x+3} = x - 1 + \dfrac{2}{x+3}$ and $y_1 = y_2$.

11.
$$
\begin{array}{r}
2x + 4 \\
x + 3\overline{\big)\, 2x^2 + 10x + 12} \\
\underline{2x^2 + 6x} \\
4x + 12 \\
\underline{4x + 12} \\
0
\end{array}
$$

$$\frac{2x^2 + 10x + 12}{x+3} = 2x + 4,\ x \neq 3$$

13.
$$
\begin{array}{r}
x^2 - 3x + 1 \\
4x + 5\overline{\big)\, 4x^3 - 7x^2 - 11x + 5} \\
\underline{4x^3 + 5x^2} \\
-12x^2 - 11x \\
\underline{-12x^2 - 15x} \\
4x + 5 \\
\underline{4x + 5} \\
0
\end{array}
$$

$$\frac{4x^3 - 7x^2 - 11x + 5}{4x + 5} = x^2 - 3x + 1,\ x \neq -\frac{5}{4}$$

15.
$$
\begin{array}{r}
x^3 + 3x^2 \qquad -1 \\
x + 2\overline{\big)\, x^4 + 5x^3 + 6x^2 - x - 2} \\
\underline{x^4 + 2x^3} \\
3x^3 + 6x^2 \\
\underline{3x^3 + 6x^2} \\
-x - 2 \\
\underline{-x - 2} \\
0
\end{array}
$$

$$\frac{x^4 + 5x^3 + 6x^2 - x - 2}{x + 2} = x^3 + 3x^2 - 1,\ x \neq -2$$

17.
$$
\begin{array}{r}
6 \\
x + 1\overline{\big)\, 6x + 5} \\
\underline{6x + 6} \\
-1
\end{array}
$$

$$\frac{6x + 5}{x + 1} = 6 - \frac{1}{x + 1}$$

19.
$$
\begin{array}{r}
x \\
x^2 + 0x + 1\overline{\big)\, x^3 + 0x^2 + 0x - 9} \\
\underline{x^3 + 0x^2 + x} \\
-x - 9
\end{array}
$$

$$\frac{x^3 - 9}{x^2 + 1} = x - \frac{x + 9}{x^2 + 1}$$

21.
$$
\begin{array}{r}
2x - 8 \\
x^2 + 0x + 1\overline{\big)\, 2x^3 - 8x^2 + 3x - 9} \\
\underline{2x^3 + 0x^2 + 2x} \\
-8x^2 + x - 9 \\
\underline{-8x^2 - 0x - 8} \\
x - 1
\end{array}
$$

$$\frac{2x^3 - 8x^2 + 3x - 9}{x^2 + 1} = 2x - 8 + \frac{x - 1}{x^2 + 1}$$

23.
$$
\begin{array}{r}
x + 3 \\
x^3 - 3x^2 + 3x - 1\overline{\big)\, x^4 + 0x^3 + 0x^2 + 0x + 0} \\
\underline{x^4 - 3x^3 + 3x^2 - x} \\
3x^3 - 3x^2 + x + 0 \\
\underline{3x^3 - 9x^2 + 9x - 3} \\
6x^2 - 8x + 3
\end{array}
$$

$$\frac{x^4}{(x-1)^3} = x + 3 + \frac{6x^2 - 8x + 3}{(x-1)^3}$$

25.

$$\begin{array}{r|rrrr} 4 & 2 & -10 & 14 & -24 \\ & & 8 & -8 & 24 \\ \hline & 2 & -2 & 6 & 0 \end{array}$$

$$\frac{2x^3 - 10x^2 + 14x - 24}{x - 4} = 2x^2 - 2x + 6, x \neq 4$$

27.

$$\begin{array}{r|rrrr} 3 & 6 & 7 & -1 & 26 \\ & & 18 & 75 & 222 \\ \hline & 6 & 25 & 74 & 248 \end{array}$$

$$\frac{6x^3 + 7x^2 - x + 26}{x - 3} = 6x^2 + 25x + 74 + \frac{248}{x - 3}$$

29.

$$\begin{array}{r|rrrr} -2 & 4 & 8 & -9 & -18 \\ & & -8 & 0 & 18 \\ \hline & 4 & 0 & -9 & 0 \end{array}$$

$$\frac{4x^3 + 8x^2 - 9x - 18}{x + 2} = 4x^2 - 9, x \neq -2$$

31.

$$\begin{array}{r|rrrr} -10 & -1 & 0 & 75 & -250 \\ & & 10 & -100 & 250 \\ \hline & -1 & 10 & -25 & 0 \end{array}$$

$$\frac{-x^3 + 75x - 250}{x + 10} = -x^2 + 10x - 25, x \neq -10$$

33.

$$\begin{array}{r|rrrr} 4 & 1 & -3 & 0 & 5 \\ & & 4 & 4 & 16 \\ \hline & 1 & 1 & 4 & 21 \end{array}$$

$$\frac{x^3 - 3x^2 + 5}{x - 4} = x^2 + x + 4 + \frac{21}{x - 4}$$

35.

$$\begin{array}{r|rrrrr} 6 & 10 & -50 & 0 & 0 & -800 \\ & & 60 & 60 & 360 & 2160 \\ \hline & 10 & 10 & 60 & 360 & 1360 \end{array}$$

$$\frac{10x^4 - 50x^3 - 800}{x - 6} = 10x^3 + 10x^2 + 60x + 360 + \frac{1360}{x - 6}$$

37.

$$\begin{array}{r|rrrr} -8 & 1 & 0 & 0 & 512 \\ & & -8 & 64 & -512 \\ \hline & 1 & -8 & 64 & 0 \end{array}$$

$$\frac{x^3 + 512}{x + 8} = x^2 - 8x + 64, x \neq -8$$

39.

$$\begin{array}{r|rrrrr} 2 & -3 & 0 & 0 & 0 & 0 \\ & & -6 & -12 & -24 & -48 \\ \hline & -3 & -6 & -12 & -24 & -48 \end{array}$$

$$\frac{-3x^4}{x - 2} = -3x^3 - 6x^2 - 12x - 24 - \frac{48}{x - 2}$$

41.

$$\begin{array}{r|rrrrr} 6 & -1 & 0 & 0 & 180 & 0 \\ & & -6 & -36 & -216 & -216 \\ \hline & -1 & -6 & -36 & -36 & -216 \end{array}$$

$$\frac{180x - x^4}{x - 6} = -x^3 - 6x^2 - 36x - 36 - \frac{216}{x - 6}$$

43.

$$\begin{array}{r|rrrr} -\frac{1}{2} & 4 & 16 & -23 & -15 \\ & & -2 & -7 & 15 \\ \hline & 4 & 14 & -30 & 0 \end{array}$$

$$\frac{4x^3 + 16x^2 - 23x - 15}{x + \frac{1}{2}} = 4x^2 + 14x - 30, x \neq -\frac{1}{2}$$

45. $f(x) = x^3 - x^2 - 10x + 7, k = 3$

$$\begin{array}{r|rrrr} 3 & 1 & -1 & -10 & 7 \\ & & 3 & 6 & -12 \\ \hline & 1 & 2 & -4 & -5 \end{array}$$

$$f(x) = (x - 3)(x^2 + 2x - 4) - 5$$

$$f(3) = 3^3 - 3^2 - 10(3) + 7 = -5$$

47. $f(x) = 15x^4 + 10x^3 - 6x^2 + 14, k = -\frac{2}{3}$

$$\begin{array}{r|rrrrr} -\frac{2}{3} & 15 & 10 & -6 & 0 & 14 \\ & & -10 & 0 & 4 & -\frac{8}{3} \\ \hline & 15 & 0 & -6 & 4 & \frac{34}{3} \end{array}$$

$$f(x) = \left(x + \tfrac{2}{3}\right)\left(15x^3 - 6x + 4\right) + \tfrac{34}{3}$$

$$f\left(-\tfrac{2}{3}\right) = 15\left(-\tfrac{2}{3}\right)^4 + 10\left(-\tfrac{2}{3}\right)^3 - 6\left(-\tfrac{2}{3}\right)^2 + 14 = \tfrac{34}{3}$$

49. $f(x) = -4x^3 + 6x^2 + 12x + 4,\ k = 1 - \sqrt{3}$

$$\begin{array}{r|rrrr} 1-\sqrt{3} & -4 & 6 & 12 & 4 \\ & & -4+4\sqrt{3} & -10+2\sqrt{3} & -4 \\ \hline & -4 & 2+4\sqrt{3} & 2+2\sqrt{3} & 0 \end{array}$$

$$f(x) = \left(x - 1 + \sqrt{3}\right)\left[-4x^2 + \left(2 + 4\sqrt{3}\right)x + \left(2 + 2\sqrt{3}\right)\right]$$

$$f\left(1 - \sqrt{3}\right) = -4\left(1 - \sqrt{3}\right)^3 + 6\left(1 - \sqrt{3}\right)^2 + 12\left(1 - \sqrt{3}\right) + 4 = 0$$

51. $f(x) = 2x^3 - 7x + 3$

(a) Using the Remainder Theorem:

$f(1) = 2(1)^3 - 7(1) + 3 = -2$

Using synthetic division:

$$\begin{array}{r|rrrr} 1 & 2 & 0 & -7 & 3 \\ & & 2 & 2 & -5 \\ \hline & 2 & 2 & -5 & -2 \end{array}$$

Verify using long division:

$$\begin{array}{r} 2x^2 + 2x - 5 \\ x - 1\overline{)2x^3 + 0x^2 - 7x + 3} \\ \underline{2x^3 - 2x^2} \\ 2x^2 - 7x \\ \underline{2x^2 - 2x} \\ -5x + 3 \\ \underline{-5x + 5} \\ -2 \end{array}$$

(b) Using the Remainder Theorem:

$f(-2) = 2(-2)^3 - 7(-2) + 3 = 1$

Using synthetic division:

$$\begin{array}{r|rrrr} -2 & 2 & 0 & -7 & 3 \\ & & -4 & 8 & -2 \\ \hline & 2 & -4 & 1 & 1 \end{array}$$

Verify using long division:

$$\begin{array}{r} 2x^2 - 4x + 1 \\ x + 2\overline{)2x^3 + 0x^2 - 7x + 3} \\ \underline{2x^3 + 4x^2} \\ -4x^2 - 7x \\ \underline{-4x^2 - 8x} \\ x + 3 \\ \underline{x + 2} \\ 1 \end{array}$$

(c) Using the Remainder Theorem:

$f(3) = 2(3)^3 - 7(3) + 3 = 36$

Using synthetic division:

$$\begin{array}{r|rrrr} 3 & 2 & 0 & -7 & 3 \\ & & 6 & 18 & 33 \\ \hline & 2 & 6 & 11 & 36 \end{array}$$

Verify using long division:

$$\begin{array}{r} 2x^2 + 6x + 11 \\ x - 3\overline{)2x^3 + 0x^2 - 7x + 3} \\ \underline{2x^3 - 6x^2} \\ 6x^2 - 7x + 3 \\ \underline{6x^2 - 18x} \\ 11x + 3 \\ \underline{11x - 33} \\ 36 \end{array}$$

(d) Using the Remainder Theorem:

$f(2) = 2(2)^3 - 7(2) + 3 = 5$

Using synthetic division:

$$\begin{array}{r|rrrr} 2 & 2 & 0 & -7 & 3 \\ & & 4 & 8 & 2 \\ \hline & 2 & 4 & 1 & 5 \end{array}$$

Verify using long division:

$$\begin{array}{r} 2x^2 + 4x + 1 \\ x - 2\overline{)2x^3 + 0x^2 - 7x + 3} \\ \underline{2x^3 - 4x^2} \\ 4x^2 - 7x \\ \underline{4x^2 - 8x} \\ x + 3 \\ \underline{x - 2} \\ 5 \end{array}$$

53. $h(x) = x^3 - 5x^2 - 7x + 4$

(a) Using the Remainder Theorem:

$$h(3) = (3)^3 - 5(3)^2 - 7(3) + 4 = -35$$

Using synthetic division:

$$\begin{array}{r|rrrr} 3 & 1 & -5 & -7 & 4 \\ & & 3 & -6 & -39 \\ \hline & 1 & -2 & -13 & -35 \end{array}$$

Verify using long division:

$$\begin{array}{r} x^2 - 2x - 13 \\ x - 3\overline{)x^3 - 5x^2 - 7x + 4} \\ \underline{x^3 - 3x^2} \qquad\qquad \\ -2x^2 - 7x \qquad \\ \underline{-2x^2 + 6x} \qquad \\ -13x + 4 \\ \underline{-13x + 39} \\ -35 \end{array}$$

(b) Using the Remainder Theorem:

$$h\left(\tfrac{1}{2}\right) = \left(\tfrac{1}{2}\right)^3 - 5\left(\tfrac{1}{2}\right)^2 - 7\left(\tfrac{1}{2}\right) + 4 = -\frac{5}{8}$$

Using synthetic division:

$$\begin{array}{r|rrrr} \frac{1}{2} & 1 & -5 & -7 & 4 \\ & & \frac{1}{2} & -\frac{9}{4} & -\frac{37}{8} \\ \hline & 1 & -\frac{9}{2} & -\frac{37}{4} & -\frac{5}{8} \end{array}$$

Verify using long division:

$$\begin{array}{r} x^2 - \frac{9}{2}x - \frac{37}{4} \\ x - \frac{1}{2}\overline{)x^3 - 5x^2 - 7x + 4} \\ \underline{x^3 - \frac{1}{2}x^2} \qquad\qquad \\ -\frac{9}{2}x^2 - 7x + 4 \\ \underline{-\frac{9}{2}x^2 + \frac{9}{4}x} \qquad \\ -\frac{37}{4}x + 4 \\ \underline{-\frac{37}{4}x + \frac{37}{8}} \\ -\frac{5}{8} \end{array}$$

(c) Using the Remainder Theorem:

$$h(-2) = (-2)^3 - 5(-2)^2 - 7(-2) + 4 = -10$$

Using synthetic division:

$$\begin{array}{r|rrrr} -2 & 1 & -5 & -7 & 4 \\ & & -2 & 14 & -14 \\ \hline & 1 & -7 & 7 & -10 \end{array}$$

Verify using long division:

$$\begin{array}{r} x^2 - 7x + 7 \\ x + 2\overline{)x^3 - 5x^2 - 7x + 4} \\ \underline{x^3 + 2x^2} \qquad\qquad \\ -7x^2 - 7x \qquad \\ \underline{-7x^2 - 14x} \qquad \\ 7x + 4 \\ \underline{7x + 14} \\ -10 \end{array}$$

(d) Using the Remainder Theorem:

$$h(-5) = (-5)^3 - 5(-5)^2 - 7(-5) + 4 = -211$$

Using synthetic division:

$$\begin{array}{r|rrrr} -5 & 1 & -5 & -7 & 4 \\ & & -5 & 50 & -215 \\ \hline & 1 & -10 & 43 & -211 \end{array}$$

Verify using long division:

$$\begin{array}{r} x^2 - 10x + 43 \\ x + 5\overline{)x^3 - 5x^2 - 7x + 4} \\ \underline{x^3 + 5x^2} \qquad\qquad \\ -10x^2 - 7x \qquad \\ \underline{-10x^2 - 50x} \qquad \\ 43x + 4 \\ \underline{43x + 215} \\ -211 \end{array}$$

55.

$$\begin{array}{r|rrrr} -3 & 1 & 6 & 11 & 6 \\ & & -3 & -9 & -6 \\ \hline & 1 & 3 & 2 & 0 \end{array}$$

$$\begin{aligned} x^3 + 6x^2 + 11x + 6 &= (x + 3)(x^2 + 3x + 2) \\ &= (x + 3)(x + 2)(x + 1) \end{aligned}$$

Zeros: $-3, -2, -1$

57.

$$\begin{array}{r|rrrr} \frac{1}{2} & 2 & -15 & 27 & -10 \\ & & 1 & -7 & 10 \\ \hline & 2 & -14 & 20 & 0 \end{array}$$

$$\begin{aligned} 2x^3 - 15x^2 + 27x - 10 &= \left(x - \tfrac{1}{2}\right)(2x^2 - 14x + 20) \\ &= (2x - 1)(x - 2)(x - 5) \end{aligned}$$

Zeros: $\frac{1}{2}, 2, 5$

59.

$$\begin{array}{r|rrrr} \sqrt{3} & 1 & 2 & -3 & -6 \\ & & \sqrt{3} & 3 + 2\sqrt{3} & 6 \\ \hline & 1 & 2 + \sqrt{3} & 2\sqrt{3} & 0 \end{array}$$

$$\begin{array}{r|rrr} -\sqrt{3} & 1 & 2 + \sqrt{3} & 2\sqrt{3} \\ & & -\sqrt{3} & -2\sqrt{3} \\ \hline & 1 & 2 & 0 \end{array}$$

$$x^3 + 2x^2 - 3x - 6 = \left(x - \sqrt{3}\right)\left(x + \sqrt{3}\right)(x + 2)$$

Zeros: $-\sqrt{3}, \sqrt{3}, -2$

61. $$\begin{array}{r|rrrr} 1+\sqrt{3} & 1 & -3 & 0 & 2 \\ & & 1+\sqrt{3} & 1-\sqrt{3} & -2 \\ \hline & 1 & -2+\sqrt{3} & 1-\sqrt{3} & 0 \end{array}$$

$$\begin{array}{r|rrr} 1-\sqrt{3} & 1 & -2+\sqrt{3} & 1-\sqrt{3} \\ & & 1-\sqrt{3} & -1+\sqrt{3} \\ \hline & 1 & -1 & 0 \end{array}$$

$$x^3 - 3x^2 + 2 = \left[x - \left(1 + \sqrt{3}\right)\right]\left[x - \left(1 - \sqrt{3}\right)\right](x - 1)$$
$$= (x - 1)\left(x - 1 - \sqrt{3}\right)\left(x - 1 + \sqrt{3}\right)$$

Zeros: $1, 1 - \sqrt{3}, 1 + \sqrt{3}$

63. $f(x) = 2x^3 + x^2 - 5x + 2$; Factors: $(x + 2), (x - 1)$

(a) $$\begin{array}{r|rrrr} -2 & 2 & 1 & -5 & 2 \\ & & -4 & 6 & -2 \\ \hline & 2 & -3 & 1 & 0 \end{array}$$

$$\begin{array}{r|rrr} 1 & 2 & -3 & 1 \\ & & 2 & -1 \\ \hline & 2 & -1 & 0 \end{array}$$

Both are factors of $f(x)$ because the remainders are zero.

(b) The remaining factor of $f(x)$ is $(2x - 1)$.

(c) $f(x) = (2x - 1)(x + 2)(x - 1)$

(d) Zeros: $\frac{1}{2}, -2, 1$

(e)

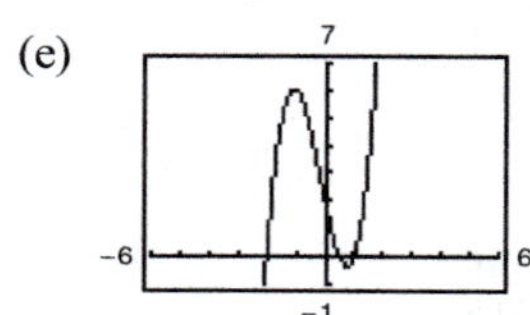

65. $f(x) = x^4 - 8x^3 + 9x^2 + 38x - 40$;

Factors: $(x - 5), (x + 2)$

(a) $$\begin{array}{r|rrrrr} 5 & 1 & -8 & 9 & 38 & -40 \\ & & 5 & -15 & -30 & 40 \\ \hline & 1 & -3 & -6 & 8 & 0 \end{array}$$

$$\begin{array}{r|rrrr} -2 & 1 & -3 & -6 & 8 \\ & & -2 & 10 & -8 \\ \hline & 1 & -5 & 4 & 0 \end{array}$$

Both are factors of $f(x)$ because the remainders are zero.

(b) $x^2 - 5x + 4 = (x - 1)(x - 4)$

The remaining factors are $(x - 1)$ and $(x - 2)$.

(c) $f(x) = (x - 5)(x + 2)(x - 1)(x - 4)$

(d) Zeros: $-2, 1, 4, 5$

(e)

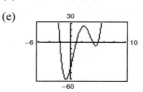

67. $f(x) = 6x^3 + 41x^2 - 9x - 14$;

Factors: $(2x + 1), (3x - 2)$

(a) $$\begin{array}{r|rrrr} -\frac{1}{2} & 6 & 41 & -9 & -14 \\ & & -3 & -19 & 14 \\ \hline & 6 & 38 & -28 & 0 \end{array}$$

$$\begin{array}{r|rrr} \frac{2}{3} & 6 & 38 & -28 \\ & & 4 & 28 \\ \hline & 6 & 42 & 0 \end{array}$$

Both are factors of $f(x)$ because the remainders are zero.

(b) $6x + 42 = 6(x + 7)$

This shows that $\dfrac{f(x)}{\left(x + \frac{1}{2}\right)\left(x - \frac{2}{3}\right)} = 6(x + 7)$,

so $\dfrac{f(x)}{(2x + 1)(3x - 2)} = x + 7$.

The remaining factor is $(x + 7)$.

(c) $f(x) = (x + 7)(2x + 1)(3x - 2)$

(d) Zeros: $-7, -\frac{1}{2}, \frac{2}{3}$

(e)

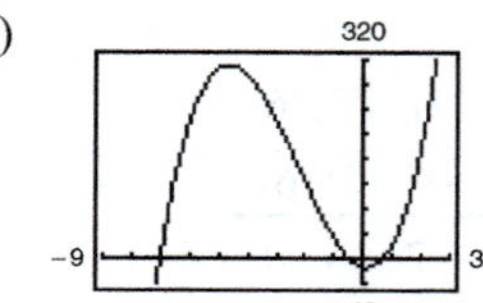

69. $f(x) = 2x^3 - x^2 - 10x + 5;$

Factors: $(2x - 1), \left(x + \sqrt{5}\right)$

(a)
$$\begin{array}{r|rrrr} \frac{1}{2} & 2 & -1 & -10 & 5 \\ & & 1 & 0 & -5 \\ \hline & 2 & 0 & -10 & 0 \end{array}$$

$$\begin{array}{r|rrr} -\sqrt{5} & 2 & 0 & -10 \\ & & -2\sqrt{5} & 10 \\ \hline & 2 & -2\sqrt{5} & 0 \end{array}$$

Both are factors of $f(x)$ because the remainders are zero.

(b) $2x - 2\sqrt{5} = 2\left(x - \sqrt{5}\right)$

This shows that $\dfrac{f(x)}{\left(x - \frac{1}{2}\right)\left(x + \sqrt{5}\right)} = 2\left(x - \sqrt{5}\right),$

so $\dfrac{f(x)}{(2x - 1)\left(x + \sqrt{5}\right)} = x - \sqrt{5}.$

The remaining factor is $\left(x - \sqrt{5}\right)$.

(c) $f(x) = \left(x + \sqrt{5}\right)\left(x - \sqrt{5}\right)(2x - 1)$

(d) Zeros: $-\sqrt{5}, \sqrt{5}, \dfrac{1}{2}$

(e)

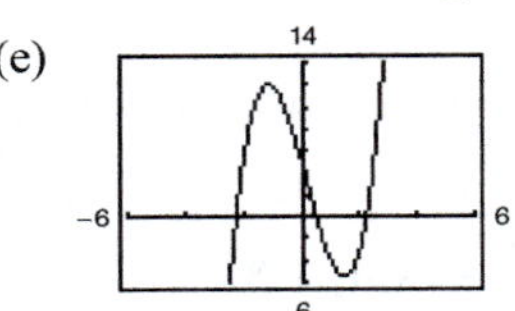

71. $f(x) = x^3 - 2x^2 - 5x + 10$

(a) The zeros of f are $x = 2$ and $x \approx \pm 2.236$.

(b) An exact zero is $x = 2$.

(c)
$$\begin{array}{r|rrrr} 2 & 1 & -2 & -5 & 10 \\ & & 2 & 0 & -10 \\ \hline & 1 & 0 & -5 & 0 \end{array}$$

$$\begin{aligned} f(x) &= (x - 2)\left(x^2 - 5\right) \\ &= (x - 2)\left(x - \sqrt{5}\right)\left(x + \sqrt{5}\right) \end{aligned}$$

73. $h(t) = t^3 - 2t^2 - 7t + 2$

(a) The zeros of h are $t = -2, t \approx 3.732, t \approx 0.268$.

(b) An exact zero is $t = -2$.

(c)
$$\begin{array}{r|rrrr} -2 & 1 & -2 & -7 & 2 \\ & & -2 & 8 & -2 \\ \hline & 1 & -4 & 1 & 0 \end{array}$$

$$h(t) = (t + 2)\left(t^2 - 4t + 1\right)$$

By the Quadratic Formula, the zeros of $t^2 - 4t + 1$ are $2 \pm \sqrt{3}$. Thus,

$$h(t) = (t + 2)\left[t - \left(2 + \sqrt{3}\right)\right]\left[t - \left(2 - \sqrt{3}\right)\right].$$

75. $h(x) = x^5 - 7x^4 + 10x^3 + 14x^2 - 24x$

(a) The zeros of h are $x = 0, x = 3, x = 4,$ $x \approx 1.414, x \approx -1.414$.

(b) An exact zero is $x = 4$.

(c)
$$\begin{array}{r|rrrrr} 4 & 1 & -7 & 10 & 14 & -24 \\ & & 4 & -12 & -8 & 24 \\ \hline & 1 & -3 & -2 & 6 & 0 \end{array}$$

$$\begin{aligned} h(x) &= (x - 4)\left(x^4 - 3x^3 - 2x^2 + 6x\right) \\ &= x(x - 4)(x - 3)\left(x + \sqrt{2}\right)\left(x - \sqrt{2}\right) \end{aligned}$$

77. $\dfrac{x^3 + x^2 - 64x - 64}{x + 8}$

$$\begin{array}{r|rrrr} -8 & 1 & 1 & -64 & -64 \\ & & -8 & 56 & 64 \\ \hline & 1 & -7 & -8 & 0 \end{array}$$

$$\frac{x^3 + x^2 - 64x - 64}{x + 8} = x^2 - 7x - 8, x \neq -8$$

79. $\dfrac{x^4 + 6x^3 + 11x^2 + 6x}{x^2 + 3x + 2} = \dfrac{x^4 + 6x^3 + 11x^2 + 6x}{(x + 1)(x + 2)}$

$$\begin{array}{r|rrrrr} -1 & 1 & 6 & 11 & 6 & 0 \\ & & -1 & -5 & -6 & 0 \\ \hline & 1 & 5 & 6 & 0 & 0 \end{array}$$

$$\begin{array}{r|rrrr} -2 & 1 & 5 & 6 & 0 \\ & & -2 & -6 & 0 \\ \hline & 1 & 3 & 0 & 0 \end{array}$$

$$\frac{x^4 + 6x^3 + 11x^2 + 6x}{(x + 1)(x + 2)} = x^2 + 3x, x \neq -2, -1$$

81. (a)

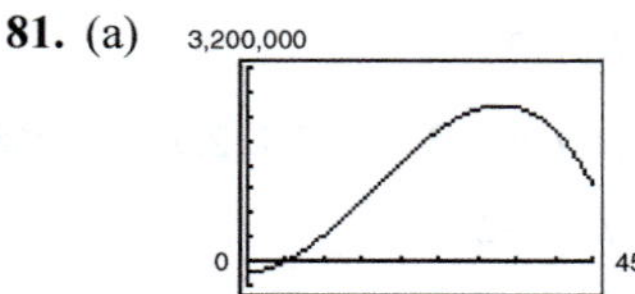

(b) Using the trace and zoom features, when $x = 25$, an advertising expense of about $250,000 would produce the same profit of $2,174,375.

(c) $x = 25$

$$\begin{array}{r|rrrr} 25 & -152 & 7545 & 0 & -169{,}625 \\ & & -3800 & 93{,}625 & 2{,}340{,}625 \\ \hline & -152 & 3745 & 93{,}625 & 2{,}171{,}000 \end{array}$$

So, an advertising expense of $250,000 yields a profit of $2,171,000, which is close to $2,174,375.

83. False. If $(7x + 4)$ is a factor of f, then $-\frac{4}{7}$ is a zero of f.

85. True. The degree of the numerator is greater than the degree of the denominator.

87.

$$\begin{array}{r} x^{2n} + 6x^n + 9 \\ x^n + 3 \overline{\smash{)}\, x^{3n} + 9x^{2n} + 27x^n + 27} \\ \underline{x^{3n} + 3x^{2n}} \qquad\qquad\quad \\ 6x^{2n} + 27x^n \qquad \\ \underline{6x^{2n} + 18x^n} \qquad \\ 9x^n + 27 \\ \underline{9x^n + 27} \\ 0 \end{array}$$

$$\frac{x^{3n} + 9x^{2n} + 27x^n + 27}{x^n + 3} = x^{2n} + 6x^n + 9,\ x^n \neq -3$$

89. To divide $x^2 + 3x - 5$ by $x + 1$ using synthetic division, the value of k is $k = -1$ not $k = 1$ as shown.

$$\begin{array}{r|rrr} -1 & 1 & 3 & -5 \\ & & -1 & -2 \\ \hline & 1 & 2 & -7 \end{array} \leftarrow \text{Remainder: } -7$$

91.

$$\begin{array}{r|rrrr} 5 & 1 & 4 & -3 & c \\ & & 5 & 45 & 210 \\ \hline & 1 & 9 & 42 & c + 210 \end{array}$$

To divide evenly, $c + 210$ must equal zero. So, c must equal -210.

93. If $x - 4$ is a factor of $f(x) = x^3 - kx^2 + 2kx - 8$, then $f(4) = 0$.

$$\begin{aligned} f(4) &= (4)^3 - k(4)^2 + 2k(4) - 8 \\ 0 &= 64 - 16k + 8k - 8 \\ -56 &= -8k \\ 7 &= k \end{aligned}$$

Section 2.4 Complex Numbers

1. real

3. pure imaginary

5. principal square

7. $a + bi = 9 + 8i$

$a = 9$

$b = 8$

9. $(a - 2) + (b + 1)i = 6 + 5i$

$a - 2 = 6 \Rightarrow a = 8$

$b + 1 = 5 \Rightarrow b = 4$

11. $2 + \sqrt{-25} = 2 + 5i$

13. $1 - \sqrt{-12} = 1 - 2\sqrt{3}i$

15. $\sqrt{-40} = 2\sqrt{10}\,i$

17. 23

19. $-6i + i^2 = -6i + (-1)$

$= -1 - 6i$

21. $\sqrt{-0.04} = \sqrt{0.04}\,i$

$= 0.2i$

23. $(5 + i) + (2 + 3i) = 5 + i + 2 + 3i$
$= 7 + 4i$

25. $(9 - i) - (8 - i) = 1$

27. $\left(-2 + \sqrt{-8}\right) + \left(5 - \sqrt{-50}\right) = -2 + 2\sqrt{2}i + 5 - 5\sqrt{2}i$
$= 3 - 3\sqrt{2}i$

29. $13i - (14 - 7i) = 13i - 14 + 7i$
$= -14 + 20i$

31. $(1 + i)(3 - 2i) = 3 - 2i + 3i - 2i^2$
$= 3 + i + 2 = 5 + i$

33. $12i(1 - 9i) = 12i - 108i^2$
$= 12i + 108$
$= 108 + 12i$

35. $\left(\sqrt{2} + 3i\right)\left(\sqrt{2} - 3i\right) = 2 - 9i^2$
$= 2 + 9 = 11$

37. $(6 + 7i)^2 = 36 + 84i + 49i^2$
$= 36 + 84i - 49$
$= -13 + 84i$

39. The complex conjugate of $9 + 2i$ is $9 - 2i$.

$(9 + 2i)(9 - 2i) = 81 - 4i^2$
$= 81 + 4$
$= 85$

41. The complex conjugate of $-1 - \sqrt{5}i$ is $-1 + \sqrt{5}i$.

$\left(-1 - \sqrt{5}i\right)\left(-1 + \sqrt{5}i\right) = 1 - 5i^2$
$= 1 + 5 = 6$

43. The complex conjugate of $\sqrt{-20} = 2\sqrt{5}i$ is $-2\sqrt{5}i$.

$\left(2\sqrt{5}i\right)\left(-2\sqrt{5}i\right) = -20i^2 = 20$

45. The complex conjugate of $\sqrt{6}$ is $\sqrt{6}$.

$\left(\sqrt{6}\right)\left(\sqrt{6}\right) = 6$

47. $\dfrac{2}{4 - 5i} = \dfrac{2}{4 - 5i} \cdot \dfrac{4 + 5i}{4 + 5i}$
$= \dfrac{2(4 + 5i)}{16 + 25} = \dfrac{8 + 10i}{41} = \dfrac{8}{41} + \dfrac{10}{41}i$

49. $\dfrac{5 + i}{5 - i} \cdot \dfrac{(5 + i)}{(5 + i)} = \dfrac{25 + 10i + i^2}{25 - i^2}$
$= \dfrac{24 + 10i}{26} = \dfrac{12}{13} + \dfrac{5}{13}i$

51. $\dfrac{9 - 4i}{i} \cdot \dfrac{-i}{-i} = \dfrac{-9i + 4i^2}{-i^2} = -4 - 9i$

53. $\dfrac{3i}{(4 - 5i)^2} = \dfrac{3i}{16 - 40i + 25i^2} = \dfrac{3i}{-9 - 40i} \cdot \dfrac{-9 + 40i}{-9 + 40i}$
$= \dfrac{-27i + 120i^2}{81 + 1600} = \dfrac{-120 - 27i}{1681}$
$= -\dfrac{120}{1681} - \dfrac{27}{1681}i$

55. $\dfrac{2}{1 + i} - \dfrac{3}{1 - i} = \dfrac{2(1 - i) - 3(1 + i)}{(1 + i)(1 - i)}$
$= \dfrac{2 - 2i - 3 - 3i}{1 + 1}$
$= \dfrac{-1 - 5i}{2}$
$= -\dfrac{1}{2} - \dfrac{5}{2}i$

57. $\dfrac{i}{3 - 2i} + \dfrac{2i}{3 + 8i} = \dfrac{i(3 + 8i) + 2i(3 - 2i)}{(3 - 2i)(3 + 8i)}$
$= \dfrac{3i + 8i^2 + 6i - 4i^2}{9 + 24i - 6i - 16i^2}$
$= \dfrac{4i^2 + 9i}{9 + 18i + 16}$
$= \dfrac{-4 + 9i}{25 + 18i} \cdot \dfrac{25 - 18i}{25 - 18i}$
$= \dfrac{-100 + 72i + 225i - 162i^2}{625 + 324}$
$= \dfrac{62 + 297i}{949} = \dfrac{62}{949} + \dfrac{297}{949}i$

59. $\sqrt{-6} \cdot \sqrt{-2} = \left(\sqrt{6}i\right)\left(\sqrt{2}i\right) = \sqrt{12}i^2 = \left(2\sqrt{3}\right)(-1)$
$= -2\sqrt{3}$

61. $\left(\sqrt{-15}\right)^2 = \left(\sqrt{15}i\right)^2 = 15i^2 = -15$

63. $\sqrt{-8} + \sqrt{-50} = \sqrt{8}i + \sqrt{50}i$
$= 2\sqrt{2}i + 5\sqrt{2}i$
$= 7\sqrt{2}i$

65. $\left(3 + \sqrt{-5}\right)\left(7 - \sqrt{-10}\right) = \left(3 + \sqrt{5}i\right)\left(7 - \sqrt{10}i\right)$

$$= 21 - 3\sqrt{10}i + 7\sqrt{5}i - \sqrt{50}i^2$$
$$= \left(21 + \sqrt{50}\right) + \left(7\sqrt{5} - 3\sqrt{10}\right)i$$
$$= \left(21 + 5\sqrt{2}\right) + \left(7\sqrt{5} - 3\sqrt{10}\right)i$$

67. $x^2 - 2x + 2 = 0; a = 1, b = -2, c = 2$

$$x = \frac{-(-2) \pm \sqrt{(-2)^2 - 4(1)(2)}}{2(1)}$$
$$= \frac{2 \pm \sqrt{-4}}{2}$$
$$= \frac{2 \pm 2i}{2}$$
$$= 1 \pm i$$

69. $4x^2 + 16x + 17 = 0; a = 4, b = 16, c = 17$

$$x = \frac{-16 \pm \sqrt{(16)^2 - 4(4)(17)}}{2(4)}$$
$$= \frac{-16 \pm \sqrt{-16}}{8}$$
$$= \frac{-16 \pm 4i}{8}$$
$$= -2 \pm \frac{1}{2}i$$

71. $4x^2 + 16x + 21 = 0; a = 4, b = 16, c = 21$

$$x = \frac{-16 \pm \sqrt{(16)^2 - 4(4)(21)}}{2(4)}$$
$$= \frac{-16 \pm \sqrt{-80}}{8}$$
$$= \frac{-16 \pm \sqrt{80}\, i}{8}$$
$$= \frac{-16 \pm 4\sqrt{5}\, i}{8}$$
$$= -2 \pm \frac{\sqrt{5}}{2}i$$

73. $\frac{3}{2}x^2 - 6x + 9 = 0$ Multiply both sides by 2.

$3x^2 - 12x + 18 = 0; a = 3, b = -12, c = 18$

$$x = \frac{-(-12) \pm \sqrt{(-12)^2 - 4(3)(18)}}{2(3)}$$
$$= \frac{12 \pm \sqrt{-72}}{6}$$
$$= \frac{12 \pm 6\sqrt{2}i}{6}$$
$$= 2 \pm \sqrt{2}i$$

75. $1.4x^2 - 2x + 10 = 0 \Rightarrow 14x^2 - 20x + 100 = 0;$

$a = 14, b = -20, c = 100$

$$x = \frac{-(-20) \pm \sqrt{(-20)^2 - 4(14)(100)}}{2(14)}$$
$$= \frac{20 \pm \sqrt{-5200}}{28}$$
$$= \frac{20 \pm 20\sqrt{13}\, i}{28}$$
$$= \frac{20}{28} \pm \frac{20\sqrt{13}\, i}{28}$$
$$= \frac{5}{7} \pm \frac{5\sqrt{13}}{7}i$$

77. $-6i^3 + i^2 = -6i^2 i + i^2$

$$= -6(-1)i + (-1)$$
$$= 6i - 1$$
$$= -1 + 6i$$

79. $-14i^5 = -14i^2 i^2 i = -14(-1)(-1)(i) = -14i$

81. $\left(\sqrt{-72}\right)^3 = \left(6\sqrt{2}i\right)^3$

$$= 6^3\left(\sqrt{2}\right)^3 i^3$$
$$= 216\left(2\sqrt{2}\right)i^2 i$$
$$= 432\sqrt{2}(-1)i$$
$$= -432\sqrt{2}i$$

83. $\frac{1}{i^3} = \frac{1}{i^2 i} = \frac{1}{-i} = \frac{1}{-i} \cdot \frac{i}{i} = \frac{i}{-i^2} = i$

85. $(3i)^4 = 81i^4 = 81i^2i^2 = 81(-1)(-1) = 81$

87. (a) $z_1 = 9 + 16i, z_2 = 20 - 10i$

(b) $\frac{1}{z} = \frac{1}{z_1} + \frac{1}{z_2} = \frac{1}{9 + 16i} + \frac{1}{20 - 10i} = \frac{20 - 10i + 9 + 16i}{(9 + 16i)(20 - 10i)} = \frac{29 + 6i}{340 + 230i}$

$z = \left(\frac{340 + 230i}{29 + 6i}\right)\left(\frac{29 - 6i}{29 - 6i}\right) = \frac{11{,}240 + 4630i}{877} = \frac{11{,}240}{877} + \frac{4630}{877}i$

89. False.

Sample answer: $(1 + i) + (3 + i) = 4 + 2i$ which is not a real number.

91. True.

$$x^4 - x^2 + 14 = 56$$
$$\left(-i\sqrt{6}\right)^4 - \left(-i\sqrt{6}\right)^2 + 14 \stackrel{?}{=} 56$$
$$36 + 6 + 14 \stackrel{?}{=} 56$$
$$56 = 56$$

93. $i = i$

$i^2 = -1$

$i^3 = -i$

$i^4 = 1$

$i^5 = i^4 i = i$

$i^6 = i^4 i^2 = -1$

$i^7 = i^4 i^3 = -i$

$i^8 = i^4 i^4 = 1$

$i^9 = i^4 i^4 i = i$

$i^{10} = i^4 i^4 i^2 = -1$

$i^{11} = i^4 i^4 i^3 = -i$

$i^{12} = i^4 i^4 i^4 = 1$

The pattern $i, -1, -i, 1$ repeats. Divide the exponent by 4.

If the remainder is 1, the result is i.

If the remainder is 2, the result is -1.

If the remainder is 3, the result is $-i$.

If the remainder is 0, the result is 1.

95. $\sqrt{-6}\sqrt{-6} = \sqrt{6}i\sqrt{6}i = 6i^2 = -6$

97. $(a_1 + b_1 i) + (a_2 + b_2 i) = (a_1 + a_2) + (b_1 + b_2)i$

The complex conjugate of this sum is $(a_1 + a_2) - (b_1 + b_2)i$.

The sum of the complex conjugates is $(a_1 - b_1 i) + (a_2 - b_2 i) = (a_1 + a_2) - (b_1 + b_2)i$.

So, the complex conjugate of the sum of two complex numbers is the sum of their complex conjugates.

Section 2.5 Zeros of Polynomial Functions

1. Fundamental Theorem of Algebra

3. Rational Zero

5. linear; quadratic; quadratic

7. Descartes's Rule of Signs

9. f is a 3rd degree polynomial function, so there are three zeros.

11. f is a 5th degree polynomial function, so there are five zeros.

13. f is a 2nd degree polynomial function, so there are two zeros.

15. $f(x) = x^3 + 2x^2 - x - 2$

Possible rational zeros: $\pm 1, \pm 2$

Zeros shown on graph: $-2, -1, 1, 2$

17. $f(x) = 2x^4 - 17x^3 + 35x^2 + 9x - 45$

Possible rational zeros: $\pm1, \pm3, \pm5, \pm9, \pm15, \pm45, \pm\frac{1}{2}, \pm\frac{3}{2}, \pm\frac{5}{2}, \pm\frac{9}{2}, \pm\frac{15}{2}, \pm\frac{45}{2}$

Zeros shown on graph: $-1, \frac{3}{2}, 3, 5$

19. $f(x) = x^3 - 7x - 6$

Possible rational zeros: $\pm1, \pm2, \pm3, \pm6$

$$\begin{array}{r|rrrr} 3 & 1 & 0 & -7 & -6 \\ & & 3 & 9 & 6 \\ \hline & 1 & 3 & 2 & 0 \end{array}$$

$$\begin{aligned} f(x) &= (x-3)(x^2+3x+2) \\ &= (x-3)(x+2)(x+1) \end{aligned}$$

So, the rational zeros are -2, -1, and 3.

21. $g(t) = t^3 - 4t^2 + 4$

Possible rational zeros: $\pm1, \pm 2, \pm 4$

After testing all six possible rational zeros by synthetic division, you can conclude there are no rational zeros.

23. $h(t) = t^3 + 8t^2 + 13t + 6$

Possible rational zeros: $\pm1, \pm2, \pm3, \pm6$

$$\begin{array}{r|rrrr} -6 & 1 & 8 & 13 & 6 \\ & & -6 & -12 & -6 \\ \hline & 1 & 2 & 1 & 0 \end{array}$$

$$\begin{aligned} t^3 + 8t^2 + 13t + 6 &= (t+6)(t^2+2t+1) \\ &= (t+6)(t+1)(t+1) \end{aligned}$$

So, the rational zeros are -1 and -6.

25. $C(x) = 2x^3 + 3x^2 - 1$

Possible rational zeros: $\pm1, \pm\frac{1}{2}$

$$\begin{array}{r|rrrr} -1 & 2 & 3 & 0 & -1 \\ & & -2 & -1 & 1 \\ \hline & 2 & 1 & -1 & 0 \end{array}$$

$$\begin{aligned} 2x^3 + 3x^2 - 1 &= (x+1)(2x^2+x-1) \\ &= (x+1)(x+1)(2x-1) \\ &= (x+1)^2(2x-1) \end{aligned}$$

So, the rational zeros are -1 and $\frac{1}{2}$.

27. $f(x) = 9x^4 - 9x^3 - 58x^2 + 4x + 24$

Possible rational zeros: $\pm1, \pm2, \pm3, \pm4, \pm6, \pm8, \pm12, \pm24, \pm\frac{1}{3}, \pm\frac{2}{3}, \pm\frac{4}{3}, \pm\frac{8}{3}, \pm\frac{1}{9}, \pm\frac{2}{9}, \pm\frac{4}{9}, \pm\frac{8}{9}$

$$\begin{array}{r|rrrrr} -2 & 9 & -9 & -58 & 4 & 24 \\ & & -18 & 54 & 8 & -24 \\ \hline & 9 & -27 & -4 & 12 & 0 \end{array}$$

$$\begin{array}{r|rrrr} 3 & 9 & -27 & -4 & 12 \\ & & 27 & 0 & -12 \\ \hline & 9 & 0 & -4 & 0 \end{array}$$

$$\begin{aligned} f(x) &= (x+2)(x-3)(9x^2-4) \\ &= (x+2)(x-3)(3x-2)(3x+2) \end{aligned}$$

So, the rational zeros are $-2, 3, \frac{2}{3}$, and $-\frac{2}{3}$.

29. $-5x^3 + 11x^2 - 4x - 2 = 0$

Possible rational zeros: $\dfrac{\pm1, \pm2}{\pm1, \pm5} = \pm\dfrac{1}{5}, \pm\dfrac{2}{5}, \pm 1, \pm 2$

$$\begin{array}{r|rrrr} 1 & -5 & 11 & -4 & -2 \\ & & -5 & 6 & 2 \\ \hline & -5 & 6 & 2 & 0 \end{array}$$

$$\begin{aligned} (x-1)(-5x^2+6x+2) &= 0 \\ -5x^2 + 6x + 2 &= 0 \\ 5x^2 - 6x - 2 &= 0 \end{aligned}$$

$$x = \frac{-b \pm \sqrt{b^2 - 4ac}}{2a}$$

$$x = \frac{-(-6) \pm \sqrt{(-6)^2 - 4(5)(-2)}}{2(5)}$$

$$x = \frac{6 \pm \sqrt{76}}{10}$$

$$x = \frac{2(3 \pm \sqrt{19})}{10}$$

$$x = \frac{3 \pm \sqrt{19}}{5}$$

So, the real zeros are $x = 1, x = \dfrac{3}{5} \pm \dfrac{\sqrt{19}}{5}$.

31. $x^4 + 6x^3 + 3x^2 - 16x + 6 = 0$

Possible rational zeros: $\pm1, \pm2, \pm3, \pm6$

$$\begin{array}{r|rrrrr} 1 & 1 & 6 & 3 & -16 & 6 \\ & & 1 & 7 & 10 & -6 \\ \hline & 1 & 7 & 10 & -6 & 0 \end{array}$$

$$\begin{array}{r|rrrr} -3 & 1 & 7 & 10 & -6 \\ & & -3 & -12 & 6 \\ \hline & 1 & 4 & -2 & 0 \end{array}$$

$$(x-1)(x+3)(x^2+4x-2) = 0$$
$$x^2 + 4x - 2 = 0$$
$$x^2 + 4x + 4 = 2 + 4$$
$$(x+2)^2 = 6$$
$$x + 2 = \pm\sqrt{6}$$
$$x = -2 \pm \sqrt{6}$$

So the real zeros are $x = 1, -3, -2 \pm 2\sqrt{6}$.

33. $f(x) = x^3 + x^2 - 4x - 4$

(a) Possible rational zeros: $\pm1, \pm2, \pm4$

(b)

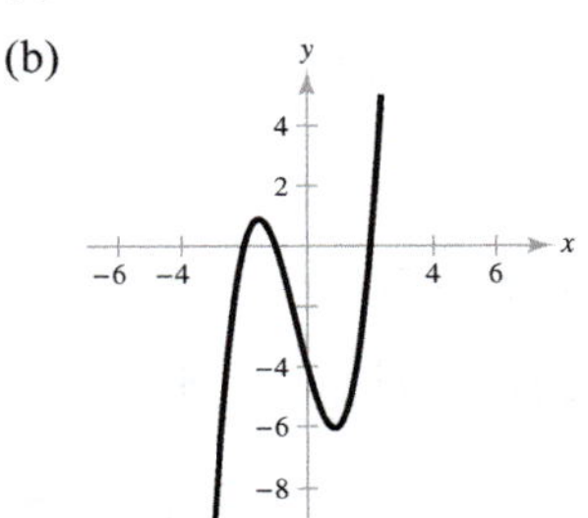

(c) Real zeros: $-2, -1, 2$

35. $f(x) = -4x^3 + 15x^2 - 8x - 3$

(a) Possible rational zeros: $\pm1, \pm3, \pm\frac{1}{2}, \pm\frac{3}{2}, \pm\frac{1}{4}, \pm\frac{3}{4}$

(b)

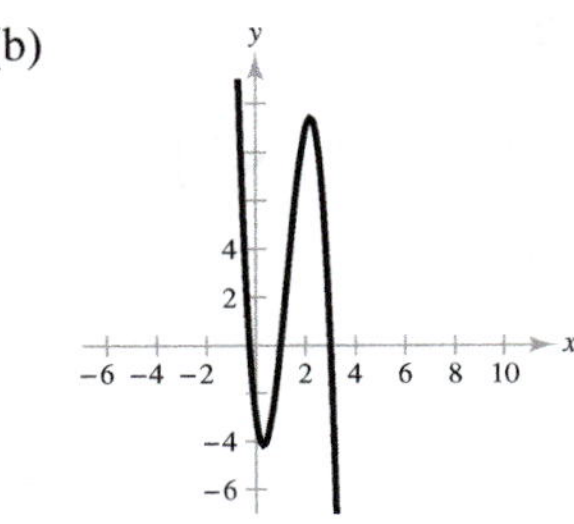

(c) Real zeros: $-\frac{1}{4}, 1, 3$

37. $f(x) = -2x^4 + 13x^3 - 21x^2 + 2x + 8$

(a) Possible rational zeros: $\pm1, \pm2, \pm4, \pm8, \pm\frac{1}{2}$

(b)

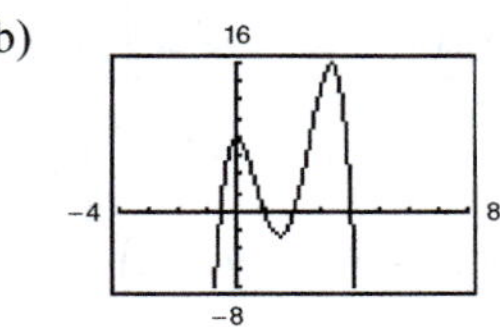

(c) Real zeros: $-\frac{1}{2}, 1, 2, 4$

39. $f(x) = 32x^3 - 52x^2 + 17x + 3$

(a) Possible rational zeros: $\pm1, \pm3, \pm\frac{1}{2}, \pm\frac{3}{2}, \pm\frac{1}{4}, \pm\frac{3}{4}, \pm\frac{1}{8}, \pm\frac{3}{8}, \pm\frac{1}{16}, \pm\frac{3}{16}, \pm\frac{1}{32}, \pm\frac{3}{32}$

(b)

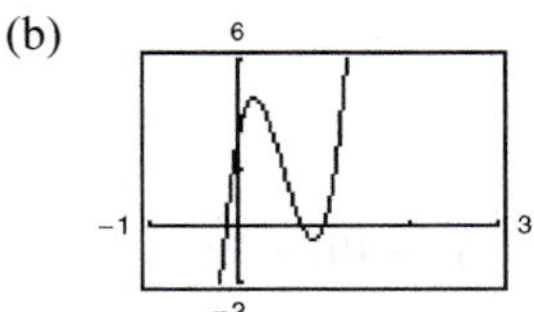

(c) Real zeros: $-\frac{1}{8}, \frac{3}{4}, 1$

41. $f(x) = (x-1)(x-5i)(x+5i)$
$= (x-1)(x^2+25)$
$= x^3 - x^2 + 25x - 25$

Note: $f(x) = a(x^3 - x^2 + 25x - 25)$, where a is any nonzero real number, has the zeros 1 and $\pm5i$.

43. If $1 + i$ is a zero, so is its conjugate, $1 - i$.

$f(x) = (x-2)(x-2)(x-(1+i))(x-(1-i))$
$= (x^2 - 4x + 4)(x^2 - 2x + 2)$
$= x^4 - 6x^3 + 14x^2 - 16x + 8$

Note: $f(x) = a(x^4 - 6x^3 + 14x^2 - 16x + 8)$, where a is any nonzero real number, has the zeros 2, 2 and $1 \pm i$.

45. If $3 + \sqrt{2}i$ is a zero, so is its conjugate, $3 - \sqrt{2}i$.

$$\begin{aligned} f(x) &= (3x - 2)(x + 1)\left[x - \left(3 + \sqrt{2}i\right)\right]\left[x - \left(3 - \sqrt{2}i\right)\right] \\ &= (3x - 2)(x + 1)\left[(x - 3) - \sqrt{2}i\right]\left[(x - 3) + \sqrt{2}i\right] \\ &= \left(3x^2 + x - 2\right)\left[(x - 3)^2 - \left(\sqrt{2}i\right)^2\right] \\ &= \left(3x^2 + x - 2\right)\left(x^2 - 6x + 9 + 2\right) \\ &= \left(3x^2 + x - 2\right)\left(x^2 - 6x + 11\right) \\ &= 3x^4 - 17x^3 + 25x^2 + 23x - 22 \end{aligned}$$

Note: $f(x) = a\left(3x^4 - 17x^3 + 25x^2 + 23x - 22\right)$, where a is any nonzero real number, has the zeros $\frac{2}{3}$, -1, and $3 \pm \sqrt{2}i$.

47. $f(x) = a(x + 2)(x - 1)(x - i)(x + i)$

$$\begin{aligned} &= a\left(x^2 + x - 2\right)\left(x^2 + 1\right) \\ &= a\left(x^4 + x^3 - x^2 + x - 2\right) \end{aligned}$$

Since $f(0) = -4$

$$\begin{aligned} -4 &= a\left((0)^4 + (0)^3 - (0)^2 + (0) - 2\right) \\ -4 &= -2a \\ a &= 2 \end{aligned}$$

So, $f(x) = 2\left(x^4 + x^3 - x^2 + x - 2\right)$

$$= 2x^4 + 2x^3 - 2x^2 + 2x - 4.$$

49. $f(x) = a(x + 3)\left(x - \left(1 + \sqrt{3}i\right)\right)\left(x - \left(1 - \sqrt{3}i\right)\right)$

$$\begin{aligned} &= a(x + 3)\left(x^2 - 2x + 4\right) \\ &= a\left(x^3 + x^2 - 2x + 12\right) \end{aligned}$$

Since $f(-2) = 12$

$$\begin{aligned} 12 &= a\left((-2)^3 + (-2)^2 - 2(-2) + 12\right) \\ 12 &= 12a \\ a &= 1 \end{aligned}$$

So, $f(x) = (1)\left(x^4 - x^3 - 2x - 4\right)$

$$= x^3 + x^2 - 2x + 12.$$

51. $f(x) = x^4 + 2x^2 - 8$

(a) $f(x) = \left(x^2 + 4\right)\left(x^2 - 2\right)$

(b) $f(x) = \left(x^2 + 4\right)\left(x + \sqrt{2}\right)\left(x - \sqrt{2}\right)$

(c) $f(x) = (x + 2i)(x - 2i)\left(x + \sqrt{2}\right)\left(x - \sqrt{2}\right)$

53. $f(x) = x^4 - 2x^3 - 3x^2 + 12x - 18$

$$\begin{array}{r} x^2 - 2x + 3 \\ x^2 - 6 \overline{\big) x^4 - 2x^3 - 3x^2 + 12x - 18} \\ \underline{x^4 \qquad\quad - 6x^2 \qquad\qquad\quad} \\ -2x^3 + 3x^2 + 12x \qquad \\ \underline{-2x^3 \qquad\quad + 12x \qquad} \\ 3x^2 \qquad\quad -18 \\ \underline{3x^2 \qquad\quad -18} \\ 0 \end{array}$$

(a) $f(x) = \left(x^2 - 6\right)\left(x^2 - 2x + 3\right)$

(b) $f(x) = \left(x + \sqrt{6}\right)\left(x - \sqrt{6}\right)\left(x^2 - 2x + 3\right)$

(c) $f(x) = \left(x + \sqrt{6}\right)\left(x - \sqrt{6}\right)\left(x - 1 - \sqrt{2}i\right)\left(x - 1 + \sqrt{2}i\right)$

Note: Use the Quadratic Formula for (c).

55. $f(x) = x^3 - x^2 + 4x - 4$

Because $2i$ is a zero, so is $-2i$.

$$\begin{array}{r|rrrr} 2i & 1 & -1 & 4 & -4 \\ & & 2i & -4-2i & 4 \\ \hline & 1 & 2i-1 & -2i & 0 \end{array}$$

$$\begin{array}{r|rrr} -2i & 1 & 2i-1 & -2i \\ & & -2i & 2i \\ \hline & 1 & -1 & 0 \end{array}$$

$f(x) = (x - 2i)(x + 2i)(x - 1)$

The zeros of $f(x)$ are $x = 1, \pm 2i$.

Alternate Solution:

Because $x = \pm 2i$ are zeros of $f(x)$,

$(x + 2i)(x - 2i) = x^2 + 4$ is a factor of $f(x)$.

By long division, you have:

$$\begin{array}{r} x - 1 \\ x^2 + 0x + 4 \overline{\big) x^3 - x^2 + 4x - 4} \\ \underline{x^3 + 0x^2 + 4x} \\ -x^2 + 0x - 4 \\ \underline{-x^2 + 0x - 4} \\ 0 \end{array}$$

$f(x) = (x^2 + 4)(x - 1)$

The zeros of $f(x)$ are $x = 1, \pm 2i$.

57. $f(x) = x^3 - 8x^2 + 25x - 26$

Because $3 + 2i$ is a zero, so is $3 - 2i$.

$$\begin{array}{r|rrrr} 3+2i & 1 & -8 & 25 & -26 \\ & & 3+2i & -19-4i & 26 \\ \hline & 1 & -5+2i & 6-4i & 0 \end{array}$$

$$\begin{array}{r|rrr} 3-2i & 1 & -5+2i & 6-4i \\ & & 3-2i & -6+4i \\ \hline & 1 & -2 & 0 \end{array}$$

$f(x) = (x - (3 + 2i))(x - (3 - 2i))(x - 2)$

The zeros of $f(x)$ are $x = 3 \pm 2i, 2$.

Alternate Solution:

Because $x = 3 \pm 2i$ are zeros of $f(x)$, $(x - (3 + 2i))(x - (3 - 2i)) = x^2 - 6x + 13$ is a factor of $f(x.)$

By long division, you have:

$$\begin{array}{r} x - 2 \\ x^2 - 6x + 13 \overline{\big) x^3 - 8x^2 + 25x - 26} \\ \underline{x^3 - 6x^2 + 13x} \\ -2x^2 + 12x - 26 \\ \underline{-2x^2 + 12x^2 - 26} \\ 0 \end{array}$$

$f(x) = (x^2 - 6x + 13)(x - 2)$

The zeros of $f(x)$ are $x = 3 \pm 2i, 2$.

59. $f(x) = x^4 - 6x^3 + 14x^2 - 18x + 9$

Because $1 - \sqrt{2}i$ is a zero, so is $1 + \sqrt{2}i$, and

$$\begin{aligned} \left[x - \left(1 - \sqrt{2}i\right)\right]\left[x - \left(1 + \sqrt{2}i\right)\right] &= \left[(x - 1) - \sqrt{2}i\right]\left[(x - 1) - \sqrt{2}i\right]\left[(x - 1) + \sqrt{2}i\right] \\ &= (x - 1)^2 - \left(\sqrt{2}i\right)^2 \\ &= x^2 - 2x + 1 - 2i^2 \\ &= x^2 - 2x + 3 \end{aligned}$$

is a factor of $f(x)$. By long division, you have:

$$\begin{array}{r} x^2 - 4x + 3 \\ x^2 - 2x + 3 \overline{\big) x^4 - 6x^3 + 14x^2 - 18x + 9} \\ \underline{x^4 - 2x^3 + 3x^2} \\ -4x^3 + 11x^2 - 18x + 9 \\ \underline{-4x^3 + 8x^2 - 12x} \\ 3x^2 - 6x + 9 \\ \underline{3x^2 - 6x + 9} \\ 0 \end{array}$$

$$\begin{aligned} f(x) &= (x^2 - 2x + 3)(x^2 - 4x + 3) \\ &= (x^2 - 2x + 3)(x - 1)(x - 3) \end{aligned}$$

The zeros of $f(x)$ are $x = 1 \pm \sqrt{2}i, 1, 3$.

61. $f(x) = x^2 + 36$
$= (x + 6i)(x - 6i)$

The zeros of $f(x)$ are $x = \pm 6i$.

63. $h(x) = x^2 - 2x + 17$

By the Quadratic Formula, the zeros of $f(x)$ are

$$x = \frac{2 \pm \sqrt{4 - 68}}{2} = \frac{2 \pm \sqrt{-64}}{2} = 1 \pm 4i.$$

$$f(x) = (x - (1 + 4i))(x - (1 - 4i))$$
$$= (x - 1 - 4i)(x - 1 + 4i)$$

65. $f(x) = x^4 - 16$
$= (x^2 - 4)(x^2 + 4)$
$= (x - 2)(x + 2)(x - 2i)(x + 2i)$

Zeros: $\pm 2, \pm 2i$

67. $f(z) = z^2 - 2z + 2$

By the Quadratic Formula, the zeros of $f(z)$ are

$$z = \frac{2 \pm \sqrt{4 - 8}}{2} = 1 \pm i.$$

$$f(z) = [z - (1 + i)][z - (1 - i)]$$
$$= (z - 1 - i)(z - 1 + i)$$

69. $g(x) = x^3 - 3x^2 + x + 5$

Possible rational zeros: $\pm 1, \pm 5$

−1	1	−3	1	5
		−1	4	−5
	1	−4	5	0

By the Quadratic Formula, the zeros of $x^2 - 4x + 5$ are: $x = \frac{4 \pm \sqrt{16 - 20}}{2} = 2 \pm i$

Zeros: $-1, 2 \pm i$

$g(x) = (x + 1)(x - 2 - i)(x - 2 + i)$

71. $g(x) = x^4 - 4x^3 + 8x^2 - 16x + 16$

Possible rational zeros: $\pm 1, \pm 2, \pm 4, \pm 8, \pm 16$

2	1	−4	8	−16	16
		2	−4	8	−16
	1	−2	4	−8	0

2	1	−2	4	−8
		2	0	8
	1	0	4	0

$g(x) = (x - 2)(x - 2)(x^2 + 4)$
$= (x - 2)^2(x + 2i)(x - 2i)$

Zeros: $2, \pm 2i$

73. $f(x) = x^3 + 24x^2 + 214x + 740$

Possible rational zeros: $\pm 1, \pm 2, \pm 4, \pm 5, \pm 10, \pm 20, \pm 37, \pm 74, \pm 148, \pm 185, \pm 370, \pm 740$

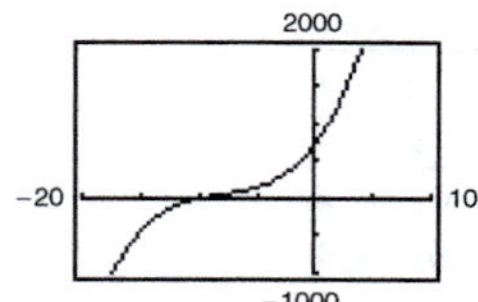

Based on the graph, try $x = -10$.

−10	1	24	214	740
		−10	−140	−740
	1	14	74	0

By the Quadratic Formula, the zeros of $x^2 + 14x + 74$ are $x = \frac{-14 \pm \sqrt{196 - 296}}{2} = -7 \pm 5i$.

The zeros of $f(x)$ are $x = -10$ and $x = -7 \pm 5i$.

75. $f(x) = 16x^3 - 20x^2 - 4x + 15$

Possible rational zeros:

$\pm1, \pm3, \pm5, \pm15, \pm\frac{1}{2}, \pm\frac{3}{2}, \pm\frac{5}{2}, \pm\frac{15}{2}, \pm\frac{1}{4}, \pm\frac{3}{4},$

$\pm\frac{5}{4}, \pm\frac{15}{4}, \pm\frac{1}{8}, \pm\frac{3}{8}, \pm\frac{5}{8}, \pm\frac{15}{8}, \pm\frac{1}{16}, \pm\frac{3}{16}, \pm\frac{5}{16}, \pm\frac{15}{16}$

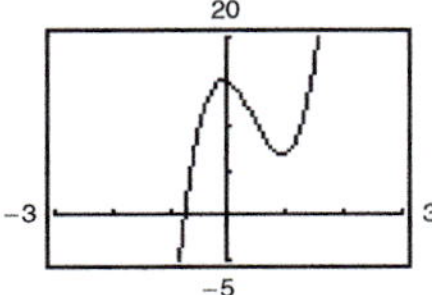

Based on the graph, try $x = -\frac{3}{4}$.

$$\begin{array}{r|rrrr} -\frac{3}{4} & 16 & -20 & -4 & 15 \\ & & -12 & 24 & -15 \\ \hline & 16 & -32 & 20 & 0 \end{array}$$

By the Quadratic Formula, the zeros of $16x^2 - 32x + 20 = 4(4x^2 - 8x + 5)$ are

$$x = \frac{8 \pm \sqrt{64 - 80}}{8} = 1 \pm \frac{1}{2}i.$$

The zeros of $f(x)$ are $x = -\frac{3}{4}$ and $x = 1 \pm \frac{1}{2}i$.

77. $f(x) = 2x^4 + 5x^3 + 4x^2 + 5x + 2$

Possible rational zeros: $\pm1, \pm2, \pm\frac{1}{2}$

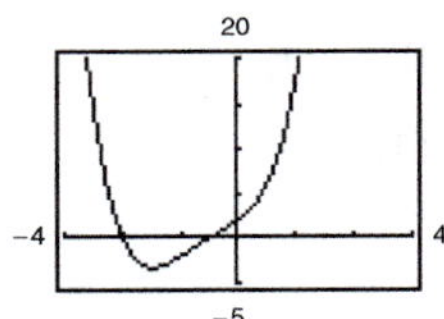

Based on the graph, try $x = -2$ and $x = -\frac{1}{2}$.

$$\begin{array}{r|rrrrr} -2 & 2 & 5 & 4 & 5 & 2 \\ & & -4 & -2 & -4 & -2 \\ \hline & 2 & 1 & 2 & 1 & 0 \end{array}$$

$$\begin{array}{r|rrrr} -\frac{1}{2} & 2 & 1 & 2 & 1 \\ & & -1 & 0 & -1 \\ \hline & 2 & 0 & 2 & 0 \end{array}$$

The zeros of $2x^2 + 2 = 2(x^2 + 1)$ are $x = \pm i$.

The zeros of $f(x)$ are $x = -2$, $x = -\frac{1}{2}$, and $x = \pm i$.

79. $g(x) = 2x^3 - 3x^2 - 3$

Sign variations: 1, positive zeros: 1

$g(-x) = -2x^3 - 3x^2 - 3$

Sign variations: 0, negative zeros: 0

81. $h(x) = 2x^3 + 3x^2 + 1$

Sign variations: 0, positive zeros: 0

$h(-x) = -2x^3 + 3x^2 + 1$

Sign variations: 1, negative zeros: 1

83. $g(x) = 6x^4 + 2x^3 - 3x^2 + 2$

Sign variations: 2, positive zeros: 2 or 0

$g(-x) = 6x^4 - 2x^3 - 3x^2 + 2$

Sign variations: 2, negative zeros: 2 or 0

85. $f(x) = 5x^3 + x^2 - x + 5$

Sign variations: 2, positive zeros: 2 or 0

$f(-x) = -5x^3 + x^2 + x + 5$

Sign variations: 1, negative zeros: 1.

87. $f(x) = x^3 + 3x^2 - 2x + 1$

(a)
$$\begin{array}{r|rrrr} 1 & 1 & 3 & -2 & 1 \\ & & 1 & 4 & 2 \\ \hline & 1 & 4 & 2 & 3 \end{array}$$

1 is an upper bound.

(b)
$$\begin{array}{r|rrrr} -4 & 1 & 3 & -2 & 1 \\ & & -4 & 4 & -8 \\ \hline & 1 & -1 & 2 & -7 \end{array}$$

−4 is a lower bound.

89. $f(x) = x^4 - 4x^3 + 16x - 16$

(a)
$$\begin{array}{r|rrrrr} 5 & 1 & -4 & 0 & 16 & -16 \\ & & 5 & 5 & 25 & 205 \\ \hline & 1 & 1 & 5 & 41 & 189 \end{array}$$

5 is an upper bound.

(b)
$$\begin{array}{r|rrrrr} -3 & 1 & -4 & 0 & 16 & -16 \\ & & -3 & 21 & -63 & 141 \\ \hline & 1 & -7 & 21 & -47 & 125 \end{array}$$

−3 is a lower bound.

91. $f(x) = 16x^3 - 12x^2 - 4x + 3$

Possible rational zeros: $\dfrac{\pm 1, \pm 3}{\pm, \pm 2, \pm 4, \pm 8, \pm 16} = \pm\frac{1}{16}, \pm\frac{1}{8}, \pm\frac{3}{16}, \pm\frac{1}{4}, \pm\frac{3}{8}, \pm\frac{3}{4}, \pm\frac{1}{2}, \pm 1, \pm\frac{3}{2}, \pm 3$

However, the function factors by grouping.

$$\begin{aligned} 16x^2 - 12x^2 - 4x + 3 &= 0 \\ 4x^2(4x - 3) - (4x - 3) &= 0 \\ (4x - 3)(4x^2 - 1) &= 0 \\ (4x - 3)(2x - 1)(2x + 1) &= 0 \end{aligned}$$

$4x - 3 = 0 \rightarrow x = \frac{3}{4}$

$2x - 1 = 0 \rightarrow x = \frac{1}{2}$

$2x + 1 = 0 \rightarrow x = -\frac{1}{2}$

So, the zeros are $x = \frac{3}{4}, \pm\frac{1}{2}$.

93. $f(y) = 4y^3 + 3y^2 + 8y + 6$

Possible rational zeros: $\pm 1, \pm 2, \pm 3, \pm 6, \pm\frac{1}{2}, \pm\frac{3}{2}, \pm\frac{1}{4}, \pm\frac{3}{4}$

$$\begin{array}{r|rrrr} -\frac{3}{4} & 4 & 3 & 8 & 6 \\ & & -3 & 0 & -6 \\ \hline & 4 & 0 & 8 & 0 \end{array}$$

$$\begin{aligned} 4y^3 + 3y^2 + 8y + 6 &= \left(y + \tfrac{3}{4}\right)\left(4y^2 + 8\right) \\ &= \left(y + \tfrac{3}{4}\right)4\left(y^2 + 2\right) \\ &= (4y + 3)\left(y^2 + 2\right) \end{aligned}$$

So, the only real zero is $-\frac{3}{4}$.

95.
$$\begin{aligned} P(x) &= x^4 - \tfrac{25}{4}x^2 + 9 \\ &= \tfrac{1}{4}\left(4x^4 - 25x^2 + 36\right) \\ &= \tfrac{1}{4}\left(4x^2 - 9\right)\left(x^2 - 4\right) \\ &= \tfrac{1}{4}(2x + 3)(2x - 3)(x + 2)(x - 2) \end{aligned}$$

The rational zeros are $\pm\frac{3}{2}$ and ± 2.

97.
$$\begin{aligned} f(x) &= x^3 - \tfrac{1}{4}x^2 - x + \tfrac{1}{4} \\ &= \tfrac{1}{4}\left(4x^3 - x^2 - 4x + 1\right) \\ &= \tfrac{1}{4}\left[x^2(4x - 1) - 1(4x - 1)\right] \\ &= \tfrac{1}{4}(4x - 1)\left(x^2 - 1\right) \\ &= \tfrac{1}{4}(4x - 1)(x + 1)(x - 1) \end{aligned}$$

The rational zeros are $\frac{1}{4}$ and ± 1.

99. $f(x) = x^3 - 1 = (x - 1)\left(x^2 + x + 1\right)$

Rational zeros: $1\ (x = 1)$

Irrational zeros: 0

Matches (d).

100.
$$\begin{aligned} f(x) &= x^3 - 2 \\ &= \left(x - \sqrt[3]{2}\right)\left(x^2 + \sqrt[3]{2}x + \sqrt[3]{4}\right) \end{aligned}$$

Rational zeros: 0

Irrational zeros: $1\left(x = \sqrt[3]{2}\right)$

Matches (a).

101. $f(x) = x^3 - x = x(x + 1)(x - 1)$

Rational zeros: $3\ (x = 0, \pm 1)$

Irrational zeros: 0

Matches (b).

102.
$$\begin{aligned} f(x) &= x^3 - 2x \\ &= x\left(x^2 - 2\right) \\ &= x\left(x + \sqrt{2}\right)\left(x - \sqrt{2}\right) \end{aligned}$$

Rational zeros: $1\ (x = 0)$

Irrational zeros: $2\left(x = \pm\sqrt{2}\right)$

Matches (c).

103. (a)

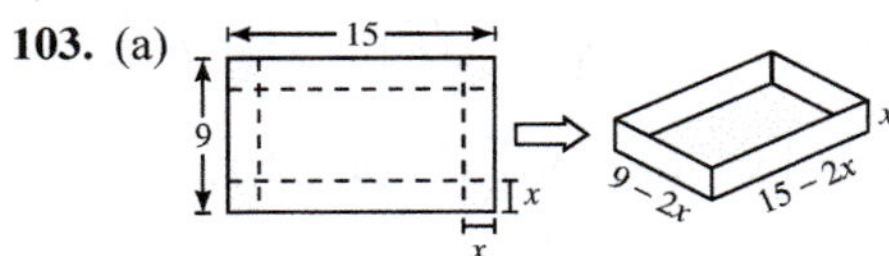

(b) $V = l \cdot w \cdot h = (15 - 2x)(9 - 2x)x$

$= x(9 - 2x)(15 - 2x)$

Because length, width, and height must be positive, you have $0 < x < \frac{9}{2}$ for the domain.

(c)

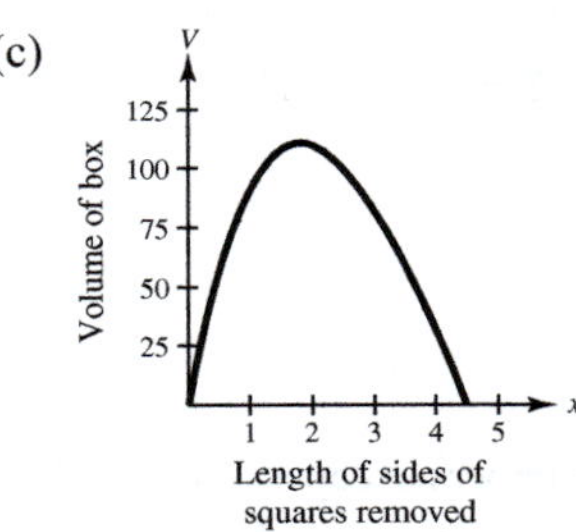

The volume is maximum when $x \approx 1.82$.

The dimensions are: $r = ak^3 + bk^2 + ck + d, f(k) = r.$

1.82 cm × 5.36 cm × 11.36 cm

(d) $56 = x(9 - 2x)(15 - 2x)$

$56 = 135x - 48x^2 + 4x^3$

$0 = 4x^3 - 48x^2 + 135x - 56$

The zeros of this polynomial are $\frac{1}{2}, \frac{7}{2},$ and 8.

x cannot equal 8 because it is not in the domain of V. [The length cannot equal -1 and the width cannot equal -7. The product of $(8)(-1)(-7) = 56$ so it showed up as an extraneous solution.]

So, the volume is 56 cubic centimeters when $x = \frac{1}{2}$ centimeter or $x = \frac{7}{2}$ centimeters.

105. (a) Current bin: $V = 2 \times 3 \times 4 = 24$ cubic feet

New bin: $V = 5(24) = 120$ cubic feet

$V(x) = (2 + x)(3 + x)(4 + x) = 120$

(b) $x^3 + 9x^2 + 26x + 24 = 120$

$x^3 + 9x^2 + 26x - 96 = 0$

The only real zero of this polynomial is $x = 2$. All the dimensions should be increased by 2 feet, so the new bin will have dimensions of 4 feet by 5 feet by 6 feet.

107. False. The most complex zeros it can have is two, and the Linear Factorization Theorem guarantees that there are three linear factors, so one zero must be real.

109. $g(x) = -f(x)$. This function would have the same zeros as $f(x)$, so $r_1, r_2,$ and r_3 are also zeros of $g(x)$.

111. $g(x) = f(x - 5)$. The graph of $g(x)$ is a horizontal shift of the graph of $f(x)$ five units of the right, so the zeros of $g(x)$ are $5 + r_1, 5 + r_2,$ and $5 + r_3$.

113. $g(x) = 3 + f(x)$. Because $g(x)$ is a vertical shift of the graph of $f(x)$, the zeros of $g(x)$ cannot be determined.

115. Zeros: $-2, \frac{1}{2}, 3$

$f(x) = -(x + 2)(2x - 1)(x - 3)$

$= -2x^3 + 3x^2 + 11x - 6$

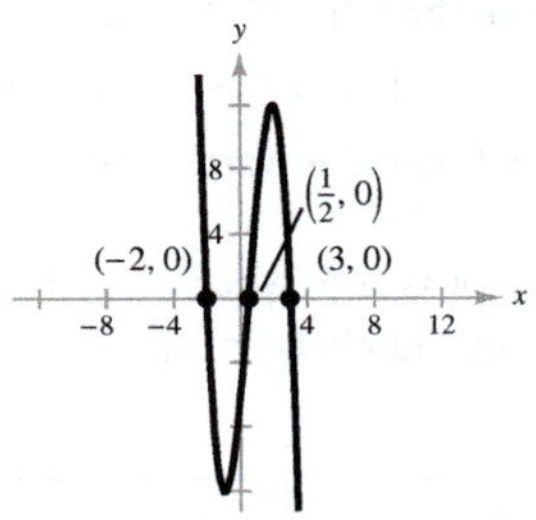

Any nonzero scalar multiple of f would have the same three zeros. Let $g(x) = af(x), a > 0$. There are infinitely many possible functions for f.

117. Because $1 + i$ is a zero of f, so is $1 - i$. From the graph, 1 is also a zero.

$f(x) = (x - (1 + i))(x - (1 - i))(x - 1)$

$= (x^2 - 2x + 2)(x - 1)$

$= x^3 - 3x^2 + 4x - 2$

119.

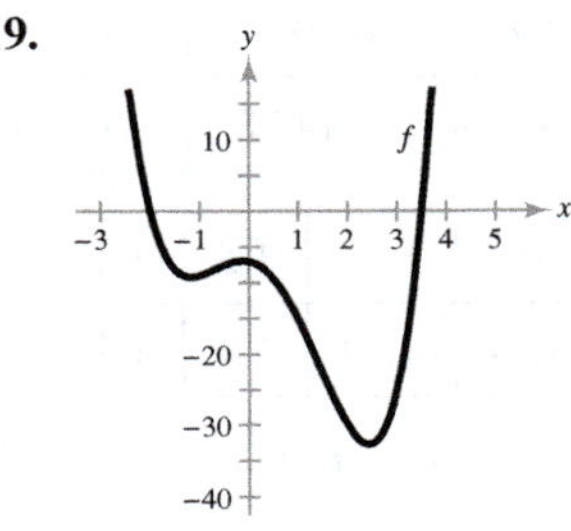

If $x = i$ is a zero, then $x = -i$ is also a zero. So the function is $f(x) = (x - 2)(x - 3.5)(x - i)(x + i)$.

121. Because $f(i) = f(2i) = 0$, then i and $2i$ are zeros of f. Because i and $2i$ are zeros of f, so are $-i$ and $-2i$.

$$f(x) = (x - i)(x + i)(x - 2i)(x + 2i)$$
$$= (x^2 + 1)(x^2 + 4)$$
$$= x^4 + 5x^2 + 4$$

123. (a) $f(x) = \left(x - \sqrt{b}i\right)\left(x + \sqrt{b}i\right) = x^2 + b$

(b) $f(x) = \left[x - (a + bi)\right]\left[x - (a - bi)\right]$
$$= \left[(x - a) - bi\right]\left[(x - a) + bi\right]$$
$$= (x - a)^2 - (bi)^2$$
$$= x^2 - 2ax + a^2 + b^2$$

Section 2.6 Rational Functions

1. rational functions

3. horizontal asymptote

5. Because the denominator is zero when $x - 1 = 0$, the domain of f is all real numbers except $x = 1$.

x	0	0.5	0.9	0.99	$\to 1$
$f(x)$	-1	-2	-10	-100	$\to -\infty$

x	$1 \leftarrow$	1.01	1.1	1.5	2
$f(x)$	$\infty \leftarrow$	100	10	2	1

As x approaches 1 from the left, $f(x)$ decreases without bound towards $-\infty$. As x approaches 1 from the right, $f(x)$ increases without bound towards $+\infty$.

7. Because the denominator is zero when $x^2 - 1 = 0$, the domain of f is all real numbers except $x = -1$ and $x = 1$.

x	-2	-1.5	-1.1	-1.01	$\to -1$
$f(x)$	4	5.4	17.3	152.3	$\to \infty$

x	$-1 \leftarrow$	-0.99	-0.9	-0.5	0
$f(x)$	$-\infty \leftarrow$	-147.8	-12.8	-1	0

As x approaches -1 from the left, $f(x)$ increases without bound (∞). As x approaches -1 from the right, $f(x)$ decreases without bound $(-\infty)$.

x	0	0.5	0.9	0.99	$\to 1$
$f(x)$	0	-1	-12.8	-147.8	$\to -\infty$

x	$1 \leftarrow$	1.01	1.1	1.5	2
$f(x)$	$\infty \leftarrow$	152.3	17.3	5.4	4

As x approaches 1 from the left, $f(x)$ decreases without bound $(-\infty)$. As x approaches 1 from the right, $f(x)$ increases without bound (∞).

9. $f(x) = \dfrac{4}{x^2}$

Domain: all real numbers except $x = 0$

Vertical asymptote: $x = 0$

Horizontal asymptote: $y = 0$

$\left[\text{Degree of } N(x) < \text{degree of } D(x)\right]$

11. $f(x) = \dfrac{5 + x}{5 - x} = \dfrac{x + 5}{-x + 5}$

Domain: all real numbers except $x = 5$

Vertical asymptote: $x = 5$

Horizontal asymptote: $y = -1$

$\left[\text{Degree of } N(x) = \text{degree of } D(x)\right]$

13. $f(x) = \dfrac{x^3}{x^2 - 1}$

Domain: all real numbers except $x = \pm 1$

Vertical asymptotes: $x = \pm 1$

Horizontal asymptote: None

$\left[\text{Degree of } N(x) > \text{degree of } D(x)\right]$

15. $f(x) = \dfrac{x^2 - 3x - 4}{2x^2 + x - 1}$
$$= \frac{(x + 1)(x - 4)}{(2x - 1)(x + 1)}$$
$$= \frac{x - 4}{2x - 1}, x \neq -1$$

Horizontal asymptote: $y = \dfrac{1}{2}$

$(\text{Degree of } N(x) = \text{degree of } D(x))$

Vertical asymptote: $x = \dfrac{1}{2}$

(Because $x + 1$ is a common factor of $N(x)$ and $D(x)$, $x = -1$ is not a vertical asymptote of $f(x)$.)

17. $f(x) = \dfrac{1}{x + 1}$

(a) Domain: all real numbers x except $x = -1$

(b) y-intercept: $(0, 1)$

(c) Vertical asymptote: $x = -1$

Horizontal asymptote: $y = 0$

(d)

x	-4	-3	0	1	2	3
$f(x)$	$-\frac{1}{3}$	$-\frac{1}{2}$	1	$\frac{1}{2}$	$\frac{1}{3}$	$\frac{1}{4}$

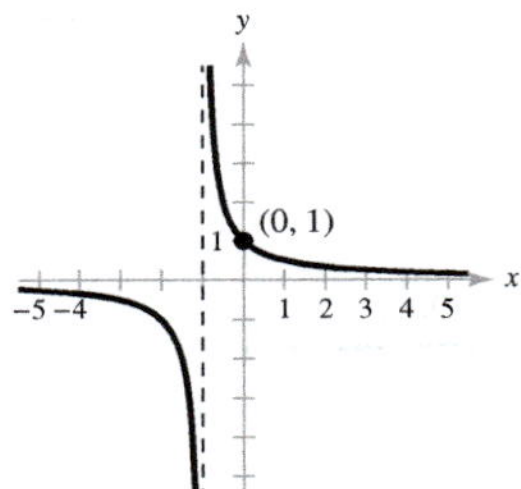

19. $h(x) = \dfrac{-1}{x + 4}$

(a) Domain: all real numbers x except $x = -4$

(b) y-intercept: $\left(0, -\dfrac{1}{4}\right)$

(c) Vertical asymptote: $x = -4$

Horizontal asymptote: $y = 0$

(d)

x	-6	-5	-3	-2	-1	0
$h(x)$	$\frac{1}{2}$	1	-1	$-\frac{1}{2}$	$-\frac{1}{3}$	$-\frac{1}{4}$

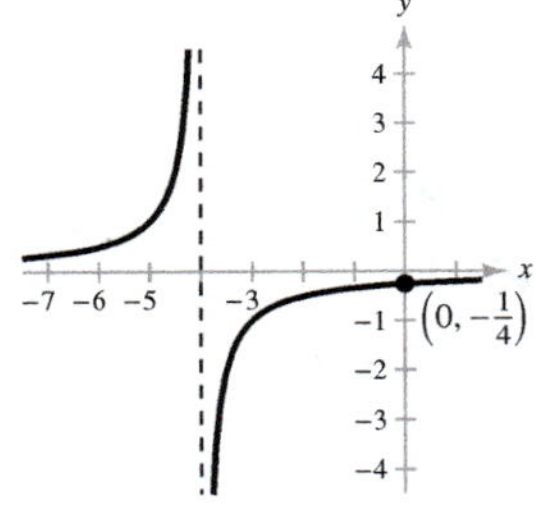

21. $C(x) = \dfrac{2x + 3}{x + 2}$

(a) Domain: all real numbers x except $x = -2$

(b) x-intercept: $\left(-\dfrac{3}{2}, 0\right)$

y-intercept: $\left(0, \dfrac{3}{2}\right)$

(c) Vertical asymptote: $x = -2$

Horizontal asymptote: $y = 2$

(d)

x	-4	-3	-1	0	1	2
$C(x)$	$\frac{5}{2}$	3	1	$\frac{3}{2}$	$\frac{5}{3}$	$\frac{7}{4}$

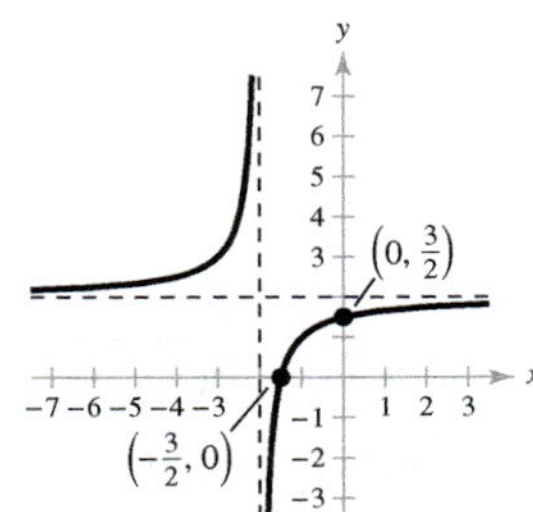

23. $f(x) = \dfrac{x^2}{x^2 + 9}$

(a) Domain: all real numbers x

(b) Intercept: $(0, 0)$

(c) Horizontal asymptote: $y = 1$

(d)

x	±1	±2	±3
$f(x)$	$\frac{1}{10}$	$\frac{4}{13}$	$\frac{1}{2}$

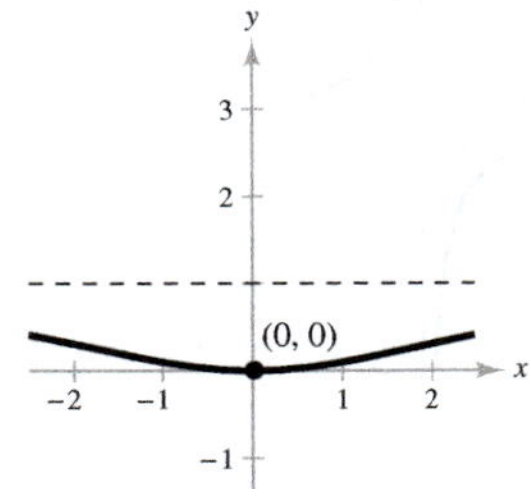

25. $g(s) = \dfrac{4s}{s^2 + 4}$

(a) Domain: all real numbers s

(b) Intercept: $(0, 0)$

(c) Horizontal asymptote: $y = 0$

(d)

s	-2	-1	0	1	2
$g(s)$	-1	$-\frac{4}{5}$	0	$\frac{4}{5}$	1

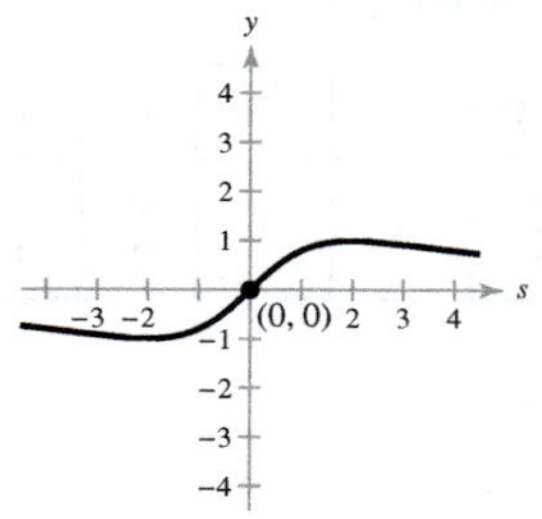

27. $f(x) = \dfrac{2x}{x^2 - 3x - 4} = \dfrac{2x}{(x - 4)(x + 1)}$

(a) Domain: all real numbers x except $x = 4$ and $x = -1$

(b) Intercept: $(0, 0)$

(c) Vertical asymptotes: $x = 4, x = -1$

Horizontal asymptote: $y = 0$

(d)

x	-3	-2	0	1	2	3	5
$f(x)$	$-\frac{3}{7}$	$-\frac{2}{3}$	0	$-\frac{1}{3}$	$-\frac{2}{3}$	$-\frac{3}{2}$	$\frac{5}{3}$

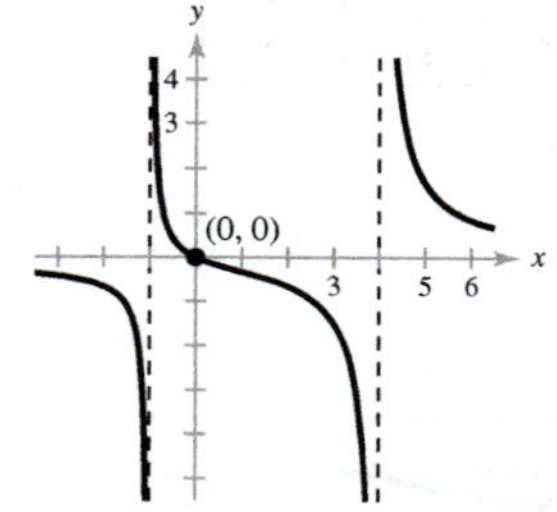

29. $f(x) = \dfrac{x - 4}{x^2 - 16} = \dfrac{x - 4}{(x + 4)(x - 4)} = \dfrac{1}{x + 4}, x \neq 4$

(a) Domain: all real numbers except $x = \pm 4$

(b) y-intercept: $\left(0, \dfrac{1}{4}\right)$

(c) Vertical asymptote: $x = -4$

Horizontal asymptote: $y = 0$

(d)

x	-5	-4	-3	-2	-1
$f(x)$	-1	Undef.	1	$\frac{1}{2}$	$\frac{1}{3}$

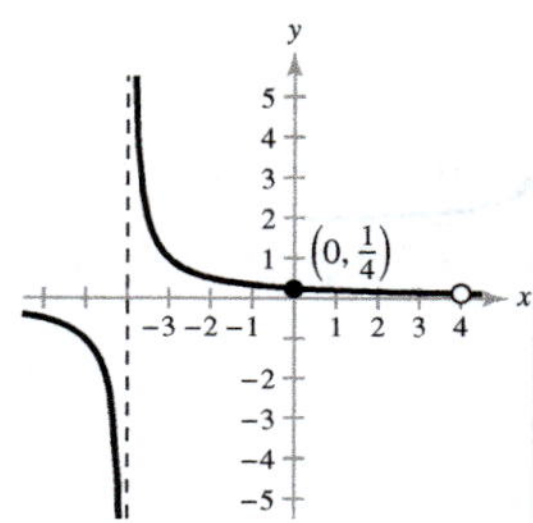

31. $f(t) = \dfrac{t^2 - 1}{t - 1} = \dfrac{(t + 1)(t - 1)}{t - 1} = t + 1, \quad t \neq 1$

(a) Domain: all real numbers t except $t = 1$

(b) t-intercept: $(-1, 0)$

y-intercept: $(0, 1)$

(c) No asymptotes

(d)

t	-3	-2	-1	0	1	2
$f(t)$	-2	-1	0	1	Undef.	3

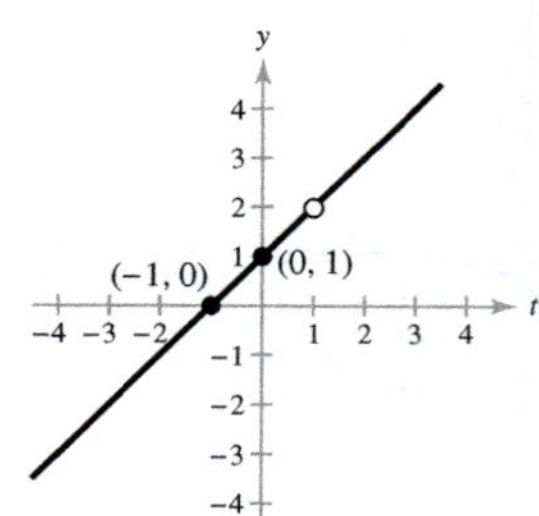

33. $f(x) = \frac{x^2 - 25}{x^2 - 4x - 5} = \frac{(x + 5)(x - 5)}{(x - 5)(x + 1)} = \frac{x + 5}{x + 1}, x \neq 5$

(a) Domain: all real numbers x except $x = 5$ and $x = -1$

(b) x-intercept: $(-5, 0)$

y-intercept: $(0, 5)$

(c) Vertical asymptote: $x = -1$

Horizontal asymptote: $y = 1$

(d)

x	-5	-3	0	3	5
$f(x)$	0	-1	5	2	Undef.

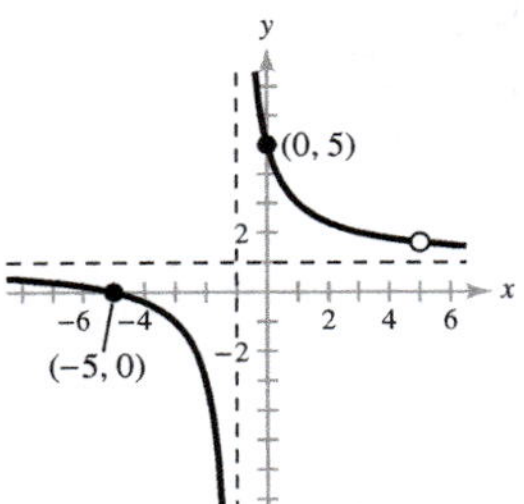

35. $f(x) = \frac{x^2 + 3x}{x^2 + x - 6}$

$= \frac{x(x + 3)}{(x + 3)(x - 2)}$

$= \frac{x}{x - 2}, x \neq -3$

(a) Domain: all real numbers x except $x = -3$ and $x = 2$

(b) Intercept: $(0, 0)$

(c) Vertical asymptote: $x = 2$

Horizontal asymptote: $y = 1$

(d)

x	-1	0	1	3	4
$f(x)$	$\frac{1}{3}$	0	-1	3	2

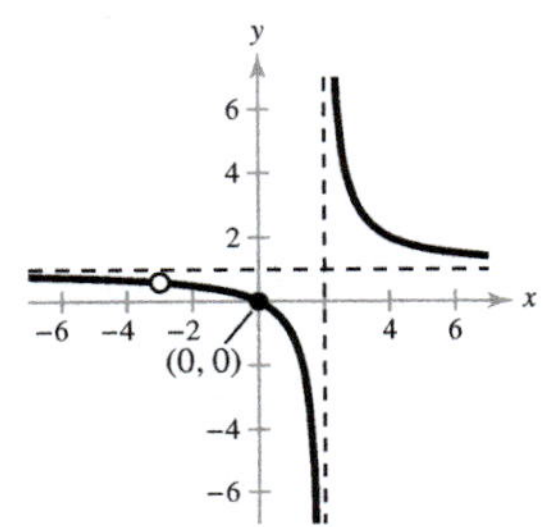

37. $f(x) = \frac{2x^2 - 5x - 3}{x^3 - 2x^2 - x + 2} = \frac{(2x + 1)(x - 3)}{(x - 2)(x + 1)(x - 1)}$

(a) Domain: all real numbers x except $x = 2, x = \pm 1$

(b) x-intercepts: $\left(-\frac{1}{2}, 0\right), (3, 0)$

y-intercept: $\left(0, -\frac{3}{2}\right)$

(c) Vertical asymptotes: $x = 2, x = -1,$ and $x = 1$

Horizontal asymptote: $y = 0$

(d)

x	-3	-2	0	$\frac{3}{2}$	3	4
$f(x)$	$-\frac{3}{4}$	$-\frac{5}{4}$	$-\frac{3}{2}$	$\frac{48}{5}$	0	$\frac{3}{10}$

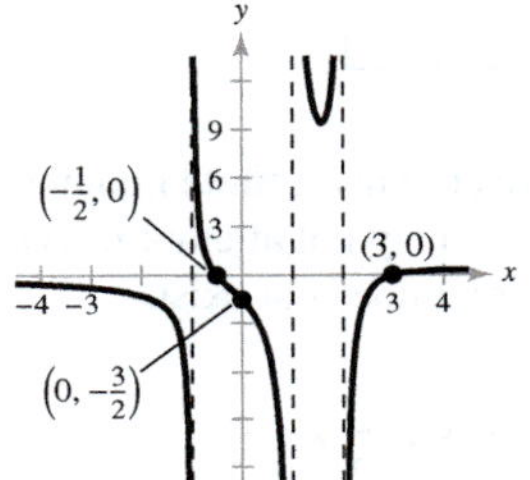

39. Because the function has a vertical asymptote at $x = -2$ and a horizontal asymptote at $y = 0$,

$f(x) = \frac{4}{x + 2}$ matches graph (d).

40. Because the function has a vertical asymptote at $x = 2$ and a horizontal asymptote at $y = 0$,

$f(x) = \frac{5}{x - 2}$ matches graph (a).

41. Because the function has vertical asymptotes at $x = \pm 2$ and a horizontal asymptote at $y = 2$,

$f(x) = \frac{2x^2}{x^2 - 4}$ matches graph (c).

42. Because the function has a vertical asymptote at $x = -2$ and a horizontal asymptote at $y = 0$,

$f(x) = \frac{3x}{(x + 2)^2}$ matches graph (b).

43. (a) Domain of f: all real numbers x except $x = -1$

Domain of g: all real numbers x

(b)

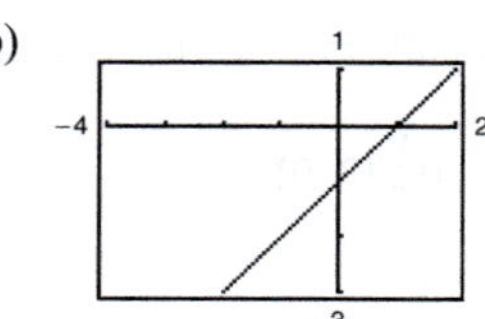

(c) Because there are only a finite number of pixels, the graphing utility may not attempt to evaluate the function where it does not exist.

45. (a) Domain of f: all real numbers x except $x = 0, 2$

Domain of g: all real numbers $x = 0$

(b)

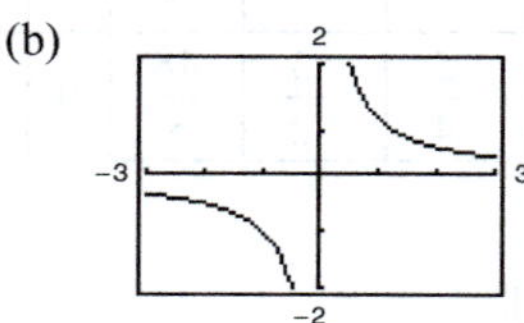

(c) Because there are only a finite number of pixels, the graphing utility may not attempt to evaluate the function where it does not exist.

47. $h(x) = \dfrac{x^2 - 4}{x} = x - \dfrac{4}{x}$

(a) Domain: all real numbers x except $x = 0$

(b) x-intercepts: $(-2, 0)$, $(2, 0)$

(c) Vertical asymptote: $x = 0$

Slant asymptote: $y = x$

(d)

x	-4	-3	-1	1	3	4
$h(x)$	-3	$\dfrac{-5}{3}$	3	-3	$\dfrac{5}{3}$	3

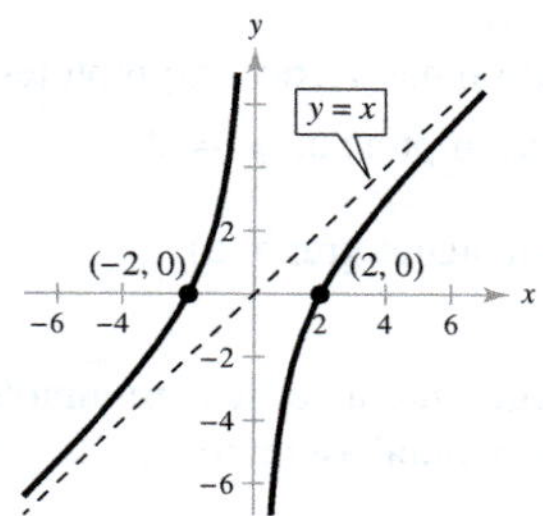

49. $f(x) = \dfrac{2x^2 + 1}{x} = 2x + \dfrac{1}{x}$

(a) Domain: all real numbers x except $x = 0$

(b) No intercepts

(c) Vertical asymptote: $x = 0$

Slant asymptote: $y = 2x$

(d)

x	-4	-2	2	4	6
$f(x)$	$-\dfrac{33}{4}$	$-\dfrac{9}{2}$	$\dfrac{9}{2}$	$\dfrac{33}{4}$	$\dfrac{73}{6}$

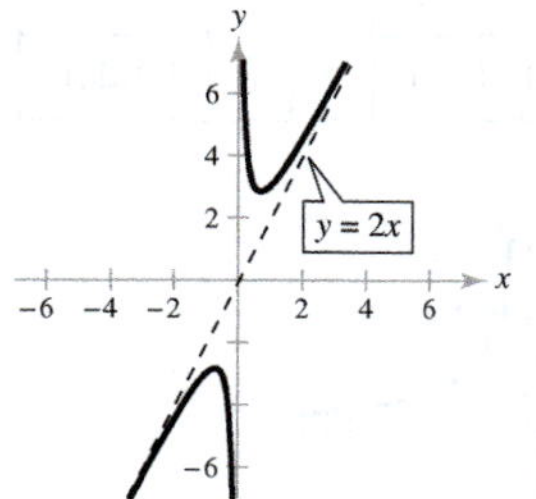

51. $g(x) = \dfrac{x^2 + 1}{x} = x + \dfrac{1}{x}$

(a) Domain: all real numbers x except $x = 0$

(b) No intercepts

(c) Vertical asymptote: $x = 0$

Slant asymptote: $y = x$

(d)

x	-4	-2	2	4	6
$g(x)$	$-\dfrac{17}{4}$	$-\dfrac{5}{2}$	$\dfrac{5}{2}$	$\dfrac{17}{4}$	$\dfrac{37}{6}$

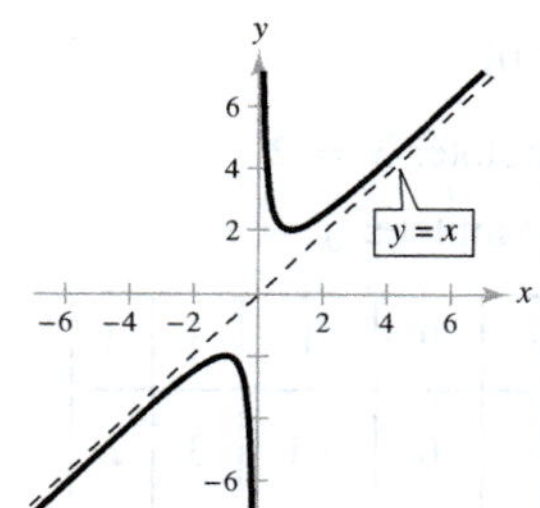

53. $f(t) = \frac{t^2 + 1}{t + 5} = -t + 5 - \frac{26}{t + 5}$

(a) Domain: all real numbers t except $t = -5$

(b) Intercept: $\left(0, -\frac{1}{5}\right)$

(c) Vertical asymptote: $t = -5$

Slant asymptote: $y = -t + 5$

(d)

t	-7	-6	-4	-3	0
$f(t)$	25	37	-17	-5	$-\frac{1}{5}$

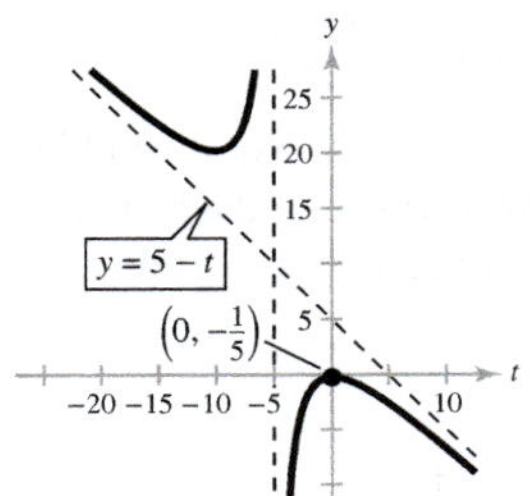

55. $f(x) = \frac{x^3}{x^2 - 4} = x + \frac{4x}{x^2 - 4}$

(a) Domain: all real numbers x except $x = \pm 2$

(b) Intercept: $(0, 0)$

(c) Vertical asymptotes: $x = \pm 2$

Slant asymptote: $y = x$

(d)

x	-6	-4	-1	0	1	4	6
$f(x)$	$-\frac{27}{4}$	$-\frac{16}{3}$	$\frac{1}{3}$	0	$-\frac{1}{3}$	$\frac{16}{3}$	$\frac{27}{4}$

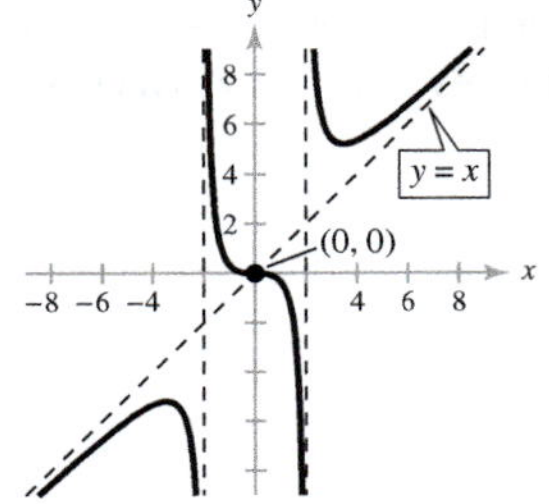

57. $f(x) = \frac{x^2 - x + 1}{x - 1} = x + \frac{1}{x - 1}$

(a) Domain: all real numbers x except $x = 1$

(b) y-intercept: $(0, -1)$

(c) Vertical asymptote: $x = 1$

Slant asymptote: $y = x$

(d)

x	-4	-2	0	2	4
$f(x)$	$-\frac{21}{5}$	$-\frac{7}{3}$	-1	3	$\frac{13}{3}$

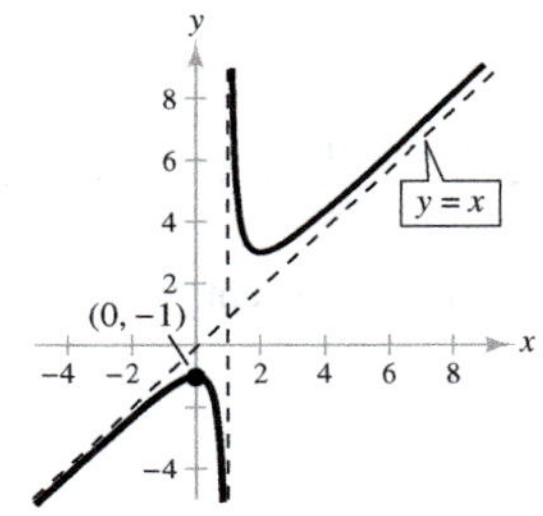

59. $f(x) = \frac{2x^3 - x^2 - 2x + 1}{x^2 + 3x + 2} = \frac{(2x - 1)(x + 1)(x - 1)}{(x + 1)(x + 2)}$

$= \frac{(2x - 1)(x - 1)}{x + 2}, \quad x \neq -1$

$= \frac{2x^2 - 3x + 1}{x + 2} = 2x - 7 + \frac{15}{x + 2}, \quad x \neq -1$

(a) Domain: all real numbers x except $x = -1$ and $x = -2$

(b) y-intercept: $\left(0, \frac{1}{2}\right)$

x-intercepts: $\left(\frac{1}{2}, 0\right), (1, 0)$

(c) Vertical asymptote: $x = -2$

Slant asymptote: $y = 2x - 7$

(d)

x	-4	-3	$-\frac{3}{2}$	0	1
$f(x)$	$-\frac{45}{2}$	-28	20	$\frac{1}{2}$	0

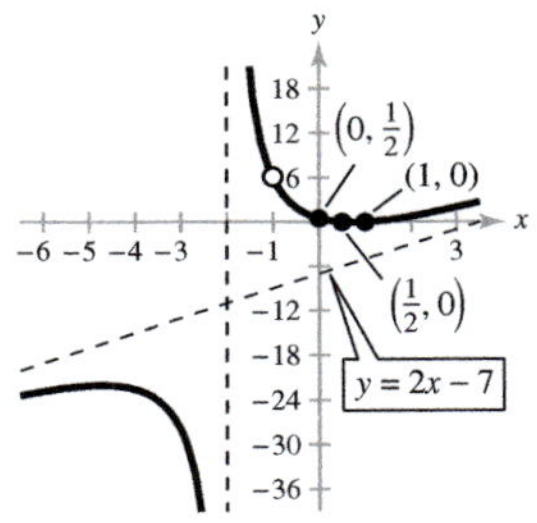

61. $f(x) = \dfrac{x^2 + 2x - 8}{x + 2} = x - \dfrac{8}{x + 2}$

Domain: all real numbers x except $x = -2$

Vertical asymptote: $x = -2$

Slant asymptote: $y = x$

Line: $y = x$

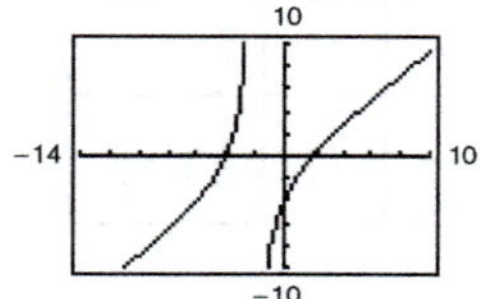

63. $g(x) = \dfrac{1 + 3x^2 - x^3}{x^2} = \dfrac{1}{x^2} + 3 - x = -x + 3 + \dfrac{1}{x^2}$

Domain: all real numbers x except $x = 0$

Vertical asymptote: $x = 0$

Slant asymptote: $y = -x + 3$

Line: $y = -x + 3$

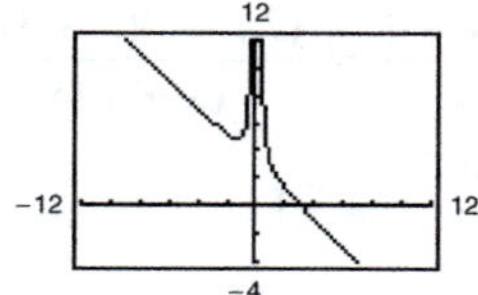

65. $y = \dfrac{x + 1}{x - 3}$

(a) x-intercept: $(-1, 0)$

(b) $0 = \dfrac{x + 1}{x - 3}$

$0 = x + 1$

$-1 = x$

67. $y = \dfrac{1}{x} - x$

(a) x-intercepts: $(-1, 0), (1, 0)$

(b) $0 = \dfrac{1}{x} - x$

$x = \dfrac{1}{x}$

$x^2 = 1$

$x = \pm 1$

69. $C = \dfrac{25{,}000p}{100 - p}, \; 0 \le p < 100$

(a)

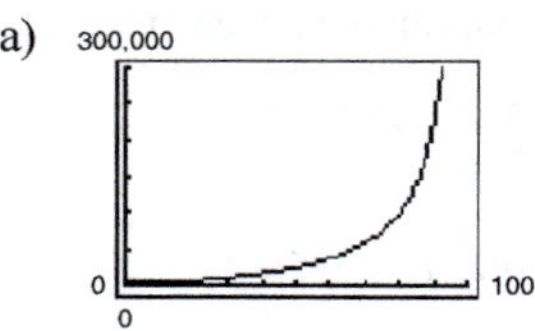

(b) $C = \dfrac{25{,}000(15)}{100 - 15} \approx \$4411.76$

$C = \dfrac{25{,}000(50)}{100 - 50} = \$25{,}000$

$C = \dfrac{25{,}000(90)}{100 - 90} = \$225{,}000$

(c) $C \to \infty$ as $x \to 100$. No, it would not be possible to supply bins to 100% of the residents because the model is undefined for $p = 100$.

71. $A = xy$ and

$(x - 3)(y - 2) = 64$

$y - 2 = \dfrac{64}{x - 3}$

$y = 2 + \dfrac{64}{x - 3} = \dfrac{2x + 58}{x - 3}$

Thus, $A = xy = x\left(\dfrac{2x + 58}{x - 3}\right) = \dfrac{2x(x + 29)}{x - 3}, x > 3.$

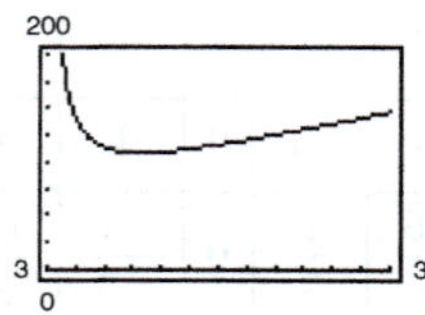

By graphing the area function, we see that A is minimum when $x \approx 12.8$ inches and $y \approx 8.5$ inches.

73. (a) Let t_1 = time from Akron to Columbus and t_2 = time from Columbus back to Akron.

$$xt_1 = 100 \Rightarrow t_1 = \frac{100}{x}$$

$$yt_2 = 100 \Rightarrow t_2 = \frac{100}{y}$$

$$50(t_1 + t_2) = 200$$
$$t_1 + t_2 = 4$$
$$\frac{100}{x} + \frac{100}{y} = 4$$
$$100y + 100x = 4xy$$
$$25y + 25x = xy$$
$$25x = xy - 25y$$
$$25x = y(x - 25)$$

Thus, $y = \dfrac{25x}{x - 25}$.

(b) Vertical asymptote: $x = 25$

Horizontal asymptote: $y = 25$

(c)

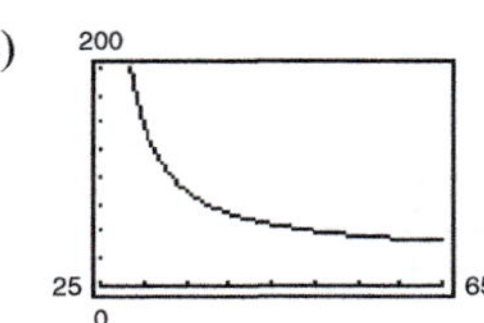

(d)

x	30	35	40	45	50	55	60
y	150	87.5	66.7	56.3	50	45.8	42.9

(e) *Sample answer:* No. You might expect the average speed for the round trip to be the average of the average speeds for the two parts of the trip.

(f) No. At 20 miles per hour you would use more time in one direction than is required for the round trip at an average speed of 50 miles per hour.

75. False. Polynomial functions do not have vertical asymptotes.

77. False. A graph can have a vertical asymptote and a horizontal asymptote or a vertical asymptote and a slant asymptote, but a graph cannot have both a horizontal asymptote and a slant asymptote.

A horizontal asymptote occurs when the degree of $N(x)$ is equal to the degree of $D(x)$ or when the degree of $N(x)$ is less than the degree of $D(x)$. A slant asymptote occurs when the degree of $N(x)$ is greater than the degree of $D(x)$ by one. Because the degree of a polynomial is constant, it is impossible to have both relationships at the same time.

79. Yes. No. Every rational function is the ratio of two polynomial functions of the form $f(x) = \dfrac{N(x)}{D(x)}$.

81. Vertical asymptotes: $x = -2, x = 1 \Rightarrow (x + 2)(x - 1)$ are factors of the denominator.

Horizontal asymptotes: None $\Rightarrow$ The degree of the numerator is greater than the degree of the denominator.

$f(x) = \dfrac{x^3}{(x + 2)(x - 1)}$ is one possible function. There are many correct answers.

Section 2.7 Nonlinear Inequalities

1. positive; negative

3. zeros; undefined values

5. $x^2 - 3 < 0$

(a) $x = 3$

$(3)^2 - 3 \stackrel{?}{<} 0$

$6 \not< 0$

No, $x = 3$ *is not* a solution.

(b) $x = 0$

$(0)^2 - 3 \stackrel{?}{<} 0$

$-3 < 0$

Yes, $x = 0$ *is* a solution.

(c) $x = \frac{3}{2}$

$\left(\frac{3}{2}\right)^2 - 3 \stackrel{?}{<} 0$

$-\frac{3}{4} < 0$

Yes, $x = \frac{3}{2}$ *is* a solution.

(d) $x = -5$

$(-5)^2 - 3 \stackrel{?}{<} 0$

$22 \not< 0$

No, $x = -5$ *is not* a solution

7. $\dfrac{x+2}{x-4} \geq 3$

(a) $x = 5$

$$\frac{5+2}{5-4} \stackrel{?}{\geq} 3$$
$$7 \geq 3$$

Yes, $x = 5$ *is* a solution.

(b) $x = 4$

$$\frac{4+2}{4-4} \stackrel{?}{\geq} 3$$

$\dfrac{6}{0}$ is undefined.

No, $x = 4$ *is not* a solution.

(c) $x = -\dfrac{9}{2}$

$$\frac{-\frac{9}{2}+2}{-\frac{9}{2}-4} \stackrel{?}{\geq} 3$$
$$\frac{5}{17} \not\geq 3$$

No, $x = -\dfrac{9}{2}$ *is not* a solution.

(d) $x = \dfrac{9}{2}$

$$\frac{\frac{9}{2}+2}{\frac{9}{2}-4} \stackrel{?}{\geq} 3$$
$$13 \geq 3$$

Yes, $x = \dfrac{9}{2}$ *is* a solution.

9. $x^2 - 3x - 18 = (x+3)(x-6)$

$x + 3 = 0 \Rightarrow x = -3$

$x - 6 = 0 \Rightarrow x = 6$

The key numbers are -3 and 6.

11. $\dfrac{1}{x-5} + 1 = \dfrac{1 + 1(x-5)}{x-5}$

$= \dfrac{x-4}{x-5}$

$x - 4 = 0 \Rightarrow x = 4$

$x - 5 = 0 \Rightarrow x = 5$

The key numbers are 4 and 5.

13. $2x^2 + 4x < 0$

$2x(x+2) < 0$

Key numbers: $x = 0, \ -2$

Test intervals: $(-\infty, -2), (-2, 0), (0, \infty)$

Test: Is $2x(x+2) < 0$?

Interval	x-Value	Value of $2x(x+2)$	Conclusion
$(-\infty, -2)$	-3	6	Positive
$(-2, 0)$	-1	-2	Negative
$(0, \infty)$	1	3	Positive

Solution Set: $(-2, 0)$

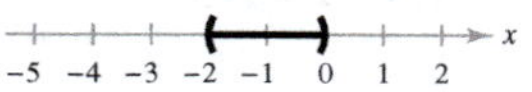

15. $x^2 < 9$

$x^2 - 9 < 0$

$(x+3)(x-3) < 0$

Key numbers: $x = \pm 3$

Test intervals: $(-\infty, -3), (-3, 3), (3, \infty)$

Test: Is $(x+3)(x-3) < 0$?

Interval	x-Value	Value of $x^2 - 9$	Conclusion
$(-\infty, -3)$	-4	7	Positive
$(-3, 3)$	0	-9	Negative
$(3, \infty)$	4	7	Positive

Solution set: $(-3, 3)$

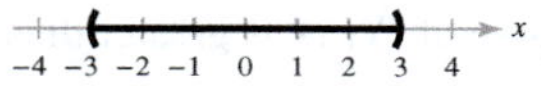

17. $(x+2)^2 \leq 25$

$x^2 + 4x + 4 \leq 25$

$x^2 + 4x - 21 \leq 0$

$(x+7)(x-3) \leq 0$

Key numbers: $x = -7, x = 3$

Test intervals: $(-\infty, -7), (-7, 3), (3, \infty)$

Test: Is $(x+7)(x-3) \leq 0$?

Interval	x-Value	Value of $(x+7)(x-3)$	Conclusion
$(-\infty, -7)$	-8	$(-1)(-11) = 11$	Positive
$(-7, 3)$	0	$(7)(-3) = -21$	Negative
$(3, \infty)$	4	$(11)(1) = 11$	Positive

Solution set: $[-7, 3]$

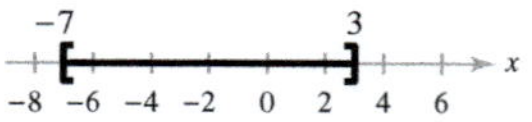

19. $x^2 + 6x + 1 \geq -7$

$x^2 + 6x + 8 \geq 0$

$(x + 2)(x + 4) \geq 0$

Key numbers: $x = -2, x = -4$

Test Intervals: $(-\infty, -4), (-4, -2), (-2, \infty)$

Test: Is $(x + 2)(x + 4) > 0$?

Interval	x-Value	Value of $(x + 2)(x + 4)$	Conclusion
$(-\infty, -4)$	-6	8	Positive
$(-4, -2)$	-3	-1	Negative
$(-2, \infty)$	0	8	Positive

Solution set: $(-\infty, -4] \cup [-2, \infty)$

21. $x^2 + x < 6$

$x^2 + x - 6 < 0$

$(x + 3)(x - 2) < 0$

Key numbers: $x = -3, x = 2$

Test intervals: $(-\infty, -3), (-3, 2), (2, \infty)$

Test: Is $(x + 3)(x - 2) < 0$?

Interval	x-Value	Value of $(x + 3)(x - 2)$	Conclusion
$(-\infty, -3)$	-4	$(-1)(-6) = 6$	Positive
$(-3, 2)$	0	$(3)(-2) = -6$	Negative
$(2, \infty)$	3	$(6)(1) = 6$	Positive

Solution set: $(-3, 2)$

23. $x^2 < 3 - 2x$

$x^2 + 2x - 3 < 0$

$(x + 3)(x - 1) < 0$

Key numbers: $x = -3, x = 1$

Test intervals: $(-\infty, -3), (-3, 1), (1, \infty)$

Test: Is $(x + 3)(x - 1) < 0$?

Interval	x-Value	Value of $(x + 3)(x - 1)$	Conclusion
$(-\infty, -3)$	-4	$(-1)(-5) = 5$	Positive
$(-3, 1)$	0	$(3)(-1) = -3$	Negative
$(1, \infty)$	2	$(5)(1) = 5$	Positive

Solution set: $(-3, 1)$

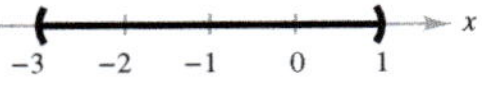

25. $3x^2 - 11x > 20$

$3x^2 - 11x - 20 > 0$

$(3x + 4)(x - 5) > 0$

Key numbers: $x = 5, x = -\frac{4}{3}$

Test intervals: $\left(-\infty, -\frac{4}{5}\right), \left(-\frac{4}{3}, 5\right), (5, \infty)$

Test: Is $(3x + 4)(x - 5) > 0$?

Interval	x-Value	Value of $(3x + 4)(x - 5)$	Conclusion
$\left(-\infty, -\frac{4}{3}\right)$	-3	$(-5)(-8) = 40$	Positive
$\left(-\frac{4}{3}, 5\right)$	0	$(4)(-5) = -20$	Negative
$(5, \infty)$	6	$(22)(1) = 22$	Positive

Solution set: $\left(-\infty, -\frac{4}{3}\right) \cup (5, \infty)$

27. $x^3 - 3x^2 - x + 3 > 0$

$x^2(x - 3) - (x - 3) > 0$

$(x - 3)(x^2 - 1) > 0$

$(x - 3)(x + 1)(x - 1) > 0$

Key numbers: $x = -1, x = 1, x = 3$

Test intervals: $(-\infty -1), (-1, 1), (1, 3), (3, \infty)$

Test: Is $(x - 3)(x + 1)(x - 1) > 0$?

Interval	x-Value	Value of $(x - 3)(x + 1)(x - 1)$	Conclusion
$(-\infty, -1)$	-2	$(-5)(-1)(-3) = -15$	Negative
$(-1, 1)$	0	$(-3)(1)(-1) = 3$	Positive
$(1, 3)$	2	$(-1)(3)(1) = -3$	Negative
$(3, \infty)$	4	$(1)(5)(3) = 15$	Positive

Solution set: $(-1, 1) \cup (3, \infty)$

−3 −2 −1 0 1 2 3 4 5 6 7 x

29. $-x^3 + 7x^2 + 9x > 63$

$x^3 - 7x^2 - 9x < -63$

$x^3 - 7x^2 - 9x + 63 < 0$

$x^2(x - 7) - 9(x - 7) < 0$

$(x - 7)(x^2 - 9) < 0$

$(x - 7)(x + 3)(x - 3) < 0$

Key numbers: $x = -3, x = 3, x = 7$

Test intervals: $(-\infty, -3), (-3, 3), (3, 7), (7, \infty)$

Test: Is $(x - 7)(x + 3)(x - 3) < 0$?

Interval	x-Value	Value of $(x - 7)(x + 3)(x - 3)$	Conclusion
$(-\infty, -3)$	-4	$(-11)(-1)(-7) = -77$	Negative
$(-3, 3)$	0	$(-7)(3)(-3) = 63$	Positive
$(3, 7)$	4	$(-3)(7)(1) = -21$	Negative
$(7, \infty)$	8	$(1)(11)(5) = 55$	Positive

Solution set: $(-\infty, -3) \cup (3, 7)$

−3 3 7

−4 −2 0 2 4 6 8 x

31. $4x^3 - 6x^2 < 0$

$2x^2(2x - 3) < 0$

Key numbers: $x = 0, x = \frac{3}{2}$

Test intervals: $(-\infty, 0) \Rightarrow 2x^2(2x - 3) < 0$

$\left(0, \frac{3}{2}\right) \Rightarrow 2 \Rightarrow 2x^2(2x - 3) < 0$

$\left(\frac{3}{2}, \infty\right) \Rightarrow 2x^2(2x - 3) > 0$

Solution set: $(-\infty, 0) \cup \left(0, \frac{3}{2}\right)$

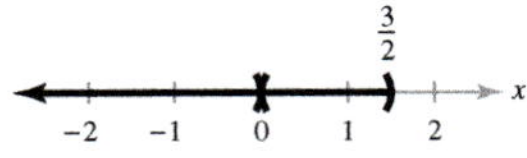

33. $x^3 - 4x \geq 0$

$x(x + 2)(x - 2) \geq 0$

Key numbers: $x = 0, x = \pm 2$

Test intervals: $(-\infty, -2) \Rightarrow x(x + 2)(x - 2) < 0$

$(-2, 0) \Rightarrow x(x + 2)(x - 2) > 0$

$(0, 2) \Rightarrow x(x + 2)(x - 2) < 0$

$(2, \infty) \Rightarrow x(x + 2)(x - 2) > 0$

Solution set: $[-2, 0] \cup [2, \infty)$

35. $(x - 1)^2(x + 2)^3 \geq 0$

Key numbers: $x = 1, x = -2$

Test intervals: $(-\infty, -2) \Rightarrow (x - 1)^2(x + 2)^3 < 0$

$(-2, 1) \Rightarrow (x - 1)^2(x + 2)^3 > 0$

$(1, \infty) \Rightarrow (x - 1)^2(x + 2)^3 > 0$

Solution set: $[-2, \infty)$

37. $4x^2 - 4x + 1 \leq 0$

$(2x - 1)^2 \leq 0$

Key number: $x = \frac{1}{2}$

Test Interval	x-Value	Polynomial Value	Conclusion
$\left(-\infty, \frac{1}{2}\right)$	$x = 0$	$[2(0) - 1]^2 = 1$	Positive
$\left(\frac{1}{2}, \infty\right)$	$x = 1$	$[2(1) - 1]^2 = 1$	Positive

The solution set consists of the single real number $\frac{1}{2}$.

39. $x^2 - 6x + 12 \leq 0$

Using the Quadratic Formula, you can determine that the key numbers are $x = 3 \pm \sqrt{3}i$.

Test Interval	x-Value	Polynomial Value	Conclusion
$(-\infty, \infty)$	$x = 0$	$(0)^2 - 6(0) + 12 = 12$	Positive

The solution set is empty, that is there are no real solutions.

41. $\dfrac{4x - 1}{x} > 0$

Key numbers: $x = 0, x = \dfrac{1}{4}$

Test intervals: $(-\infty, 0), \left(0, \frac{1}{4}\right), \left(\frac{1}{4}, \infty\right)$

Test: Is $\dfrac{4x - 1}{x} > 0$?

Interval	x-Value	Value of $\dfrac{4x - 1}{x}$	Conclusion
$(-\infty, 0)$	-1	$\dfrac{-5}{-1} = 5$	Positive
$\left(0, \dfrac{1}{4}\right)$	$\dfrac{1}{8}$	$\dfrac{-\frac{1}{2}}{\frac{1}{8}} = -4$	Negative
$\left(\dfrac{1}{4}, \infty\right)$	1	$\dfrac{3}{1} = 3$	Positive

Solution set: $(-\infty, 0) \cup \left(\dfrac{1}{4}, \infty\right)$

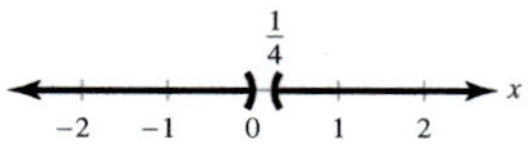

43.
$$\frac{3x + 5}{x - 1} < 2$$
$$\frac{3x + 5}{x - 1} - 2 < 0$$
$$\frac{3x + 5 - 2(x - 1)}{x - 1} < 0$$
$$\frac{x + 7}{x - 1} < 0$$

Key numbers: $x = -7, x = 1$

Test intervals: $(-\infty, -7), (-7, 1), (1, \infty)$

Test: Is $\dfrac{x + 7}{x - 1} < 0$?

Interval	x-Value	Value of $\dfrac{x + 7}{x - 1}$	Conclusion
$(-\infty, -7)$	-8	$\dfrac{-1}{-9} = \dfrac{1}{9}$	Positive
$(-7, 1)$	0	$\dfrac{0 + 7}{0 - 1} = -7$	Negative
$(1, \infty)$	2	$\dfrac{2 + 9}{2 - 1} = 11$	Positive

Solution set: $(-7, 1)$

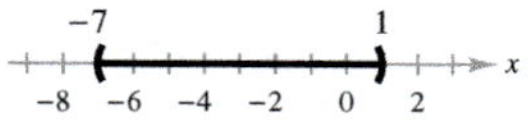

45.
$$\frac{2}{x+5} > \frac{1}{x-3}$$
$$\frac{2}{x+5} - \frac{1}{x-3} > 0$$
$$\frac{2(x-3) - 1(x+5)}{(x+5)(x-3)} > 0$$
$$\frac{x-11}{(x+5)(x-3)} > 0$$

Key numbers: $x = -5, x = 3, x = 11$

Test intervals: $(-\infty, -5) \Rightarrow \dfrac{x-11}{(x+5)(x-3)} < 0$

$(-5, 3) \Rightarrow \dfrac{x-11}{(x+5)(x-3)} > 0$

$(3, 11) \Rightarrow \dfrac{x-11}{(x+5)(x-3)} < 0$

$(11, \infty) \Rightarrow \dfrac{x-11}{(x+5)(x-3)} > 0$

Solution set: $(-5, 3) \cup (11, \infty)$

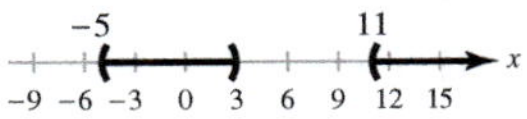

47.
$$\frac{1}{x-3} \le \frac{9}{4x+3}$$
$$\frac{1}{x-3} - \frac{9}{4x+3} \le 0$$
$$\frac{4x+3-9(x-3)}{(x-3)(4x+3)} \le 0$$
$$\frac{30-5x}{(x-3)(4x+3)} \le 0$$

Key numbers: $x = 3, x = -\dfrac{3}{4}, x = 6$

Test intervals: $\left(-\infty, -\dfrac{3}{4}\right) \Rightarrow \dfrac{30-5x}{(x-3)(4x+3)} > 0$

$\left(-\dfrac{3}{4}, 3\right) \Rightarrow \dfrac{30-5x}{(x-3)(4x+3)} < 0$

$(3, 6) \Rightarrow \dfrac{30-5x}{(x-3)(4x+3)} > 0$

$(6, \infty) \Rightarrow \dfrac{30-5x}{(x-3)(4x+3)} < 0$

Solution set: $\left(-\dfrac{3}{4}, 3\right) \cup [6, \infty)$

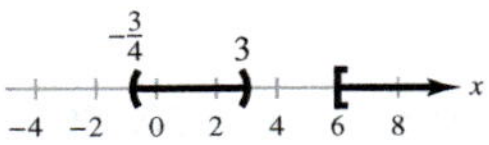

49.
$$\frac{x^2+2x}{x^2-9} \le 0$$
$$\frac{x(x+2)}{(x+3)(x-3)} \le 0$$

Key numbers: $x = 0, x = -2, x = \pm 3$

Test intervals: $(-\infty, -3) \Rightarrow \dfrac{x(x+2)}{(x+3)(x-3)} > 0$

$(-3, -2) \Rightarrow \dfrac{x(x+2)}{(x+3)(x-3)} < 0$

$(-2, 0) \Rightarrow \dfrac{x(x+2)}{(x+3)(x-3)} > 0$

$(0, 3) \Rightarrow \dfrac{x(x+2)}{(x+3)(x-3)} < 0$

$(3, \infty) \Rightarrow \dfrac{x(x+2)}{(x+3)(x-3)} > 0$

Solution set: $(-3, -2] \cup [0, 3)$

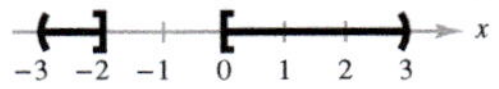

51.
$$\frac{3}{x-1} + \frac{2x}{x+1} > -1$$
$$\frac{3(x+1) + 2x(x-1) + 1(x+1)(x-1)}{(x-1)(x+1)} > 0$$
$$\frac{3x^2+x+2}{(x-1)(x+1)} > 0$$

Key numbers: $x = -1, x = 1$

Test intervals: $(-\infty, -1) \Rightarrow \dfrac{3x^2+x+2}{(x-1)(x+1)} > 0$

$(-1, 1) \Rightarrow \dfrac{3x^2+x+2}{(x-1)(x+1)} < 0$

$(1, \infty) \Rightarrow \dfrac{3x^2+x+2}{(x-1)(x+1)} > 0$

Solution set: $(-\infty, -1) \cup (1, \infty)$

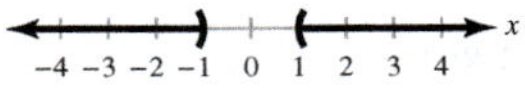

53. $y = -x^2 + 2x + 3$

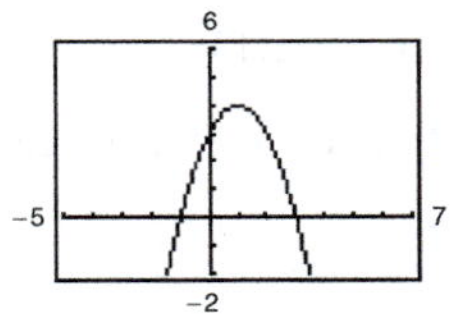

(a) $y \le 0$ when $x \le -1$ or $x \ge 3$.

(b) $y \ge 3$ when $0 \le x \le 2$.

55. $y = \frac{1}{8}x^3 - \frac{1}{2}x$

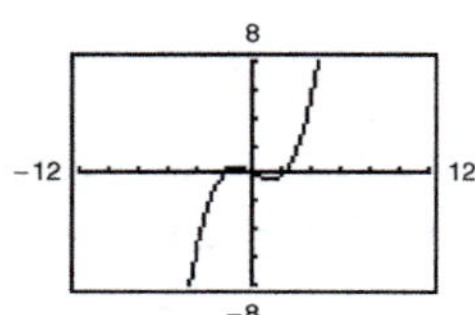

(a) $y \geq 0$ when $-2 \leq x \leq 0$ or $2 \leq x < \infty$.

(b) $y \leq 6$ when $x \leq 4$.

57. $y = \dfrac{3x}{x - 2}$

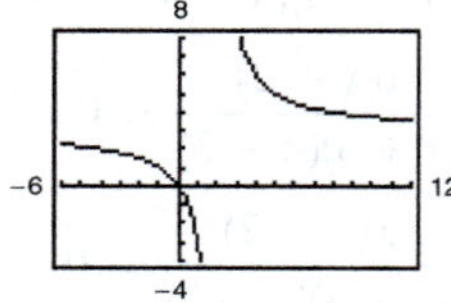

(a) $y \leq 0$ when $0 \leq x < 2$.

(b) $y \geq 6$ when $2 < x \leq 4$.

59. $y = \dfrac{2x^2}{x^2 + 4}$

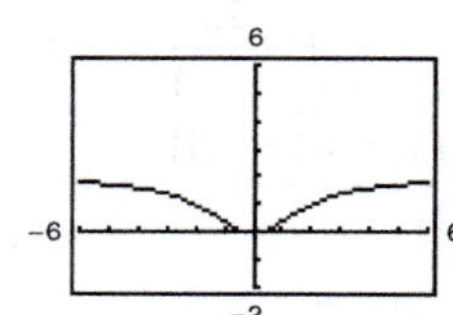

(a) $y \geq 1$ when $x \leq -2$ or $x \geq 2$.

This can also be expressed as $|x| \geq 2$.

(b) $y \leq 2$ for all real numbers x.

This can also be expressed as $-\infty < x < \infty$.

61. $0.3x^2 + 6.26 < 10.8$

$0.3x^2 + 4.54 < 0$

Key numbers: $x \approx \pm 3.89$

Test intervals: $(-\infty, -3.89), (-3.89, 3.89), (3.89, \infty)$

Solution set: $(-3.89, 3.89)$

63. $-0.5x^2 + 12.5x + 1.6 > 0$

Key numbers: $x \approx -0.13, x \approx 25.13$

Test intervals: $(-\infty, -0.13), (-0.13, 25.13), (25.13, \infty)$

Solution set: $(-0.13, 25.13)$

65.
$$\frac{1}{2.3x - 5.2} > 3.4$$
$$\frac{1}{2.3x - 5.2} - 3.4 > 0$$
$$\frac{1 - 3.4(2.3x - 5.2)}{2.3x - 5.2} > 0$$
$$\frac{-7.82x + 18.68}{2.3x - 5.2} > 0$$

Key numbers: $x \approx 2.39, x \approx 2.26$

Test intervals: $(-\infty, 2.26), (2.26, 2.39), (2.39, \infty)$

Solution set: $(2.26, 2.39)$

67. $s = -16t^2 + v_0t + s_0 = -16t^2 + 160t$

(a)
$$-16t^2 + 160t = 0$$
$$-16t(t - 10) = 0$$
$$t = 0, t = 10$$

It will be back on the ground in 10 seconds.

(b)
$$-16t^2 + 160t > 384$$
$$-16t^2 + 160t - 384 > 0$$
$$-16(t^2 - 10t + 24) > 0$$
$$t^2 - 10t + 24 < 0$$
$$(t - 4)(t - 6) < 0$$

Key numbers: $t = 4, t = 6$

Test intervals: $(-\infty, 4), (4, 6), (6, \infty)$

Solution set: 4 seconds $< t <$ 6 seconds

69. $R = x(75 - 0.0005x)$ and $C = 30x + 250{,}000$

$$P = R - C$$
$$= (75x - 0.0005x^2) - (30x + 250{,}000)$$
$$= -0.0005x^2 + 45x - 250{,}000$$
$$P \geq 750{,}000$$
$$-0.0005x^2 + 45x - 250{,}000 \geq 750{,}000$$
$$-0.0005x^2 + 45x - 1{,}000{,}000 \geq 0$$

Key numbers: $x = 40{,}000, x = 50{,}000$

(These were obtained by using the Quadratic Formula.)

Test intervals:

$(0, 40{,}000), (40{,}000, 50{,}000), (50{,}000, \infty)$

The solution set is $[40{,}000, 50{,}000]$ or $40{,}000 \leq x \leq 50{,}000$. The price per unit is

$$p = \frac{R}{x} = 75 - 0.0005x.$$

For $x = 40{,}000$, $p = \$55$. For $x = 50{,}000$, $p = \$50$. So, for $40{,}000 \leq x \leq 50{,}000$, $\$50.00 \leq p \leq \$55.00$.

71. $4 - x^2 \ge 0$

$(2 + x)(2 - x) \ge 0$

Key numbers: $x = \pm 2$

Test intervals: $(-\infty, -2) \Rightarrow 4 - x^2 < 0$

$(-2, 2) \Rightarrow 4 - x^2 > 0$

$(2, \infty) \Rightarrow 4 - x^2 < 0$

Domain: $[-2, 2]$

73. $x^2 - 9x + 20 \ge 0$

$(x - 4)(x - 5) \ge 0$

Key numbers: $x = 4, x = 5$

Test intervals: $(-\infty, 4), (4, 5), (5, \infty)$

Interval	x-Value	Value of $(x - 4)(x - 5)$	Conclusion
$(-\infty, 4)$	0	$(-4)(-5) = 20$	Positive
$(4, 5)$	$\frac{9}{2}$	$\left(\frac{1}{2}\right)\left(-\frac{1}{2}\right) = -\frac{1}{4}$	Negative
$(5, \infty)$	6	$(2)(1) = 2$	Positive

Domain: $(-\infty, 4] \cup [5, \infty)$

75. $\dfrac{x}{x^2 - 2x - 35} \ge 0$

$\dfrac{x}{(x + 5)(x - 7)} \ge 0$

Key numbers: $x = 0, x = -5, x = 7$

Test intervals: $(-\infty, -5) \Rightarrow \dfrac{x}{(x + 5)(x - 7)} < 0$

$(-5, 0) \Rightarrow \dfrac{x}{(x + 5)(x - 7)} > 0$

$(0, 7) \Rightarrow \dfrac{x}{(x + 5)(x - 7)} < 0$

$(7, \infty) \Rightarrow \dfrac{x}{(x + 5)(x - 7)} > 0$

Domain: $(-5, 0] \cup (7, \infty)$

77. (a) and (c)

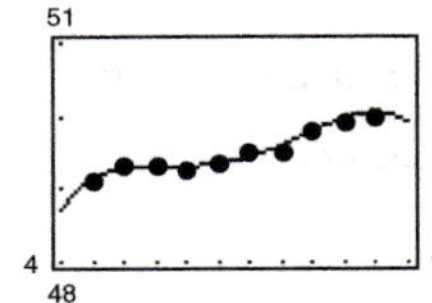

(b) $N = -0.001231t^4 + 0.04723t^3 - 0.6452t^2 + 3.783t + 41.21$

The model fits the data well.

(d) Using the zoom and trace features, the number of students enrolled in elementary and secondary schools fell below 48 million in the year 2017.

(e) No. The model can be used to predict enrollments for years close to those in its domain, $5 \le t \le 14$, but when you project too far into the future, the numbers predicted by the model decrease too rapidly to be considered reasonable.

79. $2L + 2W = 100 \Rightarrow W = 50 - L$

$LW \ge 500$

$L(50 - L) \ge 500$

$-L^2 + 50L - 500 \ge 0$

By the Quadratic Formula you have:

Key numbers: $L = 25 \pm 5\sqrt{5}$

Test: Is $-L^2 + 50L - 500 \ge 0$?

Solution set: $25 - 5\sqrt{5} \le L \le 25 + 5\sqrt{5}$

13.8 meters $\le L \le$ 36.2 meters

81. $\dfrac{1}{R} = \dfrac{1}{R_1} + \dfrac{1}{2}$

$2R_1 = 2R + RR_1$

$2R_1 = R(2 + R_1)$

$\dfrac{2R_1}{2 + R_1} = R$

Because $R \ge 1$,

$\dfrac{2R_1}{2 + R_1} \ge 1$

$\dfrac{2R_1}{2 + R_1} - 1 \ge 0$

$\dfrac{R_1 - 2}{2 + R_1} \ge 0.$

Because $R_1 > 0$, the only key number is $R_1 = 2$. The inequality is satisfied when $R_1 \ge 2$ ohms.

83. False.

There are four test intervals. The test intervals are $(-\infty, -3)$, $(-3, 1)$, $(1, 4)$, and $(4, \infty)$.

85.

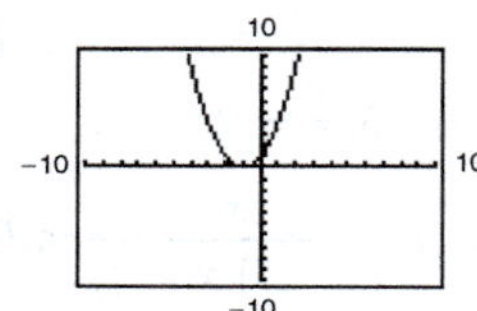

For part (b), the y-values that are less than or equal to 0 occur only at $x = -1$.

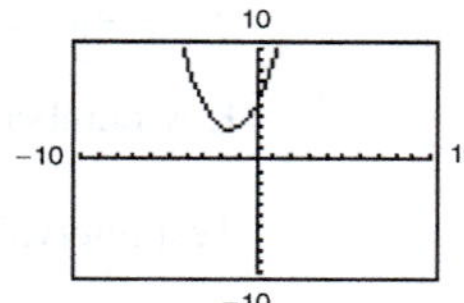

For part (c), there are no y-values that are less than 0.

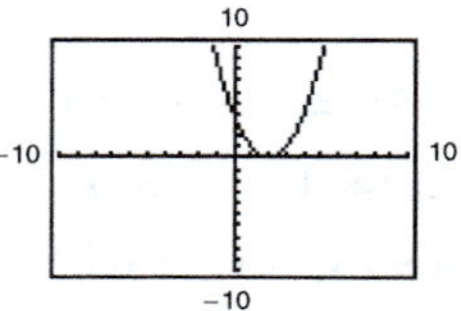

For part (d), the y-value that are greater than 0 occur for all values of x except 2.

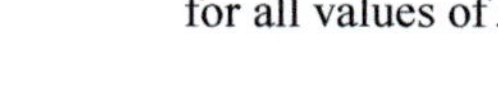

87. $x^2 + bx + 9 = 0$

(a) To have at least one real solution, $b^2 - 4ac \geq 0$.

$$b^2 - 4(1)(9) \geq 0$$
$$b^2 - 36 \geq 0$$

Key numbers: $b = -6, b = 6$

Test intervals: $(-\infty, -6) \Rightarrow b^2 - 36 > 0$

$(-6, 6) \Rightarrow b^2 - 36 < 0$

$(6, \infty) \Rightarrow b^2 - 36 > 0$

Solution set: $(-\infty, -6] \cup [6, \infty)$

(b) $b^2 - 4ac \geq 0$

Key numbers: $b = -2\sqrt{ac}, b = 2\sqrt{ac}$

Similar to part (a), if $a > 0$ and $c > 0$, $b \leq -2\sqrt{ac}$ or $b \geq 2\sqrt{ac}$.

89. $3x^2 + bx + 10 = 0$

(a) To have at least one real solution, $b^2 - 4ac \geq 0$.

$$b^2 - 4(3)(10) \geq 0$$
$$b^2 - 120 \geq 0$$

Key numbers: $b = -2\sqrt{30}, b = 2\sqrt{30}$

Test intervals: $\left(-\infty, -2\sqrt{30}\right) \Rightarrow b^2 - 120 > 0$

$\left(-2\sqrt{30}, 2\sqrt{30}\right) \Rightarrow b^2 - 120 < 0$

$\left(2\sqrt{30}, \infty\right) \Rightarrow b^2 - 120 > 0$

Solution set: $\left(-\infty, -2\sqrt{30}\right] \cup \left[2\sqrt{30}, \infty\right)$

(b) $b^2 - 4ac \geq 0$

Similar to part (a), if $a > 0$ and $c > 0$, $b \leq -2\sqrt{ac}$ or $b \geq 2\sqrt{ac}$.

Review Exercises for Chapter 2

1. (a) $y = -2x^2$

Vertical stretch and a reflection in the x-axis

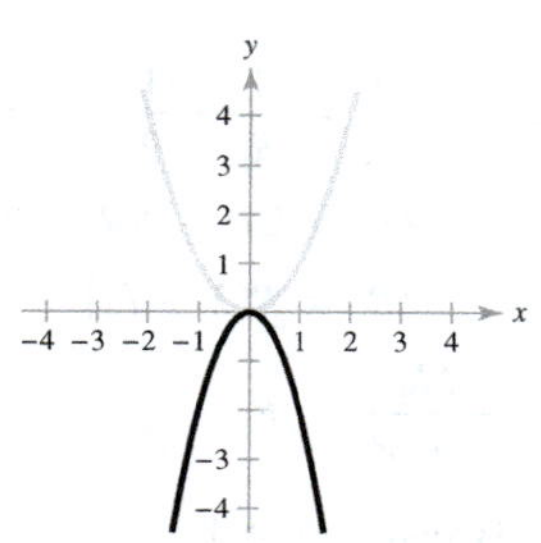

(b) $y = x^2 + 2$

Upward shift of two units

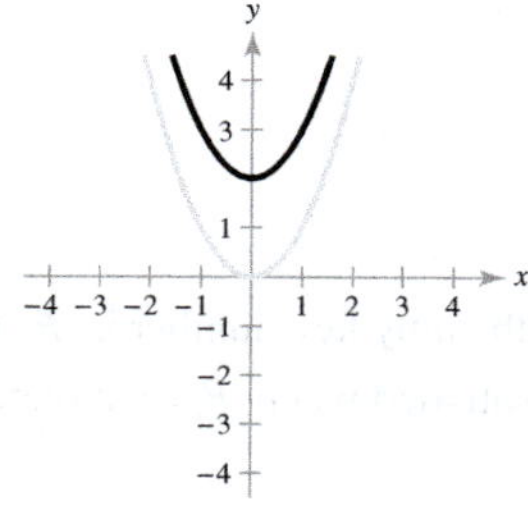

3. $g(x) = x^2 - 2x$

$$= x^2 - 2x + 1 - 1$$
$$= (x - 1)^2 - 1$$

Vertex: $(1, -1)$

Axis of symmetry: $x = 1$

$0 = x^2 - 2x = x(x - 2)$

x-intercepts: $(0, 0)$, $(2, 0)$

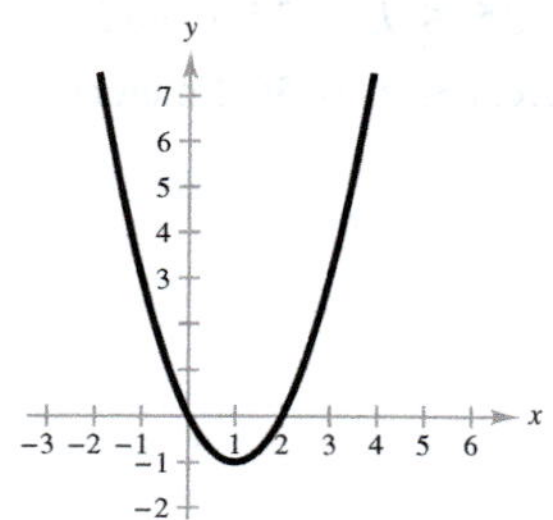

5. $h(x) = 3 + 4x - x^2$

$= -(x^2 - 4x - 3)$

$= -(x^2 - 4x + 4 - 4 - 3)$

$= -\left[(x - 2)^2 - 7\right]$

$= -(x - 2)^2 + 7$

Vertex: $(2, 7)$

Axis of symmetry: $x = 2$

$0 = 3 + 4x - x^2$

$0 = x^2 - 4x - 3$

$$x = \frac{-(-4) \pm \sqrt{(-4)^2 - 4(1)(-3)}}{2(1)}$$

$$= \frac{4 \pm \sqrt{28}}{2} = 2 \pm \sqrt{7}$$

x-intercepts: $\left(2 \pm \sqrt{7}, 0\right)$

7. $h(x) = 4x^2 + 4x + 13$

$= 4(x^2 + x) + 13$

$= 4\left(x^2 + x + \frac{1}{4} - \frac{1}{4}\right) + 13$

$= 4\left(x^2 + x + \frac{1}{4}\right) - 1 + 13$

$= 4\left(x + \frac{1}{2}\right)^2 + 12$

Vertex: $\left(-\frac{1}{2}, 12\right)$

Axis of symmetry: $x = -\frac{1}{2}$

$0 = 4\left(x + \frac{1}{2}\right)^2 + 12$

$\left(x + \frac{1}{2}\right)^2 = -3$

No real zeros

x-intercepts: none

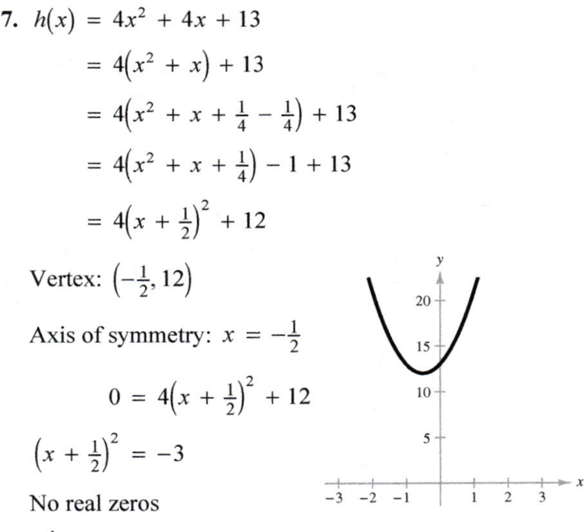

9. (a) $x + x + y + y = P$

$2x + 2y = 1000$

$y = 500 - x$

$A = xy = x(500 - x) = 500x - x^2$

(b) $A = 500x - x^2 = -(x^2 - 500x + 62{,}500) + 62{,}500 = -(x - 250)^2 + 62{,}500$

The maximum area occurs at the vertex when $x = 250$ and $y = 500 - 250 = 250$.

The dimensions with the maximum area are $x = 250$ meters and $y = 250$ meters.

11. $y = x^4$, $f(x) = 6 - x^4$

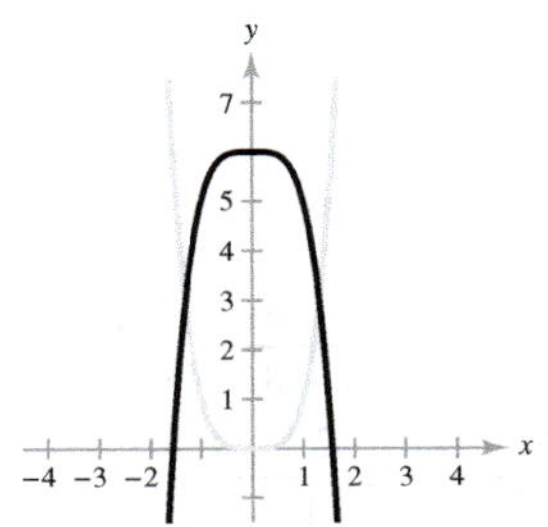

Transformation: Reflection in the x-axis and a vertical shift six units upward

13. $f(x) = -2x^2 - 5x + 12$

The degree is even and the leading coefficient is negative. The graph falls to the left and falls to the right.

15. $g(x) = -3x^3 - 8x^4 + x^5$

The degree is odd and the leading coefficient is positive. The graph falls to the left and rises to the right.

17. $g(x) = 2x^3 + 4x^2$

(a) The degree is odd and the leading coefficient is positive. The graph falls to the left and rises to the right.

(b) $g(x) = 2x^3 + 4x^2$

$0 = 2x^3 + 4x^2$

$0 = 2x^2(x + 2)$

$0 = x^2(x + 2)$

Zeros: $x = -2, 0$

(c)

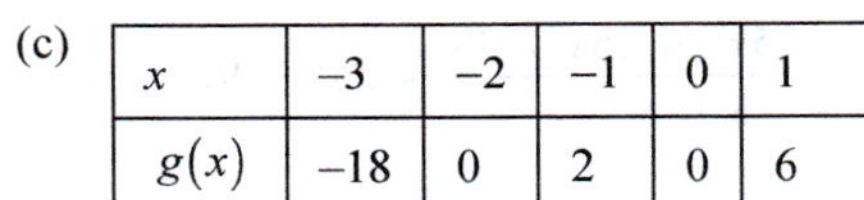

x	-3	-2	-1	0	1
$g(x)$	-18	0	2	0	6

(d)

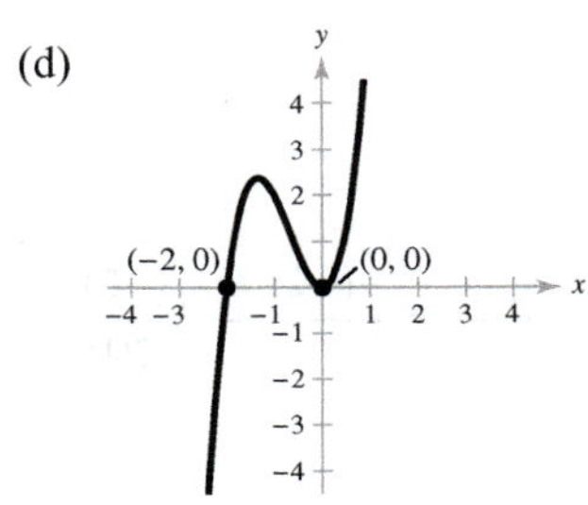

19. $f(x) = -x^3 + x^2 - 2$

(a) The degree is odd and the leading coefficient is negative. The graph rises to the left and falls to the right.

(b) Zero: $x = -1$

(c)

x	-3	-2	-1	0	1	2
$f(x)$	34	10	0	-2	-2	-6

(d)

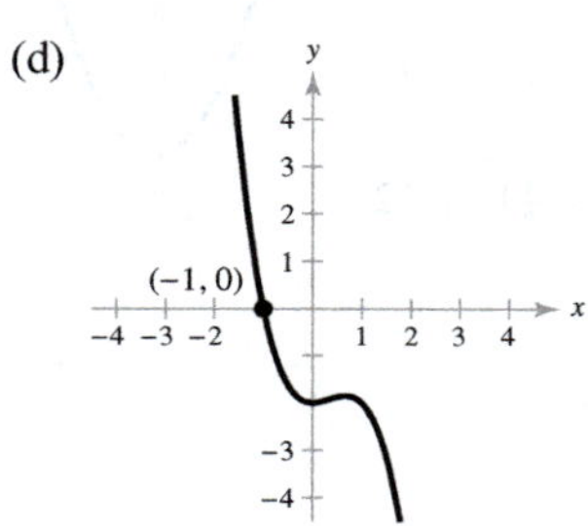

21. (a) $f(x) = 3x^3 - x^2 + 3$

x	-3	-2	-1	0	1	2	3
$f(x)$	-87	-25	-1	3	5	23	75

The zero is in the interval $[-1, 0]$.

(b) Zero: $x \approx -0.900$

23.
$$\begin{array}{r} 6x + 3 \\ 5x - 3\overline{)30x^2 - 3x + 8} \\ \underline{30x^2 - 18x} \\ 15x + 8 \\ \underline{15x - 9} \\ 17 \end{array}$$

$$\frac{30x^2 - 3x + 8}{5x - 3} = 6x + 3 + \frac{17}{5x - 3}$$

25.
$$\begin{array}{r|rrrr} 8 & 2 & -25 & 66 & 48 \\ & & 16 & -72 & -48 \\ \hline & 2 & -9 & -6 & 0 \end{array}$$

$$\frac{2x^3 - 25x^2 + 66x + 48}{x - 8} = 2x^2 - 9x - 6, \quad x \neq 8$$

27. $f(x) = 2x^3 + 11x^2 - 21x - 90$; Factor: $(x + 6)$

(a)
$$\begin{array}{r|rrrr} -6 & 2 & 11 & -21 & -90 \\ & & -12 & 6 & 90 \\ \hline & 2 & -1 & -15 & 0 \end{array}$$

Yes, $(x + 6)$ is a factor of $f(x)$.

(b) $2x^2 - x - 15 = (2x + 5)(x - 3)$

The remaining factors are $(2x + 5)$ and $(x - 3)$.

(c) $f(x) = (2x + 5)(x - 3)(x + 6)$

(d) Zeros: $x = -\frac{5}{2}, 3, -6$

(e)

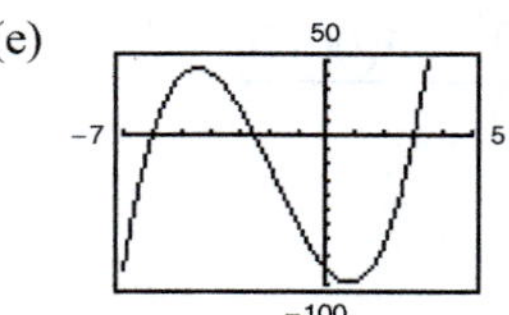

29. $4 + \sqrt{-9} = 4 + 3i$

31. $(6 - 4i) + (-9 + i) = (6 + (-9)) + (-4i + i)$
$= -3 - 3i$

33. $-3i(-2 + 5i) = 6i - 15i^2 = 6i - 15(-1) = 15 + 6i$

35.
$$\begin{aligned} \frac{4}{1 - 2i} &= \frac{4}{1 - 2i} \cdot \frac{1 + 2i}{1 + 2i} \\ &= \frac{4 + 8i}{1 - 4i^2} \\ &= \frac{4 + 8i}{5} \\ &= \frac{4}{5} + \frac{8}{5}i \end{aligned}$$

37.
$$\begin{aligned} \frac{4}{2 - 3i} + \frac{2}{1 + i} &= \frac{4}{2 - 3i} \cdot \frac{2 + 3i}{2 + 3i} + \frac{2}{1 + i} \cdot \frac{1 - i}{1 - i} \\ &= \frac{8 + 12i}{4 + 9} + \frac{2 - 2i}{1 + 1} \\ &= \frac{8}{13} + \frac{12}{13}i + 1 - i \\ &= \left(\frac{8}{13} + 1\right) + \left(\frac{12}{13}i - i\right) \\ &= \frac{21}{13} - \frac{1}{13}i \end{aligned}$$

39. $x^2 - 2x + 10 = 0$

$$x = \frac{-b \pm \sqrt{b^2 - 4ac}}{2a} = \frac{-(-2) \pm \sqrt{(-2)^2 - 4(1)(10)}}{2(1)} = \frac{2 \pm \sqrt{-36}}{2} = \frac{2 \pm 6i}{2} = 1 \pm 3i$$

41. Since $g(x) = x^2 - 2x - 8$ is a 2nd degree polynomial function, it has two zeros.

43. $f(x) = 4x^3 - 27x^2 + 11x + 42$

Possible rational zeros: $\pm\frac{1}{4}, \pm\frac{1}{2}, \pm\frac{3}{4}, \pm1, \pm\frac{3}{2}, \pm\frac{7}{4},$

$\pm2, \pm3, \pm\frac{7}{2}, \pm\frac{21}{4}, \pm6, \pm7, \pm\frac{21}{2}, \pm14, \pm21, \pm42$

$$\begin{array}{r|rrrr} -1 & 4 & -27 & 11 & 42 \\ & & -4 & 31 & -42 \\ \hline & 4 & -31 & 42 & 0 \end{array}$$

$$\begin{aligned} 4x^3 - 27x^2 + 11x + 42 &= (x + 1)(4x^2 - 31x + 42) \\ &= (x + 1)(x - 6)(4x - 7) \end{aligned}$$

The zeros of $f(x)$ are $x = -1, x = \frac{7}{4},$ and $x = 6$.

45. $g(x) = x^3 - 7x^2 + 36$

$$\begin{array}{r|rrrr} -2 & 1 & -7 & 0 & 36 \\ & & -2 & 18 & -36 \\ \hline & 1 & -9 & 18 & 0 \end{array}$$

The zeros of $x^2 - 9x + 18 = (x - 3)(x - 6)$ are $x = 3, 6$. The zeros of $g(x)$ are $x = -2, 3, 6$.

$g(x) = (x + 2)(x - 3)(x - 6)$

47. $h(x) = -2x^5 + 4x^3 - 2x^2 + 5$

$h(x)$ has three variations in sign, so h has either three or one positive real zeros.

$$\begin{aligned} h(-x) &= -2(-x)^5 + 4(-x)^3 - 2(-x)^2 + 5 \\ &= 2x^5 - 4x^3 - 2x^2 + 5 \end{aligned}$$

$h(-x)$ has two variations in sign, so h has either two or no negative real zeros.

49. Because the denominator is zero when $x + 10 = 0$, the domain of f is all real numbers except $x = -10$.

x	-11	-10.5	-10.1	-10.01	-10.001	$\to -10$
$f(x)$	33	63	303	3003	30,003	$\to \infty$

x	$-10 \leftarrow$	-9.999	-9.99	-9.9	-9.5	-9
$f(x)$	$-\infty \leftarrow$	$-29{,}997$	-2997	-297	-57	-27

As x approaches -10 from the left, $f(x)$ increases without bound.

As x approaches -10 from the right, $f(x)$ decreases without bound.

Vertical asymptote: $x = -10$

Horizontal asymptote: $y = 3$

51. $f(x) = \dfrac{4}{x}$

(a) Domain: all real numbers x except $x = 0$

(b) No intercepts

(c) Vertical asymptote: $x = 0$

Horizontal asymptote: $y = 0$

(d)

x	-3	-2	-1	1	2	3
$f(x)$	$-\frac{4}{3}$	-2	-4	4	2	$\frac{4}{3}$

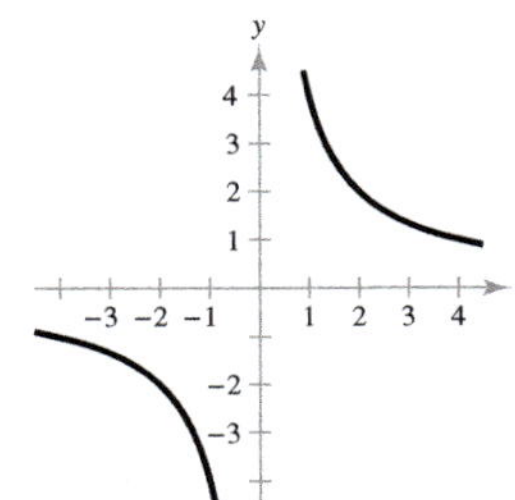

53. $f(x) = \dfrac{x}{x^2 - 16}$

(a) Domain: all real numbers x except $x \neq \pm 4$

(b) Intercept: $(0, 0)$

(c) Vertical asymptotes: $x = \pm 4$

Horizontal asymptote: $y = 0$

(d)

x	-5	-3	-2	-1	0	1	2	3	5
$f(x)$	$-\frac{5}{9}$	$\frac{3}{7}$	$\frac{1}{6}$	$\frac{1}{15}$	0	$-\frac{1}{15}$	$-\frac{1}{6}$	$-\frac{3}{7}$	$\frac{5}{9}$

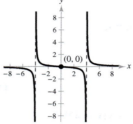

55. $f(x) = \dfrac{6x^2 - 11x + 3}{3x^2 - x}$

$= \dfrac{(3x - 1)(2x - 3)}{x(3x - 1)}$

$= \dfrac{2x - 3}{x}, \ x \neq \dfrac{1}{3}$

(a) Domain: all real numbers x except $x = 0$ and $x = \dfrac{1}{3}$

(b) x-intercept: $\left(\dfrac{3}{2}, 0\right)$

(c) Vertical asymptote: $x = 0$

Horizontal asymptote: $y = 2$

(d)

x	-2	-1	1	2	3	4
$f(x)$	$\frac{7}{2}$	5	-1	$\frac{1}{2}$	1	$\frac{5}{4}$

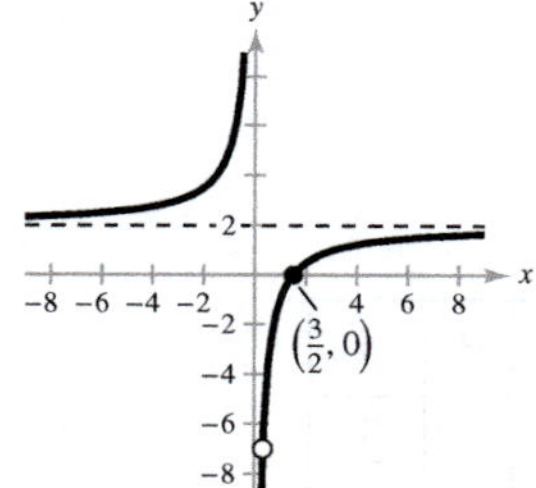

57. $f(x) = \dfrac{2x^3}{x^2 + 1} = 2x - \dfrac{2x}{x^2 + 1}$

(a) Domain: all real numbers x

(b) Intercept: $(0, 0)$

(c) Slant asymptote: $y = 2x$

(d)

x	-2	-1	0	1	2
$f(x)$	$-\frac{16}{5}$	-1	0	1	$\frac{16}{5}$

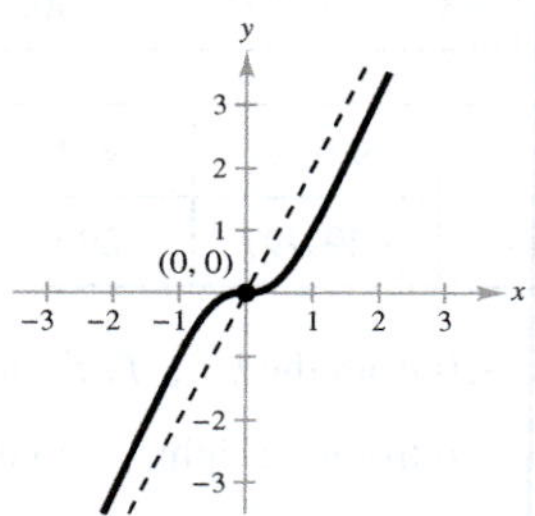

59. $C = \dfrac{528p}{100 - p}, \ 0 \leq p < 100$

(a)

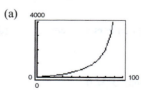

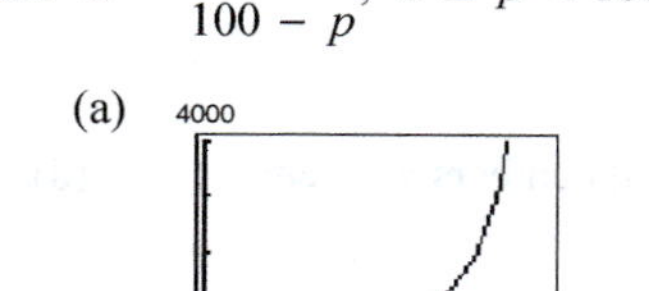

(b) When $p = 25$, $C = \dfrac{528(25)}{100 - 25} = \$176$ million.

When $p = 50$, $C = \dfrac{528(50)}{100 - 50} = \$528$ million.

When $p = 75$, $C = \dfrac{528(75)}{100 - 75} = \$1584$ million.

(c) As $p \to 100$, $C \to \infty$. No, the function is undefined when $p = 100$.

61. $12x^2 + 5x < 2$

$12x^2 + 5x - 2 < 0$

$(4x - 1)(3x + 2) < 0$

Key numbers: $x = -\frac{2}{3}, x = \frac{1}{4}$

Test intervals: $\left(-\infty, -\frac{2}{3}\right), \left(-\frac{2}{3}, \frac{1}{4}\right), \left(\frac{1}{4}, \infty\right)$

Test: Is $(4x - 1)(3x + 2) < 0$?

By testing an x-value in each test interval in the inequality, you see that the solution set is $\left(-\frac{2}{3}, \frac{1}{4}\right)$.

63. $\frac{2}{x + 1} \le \frac{3}{x - 1}$

$\frac{2(x - 1) - 3(x + 1)}{(x + 1)(x - 1)} \le 0$

$\frac{2x - 2 - 3x - 3}{(x + 1)(x - 1)} \le 0$

$\frac{-(x + 5)}{(x + 1)(x - 1)} \le 0$

Key numbers: $x = -5, x = \pm 1$

Test intervals: $(-\infty, -5), (-5, -1), (-1, 1), (1, \infty)$

Test: Is $\frac{-(x + 5)}{(x + 1)(x - 1)} \le 0$?

By testing an x-value in each test interval in the inequality, you see that the solution set is $[-5, -1) \cup (1, \infty)$.

65. $P = \frac{1000(1 + 3t)}{5 + t}$

$2000 \le \frac{1000(1 + 3t)}{5 + t}$

$2000(5 + t) \le 1000(1 + 3t)$

$10{,}000 + 2000t \le 1000 + 3000t$

$-1000t \le -9000$

$t \ge 9$ days

67. False. The domain of $f(x) = \frac{1}{x^2 + 1}$ is the set of all real numbers x.

Problem Solving for Chapter 2

1. $f(x) = ax^3 + bx^2 + cx + d$

$$\begin{array}{r}
ax^2 + (ak + b)x + (ak^2 + bk + c) \\
x - k\overline{)\,ax^3 + bx^2 \qquad + cx \qquad + d} \\
\underline{ax^3 - akx^2} \qquad\qquad\qquad\qquad \\
(ak + b)x^2 + cx \qquad\qquad \\
\underline{(ak + b)x^2 - (ak^2 + bk)x} \qquad \\
(ak^2 + bk + c)x + d \\
\underline{(ak^2 + bk + c)x - (ak^3 + bk^2 + ck)} \\
(ak^3 + bk^2 + ck + d)
\end{array}$$

So, $f(x) = ax^3 + bx^2 + cx + d = (x - k)\left[ax^2 + (ak + b)x + (ak^2 + bx + c)\right] + ak^3 + bk^2 + ck + d$ and $f(x) = ak^3 + bk^2 + ck + d$. Because the remainder is $r = ak^3 + bk^2 + ck + d$, $f(k) = r$.

3. $V = l \cdot w \cdot h = x^2(x + 3)$

$$x^2(x + 3) = 20$$

$$x^3 + 3x^2 - 20 = 0$$

Possible rational zeros: $\pm1, \pm2, \pm4, \pm5, \pm10, \pm20$

$$\begin{array}{r|rrrr} 2 & 1 & 3 & 0 & -20 \\ & & 2 & 10 & 20 \\ \hline & 1 & 5 & 10 & 0 \end{array}$$

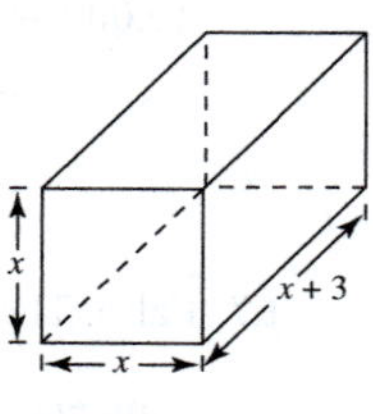

$$(x - 2)(x^2 + 5x + 10) = 0$$

$$x = 2 \text{ or } x = \frac{-5 \pm \sqrt{15}i}{2}$$

Choosing the real positive value for x we have: $x = 2$ and $x + 3 = 5$.

The dimensions of the mold are 2 inches × 2 inches × 5 inches.

5. (a) $y = ax^2 + bx + c$

$(0, -4)$: $-4 = a(0)^2 + b(0) + c$

$-4 = c$

$(4, 0)$: $0 = a(4)^2 + b(4) - 4$

$0 = 16a + 4b - 4 = 4(4a + b - 1)$

$0 = 4a + b - 1$ or $b = 1 - 4a$

$(1, 0)$: $0 = a(1)^2 + b(1) - 4$

$4 = a + b$

$4 = a + (1 - 4a)$

$4 = 1 - 3a$

$3 = -3a$

$a = -1$

$b = 1 - 4(-1) = 5$

$y = -x^2 + 5x - 4$

(b) Enter the data points $(0, -4)$, $(1, 0)$, $(2, 2)$, $(4, 0)$, $(6, -10)$ and use the regression feature to obtain $y = -x^2 + 5x - 4$.

7. $f(x) = (x - k)q(x) + r$

(a) Cubic, passes through $(2, 5)$, rises to the right

One possibility:

$$f(x) = (x - 2)x^2 + 5$$
$$= x^3 - 2x^2 + 5$$

(b) Cubic, passes through $(-3, 1)$, falls to the right

One possibility:

$$f(x) = -(x + 3)x^2 + 1$$
$$= -x^3 - 3x^2 + 1$$

9. $(a + bi)(a - bi) = a^2 - abi + abi - b^2i^2 = a^2 + b^2$

Since a and b are real numbers, $a^2 + b^2$ is also a real number.

11. $f(x) = \dfrac{ax}{(x - b)^2}$

(a) $b \neq 0 \Rightarrow x = b$ is a vertical asymptote. a causes a vertical stretch if $|a| > 1$ and a vertical shrink if $0 < |a| < 1$. For $|a| > 1$, the graph becomes wider as $|a|$ increases. When a is negative, the graph is reflected about the x-axis.

(b) $a \neq 0$. Varying the value of b varies the vertical asymptote of the graph of f. For $b > 0$, the graph is translated to the right. For $b < 0$, the graph is reflected in the x-axis and is translated to the left.

13. Because complex zeros always occur in conjugate pairs, and a cubic function has three zeros and not four, a cubic function with real coefficients cannot have two real zeros and one complex zero.

Practice Test for Chapter 2

1. Sketch the graph of $f(x) = x^2 - 6x + 5$ and identify the vertex and the intercepts.

2. Find the number of units x that produce a minimum cost C if

$C = 0.01x^2 - 90x + 15{,}000$.

3. Find the quadratic function that has a maximum at $(1, 7)$ and passes through the point $(2, 5)$.

4. Find two quadratic functions that have x-intercepts $(2, 0)$ and $\left(\frac{4}{3}, 0\right)$.

5. Use the leading coefficient test to determine the right and left end behavior of the graph of the polynomial function $f(x) = -3x^5 + 2x^3 - 17$.

6. Find all the real zeros of $f(x) = x^5 - 5x^3 + 4x$.

7. Find a polynomial function with 0, 3, and -2 as zeros.

8. Sketch $f(x) = x^3 - 12x$.

9. Divide $3x^4 - 7x^2 + 2x - 10$ by $x - 3$ using long division.

10. Divide $x^3 - 11$ by $x^2 + 2x - 1$.

11. Use synthetic division to divide $3x^5 + 13x^4 + 12x - 1$ by $x + 5$.

12. Use synthetic division to find $f(-6)$ given $f(x) = 7x^3 + 40x^2 - 12x + 15$.

13. Find the real zeros of $f(x) = x^3 - 19x - 30$.

14. Find the real zeros of $f(x) = x^4 + x^3 - 8x^2 - 9x - 9$.

15. List all possible rational zeros of the function $f(x) = 6x^3 - 5x^2 + 4x - 15$.

16. Find the rational zeros of the polynomial $f(x) = x^3 - \frac{20}{3}x^2 + 9x - \frac{10}{3}$.

17. Write $f(x) = x^4 + x^3 + 5x - 10$ as a product of linear factors.

18. Find a polynomial with real coefficients that has $2, 3 + i,$ and $3 - 2i$ as zeros.

19. Use synthetic division to show that $3i$ is a zero of $f(x) = x^3 + 4x^2 + 9x + 36$.

20. Sketch the graph of $f(x) = \dfrac{x - 1}{2x}$ and label all intercepts and asymptotes.

21. Find all the asymptotes of $f(x) = \dfrac{8x^2 - 9}{x^2 + 1}$.

22. Find all the asymptotes of $f(x) = \dfrac{4x^2 - 2x + 7}{x - 1}$.

23. Given $z_1 = 4 - 3i$ and $z_2 = -2 + i$, find the following:

(a) $z_1 - z_2$

(b) $z_1 z_2$

(c) z_1/z_2

24. Solve the inequality: $x^2 - 49 \leq 0$

25. Solve the inequality: $\dfrac{x + 3}{x - 7} \geq 0$

CHAPTER 3
Exponential and Logarithmic Functions

CHAPTER 3
Exponential and Logarithmic Functions

Section 3.1 Exponential Functions and Their Graphs

1. algebraic

3. One-to-One

5. $A = P\left(1 + \frac{r}{n}\right)^{nt}$

7. $f(1.4) = (0.9)^{1.4} \approx 0.863$

9. $f\left(\frac{2}{5}\right) = 3^{2/5} \approx 1.552$

11. $f(-1.5) = 5000(2^{-1.5})$
    ≈ 1767.767

13. $f(x) = 2^x$

    Increasing

    Asymptote: $y = 0$

    Intercept: $(0, 1)$

    Matches graph (d).

14. $f(x) = 2^x + 1$

    Increasing

    Asymptote: $y = 1$

    Intercept: $(0, 2)$

    Matches graph (c).

15. $f(x) = 2^{-x}$

    Decreasing

    Asymptote: $y = 0$

    Intercept: $(0, 1)$

    Matches graph (a).

16. $f(x) = 2^{x-2}$

    Increasing

    Asymptote: $y = 0$

    Intercept: $\left(0, \frac{1}{4}\right)$

    Matches graph (b).

17. $f(x) = 7^x$

x	-2	-1	0	1	2
$f(x)$	0.020	0.143	1	7	49

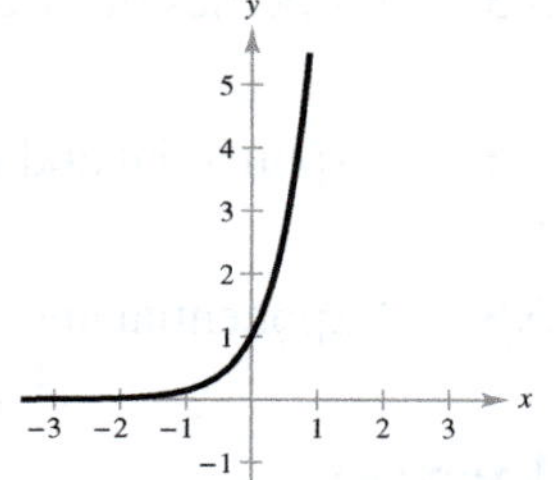

19. $f(x) = \left(\frac{1}{4}\right)^{-x}$

x	-2	-1	0	1	2
$f(x)$	0.063	0.25	1	4	16

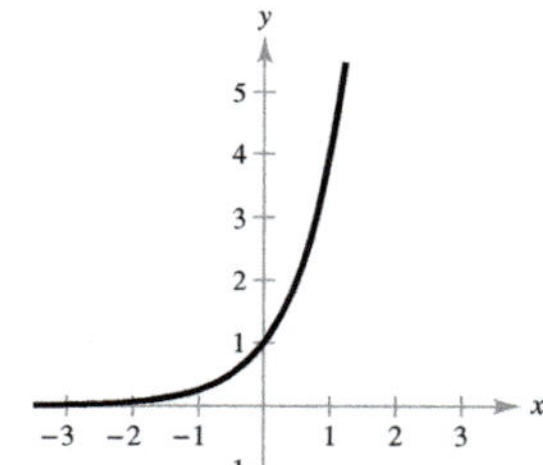

21. $f(x) = 4^{x-1}$

x	-2	-1	0	1	2
$f(x)$	0.016	0.063	0.25	1	4

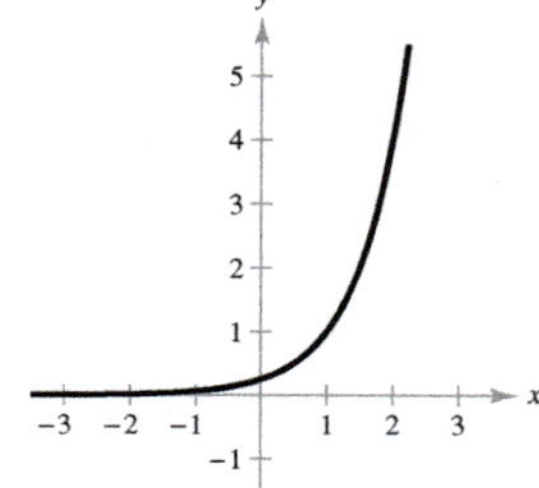

23. $f(x) = 2^{x+1} + 3$

x	-3	-2	-1	0	1
$f(x)$	3.25	3.5	4	5	7

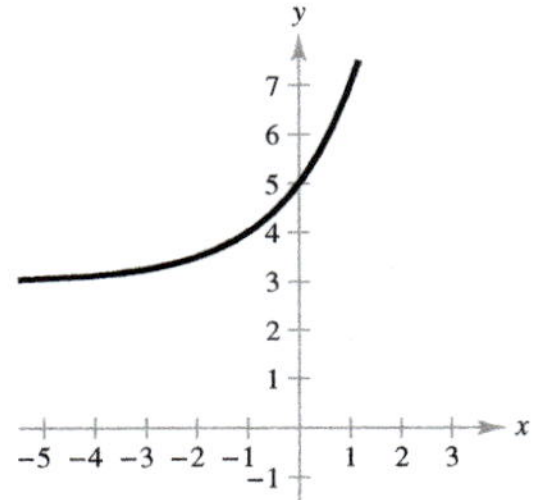

25.
$$3^{x+1} = 27$$
$$3^{x+1} = 3^3$$
$$x + 1 = 3$$
$$x = 2$$

27.
$$\left(\frac{1}{2}\right)^x = 32$$
$$\left(\frac{1}{2}\right)^x = \left(\frac{1}{2}\right)^{-5}$$
$$x = -5$$

29. $f(x) = 3^x$, $g(x) = 3^x + 1$

Because $g(x) = f(x) + 1$, the graph of g can be obtained by shifting the graph of f one unit upward.

31. $f(x) = 10^x$, $g(x) = 10^{-x+3}$

Because $g(x) = f(-x + 3)$, the graph of g can be obtained by reflecting the graph of f in the y-axis and shifting f three units to the right. (**Note:** This is equivalent to shifting f three units to the left and then reflecting the graph in the y-axis.)

33.
$$f(x) = e^x$$
$$f(1.9) = e^{1.9} \approx 6.686$$

35. $f(6) = 5000e^{0.06(6)} \approx 7166.647$

37. $f(x) = 3e^{x+4}$

x	-8	-7	-6	-5	-4
$f(x)$	0.055	0.149	0.406	1.104	3

Asymptote: $y = 0$

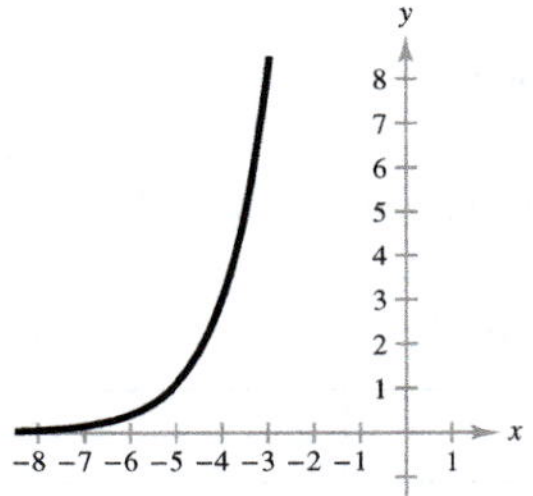

39. $f(x) = 2e^{x-2} + 4$

x	-2	-1	0	1	2
$f(x)$	4.037	4.100	4.271	4.736	6

Asymptote: $y = 4$

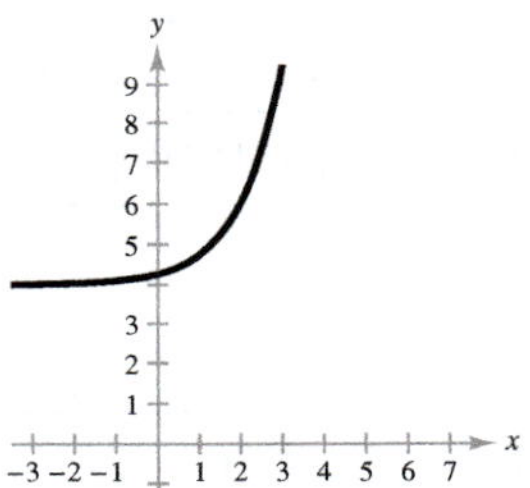

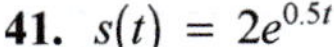

41. $s(t) = 2e^{0.5t}$

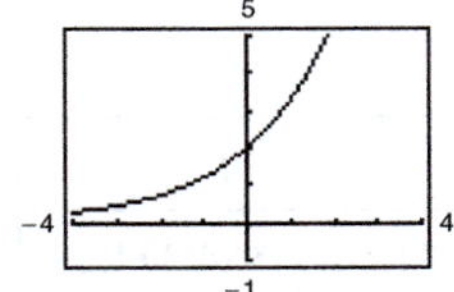

43. $g(x) = 1 + e^{-x}$

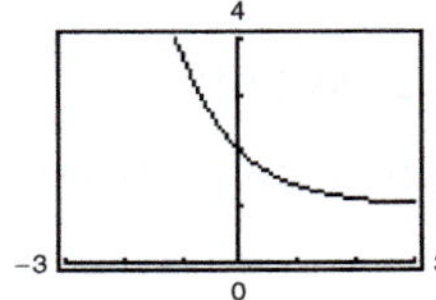

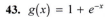

45. $e^{3x+2} = e^3$

$3x + 2 = 3$

$3x = 1$

$x = \frac{1}{3}$

47. $e^{x^2-3} = e^{2x}$

$x^2 - 3 = 2x$

$x^2 - 2x - 3 = 0$

$(x - 3)(x + 1) = 0$

$x = 3$ or $x = -1$

49. $P = \$1500, r = 2\%, t = 10$ years

Compounded n times per year: $A = P\left(1 + \frac{r}{n}\right)^{nt} = 1500\left(1 + \frac{0.02}{n}\right)^{10n}$

Compounded continuously: $A = Pe^{rt} = 1500e^{0.02(10)}$

n	1	2	4	12	365	Continuous
A	\$1828.49	\$1830.29	\$1831.19	\$1831.80	\$1832.09	\$1832.10

51. $P = \$2500, r = 4\%, t = 20$ years

Compounded n times per year: $A = P\left(1 + \frac{r}{n}\right)^{nt} = 2500\left(1 + \frac{0.04}{n}\right)^{20n}$

Compounded continuously: $A = Pe^{rt} = 2500e^{0.04(20)}$

n	1	2	4	12	365	Continuous
A	\$5477.81	\$5520.10	\$5541.79	\$5556.46	\$5563.61	\$5563.85

53. $A = Pe^{rt} = 12{,}000e^{0.04t}$

t	10	20	30	40	50
A	\$17,901.90	\$26,706.49	\$39,841.40	\$59,436.39	\$88,668.67

55. $A = Pe^{rt} = 12{,}000e^{0.065t}$

t	10	20	30	40	50
A	\$22,986.49	\$44,031.56	\$84,344.25	\$161,564.86	\$309,484.08

57. $A = 30{,}000e^{(0.05)(25)} \approx \$104{,}710.29$

59. $C(t) = 29.88(1.04)^t$

Ten years from today, $t = 10$: $C(10) = 29.88(1.04)^{10} \approx \$44.23$

61. (a)

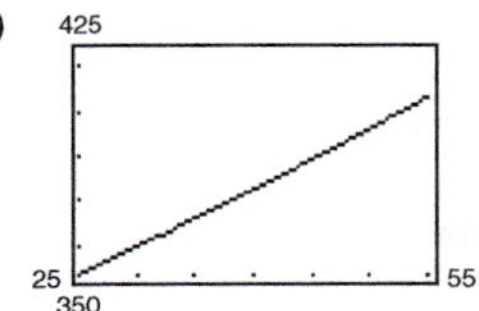

(b)

t	25	26	27	28
P (in millions)	350.281	352.107	353.943	355.788

t	29	30	31	32
P (in millions)	357.643	359.508	361.382	363.266

t	33	34	35	36
P (in millions)	365.160	367.064	368.977	370.901

t	37	38	39	40
P (in millions)	372.835	374.779	376.732	378.697

t	41	42	43	44
P (in millions)	380.671	382.656	384.651	386.656

t	45	46	47	48
P (in millions)	388.672	390.698	392.735	394.783

t	49	50	51	52
P (in millions)	396.841	398.910	400.989	403.080

t	53	54	55
P (in millions)	405.182	407.294	409.417

(c) Using the model and extending the table beyond the year 2055, the population will exceed 430 million in 2064.

t	55	56	57	58	59	60	61	62	63	64	65
P (in millions)	409.417	411.552	413.698	415.854	418.022	420.202	422.393	424.595	426.808	429.034	431.270

63. $Q = 16\left(\frac{1}{2}\right)^{t/24,100}$

(a) $Q(0) = 16$ grams

(b) $Q(75{,}000) \approx 1.85$ grams

(c)

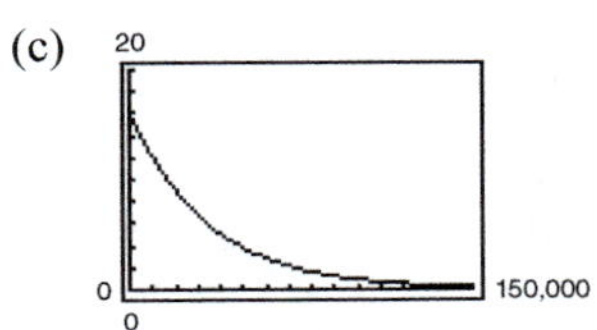

65. (a) $V(t) = 49{,}810\left(\frac{7}{8}\right)^t$ where t is the number of years since it was purchased.

(b) $V(4) = 49{,}810\left(\frac{7}{8}\right)^4 \approx 29,197.71$

After 4 years, the value of the van is about $29,198.

67. True. The line $y = -2$ is a horizontal asymptote for the graph of $f(x) = 10^x - 2$. As $x \to -\infty$, $f(x) \to -2$ but never reaches -2.

69. $f(x) = 3^{x-2}$

$= 3^x 3^{-2}$

$= 3^x\left(\dfrac{1}{3^2}\right)$

$= \dfrac{1}{9}(3^x)$

$= h(x)$

So, $f(x) \neq g(x)$, but $f(x) = h(x)$.

71. $f(x) = 16(4^{-x})$ and $f(x) = 16(4^{-x})$

$= 4^2(4^{-x})$ $\qquad = 16(2^2)^{-x}$

$= 4^{2-x}$ $\qquad = 16(2^{-2x})$

$= \left(\frac{1}{4}\right)^{-(2-x)}$ $\qquad = h(x)$

$= \left(\frac{1}{4}\right)^{x-2}$

$= g(x)$

So, $f(x) = g(x) = h(x)$.

73. $y = 3^x$ and $y = 4^x$

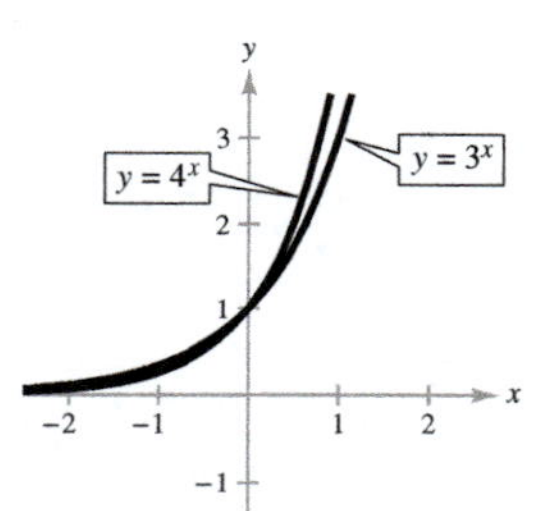

x	-2	-1	0	1	2
3^x	$\frac{1}{9}$	$\frac{1}{3}$	1	3	9
4^x	$\frac{1}{16}$	$\frac{1}{4}$	1	4	16

(a) $4^x < 3^x$ when $x < 0$.

(b) $4^x > 3^x$ when $x > 0$.

75.

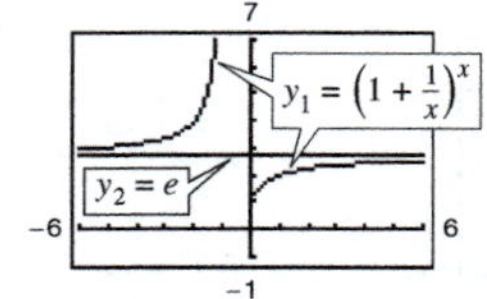

As x increases, the graph of y_1 approaches e, which is y_2.

77. (a)

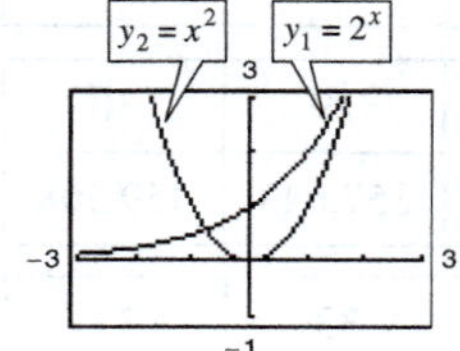

At $x = 2$, both functions have a value of 4. The function y_1 increases for all values of x. The function y_2 is symmetric with respect to the y-axis.

(b)

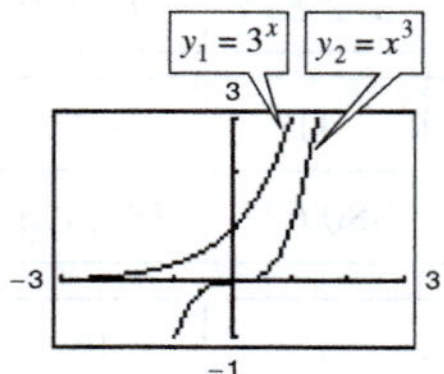

Both functions are increasing for all values of x. For $x > 0$, both functions have a similar shape. The function y_2 is symmetric with respect to the origin.

In both viewing windows, the constant raised to a variable power increases more rapidly than the variable raised to a constant power.

79. The functions (c) $h(x) = 3^x$ and (d) $k(x) = 2^{-x}$ are exponential.

Section 3.2 Logarithmic Functions and Their Graphs

1. logarithmic

3. natural; e

5. $x = y$

7. $\log_4 16 = 2 \Rightarrow 4^2 = 16$

9. $\log_{12} 12 = 1 \Rightarrow 12^1 = 12$

11. $5^3 = 125 \Rightarrow \log_5 125 = 3$

13. $4^{-3} = \frac{1}{64} \Rightarrow \log_4 \frac{1}{64} = -3$

15. $f(x) = \log_2 x$

$f(64) = \log_2 64 = 6$ because $2^6 = 64$

17. $f(x) = \log_8 x$

$f(1) = \log_8 1 = 0$ because $8^0 = 1$

19. $g(x) = \log_a x$

$g(a^2) = \log_a a^{-2}$

$= -2$

21. $f(x) = \log x$

$f\left(\frac{7}{8}\right) = \log\left(\frac{7}{8}\right) \approx -0.058$

23. $f(x) = \log x$

$f(12.5) = \log 12.5 \approx 1.097$

25. $\log_8 8 = 1$ because $8^1 = 8$

27. $\log_{7.5} 1 = 0$ because $7.5^0 = 1$

29. $\log_5(x + 1) = \log_5 6$

$x + 1 = 6$

$x = 5$

31. $\log 11 = \log(x^2 + 7)$

$11 = x^2 + 7$

$x^2 = 4$

$x = \pm 2$

33.

x	-2	-1	0	1	2
$f(x) = 7^x$	$\frac{1}{49}$	$\frac{1}{7}$	1	7	49

x	$\frac{1}{49}$	$\frac{1}{7}$	1	7	49
$g(x) = \log_7 x$	-2	-1	0	1	2

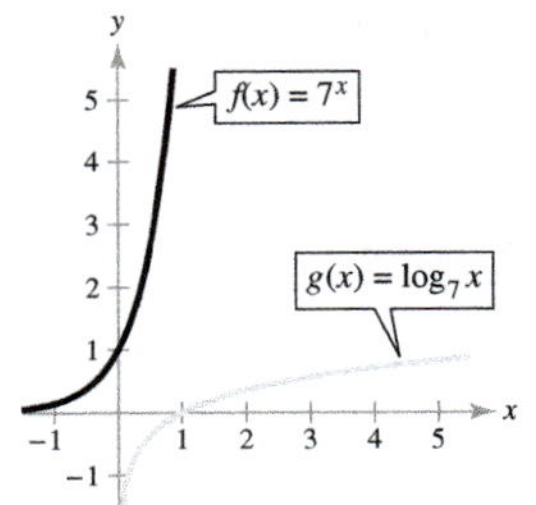

35.

x	-2	-1	0	1	2
$f(x) = 6^x$	$\frac{1}{36}$	$\frac{1}{6}$	1	6	36

x	$\frac{1}{36}$	$\frac{1}{6}$	1	6	36
$g(x) = \log_6 x$	-2	-1	0	1	2

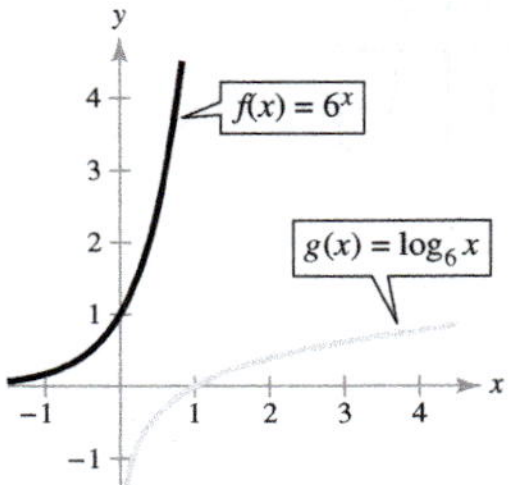

37. $f(x) = \log_3 x + 2$

Asymptote: $x = 0$

Point on graph: $(1, 2)$

Matches graph (a).

The graph of $f(x)$ is obtained from $g(x)$ by shifting the graph two units upward.

38. $f(x) = \log_3(x - 1)$

Asymptote: $x = 1$

Point on graph: $(2, 0)$

Matches graph (d).

$f(x)$ shifts $g(x)$ one unit to the right.

39. $f(x) = \log_3(1 - x) = \log_3[-(x - 1)]$

Asymptote: $x = 1$

Point on graph: $(0, 0)$

Matches graph (b).

The graph of $f(x)$ is obtained by reflecting the graph of $g(x)$ in the y-axis and shifting the graph one unit to the right.

40. $f(x) = -\log_3 x$

Asymptote: $x = 0$

Point on graph: $(1, 0)$

Matches graph (c).

$f(x)$ reflects $g(x)$ in the x-axis.

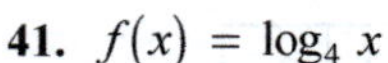

41. $f(x) = \log_4 x$

Domain: $(0, \infty)$

x-intercept: $(1, 0)$

Vertical asymptote: $x = 0$

$y = \log_4 x \Rightarrow 4^y = x$

x	$\frac{1}{4}$	1	4	2
$f(x)$	-1	0	1	$\frac{1}{2}$

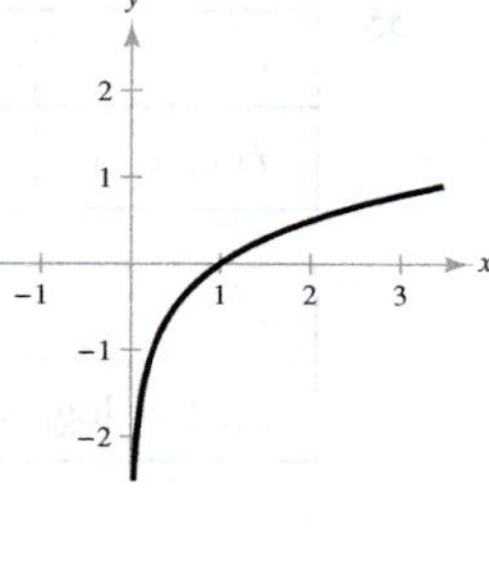

43. $y = \log_3 x + 1$

Domain: $(0, \infty)$

x-intercept:

$$\log_3 x + 1 = 0$$
$$\log_3 x = -1$$
$$3^{-1} = x$$
$$\frac{1}{3} = x$$

The x-intercept is $\left(\frac{1}{3}, 0\right)$.

Vertical asymptote: $x = 0$

$y = \log_3 x + 1$

$\log_3 x = y - 1 \Rightarrow 3^{y-1} = x$

x	$\frac{1}{9}$	$\frac{1}{3}$	0	3	9
y	-1	0	1	2	3

45. $f(x) = -\log_6(x + 2)$

Domain: $x + 2 > 0 \Rightarrow x > -2$

The domain is $(-2, \infty)$.

x-intercept:

$$0 = -\log_6(x + 2)$$
$$0 = \log_6(x + 2)$$
$$6^0 = x + 2$$
$$1 = x + 2$$
$$-1 = x$$

The x-intercept is $(-1, 0)$.

Vertical asymptote: $x + 2 = 0 \Rightarrow x = -2$

$$y = -\log_6(x + 2)$$
$$-y = \log_6(x + 2)$$
$$6^{-y} - 2 = x$$

x	4	-1	$-1\frac{5}{6}$	$-1\frac{35}{36}$
$f(x)$	-1	0	1	2

47. $y = \log\left(\frac{x}{7}\right)$

Domain: $\frac{x}{7} > 0 \Rightarrow x > 0$

The domain is $(0, \infty)$.

x-intercept: $\log\left(\frac{x}{7}\right) = 0$

$$\frac{x}{7} = 10^0$$
$$\frac{x}{7} = 1$$
$$x = 7$$

The x-intercept is $(7, 0)$.

Vertical asymptote: $\frac{x}{7} = 0 \Rightarrow x = 0$

The vertical asymptote is the y-axis.

x	1	2	3	4	5
y	-0.85	-0.54	-0.37	-0.24	-0.15

x	6	7	8
y	-0.069	0	0.06

49. $\ln \frac{1}{2} = -0.693\ldots \Rightarrow e^{-0.693\ldots} = \frac{1}{2}$

51. $\ln 250 = 5.521\ldots \Rightarrow e^{5.521\ldots} = 250$

53. $e^2 = 7.3890\ldots \Rightarrow \ln 7.3890\ldots = 2$

55. $e^{-4x} = \frac{1}{2} \Rightarrow \ln \frac{1}{2} = -4x$

57. $f(x) = \ln x$

$f(18.42) = \ln 18.42 \approx 2.913$

59. $g(x) = 8 \ln x$

$g(\sqrt{5}) = 8 \ln \sqrt{5} \approx 6.438$

61. $e^{\ln 4} = 4$

63. $2.5 \ln 1 = 2.5(0) = 0$

65. $\ln e^{\ln e} = \ln e^1 = 1$

67. $f(x) = \ln(x - 4)$

Domain: $x - 4 > 0 \Rightarrow x > 4$

The domain is $(4, \infty)$.

x-intercept: $0 = \ln(x - 4)$

$e^0 = x - 4$

$5 = x$

The x-intercept is $(5, 0)$.

Vertical asymptote: $x - 4 = 0 \Rightarrow x = 4$

x	4.5	5	6	7
$f(x)$	−0.69	0	0.69	1.10

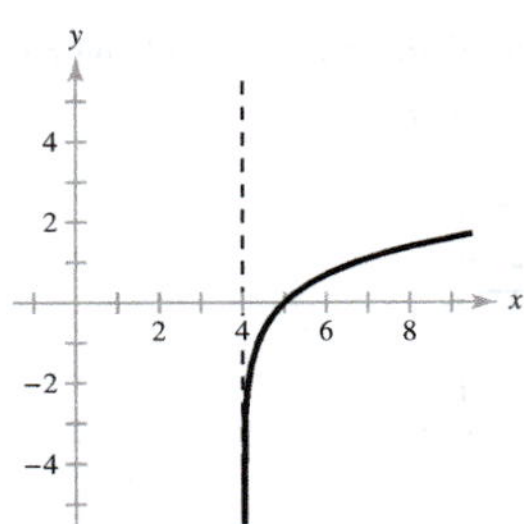

69. $g(x) = \ln(-x)$

Domain: $-x > 0 \Rightarrow x < 0$

The domain is $(-\infty, 0)$.

x-intercept:

$0 = \ln(-x)$

$e^0 = -x$

$-1 = x$

The x-intercept is $(-1, 0)$.

Vertical asymptote: $-x = 0 \Rightarrow x = 0$

x	−0.5	−1	−2	−3
$g(x)$	−0.69	0	0.69	1.10

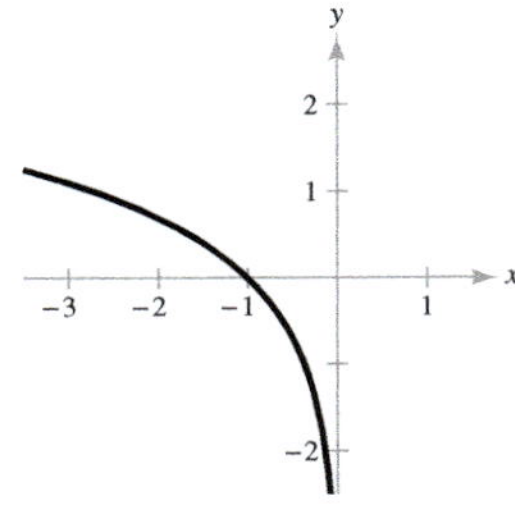

71. $f(x) = \ln(x - 1)$

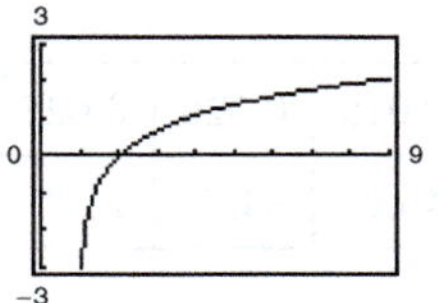

73. $f(x) = -\ln x + 8$

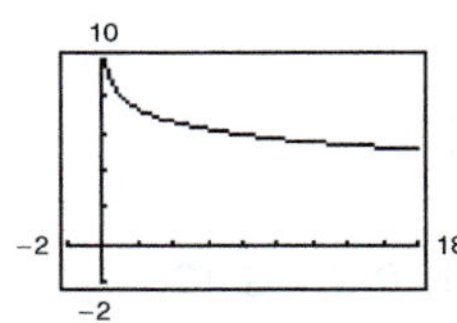

75. $\ln(x + 4) = \ln 12$

$x + 4 = 12$

$x = 8$

77. $\ln(x^2 - x) = \ln 6$

$x^2 - x = 6$

$x^2 - x - 6 = 0$

$(x - 3)(x + 2) = 0$

$x = -2 \text{ or } x = 3$

79. $t = 16.625 \ln\left(\dfrac{x}{x - 750}\right), x > 750$

(a) When $x = \$897.72$:

$$t = 16.625 \ln\left(\frac{897.72}{897.72 - 750}\right) \approx 30 \text{ years}$$

When $x = \$1659.24$:

$$t = 16.625 \ln\left(\frac{1659.24}{1659.24 - 750}\right) \approx 10 \text{ years}$$

(b) Total amounts:

$(897.72)(12)(30) = \$323{,}179.20 \approx \$323{,}179$

$(1659.24)(12)(10) = \$199{,}108.80 \approx \$199{,}109$

Interest charges:

$323{,}179.20 - 150{,}000 = \$173{,}179.20 \approx \$173{,}179$

$199{,}108.80 - 150{,}000 = \$49{,}108.80 \approx \$49{,}109$

(c) The vertical asymptote is $x = 750$. The closer the payment is to \$750 per month, the longer the length of the mortgage will be. Also, the monthly payment must be greater than \$750.

81. $t = \dfrac{\ln 2}{r}$

(a)

r	0.005	0.010	0.015	0.020	0.025	0.030
t	138.6	69.3	46.2	34.7	27.7	23.1

(b)

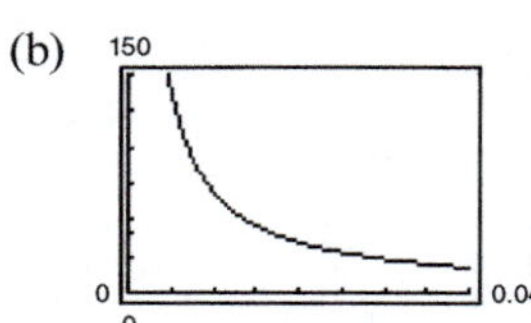

83. $f(t) = 80 - 17\log(t + 1), 0 \le t \le 12$

(a)

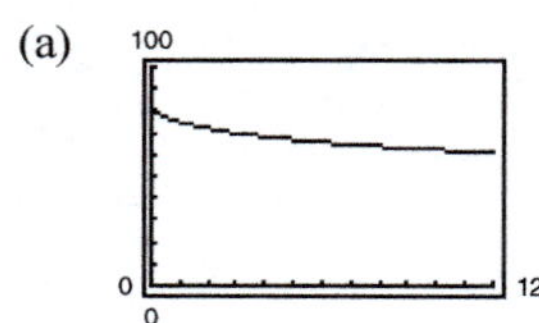

(b) $f(0) = 80 - 17 \log 1 = 80.0$

(c) $f(4) = 80 - 17 \log 5 \approx 68.1$

(d) $f(10) = 80 - 17 \log 11 \approx 62.3$

85. False. Reflecting $g(x)$ about the line $y = x$ will determine the graph of $f(x)$.

87. (a) $f(x) = \ln x, g(x) = \sqrt{x}$

The natural log function grows at a slower rate than the square root function.

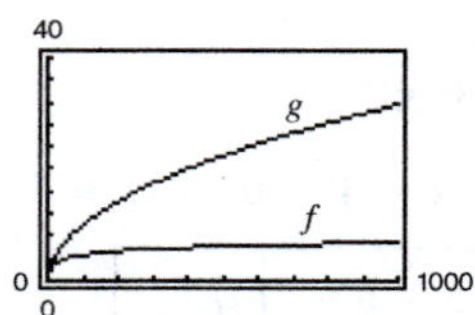

(b) $f(x) = \ln x, g(x) = \sqrt[4]{x}$

The natural log function grows at a slower rate than the fourth root function.

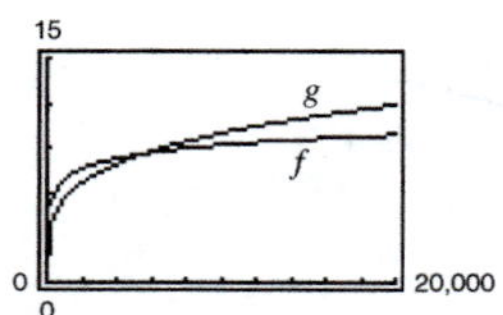

89.

x	1	2	8
y	0	1	3

y is not an exponential function of x, but it is a logarithmic function of x, $y = \log_2 x$.

91. $f(x) = \dfrac{\ln x}{x}$

(a)

x	1	5	10	10^2	10^4	10^6
$f(x)$	0	0.322	0.230	0.046	0.00092	0.0000138

(b) As $x \to \infty$, $f(x) \to 0$.

(c)

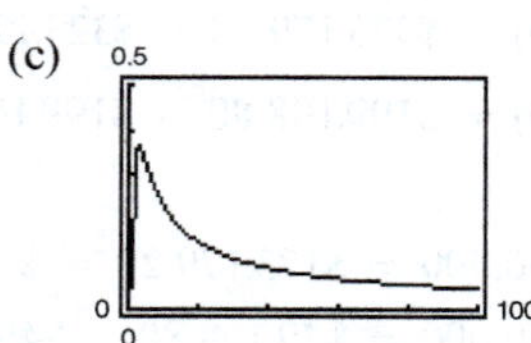

Section 3.3 Properties of Logarithms

1. change-of-base

3. $\dfrac{1}{\log_b a}$

5. (a) $\log_5 16 = \dfrac{\log 16}{\log 5}$

(b) $\log_5 16 = \dfrac{\ln 16}{\ln 5}$

7. (a) $\log_x \dfrac{3}{10} = \dfrac{\log(3/10)}{\log x}$

(b) $\log_x \dfrac{3}{10} = \dfrac{\ln(3/10)}{\ln x}$

9. $\log_3 17 = \dfrac{\log 17}{\log 3} = \dfrac{\ln 17}{\ln 3} \approx 2.579$

11. $\log_\pi 0.5 = \dfrac{\log 0.5}{\log \pi} = \dfrac{\ln 0.5}{\ln \pi} \approx -0.606$

13. $\log_3 35 = \log_3(5 \cdot 7) = \log_3 5 + \log_3 7$

15. $\log_3\left(\frac{7}{25}\right) = \log_3 7 - \log_3 25$
$= \log_3 7 - \log_3 5^2$
$= \log_3 7 - 2\log_3 5$

17. $\log_3\left(\frac{21}{5}\right) = \log_3 21 - \log_3 5$
$= \log_3(3 \cdot 7) - \log_3 5$
$= \log_3 3 + \log_3 7 - \log_3 5$
$= 1 + \log_3 7 - \log_3 5$

19. $\log_3 9 = 2\log_3 3 = 2$

21. $\log_6 \sqrt[3]{\frac{1}{6}} = \log_6\left(\frac{1}{6}\right)^{1/3}$
$= \frac{1}{3}\log_6\left(\frac{1}{6}\right)$
$= \frac{1}{3}\log_6 6^{-1}$
$= \frac{1}{3}(-1)$
$= -\frac{1}{3}$

23. $\log_2(-2)$ is undefined. -2 is not in the domain of $\log_2 x$.

25. $\ln \sqrt[4]{e^3} = \ln e^{3/4} = \frac{3}{4}\ln e = \frac{3}{4}(1) = \frac{3}{4}$

27. $\ln e^2 + \ln e^5 = 2 + 5 = 7$

29. $\log_5 75 - \log_5 3 = \log_5 \frac{75}{3}$
$= \log_5 25$
$= \log_5 5^2$
$= 2\log_5 5$
$= 2$

31. $\log_4 8 = \log_4(4 \cdot 2)$
$= \log_4 4 + \log_4 2$
$= \log_4 4 + \log_4 4^{1/2}$
$= 1 + \frac{1}{2}$
$= \frac{3}{2}$

33. $\log_b 10 = \log_b 2.5$
$= \log_b 2 + \log_b 5$
$\approx 0.3562 + 0.8271$
$= 1.1833$

35. $\log_b 0.04 = \log_b \frac{4}{100} = \log_b \frac{1}{25}$
$= \log_b 1 - \log_b 25$
$= \log_b 1 - \log_b 5^2$
$= 0 - 2\log_b 5$
$\approx -2(0.8271)$
$= -1.6542$

37. $\log_b 45 = \log_b 9.5$
$= \log_b 9 + \log_b 5$
$= \log_b 3^2 + \log_b 5$
$= 2\log_b 3 + \log_b 5$
$\approx 2(0.5646) + 0.8271$
$= 1.9563$

39. $\log_b (2b)^{-2} = -2\log_b 2b$
$= -2(\log_b 2 + \log_b b)$
$\approx -2(0.3562 + 1)$
$= -2.7124$

41. $\ln 7x = \ln 7 + \ln x$

43. $\log_8 x^4 = 4\log_8 x$

45. $\log_5 \dfrac{5}{x} = \log_5 5 - \log_5 x = 1 - \log_5 x$

47. $\ln\sqrt{z} = \ln z^{1/2} = \frac{1}{2}\ln z$

49. $\ln xyz^2 = \ln x + \ln y + \ln z^2 = \ln x + \ln y + 2\ln z$

51. $\ln z(z-1)^2 = \ln z + \ln(z-1)^2 = \ln z + 2\ln(z-1),\ z > 1$

53. $\log_2\left(\dfrac{\sqrt{a^2-4}}{7}\right) = \log_2\sqrt{a^2-4} - \log_2 7$

$$= \log_2\left(a^2-4\right)^{1/2} - \log_2 7$$
$$= \tfrac{1}{2}\log_2\left(a^2-4\right) - \log_2 7$$
$$= \tfrac{1}{2}\log_2\left[(a-2)(a+2)\right] - \log_2 7$$
$$= \tfrac{1}{2}\left[\log_2(a-2) + \log_2(a+2)\right] - \log_2 7$$
$$= \tfrac{1}{2}\log_2(a-2) + \tfrac{1}{2}\log_2(a+2) - \log_2 7$$

55. $\log_5\left(\dfrac{x^2}{y^2z^3}\right) = \log_5 x^2 - \log_5 y^2z^3$

$$= \log_5 x^2 - \left(\log_5 y^2 + \log_5 z^3\right)$$
$$= 2\log_5 x - 2\log_5 y - 3\log_5 z$$

57. $\ln\sqrt[3]{\dfrac{yz}{x^2}} = \ln\left(\dfrac{yz}{x^2}\right)^{1/3}$

$$= \frac{1}{3}\ln\left(\frac{yz}{x^2}\right)$$
$$= \frac{1}{3}\left[\ln(yz) - \ln x^2\right]$$
$$= \frac{1}{3}\left[\ln(yz) - 2\ln x\right]$$
$$= \frac{1}{3}\left[\ln y + \ln z - 2\ln x\right]$$
$$= \frac{1}{3}\ln y + \frac{1}{3}\ln z - \frac{2}{3}\ln x$$

59. $\ln\sqrt[4]{x^3\left(x^2+3\right)} = \tfrac{1}{4}\ln x^3\left(x^2+3\right)$

$$= \tfrac{1}{4}\left[\ln x^3 + \ln\left(x^2+3\right)\right]$$
$$= \tfrac{1}{4}\left[3\ln x + \ln\left(x^2+3\right)\right]$$
$$= \tfrac{3}{4}\ln x + \tfrac{1}{4}\ln\left(x^2+3\right)$$

61. $\ln 3 + \ln x = \ln(3x)$

63. $\tfrac{2}{3}\log_7(z-2) = \log_7(z-2)^{2/3}$

65. $\log_3 5x - 4\log_3 x = \log_3 5x - \log_3 x^4$

$$= \log_3\left(\frac{5x}{x^4}\right)$$
$$= \log_3\left(\frac{5}{x^3}\right)$$

67. $\log x + 2\log(x+1) = \log x + \log(x+1)^2$

$$= \log\left[x(x+1)^2\right]$$

69. $\log x - 2\log y + 3\log z = \log x - \log y^2 + \log z^3$

$$= \log\frac{x}{y^2} + \log z^3$$
$$= \log\frac{xz^3}{y^2}$$

71. $\ln x - \left[\ln(x+1) + \ln(x-1)\right] = \ln x - \ln(x+1)(x-1)$

$$= \ln\frac{x}{(x+1)(x-1)}$$

73. $\dfrac{1}{2}\left[2\ln(x+3) + \ln x - \ln\left(x^2-1\right)\right] = \dfrac{1}{2}\left[\ln(x+3)^2 + \ln x - \ln\left(x^2-1\right)\right]$

$$= \frac{1}{2}\left[\ln\left[x(x+3)^2\right] - \ln\left(x^2-1\right)\right]$$
$$= \frac{1}{2}\left[\ln\left(\frac{x(x+3)^2}{x^2-1}\right)\right]$$
$$= \frac{1}{2}\ln\left[\frac{x(x+3)^2}{x^2-1}\right]$$
$$= \ln\sqrt{\frac{x(x+3)^2}{x^2-1}}$$

75. $\frac{1}{3}\left[\log_8 y + 2\log_8(y+4)\right] - \log_8(y-1) = \frac{1}{3}\left[\log_8 y + \log_8(y+4)^2\right] - \log_8(y-1)$

$$= \frac{1}{3}\log_8 y(y+4)^2 - \log_8(y-1)$$

$$= \log_8 \sqrt[3]{y(y+4)^2} - \log_8(y-1)$$

$$= \log_8\left(\frac{\sqrt[3]{y(y+4)^2}}{y-1}\right)$$

77. $\log_2 \frac{32}{4} = \log_2 32 - \log_2 4 \neq \frac{\log_2 32}{\log_2 4}$

The second and third expressions are equal by Property 2.

79. $\beta = 10\log\left(\frac{I}{10^{-12}}\right) = 10\left[\log I - \log 10^{-12}\right] = 10\left[\log I + 12\right] = 120 + 10\log I$

When $I = 10^{-6}$:

$\beta = 120 + 10\log 10^{-6} = 120 + 10(-6) = 60$ decibels

81. $\beta = 10\log\left(\frac{I}{10^{-12}}\right)$

Difference $= 10\log\left(\frac{10^{-4}}{10^{-12}}\right) - 10\log\left(\frac{10^{-11}}{10^{-12}}\right) = 10\left[\log 10^8 - \log 10\right] = 10(8-1) = 10(7) = 70$ dB

83.

x	1	2	3	4	5	6
y	1.000	1.189	1.316	1.414	1.495	1.565
$\ln x$	0	0.693	1.099	1.386	1.609	1.792
$\ln y$	0	0.173	0.275	0.346	0.402	0.448

The slope of the line is $\frac{1}{4}$. So, $\ln y = \frac{1}{4}\ln x$

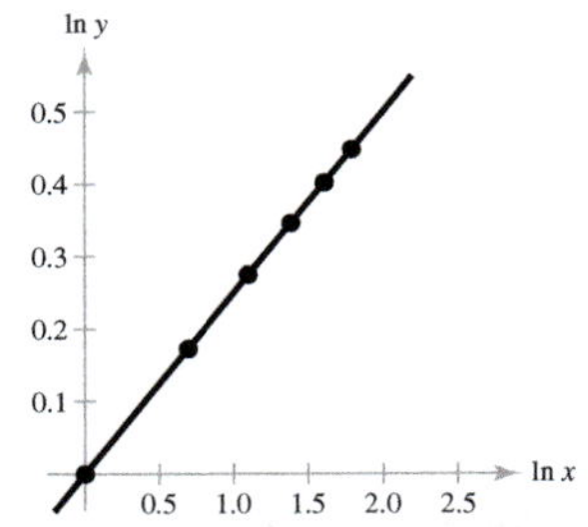

The slope of the line is $-\frac{2}{3}$. So, $\ln y = -\frac{2}{3}\ln x$.

85.

x	1	2	3	4	5	6
y	2.500	2.102	1.900	1.768	1.672	1.597
$\ln x$	0	0.693	1.099	1.386	1.609	1.792
$\ln y$	0.916	0.743	0.642	0.570	0.514	0.468

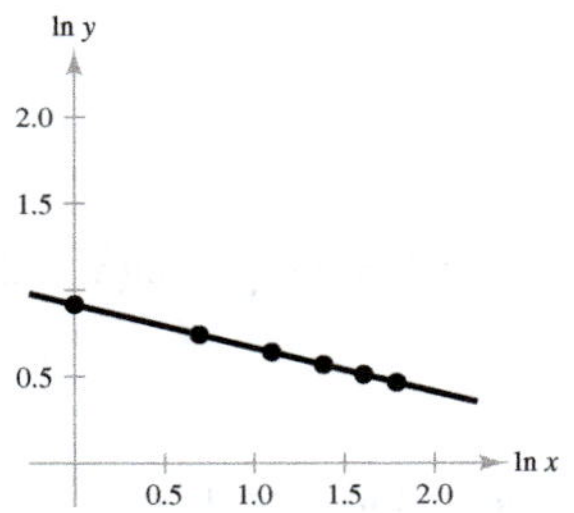

The slope of the line is $-\frac{1}{4}$. So, $\ln y = -\frac{1}{4}\ln x + \ln\frac{5}{2}$.

87.

Weight, x	25	35	50	75	500	1000
Galloping Speed, y	191.5	182.7	173.8	164.2	125.9	114.2
$\ln x$	3.219	3.555	3.912	4.317	6.215	6.908
$\ln y$	5.255	5.208	5.158	5.101	4.835	4.738

$\ln y = -0.14\ln x + 5.7$

89. (a)

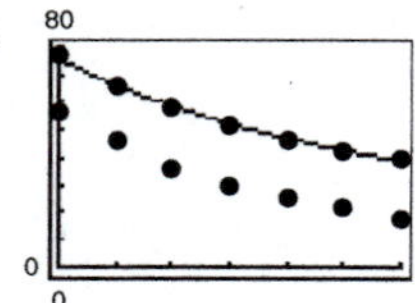

(b) $T - 21 = 54.4(0.964)^t$

$T = 54.4(0.964)^t + 21$

See graph in (a).

(c)

t (in minutes)	T (°C)	$T - 21$ (°C)	$\ln(T - 21)$	$1/(T - 21)$
0	78	57	4.043	0.0175
5	66	45	3.807	0.0222
10	57.5	36.5	3.597	0.0274
15	51.2	30.2	3.408	0.0331
20	46.3	25.3	3.231	0.0395
25	42.5	21.5	3.068	0.0465
30	39.6	18.6	2.923	0.0538

$\ln(T - 21) = -0.037t + 4$

$T = e^{-0.037t+3.997} + 21$

This graph is identical to T in (b).

(d) $\dfrac{1}{T - 21} = 0.0012t + 0.016$

$T = \dfrac{1}{0.001t + 0.016} + 21$

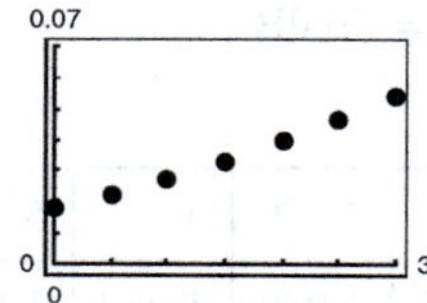

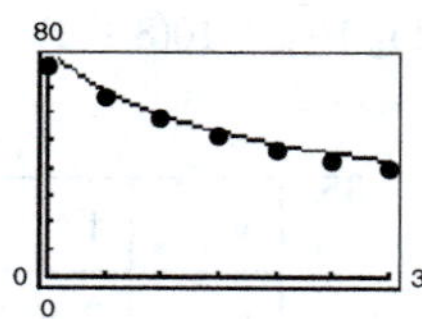

91. $f(x) = \ln x$

False, $f(0) \neq 0$ because 0 is not in the domain of $f(x)$.

$f(1) = \ln 1 = 0$

93. False.

$f(x) - f(2) = \ln x - \ln 2 = \ln \dfrac{x}{2} \neq \ln(x - 2)$

95. False.

$f(u) = 2f(v) \Rightarrow \ln u = 2 \ln v \Rightarrow \ln u = \ln v^2 \Rightarrow u = v^2$

97. $f(x) = \log_2 x = \dfrac{\log x}{\log 2} = \dfrac{\ln x}{\ln 2}$

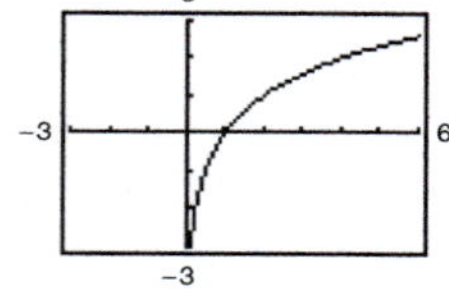

99. $f(x) = \log_{1/4} x = \dfrac{\log x}{\log(1/4)} = \dfrac{\ln x}{\ln(1/4)}$

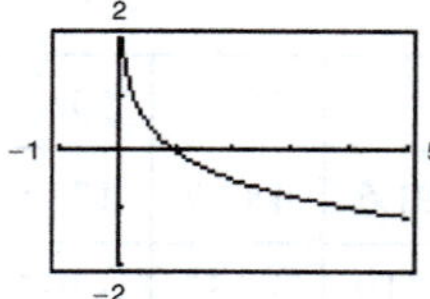

101. The power property cannot be used because $\ln e$ is raised to the second power, not just e.

A correct statement is $(\ln e)^2 = (1)^2 = 1.$

103.

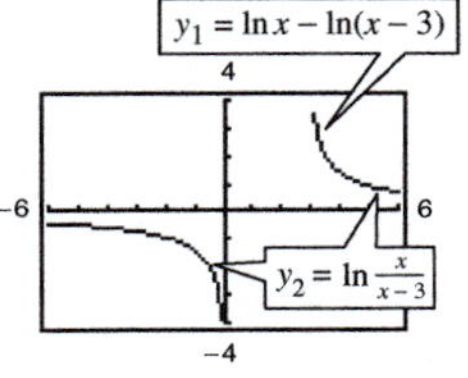

The graphing utility does not show the functions with the same domain. The domain of $y_1 = \ln x - \ln(x - 3)$ is $(3, \infty)$ and the domain of $y_2 = \ln \dfrac{x}{x-3}$ is $(-\infty, 0) \cup (3, \infty)$.

105. $\ln 2 \approx 0.6931, \ln 3 \approx 1.0986, \ln 5 \approx 1.6094$

$\ln 1 = 0$

$\ln 2 \approx 0.6931$

$\ln 3 \approx 1.0986$

$\ln 4 = \ln(2 \cdot 2) = \ln 2 + \ln 2 \approx 0.6931 + 0.6931 = 1.3862$

$\ln 5 \approx 1.6094$

$\ln 6 = \ln(2 \cdot 3) = \ln 2 + \ln 3 \approx 0.6931 + 1.0986 = 1.7917$

$\ln 8 = \ln 2^3 = 3 \ln 2 \approx 3(0.6931) = 2.0793$

$\ln 9 = \ln 3^2 = 2 \ln 3 \approx 2(1.0986) = 2.1972$

$\ln 10 = \ln(5 \cdot 2) = \ln 5 + \ln 2 \approx 1.6094 + 0.6931 = 2.3025$

$\ln 12 = \ln(2^2 \cdot 3) = \ln 2^2 + \ln 3 = 2 \ln 2 + \ln 3 \approx 2(0.6931) + 1.0986 = 2.4848$

$\ln 15 = \ln(5 \cdot 3) = \ln 5 + \ln 3 \approx 1.6094 + 1.0986 = 2.7080$

$\ln 16 = \ln 2^4 = 4 \ln 2 \approx 4(0.6931) = 2.7724$

$\ln 18 = \ln(3^2 \cdot 2) = \ln 3^2 + \ln 2 = 2 \ln 3 + \ln 2 \approx 2(1.0986) + 0.6931 = 2.8903$

$\ln 20 = \ln(5 \cdot 2^2) = \ln 5 + \ln 2^2 = \ln 5 + 2 \ln 2 \approx 1.6094 + 2(0.6931) = 2.9956$

Section 3.4 Exponential and Logarithmic Equations

1. (a) $x = y$

(b) $x = y$

(c) x

(d) x

3. $4^{2x-7} = 64$

(a) $x = 5$

$4^{2(5)-7} = 4^3 = 64$

Yes, $x = 5$ *is* a solution.

(b) $x = 2$

$4^{2(2)-7} = 4^{-3} = \frac{1}{64} \neq 64$

No, $x = 2$ *is not* a solution.

(c) $x = \frac{1}{2}(\log_4 64 + 7)$

$4^{2(1/2(\log_4 64+7))-7} = 64$

$4^{(\log_4 64+7)-7} = 64$

$4^{(3+7)-7} = 64$

$4^3 = 64$

Yes, $x = \frac{1}{2}(\log_4 64 + 7)$ *is* a solution.

5. $\log_2(x + 3) = 10$

(a) $x = 1021$

$\log_2(1021 + 3) = \log_2(1024)$

Because $2^{10} = 1024$, $x = 1021$ *is* a solution.

(b) $x = 17$

$\log_2(17 + 3) = \log_2(20)$

Because $2^{10} \neq 20$, $x = 17$ *is not* a solution.

(c) $x = 10^2 - 3 = 97$

$\log_2(97 + 3) = \log_2(100)$

Because $2^{10} \neq 100$, $10^2 - 3$ *is not* a solution.

7. $4^x = 16$

$4^x = 4^2$

$x = 2$

9. $\ln x - \ln 2 = 0$

$\ln x = \ln 2$

$x = 2$

11. $e^x = 2$

$\ln e^x = \ln 2$

$x = \ln 2$

$x \approx 0.693$

13. $\ln x = -1$

$e^{\ln x} = e^{-1}$

$x = e^{-1}$

$x \approx 0.368$

15. $\log_4 x = 3$

$4^{\log_4 x} = 4^3$

$x = 4^3$

$x = 64$

17. $f(x) = g(x)$

$2^x = 8$

$2^x = 2^3$

$x = 3$

Point of intersection:

$(3, 8)$

19. $e^x = e^{x^2 - 2}$

$x = x^2 - 2$

$0 = x^2 - x - 2$

$0 = (x + 1)(x - 2)$

$x = -1, x = 2$

21. $4(3^x) = 20$

$3^x = 5$

$\log_3 3^x = \log_3 5$

$x = \log_3 5 = \dfrac{\log 5}{\log 3}$ or $\dfrac{\ln 5}{\ln 3}$

$x \approx 1.465$

23. $e^x - 8 = 31$

$e^x = 39$

$\ln e^x = \ln 39$

$x = \ln 39 \approx 3.664$

25. $3^{2x} = 80$

$\ln 3^{2x} = \ln 80$

$2x \ln 3 = \ln 80$

$x = \dfrac{\ln 80}{2 \ln 3} \approx 1.994$

27. $3^{2-x} = 400$

$\ln 3^{2-x} = \ln 400$

$(2 - x) \ln 3 = \ln 400$

$2 \ln 3 - x \ln 3 = \ln 400$

$-x \ln 3 = \ln 400 - 2 \ln 3$

$x \ln 3 = 2 \ln 3 - \ln 400$

$x = \dfrac{2 \ln 3 - \ln 400}{\ln 3}$

$x = 2 - \dfrac{\ln 400}{\ln 3} \approx -3.454$

29. $8(10^{3x}) = 12$

$10^{3x} = \frac{12}{8}$

$\log 10^{3x} = \log\left(\frac{3}{2}\right)$

$3x = \log\left(\frac{3}{2}\right)$

$x = \frac{1}{3} \log\left(\frac{3}{2}\right)$

$x \approx 0.059$

31. $e^{3x} = 12$

$3x = \ln 12$

$x = \dfrac{\ln 12}{3} \approx 0.828$

33. $7 - 2e^x = 5$

$-2e^x = -2$

$e^x = 1$

$x = \ln 1 = 0$

35. $6(2^{3x-1}) - 7 = 9$

$6(2^{3x-1}) = 16$

$2^{3x-1} = \dfrac{8}{3}$

$\log_2 2^{3x-1} = \log_2\left(\dfrac{8}{3}\right)$

$3x - 1 = \log_2\left(\dfrac{8}{3}\right) = \dfrac{\log(8/3)}{\log 2}$ or $\dfrac{\ln(8/3)}{\ln 2}$

$x = \dfrac{1}{3}\left[\dfrac{\log(8/3)}{\log 2} + 1\right] \approx 0.805$

37. $3^x = 2^{x-1}$

$\ln 3^x = \ln 2^{x-1}$

$x \ln 3 = (x - 1)\ln 2$

$x \ln 3 = x \ln 2 - \ln 2$

$x \ln 3 - x \ln 2 = -\ln 2$

$x(\ln 3 - \ln 2) = -\ln 2$

$x = \dfrac{\ln 2}{\ln 2 - \ln 3} \approx -1.710$

39. $4^x = 5^{x^2}$

$\ln 4^x = \ln 5^{x^2}$

$x \ln 4 = x^2 \ln 5$

$x^2 \ln 5 - x \ln 4 = 0$

$x(x \ln 5 - \ln 4) = 0$

$x = 0$

$x \ln 5 - \ln 4 = 0 \Rightarrow x = \dfrac{\ln 4}{\ln 5} \approx 0.861$

41. $e^{2x} - 4e^x - 5 = 0$

$(e^x + 1)(e^x - 5) = 0$

$e^x = -1$ or $e^x = 5$

(No solution) $x = \ln 5 \approx 1.609$

43. $\dfrac{1}{1 - e^x} = 5$

$1 = 5(1 - e^x)$

$\dfrac{1}{5} = 1 - e^x$

$\dfrac{1}{5} - 1 = -e^x$

$-\dfrac{4}{5} = -e^x$

$\dfrac{4}{5} = e^x$

$\ln \dfrac{4}{5} = \ln e^x$

$\ln \dfrac{4}{5} = x$

$x \approx -0.223$

45. $\left(1 + \dfrac{0.065}{365}\right)^{365t} = 4$

$\ln\left(1 + \dfrac{0.065}{365}\right)^{365t} = \ln 4$

$365t \ln\left(1 + \dfrac{0.065}{365}\right) = \ln 4$

$t = \dfrac{\ln 4}{365 \ln\left(1 + \dfrac{0.065}{365}\right)} \approx 21.330$

47. $\ln x = -3$

$x = e^{-3} \approx 0.050$

49. $2.1 = \ln 6x$

$e^{2.1} = 6x$

$\dfrac{e^{2.1}}{6} = x$

$1.361 \approx x$

51. $3 - 4 \ln x = 11$

$-4 \ln x = 8$

$\ln x = -2$

$x = e^{-2} = \dfrac{1}{e^2} \approx 0.135$

53. $6 \log_3(0.5x) = 11$

$\log_3(0.5x) = \frac{11}{6}$

$3^{\log_3(0.5x)} = 3^{11/6}$

$0.5x = 3^{11/6}$

$x = 2(3^{11/6}) \approx 14.988$

55. $\ln x - \ln(x + 1) = 2$

$$\ln\left(\frac{x}{x+1}\right) = 2$$

$$\frac{x}{x+1} = e^2$$

$$x = e^2(x + 1)$$

$$x = e^2x + e^2$$

$$x - e^2x = e^2$$

$$x(1 - e^2) = e^2$$

$$x = \frac{e^2}{1 - e^2} \approx -1.157$$

This negative value is extraneous. The equation has no solution.

57. $\ln(x + 5) = \ln(x - 1) - \ln(x + 1)$

$$\ln(x + 5) = \ln\left(\frac{x-1}{x+1}\right)$$

$$x + 5 = \frac{x-1}{x+1}$$

$$(x + 5)(x + 1) = x - 1$$

$$x^2 + 6x + 5 = x - 1$$

$$x^2 + 5x + 6 = 0$$

$$(x + 2)(x + 3) = 0$$

$$x = -2 \quad \text{or} \quad x = -3$$

Both of these solutions are extraneous, so the equation has no solution.

59. $\log(3x + 4) = \log(x - 10)$

$$3x + 4 = x - 10$$

$$2x = -14$$

$$x = -7$$

The negative value is extraneous.

The equation has no solution.

61. $\log_4 x - \log_4(x - 1) = \frac{1}{2}$

$$\log_4\left(\frac{x}{x-1}\right) = \frac{1}{2}$$

$$4^{\log_4[x/(x-1)]} = 4^{1/2}$$

$$\frac{x}{x-1} = 4^{1/2}$$

$$x = 2(x - 1)$$

$$x = 2x - 2$$

$$-x = -2$$

$$x = 2$$

63. $f(x) = 5^x - 212$

Algebraically:

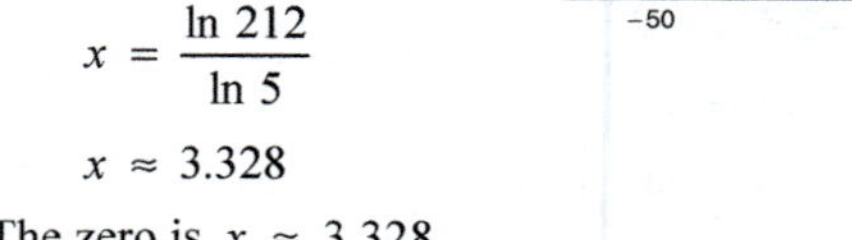

$$5^x = 212$$

$$\ln 5^x = \ln 212$$

$$x \ln 5 = \ln 212$$

$$x = \frac{\ln 212}{\ln 5}$$

$$x \approx 3.328$$

The zero is $x \approx 3.328$.

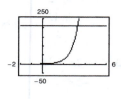

65. $g(x) = 8e^{-2x/3} - 11$

Algebraically:

$$8e^{-2x/3} = 11$$

$$e^{-2x/3} = 1.375$$

$$-\frac{2x}{3} = \ln 1.375$$

$$x = -1.5 \ln 1.375$$

$$x \approx -0.478$$

The zero is $x \approx -0.478$.

67. $y_1 = 3$

$y_2 = \ln x$

From the graph,

$x \approx 20.086$ when $y = 3$.

Algebraically:

$$3 - \ln x = 0$$

$$\ln x = 3$$

$$x = e^3 \approx 20.086$$

69. $y_1 = 2\ln(x + 3)$

$y_2 = 3$

From the graph, $x \approx 1.482$ when $y = 3$.

Algebraically:

$$2\ln(x + 3) = 3$$

$$\ln(x + 3) = \tfrac{3}{2}$$

$$x + 3 = e^{3/2}$$

$$x = e^{3/2} - 3 \approx 1.482$$

71. (a)
$$\begin{aligned} r &= 0.025 \\ A &= Pe^{rt} \\ 5000 &= 2500e^{0.025t} \\ 2 &= e^{0.025t} \\ \ln 2 &= 0.025t \\ \frac{\ln 2}{0.025} &= t \\ t &\approx 27.73 \text{ years} \end{aligned}$$

(b)
$$\begin{aligned} r &= 0.025 \\ A &= Pe^{rt} \\ 7500 &= 2500e^{0.025t} \\ 3 &= e^{0.025t} \\ \ln 3 &= 0.025t \\ \frac{\ln 3}{0.025} &= t \\ t &\approx 43.94 \text{ years} \end{aligned}$$

73.
$$\begin{aligned} 2x^2e^{2x} + 2xe^{2x} &= 0 \\ \left(2x^2 + 2x\right)e^{2x} &= 0 \\ 2x^2 + 2x &= 0 \qquad \left(\text{because } e^{2x} \neq 0\right) \\ 2x(x + 1) &= 0 \\ x &= 0, -1 \end{aligned}$$

75.
$$\begin{aligned} -xe^{-x} + e^{-x} &= 0 \\ (-x + 1)e^{-x} &= 0 \\ -x + 1 &= 0 \qquad \left(\text{because } e^{-x} \neq 0\right) \\ x &= 1 \end{aligned}$$

77.
$$\begin{aligned} \frac{1 + \ln x}{2} &= 0 \\ 1 + \ln x &= 0 \\ \ln x &= -1 \\ x &= e^{-1} = \frac{1}{e} \approx 0.368 \end{aligned}$$

79.
$$\begin{aligned} 2x \ln x + x &= 0 \\ x(2 \ln x + 1) &= 0 \\ 2 \ln x + 1 &= 0 \qquad (\text{because } x > 0) \\ \ln x &= -\tfrac{1}{2} \\ x &= e^{-1/2} \approx 0.607 \end{aligned}$$

81. (a)

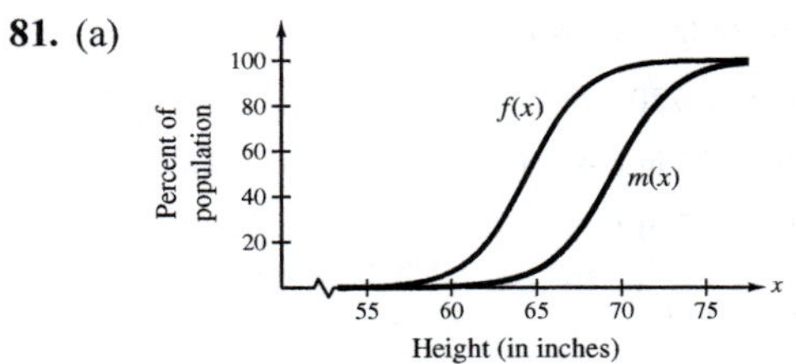

From the graph you see horizontal asymptotes at $y = 0$ and $y = 100$.

These represent the lower and upper percent bounds; the range falls between 0% and 100%.

(b) Males:
$$\begin{aligned} 50 &= \frac{100}{1 + e^{-0.5536(x-69.51)}} \\ 1 + e^{-0.5536(x-69.51)} &= 2 \\ e^{-0.5536(x-69.51)} &= 1 \\ -0.5536(x - 69.51) &= \ln 1 \\ -0.5536(x - 69.51) &= 0 \\ x &= 69.51 \end{aligned}$$

The average height of an American male is 69.51 inches.

Females:
$$\begin{aligned} 50 &= \frac{100}{1 + e^{-0.5834(x-64.49)}} \\ 1 + e^{-0.5834(x-64.49)} &= 2 \\ e^{-0.5834(x-64.49)} &= 1 \\ -0.5834(x - 64.49) &= \ln 1 \\ -0.5834(x - 64.49) &= 0 \\ x &= 64.49 \end{aligned}$$

The average height of an American female is 64.49 inches.

83. $N = 5.5 \cdot 10^{0.23x}$

When $N = 78$:
$$\begin{aligned} 78 &= 5.5 \cdot 10^{0.23x} \\ \frac{78}{5.5} &= 10^{0.23x} \\ \log_{10} \frac{78}{5.5} &= 0.23x \\ x &= \frac{\log_{10}(78/5.5)}{0.23} \approx 5.008 \text{ years} \end{aligned}$$

The beaver population will reach 78 in about 5 years.

85. $P = 75 \ln t + 540$

Let $P = 720$

$720 = 75 \ln t + 540$

$180 = 75 \ln t$

$\frac{180}{75} = \ln t$

$\ln t = 2.4$

$t = e^{2.4} \approx 11.02$ or 2011

87. $T = 20 + 60e^{-0.06m}$

Let $T = 70$

$70 = 20 + 60e^{-0.06m}$

$50 = 60e^{-0.06m}$

$\frac{5}{6} = e^{-0.06m}$

$\ln \frac{5}{6} = -0.06m$

$m = -\frac{1}{0.06} \ln \frac{5}{6}$

$m \approx 3.039$ minutes

89. $\log_a(uv) = \log_a u + \log_a v$

True by Property 1 in Section 3.3.

91. $\log_a(u - v) = \log_a u - \log_a v$

False.

$1.95 = \log(100 - 10)$

$\neq \log 100 - \log 10 = 1$

93. Yes, a logarithmic equation can have more than one extraneous solution. See Exercise 57.

95. $A = Pe^{rt}$

(a) $A = (2P)e^{rt} = 2(Pe^{rt})$ This doubles your money.

(b) $A = Pe^{(2r)t} = Pe^{rt}e^{rt} = e^{rt}(Pe^{rt})$

(c) $A = Pe^{r(2t)} = Pe^{rt}e^{rt} = e^{rt}(Pe^{rt})$

Doubling the interest rate yields the same result as doubling the number of years.

If $2 > e^{rt}$ (i.e., $rt < \ln 2$), then doubling your investment would yield the most money. If $rt > \ln 2$, then doubling either the interest rate or the number of years would yield more money.

97. (a) $P = 1000$, $r = 0.07$, compounded annually, $n = 1$

Effective yield:

$$A = P\left(1 + \frac{r}{n}\right)^{nt} = 1000\left(1 + \frac{0.07}{1}\right)^{1} = \$1070$$

$$\frac{1070 - 1000}{1000} = 7\%$$

The effective yield is 7%.

Balance after 5 years:

$$A = P\left(1 + \frac{r}{n}\right)^{nt} = 1000\left(1 + \frac{0.07}{1}\right)^{1(5)} \approx \$1402.55$$

(b) $P = 1000$, $r = 0.07$, compounded continuously

Effective yield:

$$A = Pe^{rt} = 1000e^{0.07(1)} \approx \$1072.51$$

$$\frac{1072.51 - 1000}{1000} = 7.25\%$$

The effective yield is about 7.25%.

Balance after 5 years:

$$A = Pe^{rt} = 1000e^{0.07(5)} \approx \$1419.07$$

(c) $P = 1000$, $r = 0.07$, compounded quarterly, $n = 4$

Effective yield:

$$A = P\left(1 + \frac{r}{n}\right)^{nt} = 1000\left(1 + \frac{0.07}{4}\right)^{4(1)} \approx \$1071.86$$

$$\frac{1071.86 - 1000}{1000} = 7.19\%$$

The effective yield is about 7.19%.

Balance after 5 years:

$$A = P\left(1 + \frac{r}{n}\right)^{nt} = 1000\left(1 + \frac{0.07}{4}\right)^{4(5)} \approx \$1414.78$$

(d) $P = 1000$, $r = 0.0725$, compounded quarterly, $n = 4$

Effective yield:

$$A = P\left(1 + \frac{r}{n}\right)^{nt} = 1000\left(1 + \frac{0.0725}{4}\right)^{4(1)} \approx \$1074.50$$

$$\frac{1074.50 - 1000}{1000} \approx 7.45\%$$

The effective yield is about 7.45%.

Balance after 5 years:

$$A = P\left(1 + \frac{r}{n}\right)^{nt} = 1000\left(1 + \frac{0.0725}{4}\right)^{4(5)} \approx \$1432.26$$

Savings plan (d) has the greatest effective yield and the highest balance after 5 years.

Section 3.5 Exponential and Logarithmic Models

1. $y = ae^{bx}$; $y = ae^{-bx}$

3. normally distributed

5. (a) $A = Pe^{rt}$

$$\frac{A}{e^{rt}} = P$$

(b)
$$\begin{aligned} A &= Pe^{rt} \\ \frac{A}{P} &= e^{rt} \\ \ln\frac{A}{P} &= \ln e^{rt} \\ \ln\frac{A}{P} &= rt \\ \frac{\ln(A/P)}{r} &= t \end{aligned}$$

7. Because $A = 1000e^{0.035t}$, the time to double is given by $2000 = 1000e^{0.035t}$ and you have

$$\begin{aligned} 2 &= e^{0.035t} \\ \ln 2 &= \ln e^{0.035t} \\ \ln 2 &= 0.035t \\ t &= \frac{\ln 2}{0.035} \approx 19.8 \text{ years.} \end{aligned}$$

Amount after 10 years: $A = 1000e^{0.35} \approx \$1419.07$

9. Because $A = 750e^{rt}$ and $A = 1500$ when $t = 7.75$, you have

$$\begin{aligned} 1500 &= 750e^{7.75r} \\ 2 &= e^{7.75r} \\ \ln 2 &= \ln e^{7.75r} \\ \ln 2 &= 7.75r \\ r &= \frac{\ln 2}{7.75} \approx 0.089438 = 8.9438\%. \end{aligned}$$

Amount after 10 years: $A = 750e^{0.089438(10)} \approx \$1834.37$

11. Because $A = Pe^{0.045t}$ and $A = 10{,}000.00$ when $t = 10$, you have

$$\begin{aligned} 10{,}000.00 &= Pe^{0.045(10)} \\ \frac{10{,}000.00}{e^{0.045(10)}} &= P \approx \$6376.28. \end{aligned}$$

The time to double is given by

$$t = \frac{\ln 2}{0.045} \approx 15.40 \text{ years.}$$

13. $A = 500{,}000$, $r = 0.05$, $n = 12$, $t = 10$

$$\begin{aligned} A &= P\left(1 + \frac{r}{n}\right)^{nt} \\ 500{,}000 &= P\left(1 + \frac{0.05}{12}\right)^{12(10)} \\ P &= \frac{500{,}000}{\left(1 + \frac{0.05}{12}\right)^{12(10)}} \approx \$303{,}580.52 \end{aligned}$$

15. $P = 1000, r = 0.1, A = 2000$

$$A = P\left(1 + \frac{r}{n}\right)^{nt}$$

$$2000 = 1000\left(1 + \frac{0.1}{n}\right)^{nt}$$

$$2 = \left(1 + \frac{0.1}{n}\right)^{nt}$$

(a) $n = 1$

$$(1 + 0.1)^t = 2$$
$$(1.1)^t = 2$$
$$\ln(1.1)^t = \ln 2$$
$$t \ln 1.1 = \ln 2$$
$$t = \frac{\ln 2}{\ln 1.1}$$
$$\approx 7.27 \text{ years}$$

(b) $n = 12$

$$\left(1 + \frac{0.1}{12}\right)^{12t} = 2$$
$$\ln\left(\frac{12.1}{12}\right)^{12t} = \ln 2$$
$$12t \ln\left(\frac{12.1}{12}\right) = \ln 2$$
$$12t = \frac{\ln 2}{\ln(12.1/12)}$$
$$t = \frac{\ln 2}{12 \ln(12.1/12)}$$
$$\approx 6.96 \text{ years}$$

(c) $n = 365$

$$\left(1 + \frac{0.065}{365}\right)^{365t} = 2$$
$$\ln\left(\frac{365.065}{365}\right)^{365t} = \ln 2$$
$$365t \ln\left(\frac{365.065}{365}\right) = \ln 2$$
$$365t = \frac{\ln 2}{\ln(365.065/365)}$$
$$t = \frac{\ln 2}{365 \ln(365.065/365)}$$
$$\approx 10.66 \text{ years}$$

(d) Compounded continuously

$$A = Pe^{rt}$$
$$2000 = 1000e^{0.065t}$$
$$2 = e^{0.065t}$$
$$\ln 2 = \ln e^{0.065t}$$
$$0.065t = \ln 2$$
$$t = \frac{\ln 2}{0.065}$$
$$\approx 10.66 \text{ years}$$

17. (a)

$$3P = Pe^{rt}$$
$$3 = e^{rt}$$
$$\ln 3 = rt$$
$$\frac{\ln 3}{r} = t$$

r	2%	4%	6%	8%	10%	12%
$t = \frac{\ln 3}{r}$ (years)	54.93	27.47	18.31	13.73	10.99	9.16

(b)

$$3P = P(1 + r)^t$$
$$3 = (1 + r)^t$$
$$\ln 3 = \ln (1 + r)^t$$
$$\frac{\ln 3}{\ln (1 + r)} = t$$

r	2%	4%	6%	8%	10%	12%
$t = \frac{\ln 3}{\ln (1 + r)}$ (years)	55.48	28.01	18.85	14.27	11.53	9.69

19. Continuous compounding results in faster growth.

$A = 1 + 0.075[\![t]\!]$ and $A = e^{0.07t}$

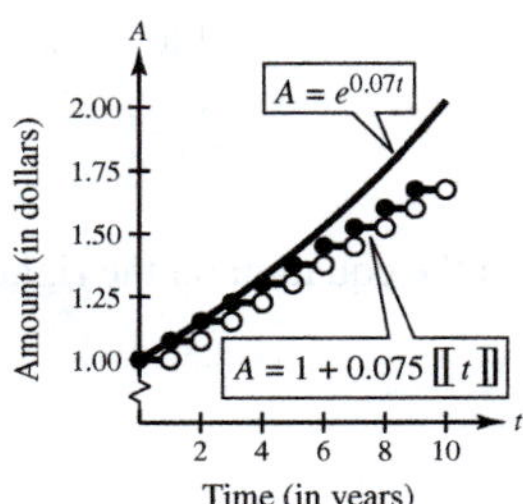

21. $a = 10, y = \frac{1}{2}(10) = 5, t = 1599$

$$\begin{aligned} y &= ae^{-bt} \\ 5 &= 10e^{-b(1599)} \\ 0.5 &= e^{-1599b} \\ \ln 0.5 &= \ln e^{-1599b} \\ \ln 0.5 &= -1599b \\ b &= -\frac{\ln 0.5}{1599} \end{aligned}$$

Given an initial quantity of 10 grams, after 1000 years, you have

$y = 10e^{-[-(\ln 0.5)/1599](1000)} \approx 6.48$ grams.

23. $y = 2, a = 2(2) = 4, t = 5715$

$$\begin{aligned} y &= ae^{-bt} \\ 2 &= 4e^{-b(5715)} \\ 0.5 &= e^{-5715b} \\ \ln 0.5 &= \ln e^{-5715b} \\ \ln 0.5 &= -5715b \\ b &= -\frac{\ln 0.5}{5715} \end{aligned}$$

Given 2 grams after 1000 years, the initial amount is

$2 = ae^{-[-(\ln 0.5)/5715](1000)}$

$a \approx 2.26$ grams.

25.

$$\begin{aligned} y &= ae^{bx} \\ 1 &= ae^{b(0)} \Rightarrow 1 = a \\ 10 &= e^{b(3)} \\ \ln 10 &= 3b \\ \frac{\ln 10}{3} &= b \Rightarrow b \approx 0.7675 \end{aligned}$$

So, $y = e^{0.7675x}$.

27.

$$\begin{aligned} y &= ae^{bx} \\ 5 &= ae^{b(0)} \Rightarrow 5 = a \\ 1 &= 5e^{b(4)} \\ \frac{1}{5} &= e^{4b} \\ \ln\left(\frac{1}{5}\right) &= 4b \\ \frac{\ln(1/5)}{4} &= b \Rightarrow b \approx -0.4024 \end{aligned}$$

So, $y = 5e^{-0.4024x}$.

29. (a) $P = 76.6e^{0.0313t}$

Year	1980	1990	2000	2010
P	104.752	143.251	195.899	267.896
Population	104,752	143,251	195,899	267,896

(b) Let $P = 360$, and solve for t.

$$\begin{aligned} 360 &= 76.6e^{0.0313t} \\ \frac{360}{76.6} &= e^{0.0313t} \\ \ln\left(\frac{360}{76.6}\right) &= 0.0313t \\ \frac{1}{0.0313}\ln\left(\frac{360}{76.6}\right) &= t \\ 49.4 &\approx t \end{aligned}$$

According to the model, the population will reach 360,000 in 2019.

(c) No; As t increases, the population increases rapidly.

31. $y = 4080e^{kt}$

When $t = 3$, $y = 10{,}000$:

$$10{,}000 = 4080e^{k(3)}$$

$$\frac{10{,}000}{4080} = e^{3k}$$

$$\ln\left(\frac{10{,}000}{4080}\right) = 3k$$

$$k = \frac{\ln(10{,}000/4080)}{3} \approx 0.2988$$

When $t = 24$: $y = 4080e^{0.2988(24)} \approx 5{,}309{,}734$ hits

33. $y = ae^{bt}$

When $t = 3$, $y = 100$:

$$100 = ae^{3b}$$

$$\frac{100}{e^{3b}} = a$$

When $t = 5$, $y = 400$:

$$400 = ae^{5b}$$

Substitute $\frac{100}{e^{3b}}$ for a in the equation on the right.

$$400 = \frac{100}{e^{3b}}e^{5b}$$

$$400 = 100e^{2b}$$

$$4 = e^{2b}$$

$$\ln 4 = 2b$$

$$\ln 2^2 = 2b$$

$$2\ln 2 = 2b$$

$$\ln 2 = b$$

$$a = \frac{100}{e^{3b}} = \frac{100}{e^{3\ln 2}} = \frac{100}{e^{\ln 2^3}} = \frac{100}{2^3} = \frac{100}{8} = 12.5$$

$$y = 12.5e^{(\ln 2)t}$$

After 6 hours, there are $y = 12.5e^{(\ln 2)(6)} = 800$ bacteria.

35. $(0, 575), (2, 275)$

(a) $m = \dfrac{275 - 575}{2 - 0} = -150$

$V = -150t + 575$

(b) Since $V = 575$, when $t = 0$, $575 = ae^{(b)(0)} \rightarrow a = 575$

Then $275 = 575e^{k(2)}$

$\ln\left(\dfrac{275}{575}\right) = 2k \Rightarrow k \approx -0.3688$

$V = 575e^{-0.3688t}$

(c)

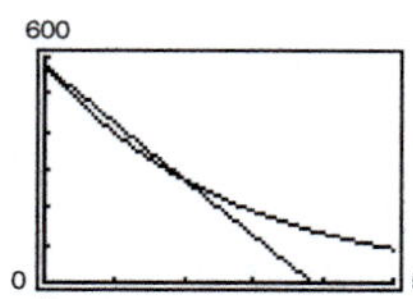

The exponential model depreciates faster in the first two years.

(d)

t	1	3
$V = -150t + 575$	\$425	\$125
$V = 575e^{-0.3688t}$	\$397.65	\$190.18

(e) Answers will vary. Sample Answer: The slope of the linear model means that the laptop depreciates \$150 per year, then loses all value late in the third year. The exponential model depreciates faster in the first three years but maintains value longer.

37. $R = \dfrac{1}{10^{12}}e^{-t/8223}$

$$R = \frac{1}{8^{14}}$$

$$\frac{1}{10^{12}}e^{-t/8223} = \frac{1}{8^{14}}$$

$$e^{-t/8223} = \frac{10^{12}}{8^{14}}$$

$$-\frac{t}{8223} = \ln\left(\frac{10^{12}}{8^{14}}\right)$$

$$t = -8223\ln\left(\frac{10^{12}}{8^{14}}\right) \approx 12{,}180 \text{ years old}$$

39. $y = 0.0266e^{-(x-100)^2/450}, \; 70 \le x \le 116$

(a)

(b) The average IQ score of an adult student is 100.

41. (a) 1998: $t = 18$, $y = \dfrac{320{,}110}{1 + 374e^{-0.252(18)}}$

$\approx 63{,}992$ sites

2003: $t = 23$, $y = \dfrac{320{,}110}{1 + 374e^{-0.252(23)}}$

$\approx 149{,}805$ sites

2006: $t = 26$, $y = \dfrac{320{,}110}{1 + 374e^{-0.252(26)}}$

$\approx 208{,}705$ sites

(b)

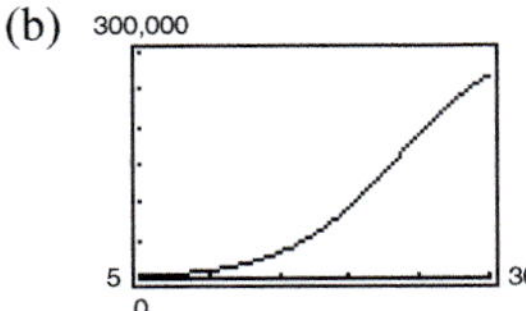

(c) When $y = 270{,}000$, $t \approx 30.2$. So, the number of cell sites will reach 270,000 in the year 2010.

(d) Let $y = 270{,}000$ and solve for t.

$$270{,}000 = \frac{320{,}110}{1 + 374e^{-0.252t}}$$

$$1 + 374e^{-0.252t} = \frac{320{,}110}{270{,}000}$$

$$374e^{-0.252t} = 0.1855926$$

$$e^{-0.252t} \approx 0.000496237$$

$$-0.252t \approx \ln(0.000496237)$$

$$t \approx 30.2$$

The number of cell sites will reach 270,000 during the year 2010.

43. $p(t) = \dfrac{1000}{1 + 9e^{-0.1656t}}$

(a) $p(5) = \dfrac{1000}{1 + 9e^{-0.1656(5)}} \approx 203$ animals

(b)

$$500 = \frac{1000}{1 + 9e^{-0.1656t}}$$

$$1 + 9e^{-0.1656t} = 2$$

$$9e^{-0.1656t} = 1$$

$$e^{-0.1656t} = \frac{1}{9}$$

$$t = -\frac{\ln(1/9)}{0.1656} \approx 13 \text{ months}$$

(c)

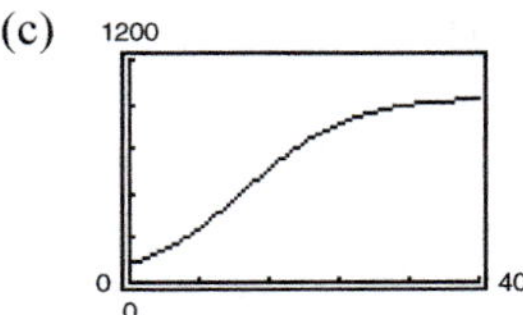

The horizontal asymptotes are $p = 0$ and $p = 1000$. The asymptote with the larger p-value, $p = 1000$, indicates that the population size will approach 1000 as time increases.

45. $R = \log\dfrac{I}{I_0} = \log I$ because $I_0 = 1$.

(a)

$$R = 7.6$$

$$7.6 = \log I$$

$$10^{7.6} = 10^{\log I}$$

$$39{,}810{,}717 \approx I$$

(b)

$$R = 5.6$$

$$5.6 = \log I$$

$$10^{5.6} = 10^{\log I}$$

$$10^{5.6} = I$$

$$398{,}107 \approx I$$

(c)

$$R = 6.6$$

$$6.6 = \log I$$

$$10^{6.6} = 10^{\log I}$$

$$3{,}981{,}072 \approx I$$

47. $\beta = 10\log\dfrac{I}{I_0}$ where $I_0 = 10^{-12}$ watt/m^2.

(a) $\beta = 10\log\dfrac{10^{-10}}{10^{-12}} = 10\log 10^2 = 20$ decibels

(b) $\beta = 10\log\dfrac{10^{-5}}{10^{-12}} = 10\log 10^7 = 70$ decibels

(c) $\beta = 10\log\dfrac{10^{-8}}{10^{-12}} = 10\log 10^4 = 40$ decibels

(d) $\beta = 10\log\dfrac{10^{-3}}{10^{-12}} = 10\log 10^9 = 90$ decibels

49.
$$\beta = 10 \log \frac{I}{I_0}$$
$$\frac{\beta}{10} = \log \frac{I}{I_0}$$
$$10^{\beta/10} = 10^{\log I/I_0}$$
$$10^{\beta/10} = \frac{I}{I_0}$$
$$I = I_0 10^{\beta/10}$$
$$\% \text{ decrease} = \frac{I_0 10^{9.3} - I_0 10^{8.0}}{I_0 10^{9.3}} \times 100 \approx 95\%$$

51. $\text{pH} = -\log[\text{H}^+]$

$-\log(2.3 \times 10^{-5}) \approx 4.64$

53.
$$5.8 = -\log[\text{H}^+]$$
$$-5.8 = \log[\text{H}^+]$$
$$10^{-5.8} = 10^{\log[\text{H}^+]}$$
$$10^{-5.8} = [\text{H}^+]$$
$$[\text{H}^+] \approx 1.58 \times 10^{-6} \text{ moles per liter}$$

55.
$$2.9 = -\log[\text{H}^+]$$
$$-2.9 = \log[\text{H}^+]$$
$[\text{H}^+] = 10^{-2.9}$ for the apple juice
$$8.0 = -\log[\text{H}^+]$$
$$-8.0 = \log[\text{H}^+]$$
$[\text{H}^+] = 10^{-8}$ for the drinking water

$\frac{10^{-2.9}}{10^{-8}} = 10^{5.1}$ times the hydrogen ion concentration of drinking water

57. $t = -10 \ln \frac{T - 70}{98.6 - 70}$

At 9:00 A.M. you have:

$t = -10 \ln \frac{85.7 - 70}{98.6 - 70} \approx 6$ hours

From this you can conclude that the person died at 3:00 A.M.

59. $u = 120{,}000\left[\frac{0.075t}{1 - \left(\frac{1}{1 + 0.075/12}\right)^{12t}} - 1\right]$

(a)

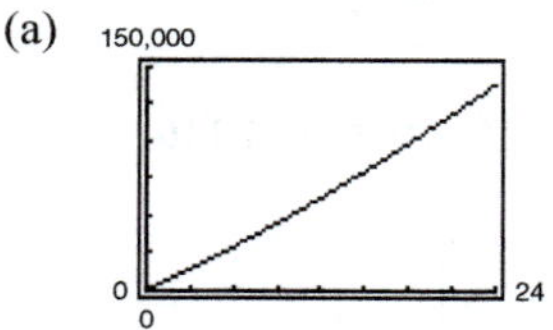

(b) From the graph, $u = \$120{,}000$ when $t \approx 21$ years. It would take approximately 37.6 years to pay \$240,000 in interest. Yes, it is possible to pay twice as much in interest charges as the size of the mortgage. It is especially likely when the interest rates are higher.

61. False. The domain can be the set of real numbers for a logistic growth function.

63. False. The graph of $f(x)$ is the graph of $g(x)$ shifted upward five units.

65. Answers will vary.

Review Exercises for Chapter 3

1. $f(x) = 0.3^x$

$f(1.5) = 0.3^{1.5} \approx 0.164$

3. $f(x) = 2^x$

$f\left(\frac{2}{3}\right) = 2^{2/3} \approx 1.587$

5.
$$f(x) = 7(0.2^x)$$
$$f(-\sqrt{11}) = 7(0.2^{-\sqrt{11}})$$
$$\approx 1456.529$$

7. $f(x) = 4^{-x} + 4$

Horizontal asymptote: $y = 4$

x	−1	0	1	2	3
$f(x)$	8	5	4.25	4.063	4.016

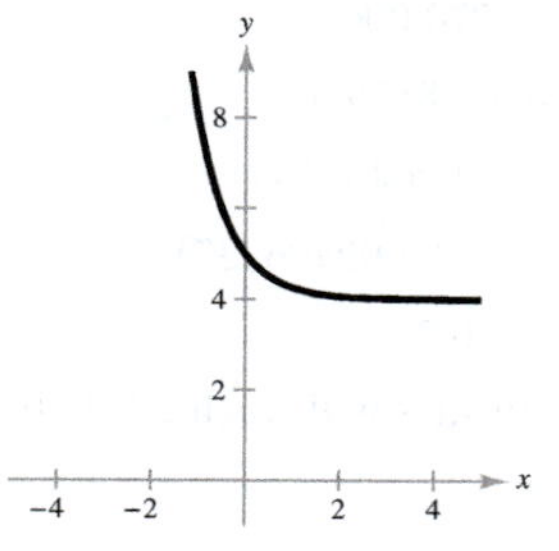

9. $f(x) = 5^{x-2} + 4$

Horizontal asymptote: $y = 4$

x	-1	0	1	2	3
$f(x)$	4.008	4.04	4.2	5	9

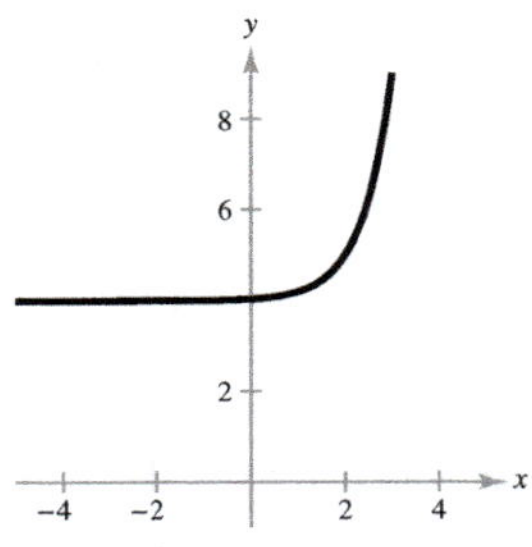

11. $f(x) = \left(\frac{1}{2}\right)^{-x} + 3 = 2^x + 3$

Horizontal asymptote: $y = 3$

x	-2	-1	0	1	2
$f(x)$	3.25	3.5	4	5	7

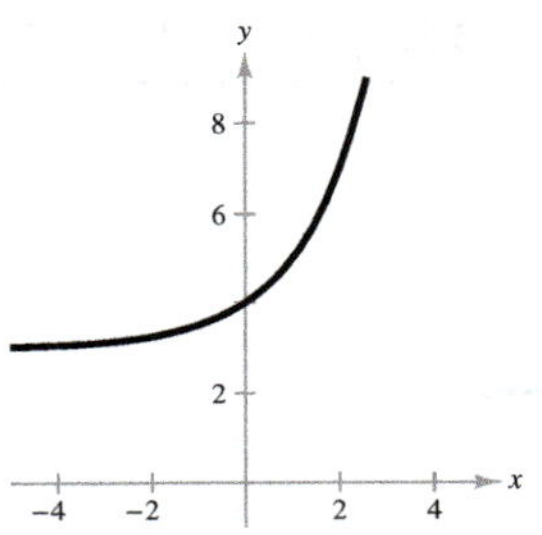

13. $\left(\frac{1}{3}\right)^{x-3} = 9$

$\left(\frac{1}{3}\right)^{x-3} = 3^2$

$\left(\frac{1}{3}\right)^{x-3} = \left(\frac{1}{3}\right)^{-2}$

$x - 3 = -2$

$x = 1$

15. $e^{3x-5} = e^7$

$3x - 5 = 7$

$3x = 12$

$x = 4$

17. $f(x) = 5^x$, $g(x) = 5^x + 1$

Because $g(x) = f(x) + 1$, the graph of g can be obtained by shifting the graph of f one unit upward.

19. $f(x) = 3^x$, $g(x) = 1 - 3^x$

Because $g(x) = 1 - f(x)$, the graph of g can be obtained by reflecting the graph of f in the x-axis and shifting the graph one unit upward. (**Note:** This is equivalent to shifting the graph of f one unit upward and then reflecting the graph in the x-axis.)

21. $f(x) = e^x$

$f(3.4) = e^{3.4} \approx 29.964$

23. $f(x) = e^x$

$f\left(\frac{3}{5}\right) = e^{3/5} \approx 1.822$

25. $h(x) = e^{-x/2}$

x	-2	-1	0	1	2
$h(x)$	2.72	1.65	1	0.61	0.37

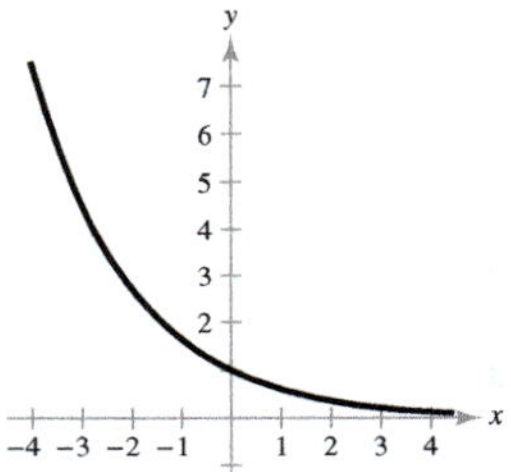

27. $f(x) = e^{x+2}$

x	-3	-2	-1	0	1
$f(x)$	0.37	1	2.72	7.39	20.09

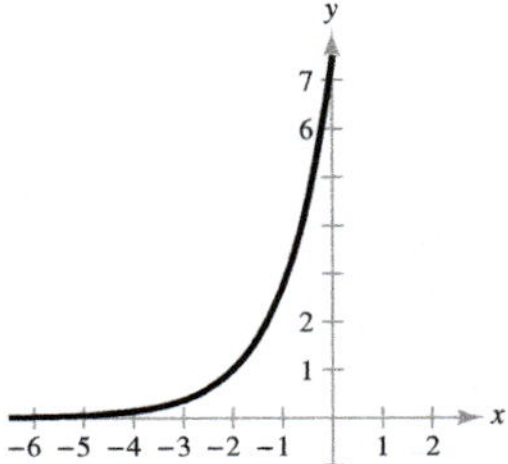

29. $F(t) = 1 - e^{-t/3}$

(a) $F(1) \approx 0.283$

(b) $F(2) \approx 0.487$

(c) $F(5) \approx 0.811$

31. $P = \$5000, r = 3\%, t = 10$ years

Compounded n times per year: $A = P\left(1 + \frac{r}{n}\right)^{nt} = 5000\left(1 + \frac{0.03}{n}\right)^{10n}$

Compounded continuously: $A = Pe^{rt} = 5000e^{0.03(10)}$

n	1	2	4	12	365	Continuous
A	\$6719.58	\$6734.28	\$6741.74	\$6746.77	\$6749.21	\$6749.29

33. $3^3 = 27$

$\log_3 27 = 3$

35. $e^{0.8} = 2.2255\ldots$

$\ln 2.2255\ldots = 0.8$

37. $f(x) = \log x$

$f(1000) = \log 1000$

$= \log 10^3 = 3$

39. $g(x) = \log_2 x$

$g\left(\frac{1}{4}\right) = \log_2 \frac{1}{4}$

$= \log_2 2^{-2} = -2$

41. $\log_4(x + 7) = \log_4 14$

$x + 7 = 14$

$x = 7$

43. $\ln(x + 9) = \ln 4$

$x + 9 = 4$

$x = -5$

45. $g(x) = \log_7 x \Rightarrow x = 7^y$

Domain: $(0, \infty)$

x-intercept: $(1, 0)$

Vertical asymptote: $x = 0$

x	$\frac{1}{7}$	1	7	49
$g(x)$	−1	0	1	2

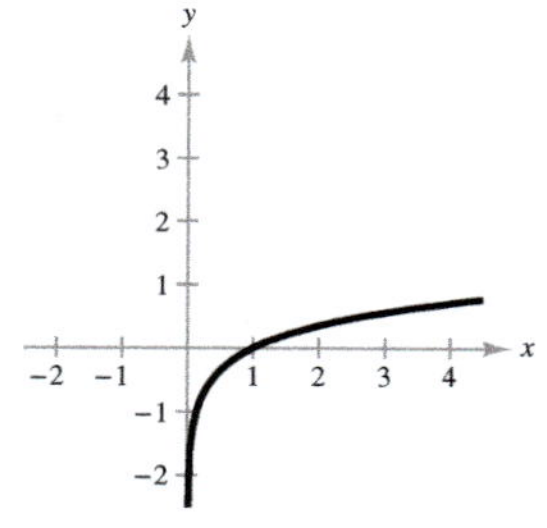

47. $f(x) = 4 - \log(x + 5)$

Domain: $(-5, \infty)$

Because

$4 - \log(x + 5) = 0 \Rightarrow \log(x + 5) = 4$

$x + 5 = 10^4$

$x = 10^4 - 5$

$= 9995.$

x-intercept: $(9995, 0)$

Vertical asymptote: $x = -5$

x	−4	−3	−2	−1	0	1
$f(x)$	4	3.70	3.52	3.40	3.30	3.22

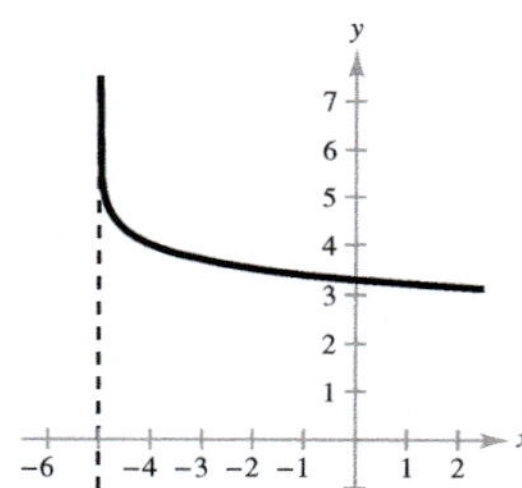

49. $f(22.6) = \ln 22.6 \approx 3.118$

51. $f\left(\sqrt{e}\right) = \frac{1}{2} \ln \sqrt{e} = 0.25$

53. $f(x) = \ln x + 6 = 6 + \ln x$

Domain: $(0, \infty)$

$$\ln x + 6 = 0$$
$$\ln x = -6$$
$$x = e^{-6}$$

x-intercept: $(e^{-6}, 0)$

Vertical asymptote: $x = 0$

x	$\frac{1}{4}$	$\frac{1}{2}$	1	2	3
$f(x)$	4.613	5.037	6	6.693	7.098

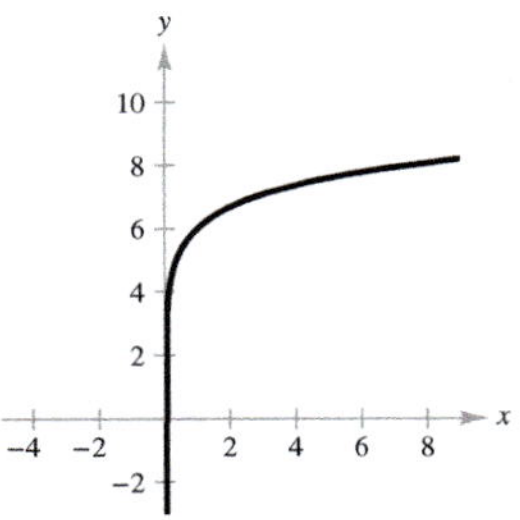

55. $f(x) = \ln(x - 6)$

Domain: $(6, \infty)$

$$\ln(x - 6) = 0$$
$$x - 6 = e^0$$
$$x - 6 = 1$$
$$x = 7$$

x-intercept: $(7, 0)$

Vertical asymptote: $x = 6$

x	6.5	7	8	9	10
$f(x)$	−0.693	0	0.693	1.099	1.386

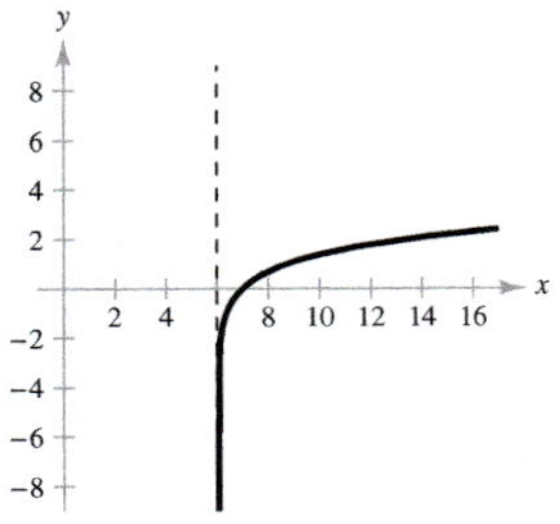

57. $M = m - 5\log\left(\frac{d}{10}\right)$

Let $m = 2.08$ and $M = 1.3$ and solve for d.

$$1.3 = 2.08 - 5\log\left(\frac{d}{10}\right)$$
$$-0.78 = -5\log\left(\frac{d}{10}\right)$$
$$0.156 = \log\left(\frac{d}{10}\right)$$
$$10^{0.156} = 10^{\log(d/10)}$$
$$10^{0.156} = \frac{d}{10}$$
$$10 \cdot 10^{0.156} = d$$
$$d = 10^{1.156} \approx 14.32 \text{ parsecs}$$

59. (a) $\log_2 6 = \frac{\log 6}{\log 2} \approx 2.585$

(b) $\log_2 6 = \frac{\ln 6}{\ln 2} \approx 2.585$

61. (a) $\log_{1/2} 5 = \frac{\log 5}{\log(1/2)} \approx -2.322$

(b) $\log_{1/2} 5 = \frac{\ln 5}{\ln(1/2)} \approx -2.322$

63. $\log_2 \frac{5}{3} = \log_2 5 - \log_2 3$

65. $\log_2 \frac{9}{5} = \log_2 9 - \log_2 5$

$$= \log_2 3^2 - \log_2 9$$
$$= 2\log_2 3 - \log_2 5$$

67. $\log 7x^2 = \log 7 + \log x^2$

$$= \log 7 + 2\log x$$

69. $\log_3 \frac{9}{\sqrt{x}} = \log_3 9 - \log_3 \sqrt{x}$

$$= \log_3 3^2 - \log_3 x^{1/2}$$
$$= 2 - \frac{1}{2}\log_3 x$$

71. $\ln x^2y^2z = \ln x^2 + \ln y^2 + \ln z$

$$= 2\ln x + 2\ln y + \ln z$$

73. $\ln 7 + \ln x = \ln(7x)$

75. $\log x - \frac{1}{2}\log y = \log x - \log y^{1/2}$

$= \log\left(\frac{x}{\sqrt{y}}\right)$

77. $\frac{1}{2}\log_3 x - 2\log_3(y+8) = \log_3 x^{1/2} - \log_3(y+8)^2$

$= \log_3 \sqrt{x} - \log_3(y+8)^2$

$= \log_3 \frac{\sqrt{x}}{(y+8)^2}$

79. $t = 50 \log \frac{18{,}000}{18{,}000 - h}$

(a) Domain: $0 \le h < 18{,}000$

(b)

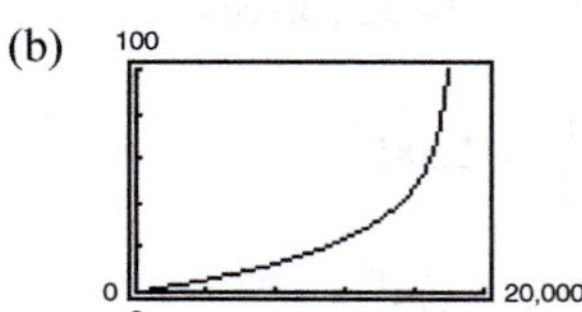

Vertical asymptote: $h = 18{,}000$

(c) As the plane approaches its absolute ceiling, it climbs at a slower rate, so the time required increases.

(d) $50 \log \frac{18{,}000}{18{,}000 - 4000} \approx 5.46$ minutes

81. $5^x = 125$

$5^x = 5^3$

$x = 3$

83. $e^x = 3$

$x = \ln 3 \approx 1.099$

85. $\ln x = 4$

$x = e^4 \approx 54.598$

87. $e^{4x} = e^{x^2+3}$

$4x = x^2 + 3$

$0 = x^2 - 4x + 3$

$0 = (x-1)(x-3)$

$x = 1, x = 3$

89. $2^x - 3 = 29$

$2^x = 32$

$2^x = 2^5$

$x = 5$

91. $\ln 3x = 8.2$

$e^{\ln 3x} = e^{8.2}$

$3x = e^{8.2}$

$x = \frac{e^{8.2}}{3} \approx 1213.650$

93. $\ln x + \ln(x-3) = 1$

$\ln[x(x-3)] = 1$

$\ln(x^2 - 3x) = 1$

$e^{\ln(x^2-3x)} = e^1$

$x^2 - 3x - e = 0$

$x = \frac{-b \pm \sqrt{b^2 - 4ac}}{2a}$

$x = \frac{-(-3) \pm \sqrt{(-3)^2 - 4(1)(-e)}}{2(1)}$

$x = \frac{3 \pm \sqrt{9 + 4e}}{2}$

$x = \frac{3 + \sqrt{9 + 4e}}{2} \approx 3.729$

$x = \frac{3 - \sqrt{9 + 4e}}{2}$ is extraneous since the domain of the $\ln x$ term is $x > 0$.

95.
$$\begin{aligned}
\log_8(x-1) &= \log_8(x-2) - \log_8(x+2)\\
\log_8(x-1) &= \log_8\left(\frac{x-2}{x+2}\right)\\
x-1 &= \frac{x-2}{x+2}\\
(x-1)(x+2) &= x-2\\
x^2+x-2 &= x-2\\
x^2 &= 0\\
x &= 0
\end{aligned}$$

Because $x = 0$ is not in the domain of $\log_8(x-1)$ or of $\log_8(x-2)$, it is an extraneous solution. The equation has no solution.

97.
$$\begin{aligned}
\log(1-x) &= -1\\
1-x &= 10^{-1}\\
1-\tfrac{1}{10} &= x\\
x &= 0.900
\end{aligned}$$

99. $25e^{-0.3x} = 12$

Graph $y_1 = 25e^{-0.3x}$ and $y_2 = 12$.

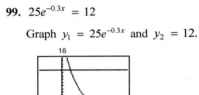

The graphs intersect at $x \approx 2.447$.

101. $2\ln(x+3) - 3 = 0$

Graph $y_1 = 2\ln(x+3) - 3$.

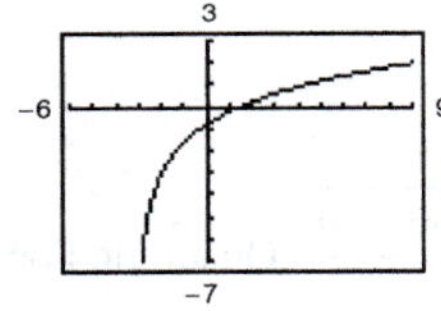

The x-intercept is at $x \approx 1.482$.

103. $P = 8500,\ A = 3(8500) = 25{,}500,\ r = 1.5\%$

$$\begin{aligned}
A &= Pe^{rt}\\
25{,}500 &= 8500e^{0.015t}\\
3 &= e^{0.015t}\\
\ln 3 &= 0.015t\\
t &= \frac{\ln 3}{0.015} \approx 73.2 \text{ years}
\end{aligned}$$

105. $y = 3e^{-2x/3}$

Exponential decay model

Matches graph (e).

106. $y = 4e^{2x/3}$

Exponential growth model

Matches graph (b).

107. $y = \ln(x+3)$

Logarithmic model

Vertical asymptote: $x = -3$

Graph includes $(-2, 0)$

Matches graph (f).

108. $y = 7 - \log(x+3)$

Logarithmic model

Vertical asymptote: $x = -3$

Matches graph (d).

109. $y = 2e^{-(x+4)^2/3}$

Gaussian model

Matches graph (a).

110. $y = \dfrac{6}{1 + 2e^{-2x}}$

Logistics growth model

Matches graph (c).

111. $y = ae^{bx}$

Using the point $(0, 2)$, you have

$$\begin{aligned}
2 &= ae^{b(0)}\\
2 &= ae^0\\
2 &= a(1)\\
2 &= a
\end{aligned}$$

Then, using the point $(4, 3)$, you have

$$\begin{aligned}
3 &= 2e^{b(4)}\\
3 &= 2e^{4b}\\
\tfrac{3}{2} &= e^{4b}\\
\ln\tfrac{3}{2} &= 4b\\
\tfrac{1}{4}\ln\left(\tfrac{3}{2}\right) &= b
\end{aligned}$$

So, $y = 2e^{\frac{1}{4}\ln\left(\frac{3}{2}\right)x}$

or

$y = 2e^{0.1014x}$

113. $y = 0.0499e^{-(x-71)^2/128},\ 40 \le x \le 100$

Graph $y_1 = 0.0499e^{-(x-71)^2/128}$.

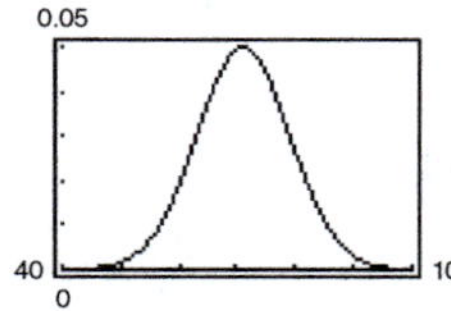

The average test score is 71.

115. $\beta = 10\log\left(\dfrac{I}{10^{-12}}\right)$

$\dfrac{\beta}{10} = \log\left(\dfrac{I}{10^{-12}}\right)$

$10^{\beta/10} = \dfrac{I}{10^{-12}}$

$I = 10^{\beta/10-12}$

(a) $\beta = 60$

$I = 10^{60/10-12}$

$= 10^{-6}$ watt/m^2

(b) $\beta = 135$

$I = 10^{135/10-12}$

$= 10^{1.5}$

$= 10\sqrt{10}$ watts/m^2

(c) $\beta = 1$

$I = 10^{1/10-12}$

$= 10^{\frac{1}{10}} \times 10^{-12}$

$\approx 1.259 \times 10^{-12}$ watt/m^2

117. True. By the inverse properties, $\log_b b^{2x} = 2x$.

Problem Solving for Chapter 3

1. $y = a^x$

$y_1 = 0.5^x$

$y_2 = 1.2^x$

$y_3 = 2.0^x$

$y_4 = x$

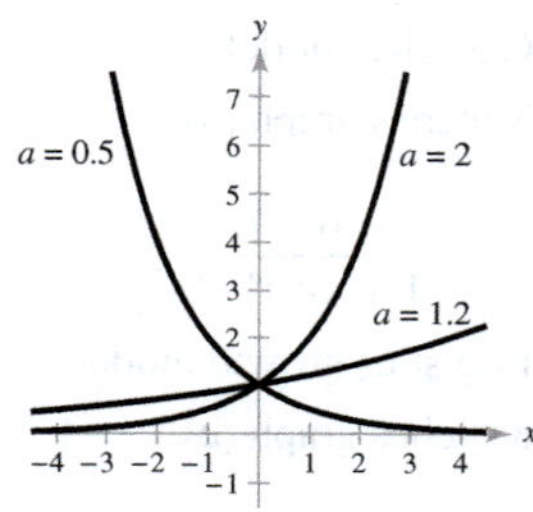

The curves $y = 0.5^x$ and $y = 1.2^x$ cross the line $y = x$. From checking the graphs it appears that $y = x$ will cross $y = a^x$ for $0 \le a \le 1.44$.

3. The exponential function, $y = e^x$, increases at a faster rate than the polynomial $y = x^n$.

5. (a) $f(u + v) = a^{u+v} = a^u \cdot a^v = f(u) \cdot f(v)$

(b) $f(2x) = a^{2x} = (a^x)^2 = [f(x)]^2$

7. (a)

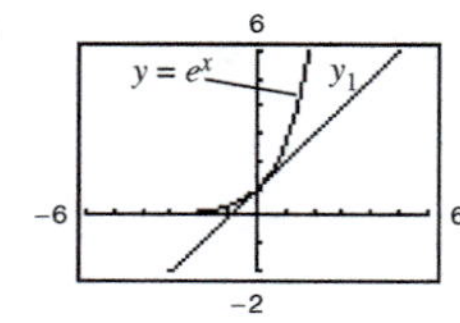

(b)

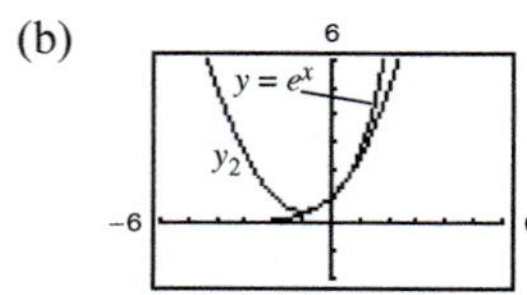

(c)

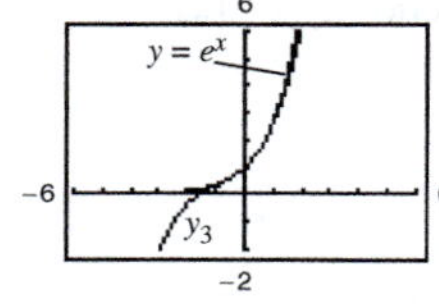

9.

$f(x) = e^x - e^{-x}$

$y = e^x - e^{-x}$

$x = e^y - e^{-y}$

$x = \dfrac{e^{2y} - 1}{e^y}$

$xe^y = e^{2y} - 1$

$e^{2y} - xe^y - 1 = 0$

$e^y = \dfrac{x \pm \sqrt{x^2 + 4}}{2}$ Quadratic Formula

Choosing the positive quantity for e^y you have

$y = \ln\left(\dfrac{x + \sqrt{x^2 + 4}}{2}\right)$. So,

$f^{-1}(x) = \ln\left(\dfrac{x + \sqrt{x^2 + 4}}{2}\right)$.

11. Answer (c). $y = 6\left(1 - e^{-x^2/2}\right)$

The graph passes through (0, 0) and neither (a) nor (b) pass through the origin. Also, the graph has y-axis symmetry and a horizontal asymptote at $y = 6$.

13. $y_1 = c_1\left(\frac{1}{2}\right)^{t/k_1}$ and $y_2 = c_2\left(\frac{1}{2}\right)^{t/k_2}$

$$c_1\left(\frac{1}{2}\right)^{t/k_1} = c_2\left(\frac{1}{2}\right)^{t/k_2}$$

$$\frac{c_1}{c_2} = \left(\frac{1}{2}\right)^{(t/k_2 - t/k_1)}$$

$$\ln\left(\frac{c_1}{c_2}\right) = \left(\frac{t}{k_2} - \frac{t}{k_1}\right)\ln\left(\frac{1}{2}\right)$$

$$\ln c_1 - \ln c_2 = t\left(\frac{1}{k_2} - \frac{1}{k_1}\right)\ln\left(\frac{1}{2}\right)$$

$$t = \frac{\ln c_1 - \ln c_2}{\left[(1/k_2) - (1/k_1)\right]\ln(1/2)}$$

15. (a) $y_1 \approx 252{,}606(1.0310)^t$

(b) $y_2 \approx 400.88t^2 - 1464.6t + 291{,}782$

(c)

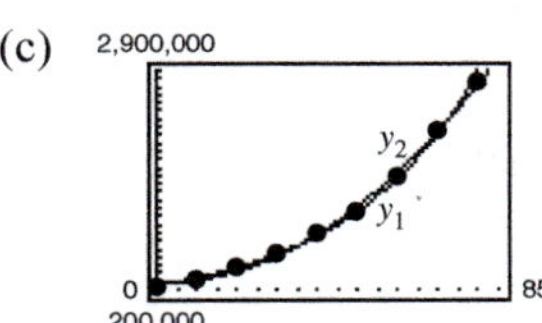

(d) The exponential model is a better fit for the data, but neither would be reliable to predict the population of the United States in 2020. The exponential model approaches infinity rapidly.

17.

$$(\ln x)^2 = \ln x^2$$

$$(\ln x)^2 - 2\ln x = 0$$

$$\ln x(\ln x - 2) = 0$$

$$\ln x = 0 \text{ or } \ln x = 2$$

$$x = 1 \text{ or } x = e^2$$

19. $y_4 = (x-1) - \frac{1}{2}(x-1)^2 + \frac{1}{3}(x-1)^3 - \frac{1}{4}(x-1)^4$

The pattern implies that $\ln x = (x-1) - \frac{1}{2}(x-1)^2 + \frac{1}{3}(x-1)^3 - \frac{1}{4}(x-1)^4 + \dots$.

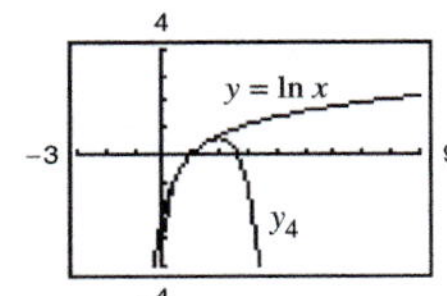

21. $y = 80.4 - 11\ln x$

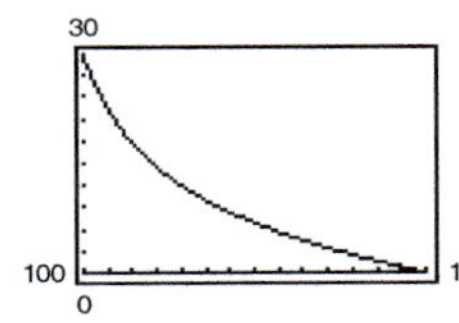

$$y(300) = 80.4 - 11\ln 300$$

$$\approx 17.7 \text{ ft}^3/\text{min}$$

23. (a)

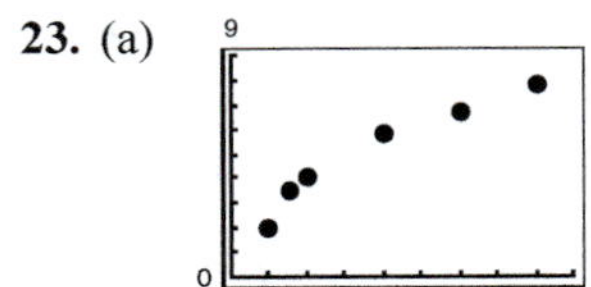

(b) The data could best be modeled by a logarithmic model.

(c) The shape of the curve looks much more logarithmic than linear or exponential.

(d) $y \approx 2.1518 + 2.7044\ln x$

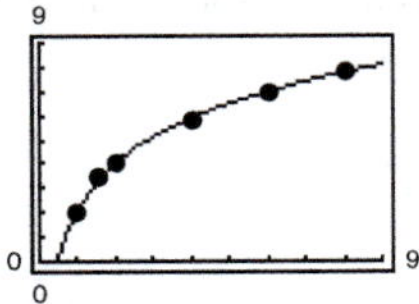

(e) The model is a good fit to the actual data.

25. (a)

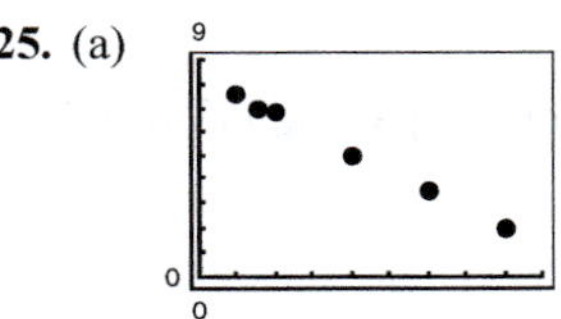

(b) The data could best be modeled by a linear model.

(c) The shape of the curve looks much more linear than exponential or logarithmic.

(d) $y \approx -0.7884x + 8.2566$

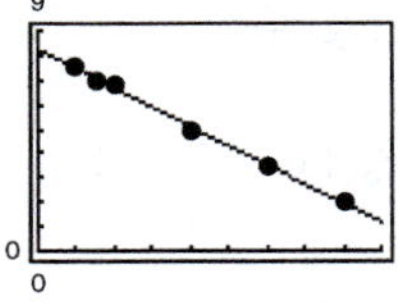

(e) The model is a good fit to the actual data.

Practice Test for Chapter 3

1. Solve for x: $x^{3/5} = 8$.

2. Solve for x: $3^{x-1} = \frac{1}{81}$.

3. Graph $f(x) = 2^{-x}$.

4. Graph $g(x) = e^x + 1$.

5. If $5000 is invested at 9% interest, find the amount after three years if the interest is compounded

(a) monthly.

(b) quarterly.

(c) continuously.

6. Write the equation in logarithmic form: $7^{-2} = \frac{1}{49}$.

7. Solve for x: $x - 4 = \log_2 \frac{1}{64}$.

8. Given $\log_b 2 = 0.3562$ and $\log_b 5 = 0.8271$, evaluate $\log_b \sqrt[4]{8/25}$.

9. Write $5 \ln x - \frac{1}{2} \ln y + 6 \ln z$ as a single logarithm.

10. Using your calculator and the change of base formula, evaluate $\log_9 28$.

11. Use your calculator to solve for N: $\log_{10} N = 0.6646$

12. Graph $y = \log_4 x$.

13. Determine the domain of $f(x) = \log_3(x^2 - 9)$.

14. Graph $y = \ln(x - 2)$.

15. True or false: $\dfrac{\ln x}{\ln y} = \ln(x - y)$

16. Solve for x: $5^x = 41$

17. Solve for x: $x - x^2 = \log_5 \frac{1}{25}$

18. Solve for x: $\log_2 x + \log_2(x - 3) = 2$

19. Solve for x: $\dfrac{e^x + e^{-x}}{3} = 4$

20. Six thousand dollars is deposited into a fund at an annual interest rate of 13%. Find the time required for the investment to double if the interest is compounded continuously.

CHAPTER 4
Trigonometry

CHAPTER 4
Trigonometry

Section 4.1 Radian and Degree Measure

1. coterminal

3. complementary; supplementary

5. linear; angular

7.

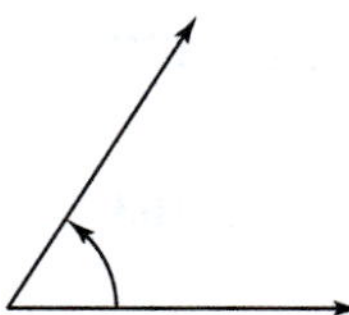

The angle shown is approximately 1 radian.

9.

The angle shown is approximately -3 radians.

11. (a) Because $0 < \frac{\pi}{4} < \frac{\pi}{2}, \frac{\pi}{4}$ lies in Quadrant I.

(b) Because $-\frac{5\pi}{4}$ is coterminal with $\frac{3\pi}{4}$ and $\frac{\pi}{2} < \frac{3\pi}{4} < \pi, -\frac{5\pi}{4}$ lies in Quadrant II.

13. (a) $\frac{\pi}{3}$

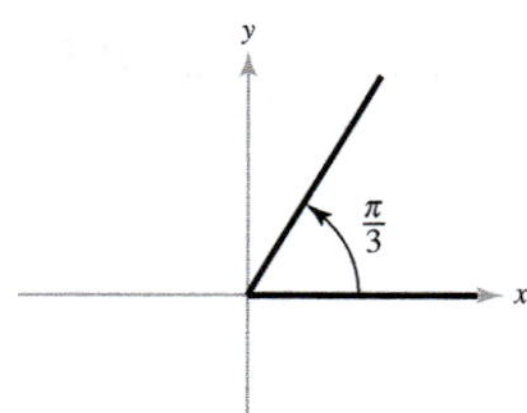

(b) $-\frac{2\pi}{3}$

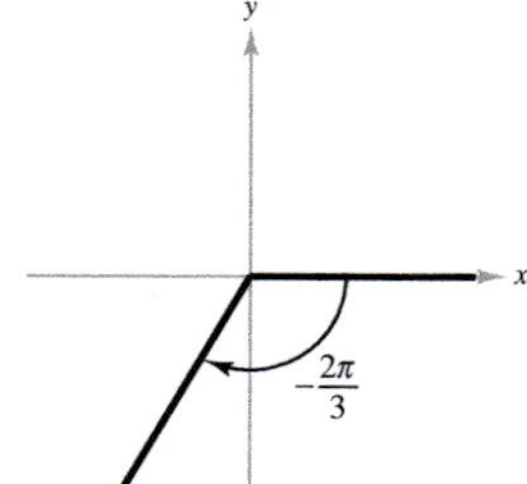

15. *Sample answers*:

(a) $\frac{\pi}{6} + 2\pi = \frac{13\pi}{6}$

$\frac{\pi}{6} - 2\pi = -\frac{11\pi}{6}$

(b) $-\frac{5\pi}{6} + 2\pi = \frac{7\pi}{6}$

$-\frac{5\pi}{6} - 2\pi = -\frac{17\pi}{6}$

17. (a) Complement: $\frac{\pi}{2} - \frac{\pi}{12} = \frac{5\pi}{12}$

Supplement: $\pi - \frac{\pi}{12} = \frac{11\pi}{12}$

(b) Complement: Not possible, $\frac{11\pi}{12}$ is greater than $\frac{\pi}{2}$.

Supplement: $\pi - \frac{11\pi}{12} = \frac{\pi}{12}$

19. (a) Complement: $\frac{\pi}{2} - 1 \approx 0.57$

Supplement: $\pi - 1 \approx 2.14$

(b) Complement: Not possible, 2 is greater than $\frac{\pi}{2}$.

Supplement: $\pi - 2 \approx 1.14$

21.

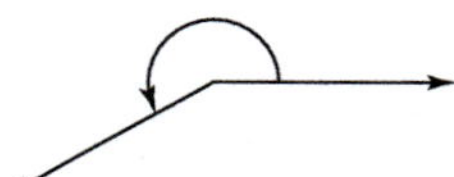

The angle shown is approximately $210°$.

23.

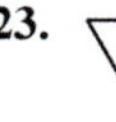
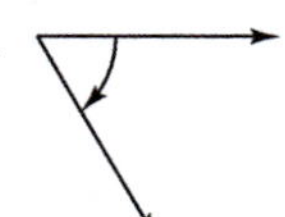

The angle shown is approximately $-60°$.

25. (a) Because $90° < 130° < 180°, 130°$ lies in Quadrant II.

(b) Because $-8.3°$ is coterminal with $351.7°$ and $270° < 351.7° < 360°, -8.3°$ lies in Quadrant IV.

27. (a) $-270°$

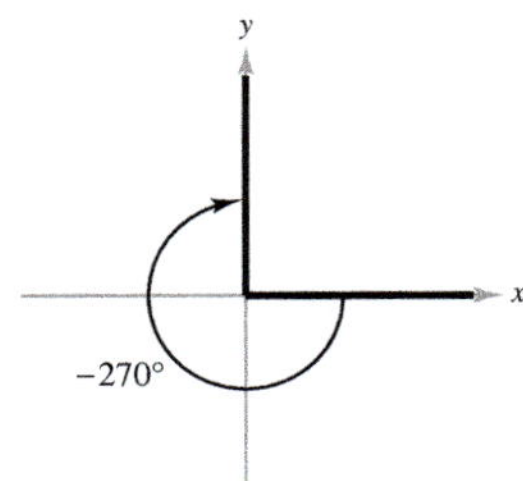

(b) $-120°$

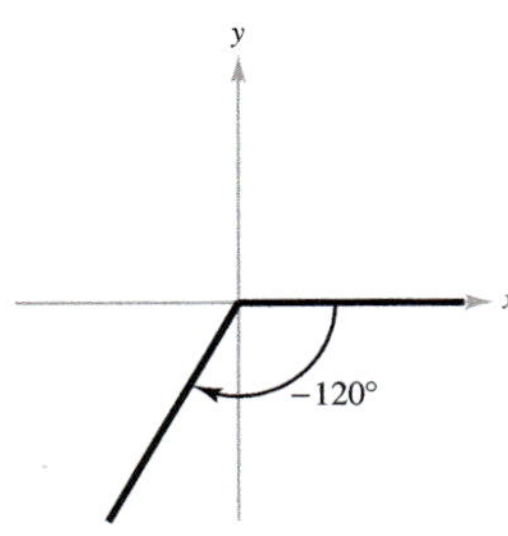

29. (a) Coterminal angles for $120°$

$120° + 360° = 480°$

$120° - 360° = -240°$

(b) Coterminal angles for $-210°$

$-210° + 360° = 150°$

$-210° - 360° = -570°$

31. (a) Complement: $90° - 18° = 72°$

Supplement: $180° - 18° = 162°$

(b) Complement: $90° - 85° = 5°$

Supplement: $180° - 85° = 95°$

33. (a) Complement: $90° - 24° = 66°$

Supplement: $180° - 24° = 156°$

(b) Complement: Not possible. $126°$ is greater than $90°$.

Supplement: $180° - 126° = 54°$

35. (a) $120° = 120°\left(\dfrac{\pi}{180°}\right) = -\dfrac{2\pi}{3}$

(b) $-20° = -20\left(\dfrac{\pi}{180°}\right) = -\dfrac{\pi}{9}$

37. (a) $\dfrac{3\pi}{2} = \dfrac{3\pi}{2}\left(\dfrac{180°}{\pi}\right) = 270°$

(b) $-\dfrac{7\pi}{6} = -\dfrac{7\pi}{6}\left(\dfrac{180°}{\pi}\right) = -210°$

39. $45° = 45\left(\dfrac{\pi}{180°}\right) \approx 0.785$ radian

41. $-0.54° = -0.54°\left(\dfrac{\pi}{180°}\right) \approx -0.009$ radian

43. $\dfrac{5\pi}{11} = \dfrac{5\pi}{11}\left(\dfrac{180°}{\pi}\right) \approx 81.818°$

45. $-4.2\pi = -4.2\pi\left(\dfrac{180°}{\pi}\right) = -756°$

47. (a) $54°45' = 54° + \left(\frac{45}{60}\right)^{\circ} = 54.75°$

(b) $-128°30' = -128^{\circ} - \left(\frac{30}{60}\right)^{\circ} = -128.5°$

49. (a) $240.6° = 240° + 0.6(60)' = 240°36'$

(b) $-145.8° = -\left[145° + 0.8(60')\right] = -145°48'$

51. $r = 15$ inches, $\theta = 120°$

$s = r\theta$

$s = 15(120°)\left(\dfrac{\pi}{180°}\right) = 10\pi$ inches

≈ 31.42 inches

53. $r = 80$ kilometers, $s = 150$ kilometers

$s = r\theta$

$150 = 80\theta$

$\theta = \frac{150}{80} = \frac{15}{8}$ radians

55. $s = r\theta$

$28 = 7\theta$

$\theta = 4$ radians

57. $r = 6$ inches, $\theta = \dfrac{\pi}{3}$

$A = \dfrac{1}{2}r^2\theta = \dfrac{1}{2}(6)^2\left(\dfrac{\pi}{3}\right) = 6\pi \text{ in.}^2 \approx 18.85 \text{ in.}^2$

59. The angle in degrees should be multiplied by $\dfrac{\pi}{180°}$.

$20° = (20°)\left(\dfrac{\pi \text{ rad}}{180°}\right) = \dfrac{\pi}{9}$ radians.

61. $\theta = 41°\,15'\,50'' - 32°\,47'\,9'' \approx 8.47806° \approx 0.14797$ radian

$s = r\theta \approx 4000(0.14782) \approx 592$ miles

63. $\theta = \dfrac{s}{r} = \dfrac{2.5}{6} = \dfrac{25}{60} = \dfrac{5}{12}$ radian $\approx 23.87°$

65. (a) 4 rpm $= 4(2\pi)$ radians/minute $= 8\pi \approx 25$ radians/minute

(b) $r = 25$ ft

$\dfrac{r\theta}{t} = 200\pi$ ft/minute

Linear speed $\approx 25(25.13)$ ft/minute ≈ 628.3 ft/minute

67. (a) Road speed (linear speed) $= \dfrac{\left(\dfrac{25}{2}\text{ in.}\right)\left(\dfrac{1\text{ ft}}{12\text{ in.}}\right)\left(\dfrac{1\text{ mi}}{5280\text{ ft}}\right)(480)(2\pi)}{1\text{ minute}\left(\dfrac{1\text{ hour}}{60\text{ minutes}}\right)} \approx 35.70$ mi/h

(b) $\dfrac{55\text{ mi}}{1\text{ h}} \times \dfrac{5280\text{ ft}}{1\text{ mi}} \times \dfrac{12\text{ in.}}{1\text{ ft}} \times \dfrac{1\text{ h}}{60\text{ min}} = \dfrac{58{,}080\text{ in.}}{1\text{ min}}$

The circumference of the machine is $C = 2\pi\left(\dfrac{25}{2}\right) = 25\pi$ inches.

The number of revolutions per minute is

$r = 58{,}080/25\pi \approx 739.50$ revolutions/min.

69. $A = \dfrac{1}{2}r^2\theta$

$= \dfrac{1}{2}(15)^2(150°)\left(\dfrac{\pi}{180°}\right)$

$= 93.75\pi$ m^2

≈ 294.52 m^2

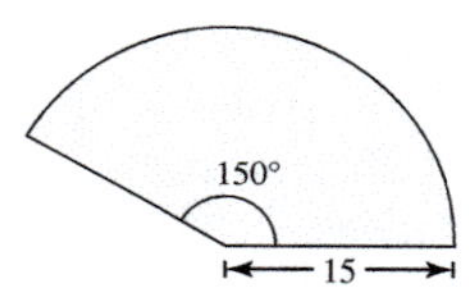

71. False. $\dfrac{180°}{\pi}$ is in degree measure.

73. True. If α and β are coterminal angles, then $\alpha = \beta + n(360°)$ or $\alpha = \beta + n(2\pi)$, where n is an integer. The difference between α and β is $\alpha - \beta = n(360°)$, or $\alpha - \beta = n(2\pi)$ if expressed in radians.

75. Since the arc length s is given by $s = r\theta$, if the central angle θ is fixed while the radius r increases, then s increases in proportion to r.

77. The speed increases. The linear speed is proportional to the radius.

79. Area of circle $= \pi r^2$

$\dfrac{\text{Area of sector}}{\text{Area of circle}} = \dfrac{\text{Measure of central angle of sector}}{\text{Measure of central angle of circle}}$

$\dfrac{\text{Area of sector}}{\pi r^2} = \dfrac{\theta}{2\pi}$

Area of sector $= \left(\pi r^2\right)\left(\dfrac{\theta}{2\pi}\right) = \dfrac{1}{2}r^2\theta$

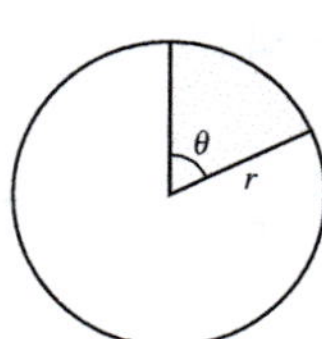

Section 4.2 Trigonometric Functions: The Unit Circle

1. unit circle

3. period

5. $x = \frac{12}{13}, y = \frac{5}{13}$

$\sin t = y = \frac{5}{13}$ $\quad \csc t = \frac{1}{y} = \frac{13}{5}$

$\cos t = x = \frac{12}{13}$ $\quad \sec t = \frac{1}{x} = \frac{13}{12}$

$\tan t = \frac{y}{x} = \frac{5}{12}$ $\quad \cot t = \frac{x}{y} = \frac{12}{5}$

7. $x = -\frac{4}{5}, y = -\frac{3}{5}$

$\sin t = y = -\frac{3}{5}$ $\quad \csc t = \frac{1}{y} = -\frac{5}{3}$

$\cos t = x = -\frac{4}{5}$ $\quad \sec t = \frac{1}{x} = -\frac{5}{4}$

$\tan t = \frac{y}{x} = \frac{3}{4}$ $\quad \cot t = \frac{x}{y} = \frac{4}{3}$

9. $t = \frac{\pi}{2}$ corresponds to the point $(x, y) = (0, 1)$.

11. $t = \frac{5\pi}{6}$ corresponds to the point $(x, y) = \left(-\frac{\sqrt{3}}{2}, \frac{1}{2}\right)$.

13. $t = \frac{\pi}{4}$ corresponds to the point $(x, y) = \left(\frac{\sqrt{2}}{2}, \frac{\sqrt{2}}{2}\right)$.

$\sin \frac{\pi}{4} = y = \frac{\sqrt{2}}{2}$

$\cos \frac{\pi}{4} = x = \frac{\sqrt{2}}{2}$

$\tan \frac{\pi}{4} = \frac{y}{x} = 1$

15. $t = -\frac{\pi}{6}$ corresponds to $\left(\frac{\sqrt{3}}{2}, -\frac{1}{2}\right)$.

$\sin -\frac{\pi}{6} = y = -\frac{1}{2}$

$\cos -\frac{\pi}{6} = x = \frac{\sqrt{3}}{2}$

$\tan -\frac{\pi}{6} = \frac{y}{x} = -\frac{1}{\sqrt{3}} = -\frac{\sqrt{3}}{3}$

17. $t = -\frac{7\pi}{4}$ corresponds to the point

$(x, y) = \left(\frac{\sqrt{2}}{2}, \frac{\sqrt{2}}{2}\right)$.

$\sin\left(-\frac{7\pi}{4}\right) = y = \frac{\sqrt{2}}{2}$

$\cos\left(-\frac{7\pi}{4}\right) = x = \frac{\sqrt{2}}{2}$

$\tan\left(-\frac{7\pi}{4}\right) = \frac{y}{x} = 1$

19. $t = \frac{11\pi}{6}$ corresponds to the point $(x, y) = \left(\frac{\sqrt{3}}{2}, -\frac{1}{2}\right)$.

$\sin \frac{11\pi}{6} = y = -\frac{1}{2}$

$\cos \frac{11\pi}{6} = x = \frac{\sqrt{3}}{2}$

$\tan \frac{11\pi}{6} = \frac{y}{x} = -\frac{1}{\sqrt{3}} = -\frac{\sqrt{3}}{3}$

21. $t = -\frac{3\pi}{2}$ corresponds to the point $(x, y) = (0, 1)$.

$\sin\left(-\frac{3\pi}{2}\right) = y = 1$

$\cos\left(-\frac{3\pi}{2}\right) = x = 0$

$\tan\left(-\frac{3\pi}{2}\right) = \frac{y}{x}$ is undefined.

23. $t = \frac{2\pi}{3}$ corresponds to the point $(x, y) = \left(-\frac{1}{2}, \frac{\sqrt{3}}{2}\right)$.

$\sin\frac{2\pi}{3} = y = \frac{\sqrt{3}}{2}$ $\qquad \csc\frac{2\pi}{3} = \frac{1}{y} = \frac{2\sqrt{3}}{3}$

$\cos\frac{2\pi}{3} = x = -\frac{1}{2}$ $\qquad \sec\frac{2\pi}{3} = \frac{1}{x} = -2$

$\tan\frac{2\pi}{3} = \frac{y}{x} = \frac{\frac{\sqrt{3}}{2}}{-\frac{1}{2}} = -\sqrt{3}$ $\qquad \cot\frac{2\pi}{3} = \frac{x}{y} = \frac{-\frac{1}{2}}{\frac{\sqrt{3}}{2}} = -\frac{\sqrt{3}}{3}$

25. $t = \frac{4\pi}{3}$ corresponds to the point $(x, y) = \left(-\frac{1}{2}, -\frac{\sqrt{3}}{2}\right)$.

$\sin\frac{4\pi}{3} = y = -\frac{\sqrt{3}}{2}$ $\qquad \csc\frac{4\pi}{3} = \frac{1}{y} = -\frac{2\sqrt{3}}{3}$

$\cos\frac{4\pi}{3} = x = -\frac{1}{2}$ $\qquad \sec\frac{4\pi}{3} = \frac{1}{x} = -2$

$\tan\frac{4\pi}{3} = \frac{y}{x} = \sqrt{3}$ $\qquad \cot\frac{4\pi}{3} = \frac{x}{y} = \frac{\sqrt{3}}{3}$

27. $t = -\frac{5\pi}{3}$ corresponds to the point $\left(\frac{1}{2}, \frac{\sqrt{3}}{2}\right)$.

$\sin\left(-\frac{5\pi}{3}\right) = y = \frac{\sqrt{3}}{2}$ $\qquad \csc\left(-\frac{5\pi}{3}\right) = \frac{1}{y} = \frac{2}{\sqrt{3}}$

$\cos\left(-\frac{5\pi}{3}\right) = x = \frac{1}{2}$ $\qquad \sec\left(-\frac{5\pi}{3}\right) = \frac{1}{x} = 2$

$\tan\left(-\frac{5\pi}{3}\right) = \frac{y}{x} = \sqrt{3}$ $\qquad \cot\left(-\frac{5\pi}{3}\right) = \frac{x}{y} = \frac{\sqrt{3}}{3}$

29. $t = -\frac{\pi}{2}$ corresponds to the point $(x, y) = (0, -1)$.

$\sin\left(-\frac{\pi}{2}\right) = y = -1$ $\qquad \csc\left(-\frac{\pi}{2}\right) = \frac{1}{y} = -1$

$\cos\left(-\frac{\pi}{2}\right) = x = 0$ $\qquad \sec\left(-\frac{\pi}{2}\right) = \frac{1}{x}$ is undefined.

$\tan\left(-\frac{\pi}{2}\right) = \frac{y}{x}$ is undefined. $\qquad \cot\left(-\frac{\pi}{2}\right) = \frac{x}{y} = 0$

31. $\sin 4\pi = \sin 0 = 0$

33. $\cos\frac{7\pi}{3} = \cos\frac{\pi}{3} = \frac{1}{2}$

35. $\sin\frac{19\pi}{6} = \sin\frac{7\pi}{6} = -\frac{1}{2}$

37. $\sin t = \frac{1}{2}$

(a) $\sin(-t) = -\sin t = -\frac{1}{2}$

(b) $\csc(-t) = -\csc t = -2$

39. $\cos(-t) = -\frac{1}{5}$

(a) $\cos t = \cos(-t) = -\frac{1}{5}$

(b) $\sec(-t) = \frac{1}{\cos(-t)} = -5$

41. $\sin t = \frac{4}{5}$

(a) $\sin(\pi - t) = \sin t = \frac{4}{5}$

(b) $\sin(t + \pi) = -\sin t = -\frac{4}{5}$

43. $\sin 0.6 \approx 0.5646$

45. $\tan\frac{\pi}{8} \approx 0.4142$

47. $\sec 3.1 = \frac{1}{\cos 3.1} \approx -1.0009$

49. $y(t) = \frac{1}{2}\cos 6t$

(a) $y(0) = \frac{1}{2}\cos 0 = 0.5$ foot

(b) $y\left(\frac{1}{4}\right) = \frac{1}{2}\cos\frac{3}{2} \approx 0.04$ foot

(c) $y\left(\frac{1}{2}\right) = \frac{1}{2}\cos 3 \approx -0.49$ foot

51. False. $\sin(-t) = -\sin t$ means the function is odd, not that the sine of a negative angle is a negative number.

For example: $\sin\left(-\frac{3\pi}{2}\right) = -\sin\left(\frac{3\pi}{2}\right) = -(-1) = 1.$

Even though the angle is negative, the sine value is positive.

53. True. $\tan a = \tan(a - 6\pi)$ because the period of the tangent function is π.

55. (a) The points have y-axis symmetry.

(b) $\sin t_1 = \sin(\pi - t_1)$ because they have the same y-value.

(c) $\cos(\pi - t_1) = -\cos t_1$ because the x-values have the opposite signs.

57. The calculator was in degree mode instead of radian mode. $\tan(\pi/2)$ is undefined.

59. (a)

Circle of radius 1 centered at $(0, 0)$

(b) The t-values represent the central angle in radians. The x- and y-values represent the location in the coordinate plane.

(c) $-1 \le x \le 1, -1 \le y \le 1$

61. Let $h(t) = f(t)g(t) = \sin t \cos t.$

Then, $h(-t) = \sin(-t)\cos(-t)$
$= -\sin t \cos t$
$= -h(t).$

So, $h(t)$ is odd.

Section 4.3 Right Triangle Trigonometry

1. (a) $\frac{\text{opposite}}{\text{hypotenuse}} = \sin\theta$ (v) (b) $\frac{\text{adjacent}}{\text{hypotenuse}} = \cos\theta$ (iv) (c) $\frac{\text{opposite}}{\text{adjacent}} = \tan\theta$ (vi)

(d) $\frac{\text{hypotenuse}}{\text{opposite}} = \csc\theta$ (iii) (e) $\frac{\text{hypotenuse}}{\text{adjacent}} = \sec\theta$ (i) (f) $\frac{\text{adjacent}}{\text{opposite}} = \cot\theta$ (ii)

3. Complementary

5. $\text{hyp} = \sqrt{6^2 + 8^2} = \sqrt{36 + 64} = \sqrt{100} = 10$

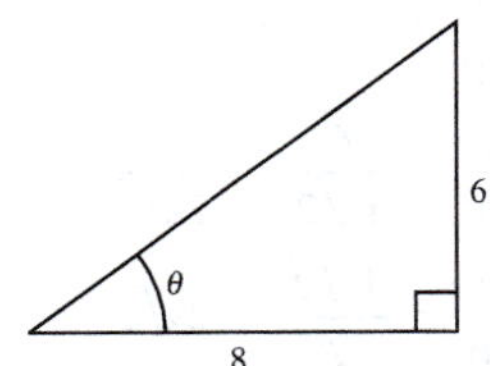

$\sin\theta = \frac{\text{opp}}{\text{hyp}} = \frac{6}{10} = \frac{3}{5}$ $\csc\theta = \frac{\text{hyp}}{\text{opp}} = \frac{10}{6} = \frac{5}{3}$

$\cos\theta = \frac{\text{adj}}{\text{hyp}} = \frac{8}{10} = \frac{4}{5}$ $\sec\theta = \frac{\text{hyp}}{\text{adj}} = \frac{10}{8} = \frac{5}{4}$

$\tan\theta = \frac{\text{opp}}{\text{adj}} = \frac{6}{8} = \frac{3}{4}$ $\cot\theta = \frac{\text{adj}}{\text{opp}} = \frac{8}{6} = \frac{4}{3}$

7. $\text{adj} = \sqrt{41^2 - 9^2} = \sqrt{1681 - 81} = \sqrt{1600} = 40$

$$\sin\theta = \frac{\text{opp}}{\text{hyp}} = \frac{9}{41} \qquad \csc\theta = \frac{\text{hyp}}{\text{opp}} = \frac{41}{9}$$

$$\cos\theta = \frac{\text{adj}}{\text{hyp}} = \frac{40}{41} \qquad \sec\theta = \frac{\text{hyp}}{\text{adj}} = \frac{41}{40}$$

$$\tan\theta = \frac{\text{opp}}{\text{adj}} = \frac{9}{40} \qquad \cot\theta = \frac{\text{adj}}{\text{opp}} = \frac{40}{9}$$

9. $\text{hyp} = \sqrt{4^2 + 4^2} = \sqrt{32} = 4\sqrt{2}$

$$\sin\theta = \frac{\text{opp}}{\text{hyp}} = \frac{4}{4\sqrt{2}} = \frac{1}{\sqrt{2}} = \frac{\sqrt{2}}{2} \qquad \csc\theta = \frac{\text{hyp}}{\text{opp}} = \frac{4\sqrt{2}}{4} = \sqrt{2}$$

$$\cos\theta = \frac{\text{adj}}{\text{hyp}} = \frac{4}{4\sqrt{2}} = \frac{1}{\sqrt{2}} = \frac{\sqrt{2}}{2} \qquad \sec\theta = \frac{\text{hyp}}{\text{adj}} = \frac{4\sqrt{2}}{4} = \sqrt{2}$$

$$\tan\theta = \frac{\text{opp}}{\text{adj}} = \frac{4}{4} = 1 \qquad \cot\theta = \frac{\text{adj}}{\text{opp}} = \frac{4}{4} = 1$$

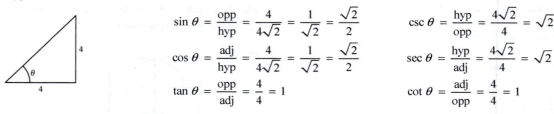

11. $\text{hyp} = \sqrt{15^2 + 8^2} = \sqrt{289} = 17$

$$\sin\theta = \frac{\text{opp}}{\text{hyp}} = \frac{8}{17} \qquad \csc\theta = \frac{\text{hyp}}{\text{opp}} = \frac{17}{8}$$

$$\cos\theta = \frac{\text{adj}}{\text{hyp}} = \frac{15}{17} \qquad \sec\theta = \frac{\text{hyp}}{\text{adj}} = \frac{17}{15}$$

$$\tan\theta = \frac{\text{opp}}{\text{adj}} = \frac{8}{15} \qquad \cot\theta = \frac{\text{adj}}{\text{opp}} = \frac{15}{8}$$

$\text{hyp} = \sqrt{7.5^2 + 4^2} = \frac{17}{2}$

$$\sin\theta = \frac{\text{opp}}{\text{hyp}} = \frac{4}{(17/2)} = \frac{8}{17} \qquad \csc\theta = \frac{\text{hyp}}{\text{opp}} = \frac{(17/2)}{4} = \frac{17}{8}$$

$$\cos\theta = \frac{\text{adj}}{\text{hyp}} = \frac{7.5}{(17/2)} = \frac{15}{17} \qquad \sec\theta = \frac{\text{hyp}}{\text{adj}} = \frac{(17/2)}{7.5} = \frac{17}{15}$$

$$\tan\theta = \frac{\text{opp}}{\text{adj}} = \frac{4}{7.5} = \frac{8}{15} \qquad \cot\theta = \frac{\text{adj}}{\text{opp}} = \frac{7.5}{4} = \frac{15}{8}$$

The function values are the same because the triangles are similar, and corresponding sides are proportional.

13. $\text{adj} = \sqrt{3^2 - 1^2} = \sqrt{8} = 2\sqrt{2}$

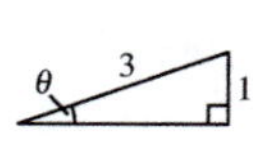

$$\sin\theta = \frac{\text{opp}}{\text{hyp}} = \frac{1}{3} \qquad \csc\theta = \frac{\text{hyp}}{\text{opp}} = 3$$

$$\cos\theta = \frac{\text{adj}}{\text{hyp}} = \frac{2\sqrt{2}}{3} \qquad \sec\theta = \frac{\text{hyp}}{\text{adj}} = \frac{3}{2\sqrt{2}} = \frac{3\sqrt{2}}{4}$$

$$\tan\theta = \frac{\text{opp}}{\text{adj}} = \frac{1}{2\sqrt{2}} = \frac{\sqrt{2}}{4} \qquad \cot\theta = \frac{\text{adj}}{\text{opp}} = 2\sqrt{2}$$

$\text{adj} = \sqrt{6^2 - 2^2} = \sqrt{32} = 4\sqrt{2}$

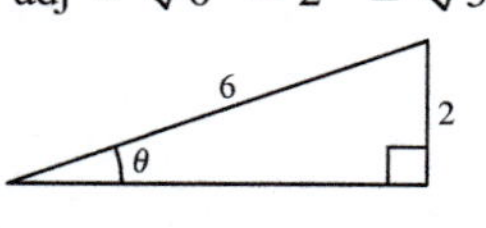

$$\sin\theta = \frac{\text{opp}}{\text{hyp}} = \frac{2}{6} = \frac{1}{3} \qquad \csc\theta = \frac{\text{hyp}}{\text{opp}} = \frac{6}{2} = 3$$

$$\cos\theta = \frac{\text{adj}}{\text{hyp}} = \frac{4\sqrt{2}}{6} = \frac{2\sqrt{2}}{3} \qquad \sec\theta = \frac{\text{hyp}}{\text{adj}} = \frac{6}{4\sqrt{2}} = \frac{3}{2\sqrt{2}} = \frac{3\sqrt{2}}{4}$$

$$\tan\theta = \frac{\text{opp}}{\text{adj}} = \frac{5\sqrt{2}}{4} = \frac{1}{2\sqrt{2}} = \frac{\sqrt{2}}{4} \qquad \cot\theta = \frac{\text{adj}}{\text{opp}} = \frac{4\sqrt{2}}{2} = 2\sqrt{2}$$

The function values are the same since the triangles are similar and the corresponding sides are proportional.

15. Given: $\cos\theta = \dfrac{15}{17} = \dfrac{\text{adj}}{\text{hyp}}$

$$(\text{opp})^2 + 15^2 = 17^2$$
$$\text{opp} = \sqrt{289 - 225}$$
$$\text{opp} = \sqrt{64} = 8$$

$$\sin\theta = \frac{\text{opp}}{\text{hyp}} = \frac{8}{17}$$
$$\tan\theta = \frac{\text{opp}}{\text{adj}} = \frac{8}{15}$$
$$\csc\theta = \frac{\text{hyp}}{\text{opp}} = \frac{17}{8}$$
$$\sec\theta = \frac{\text{hyp}}{\text{adj}} = \frac{17}{15}$$
$$\cot\theta = \frac{\text{adj}}{\text{opp}} = \frac{15}{8}$$

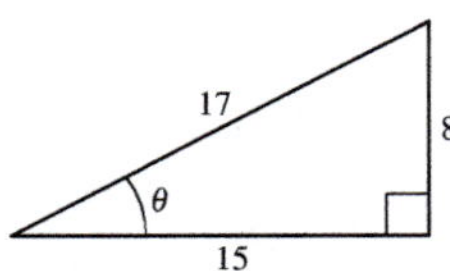

17. Given: $\sec\theta = \dfrac{6}{5} = \dfrac{\text{hyp}}{\text{adj}}$

$$(\text{opp})^2 + 5^2 = 6^2$$
$$\text{opp} = \sqrt{36 - 25} = \sqrt{11}$$

$$\sin\theta = \frac{\text{opp}}{\text{hyp}} = \frac{\sqrt{11}}{6}$$
$$\cos\theta = \frac{\text{adj}}{\text{hyp}} = \frac{5}{6}$$
$$\tan\theta = \frac{\text{opp}}{\text{adj}} = \frac{\sqrt{11}}{5}$$
$$\csc\theta = \frac{\text{hyp}}{\text{opp}} = \frac{6\sqrt{11}}{11}$$
$$\cot\theta = \frac{\text{adj}}{\text{opp}} = \frac{5\sqrt{11}}{11}$$

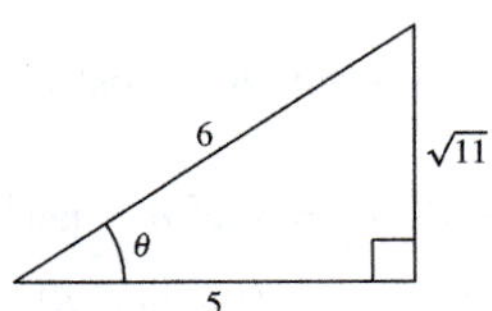

19. Given: $\sin\theta = \dfrac{1}{5} = \dfrac{\text{opp}}{\text{hyp}}$

$$1^2 + (\text{adj})^2 = 5^2$$
$$\text{adj} = \sqrt{24} = 2\sqrt{6}$$

$$\cos\theta = \frac{\text{adj}}{\text{hyp}} = \frac{2\sqrt{6}}{5}$$
$$\tan\theta = \frac{\text{opp}}{\text{adj}} = \frac{\sqrt{6}}{12}$$
$$\csc\theta = \frac{\text{hyp}}{\text{opp}} = 5$$
$$\sec\theta = \frac{\text{hyp}}{\text{adj}} = \frac{5\sqrt{6}}{12}$$
$$\cot\theta = \frac{\text{adj}}{\text{opp}} = 2\sqrt{6}$$

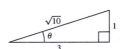

21. Given: $\cot\theta = 3 = \dfrac{3}{1} = \dfrac{\text{adj}}{\text{opp}}$

$$1^2 + 3^2 = (\text{hyp})^2$$
$$\text{hyp} = \sqrt{10}$$

$$\sin\theta = \frac{\text{opp}}{\text{hyp}} = \frac{\sqrt{10}}{10}$$
$$\cos\theta = \frac{\text{adj}}{\text{hyp}} = \frac{3\sqrt{10}}{10}$$
$$\tan\theta = \frac{\text{opp}}{\text{adj}} = \frac{1}{3}$$
$$\csc\theta = \frac{\text{hyp}}{\text{opp}} = \sqrt{10}$$
$$\sec\theta = \frac{\text{hyp}}{\text{adj}} = \frac{\sqrt{10}}{3}$$

23.

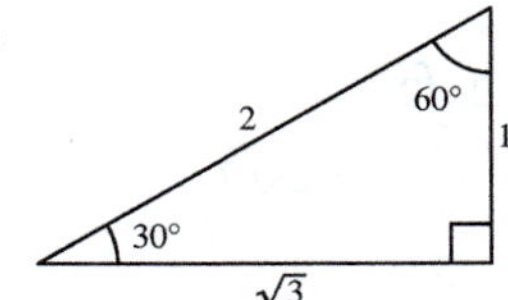

$$30° = 30°\left(\frac{\pi}{180°}\right) = \frac{\pi}{6} \text{ radian}$$

$$\tan 30° = \frac{\text{opp}}{\text{adj}} = \frac{1}{\sqrt{3}} = \frac{\sqrt{3}}{3}$$

25. 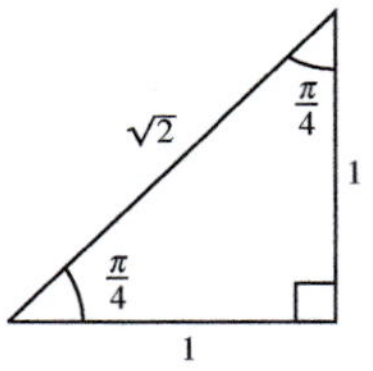

$$\frac{\pi}{4} = \frac{\pi}{4}\left(\frac{180°}{\pi}\right) = 45°$$

$$\sin\frac{\pi}{4} = \frac{\text{opp}}{\text{hyp}} = \frac{1}{\sqrt{2}} = \frac{\sqrt{2}}{2}$$

27. 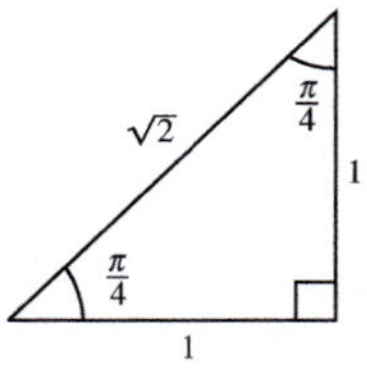

$$\frac{\pi}{4} = \frac{\pi}{4}\left(\frac{180°}{\pi}\right) = 45°$$

$$\sec\frac{\pi}{4} = \frac{\text{hyp}}{\text{adj}} = \frac{\sqrt{2}}{1} = \sqrt{2}$$

29. (a) $\sin 20° \approx 0.3420$

(b) $\cos 70° \approx 0.3420$

31. (a) $\sin 14.21° \approx 0.2455$

(b) $\csc 14.21° = \dfrac{1}{\sin 14.21°} \approx 4.0737$

33. (a) $\cos 4° 50' 15'' = \cos\left(4 + \dfrac{50}{60} + \dfrac{15}{3600}\right)^{\circ} \approx 0.9964$

(b) $\sec 4° 50' 15'' = \dfrac{1}{\cos 4° 50' 15''} \approx 1.0036$

35. (a) $\cot 17° 15' = \cot\left(17 + \dfrac{15}{60}\right)^{\circ} = \dfrac{1}{\tan 17.25°} \approx 3.2205$

(b) $\tan 17° 15' = \tan\left(17 + \dfrac{15}{60}\right)^{\circ} = \tan 17.25° \approx 0.3105$

37. $\sin 60° = \dfrac{\sqrt{3}}{2}$, $\cos 60° = \dfrac{1}{2}$

(a) $\sin 30° = \cos 60° = \dfrac{1}{2}$

(b) $\cos 30° = \sin 60° = \dfrac{\sqrt{3}}{2}$

(c) $\tan 60° = \dfrac{\sin 60°}{\cos 60°} = \sqrt{3}$

(d) $\cot 60° = \dfrac{\cos 60°}{\sin 60°} = \dfrac{1}{\sqrt{3}} = \dfrac{\sqrt{3}}{3}$

39. $\cos\theta = \dfrac{1}{3}$

(a)
$$\sin^2\theta + \cos^2\theta = 1$$
$$\sin^2\theta + \left(\frac{1}{3}\right)^2 = 1$$
$$\sin^2\theta = \frac{8}{9}$$
$$\sin\theta = \frac{2\sqrt{2}}{3}$$

(b) $\tan\theta = \dfrac{\sin\theta}{\cos\theta} = \dfrac{\frac{2\sqrt{2}}{3}}{\frac{1}{3}} = 2\sqrt{2}$

(c) $\sec\theta = \dfrac{1}{\cos\theta} = 3$

(d) $\csc(90° - \theta) = \sec\theta = 3$

41. $\cot\alpha = 3$

(a) $\tan\alpha = \dfrac{1}{\cot\alpha} = \dfrac{1}{3}$

(b)
$$\csc^2\alpha = 1 + \cot^2\alpha$$
$$\csc^2\alpha = 1 + 3^2$$
$$\csc^2\alpha = 10$$
$$\csc\alpha = \sqrt{10}$$

(c) $\cot(90° - \alpha) = \tan\alpha = \dfrac{1}{3}$

(d) $\csc\alpha = \sqrt{10}$

$$\sin\alpha = \frac{1}{\csc\alpha} = \frac{1}{\sqrt{10}} = \frac{\sqrt{10}}{10}$$

43. $\tan\theta\cot\theta = \tan\theta\left(\dfrac{1}{\tan\theta}\right) = 1$

45. $\tan\alpha\cos\alpha = \left(\dfrac{\sin\alpha}{\cos\alpha}\right)\cos\alpha = \sin\alpha$

47. $(1 + \sin\theta)(1 - \sin\theta) = 1 - \sin^2\theta = \cos^2\theta$

49.
$$\begin{aligned}(\sec\theta + \tan\theta)(\sec\theta - \tan\theta) &= \sec^2\theta - \tan^2\theta \\ &= (1 + \tan^2\theta) - \tan^2\theta \\ &= 1\end{aligned}$$

51. $$\frac{\sin\theta}{\cos\theta} + \frac{\cos\theta}{\sin\theta} = \frac{\sin^2\theta + \cos^2\theta}{\sin\theta\cos\theta}$$
$$= \frac{1}{\sin\theta\cos\theta}$$
$$= \frac{1}{\sin\theta}\cdot\frac{1}{\cos\theta}$$
$$= \csc\theta\sec\theta$$

53. (a) $\sin\theta = \frac{1}{2} \Rightarrow \theta = 30° = \frac{\pi}{6}$

(b) $\csc\theta = 2 \Rightarrow \theta = 30° = \frac{\pi}{6}$

55. (a) $\sec\theta = 2 \Rightarrow \theta = 60° = \frac{\pi}{3}$

(b) $\cot\theta = 1 \Rightarrow \theta = 45° = \frac{\pi}{4}$

57. (a) $\csc\theta = \frac{2\sqrt{3}}{3} \Rightarrow \theta = 60° = \frac{\pi}{3}$

(b) $\sin\theta = \frac{\sqrt{2}}{2} \Rightarrow \theta = 45° = \frac{\pi}{4}$

59. $$\cos 60° = \frac{x}{18}$$
$$x = 18\cos 60° = 18\left(\frac{1}{2}\right) = 9$$
$$\sin 60° = \frac{y}{18}$$
$$y = 18\sin 60° = 18\frac{\sqrt{3}}{2} = 9\sqrt{3}$$

61. $$\tan 60° = \frac{32}{x}$$
$$\sqrt{3} = \frac{32}{x}$$
$$\sqrt{3}x = 32$$
$$x = \frac{32}{\sqrt{3}} = \frac{32\sqrt{3}}{3}$$
$$\sin 60° = \frac{32}{r}$$
$$r = \frac{32}{\sin 60°}$$
$$r = \frac{32}{\frac{\sqrt{3}}{2}} = \frac{64\sqrt{3}}{3}$$

63. $$\tan 82° = \frac{x}{45}$$
$$x = 45\tan 82°$$

Height of the building:

$123 + 45\tan 82° \approx 443.2$ meters

Distance between friends:

$$\cos 82° = \frac{45}{y} \Rightarrow y = \frac{45}{\cos 82°}$$
$$\approx 323.34 \text{ meters}$$

65. $$\sin\theta = \frac{1250}{2500} = \frac{1}{2}$$
$$\theta = 30° = \frac{\pi}{6}$$

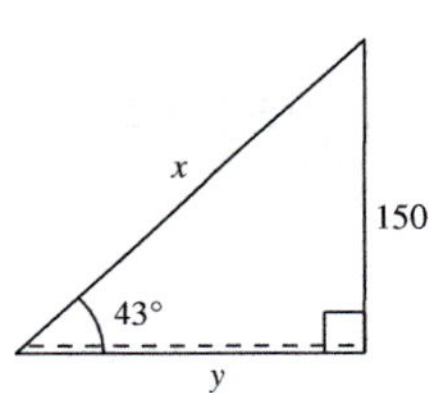

67. (a) $$\sin 43° = \frac{150}{x}$$
$$x = \frac{150}{\sin 43°} \approx 219.9 \text{ ft}$$

(b) $$\tan 43° = \frac{150}{y}$$
$$y = \frac{150}{\tan 43°} \approx 160.9 \text{ ft}$$

69.

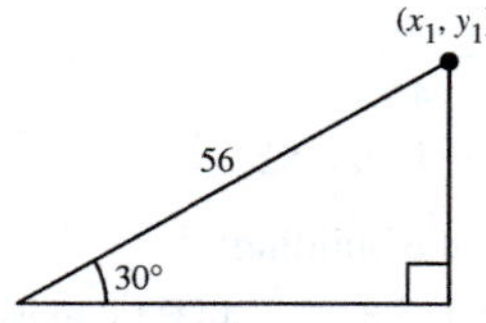

$\sin 30° = \frac{y_1}{56}$

$y_1 = (\sin 30°)(56) = \left(\frac{1}{2}\right)(56) = 28$

$\cos 30° = \frac{x_1}{56}$

$x_1 = \cos 30°(56) = \frac{\sqrt{3}}{2}(56) = 28\sqrt{3}$

$(x_1, y_1) = \left(28\sqrt{3}, 28\right)$

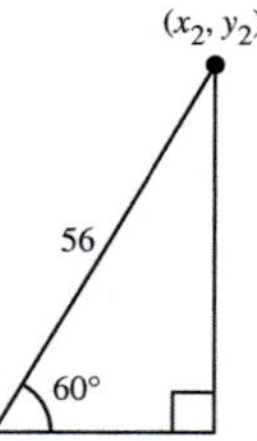

$\sin 60° = \frac{y_2}{56}$

$y_2 = \sin 60°(56) = \left(\frac{\sqrt{3}}{2}\right)(56) = 28\sqrt{3}$

$\cos 60° = \frac{x_2}{56}$

$x_2 = (\cos 60°)(56) = \left(\frac{1}{2}\right)(56) = 28$

$(x_2, y_2) = \left(28, 28\sqrt{3}\right)$

71. $x \approx 9.397, y \approx 3.420$

$\sin 20° = \frac{y}{10} \approx 0.34$

$\cos 20° = \frac{x}{10} \approx 0.94$

$\tan 20° = \frac{y}{x} \approx 0.36$

$\cot 20° = \frac{x}{y} \approx 2.75$

$\sec 20° = \frac{10}{x} \approx 1.06$

$\csc 20° = \frac{10}{y} \approx 2.92$

73. (a)

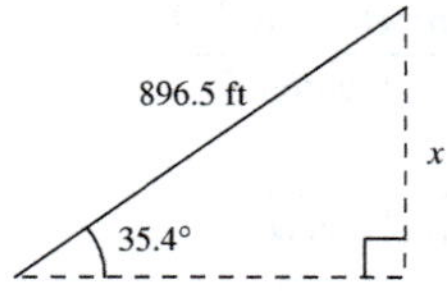

$\sin 35.4° = \frac{x}{896.5}$

$x = 896.5 \sin 35.4° \approx 519.33$ feet

(b) Because the top of the incline is 1693.5 feet above sea level and the vertical rise of the inclined plane is 519.33 feet, the elevation of the lower end of the inclined plan is about $1693.5 - 519.33 = 1174.17$ feet.

(c) Ascent time: $d = rt$

$896.5 = 300t$

$3 \approx t$

It takes about 3 minutes for the cars to get from the bottom to the top.

Vertical rate: $d = rt$

$519.33 \approx r(3)$

$r \approx 173.11$ ft/min

75. $\sin 60° \csc 60° = 1$

True.

$\csc x = \frac{1}{\sin x} \Rightarrow \sin 60° \csc 60° = \sin 60°\left(\frac{1}{\sin 60°}\right)$

$= 1$

77. False, $\frac{\sqrt{2}}{2} + \frac{\sqrt{2}}{2} = \sqrt{2} \neq 1$

79. False, $\frac{\sin 60°}{\sin 30°} = \frac{\cos 30°}{\sin 30°} = \cot 30° \approx 1.7321$

$\sin 2° \approx 0.0349$

81. Yes. Given $\tan\theta$, $\sec\theta$ can be found from the identity $1 + \tan^2\theta = \sec^2\theta$.

83.

θ	0.1	0.2	0.3	0.4	0.5
$\sin\theta$	0.0998	0.1987	0.2955	0.3894	0.4794

(a) In the interval $(0, 0.5]$, $\theta > \sin\theta$.

(b) As $\theta \to 0$, $\sin\theta \to 0$, and $\dfrac{\theta}{\sin\theta} \to 1$.

Section 4.4 Trigonometric Functions of Any Angle

1. $\dfrac{y}{r}$

3. $\dfrac{y}{x}$

5. $\cos\theta$

7. zero; defined

9. (a) $(x, y) = (4, 3)$

$r = \sqrt{16 + 9} = 5$

$\sin\theta = \dfrac{y}{r} = \dfrac{3}{5}$ $\quad\csc\theta = \dfrac{r}{y} = \dfrac{5}{3}$

$\cos\theta = \dfrac{x}{r} = \dfrac{4}{5}$ $\quad\sec\theta = \dfrac{r}{x} = \dfrac{5}{4}$

$\tan\theta = \dfrac{y}{x} = \dfrac{3}{4}$ $\quad\cot\theta = \dfrac{x}{y} = \dfrac{4}{3}$

(b) $(x, y) = (-8, 15)$

$r = \sqrt{64 + 225} = 17$

$\sin\theta = \dfrac{y}{r} = \dfrac{15}{17}$ $\quad\csc\theta = \dfrac{r}{y} = \dfrac{17}{15}$

$\cos\theta = \dfrac{x}{r} = -\dfrac{8}{17}$ $\quad\sec\theta = \dfrac{r}{x} = -\dfrac{17}{8}$

$\tan\theta = \dfrac{y}{x} = -\dfrac{15}{8}$ $\quad\cot\theta = \dfrac{x}{y} = -\dfrac{8}{15}$

11. (a) $(x, y) = \left(-\sqrt{3}, -1\right)$

$r = \sqrt{3 + 1} = 2$

$\sin\theta = \dfrac{y}{r} = -\dfrac{1}{2}$ $\quad\csc\theta = \dfrac{r}{y} = -2$

$\cos\theta = \dfrac{x}{r} = -\dfrac{\sqrt{3}}{2}$ $\quad\sec\theta = \dfrac{r}{x} = -\dfrac{2\sqrt{3}}{3}$

$\tan\theta = \dfrac{y}{x} = \dfrac{\sqrt{3}}{3}$ $\quad\cot\theta = \dfrac{x}{y} = \sqrt{3}$

(b) $(x, y) = (4, -1)$

$r = \sqrt{16 + 1} = \sqrt{17}$

$\sin\theta = \dfrac{y}{r} = -\dfrac{1}{\sqrt{17}} = -\dfrac{\sqrt{17}}{17}$ $\quad\csc\theta = \dfrac{r}{y} = -\sqrt{17}$

$\cos\theta = \dfrac{x}{r} = \dfrac{4}{\sqrt{17}} = \dfrac{4\sqrt{17}}{17}$ $\quad\sec\theta = \dfrac{r}{x} = \dfrac{\sqrt{17}}{4}$

$\tan\theta = \dfrac{y}{x} = -\dfrac{1}{4}$ $\quad\cot\theta = \dfrac{x}{y} = -4$

13. $(x, y) = (5, 12)$

$r = \sqrt{25 + 144} = 13$

$\sin\theta = \dfrac{y}{r} = \dfrac{12}{13}$ $\quad\csc\theta = \dfrac{r}{y} = \dfrac{13}{12}$

$\cos\theta = \dfrac{x}{r} = \dfrac{5}{13}$ $\quad\sec\theta = \dfrac{r}{x} = \dfrac{13}{5}$

$\tan\theta = \dfrac{y}{x} = \dfrac{12}{5}$ $\quad\cot\theta = \dfrac{x}{y} = \dfrac{5}{12}$

15. $x = -5, y = -2$

$r = \sqrt{(-5)^2 + (-2)^2} = \sqrt{29}$

$\sin\theta = \dfrac{y}{r} = \dfrac{-2}{\sqrt{29}} = -\dfrac{2\sqrt{29}}{29}$

$\cos\theta = \dfrac{x}{r} = \dfrac{-5}{\sqrt{29}} = -\dfrac{5\sqrt{29}}{29}$

$\tan\theta = \dfrac{y}{x} = \dfrac{-2}{-5} = \dfrac{2}{5}$

$\csc\theta = \dfrac{r}{y} = \dfrac{\sqrt{29}}{-2} = -\dfrac{\sqrt{29}}{2}$

$\sec\theta = \dfrac{r}{x} = \dfrac{\sqrt{29}}{-5} = -\dfrac{\sqrt{29}}{5}$

$\cot\theta = \dfrac{x}{y} = \dfrac{-5}{-2} = \dfrac{5}{2}$

17. $(x, y) = (-5.4, 7.2)$

$r = \sqrt{29.16 + 51.84} = 9$

$\sin\theta = \frac{y}{r} = \frac{7.2}{9} = \frac{4}{5}$ $\qquad \csc\theta = \frac{r}{y} = \frac{9}{7.2} = \frac{5}{4}$

$\cos\theta = \frac{x}{r} = -\frac{5.4}{9} = -\frac{3}{5}$ $\qquad \sec\theta = \frac{r}{x} = -\frac{9}{5.4} = -\frac{5}{3}$

$\tan\theta = \frac{y}{x} = -\frac{7.2}{5.4} = -\frac{4}{3}$ $\qquad \tan\theta = \frac{x}{y} = -\frac{5.4}{7.2} = -\frac{3}{4}$

19. $\sin\theta > 0 \Rightarrow \theta$ lies in Quadrant I or in Quadrant II.

$\cos\theta > 0 \Rightarrow \theta$ lies in Quadrant I or in Quadrant IV.

$\sin\theta > 0$ and $\cos\theta > 0 \Rightarrow \theta$ lies in Quadrant I.

21. $\sin\theta > 0 \Rightarrow \theta$ lies in Quadrant I or in Quadrant II.

$\cos\theta < 0 \Rightarrow \theta$ lies in Quadrant II or in Quadrant III.

$\sin\theta > 0$ and $\cos\theta < 0 \Rightarrow \theta$ lies in Quadrant II.

23. $\tan\theta > 0$ and $\sin\theta > 0 \Rightarrow \theta$ is in Quadrant I $\Rightarrow x > 0$ and $y > 0$.

$\tan\theta = \frac{y}{x} = \frac{15}{8} \Rightarrow r = 17$

$\sin\theta = \frac{y}{r} = \frac{15}{7}$ $\qquad \sec\theta = \frac{r}{x} = \frac{17}{8}$

$\cos\theta = \frac{x}{r} = \frac{8}{17}$ $\qquad \cot\theta = \frac{x}{y} = \frac{8}{15}$

$\csc\theta = \frac{r}{y} = \frac{17}{15}$

25. $\sin\theta = 0.6 = \frac{3}{5}$ and θ in Quadrant II

$x^2 + y^2 = r^2$

$\sin\theta = \frac{3}{5}$, $\qquad x = -\sqrt{r^2 - y^2}$

$x = -\sqrt{5^2 - 3^2} = -4$

$\cos\theta = \frac{x}{r} = -\frac{4}{5}$ $\qquad \sec\theta = \frac{r}{x} = -\frac{5}{4}$

$\tan\theta = \frac{y}{x} = -\frac{3}{4}$ $\qquad \cot\theta = \frac{x}{y} = -\frac{4}{3}$

$\csc\theta = \frac{r}{y} = \frac{5}{3}$

27. $\cot\theta = \frac{x}{y} = -\frac{3}{1} = \frac{3}{-1}$

$\cos\theta > 0 \Rightarrow \theta$ is in Quadrant IV $\Rightarrow x$ is positive;

$x = 3, y = -1, r = \sqrt{10}$

$\sin\theta = \frac{y}{r} = -\frac{\sqrt{10}}{10}$ $\qquad \csc\theta = \frac{r}{y} = -\sqrt{10}$

$\cos\theta = \frac{x}{r} = \frac{3\sqrt{10}}{10}$ $\qquad \sec\theta = \frac{r}{x} = \frac{\sqrt{10}}{3}$

$\tan\theta = \frac{y}{x} = -\frac{1}{3}$

29. $\cos\theta = 0 \Rightarrow \theta = \frac{\pi}{2} + \pi n$

$\csc\theta = 1 \Rightarrow \theta = \frac{\pi}{2} + 2\pi n$

$y = 1, x = 0, r = 1$

$\sin\theta = \frac{y}{r} = 1$

$\tan\theta = \frac{y}{x}$ is undefined

$\sec\theta$ is undefined

$\cot\theta = 0.$

31. $\cot\theta$ is undefined,

$\frac{\pi}{2} \le \theta \le \frac{3\pi}{2} \Rightarrow y = 0 \Rightarrow \theta = \pi$

$\sin\theta = 0$ $\qquad \csc\theta$ is undefined.

$\cos\theta = -1$ $\qquad \sec\theta = -1$

$\tan\theta = 0$ $\qquad \cot\theta$ is undefined.

33. To find a point on the terminal side of θ, use any point on the line $y = -x$ that lies in Quadrant II. $(-1, 1)$ is one such point.

$x = -1, y = 1, r = \sqrt{2}$

$\sin\theta = \frac{1}{\sqrt{2}} = \frac{\sqrt{2}}{2}$ $\qquad \csc\theta = \sqrt{2}$

$\cos\theta = -\frac{1}{\sqrt{2}} = -\frac{\sqrt{2}}{2}$ $\qquad \sec\theta = -\sqrt{2}$

$\tan\theta = -1$ $\qquad \cot\theta = -1$

35. To find a point on the terminal side of θ, use any point on the line $y = 2x$ that lies in Quadrant I. (1, 2) is one such point.

$x = 1, y = 2, y = \sqrt{5}$

$\sin\theta = \dfrac{2}{\sqrt{5}} = \dfrac{2\sqrt{5}}{5}$ $\qquad \csc\theta = \dfrac{\sqrt{5}}{2} = \dfrac{\sqrt{5}}{2}$

$\cos\theta = \dfrac{1}{\sqrt{5}} = \dfrac{\sqrt{5}}{5}$ $\qquad \sec\theta = \dfrac{\sqrt{5}}{1} = \sqrt{5}$

$\tan\theta = \dfrac{2}{1} = 2$ $\qquad \cot\theta = \dfrac{1}{2}$

37. $(x, y) = (-1, 0), r = 1$

$\sin\pi = \dfrac{y}{r} = \dfrac{0}{1} = 0$

39. $(x, y) = (0, -1), r = 1$

$\sec\dfrac{3\pi}{2} = \dfrac{r}{x} = \dfrac{1}{0} \Rightarrow$ undefined

41. $(x, y) = (0, 1), r = 1$

$\sin\dfrac{\pi}{2} = \dfrac{y}{r} = \dfrac{1}{1} = 1$

43. $(x, y) = (-1, 0), r = 1$

$\csc\pi = \dfrac{r}{y} = \dfrac{1}{0} \Rightarrow$ undefined

45. $(x, y) = (0, 1)$

$\cot\dfrac{9\pi}{2} = \dfrac{x}{y} = \dfrac{0}{1} = 0$

47. $\theta = 160°$

$\theta' = 180° - 160° = 20°$

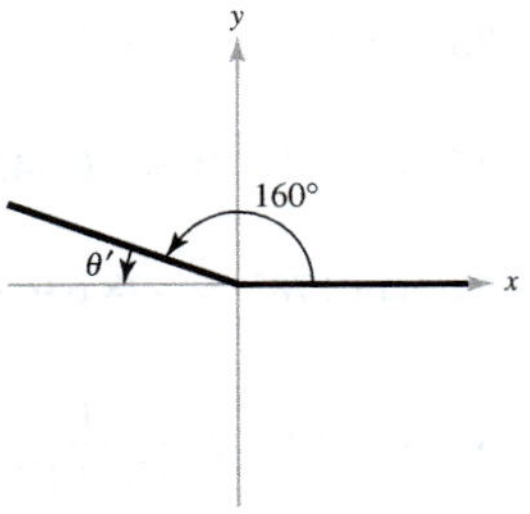

y
160°
θ′
x

49. $\theta = -125°$

$360° - 125° = 235°$ (coterminal angle)

$\theta' = 235° - 180° = 55°$

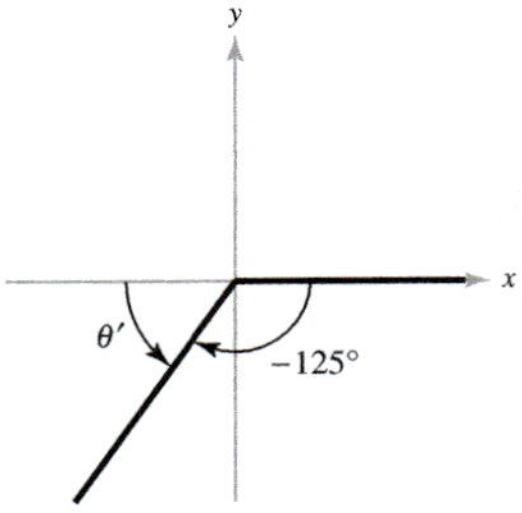

y
θ′
−125°
x

51. $\theta = \dfrac{2\pi}{3}$

$\theta' = \pi - \dfrac{2\pi}{3} = \dfrac{\pi}{3}$

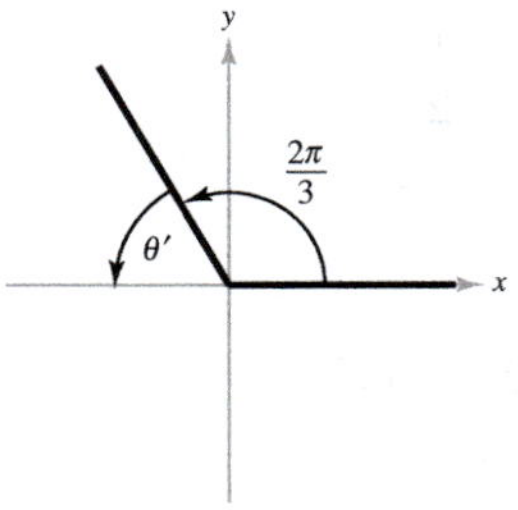

y
$\frac{2\pi}{3}$
θ′
x

53. $\theta = 4.8$

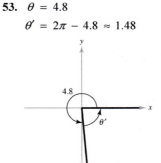

$\theta' = 2\pi - 4.8 \approx 1.48$

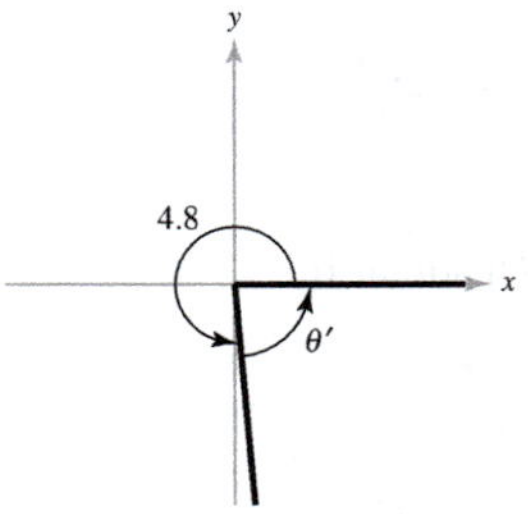

y
4.8
θ′
x

55. $\theta = 225°, \theta' = 45°$, Quadrant III

$\sin 225° = -\sin 45° = -\dfrac{\sqrt{2}}{2}$

$\cos 225° = -\cos 45° = -\dfrac{\sqrt{2}}{2}$

$\tan 225° = \tan 45° = 1$

57. $\theta = 750°, \theta' = 30°$, Quadrant I

$\sin 750° = \sin 30° = \dfrac{1}{2}$

$\cos 750° = \cos 30° = \dfrac{\sqrt{3}}{2}$

$\tan 750° = \tan 30° = \dfrac{\sqrt{3}}{3}$

59. $\theta = -120°, \theta' = 60°$, Quadrant III

$\sin(-120°) = -\sin 60° = -\dfrac{\sqrt{3}}{2}$

$\cos(-120°) = -\cos 60° = -\dfrac{1}{2}$

$\tan(-120°) = \tan 60° = \sqrt{3}$

61. $\theta = \frac{2\pi}{3}, \theta' = \frac{\pi}{3}$ in Quadrant II

$$\sin\frac{2\pi}{3} = \sin\frac{\pi}{3} = \frac{\sqrt{3}}{2}$$

$$\cos\frac{2\pi}{3} = -\cos\frac{\pi}{3} = -\frac{1}{2}$$

$$\tan\frac{2\pi}{3} = -\tan\frac{\pi}{3} = -\sqrt{3}$$

63. $\theta = -\frac{\pi}{6}, \theta' = \frac{\pi}{6}$, Quadrant IV

$$\sin\left(-\frac{\pi}{6}\right) = -\sin\frac{\pi}{6} = -\frac{1}{2}$$

$$\cos\left(-\frac{\pi}{6}\right) = \cos\frac{\pi}{6} = \frac{\sqrt{3}}{2}$$

$$\tan\left(-\frac{\pi}{6}\right) = -\tan\frac{\pi}{6} = -\frac{\sqrt{3}}{3}$$

65. $\theta = \frac{11\pi}{4}, \theta' = \frac{\pi}{4}$, Quadrant II

$$\sin\frac{11\pi}{4} = \sin\frac{\pi}{4} = \frac{\sqrt{2}}{2}$$

$$\cos\frac{11\pi}{4} = -\cos\frac{\pi}{4} = -\frac{\sqrt{2}}{2}$$

$$\tan\frac{11\pi}{4} = -\tan\frac{\pi}{4} = -1$$

67. $\theta = -\frac{17\pi}{6}, \theta' = \frac{\pi}{6}$ in Quadrant III

$$\sin\left(-\frac{17\pi}{6}\right) = -\sin\frac{\pi}{6} = -\frac{1}{2}$$

$$\cos\left(-\frac{17\pi}{6}\right) = -\cos\frac{\pi}{6} = -\frac{\sqrt{3}}{2}$$

$$\tan\left(-\frac{17\pi}{6}\right) = \tan\frac{\pi}{6} = \frac{\sqrt{3}}{3}$$

69. $\sin\theta = -\frac{3}{5}$

$$\sin^2\theta + \cos^2\theta = 1$$

$$\cos^2\theta = 1 - \sin^2\theta$$

$$\cos^2\theta = 1 - \left(-\frac{3}{5}\right)^2$$

$$\cos^2\theta = 1 - \frac{9}{25}$$

$$\cos^2\theta = \frac{16}{25}$$

$\cos\theta > 0$ in Quadrant IV.

$$\cos\theta = \frac{4}{5}$$

71. $\tan\theta = \frac{3}{2}$

$$\sec^2\theta = 1 + \tan^2\theta$$

$$\sec^2\theta = 1 + \left(\frac{3}{2}\right)^2$$

$$\sec^2\theta = 1 + \frac{9}{4}$$

$$\sec^2\theta = \frac{13}{4}$$

$\sec\theta < 0$ in Quadrant III.

$$\sec\theta = -\frac{\sqrt{13}}{2}$$

73. $\cos\theta = \frac{5}{8}$

$$\sin^2\theta + \cos^2\theta = 1$$

$$\sin^2\theta + \left(\frac{5}{8}\right)^2 = 1$$

$$\sin^2\theta = 1 - \frac{25}{64}$$

$$\sin^2\theta = \frac{39}{64}$$

$\sin\theta > 0$ in Quadrant I.

$$\sin\theta = \frac{\sqrt{39}}{8}$$

$$\csc\theta = \frac{1}{\sin\theta}$$

$$\csc\theta = \frac{8\sqrt{39}}{39}$$

75. $\sin 10^\circ \approx 0.1736$

77. $\cos(-110^\circ) \approx -0.3420$

79. $\cot 178^\circ \approx -28.6363$

81. $\csc 405^\circ = \frac{1}{\sin 405^\circ} \approx 1.4142$

83. $\tan\left(\frac{\pi}{9}\right) \approx 0.3640$

85. $\sec\frac{11\pi}{8} = \frac{1}{\cos\frac{11\pi}{8}} \approx -2.6131$

87. $\sin(-0.65) \approx -0.6052$

89. $\csc(-10) = \dfrac{1}{\sin(-10)} \approx 1.8382$

91. (a) $\sin\theta = \dfrac{1}{2} \Rightarrow$ reference angle is $30°$ or $\dfrac{\pi}{6}$ and θ is in Quadrant I or Quadrant II.

Values in degrees: $30°, 150°$

Values in radian: $\dfrac{\pi}{6}, \dfrac{5\pi}{6}$

(b) $\sin\theta = \dfrac{1}{2} \Rightarrow$ reference angle is $30°$ or $\dfrac{\pi}{6}$ and θ is in Quadrant III or Quadrant IV.

Values in degrees: $210°, 330°$

Values in radians: $\dfrac{7\pi}{6}, \dfrac{11\pi}{6}$

93. (a) $\cos\theta = \dfrac{1}{2} \Rightarrow$ reference angle is $60°$ or $\dfrac{\pi}{3}$ and θ is in Quadrant I or IV.

Values in degrees: $60°, 300°$

Values in radians: $\dfrac{\pi}{3}, \dfrac{5\pi}{3}$

(b) $\sec\theta = 2 \Rightarrow \cos\theta = \dfrac{1}{2} \Rightarrow$ reference angle is $60°$ or $\dfrac{\pi}{3}$ and θ is in Quadrant I or IV.

Values in degrees: $60°, 300°$

Values in radians: $\dfrac{\pi}{3}, \dfrac{5\pi}{3}$

95. (a) $\tan\theta = 1 \Rightarrow$ reference angle is $45°$ or $\dfrac{\pi}{4}$ and θ is in Quadrant I or Quadrant III.

Values in degrees: $45°, 225°$

Values in radians: $\dfrac{\pi}{4}, \dfrac{5\pi}{4}$

(b) $\cot\theta = -\sqrt{3} \Rightarrow$ reference angle is $30°$ or $\dfrac{\pi}{6}$ and θ is in Quadrant II or Quadrant IV.

Values in degrees: $150°, 330°$

Values in radians: $\dfrac{5\pi}{6}, \dfrac{11\pi}{6}$

97. $\sin\theta = \dfrac{6}{d} \Rightarrow d = \dfrac{6}{\sin\theta}$

(a) $\theta = 30°$

$$d = \frac{6}{\sin 30°} = \frac{6}{1/2} = 12 \text{ miles}$$

(b) $\theta = 90°$

$$d = \frac{6}{\sin 90°} = \frac{6}{1} = 6 \text{ miles}$$

(c) $\theta = 120°$

$$d = \frac{6}{\sin 120°} = \frac{6}{\sqrt{3}/2} \approx 6.9 \text{ miles}$$

99. (a) Boston: $B \approx 24.593\sin(0.495t - 2.262) + 57.387$

Fairbanks:
$F \approx 39.071\sin(0.448t - 1.366) + 32.204$

(b)

Month	*Boston, B*	*Fairbanks, F*
February	33.9°	14.5°
April	50.5°	48.3°
May	62.6°	62.2°
July	80.3°	70.5°
September	77.4°	50.1°
October	68.2°	33.3°
December	44.8°	2.4°

(c)

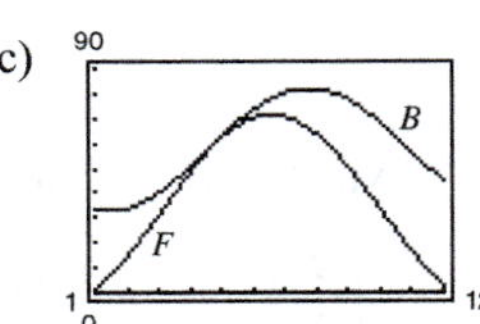

Answers will vary.

101. $I = 5e^{-2t}\sin t$

$I(0.7) = 5e^{-1.4}\sin 0.7 \approx 0.79$ amp

103. False. In each of the four quadrants, the sign of the secant function and the cosine function will be the same because they are reciprocals of each other.

105. Answers will vary.

Section 4.5 Graphs of Sine and Cosine Functions

1. cycle

3. phase shift

5. $y = 2\sin 5x$

Period: $\frac{2\pi}{5}$

Amplitude: $|2| = 2$

7. $y = \frac{3}{4}\cos\frac{\pi x}{2}$

Period: $\frac{2\pi}{b} = \frac{2\pi}{\pi/2} = 4$

Amplitude: $\left|\frac{3}{4}\right| = \frac{3}{4}$

9. $y = -\frac{1}{2}\sin\frac{5x}{4}$

Period: $\frac{2\pi}{b} = \frac{2\pi}{5/4} = \frac{8\pi}{5}$

Amplitude: $\left|-\frac{1}{2}\right| = \frac{1}{2}$

11. $y = -\frac{5}{3}\sin\frac{\pi x}{12}$

Period: $\frac{2\pi}{b} = \frac{2\pi}{\pi/12} = 24$

Amplitude: $\left|-\frac{5}{3}\right| = \frac{5}{3}$

13. $f(x) = \cos x$

$g(x) = \cos 5x$

The period of g is one-fifth the period of f.

15. $f(x) = \cos 2x$

$g(x) = -\cos 2x$

g is a reflection of the graph of f in the x-axis.

17. $f(x) = \sin x$

$g(x) = \sin(x - \pi)$

g is a horizontal shift to the right π units of the graph of f.

19. $f(x) = \sin 2x$

$f(x) = 3 + \sin 2x$

g is a vertical shift three units upward of the graph of f.

21. The graph of g has twice the amplitude as the graph of f. The period is the same.

23. The graph of g is a horizontal shift π units to the right of the graph of f.

25. $f(x) = \sin x$

Period: $\frac{2\pi}{b} = \frac{2\pi}{1} = 2\pi$

Amplitude: 1

Symmetry: origin

Key points: Intercept	Maximum	Intercept	Minimum	Intercept
$(0, 0)$	$\left(\frac{\pi}{2}, 1\right)$	$(\pi, 0)$	$\left(\frac{3\pi}{2}, -1\right)$	$(2\pi, 0)$

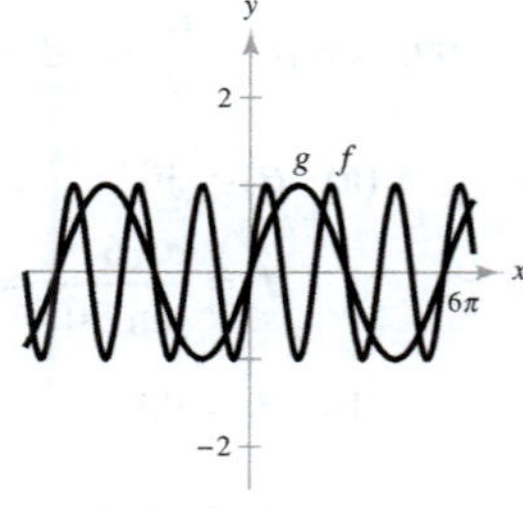

Because $g(x) = \sin\left(\frac{x}{3}\right) = f\left(\frac{x}{3}\right)$, the graph of $g(x)$ is the graph of $f(x)$, but stretched horizontally by a factor of 3.

Generate key points for the graph of $g(x)$ by multiplying the x-coordinate of each key point of $f(x)$ by 3.

27. $f(x) = \cos x$

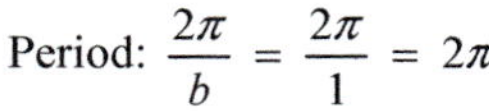

Period: $\dfrac{2\pi}{b} = \dfrac{2\pi}{1} = 2\pi$

Amplitude: $|1| = 1$

Symmetry: y-axis

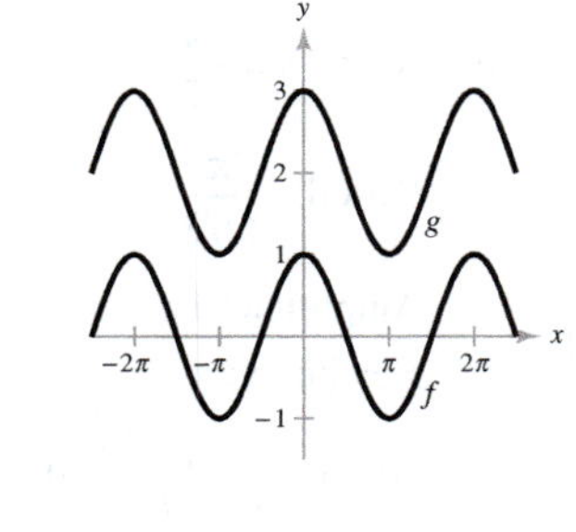

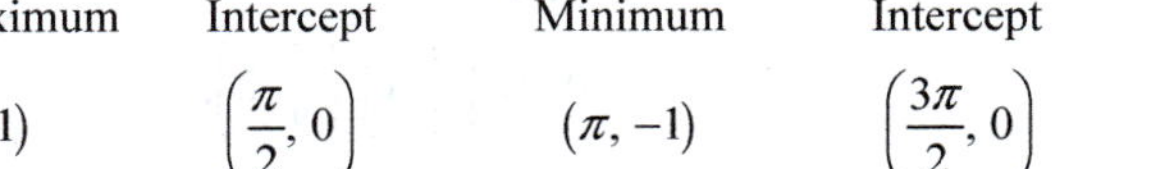

Key points: Maximum	Intercept	Minimum	Intercept	Maximum
$(0, 1)$	$\left(\dfrac{\pi}{2}, 0\right)$	$(\pi, -1)$	$\left(\dfrac{3\pi}{2}, 0\right)$	$(2\pi, 1)$

Because $g(x) = 2 + \cos x = f(x) + 2$, the graph of $g(x)$ is the graph of $f(x)$, but translated upward by two units. Generate key points of $g(x)$ by adding 2 to the y-coordinate of each key point of $f(x)$.

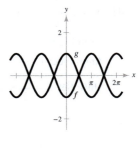

29. $f(x) = -\cos x$

Period: $\dfrac{2\pi}{b} = \dfrac{2\pi}{1} = 2\pi$

Amplitude: 1

Symmetry: y-axis

Key points: Minimum	Intercept	Maximum	Intercept	Minimum
$(0, -1)$	$\left(\dfrac{\pi}{2}, 0\right)$	$(\pi, 1)$	$\left(\dfrac{3\pi}{2}, 0\right)$	$(2\pi, -1)$

Because $g(x) = -\cos(x - \pi) = f(x - \pi)$, the graph of $g(x)$ is the graph of $f(x)$, but with a phase shift (horizontal translation) of π. Generate key points for the graph of $g(x)$ by shifting each key point of $f(x)$ π units to the right.

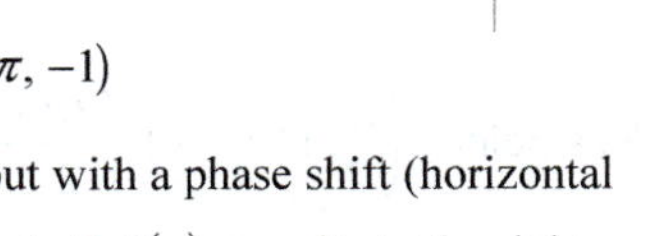

31. $y = 5 \sin x$

Period: 2π

Amplitude: 5

Key points:

$(0, 0), \left(\dfrac{\pi}{2}, 5\right), (\pi, 0),$

$\left(\dfrac{3\pi}{2}, -5\right), (2\pi, 0)$

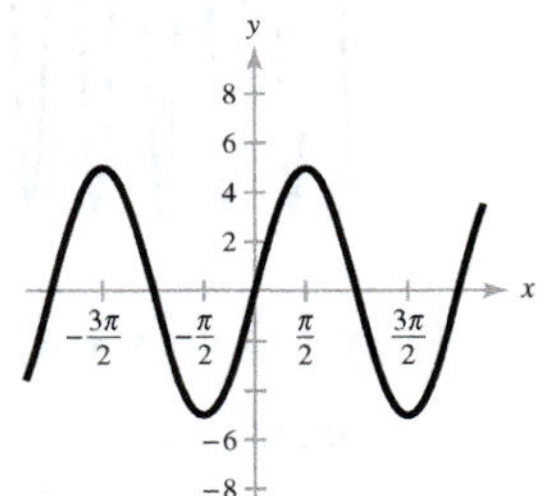

33. $y = \dfrac{1}{3} \cos x$

Period: 2π

Amplitude: $\dfrac{1}{3}$

Key points:

$\left(0, \dfrac{1}{3}\right), \left(\dfrac{\pi}{2}, 0\right), \left(\pi, -\dfrac{1}{3}\right),$

$\left(\dfrac{3\pi}{2}, 0\right), \left(2\pi, \dfrac{1}{3}\right)$

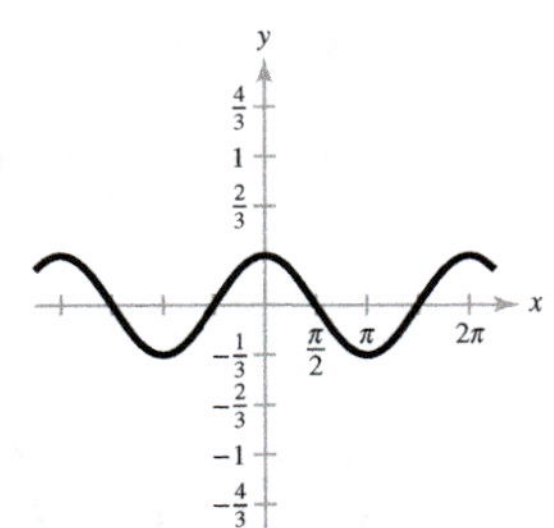

35. $y = \cos \dfrac{x}{2}$

Period $\dfrac{2\pi}{1/2} = 4\pi$

Amplitude: 1

Key points:

$(0, 1), (\pi, 0), (2\pi, -1),$

$(3\pi, 0), (4\pi, 1)$

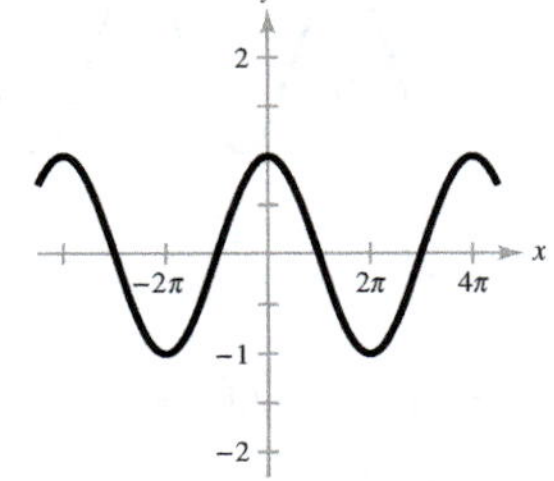

37. $y = \cos 2\pi x$

Period: $\dfrac{2\pi}{2\pi} = 1$

Amplitude: 1

Key points:

$(0, 1), \left(\dfrac{1}{4}, 0\right), \left(\dfrac{1}{2}, -1\right), \left(\dfrac{3}{4}, 0\right)$

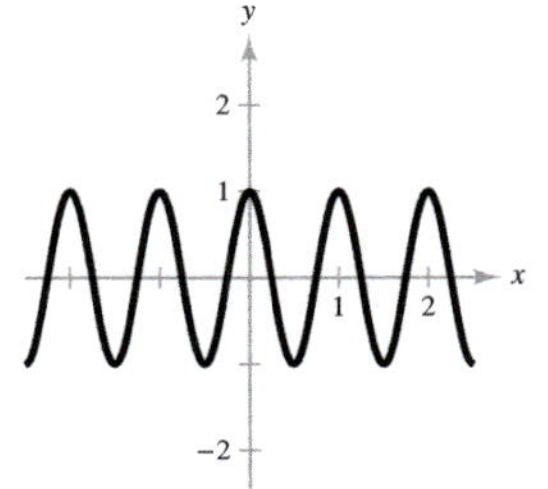

39. $y = -\sin \dfrac{2\pi x}{3}$

Period: $\dfrac{2\pi}{2\pi/3} = 3$

Amplitude: 1

Key points:

$(0, 0), \left(\dfrac{3}{4}, -1\right), \left(\dfrac{3}{2}, 0\right),$

$\left(\dfrac{9}{4}, 1\right), (3, 0)$

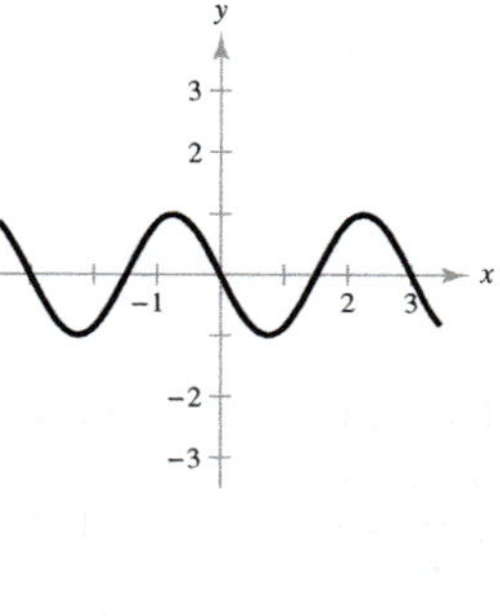

41. $y = \cos\left(x - \dfrac{\pi}{2}\right)$

Period: 2π

Amplitude: 1

Shift: Set $x - \dfrac{\pi}{2} = 0$ and $x - \dfrac{\pi}{2} = 2\pi$

$x = \dfrac{\pi}{2} \qquad x = \dfrac{5\pi}{2}$

Key points: $\left(\dfrac{\pi}{2}, 1\right), (\pi, 0), \left(\dfrac{3\pi}{2}, -1\right), (2\pi, 0), \left(\dfrac{5\pi}{2}, 1\right)$

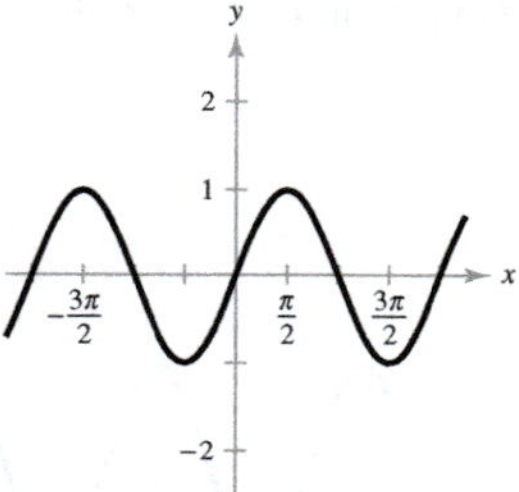

43. $y = 3\sin(x + \pi)$

Period: 2π

Amplitude: 3

Shift: Set $x + \pi = 0$ and $x + \pi = 2\pi$

$x = -\pi \qquad x = \pi$

Key points: $(-\pi, 0), \left(-\dfrac{\pi}{2}, 3\right), (0, 0), \left(\dfrac{\pi}{2}, -3\right), (\pi, 0)$

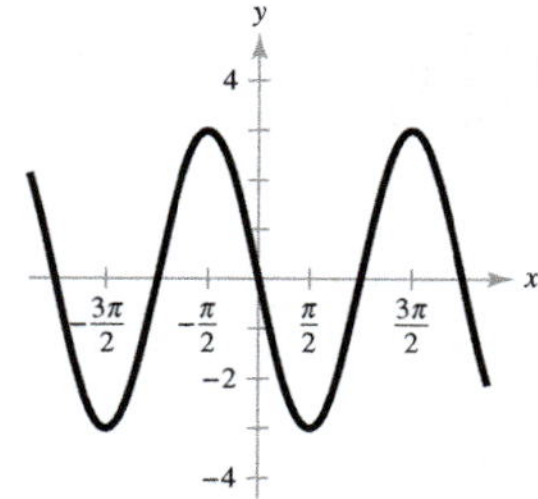

45. $y = 2 - \sin \dfrac{2\pi x}{3}$

Period: $\dfrac{2\pi}{2\pi/3} = 3$

Amplitude: 1

Key points:

$(0, 2), \left(\dfrac{3}{4}, 1\right), \left(\dfrac{3}{2}, 2\right),$

$\left(\dfrac{9}{4}, 3\right), (3, 2)$

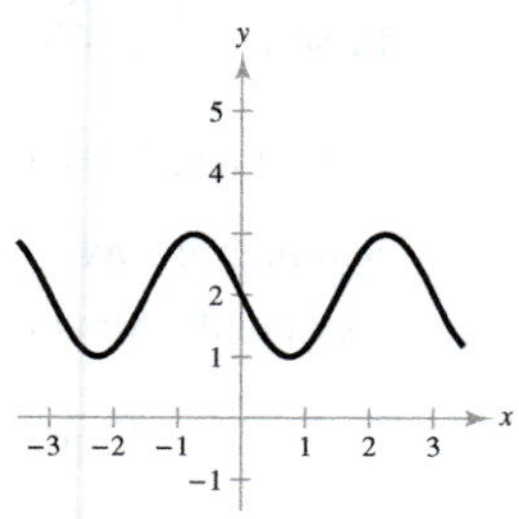

47. $y = 2 + 5\cos 6\pi x$

Period: $\dfrac{2\pi}{6\pi} = \dfrac{1}{3}$

Amplitude: 5

Shift: Set $6\pi x = 0$ and $6\pi x = 2\pi$

$x = 0 \qquad x = \dfrac{1}{3}$

Key points: $\left(\dfrac{1}{3}, 7\right), \left(\dfrac{5}{12}, 2\right), \left(\dfrac{1}{2}, -3\right), \left(\dfrac{7}{12}, 2\right), \left(\dfrac{1}{3}, 7\right)$

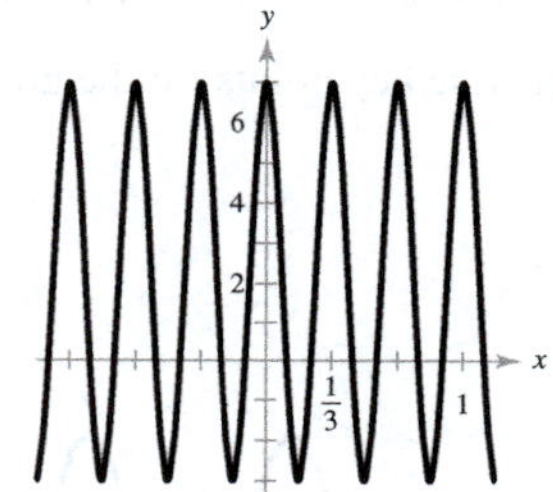

49. $y = 3\sin(x + \pi) - 3$

Period: 2π

Amplitude: 3

Shift: Set $x + \pi = 0$ and $x + \pi = 2\pi$

$x = -\pi \qquad x = \pi$

Key points:

$(-\pi, -3), \left(-\dfrac{\pi}{2}, 0\right), (0, -3), \left(\dfrac{\pi}{2}, -6\right), (\pi, -3)$

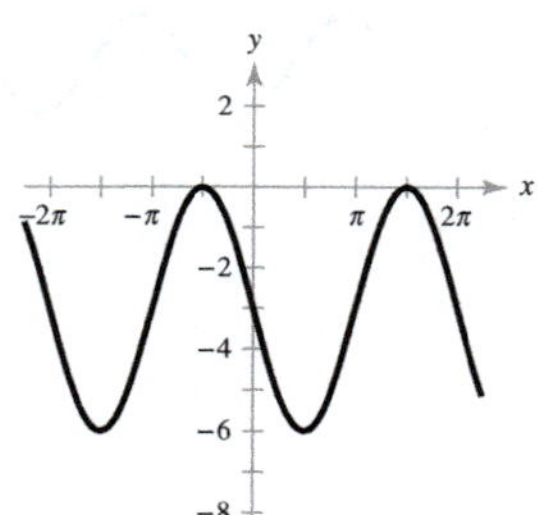

51. $y = \frac{2}{3}\cos\left(\frac{x}{2} - \frac{\pi}{4}\right)$

Period: $\frac{2\pi}{1/2} = 4\pi$

Amplitude: $\frac{2}{3}$

Shift: $\frac{x}{2} - \frac{\pi}{4} = 0$ and $\frac{\pi}{2} - \frac{\pi}{4} = 2\pi$

$x = \frac{\pi}{2}$ $\qquad x = \frac{9\pi}{2}$

Key points:

$\left(\frac{\pi}{2}, \frac{2}{3}\right), \left(\frac{3\pi}{2}, 0\right), \left(\frac{5\pi}{2}, \frac{-2}{3}\right), \left(\frac{7\pi}{2}, 0\right), \left(\frac{9\pi}{2}, \frac{2}{3}\right)$

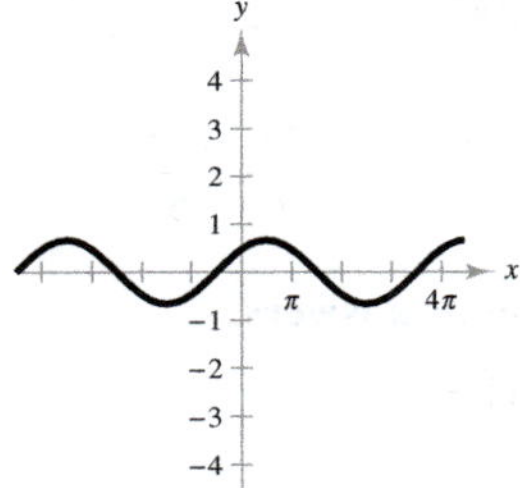

53. $g(x) = \sin(4x - \pi)$

(a) $g(x)$ is obtained by a horizontal shrink and a phase shift of $\frac{\pi}{4}$. One cycle of $g(x)$ corresponds to the interval $\left[\frac{\pi}{4}, \frac{3\pi}{4}\right]$.

(b)

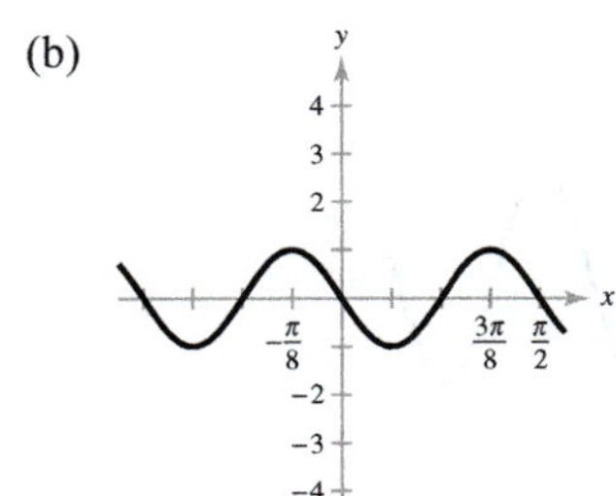

(c) $g(x) = f(4x - \pi)$ where $f(x) = \sin x$.

55. $g(x) = \cos\left(x - \frac{\pi}{2}\right) + 2$

(a) $g(x)$ is obtained by shifting $f(x)$ two units upward and a phase shift of $\frac{\pi}{2}$. One cycle of $g(x)$ corresponds to the interval $[\pi, 3\pi]$.

(b)

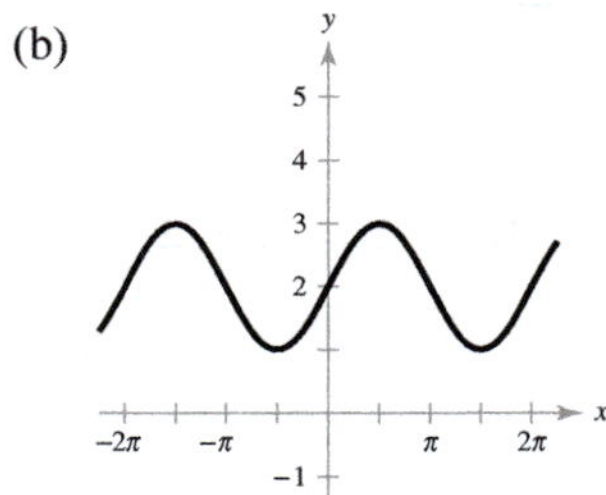

(c) $g(x) = f\left(x - \frac{\pi}{2}\right) + 2$ where $f(x) = \cos x$

57. $g(x) = 2\sin(4x - \pi) - 3$

(a) $g(x)$ is obtained by a vertical stretch, a horizontal shrink, a phase shift of $\frac{\pi}{4}$, and shifting $f(x)$ three units downward. One cycle of $g(x)$ corresponds to the interval $\left[\frac{\pi}{4}, \frac{3\pi}{4}\right]$.

(b)

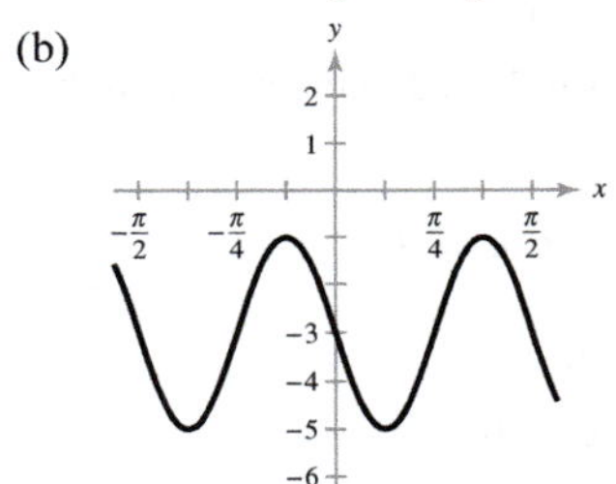

(c) $g(x) = 2f(4x - \pi) - 3$ where $f(x) = \sin x$

59. $y = -2\sin(4x + \pi)$

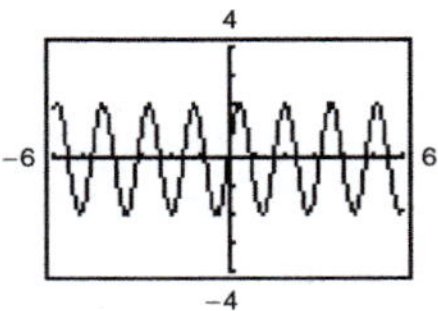

61. $y = \cos\left(2\pi x - \frac{\pi}{2}\right) + 1$

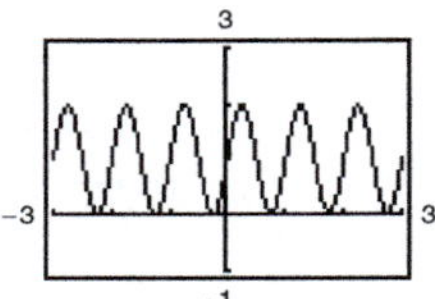

63. $y = -0.1\sin\left(\frac{\pi x}{10} + \pi\right)$

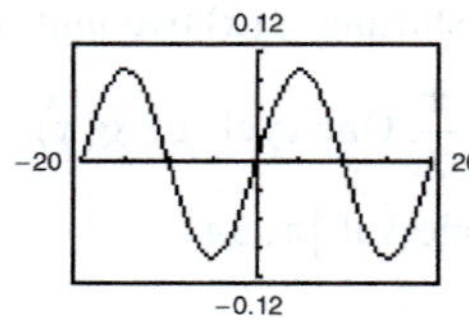

65. $f(x) = a\cos x + d$

Amplitude: $\frac{1}{2}[3 - (-1)] = 2 \Rightarrow a = 2$

$3 = 2\cos 0 + d$

$d = 3 - 2 = 1$

$a = 2, d = 1$

67. $f(x) = a\cos x + d$

Amplitude: $\frac{1}{2}[8 - 0] = 4$

Reflected in the x-axis: $a = -4$

$0 = -4\cos 0 + d$

$d = 4$

$a = -4, d = 4$

69. $y = a\sin(bx - c)$

Amplitude: $|a| = |3|$

Since the graph is reflected in the x-axis, we have $a = -3$.

Period: $\frac{2\pi}{b} = \pi \Rightarrow b = 2$

Phase shift: $c = 0$

$a = -3, b = 2, c = 0$

71. $y = a\sin(bx - c)$

Amplitude: $a = 2$

Period: $2\pi \Rightarrow b = 1$

Phase shift: $bx - c = 0$ when $x = -\frac{\pi}{4}$

$$(1)\left(-\frac{\pi}{4}\right) - c = 0 \Rightarrow c = -\frac{\pi}{4}$$

$a = 2, b = 1, c = -\frac{\pi}{4}$

73. $y_1 = \sin x$

$y_2 = -\frac{1}{2}$

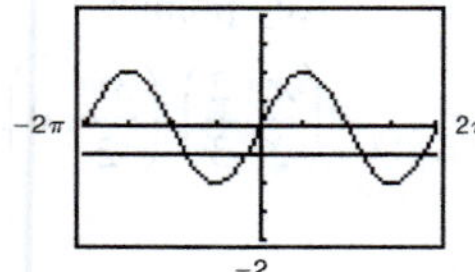

In the interval $[-2\pi, 2\pi]$,

$y_1 = y_2$ when $x = -\frac{5\pi}{6}, -\frac{\pi}{6}, \frac{7\pi}{6}, \frac{11\pi}{6}$.

Answers for 75–77 are sample answers.

75. $f(x) = 2\sin(2x - \pi) + 1$

77. $f(x) = \cos(2x + 2\pi) - \frac{3}{2}$

79. $v = 1.75\sin\frac{\pi t}{2}$

(a) Period $= \frac{2\pi}{\pi/2} = 4$ seconds

(b) $\frac{1 \text{ cycle}}{4 \text{ seconds}} \cdot \frac{60 \text{ seconds}}{1 \text{ minute}} = 15$ cycles per minute

(c)

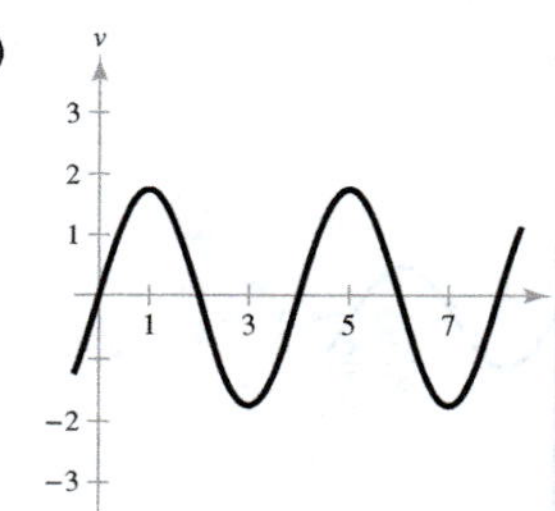

81. $P = 100 - 20\cos\frac{5\pi t}{3}$

(a) Period: $\frac{2\pi}{(5\pi)/3} = \frac{6}{5}$ seconds

(b) $\frac{1 \text{ heartbeat}}{6/5 \text{ seconds}} \cdot \frac{60 \text{ seconds}}{1 \text{ minute}} = 50$ heartbeats per minute

83. (a) and (c)

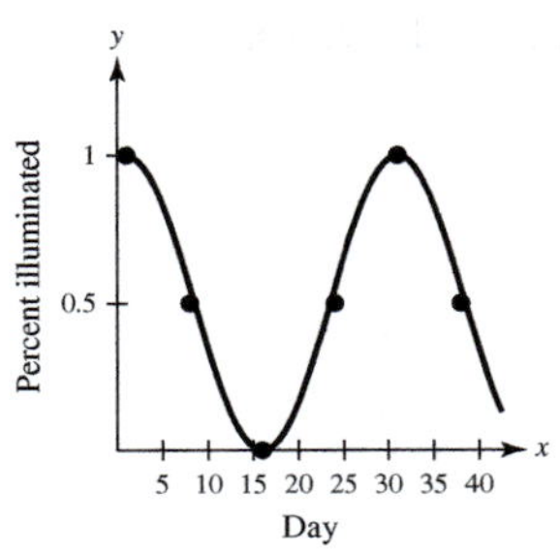

The model fits the data well.

(b) Amplitude $a = \frac{1}{2}[\text{max. percent} - \text{min. percent}] = \frac{1}{2}[1.0 - 0] = 0.5$

Period: $p = \left(2^{\text{nd}} \text{ day of } 1.0 \text{ percent} - 1^{\text{st}} \text{ day of } 1.0 \text{ percent}\right) = 31 - 1 = 30$

$$b = \frac{2\pi}{p} = \frac{2\pi}{30} = \frac{\pi}{15}$$

Because the zero percent occurs at day 16, $x = 16$.

$$bx - c = \pi$$

$$\left(\frac{\pi}{15}\right)(16) - c = \pi$$

$$\frac{16\pi}{15} - \pi = c$$

$$\frac{\pi}{15} = c$$

The average percent of illumination is $\frac{1}{2}(1.0 - 0) = 0.5$. So, $d = 0.5$.

So the model is $y = 0.5 \cos\left(\frac{\pi x}{15} - \frac{\pi}{15}\right) + 0.5$.

(d) The period of the model is $p = 30$ days.

(e) Because March 12, 2018 is the 76th day of the year, $x = 76$. So, the percent of the moon's face illuminated is $y = 0.5 \cos\left(\frac{\pi(71)}{15} - \frac{\pi}{15}\right) + 0.5 = 0.25 = 25\%$.

85. (a) Period $= \dfrac{2\pi}{\left(\frac{\pi}{10}\right)} = 20$ seconds

The wheel takes 20 seconds to revolve once.

(b) Amplitude: 50 feet

The radius of the wheel is 50 feet.

(c)

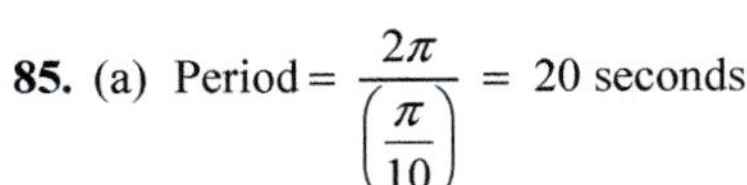
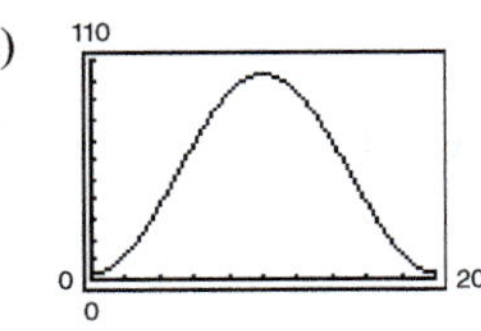

87. False. The graph of $g(x) = \sin(x + 2\pi)$ is the graph of $f(x) = \sin(x)$ translated to the *left* by one period, and the graphs are identical.

89. True. $y = -\cos x$ is a reflection of $y = \sin\left(x + \frac{\pi}{2}\right)$ in the x-axis. So, $-\cos x = -\sin\left(x + \frac{\pi}{2}\right)$.

91.

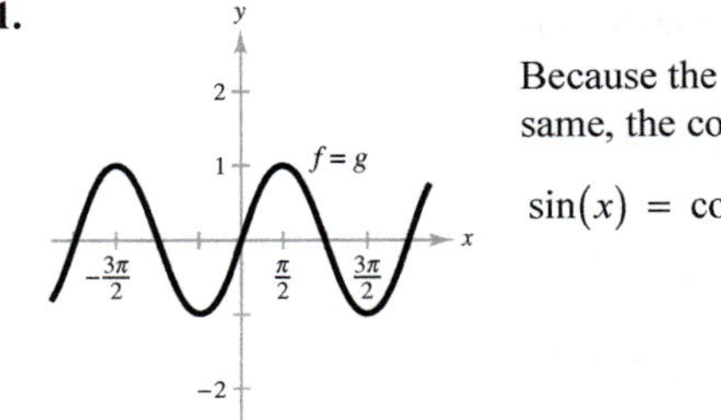

Because the graphs are the same, the conjecture is that $\sin(x) = \cos\left(x - \frac{\pi}{2}\right)$.

93.

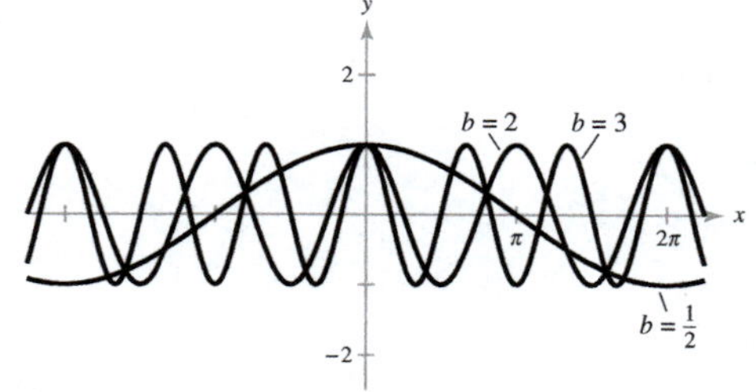

As the value of b increases, the period decreases.

$b = \frac{1}{2} \to \frac{1}{2}$ cycle

$b = 2 \to 2$ cycles

$b = 3 \to 3$ cycles

95. (a) $\sin\frac{1}{2} \approx \frac{1}{2} - \frac{(1/2)^3}{3!} + \frac{(1/2)^5}{5!} \approx 0.4794$

$\sin\frac{1}{2} \approx 0.4794$ (by calculator)

(b) $\sin 1 \approx 1 - \frac{1}{3!} + \frac{1}{5!} \approx 0.8417$

$\sin 1 \approx 0.8415$ (by calculator)

(c) $\sin\frac{\pi}{6} \approx 1 - \frac{(\pi/6)^3}{3!} + \frac{(\pi/6)^5}{5!} \approx 0.5000$

$\sin\frac{\pi}{6} = 0.5$ (by calculator)

(d) $\cos(-0.5) \approx 1 - \frac{(-0.5)^2}{2!} + \frac{(-0.5)^4}{4!} \approx 0.8776$

$\cos(-0.5) \approx 0.8776$ (by calculator)

(e) $\cos 1 \approx 1 - \frac{1}{2!} + \frac{1}{4!} \approx 0.5417$

$\cos 1 \approx 0.5403$ (by calculator)

(f) $\cos\frac{\pi}{4} \approx 1 - \frac{(\pi/4)^2}{2!} + \frac{(\pi/4)^2}{4!} = 0.7074$

$\cos\frac{\pi}{4} \approx 0.7071$ (by calculator)

The error in the approximation is not the same in each case. The error appears to increase as x moves farther away from 0.

Section 4.6 Graphs of Other Trigonometric Functions

1. odd; origin

3. reciprocal

5. π

7. $(-\infty, -1] \cup [1, \infty)$

9. $y = \sec 2x$

Period: $\frac{2\pi}{2} = \pi$

Matches graph (e).

10. $y = \tan\frac{x}{2}$

Period: $\frac{\pi}{b} = \frac{\pi}{1/2} = 2\pi$

Asymptotes: $x = -\pi, x = \pi$

Matches graph (c).

11. $y = \frac{1}{2}\cot \pi x$

Period: $\frac{\pi}{\pi} = 1$

Matches graph (a).

12. $y = -\csc x$

Period: 2π

Matches graph (d).

13. $y = \frac{1}{2}\sec\frac{\pi x}{2}$

Period: $\frac{2\pi}{b} = \frac{2\pi}{\pi/2} = 4$

Asymptotes: $x = -1, x = 1$

Matches graph (f).

14. $y = -2\sec\frac{\pi x}{2}$

Period: $\frac{2\pi}{b} = \frac{2\pi}{\pi/2} = 4$

Asymptotes: $x = -1, x = 1$

Reflected in x-axis

Matches graph (b).

15. $y = \frac{1}{3}\tan x$

Period: π

Two consecutive asymptotes:

$x = -\frac{\pi}{2}$ and $x = \frac{\pi}{2}$

x	$-\frac{\pi}{4}$	0	$\frac{\pi}{4}$
y	$-\frac{1}{3}$	0	$\frac{1}{3}$

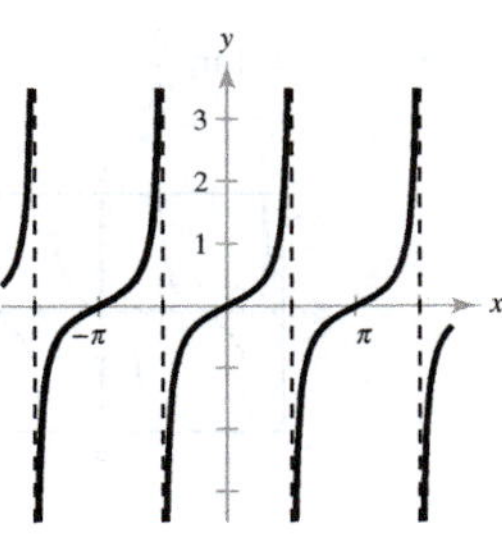

17. $y = -\frac{1}{2}\sec x$

Period: 2π

Two consecutive asymptotes:

$x = -\frac{\pi}{2}, x = \frac{\pi}{2}$

x	$-\frac{\pi}{3}$	0	$\frac{\pi}{3}$
y	-1	$-\frac{1}{2}$	-1

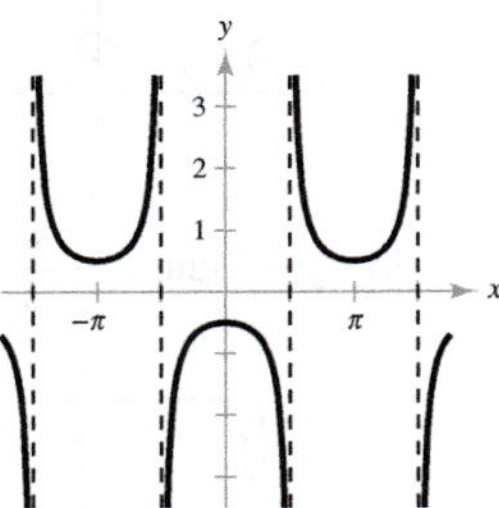

19. $y = -2\tan 3x$

Period: $\frac{\pi}{3}$

Two consecutive asymptotes:

$x = -\frac{\pi}{6}, x = \frac{\pi}{6}$

x	$-\frac{\pi}{3}$	0	$\frac{\pi}{3}$
y	0	0	0

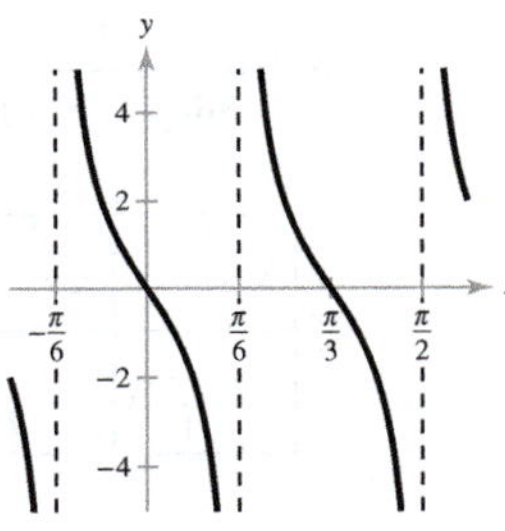

21. $y = \csc \pi x$

Period: $\frac{2\pi}{\pi} = 2$

Two consecutive asymptotes:

$x = 0, x = 1$

x	$\frac{1}{6}$	$\frac{1}{2}$	$\frac{5}{6}$
y	2	1	2

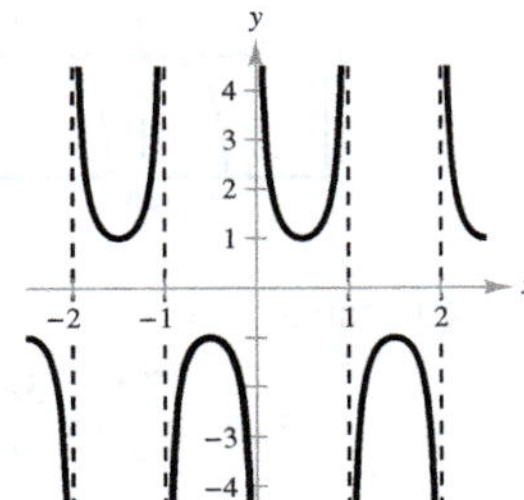

23. $y = \frac{1}{2}\sec \pi x$

Period: 2

Two consecutive asymptotes:

$x = -\frac{1}{2}, x = \frac{1}{2}$

x	-1	0	1
y	$-\frac{1}{2}$	$\frac{1}{2}$	$-\frac{1}{2}$

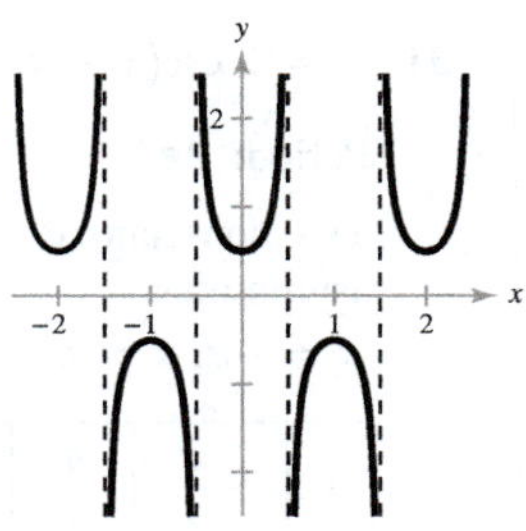

25. $y = \csc \frac{x}{2}$

Period: $\frac{2\pi}{1/2} = 4\pi$

Two consecutive asymptotes:

$x = 0, x = 2\pi$

x	$\frac{\pi}{3}$	π	$\frac{5\pi}{3}$
y	2	1	2

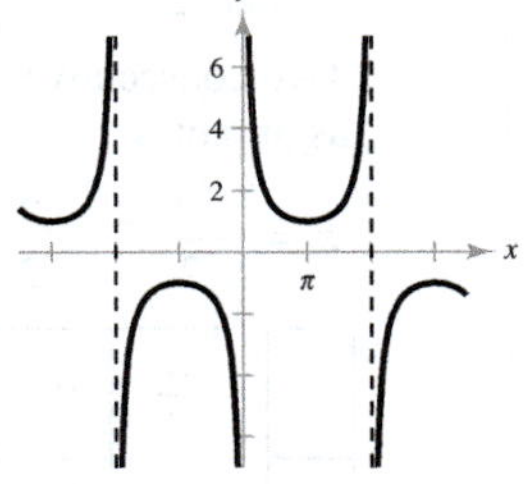

27. $y = 3\cot 2x$

Period: $\frac{\pi}{2}$

Two consecutive asymptotes:

$x = -\frac{\pi}{2}, x = \frac{\pi}{2}$

x	$-\frac{\pi}{6}$	$-\frac{\pi}{8}$	$\frac{\pi}{8}$	$\frac{\pi}{6}$
y	$-3\sqrt{3}$	-3	3	$3\sqrt{3}$

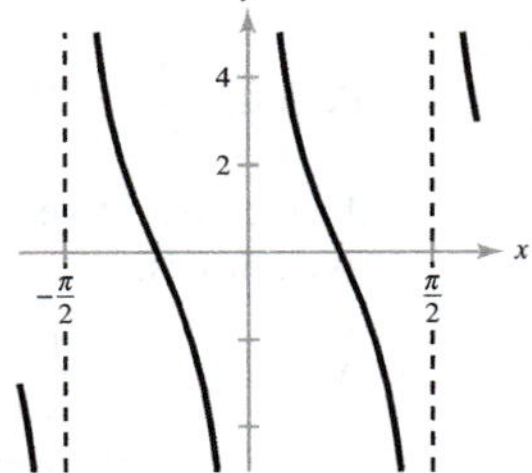

29. $y = \tan \frac{\pi x}{4}$

Period: $\frac{\pi}{\pi/4} = 4$

Two consecutive asymptotes:

$\frac{\pi x}{4} = -\frac{\pi}{2} \Rightarrow x = -2$

$\frac{\pi x}{4} = \frac{\pi}{2} \Rightarrow x = 2$

x	-1	0	1
y	-1	0	1

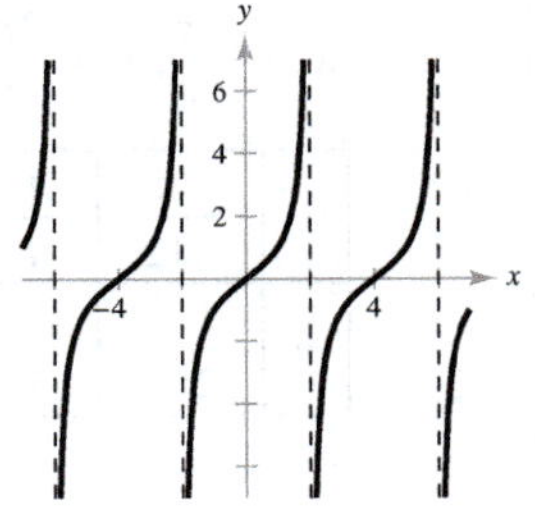

31. $y = 2\csc(x - \pi)$

Period: 2π

Two consecutive asymptotes:

$x = -\pi,\ x = \pi$

x	$-\dfrac{\pi}{2}$	$\dfrac{\pi}{2}$	$\dfrac{3\pi}{2}$
y	2	-2	-2

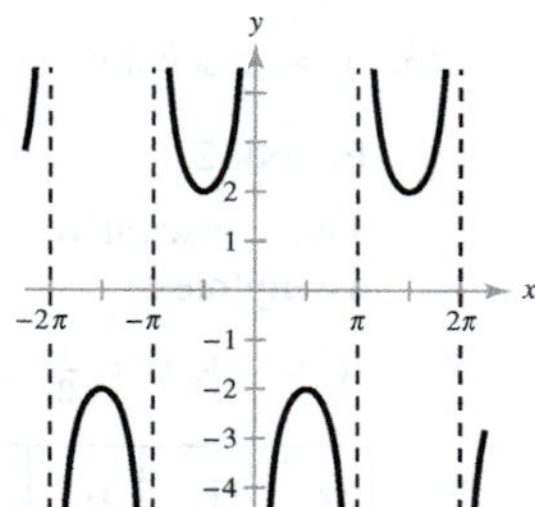

33. $y = 2\sec(x + \pi)$

Period: 2π

Two consecutive asymptotes:

$x = -\dfrac{\pi}{2},\ x = \dfrac{\pi}{2}$

x	$-\dfrac{\pi}{3}$	0	$\dfrac{\pi}{3}$
y	-4	-2	-4

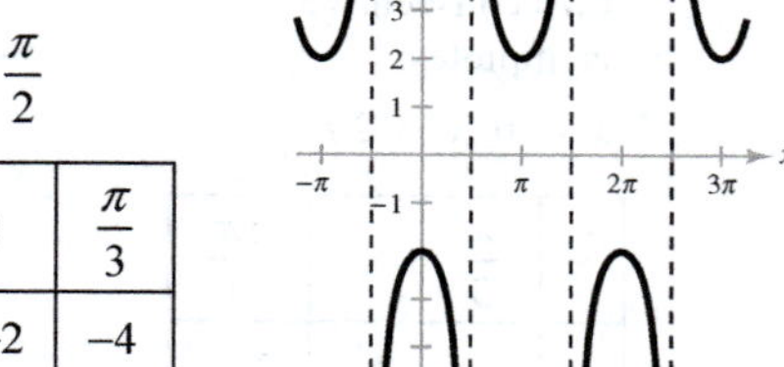

35. $y = -\sec \pi x + 1$

Period: $\dfrac{2\pi}{\pi} = 2$

Two consecutive asymptotes:

$x = -\dfrac{1}{2},\ x = \dfrac{1}{2}$

x	$-\dfrac{1}{3}$	0	$\dfrac{1}{3}$
y	-1	0	1

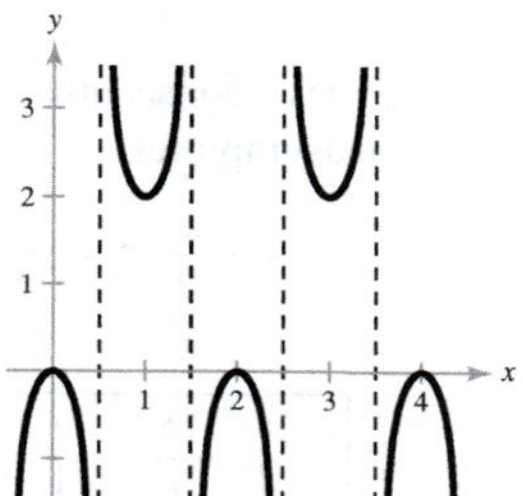

37. $y = \dfrac{1}{4}\csc\left(x + \dfrac{\pi}{4}\right)$

Period: 2π

Two consecutive asymptotes:

$x = -\dfrac{\pi}{4},\ x = \dfrac{3\pi}{4}$

x	$-\dfrac{\pi}{12}$	$\dfrac{\pi}{4}$	$\dfrac{7\pi}{12}$
y	$\dfrac{1}{2}$	$\dfrac{1}{4}$	$\dfrac{1}{2}$

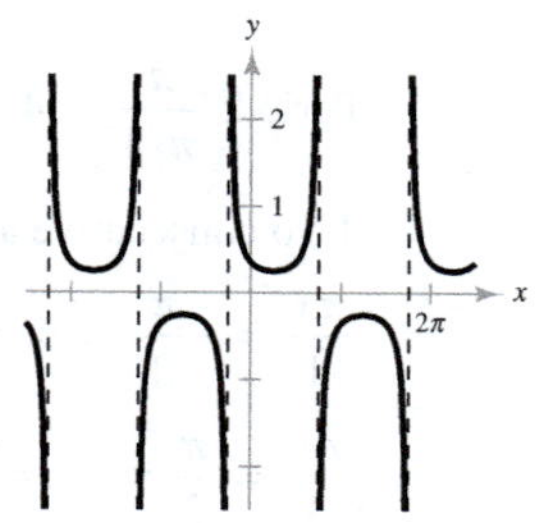

39. $y = \tan \dfrac{x}{3}$

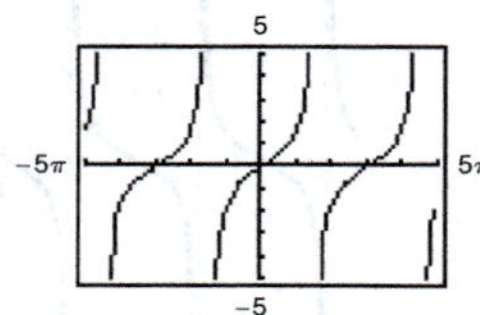

41. $y = -2\sec 4x = \dfrac{-2}{\cos 4x}$

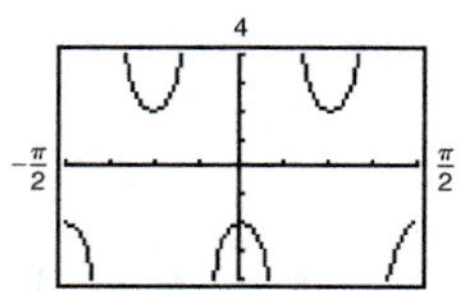

43. $y = \tan\left(x - \dfrac{\pi}{4}\right)$

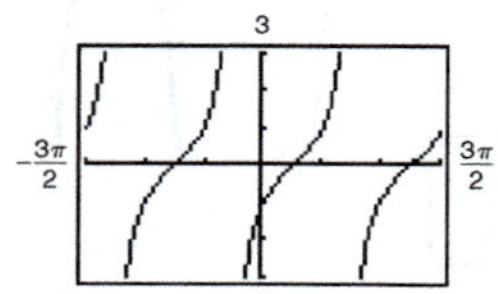

45. $y = -\csc(4x - \pi)$

$y = \dfrac{-1}{\sin(4x - \pi)}$

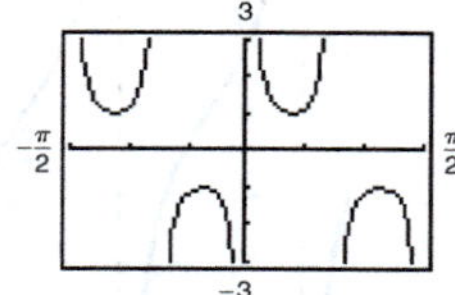

47. $y = 0.1\tan\left(\dfrac{\pi x}{4} + \dfrac{\pi}{4}\right)$

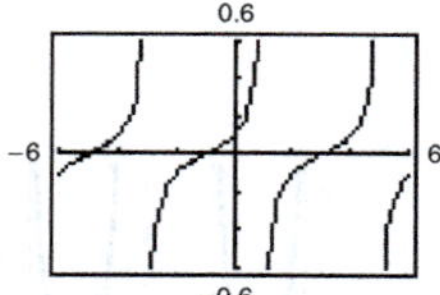

49. $\tan x = 1$

$x = -\dfrac{7\pi}{4}, -\dfrac{3\pi}{4}, \dfrac{\pi}{4}, \dfrac{5\pi}{4}$

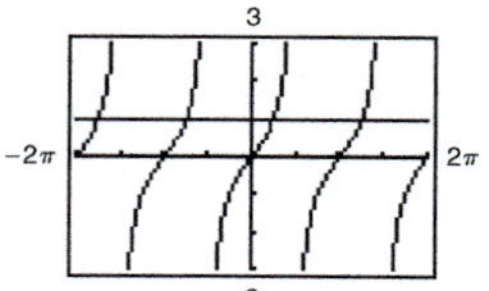

51. $\cot x = -\sqrt{3}$

$$x = -\frac{7\pi}{6}, -\frac{\pi}{6}, \frac{5\pi}{6}, \frac{11\pi}{6}$$

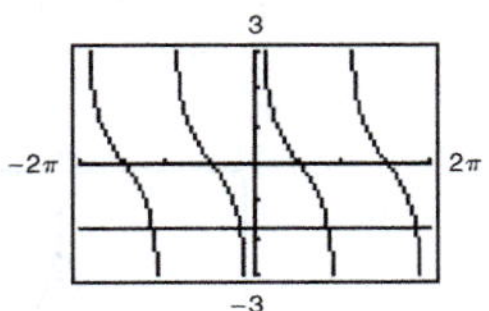

53. $\sec x = -2$

$$x = \frac{2\pi}{3}, \frac{4\pi}{3}, -\frac{2\pi}{3}, -\frac{4\pi}{3}$$

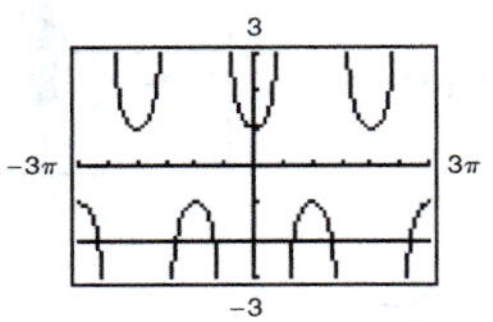

55. $\csc x = \sqrt{2}$

$$x = -\frac{7\pi}{4}, -\frac{5\pi}{4}, \frac{\pi}{4}, \frac{3\pi}{4}$$

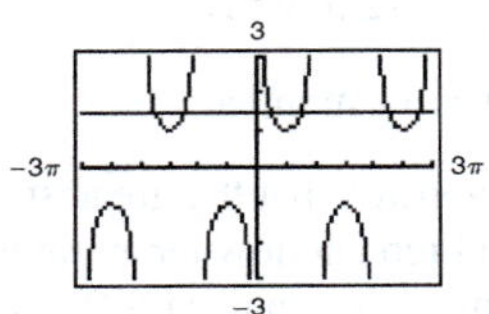

57. $f(x) = \sec x = \dfrac{1}{\cos x}$

$$f(-x) = \sec(-x) = \frac{1}{\cos(-x)} = \frac{1}{\cos x} = f(x)$$

So, $f(x) = \sec x$ is an even function and the graph has y-axis symmetry.

59. $g(x) = \cot x = \dfrac{1}{\tan x}$

$$g(-x) = \cot(-x) = \frac{1}{\tan(-x)} = -\frac{1}{\tan x} = -g(x)$$

So, $g(x) = \cot x$ is an odd function and the graph has origin symmetry.

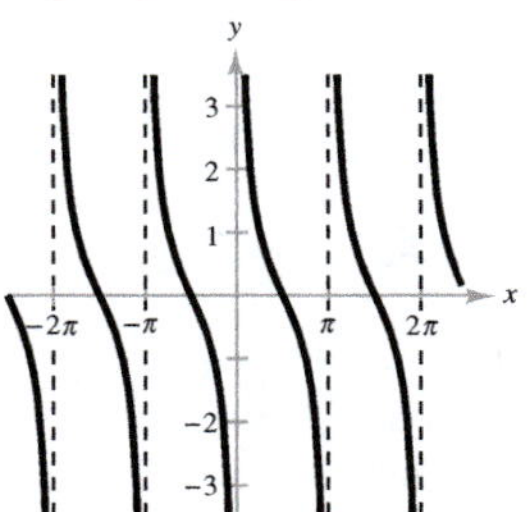

61. $f(x) = x + \tan x$

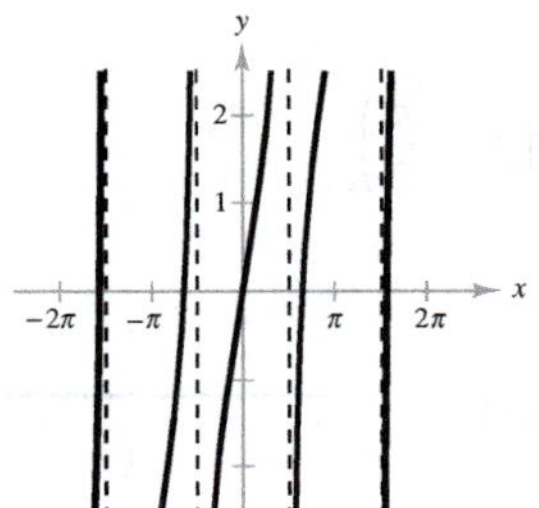

$$f(-x) = (-x) + \tan(-x) = -x - \tan x = -(x + \tan x) = -f(x)$$

So, $f(x) = x + \tan x$ is an odd function and the graph has origin symmetry.

63. $g(x) = x \csc x = \dfrac{x}{\sin x}$

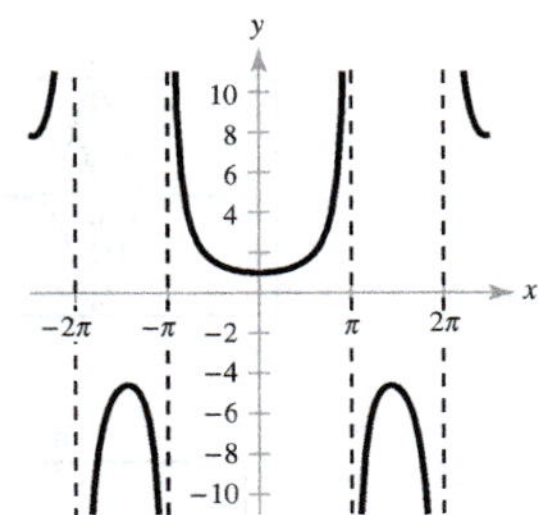

$$g(-x) = (-x)\csc(-x) = \frac{-x}{\sin(-x)} = \frac{-x}{-\sin x} = \frac{x}{\sin x} = x\csc x = g(x)$$

So, $g(x) = x \csc x$ is an even function and the graph has y-axis symmetry.

65. $f(x) = |x \cos x|$

Matches graph (d).

As $x \to 0$, $f(x) \to 0$.

66. $f(x) = x \sin x$

Matches graph (a)

As $x \to 0$, $f(x) \to 0$.

67. $g(x) = |x| \sin x$

Matches graph (b).

As $x \to 0$, $g(x) \to 0$.

68. $g(x) = |x| \cos x$

Matches graph (c).

As $x \to 0$, $g(x) \to 0$.

69. $f(x) = \sin x + \cos\left(x + \dfrac{\pi}{2}\right)$

$g(x) = 0$

$f(x) = g(x)$

The functions are equal.

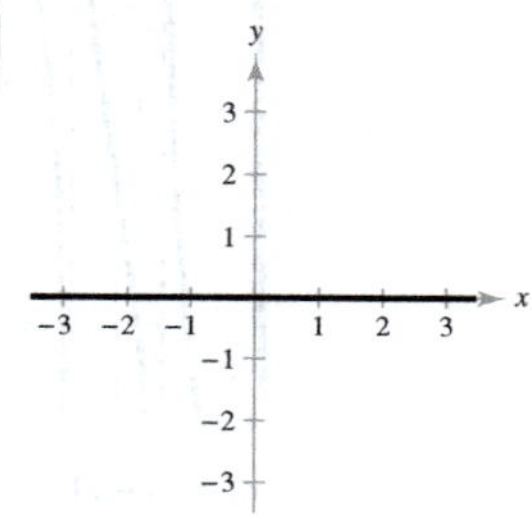

71. $f(x) = \sin^2 x$

$g(x) = \frac{1}{2}(1 - \cos 2x)$

$f(x) = g(x)$

The functions are equal.

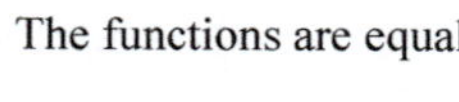

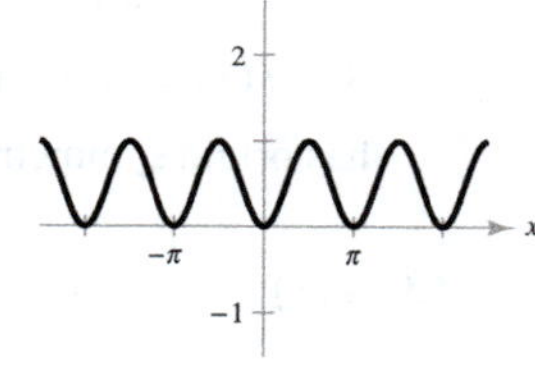

73. $g(x) = e^{-x^2/2} \sin x$

Damping factor: $e^{-x^2/2}$

As $x \to \infty$, $g(x) \to 0$.

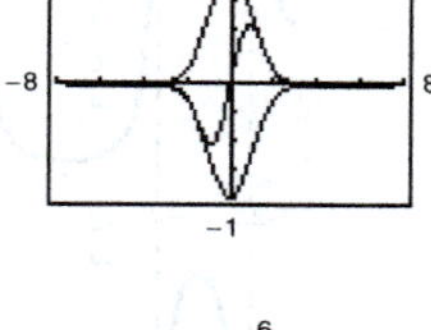

75. $f(x) = 2^{-x/4} \cos \pi x$

Damping factor: $y = 2^{-x/4}$

As $x \to \infty$, $f(x) \to 0$.

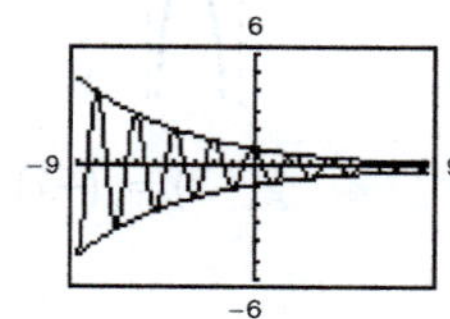

77. $y = \dfrac{6}{x} + \cos x,\ x > 0$

As $x \to 0$, $y \to \infty$.

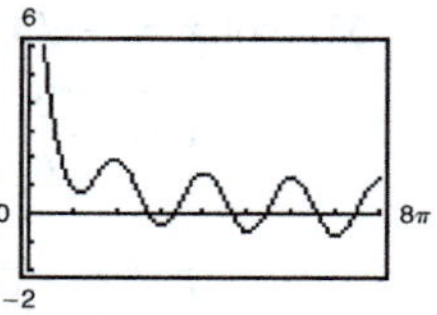

79. $g(x) = \dfrac{\sin x}{x}$

As $x \to 0$, $g(x) \to 1$.

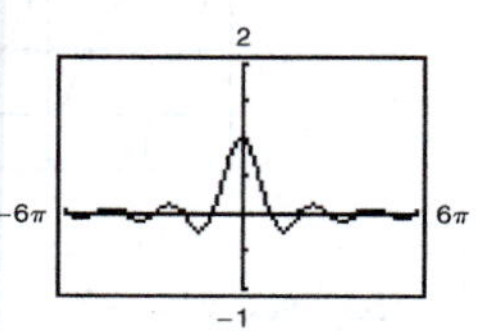

81. $f(x) = \sin \dfrac{1}{x}$

As $x \to 0$, $f(x)$ oscillates between -1 and 1.

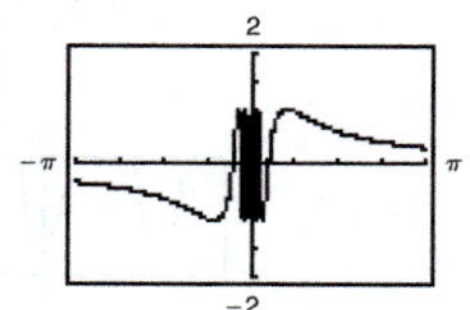

83. (a) Period of $\cos \dfrac{\pi t}{6} = \dfrac{2\pi}{\pi/6} = 12$

Period of $\sin \dfrac{\pi t}{6} = \dfrac{2\pi}{\pi/6} = 12$

The period of $H(t)$ is 12 months.

The period of $L(t)$ is 12 months.

(b) From the graph, it appears that the greatest difference between high and low temperatures occurs in the summer. The smallest difference occurs in the winter.

(c) The highest high and low temperatures appear to occur about half of a month after the time when the sun is northernmost in the sky.

85. $\cos x = \dfrac{27}{d}$

$d = \dfrac{27}{\cos x} = 27 \sec x,\ -\dfrac{\pi}{2} < x < \dfrac{\pi}{2}$

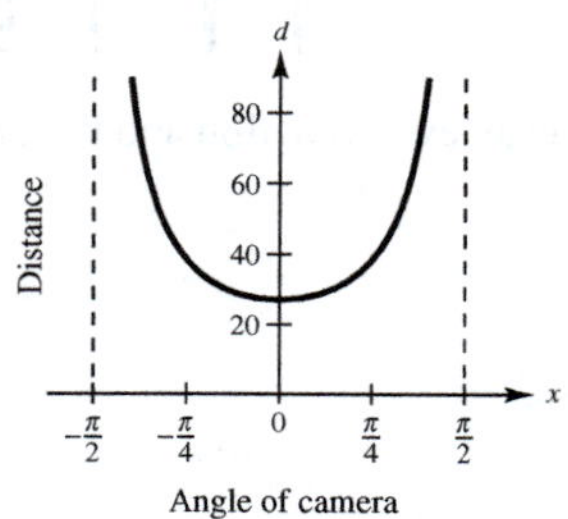

87. True. Because

$y = \csc x = \dfrac{1}{\sin x},$

for a given value of x, the y-coordinate of $\csc x$ is the reciprocal of the y-coordinate of $\sin x$.

89. $f(x) = x - \cos x$

(a)

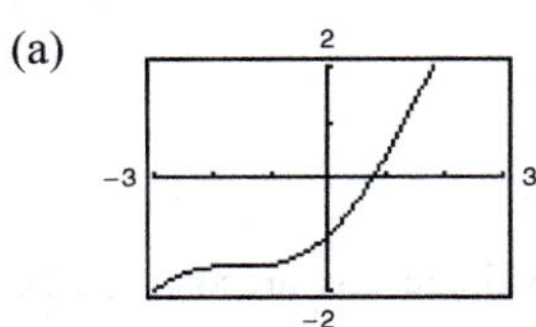

The zero between 0 and 1 occurs at $x \approx 0.7391$.

(b) $x_n = \cos(x_{n-1})$

$x_0 = 1$

$x_1 = \cos 1 \approx 0.5403$

$x_2 = \cos 0.5403 \approx 0.8576$

$x_3 = \cos 0.8576 \approx 0.6543$

$x_4 = \cos 0.6543 \approx 0.7935$

$x_5 = \cos 0.7935 \approx 0.7014$

$x_6 = \cos 0.7014 \approx 0.7640$

$x_7 = \cos 0.7640 \approx 0.7221$

$x_8 = \cos 0.7221 \approx 0.7504$

$x_9 = \cos 0.7504 \approx 0.7314$

$\vdots$

This sequence appears to be approaching the zero of f: $x \approx 0.7391$.

91. $f(x) = \cot x$

(a) $x \to 0^+, f(x) \to \infty$

(b) $x \to 0^-, f(x) \to -\infty$

(c) $x \to \pi^+, f(x) \to \infty$

(d) $x \to \pi^-, f(x) \to -\infty$

93. $f(x) = \tan x$

(a) $x \to \dfrac{\pi}{2}^+, f(x) \to -\infty$

(b) $x \to \dfrac{\pi}{2}^-, f(x) \to \infty$

(c) $x \to -\dfrac{\pi}{2}^+, f(x) \to -\infty$

(d) $x \to -\dfrac{\pi}{2}^-, f(x) \to \infty$

Section 4.7 Inverse Trigonometric Functions

Function	*Alternative Notation*	*Domain*	*Range*
1. $y = \arcsin x$	$y = \sin^{-1} x$	$-1 \le x \le 1$	$-\dfrac{\pi}{2} \le y \le \dfrac{\pi}{2}$
3. $y = \arctan x$	$y = \tan^{-1} x$	$-\infty < x < \infty$	$-\dfrac{\pi}{2} < y < \dfrac{\pi}{2}$

5. $y = \arcsin \dfrac{1}{2} \Rightarrow \sin y = \dfrac{1}{2}$ for $-\dfrac{\pi}{2} \le y \le \dfrac{\pi}{2} \Rightarrow y = \dfrac{\pi}{6}$

7. $y = \arccos \dfrac{1}{2} \Rightarrow \cos y = \dfrac{1}{2}$ for $0 \le y \le \pi \Rightarrow y = \dfrac{\pi}{3}$

9. $y = \arctan \dfrac{\sqrt{3}}{3} \Rightarrow \tan y = \dfrac{\sqrt{3}}{3}$ for $-\dfrac{\pi}{2} < y < \dfrac{\pi}{2} \Rightarrow y = \dfrac{\pi}{6}$

11. It is not possible to evaluate arcsin 3. The domain of the inverse sine function is $[-1, 1]$.

13. $y = \arctan(-\sqrt{3}) \Rightarrow \tan y = -\sqrt{3}$ for $-\dfrac{\pi}{2} < y < \dfrac{\pi}{2} \Rightarrow y = -\dfrac{\pi}{3}$

15. $y = \arccos\left(-\dfrac{1}{2}\right) \Rightarrow \cos y = -\dfrac{1}{2}$ for $0 \le y \le \pi \Rightarrow y = \dfrac{2\pi}{3}$

17. $y = \sin^{-1} -\dfrac{\sqrt{3}}{2} \Rightarrow \sin y = -\dfrac{\sqrt{3}}{2}$ for $-\dfrac{\pi}{2} \le y \le \dfrac{\pi}{2} \Rightarrow y = -\dfrac{\pi}{3}$

19. $f(x) = \cos x$

$g(x) = \arccos x$

$y = x$

21. $\arccos 0.37 = \cos^{-1}(0.37) \approx 1.19$

23. $\arcsin(-0.75) = \sin^{-1}(-0.75) \approx -0.85$

25. $\arctan(-3) = \tan^{-1}(-3) \approx -1.25$

27. It is not possible to evaluate $\sin^{-1} 1.36$. The domain of the inverse sine function is $[-1, 1]$.

29. $\arccos(-0.41) = \cos^{-1}(-0.41) \approx 1.99$

31. $\arctan 0.92 = \tan^{-1} 0.92 \approx 0.74$

33. $\arcsin \frac{7}{8} = \sin^{-1}\left(\frac{7}{8}\right) \approx 1.07$

35. $\tan^{-1}\left(-\frac{95}{7}\right) \approx -1.50$

37. $\arctan\left(-\sqrt{3}\right) = -\dfrac{\pi}{3}$

$\tan\left(-\dfrac{\pi}{6}\right) = -\dfrac{\sqrt{3}}{3}$

$\tan\left(\dfrac{\pi}{4}\right) = 1$

39. $\tan \theta = \dfrac{x}{4}$

$\theta = \arctan \dfrac{x}{4}$

41. $\sin \theta = \dfrac{x+2}{5}$

$\theta = \arcsin\left(\dfrac{x+2}{5}\right)$

43. $\cos \theta = \dfrac{x+3}{2x}$

$\theta = \arccos \dfrac{x+3}{2x}$

45. $\sin(\arcsin 0.3) = 0.3$

47. It is not possible to evaluate $\cos\left[\arccos\left(-\sqrt{3}\right)\right]$. The domain of the inverse cosine function is $[-1, 1]$.

49. $\arcsin\left[\sin\left(\dfrac{9\pi}{4}\right)\right] = \arcsin\left(\dfrac{\sqrt{2}}{2}\right) = \dfrac{\pi}{4}.$

Note: $\dfrac{9\pi}{4}$ is not in the range of the arcsin function.

51. Let $u = \arctan \dfrac{3}{4}$.

$\tan u = \dfrac{3}{4}, 0 < u < \dfrac{\pi}{2},$

$\sin\left(\arctan \dfrac{3}{4}\right) = \sin u = \dfrac{3}{5}$

53. Let $u = \tan^{-1} 2,$

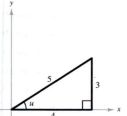

$\tan u = 2 = \dfrac{2}{1}, 0 < u < \dfrac{\pi}{2},$

$\cos\left(\tan^{-1} 2\right) = \cos u = \dfrac{1}{\sqrt{5}} = \dfrac{\sqrt{5}}{5}.$

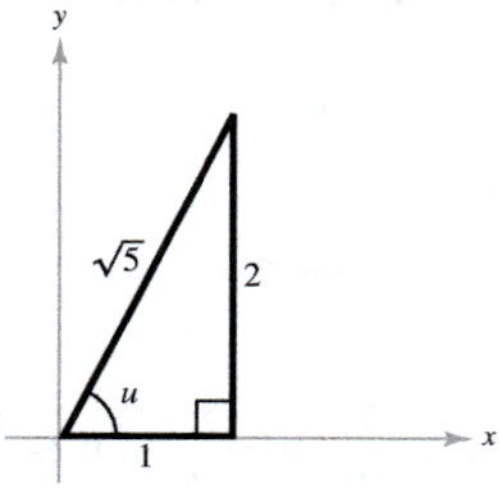

55. Let $u = \arcsin \dfrac{5}{13},$

$\sin u = \dfrac{5}{13}, 0 < u < \dfrac{\pi}{2},$

$\sec\left(\arcsin \dfrac{5}{13}\right) = \sec u = \dfrac{13}{12}.$

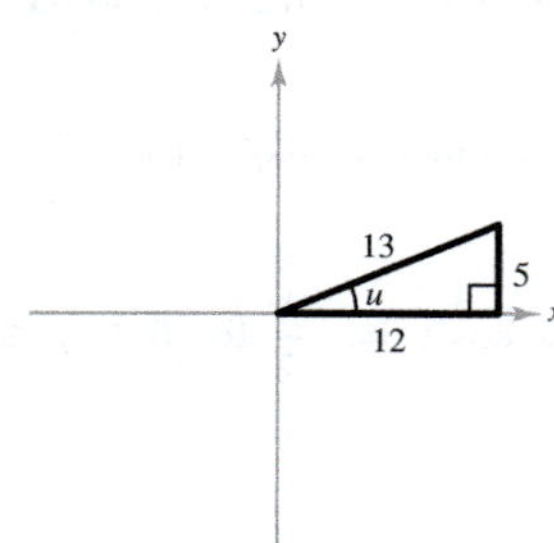

57. Let $u = \arctan\left(-\frac{3}{5}\right)$,

$\tan u = -\frac{3}{5}, -\frac{\pi}{2} < y < 0,$

$\cot\left[\arctan\left(-\frac{3}{5}\right)\right] = \cot u = -\frac{5}{3}.$

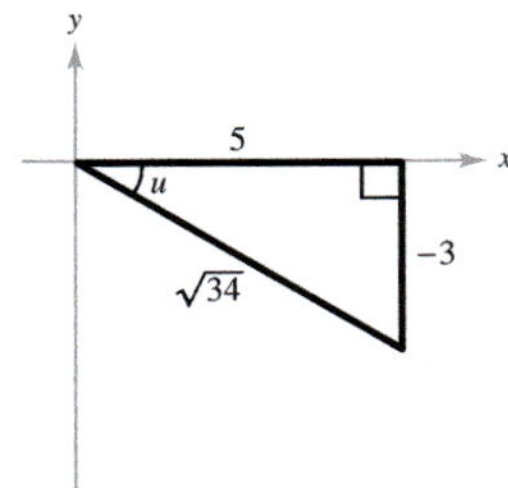

59. Let $u = \arccos\left(-\frac{2}{3}\right)$.

$\cos u = -\frac{2}{3}, \frac{\pi}{2} < u < \pi,$

$\tan\left[\arccos\left(-\frac{2}{3}\right)\right] = \tan u = -\frac{\sqrt{5}}{2}$

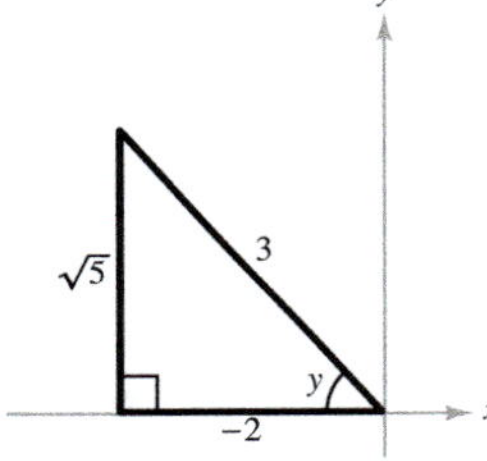

61. Let $u = \cos^{-1}\frac{\sqrt{3}}{2}$.

$\cos u = \frac{\sqrt{3}}{2}, 0 < u < \frac{\pi}{2},$

$\csc\left[\cos^{-1}\frac{\sqrt{3}}{2}\right] = \csc u = 2.$

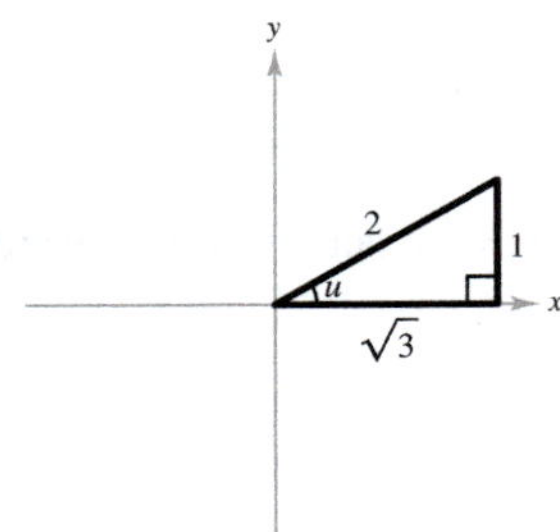

63. Let $u = \arcsin(2x)$.

$\sin u = 2x = \frac{2x}{1},$

$\cos(\arcsin 2x) = \cos u = \sqrt{1 - 4x^2}$

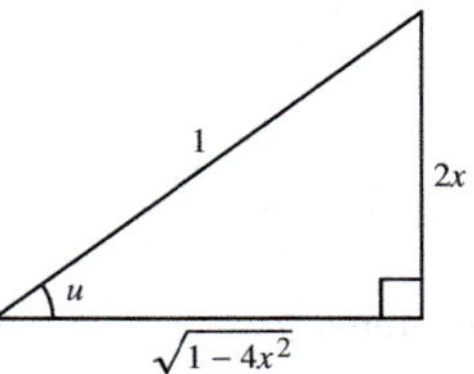

65. Let $u = \arctan x$.

$\tan u = x = \frac{x}{1},$

$\cot(\arctan x) = \cot u = \frac{1}{x}$

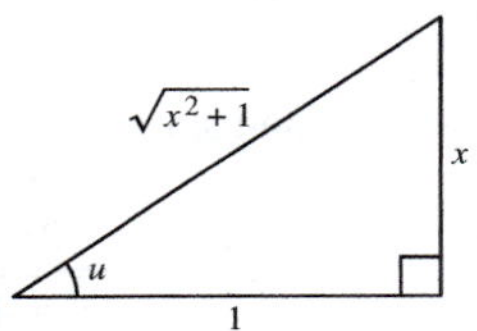

67. Let $u = \arccos x$.

$\cos u = x = \frac{x}{1},$

$\sin(\arccos x) = \sin u = \sqrt{1 - x^2}$

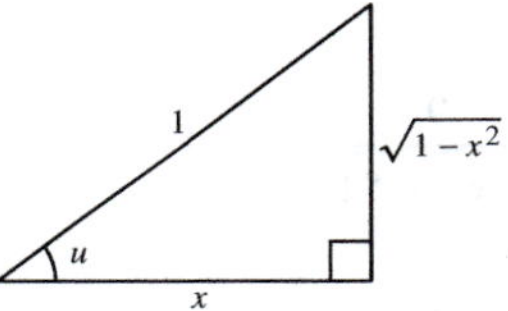

69. Let $u = \arccos\left(\frac{x}{3}\right)$.

$\cos u = \frac{x}{3},$

$\tan\left(\arccos\frac{x}{3}\right) = \tan u = \frac{\sqrt{9 - x^2}}{x}$

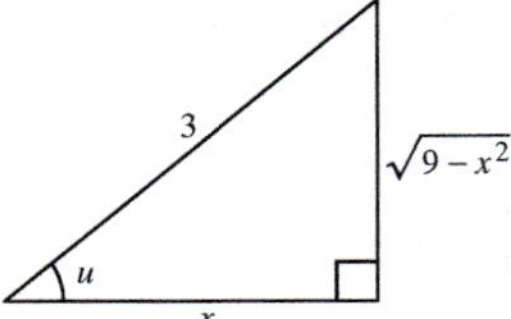

71. Let $u = \arctan \dfrac{x}{a}$.

$\tan u = \dfrac{x}{a}$,

$\csc\left(\arctan \dfrac{x}{a}\right) = \csc u = \dfrac{\sqrt{x^2 + a^2}}{a}$

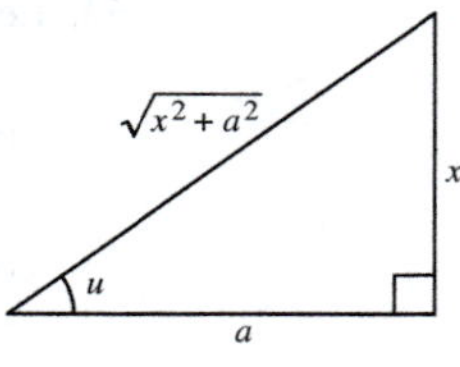

73. $f(x) = \sin(\arctan 2x)$, $g(x) = \dfrac{2x}{\sqrt{1 + 4x^2}}$

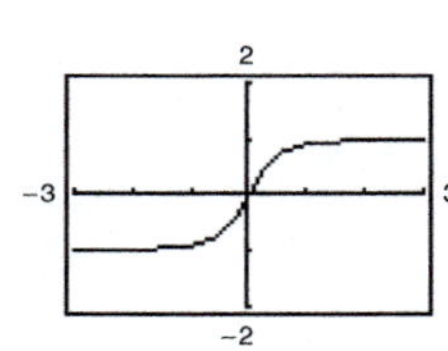

They are equal. Let $u = \arctan 2x$,

$\tan u = 2x = \dfrac{2x}{1}$,

and $\sin u = \dfrac{2x}{\sqrt{1 + 4x^2}}$.

$g(x) = \dfrac{2x}{\sqrt{1 + 4x^2}} = f(x)$

The graph has horizontal asymptotes at $y = \pm 1$.

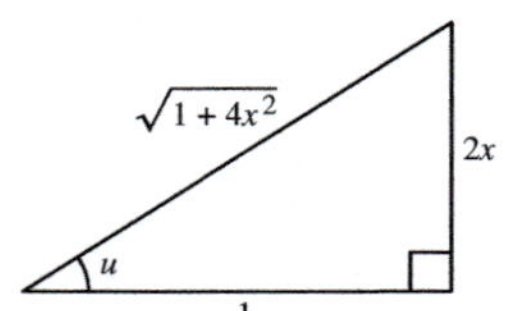

75. Let $u = \arctan \dfrac{9}{x}$.

$\tan u = \dfrac{9}{x}$ and $\sin u = \dfrac{9}{\sqrt{x^2 + 81}}$, $x > 0$

So,

$\arctan \dfrac{9}{x} = \arcsin \dfrac{9}{\sqrt{x^2 + 81}}$, $x > 0$.

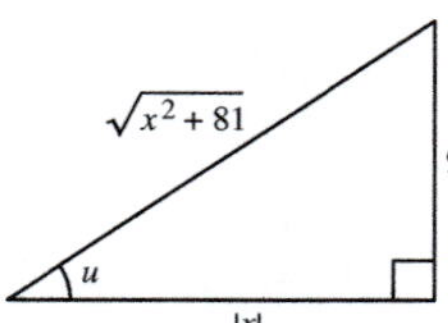

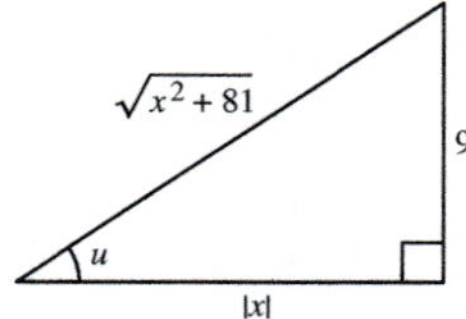

77. Let $u = \arccos \dfrac{3}{\sqrt{x^2 - 2x + 10}}$. Then,

$\cos u = \dfrac{3}{\sqrt{x^2 - 2x + 10}} = \dfrac{3}{\sqrt{(x - 1)^2 + 9}}$

and $\sin u = \dfrac{|x - 1|}{\sqrt{(x - 1)^2 + 9}}$.

So, $u = \arcsin \dfrac{|x - 1|}{\sqrt{x^2 - 2x + 10}}$.

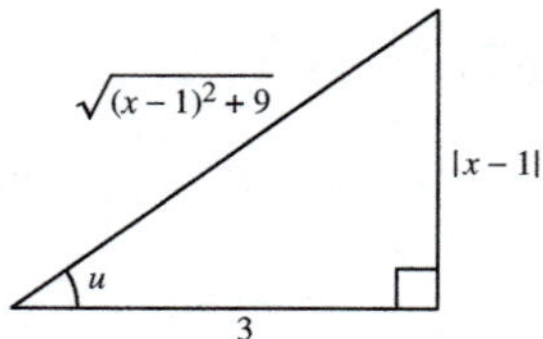

79. $g(x) = 2 \arcsin x$

Domain: $-1 \le x \le 1$

Range: $-\pi \le y \le \pi$

This is the graph of $f(x) = \arcsin(x)$ with a vertical stretch.

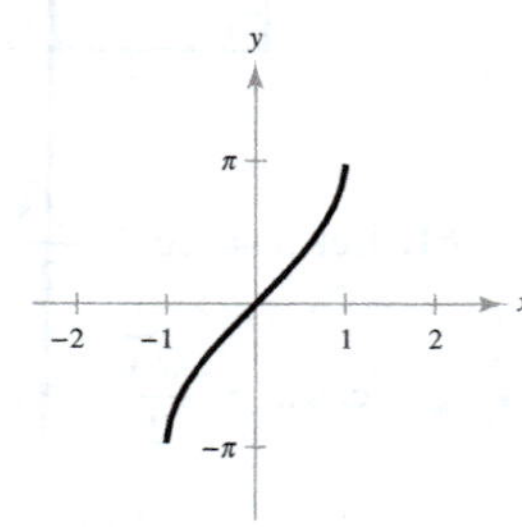

81. $f(x) = \dfrac{\pi}{2} + \arctan x$

Domain: all real numbers

Range: $0 < y \le \pi$

This is the graph of $y = \arctan x$ shifted upward $\pi/2$ units.

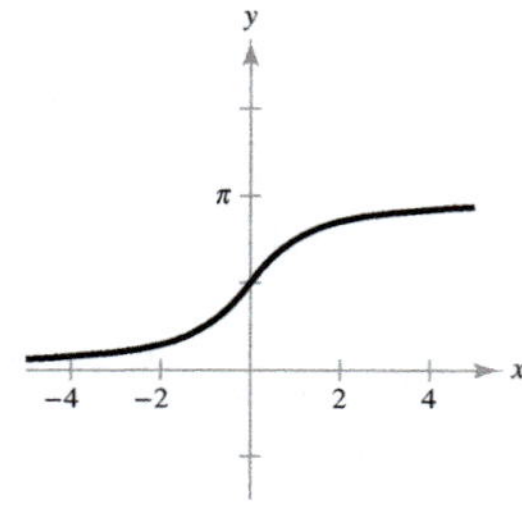

83. $h(v) = \arccos \dfrac{v}{2}$

Domain: $-2 \le v \le 2$

Range: $0 \le y \le \pi$

This is the graph of $h(v) = \arccos v$ with a horizontal stretch.

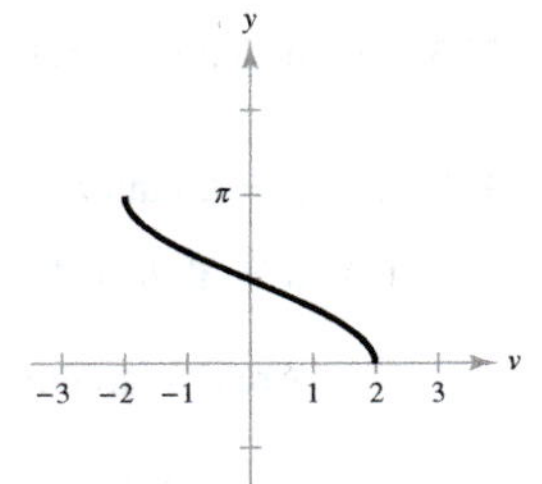

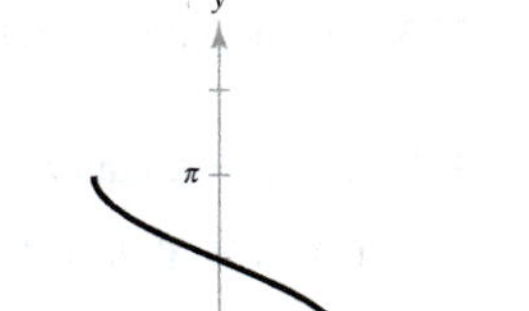

85. $f(x) = 2 \arccos(2x)$

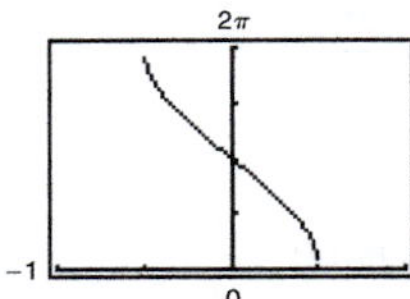

87. $f(x) = \arctan(2x - 3)$

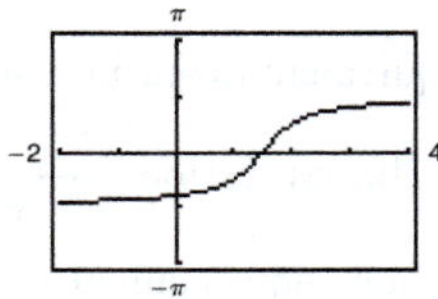

89. $f(x) = \pi - \sin^{-1}\left(\dfrac{2}{3}\right) \approx 2.412$

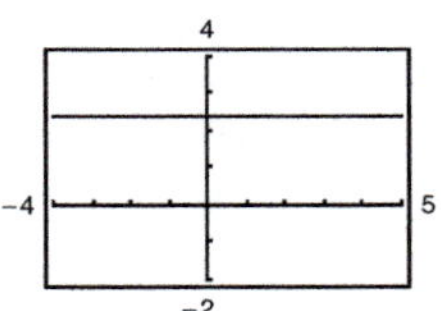

91. $f(t) = 3 \cos 2t + 3 \sin 2t = \sqrt{3^2 + 3^2} \sin\left(2t + \arctan \dfrac{3}{3}\right)$

$= 3\sqrt{2} \sin(2t + \arctan 1)$

$= 3\sqrt{2} \sin\left(2t + \dfrac{\pi}{4}\right)$

The graph implies that the identity is true.

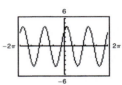

93. $\dfrac{\pi}{2}$

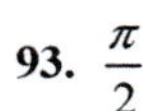

95. $\dfrac{\pi}{2}$

97. π

99. (a) $\sin \theta = \dfrac{5}{s}$

$\theta = \arcsin \dfrac{5}{s}$

(b) $s = 40$: $\theta = \arcsin \dfrac{5}{40} \approx 0.13$

$s = 20$: $\theta = \arcsin \dfrac{5}{20} \approx 0.25$

101. (a) $\tan \theta = \dfrac{5.5}{8.5}$

$\theta \approx 32.9°$

(b) $\tan 32.9° = \dfrac{h}{10}$

$h = 10 \tan 32.9°$

$h \approx 6.49$ meters

The height is about 6.5 meters.

103. $\beta = \arctan \dfrac{3x}{x^2 + 4}$

(a)

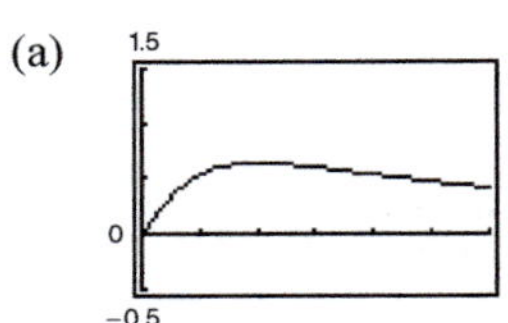

(b) β is maximum when $x = 2$ feet.

(c) The graph has a horizontal asymptote at $\beta = 0$. As x increases, β decreases.

105. (a) $\tan \theta = \dfrac{x}{20}$

$\theta = \arctan \dfrac{x}{20}$

(b) $x = 5$: $\theta = \arctan \dfrac{5}{20} \approx 14.0°$

$x = 12$: $\theta = \arctan \dfrac{12}{20} \approx 31.0°$

107. True. $-\dfrac{\pi}{4}$ is in the range of the arctangent function.

$\tan\left(-\dfrac{\pi}{4}\right) = -1 \Leftrightarrow \arctan(-1) = -\dfrac{\pi}{4}.$

109. False. $\sin^{-1} x \neq \dfrac{1}{\sin x}$

The function $\sin^{-1} x$ is equivalent to arcsin x, which is the inverse sine function. The expression, $\dfrac{1}{\sin x}$ is the reciprocal of the sine function and is equivalent to csc x.

111. $y = \operatorname{arccot} x$ if and only if $\cot y = x$.

Domain: $(-\infty, \infty)$

Range: $(0, \pi)$

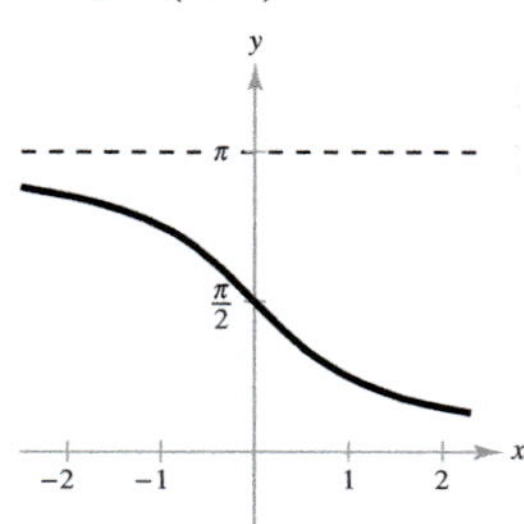

113. $y = \operatorname{arccsc} x$ if and only if $\csc y = x$.

Domain: $(-\infty, -1] \cup [1, \infty)$

Range: $\left[-\dfrac{\pi}{2}, 0\right) \cup \left(0, \dfrac{\pi}{2}\right]$

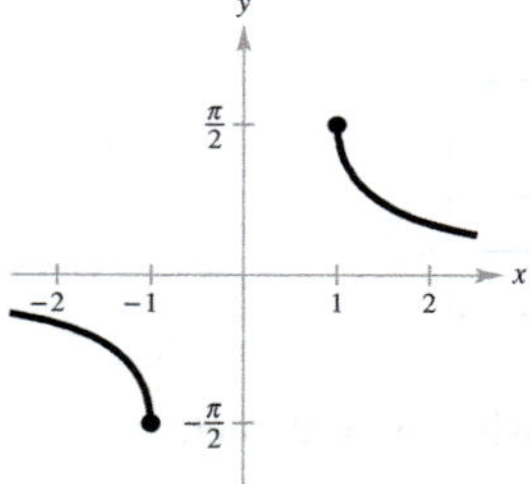

115. $y = \operatorname{arcsec} \sqrt{2} \Rightarrow \sec y = \sqrt{2}$ and

$0 \leq y < \dfrac{\pi}{2} \cup \dfrac{\pi}{2} < y \leq \pi \Rightarrow y = \dfrac{\pi}{4}$

117. $y = \operatorname{arccot}(-1) \Rightarrow \cot y = -1$ and

$0 < y < \pi \Rightarrow y = \dfrac{3\pi}{4}$

119. $y = \operatorname{arccsc}(-1) \Rightarrow \csc y = -1$ and

$-\dfrac{\pi}{2} \leq y < 0 \cup 0 < y \leq \dfrac{\pi}{2} \Rightarrow y = -\dfrac{\pi}{2}$

121. $\operatorname{arcsec} 2.54 = \arccos\left(\dfrac{1}{2.54}\right) \approx 1.17$

123. $\operatorname{arccsc}\left(-\frac{25}{3}\right) = \arcsin\left(-\frac{3}{25}\right) \approx -0.12$

125. $\operatorname{arccot} 5.25 = \arctan\left(\dfrac{1}{5.25}\right) \approx 0.19$

127. Area $= \arctan b - \arctan a$

(a) $a = 0, b = 1$

Area $= \arctan 1 - \arctan 0 = \dfrac{\pi}{4} - 0 = \dfrac{\pi}{4}$

(b) $a = -1, b = 1$

Area $= \arctan 1 - \arctan(-1)$

$= \dfrac{\pi}{4} - \left(-\dfrac{\pi}{4}\right) = \dfrac{\pi}{2}$

(c) $a = 0, b = 3$

Area $= \arctan 3 - \arctan 0$

$\approx 1.25 - 0 = 1.25$

(d) $a = -1, b = 3$

Area $= \arctan 3 - \arctan(-1)$

$\approx 1.25 - \left(-\dfrac{\pi}{4}\right) \approx 2.03$

129. $f(x) = \sin(x)$, $f^{-1}(x) = \arcsin(x)$

(a) $f \circ f^{-1} = \sin(\arcsin x)$

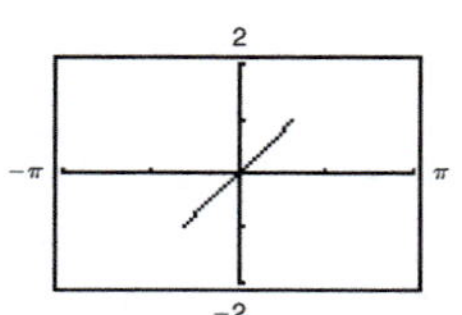

$f^{-1} \circ f = \arcsin(\sin x)$

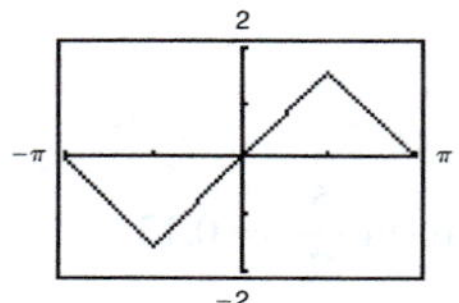

(b) The graphs coincide with the graph of $y = x$ only for certain values of x.

$f \circ f^{-1} = x$ over its entire domain, $-1 \leq x \leq 1$.

$f^{-1} \circ f = x$ over the region $-\dfrac{\pi}{2} \leq x \leq \dfrac{\pi}{2}$, corresponding to the region where $\sin x$ is one-to-one and has an inverse.

Section 4.8 Applications and Models

1. bearing

3. period

5. Given: $A = 60°, c = 12$

$$\sin A = \frac{a}{c} \Rightarrow a = c \sin A = 12 \sin 60° = \frac{12\sqrt{3}}{2} = 6\sqrt{3} \approx 10.39$$

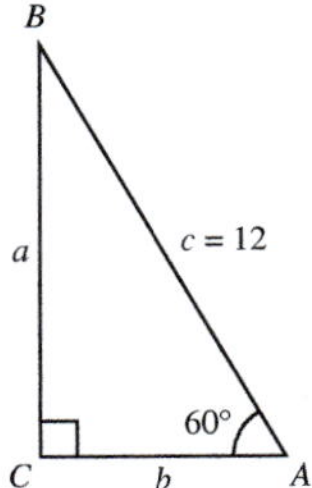

7. Given: $B = 72.8°, a = 4.4$

$$\cos B = \frac{a}{c} \Rightarrow c = \frac{a}{\cos B} = \frac{4.4}{\cos 72.8°} \approx 14.88$$

$$\tan B = \frac{b}{a} \Rightarrow b = a \tan B = 4.4 \tan 72.8° \approx 14.21$$

$$A = 90° - 72.8° = 17.2°$$

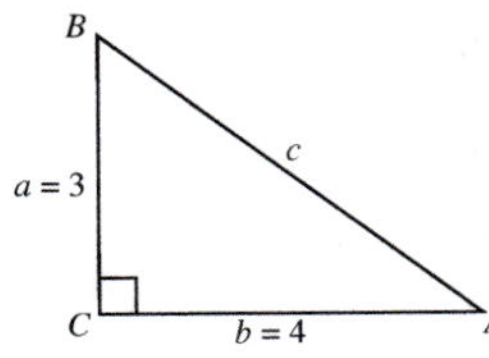

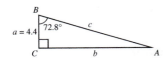

9. Given: $a = 3, b = 4$

$$a^2 + b^2 = c^2 \Rightarrow c^2 = (3)^2 + (4)^2 \Rightarrow c = 5$$

$$\tan A = \frac{a}{b} \Rightarrow A = \tan^{-1}\left(\frac{a}{b}\right) = \tan^{-1}\left(\frac{3}{4}\right) \approx 36.87°$$

$$B = 90° - 36.87° = 53.13°$$

11. Given: $b = 15.70, c = 55.16$

$$a = \sqrt{55.16^2 - 15.7^2} \approx 52.88$$

$$\cos A = \frac{b}{c}$$

$$\cos A = \frac{15.7}{55.15}$$

$$A = \arccos \frac{15.7}{55.15} \approx 73.46°$$

$$B = 90° - 73.46° \approx 16.54°$$

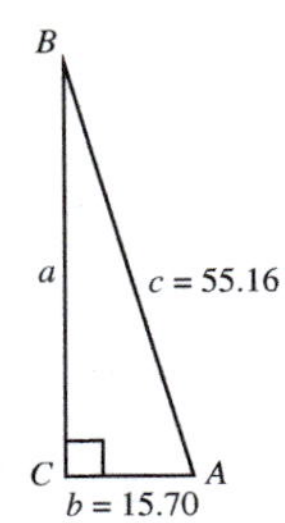

13. $\theta = 45°, b = 6$

$$\tan \theta = \frac{h}{(1/2)b} \Rightarrow h = \frac{1}{2}b \tan \theta$$

$$h = \frac{1}{2}(6) \tan 45° = 3.00 \text{ units}$$

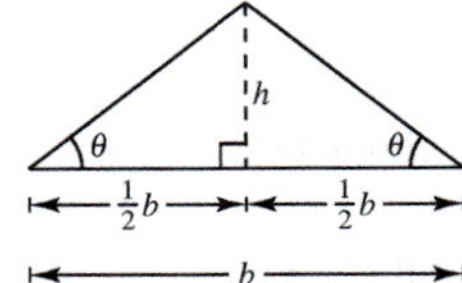

15. $\tan \theta = \dfrac{h}{(1/2)b} \Rightarrow h = \dfrac{1}{2}b \tan \theta$

$$h = \frac{1}{2}(8) \tan 32° \approx 2.50 \text{ units}$$

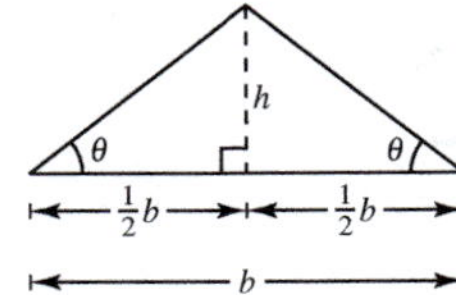

17. $\tan 25° = \dfrac{100}{x}$

$$x = \frac{100}{\tan 25°}$$

$$\approx 214.45 \text{ feet}$$

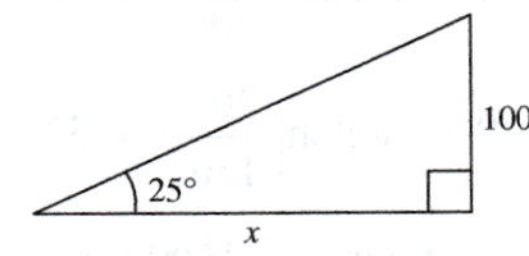

19. $\sin 80° = \dfrac{h}{20}$

$20 \sin 80° = h$

$h \approx 19.7$ feet

21. Let the height of the church $= x$ and the height of the church and steeple $= y$.

$\tan 35° = \dfrac{x}{50}$ and $\tan 48° = \dfrac{y}{50}$

$x = 50 \tan 35° \approx 35.01$ and $y = 50 \tan 48° \approx 55.53$

$h = y - x = 55.53 - 35.01 = 20.52.$

$h \approx 20.5$ feet

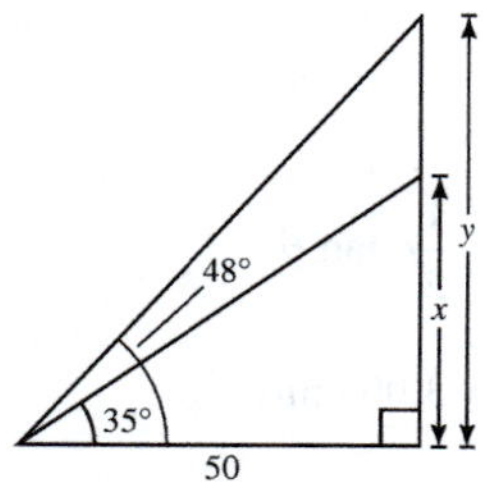

23. $\cot 55 = \dfrac{d}{10} \Rightarrow d \approx 7$ kilometers

$\cot 28° = \dfrac{D}{10} \Rightarrow D \approx 18.8$ kilometers

Distance between towns:

$D - d = 18.8 - 7 = 11.8$ kilometers

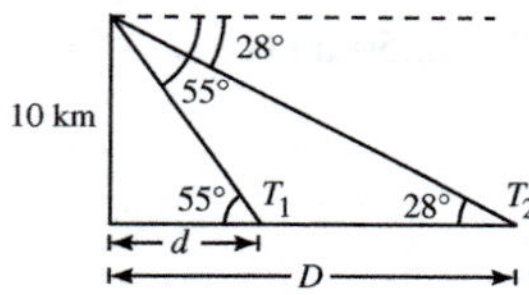

25. $\tan \theta = \frac{75}{50}$

$\theta = \arctan \frac{3}{2} \approx 56.3°$

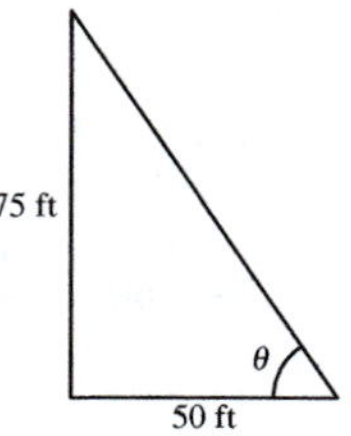

27. $12{,}500 + 4000 = 16{,}500$

$\sin \theta = \dfrac{4000}{16{,}500}$

$\theta = \arcsin\left(\dfrac{4000}{16{,}500}\right)$

$\theta \approx 14.03°$

Angle of depression $= \alpha \approx 90° - 14.03° = 75.97°$

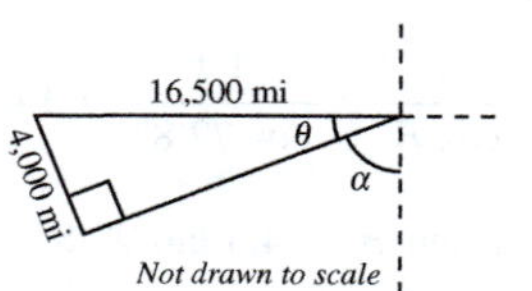

29. $\tan 57° = \dfrac{a}{x} \Rightarrow x = a \cot 57°$

$\tan 16° = \dfrac{a}{x + (55/6)}$

$\tan 16° = \dfrac{a}{a \cot 57° + (55/6)}$

$\cot 16° = \dfrac{a \cot 57° + (55/6)}{a}$

$a \cot 16° - a \cot 57° = \dfrac{55}{6} \Rightarrow a \approx 3.23$ miles

$\approx 17{,}054$ feet

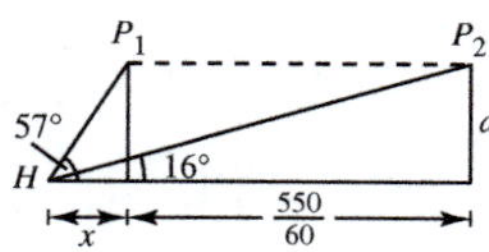

31. (a) $l^2 = (200)^2 + (150)^2$

$l = 250$ feet

$\tan A = \dfrac{150}{200} \Rightarrow A = \arctan\left(\dfrac{150}{200}\right) \approx 36.87°$

$\tan B = \dfrac{200}{150} \Rightarrow B = \arctan\left(\dfrac{200}{150}\right) \approx 53.13°$

(b) $250 \text{ ft} \times \dfrac{\text{mile}}{5280 \text{ ft}} \times \dfrac{\text{hour}}{35 \text{ miles}} \times \dfrac{3600 \text{ sec}}{\text{hour}} \approx 4.87$ seconds

33. The plane has traveled $1.5(550) = 825$ miles.

$$\sin 38° = \frac{a}{825} \Rightarrow a \approx 508 \text{ miles north}$$

$$\cos 38° = \frac{b}{825} \Rightarrow b \approx 650 \text{ miles east}$$

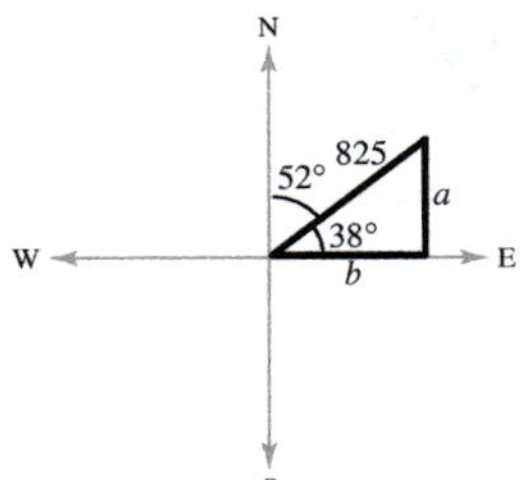

35.

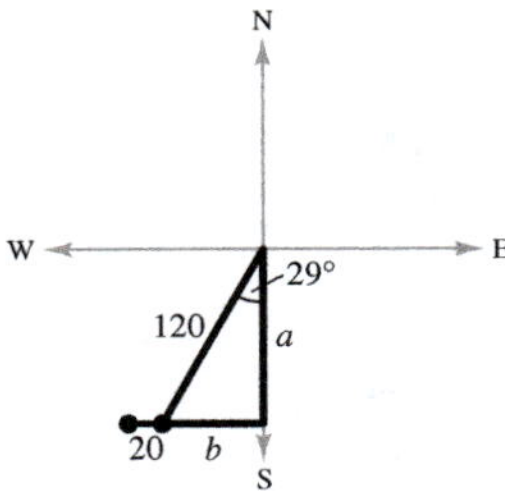

(a) $\cos 29° = \dfrac{a}{120} \Rightarrow a \approx 104.95$ nautical miles south

$\sin 29° = \dfrac{b}{120} \Rightarrow b \approx 58.18$ nautical miles west

(b) $\tan \theta = \dfrac{20 + b}{a} \approx \dfrac{78.18}{104.95} \Rightarrow \theta \approx 36.7°$

Bearing: S 36.7° W

Distance: $d \approx \sqrt{104.95^2 + 78.18^2}$

≈ 130.9 nautical miles from port

37. $\tan \theta = \frac{45}{30} \Rightarrow \theta \approx 56.3°$

Bearing: N 56.31°

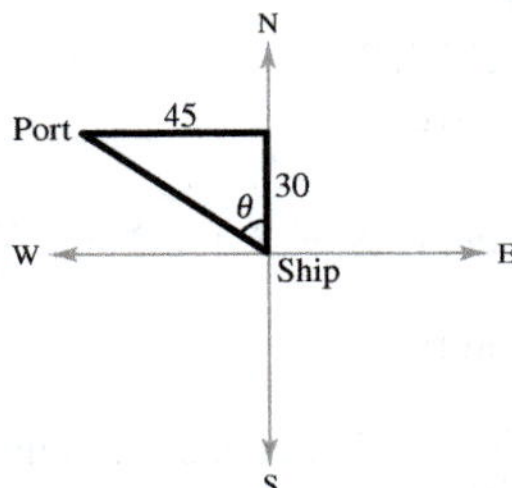

39. $\theta = 32°, \phi = 68°$

(a) $\alpha = 90° - 32° = 58°$

Bearing from A to C: N 58° E

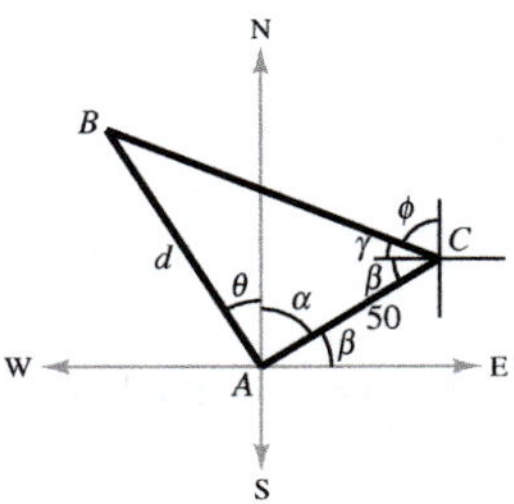

(b) $\beta = \theta = 32°$

$\gamma = 90° - \phi = 22°$

$C = \beta + \gamma = 54°$

$$\tan C = \frac{d}{50} \Rightarrow \tan 54°$$

$$= \frac{d}{50} \Rightarrow d \approx 68.82 \text{ meters}$$

41. The diagonal of the base has a length of $\sqrt{a^2 + a^2} = \sqrt{2}a$. Now, you have

$$\tan \theta = \frac{a}{\sqrt{2}a} = \frac{1}{\sqrt{2}}$$

$$\theta = \arctan \frac{1}{\sqrt{2}}$$

$$\theta \approx 35.3°.$$

43. $\sin 36° = \dfrac{d}{25} \Rightarrow d \approx 14.69$

Length of side: $2d \approx 29.4$ inches

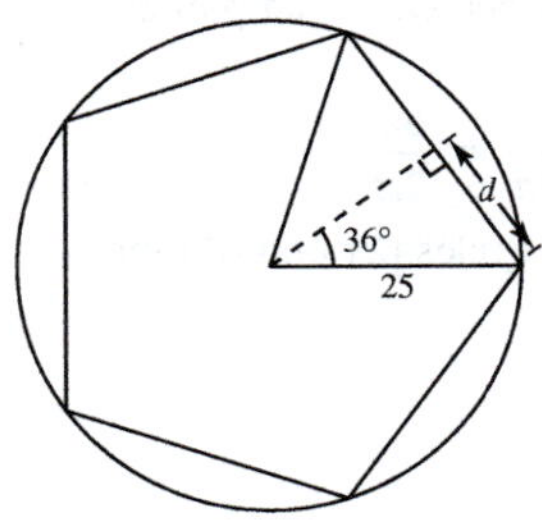

45. Use $d = a \sin \omega t$ because $d = 0$ when $t = 0$.

Period: $\dfrac{2\pi}{\omega} = 2 \Rightarrow \omega = \pi$

So, $d = 4 \sin(\pi t)$.

47. Use $d = a \cos \omega t$ because $d = 3$ when $t = 0$.

Period: $\dfrac{2\pi}{\omega} = 1.5 \Rightarrow \omega = \dfrac{4\pi}{3}$

So, $d = 3 \cos\left(\dfrac{4\pi}{3}t\right) = 3 \cos\left(\dfrac{4\pi t}{3}\right)$.

49. $d = a \sin \omega t$

$$\text{Frequency} = \frac{\omega}{2\pi}$$

$$262 = \frac{\omega}{2\pi}$$

$$\omega = 2\pi(262) = 524\pi$$

51. $d = 9 \cos \frac{6\pi}{5}t$

(a) Maximum displacement = amplitude = 9

(b) $\text{Frequency} = \frac{\omega}{2\pi} = \frac{\frac{6\pi}{5}}{2\pi}$

$= \frac{3}{5}$ cycle per unit of time

(c) $d = 9 \cos \frac{6\pi}{5}(5) = 9$

(d) $9 \cos \frac{6\pi}{5}t = 0$

$$\cos \frac{6\pi}{5}t = 0$$

$$\frac{6\pi}{5}t = \arccos 0$$

$$\frac{6\pi}{5}t = \frac{\pi}{2}$$

$$t = \frac{5}{12}$$

53. $d = \frac{1}{4} \sin 6\pi t$

(a) Maximum displacement = amplitude = $\frac{1}{4}$

(b) $\text{Frequency} = \frac{\omega}{2\pi} = \frac{6\pi}{2\pi}$

= 3 cycles per unit of time

(c) $d = \frac{1}{4} \sin 30\pi \approx 0$

(d) $\frac{1}{4} \sin 6\pi t = 0$

$$\sin 6\pi t = 0$$

$$6\pi t = \arcsin 0$$

$$6\pi t = \pi$$

$$t = \frac{1}{6}$$

55. $y = \frac{1}{4} \cos 16t, t > 0$

(a)

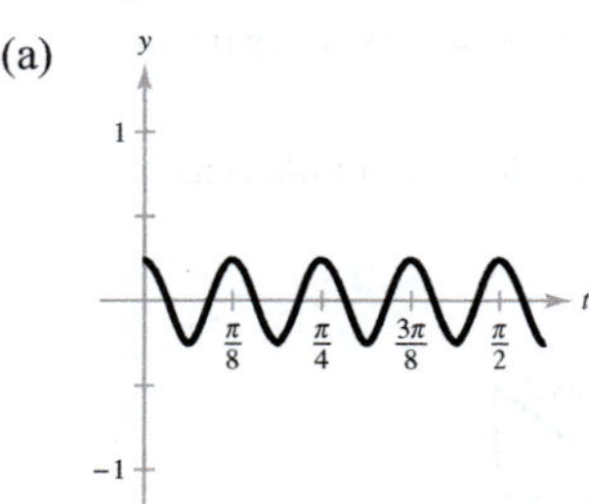

(b) Period: $\frac{2\pi}{16} = \frac{\pi}{8}$

(c) $\frac{1}{4} \cos 16t = 0$ when $16t = \frac{\pi}{2} \Rightarrow t = \frac{\pi}{32}$

57. (a)

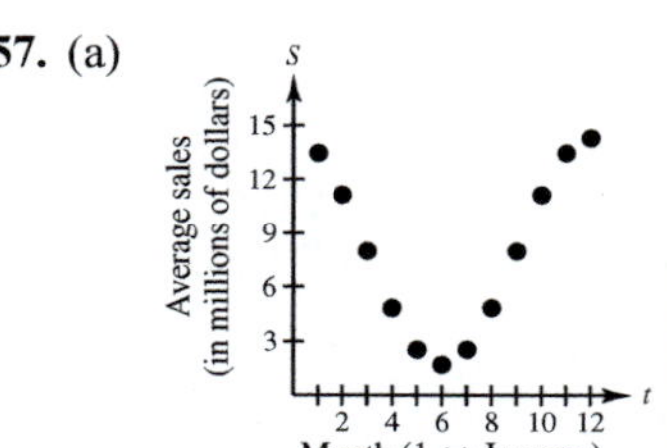

(b) $a = \frac{1}{2}(14.3 - 1.7) = 6.3$

$$\frac{2\pi}{b} = 12 \Rightarrow b = \frac{\pi}{6}$$

Shift: $d = 14.3 - 6.3 = 8$

$$S = d + a \cos bt$$

$$S = 8 + 6.3 \cos\left(\frac{\pi t}{6}\right)$$

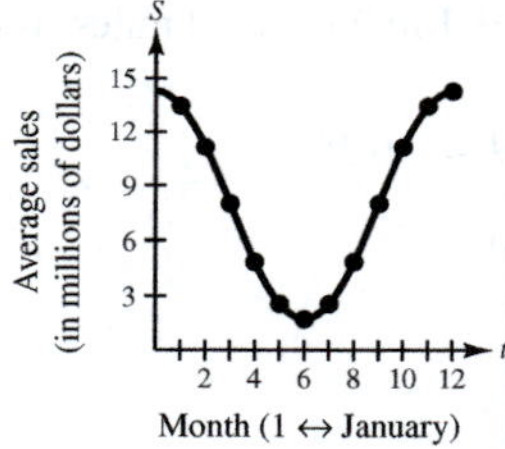

Note: Another model is $S = 8 + 6.3 \sin\left(\frac{\pi t}{6} + \frac{\pi}{2}\right)$.

The model is a good fit.

(c) The period is $\frac{2\pi}{(\pi/6)} = 12$. Yes, sales of outwear are seasonal.

(d) The amplitude is the maximum displacement from average sales of $8 million

59. False. The tower isn't vertical and so the triangle formed is not a right triangle.

Review Exercises for Chapter 4

1. $\theta = \dfrac{15\pi}{4}$

(a)

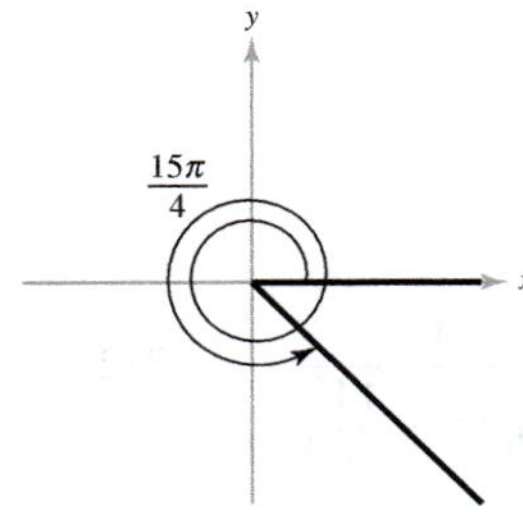

(b) Quadrant IV

(c) $\dfrac{15\pi}{4} - 2\pi = \dfrac{7\pi}{4}$

$\dfrac{7\pi}{4} - 2\pi = -\dfrac{\pi}{4}$

3. $\theta = -110°$

(a)

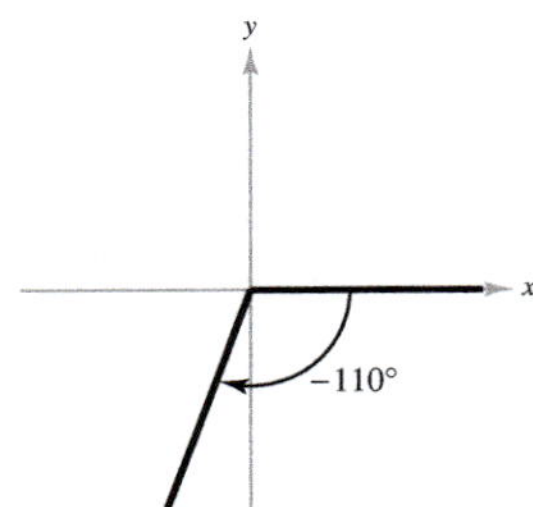

(b) Quadrant III

(c) Coterminal angles:

$-110° + 360° = 250°$

$-110° - 360° = -470°$

5. $450° = 450° \cdot \dfrac{\pi \text{ rad}}{180°} = \dfrac{5\pi}{2} \approx 7.854$ radians

7. $-16° = -16° \cdot \dfrac{\pi \text{ rad}}{180°} \approx -0.279$ radians

9. $\dfrac{3\pi}{10} = \dfrac{3\pi}{10} \cdot \dfrac{180°}{\pi \text{ rad}} = 54°$

11. $-3.5 \text{ rad} = -3.5 \text{ rad} \cdot \dfrac{180°}{\pi \text{ rad}} \approx -200.535°$

13. $198.4° = 198° + 0.4(60)' = 198°24'$

15. $138° = \dfrac{138\pi}{180} = \dfrac{23\pi}{30}$ radians

$s = r\theta = 20\left(\dfrac{23\pi}{30}\right) \approx 48.17$ inches

17. $150° = \dfrac{150\pi}{180} = \dfrac{5\pi}{6}$ radians

$A = \dfrac{1}{2}r^2\theta = \dfrac{1}{2}(20)^2\left(\dfrac{5\pi}{6}\right) = \dfrac{500\pi}{3}$

≈ 523.6 square inches

19. $t = \dfrac{2\pi}{3}$ corresponds to the point $\left(-\dfrac{1}{2}, \dfrac{\sqrt{3}}{2}\right)$.

21. $t = \dfrac{7\pi}{6}$ corresponds to the point

$(x, y) = \left(-\dfrac{\sqrt{3}}{2}, -\dfrac{1}{2}\right)$.

23. $t = \dfrac{3\pi}{4}$ corresponds to the point $(x, y) = \left(-\dfrac{\sqrt{2}}{2}, \dfrac{\sqrt{2}}{2}\right)$.

$\sin\dfrac{3\pi}{4} = y = \dfrac{\sqrt{2}}{2}$ $\qquad \csc\dfrac{3\pi}{4} = \dfrac{1}{y} = \sqrt{2}$

$\cos\dfrac{3\pi}{4} = x = -\dfrac{\sqrt{2}}{2}$ $\qquad \sec\dfrac{3\pi}{4} = \dfrac{1}{x} = -\sqrt{2}$

$\tan\dfrac{3\pi}{4} = \dfrac{y}{x} = -1$ $\qquad \cot\dfrac{3\pi}{4} = \dfrac{x}{y} = -1$

25. $\sin\dfrac{11\pi}{4} = \sin\dfrac{3\pi}{4} = \dfrac{\sqrt{2}}{2}$

27. $\cos\left(-\dfrac{17\pi}{6}\right) = \cos\dfrac{7\pi}{6} = -\dfrac{\sqrt{3}}{2}$

29. $\sec\left(\dfrac{12\pi}{5}\right) = \dfrac{1}{\cos\left(\dfrac{12\pi}{5}\right)} \approx 3.2361$

31. $\tan 33 \approx -75.3130$

33. opp $= 4$, adj $= 5$, hyp $= \sqrt{4^2 + 5^2} = \sqrt{41}$

$$\sin\theta = \frac{\text{opp}}{\text{hyp}} = \frac{4}{\sqrt{41}} = \frac{4\sqrt{41}}{41} \qquad \csc\theta = \frac{\text{hyp}}{\text{opp}} = \frac{\sqrt{41}}{4}$$

$$\cos\theta = \frac{\text{adj}}{\text{hyp}} = \frac{5}{\sqrt{41}} = \frac{5\sqrt{41}}{41} \qquad \sec\theta = \frac{\text{hyp}}{\text{adj}} = \frac{\sqrt{41}}{5}$$

$$\tan\theta = \frac{\text{opp}}{\text{adj}} = \frac{4}{5} \qquad \cot\theta = \frac{\text{adj}}{\text{opp}} = \frac{5}{4}$$

35. $\tan 33° \approx 0.6494$

37. $\cot 15°\,14' = \dfrac{1}{\tan\left(15 + \frac{14}{60}\right)} \approx 3.6722$

39. $\sin\theta = \dfrac{1}{3}$

(a) $\csc\theta = \dfrac{1}{\sin\theta} = 3$

(b) $\sin^2\theta + \cos^2\theta = 1$

$$\left(\frac{1}{3}\right)^2 + \cos^2\theta = 1$$

$$\cos^2\theta = 1 - \frac{1}{9}$$

$$\cos^2\theta = \frac{8}{9}$$

$$\cos\theta = \sqrt{\frac{8}{9}}$$

$$\cos\theta = \frac{2\sqrt{2}}{3}$$

(c) $\sec\theta = \dfrac{1}{\cos\theta} = \dfrac{3}{2\sqrt{2}} = \dfrac{3\sqrt{2}}{4}$

(d) $\tan\theta = \dfrac{\sin\theta}{\cos\theta} = \dfrac{1/3}{\left(2\sqrt{2}\right)/3} = \dfrac{1}{2\sqrt{2}} = \dfrac{\sqrt{2}}{4}$

41. $\sin 1.2° = \dfrac{x}{3.5}$

$x = 3.5 \sin 1.2°$

≈ 0.0733 kilometer or 73.3 meters

Not drawn to scale

43. $x = 12$, $y = 16$, $r = \sqrt{144 + 256} = \sqrt{400} = 20$

$$\sin\theta = \frac{y}{r} = \frac{4}{5} \qquad \csc\theta = \frac{r}{y} = \frac{5}{4}$$

$$\cos\theta = \frac{x}{r} = \frac{3}{5} \qquad \sec\theta = \frac{r}{x} = \frac{5}{3}$$

$$\tan\theta = \frac{y}{x} = \frac{4}{3} \qquad \cot\theta = \frac{x}{y} = \frac{3}{4}$$

45. $x = 0.3$, $y = 0.4$

$$r = \sqrt{(0.3)^2 + (0.4)^2} = 0.5$$

$$\sin\theta = \frac{y}{r} = \frac{0.4}{0.5} = \frac{4}{5} = 0.8 \qquad \csc\theta = \frac{r}{y} = \frac{0.5}{0.4} = \frac{5}{4} = 1.25$$

$$\cos\theta = \frac{x}{r} = \frac{0.3}{0.5} = \frac{3}{5} = 0.6 \qquad \sec\theta = \frac{r}{x} = \frac{0.5}{0.3} = \frac{5}{3} \approx 1.67$$

$$\tan\theta = \frac{y}{x} = \frac{0.4}{0.3} = \frac{4}{3} \approx 1.33 \qquad \cot\theta = \frac{x}{y} = \frac{0.3}{0.4} = \frac{3}{4} = 0.75$$

47. $\sec\theta = \frac{6}{5}$, $\tan\theta < 0 \Rightarrow \theta$ is in Quadrant IV.

$r = 6, x = 5, y = -\sqrt{36 - 25} = -\sqrt{11}$

$\sin\theta = \frac{y}{r} = -\frac{\sqrt{11}}{6}$

$\cos\theta = \frac{x}{r} = \frac{5}{6}$

$\tan\theta = \frac{y}{x} = -\frac{\sqrt{11}}{5}$

$\csc\theta = \frac{r}{y} = -\frac{6\sqrt{11}}{11}$

$\cot\theta = -\frac{5\sqrt{11}}{11}$

49. $\cos\theta = \frac{x}{r} = \frac{-2}{5} \Rightarrow y^2 = 21$

$\sin\theta > 0 \Rightarrow \theta$ is in Quadrant II $\Rightarrow y = \sqrt{21}$

$\sin\theta = \frac{y}{r} = \frac{\sqrt{21}}{5}$

$\tan\theta = \frac{y}{x} = -\frac{\sqrt{21}}{2}$

$\csc\theta = \frac{r}{y} = \frac{5}{\sqrt{21}} = \frac{5\sqrt{21}}{21}$

$\sec\theta = \frac{r}{x} = \frac{5}{-2} = -\frac{5}{2}$

$\cot\theta = \frac{x}{y} = \frac{-2}{\sqrt{21}} = -\frac{2\sqrt{21}}{21}$

51. $\theta = 264°$

$= 264° - 180° = 84°$

53. $\theta = -\frac{6\pi}{5}$

$-\frac{6\pi}{5} + 2\pi = \frac{4\pi}{5}$

$\theta' = \pi - \frac{4\pi}{5}$

$= \frac{\pi}{5}$

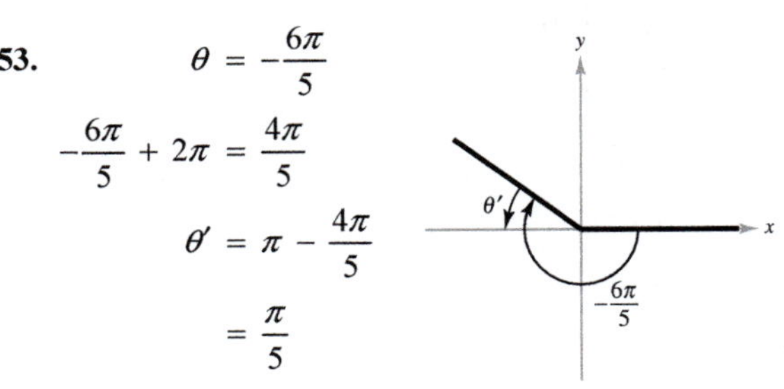

55. $\sin(-150°) = -\frac{1}{2}$

$\cos(-150°) = -\frac{\sqrt{3}}{2}$

$\tan(-150°) = \frac{-1/2}{-\sqrt{3}/2} = \frac{\sqrt{3}}{3}$

57. $\sin\frac{\pi}{3} = \frac{\sqrt{3}}{2}$

$\cos\frac{\pi}{3} = \frac{1}{2}$

$\tan\frac{\pi}{3} = \sqrt{3}$

59. $\sin 106° \approx 0.9613$

61. $\tan\left(-\frac{17\pi}{15}\right) \approx -0.4452$

63. $y = \sin 6x$

Amplitude: 1

Period: $\frac{2\pi}{6} = \frac{\pi}{3}$

65. $y = 5 + \sin \pi x$

Amplitude: 1

Period: $\frac{2\pi}{\pi} = 2$

67. $g(t) = \frac{5}{2}\sin(t - \pi)$

Amplitude: $\frac{5}{2}$

Period: 2π

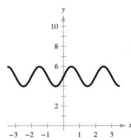

69. $y = a \sin bx$

(a) $a = 2,$

$$\frac{2\pi}{b} = \frac{1}{264} \Rightarrow b = 528\pi$$

$$y = 2 \sin 528\pi x$$

(b) $f = \dfrac{1}{1/264} = 264$ cycles per second

71. $f(t) = \tan\left(t + \dfrac{\pi}{2}\right)$

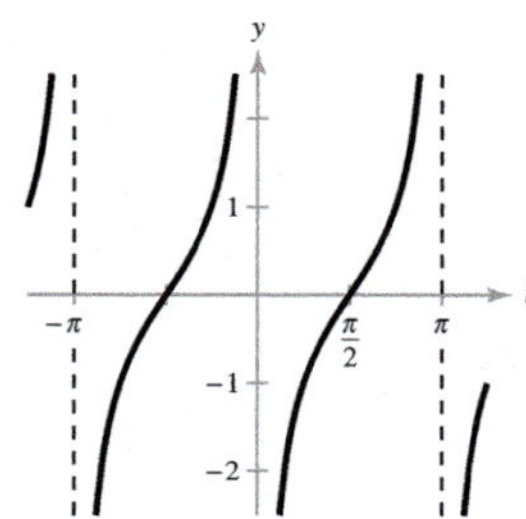

73. $f(x) = \dfrac{1}{2} \csc \dfrac{x}{2}$

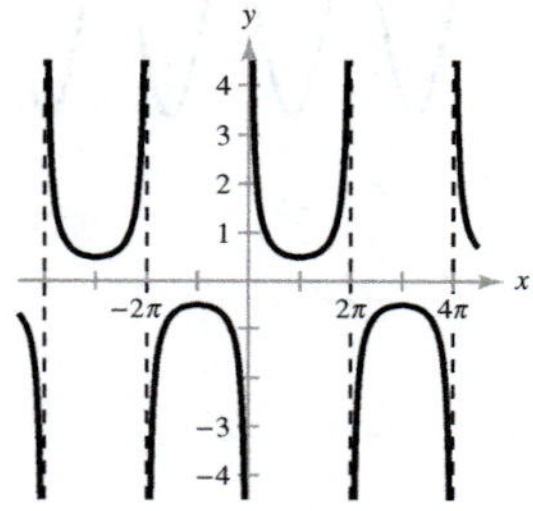

75. $f(x) = x \cos x$

Damping factor: x

As $x \to \infty$, $f(x)$ oscillates.

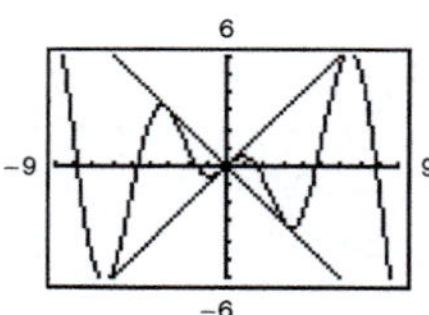

77. $\arcsin(-1) = -\dfrac{\pi}{2}$

79. $\operatorname{arccot} \sqrt{3} = \dfrac{\pi}{6}$

81. $\tan^{-1}(-1.3) \approx -0.92$ radian

83. $\operatorname{arccot} 15.5 = \arctan \dfrac{1}{15.5} \approx 0.06$

85. $f(x) = \arctan\left(\dfrac{x}{2}\right) = \tan^{-1}\left(\dfrac{x}{2}\right)$

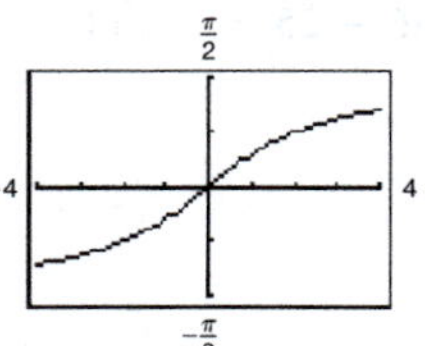

87. Let $u = \arctan \frac{3}{4}$ then

$\tan u = \frac{3}{4}.$

$\cos\left(\arctan \frac{3}{4}\right) = \frac{4}{5}$

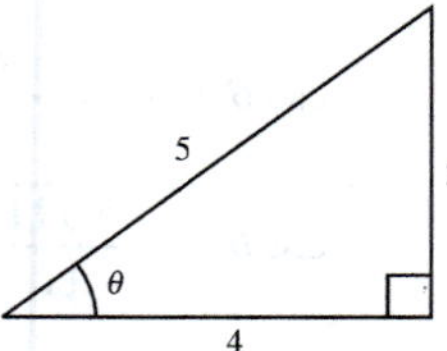

89. Let $u = \arctan \frac{12}{5}$

then $\tan u = \frac{12}{5}.$

$\sec\left(\arctan \frac{12}{5}\right) = \frac{13}{5}$

13
12
θ
5

91. Let $y = \arccos\left(\dfrac{x}{2}\right)$. Then

$\cos y = \dfrac{x}{2}$ and $\tan y = \tan\left(\arccos\left(\dfrac{x}{2}\right)\right) = \dfrac{\sqrt{4 - x^2}}{x}.$

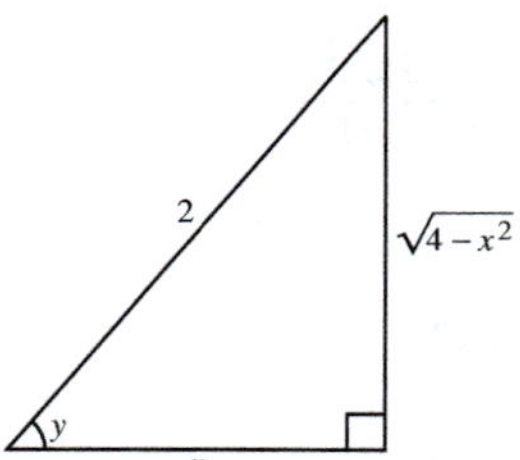

93. $\tan \theta = \frac{70}{30}$

$\theta = \arctan\left(\frac{70}{30}\right) \approx 66.8°$

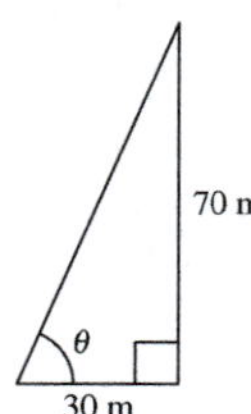

95. $\left.\begin{aligned} \sin 48° &= \frac{d_1}{650} \Rightarrow d_1 \approx 483 \\ \cos 25° &= \frac{d_2}{810} \Rightarrow d_2 \approx 734 \end{aligned}\right\} d_1 + d_2 \approx 1217$

$\left.\begin{aligned} \cos 48° &= \frac{d_3}{650} \Rightarrow d_3 \approx 435 \\ \sin 25° &= \frac{d_4}{810} \Rightarrow d_4 \approx 342 \end{aligned}\right\} d_3 - d_4 \approx 93$

$\tan \theta \approx \frac{93}{1217} \Rightarrow \theta \approx 4.4°$

$\sec 4.4° \approx \frac{D}{1217} \Rightarrow D \approx 1217 \sec 4.4° \approx 1221$

The distance is 1221 miles and the bearing is 85.6°.

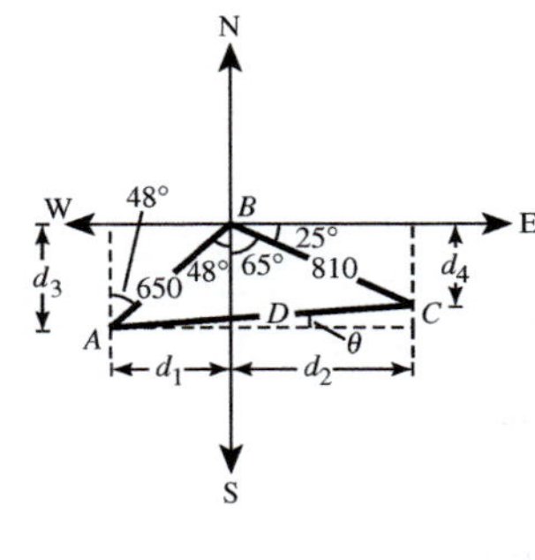

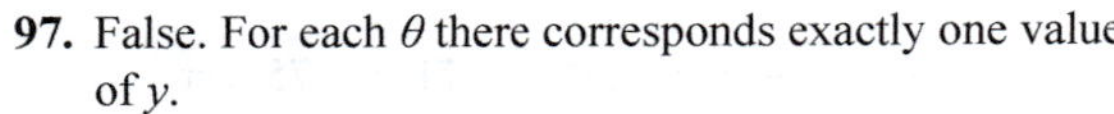

97. False. For each θ there corresponds exactly one value of y.

99. $f(\theta) = \sec \theta$ is undefined at the zeros of $g(\theta) = \cos \theta$ because $\sec \theta = \frac{1}{\cos \theta}$.

101. The ranges of the other four trigonometric functions are not bounded.

For $y = \tan x$ and $y = \cot x$, the range is $(-\infty, \infty)$.

For $y = \sec x$ and $y = \csc x$, the range is $(-\infty, -1] \cup [1, \infty)$.

Problem Solving for Chapter 4

1. (a) $8{:}57 - 6{:}45 = 2$ hours 12 minutes $= 132$ minutes

$\frac{132}{48} = \frac{11}{4}$ revolutions

$\theta = \left(\frac{11}{4}\right)(2\pi) = \frac{11\pi}{2}$ radians or 990°

(b) $s = r\theta = 47.25(5.5\pi) \approx 816.42$ feet

3. If you alter the model so that $h = 1$ when $t = 0$, you can use either a sine or a cosine model.

$a = \frac{1}{2}[\max - \min] = \frac{1}{2}[101 - 1] = 50$

$d = \frac{1}{2}[\max + \min] = \frac{1}{2}[101 + 1] = 51$

$b = 8\pi$

Cosine model: $h = 51 - 50\cos(8\pi t)$

Sine model: $h = 51 - 50\sin\left(8\pi t + \frac{\pi}{2}\right)$

Notice that you needed the horizontal shift so that the sine value was one when $t = 0$.

Another model would be: $h = 51 + 50\sin\left(8\pi t + \frac{3\pi}{2}\right)$

Here you wanted the sine value to be 1 when $t = 0$.

5. (a) $\sin 39° = \frac{3000}{d}$

$d = \frac{3000}{\sin 39°} \approx 4767$ feet

(b) $\tan 39° = \frac{3000}{x}$

$x = \frac{3000}{\tan 39°} \approx 3705$ feet

(c) $\tan 63° = \frac{w + 3705}{3000}$

$3000 \tan 63° = w + 3705$

$w = 3000 \tan 63° - 3705 \approx 2183$ feet

7. (a) $h(x) = \cos^2 x$

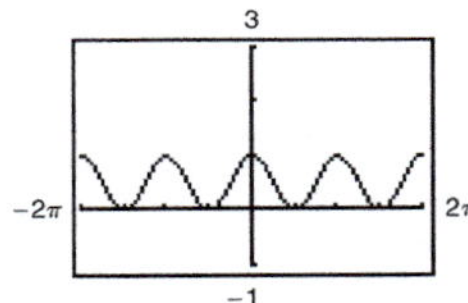

h is even.

(b) $h(x) = \sin^2 x$

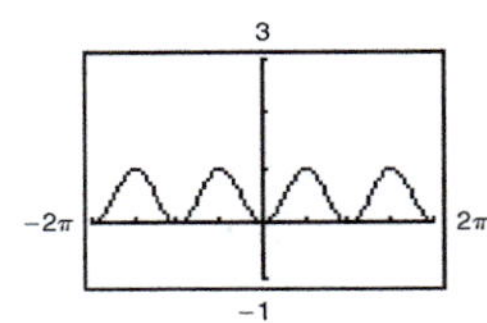

h is even.

9. $P = 100 - 20\cos\left(\frac{8\pi}{3}t\right)$

(a) 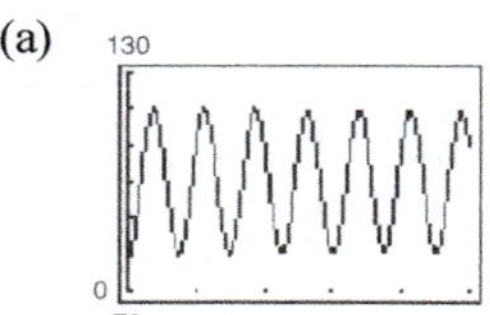

(b) Period $= \frac{2\pi}{8\pi/3} = \frac{6}{8} = \frac{3}{4}$ sec

This is the time between heartbeats.

(c) Amplitude: 20

The blood pressure ranges between $100 - 20 = 80$ and $100 + 20 = 120$.

(d) Pulse rate $= \frac{60 \text{ sec/min}}{\frac{3}{4} \text{ sec/beat}} = 80$ beats/min

(e) Period $= \frac{60}{64} = \frac{15}{16}$ sec

$64 = \frac{60}{2\pi/b} \Rightarrow b = \frac{64}{60} \cdot 2\pi = \frac{32}{15}\pi$

11. $f(x) = 2\cos 2x + 3\sin 3x$

$g(x) = 2\cos 2x + 3\sin 4x$

(a) 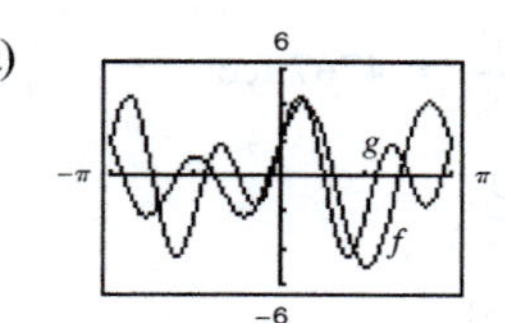

(b) The period of $f(x)$ is 2π.

The period of $g(x)$ is π.

(c) $h(x) = A\cos\alpha x + B\sin\beta x$ is periodic because the sine and cosine functions are periodic.

13. 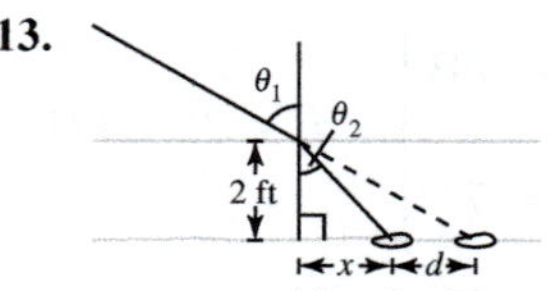

(a) $\frac{\sin\theta_1}{\sin\theta_2} = 1.333$

$\sin\theta_2 = \frac{\sin\theta_1}{1.333} = \frac{\sin 60°}{1.333} \approx 0.6497$

$\theta_2 = 40.5°$

(b) $\tan\theta_2 = \frac{x}{2} \Rightarrow x = 2\tan 40.52° \approx 1.71$ feet

$\tan\theta_1 = \frac{y}{2} \Rightarrow y = 2\tan 60° \approx 3.46$ feet

(c) $d = y - x = 3.46 - 1.71 = 1.75$ feet

(d) As you move closer to the rock, θ_1 decreases, which causes y to decrease, which in turn causes d to decrease.

Practice Test for Chapter 4

1. Express 350° in radian measure.

2. Express $(5\pi)/9$ in degree measure.

3. Convert $135° 14' 12''$ to decimal form.

4. Convert $-22.569°$ to $D° M' S''$ form.

5. If $\cos\theta = \frac{2}{3}$, use the trigonometric identities to find $\tan\theta$.

6. Find θ given $\sin\theta = 0.9063$.

7. Solve for x in the figure below.

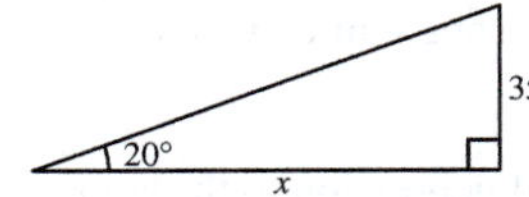

8. Find the magnitude of the reference angle for $\theta = (6\pi)/5$.

9. Evaluate $\csc 3.92$.

10. Find $\sec\theta$ given that θ lies in Quadrant III and $\tan\theta = 6$.

11. Graph $y = 3\sin\frac{x}{2}$.

12. Graph $y = -2\cos(x - \pi)$.

13. Graph $y = \tan 2x$.

14. Graph $y = -\csc\left(x + \frac{\pi}{4}\right)$.

15. Graph $y = 2x + \sin x$, using a graphing calculator.

16. Graph $y = 3x\cos x$, using a graphing calculator.

17. Evaluate $\arcsin 1$.

18. Evaluate $\arctan(-3)$.

19. Evaluate $\sin\left(\arccos\frac{4}{\sqrt{35}}\right)$.

20. Write an algebraic expression for $\cos\left(\arcsin\frac{x}{4}\right)$.

For Exercises 21–23, solve the right triangle.

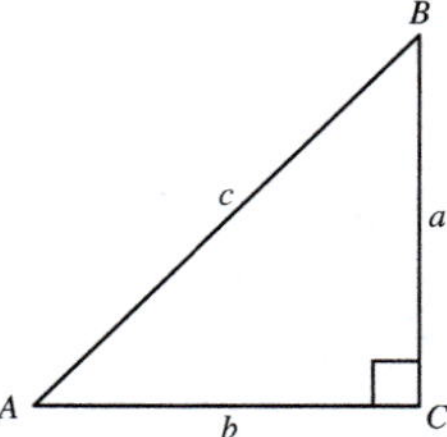

21. $A = 40°, c = 12$

22. $B = 6.84°, a = 21.3$

23. $a = 5, b = 9$

24. A 20-foot ladder leans against the side of a barn. Find the height of the top of the ladder if the angle of elevation of the ladder is 67°.

25. An observer in a lighthouse 250 feet above sea level spots a ship off the shore. If the angle of depression to the ship is 5°, how far out is the ship?

CHAPTER 5
Analytic Trigonometry

CHAPTER 5
Analytic Trigonometry

Section 5.1 Using Fundamental Identities

1. $\tan u$

3. $\cot u$

5. 1

7. $\sec x = -\frac{5}{2}$, $\tan x < 0 \Rightarrow x$ is in Quadrant II.

$$\cos x = \frac{1}{\sec x} = \frac{1}{-\frac{5}{2}} = -\frac{2}{5}$$

$$\sin x = \sqrt{1 - \left(-\frac{2}{5}\right)^2} = \sqrt{1 - \frac{4}{25}} = \frac{\sqrt{21}}{5}$$

$$\tan x = \frac{\sin x}{\cos x} = \frac{\frac{\sqrt{21}}{5}}{-\frac{2}{5}} = -\frac{\sqrt{21}}{2}$$

$$\csc x = \frac{1}{\sin x} = \frac{5}{\sqrt{21}} = \frac{5\sqrt{21}}{21}$$

$$\cot x = \frac{1}{\tan x} = -\frac{2}{\sqrt{21}} = -\frac{2\sqrt{21}}{21}$$

9. $\sin\theta = -\frac{3}{4}$, $\cos\theta > 0 \Rightarrow \theta$ is in Quadrant IV.

$$\cos\theta = \sqrt{1 - \left(-\frac{3}{4}\right)^2} = \sqrt{1 - \frac{9}{16}} = \frac{\sqrt{7}}{4}$$

$$\tan\theta = \frac{\sin\theta}{\cos\theta} = \frac{-\frac{3}{4}}{\frac{\sqrt{7}}{4}} = -\frac{3}{\sqrt{7}} = -\frac{3\sqrt{7}}{7}$$

$$\sec\theta = \frac{1}{\cos\theta} = \frac{1}{\frac{\sqrt{7}}{4}} = \frac{4}{\sqrt{7}} = \frac{4\sqrt{7}}{7}$$

$$\cot\theta = \frac{1}{\tan\theta} = \frac{1}{-\frac{3}{\sqrt{7}}} = -\frac{\sqrt{7}}{3}$$

$$\csc\theta = \frac{1}{\sin\theta} = \frac{1}{-\frac{3}{4}} = -\frac{4}{3}$$

11. $\tan x = \frac{2}{3}$, $\cos x > 0 \Rightarrow x$ is in Quadrant I.

$$\cot x = \frac{1}{\tan x} = \frac{1}{\frac{2}{3}} = \frac{3}{2}$$

$$\sec x = \sqrt{1 + \left(\frac{2}{3}\right)^2} = \sqrt{1 + \frac{4}{9}} = \frac{\sqrt{13}}{3}$$

$$\csc x = \sqrt{1 + \left(\frac{3}{2}\right)^2} = \sqrt{1 + \frac{9}{4}} = \frac{\sqrt{13}}{2}$$

$$\sin x = \frac{1}{\csc x} = \frac{1}{\frac{\sqrt{13}}{2}} = \frac{2}{\sqrt{13}} = \frac{2\sqrt{13}}{13}$$

$$\cos x = \frac{1}{\sec x} = \frac{1}{\frac{\sqrt{13}}{3}} = \frac{3}{\sqrt{13}} = \frac{3\sqrt{13}}{13}$$

$$\cot x = \frac{1}{\tan x} = \frac{1}{\frac{2}{3}} = \frac{3}{2}$$

13. $\sec x \cos x = \left(\frac{1}{\cancel{\cos x}}\right)\cancel{\cos x} = 1$

Matches (c).

14. $\cot^2 x - \csc^2 x = \left(\csc^2 x - 1\right) - \csc^2 x = -1$

Matches (b).

15. $\cos x\left(1 + \tan^2 x\right) = \cos x\left(\sec^2 x\right)$

$$= \cos x\left(\frac{1}{\cos^2 x}\right)$$

$$= \frac{1}{\cos x}$$

$$= \sec x$$

Matches (f).

16. $\cot x \sec x = \frac{\cos x}{\sin x} \cdot \frac{1}{\cos x} = \frac{1}{\sin x} = \csc x$

Matches (a).

17. $\frac{\sec^2 x - 1}{\sin^2 x} = \frac{\tan^2 x}{\sin^2 x} = \frac{\sin^2 x}{\cos^2 x} \cdot \frac{1}{\sin^2 x} = \sec^2 x$

Matches (e).

18. $\frac{\cos^2\left[(\pi/2) - x\right]}{\cos x} = \frac{\sin^2 x}{\cos x} = \frac{\sin x}{\cos x}\sin x = \tan x \sin x$

Matches (d).

19. $\dfrac{\tan\theta\cot\theta}{\sec\theta} = \dfrac{\tan\theta\left(\dfrac{1}{\tan\theta}\right)}{\dfrac{1}{\cos\theta}} = \dfrac{1}{\dfrac{1}{\cos\theta}} = \cos\theta$

21. $\tan^2 x - \tan^2 x \sin^2 x = \tan^2 x(1 - \sin^2 x)$

$= \tan^2 x \cos^2 x$

$= \dfrac{\sin^2 x}{\cos^2 x} \cdot \cos^2 x$

$= \sin^2 x$

23. $\dfrac{\sec^2 x - 1}{\sec x - 1} = \dfrac{(\sec x + 1)(\sec x - 1)}{\sec x - 1}$

$= \sec x + 1$

25. $1 - 2\cos^2 x + \cos^4 x = (1 - \cos^2 x)^2$

$= (\sin^2 x)^2$

$= \sin^4 x$

27. $\cot^3 x + \cot^2 x + \cot x + 1 = \cot^2 x(\cot x + 1) + (\cot x + 1)$

$= (\cot x + 1)(\cot^2 x + 1)$

$= (\cot x + 1)\csc^2 x$

29. $3\sin^2 x - 5\sin x - 2 = (3\sin x + 1)(\sin x - 2)$

31. $\cot^2 x + \csc x - 1 = (\csc^2 x - 1) + \csc x - 1$

$= \csc^2 x + \csc x - 2$

$= (\csc x - 1)(\csc x + 2)$

33. $\tan\theta\csc\theta = \dfrac{\sin\theta}{\cos\theta} \cdot \dfrac{1}{\sin\theta} = \dfrac{1}{\cos\theta} = \sec\theta$

35. $\sin\phi(\csc\phi - \sin\phi) = (\sin\phi)\dfrac{1}{\sin\phi} - \sin^2\phi$

$= 1 - \sin^2\phi = \cos^2\phi$

37. $\sin\beta\tan\beta + \cos\beta = (\sin\beta)\dfrac{\sin\beta}{\cos\beta} + \cos\beta$

$= \dfrac{\sin^2\beta}{\cos\beta} + \dfrac{\cos^2\beta}{\cos\beta}$

$= \dfrac{\sin^2\beta + \cos^2\beta}{\cos\beta}$

$= \dfrac{1}{\cos\beta}$

$= \sec\beta$

39. $\dfrac{1 - \sin^2 x}{\csc^2 x - 1} = \dfrac{\cos^2 x}{\cot^2 x} = \cos^2 x\tan^2 x = (\cos^2 x)\dfrac{\sin^2 x}{\cos^2 x}$

$= \sin^2 x$

41. $(\sin x + \cos x)^2 = \sin^2 x + 2\sin x\cos x + \cos^2 x$

$= (\sin^2 x + \cos^2 x) + 2\sin x\cos x$

$= 1 + 2\sin x\cos x$

43. $\dfrac{1}{1 + \cos x} + \dfrac{1}{1 - \cos x} = \dfrac{1 - \cos x + 1 + \cos x}{(1 + \cos x)(1 - \cos x)}$

$= \dfrac{2}{1 - \cos^2 x}$

$= \dfrac{2}{\sin^2 x}$

$= 2\csc^2 x$

45. $\dfrac{\cos x}{1 + \sin x} - \dfrac{\cos x}{1 - \sin x} = \dfrac{\cos x(1 - \sin x) - \cos x(1 + \sin x)}{(1 + \sin x)(1 - \sin x)}$

$= \dfrac{\cos x - \sin x\cos x - \cos x - \sin x\cos x}{(1 + \sin x)(1 - \sin x)}$

$= \dfrac{-2\sin x\cos x}{1 - \sin^2 x}$

$= \dfrac{-2\sin x\cos x}{\cos^2 x}$

$= \dfrac{-2\sin x}{\cos x}$

$= -2\tan x$

47. $\tan x - \dfrac{\sec^2 x}{\tan x} = \dfrac{\tan^2 x - \sec^2 x}{\tan x}$

$= \dfrac{-1}{\tan x} = -\cot x$

49. $\dfrac{\sin^2 y}{1 - \cos y} = \dfrac{1 - \cos^2 y}{1 - \cos y}$

$= \dfrac{(1 + \cos y)(1 - \cos y)}{1 - \cos y} = 1 + \cos y$

51. $y_1 = \dfrac{\tan x + 1}{\sec x + \csc x}$

$= \dfrac{\dfrac{\sin x}{\cos x} + 1}{\dfrac{1}{\cos x} + \dfrac{1}{\sin x}}$

$= \dfrac{\dfrac{\sin x + \cos x}{\cos x}}{\dfrac{\sin x + \cos x}{\sin x \cos x}}$

$= \left(\dfrac{\sin x + \cos x}{\cos x}\right)\left(\dfrac{\sin x \cos x}{\sin x + \cos x}\right)$

$= \sin x$

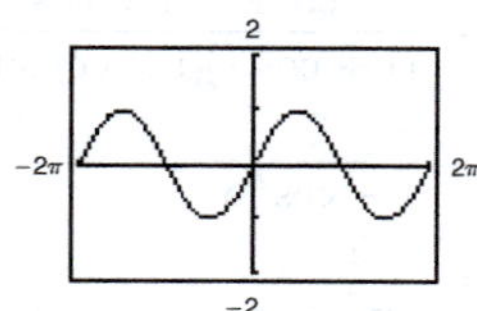

53. Let $x = 3 \cos \theta$.

$\sqrt{9 - x^2} = \sqrt{9 - (3 \cos \theta)^2}$

$= \sqrt{9 - 9 \cos^2 \theta}$

$= \sqrt{9(1 - \cos^2 \theta)}$

$= \sqrt{9 \sin^2 \theta} = 3 \sin \theta$

55. Let $x = 2 \sec \theta$.

$\sqrt{x^2 - 4} = \sqrt{(2 \sec \theta)^2 - 4}$

$= \sqrt{4(\sec^2 \theta - 1)}$

$= \sqrt{4 \tan^2 \theta}$

$= 2 \tan \theta$

57. Let $x = 2 \sin \theta$.

$\sqrt{4 - x^2} = \sqrt{2}$

$\sqrt{4 - (2 \sin \theta)^2} = \sqrt{2}$

$\sqrt{4 - 4 \sin^2 \theta} = \sqrt{2}$

$\sqrt{4(1 - \sin^2 \theta)} = \sqrt{2}$

$\sqrt{4 \cos^2 \theta} = \sqrt{2}$

$2 \cos \theta = \sqrt{2}$

$\cos \theta = \dfrac{\sqrt{2}}{2}$

$\sin \theta = \sqrt{1 - \cos^2 \theta}$

$= \sqrt{1 - \left(\dfrac{\sqrt{2}}{2}\right)^2}$

$= \pm\dfrac{\sqrt{2}}{2}$

59. $\sin \theta = \sqrt{1 - \cos^2 \theta}$

Let $y_1 = \sin x$ and $y_2 = \sqrt{1 - \cos^2 x}, 0 \le x \le 2\pi$.

$y_1 = y_2$ for $0 \le x \le \pi$.

So, $\sin \theta = \sqrt{1 - \cos^2 \theta}$ for $0 \le \theta \le \pi$.

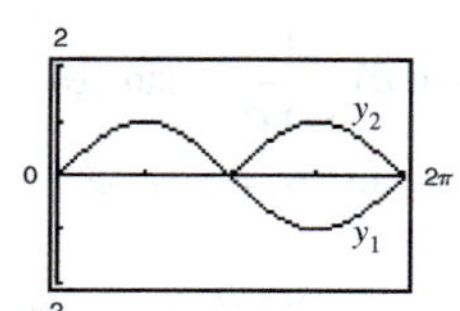

61. $\ln|\sin x| + \ln|\cot x| = \ln|\sin x \cot x|$

$= \ln\left|\sin x \cdot \dfrac{\cos x}{\sin x}\right|$

$= \ln|\cos x|$

63. $\ln|\tan t| - \ln(1 - \cos^2 t) = \ln\left[\dfrac{|\tan t|}{1 - \cos^2 t}\right]$

$= \ln\left|\dfrac{\tan t}{\sin^2 t}\right|$

$= \ln\left|\dfrac{\sin t}{\cos t} \cdot \dfrac{1}{\sin^2 t}\right|$

$= \ln\left|\dfrac{1}{\cos t \sin t}\right|$

$= \ln|\sec t \csc t|$

65. $\mu W \cos \theta = W \sin \theta$

$\mu = \dfrac{W \sin \theta}{W \cos \theta} = \tan \theta$

67. True.

$$\tan u = \frac{\sin u}{\cos u}$$

$$\cot u = \frac{\cos u}{\sin u}$$

$$\sec u = \frac{1}{\cos u}$$

$$\csc u = \frac{1}{\sin u}$$

69. As $x \to \frac{\pi^-}{2}$, $\tan x \to \infty$ and $\cot x \to 0$.

71. $\cos(-\theta) \neq -\cos\theta$

$\cos(-\theta) = \cos\theta$

The correct identity is $\dfrac{\sin\theta}{\cos(-\theta)} = \dfrac{\sin\theta}{\cos\theta}$

$= \tan\theta$

73. Because $\sin^2\theta + \cos^2\theta = 1$, then

$\cos^2\theta = 1 - \sin^2\theta.$

$$\cos\theta = \pm\sqrt{1 - \sin\theta}$$

$$\tan\theta = \frac{\sin\theta}{\cos\theta} = \frac{\sin\theta}{\pm\sqrt{1 - \sin^2\theta}}$$

$$\cot\theta = \frac{\cos\theta}{\sin\theta} = \frac{\pm\sqrt{1 - \sin^2\theta}}{\sin\theta}$$

$$\sec\theta = \frac{1}{\cos\theta} = \frac{1}{\pm\sqrt{1 - \sin^2\theta}}$$

$$\csc\theta = \frac{1}{\sin\theta}$$

75.
$$\frac{\sec\theta(1 + \tan\theta)}{\sec\theta + \csc\theta} = \frac{\left(\frac{1}{\cos\theta}\right)\left(1 + \frac{\sin\theta}{\cos\theta}\right)}{\frac{1}{\cos\theta} + \frac{1}{\sin\theta}}$$

$$= \frac{\frac{\cos\theta + \sin\theta}{\cos^2\theta}}{\frac{\sin\theta + \cos\theta}{\sin\theta\cos\theta}}$$

$$= \left(\frac{\sin\theta + \cos\theta}{\cos^2\theta}\right)\left(\frac{\sin\theta\cos\theta}{\sin\theta + \cos\theta}\right)$$

$$= \frac{\sin\theta}{\cos\theta}$$

Section 5.2 Verifying Trigonometric Identities

1. identity

3. $\tan u$

5. $\sin u$

7. $-\csc u$

9. $\tan t \cot t = \dfrac{\sin t}{\cos t} \cdot \dfrac{\cos t}{\sin t} = 1$

11. $(1 + \sin\alpha)(1 - \sin\alpha) = 1 - \sin^2\alpha = \cos^2\alpha$

13.
$$\cos^2\beta - \sin^2\beta = (1 - \sin^2\beta) - \sin^2\beta$$
$$= 1 - 2\sin^2\beta$$

15.
$$\tan\left(\frac{\pi}{2} - \theta\right)\tan\theta = \cot\theta\tan\theta$$
$$= \left(\frac{1}{\tan\theta}\right)\tan\theta$$
$$= 1$$

17.
$$\sin t \csc\left(\frac{\pi}{2} - t\right) = \sin t \sec t = \sin t\left(\frac{1}{\cos t}\right)$$
$$= \frac{\sin t}{\cos t} = \tan t$$

19.
$$\frac{1}{\tan x} + \frac{1}{\cot x} = \frac{\cot x + \tan x}{\tan x \cot x}$$
$$= \frac{\cot x + \tan x}{1}$$
$$= \tan x + \cot x$$

21. $\dfrac{1+\sin\theta}{\cos\theta}+\dfrac{\cos\theta}{1+\sin\theta}=\dfrac{(1+\sin\theta)^2+\cos^2\theta}{\cos\theta(1+\sin\theta)}$

$$=\frac{1+2\sin\theta+\sin^2\theta+\cos^2\theta}{\cos\theta(1+\sin\theta)}$$

$$=\frac{2+2\sin\theta}{\cos\theta(1+\sin\theta)}$$

$$=\frac{2(1+\sin\theta)}{\cos\theta(1+\sin\theta)}$$

$$=\frac{2}{\cos\theta}$$

$$=2\sec\theta$$

23. $\dfrac{1}{\cos x+1}+\dfrac{1}{\cos x-1}=\dfrac{\cos x-1+\cos x+1}{(\cos x+1)(\cos x-1)}$

$$=\frac{2\cos x}{\cos^2 x-1}$$

$$=\frac{2\cos x}{-\sin^2 x}$$

$$=-2\cdot\frac{1}{\sin x}\cdot\frac{\cos x}{\sin x}$$

$$=-2\csc x\cot x$$

25. $\sec y\cos y=\left(\dfrac{1}{\cos y}\right)\cos y=1$

27. $\dfrac{\tan^2\theta}{\sec\theta}=\dfrac{(\sin\theta/\cos\theta)\tan\theta}{1/\cos\theta}=\sin\theta\tan\theta$

29. $\dfrac{1}{\tan\beta}+\tan\beta=\dfrac{1+\tan^2\beta}{\tan\beta}$

$$=\frac{\sec^2\beta}{\tan\beta}$$

31. $\dfrac{\cot^2 t}{\csc t}=\dfrac{\cos^2 t/\sin^2 t}{1/\sin t}=\dfrac{\cos^2 t}{\sin t}=\dfrac{1-\sin^2 t}{\sin t}$

33. $\sec x-\cos x=\dfrac{1}{\cos x}-\cos x$

$$=\frac{1-\cos^2 x}{\cos x}$$

$$=\frac{\sin^2 x}{\cos x}$$

$$=\sin x\cdot\frac{\sin x}{\cos x}$$

$$=\sin x\tan x$$

35. $\dfrac{\cot x}{\sec x}=\dfrac{\cos x/\sin x}{1/\cos x}=\dfrac{\cos^2 x}{\sin x}=\dfrac{1-\sin^2 x}{\sin x}=\dfrac{1}{\sin x}-\dfrac{\sin^2 x}{\sin x}=\csc x-\sin x$

37. $\sin^{1/2}x\cos x-\sin^{5/2}x\cos x=\sin^{1/2}x\cos x\left(1-\sin^2 x\right)=\sin^{1/2}x\cos x\cdot\cos^2 x=\cos^3 x\sqrt{\sin x}$

39. $(1+\sin y)[1+\sin(-y)]=(1+\sin y)(1-\sin y)$

$$=1-\sin^2 y$$

$$=\cos^2 y$$

41.

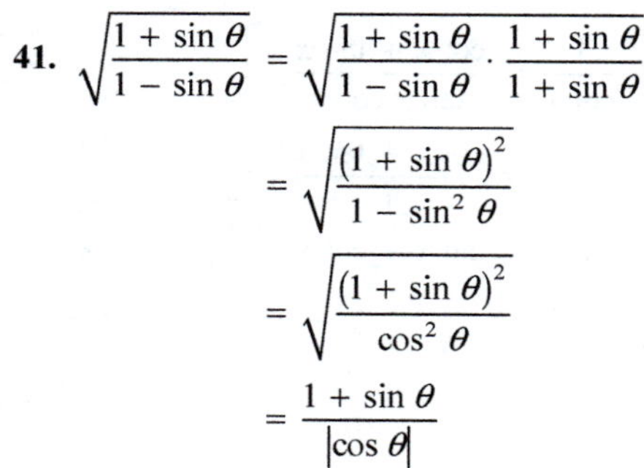

$$\sqrt{\frac{1+\sin\theta}{1-\sin\theta}}=\sqrt{\frac{1+\sin\theta}{1-\sin\theta}\cdot\frac{1+\sin\theta}{1+\sin\theta}}$$

$$=\sqrt{\frac{(1+\sin\theta)^2}{1-\sin^2\theta}}$$

$$=\sqrt{\frac{(1+\sin\theta)^2}{\cos^2\theta}}$$

$$=\frac{1+\sin\theta}{|\cos\theta|}$$

43. $\cot(-x)\neq\cot x$

The correct substitution is $\cot(-x)=-\cot x$.

$$\frac{1}{\tan x}+\cot(-x)=\cot x-\cot x=0$$

45. (a)

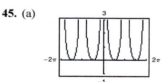

Identity

(b)

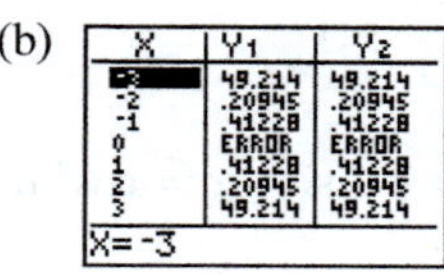

X	Y1	Y2
-3	49.214	49.214
-2	.20945	.20945
-1	.41228	.41228
0	ERROR	ERROR
1	.41228	.41228
2	.20945	.20945
3	49.214	49.214

X=-3

Identity

(c) $\left(1+\cot^2 x\right)\left(\cos^2 x\right)=\csc^2 x\cos^2 x$

$$=\frac{1}{\sin^2 x}\cdot\cos^2 x$$

$$=\cot^2 x$$

47. (a)

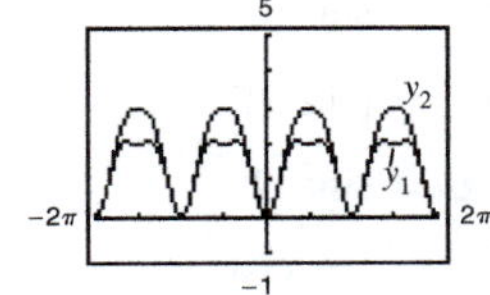

Not an identity

(b)

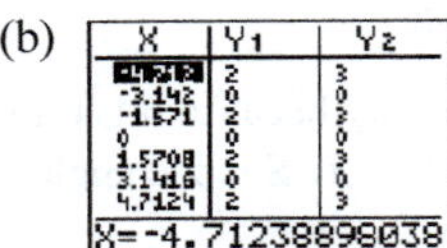

X	Y_1	Y_2
-4.712	2	3
-3.142	0	0
-1.571	2	3
0	0	0
1.5708	2	3
3.1416	0	0
4.7124	2	3

X=-4.71238898038

Not an identity

(c) $2 + \cos^2 x - 3\cos^4 x = (1 - \cos^2 x)(2 + 3\cos^2 x)$

$= \sin^2 x(2 + 3\cos^2 x)$

$\neq \sin^2 x(3 + 2\cos^2 x)$

49. (a)

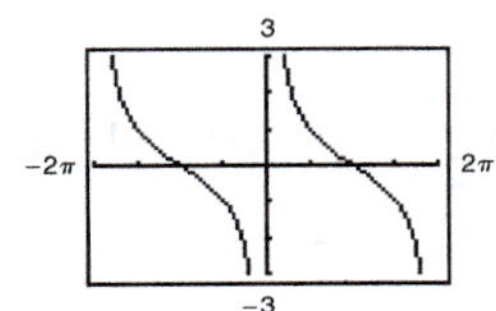

Identity

(b)

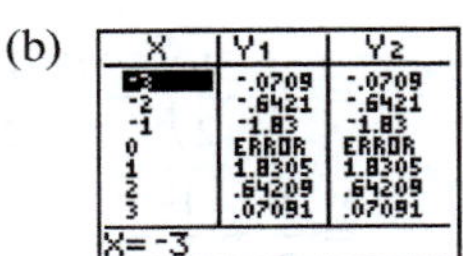

X	Y_1	Y_2
-3	-.0709	-.0709
-2	-.6421	-.6421
-1	-1.83	-1.83
0	ERROR	ERROR
1	1.8305	1.8305
2	.64209	.64209
3	.07091	.07091

X=-3

Identity

(c) $$\frac{1 + \cos x}{\sin x} = \frac{(1 + \cos x)(1 - \cos x)}{\sin x(1 - \cos x)}$$

$$= \frac{1 - \cos^2 x}{\sin x(1 - \cos x)}$$

$$= \frac{\sin^2 x}{\sin x(1 - \cos x)}$$

$$= \frac{\sin x}{1 - \cos x}$$

51. $\tan^3 x \sec^2 x - \tan^3 x = \tan^3 x(\sec^2 x - 1)$

$= \tan^3 x \tan^2 x$

$= \tan^5 x$

53. $(\sin^2 x - \sin^4 x)\cos x = \sin^2 x(1 - \sin^2 x)\cos x$

$= \sin^2 x \cos^2 x \cos x$

$= \sin^2 x \cos^3 x$

55. $\sin^2 25° + \sin^2 65° = \sin^2 25° + \cos^2(90° - 65°)$

$= \sin^2 25° + \cos^2 25°$

$= 1$

57. Let $\theta = \sin^{-1} x \Rightarrow \sin\theta = x = \dfrac{x}{1}$.

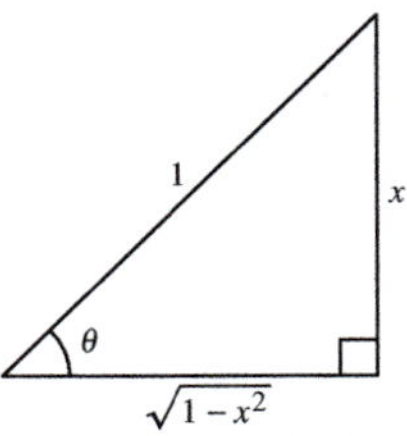

From the diagram,

$$\tan(\sin^{-1} x) = \tan\theta = \frac{x}{\sqrt{1 - x^2}}.$$

59. Let $\theta = \sin^{-1}\dfrac{x - 1}{4} \Rightarrow \sin\theta = \dfrac{x - 1}{4}$.

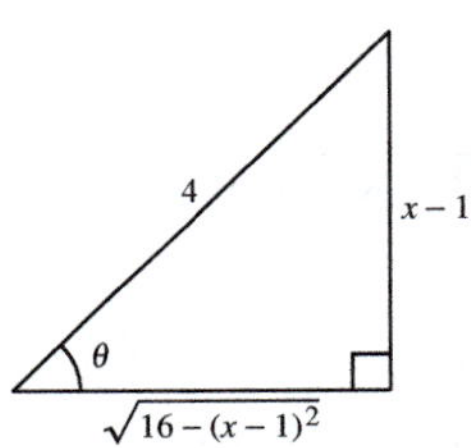

From the diagram,

$$\tan\left(\sin^{-1}\frac{x - 1}{4}\right) = \tan\theta = \frac{x - 1}{\sqrt{16 - (x - 1)^2}}.$$

61. $\cos x - \csc x \cot x = \cos x - \dfrac{1}{\sin x}\dfrac{\cos x}{\sin x}$

$= \cos x\left(1 - \dfrac{1}{\sin^2 x}\right)$

$= \cos x(1 - \csc^2 x)$

$= -\cos x(\csc^2 x - 1)$

$= -\cos x \cot^2 x$

63. False. $\tan x^2 = \tan(x \cdot x)$ and $\tan^2 x = (\tan x)(\tan x)$, $\tan x^2 \neq \tan^2 x$.

65. False. For the equation to be an identity, it must be true for all values of θ in the domain.

67. Because $\sin^2 \theta = 1 - \cos^2 \theta$, then

$\sin \theta = \pm\sqrt{1 - \cos^2 \theta}$; $\sin \theta \neq \sqrt{1 - \cos^2 \theta}$ if θ lies in Quadrant III or IV.

One such angle is $\theta = \dfrac{7\pi}{4}$.

69.

$$\begin{aligned} 1 - \cos \theta &= \sin \theta \\ (1 - \cos \theta)^2 &= (\sin \theta)^2 \\ 1 - 2\cos \theta + \cos^2 \theta &= \sin^2 \theta \\ 1 - 2\cos \theta + \cos^2 \theta &= 1 - \cos^2 \theta \\ 2\cos^2 \theta - 2\cos \theta &= 0 \\ 2\cos \theta(\cos \theta - 1) &= 0 \end{aligned}$$

The equation is not an identity because it is only true when $\cos \theta = 0$ or $\cos \theta = 1$. So, one angle for which the equation is not true is $-\dfrac{\pi}{2}$.

Section 5.3 Solving Trigonometric Equations

1. isolate

3. quadratic

5. $\tan x - \sqrt{3} = 0$

(a) $x = \dfrac{\pi}{3}$

$$\tan \frac{\pi}{3} - \sqrt{3} = \sqrt{3} - \sqrt{3} = 0$$

(b) $x = \dfrac{4\pi}{3}$

$$\tan \frac{4\pi}{3} - \sqrt{3} = \sqrt{3} - \sqrt{3} = 0$$

7. $3\tan^2 2x - 1 = 0$

(a) $x = \dfrac{\pi}{12}$

$$\begin{aligned} 3\left[\tan 2\left(\frac{\pi}{12}\right)\right]^2 - 1 &= 3\tan^2 \frac{\pi}{6} - 1 \\ &= 3\left(\frac{1}{\sqrt{3}}\right)^2 - 1 \\ &= 0 \end{aligned}$$

(b) $x = \dfrac{5\pi}{12}$

$$\begin{aligned} 3\left[\tan 2\left(\frac{5\pi}{12}\right)\right]^2 - 1 &= 3\tan^2 \frac{5\pi}{6} - 1 \\ &= 3\left(-\frac{1}{\sqrt{3}}\right)^2 - 1 \\ &= 0 \end{aligned}$$

9. $2\sin^2 x - \sin x - 1 = 0$

(a) $x = \dfrac{\pi}{2}$

$$\begin{aligned} 2\sin^2 \frac{\pi}{2} - \sin \frac{\pi}{2} - 1 &= 2(1)^2 - 1 - 1 \\ &= 0 \end{aligned}$$

(b) $x = \dfrac{7\pi}{6}$

$$\begin{aligned} 2\sin^2 \frac{7\pi}{6} - \sin \frac{7\pi}{6} - 1 &= 2\left(-\frac{1}{2}\right)^2 - \left(-\frac{1}{2}\right) - 1 \\ &= \frac{1}{2} + \frac{1}{2} - 1 \\ &= 0 \end{aligned}$$

11.

$$\begin{aligned} \sqrt{3}\csc x - 2 &= 0 \\ \sqrt{3}\csc x &= 2 \\ \csc x &= \frac{2}{\sqrt{3}} \\ x &= \frac{\pi}{3} + 2n\pi \\ \text{or } x &= \frac{2\pi}{3} + 2n\pi \end{aligned}$$

13.

$$\begin{aligned} \cos x + 1 &= -\cos x \\ 2\cos x + 1 &= 0 \\ \cos x &= -\frac{1}{2} \\ x &= \frac{2\pi}{3} + 2n\pi \text{ or } x = \frac{4\pi}{3} + 2n\pi \end{aligned}$$

15. $3\sec^2 x - 4 = 0$

$$\sec^2 x = \frac{4}{3}$$

$$\sec x = \pm\frac{2}{\sqrt{3}}$$

$$x = \frac{\pi}{6} + n\pi$$

$$\text{or } x = \frac{5\pi}{6} + n\pi$$

17. $4\cos^2 x - 1 = 0$

$$\cos^2 x = \frac{1}{4}$$

$$\cos x = \pm\frac{1}{2}$$

$$x = \frac{\pi}{3} + n\pi \quad \text{or} \quad x = \frac{2\pi}{3} + n\pi$$

19. $\sin x(\sin x + 1) = 0$

$$\sin x = 0 \quad \text{or} \quad \sin x = -1$$

$$x = n\pi \qquad x = \frac{3\pi}{2} + 2n\pi$$

21. $\cos^3 x - \cos x = 0$

$$\cos x(\cos^2 x - 1) = 0$$

$$\cos x = 0 \quad \text{or } \cos^2 x - 1 = 0$$

$$x = \frac{\pi}{2} + n\pi \qquad \cos x = \pm 1$$

$$x = n\pi$$

Both of these answers can be represented as $x = \frac{n\pi}{2}$.

23. $3\tan^3 x = \tan x$

$$3\tan^3 x - \tan x = 0$$

$$\tan x(3\tan^2 x - 1) = 0$$

$$\tan x = 0 \quad \text{or } 3\tan^2 x - 1 = 0$$

$$x = n\pi \qquad \tan x = \pm\frac{\sqrt{3}}{3}$$

$$x = \frac{\pi}{6} + n\pi, \frac{5\pi}{6} + n\pi$$

25. $2\cos^2 x + \cos x - 1 = 0$

$$(2\cos x - 1)(\cos x + 1) = 0$$

$$2\cos x - 1 = 0 \quad \text{or} \quad \cos x + 1 = 0$$

$$\cos x = \frac{1}{2} \qquad \cos x = -1$$

$$x = \frac{\pi}{3} + 2n\pi, \frac{5\pi}{3} + 2n\pi \qquad x = \pi + 2n\pi$$

27. $\sec^2 x - \sec x = 2$

$$\sec^2 x - \sec x - 2 = 0$$

$$(\sec x - 2)(\sec x + 1) = 0$$

$$\sec x - 2 = 0 \quad \text{or} \quad \sec x + 1 = 0$$

$$\sec x = 2 \qquad \sec x = -1$$

$$x = \frac{\pi}{3} + 2n\pi, \frac{5\pi}{3} + 2n\pi \qquad x = \pi + 2n\pi$$

29. $\sin x - 2 = \cos x - 2$

$$\sin x = \cos x$$

$$\frac{\sin x}{\cos x} = 1$$

$$\tan x = 1$$

$$x = \tan^{-1} 1$$

$$x = \frac{\pi}{4}, \frac{5\pi}{4}$$

31. $2\sin^2 x = 2 + \cos x$

$2 - 2\cos^2 x = 2 + \cos x$

$2\cos^2 x + \cos x = 0$

$\cos x(2\cos x + 1) = 0$

$\cos x = 0$ or $2\cos x + 1 = 0$

$x = \frac{\pi}{2}, \frac{3\pi}{2}$ $\quad 2\cos x = -1$

$\cos x = -\frac{1}{2}$

$x = \frac{2\pi}{3}, \frac{4\pi}{3}$

33. $\sin^2 x = 3\cos^2 x$

$\sin^2 x - 3\cos^2 x = 0$

$\sin^2 x - 3(1 - \sin^2 x) = 0$

$4\sin^2 x = 3$

$\sin x = \pm\frac{\sqrt{3}}{2}$

$x = \frac{\pi}{3}, \frac{2\pi}{3}, \frac{4\pi}{3}, \frac{5\pi}{3}$

35. $2\sin x + \csc x = 0$

$2\sin x + \frac{1}{\sin x} = 0$

$2\sin^2 x + 1 = 0$

$\sin^2 x = -\frac{1}{2} \Rightarrow$ No solution

37. $\csc x + \cot x = 1$

$(\csc x + \cot x)^2 = 1^2$

$\csc^2 x + 2\csc x\cot x + \cot^2 x = 1$

$\cot^2 x + 1 + 2\csc x\cot x + \cot^2 x = 1$

$2\cot^2 x + 2\csc x\cot x = 0$

$2\cot x(\cot x + \csc x) = 0$

$2\cot x = 0$ or $\cot x + \csc x = 0$

$x = \frac{\pi}{2}, \frac{3\pi}{2}$ $\quad \frac{\cos x}{\sin x} = -\frac{1}{\sin x}$

$\left(\frac{3\pi}{2} \text{ is extraneous.}\right)$ $\quad \cos x = -1$

$x = \pi$

(π is extraneous.)

$x = \pi/2$ is the only solution.

39. $2\cos 2x - 1 = 0$

$\cos 2x = \frac{1}{2}$

$2x = \frac{\pi}{3} + 2n\pi$ or $2x = \frac{5\pi}{3} + 2n\pi$

$x = \frac{\pi}{6} + n\pi$ $\quad x = \frac{5\pi}{6} + n\pi$

41. $\tan 3x - 1 = 0$

$\tan 3x = 1$

$3x = \frac{\pi}{4} + n\pi$

$x = \frac{\pi}{12} + \frac{n\pi}{3}$

43. $2\cos\frac{x}{2} = \sqrt{2} = 0$

$\cos\frac{x}{2} = \frac{\sqrt{2}}{2}$

$\frac{x}{2} = \frac{\pi}{4} + 2n\pi$ or $\frac{x}{2} = \frac{7\pi}{4} + 2n\pi$

$x = \frac{\pi}{2} + 4n\pi$ $\quad x = \frac{7\pi}{2} + 4n\pi$

45. $3\tan\frac{x}{2} - \sqrt{3} = 0$

$\tan\frac{x}{2} = \frac{\sqrt{3}}{3}$

$\frac{x}{2} = \frac{\pi}{6} + n\pi \Rightarrow x = \frac{\pi}{3} + 2n\pi$

47. $y = \sin\frac{\pi x}{2} + 1$

$\sin\left(\frac{\pi x}{2}\right) + 1 = 0$

$\sin\left(\frac{\pi x}{2}\right) = -1$

$\frac{\pi x}{2} = \frac{3\pi}{2} + 2n\pi$

$x = 3 + 4n$

For $-2 < x < 4$, the intercepts are -1 and 3.

49. $5\sin x + 2 = 0$

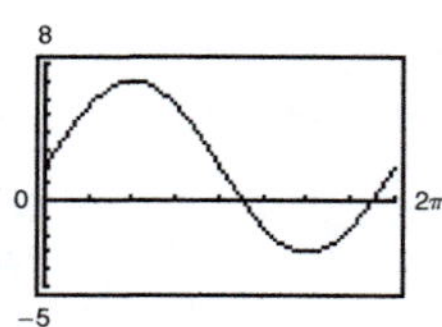

$x \approx 3.553$ and $x \approx 5.872$

51. $\sin x - 3\cos x = 0$

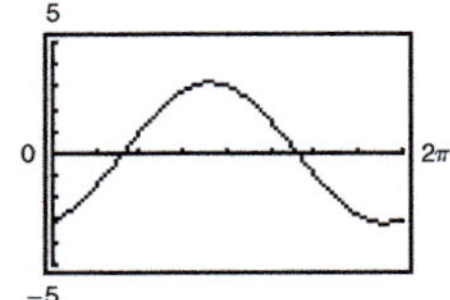

$x \approx 1.249$ and $x \approx 4.391$

53. $\cos x = x$

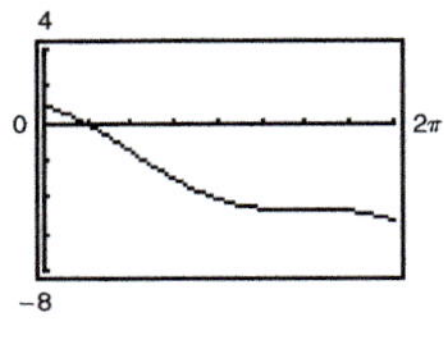

$x \approx 0.739$

55. $\sec^2 x - 3 = 0$

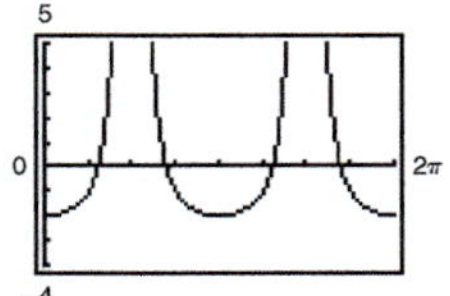

$x \approx 0.955$, $x \approx 2.186$, $x \approx 4.097$ and $x \approx 5.328$

57. $2\tan^2 x = 15$

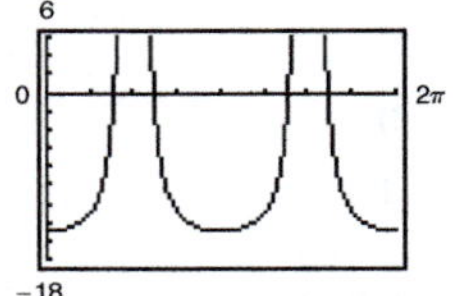

$x \approx 1.221$, $x \approx 1.921$, $x \approx 4.362$ and $x \approx 5.062$

59.
$$\tan^2 x + \tan x - 12 = 0$$
$$(\tan x + 4)(\tan x - 3) = 0$$

$\tan x + 4 = 0$ or $\tan x - 3 = 0$

$\tan x = -4$ $\qquad$ $\tan x = 3$

$x = \arctan(-4) + n\pi$ $\qquad$ $x = \arctan 3 + n\pi$

61.
$$\sec^2 x - 6\tan x = -4$$
$$1 + \tan^2 x - 6\tan x + 4 = 0$$
$$\tan^2 x - 6\tan x + 5 = 0$$
$$(\tan x - 1)(\tan x - 5) = 0$$

$\tan x - 1 = 0$ $\qquad$ $\tan x - 5 = 0$

$\tan x = 1$ $\qquad$ $\tan x = 5$

$x = \dfrac{\pi}{4} + n\pi$ $\qquad$ $x = \arctan 5 + n\pi$

63.
$$2\sin^2 x + 5\cos x = 4$$
$$2(1 - \cos^2 x) + 5\cos x - 4 = 0$$
$$-2\cos^2 x + 5\cos x - 2 = 0$$
$$-(2\cos x - 1)(\cos x - 2) = 0$$

$2\cos x - 1 = 0$ or $\cos x - 2 = 0$

$\cos x = \dfrac{1}{2}$ $\qquad$ $\cos x = 2$

$x = \dfrac{\pi}{3} + 2n\pi, \dfrac{5\pi}{3} + 2n\pi$ $\qquad$ No solution

65. $\cot^2 x - 9 = 0$

$\cot^2 x = 9$

$\frac{1}{9} = \tan^2 x$

$\pm\frac{1}{3} = \tan x$

$x = \arctan \frac{1}{3} + n\pi, \arctan\left(-\frac{1}{3}\right) + n\pi$

67. $\sec^2 x - 4 \sec x = 0$

$\sec x(\sec x - 4) = 0$

$\sec x = 0$ $\sec x - 4 = 0$

No solution $\sec x = 4$

$\frac{1}{4} = \cos x$

$x = \arccos \frac{1}{4} + 2n\pi, -\arccos \frac{1}{4} + 2n\pi$

69. $\csc^2 x + 3 \csc x - 4 = 0$

$(\csc x + 4)(\csc x - 1) = 0$

$\csc x + 4 = 0$ or $\csc x - 1 = 0$

$\csc x = -4$ $\csc x = 1$

$-\frac{1}{4} = \sin x$ $1 = \sin x$

$x = \arcsin\left(\frac{1}{4}\right) + 2n\pi, \arcsin\left(-\frac{1}{4}\right) + 2n\pi$ $x = \frac{\pi}{2} + 2n\pi$

71. $12 \sin^2 x - 13 \sin x + 3 = 0$

$$\sin x = \frac{-(-13) \pm \sqrt{(-13)^2 - 4(12)(3)}}{2(12)} = \frac{13 \pm 5}{24}$$

$\sin x = \frac{1}{3}$ or $\sin x = \frac{3}{4}$

$x \approx 0.3398, 2.8018$ $x \approx 0.8481, 2.2935$

30

0   2π

−10

The x-intercepts occur at $x \approx 0.3398$, $x \approx 0.8481$, $x \approx 2.2935$, and $x \approx 2.8018$.

73. $\tan^2 x + 3 \tan x + 1 = 0$

$$\tan x = \frac{-3 \pm \sqrt{3^2 - 4(1)(1)}}{2(1)} = \frac{-3 \pm \sqrt{5}}{2}$$

$\tan x = \dfrac{-3 - \sqrt{5}}{2}$ or $\tan x = \dfrac{-3 + \sqrt{5}}{2}$

$x \approx 1.9357, 5.0773$ $\quad$ $x \approx 2.7767, 5.9183$

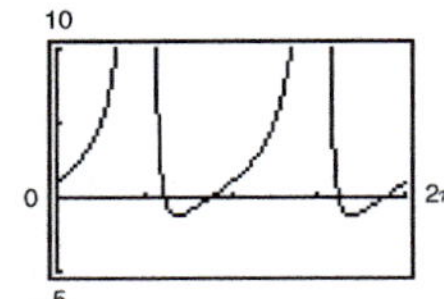

The x-intercepts occur at $x \approx 1.9357$, $x \approx 2.7767$, $x \approx 5.0773$, and $x \approx 5.9183$.

75. $3 \tan^2 x + 5 \tan x - 4 = 0, \left[-\dfrac{\pi}{2}, \dfrac{\pi}{2}\right]$

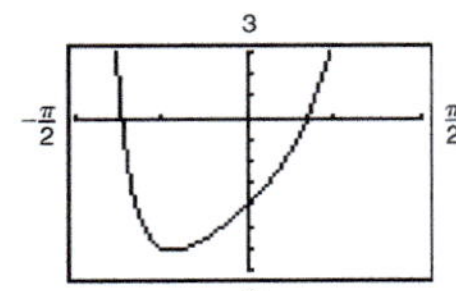

$x \approx -1.154, 0.534$

77. $4 \cos^2 x - 2 \sin x + 1 = 0, \left[-\dfrac{\pi}{2}, \dfrac{\pi}{2}\right]$

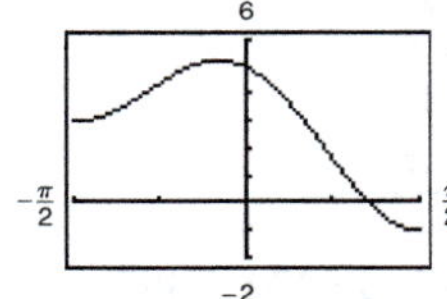

$x \approx 1.110$

79. (a) $f(x) = \sin^2 x + \cos x$

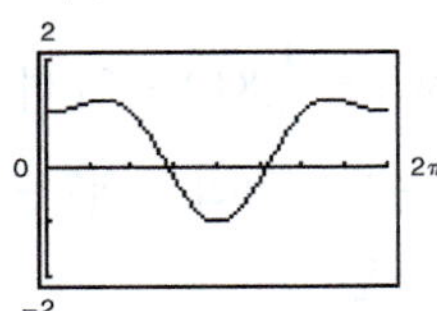

Maximum: $(1.0472, 1.25)$

Maximum: $(5.2360, 1.25)$

Minimum: $(0, 1)$

Minimum: $(3.1416, -1)$

(b) $2 \sin x \cos x - \sin x = 0$

$\sin x(2 \cos x - 1) = 0$

$\sin x = 0$ or $2 \cos x - 1 = 0$

$x = 0, \pi$

$\approx 0, 3.1416$

$\cos x = \dfrac{1}{2}$

$x = \dfrac{\pi}{3}, \dfrac{5\pi}{3}$

$\approx 1.0472, 5.2360$

81. (a) $f(x) = \sin x + \cos x$

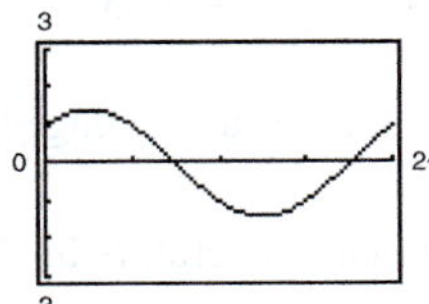

Maximum: $(0.7854, 1.4142)$

Minimum: $(3.9270, -1.4142)$

(b) $\cos x - \sin x = 0$

$\cos x = \sin x$

$1 = \dfrac{\sin x}{\cos x}$

$\tan x = 1$

$x = \dfrac{\pi}{4}, \dfrac{5\pi}{4}$

$\approx 0.7854, 3.9270$

83. (a) $f(x) = \sin x \cos x$

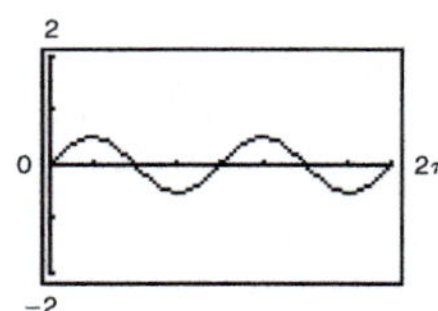

Maximum: $(0.7854, 0.5)$

Maximum: $(3.9270, 0.5)$

Minimum: $(2.3562, -0.5)$

Minimum: $(5.4978, -0.5)$

(b)
$$-\sin^2 x + \cos^2 x = 0$$
$$-\sin^2 x + 1 - \sin^2 x = 0$$
$$-2\sin^2 x + 1 = 0$$
$$\sin^2 x = \frac{1}{2}$$
$$\sin x = \pm\sqrt{\frac{1}{2}} = \pm\frac{\sqrt{2}}{2}$$
$$x = \frac{\pi}{4}, \frac{3\pi}{4}, \frac{5\pi}{4}, \frac{7\pi}{4}$$
$$\approx 0.7854, 2.3562, 3.9270, 5.4978$$

85. The graphs of $y_1 = 2\sin x$ and $y_2 = 3x + 1$ appear to have one point of intersection. This implies there is one solution to the equation $2\sin x = 3x + 1$.

87. $f(x) = \dfrac{\sin x}{x}$

(a) Domain: all real numbers except $x = 0$.

(b) The graph has y-axis symmetry and a horizontal asymptote at $y = 0$.

(c) As $x \to 0$, $f(x) \to 1$.

(d) $\dfrac{\sin x}{x} = 0$ has four solutions in the interval $[-8, 8]$.

$$\sin x\left(\frac{1}{x}\right) = 0$$
$$\sin x = 0$$
$$x = -2\pi, -\pi, \pi, 2\pi$$

89.
$$y = \frac{1}{12}(\cos 8t - 3\sin 8t)$$
$$\frac{1}{12}(\cos 8t - 3\sin 8t) = 0$$
$$\cos 8t = 3\sin 8t$$
$$\frac{1}{3} = \tan 8t$$
$$8t \approx 0.32175 + n\pi$$
$$t \approx 0.04 + \frac{n\pi}{8}$$

In the interval $0 \le t \le 1$, $t \approx 0.04, 0.43,$ and 0.83.

91. Graph $y_1 = 58.3 + 32\cos\left(\dfrac{\pi t}{6}\right)$

$y_2 = 75.$

Left point of intersection: $(1.95, 75)$

Right point of intersection: $(10.05, 75)$

So, sales exceed 7500 in January, November, and December.

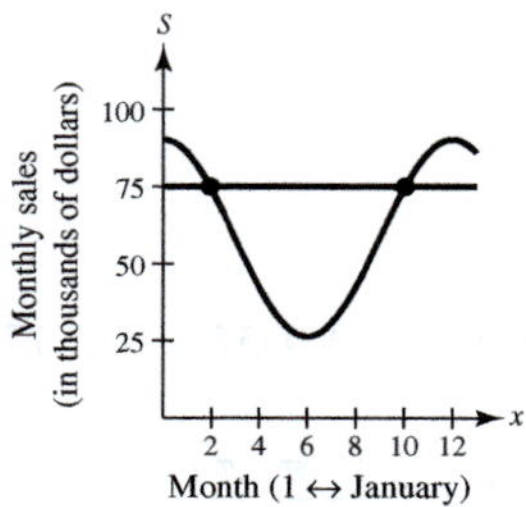

93. (a) and (c)

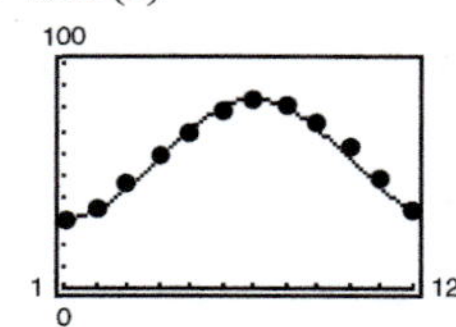

The model fits the data well.

(b) $C = a\cos(bt - c) + d$

$$a = \frac{1}{2}[\text{high} - \text{low}] = \frac{1}{2}[84.1 - 31.0] = 26.55$$
$$p = 2[\text{high time} - \text{low time}] = 2[7 - 1] = 12$$
$$b = \frac{2\pi}{p} = \frac{2\pi}{12} = \frac{\pi}{6}$$

The maximum occurs at 7, so the left end point is

$$\frac{c}{b} = 7 \Rightarrow c = 7\left(\frac{\pi}{6}\right) = \frac{7\pi}{6}$$
$$d = \frac{1}{2}[\text{high} + \text{low}] = \frac{1}{2}[93.6 + 62.3] = 57.55$$
$$C = 26.55\cos\left(\frac{\pi}{6}t - \frac{7\pi}{6}\right) + 57.55$$

(d) The constant term, d, gives the average maximum temperature.

The average maximum temperature in Chicago is 57.55°F.

(e) The average maximum temperature is above 72°F from June through September. The average maximum temperature is below 70°F from October through May.

95. $A = 2x\cos x,\ 0 < x < \dfrac{\pi}{2}$

(a)

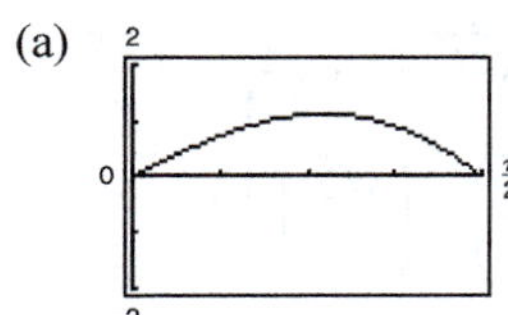

The maximum area of $A \approx 1.12$ occurs when $x \approx 0.86$.

(b) $A \ge 1$ for $0.6 < x < 1.1$

97. $f(x) = \tan \dfrac{\pi x}{4}$

Because $\tan \pi/4 = 1,\ x = 1$ is the smallest nonnegative fixed point.

99. True. The period of $2\sin 4t - 1$ is $\dfrac{\pi}{2}$ and the period of $2\sin t - 1$ is 2π.

In the interval $[0, 2\pi)$ the first equation has four cycles whereas the second equation has only one cycle, so the first equation has four times the x-intercepts (solutions) as the second equation.

101. $\cot x\cos^2 x = 2\cot x$

$\cos^2 x = 2$

$\cos x = \pm\sqrt{2}$

No solution

Because you solved this problem by first dividing by $\cot x$, you do not get the same solution as Example 3.

When solving equations, you do not want to divide each side by a variable expression that will cancel out because you may accidentally remove one of the solutions.

103. (a)

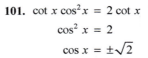

The graphs intersect when $x = \dfrac{\pi}{2}$ and $x = \pi$.

(b)

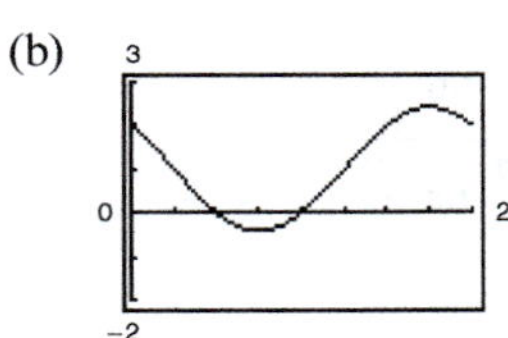

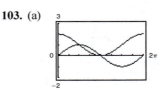

The x-intercepts are $\left(\dfrac{\pi}{2}, 0\right)$ and $(\pi, 0)$.

(c) Both methods produce the same x-values. Answers will vary on which method is preferred.

Section 5.4 Sum and Difference Formulas

1. $\sin u\cos v - \cos u\sin v$

3. $\dfrac{\tan u + \tan v}{1 - \tan u\tan v}$

5. $\cos u\cos v + \sin u\sin v$

7. (a) $\cos\left(\dfrac{\pi}{4} + \dfrac{\pi}{3}\right) = \cos\dfrac{\pi}{4}\cos\dfrac{\pi}{3} - \sin\dfrac{\pi}{4}\sin\dfrac{\pi}{3}$

$$= \frac{\sqrt{2}}{2}\cdot\frac{1}{2} - \frac{\sqrt{2}}{2}\cdot\frac{\sqrt{3}}{2}$$

$$= \frac{\sqrt{2} - \sqrt{6}}{4}$$

(b) $\cos\dfrac{\pi}{4} + \cos\dfrac{\pi}{3} = \dfrac{\sqrt{2}}{2} + \dfrac{1}{2} = \dfrac{\sqrt{2} + 1}{2}$

9. (a) $\sin(135^\circ - 30^\circ) = \sin 135^\circ\cos 30^\circ - \cos 135^\circ\sin 30^\circ$

$$= \left(\frac{\sqrt{2}}{2}\right)\left(\frac{\sqrt{3}}{2}\right) - \left(-\frac{\sqrt{2}}{2}\right)\left(\frac{1}{2}\right) = \frac{\sqrt{6} + \sqrt{2}}{4}$$

(b) $\sin 135^\circ - \cos 30^\circ = \dfrac{\sqrt{2}}{2} - \dfrac{\sqrt{3}}{2} = \dfrac{\sqrt{2} - \sqrt{3}}{2}$

11. $\sin\frac{11\pi}{12} = \sin\left(\frac{3\pi}{4} + \frac{\pi}{6}\right)$

$= \sin\frac{3\pi}{4}\cos\frac{\pi}{6} + \cos\frac{3\pi}{4}\sin\frac{\pi}{6}$

$= \frac{\sqrt{2}}{2}\cdot\frac{\sqrt{3}}{2} + \left(-\frac{\sqrt{2}}{2}\right)\frac{1}{2}$

$= \frac{\sqrt{2}}{4}\left(\sqrt{3} - 1\right)$

$\cos\frac{11\pi}{12} = \cos\left(\frac{3\pi}{4} + \frac{\pi}{6}\right)$

$= \cos\frac{3\pi}{4}\cos\frac{\pi}{6} - \sin\frac{3\pi}{4}\sin\frac{\pi}{6}$

$= -\frac{\sqrt{2}}{2}\cdot\frac{\sqrt{3}}{2} - \frac{\sqrt{2}}{2}\cdot\frac{1}{2} = -\frac{\sqrt{2}}{4}\left(\sqrt{3} + 1\right)$

$\tan\frac{11\pi}{4} = \tan\left(\frac{3\pi}{4} + \frac{\pi}{6}\right)$

$= \frac{\tan\frac{3\pi}{4} + \tan\frac{\pi}{6}}{1 - \tan\frac{3\pi}{4}\tan\frac{\pi}{6}}$

$= \frac{-1 + \frac{\sqrt{3}}{3}}{1 - (-1)\frac{\sqrt{3}}{3}}$

$= \frac{-3 + \sqrt{3}}{3 + \sqrt{3}}\cdot\frac{3 - \sqrt{3}}{3 - \sqrt{3}}$

$= \frac{-12 + 6\sqrt{3}}{6} = -2 + \sqrt{3}$

13. $\sin\frac{17\pi}{12} = \sin\left(\frac{9\pi}{4} - \frac{5\pi}{6}\right)$

$= \sin\frac{9\pi}{4}\cos\frac{5\pi}{6} - \cos\frac{9\pi}{4}\sin\frac{5\pi}{6}$

$= \frac{\sqrt{2}}{2}\left(-\frac{\sqrt{3}}{2}\right) - \left(\frac{\sqrt{2}}{2}\right)\left(\frac{1}{2}\right)$

$= -\frac{\sqrt{2}}{4}\left(\sqrt{3} + 1\right)$

$\cos\frac{17\pi}{12} = \cos\left(\frac{9\pi}{4} - \frac{5\pi}{6}\right)$

$= \cos\frac{9\pi}{4}\cos\frac{5\pi}{6} + \sin\frac{9\pi}{4}\sin\frac{5\pi}{6}$

$= \frac{\sqrt{2}}{2}\left(-\frac{\sqrt{3}}{2}\right) + \frac{\sqrt{2}}{2}\left(\frac{1}{2}\right)$

$= \frac{\sqrt{2}}{4}\left(1 - \sqrt{3}\right)$

$\tan\frac{17\pi}{12} = \tan\left(\frac{9\pi}{4} - \frac{5\pi}{6}\right)$

$= \frac{\tan(9\pi/4) - \tan(5\pi/6)}{1 + \tan(9\pi/4)\tan(5\pi/6)}$

$= \frac{1 - \left(-\sqrt{3}/3\right)}{1 + \left(-\sqrt{3}/3\right)}$

$= \frac{3 + \sqrt{3}}{3 - \sqrt{3}}\cdot\frac{3 + \sqrt{3}}{3 + \sqrt{3}}$

$= \frac{12 + 6\sqrt{3}}{6} = 2 + \sqrt{3}$

15. $\sin 105° = \sin(60° + 45°) = \sin 60°\cos 45° + \cos 60°\sin 45°$

$= \frac{\sqrt{3}}{2}\cdot\frac{\sqrt{2}}{2} + \frac{1}{2}\cdot\frac{\sqrt{2}}{2} = \frac{\sqrt{2}}{4}\left(\sqrt{3} + 1\right)$

$\cos 105° = \cos(60° + 45°) = \cos 60°\cos 45° - \sin 60°\sin 45°$

$= \frac{1}{2}\cdot\frac{\sqrt{2}}{2} - \frac{\sqrt{3}}{2}\cdot\frac{\sqrt{2}}{2} = \frac{\sqrt{2}}{4}\left(1 - \sqrt{3}\right)$

$\tan 105° = \tan(60° + 45°) = \frac{\tan 60° + \tan 45°}{1 - \tan 60°\tan 45°}$

$= \frac{\sqrt{3} + 1}{1 - \sqrt{3}} = \frac{\sqrt{3} + 1}{1 - \sqrt{3}}\cdot\frac{1 + \sqrt{3}}{1 + \sqrt{3}} = \frac{4 + 2\sqrt{3}}{-2} = -2 - \sqrt{3}$

17. $\sin(-195°) = \sin(30° - 225°)$

$= \sin 30° \cos 225° - \cos 30° \sin 225°$

$= \sin 30°(-\cos 45°) - \cos 30°(-\sin 45°)$

$= \frac{1}{2}\left(-\frac{\sqrt{2}}{2}\right) - \frac{\sqrt{3}}{2}\left(-\frac{\sqrt{2}}{2}\right)$

$= -\frac{\sqrt{2}}{4}\left(1 - \sqrt{3}\right)$

$= \frac{\sqrt{2}}{4}\left(\sqrt{3} - 1\right)$

$\cos(-195°) = \cos(30° - 225°)$

$= \cos 30° \cos 225° + \sin 30° \sin 225°$

$= \cos 30°(-\cos 45°) + \sin 30°(-45°)$

$= \frac{\sqrt{3}}{2}\left(-\frac{\sqrt{2}}{2}\right) + \frac{1}{2}\left(-\frac{\sqrt{2}}{2}\right)$

$= -\frac{\sqrt{2}}{4}\left(\sqrt{3} + 1\right)$

$\tan(-195°) = \tan(30° - 225°)$

$= \frac{\tan 30° - \tan 225°}{1 + \tan 30° \tan 225°}$

$= \frac{\tan 30° - \tan 45°}{1 + \tan 30° \tan 45°}$

$= \frac{\left(\frac{\sqrt{3}}{3}\right) - 1}{1 + \left(\frac{\sqrt{3}}{3}\right)} = \frac{\sqrt{3} - 3}{3 + \sqrt{3}} \cdot \frac{3 - \sqrt{3}}{3 - \sqrt{3}}$

$= \frac{-12 + 6\sqrt{3}}{6} = -2 + \sqrt{3}$

19. $\frac{13\pi}{12} = \frac{3\pi}{4} + \frac{\pi}{3}$

$\sin \frac{13\pi}{12} = \sin\left(\frac{3\pi}{4} + \frac{\pi}{3}\right)$

$= \sin \frac{3\pi}{4} \cos \frac{\pi}{3} + \cos \frac{3\pi}{4} \sin \frac{\pi}{3}$

$= \frac{\sqrt{2}}{2} \cdot \frac{1}{2} + \left(-\frac{\sqrt{2}}{2}\right)\left(\frac{\sqrt{3}}{2}\right)$

$= \frac{\sqrt{2}}{4}\left(1 - \sqrt{3}\right)$

$\cos \frac{13\pi}{12} = \cos\left(\frac{3\pi}{4} + \frac{\pi}{3}\right)$

$= \cos \frac{3\pi}{4} \cos \frac{\pi}{3} - \sin \frac{3\pi}{4} \sin \frac{\pi}{3}$

$= -\frac{\sqrt{2}}{2} \cdot \frac{1}{2} - \frac{\sqrt{2}}{2} \cdot \frac{\sqrt{3}}{2} = -\frac{\sqrt{2}}{4}\left(1 + \sqrt{3}\right)$

$\tan \frac{13\pi}{12} = \tan\left(\frac{3\pi}{4} + \frac{\pi}{3}\right)$

$= \frac{\tan\left(\frac{3\pi}{4}\right) + \tan\left(\frac{\pi}{3}\right)}{1 - \tan\left(\frac{3\pi}{4}\right)\tan\left(\frac{\pi}{3}\right)}$

$= \frac{-1 + \sqrt{3}}{1 - (-1)\left(\sqrt{3}\right)}$

$= -\frac{1 - \sqrt{3}}{1 + \sqrt{3}} \cdot \frac{1 - \sqrt{3}}{1 - \sqrt{3}}$

$= -\frac{4 - 2\sqrt{3}}{-2}$

$= 2 - \sqrt{3}$

21. $-\frac{5\pi}{12} = -\frac{\pi}{4} - \frac{\pi}{6}$

$\sin\left(-\frac{\pi}{4} - \frac{\pi}{6}\right) = \sin\left(-\frac{\pi}{4}\right)\cos \frac{\pi}{6} - \cos\left(-\frac{\pi}{4}\right)\sin \frac{\pi}{6}$

$= \left(-\frac{\sqrt{2}}{2}\right)\left(\frac{\sqrt{3}}{2}\right) - \left(\frac{\sqrt{2}}{2}\right)\left(\frac{1}{2}\right) = -\frac{\sqrt{2}}{4}\left(\sqrt{3} + 1\right)$

$\cos\left(-\frac{\pi}{4} - \frac{\pi}{6}\right) = \cos\left(-\frac{\pi}{4}\right)\cos \frac{\pi}{6} + \sin\left(-\frac{\pi}{4}\right)\sin \frac{\pi}{6}$

$= \left(\frac{\sqrt{2}}{2}\right)\left(\frac{\sqrt{3}}{2}\right) + \left(-\frac{\sqrt{2}}{2}\right)\left(\frac{1}{2}\right) = \frac{\sqrt{2}}{4}\left(\sqrt{3} - 1\right)$

$\tan\left(-\frac{\pi}{4} - \frac{\pi}{6}\right) = \frac{\tan\left(-\frac{\pi}{4}\right) - \tan \frac{\pi}{6}}{1 + \tan\left(-\frac{\pi}{4}\right)\tan \frac{\pi}{6}} = \frac{-1 - \frac{\sqrt{3}}{3}}{1 + (-1)\left(\frac{\sqrt{3}}{3}\right)} = \frac{-3 - \sqrt{3}}{3 - \sqrt{3}}$

$= \frac{-3 - \sqrt{3}}{3 - \sqrt{3}} \cdot \frac{3 + \sqrt{3}}{3 + \sqrt{3}} = \frac{-12 - 6\sqrt{3}}{6} = -2 - \sqrt{3}$

23. $285° = 225° + 60°$

$\sin 285° = \sin(225° + 60°) = \sin 225° \cos 60° + \cos 225° \sin 60°$

$= -\frac{\sqrt{2}}{2}\left(\frac{1}{2}\right) - \frac{\sqrt{2}}{2}\left(\frac{\sqrt{3}}{2}\right) = -\frac{\sqrt{2}}{4}\left(\sqrt{3} + 1\right)$

$\cos 285° = \cos(225° + 60°) = \cos 225° \cos 60° - \sin 225° \sin 60°$

$= -\frac{\sqrt{2}}{2}\left(\frac{1}{2}\right) - \left(-\frac{\sqrt{2}}{2}\right)\left(\frac{\sqrt{3}}{2}\right) = \frac{\sqrt{2}}{4}\left(\sqrt{3} - 1\right)$

$\tan 285° = \tan(225° + 60°) = \frac{\tan 225° + \tan 60°}{1 - \tan 225° \tan 60°}$

$= \frac{1 + \sqrt{3}}{1 - \sqrt{3}} \cdot \frac{1 + \sqrt{3}}{1 + \sqrt{3}} = \frac{4 + 2\sqrt{3}}{-2} = -2 - \sqrt{3} = -\left(2 + \sqrt{3}\right)$

25. $-165° = -(120° + 45°)$

$\sin(-165°) = \sin\left[-(120° + 45°)\right] = -\sin(120° + 45°) = -[\sin 120° \cos 45° + \cos 120° \sin 45°]$

$= -\left[\frac{\sqrt{3}}{2} \cdot \frac{\sqrt{2}}{2} - \frac{1}{2} \cdot \frac{\sqrt{2}}{2}\right] = -\frac{\sqrt{2}}{4}\left(\sqrt{3} - 1\right)$

$\cos(-165°) = \cos\left[-(120° + 45°)\right] = \cos(120° + 45°) = \cos 120° \cos 45° - \sin 120° \sin 45°$

$= -\frac{1}{2} \cdot \frac{\sqrt{2}}{2} - \frac{\sqrt{3}}{2} \cdot \frac{\sqrt{2}}{2} = -\frac{\sqrt{2}}{4}\left(1 + \sqrt{3}\right)$

$\tan(-165°) = \tan\left[-(120° + 45°)\right] = -\tan(120° + \tan 45°) = -\frac{\tan 120° + \tan 45°}{1 - \tan 120° \tan 45°}$

$= -\frac{-\sqrt{3} + 1}{1 - \left(-\sqrt{3}\right)(1)} = -\frac{1 - \sqrt{3}}{1 + \sqrt{3}} \cdot \frac{1 - \sqrt{3}}{1 - \sqrt{3}} = -\frac{4 - 2\sqrt{3}}{-2} = 2 - \sqrt{3}$

27. $\sin 3 \cos 1.2 - \cos 3 \sin 1.2 = \sin(3 - 1.2) = \sin 1.8$

29. $\sin 60° \cos 15° + \cos 60° \sin 15° = \sin(60° + 15°)$

$= \sin 75°$

31. $\frac{\tan(\pi/15) + \tan(2\pi/5)}{1 - \tan(\pi/15)\tan(2\pi/5)} = \tan(\pi/15 + 2\pi/5)$

$= \tan(7\pi/15)$

33. $\cos 3x \cos 2y + \sin 3x \sin 2y = \cos(3x - 2y)$

35. $\sin \frac{\pi}{12} \cos \frac{\pi}{4} + \cos \frac{\pi}{12} \sin \frac{\pi}{4} = \sin\left(\frac{\pi}{12} + \frac{\pi}{4}\right)$

$= \sin \frac{\pi}{3}$

$= \frac{\sqrt{3}}{2}$

37. $\cos 130° \cos 10° + \sin 130° \sin 10° = \cos(130° - 10°)$

$= \cos 120°$

$= -\frac{1}{2}$

39. $\frac{\tan(9\pi/8) - \tan(\pi/8)}{1 + \tan(9\pi/8)\tan(\pi/8)} = \tan\left(\frac{9\pi}{8} - \frac{\pi}{8}\right)$

$= \tan \pi$

$= 0$

For Exercises 41–45, you have:

$\sin u = -\frac{3}{5}$, u in Quadrant IV $\Rightarrow$ $\cos u = \frac{4}{5}$, $\tan u = -\frac{4}{3}$

$\cos v = \frac{15}{17}$, v in Quadrant I $\Rightarrow$ $\sin v = \frac{8}{17}$, $\tan v = \frac{8}{15}$

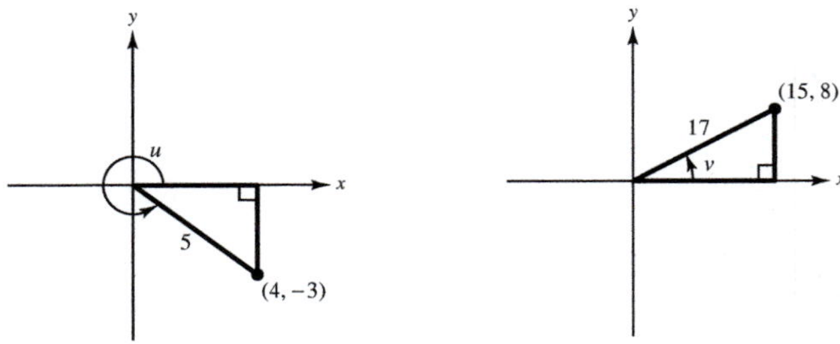

Figures for Exercises 41–45

41. $\sin(u + v) = \sin u \cos v + \cos u \sin v = \left(-\frac{3}{5}\right)\left(\frac{15}{17}\right) + \left(\frac{4}{5}\right)\left(\frac{8}{17}\right) = -\frac{13}{85}$

43. $\tan(u + v) = \dfrac{\tan u + \tan v}{1 - \tan u \tan v} = \dfrac{-\frac{3}{4} + \left(\frac{8}{15}\right)}{1 - \left(-\frac{3}{4}\right)\left(\frac{8}{15}\right)} = \dfrac{-\frac{13}{60}}{1 + \frac{32}{60}} = \left(-\frac{13}{60}\right)\left(\frac{5}{7}\right) = -\frac{13}{84}$

45. $\sec(v - u) = \dfrac{1}{\cos(v - u)} = \dfrac{1}{\cos v \cos u + \sin v \sin u} = \dfrac{1}{\left(\frac{15}{17}\right)\left(\frac{4}{5}\right) + \left(\frac{8}{17}\right)\left(-\frac{3}{5}\right)} = \dfrac{1}{\left(\frac{60}{85}\right) + \left(-\frac{24}{85}\right)} = \dfrac{1}{\frac{36}{85}} = \dfrac{85}{36}$

For Exercises 47–51, you have:

$\sin u = -\frac{7}{25}$, u in Quadrant III $\Rightarrow$ $\cos u = -\frac{24}{25}$, $\tan u = \frac{7}{24}$

$\cos v = -\frac{4}{5}$, v in Quadrant III $\Rightarrow$ $\sin v = -\frac{3}{5}$, $\tan v = \frac{3}{4}$

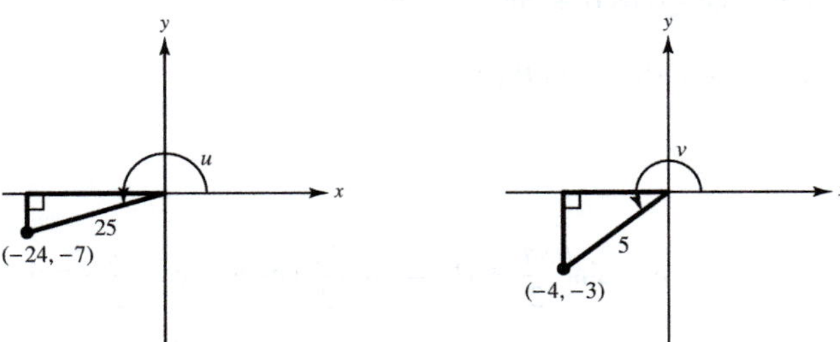

Figures for Exercises 47–51

47. $\cos(u + v) = \cos u \cos v - \sin u \sin v = \left(-\frac{24}{25}\right)\left(-\frac{4}{5}\right) - \left(-\frac{7}{25}\right)\left(-\frac{3}{5}\right) = \frac{3}{5}$

49. $\tan(u - v) = \dfrac{\tan u - \tan v}{1 + \tan u \tan v} = \dfrac{\frac{7}{24} - \frac{3}{4}}{1 + \left(\frac{7}{24}\right)\left(\frac{3}{4}\right)} = \dfrac{-\frac{11}{24}}{\frac{39}{32}} = -\frac{44}{117}$

51. $\csc(u - v) = \dfrac{1}{\sin(u - v)} = \dfrac{1}{\sin u \cos v - \cos u \sin v} = \dfrac{1}{\left(-\frac{7}{25}\right)\left(-\frac{4}{5}\right) - \left(-\frac{24}{25}\right)\left(-\frac{3}{5}\right)} = \dfrac{1}{-\frac{44}{125}} = -\frac{125}{44}$

53. $\sin(\arcsin x + \arccos x) = \sin(\arcsin x)\cos(\arccos x) + \sin(\arccos x)\cos(\arcsin x)$

$$= x \cdot x + \sqrt{1 - x^2} \cdot \sqrt{1 - x^2}$$
$$= x^2 + 1 - x^2$$
$$= 1$$

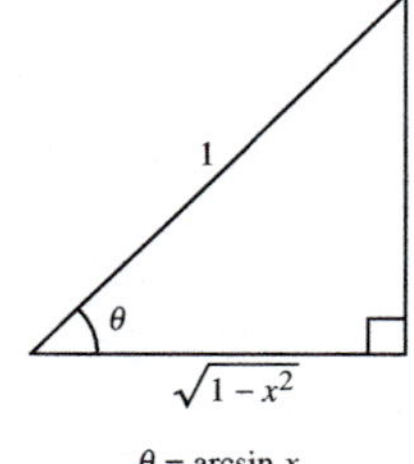

$\theta = \arcsin x$

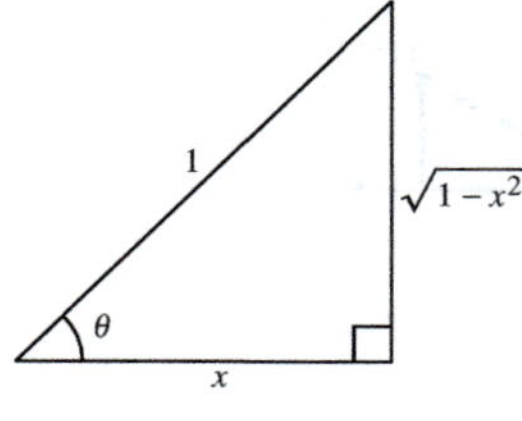

$\theta = \arccos x$

55. $\cos(\arccos x + \arcsin x) = \cos(\arccos x)\cos(\arcsin x) - \sin(\arccos x)\sin(\arcsin x)$

$$= x \cdot \sqrt{1 - x^2} - \sqrt{1 - x^2} \cdot x$$
$$= 0$$

(Use the triangles in Exercise 53.)

57. $\sin\left(\frac{\pi}{2} - x\right) = \sin\frac{\pi}{2}\cos x - \cos\frac{\pi}{2}\sin x$

$$= (1)(\cos x) - (0)(\sin x)$$
$$= \cos x$$

59. $\sin\left(\frac{\pi}{6} + x\right) = \sin\frac{\pi}{6}\cos x + \cos\frac{\pi}{6}\sin x$

$$= \frac{1}{2}\left(\cos x + \sqrt{3}\sin x\right)$$

61. $\tan(\theta + \pi) = \dfrac{\tan\theta + \tan\pi}{1 - \tan\theta\tan\pi} = \dfrac{\tan\theta + 0}{1 - (\tan\theta)(0)} = \dfrac{\tan\theta}{1} = \tan\theta$

63. $\cos(\pi - \theta) + \sin\left(\frac{\pi}{2} + \theta\right) = \cos\pi\cos\theta + \sin\pi\sin\theta + \sin\frac{\pi}{2}\cos\theta + \cos\frac{\pi}{2}\sin\theta$

$$= (-1)(\cos\theta) + (0)(\sin\theta) + (1)(\cos\theta) + (\sin\theta)(0)$$
$$= -\cos\theta + \cos\theta$$
$$= 0$$

65. $\cos\left(\frac{3\pi}{2} - \theta\right) = \cos\frac{3\pi}{2}\cos\theta + \sin\frac{3\pi}{2}\sin\theta$

$$= (0)(\cos\theta) + (-1)(\sin\theta)$$
$$= -\sin\theta$$

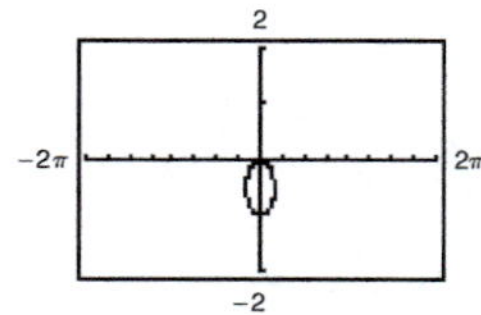

The graphs appear to coincide, so

$\cos\left(\frac{3\pi}{2} - \theta\right) = -\sin\theta.$

67. $\sin\left(\frac{3\pi}{2} + \theta\right) = \sin\frac{3\pi}{2}\cos\theta + \cos\frac{3\pi}{2}\sin\theta$

$$= (-1)(\cos\theta) + (0)(\sin\theta)$$
$$= -\cos\theta$$

$$\csc\left(\frac{3\pi}{2} + \theta\right) = \frac{1}{\sin\left(\frac{3\pi}{2} + \theta\right)} = \frac{1}{-\cos\theta} = -\sec\theta$$

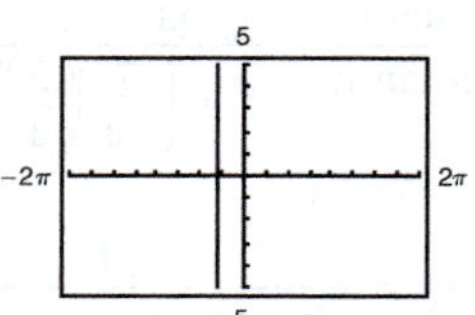

The graphs appear to coincide, so

$\csc\left(\frac{3\pi}{2} + \theta\right) = -\sec\theta.$

69.

$$\sin(x + \pi) - \sin x + 1 = 0$$
$$\sin x \cos \pi + \cos x \sin \pi - \sin x + 1 = 0$$
$$(\sin x)(-1) + (\cos x)(0) - \sin x + 1 = 0$$
$$-2 \sin x + 1 = 0$$
$$\sin x = \frac{1}{2}$$
$$x = \frac{\pi}{6}, \frac{5\pi}{6}$$

71.

$$\cos\left(x + \frac{\pi}{4}\right) - \cos\left(x - \frac{\pi}{4}\right) = 1$$
$$\cos x \cos \frac{\pi}{4} - \sin x \sin \frac{\pi}{4} - \left(\cos x \cos \frac{\pi}{4} + \sin x \sin \frac{\pi}{4}\right) = 1$$
$$-2 \sin x\left(\frac{\sqrt{2}}{2}\right) = 1$$
$$-\sqrt{2} \sin x = 1$$
$$\sin x = -\frac{1}{\sqrt{2}}$$
$$\sin x = -\frac{\sqrt{2}}{2}$$
$$x = \frac{5\pi}{4}, \frac{7\pi}{4}$$

73.

$$\tan(x + \pi) + 2 \sin(x + \pi) = 0$$
$$\frac{\tan x + \tan \pi}{1 - \tan x \tan \pi} + 2(\sin x \cos \pi + \cos x \sin \pi) = 0$$
$$\frac{\tan x + 0}{1 - \tan x(0)} + 2\left[\sin x(-1) + \cos x(0)\right] = 0$$
$$\frac{\tan x}{1} - 2 \sin x = 0$$
$$\frac{\sin x}{\cos x} = 2 \sin x$$
$$\sin x = 2 \sin x \cos x$$
$$\sin x(1 - 2 \cos x) = 0$$
$$\sin x = 0 \quad \text{or} \quad \cos x = \frac{1}{2}$$
$$x = 0, \pi \qquad x = \frac{\pi}{3}, \frac{5\pi}{3}$$

75. $\cos\left(x + \frac{\pi}{4}\right) + \cos\left(x - \frac{\pi}{4}\right) = 1$

Graph $y_1 = \cos\left(x + \frac{\pi}{4}\right) + \cos\left(x - \frac{\pi}{4}\right)$ and $y_2 = 1$.

$x = \frac{\pi}{4}, \frac{7\pi}{4}$

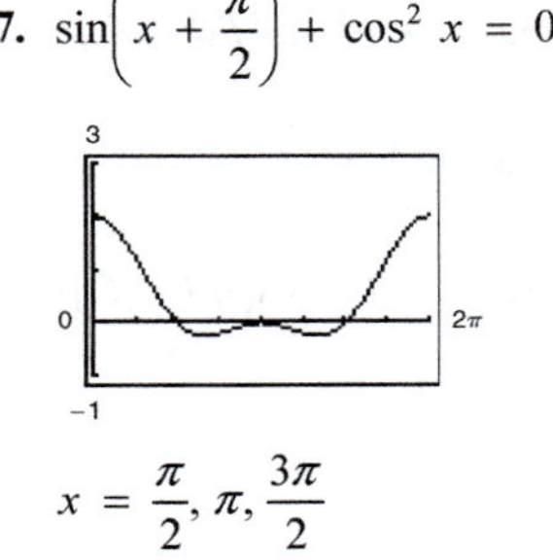

77. $\sin\left(x + \frac{\pi}{2}\right) + \cos^2 x = 0$

$x = \frac{\pi}{2}, \pi, \frac{3\pi}{2}$

79. $y = \frac{1}{3}\sin 2t + \frac{1}{4}\cos 2t$

(a) $a = \frac{1}{3},\ b = \frac{1}{4},\ B = 2$

$C = \arctan\frac{b}{a} = \arctan\frac{3}{4} \approx 0.6435$

$y \approx \sqrt{\left(\frac{1}{3}\right)^2 + \left(\frac{1}{4}\right)^2}\sin(2t + 0.6435) = \frac{5}{12}\sin(2t + 0.6435)$

(b) Amplitude: $\frac{5}{12}$ feet

(c) Frequency: $\frac{1}{\text{period}} = \frac{B}{2\pi} = \frac{2}{2\pi} = \frac{1}{\pi}$ cycle per second

81. True.

$\sin(u + v) = \sin u\cos v + \cos u\sin v$

$\sin(u - v) = \sin u\cos v - \cos u\sin v$

So, $\sin(u \pm v) = \sin u\cos v \pm \cos u\sin v$.

83. $\sin(\alpha + \beta) = \sin\alpha\cos\beta + \sin\beta\cos\alpha = 0$

$\sin\alpha\cos\beta + \sin\beta\cos\alpha = 0$

$\sin\alpha\cos\beta = -\sin\beta\cos\alpha$

False. When α and β are supplementary, $\sin\alpha\cos\beta = -\cos\alpha\sin\beta$.

85. The denominator should be $1 + \tan x\tan(\pi/4)$.

$\tan\left(x - \frac{\pi}{4}\right) = \frac{\tan x - \tan(\pi/4)}{1 + \tan x\tan(\pi/4)} = \frac{\tan x - 1}{1 + \tan x}$

87. $\cos(n\pi + \theta) = \cos n\pi\cos\theta - \sin n\pi\sin\theta = (-1)^n(\cos\theta) - (0)(\sin\theta) = (-1)^n(\cos\theta)$, where n is an integer.

89. $C = \arctan\frac{b}{a} \Rightarrow \sin C = \frac{b}{\sqrt{a^2 + b^2}},\ \cos C = \frac{a}{\sqrt{a^2 + b^2}}$

$\sqrt{a^2 + b^2}\sin(B\theta + C) = \sqrt{a^2 + b^2}\left(\sin B\theta \cdot \frac{a}{\sqrt{a^2 + b^2}} + \frac{b}{\sqrt{a^2 + b^2}} \cdot \cos B\theta\right) = a\sin B\theta + b\cos B\theta$

91. $\sin\theta + \cos\theta$

$a = 1,\ b = 1,\ B = 1$

(a) $C = \arctan\frac{b}{a} = \arctan 1 = \frac{\pi}{4}$

$\sin\theta + \cos\theta = \sqrt{a^2 + b^2}\sin(B\theta + C) = \sqrt{2}\sin\left(\theta + \frac{\pi}{4}\right)$

(b) $C = \arctan\frac{a}{b} = \arctan 1 = \frac{\pi}{4}$

$\sin\theta + \cos\theta = \sqrt{a^2 + b^2}\cos(B\theta - C) = \sqrt{2}\cos\left(\theta - \frac{\pi}{4}\right)$

93. $12 \sin 3\theta + 5 \cos 3\theta$

$a = 12, b = 5, B = 3$

(a) $C = \arctan \dfrac{b}{a} = \arctan \dfrac{5}{12} \approx 0.3948$

$$12 \sin 3\theta + 5 \cos 3\theta = \sqrt{a^2 + b^2} \sin(B\theta + C)$$
$$\approx 13 \sin(3\theta + 0.3948)$$

(b) $C = \arctan \dfrac{a}{b} = \arctan \dfrac{12}{5} \approx 1.1760$

$$12 \sin 3\theta + 5 \cos 3\theta = \sqrt{a^2 + b^2} \cos(B\theta - C)$$
$$\approx 13 \cos(3\theta - 1.1760)$$

95. $C = \arctan \dfrac{b}{a} = \dfrac{\pi}{4} \Rightarrow a = b, a > 0, b > 0$

$\sqrt{a^2 + b^2} = 2 \Rightarrow a = b = \sqrt{2}$

$B = 1$

$$2 \sin\left(\theta + \frac{\pi}{4}\right) = \sqrt{2} \sin \theta + \sqrt{2} \cos \theta$$

97.

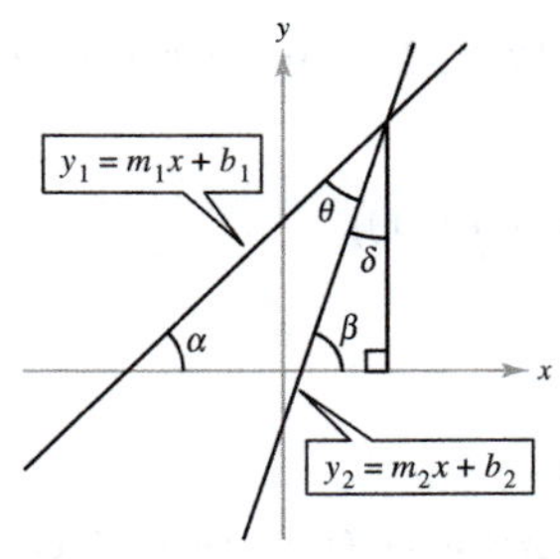

$m_1 = \tan \alpha$ and $m_2 = \tan \beta$

$\beta + \delta = 90° \Rightarrow \delta = 90° - \beta$

$\alpha + \theta + \delta = 90° \Rightarrow \alpha + \theta + (90° - \beta)$
$= 90° \Rightarrow \theta = \beta - \alpha$

So, $\theta = \arctan m_2 - \arctan m_1$. For $y = x$ and $y = \sqrt{3}x$ you have $m_1 = 1$ and $m_2 = \sqrt{3}$.

$\theta = \arctan\sqrt{3} - \arctan 1 = 60° - 45° = 15°$

99. $y_1 = \cos(x + 2), y_2 = \cos x + \cos 2$

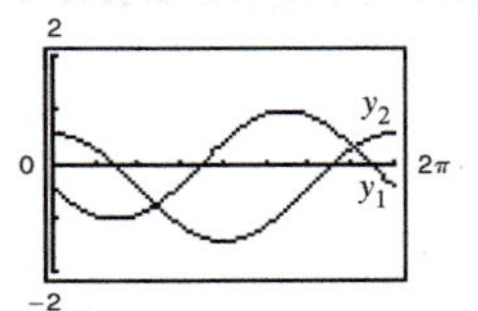

No, $y_1 \neq y_2$ because their graphs are different.

101. (a) To prove the identity for $\sin(u + v)$ you first need to prove the identity for $\cos(u - v)$. Assume $0 < v < u < 2\pi$ and locate u, v, and $u - v$ on the unit circle.

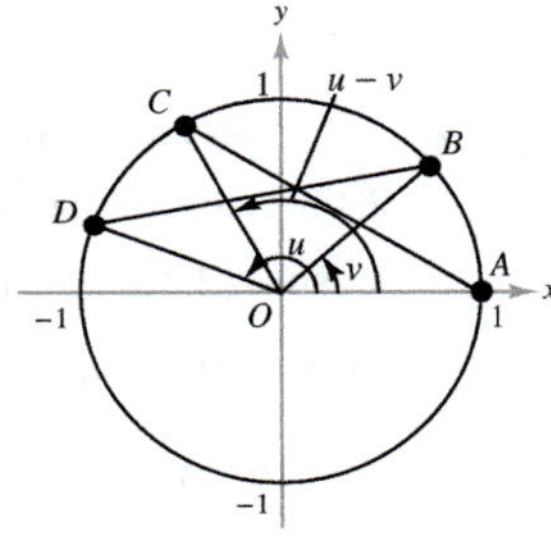

The coordinates of the points on the circle are:

$A = (1, 0), B = (\cos v, \sin v), C = (\cos(u - v), \sin(u - v))$, and $D = (\cos u, \sin u)$.

Because $\angle DOB = \angle COA$, chords AC and BD are equal. By the Distance Formula:

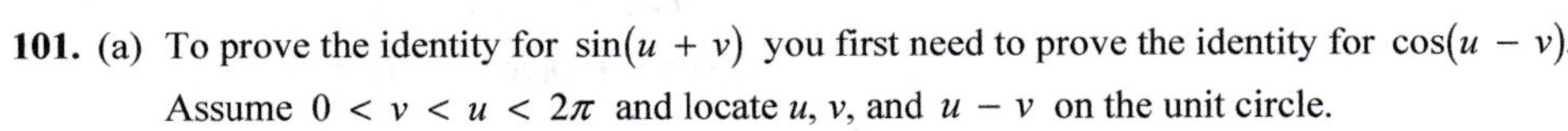

$$\sqrt{[\cos(u - v) - 1]^2 + [\sin(u - v) - 0]^2} = \sqrt{(\cos u - \cos v)^2 + (\sin u - \sin v)^2}$$

$$\cos^2(u - v) - 2\cos(u - v) + 1 + \sin^2(u - v) = \cos^2 u - 2 \cos u \cos v + \cos^2 v + \sin^2 u - 2 \sin u \sin v + \sin^2 v$$

$$[\cos^2(u - v) + \sin^2(u - v)] + 1 - 2\cos(u - v) = (\cos^2 u + \sin^2 u) + (\cos^2 v + \sin^2 v) - 2 \cos u \cos v - 2 \sin u \sin v$$

$$2 - 2\cos(u - v) = 2 - 2 \cos u \cos v - 2 \sin u \sin v$$

$$-2\cos(u - v) = -2(\cos u \cos v + \sin u \sin v)$$

$$\cos(u - v) = \cos u \cos v + \sin u \sin v$$

Now, to prove the identity for $\sin(u + v)$, use cofunction identities.

$$\sin(u + v) = \cos\left[\frac{\pi}{2} - (u + v)\right] = \cos\left[\left(\frac{\pi}{2} - u\right) - v\right]$$
$$= \cos\left(\frac{\pi}{2} - u\right)\cos v + \sin\left(\frac{\pi}{2} - u\right)\sin v$$
$$= \sin u \cos v + \cos u \sin v$$

(b) First, prove $\cos(u - v) = \cos u \cos v + \sin u \sin v$ using the figure containing points

$A(1, 0)$

$B(\cos(u - v), \sin(u - v))$

$C(\cos v, \sin v)$

$D(\cos u, \sin u)$

on the unit circle.

Because chords AB and CD are each subtended by angle $u - v$, their lengths are equal. Equating $[d(A, B)]^2 = [d(C, D)]^2$ you have $(\cos(u - v) - 1)^2 + \sin^2(u - v) = (\cos u - \cos v)^2 + (\sin u - \sin v)^2$.

Simplifying and solving for $\cos(u - v)$, you have $\cos(u - v) = \cos u \cos v + \sin u \sin v$.

Using $\sin\theta = \cos\left(\frac{\pi}{2} - \theta\right)$,

$$\sin(u - v) = \cos\left[\frac{\pi}{2} - (u - v)\right] = \cos\left[\left(\frac{\pi}{2} - u\right) - (-v)\right] = \cos\left(\frac{\pi}{2} - u\right)\cos(-v) + \sin\left(\frac{\pi}{2} - u\right)\sin(-v)$$

$$= \sin u \cos v - \cos u \sin v$$

Section 5.5 Multiple-Angle and Product-to-Sum Formulas

1. $2 \sin u \cos u$

3. $\frac{1}{2}\left[\sin(u + v) + \sin(u - v)\right]$

5. $\pm\sqrt{\frac{1 - \cos u}{2}}$

7.
$$\sin 2x - \sin x = 0$$
$$2 \sin x \cos x - \sin x = 0$$
$$\sin x(2 \cos x - 1) = 0$$

$\sin x = 0$ or $2 \cos x - 1 = 0$

$x = n\pi$ $\qquad$ $\cos x = \frac{1}{2}$

$x = \frac{\pi}{3} + 2n\pi, \frac{5\pi}{3} + 2n\pi$

9.
$$\cos 2x - \cos x = 0$$
$$\cos 2x = \cos x$$
$$\cos^2 x - \sin^2 x = \cos x$$
$$\cos^2 x - (1 - \cos^2 x) - \cos x = 0$$
$$2 \cos^2 x - \cos x - 1 = 0$$
$$(2 \cos x + 1)(\cos x - 1) = 0$$

$2 \cos x + 1 = 0$ or $\cos x - 1 = 0$

$\cos x = -\frac{1}{2}$ $\qquad$ $\cos x = 1$

$x = \frac{2n\pi}{3}$ $\qquad$ $x = 0$

11.
$$\sin 4x = -2 \sin 2x$$
$$\sin 4x + 2 \sin 2x = 0$$
$$2 \sin 2x \cos 2x + 2 \sin 2x = 0$$
$$2 \sin 2x(\cos 2x + 1) = 0$$

$2 \sin 2x = 0$ or $\cos 2x + 1 = 0$

$\sin 2x = 0$ $\qquad$ $\cos 2x = -1$

$2x = n\pi$ $\qquad$ $2x = \pi + 2n\pi$

$x = \frac{n}{2}\pi$ $\qquad$ $x = \frac{\pi}{2} + n\pi$

13. $$\tan 2x - \cot x = 0$$
$$\frac{2\tan x}{1 - \tan^2 x} = \cot x$$
$$2\tan x = \cot x(1 - \tan^2 x)$$
$$2\tan x = \cot x - \cot x \tan^2 x$$
$$2\tan x = \cot x - \tan x$$
$$3\tan x = \cot x$$
$$3\tan x - \cot x = 0$$
$$3\tan x - \frac{1}{\tan x} = 0$$
$$\frac{3\tan^2 x - 1}{\tan x} = 0$$
$$\frac{1}{\tan x}(3\tan^2 x - 1) = 0$$
$$\cot x(3\tan^2 x - 1) = 0$$

$\cot x = 0$ or $3\tan^2 x - 1 = 0$

$x = \frac{\pi}{2} + n\pi$ $\qquad \tan^2 x = \frac{1}{3}$

$\tan x = \pm\frac{\sqrt{3}}{3}$

$x = \frac{\pi}{6} + n\pi, \frac{5\pi}{6} + n\pi$

15. $6\sin x\cos x = 3(2\sin x\cos x) = 3\sin 2x$

17. $6\cos^2 x - 3 = 3(2\cos^2 x - 1) = 3\cos 2x$

19. $4 - 8\sin^2 x = 4(1 - 2\sin^2 x) = 4\cos 2x$

21. $\sin u = -\frac{3}{5}, \frac{3\pi}{2} < u < 2\pi$

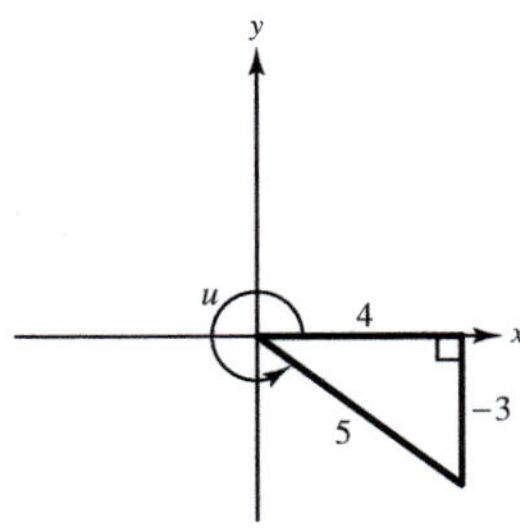

$$\sin 2u = 2\sin u\cos u = 2\left(-\frac{3}{5}\right)\left(\frac{4}{5}\right) = -\frac{24}{25}$$
$$\cos 2u = \cos^2 u - \sin^2 u = \frac{16}{25} - \frac{9}{25} = \frac{7}{25}$$
$$\tan 2u = \frac{2\tan u}{1 - \tan^2 u} = \frac{2\left(-\frac{3}{4}\right)}{1 - \frac{9}{16}} = -\frac{3}{2}\left(\frac{16}{7}\right) = -\frac{24}{7}$$

23. $\tan u = \frac{3}{5}, 0 < u < \frac{\pi}{2}$

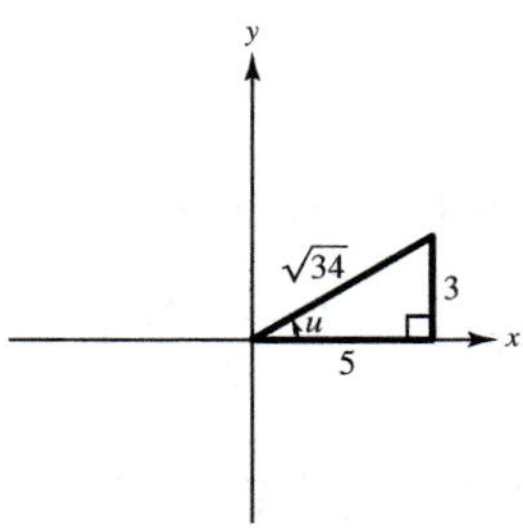

$$\sin 2u = 2\sin u\cos u = 2\left(\frac{3}{\sqrt{34}}\right)\left(\frac{5}{\sqrt{34}}\right) = \frac{15}{17}$$
$$\cos 2u = \cos^2 u - \sin^2 u = \frac{25}{34} - \frac{9}{34} = \frac{8}{17}$$
$$\tan 2u = \frac{2\tan u}{1 - \tan^2 u} = \frac{2\left(\frac{3}{5}\right)}{1 - \frac{9}{25}} = \frac{6}{5}\left(\frac{25}{16}\right) = \frac{15}{8}$$

25. $$\cos 4x = \cos(2x + 2x)$$
$$= \cos 2x\cos 2x - \sin 2x\sin 2x$$
$$= \cos^2 2x - \sin^2 2x$$
$$= \cos^2 2x - (1 - \cos^2 2x)$$
$$= 2\cos^2 2x - 1$$
$$= 2(\cos 2x)^2 - 1$$
$$= 2(2\cos^2 x - 1)^2 - 1$$
$$= 2(4\cos^4 x - 4\cos x + 1) - 1$$
$$= 8\cos^4 x - 8\cos x + 1$$

27. $$\cos^4 x = (\cos^2 x)(\cos^2 x)$$
$$= \left(\frac{1 + \cos 2x}{2}\right)\left(\frac{1 + \cos 2x}{2}\right)$$
$$= \frac{1 + 2\cos 2x + \cos^2 2x}{4}$$
$$= \frac{1 + 2\cos 2x + \frac{1 + \cos 4x}{2}}{4}$$
$$= \frac{2 + 4\cos 2x + 1 + \cos 4x}{8}$$
$$= \frac{3 + 4\cos 2x + \cos 4x}{8}$$
$$= \frac{1}{8}(3 + 4\cos 2x + \cos 4x)$$

29. $\sin^4 2x = \left(\sin^2 2x\right)^2$

$$= \left(\frac{1 - \cos 4x}{2}\right)^2$$

$$= \frac{1}{4}\left(1 - 2\cos 4x + \cos^2 4x\right)$$

$$= \frac{1}{4}\left(1 - 2\cos 4x + \frac{1 + \cos 8x}{2}\right)$$

$$= \frac{1}{4} - \frac{1}{2}\cos 4x + \frac{1}{8} + \frac{1}{8}\cos 8x$$

$$= \frac{3}{8} - \frac{1}{2}\cos 4x + \frac{1}{8}\cos 8x$$

$$= \frac{1}{8}(3 - 4\cos 4x + \cos 8x)$$

31. $\tan^4 2x = \left(\tan^2 2x\right)^2$

$$= \left(\frac{1 - \cos 4x}{1 + \cos 4x}\right)^2$$

$$= \frac{1 - 2\cos 4x + \cos^2 4x}{1 + 2\cos 4x + \cos^2 4x}$$

$$= \frac{1 - 2\cos 4x + \dfrac{1 + \cos 8x}{2}}{1 + 2\cos 4x + \dfrac{1 + \cos 8x}{2}}$$

$$= \frac{\frac{1}{2}(2 - 4\cos 4x + 1 + \cos 8x)}{\frac{1}{2}(2 + 4\cos 4x + 1 + \cos 8x)}$$

$$= \frac{3 - 4\cos 4x + \cos 8x}{3 + 4\cos 4x + \cos 8x}$$

33. $\sin^2 2x \cos^2 2x = \left(\frac{1 - \cos 4x}{2}\right)\left(\frac{1 + \cos 4x}{2}\right)$

$$= \frac{1}{4}\left(1 - \cos^2 4x\right)$$

$$= \frac{1}{4}\left(1 - \frac{1 + \cos 8x}{2}\right)$$

$$= \frac{1}{4} - \frac{1}{8} - \frac{1}{8}\cos 8x$$

$$= \frac{1}{8} - \frac{1}{8}\cos 8x$$

$$= \frac{1}{8}(1 - \cos 8x)$$

35. $\sin 75^\circ = \sin\left(\frac{1}{2}\cdot 150^\circ\right) = \sqrt{\frac{1 - \cos 150^\circ}{2}}$

$$= \sqrt{\frac{1 + \left(\sqrt{3}/2\right)}{2}} = \frac{1}{2}\sqrt{2 + \sqrt{3}}$$

$\cos 75^\circ = \cos\left(\frac{1}{2}\cdot 150^\circ\right) = \sqrt{\frac{1 + \cos 150^\circ}{2}}$

$$= \sqrt{\frac{1 - \left(\sqrt{3}/2\right)}{2}} = \frac{1}{2}\sqrt{2 - \sqrt{3}}$$

$\tan 75^\circ = \tan\left(\frac{1}{2}\cdot 150^\circ\right) = \frac{\sin 150^\circ}{1 + \cos 150^\circ}$

$$= \frac{1/2}{1 - \left(\sqrt{3}/2\right)} = \frac{1}{2 - \sqrt{3}}\cdot\frac{2 + \sqrt{3}}{2 + \sqrt{3}}$$

$$= \frac{2 + \sqrt{3}}{4 - 3} = 2 + \sqrt{3}$$

37. $\sin 112^\circ 30' = \sin\left(\frac{1}{2}\cdot 225^\circ\right) = \sqrt{\frac{1 - \cos 225^\circ}{2}} = \sqrt{\frac{1 - \left(-\sqrt{2}/2\right)}{2}} = \frac{1}{2}\sqrt{2 + \sqrt{2}}$

$\cos 112^\circ 30' = \cos\left(\frac{1}{2}\cdot 225^\circ\right) = -\sqrt{\frac{1 + \cos 225^\circ}{2}} = -\sqrt{\frac{1 + \left(-\sqrt{2}/2\right)}{2}} = \frac{1}{2} - \sqrt{2 - 2}$

$\tan 112^\circ 30' = \tan\left(\frac{1}{2}\cdot 225^\circ\right) = \frac{\sin 225^\circ}{1 + \cos 225^\circ} = \frac{-\sqrt{2}/2}{1 + \left(-\sqrt{2}/2\right)} = \frac{-\sqrt{2}}{2 - \sqrt{2}}\cdot\frac{2 + \sqrt{2}}{2 + \sqrt{2}} = \frac{-2\sqrt{2} - 2}{2} = -1 - \sqrt{2}$

39. $\sin\frac{\pi}{8} = \sin\left[\frac{1}{2}\left(\frac{\pi}{4}\right)\right] = \sqrt{\frac{1 - \cos\frac{\pi}{4}}{2}} = \frac{1}{2}\sqrt{2 - \sqrt{2}}$

$\cos\frac{\pi}{8} = \cos\left[\frac{1}{2}\left(\frac{\pi}{4}\right)\right] = \sqrt{\frac{1 + \cos\frac{\pi}{4}}{2}} = \frac{1}{2}\sqrt{2 + \sqrt{2}}$

$\tan\frac{\pi}{8} = \tan\left[\frac{1}{2}\left(\frac{\pi}{4}\right)\right] = \frac{\sin\frac{\pi}{4}}{1 + \cos\frac{\pi}{4}} = \frac{\frac{\sqrt{2}}{2}}{1 + \frac{\sqrt{2}}{2}} = \sqrt{2} - 1$

41. $\cos u = \dfrac{7}{25},\ 0 < u < \dfrac{\pi}{2}$

(a) Because u is in Quadrant I, $\dfrac{u}{2}$ is also in Quadrant I.

(b) $$\sin\frac{u}{2} = \sqrt{\frac{1-\cos u}{2}} = \sqrt{\frac{1-\frac{7}{25}}{2}} = \sqrt{\frac{9}{25}} = \frac{3}{5}$$

$$\cos\frac{u}{2} = \sqrt{\frac{1+\cos u}{2}} = \sqrt{\frac{1+\frac{7}{25}}{2}} = \sqrt{\frac{16}{25}} = \frac{4}{5}$$

$$\tan\frac{u}{2} = \frac{1-\cos u}{\sin u} = \frac{1-\frac{7}{25}}{\frac{24}{25}} = \frac{3}{4}$$

43. $\tan u = -\dfrac{5}{12},\ \dfrac{3\pi}{2} < u < 2\pi$

(a) Because u is in Quadrant IV, $\dfrac{u}{2}$ is in Quadrant II.

(b) $$\sin\frac{u}{2} = \sqrt{\frac{1-\cos u}{2}} = \sqrt{\frac{1-\frac{12}{13}}{2}} = \sqrt{\frac{1}{26}} = \frac{\sqrt{26}}{26}$$

$$\cos\frac{u}{2} = -\sqrt{\frac{1+\cos u}{2}} = -\sqrt{\frac{1+\frac{12}{13}}{2}} = -\sqrt{\frac{25}{26}} = -\frac{5\sqrt{26}}{26}$$

$$\tan\frac{u}{2} = \frac{1-\cos u}{\sin u} = \frac{1-\frac{12}{13}}{\left(-\frac{5}{13}\right)} = -\frac{1}{5}$$

45. $$\sin\frac{x}{2} + \cos x = 0$$

$$\pm\sqrt{\frac{1-\cos x}{2}} = -\cos x$$

$$\frac{1-\cos x}{2} = \cos^2 x$$

$$0 = 2\cos^2 x + \cos x - 1$$

$$= (2\cos x - 1)(\cos x + 1)$$

$\cos x = \dfrac{1}{2}$ or $\cos x = -1$

$x = \dfrac{\pi}{3}, \dfrac{5\pi}{3}$ $\qquad x = \pi$

By checking these values in the original equation, $x = \pi/3$ and $x = 5\pi/3$ are extraneous, and $x = \pi$ is the only solution.

47. $$\cos\frac{x}{2} - \sin x = 0$$

$$\pm\sqrt{\frac{1+\cos x}{2}} = \sin x$$

$$\frac{1+\cos x}{2} = \sin^2 x$$

$$1 + \cos x = 2\sin^2 x$$

$$1 + \cos x = 2 - 2\cos^2 x$$

$$2\cos^2 x + \cos x - 1 = 0$$

$$(2\cos x - 1)(\cos x + 1) = 0$$

$2\cos x - 1 = 0$ or $\cos x + 1 = 0$

$\cos x = \dfrac{1}{2}$ $\qquad \cos x = -1$

$x = \dfrac{\pi}{3}, \dfrac{5\pi}{3}$ $\qquad x = \pi$

$x = \dfrac{\pi}{3}, \pi, \dfrac{5\pi}{3}$

$\pi/3$, π, and $5\pi/3$ are all solutions to the equation.

49. $\sin 5\theta \sin 3\theta = \frac{1}{2}\left[\cos(5\theta - 3\theta) - \cos(5\theta + 3\theta)\right] = \frac{1}{2}(\cos 2\theta - \cos 8\theta)$

51. $\cos 2\theta \cos 4\theta = \frac{1}{2}\left[\cos(2\theta - 4\theta) + \cos(2\theta + 4\theta)\right] = \frac{1}{2}\left[\cos(-2\theta) + \cos 6\theta\right]$

53.
$$\sin 5\theta - \sin 3\theta = 2\cos\left(\frac{5\theta + 3\theta}{2}\right)\sin\left(\frac{5\theta - 3\theta}{2}\right)$$
$$= 2\cos 4\theta \sin \theta$$

55.
$$\cos 6x + \cos 2x = 2\cos\left(\frac{6x + 2x}{2}\right)\cos\left(\frac{6x - 2x}{2}\right)$$
$$= 2\cos 4x \cos 2x$$

57.
$$\sin 75° + \sin 15° = 2\sin\left(\frac{75° + 15°}{2}\right)\cos\left(\frac{75° - 15°}{2}\right)$$
$$= 2\sin 45° \cos 30°$$
$$= 2\left(\frac{\sqrt{2}}{2}\right)\left(\frac{\sqrt{3}}{2}\right)$$
$$= \frac{\sqrt{6}}{2}$$

59.
$$\cos\frac{3\pi}{4} - \cos\frac{\pi}{4} = -2\sin\left(\frac{\frac{3\pi}{4} + \frac{\pi}{4}}{2}\right)\sin\left(\frac{\frac{3\pi}{4} - \frac{\pi}{4}}{2}\right)$$
$$= -2\sin\frac{\pi}{2}\sin\frac{\pi}{4}$$
$$\cos\frac{3\pi}{4} - \cos\frac{\pi}{4} = -\frac{\sqrt{2}}{2} - \frac{\sqrt{2}}{2} = -\sqrt{2}$$

61.
$$\sin 6x + \sin 2x = 0$$
$$2\sin\left(\frac{6x + 2x}{2}\right)\cos\left(\frac{6x - 2x}{2}\right) = 0$$
$$2(\sin 4x)\cos 2x = 0$$

$\sin 4x = 0$ or $\cos 2x = 0$

$4x = n\pi$ $\quad$ $2x = \frac{\pi}{2} + n\pi$

$x = \frac{n\pi}{4}$ $\quad$ $x = \frac{\pi}{4} + \frac{n\pi}{2}$

In the interval $[0, 2\pi)$

$x = 0, \frac{\pi}{4}, \frac{\pi}{2}, \frac{3\pi}{4}, \pi, \frac{5\pi}{4}, \frac{3\pi}{2}, \frac{7\pi}{4}.$

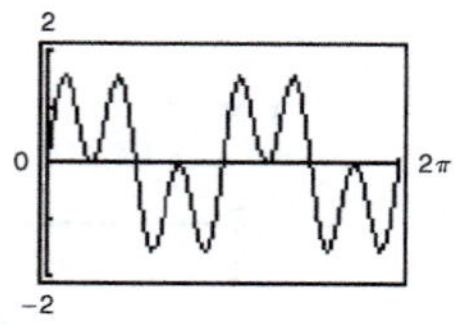

63.
$$\frac{\cos 2x}{\sin 3x - \sin x} - 1 = 0$$
$$\frac{\cos 2x}{\sin 3x - \sin x} = 1$$
$$\frac{\cos 2x}{2\cos 2x \sin x} = 1$$
$$2\sin x = 1$$
$$\sin x = \frac{1}{2}$$
$$x = \frac{\pi}{6}, \frac{5\pi}{6}$$

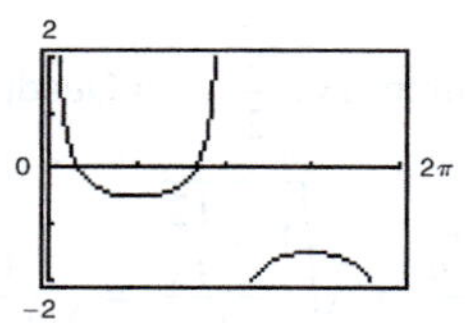

65.
$$\csc 2\theta = \frac{1}{\sin 2\theta}$$
$$= \frac{1}{2\sin\theta\cos\theta}$$
$$= \frac{1}{\sin\theta}\cdot\frac{1}{2\cos\theta}$$
$$= \frac{\csc\theta}{2\cos\theta}$$

67.
$$(\sin x + \cos x)^2 = \sin^2 x + 2\sin x\cos x + \cos^2 x$$
$$= \left(\sin^2 x + \cos^2 x\right) + 2\sin x\cos x$$
$$= 1 + \sin 2x$$

69.
$$\frac{\sin x \pm \sin y}{\cos x + \cos y} = \frac{2\sin\left(\frac{x \pm y}{2}\right)\cos\left(\frac{x \mp y}{2}\right)}{2\cos\left(\frac{x + y}{2}\right)\cos\left(\frac{x - y}{2}\right)}$$
$$= \tan\left(\frac{x \pm y}{2}\right)$$

71. (a)
$$\sin\left(\frac{\theta}{2}\right) = \pm\sqrt{\frac{1-\cos\theta}{2}} = \frac{1}{M}$$
$$\left(\pm\sqrt{\frac{1-\cos\theta}{2}}\right)^2 = \left(\frac{1}{M}\right)^2$$
$$\frac{1-\cos\theta}{2} = \frac{1}{M^2}$$
$$M^2(1-\cos\theta) = 2$$
$$1-\cos\theta = \frac{2}{M^2}$$
$$-\cos\theta = \frac{2}{M^2} - 1$$
$$\cos\theta = 1 - \frac{2}{M^2}$$
$$\cos\theta = \frac{M^2-2}{M^2}$$

(b) When $M = 2$, $\cos\theta = \dfrac{2^2-2}{2^2} = \dfrac{1}{2}$. So, $\theta = \dfrac{\pi}{3}$.

(c) When $M = 4.5$, $\cos\theta = \dfrac{(4.5)^2-2}{(4.5)^2}$

$$\cos\theta \approx 0.901235.$$

So, $\theta \approx 0.4482$ radian.

(d) When $M = 2$, $\dfrac{\text{speed of object}}{\text{speed of sound}} = M$

$$\frac{\text{speed of object}}{760 \text{ mph}} = 2$$
$$\text{speed of object} = 1520 \text{ mph}.$$

When $M = 4.5$, $\dfrac{\text{speed of object}}{\text{speed of sound}} = M$

$$\frac{\text{speed of object}}{760 \text{ mph}} = 4.5$$
$$\text{speed of object} = 3420 \text{ mph}.$$

73. $\dfrac{x}{2} = 2r\sin^2\dfrac{\theta}{2} = 2r\left(\dfrac{1-\cos\theta}{2}\right)$

$= r(1-\cos\theta)$

So, $x = 2r(1-\cos\theta)$.

75. True. Using the double angle formula and that sine is an odd function and cosine is an even function,

$$\begin{aligned}\sin(-2x) &= \sin[2(-x)]\\ &= 2\sin(-x)\cos(-x)\\ &= 2(-\sin x)\cos x\\ &= -2\sin x\cos x.\end{aligned}$$

77. Because ϕ and θ are complementary angles, $\sin\phi = \cos\theta$ and $\cos\phi = \sin\theta$.

(a) $$\begin{aligned}\sin(\phi-\theta) &= \sin\phi\cos\theta - \sin\theta\cos\phi\\ &= (\cos\theta)(\cos\theta) - (\sin\theta)(\sin\theta)\\ &= \cos^2\theta - \sin^2\theta\\ &= \cos 2\theta\end{aligned}$$

(b) $$\begin{aligned}\cos(\phi-\theta) &= \cos\phi\cos\theta + \sin\phi\sin\theta\\ &= (\sin\theta)(\cos\theta) + (\cos\theta)(\sin\theta)\\ &= 2\sin\theta\cos\theta\\ &= \sin 2\theta\end{aligned}$$

Review Exercises for Chapter 5

1. $\cot x$

3. $\cos x$

5. $\cos\theta = -\dfrac{2}{5}$, $\tan\theta > 0$, θ is in Quadrant III.

$$\sec\theta = \frac{1}{\cos\theta} = -\frac{5}{2}$$
$$\sin\theta = -\sqrt{1-\cos^2\theta} = -\sqrt{1-\frac{4}{25}} = -\sqrt{\frac{21}{25}} = -\frac{\sqrt{21}}{5}$$
$$\csc\theta = \frac{1}{\sin\theta} = -\frac{5}{\sqrt{21}} = -\frac{5\sqrt{21}}{21}$$
$$\tan\theta = \frac{\sin}{\cos\theta} = \frac{-\frac{\sqrt{21}}{5}}{-\frac{2}{5}} = \frac{\sqrt{21}}{2}$$
$$\cot\theta = \frac{1}{\tan\theta} = \frac{2}{\sqrt{21}} = \frac{2\sqrt{21}}{21}$$

7. $\dfrac{1}{\cot^2 x + 1} = \dfrac{1}{\csc^2 x} = \sin^2 x$

9. $\tan^2 x(\csc^2 x - 1) = \tan^2 x(\cot^2 x)$
$$= \tan^2 x\left(\frac{1}{\tan^2 x}\right)$$
$$= 1$$

11. $\dfrac{\cot\left(\frac{\pi}{2} - u\right)}{\cos u} = \dfrac{\tan u}{\cos u} = \tan u \sec u$

13. $\cos^2 x + \cos^2 x \cot^2 x = \cos^2 x(1 + \cot^2 x)$
$$= \cos^2 x(\csc^2 x)$$
$$= \cos^2 x\left(\frac{1}{\sin^2 x}\right)$$
$$= \frac{\cos^2 x}{\sin^2 x}$$
$$= \cot^2 x$$

15. $\dfrac{1}{\csc\theta + 1} - \dfrac{1}{\csc\theta - 1} = \dfrac{(\csc\theta - 1) - (\csc\theta + 1)}{(\csc\theta + 1)(\csc\theta - 1)}$
$$= \frac{-2}{\csc^2\theta - 1}$$
$$= \frac{-2}{\cot^2\theta}$$
$$= -2\tan^2\theta$$

17. Let $x = 5\sin\theta$, then
$$\sqrt{25 - x^2} = \sqrt{25 - (5\sin\theta)^2}$$
$$= \sqrt{25 - 25\sin^2\theta}$$
$$= \sqrt{25(1 - \sin^2\theta)}$$
$$= \sqrt{25\cos^2\theta}$$
$$= 5\cos\theta.$$

19. $\cos x(\tan^2 x + 1) = \cos x \sec^2 x$
$$= \frac{1}{\sec x}\sec^2 x$$
$$= \sec x$$

21. $\sin\left(\dfrac{\pi}{2} - \theta\right)\tan\theta = \cos\theta\tan\theta$
$$= \cos\theta\left(\frac{\sin\theta}{\cos\theta}\right)$$
$$= \sin\theta$$

23. $\dfrac{1}{\tan\theta\csc\theta} = \dfrac{1}{\dfrac{\sin\theta}{\cos\theta}\cdot\dfrac{1}{\sin\theta}} = \cos\theta$

25. $\sin^5 x\cos^2 x = \sin^4 x\cos^2 x\sin x$
$$= (1 - \cos^2 x)^2\cos^2 x\sin x$$
$$= (1 - 2\cos^2 x + \cos^4 x)\cos^2 x\sin x$$
$$= (\cos^2 x - 2\cos^4 x + \cos^6 x)\sin x$$

27. $\sin x = \sqrt{3} - \sin x$
$$\sin x = \frac{\sqrt{3}}{2}$$
$$x = \frac{\pi}{3} + 2\pi n, \frac{2\pi}{3} + 2\pi n$$

29. $3\sqrt{3}\tan u = 3$
$$\tan u = \frac{1}{\sqrt{3}}$$
$$u = \frac{\pi}{6} + n\pi$$

31. $3\csc^2 x = 4$
$$\csc^2 x = \frac{4}{3}$$
$$\sin x = \pm\frac{\sqrt{3}}{2}$$
$$x = \frac{\pi}{3} + 2\pi n, \frac{2\pi}{3} + 2\pi n, \frac{4\pi}{3} + 2\pi n, \frac{5\pi}{3} + 2\pi n$$

These can be combined as:
$$x = \frac{\pi}{3} + n\pi \text{ or } x = \frac{2\pi}{3} + n\pi$$

33. $\sin^3 x = \sin x$
$$\sin^3 x - \sin x = 0$$
$$\sin x(\sin^2 x - 1) = 0$$
$$\sin x = 0 \Rightarrow x = 0, \pi$$
$$\sin^2 x = 1$$
$$\sin x = \pm 1 \Rightarrow x = \frac{\pi}{2}, \frac{3\pi}{2}$$

35. $\cos^2 x + \sin x = 1$
$$1 - \sin^2 x + \sin x - 1 = 0$$
$$-\sin x(\sin x - 1) = 0$$
$$\sin x = 0 \qquad \sin x - 1 = 0$$
$$x = 0, \pi \qquad \sin x = 1$$
$$x = \frac{\pi}{2}$$

37. $2\sin 2x - \sqrt{2} = 0$

$$\sin 2x = \frac{\sqrt{2}}{2}$$

$$2x = \frac{\pi}{4} + 2\pi n, \frac{3\pi}{4} + 2\pi n$$

$$x = \frac{\pi}{8} + \pi n, \frac{3\pi}{8} + \pi n$$

$$x = \frac{\pi}{8}, \frac{3\pi}{8}, \frac{9\pi}{8}, \frac{11\pi}{8}$$

39. $3\tan^2\left(\frac{x}{3}\right) - 1 = 0$

$$\tan^2\left(\frac{x}{3}\right) = \frac{1}{3}$$

$$\tan\frac{x}{3} = \pm\sqrt{\frac{1}{3}}$$

$$\tan\frac{x}{3} = \pm\frac{\sqrt{3}}{3}$$

$$\frac{x}{3} = \frac{\pi}{6}, \frac{5\pi}{6}, \frac{7\pi}{6}$$

$$x = \frac{\pi}{2}, \frac{5\pi}{2}, \frac{7\pi}{2}$$

$\frac{5\pi}{2}$ and $\frac{7\pi}{2}$ are greater than 2π, so they are not solutions. The solution is $x = \frac{\pi}{2}$.

41. $\cos 4x(\cos x - 1) = 0$

$\cos 4x = 0$

$$4x = \frac{\pi}{2} + 2\pi n, \frac{3\pi}{2} + 2\pi n$$

$$x = \frac{\pi}{8} + \frac{\pi}{2}n, \frac{3\pi}{8} + \frac{\pi}{2}n$$

$\cos x - 1 = 0$

$\cos x = 1$

$x = 0$

$$x = 0, \frac{\pi}{8}, \frac{3\pi}{8}, \frac{5\pi}{8}, \frac{7\pi}{8}, \frac{9\pi}{8}, \frac{11\pi}{8}, \frac{13\pi}{8}, \frac{15\pi}{8}$$

43. $\tan^2 x - 2\tan x = 0$

$\tan x(\tan x - 2) = 0$

$\tan x = 0$ or $\tan x - 2 = 0$

$x = n\pi$ $\tan x = 2$

$x = \arctan 2 + n\pi$

45. $\tan^2\theta + \tan\theta - 6 = 0$

$(\tan\theta + 3)(\tan\theta - 2) = 0$

$\tan\theta + 3 = 0$ or $\tan\theta - 2 = 0$

$\tan\theta = -3$ $\tan\theta = 2$

$\theta = \arctan(-3) + n\pi$ $\theta = \arctan 2 + n\pi$

47. $\sin 75° = \sin(120° - 45°) = \sin 120°\cos 45° - \cos 120°\sin 45° = \left(\frac{\sqrt{3}}{2}\right)\left(\frac{\sqrt{2}}{2}\right) - \left(-\frac{1}{2}\right)\left(\frac{\sqrt{2}}{2}\right) = \frac{\sqrt{2}}{4}(\sqrt{3} + 1)$

$\cos 75° = \cos(120° - 45°) = \cos 120°\cos 45° + \sin 120°\sin 45° = \left(-\frac{1}{2}\right)\left(\frac{\sqrt{2}}{2}\right) + \left(\frac{\sqrt{3}}{2}\right)\left(\frac{\sqrt{2}}{2}\right) = \frac{\sqrt{2}}{4}(\sqrt{3} - 1)$

$$\tan 75° = \tan(120° - 45°) = \frac{\tan 120° - \tan 45°}{1 + \tan 120°\tan 45°} = \frac{-\sqrt{3} - 1}{1 + (-\sqrt{3})(1)} = \frac{-\sqrt{3} - 1}{1 - \sqrt{3}}$$

$$= \frac{-\sqrt{3} - 1}{1 - \sqrt{3}} \cdot \frac{1 + \sqrt{3}}{1 + \sqrt{3}} = \frac{-4 - 2\sqrt{3}}{-2} = 2 + \sqrt{3}$$

49. $\sin\frac{25\pi}{12} = \sin\left(\frac{11\pi}{6} + \frac{\pi}{4}\right) = \sin\frac{11\pi}{6}\cos\frac{\pi}{4} + \cos\frac{11\pi}{6}\sin\frac{\pi}{4}$

$= \left(-\frac{1}{2}\right)\left(\frac{\sqrt{2}}{2}\right) + \left(\frac{\sqrt{3}}{2}\right)\left(\frac{\sqrt{2}}{2}\right) = \frac{\sqrt{2}}{4}\left(\sqrt{3} - 1\right)$

$\cos\frac{25\pi}{12} = \cos\left(\frac{11\pi}{6} + \frac{\pi}{4}\right) = \cos\frac{11\pi}{6}\cos\frac{\pi}{4} - \sin\frac{11\pi}{6}\sin\frac{\pi}{4}$

$= \left(\frac{\sqrt{3}}{2}\right)\left(\frac{\sqrt{2}}{2}\right) - \left(-\frac{1}{2}\right)\left(\frac{\sqrt{2}}{2}\right) = \frac{\sqrt{2}}{4}\left(\sqrt{3} + 1\right)$

$\tan\frac{25\pi}{12} = \tan\left(\frac{11\pi}{6} + \frac{\pi}{4}\right) = \frac{\tan\frac{11\pi}{6} + \tan\frac{\pi}{4}}{1 - \tan\frac{11\pi}{6}\tan\frac{\pi}{4}} = \frac{\left(-\frac{\sqrt{3}}{3}\right) + 1}{1 - \left(-\frac{\sqrt{3}}{3}\right)(1)} = 2 - \sqrt{3}$

51. $\sin 60^\circ \cos 45^\circ - \cos 60^\circ \sin 45^\circ = \sin(60^\circ - 45^\circ) = \sin 15^\circ$

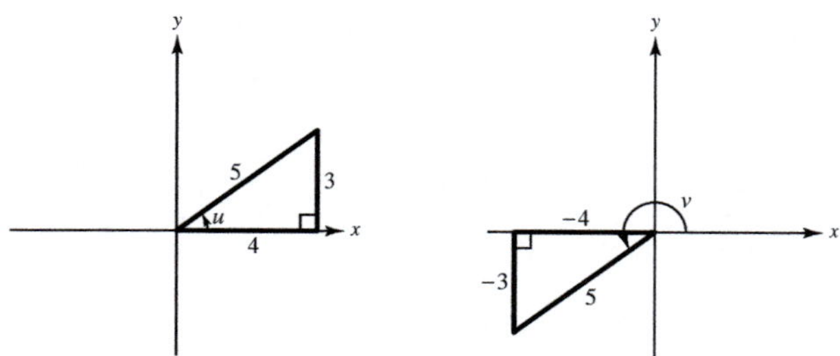

Figures for Exercises 53–55

53. $\sin(u + v) = \sin u \cos v + \cos u \sin v = \frac{3}{5}\left(-\frac{4}{5}\right) + \frac{4}{5}\left(-\frac{3}{5}\right) = -\frac{24}{25}$

55. $\cos(u - v) = \cos u \cos v + \sin u \sin v = \frac{4}{5}\left(-\frac{4}{5}\right) + \frac{3}{5}\left(-\frac{3}{5}\right) = -1$

57. $\cos\left(x + \frac{\pi}{2}\right) = \cos x \cos\frac{\pi}{2} - \sin x \sin\frac{\pi}{2}$

$= \cos x(0) - \sin x(1)$

$= -\sin x$

59. $\tan(\pi - x) = \frac{\tan\pi - \tan x}{1 - \tan\pi\tan x} = -\tan x$

61. $\sin\left(x + \frac{\pi}{4}\right) - \sin\left(x - \frac{\pi}{4}\right) = 1$

$2\cos x \sin\frac{\pi}{4} = 1$

$\cos x = \frac{\sqrt{2}}{2}$

$x = \frac{\pi}{4}, \frac{7\pi}{4}$

63. $\sin u = -\frac{4}{5}, \pi < u < \frac{3\pi}{2}$

$\cos u = -\sqrt{1 - \sin^2 u} = \frac{-3}{5}$

$\tan u = \frac{\sin u}{\cos u} = \frac{4}{3}$

$\sin 2u = 2\sin u \cos u = 2\left(-\frac{4}{5}\right)\left(-\frac{3}{5}\right) = \frac{24}{25}$

$\cos 2u = \cos^2 u - \sin^2 u = \left(-\frac{3}{5}\right)^2 - \left(-\frac{4}{5}\right)^2 = -\frac{7}{25}$

$\tan 2u = \frac{2\tan u}{1 - \tan^2 u} = \frac{2\left(\frac{4}{3}\right)}{1 - \left(\frac{4}{3}\right)^2} = -\frac{24}{7}$

65. $\sin 4x = 2\sin 2x\cos 2x$
$= 2\left[2\sin x\cos x\left(\cos^2 x - \sin^2 x\right)\right]$
$= 4\sin x\cos x\left(2\cos^2 x - 1\right)$
$= 8\cos^3 x\sin x - 4\cos x\sin x$

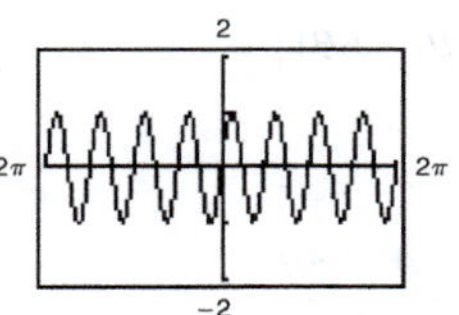

67. $\tan^2 3x = \dfrac{\sin^2 3x}{\cos^2 3x} = \dfrac{\frac{1-\cos 6x}{2}}{\frac{1+\cos 6x}{2}} = \dfrac{1-\cos 6x}{1+\cos 6x}$

69. $\sin(-75°) = -\sqrt{\dfrac{1-\cos 150°}{2}} = -\sqrt{\dfrac{1-\left(-\frac{\sqrt{3}}{2}\right)}{2}} = -\dfrac{\sqrt{2+\sqrt{3}}}{2} = -\dfrac{1}{2}\sqrt{2+\sqrt{3}}$

$\cos(-75°) = -\sqrt{\dfrac{1+\cos 150°}{2}} = \sqrt{\dfrac{1+\left(-\frac{\sqrt{3}}{2}\right)}{2}} = \dfrac{\sqrt{2-\sqrt{3}}}{2} = \dfrac{1}{2}\sqrt{2-\sqrt{3}}$

$\tan(-75°) = -\left(\dfrac{1-\cos 150°}{\sin 150°}\right) = -\left(\dfrac{1-\left(-\frac{\sqrt{3}}{2}\right)}{\frac{1}{2}}\right) = -\left(2+\sqrt{3}\right) = -2-\sqrt{3}$

71. $\tan u = \dfrac{4}{3}, \pi < u < \dfrac{3\pi}{2}$

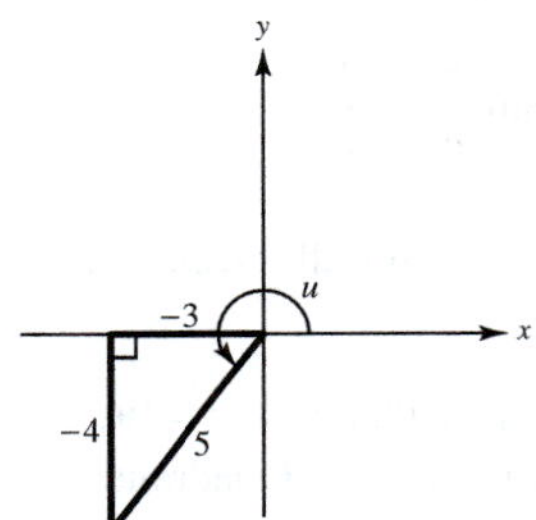

(a) Because u is in Quadrant III, $\dfrac{u}{2}$ is in Quadrant II.

(b) $\sin\dfrac{u}{2} = \sqrt{\dfrac{1-\cos u}{2}} = \sqrt{\dfrac{1-\left(-\frac{3}{5}\right)}{2}} = \sqrt{\dfrac{4}{5}}$
$= \dfrac{2\sqrt{5}}{5}$

$\cos\dfrac{u}{2} = -\sqrt{\dfrac{1+\cos u}{2}} = -\sqrt{\dfrac{1+\left(-\frac{3}{5}\right)}{2}} = -\sqrt{\dfrac{1}{5}}$
$= -\dfrac{\sqrt{5}}{5}$

$\tan\dfrac{u}{2} = \dfrac{1-\cos u}{\sin u} = \dfrac{1-\left(-\frac{3}{5}\right)}{\left(-\frac{4}{5}\right)} = -2$

73. $\cos u = -\dfrac{2}{7}, \dfrac{\pi}{2} < u < \pi$

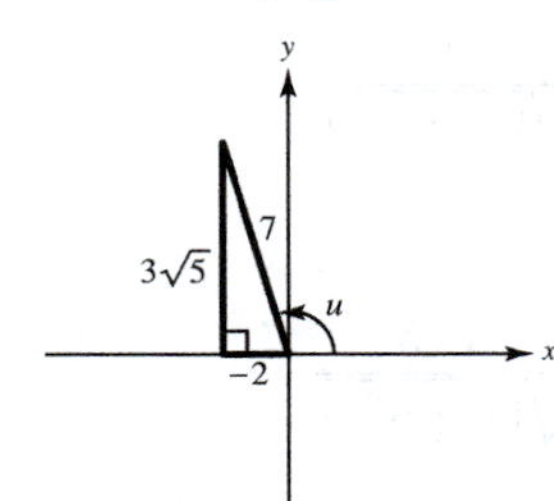

(a) Because u is in Quadrant II, $\dfrac{u}{2}$ is in Quadrant I.

(b) $\sin\dfrac{u}{2} = \sqrt{\dfrac{1-\cos u}{2}} = \sqrt{\dfrac{1-\left(-\frac{2}{7}\right)}{2}} = \sqrt{\dfrac{9}{14}}$
$= \dfrac{3\sqrt{14}}{14}$

$\cos\dfrac{u}{2} = \sqrt{\dfrac{1+\cos u}{2}} = \sqrt{\dfrac{1+\left(-\frac{2}{7}\right)}{2}} = \sqrt{\dfrac{5}{14}}$
$= \dfrac{\sqrt{70}}{14}$

$\tan\dfrac{u}{2} = \dfrac{1-\cos u}{\sin u} = \dfrac{1-\left(-\frac{2}{7}\right)}{\frac{3\sqrt{5}}{7}} = \dfrac{3\sqrt{5}}{5}$

75. $\cos 4\theta \sin 6\theta = \frac{1}{2}\left[\sin(4\theta + 6\theta) - \sin(4\theta - 6\theta)\right]$
$= \frac{1}{2}\left[\sin 10\theta - \sin(-2\theta)\right]$

77. $\cos 6\theta + \cos 5\theta = 2\cos\left(\frac{6\theta + 5\theta}{2}\right)\cos\left(\frac{6\theta - 5\theta}{2}\right)$
$= 2\cos\frac{11\theta}{2}\cos\frac{\theta}{2}$

79. $r = \frac{1}{32}v_0^2 \sin 2\theta$

range $= 100$ feet

$v_0 = 80$ feet per second

$r = \frac{1}{32}(80)^2 \sin 2\theta = 100$

$\sin 2\theta = 0.5$

$2\theta = 30°$

$\theta = 15°$ or $\frac{\pi}{12}$

81. False. If $\frac{\pi}{2} < \theta < \pi$, then $\frac{\pi}{4} < \frac{\theta}{2} < \frac{\pi}{2}$, and $\frac{\theta}{2}$ is in Quadrant I. $\cos\frac{\theta}{2} > 0$

83. True. $4\sin(-x)\cos(-x) = 4(-\sin x)\cos x$
$= -4\sin x\cos x$
$= -2(2\sin x\cos x)$
$= -2\sin 2x$

85. Yes. *Sample Answer*. When the domain is all real numbers, the solutions of $\sin x = \frac{1}{2}$ are $x = \frac{\pi}{6} + 2n\pi$ and $x = \frac{5\pi}{6} + 2n\pi$, so there are infinitely many solutions.

Problem Solving for Chapter 5

1. $\sin\theta = \pm\sqrt{1 - \cos^2\theta}$

$\tan\theta = \frac{\sin\theta}{\cos\theta} = \pm\frac{\sqrt{1 - \cos^2\theta}}{\cos\theta}$

$\csc\theta = \frac{1}{\sin\theta} = \pm\frac{1}{\sqrt{1 - \cos^2\theta}}$

$\sec\theta = \frac{1}{\cos\theta}$

$\cot\theta = \frac{1}{\tan\theta} = \pm\frac{\cos\theta}{\sqrt{1 - \cos^2\theta}}$

You also have the following relationships:

$\sin\theta = \cos\left(\frac{\pi}{2} - \theta\right)$

$\tan\theta = \frac{\cos\left[(\pi/2) - \theta\right]}{\cos\theta}$

$\csc\theta = \frac{1}{\cos\left[(\pi/2) - \theta\right]}$

$\sec\theta = \frac{1}{\cos\theta}$

$\cot\theta = \frac{\cos\theta}{\cos\left[(\pi/2) - \theta\right]}$

3. $\sin\left[\frac{(12n + 1)\pi}{6}\right] = \sin\left[\frac{1}{6}(12n\pi + \pi)\right]$
$= \sin\left(2n\pi + \frac{\pi}{6}\right)$
$= \sin\frac{\pi}{6} = \frac{1}{2}$

So, $\sin\left[\frac{(12n + 1)\pi}{6}\right] = \frac{1}{2}$ for all integers n.

5. From the figure, it appears that $u + v = w$. Assume that u, v, and w are all in Quadrant I.

From the figure:

$\tan u = \frac{s}{3s} = \frac{1}{3}$

$\tan v = \frac{s}{2s} = \frac{1}{2}$

$\tan w = \frac{s}{s} = 1$

$\tan(u + v) = \frac{\tan u + \tan v}{1 - \tan u \tan v}$
$= \frac{1/3 + 1/2}{1 - (1/3)(1/2)}$
$= \frac{5/6}{1 - (1/6)} = 1 = \tan w.$

So, $\tan(u + v) = \tan w$. Because u, v, and w are all in Quadrant I, you have

$\arctan\left[\tan(u + v)\right] = \arctan[\tan w]u + v = w.$

7. (a)

$$\sin\frac{\theta}{2} = \frac{\frac{1}{2}b}{10} \qquad \text{and} \qquad \cos\frac{\theta}{2} = \frac{h}{10}$$

$$b = 20\sin\frac{\theta}{2} \qquad\qquad h = 10\cos\frac{\theta}{2}$$

$$A = \frac{1}{2}bh$$

$$= \frac{1}{2}\left(20\sin\frac{\theta}{2}\right)\left(10\cos\frac{\theta}{2}\right)$$

$$= 100\sin\frac{\theta}{2}\cos\frac{\theta}{2}$$

(b) $$A = 50\left(2\sin\frac{\theta}{2}\cos\frac{\theta}{2}\right)$$

$$= 50\sin\left(2\left(\frac{\theta}{2}\right)\right)$$

$$= 50\sin\theta$$

Because $\sin\frac{\pi}{2} = 1$ is a maximum, $\theta = \frac{\pi}{2}$. So, the area is a maximum at $A = 50\sin\frac{\pi}{2} = 50$ square meters.

9. $$F = \frac{0.6W\sin(\theta + 90°)}{\sin 12°}$$

(a) $$F = \frac{0.6W(\sin\theta\cos 90° + \cos\theta\sin 90°)}{\sin 12°}$$

$$= \frac{0.6W\left[(\sin\theta)(0) + (\cos\theta)(1)\right]}{\sin 12°}$$

$$= \frac{0.6W\cos\theta}{\sin 12°}$$

(b) Let $y_1 = \dfrac{0.6(185)\cos x}{\sin 12°}$.

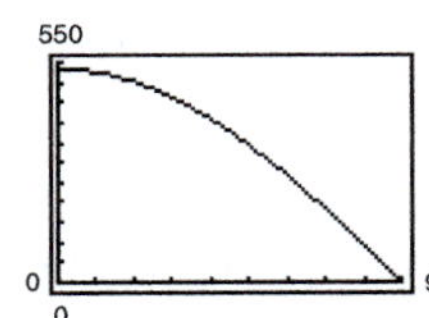

(c) The force is maximum (533.88 pounds) when $\theta = 0°$.

The force is minimum (0 pounds) when $\theta = 90°$.

11. $d = 35 - 28\cos\dfrac{\pi}{6.2}t$ when $t = 0$ corresponds to 12:00 A.M.

(a) The high tides occur when $\cos\dfrac{\pi}{6.2}t = -1$. Solving yields $t = 6.2$ or $t = 18.6$.

These t-values correspond to 6:12 A.M. and 6:36 P.M.

The low tide occurs when $\cos\dfrac{\pi}{6.2}t = 1$. Solving yields $t = 0$ and $t = 12.4$ which corresponds to 12:00 A.M. and 12:24 P.M.

(b) The water depth is never 3.5 feet. At low tide, the depth is $d = 35 - 28 = 7$ feet.

(c)

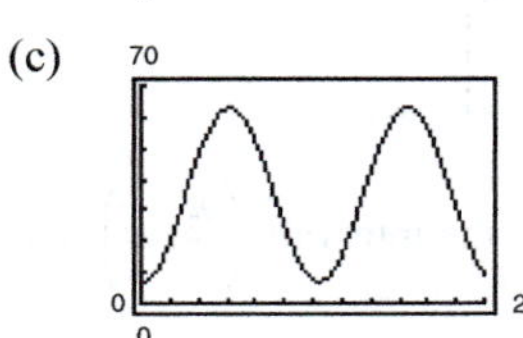

13. (a) $$n = \frac{\sin\left(\frac{\theta}{2} + \frac{\alpha}{2}\right)}{\sin\frac{\theta}{2}}$$

$$= \frac{\sin\left(\frac{\theta}{2}\right)\cos\left(\frac{\alpha}{2}\right) + \cos\left(\frac{\theta}{2}\right)\sin\left(\frac{\alpha}{2}\right)}{\sin\left(\frac{\theta}{2}\right)}$$

$$= \cos\left(\frac{\alpha}{2}\right) + \cot\left(\frac{\theta}{2}\right)\sin\left(\frac{\alpha}{2}\right)$$

For $\alpha = 60°$, $n = \cos 30° + \cot\left(\frac{\theta}{2}\right)\sin 30°$

$$n = \frac{\sqrt{3}}{2} + \frac{1}{2}\cot\left(\frac{\theta}{2}\right).$$

(b) For glass, $n = 1.50$.

$$1.50 = \frac{\sqrt{3}}{2} + \frac{1}{2}\cot\left(\frac{\theta}{2}\right)$$

$$2\left(1.50 - \frac{\sqrt{3}}{2}\right) = \cot\left(\frac{\theta}{2}\right)$$

$$\frac{1}{3 - \sqrt{3}} = \tan\left(\frac{\theta}{2}\right)$$

$$\theta = 2\tan^{-1}\left(\frac{1}{3 - \sqrt{3}}\right)$$

$$\theta \approx 76.5°$$

15. (a) Let $y_1 = \sin x$ and $y_2 = 0.5$.

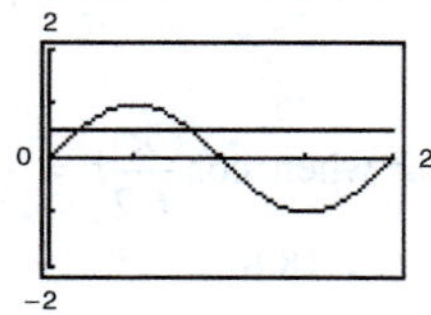

$\sin x \geq 0.5$ on the interval $\left[\frac{\pi}{6}, \frac{5\pi}{6}\right]$.

(b) Let $y_1 = \cos x$ and $y_2 = -0.5$.

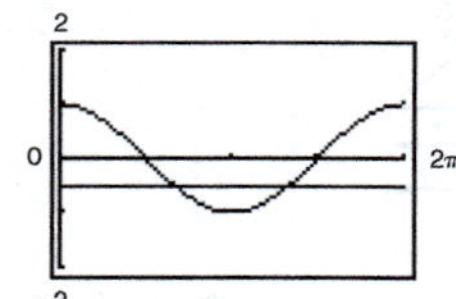

$\cos x \leq -0.5$ on the interval $\left[\frac{2\pi}{3}, \frac{4\pi}{3}\right]$.

(c) Let $y_1 = \tan x$ and $y_2 = \sin x$.

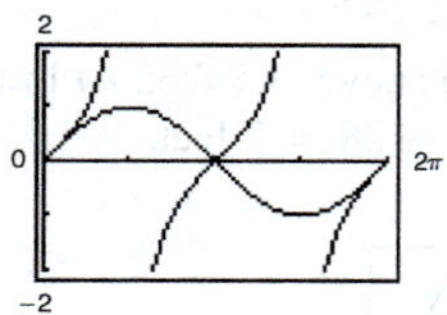

$\tan x < \sin x$ on the intervals $\left(\frac{\pi}{2}, \pi\right)$ and $\left(\frac{3\pi}{2}, 2\pi\right)$.

(d) Let $y_1 = \cos x$ and $y_2 = \sin x$.

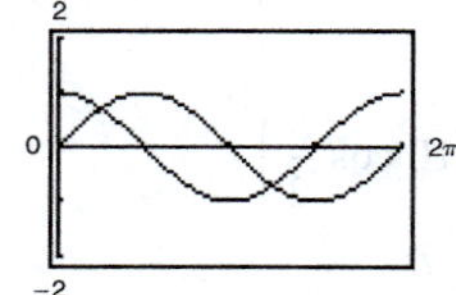

$\cos x \geq \sin x$ on the intervals $\left[0, \frac{\pi}{4}\right]$ and $\left[\frac{5\pi}{4}, 2\pi\right)$.

Practice Test for Chapter 5

1. Find the value of the other five trigonometric functions, given $\tan x = \frac{4}{11}$, $\sec x < 0$.

2. Simplify $\dfrac{\sec^2 x + \csc^2 x}{\csc^2 x(1 + \tan^2 x)}$.

3. Rewrite as a single logarithm and simplify $\ln|\tan \theta| - \ln|\cot \theta|$.

4. True or false:

$$\cos\left(\frac{\pi}{2} - x\right) = \frac{1}{\csc x}$$

5. Factor and simplify: $\sin^4 x + (\sin^2 x)\cos^2 x$

6. Multiply and simplify: $(\csc x + 1)(\csc x - 1)$

7. Rationalize the denominator and simplify:

$$\frac{\cos^2 x}{1 - \sin x}$$

8. Verify:

$$\frac{1 + \cos \theta}{\sin \theta} + \frac{\sin \theta}{1 + \cos \theta} = 2 \csc \theta$$

9. Verify:

$$\tan^4 x + 2\tan^2 x + 1 = \sec^4 x$$

10. Use the sum or difference formulas to determine:

(a) $\sin 105°$

(b) $\tan 15°$

11. Simplify: $(\sin 42°)\cos 38° - (\cos 42°)\sin 38°$

12. Verify $\tan\left(\theta + \dfrac{\pi}{4}\right) = \dfrac{1 + \tan \theta}{1 - \tan \theta}$.

13. Write $\sin(\arcsin x - \arccos x)$ as an algebraic expression in x.

14. Use the double-angle formulas to determine:

(a) $\cos 120°$

(b) $\tan 300°$

15. Use the half-angle formulas to determine:

(a) $\sin 22.5°$

(b) $\tan \dfrac{\pi}{12}$

16. Given $\sin \theta = 4/5$, θ lies in Quadrant II, find $\cos(\theta/2)$.

17. Use the power-reducing identities to write $(\sin^2 x)\cos^2 x$ in terms of the first power of cosine.

18. Rewrite as a sum: $6(\sin 5\theta)\cos 2\theta$.

19. Rewrite as a product: $\sin(x + \pi) + \sin(x - \pi)$.

20. Verify $\dfrac{\sin 9x + \sin 5x}{\cos 9x - \cos 5x} = -\cot 2x$.

21. Verify:

$(\cos u)\sin v = \frac{1}{2}[\sin(u + v) - \sin(u - v)]$.

22. Find all solutions in the interval $[0, 2\pi)$:

$4\sin^2 x = 1$

23. Find all solutions in the interval $[0, 2\pi)$:

$\tan^2\theta + (\sqrt{3} - 1)\tan\theta - \sqrt{3} = 0$

24. Find all solutions in the interval $[0, 2\pi)$:

$\sin 2x = \cos x$

25. Use the quadratic formula to find all solutions in the interval $[0, 2\pi)$:

$\tan^2 x - 6\tan x + 4 = 0$

CHAPTER 6
Additional Topics in Trigonometry

CHAPTER 6
Additional Topics in Trigonometry

Section 6.1 Law of Sines

1. oblique

3. angles; side

5.

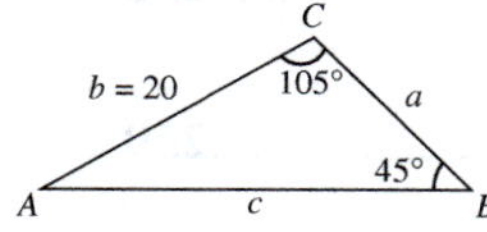

Given: $B = 45°, C = 105°, b = 20$

$A = 180° - B - C = 30°$

$$a = \frac{b}{\sin B}(\sin A) = \frac{20 \sin 30°}{\sin 45°} = 10\sqrt{2} \approx 14.14$$

$$C = \frac{b}{\sin B}(\sin C) = \frac{20 \sin 105°}{\sin 45°} \approx 27.32$$

7.

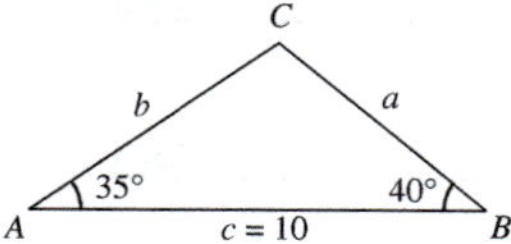

Given: $A = 35°, B = 40°, c = 10$

$C = 180° - A - B = 105°$

$$a = \frac{c}{\sin C}(\sin A) = \frac{10 \sin 35°}{\sin 105°} \approx 5.94$$

$$b = \frac{c}{\sin C}(\sin B) = \frac{10 \sin 40°}{\sin 105°} \approx 6.65$$

9. Given: $A = 102.4°, C = 16.7°, a = 21.6$

$B = 180° - A - C = 60.9°$

$$b = \frac{a}{\sin A}(\sin B) = \frac{21.6}{\sin 102.4°}(\sin 60.9°) \approx 19.32$$

$$c = \frac{a}{\sin A}(\sin C) = \frac{21.6}{\sin 102.4°}(\sin 16.7°) \approx 6.36$$

11. Given: $A = 83°20', C = 54.6°, c = 18.1$

$B = 180° - A - C = 180° - 83°20' - 54°36' = 42°4'$

$$a = \frac{c}{\sin C}(\sin A) = \frac{18.1}{\sin 54.6°}(\sin 83°20') \approx 22.05$$

$$b = \frac{c}{\sin C}(\sin B) = \frac{18.1}{\sin 54.6°}(\sin 42°4') \approx 14.88$$

13. Given: $A = 35°, B = 65°, c = 10$

$C = 180° - A - B = 80°$

$$a = \frac{c}{\sin C}(\sin A) = \frac{10 \sin 35°}{\sin 80°} \approx 5.82$$

$$b = \frac{c}{\sin C}(\sin B) = \frac{10 \sin 65°}{\sin 80°} \approx 9.20$$

15. Given: $A = 55°, B = 42°, c = \frac{3}{4}$

$C = 180° - A - B = 83°$

$$a = \frac{c}{\sin C}(\sin A) = \frac{0.75}{\sin 83°}(\sin 55°) \approx 0.62$$

$$b = \frac{c}{\sin C}(\sin B) = \frac{0.75}{\sin 83°}(\sin 42°) \approx 0.51$$

17. Given: $A = 36°, a = 8, b = 5$

$$\sin B = \frac{b \sin A}{a} = \frac{5 \sin 36°}{8} \approx 0.36737 \Rightarrow B \approx 21.55°$$

$C = 180° - A - B \approx 180° - 36° - 21.55 = 122.45°$

$$c = \frac{a}{\sin A}(\sin C) = \frac{8}{\sin 36°}(\sin 122.45°) \approx 11.49$$

19. Given: $A = 145°, a = 14, b = 4$

$$\sin B = \frac{b \sin A}{a} = \frac{4 \sin 145°}{14} \approx 0.1639 \Rightarrow B \approx 9.43°$$

$C = 180° - A - B \approx 25.57°$

$$c = \frac{a}{\sin A}(\sin C) \approx \frac{14 \sin 25.57°}{\sin 145°} \approx 10.53$$

21. Given: $B = 15°30'$, $a = 4.5$, $b = 6.8$

$$\sin A = \frac{a \sin B}{b} = \frac{4.5 \sin 15°30'}{6.8} \approx 0.17685 \Rightarrow A \approx 10°11'$$

$$C = 180° - A - B \approx 180° - 10°11' - 15°30' = 154°19'$$

$$c = \frac{b}{\sin B}(\sin C) = \frac{6.8}{\sin 15°30'}(\sin 154°19') \approx 11.03$$

23. Given: $A = 110°$, $a = 125$, $b = 100$

$$\sin B = \frac{b \sin A}{a} = \frac{100 \sin 110°}{125} \approx 0.75175 \Rightarrow B \approx 48.74°$$

$$C = 180° - A - B \approx 21.26°$$

$$c = \frac{a \sin C}{\sin A} \approx \frac{125 \sin 21.26°}{\sin 110°} \approx 48.23$$

25. Given: $a = 18$, $b = 20$, $A = 76°$

$h = 20 \sin 76° \approx 19.41$

Because $a < h$, no triangle is formed.

27. Given: $A = 58°$, $a = 11.4$, $c = 12.8$

$$\sin B = \frac{b \sin A}{a} = \frac{12.8 \sin 58°}{11.4} \approx 0.9522 \Rightarrow B \approx 72.21° \text{ or } B \approx 107.79°$$

Case 1

$B \approx 72.21°$

$C = 180° - A - B \approx 49.79°$

$$c = \frac{a}{\sin A}(\sin C) \approx \frac{11.4 \sin 49.79°}{\sin 58°} \approx 10.27$$

Case 2

$B \approx 107.79°$

$C = 180° - A - B \approx 14.21°$

$$c = \frac{a}{\sin A}(\sin C) \approx \frac{11.4 \sin 14.21°}{\sin 58°} \approx 3.30$$

29. Given: $A = 120°$, $a = b = 25$

No triangle is formed because A is obtuse and $a = b$.

31. Given: $A = 45°$, $a = b = 1$

Because $a = b = 1$, $B = 45°$.

$C = 180° - A - B = 90°$

$$c = \frac{a}{\sin A}(\sin C) = \frac{1 \sin 90°}{\sin 45°} = \sqrt{2} \approx 1.41$$

33. Given: $A = 36°$, $a = 5$

(a) One solution if $b \le 5$ or $b = \dfrac{5}{\sin 36°}$.

(b) Two solutions if $5 < b < \dfrac{5}{\sin 36°}$.

(c) No solution if $b > \dfrac{5}{\sin 36°}$.

35. Given: $A = 105°$, $a = 80$

(a) One solution if $b < 80$.

(b) Not possible for two solutions.

(c) No solution if $b \ge 80$.

37. $A = 125°$, $b = 9$, $c = 6$

$$\text{Area} = \frac{1}{2}bc \sin A = \frac{1}{2}(9)(6) \sin 125° \approx 22.1$$

39. $B = 39°$, $a = 25$, $c = 12$

$$\text{Area} = \frac{1}{2}ac \sin B = \frac{1}{2}(25)(12) \sin 39° = 94.4$$

41. $C = 103°\,15'$, $a = 16$, $b = 28$

$$\text{Area} = \frac{1}{2}ab \sin C = \frac{1}{2}(16)(28) \sin 103°\,15' \approx 218.0$$

43. $A = 67°, B = 43°, a = 8$

$$b = \frac{a}{\sin A}(\sin B) = \frac{8 \sin 43°}{\sin 67°} \approx 5.927$$

$$C = 180° - A - B = 70°$$

$$\text{Area} = \frac{1}{2}ab \sin C = \frac{1}{2}(8)(5.927) \sin 70° \approx 22.3$$

45. (a) $C = 180° - 94° - 30° = 56°$

$$\frac{h}{\sin 30°} = \frac{40}{\sin 56°}$$

$$h = \frac{40}{\sin 56°}(\sin 30°)$$

(b) $h = \dfrac{40}{\sin 56°}(\sin 30°) \approx 24.1$ meters

47. Given: $A = 15°, B = 135°, c = 30$

$$C = 180° - A - B = 30°$$

From Pine Knob:

$$b = \frac{c \sin B}{\sin C} = \frac{30 \sin 135°}{\sin 30°} \approx 42.4 \text{ kilometers}$$

From Colt Station:

$$a = \frac{c \sin A}{\sin C} = \frac{30 \sin 15°}{\sin 30°} \approx 15.5 \text{ kilometers}$$

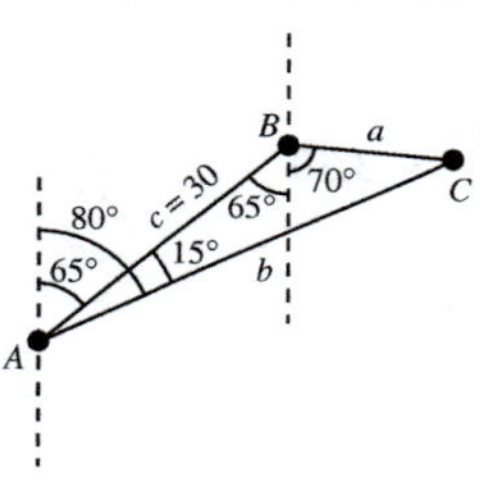

49. $\dfrac{\sin(42° - \theta)}{10} = \dfrac{\sin 48°}{17}$

$$\sin(42° - \theta) \approx 0.43714$$

$$42° - \theta \approx 25.9°$$

$$\theta \approx 16.1°$$

51. Given: $A = 55°, c = 2.2$

(a)

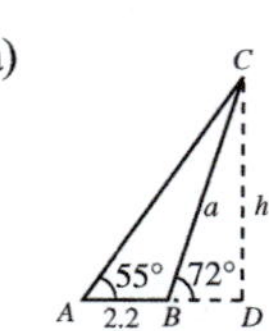

(b) $B = 180° - 72° = 108°$

$$C = 180° - 55° - 108° = 17°$$

$$\frac{a}{\sin A} = \frac{c}{\sin c}$$

$$a = \frac{c}{\sin c}(\sin A) = \frac{2.2 \sin 55°}{\sin 17°} \approx 6.16$$

(c) $h = a \sin 72° \approx 6.16 \sin 72° \approx 5.86$ miles

(d) The plane must travel a horizontal distance d to be directly above point A.

$$\angle ACD = \angle ACB + \angle BCD$$
$$= 17° + (180° - 72° - 90°)$$
$$= 17° + 18° = 35°$$

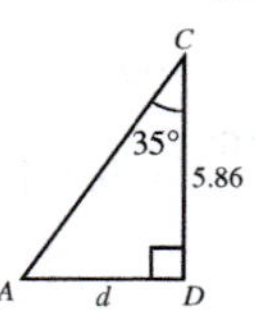

$$\tan 35° = \frac{d}{5.86}$$

$$d = 5.86 \tan 35° \approx 4.10 \text{ miles}$$

53. $\alpha = 180 - \theta - (180 - \phi) = \phi - \theta$

$$\frac{d}{\sin \theta} = \frac{2}{\sin \alpha}$$

$$d = \frac{2 \sin \theta}{\sin(\phi - \theta)}$$

55. True. If one angle of a triangle is obtuse, then there is less than 90° left for the other two angles, so it cannot contain a right angle. It must be oblique.

57. False. To solve an oblique triangle using the Law of Sines, you need to know two angles and any side, or two sides and an angle opposite one of them.

59. To find the area using angle C, the formula should be $A = \frac{1}{2}ab \sin C$ and not $A = \frac{1}{2}bc \sin C$. So first find angle B, to find side a. Then the area can be calculated.

61. Yes.

$$A = 180° - B - C = 40°$$

$$\frac{a}{\sin A} = \frac{c}{\sin C}$$

$$c = \frac{a}{\sin A} \sin C$$
$$\approx 15.6$$

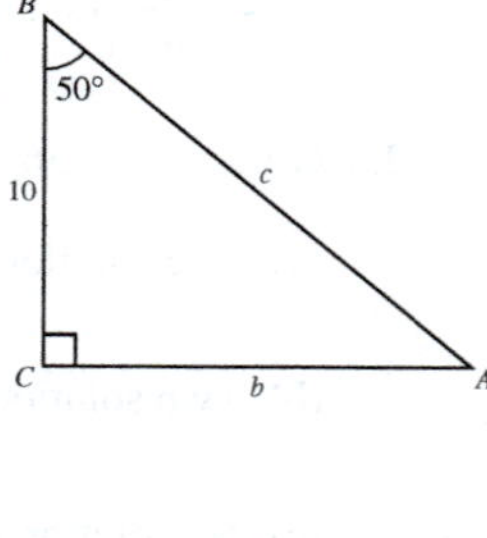

$$\frac{b}{\sin B} = \frac{c}{\sin C}$$

$$b = \frac{c}{\sin C} \sin B$$
$$\approx 11.9$$

An alternative method is to use the trigonometric ratios of a right triangle.

That is $\sin A = \dfrac{opp}{hyp} = \dfrac{a}{c}$ and $\tan A = \dfrac{opp}{hyp} = \dfrac{a}{b}$.

Section 6.2 Law of Cosines

1. $b^2 = a^2 + c^2 - 2ac\cos B$

3. standard

5. Given: $a = 10, b = 12, c = 16$

$$\cos C = \frac{a^2 + b^2 - c^2}{2ab} = \frac{100 + 144 - 256}{2(10)(12)} = -0.05 \Rightarrow C \approx 92.87°$$

$$\sin B = \frac{b\sin C}{c} \approx \frac{12\sin 92.87°}{16} \approx 0.749059 \Rightarrow B \approx 48.51°$$

$$A \approx 180° - 48.51° - 92.87° = 38.62°$$

7. Given: $a = 6, b = 8, c = 12$

$$\cos C = \frac{a^2 + b^2 + c^2}{2ab} = \frac{6^2 + 8^2 - 12^2}{2(6)(8)} \approx -0.458333 \Rightarrow C \approx 117.28°$$

$$\sin B = b\left(\frac{\sin C}{c}\right) = 8\left(\frac{\sin 117.28°}{12}\right) \approx 0.592518 \Rightarrow B \approx 36.34°$$

$$A = 180° - B - C \approx 180° - 36.34° - 117.28° = 26.38°$$

9. Given: $A = 30°, b = 15, c = 30$

$$a^2 = b^2 + c^2 - 2bc\cos A$$
$$= 225 + 900 - 2(15)(30)\cos 30° \approx 345.5771$$
$$a \approx 18.59$$

$$\cos C = \frac{a^2 + b^2 - c^2}{2ab} \approx \frac{(345.5771)^2 + 15^2 - 30^2}{2(18.59)(15)} \approx -0.590681 \Rightarrow C \approx 126.21°$$

$$B \approx 180° - 30° - 126.21° = 23.79°$$

11. Given: $A = 50°, b = 15, c = 30$

$$a^2 = b^2 + c^2 - 2bc\cos A = 15^2 + 30^2 - 2(15)(30)\cos 50°$$
$$\approx 546.4912 \Rightarrow a \approx 23.38$$

$$\sin C = c\left(\frac{\sin A}{a}\right) \approx 30\left(\frac{\sin 50°}{23.3772}\right) \approx 0.983066$$

There are two angles between $0°$ and $180°$ whose sine is 0.983066, $C_1 \approx 79.4408°$ and $C_2 \approx 180° - 79.4408° \approx 100.56°$.

Because side c is the longest side of the triangle, C must be the largest angle of the triangle. So, $C \approx 100.56°$ and $B = 180° - A - C \approx 180° - 50° - 100.56° = 29.44°$.

13. Given: $a = 11, b = 15, c = 21$

$$\cos C = \frac{a^2 + b^2 - c^2}{2ab} = \frac{121 + 225 - 441}{2(11)(15)} \approx -0.287879 \Rightarrow C \approx 106.73°$$

$$\sin B = \frac{b\sin C}{c} = \frac{15\sin 106.73°}{21} \approx 0.684051 \Rightarrow B \approx 43.16°$$

$$A \approx 180° - 43.16° - 106.73° = 30.11°$$

15. Given: $a = 2.5, b = 1.8, c = 0.9$

$$\cos A = \frac{b^2 + c^2 - a^2}{2bc} = \frac{(1.8)^2 + (0.9)^2 - (2.5)^2}{2(1.8)(0.9)} = -0.679012 \Rightarrow A \approx 132.77°$$

$$\cos B = \frac{a^2 + c^2 - b^2}{2ac} = \frac{(2.5)^2 + (0.9)^2 - (1.8)^2}{2(2.5)(0.9)} \approx 0.848889 \Rightarrow B \approx 31.91°$$

$$C = 180° - 132.77° - 31.91° = 15.32°$$

17. Given: $A = 120°, b = 6, c = 7$

$$a^2 = b^2 + c^2 - 2bc\cos A = 36 + 49 - 2(6)(7)\cos 120° = 127 \Rightarrow a \approx 11.27$$

$$\sin B = \frac{b\sin A}{a} \approx \frac{6\sin 120°}{11.27} \approx 0.461061 \Rightarrow B \approx 27.46°$$

$$C \approx 180° - 120° - 27.46° = 32.54°$$

19. Given: $B = 10° 35', a = 40, c = 30$

$$b^2 = a^2 + c^2 - 2ac\cos B = 1600 + 900 - 2(40)(30)\cos 10° 35' \approx 140.8268 \Rightarrow b \approx 11.87$$

$$\sin C = \frac{c\sin B}{b} = \frac{30\sin 10° 35'}{11.87} \approx 0.464192 \Rightarrow C \approx 27.66° \approx 27° 40'$$

$$A \approx 180° - 10° 35' - 27° 40' = 141° 45'$$

21. Given: $B = 125° 40', a = 37, c = 37$

$$b^2 = a^2 + c^2 - 2ac\cos B = 1369 + 1369 - 2(37)(37)\cos 125° 40' \approx 4334.4420 \Rightarrow b \approx 65.84$$

$$A = C \Rightarrow 2A = 180 - 125° 40' = 54° 20' \Rightarrow A = C = 27° 10'$$

23. $C = 43°, a = \frac{4}{9}, b = \frac{7}{9}$

$$c^2 = a^2 + b^2 - 2ab\cos C = \left(\frac{4}{9}\right)^2 + \left(\frac{7}{9}\right)^2 - 2\left(\frac{4}{9}\right)\left(\frac{7}{9}\right)\cos 43° \approx 0.296842 \Rightarrow c \approx 0.54$$

$$\sin A = \frac{a\sin C}{c} = \frac{(4/9)\sin 43°}{0.544832} \approx 0.556337 \Rightarrow A \approx 33.80°$$

$$B \approx 180° - 43° - 33.8° = 103.20°$$

25. $d^2 = 5^2 + 8^2 - 2(5)(8)\cos 45° \approx 32.4315 \Rightarrow d \approx 5.69$

$2\phi = 360° - 2(45°) = 270° \Rightarrow \phi = 135°$

$c^2 = 5^2 + 8^2 - 2(5)(8)\cos 135° \approx 145.5685 \Rightarrow c \approx 12.07$

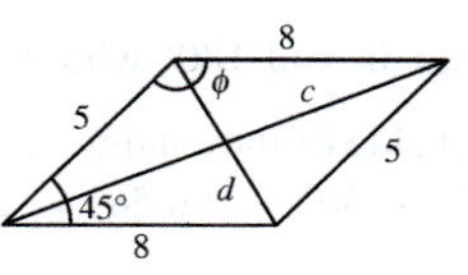

27.

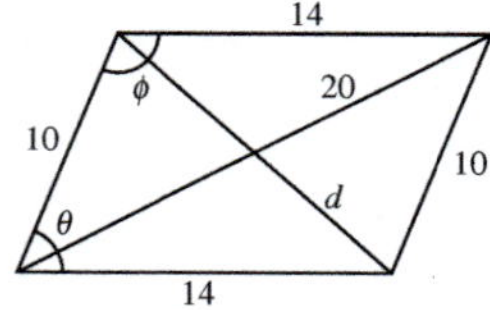

$$\cos\phi = \frac{10^2 + 14^2 - 20^2}{2(10)(14)}$$

$$\phi \approx 111.8°$$

$$2\theta \approx 360° - 2(111.8°)$$

$$\theta = 68.2°$$

$$d^2 = 10^2 + 14^2 - 2(10)(14)\cos 68.2°$$

$$d \approx 13.86$$

29. $\cos \alpha = \dfrac{(12.5)^2 + (15)^2 - 10^2}{2(12.5)(15)} = 0.75 \Rightarrow \alpha \approx 41.41°$

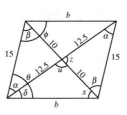

$\cos \beta = \dfrac{10^2 + 15^2 - (12.5)^2}{2(10)(15)} = 0.5625 \Rightarrow \beta \approx 55.77°$

$z = 180° - \alpha - \beta = 82.82°$

$u = 180° - z = 97.18°$

$b^2 = 12.5^2 + 10^2 - 2(12.5)(10) \cos 97.18° \approx 287.4967 \Rightarrow b \approx 16.96$

$\cos \delta = \dfrac{12.5^2 + 16.96^2 - 10^2}{2(12.5)(16.96)} \approx 0.8111 \Rightarrow \delta \approx 35.80°$

$\theta = \alpha + \delta = 41.41° + 35.80° = 77.2°$

$2\phi = 360° - 2\theta \Rightarrow \phi = \dfrac{360° - 2(77.21°)}{2} = 102.8°$

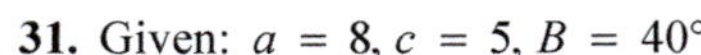

31. Given: $a = 8, c = 5, B = 40°$

Given two sides and included angle, use the Law of Cosines.

$b^2 = a^2 + c^2 - 2ac \cos B = 64 + 25 - 2(8)(5) \cos 40° \approx 27.7164 \Rightarrow b \approx 5.26$

$\cos A = \dfrac{b^2 + c^2 - a^2}{2bc} \approx \dfrac{(5.26)^2 + 25 - 64}{2(5.26)(5)} \approx -0.2154 \Rightarrow A \approx 102.44°$

$C \approx 180° - 102.44° - 40° = 37.56°$

33. Given: $A = 24°, a = 4, b = 18$

Given two sides and an angle opposite one of them, use the Law of Sines.

$h = b \sin A = 18 \sin 24° \approx 7.32$

Because $a < h$, no triangle is formed.

35. Given: $A = 42°, B = 35°, c = 1.2$

Given two angles and a side, use the Law of Sines.

$C = 180° - 42° - 35° = 103°$

$a = \dfrac{c \sin A}{\sin C} = \dfrac{1.2 \sin 42°}{\sin 103°} \approx 0.82$

$b = \dfrac{c \sin B}{\sin C} = \dfrac{1.2 \sin 35°}{\sin 103°} \approx 0.71$

37. $a = 6, b = 12, c = 17$

$s = \dfrac{a + b + c}{2} = \dfrac{6 + 12 + 17}{2} = 17.5$

$\text{Area} = \sqrt{s(s - a)(s - b)(s - c)}$

$= \sqrt{17.5(11.5)(5.5)(0.5)}$

≈ 23.53

39. $a = 2.5, b = 10.2, c = 8$

$s = \dfrac{a + b + c}{2} = \dfrac{2.5 + 10.2 + 8}{2} = 10.35$

$\text{Area} = \sqrt{s(s - a)(s - b)(s - c)}$

$= \sqrt{10.35(7.85)(0.15)(2.35)}$

≈ 5.35

41. Given: $a = 1, b = \dfrac{1}{2}, c = \dfrac{5}{4}$

$s = \dfrac{a + b + c}{2} = \dfrac{1 + \frac{1}{2} + \frac{5}{4}}{2} = \dfrac{11}{8}$

$\text{Area} = \sqrt{s(s - a)(s - b)(s - c)}$

$= \sqrt{\dfrac{11}{8}\left(\dfrac{3}{8}\right)\left(\dfrac{7}{8}\right)\left(\dfrac{1}{8}\right)}$

≈ 0.24

43. $\text{Area} = \dfrac{1}{2}bc \sin A$

$= \dfrac{1}{2}(75)(41) \sin 80°$

≈ 1514.14

45. $b^2 = 220^2 + 250^2 - 2(220)(250)\cos 105^\circ \Rightarrow b \approx 373.3$ meters

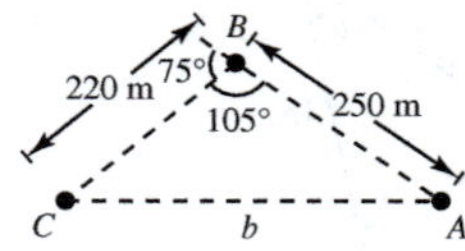

47. $d = \sqrt{330^2 + 420^2 - 2(330)(420)\cos 8^\circ} \approx 103.9$ feet

49. The angles at the base of the tower are 96° and 84°.

The longer guy wire g_1 is given by:

$$g_1^2 = 75^2 + 100^2 - 2(75)(100)\cos 96^\circ \approx 17{,}192.9 \Rightarrow g_1 \approx 131.1 \text{ feet}$$

The shorter guy wire g_2 is given by:

$$g_2^2 = 75^2 + 100^2 - 2(75)(100)\cos 84 \approx 14{,}057.1 \Rightarrow g_2 \approx 118.6 \text{ feet}$$

51. $\cos B = \dfrac{1700^2 + 3700^2 - 3000^2}{2(1700)(3700)} \Rightarrow B \approx 52.9^\circ$

Bearing: $90^\circ - 52.9^\circ =$ N 37.1° E

$\cos C = \dfrac{1700^2 + 3000^2 - 3700^2}{2(1700)(3000)} \Rightarrow C \approx 100.2^\circ$

Bearing: $90^\circ - 26.9^\circ =$ S 63.1° E

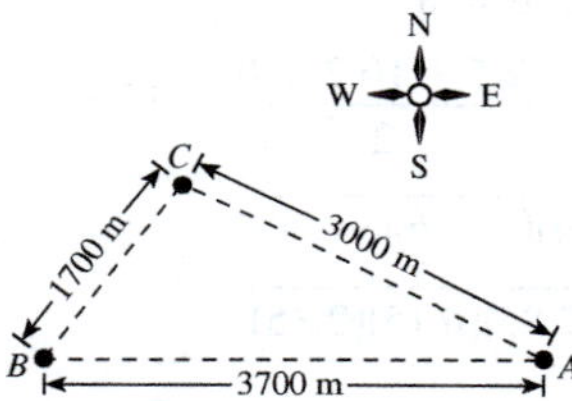

53.

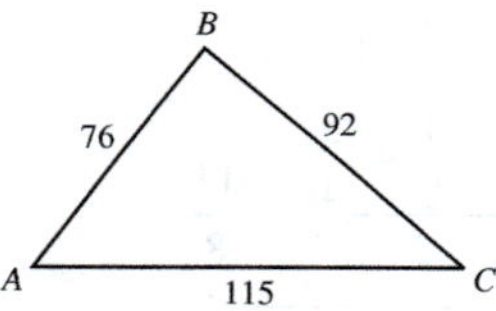

$\cos A = \dfrac{115^2 + 76^2 - 92^2}{2(115)(76)} \approx 0.6028 \Rightarrow A \approx 52.9^\circ$

$\cos C = \dfrac{115^2 + 92^2 - 76^2}{2(115)(92)} \approx 0.75203 \Rightarrow c \approx 41.2^\circ$

55. (a) $C = 180^\circ - 53^\circ - 67^\circ = 60^\circ$

$$\begin{aligned} d^2 &= a^2 + (3s)^2 - 2ab\cos C \\ &= 36^2 + 9s^2 - 2(36)(3s)(0.5) \\ d &= \sqrt{9s^2 - 108s + 1296} \end{aligned}$$

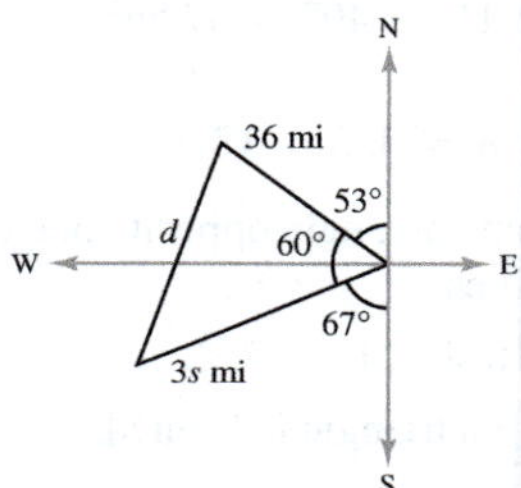

(b) $43 = \sqrt{9s^2 - 108s + 1296}$

$9s^2 - 108s - 553 = 0$

Using the quadratic formula, $s \approx 15.87$ mph.

57. $a = 200$

$b = 500$

$c = 600 \Rightarrow s = \dfrac{200 + 500 + 600}{2} = 650$

Area $= \sqrt{650(450)(150)(50)} \approx 46{,}837.5$ square feet

59. $s = \dfrac{510 + 840 + 1120}{2} = 1235$

Area $= \sqrt{1235(1235 - 510)(1235 - 840)(1235 - 1120)}$

$\approx 201{,}674$ square yards

Cost $\approx \left(\dfrac{201{,}674.02}{4840}\right)(2000) \approx \$83{,}336.37$

61. False. The average of the three sides of a triangle is

$$\frac{a+b+c}{3}, \text{ not } \frac{a+b+c}{2} = s.$$

63. $c^2 = a^2 + b^2 - 2ab\cos C$

$= a^2 + b^2 - 2ab\cos 90°$

$= a^2 + b^2 - 2ab(0)$

$= a^2 + b^2$

When $C = 90°$, you obtain the Pythagorean Theorem. The Pythagorean Theorem is a special case of the Law of Cosines.

65. There is no method that can be used to solve the no-solution case of SSA.

The Law of Cosines can be used to solve the single-solution case of SSA. You can substitute values into $a^2 = b^2 + c^2 - 2bc\cos A$. The simplified quadratic equation in terms of c can be solved, with one positive solution and one negative solution. The negative solution can be discarded because length is positive. You can use the positive solution to solve the triangle.

67. (a)
$$\begin{aligned}\frac{1}{2}bc(1+\cos A) &= \frac{1}{2}bc\left[1 + \frac{b^2+c^2-a^2}{2bc}\right]\\ &= \frac{1}{2}bc\left[\frac{2bc+b^2+c^2-a^2}{2bc}\right]\\ &= \frac{1}{4}\left[(b+c)^2 - a^2\right]\\ &= \frac{1}{4}\left[(b+c)+a\right]\left[(b+c)-a\right]\\ &= \frac{b+c+a}{2}\cdot\frac{b+c-a}{2}\\ &= \frac{a+b+c}{2}\cdot\frac{-a+b+c}{2}\end{aligned}$$

(b)
$$\begin{aligned}\frac{1}{2}bc(1-\cos A) &= \frac{1}{2}bc\left[1 + \frac{a^2-(b^2+c^2)}{2bc}\right]\\ &= \frac{1}{2}bc\left[\frac{2bc+a^2-b^2-c^2}{2bc}\right]\\ &= \frac{a^2-(b^2-2bc+c^2)}{4}\\ &= \frac{a^2-(b-c)^2}{4}\\ &= \left(\frac{a-(b-c)}{2}\right)\left(\frac{a+(b-c)}{2}\right)\\ &= \frac{a-b+c}{2}\cdot\frac{a+b-c}{2}\end{aligned}$$

Section 6.3 Vectors in the Plane

1. directed line segment

3. vector

5. standard position

7. multiplication; addition

9. $\|\mathbf{u}\| = \sqrt{(6-2)^2 + (5-4)^2} = \sqrt{17}$

$\|\mathbf{v}\| = \sqrt{(4-0)^2 + (1-0)^2} = \sqrt{17}$

$\text{slope}_{\mathbf{u}} = \dfrac{5-4}{6-2} = \dfrac{1}{4}$

$\text{slope}_{\mathbf{v}} = \dfrac{1-0}{4-0} = \dfrac{1}{4}$

u and **v** have the same magnitude and direction so they are equivalent.

11. $\|\mathbf{u}\| = \sqrt{(-1-2)^2 + (4-2)^2} = \sqrt{13}$

$\|\mathbf{v}\| = \sqrt{(-5-(-3))^2 + (2-(-1))^2} = \sqrt{13}$

$\text{slope}_{\mathbf{u}} = \dfrac{4-2}{-1-2} = -\dfrac{2}{3}$

$\text{slope}_{\mathbf{v}} = \dfrac{2-(-1)}{-5-(-3)} = -\dfrac{3}{2}$

u and **v** have the same magnitude but not the same direction so they are not equivalent.

13. $\|\mathbf{u}\| = \sqrt{(5-2)^2 + (-10-(-1))^2} = \sqrt{90} = 3\sqrt{10}$

$\|\mathbf{v}\| = \sqrt{(9-6)^2 + (-8-1)^2} = \sqrt{90} = 3\sqrt{10}$

$\text{slope}_{\mathbf{u}} = \dfrac{-10-(-1)}{5-2} = -3$

$\text{slope}_{\mathbf{v}} = \dfrac{-8-1}{9-6} = -3$

u and **v** have the same magnitude and direction so they are equivalent.

15. Initial point: $(0, 0)$

Terminal point: $(1, 3)$

$\mathbf{v} = \langle 1 - 0, 3 - 0 \rangle = \langle 1, 3 \rangle$

$\|\mathbf{v}\| = \sqrt{1^2 + 3^2} = \sqrt{10}$

17. Initial point: $(3, -2)$

Terminal point: $(3, 3)$

$\mathbf{v} = \langle 3 - 3, 3 - (-2) \rangle = \langle 0, 5 \rangle$

$\|\mathbf{v}\| = \sqrt{0^2 + 5^2} = \sqrt{25} = 5$

19. Initial point: $(-3, -5)$

Terminal point: $(-11, 1)$

$\mathbf{v} = \langle -11 - (-3), 1 - (-5) \rangle = \langle -8, 6 \rangle$

$\|\mathbf{v}\| = \sqrt{(-8)^2 + 6^2} = \sqrt{100} = 10$

21. Initial point: $(1, 3)$

Terminal point: $(-8, -9)$

$\mathbf{v} = \langle -8 - 1, -9 - 3 \rangle = \langle -9, -12 \rangle$

$\|\mathbf{v}\| = \sqrt{(-9)^2 + (-12)^2} = \sqrt{225} = 15$

23. Initial point: $(-1, 5)$

Terminal point: $(15, -21)$

$\mathbf{v} = \langle 15 - (-1), -21 - 5 \rangle = \langle 16, -26 \rangle$

$\|\mathbf{v}\| = \sqrt{(16)^2 + (-26)^2} = \sqrt{932} = 2\sqrt{233}$

25. $-\mathbf{v}$

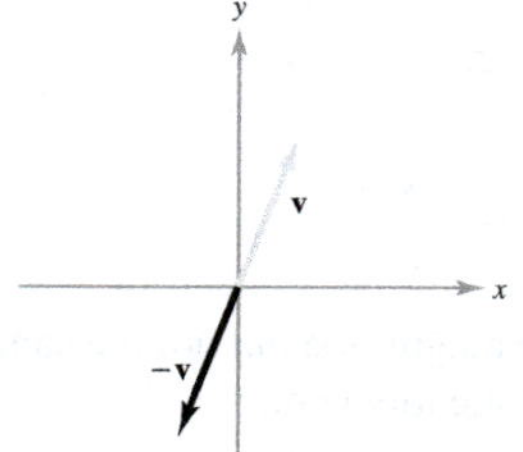

27. $\mathbf{u} + \mathbf{v}$

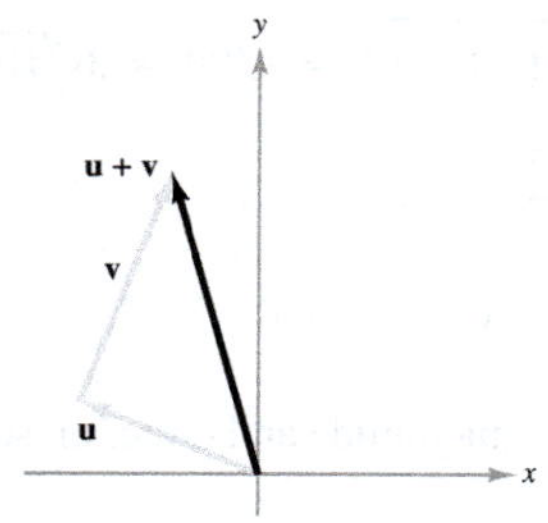

29. $\mathbf{u} - \mathbf{v}$

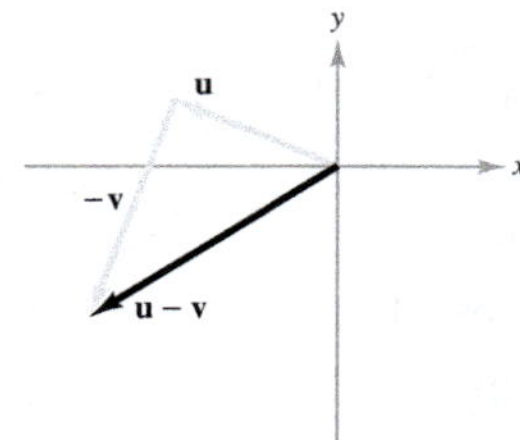

31. $\mathbf{u} = \langle 2, 1 \rangle, \mathbf{v} = \langle 1, 3 \rangle$

(a) $\mathbf{u} + \mathbf{v} = \langle 3, 4 \rangle$

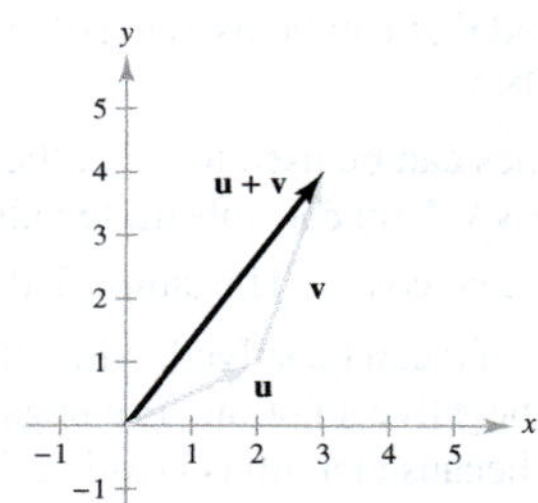

(b) $\mathbf{u} - \mathbf{v} = \langle 1, -2 \rangle$

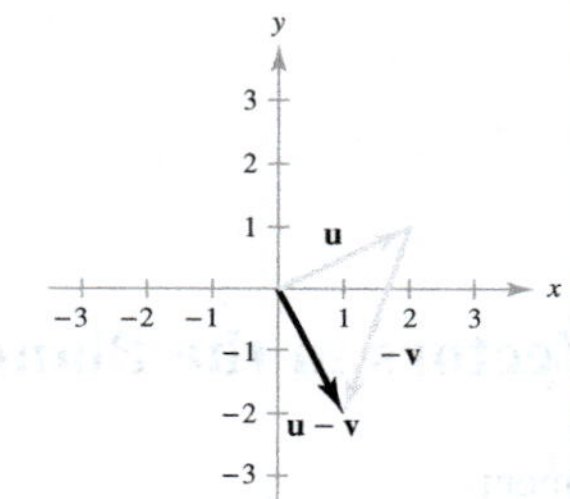

(c) $2\mathbf{u} - 3\mathbf{v} = \langle 4, 2 \rangle - \langle 3, 9 \rangle = \langle 1, -7 \rangle$

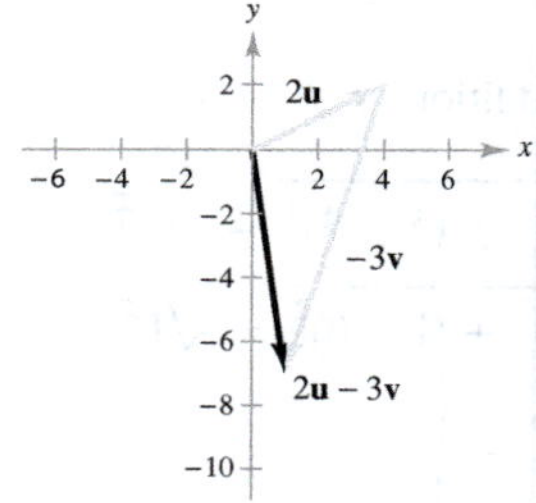

33. $\mathbf{u} = \langle -5, 3\rangle, \mathbf{v} = \langle 0, 0\rangle$

(a) $\mathbf{u} + \mathbf{v} = \langle -5, 3\rangle = \mathbf{u}$

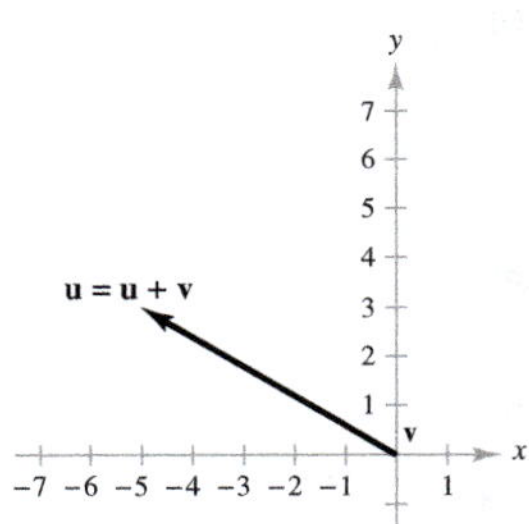

(b) $\mathbf{u} - \mathbf{v} = \langle -5, 3\rangle = \mathbf{u}$

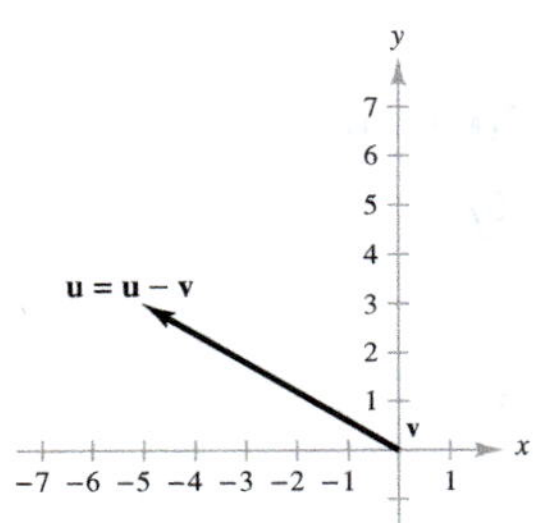

(c) $2\mathbf{u} - 3\mathbf{v} = \langle -10, 6\rangle = 2\mathbf{u}$

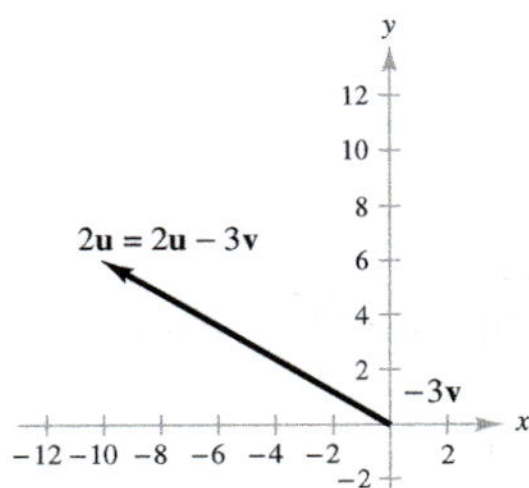

35. $\mathbf{u} = -7\mathbf{j}, \mathbf{v} = \mathbf{i} - 2\mathbf{j}$

(a) $\mathbf{u} + \mathbf{v} = \mathbf{i} - 9\mathbf{j}$

$\langle 1, -9\rangle$

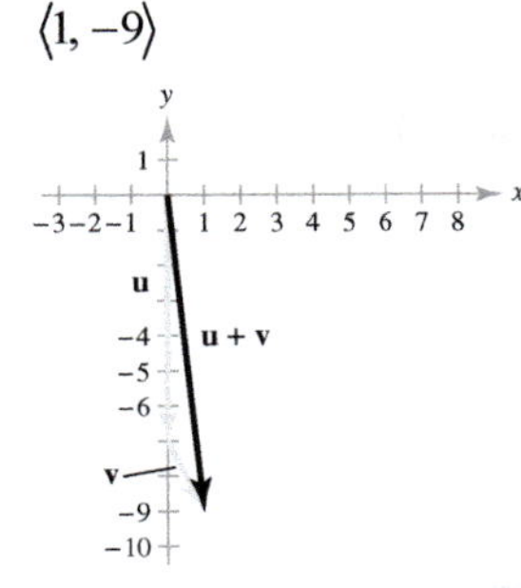

(b) $\mathbf{u} - \mathbf{v} = -\mathbf{i} - 5\mathbf{j}$

$\langle -1, -5\rangle$

(c) $2\mathbf{u} - 3\mathbf{v} = (-14\mathbf{j}) - (3\mathbf{i} - 6\mathbf{j}) = -3\mathbf{i} - 8\mathbf{j}$

$\langle -3, -8\rangle$

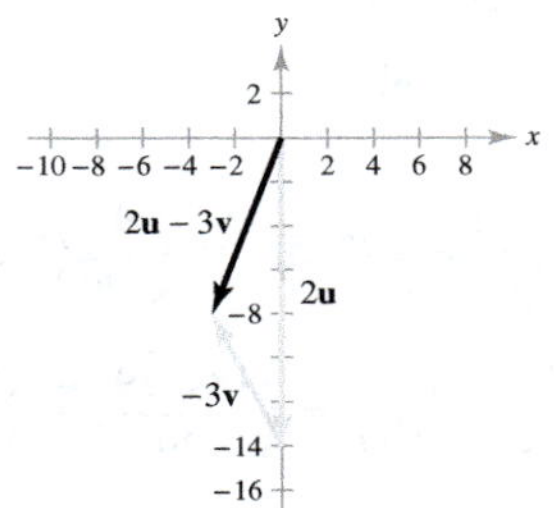

37.

$$\mathbf{u} = \langle 2, 0\rangle$$
$$5\mathbf{u} = \langle 10, 0\rangle$$
$$\|5\mathbf{u}\| = \sqrt{(10)^2 + 0^2} = 10$$

39. $\mathbf{v} = \langle -3, 6\rangle$

$$-3\mathbf{v} = \langle 9, -18\rangle$$
$$\|4\mathbf{v}\| = \sqrt{9^2 + (-18)^2} = \sqrt{405} = 9\sqrt{5}$$

41. $\mathbf{v} = \langle 3, 0\rangle$

$$\mathbf{u} = \frac{1}{\|\mathbf{v}\|}\mathbf{v} = \frac{1}{\sqrt{3^2 + 0^2}}\langle 3, 0\rangle = \frac{1}{3}\langle 3, 0\rangle = \langle 1, 0\rangle$$
$$\|\mathbf{u}\| = \sqrt{1^2 + 0^2} = 1$$

43. $\mathbf{v} = \langle -2, 2\rangle$

$$\mathbf{u} = \frac{1}{\|\mathbf{v}\|}\mathbf{v} = \frac{1}{\sqrt{(-2)^2 + 2^2}}\langle -2, 2\rangle = \frac{1}{2\sqrt{2}}\langle -2, 2\rangle$$
$$= \left\langle -\frac{1}{\sqrt{2}}, \frac{1}{\sqrt{2}}\right\rangle$$
$$= \left\langle -\frac{\sqrt{2}}{2}, \frac{\sqrt{2}}{2}\right\rangle$$
$$\|\mathbf{u}\| = \sqrt{\left(\frac{-\sqrt{2}}{2}\right)^2 + \left(\frac{\sqrt{2}}{2}\right)^2} = 1$$

45. $\mathbf{v} = \langle 1, -6 \rangle$

$$\mathbf{u} = \frac{1}{\|\mathbf{v}\|}\mathbf{v} = \frac{1}{\sqrt{1^2 + (-6)^2}}\langle 1, -6 \rangle = \frac{1}{\sqrt{37}}\langle 1, -6 \rangle$$

$$= \frac{1}{\sqrt{37}}\langle 1, -6 \rangle = \left\langle \frac{\sqrt{37}}{37}, -\frac{6\sqrt{37}}{37} \right\rangle$$

$$\|\mathbf{u}\| = \sqrt{\left(\frac{\sqrt{37}}{37}\right)^2 + \left(\frac{-6\sqrt{37}}{37}\right)^2} = 1$$

47. $\mathbf{v} = 10\left(\frac{1}{\|\mathbf{u}\|}\mathbf{u}\right) = 10\left(\frac{1}{\sqrt{(-3)^2 + 4^2}}\langle -3, 4 \rangle\right)$

$$= 2\langle -3, 4 \rangle$$

$$= \langle -6, 8 \rangle$$

49. $9\left(\frac{1}{\|\mathbf{u}\|}\mathbf{u}\right) = 9\left(\frac{1}{\sqrt{2^2 + 5^2}}\langle 2, 5 \rangle\right) = \frac{9}{\sqrt{29}}\langle 2, 5 \rangle$

$$= \left\langle \frac{18}{\sqrt{29}}, \frac{45}{\sqrt{29}} \right\rangle = \left\langle \frac{18\sqrt{29}}{29}, \frac{45\sqrt{29}}{29} \right\rangle$$

51. $\mathbf{u} = \langle 3 - (-2), -2 - 1 \rangle$

$$= \langle 5, -3 \rangle$$

$$= 5\mathbf{i} - 3\mathbf{j}$$

53. $\mathbf{u} = \langle -6 - 0, 4 - 1 \rangle$

$\mathbf{u} = \langle -6, 3 \rangle$

$\mathbf{u} = -6\mathbf{i} + 3\mathbf{j}$

55. $\mathbf{v} = \frac{3}{2}\mathbf{u}$

$$= \frac{3}{2}(2\mathbf{i} - \mathbf{j})$$

$$= 3\mathbf{i} - \frac{3}{2}\mathbf{j} = \left\langle 3, -\frac{3}{2} \right\rangle$$

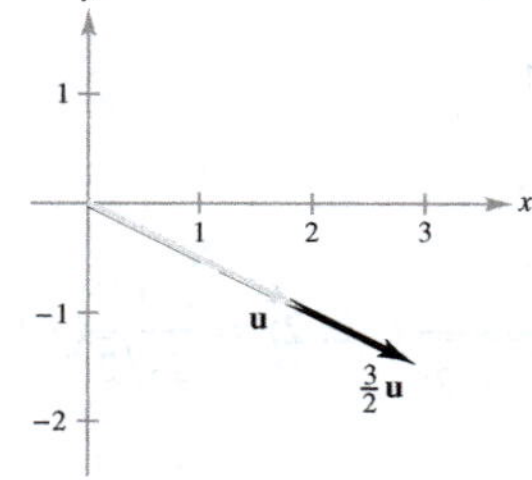

57. $\mathbf{v} = \mathbf{u} + 2\mathbf{w}$

$$= (2\mathbf{i} - \mathbf{j}) + 2(\mathbf{i} + 2\mathbf{j})$$

$$= 4\mathbf{i} + 3\mathbf{j} = \langle 4, 3 \rangle$$

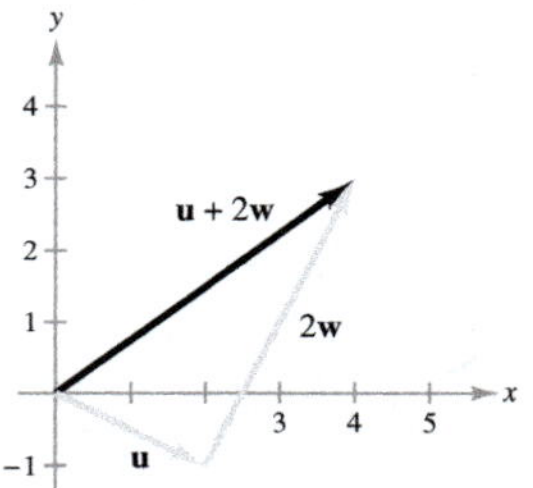

59. $\mathbf{v} = \mathbf{u} - 2\mathbf{w}$

$$= (2\mathbf{i} - \mathbf{j}) - 2(\mathbf{i} + 2\mathbf{j})$$

$$= -5\mathbf{j} = \langle 0, -5 \rangle$$

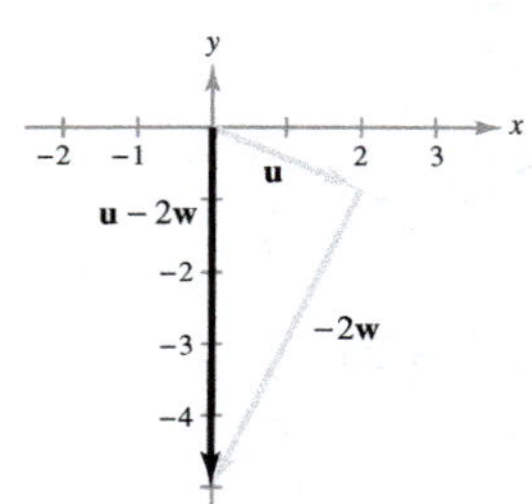

61. $\mathbf{v} = 6\mathbf{i} - 6\mathbf{j}$

$$\|\mathbf{v}\| = \sqrt{6^2 + (-6)^2} = \sqrt{72} = 6\sqrt{2}$$

$$\tan\theta = \frac{-6}{6} = -1$$

Since $\mathbf{v}$ lies in Quadrant IV, $\theta = 315°$.

63. $\mathbf{v} = 3(\cos 60°\mathbf{i} + \sin 60°\mathbf{j})$

$\|\mathbf{v}\| = 3, \theta = 60°$

65. $\mathbf{v} = \langle 3\cos 0°, 3\sin 0° \rangle$

$$= \langle 3, 0 \rangle$$

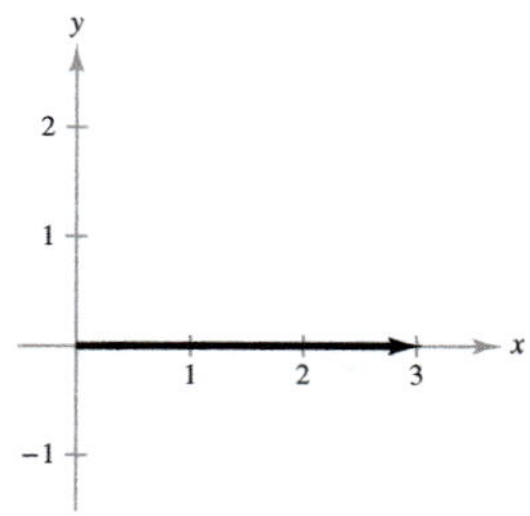

67. $\mathbf{v} = \left\langle \frac{7}{2}\cos 150°, \frac{7}{2}\sin 150° \right\rangle$

$= \left\langle -\frac{7\sqrt{3}}{4}, \frac{7}{4} \right\rangle$

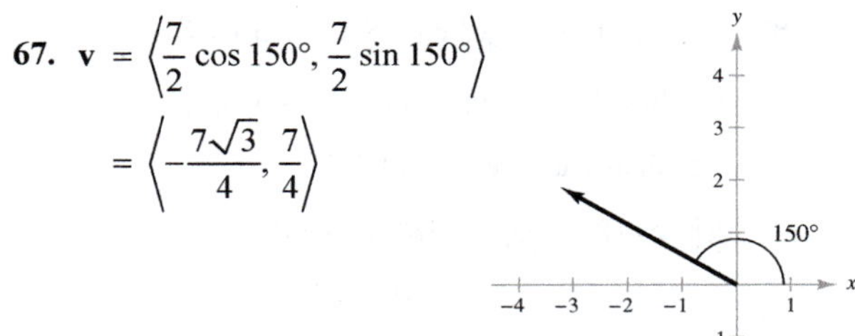

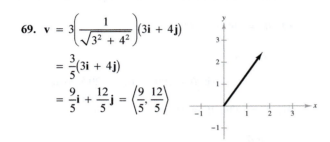

69. $\mathbf{v} = 3\left(\frac{1}{\sqrt{3^2 + 4^2}}\right)(3\mathbf{i} + 4\mathbf{j})$

$= \frac{3}{5}(3\mathbf{i} + 4\mathbf{j})$

$= \frac{9}{5}\mathbf{i} + \frac{12}{5}\mathbf{j} = \left\langle \frac{9}{5}, \frac{12}{5} \right\rangle$

71. $\mathbf{u} = \langle 4\cos 60°, 4\sin 60° \rangle = \langle 2, 2\sqrt{3} \rangle$

$\mathbf{v} = \langle 4\cos 90°, 4\sin 90° \rangle = \langle 0, 4 \rangle$

$\mathbf{u} + \mathbf{v} = \langle 2, 4 + 2\sqrt{3} \rangle$

73. $\mathbf{v} = \mathbf{i} + \mathbf{j}$

$\mathbf{w} = 2\mathbf{i} - 2\mathbf{j}$

$\mathbf{u} = \mathbf{v} - \mathbf{w} = -\mathbf{i} + 3\mathbf{j}$

$\|\mathbf{v}\| = \sqrt{2}$

$\|\mathbf{w}\| = 2\sqrt{2}$

$\|\mathbf{v} - \mathbf{w}\| = \sqrt{10}$

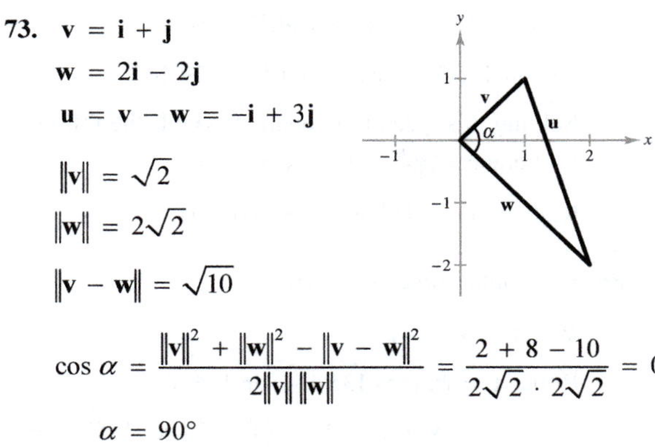

$$\cos\alpha = \frac{\|\mathbf{v}\|^2 + \|\mathbf{w}\|^2 - \|\mathbf{v} - \mathbf{w}\|^2}{2\|\mathbf{v}\|\,\|\mathbf{w}\|} = \frac{2 + 8 - 10}{2\sqrt{2}\cdot 2\sqrt{2}} = 0$$

$\alpha = 90°$

75. Force One: $\mathbf{u} = 45\mathbf{i}$

Force Two: $\mathbf{v} = 60\cos\theta\mathbf{i} + 60\sin\theta\mathbf{j}$

Resultant Force: $\mathbf{u} + \mathbf{v} = (45 + 60\cos\theta)\mathbf{i} + 60\sin\theta\mathbf{j}$

$$\|\mathbf{u} + \mathbf{v}\| = \sqrt{(45 + 60\cos\theta)^2 + (60\sin\theta)^2} = 90$$

$$2025 + 5400\cos\theta + 3600 = 8100$$

$$5400\cos\theta = 2475$$

$$\cos\theta = \tfrac{2475}{5400} \approx 0.4583$$

$$\theta \approx 62.7°$$

77. Horizontal component of velocity:
$1200\cos 6° \approx 1193.4$ ft/sec

Vertical component of velocity:
$1200\sin 6° \approx 125.4$ ft/sec

79. $\mathbf{u} = 300\mathbf{i}$

$\mathbf{v} = (125\cos 45°)\mathbf{i} + (125\sin 45°)\mathbf{j}$

$= \frac{125}{\sqrt{2}}\mathbf{i} + \frac{125}{\sqrt{2}}\mathbf{j}$

$\mathbf{u} + \mathbf{v} = \left(300 + \frac{125}{\sqrt{2}}\right)\mathbf{i} + \frac{125}{\sqrt{2}}\mathbf{j}$

$\|\mathbf{u} + \mathbf{v}\| = \sqrt{\left(300 + \frac{125}{\sqrt{2}}\right)^2 + \left(\frac{125}{\sqrt{2}}\right)^2}$

≈ 398.32 newtons

$$\tan\theta = \frac{\frac{125}{\sqrt{2}}}{300 + \left(\frac{125}{\sqrt{2}}\right)} \Rightarrow \theta \approx 12.8°$$

81. $\mathbf{u} = (75\cos 30°)\mathbf{i} + (75\sin 30°)\mathbf{j}$

$\approx 64.95\mathbf{i} + 37.5\mathbf{j}$

$\mathbf{v} = (100\cos 45°)\mathbf{i} + (100\sin 45°)\mathbf{j}$

$\approx 70.71\mathbf{i} + 70.71\mathbf{j}$

$\mathbf{w} = (125\cos 120°)\mathbf{i} + (125\sin 120°)\mathbf{j}$

$\approx -62.5\mathbf{i} + 108.3\mathbf{j}$

$\mathbf{u} + \mathbf{v} + \mathbf{w} \approx 73.16\mathbf{i} + 216.5\mathbf{j}$

$\|\mathbf{u} + \mathbf{v} + \mathbf{w}\| \approx 228.5$ pounds

$\tan\theta \approx \frac{216.5}{73.16} \approx 2.9593$

$\theta \approx 71.3°$

83. Left crane: $\mathbf{u} = \|\mathbf{u}\|(\cos 155.7°\mathbf{i} + \sin 155.7°\mathbf{j})$

Right crane: $\mathbf{v} = \|\mathbf{v}\|(\cos 44.5°\mathbf{i} + \sin 44.5°\mathbf{j})$

Resultant: $\mathbf{u} + \mathbf{v} = -20{,}240\mathbf{j}$

System of equations:

$\|\mathbf{u}\| \cos 155.7° + \|\mathbf{v}\| \cos 44.5° = 0$

$\|\mathbf{u}\| \sin 155.7° + \|\mathbf{v}\| \sin 44.5° = 20{,}240$

Solving this system of equations yields the following:

Left crane $= \|\mathbf{u}\| \approx 15{,}484$ pounds

Right crane $= \|\mathbf{v}\| \approx 19{,}786$ pounds

85. Horizontal force: $\mathbf{u} = \|\mathbf{u}\|\mathbf{i}$

Weight: $\mathbf{w} = -\mathbf{j}$

Rope: $\mathbf{t} = \|\mathbf{t}\|(\cos 135°\mathbf{i} + \sin 135°\mathbf{j})$

$$\mathbf{u} + \mathbf{w} + \mathbf{t} = \mathbf{0} \Rightarrow \|\mathbf{u}\| + \|\mathbf{t}\| \cos 135° = 0$$
$$-1 + \|\mathbf{t}\| \sin 135° = 0$$

$\|\mathbf{t}\| \approx \sqrt{2}$ pounds

$\|\mathbf{u}\| \approx 1$ pound

87. Towline 1: $\mathbf{u} = \|\mathbf{u}\|(\cos 18°\mathbf{i} + \sin 18°\mathbf{j})$

Towline 2: $\mathbf{v} = \|\mathbf{u}\|(\cos 18°\mathbf{i} - \sin 18°\mathbf{j})$

Resultant: $\mathbf{u} + \mathbf{v} = 6000\mathbf{i}$

$\|\mathbf{u}\| \cos 18° + \|\mathbf{u}\| \cos 18° = 6000$

$\|\mathbf{u}\| \approx 3154.4$

So, the tension on each towline is $\|\mathbf{u}\| \approx 3154.4$ pounds.

89. $W = 100, \theta = 12°$

$$\sin \theta = \frac{F}{W}$$
$$F = W \sin \theta = 100 \sin 12° \approx 20.8 \text{ pounds}$$

91. $F = 5000, W = 15{,}000$

$$\sin \theta = \frac{F}{W}$$
$$\sin \theta = \frac{5000}{15{,}000}$$
$$\theta = \sin^{-1} \frac{1}{3} \approx 19.5°$$

93. Airspeed: $\mathbf{u} = (875 \cos 58°)\mathbf{i} - (875 \sin 58°)\mathbf{j}$

Groundspeed: $\mathbf{v} = (800 \cos 50°)\mathbf{i} - (800 \sin 50°)\mathbf{j}$

Wind: $\mathbf{w} = \mathbf{v} - \mathbf{u} = (800 \cos 50° - 875 \cos 58°)\mathbf{i} + (-800 \sin 50° + 875 \sin 58°)\mathbf{j}$

$\approx 50.5507\mathbf{i} + 129.2065\mathbf{j}$

Wind speed: $\|\mathbf{w}\| \approx \sqrt{(50.5507)^2 + (129.2065)^2} \approx 138.7$ kilometers per hour

Wind direction: $\tan \theta \approx \dfrac{129.2065}{50.5507}$

$\theta \approx 68.6°;\ 90° - \theta = 21.4°$

Bearing: N 21.4° E

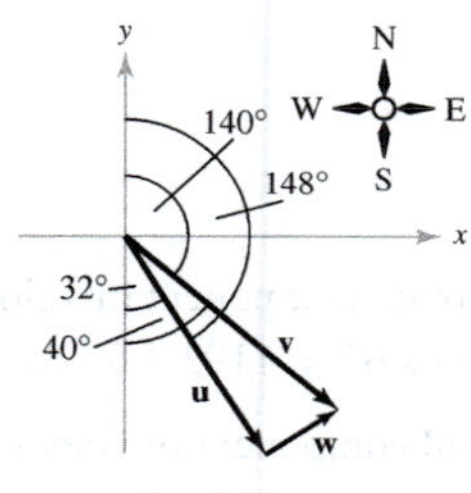

95. True. Two directed line segments that have the same magnitude and direction are equivalent (see Example 1).

97. True. If $\mathbf{v} = a\mathbf{i} + b\mathbf{j} = 0$ is the zero vector, then $a = b = 0$. So, $a = -b$.

99. The order of subtraction should be switched.

$\mathbf{u} = \langle 6 - (-3), -1 - 4 \rangle = \langle 9, -5 \rangle$

101. Let $\mathbf{v} = (\cos \theta)\mathbf{i} + (\sin \theta)\mathbf{j}$.

$\|\mathbf{v}\| = \sqrt{\cos^2 \theta + \sin^2 \theta} = \sqrt{1} = 1$

So, $\mathbf{v}$ is a unit vector for any value of θ.

103.
$$\mathbf{u} = \langle 5 - 1, 2 - 6 \rangle = \langle 4, -4 \rangle$$
$$\mathbf{v} = \langle 9 - 4, 4 - 5 \rangle = \langle 5, -1 \rangle$$
$$\mathbf{u} - \mathbf{v} = \langle -1, -3 \rangle \text{ or } \mathbf{v} - \mathbf{u} = \langle 1, 3 \rangle$$

105. $\mathbf{F}_1 = \langle 10, 0\rangle, \mathbf{F}_2 = 5\langle \cos\theta, \sin\theta\rangle$

(a) $\mathbf{F}_1 + \mathbf{F}_2 = \langle 10 + 5\cos\theta, 5\sin\theta\rangle$

$$\|\mathbf{F}_1 + \mathbf{F}_2\| = \sqrt{(10 + 5\cos\theta)^2 + (5\sin\theta)^2} = \sqrt{100 + 100\cos\theta + 25\cos^2\theta + 25\sin^2\theta}$$

$$= 5\sqrt{4 + 4\cos\theta + \cos^2\theta + \sin^2\theta} = 5\sqrt{4 + 4\cos\theta + 1} = 5\sqrt{5 + 4\cos\theta}$$

(b)

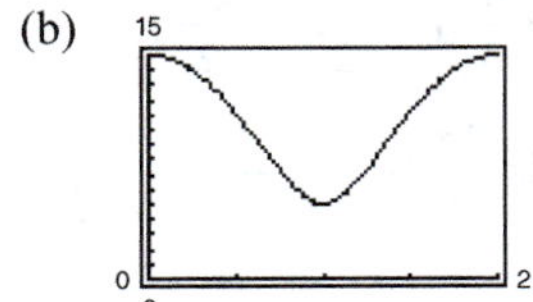

(c) Range: $[5, 15]$

Maximum is 15 when $\theta = 0$.

Minimum is 5 when $\theta = \pi$.

(d) The magnitude of the resultant is never 0 because the magnitudes of $\mathbf{F}_1$ and $\mathbf{F}_2$ are not the same.

107. (a) Answers will vary. *Sample answer:* To add two vectors **u** and **v** geometrically, first position them (without changing their lengths or directions) so that the initial point of the second vector **v** coincides with the terminal point of the first vector **u**. The sum $\mathbf{u} + \mathbf{v}$ is the vector formed by joining the initial point of the first vector **u** with the terminal point of the second vector **v**.

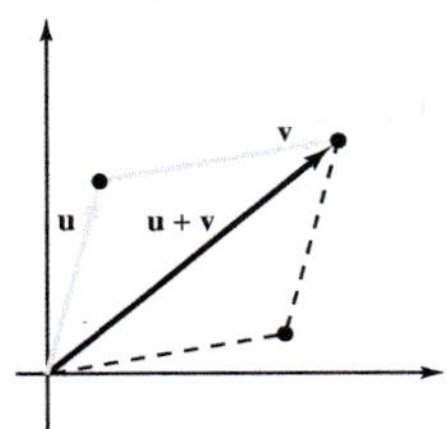

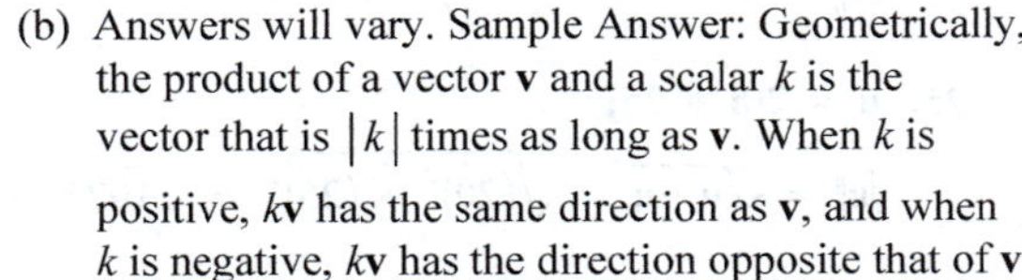
(b) Answers will vary. Sample Answer: Geometrically, the product of a vector **v** and a scalar k is the vector that is $|k|$ times as long as **v**. When k is positive, $k\mathbf{v}$ has the same direction as **v**, and when k is negative, $k\mathbf{v}$ has the direction opposite that of **v**.

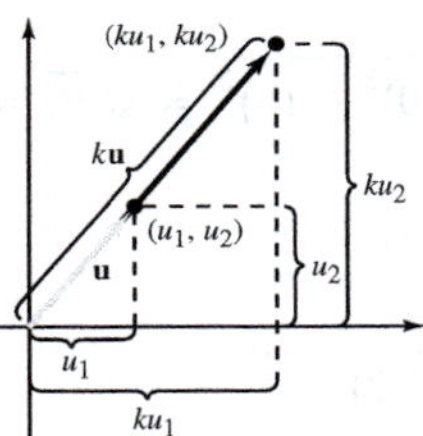

Section 6.4 Vectors and Dot Products

1. dot product

3. $\dfrac{\mathbf{u}\cdot\mathbf{v}}{\|\mathbf{u}\|\,\|\mathbf{v}\|}$

5. $\left(\dfrac{\mathbf{u}\cdot\mathbf{v}}{\|\mathbf{v}\|^2}\right)\mathbf{v}$

7. $\mathbf{u} = \langle 7, 1\rangle, \mathbf{v} = \langle -3, 2\rangle$

$\mathbf{u}\cdot\mathbf{v} = 7(-3) + 1(2) = -19$

9. $\mathbf{u} = \langle -6, 2\rangle, \mathbf{v} = \langle 1, 3\rangle$

$\mathbf{u}\cdot\mathbf{v} = -6(1) + 2(3) = 0$

11. $\mathbf{u} = 4\mathbf{i} - 2\mathbf{j}, \mathbf{v} = \mathbf{i} - \mathbf{j}$

$\mathbf{u}\cdot\mathbf{v} = 4(1) + (-2)(-1) = 6$

13. $\mathbf{u} = \langle 3, 3\rangle$

$\mathbf{u}\cdot\mathbf{u} = 3(3) + 3(3) = 18$

The result is a scalar.

15. $\mathbf{u} = \langle 3, 3\rangle, \mathbf{v} = \langle -4, 2\rangle$

$$(\mathbf{u}\cdot\mathbf{v})\mathbf{v} = [3(-4) + 3(2)]\langle -4, 2\rangle$$
$$= -6\langle -4, 2\rangle$$
$$= \langle 24, -12\rangle$$

The result is a vector.

17. $\mathbf{u} = \langle 3, 3\rangle, \mathbf{v} = \langle -4, 2\rangle, \mathbf{w} = \langle 3, -1\rangle$

$(\mathbf{v} \cdot \mathbf{0})\mathbf{w} = 0\langle 3, -1\rangle = \langle 0, 0\rangle = \mathbf{0}$

The result is a vector.

19. $\mathbf{w} = \langle 3, -1\rangle$

$\|\mathbf{w}\| - 1 = \sqrt{3^2 + (-1)^2} - 1 = \sqrt{10} - 1$

The result is a scalar.

21. $\mathbf{u} = \langle 3, 3\rangle, \mathbf{v} = \langle -4, 2\rangle, \mathbf{w} = \langle 3, -1\rangle$

$$\begin{aligned}(\mathbf{u} \cdot \mathbf{v}) - (\mathbf{u} \cdot \mathbf{w}) &= [3(-4) + 3(2)] - [3(3) + 3(-1)]\\ &= -6 - 6\\ &= -12\end{aligned}$$

The result is a scalar.

23. $\mathbf{u} = \langle -8, 15\rangle$

$\|\mathbf{u}\| = \sqrt{\mathbf{u} \cdot \mathbf{u}} = \sqrt{(-8)(-8) + 15(15)} = \sqrt{289} = 17$

25. $\mathbf{u} = 20\mathbf{i} + 25\mathbf{j}$

$\|\mathbf{u}\| = \sqrt{\mathbf{u} \cdot \mathbf{u}} = \sqrt{(20)^2 + (25)^2} = \sqrt{1025} = 5\sqrt{41}$

27. $\mathbf{u} = 6\mathbf{j}$

$\|\mathbf{u}\| = \sqrt{\mathbf{u} \cdot \mathbf{u}} = \sqrt{(0)^2 + (6)^2} = \sqrt{36} = 6$

29. $\mathbf{u} = \langle 1, 0\rangle, \mathbf{v} = \langle 0, -2\rangle$

$\cos\theta = \dfrac{\mathbf{u} \cdot \mathbf{v}}{\|\mathbf{u}\|\,\|\mathbf{v}\|} = \dfrac{0}{(1)(2)} = 0$

$\theta = \dfrac{\pi}{2}$ radians

31. $\mathbf{u} = 3\mathbf{i} + 4\mathbf{j}, \mathbf{v} = -2\mathbf{j}$

$\cos\theta = \dfrac{\mathbf{u} \cdot \mathbf{v}}{\|\mathbf{u}\|\,\|\mathbf{v}\|} = -\dfrac{8}{(5)(2)}$

$\theta = \arccos\left(-\dfrac{4}{5}\right)$

$\theta \approx 2.50$ radians

33. $\mathbf{u} = 2\mathbf{i} - \mathbf{j}, \mathbf{v} = 6\mathbf{i} - 3\mathbf{j}$

$$\begin{aligned}\cos\theta &= \frac{\mathbf{u} \cdot \mathbf{v}}{\|\mathbf{u}\|\,\|\mathbf{v}\|}\\ &= \frac{2(6) + (-1)(-3)}{\sqrt{2^2 + (-1)^2}\sqrt{6^2 + (-3)^2}}\\ &= \frac{15}{\sqrt{225}} = 1\end{aligned}$$

$\theta = 0$

35. $\mathbf{u} = -6\mathbf{i} - 3\mathbf{j}, \mathbf{v} = -8\mathbf{i} + 4\mathbf{j}$

$\cos\mathbf{u} = \dfrac{\mathbf{u} \cdot \mathbf{v}}{\|\mathbf{u}\|\,\|\mathbf{v}\|} = \dfrac{-6(-8) + (-3)(4)}{\sqrt{45}\sqrt{80}} = \dfrac{36}{60} = 0.6$

$\theta \approx 0.93$ radian

37. $\mathbf{u} = \left(\cos\dfrac{\pi}{3}\right)\mathbf{i} + \left(\sin\dfrac{\pi}{3}\right)\mathbf{j} = \dfrac{1}{2}\mathbf{i} + \dfrac{\sqrt{3}}{2}\mathbf{j}$

$\mathbf{v} = \left(\cos\dfrac{3\pi}{4}\right)\mathbf{i} + \left(\sin\dfrac{3\pi}{4}\right)\mathbf{j} = -\dfrac{\sqrt{2}}{2}\mathbf{i} + \dfrac{\sqrt{2}}{2}\mathbf{j}$

$\|\mathbf{u}\| = \|\mathbf{v}\| = 1$

$$\begin{aligned}\cos\theta &= \frac{\mathbf{u} \cdot \mathbf{v}}{\|\mathbf{u}\|\,\|\mathbf{v}\|} = \mathbf{u} \cdot \mathbf{v}\\ &= \left(\frac{1}{2}\right)\left(-\frac{\sqrt{2}}{2}\right) + \left(\frac{\sqrt{3}}{2}\right)\left(\frac{\sqrt{2}}{2}\right) = \frac{-\sqrt{2} + \sqrt{6}}{4}\end{aligned}$$

$\theta = \arccos\left(\dfrac{-\sqrt{2} + \sqrt{6}}{4}\right) = \dfrac{5\pi}{12}$

39. $\mathbf{u} = 3\mathbf{i} + 4\mathbf{j}$

$\mathbf{v} = -7\mathbf{i} + 5\mathbf{j}$

$$\begin{aligned}\cos\theta &= \frac{\mathbf{u} \cdot \mathbf{v}}{\|\mathbf{u}\|\,\|\mathbf{v}\|}\\ &= \frac{3(-7) + 4(5)}{3\sqrt{74}}\\ &= \frac{-1}{5\sqrt{74}} \approx -0.0232\end{aligned}$$

$\theta \approx 91.33°$

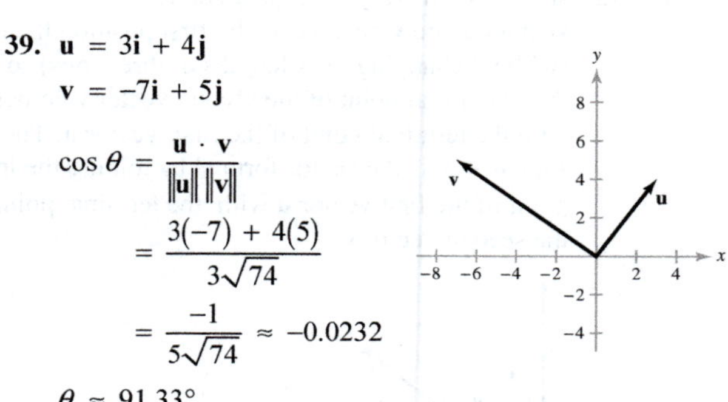

41. $\mathbf{u} = -5\mathbf{i} - 5\mathbf{j}$

$\mathbf{v} = -8\mathbf{i} + 8\mathbf{j}$

$$\begin{aligned}\cos\theta &= \frac{\mathbf{u} \cdot \mathbf{v}}{\|\mathbf{u}\|\,\|\mathbf{v}\|}\\ &= \frac{-5(-8) + (-5)(8)}{\sqrt{50}\sqrt{128}}\\ &= 0\end{aligned}$$

$\theta = 90°$

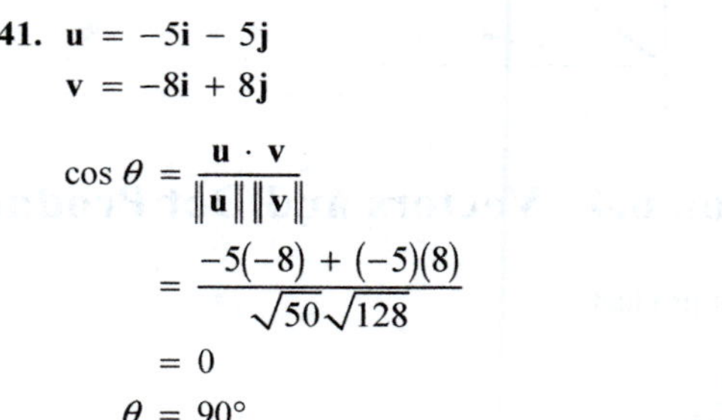

43. $P = (1, 2), Q = (3, 4), R = (2, 5)$

$\overrightarrow{PQ} = \langle 2, 2\rangle, \overrightarrow{PR} = \langle 1, 3\rangle, \overrightarrow{QR} = \langle -1, 1\rangle$

$$\cos\alpha = \frac{\overrightarrow{PQ}\cdot\overrightarrow{PR}}{\|\overrightarrow{PQ}\|\|\overrightarrow{PR}\|} = \frac{8}{(2\sqrt{2})(\sqrt{10})} \Rightarrow \alpha = \arccos\frac{2}{\sqrt{5}} \approx 26.57°$$

$$\cos\beta = \frac{\overrightarrow{PQ}\cdot\overrightarrow{QR}}{\|\overrightarrow{PQ}\|\|\overrightarrow{QR}\|} = 0 \Rightarrow \beta = 90°$$

$$\gamma = 180° - 26.57° - 90° = 63.43°$$

45. $P = (-3, 0), Q = (2, 2), R = (0, 6)$

$\overrightarrow{QP} = \langle -5, -2\rangle, \overrightarrow{PR} = \langle 3, 6\rangle, \overrightarrow{QR} = \langle -2, 4\rangle, \overrightarrow{PQ} = \langle 5, 2\rangle$

$$\cos\alpha = \frac{\overrightarrow{PQ}\cdot\overrightarrow{PR}}{\|\overrightarrow{PQ}\|\|\overrightarrow{PR}\|} = \frac{27}{\sqrt{29}\sqrt{45}} \Rightarrow \alpha \approx 41.63°$$

$$\cos\beta = \frac{\overrightarrow{QP}\cdot\overrightarrow{QR}}{\|\overrightarrow{QP}\|\|\overrightarrow{PR}\|} = \frac{2}{\sqrt{29}\sqrt{20}} \Rightarrow \beta \approx 85.24°$$

$$\delta = 180° - 41.63° - 85.24° = 53.13°$$

47. $\|\mathbf{u}\| = 4, \|\mathbf{v}\| = 10, \theta = \frac{2\pi}{3}$

$$\begin{aligned}\mathbf{u}\cdot\mathbf{v} &= \|\mathbf{u}\|\|\mathbf{v}\|\cos\theta\\ &= (4)(10)\cos\frac{2\pi}{3}\\ &= 40\left(-\frac{1}{2}\right)\\ &= -20\end{aligned}$$

49. $\|\mathbf{u}\| = 100, \|\mathbf{v}\| = 250, \theta = \frac{\pi}{6}$

$$\begin{aligned}\mathbf{u}\cdot\mathbf{v} &= \|\mathbf{u}\|\|\mathbf{v}\|\cos\theta\\ &= (100)(250)\cos\frac{\pi}{6}\\ &= 25{,}000\cdot\frac{\sqrt{3}}{2}\\ &= 12{,}500\sqrt{3}\end{aligned}$$

51. $\mathbf{u} = \langle 3, 15\rangle, \mathbf{v} = \langle -1, 5\rangle$

$\mathbf{u} \neq k\mathbf{v} \Rightarrow$ Not parallel

$\mathbf{u}\cdot\mathbf{v} \neq 0 \Rightarrow$ Not orthogonal

Neither

53. $\mathbf{u} = 2\mathbf{i} - 2\mathbf{j}, \mathbf{v} = -\mathbf{i} - \mathbf{j}$

$\mathbf{u}\cdot\mathbf{v} = 0 \Rightarrow \mathbf{u}$ and $\mathbf{v}$ are orthogonal.

55. $\mathbf{u} = \mathbf{i}, \mathbf{v} = -2\mathbf{i} + 2\mathbf{j}$

$\mathbf{u} \neq k\mathbf{v} \Rightarrow$ Not parallel

$\mathbf{u}\cdot\mathbf{v} \neq 0 \Rightarrow$ Not orthogonal

Neither

57. $\mathbf{u} = \langle 2, 2\rangle, \mathbf{v} = \langle 6, 1\rangle$

$$\mathbf{w}_1 = \text{proj}_{\mathbf{v}}\mathbf{u} = \left(\frac{\mathbf{u}\cdot\mathbf{v}}{\|\mathbf{v}\|^2}\right)\mathbf{v} = \frac{14}{37}\langle 6, 1\rangle = \frac{1}{37}\langle 84, 14\rangle$$

$$\mathbf{w}_2 = \mathbf{u} - \mathbf{w}_1 = \langle 2, 2\rangle - \frac{14}{37}\langle 6, 1\rangle = \left\langle -\frac{10}{37}, \frac{60}{37}\right\rangle = \frac{10}{37}\langle -1, 6\rangle = \frac{1}{37}\langle -10, 60\rangle$$

$$\mathbf{u} = \frac{1}{37}\langle 84, 14\rangle + \frac{1}{37}\langle -10, 60\rangle = \langle 2, 2\rangle$$

59. $\mathbf{u} = \langle 4, 2\rangle, \mathbf{v} = \langle 1, -2\rangle$

$$\mathbf{w}_1 = \text{proj}_{\mathbf{v}}\mathbf{u} = \left(\frac{\mathbf{u}\cdot\mathbf{v}}{\|\mathbf{v}\|^2}\right)\mathbf{v} = 0\langle 1, -2\rangle = \langle 0, 0\rangle$$

$$\mathbf{w}_2 = \mathbf{u} - \mathbf{w}_1 = \langle 4, 2\rangle - \langle 0, 0\rangle = \langle 4, 2\rangle$$

$$\mathbf{u} = \langle 4, 2\rangle + \langle 0, 0\rangle = \langle 4, 2\rangle$$

61. $\text{proj}_{\mathbf{v}}\mathbf{u} = \mathbf{u}$ because $\mathbf{u}$ and $\mathbf{v}$ are parallel.

$$\text{proj}_{\mathbf{v}}\mathbf{u} = \frac{\mathbf{u} \cdot \mathbf{v}}{\|\mathbf{v}\|^2}\mathbf{v} = \frac{3(6) + 2(4)}{\left(\sqrt{6^2 + 4^2}\right)^2}\langle 6, 4\rangle = \frac{1}{2}\langle 6, 4\rangle = \langle 3, 2\rangle = \mathbf{u}$$

63. Because $\mathbf{u}$ and $\mathbf{v}$ are orthogonal, $\mathbf{u} \cdot \mathbf{v} = 0$ and $\text{proj}_{\mathbf{v}}\mathbf{u} = 0$.

$$\text{proj}_{\mathbf{v}}\mathbf{u} = \frac{\mathbf{u} \cdot \mathbf{v}}{\|\mathbf{v}\|^2}\mathbf{v} = 0, \text{ because } \mathbf{u} \cdot \mathbf{v} = 0.$$

65. $\mathbf{u} = \langle 3, 5\rangle$

For $\mathbf{v}$ to be orthogonal to $\mathbf{u}$, $\mathbf{u} \cdot \mathbf{v}$ must equal 0.

Two possibilities: $\langle -5, 3\rangle$ and $\langle 5, -3\rangle$

67. $\mathbf{u} = \frac{1}{2}\mathbf{i} - \frac{2}{3}\mathbf{j}$

For $\mathbf{u}$ and $\mathbf{v}$ to be orthogonal, $\mathbf{u} \cdot \mathbf{v}$ must equal 0.

Two possibilities: $\mathbf{v} = \frac{2}{3}\mathbf{i} + \frac{1}{2}\mathbf{j}$ and $\mathbf{v} = -\frac{2}{3}\mathbf{i} - \frac{1}{2}\mathbf{j}$

69. Work $= \|\text{proj}_{\overrightarrow{PQ}}\mathbf{v}\| \|\overrightarrow{PQ}\|$ where $\overrightarrow{PQ} = \langle 4, 7\rangle$ and $\mathbf{v} = \langle 1, 4\rangle$.

$$\text{proj}_{\overrightarrow{PQ}}\mathbf{v} = \left(\frac{\mathbf{v} \cdot \overrightarrow{PQ}}{\|\overrightarrow{PQ}\|^2}\right)\overrightarrow{PQ} = \left(\frac{32}{65}\right)\langle 4, 7\rangle$$

$$\text{Work} = \|\text{proj}_{\overrightarrow{PQ}}\mathbf{v}\| \|\overrightarrow{PQ}\| = \left(\frac{32\sqrt{65}}{65}\right)\left(\sqrt{65}\right) = 32$$

71. (a) $\mathbf{u} \cdot \mathbf{v} = 1225(12.20) + 2445(8.50)$
$= 35{,}727.5$

The total amount paid to the employees is \$35,727.50.

(b) To increase wages by 2%, use scalar multiplication to multiply 1.02 by $\mathbf{v}$.

73. (a) Force due to gravity:

$\mathbf{F} = -30{,}000\mathbf{j}$

Unit vector along hill:

$\mathbf{v} = (\cos d)\mathbf{i} + (\sin d)\mathbf{j}$

Projection of $\mathbf{F}$ onto $\mathbf{v}$:

$$\mathbf{w}_1 = \text{proj}_{\mathbf{v}}\mathbf{F} = \left(\frac{\mathbf{F} \cdot \mathbf{v}}{\|\mathbf{v}\|^2}\right)\mathbf{v} = (\mathbf{F} \cdot \mathbf{v})\mathbf{v} = -30{,}000 \sin d\mathbf{v}$$

The magnitude of the force is $30{,}000 \sin d$.

(b)

d	0°	1°	2°	3°	4°	5°	6°	7°	8°	9°	10°
Force	0	523.6	1047.0	1570.1	2092.7	2614.7	3135.9	3656.1	4175.2	4693.0	5209.4

(c) Force perpendicular to the hill when $d = 5°$:

$$\text{Force} = \sqrt{(30{,}000)^2 - (2614.7)^2} \approx 29{,}885.8 \text{ pounds}$$

75. Work $= (245)(3) = 735$ newton-meters

77. Work $= (\cos 30°)(45)(20) \approx 779.4$ foot-pounds

79. Work $= (\cos 35°)(15{,}691)(800)$
$\approx 10{,}282{,}651.78$ newton-meters

81. Work $= (\cos \theta)\|\mathbf{F}\| \|\overrightarrow{PQ}\|$
$= (\cos 20°)(25 \text{ pounds})(50 \text{ feet})$
≈ 1174.62 foot-pounds

83. False. Work is represented by a scalar.

85. A dot product is a scalar, not a vector.

$\mathbf{v} \cdot \mathbf{0} = \langle -3, 5\rangle \cdot \langle 0, 0\rangle = (-3)(0) + (5)(0) = 0$

87. $\mathbf{u} \cdot \mathbf{v} = \langle 8, 4\rangle \cdot \langle 2, -k\rangle = 16 - 4k = 0$

$16 - 4k = 0$

$-4k = -16$

$k = 4$

89. $\mathbf{u} \cdot \mathbf{u} = \|\mathbf{u}\|^2$
$= 1^2 = 1$

91. (a) $\text{proj}_{\mathbf{v}}\mathbf{u} = \mathbf{u} \Rightarrow \mathbf{u}$ and $\mathbf{v}$ are parallel.

(b) $\text{proj}_{\mathbf{v}}\mathbf{u} = 0 \Rightarrow \mathbf{u}$ and $\mathbf{v}$ are orthogonal.

93. Let $\mathbf{u} = \langle u_1, u_2 \rangle$ and $\mathbf{v} = \langle v_1, v_2 \rangle$.

$$\mathbf{u} - \mathbf{v} = \langle u_1 - v_1, u_2 - v_2 \rangle$$

$$\begin{aligned}\|\mathbf{u} - \mathbf{v}\|^2 &= (u_1 - v_1)^2 + (u_2 - v_2)^2 \\ &= u_1^2 - 2u_1v_1 + v_1^2 + u_2^2 - 2u_2v_2 + v_2^2 \\ &= u_1^2 + u_2^2 + v_1^2 + v_2^2 - 2u_1v_1 - 2u_2v_2 \\ &= \|\mathbf{u}\|^2 + \|\mathbf{v}\|^2 - 2(u_1v_1 + u_2v_2) \\ &= \|\mathbf{u}\|^2 + \|\mathbf{v}\|^2 - 2\mathbf{u} \cdot \mathbf{v}\end{aligned}$$

Section 6.5 The Complex Plane

1. real

3. absolute value

5. reflections

7. $2 = 2 + 0i$ matches (c)

8. $3i = 0 + 3i$ matches (f)

9. $1 + 2i$ matches (h)

10. $2 + i$ matches (a)

11. $3 - i$ matches (b)

12. $-3 + i$ matches (g)

13. $-2 - i$ matches (e)

14. $-1 - 3i$ matches (d)

15. $|-7i| = \sqrt{0^2 + (-7)^2} = \sqrt{49} = 7$

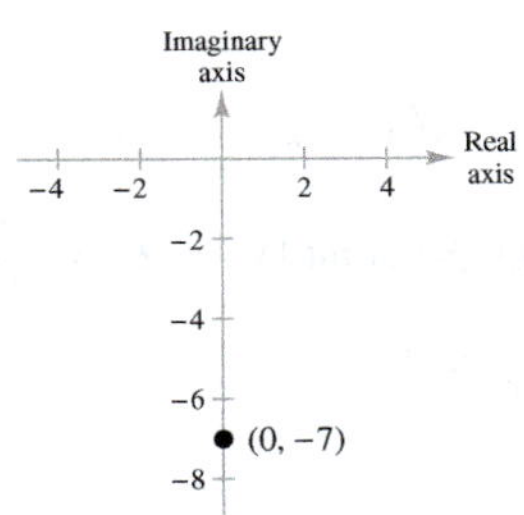

17. $|-6 + 8i| = \sqrt{(-6)^2 + 8^2} = \sqrt{100} = 10$

19. $|4 - 6i| = \sqrt{4^2 + (-6)^2} = \sqrt{52} = 2\sqrt{13}$

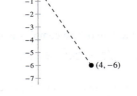

21. $(3 + i) + (2 + 5i) = 5 + 6i$

23. $(8 - 2i) + (2 + 6i) = 10 + 4i$

25. $(5 + 6i) + (1 - i) = 6 + 5i$

27. $(-3 + 4i) + (-2 + 3i) = -5 + 7i$

29. $(4 + 2i) - (6 + 4i) = -2 - 2i$

31. $(5 - i) - (-5 + 2i) = 10 - 3i$

33. $2 - (2 + 6i) = -6i$

35. $-2i - (3 - 5i) = -3 + 3i$

37.

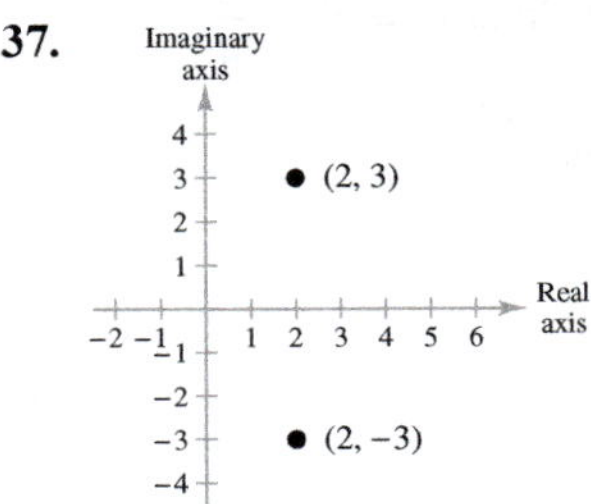

The complex conjugate of $2 + 3i$ is $2 - 3i$.

39.

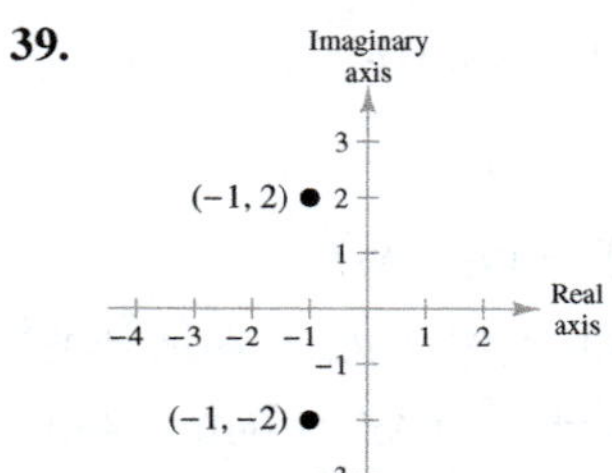

The complex conjugate of $-1 - 2i$ is $-1 + 2i$.

41. $d = \sqrt{(-1 - 1)^2 + (4 - 2)^2} = \sqrt{8} = 2\sqrt{2} \approx 2.83$

43. $d = \sqrt{(3 - 0)^2 + (-4 - 6)^2} = \sqrt{109} \approx 10.44$

45. $\text{Midpoint} = \left(\frac{2 + 6}{2}, \frac{1 + 5}{2}\right) = 4 + 3i = (4, 3)$

47. $\text{Midpoint} = \left(\frac{0 + 9}{2}, \frac{7 - 10}{2}\right)$

$= \frac{9}{2} - \frac{3}{2}i$

$= \left(\frac{9}{2}, -\frac{3}{2}\right)$

49. (a) Ship A: $3 + 4i$

Ship B: $-5 + 2i$

(b) To find the distance between the two ships using complex numbers, you can find the modulus of the difference of the two complex numbers.

$d = \sqrt{(-5 - 3)^2 + (2 - 4)^2}$

$= \sqrt{68}$

≈ 8.25 miles.

51. False. The modulus of a complex number is always real.

53. False. The modulus of the sum of two complex numbers is not equal to the sum of their moduli.

$|1 + i| + |1 - i| = \sqrt{2} + \sqrt{2} = 2\sqrt{2} \neq |(1 + i) + (1 - i)| = |2| = 2$

55. The set of all points with the same modulus represent a circle in the complex plane. The modulus represents the distance from the origin, that is the radius of the circle.

57. If two complex conjugates are plotted in the complex plane, they will form an isosceles triangle because their moduli are equal.

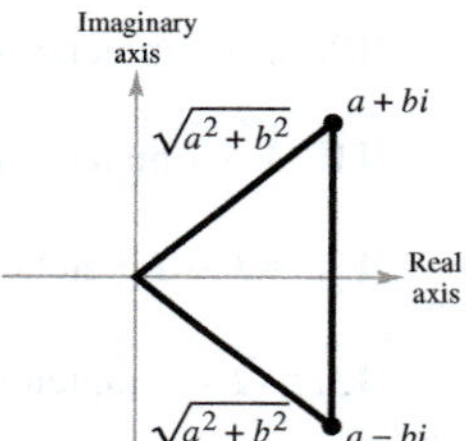

Section 6.6 Trigonometric Form of a Complex Number

1. trigonometric form; modulus; argument

3. nth root

5. $z = 1 + i$

$r = \sqrt{1^2 + 1^2} = \sqrt{2}$

$\tan\theta = 1$, θ is in Quadrant I $\Rightarrow \theta = \frac{\pi}{4}$.

$z = \sqrt{2}\left(\cos\frac{\pi}{4} + i\sin\frac{\pi}{4}\right)$

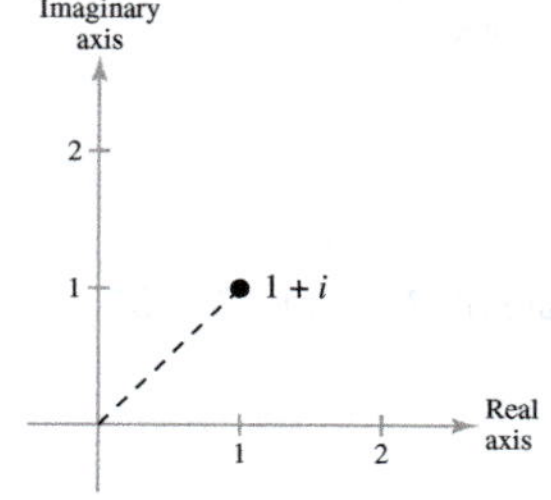

7. $z = 1 - \sqrt{3}i$

$r = \sqrt{1^2 + \left(-\sqrt{3}\right)^2} = \sqrt{4} = 2$

$\tan\theta = -\sqrt{3}$, θ is in Quadrant IV $\Rightarrow \theta = \frac{5\pi}{3}$.

$z = 2\left(\cos\frac{5\pi}{3} + i\sin\frac{5\pi}{3}\right)$

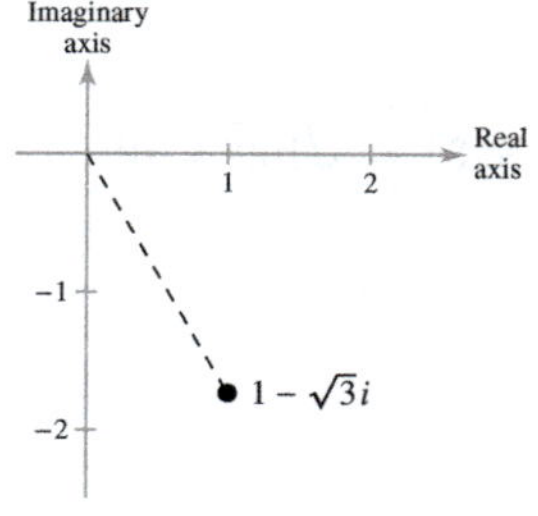

9. $z = -2(1 + \sqrt{3}i)$

$r = \sqrt{(-2)^2 + (-2\sqrt{3})^2} = \sqrt{16} = 4$

$\tan\theta = \dfrac{\sqrt{3}}{1} = \sqrt{3}$, θ is in Quadrant III $\Rightarrow \theta = \dfrac{4\pi}{3}$.

$z = 4\left(\cos\dfrac{4\pi}{3} + i\sin\dfrac{4\pi}{3}\right)$

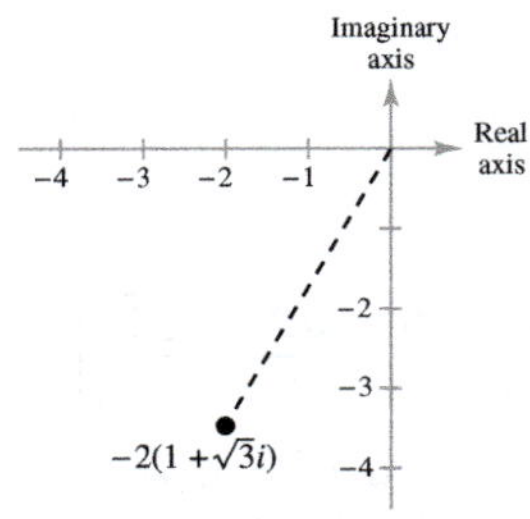

11. $z = -5i$

$r = \sqrt{0^2 + (-5)^2} = \sqrt{25} = 5$

$\tan\theta = \dfrac{-5}{0}$, undefined $\Rightarrow \theta = \dfrac{3\pi}{2}$

$z = 5\left(\cos\dfrac{3\pi}{2} + i\sin\dfrac{3\pi}{2}\right)$

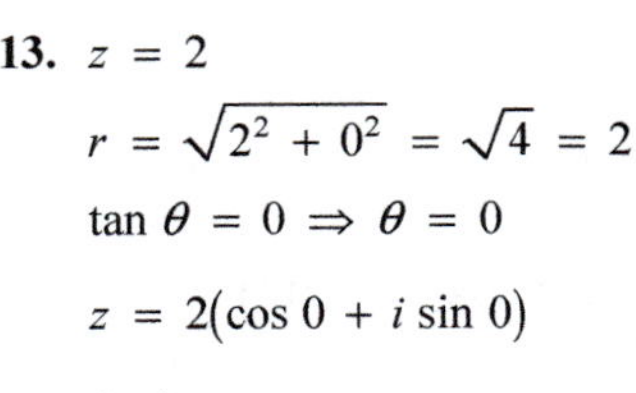

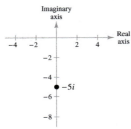

13. $z = 2$

$r = \sqrt{2^2 + 0^2} = \sqrt{4} = 2$

$\tan\theta = 0 \Rightarrow \theta = 0$

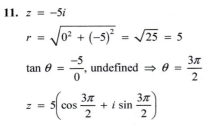

$z = 2(\cos 0 + i\sin 0)$

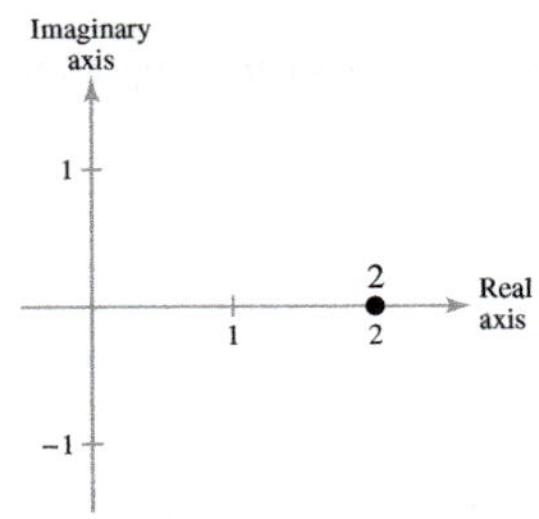

15. $z = -7 + 4i$

$r = \sqrt{(-7)^2 + (4)^2} = \sqrt{65}$

$\tan\theta = \dfrac{4}{-7}$, θ is in Quadrant II $\Rightarrow \theta \approx 2.62$.

$z \approx \sqrt{65}(\cos 2.62 + i\sin 2.62)$

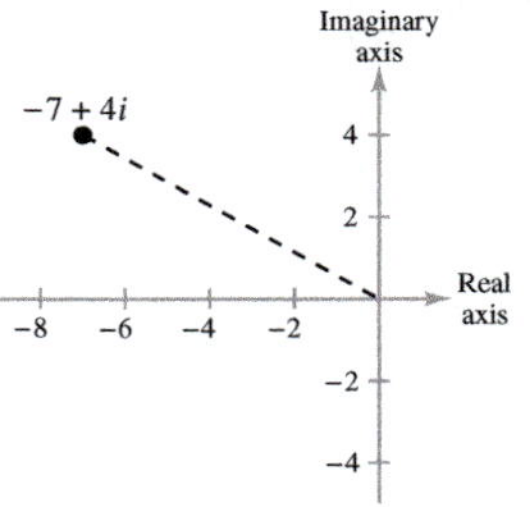

17. $z = 2\sqrt{2} - i$

$r = \sqrt{(2\sqrt{2})^2 + (-1)^2} = \sqrt{9} = 3$

$\tan\theta = \dfrac{-1}{2\sqrt{2}} = -\dfrac{\sqrt{2}}{4} \Rightarrow \theta \approx 5.94$ radians

$z = 3(\cos 5.94 + i\sin 5.94)$

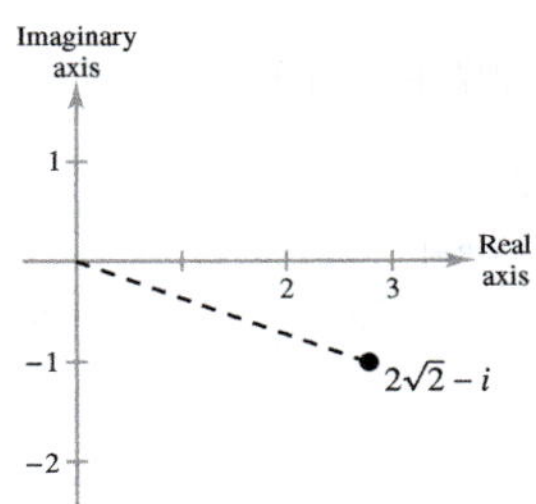

19. $z = 5 + 2i$

$r = \sqrt{5^2 + 2^2} = \sqrt{29}$

$\tan\theta = \frac{2}{5}$

$\theta \approx 0.38$

$z \approx \sqrt{29}(\cos 0.38 + i\sin 0.38)$

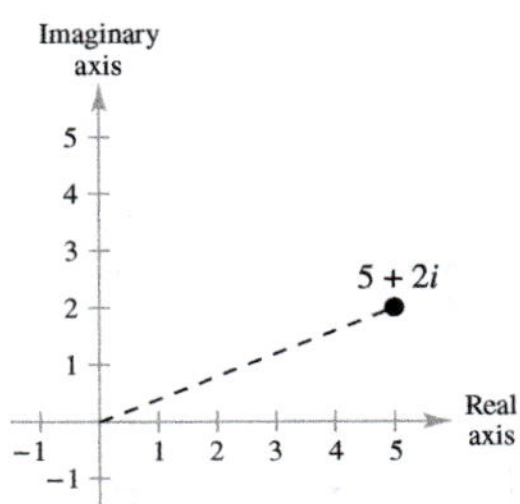

21. $z = 3 + \sqrt{3}i$

$r = \sqrt{(3)^2 + (\sqrt{3})^2} = \sqrt{12} = 2\sqrt{3}$

$\tan\theta = \dfrac{\sqrt{3}}{3} \Rightarrow \theta = \dfrac{\pi}{6}$

$z = 2\sqrt{3}\left(\cos\dfrac{\pi}{6} + i\sin\dfrac{\pi}{6}\right)$

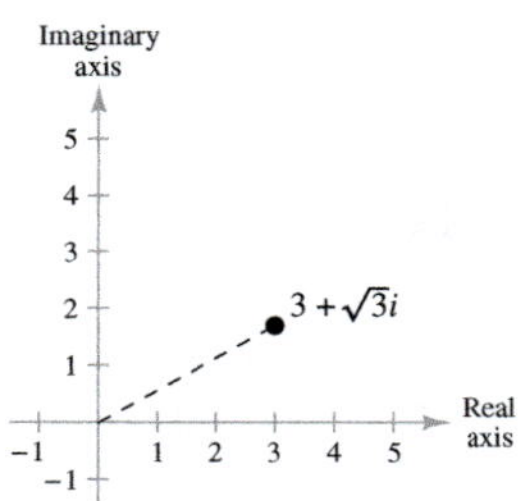

23. $z = -8 - 5\sqrt{3}i$

$r = \sqrt{(-8)^2 + (-5\sqrt{3})^2} = \sqrt{139}$

$\tan\theta = \dfrac{5\sqrt{3}}{8}$

$\theta \approx 3.97$

$z \approx \sqrt{139}(\cos 3.97 + i\sin 3.97)$

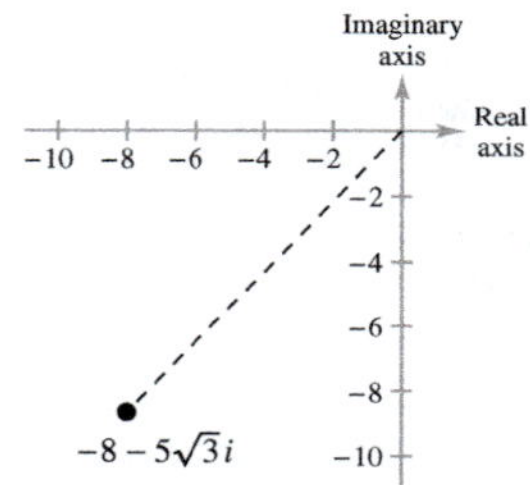

25. $2(\cos 60° + i\sin 60°) = 2\left(\dfrac{1}{2} + \dfrac{\sqrt{3}}{2}i\right) = 1 + \sqrt{3}i$

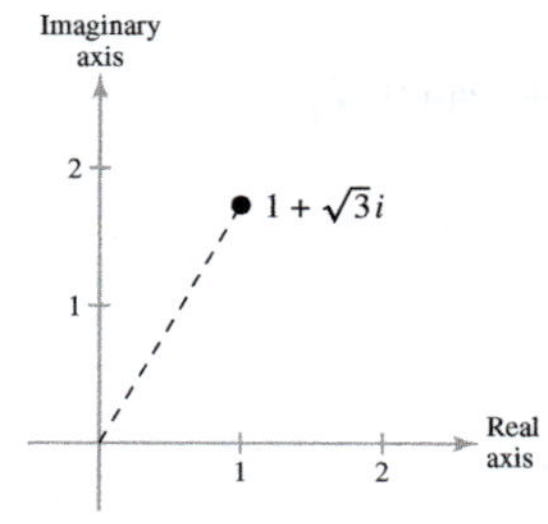

27. $\sqrt{48}\left[\cos(-30°) + i\sin(-30°)\right] = 4\sqrt{3}\left(\dfrac{\sqrt{3}}{2} - \dfrac{1}{2}i\right)$

$= 6 - 2\sqrt{3}i$

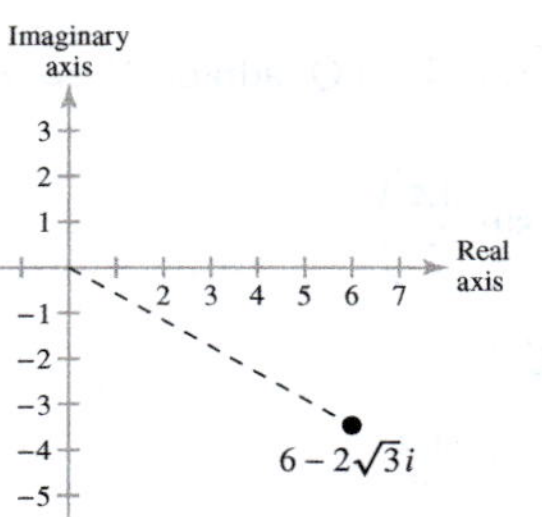

29. $\dfrac{9}{4}\left(\cos\dfrac{3\pi}{4} + i\sin\dfrac{3\pi}{4}\right) = \dfrac{9}{4}\left(-\dfrac{\sqrt{2}}{2} + \dfrac{\sqrt{2}}{2}i\right)$

$= -\dfrac{9\sqrt{2}}{8} + \dfrac{9\sqrt{2}}{8}i$

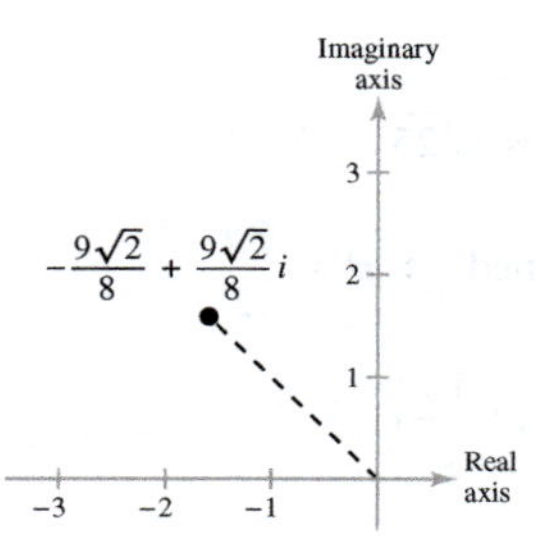

31. $5\left[\cos(198°45') + i\sin(198°45')\right] \approx -4.7347 - 1.6072i$

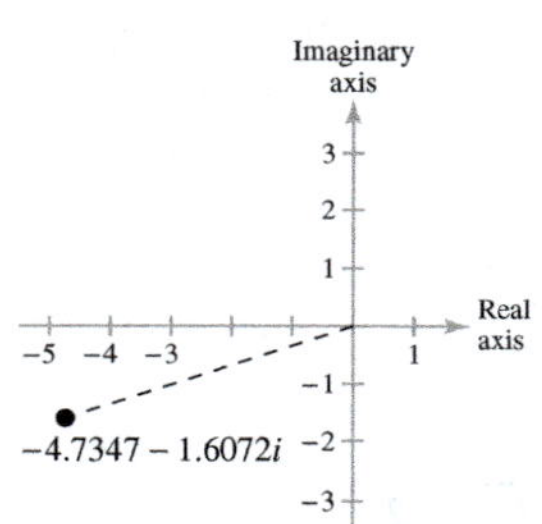

33. $5\left(\cos\dfrac{\pi}{9} + i\sin\dfrac{\pi}{9}\right) \approx 4.6985 + 1.7101i$

35. $2(\cos 155° + i\sin 155°) \approx -1.8126 + 0.8452i$

37. $\left[2\left(\cos\dfrac{\pi}{4} + i\sin\dfrac{\pi}{4}\right)\right]\left[6\left(\cos\dfrac{\pi}{12} + i\sin\dfrac{\pi}{12}\right)\right] = (2)(6)\left[\cos\left(\dfrac{\pi}{4} + \dfrac{\pi}{12}\right) + i\sin\left(\dfrac{\pi}{4} + \dfrac{\pi}{12}\right)\right] = 12\left(\cos\dfrac{\pi}{3} + i\sin\dfrac{\pi}{3}\right)$

39. $\left[\frac{5}{3}(\cos 120° + i\sin 120°)\right]\left[\frac{2}{3}(\cos 30° + i\sin 30°)\right] = \frac{5}{3}\left(\frac{2}{3}\right)\left[\cos(120° + 30°) + i\sin(120° + 30°)\right]$

$= \frac{10}{9}(\cos 150° + i\sin 150°)$

41. $\dfrac{3(\cos 50° + i\sin 50°)}{9(\cos 20° + i\sin 20°)} = \dfrac{1}{3}\left[\cos(50° - 20°) + i\sin(50° - 20°)\right] = \dfrac{1}{3}(\cos 30° + i\sin 30°)$

43. $\dfrac{\cos\pi + i\sin\pi}{\cos(\pi/3) + i\sin(\pi/3)} = \cos\left(\pi - \dfrac{\pi}{3}\right) + i\sin\left(\pi - \dfrac{\pi}{3}\right) = \cos\dfrac{2\pi}{3} + i\sin\dfrac{2\pi}{3}$

45. (a) $2 + 2i = 2\sqrt{2}\left(\cos\dfrac{\pi}{4} + i\sin\dfrac{\pi}{4}\right)$

$1 - i = \sqrt{2}\left[\cos\left(-\dfrac{\pi}{4}\right) + i\sin\left(-\dfrac{\pi}{4}\right)\right] = \sqrt{2}\left(\cos\dfrac{7\pi}{4} + i\sin\dfrac{7\pi}{4}\right)$

(b) $(2 + 2i)(1 - i) = \left[2\sqrt{2}\left(\cos\dfrac{\pi}{4} + i\sin\dfrac{\pi}{4}\right)\right]\left[\sqrt{2}\left(\cos\left(\dfrac{7\pi}{4}\right) + i\sin\left(\dfrac{7\pi}{4}\right)\right)\right] = 4(\cos 2\pi + i\sin 2\pi)$

$= 4(\cos 0 + i\sin 0) = 4$

(c) $(2 + 2i)(1 - i) = 2 - 2i + 2i - 2i^2 = 2 + 2 = 4$

47. (a) $-2i = 2\left[\cos\left(-\dfrac{\pi}{2}\right) + i\sin\left(-\dfrac{\pi}{2}\right)\right] = 2\left(\cos\dfrac{3\pi}{2} + i\sin\dfrac{3\pi}{2}\right)$

$1 + i = \sqrt{2}\left(\cos\dfrac{\pi}{4} + i\sin\dfrac{\pi}{4}\right)$

(b) $-2i(1 + i) = 2\left[\cos\left(\dfrac{3\pi}{2}\right) + i\sin\left(\dfrac{3\pi}{2}\right)\right]\left[\sqrt{2}\left(\cos\dfrac{\pi}{4} + i\sin\dfrac{\pi}{4}\right)\right]$

$= 2\sqrt{2}\left[\cos\left(\dfrac{7\pi}{4}\right) + i\sin\left(\dfrac{7\pi}{4}\right)\right]$

$= 2\sqrt{2}\left[\dfrac{1}{\sqrt{2}} - \dfrac{1}{\sqrt{2}}i\right] = 2 - 2i$

(c) $-2i(1 + i) = -2i - 2i^2 = -2i + 2 = 2 - 2i$

49. (a) $3 + 4i \approx 5(\cos 0.93 + i\sin 0.93)$

$1 - \sqrt{3}i = 2\left(\cos\dfrac{5\pi}{3} + i\sin\dfrac{5\pi}{3}\right)$

(b) $\dfrac{3 + 4i}{1 - \sqrt{3}i} \approx \dfrac{5(\cos 0.93 + i\sin 0.93)}{2\left(\cos\dfrac{5\pi}{3} + i\sin\dfrac{5\pi}{3}\right)} \approx 2.5\left[\cos(-4.31) + i\sin(-4.31)\right] = \dfrac{5}{2}(\cos 1.97 + i\sin 1.97) \approx -0.982 + 2.299i$

(c) $\dfrac{3 + 4i}{1 - \sqrt{3}i} = \dfrac{3 + 4i}{1 - \sqrt{3}i}\cdot\dfrac{1 + \sqrt{3}i}{1 + \sqrt{3}i} = \dfrac{3 + \left(4 + 3\sqrt{3}\right)i + 4\sqrt{3}i^2}{1 + 3} = \dfrac{3 - 4\sqrt{3}}{4} + \dfrac{4 + 3\sqrt{3}}{4}i \approx -0.982 + 2.299i$

51. $\left[2\left(\cos\dfrac{2\pi}{3} + i\sin\dfrac{2\pi}{3}\right)\right]\left[\dfrac{1}{2}\left(\cos\dfrac{\pi}{3} + i\sin\dfrac{\pi}{3}\right)\right] = (2)\left(\dfrac{1}{2}\right)\left[\cos\left(\dfrac{2\pi}{3} + \dfrac{\pi}{3}\right) + i\sin\left(\dfrac{2\pi}{3} + \dfrac{\pi}{3}\right)\right]$

$= (\cos\pi + i\sin\pi)$

$= -1 + 0i$

$= -1$

53. $\left[5(\cos 20° + i \sin 20°)\right]^3 = 5^3(\cos 60° + i \sin 60°)$

$= \dfrac{125}{2} + \dfrac{125\sqrt{3}}{2}i$

55. $\left(\cos \dfrac{\pi}{4} + i \sin \dfrac{\pi}{4}\right)^{12} = \cos \dfrac{12\pi}{4} + i \sin \dfrac{12\pi}{4}$

$= \cos 3\pi + i \sin 3\pi$

$= -1$

57. $\left[5(\cos 3.2 + i \sin 3.2)\right]^4 = 5^4(\cos 12.8 + i \sin 12.8)$

$\approx 608.0 + 144.7i$

59. $\left[3(\cos 15° + i \sin 15°)\right]^4 = 81(\cos 60° + i \sin 60°)$

$= \dfrac{81}{2} + \dfrac{81\sqrt{3}}{2}i$

61. $(1 + i)^5 = \left[\sqrt{2}\left(\cos \dfrac{\pi}{4} + i \sin \dfrac{\pi}{4}\right)\right]^5$

$= \left(\sqrt{2}\right)^5\left(\cos \dfrac{5\pi}{4} + i \sin \dfrac{5\pi}{4}\right)$

$= 4\sqrt{2}\left(-\dfrac{\sqrt{2}}{2} - \dfrac{\sqrt{2}}{2}i\right)$

$= -4 - 4i$

63. $(-1 + i)^6 = \left[\sqrt{2}\left(\cos \dfrac{3\pi}{4} + i \sin \dfrac{3\pi}{4}\right)\right]^6$

$= \left(\sqrt{2}\right)^6\left(\cos \dfrac{18\pi}{4} + i \sin \dfrac{18\pi}{4}\right)$

$= 8\left(\cos \dfrac{9\pi}{2} + i \sin \dfrac{9\pi}{2}\right)$

$= 8(0 + i)$

$= 8i$

65. $2\left(\sqrt{3} + i\right)^{10} = 2\left[2\left(\cos \dfrac{\pi}{6} + i \sin \dfrac{\pi}{6}\right)\right]^{16}$

$= 2\left[2^{10}\left(\cos \dfrac{10\pi}{6} + i \sin \dfrac{10\pi}{6}\right)\right]$

$= 2048\left(\cos \dfrac{5\pi}{3} + i \sin \dfrac{5\pi}{3}\right)$

$= 2048\left(\dfrac{1}{2} - \dfrac{\sqrt{3}}{2}i\right)$

$= 1024 - 1024\sqrt{3}i$

67. $(3 - 2i)^5 \approx \left[3.6056\left[\cos(-0.588) + i \sin(-0.588)\right]\right]^5$

$\approx (3.6056)^5\left[\cos(-2.94) + i \sin(-2.94)\right]$

$\approx -597 - 122i$

69. $z = \dfrac{\sqrt{2}}{2}(1 + i) = \cos 45° + i \sin 45°$

$z^2 = \cos 90° + i \sin 90° = i$

$z^3 = \cos 135° + i \sin 135° = \dfrac{\sqrt{2}}{2}(-1 + i)$

$z^4 = \cos 180° + i \sin 180° = -1$

The absolute value of each is 1, and consecutive powers of z are each $45°$ apart.

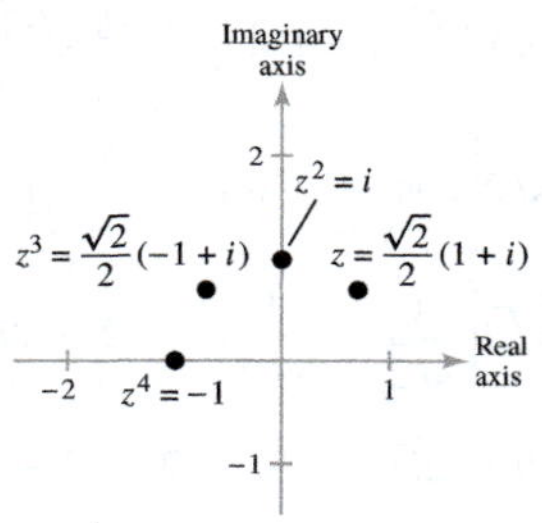

71. (a) Square roots of $5(\cos 120° + i \sin 120°)$:

$\sqrt{5}\left[\cos\left(\dfrac{120° + 360°k}{2}\right) + i \sin\left(\dfrac{120° + 360°k}{2}\right)\right], k = 0, 1$

$k = 0$: $\sqrt{5}(\cos 60° + i \sin 60°)$

$k = 1$: $\sqrt{5}(\cos 240° + i \sin 240°)$

(b) $\dfrac{\sqrt{5}}{2} + \dfrac{\sqrt{15}}{2}i, -\dfrac{\sqrt{5}}{2} - \dfrac{\sqrt{15}}{2}i$

(c)

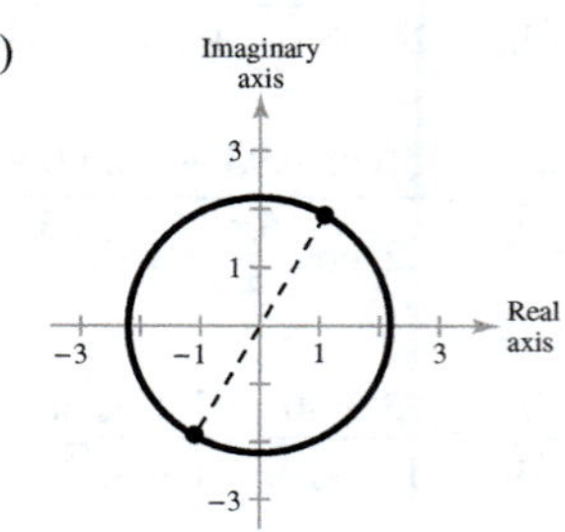

73. (a) Cube roots of $8\left(\cos \dfrac{2\pi}{3} + i \sin \dfrac{2\pi}{3}\right)$:

$$\sqrt[3]{8}\left[\cos\left(\frac{(2\pi/3) + 2\pi k}{3}\right) + i \sin\left(\frac{(2\pi/3) + 2\pi k}{3}\right)\right], k = 0, 1, 2$$

$$k = 0: 2\left(\cos \frac{2\pi}{9} + i \sin \frac{2\pi}{9}\right)$$

$$k = 1: 2\left(\cos \frac{8\pi}{9} + i \sin \frac{8\pi}{9}\right)$$

$$k = 2: 2\left(\cos \frac{14\pi}{9} + i \sin \frac{14\pi}{9}\right)$$

(b) $1.5321 + 1.2856i, -1.8794 + 0.6840i, 0.3473 - 1.9696i$

(c)

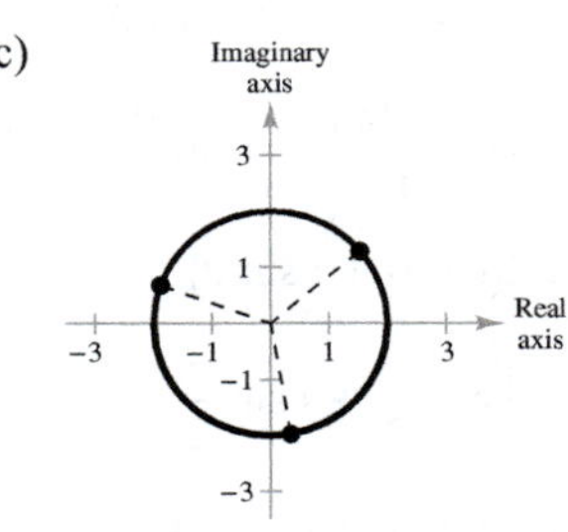

75. (a) Cube roots of $-\dfrac{125}{2}\left(1 + \sqrt{3}i\right) = 125\left(\cos \dfrac{4\pi}{3} + i \sin \dfrac{4\pi}{3}\right)$:

$$\sqrt[3]{125}\left[\cos\left(\frac{\frac{4\pi}{3} + 2k\pi}{3}\right) + i \sin\left(\frac{\frac{4\pi}{3} + 2k\pi}{3}\right)\right], k = 0, 1, 2$$

$$k = 0: 5\left(\cos \frac{4\pi}{9} + i \sin \frac{4\pi}{9}\right)$$

$$k = 1: 5\left(\cos \frac{10\pi}{9} + i \sin \frac{10\pi}{9}\right)$$

$$k = 2: 5\left(\cos \frac{16\pi}{9} + i \sin \frac{16\pi}{9}\right)$$

(b) $0.8682 + 4.9240i, -4.6985 - 1.7101i, 3.8302 - 3.2140i$

(c)

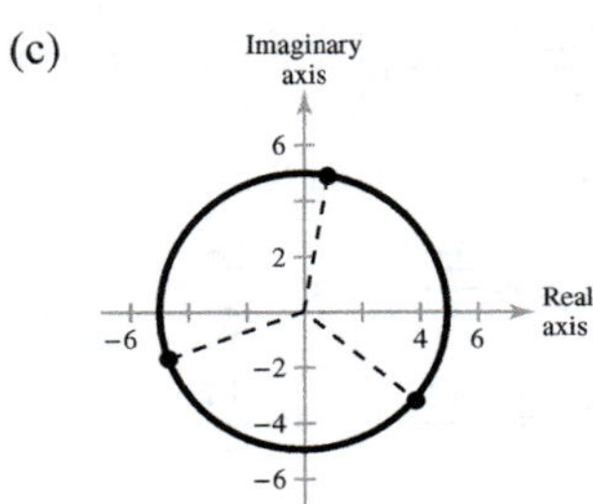

77. (a) Square roots of $-25i = 25\left(\cos \dfrac{3\pi}{2} + i \sin \dfrac{3\pi}{2}\right)$:

$$\sqrt{25}\left[\cos\left(\frac{\frac{3\pi}{2} + 2k\pi}{2}\right) + i \sin\left(\frac{\frac{3\pi}{2} + 2k\pi}{2}\right)\right], k = 0, 1$$

$$k = 0: 5\left(\cos \frac{3\pi}{4} + i \sin \frac{3\pi}{4}\right)$$

$$k = 1: 5\left(\cos \frac{7\pi}{4} + i \sin \frac{7\pi}{4}\right)$$

(b) $-\dfrac{5\sqrt{2}}{2} + \dfrac{5\sqrt{2}}{2}i, \dfrac{5\sqrt{2}}{2} - \dfrac{5\sqrt{2}}{2}i$

(c)

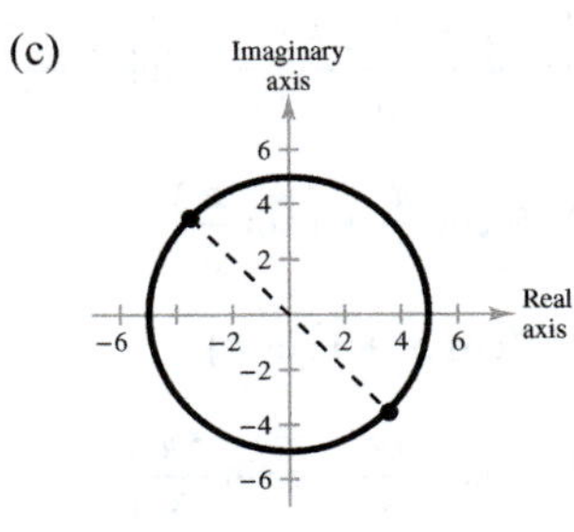

79. (a) Fourth roots of $16 = 16(\cos 0 + i \sin 0)$:

$$\sqrt[4]{16}\left[\cos \frac{0 + 2\pi k}{4} + i \sin \frac{0 + 2\pi k}{4}\right],\ k = 0, 1, 2, 3$$

$k = 0$: $2(\cos 0 + i \sin 0)$

$k = 1$: $2\left(\cos \frac{\pi}{2} + i \sin \frac{\pi}{2}\right)$

$k = 2$: $2(\cos \pi + i \sin \pi)$

$k = 3$: $2\left(\cos \frac{3\pi}{2} + i \sin \frac{3\pi}{2}\right)$

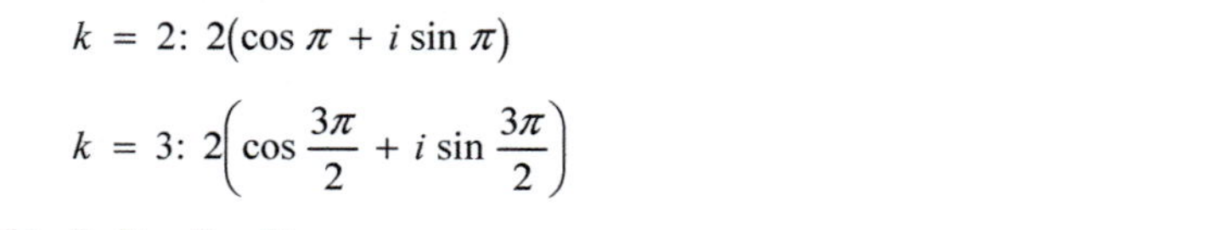

(b) $2, 2i, -2, -2i$

(c)

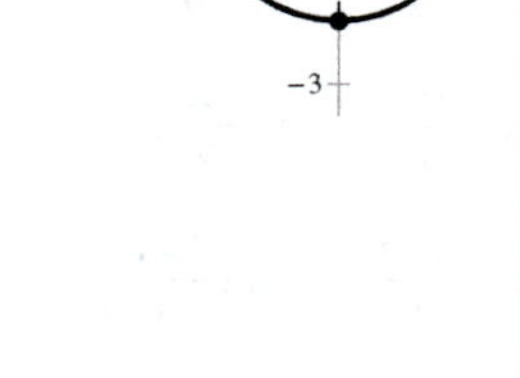

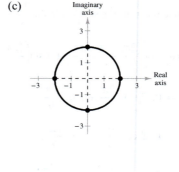

81. (a) Fifth roots of $1 = \cos 0 + i \sin 0$:

$$\cos\left(\frac{2k\pi}{5}\right) + i \sin\left(\frac{2k\pi}{5}\right),\ k = 0, 1, 2, 3, 4$$

$k = 0$: $\cos 0 + i \sin 0$

$k = 1$: $\cos \frac{2\pi}{5} + i \sin \frac{2\pi}{5}$

$k = 2$: $\cos \frac{4\pi}{5} + i \sin \frac{4\pi}{5}$

$k = 3$: $\cos \frac{6\pi}{5} + i \sin \frac{6\pi}{5}$

$k = 4$: $\cos \frac{8\pi}{5} + i \sin \frac{8\pi}{5}$

(b) $1, 0.3090 + 0.9511i, -0.8090 + 0.5878i, -0.8090 - 0.5878i, 0.3090 - 0.9511i$

(c)

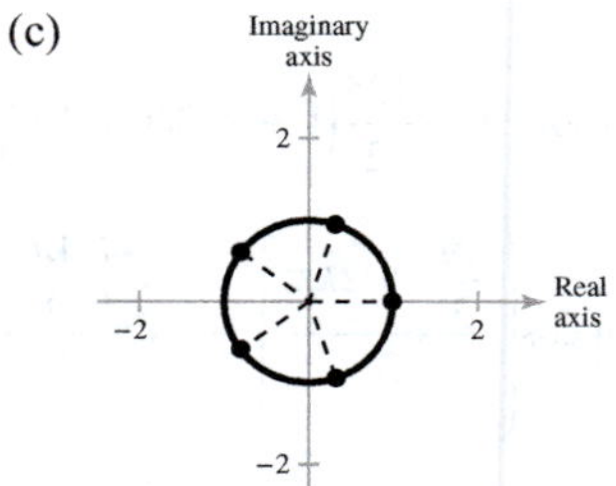

83. (a) Cube roots of $-125 = 125(\cos \pi + i \sin \pi)$:

$$\sqrt[3]{125}\left[\cos\left(\frac{\pi + 2\pi k}{3}\right) + i \sin\left(\frac{\pi + 2\pi k}{3}\right)\right],\ k = 0, 1, 2$$

$k = 0$: $5\left(\cos \frac{\pi}{3} + i \sin \frac{\pi}{3}\right)$

$k = 1$: $5(\cos \pi + i \sin \pi)$

$k = 2$: $5\left(\cos \frac{5\pi}{3} + i \sin \frac{5\pi}{3}\right)$

(b) $\frac{5}{2} + \frac{5\sqrt{3}}{2}i, -5, \frac{5}{2} - \frac{5\sqrt{3}}{2}i$

(c)

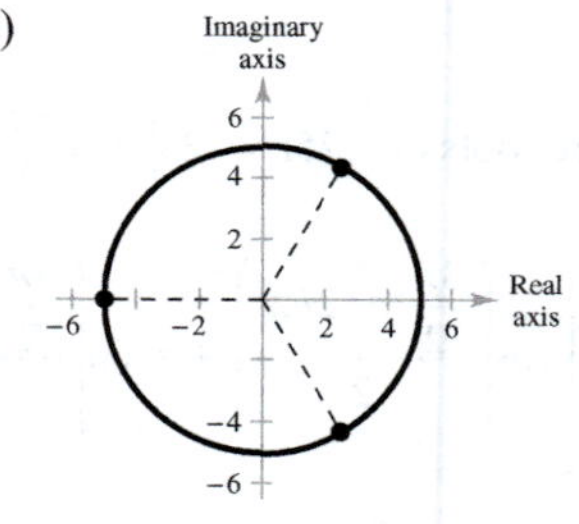

85. (a) Fifth roots of $4(1 - i) = 4\sqrt{2}\left(\cos\dfrac{7\pi}{4} + i\sin\dfrac{7\pi}{4}\right)$:

$$\sqrt[5]{4\sqrt{2}}\left[\cos\left(\frac{\frac{7\pi}{4} + 2\pi k}{5}\right) + i\sin\left(\frac{\frac{7\pi}{4} + 2\pi k}{5}\right)\right], k = 0, 1, 2, 3, 4$$

$k = 0$: $\sqrt{2}\left(\cos\dfrac{7\pi}{20} + i\sin\dfrac{7\pi}{20}\right)$

$k = 1$: $\sqrt{2}\left(\cos\dfrac{3\pi}{4} + i\sin\dfrac{3\pi}{4}\right)$

$k = 2$: $\sqrt{2}\left(\cos\dfrac{23\pi}{20} + i\sin\dfrac{23\pi}{20}\right)$

$k = 3$: $\sqrt{2}\left(\cos\dfrac{31\pi}{20} + i\sin\dfrac{31\pi}{20}\right)$

$k = 4$: $\sqrt{2}\left(\cos\dfrac{39\pi}{20} + i\sin\dfrac{39\pi}{20}\right)$

(b) $0.6420 + 1.2601i, -1 + 1i, -1.2601 - 0.6420i, 0.2212 - 1.3968i, 1.3968 - 0.2212i$

(c)

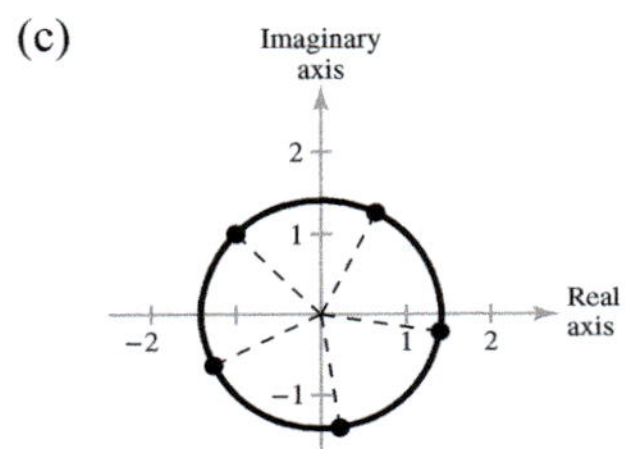

87. $x^4 + i = 0$

$x^4 = -i$

The solutions are the fourth roots of $i = \cos\dfrac{3\pi}{2} + i\sin\dfrac{3\pi}{2}$:

$$\sqrt[4]{1}\left[\cos\left(\frac{\frac{3\pi}{2} + 2k\pi}{4}\right) + i\sin\left(\frac{\frac{3\pi}{2} + 2k\pi}{4}\right)\right], k = 0, 1, 2, 3$$

$k = 0$: $\cos\dfrac{3\pi}{8} + i\sin\dfrac{3\pi}{8} \approx 0.3827 + 0.9239i$

$k = 1$: $\cos\dfrac{7\pi}{8} + i\sin\dfrac{7\pi}{8} \approx -0.9239 + 0.3827i$

$k = 2$: $\cos\dfrac{11\pi}{8} + i\sin\dfrac{11\pi}{8} \approx -0.3827 - 0.9239i$

$k = 3$: $\cos\dfrac{15\pi}{8} + i\sin\dfrac{15\pi}{8} \approx 0.9239 - 0.3827i$

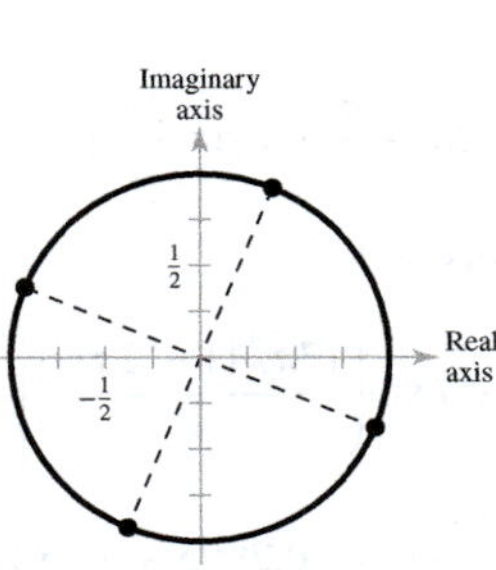

89. $x^5 + 243 = 0$

$$x^5 = -243$$

The solutions are the fifth roots of $-243 = 243(\cos \pi + i \sin \pi)$:

$$\sqrt[5]{243}\left[\cos\left(\frac{\pi + 2k\pi}{5}\right) + i \sin\left(\frac{\pi + 2k\pi}{5}\right)\right], k = 0, 1, 2, 3, 4$$

$$k = 0: 3\left(\cos \frac{\pi}{5} + i \sin \frac{\pi}{5}\right) \approx 2.4271 + 1.7634i$$

$$k = 1: 3\left(\cos \frac{3\pi}{5} + i \sin \frac{3\pi}{5}\right) \approx -0.9271 + 2.8532i$$

$$k = 2: 3(\cos \pi + i \sin \pi) = -3$$

$$k = 3: 3\left(\cos \frac{7\pi}{5} + i \sin \frac{7\pi}{5}\right) \approx -0.9271 - 2.8532i$$

$$k = 4: 3\left(\cos \frac{9\pi}{5} + i \sin \frac{9\pi}{5}\right) \approx 2.4271 - 1.7634i$$

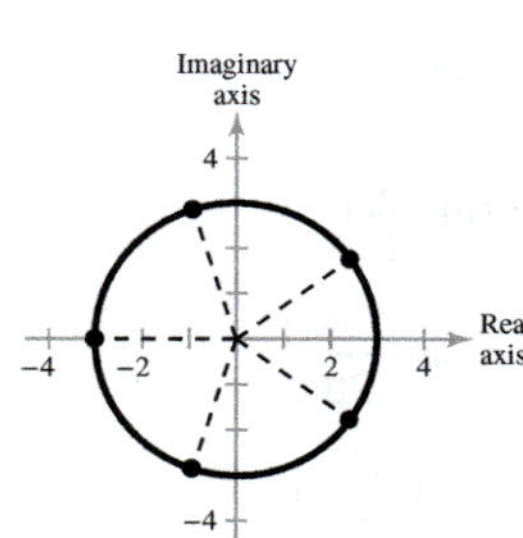

91. $x^4 + 16i = 0$

$$x^4 = -16i$$

The solutions are the fourth roots of $-16i = 16\left(\cos \frac{3\pi}{2} + i \sin \frac{3\pi}{2}\right)$:

$$\sqrt[4]{16}\left[\cos \frac{\frac{3\pi}{2} + 2\pi k}{4} + i \sin \frac{\frac{3\pi}{2} + 2\pi k}{4}\right], k = 0, 1, 2, 3$$

$$k = 0: 2\left(\cos \frac{3\pi}{8} + i \sin \frac{3\pi}{8}\right) \approx 0.7654 + 1.8478i$$

$$k = 1: 2\left(\cos \frac{7\pi}{8} + i \sin \frac{7\pi}{8}\right) \approx -1.8478 + 0.7654i$$

$$k = 2: 2\left(\cos \frac{11\pi}{8} + i \sin \frac{11\pi}{8}\right) \approx -0.7654 - 1.8478i$$

$$k = 3: 2\left(\cos \frac{15\pi}{8} + i \sin \frac{15\pi}{8}\right) \approx 1.8478 - 0.7654i$$

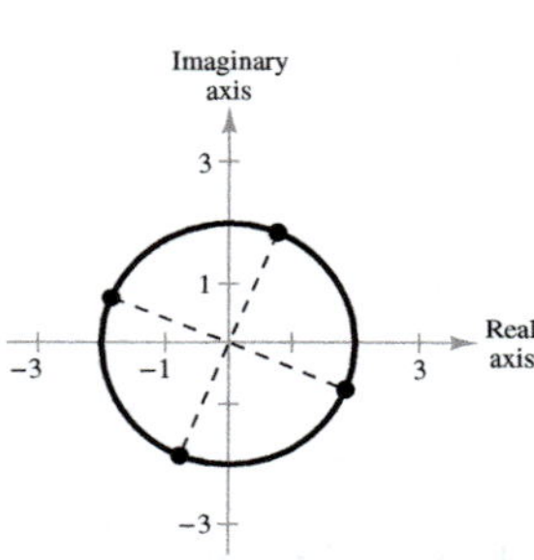

93. $x^3 - (1 - i) = 0$

$$x^3 = 1 - i = \sqrt{2}\left(\cos \frac{7\pi}{4} + i \sin \frac{7\pi}{4}\right)$$

The solutions are the cube roots of $1 - i$:

$$\sqrt[3]{\sqrt{2}}\left[\cos\left(\frac{(7\pi/4) + 2\pi k}{3}\right) + i \sin\left(\frac{(7\pi/4) + 2\pi k}{3}\right)\right], k = 0, 1, 2$$

$$k = 0: \sqrt[6]{2}\left(\cos \frac{7\pi}{12} + i \sin \frac{7\pi}{12}\right) \approx -0.2905 + 1.0842i$$

$$k = 1: \sqrt[6]{2}\left(\cos \frac{5\pi}{4} + i \sin \frac{5\pi}{4}\right) \approx -0.7937 - 0.7937i$$

$$k = 2: \sqrt[6]{2}\left(\cos \frac{23\pi}{12} + i \sin \frac{23\pi}{12}\right) \approx 1.0842 - 0.2905i$$

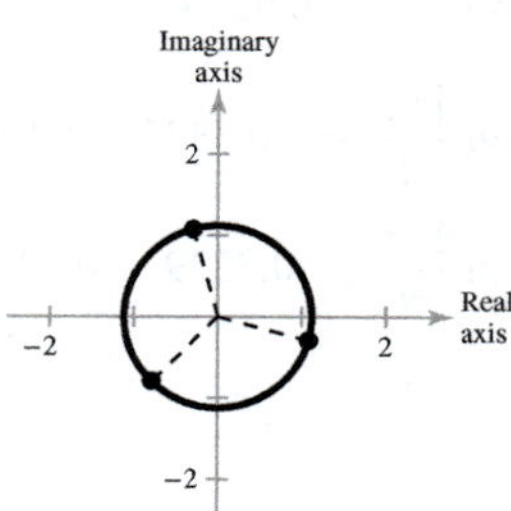

95. (a) $E = IZ$

$= [6(\cos 41° + i \sin 41°)](4[\cos(-11°) + i \sin(-11°)])$

$= 24(\cos 30° + i \sin 30°)$ volts

(b) $E = 24\left(\frac{\sqrt{3}}{2} + \frac{1}{2}i\right) = 12\sqrt{3} + 12i$ volts

(c) $|E| = \sqrt{\left(12\sqrt{3}\right)^2 + (12)^2} = \sqrt{576} = 24$ volts

97. False. They are equally spaced along the circle centered at the origin with radius $\sqrt[n]{r}$.

99. $\frac{z_1}{z_2} = \frac{r_1(\cos\theta_1 + i\sin\theta_1)}{r_2(\cos\theta_2 + i\sin\theta_2)} \cdot \frac{\cos\theta_2 - i\sin\theta_2}{\cos\theta_2 - i\sin\theta_2}$

$= \frac{r_1}{r_2(\cos^2\theta_2 + \sin^2\theta_2)}[\cos\theta_1\cos\theta_2 + \sin\theta_1\sin\theta_2 + i(\sin\theta_1\cos\theta_2 - \sin\theta_2\cos\theta_1)]$

$= \frac{r_1}{r_2}[\cos(\theta_1 - \theta_2) + i\sin(\theta_1 - \theta_2)]$

101. $\bar{z} = r[\cos(-\theta) + i\sin(-\theta)]$

$= r[\cos\theta + -i\sin\theta]$

$= r\cos\theta - ir\sin\theta$

which is the complex conjugate of $r(\cos\theta + i\sin\theta) = r\cos\theta + ir\sin\theta$.

(a) $z\bar{z} = [r(\cos\theta + i\sin\theta)][r(\cos(-\theta) + i\sin(-\theta))]$

$= r^2[\cos(\theta - \theta) + i\sin(\theta - \theta)]$

$= r^2[\cos 0 + i\sin 0]$

$= r^2$

(b) $\frac{z}{\bar{z}} = \frac{r(\cos\theta + i\sin\theta)}{r[\cos(-\theta) + i\sin(-\theta)]}$

$= \frac{r}{r}[\cos(\theta - (-\theta)) + i\sin(\theta - (-\theta))]$

$= \cos 2\theta + i\sin 2\theta$

Review Exercises for Chapter 6

1. Given: $A = 38°, B = 70°, a = 8$

$C = 180° - 38° - 70° = 72°$

$b = \frac{a\sin B}{\sin A} = \frac{8\sin 70°}{\sin 38°} \approx 12.21$

$c = \frac{a\sin C}{\sin A} = \frac{8\sin 72°}{\sin 38°} \approx 12.36$

3. Given: $B = 72°, C = 82°, b = 54$

$A = 180° - 72° - 82° = 26°$

$a = \frac{b\sin A}{\sin B} = \frac{54\sin 26°}{\sin 72°} \approx 24.89$

$c = \frac{b\sin C}{\sin B} = \frac{54\sin 82°}{\sin 72°} \approx 56.23$

5. Given: $A = 16°, B = 98°, c = 8.4$

$C = 180° - 16° - 98° = 66°$

$a = \frac{c\sin A}{\sin C} = \frac{8.4\sin 16°}{\sin 66°} \approx 2.53$

$b = \frac{c\sin B}{\sin C} = \frac{8.4\sin 98°}{\sin 66°} \approx 9.11$

7. Given: $A = 24°, C = 48°, b = 27.5$

$B = 180° - 24° - 48° = 108°$

$a = \frac{b\sin A}{\sin B} = \frac{27.5\sin 24°}{\sin 108°} \approx 11.76$

$c = \frac{b\sin C}{\sin B} = \frac{27.5\sin 48°}{\sin 108°} \approx 21.49$

9. Given: $B = 150°, b = 30, c = 10$

$$\sin C = \frac{c \sin B}{b} = \frac{10 \sin 150°}{30} \approx 0.1667 \Rightarrow C \approx 9.59°$$

$$A \approx 180° - 150° - 9.59° = 20.41°$$

$$a = \frac{b \sin A}{\sin B} = \frac{30 \sin 20.41°}{\sin 150°} \approx 20.92$$

11. $A = 75°, a = 51.2, b = 33.7$

$$\sin B = \frac{b \sin A}{a} = \frac{33.7 \sin 75°}{51.2} \approx 0.6358 \Rightarrow B \approx 39.48°$$

$$C \approx 180° - 75° - 39.48° = 65.52°$$

$$c = \frac{a \sin C}{\sin A} = \frac{51.2 \sin 65.52°}{\sin 75°} \approx 48.24$$

13. $A = 33°, b = 7, c = 10$

$$\text{Area} = \tfrac{1}{2}bc \sin A = \tfrac{1}{2}(7)(10) \sin 33° \approx 19.1$$

15. $C = 119°, a = 18, b = 6$

$$\text{Area} = \tfrac{1}{2}ab \sin C = \tfrac{1}{2}(18)(6) \sin 119° \approx 47.2$$

17. $\tan 17° = \dfrac{h}{x + 50} \Rightarrow h = (x + 50) \tan 17°$

$$h = x \tan 17° + 50 \tan 17°$$

$$\tan 31° = \frac{h}{x} \Rightarrow h = x \tan 31°$$

h
31°
17°
x
50

$$x \tan 17° + 50 \tan 17° = x \tan 31°$$

$$50 \tan 17° = x(\tan 31° - \tan 17°)$$

$$\frac{50 \tan 17°}{\tan 31° - \tan 17°} = x$$

$$x \approx 51.7959$$

$$h = x \tan 31° \approx 51.7959 \tan 31° \approx 31.1 \text{ meters}$$

The height of the building is approximately 31.1 meters.

19. Given: $a = 6, b = 9, c = 14$

$$\cos C = \frac{a^2 + b^2 - c^2}{2ab} = \frac{36 + 81 - 196}{2(6)(9)} \approx -0.7315 \Rightarrow C \approx 137.01°$$

$$\sin B = \frac{b \sin C}{c} \approx \frac{9 \sin 137.01°}{14} \approx 0.4383 \Rightarrow B \approx 26.00°$$

$$A \approx 180° - 26.00° - 137.01° = 16.99°$$

21. Given: $a = 2.5, b = 5.0, c = 4.5$

$$\cos B = \frac{a^2 + c^2 - b^2}{2ac} = 0.0667 \Rightarrow B \approx 86.18°$$

$$\cos C = \frac{a^2 + b^2 - c^2}{2ab} = 0.44 \Rightarrow C \approx 63.90°$$

$$A = 180° - B - C \approx 29.92°$$

23. Given: $B = 108°, a = 11, c = 11$

$$b^2 = a^2 + c^2 - 2ac \cos B = 11^2 + 11^2 - 2(11)(11) \cos 108° \Rightarrow b \approx 17.80$$

$$A = C = \tfrac{1}{2}(180° - 108°) = 36°$$

25. Given: $C = 43°, a = 22.5, b = 31.4$

$$c = \sqrt{a^2 + b^2 - 2ab\cos C} \approx 21.42$$

$$\cos B = \frac{a^2 + c^2 - b^2}{2ac} \approx -0.02169 \Rightarrow B \approx 91.24°$$

$$A = 180° - B - C \approx 45.76°$$

27. Given: $C = 64°, b = 9, c = 13.$

Given two sides and an angle opposite one of them, the Law of Cosines cannot be used, so use the Law of Sines.

$$\sin B = \frac{b\sin C}{c} = \frac{9\sin 64°}{13} \approx 0.62224 \Rightarrow B \approx 38.48°$$

$$A \approx 180° - 38.48° - 64° = 77.52°$$

$$a = \frac{c\sin A}{\sin C} \approx \frac{13\sin 77.52°}{\sin 64°} \approx 14.12$$

29. Given: $a = 13, b = 15, c = 24$

Given three sides, the Law of Cosines can be used.

$$\cos C = \frac{a^2 + b^2 - c^2}{2ab} = \frac{169 + 225 - 576}{2(13)(15)} \approx -0.46667 \Rightarrow C \approx 117.82°$$

$$\sin A = \frac{a\sin C}{c} \approx \frac{13\sin 117.82°}{24} \approx 0.47906 \Rightarrow A \approx 28.62°$$

$$B \approx 180° - 28.62° - 117.82° = 33.56°$$

31.

$$a^2 = 5^2 + 8^2 - 2(5)(8)\cos 28° \approx 18.364$$

$$a \approx 4.3 \text{ feet}$$

$$b^2 = 8^2 + 5^2 - 2(8)(5)\cos 152° \approx 159.636$$

$$b \approx 12.6 \text{ feet}$$

33. $a = 3, b = 6, c = 8$

$$s = \frac{a + b + c}{2} = \frac{3 + 6 + 8}{2} = 8.5$$

$$\text{Area} = \sqrt{s(s - a)(s - b)(s - c)}$$

$$= \sqrt{8.5(5.5)(2.5)(0.5)}$$

$$\approx 7.64$$

35. $a = 12.3, b = 15.8, c = 3.7$

$$s = \frac{a + b + c}{2} = \frac{12.3 + 15.8 + 3.7}{2} = 15.9$$

$$\text{Area} = \sqrt{s(s - a)(s - b)(s - c)}$$

$$= \sqrt{15.9(3.6)(0.1)(12.2)} = 8.36$$

37. $\|\mathbf{u}\| = \sqrt{(4 - (-2))^2 + (6 - 1)^2} = \sqrt{61}$

$\|\mathbf{v}\| = \sqrt{(6 - 0)^2 + (3 - (-2))^2} = \sqrt{61}$

u is directed along a line with a slope of $\frac{6 - 1}{4 - (-2)} = \frac{5}{6}$.

v is directed along a line with a slope of $\frac{3 - (-2)}{6 - 0} = \frac{5}{6}$.

Because **u** and **v** have identical magnitudes and directions, they are equivalent.

39. Initial point: $(0, 10)$

Terminal point: $(7, 3)$

$$\mathbf{v} = \langle 7 - 0, 3 - 10\rangle = \langle 7, -7\rangle$$

$$\|\mathbf{v}\| = \sqrt{7^2 + (-7)^2} = \sqrt{98} = 7\sqrt{2}$$

41. $\mathbf{u} = \langle -1, -3 \rangle, \mathbf{v} = \langle -3, 6 \rangle$

(a) $\mathbf{u} + \mathbf{v} = \langle -1, -3 \rangle + \langle -3, 6 \rangle = \langle -4, 3 \rangle$

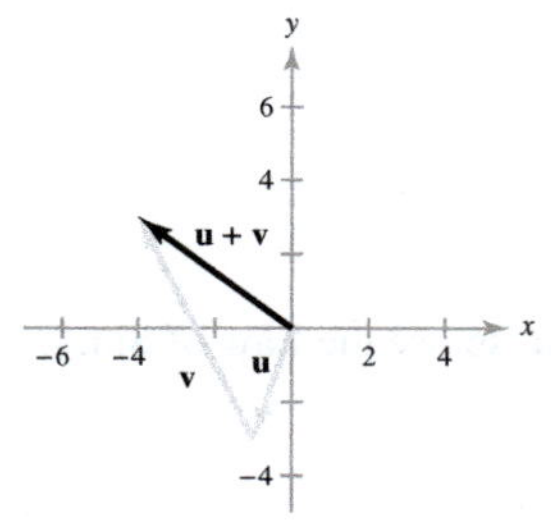

(b) $\mathbf{u} - \mathbf{v} = \langle -1, -3 \rangle - \langle -3, 6 \rangle = \langle 2, -9 \rangle$

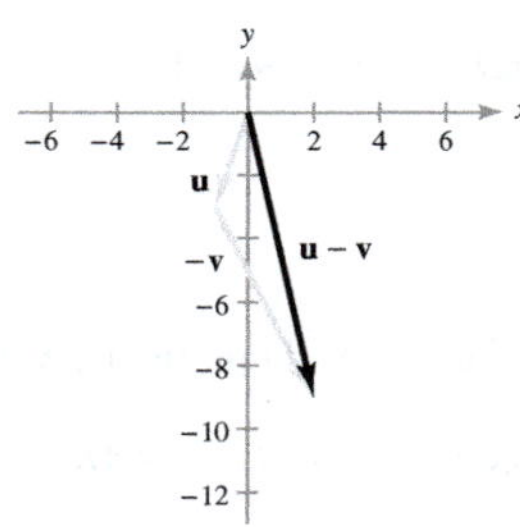

(c) $4\mathbf{u} = 4\langle -1, -3 \rangle = \langle -4, -12 \rangle$

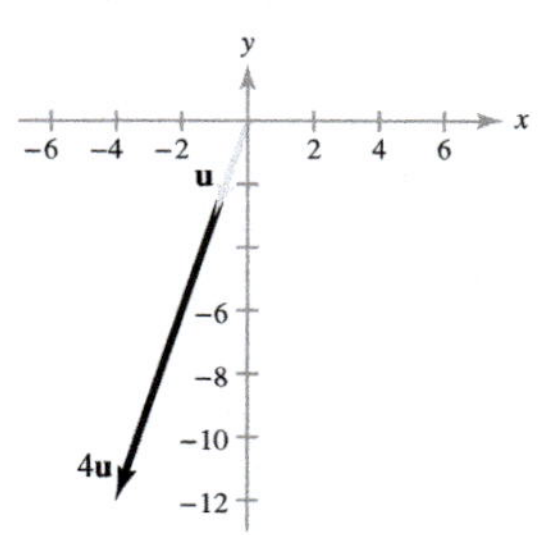

(d) $3\mathbf{v} + 5\mathbf{u} = 3\langle -3, 6 \rangle + 5\langle -1, -3 \rangle = \langle -9, 18 \rangle + \langle -5, -15 \rangle = \langle -14, 3 \rangle$

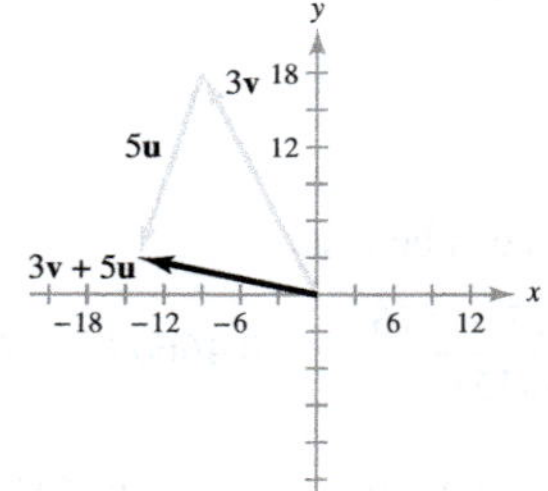

43. $\mathbf{u} = \langle -5, 2 \rangle, \mathbf{v} = \langle 4, 4 \rangle$

(a) $\mathbf{u} + \mathbf{v} = \langle -5, 2 \rangle + \langle 4, 4 \rangle = \langle -1, 6 \rangle$

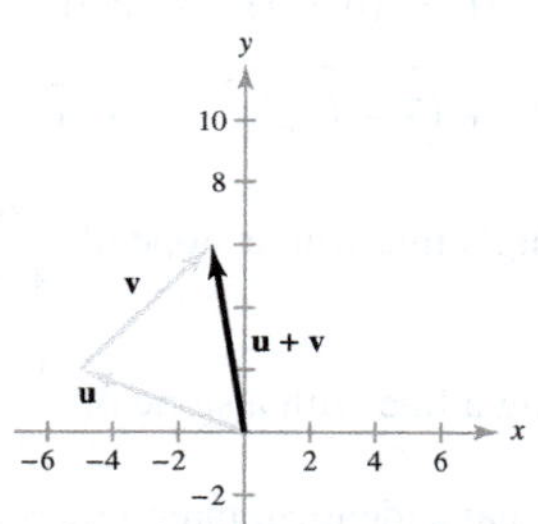

(b) $\mathbf{u} - \mathbf{v} = \langle -5, 2 \rangle - \langle 4, 4 \rangle = \langle -9, -2 \rangle$

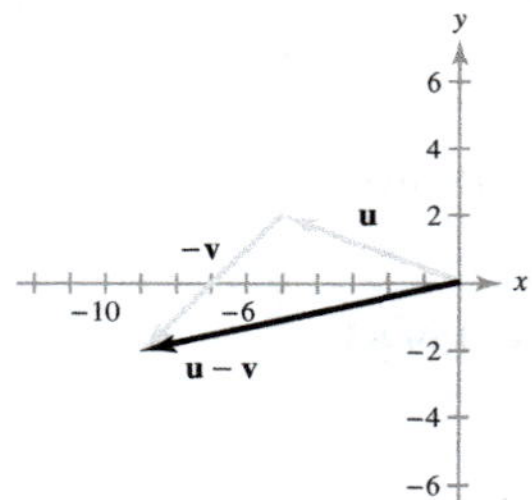

(c) $4\mathbf{u} = 4\langle -5, 2 \rangle = \langle -20, 8 \rangle$

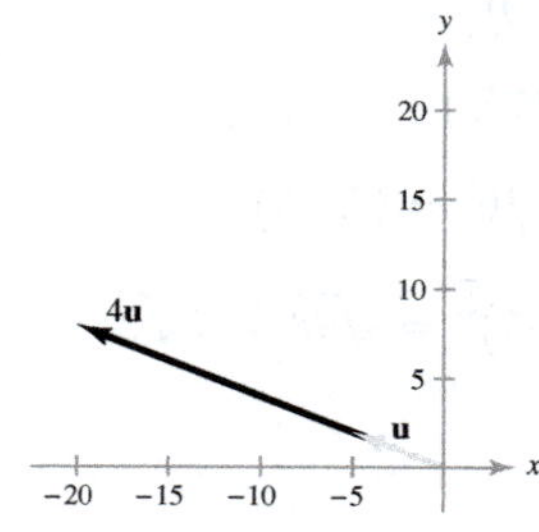

(d) $3\mathbf{v} + 5\mathbf{u} = 3\langle 4, 4 \rangle + 5\langle -5, 2 \rangle = \langle 12, 12 \rangle + \langle -25, 10 \rangle = \langle -13, 22 \rangle$

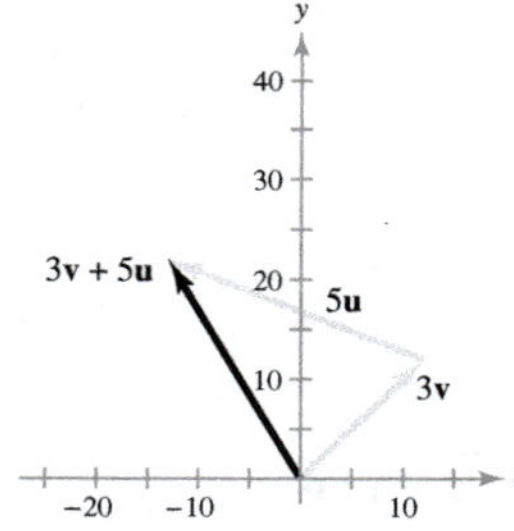

45. $\mathbf{u} = 2\mathbf{i} - \mathbf{j}, \mathbf{v} = 5\mathbf{i} + 3\mathbf{j}$

(a) $\mathbf{u} + \mathbf{v} = (2\mathbf{i} - \mathbf{j}) + (5\mathbf{i} + 3\mathbf{j}) = 7\mathbf{i} + 2\mathbf{j}$

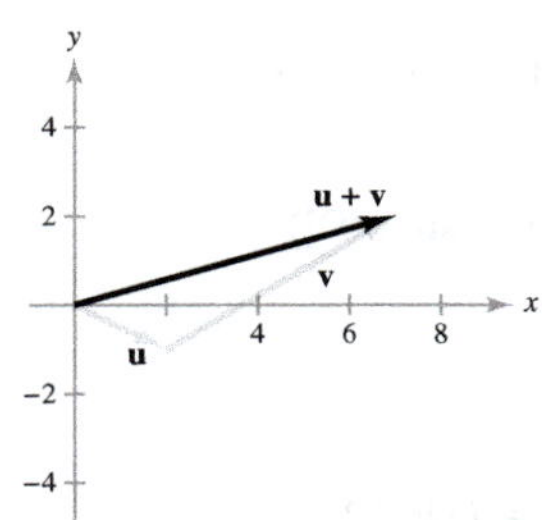

(b) $\mathbf{u} - \mathbf{v} = (2\mathbf{i} - \mathbf{j}) - (5\mathbf{i} + 3\mathbf{j}) = -3\mathbf{i} - 4\mathbf{j}$

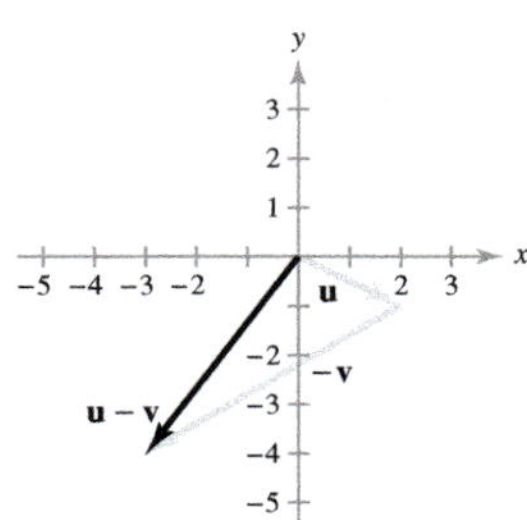

(c) $4\mathbf{u} = 4(2\mathbf{i} - \mathbf{j}) = 8\mathbf{i} - 4\mathbf{j}$

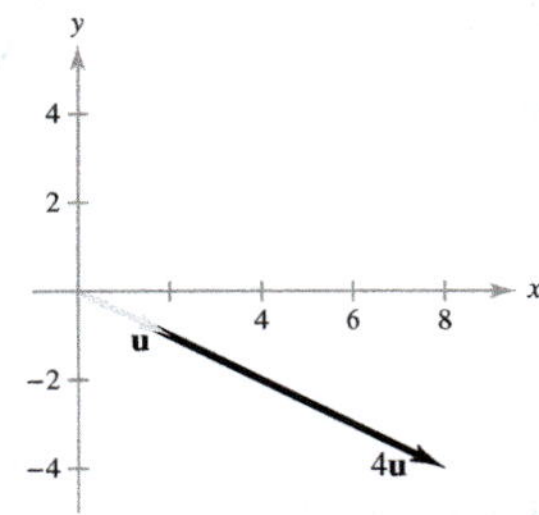

(d) $3\mathbf{v} + 5\mathbf{u} = 3(5\mathbf{i} + 3\mathbf{j}) + 5(2\mathbf{i} - \mathbf{j}) = 15\mathbf{i} + 9\mathbf{j} + 10\mathbf{i} - 5\mathbf{j} = 25\mathbf{i} + 4\mathbf{j}$

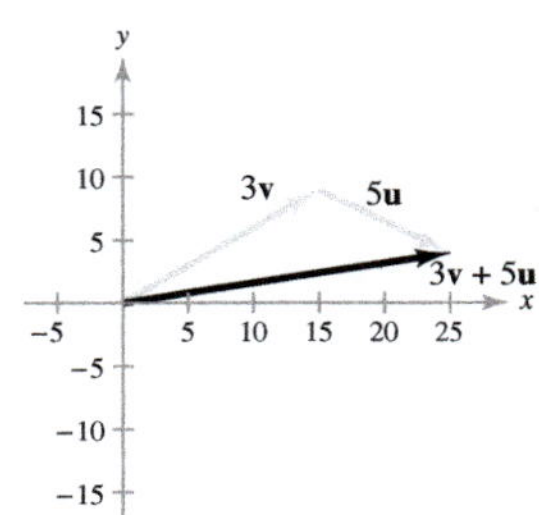

47. $\mathbf{u} = 4\mathbf{i}, \mathbf{v} = -\mathbf{i} + 6\mathbf{j}$

(a) $\mathbf{u} + \mathbf{v} = 4\mathbf{i} + (-\mathbf{i} + 6\mathbf{j}) = 3\mathbf{i} + 6\mathbf{j}$

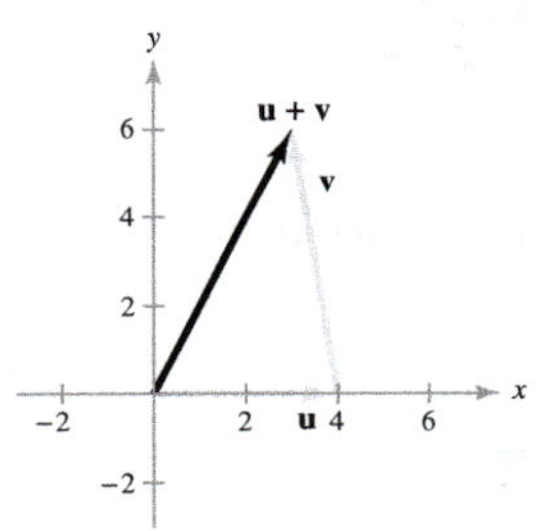

(b) $\mathbf{u} - \mathbf{v} = 4\mathbf{i} - (-\mathbf{i} + 6\mathbf{j}) = 5\mathbf{i} - 6\mathbf{j}$

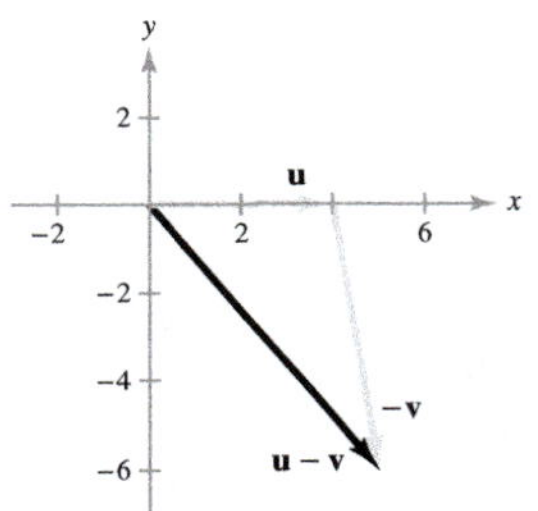

(c) $4\mathbf{u} = 4(4\mathbf{i}) = 16\mathbf{i}$

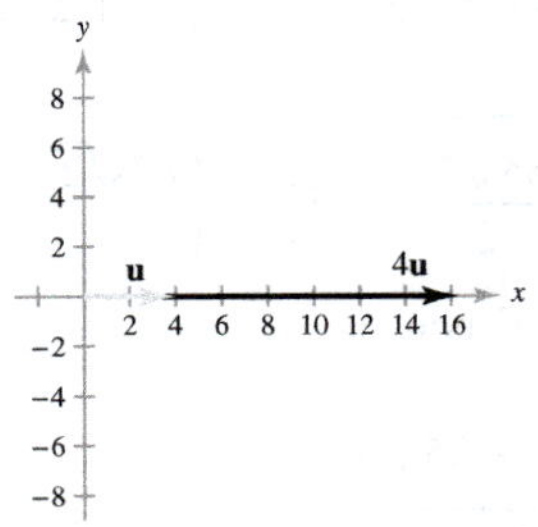

(d) $3\mathbf{v} + 5\mathbf{u} = 3(-\mathbf{i} + 6\mathbf{j}) + 5(4\mathbf{i}) = -3\mathbf{i} + 18\mathbf{j} + 20\mathbf{i} = 17\mathbf{i} + 18\mathbf{j}$

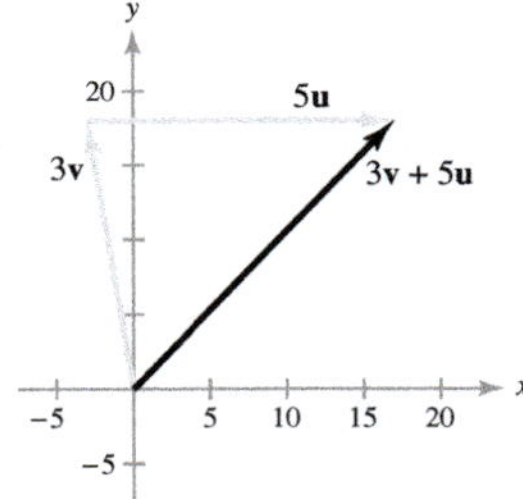

49. $P = (2, 3), Q = (1, 8)$

$\overrightarrow{PQ} = \mathbf{v} = \langle 1 - 2, 8 - 3 \rangle$

$\mathbf{v} = \langle -1, 5 \rangle$

$\mathbf{v} = -\mathbf{i} + 5\mathbf{j}$

51. $P = (3, 4), Q = (9, 8)$

$\overrightarrow{PQ} = \mathbf{v} = \langle 9 - 3, 8 - 4 \rangle$

$\mathbf{v} = \langle 6, 4 \rangle$

$\mathbf{v} = 6\mathbf{i} + 4\mathbf{j}$

53. $\mathbf{v} = 10\mathbf{i} + 3\mathbf{j}$

$$3\mathbf{v} = 3(10\mathbf{i} + 3\mathbf{j})$$
$$= 30\mathbf{i} + 9\mathbf{j}$$
$$= \langle 30, 9\rangle$$

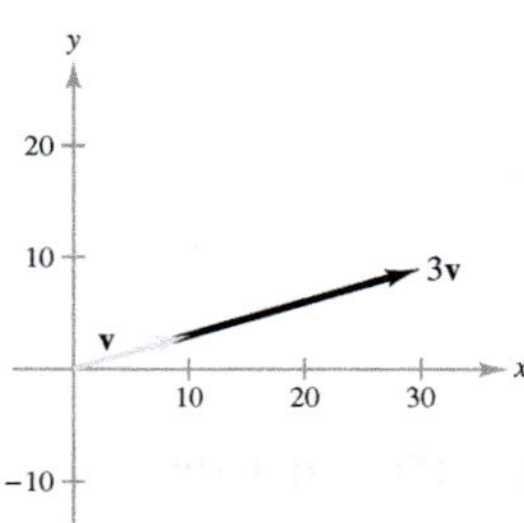

55. $\mathbf{u} = 6\mathbf{i} - 5\mathbf{j}, \mathbf{v} = 10\mathbf{i} + 3\mathbf{j}$

$$2\mathbf{u} + \mathbf{v} = 2(6\mathbf{i} - 5\mathbf{j}) + (10\mathbf{i} + 3\mathbf{j})$$
$$= 22\mathbf{i} - 7\mathbf{j}$$
$$= \langle 22, -7\rangle$$

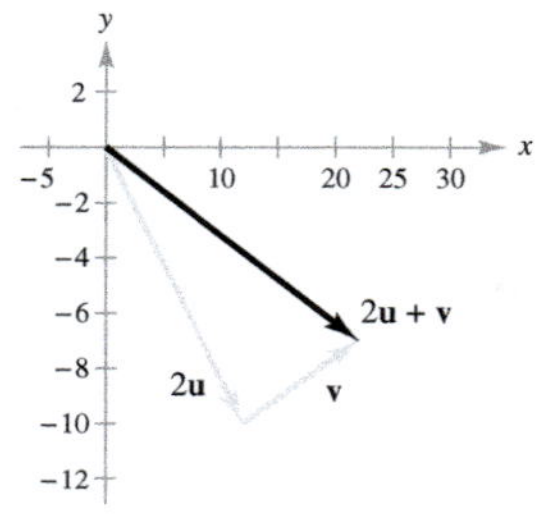

57. $\mathbf{u} = 6\mathbf{i} - 5\mathbf{j}, \mathbf{v} = 10\mathbf{i} + 3\mathbf{j}$

$$5\mathbf{u} - 4\mathbf{v} = 5(6\mathbf{i} - 5\mathbf{j}) - 4(10\mathbf{i} + 3\mathbf{j})$$
$$= 30\mathbf{i} - 25\mathbf{j} - 40\mathbf{i} - 12\mathbf{j}$$
$$= -10\mathbf{i} - 37\mathbf{j}$$
$$= \langle -10, -37\rangle$$

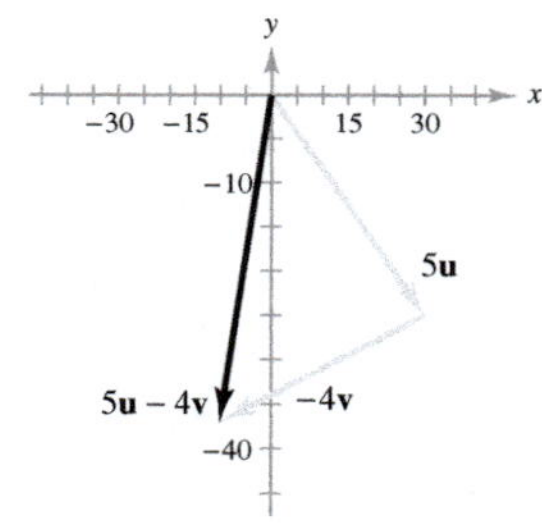

59. $\mathbf{v} = 5\mathbf{i} + 4\mathbf{j}$

$$\|\mathbf{v}\| = \sqrt{5^2 + 4^2} = \sqrt{41}$$

$$\tan\theta = \tfrac{4}{5} \Rightarrow \theta \approx 38.7^\circ$$

61. $\mathbf{v} = -3\mathbf{i} - 3\mathbf{j}$

$$\|\mathbf{v}\| = \sqrt{(-3)^2 + (-3)^2} = 3\sqrt{2}$$

$$\tan\theta = \frac{-3}{-3} = 1 \Rightarrow \theta = 225^\circ$$

63. $\mathbf{v} = 7(\cos 60^\circ\mathbf{i} + \sin 60^\circ\mathbf{j})$

$$\|\mathbf{v}\| = 7$$

$$\theta = 60^\circ$$

65. $\mathbf{v} = 8(\cos 120^\circ + i\sin 120^\circ)$

$$= 8\left(-\frac{1}{2} + \frac{\sqrt{3}}{2}i\right)$$
$$= -4 + 4\sqrt{3}i$$
$$= \langle -4, 4\sqrt{3}\rangle$$

67. Force One:

$$\mathbf{u} = 85(\cos 45^\circ\mathbf{i} + \sin 45^\circ\mathbf{j})$$
$$= 85\left(\frac{\sqrt{2}}{2}\mathbf{i} + \frac{\sqrt{2}}{2}\mathbf{j}\right)$$
$$= \frac{85\sqrt{2}}{2}\mathbf{i} + \frac{85\sqrt{2}}{2}\mathbf{j}$$

Force Two:

$$\mathbf{v} = 50(\cos 60^\circ\mathbf{i} + \sin 60^\circ\mathbf{j})$$
$$= 50\left(\frac{1}{2}\mathbf{i} + \frac{\sqrt{3}}{2}\mathbf{j}\right)$$
$$= 25\mathbf{i} + 25\sqrt{3}\mathbf{j}$$

Resultant Force:

$$\mathbf{u} + \mathbf{v} = \left(\frac{85\sqrt{2}}{2} + 25\right)\mathbf{i} + \left(\frac{85\sqrt{2}}{2} + 25\sqrt{3}\right)\mathbf{j}$$

$$\|\mathbf{u} + \mathbf{v}\| = \sqrt{\left(\frac{85\sqrt{2}}{2} + 25\right)^2 + \left(\frac{85\sqrt{2}}{2} + 25\sqrt{3}\right)^2}$$
$$\approx 133.92 \text{ pounds}$$

$$\tan\theta = \frac{\dfrac{85\sqrt{2}}{2} + 25\sqrt{3}}{\dfrac{85\sqrt{2}}{2} + 25}$$

$$\theta = 50.5^\circ$$

69. $\mathbf{u} = \langle 6, 7\rangle, \mathbf{v} = \langle -3, 9\rangle$

$$\mathbf{u}\cdot\mathbf{v} = 6(-3) + 7(9) = 45$$

71. $\mathbf{u} = 3\mathbf{i} + 7\mathbf{j}, \mathbf{v} = 11\mathbf{i} - 5\mathbf{j}$

$\mathbf{u} \cdot \mathbf{v} = 3(11) + 7(-5) = -2$

73. $\mathbf{u} = \langle -4, 2\rangle$

$2\mathbf{u} = \langle -8, 4\rangle$

$2\mathbf{u} \cdot \mathbf{u} = -8(-4) + 4(2) = 40$

The result is a scalar.

75. $\mathbf{u} = \langle -4, 2\rangle$

$4 - \|\mathbf{u}\| = 4 - \sqrt{(-4)^2 + 2^2} = 4 - \sqrt{20} = 4 - 2\sqrt{5}$

The result is a scalar.

77. $\mathbf{u} = \langle -4, 2\rangle, \mathbf{v} = \langle 5, 1\rangle$

$$\begin{aligned}\mathbf{u}(\mathbf{u} \cdot \mathbf{v}) &= \langle -4, 2\rangle[-4(5) + 2(1)]\\ &= -18\langle -4, 2\rangle\\ &= \langle 72, -36\rangle\end{aligned}$$

The result is a vector.

79. $\mathbf{u} = \langle -4, 2\rangle, \mathbf{v} = \langle 5, 1\rangle$

$$\begin{aligned}(\mathbf{u} \cdot \mathbf{u}) - (\mathbf{u} \cdot \mathbf{v}) &= [-4(-4) + 2(2)] - [-4(5) + 2(1)]\\ &= 20 - (-18)\\ &= 38\end{aligned}$$

The result is a scalar.

81. $\mathbf{u} = \langle 2\sqrt{2}, -4\rangle, \mathbf{v} = \langle -\sqrt{2}, 1\rangle$

$$\cos\theta = \frac{\mathbf{u} \cdot \mathbf{v}}{\|\mathbf{u}\|\|\mathbf{v}\|} = \frac{-8}{(\sqrt{24})(\sqrt{3})} \Rightarrow \theta \approx 160.5^\circ$$

83. $\mathbf{u} = \cos\frac{7\pi}{4}\mathbf{i} + \sin\frac{7\pi}{4}\mathbf{j} = \left\langle \frac{1}{\sqrt{2}}, -\frac{1}{\sqrt{2}}\right\rangle$

$\mathbf{v} = \cos\frac{5\pi}{6}\mathbf{i} + \sin\frac{5\pi}{6}\mathbf{j} = \left\langle -\frac{\sqrt{3}}{2}, \frac{1}{2}\right\rangle$

$$\cos\theta = \frac{\mathbf{u} \cdot \mathbf{v}}{\|\mathbf{u}\|\|\mathbf{v}\|} = \frac{-\sqrt{3} - 1}{2\sqrt{2}} \Rightarrow \theta = 165^\circ$$

85. $\mathbf{u} = \langle -3, 8\rangle$

$\mathbf{v} = \langle 8, 3\rangle$

$\mathbf{u} \cdot \mathbf{v} = -3(8) + 8(3) = 0$

u and **v** are orthogonal.

87. $\mathbf{u} = -\mathbf{i}$

$\mathbf{v} = \mathbf{i} + 2\mathbf{j}$

$\mathbf{u} \cdot \mathbf{v} \neq 0 \Rightarrow$ Not orthogonal

$\mathbf{v} \neq k\mathbf{u} \Rightarrow$ Not parallel

Neither

89. $\mathbf{u} = \langle -4, 3\rangle, \mathbf{v} = \langle -8, -2\rangle$

$$\mathbf{w}_1 = \text{proj}_{\mathbf{v}}\mathbf{u} = \left(\frac{\mathbf{u} \cdot \mathbf{v}}{\|\mathbf{v}\|^2}\right)\mathbf{v} = \left(\frac{26}{68}\right)\langle -8, -2\rangle = -\frac{13}{17}\langle 4, 1\rangle$$

$$\mathbf{w}_2 = \mathbf{u} - \mathbf{w}_1 = \langle -4, 3\rangle - \left(-\frac{13}{17}\right)\langle 4, 1\rangle = \frac{16}{17}\langle -1, 4\rangle$$

$$\mathbf{u} = \mathbf{w}_1 + \mathbf{w}_2 = -\frac{13}{17}\langle 4, 1\rangle + \frac{16}{17}\langle -1, 4\rangle$$

91. $\mathbf{u} = \langle 2, 7\rangle, \mathbf{v} = \langle 1, -1\rangle$

$$\mathbf{w}_1 = \text{proj}_{\mathbf{v}}\mathbf{u} = \left(\frac{\mathbf{u} \cdot \mathbf{v}}{\|\mathbf{v}\|^2}\right)\mathbf{v} = -\frac{5}{2}\langle 1, -1\rangle = \frac{5}{2}\langle -1, 1\rangle$$

$$\mathbf{w}_2 = \mathbf{u} - \mathbf{w}_1 = \langle 2, 7\rangle - \left(\frac{5}{2}\right)\langle -1, 1\rangle = \frac{9}{2}\langle 1, 1\rangle$$

$$\mathbf{u} = \mathbf{w}_1 + \mathbf{w}_2 = \frac{5}{2}\langle -1, 1\rangle + \frac{9}{2}\langle 1, 1\rangle$$

93. $P = (5, 3), Q = (8, 9) \Rightarrow \overrightarrow{PQ} = \langle 3, 6\rangle$

$\text{Work} = \mathbf{v} \cdot \overrightarrow{PQ} = \langle 2, 7\rangle \cdot \langle 3, 6\rangle = 48$

95. $\text{Work} = (18{,}000)\left(\frac{48}{12}\right) = 72{,}000$ foot-pounds

97. $|7i| = \sqrt{0^2 + 7^2} = 7$

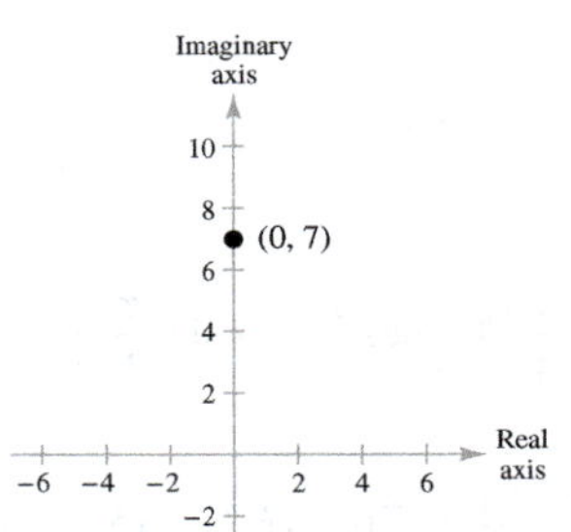

99. $|5 + 3i| = \sqrt{5^2 + 3^2} = \sqrt{34}$

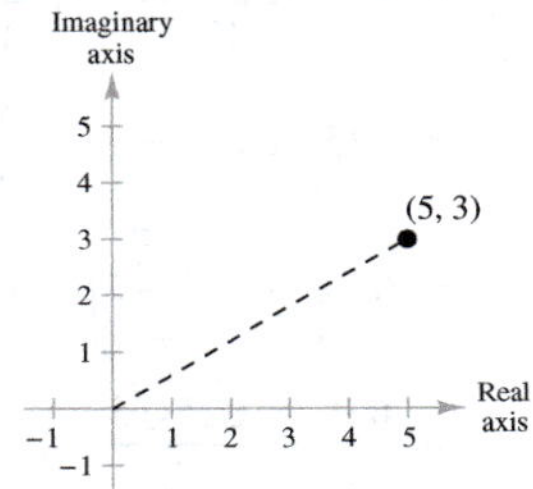

101. $(2 + 3i) + (1 - 2i) = 3 + i$

103. $(1 + 2i) - (3 + i) = -2 + i$

105. The complex conjugate of $3 + i$ is $3 - i$

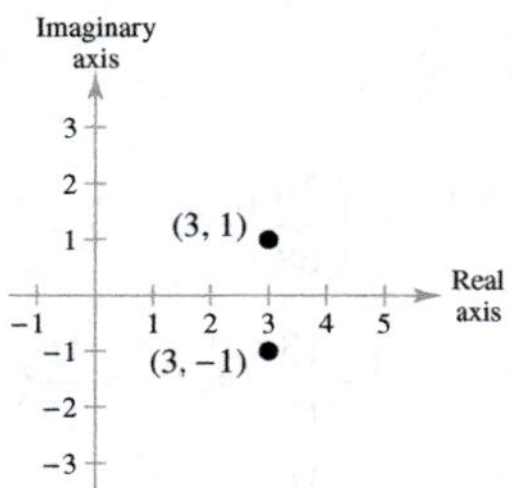

107. $d = \sqrt{(2-3)^2 + (-1-2)^2} = \sqrt{10}$

109. Midpoint $= \left(\frac{1+4}{2}, \frac{1+3}{2}i\right) = \frac{5}{2} + 2i = \left(\frac{5}{2}, 2\right)$

111. $z = 4i$

$r = \sqrt{0^2 + 4^2} = \sqrt{16} = 4$

$\tan\theta = \frac{4}{0}$, undefined $\Rightarrow \theta = \frac{\pi}{2}$

$z = 4\left(\cos\frac{\pi}{2} + i\sin\frac{\pi}{2}\right)$

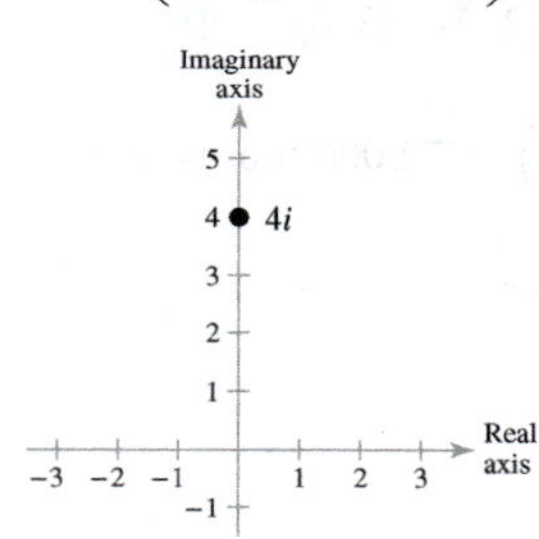

113. $z = 7 - 7i$

$r = \sqrt{(7)^2 + (-7)^2} = \sqrt{98} = 7\sqrt{2}$

$\tan\theta = \frac{-7}{7} = -1 \Rightarrow \theta = \frac{7\pi}{4}$ because the complex number lies in Quadrant IV.

$7 - 7i = 7\sqrt{2}\left(\cos\frac{7\pi}{4} + i\sin\frac{7\pi}{4}\right)$

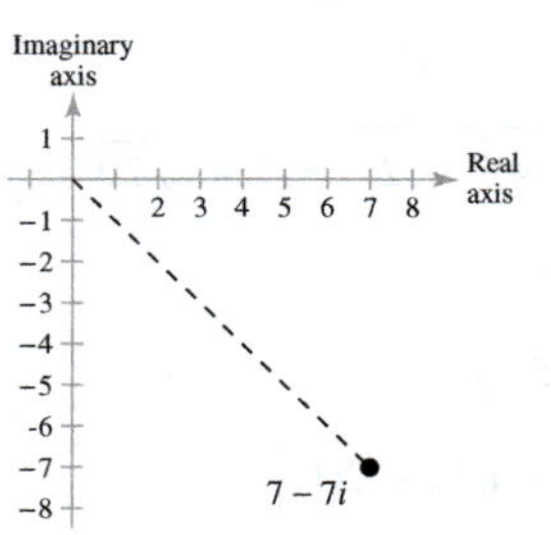

115. $z = -5 - 12i$

$r = \sqrt{(-5)^2 + (-12)^2} = \sqrt{169} = 13$

$\tan\theta = \frac{12}{5}$, θ is in Quadrant III $\Rightarrow \theta \approx 4.32$

$z = 13(\cos 4.32 + i\sin 4.32)$

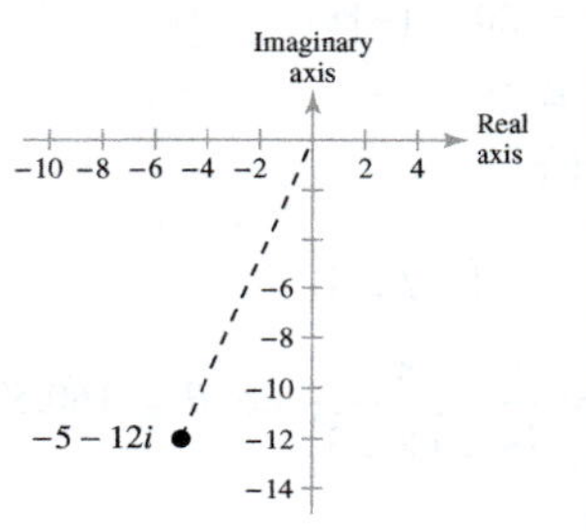

117. $\left[2\left(\cos\frac{\pi}{4} + i\sin\frac{\pi}{4}\right)\right]\left[2\left(\cos\frac{\pi}{3} + i\sin\frac{\pi}{3}\right)\right] = (2)(2)\left[\cos\left(\frac{\pi}{4} + \frac{\pi}{3}\right) + i\sin\left(\frac{\pi}{4} + \frac{\pi}{3}\right)\right] = 4\left(\cos\frac{7\pi}{12} + i\sin\frac{7\pi}{12}\right)$

119. $\frac{2[\cos 60^\circ + i\sin 60^\circ]}{3[\cos 15^\circ + i\sin 15^\circ]} = \frac{2}{3}(\cos(60^\circ - 15^\circ) + i\sin(60^\circ - 15^\circ)) = \frac{2}{3}(\cos 45^\circ + i\sin 45^\circ)$

121. $\left[5\left(\cos\frac{\pi}{12} + i\sin\frac{\pi}{12}\right)\right]^4 = 5^4\left(\cos\frac{4\pi}{12} + i\sin\frac{4\pi}{12}\right)$

$= 625\left(\cos\frac{\pi}{3} + i\sin\frac{\pi}{3}\right)$

$= 625\left(\frac{1}{2} + \frac{\sqrt{3}}{2}i\right)$

$= \frac{625}{2} + \frac{625\sqrt{3}}{2}i$

123. $(2 + 3i)^6 \approx \left[\sqrt{13}(\cos 56.3^\circ + i\sin 56.3^\circ)\right]^6$

$= 13^3(\cos 337.9^\circ + i\sin 337.9^\circ)$

$\approx 13^3(0.9263 - 0.3769i)$

$\approx 2035 - 828i$

125. Sixth roots of $-729i = 729\left(\cos \frac{3\pi}{2} + i \sin \frac{3\pi}{2}\right)$:

(a) $\sqrt[6]{729}\left[\cos\left(\frac{\frac{3\pi}{2} + 2k\pi}{6}\right) + i \sin\left(\frac{\frac{3\pi}{2} + 2k\pi}{6}\right)\right], k = 0, 1, 2, 3, 4, 5$

$k = 0$: $3\left(\cos \frac{\pi}{4} + i \sin \frac{\pi}{4}\right)$

$k = 1$: $3\left(\cos \frac{7\pi}{12} + i \sin \frac{7\pi}{12}\right)$

$k = 2$: $3\left(\cos \frac{11\pi}{12} + i \sin \frac{11\pi}{12}\right)$

$k = 3$: $3\left(\cos \frac{5\pi}{4} + i \sin \frac{5\pi}{4}\right)$

$k = 4$: $3\left(\cos \frac{19\pi}{12} + i \sin \frac{19\pi}{12}\right)$

$k = 5$: $3\left(\cos \frac{23\pi}{12} + i \sin \frac{23\pi}{12}\right)$

(b) $\frac{3\sqrt{2}}{2} + \frac{3\sqrt{2}}{2}i$

$-0.776 + 2.898i$

$-2.898 + 0.776i$

$\frac{-3\sqrt{2}}{2} - \frac{3\sqrt{2}}{2}i$

$0.776 - 2.898i$

$2.898 - 0.776i$

(c)

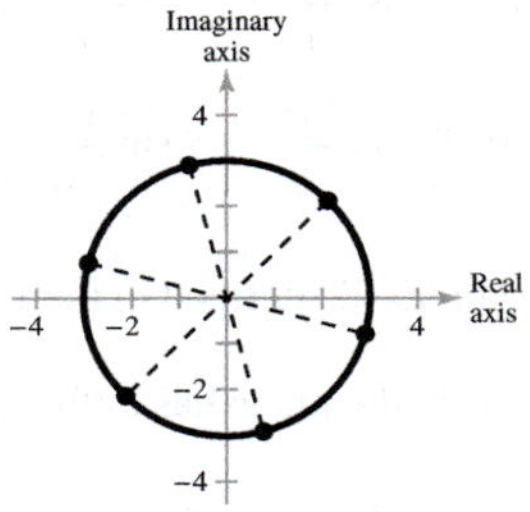
Imaginary axis
Real axis
4
−4
−2
−2
−4
4

127. Cube roots of $8 = 8(\cos 0 + i \sin 0)$, $k = 0, 1, 2$

(a) $\sqrt[3]{8}\left[\cos\left(\frac{0 + 2\pi k}{3}\right) + i \sin\left(\frac{0 + 2\pi k}{3}\right)\right]$

$k = 0$: $2(\cos 0 + i \sin 0)$

$k = 1$: $2\left(\cos \frac{2\pi}{3} + i \sin \frac{2\pi}{3}\right)$

$k = 2$: $2\left(\cos \frac{4\pi}{3} + i \sin \frac{4\pi}{3}\right)$

(b) 2

$-1 + \sqrt{3i}$

$-1 - \sqrt{3i}$

(c)

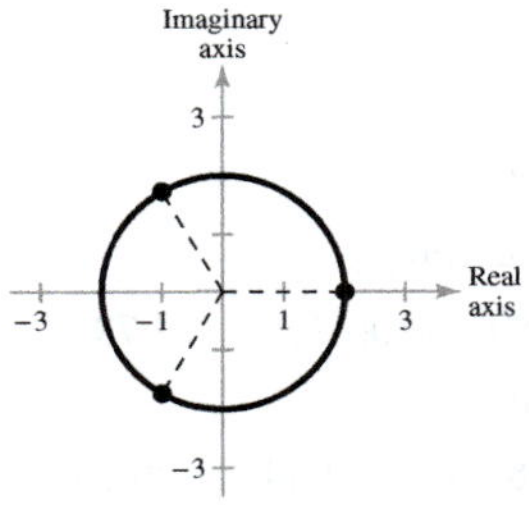
Imaginary axis
Real axis
3
−3
−1
1
3
−3

129. $x^4 + 81 = 0$

$x^4 = -81$ Solve by finding the fourth roots of -81.

$-81 = 81(\cos \pi + i \sin \pi)$

$\sqrt[4]{-81} = \sqrt[4]{81}\left[\cos\left(\frac{\pi + 2\pi k}{4}\right) + i \sin\left(\frac{\pi + 2\pi k}{4}\right)\right], k = 0, 1, 2, 3$

$k = 0$: $3\left(\cos \frac{\pi}{4} + i \sin \frac{\pi}{4}\right) = \frac{3\sqrt{2}}{2} + \frac{3\sqrt{2}}{2}i$

$k = 1$: $3\left(\cos \frac{3\pi}{4} + i \sin \frac{3\pi}{4}\right) = -\frac{3\sqrt{2}}{2} + \frac{3\sqrt{2}}{2}i$

$k = 2$: $3\left(\cos \frac{5\pi}{4} + i \sin \frac{5\pi}{4}\right) = -\frac{3\sqrt{2}}{2} - \frac{3\sqrt{2}}{2}i$

$k = 3$: $3\left(\cos \frac{7\pi}{4} + i \sin \frac{7\pi}{4}\right) = \frac{3\sqrt{2}}{2} - \frac{3\sqrt{2}}{2}i$

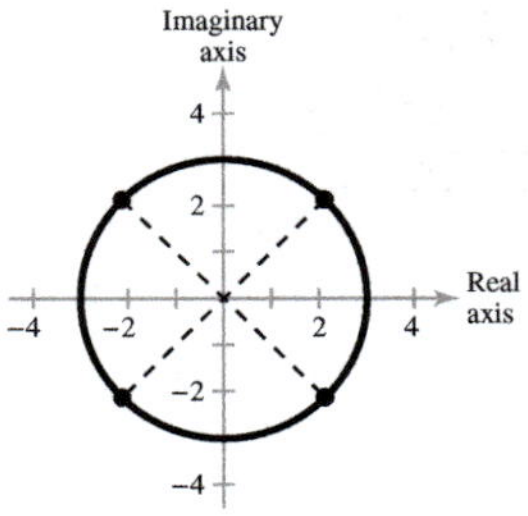
Imaginary axis
Real axis
4
2
−4
−2
2
4
−2
−4

131. $x^3 + 8i = 0$

$x^3 = -8i$ Solve by finding the cube roots of $-8i$.

$$-8i = 8\left(\cos\frac{3\pi}{2} + i\sin\frac{3\pi}{2}\right)$$

$$\sqrt[3]{-8i} = \sqrt[3]{8}\left[\cos\left(\frac{\frac{3\pi}{2} + 2\pi k}{3}\right) + i\sin\left(\frac{\frac{3\pi}{2} + 2\pi k}{3}\right)\right], k = 0, 1, 2$$

$$k = 0: 2\left(\cos\frac{\pi}{2} + i\sin\frac{\pi}{2}\right) = 2i$$

$$k = 1: 2\left(\cos\frac{7\pi}{6} + i\sin\frac{7\pi}{6}\right) = -\sqrt{3} - i$$

$$k = 2: 2\left(\cos\frac{11\pi}{6} + i\sin\frac{11\pi}{6}\right) = \sqrt{3} - i$$

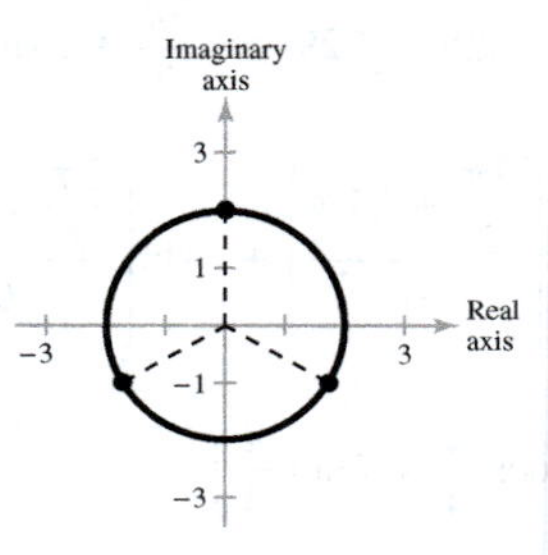

133. True. sin 90° is defined in the Law of Sines.

135. A vector in the plane has both a magnitude and a direction.

Problem Solving for Chapter 6

1. $\left(\overrightarrow{PQ}\right)^2 = 4.7^2 + 6^2 - 2(4.7)(6)\cos 25°A$

$\overrightarrow{PQ} \approx 2.6409$ feet

$\frac{\sin\alpha}{4.7} = \frac{\sin 25°}{2.6409} \Rightarrow \alpha \approx 48.78°$

$\theta + \beta = 180° - 25° - 48.78° = 106.22°$

$(\theta + \beta) + \theta = 180° \Rightarrow \theta = 180° - 106.22° = 73.78°$

$\beta = 106.22° - 73.78° = 32.44°$

$\gamma = 180° - \alpha - \beta = 180° - 48.78° - 32.44° = 98.78°$

$\phi = 180° - \gamma = 180° - 98.78° = 81.22°$

$\frac{\overrightarrow{PT}}{\sin 25°} = \frac{4.7}{\sin 81.22°}$

$\overrightarrow{PT} \approx 2.01$ feet

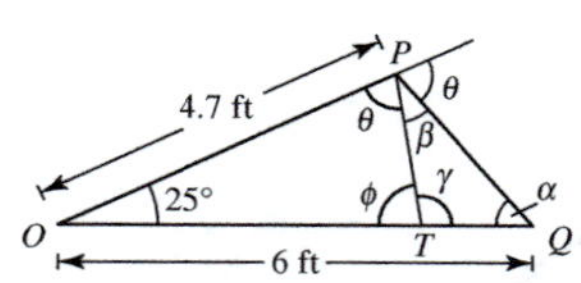

3. (a)

(b) $\frac{x}{\sin 15°} = \frac{75}{\sin 135°}$

$x \approx 27.45$ miles

and

$\frac{y}{\sin 30°} = \frac{75}{\sin 135°}$

$y \approx 53.03$ miles

(c)

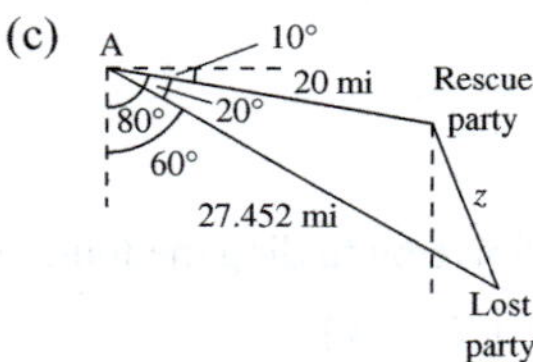

$z^2 = (27.45)^2 + (20)^2 - 2(27.45)(20)\cos 20°$

$z \approx 11.03$ miles

$\frac{\sin\theta}{27.45} = \frac{\sin 20°}{11.03}$

$\sin\theta \approx 0.8511$

$\theta = 180° - \sin^{-1}(0.8511)$

$\theta \approx 121.7°$

To find the bearing, we have $\theta - 10° - 90° \approx 21.7°$.
Bearing: S 21.7° E

5. If $\mathbf{u} \neq 0$, $\mathbf{v} \neq 0$, and $\mathbf{u} + \mathbf{v} \neq 0$, then $\left\|\frac{\mathbf{u}}{\|\mathbf{u}\|}\right\| = \left\|\frac{\mathbf{v}}{\|\mathbf{v}\|}\right\| = \left\|\frac{\mathbf{u} + \mathbf{v}}{\|\mathbf{u} + \mathbf{v}\|}\right\| = 1$ because all of these are magnitudes of unit vectors.

(a) $\mathbf{u} = \langle 1, -1 \rangle$, $\mathbf{v} = \langle -1, 2 \rangle$, $\mathbf{u} + \mathbf{v} = \langle 0, 1 \rangle$

(i) $\|\mathbf{u}\| = \sqrt{2}$ (ii) $\|\mathbf{v}\| = \sqrt{5}$ (iii) $\|\mathbf{u} + \mathbf{v}\| = 1$ (iv) $\left\|\frac{\mathbf{u}}{\|\mathbf{u}\|}\right\| = 1$ (v) $\left\|\frac{\mathbf{v}}{\|\mathbf{v}\|}\right\| = 1$ (vi) $\left\|\frac{\mathbf{u} + \mathbf{v}}{\|\mathbf{u} + \mathbf{v}\|}\right\| = 1$

(b) $\mathbf{u} = \langle 0, 1 \rangle$, $\mathbf{v} = \langle 3, -3 \rangle$, $\mathbf{u} + \mathbf{v} = \langle 3, -2 \rangle$

(i) $\|\mathbf{u}\| = 1$ (ii) $\|\mathbf{v}\| = \sqrt{18} = 3\sqrt{2}$ (iii) $\|\mathbf{u} + \mathbf{v}\| = \sqrt{13}$ (iv) $\left\|\frac{\mathbf{u}}{\|\mathbf{u}\|}\right\| = 1$ (v) $\left\|\frac{\mathbf{v}}{\|\mathbf{v}\|}\right\| = 1$ (vi) $\left\|\frac{\mathbf{u} + \mathbf{v}}{\|\mathbf{u} + \mathbf{v}\|}\right\| = 1$

(c) $\mathbf{u} = \left\langle 1, \frac{1}{2} \right\rangle$, $\mathbf{v} = \langle 2, 3 \rangle$, $\mathbf{u} + \mathbf{v} = \left\langle 3, \frac{7}{2} \right\rangle$

(i) $\|\mathbf{u}\| = \frac{\sqrt{5}}{2}$ (ii) $\|\mathbf{v}\| = \sqrt{13}$ (iii) $\|\mathbf{u} + \mathbf{v}\| = \sqrt{9 + \frac{49}{4}} = \frac{\sqrt{85}}{2}$ (iv) $\left\|\frac{\mathbf{u}}{\|\mathbf{u}\|}\right\| = 1$

(v) $\left\|\frac{\mathbf{v}}{\|\mathbf{v}\|}\right\| = 1$ (vi) $\left\|\frac{\mathbf{u} + \mathbf{v}}{\|\mathbf{u} + \mathbf{v}\|}\right\| = 1$

(d) $\mathbf{u} = \langle 2, -4 \rangle$, $\mathbf{v} = \langle 5, 5 \rangle$, $\mathbf{u} + \mathbf{v} = \langle 7, 1 \rangle$

(i) $\|\mathbf{u}\| = \sqrt{20} = 2\sqrt{5}$ (ii) $\|\mathbf{v}\| = \sqrt{50} = 5\sqrt{2}$ (iii) $\|\mathbf{u} + \mathbf{v}\| = \sqrt{50} = 5\sqrt{2}$ (iv) $\left\|\frac{\mathbf{u}}{\|\mathbf{u}\|}\right\| = 1$

(v) $\left\|\frac{\mathbf{v}}{\|\mathbf{v}\|}\right\| = 1$ (vi) $\left\|\frac{\mathbf{u} + \mathbf{v}}{\|\mathbf{u} + \mathbf{v}\|}\right\| = 1$

7. Let $\mathbf{u} \cdot \mathbf{v} = 0$ and $\mathbf{u} \cdot \mathbf{w} = 0$.

Then, $\mathbf{u} \cdot (c\mathbf{v} + d\mathbf{w}) = \mathbf{u} \cdot c\mathbf{v} + \mathbf{u} \cdot d\mathbf{w} = c(\mathbf{u} \cdot \mathbf{v}) + d(\mathbf{u} \cdot \mathbf{w}) = c(0) + d(0) = 0.$

So for all scalars c and d, $\mathbf{u}$ is orthogonal to $c\mathbf{v} + d\mathbf{w}$.

9. (a) $z_1 = 2(\cos 30° + i \sin 30°)$

$z_2 = 2(\cos 150° + i \sin 150°)$

$z_3 = 2(\cos 270° + i \sin 270°)$

(b) $z_1 = 3(\cos 45° + i \sin 45°)$

$z_2 = 3(\cos 135° + i \sin 135°)$

$z_3 = 2(\cos 225° + i \sin 225°)$

$z_4 = 2(\cos 315° + i \sin 315°)$

11. $\|\mathbf{u} + \mathbf{v}\|$ is larger in figure (a) because the angle between $\mathbf{u}$ and $\mathbf{v}$ is acute rather than obtuse as in figure (b).

As the angle between the two vectors becomes more acute the magnitude becomes greater.

Practice Test for Chapter 6

For Exercises 1 and 2, use the Law of Sines to find the remaining sides and angles of the triangle.

1. $A = 40°, B = 12°, b = 100$

2. $C = 150°, a = 5, c = 20$

3. Find the area of the triangle: $a = 3, b = 6, C = 130°$.

4. Determine the number of solutions to the triangle: $a = 10, b = 35, A = 22.5°$.

For Exercises 5 and 6, use the Law of Cosines to find the remaining sides and angles of the triangle.

5. $a = 49, b = 53, c = 38$

6. $C = 29°, a = 100, c = 300$

7. Use Heron's Formula to find the area of the triangle: $a = 4.1, b = 6.8, c = 5.5$.

8. A ship travels 40 miles due east, then adjusts its course $12°$ southward. After traveling 70 miles in that direction, how far is the ship from its point of departure?

9. $\mathbf{w} = 4\mathbf{u} - 7\mathbf{v}$ where $\mathbf{u} = 3\mathbf{i} + \mathbf{j}$ and $\mathbf{v} = -\mathbf{i} + 2\mathbf{j}$. Find $\mathbf{w}$.

10. Find a unit vector in the direction of $\mathbf{v} = 5\mathbf{i} - 3\mathbf{j}$.

11. Find the dot product and the angle between $\mathbf{u} = 6\mathbf{i} + 5\mathbf{j}$ and $\mathbf{v} = 2\mathbf{i} - 3\mathbf{j}$.

12. $\mathbf{v}$ is a vector of magnitude 4 making an angle of $30°$ with the positive x-axis. Find $\mathbf{v}$ in component form.

13. Find the projection of $\mathbf{u}$ onto $\mathbf{v}$ given $\mathbf{u} = \langle 3, -1 \rangle$ and $\mathbf{v} = \langle -2, 4 \rangle$.

14. Give the trigonometric form of $z = 5 - 5i$.

15. Give the standard form of $z = 6(\cos 225° + i \sin 225°)$.

16. Multiply $[7(\cos 23° + i \sin 23°)][4(\cos 7° + i \sin 7°)]$.

17. Divide $\dfrac{9\left(\cos \frac{5\pi}{4} + i \sin \frac{5\pi}{4}\right)}{3(\cos \pi + i \sin \pi)}$.

18. Find $(2 + 2i)^8$.

19. Find the cube roots of $8\left(\cos \frac{\pi}{3} + i \sin \frac{\pi}{3}\right)$.

20. Find all the solutions to $x^4 + i = 0$.

CHAPTER 7
Systems of Equations and Inequalities

CHAPTER 7
Systems of Equations and Inequalities

Section 7.1 Linear and Nonlinear Systems of Equations

1. solution

3. points; intersection

5. $\begin{cases} 2x - y = 4 \\ 8x + y = -9 \end{cases}$

(a) $(0, -4)$

$8(0) - 4 \neq -9$

$(0, -4)$ *is not* a solution.

(b) $(3, -1)$

$2(3) - (-1) \neq 4$

$(3, -1)$ *is not* a solution.

(c) $\left(\frac{3}{2}, -1\right)$

$8\left(\frac{3}{2}\right) - 1 \neq -9$

$\left(\frac{3}{2}, -1\right)$ *is not* a solution.

(d) $\left(-\frac{1}{2}, -5\right)$

$$2\left(-\tfrac{1}{2}\right) + 5 \stackrel{?}{=} 4$$
$$-1 + 5 = 4$$
$$8\left(-\tfrac{1}{2}\right) - 5 \stackrel{?}{=} -9$$
$$-4 - 5 = -9$$

$\left(-\frac{1}{2}, -5\right)$ *is* a solution.

7. $\begin{cases} 2x + y = 6 & \text{Equation 1} \\ -x + y = 0 & \text{Equation 2} \end{cases}$

Solve for y in Equation 1: $y = 6 - 2x$

Substitute for y in Equation 2: $-x + (6 - 2x) = 0$

Solve for x: $-3x + 6 = 0 \Rightarrow x = 2$

Back-substitute $x = 2$: $y = 6 - 2(2) = 2$

Solution: $(2, 2)$

9. $\begin{cases} x - y = -4 & \text{Equation 1} \\ x^2 - y = -2 & \text{Equation 2} \end{cases}$

Solve for y in Equation 1: $y = x + 4$

Substitute for y in Equation 2:

$x^2 - (x + 4) = -2$

Solve for x:

$x^2 - x - 2 = 0 \Rightarrow (x + 1)(x - 2) = 0 \Rightarrow x = -1, 2$

Back-substitute $x = -1$: $y = -1 + 4 = 3$

Back-substitute $x = 2$: $y = 2 + 4 = 6$

Solutions: $(-1, 3), (2, 6)$

11. $\begin{cases} x^2 + y = 0 & \text{Equation 1} \\ x^2 - 4x - y = 0 & \text{Equation 2} \end{cases}$

Solve for y in Equation 1: $y = -x^2$

Substitute for y in Equation 2: $x^2 - 4x - \left(-x^2\right) = 0$

Solve for x:

$2x^2 - 4x = 0 \Rightarrow 2x(x - 2) = 0 \Rightarrow x = 0, 2$

Back-substitute $x = 0$: $y = -0^2 = 0$

Back-substitute $x = 2$: $y = -2^2 = -4$

Solutions: $(0, 0), (2, -4)$

13. $\begin{cases} y = x^3 - 3x^2 + 1 & \text{Equation 1} \\ y = x^2 - 3x + 1 & \text{Equation 2} \end{cases}$

Substitute for y in Equation 2:

$$x^3 - 3x^2 + 1 = x^2 - 3x + 1$$
$$x^3 - 4x^2 + 3x = 0$$
$$x(x - 1)(x - 3) = 0 \Rightarrow x = 0, 1, 3$$

Back-substitute $x = 0$: $y = 0^3 - 3(0)^2 + 1 = 1$

Back-substitute $x = 1$: $y = 1^3 - 3(1)^2 + 1 = -1$

Back-substitute $x = 3$: $y = 3^3 - 3(3)^2 + 1 = 1$

Solutions: $(0, 1), (1, -1), (3, 1)$

15. $\begin{cases} x - y = 2 & \text{Equation 1} \\ 6x - 5y = 16 & \text{Equation 2} \end{cases}$

Solve for x in Equation 1: $x = y + 2$

Substitute for x in Equation 2:

$6(y + 2) - 5y = 16 \Rightarrow 6y + 12 - 5y = 16 \Rightarrow y = 4$

Back-substitute $y = 4$: $x - 4 = 2 \Rightarrow x = 6$

Solution: $(6, 4)$

17. $\begin{cases} 2x - y + 2 = 0 & \text{Equation 1} \\ 4x + y - 5 = 0 & \text{Equation 2} \end{cases}$

Solve for y in Equation 1: $y = 2x + 2$

Substitute for y in Equation 2: $4x + (2x + 2) - 5 = 0$

Solve for x: $6x - 3 = 0 \Rightarrow x = \frac{1}{2}$

Back-substitute $x = \frac{1}{2}$: $y = 2x + 2 = 2\left(\frac{1}{2}\right) + 2 = 3$

Solution: $\left(\frac{1}{2}, 3\right)$

19. $\begin{cases} 1.5x + 0.8y = 2.3 & \text{Equation 1} \\ 0.3x - 0.2y = 0.1 & \text{Equation 2} \end{cases}$

Multiply the equations by 10.

$15x + 8y = 23$ Revised Equation 1

$3x - 2y = 1$ Revised Equation 2

Solve for y in revised Equation 2: $y = \frac{3}{2}x - \frac{1}{2}$

Substitute for y in revised Equation 1:

$15x + 8\left(\frac{3}{2}x - \frac{1}{2}\right) = 23$

Solve for x:

$15x + 12x - 4 = 23 \Rightarrow 27x = 27 \Rightarrow x = 1$

Back-substitute $x = 1$: $y = \frac{3}{2}(1) - \frac{1}{2} = 1$

Solution: $(1, 1)$

21. $\begin{cases} \frac{1}{5}x + \frac{1}{2}y = 8 & \text{Equation 1} \\ x + y = 20 & \text{Equation 2} \end{cases}$

Solve for x in Equation 2: $x = 20 - y$

Substitute for x in Equation 1: $\frac{1}{5}(20 - y) + \frac{1}{2}y = 8$

Solve for y: $4 + \frac{3}{10}y = 8 \Rightarrow y = \frac{40}{3}$

Back-substitute $y = \frac{40}{3}$: $x = 20 - y = 20 - \frac{40}{3} = \frac{20}{3}$

Solution: $\left(\frac{20}{3}, \frac{40}{3}\right)$

23. $\begin{cases} 6x + 5y = -3 & \text{Equation 1} \\ -x - \frac{5}{6}y = -7 & \text{Equation 2} \end{cases}$

Solve for x in Equation 2: $x = 7 - \frac{5}{6}y$

Substitute for x in Equation 1: $6\left(7 - \frac{5}{6}y\right) + 5y = -3$

Solve for y: $42 - 5y + 5y = -3 \Rightarrow 42 = -3$ (False)

No solution

25. $\begin{cases} x + y = 12{,}000 \\ 0.02x + 0.06y = 500 \end{cases}$

Solve for y in Equation 1: $y = 12{,}000 - x$

Substitute for y in Equation 2:

$0.02x + 0.06(12{,}000 - x) = 500$

Solve for x: $0.02x + 720 - 0.06x = 500$

$-0.04x = -220$

$x = 5500$

Back-substitute $x = 5500$:

$y = 12{,}000 - 5500 = 6500$

So, \$5500 is invested at 2% and \$6500 is invested at 6%.

27. $\begin{cases} x + y = 12{,}000 \\ 0.028x + 0.038y = 396 \end{cases}$

Solve for y in Equation 1: $y = 12{,}000 - x$

Substitute for y in Equation 2:

$0.028x + 0.038(12{,}000 - x) = 396$

Solve for x: $0.028x + 456 - 0.038x = 396$

$-0.01x = -60$

$x = 6000$

Back-substitute $x = 6000$:

$y = 12{,}000 - 6000 = 6000$

So, \$6000 is invested at 2.8% and \$6000 is invested at 3.8%.

29. $\begin{cases} x^2 - y = 0 & \text{Equation 1} \\ 2x + y = 0 & \text{Equation 2} \end{cases}$

Solve for y in Equation 2: $y = -2x$

Substitute for y in Equation 1: $x^2 - (-2x) = 0$

Solve for x:

$x^2 + 2x = 0 \Rightarrow x(x + 2) = 0 \Rightarrow x = 0, -2$

Back-substitute $x = 0$: $y = -2(0) = 0$

Back-substitute $x = -2$: $y = -2(-2) = 4$

Solutions: $(0, 0), (-2, 4)$

31. $\begin{cases} x - y = -1 & \text{Equation 1} \\ x^2 - y = -4 & \text{Equation 2} \end{cases}$

Solve for y in Equation 1: $y = x + 1$

Substitute for y in Equation 2: $x^2 - (x + 1) = -4$

Solve for x: $x^2 - x - 1 = -4 \Rightarrow x^2 - x + 3 = 0$

The Quadratic Formula yields no real solutions.

33. 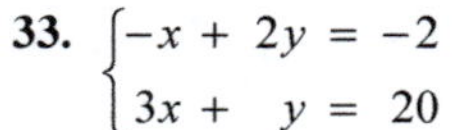

$\begin{cases} -x + 2y = -2 \\ 3x + y = 20 \end{cases}$

Point of intersection:
$(6, 2)$

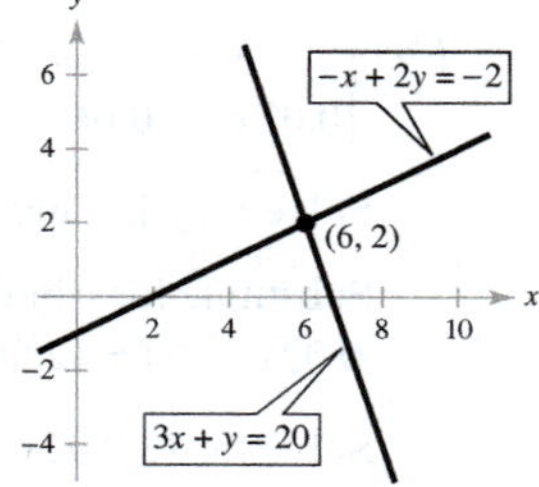

35. $\begin{cases} x - 3y = -3 \\ 5x + 3y = -6 \end{cases}$

Point of intersection:
$\left(-\frac{3}{2}, \frac{1}{2}\right)$

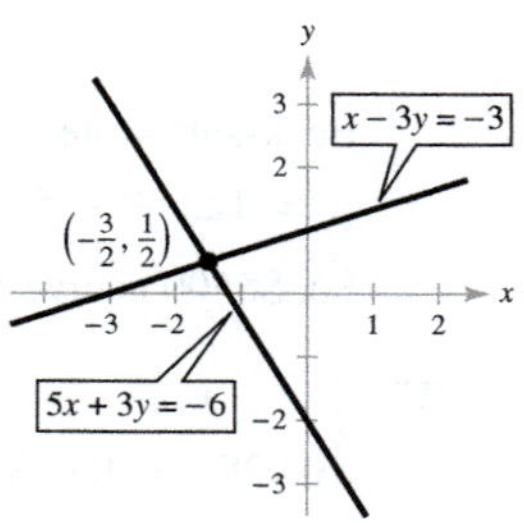

37. $\begin{cases} x + y = 4 \\ x^2 + y^2 - 4x = 0 \end{cases}$

Points of intersection:
$(2, 2), (4, 0)$

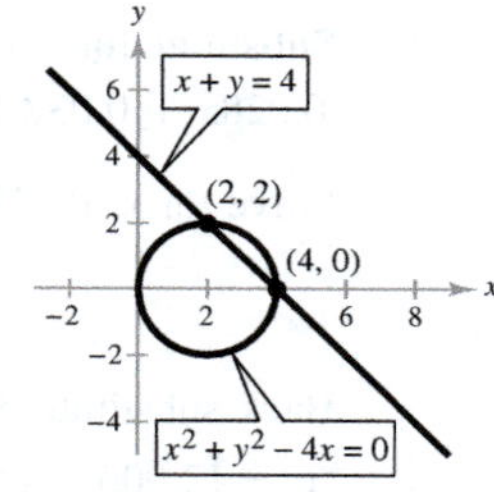

39. $\begin{cases} 3x - 2y = 0 \\ x^2 - y^2 = 4 \end{cases}$

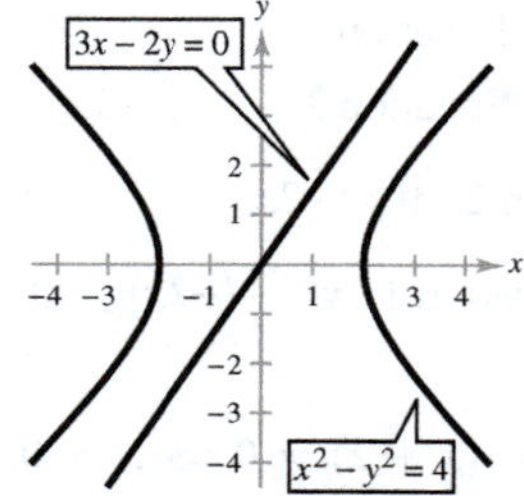

No points of intersection $\Rightarrow$ No solution

41. $\begin{cases} x^2 + y^2 = 25 \\ 3x^2 - 16y = 0 \end{cases}$

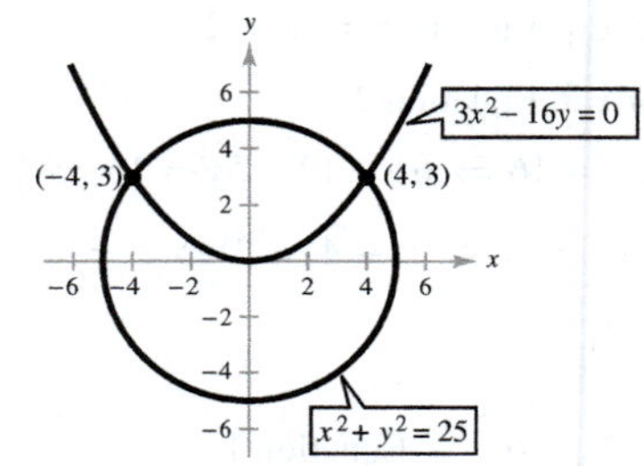

Points of intersection: $(-4, 3), (4, 3)$

43. $\begin{cases} y = e^x \\ x - y + 1 = 0 \Rightarrow y = x + 1 \end{cases}$

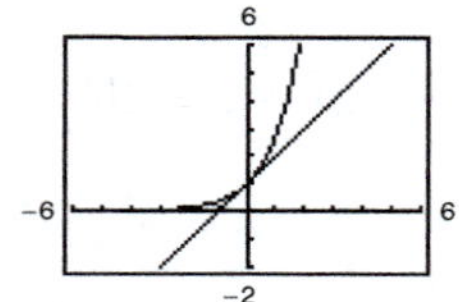

Point of intersection: $(0, 1)$

45. $\begin{cases} y = -2 + \ln(x - 1) \\ 3y + 2x = 9 \Rightarrow y = -\frac{2}{3}x + 3 \end{cases}$

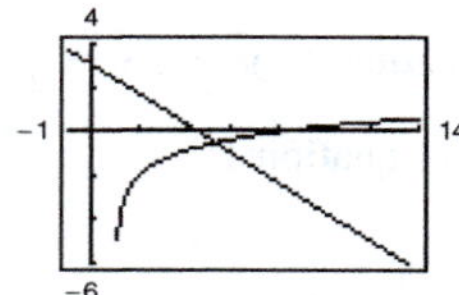

Point of intersection: $(5.31, -0.54)$

47. $\begin{cases} y = 2x & \text{Equation 1} \\ y = x^2 + 1 & \text{Equation 2} \end{cases}$

Substitute for y in Equation 2: $2x = x^2 + 1$

Solve for x: $x^2 - 2x + 1 = (x - 1)^2 = 0 \Rightarrow x = 1$

Back-substitute $x = 1$ in Equation 1: $y = 2x = 2$

Solution: $(1, 2)$

49. $\begin{cases} x - 2y = 4 & \text{Equation 1} \\ x^2 - y = 0 & \text{Equation 2} \end{cases}$

Solve for y in Equation 2: $y = x^2$

Substitute for y in Equation 1: $x - 2x^2 = 4$

Solve for x: $0 = 2x^2 - x + 4 \Rightarrow x = \dfrac{1 \pm \sqrt{1 - 4(2)(4)}}{2(2)} \Rightarrow x = \dfrac{1 \pm \sqrt{-31}}{4}$

The discriminant in the Quadratic Formula is negative.

No real solution

51. 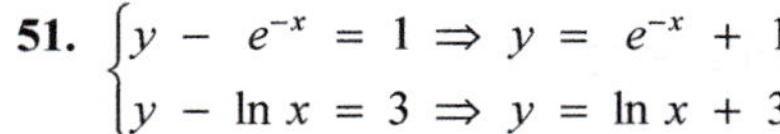

$\begin{cases} y - e^{-x} = 1 \Rightarrow y = e^{-x} + 1 \\ y - \ln x = 3 \Rightarrow y = \ln x + 3 \end{cases}$

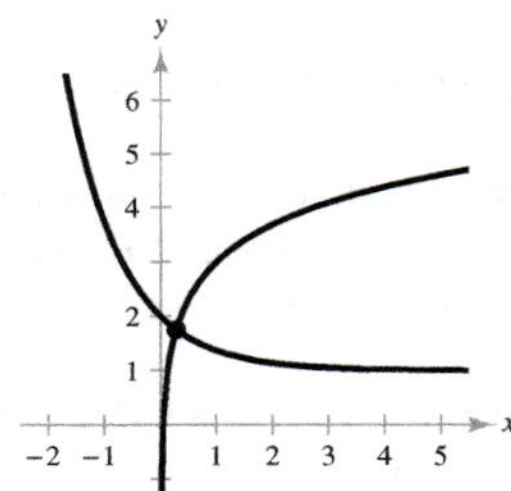

Point of intersection: approximately (0.287), (1.751)

53. $\begin{cases} xy - 1 = 0 & \text{Equation 1} \\ 2x - 4y + 7 = 0 & \text{Equation 2} \end{cases}$

Solve for y in Equation 1: $y = \dfrac{1}{x}$

Substitute for y in Equation 2: $2x - 4\left(\dfrac{1}{x}\right) + 7 = 0$

Solve for x:

$2x^2 - 4 + 7x = 0 \Rightarrow (2x - 1)(x + 4) = 0$

$\Rightarrow x = \dfrac{1}{2}, -4$

Back-substitute $x = \dfrac{1}{2}$: $y = \dfrac{1}{1/2} = 2$

Back-substitute $x = -4$: $y = \dfrac{1}{-4} = -\dfrac{1}{4}$

Solutions: $\left(\dfrac{1}{2}, 2\right), \left(-4, -\dfrac{1}{4}\right)$

55. $C = 8650x + 250{,}000$, $R = 9502x$

$$R = C$$
$$9502x = 8650x + 250{,}000$$
$$852x = 250{,}000$$
$$x \approx 293 \text{ units}$$

57. $C = 9.45x + 16{,}000$; $R = 55.95x$

(a) $R = C$

$55.95x = 9.45x + 16{,}000$

$46.5x = 16{,}000$

$x \approx 344$

About 344 units must be sold to break even.

(b) $P = R - C$

$100{,}000 = 55.95x - (9.45x + 16{,}000)$

$100{,}000 = 46.5x - 16{,}000$

$116{,}000 = 46.5x$

$x \approx 2495$

About 2495 units must be sold to earn a \$100,000 profit.

59. $\begin{cases} R = 360 - 24x & \text{Equation 1} \\ R = 24 + 18x & \text{Equation 2} \end{cases}$

(a) Substitute for R in Equation 2: $360 - 24x = 24 + 18x$

Solve for x: $336 = 42x \Rightarrow x = 8$ weeks

(b)

Weeks, x	1	2	3	4	5	6	7	8	9	10
$R = 360 - 24x$	336	312	288	264	240	216	192	168	144	120
$R = 24 + 18x$	42	60	78	96	114	132	150	168	186	204

The rentals are equal when $x = 8$ weeks.

61. The error was when the second equation was solved for y.

$$x^2 + 2x - y = 3$$
$$2x - y = 2 \Rightarrow y = 2x - 2$$
$$x^2 + 2x - (2x - 2) = 3$$
$$x^2 + 2x - 2x + 2 = 3$$
$$x^2 + 2 = 3$$
$$x^2 = 1$$
$$x = \pm 1$$

When $x = 1$, $y = 2(1) - 2 = 0$

When $x = -1$, $y = 2(-1) - 2 = -4$

Solutions: $(1, 0), (-1, -4)$.

63. $2l + 2w = 56 \quad \Rightarrow \quad l + w = 28$

$$l = w + 4 \Rightarrow (w + 4) + w = 28$$
$$2w + 4 = 28$$
$$2w = 24$$
$$w = 12 \text{ meters}$$

$l = w + 4 = 12 + 4 = 16$ meters

Dimensions: 12 meters $\times$ 16 meters

65. $44 = 2l + 2w$

$$22 = l + w \Rightarrow l = 22 - w$$
$$A = lw$$
$$120 = lw$$
$$120 = (22 - w)w$$
$$120 = 22w - w^2$$
$$w^2 - 22w + 120 = 0$$
$$(w - 10)(w - 12) = 0$$
$$w = 10, \quad w = 12$$

When $w = 10$, $l = 22 - 10 = 12$.

When $w = 12$, $l = 22 - 12 = 10$.

Dimensions: 10 kilometers $\times$ 12 kilometers

67. False. To solve a system of equations by substitution, you can solve for either variable in one of the two equations and then back-substitute.

69. *Sample answer*: If the result is a contradictory equation such as $0 = N$, then you know there are no solutions. When solving a system of equations that is a nonlinear system, there may be an equation with imaginary or extraneous solutions.

71. Answers will vary.

Section 7.2 Two-Variable Linear Systems

1. elimination

3. consistent; inconsistent

5. $\begin{cases} 2x + y = 7 & \text{Equation 1} \\ x - y = -4 & \text{Equation 2} \end{cases}$

Add to eliminate y:
$$\begin{aligned} 2x + y &= 7 \\ x - y &= -4 \\ \hline 3x \quad &= 3 \Rightarrow x = 1 \end{aligned}$$

Substitute $x = 1$ in Equation 2: $1 - y = -4 \Rightarrow y = 5$

Solution: $(1, 5)$

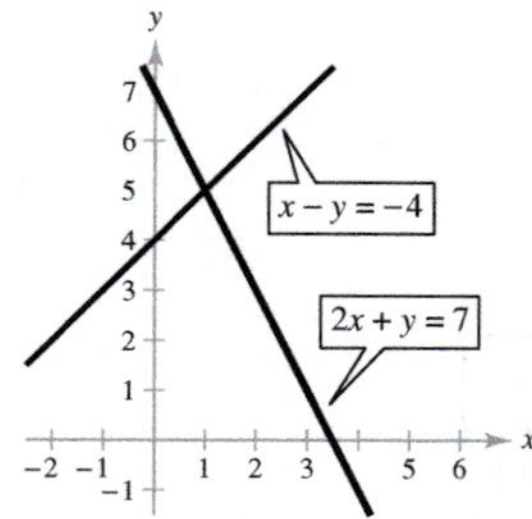

7. $\begin{cases} x + y = 0 & \text{Equation 1} \\ 3x + 2y = 1 & \text{Equation 2} \end{cases}$

Multiply Equation 1 by -2: $-2x - 2y = 0$

Add this to Equation 2 to eliminate y:
$$\begin{aligned} -2x - 2y &= 0 \\ 3x + 2y &= 1 \\ \hline x \quad &= 1 \end{aligned}$$

Substitute $x = 1$ in Equation 1: $1 + y = 0 \Rightarrow y = -1$

Solution: $(1, -1)$

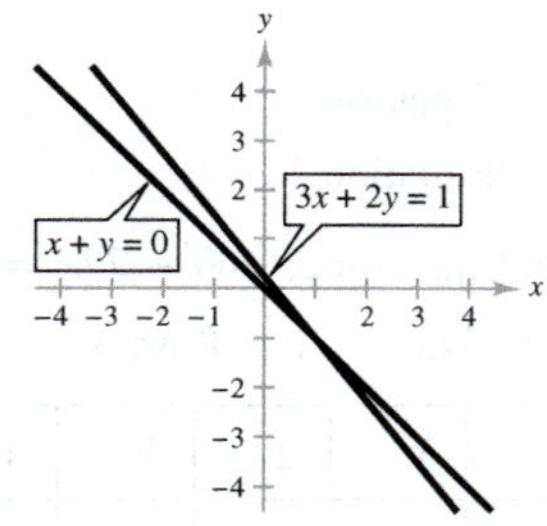

9. $\begin{cases} x - y = 2 & \text{Equation 1} \\ -2x + 2y = 5 & \text{Equation 2} \end{cases}$

Multiply Equation 1 by 2: $2x - 2y = 4$

Add this to Equation 2: $\begin{array}{r} 2x - 2y = 4 \\ -2x + 2y = 5 \\ \hline 0 = 9 \end{array}$

There are no solutions.

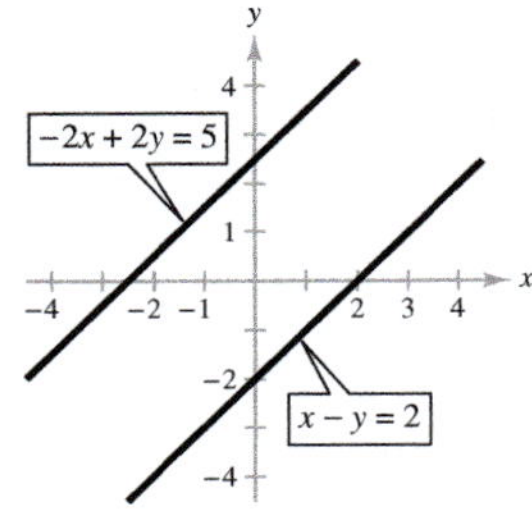

11. $\begin{cases} 3x - 2y = 5 & \text{Equation 1} \\ -6x + 4y = -10 & \text{Equation 2} \end{cases}$

Multiply Equation 1 by 2: $6x - 4y = 10$

Add this to Equation 2: $\begin{array}{r} 6x - 4y = 10 \\ -6x + 4y = -10 \\ \hline 0 = 0 \end{array}$

The equations are dependent. There are infinitely many solutions.

Let $x = a$, then $y = \dfrac{3a - 5}{2} = \dfrac{3}{2}a - \dfrac{5}{2}$.

Solution: $\left(a, \dfrac{3}{2}a - \dfrac{5}{2}\right)$, where a is any real number.

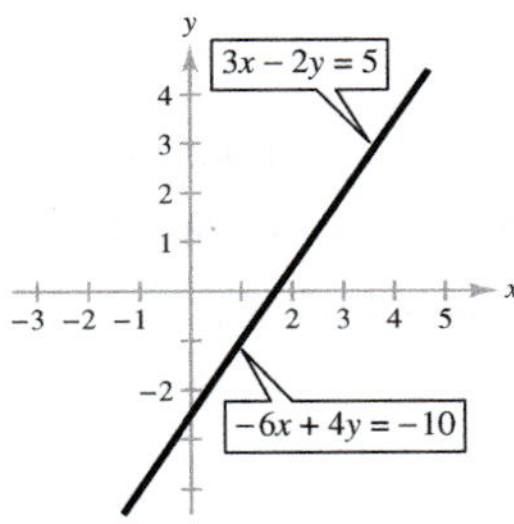

13. $\begin{cases} x + 2y = 6 & \text{Equation 1} \\ x - 2y = 2 & \text{Equation 2} \end{cases}$

Add the equations to eliminate y:

$$\begin{array}{l} x + 2y = 6 \\ x - 2y = 2 \\ \hline 2x \quad = 8 \Rightarrow x = 4 \end{array}$$

Substitute $x = 4$ into Equation 1:

$4 + 2y = 6 \Rightarrow y = 1$

Solution: $(4, 1)$

15. $\begin{cases} 5x + 3y = 6 & \text{Equation 1} \\ 3x - y = 5 & \text{Equation 2} \end{cases}$

Multiply Equation 2 by 3: $9x - 3y = 15$

Add this to Equation 1 to eliminate y:

$$\begin{array}{l} 5x + 3y = 6 \\ 9x - 3y = 15 \\ \hline 14x \quad = 21 \Rightarrow x = \frac{3}{2} \end{array}$$

Substitute $x = \frac{3}{2}$ into Equation 1:

$5\left(\frac{3}{2}\right) + 3y = 6 \Rightarrow y = -\frac{1}{2}$

Solution: $\left(\frac{3}{2}, -\frac{1}{2}\right)$

17. $\begin{cases} 2u + 3v = -1 & \text{Equation 1} \\ 7u + 15v = 4 & \text{Equation 2} \end{cases}$

Multiply Equation 1 by -5 and add to Equation 2.

$\begin{cases} -10u - 15v = 5 \\ 7u + 15v = 4 \end{cases}$

Solve for u: $-3u = 9 \Rightarrow u = -3$

Substitute $u = -3$ in Equation 1:

$2(-3) + 3v = -1 \Rightarrow v = \dfrac{5}{3}$

Solution: $\left(-3, \dfrac{5}{3}\right)$

19. $\begin{cases} 3x + 2y = 10 & \text{Equation 1} \\ 2x + 5y = 3 & \text{Equation 2} \end{cases}$

Multiply Equation 1 by 2 and Equation 2 by -3:

$\begin{cases} 6x + 4y = 20 \\ -6x - 15y = -9 \end{cases}$

Add to eliminate x: $-11y = 11 \Rightarrow y = -1$

Substitute $y = -1$ in Equation 1:

$3x - 2 = 10 \Rightarrow x = 4$

Solution: $(4, -1)$

21. $\begin{cases} 4b + 3m = 3 & \text{Equation 1} \\ 3b + 11m = 13 & \text{Equation 2} \end{cases}$

Multiply Equation 1 by 3 and Equation 2 by -4:

$\begin{cases} 12b + 9m = 9 \\ -12b - 44m = -52 \end{cases}$

Add to eliminate b: $-35m = -43 \Rightarrow m = \frac{43}{35}$

Substitute $m = \frac{43}{35}$ in Equation 1:

$4b + 3\left(\frac{43}{35}\right) = 3 \Rightarrow b = -\frac{6}{35}$

Solution: $\left(-\frac{6}{35}, \frac{43}{35}\right)$

23. $\begin{cases} 0.2x - 0.5y = -27.8 & \text{Equation 1} \\ 0.3x + 0.4y = 68.7 & \text{Equation 2} \end{cases}$

Multiply Equation 1 by 4 and Equation 2 by 5:

$\begin{cases} 0.8x - 2y = -111.2 \\ 1.5x + 2y = 343.5 \end{cases}$

Add these to eliminate y:
$$\begin{aligned} 0.8x - 2y &= -111.2 \\ 1.5x + 2y &= 343.5 \\ \hline 2.3x &= 232.3 \\ x &= 101 \end{aligned}$$

Substitute $x = 101$ in Equation 1:

$0.2(101) - 0.5y = -27.8 \Rightarrow y = 96$

Solution: $(101, 96)$

25. $\begin{cases} 3x + 2y = 4 & \text{Equation 1} \\ 9x + 6y = 3 & \text{Equation 2} \end{cases}$

Multiply Equation 1 by -3 and add to Equation 2.

$\begin{cases} -9x - 6y = -12 \\ 9x + 6y = 3 \end{cases}$

Add:
$$\begin{aligned} -9x - 6y &= -12 \\ 9x + 6y &= 3 \\ \hline 0 &\neq -9 \end{aligned}$$

No solution.

27. $\begin{cases} -5x + 6y = -3 & \text{Equation 1} \\ 20x - 24y = 12 & \text{Equation 2} \end{cases}$

Multiply Equation 1 by 4:

$\begin{cases} -20x + 24y = -12 \\ 20x - 24y = 12 \end{cases}$

Add these two together: $0 = 0$

The equations are dependent. There are infinitely many solutions.

Let $x = a$, then

$-5a + 6y = -3 \Rightarrow y = \dfrac{5a - 3}{6} = \dfrac{5}{6}a - \dfrac{1}{2}.$

Solution: $\left(a, \dfrac{5}{6}a - \dfrac{1}{2}\right)$, where a is any real number

29. $\begin{cases} \dfrac{x+3}{4} + \dfrac{y-1}{3} = 1 & \text{Equation 1} \\ 2x - y = 12 & \text{Equation 2} \end{cases}$

Multiply Equation 1 by 12 and Equation 2 by 4:

$\begin{cases} 3x + 4y = 7 \\ 8x - 4y = 48 \end{cases}$

Add to eliminate y: $11x = 55 \Rightarrow x = 5$

Substitute $x = 5$ into Equation 2:

$2(5) - y = 12 \Rightarrow y = -2$

Solution: $(5, -2)$

31. $\begin{cases} -7x + 6y = -4 \\ 14x - 12y = 8 \end{cases}$

Multiply Equation 1 by 2:

$\begin{cases} -14x + 12y = -8 \\ 14x - 12y = 8 \end{cases}$

Add this to Equation 2: $0 = 0$

The original equations are dependent.

Matches graph (a).

Number of solutions: Infinite

Consistent

33. $\begin{cases} 7x - 6y = -6 \\ -7x + 6y = -4 \end{cases}$

Add the equations: $0 = -10$

Inconsistent

Matches graph (d).

Number of solutions: None

Inconsistent

35. $\begin{cases} 3x - 5y = 7 & \text{Equation 1} \\ 2x + y = 9 & \text{Equation 2} \end{cases}$

Multiply Equation 2 by 5:

$10x + 5y = 45$

Add this to Equation 1:

$13x = 52 \Rightarrow x = 4$

Back-substitute $x = 4$ into Equation 2:

$2(4) + y = 9 \Rightarrow y = 1$

Solution: $(4, 1)$

37. $\begin{cases} -2x + 8y = 20 & \text{Equation 1} \\ y = x - 5 & \text{Equation 2} \end{cases}$

Substitute Equation 2 into Equation 1:

$$\begin{aligned} -2x + 8(x - 5) &= 20 \\ -2x + 8x - 40 &= 20 \\ 6x &= 60 \\ x &= 10 \end{aligned}$$

Back-substitute $x = 10$ into Equation 2:

$y = 10 - 5 = 5$

Solution: $(10, 5)$

39. $\begin{cases} y = -2x - 17 & \text{Equation 1} \\ y = 2 - 3x & \text{Equation 2} \end{cases}$

Use substitution because both equations are solved for y, set them equal to one another and solve for x.

$-2x - 17 = 2 - 3x$

$x = 19$

Back-substitute $x = 19$ into Equation 1:

$y = -2(19) - 17 = -55$

Solution: $(19, -15)$

41. Let and r_1 = the air speed of the plane and r_2 = the wind air speed.

$\begin{cases} 3(r_1 - r_2) = 1800 & \text{Equation 1} \\ 2.5(r_1 + r_2) = 1800 & \text{Equation 2} \end{cases}$

$r_1 - r_2 = 600$

$r_1 + r_2 = 720$

$2r_1 = 1320$

$r_1 = 660$

Back substitute into Equation 2 $660 + r_2 = 720$

$r_2 = 60$

The air speed of the plane is 660 miles per hour and the speed of the wind is 60 miles per hour.

43. Let x = the number of calories in a cheeseburger.

Let y = the number of calories in a small order of french fries.

$\begin{cases} 2x + y = 1420 & \text{Equation 1} \\ 3x + 2y = 2290 & \text{Equation 2} \end{cases}$

Multiply Equation 1 by -2 and add this to Equation 2.

$-4x - 2y = -2840$

$3x + 2y = 2290$

$-x = -550$

$x = 550$ calories

Back-substitute $x = 550$ into Equation 2:

$3(550) + 2y = 2290$

$2y = 640$

$y = 320$ calories

The cheeseburger contains 550 calories and the fries contain 320 calories.

45. $500 - 0.4x = 380 + 0.1x$

$120 = 0.5x$

$x = 240$ units

$p = \$404$

Equilibrium point: $(240, 404)$

47. $140 - 0.00002x = 80 + 0.00001x$

$60 = 0.00003x$

$x = 2{,}000{,}000$ units

$p = \$100.00$

Equilibrium point: $(2{,}000{,}000, 100)$

49. (a) Let x = the number of liters at 25%.

Let y = the number of liters at 50%.

$\begin{cases} 0.25x + 0.50y = 12 \\ x + y = 30 \end{cases}$

(b)

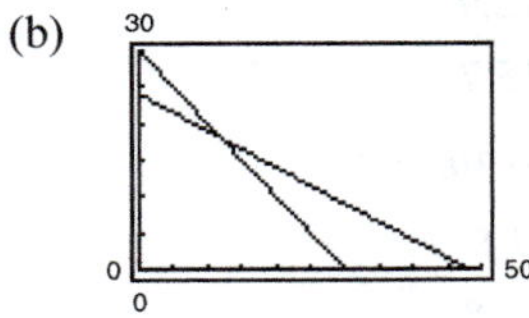

As the amount of 25% solution increases, the amount of 50% solution decreases.

(c) $\begin{cases} 0.25x + 0.50y = 12 & \text{Equation 1} \\ x + y = 30 & \text{Equation 2} \end{cases}$

Solve Equation 2 for y: $y = 30 - x$

Substitute this into Equation 1 to eliminate y:

$0.25x + 0.50(30 - x) = 12$

$0.25x + 15 - 0.50x = 12$

$-0.25x = -3$

$x = 12$ liters

Back-substitute $x = 12$ into Equation 2:

$12 + y = 30 \Rightarrow y = 18$ liters

The final mixture should contain 12 liters of the 25% solution and 18 liters of the 50% solution.

51. Let x = the amount of money invested at 3.5%.

Let y = the amount of money invested at 5%.

$\begin{cases} x + y = 24{,}000 & \text{Equation 1} \\ 0.035x + 0.05y = 930 & \text{Equation 2} \end{cases}$

Solve Equation 1 for x: $x = 24{,}000 - y$

Substitute this into Equation 2 to eliminate x:

$0.035(24{,}000 - y) + 0.05y = 930$

$840 + 0.015y = 930$

$y = \$6000$

Back-substitute $y = 6000$ into Equation 1:

$x + 6000 = 24{,}000$

$x = \$18{,}000$

$18,000 should be invested in the 3.5% bond.

53. (a) Pharmacy A: Pharmacy B:

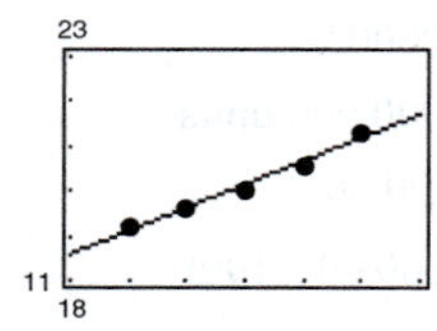

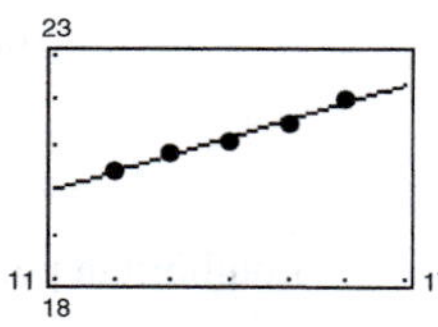

$P_A = 0.52t + 12.9$ $P_B = 0.39t + 15.7$

(b) To find when the prescriptions filled are equal, solve the system of equations consisting of $P = 0.52t + 12.9$ and $P = 0.39t + 15.7$.

$$\begin{cases} P = 0.52t + 12.9 \\ P = 0.39t + 15.7 \end{cases}$$

$$0.52t + 12.9 = 0.39t + 15.7$$
$$0.13t = 2.8$$
$$t = \frac{2.8}{013} \approx 21.5$$

The number of prescriptions filled at pharmacy A will exceed the number of prescriptions filled at pharmacy B during the year 2021.

55. $\begin{cases} 5b + 10a = 20.2 \Rightarrow & b + 2a = 4.04 \\ 10b + 30a = 50.1 \Rightarrow & -b - 3a = -5.01 \end{cases}$

$$-a = -0.97$$
$$a = 0.97$$

$$b + 2a = 4.04$$
$$b + 2(0.97) = 4.04$$
$$b = 2.1$$

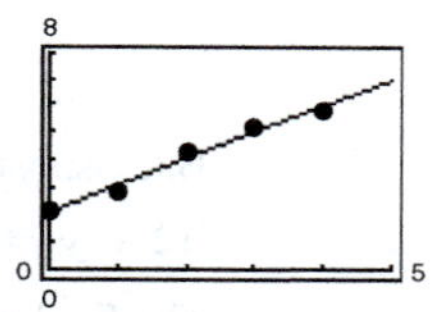

Least squares regression line:

$y = 0.97x + 2.1$

57. (a) $\begin{cases} 4b + 7.0a = 174 \Rightarrow & 28b + 49.0a = 1218 \\ 7b + 13.5a = 322 \Rightarrow & -28b - 54.0a = -1288 \end{cases}$

$$-5a = -70$$
$$a = 14$$

$$4b + 7.0a = 174$$
$$4b + 7.0(14) = 174$$
$$4b = 76$$
$$b = 19$$

Least squares regression line:

$y = 14x + 19$

(b) Substitute $x = 1.6$ into $y = 14x + 19$.

$y = 14(1.6) + 19 = 41.4$

The wheat yield is about 41.4 bushels per acre.

59. False. Two lines that coincide have infinitely many points of intersection.

61. $\begin{cases} 4x - 8y = -3 & \text{Equation 1} \\ 2x + ky = 16 & \text{Equation 2} \end{cases}$

Multiply Equation 2 by –2: $-4x - 2ky = -32$

Add this to Equation 1: $4x - 8y = -3$

$$-4x - 2ky = -32$$
$$-8y - 2ky = -35$$

The system is inconsistent if $-8y - 2ky = 0$.

This occurs when $k = -4$.

63. No, it is not possible for a consistent system of linear equations to have exactly two solutions. Either the lines will intersect once or they will coincide and then the system would have infinite solutions.

65. $\begin{cases} 3x + 2y = 4 \\ 5x - 2y = 12 \end{cases}$

$$2y = -3x + 4$$
$$y = -\tfrac{3}{2}x + 2$$

$$5x - 2\left(-\tfrac{3}{2}x + 2\right) = 12$$
$$5x + 3x - 4 = 12$$
$$8x = 16$$
$$x = 2$$

Back substitute $x = 2$: $3(2) + 2y = 4$

$$6 + 2y = 4$$
$$2y = -2$$
$$y = -1$$

Solution: $(2, -1)$

Answers will vary: *Sample answer*: If the equations can be added or subtracted without having to multiply by any coefficient, elimination of variable may be preferred. If one or both of the equations is already solved for one of the variables, the method of substitution may be more efficient.

67. $\begin{cases} 100y - x = 200 & \text{Equation 1} \\ 99y - x = -198 & \text{Equation 2} \end{cases}$

Subtract Equation 2 from Equation 1 to eliminate x:

$$100y - x = 200$$
$$-99y + x = 198$$
$$y = 398$$

Substitute $y = 398$ into Equation 1:

$100(398) - x = 200 \Rightarrow x = 39{,}600$

Solution: $(39, 600, 398)$

The lines are not parallel. The scale on the axes must be changed to see the point of intersection.

69. $\begin{cases} u \sin x + v \cos x = 0 & \text{Equation 1} \\ u \cos x - v \sin x = \sec x & \text{Equation 2} \end{cases}$

Multiply Equation 1 by $\cos x$ and multiply Equation 2 by $-\sin x$. Then add the equations to eliminate u.

$$\begin{aligned} u \sin x \cos x + v \cos^2 x &= 0 \\ -u \sin x \cos x + v \sin^2 x &= -\sin x \sec x \\ \hline v(\sin^2 x + \cos^2 x) &= -\sin x \sec x \\ v = -\sin x \sec x &= -\sin x\left(\frac{1}{\cos x}\right) = -\tan x \end{aligned}$$

Back substitute v into Equation 1

$$\begin{aligned} u \sin x + (-\tan x) \cos x &= 0 \\ u \sin x - \left(\frac{\sin x}{\cos x}\right) \cos x &= 0 \\ u \sin x - \sin x &= 0 \\ u \sin x &= \sin x \\ u &= 1 \end{aligned}$$

The solution of this system is: $u = 1$, $v = -\tan x$.

Section 7.3 Multivariable Linear Systems

1. row-echelon

3. Gaussian

5. nonsquare

7. $\begin{cases} 6x - y + z = -1 \\ 4x - 3z = -19 \\ 2y + 5z = 25 \end{cases}$

(a) $(0, 3, 1)$

$6(0) - (3) + (1) \neq 1$

$(0, 3, 1)$ *is not* a solution.

(b) $(-3, 0, 5)$

$6(-3) - 0 + 5 \neq -1$

$(-3, 0, 5)$ *is not* a solution

(c) $(0, -1, 4)$

$4(0) - 3(4) \neq -19$

$(0, -1, 4)$ *is not* a solution.

(d) $(-1, 0, 5)$

$$\begin{aligned} 6(-1) - 0 + 5 &= -1 \\ 4(-1) - 3(5) &= -19 \\ 2(0) + 5(5) &= 25 \end{aligned}$$

$(-1, 0, 5)$ *is* a solution.

9. $\begin{cases} 4x + y - z = 0 \\ -8x - 6y + z = -\frac{7}{4} \\ 3x - y = -\frac{9}{4} \end{cases}$

(a) $4\left(\frac{1}{2}\right) + \left(-\frac{3}{4}\right) - \left(-\frac{7}{4}\right) \neq 0$

$\left(\frac{1}{2}, -\frac{3}{4}, -\frac{7}{4}\right)$ *is not* a solution.

(b) $4\left(\frac{3}{2}\right) + \left(-\frac{2}{5}\right) - \left(\frac{3}{5}\right) \neq 0$

$\left(\frac{3}{2}, -\frac{2}{5}, \frac{3}{5}\right)$ *is not* a solution.

(c)
$$\begin{aligned} 4\left(-\tfrac{1}{2}\right) + \left(\tfrac{3}{4}\right) - \left(-\tfrac{5}{4}\right) &= 0 \\ -8\left(-\tfrac{1}{2}\right) - 6\left(\tfrac{3}{4}\right) + \left(-\tfrac{5}{4}\right) &= -\tfrac{7}{4} \\ 3\left(-\tfrac{1}{2}\right) - \left(\tfrac{3}{4}\right) &= -\tfrac{9}{4} \end{aligned}$$

$\left(-\frac{1}{2}, \frac{3}{4}, -\frac{5}{4}\right)$ *is* a solution.

(d) $4\left(-\frac{1}{2}\right) + \left(\frac{1}{6}\right) - \left(-\frac{3}{4}\right) \neq 0$

$\left(-\frac{1}{2}, \frac{1}{6}, -\frac{3}{4}\right)$ *is not* a solution.

11. $\begin{cases} x - y + 5z = 37 & \text{Equation 1} \\ y + 2z = 6 & \text{Equation 2} \\ z = 8 & \text{Equation 3} \end{cases}$

Back-substitute $z = 8$ into Equation 2:

$y + 2(8) = 6$

$y = -10$

Back-substitute $y = -10$ and $z = 8$ into Equation 1:

$x - (-10) + 5(8) = 37$

$x + 10 + 40 = 37$

$x = -13$

Solution: $(-13, -10, 8)$

13. $\begin{cases} x + y - 3z = 7 & \text{Equation 1} \\ y + z = 12 & \text{Equation 2} \\ z = 2 & \text{Equation 3} \end{cases}$

Back-substitute $z = 2$ into Equation 2:

$y + 2 = 12 \Rightarrow y = 10$

Back-substitute $y = 10$ and $z = 2$ into Equation 1:

$x + (10) - 3(2) = 7$

$x + 4 = 7$

$x = 3$

Solution: $(3, 10, 2)$

15. $\begin{cases} x - 2y + z = -\frac{1}{4} & \text{Equation 1} \\ y - z = -4 & \text{Equation 2} \\ z = 11 & \text{Equation 3} \end{cases}$

Back-substitute $z = 11$ into Equation 2:

$y - 11 = -4$

$y = 7$

Back-substitute $y = 7$ and $z = 11$ into Equation 1:

$x - 2(7) + (11) = -\frac{1}{4}$

$x - 3 = -\frac{1}{4}$

$x = \frac{11}{4}$

Solution: $\left(\frac{11}{4}, 7, 11\right)$

17. $\begin{cases} x - 2y + 3z = 5 & \text{Equation 1} \\ -x + 3y - 5z = 4 & \text{Equation 2} \\ 2x \qquad - 3z = 0 & \text{Equation 3} \end{cases}$

Add Equation 1 to Equation 2:

$\begin{cases} x - 2y + 3z = 5 \\ y - 2z = 9 \\ 2x \qquad - 3z = 0 \end{cases}$

This is the first step in putting the system in row-echelon form.

19. $\begin{cases} x + y = 0 \\ -2x + 3y = 10 \end{cases}$

$\begin{cases} x + y = 0 \\ 5y = 10 & 2\text{Eq.1} + \text{Eq.2} \end{cases}$

$\begin{cases} x + y = 0 \\ y = 2 & \frac{1}{5}\text{Eq.2} \end{cases}$

$x + (2) = 0$

$x = -2$

Solution: $(-2, 2)$

21. $\begin{cases} x - 2y = -2 \\ 3x - y = 9 \end{cases}$

$\begin{cases} x - 2y = -2 \\ 5y = 15 & (-3)\text{Eq.1} + \text{Eq.2} \end{cases}$

$\begin{cases} x - 2y = -2 \\ y = 3 & \frac{1}{5}\text{Eq.2} \end{cases}$

$x - 2(3) = -2$

$x = 4$

Solution: $(4, 3)$

23. $\begin{cases} x + y + z = 7 & \text{Equation 1} \\ 2x - y + z = 9 & \text{Equation 2} \\ 3x \qquad - z = 10 & \text{Equation 3} \end{cases}$

$\begin{cases} x + y + z = 7 \\ 3x \qquad + 2z = 16 & \text{Eq.2} + \text{Eq.1} \\ 3x \qquad - z = 10 \end{cases}$

$\begin{cases} x + y + z = 7 \\ 3x \qquad + 2z = 16 \\ 9x \qquad = 36 & \text{Eq.2} + 2\text{Eq.3} \end{cases}$

$\begin{cases} x + y + z = 7 \\ 3x \qquad + 2z = 16 \\ x \qquad = 4 & \frac{1}{4}\text{Eq.3} \end{cases}$

$3(4) + 2z = 16$

$2z = 4$

$z = 2$

$4 + y + 2 = 7$

$y = 1$

Solution: $(4, 1, 2)$

25. $\begin{cases} 2x + 4y - z = 7 \\ 2x - 4y + 2z = -6 \\ x + 4y + z = 0 \end{cases}$

$\begin{cases} x + 4y + z = 0 \\ 2x - 4y + 2z = -6 \\ 2x + 4y - z = 7 \end{cases}$ Interchange equations.

$\begin{cases} x + 4y + z = 0 \\ -12y = -6 & (-2)\text{Eq. } 1 + \text{Eq. } 2 \\ -4y - 3z = 7 & (-2)\text{Eq. } 1 + \text{Eq. } 3 \end{cases}$

$\begin{cases} x + 4y + z = 0 \\ y = \frac{1}{2} & -\frac{1}{12}\text{Eq. } 2 \\ -4y - 3z = 7 \end{cases}$

$$y = \tfrac{1}{2}$$

$$-4\left(\tfrac{1}{2}\right) - 3z = 7 \Rightarrow z = -3$$

$$x + 4\left(\tfrac{1}{2}\right) + (-3) = 0$$

$$x = 1$$

Solution: $\left(1, \frac{1}{2}, -3\right)$

27. $\begin{cases} x - 2y + 2z = -9 \\ 2x + y - z = 7 \\ 3x - y + z = 5 \end{cases}$ Interchange equations.

$\begin{cases} x - 2y + 2z = -9 \\ 5y - 5z = 25 & -2\text{Eq.}1 + \text{Eq.}2 \\ 5y - 5z = 32 & -3\text{Eq.}1 + \text{Eq.}3 \end{cases}$

$\begin{cases} x - 2y + 2z = -9 \\ 5y - 5z = 25 \\ 0 = 7 & -\text{Eq.}2 + \text{Eq.}3 \end{cases}$

Inconsistent, no solution

29. $\begin{cases} 3x - 5y + 5z = 1 & \text{Equation 1} \\ 2x - 2y + 3z = 0 & \text{Equation 2} \\ 7x - y + 3z = 0 & \text{Equation 3} \end{cases}$

$\begin{cases} x - 3y + 2z = 1 & \text{Eq. } 1 - \text{Eq. } 2 \\ 2x - 2y + 3z = 0 \\ 7x - y + 3z = 0 \end{cases}$

$\begin{cases} x - 3y + 2z = 1 \\ -4y + z = 2 & 2\text{Eq. } 1 - \text{Eq.}2 \\ 7x - y + 3z = 0 \end{cases}$

$\begin{cases} x - 3y + 2z = 1 \\ -4y + z = 2 \\ -20y + 11z = 7 & 7\text{Eq. } 1 - \text{Eq. } 3 \end{cases}$

$\begin{cases} x - 3y + 2z = 1 \\ -4y + z = 2 \\ 6z = -3 & -5\text{Eq. } 2 + \text{Eq. } 3 \end{cases}$

$$6z = -3 \Rightarrow \quad z = -\tfrac{1}{2}$$

$$-4y + \left(-\tfrac{1}{2}\right) = 2 \Rightarrow -4y = \tfrac{5}{2} \Rightarrow y = -\tfrac{5}{8}$$

$$x - 3\left(-\tfrac{5}{8}\right) + 2\left(-\tfrac{1}{2}\right) = 1 \Rightarrow x + \tfrac{7}{8} = 1 \Rightarrow x = \tfrac{1}{8}$$

Solution: $\left(\frac{1}{8}, -\frac{5}{8}, -\frac{1}{2}\right)$

31. $\begin{cases} 2x + 3y = 0 & \text{Equation 1} \\ 4x + 3y - z = 0 & \text{Equation 2} \\ 8x + 3y + 3z = 0 & \text{Equation 3} \end{cases}$

$\begin{cases} 2x + 3y = 0 \\ -3y - z = 0 & -2\text{Eq.}1 + \text{Eq.}2 \\ -9y + 3z = 0 & -4\text{Eq.}1 + \text{Eq.}3 \end{cases}$

$\begin{cases} 2x + 3y = 0 \\ -3y - z = 0 \\ 6z = 0 & -3\text{Eq.}2 + \text{Eq.}3 \end{cases}$

$$6z = 0 \Rightarrow z = 0$$

$$-3y - 0 = 0 \Rightarrow y = 0$$

$$2x + 3(0) = 0 \Rightarrow x = 0$$

Solution: $(0, 0, 0)$

33. $\begin{cases} x + 4z = 1 & \text{Equation 1} \\ x + y + 10z = 10 & \text{Equation 2} \\ 2x - y + 2z = -5 & \text{Equation 3} \end{cases}$

$\begin{cases} x + 4z = 1 \\ y + 6z = 9 & -\text{Eq.}1 + \text{Eq.}2 \\ -y - 6z = -7 & -2\text{Eq.}1 + \text{Eq.}3 \end{cases}$

$\begin{cases} x + 4z = 1 \\ y + 6z = 9 \\ 0 = 2 & \text{Eq.}2 + \text{Eq.}3 \end{cases}$

No solution, inconsistent

35. $\begin{cases} 3x - 3y + 6z = 6 & \text{Equation 1} \\ x + 2y - z = 5 & \text{Equation 2} \\ 5x - 8y + 13z = 7 & \text{Equation 3} \end{cases}$

$\begin{cases} x - y + 2z = 2 & \frac{1}{3}\text{Eq.1} \\ x + 2y - z = 5 \\ 5x - 8y + 13z = 7 \end{cases}$

$\begin{cases} x - y + 2z = 2 \\ 3y - 3z = 3 & -\text{Eq.1} + \text{Eq.2} \\ -3y + 3z = -3 & -5\text{Eq.1} + \text{Eq.3} \end{cases}$

$\begin{cases} x - y + 2z = 2 \\ y - z = 1 & \frac{1}{3}\text{Eq.2} \\ 0 = 0 & \text{Eq.2} + \text{Eq.3} \end{cases}$

$\begin{cases} x + z = 3 & \text{Eq.2} + \text{Eq.1} \\ y - z = 1 \end{cases}$

Let $z = a$, then:

$y = a + 1$

$x = -a + 3$

Solution: $(-a + 3, a + 1, a)$

37. $\begin{cases} x + 2y - 7z = -4 & \text{Equation 1} \\ 2x + y + z = 13 & \text{Equation 2} \\ 3x + 9y - 36z = -33 & \text{Equation 3} \end{cases}$

$\begin{cases} x + 2y - 7z = -4 \\ -3y + 15z = 21 & -2\text{Eq.1} + \text{Eq.2} \\ 3y - 15z = -21 & -3\text{Eq.1} + \text{Eq.3} \end{cases}$

$\begin{cases} x + 2y - 7z = -4 \\ -3y + 15z = 21 \\ 0 = 0 & \text{Eq.2} + \text{Eq.3} \end{cases}$

$\begin{cases} x + 2y - 7z = -4 \\ y - 5z = -7 & -\frac{1}{3}\text{Eq.2} \end{cases}$

$\begin{cases} x + 3z = 10 & -2\text{Eq.2} + \text{Eq.1} \\ y - 5z = -7 \end{cases}$

Let $z = a$, then:

$y = 5a - 7$

$x = -3a + 10$

Solution: $(-3a + 10, 5a - 7, a)$

39. $\begin{cases} x + 3w = 4 & \text{Equation 1} \\ 2y - z - w = 0 & \text{Equation 2} \\ 3y - 2w = 1 & \text{Equation 3} \\ 2x - y + 4z = 5 & \text{Equation 4} \end{cases}$

$\begin{cases} x + 3w = 4 \\ 2y - z - w = 0 \\ 3y - 2w = 1 \\ -y + 4z - 6w = -3 & -2\text{Eq.1} + \text{Eq.4} \end{cases}$

$\begin{cases} x + 3w = 4 \\ y - 4z + 6w = 3 & -\text{Eq.4 and interchange the equations.} \\ 2y - z - w = 0 \\ 3y - 2w = 1 \end{cases}$

$\begin{cases} x + 3w = 4 \\ y - 4z + 6w = 3 \\ 7z - 13w = -6 & -\text{Eq.2} + \text{Eq.3} \\ 12z - 20w = -8 & -3\text{Eq.2} + \text{Eq.4} \end{cases}$

$\begin{cases} x + 3w = 4 \\ y - 4z + 6w = 3 \\ z - 3w = -2 & -\frac{1}{2}\text{Eq.4} + \text{Eq.3} \\ 12z - 20w = -8 \end{cases}$

$\begin{cases} x + 3w = 4 \\ y - 4z + 6w = 3 \\ z - 3w = -2 \\ 16w = 16 & -12\text{Eq.3} + \text{Eq.4} \end{cases}$

$16w = 16 \Rightarrow w = 1$

$z - 3(1) = -2 \Rightarrow z = 1$

$y - 4(1) + 6(1) = 3 \Rightarrow y = 1$

$x + 3(1) = 4 \Rightarrow x = 1$

Solution: $(1, 1, 1, 1)$

41. $\begin{cases} x - 2y + 5z = 2 \\ 4x - z = 0 \end{cases}$

Let $z = a$, then: $x = \frac{1}{4}a$.

$\frac{1}{4}a - 2y + 5a = 2$

$a - 8y + 20a = 8$

$-8y = -21a + 8$

$y = \frac{21}{8}a - 1$

Answer: $\left(\frac{1}{4}a, \frac{21}{8}a - 1, a\right)$

To avoid fractions, we could go back and let $z = 8a$, then $4x - 8a = 0 \Rightarrow x = 2a$.

$2a - 2y + 5(8a) = 2$

$-2y + 42a = 2$

$y = 21a - 1$

Solution: $(2a, 21a - 1, 8a)$

43. $\begin{cases} 2x - 3y + z = -2 & \text{Equation 1} \\ -4x + 9y = 7 & \text{Equation 2} \end{cases}$

$\begin{cases} 2x - 3y + z = -2 \\ 3y + 2z = 3 & 2\text{Eq.}1 + \text{Eq.}2 \end{cases}$

$\begin{cases} 2x + 3z = 1 & \text{Eq.}2 + \text{Eq.}1 \\ 3y + 2z = 3 \end{cases}$

Let $z = a$, then:

$y = -\frac{2}{3}a + 1$

$x = -\frac{3}{2}a + \frac{1}{2}$

Solution: $\left(-\frac{3}{2}a + \frac{1}{2}, -\frac{2}{3}a + 1, a\right)$

45. $s = \frac{1}{2}at^2 + v_0t + s_0$

$(1, 128), (2, 80), (3, 0)$

$128 = \frac{1}{2}a + v_0 + s_0 \Rightarrow a + 2v_0 + 2s_0 = 256$

$80 = 2a + 2v_0 + s_0 \Rightarrow 2a + 2v_0 + s_0 = 80$

$0 = \frac{9}{2}a + 3v_0 + s_0 \Rightarrow 9a + 6v_0 + 2s_0 = 0$

Solving this system yields $a = -32, v_0 = 0, s_0 = 144$.

So, $s = \frac{1}{2}(-32)t^2 + (0)t + 144 = -16t^2 + 144$.

47. $y = ax^2 + bx + c$ passing through $(0, 0), (2, -2), (4, 0)$

$(0, 0)$: $0 = c$

$(2, -2)$: $-2 = 4a + 2b + c \Rightarrow -1 = 2a + b$

$(4, 0)$: $0 = 16a + 4b + c \Rightarrow 0 = 4a + b$

Solution: $a = \frac{1}{2}, b = -2, c = 0$

The equation of the parabola is $y = \frac{1}{2}x^2 - 2x$.

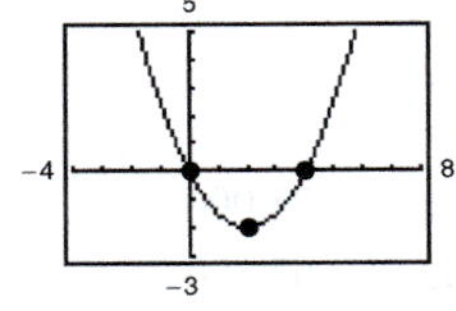

49. $y = ax^2 + bx + c$ passing through $(2, 0), (3, -1), (4, 0)$

$(2, 0)$: $0 = 4a + 2b + c$

$(3, -1)$: $-1 = 9a + 3b + c$

$(4, 0)$: $0 = 16a + 4b + c$

$\begin{cases} 0 = 4a + 2b + c \\ -1 = 5a + b & -\text{Eq.}1 + \text{Eq.}2 \\ 0 = 12a + 2b & -\text{Eq.}1 + \text{Eq.}3 \end{cases}$

$\begin{cases} 0 = 4a + 2b + c \\ -1 = 5a + b \\ 2 = 2a & -2\text{Eq.}2 + \text{Eq.}3 \end{cases}$

Solution: $a = 1, b = -6, c = 8$

The equation of the parabola is $y = x^2 - 6x + 8$.

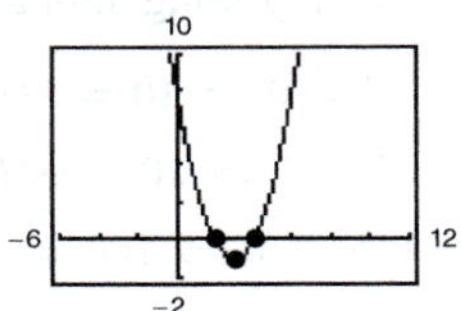

51. $y = ax^2 + bx + c$ passing through $\left(\frac{1}{2}, 1\right), (1, 3), (2, 13)$

$\left(\frac{1}{2}, 1\right)$: $1 = a\left(\frac{1}{2}\right)^2 + b\left(\frac{1}{2}\right) + c$

$(1, 3)$: $3 = a(1)^2 + b(1) + c$

$(2, 13)$: $13 = a(2)^2 + b(2) + c$

$\begin{cases} a + 2b + 4c = 4 \\ a + b + c = 3 \\ 4a + 2b + c = 13 \end{cases}$

Solution: $a = 4, b = -2, c = 1$

The equation of the parabola is $y = 4x^2 - 2x + 1$.

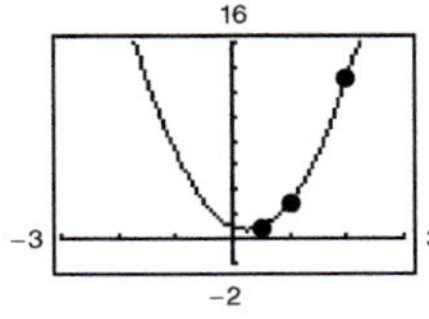

53. $x^2 + y^2 + Dx + Ey + F = 0$ passing through $(0, 0), (5, 5), (10, 0)$

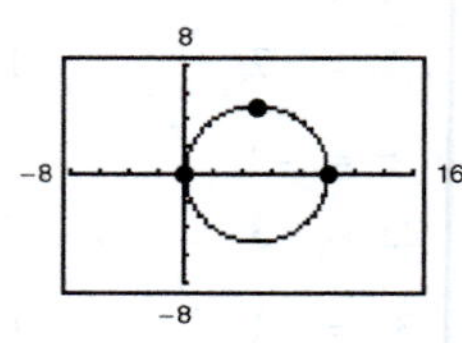

$(0, 0)$: $0^2 + 0^2 + D(0) + E(0) + F = 0 \Rightarrow F = 0$

$(5, 5)$: $5^2 + 5^2 + D(5) + E(5) + F = 0 \Rightarrow 5D + 5E + F = -50$

$(10, 0)$: $10^2 + 0^2 + D(10) + E(0) + F = 0 \Rightarrow 10D + F = -100$

Solution: $D = -10, E = 0, F = 0$

The equation of the circle is $x^2 + y^2 - 10x = 0$. To graph, complete the square first, then solve for y.

$$\left(x^2 - 10x + 25\right) + y^2 = 25$$
$$(x - 5)^2 + y^2 = 25$$
$$y^2 = 25 - (x - 5)^2$$
$$y = \pm\sqrt{25 - (x - 5)^2}$$

Let $y_1 = \sqrt{25 - (x - 5)^2}$ and $y_2 = -\sqrt{25 - (x - 5)^2}$.

55. $x^2 + y^2 + Dx + Ey + F = 0$ passing through $(-3, -1), (2, 4), (-6, 8)$

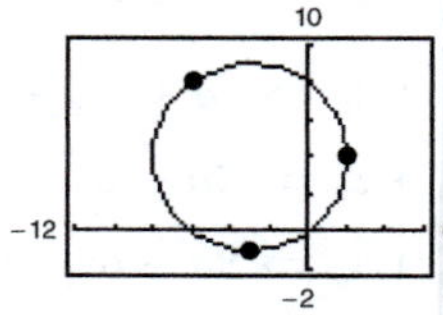

$(-3, -1)$: $10 - 3D - E + F = 0 \Rightarrow 10 = 3D + E - F$

$(2, 4)$: $20 + 2D + 4E + F = 0 \Rightarrow 20 = -2D - 4E - F$

$(-6, 8)$: $100 - 6D + 8E + F = 0 \Rightarrow 100 = 6D - 8E - F$

Solution: $D = 6, E = -8, F = 0$

The equation of the circle is $x^2 + y^2 + 6x - 8y = 0$. To graph, complete the squares first, then solve for y.

$$\left(x^2 + 6x + 9\right) + \left(y^2 - 8y + 16\right) = 0 + 9 + 16$$
$$(x + 3)^2 + (y - 4)^2 = 25$$
$$(y - 4)^2 = 25 - (x + 3)^2$$
$$y - 4 = \pm\sqrt{25 - (x + 3)^2}$$
$$y = 4 \pm \sqrt{25 - (x + 3)^2}$$

Let $y_1 = 4 + \sqrt{25 - (x + 3)^2}$ and $y_2 = 4 - \sqrt{25 - (x + 3)^2}$.

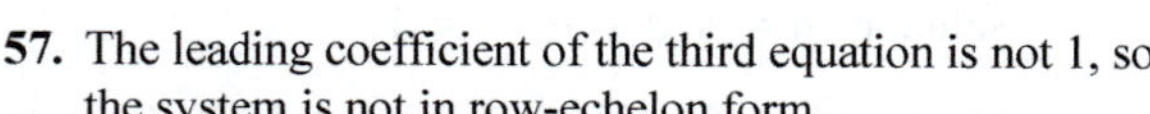

57. The leading coefficient of the third equation is not 1, so the system is not in row-echelon form.

$$\begin{cases} x - 2y + 3x = 12 \\ y + 3z = 5 \\ z = 2 \end{cases}$$

59. Let x = amount at 8%.

Let y = amount at 9%.

Let z = amount at 10%.

$$\begin{cases} x + y + z = 775{,}000 \\ 0.08x + 0.09y + 0.10z = 67{,}500 \\ x = 4z \end{cases}$$

$$\begin{cases} y + 5z = 775{,}000 \\ 0.09y + 0.42z = 67{,}500 \end{cases}$$

$z = 75{,}000$

$y = 775{,}000 - 5z = 400{,}000$

$x = 4z = 300{,}000$

\$300,000 was borrowed at 8%.

\$400,000 was borrowed at 9%.

\$75,000 was borrowed at 10%.

61. $\begin{cases} x + y + z = 180 \\ 2x + 7 + z = 180 \\ y + 2x - 7 = 180 \end{cases}$

$\begin{cases} x + y + z = 180 \\ 2x + z = 173 \\ 2x + y = 187 \end{cases}$

$\begin{cases} -x + y = 7 \\ 2x + z = 173 \\ 2x + y = 187 \end{cases}$ $\quad -\text{Eq.2} + \text{Eq.1}$

$\begin{cases} -x + y = 7 \\ 2x + z = 173 \\ 3x = 180 \end{cases}$ $\quad -\text{Eq.1} + \text{Eq.3}$

$x = 60°$

$2(60) + z = 173 \Rightarrow z = 53°$

$-60 + y = 7 \Rightarrow y = 67°$

63. Let x = the longest side (hypotenuse).

Let y = leg.

Let z = shortest leg.

$\begin{cases} x + y + z = 180 \\ x = 2z - 9 \\ y + z = 30 + x \end{cases}$

$\begin{cases} x + y + z = 180 \\ x - 2z = -9 \\ -x + y + z = 30 \end{cases}$

$\begin{cases} y + 3z = 189 \\ x - 2z = -9 \\ y - z = 21 \end{cases}$ $\quad \begin{matrix} -\text{Eq.2} + \text{Eq.1} \\ \\ \text{Eq.2} + \text{Eq.3} \end{matrix}$

$\begin{cases} 4z = 168 \\ x - 2z = -9 \\ y - z = 21 \end{cases}$ $\quad -\text{Eq.3} + \text{Eq.1}$

$z = 42$

$x - 2(42) = -9 \Rightarrow x = 75$

$y - 42 = 21 \Rightarrow y = 63$

So, the longest side measures 75 feet, the shortest side measures 42 feet, and the third side measures 63 feet.

65. $\begin{cases} I_1 - I_2 + I_3 = 0 & \text{Equation 1} \\ 3I_1 + 2I_2 = 7 & \text{Equation 2} \\ 2I_2 + 4I_3 = 8 & \text{Equation 3} \end{cases}$

$\begin{cases} I_1 - I_2 + I_3 = 0 \\ 5I_2 - 3I_3 = 7 & (-3)\text{Eq.1} + \text{Eq.2} \\ 2I_2 + 4I_3 = 8 \end{cases}$

$\begin{cases} I_1 - I_2 + I_3 = 0 \\ 10I_2 - 6I_3 = 14 & 2\text{Eq.2} \\ 10I_2 + 20I_3 = 40 & 5\text{Eq.3} \end{cases}$

$\begin{cases} I_1 - I_2 + I_3 = 0 \\ 10I_2 - 6I_3 = 14 \\ 26I_3 = 26 & (-1)\text{Eq.2} + \text{Eq.3} \end{cases}$

$26I_3 = 26 \Rightarrow I_3 = 1$

$10I_2 - 6(1) = 14 \Rightarrow I_2 = 2$

$I_1 - 2 + 1 = 0 \Rightarrow I_1 = 1$

Solution: $I_1 = 1, I_2 = 2, I_3 = 1$

67. $\begin{cases} 4c + 9b + 29a = 20 \\ 9c + 29b + 99a = 70 \\ 29c + 99b + 353a = 254 \end{cases}$

$\begin{cases} 9c + 29b + 99a = 70 \\ 4c + 9b + 29a = 20 \\ 29c + 99b + 353a = 254 \end{cases}$ Interchange equations.

$\begin{cases} c + 11b + 41a = 30 & -2\text{Eq.}2 + \text{Eq.}1 \\ -35b - 135a = -100 & -4\text{Eq.}1 + \text{Eq.}2 \\ -220b - 836a = -616 & -29\text{Eq.}1 + \text{Eq.}3 \end{cases}$

$\begin{cases} c + 11b + 41a = 30 \\ 1540b + 5940a = 4400 & -44\text{Eq.}2 \\ -1540b - 5852a = -4312 & 7\text{Eq.}3 \end{cases}$

$\begin{cases} c + 11b + 41a = 30 \\ 1540b + 5940a = 4400 \\ 88a = 88 & \text{Eq.}2 + \text{Eq.}3 \end{cases}$

$$88a = 88 \Rightarrow a = 1$$
$$1540b + 5940(1) = 4400 \Rightarrow b = -1$$
$$c + 11(-1) + 41(1) = 30 \Rightarrow c = 0$$

Least squares regression parabola: $y = x^2 - x$

69. (a) $\begin{cases} 5c + 250b + 13{,}500a = 923 \\ 250c + 13{,}500b + 775{,}000a = 52{,}170 \\ 13{,}500c + 775{,}000b + 46{,}590{,}000a = 3{,}101{,}300 \end{cases}$

$\begin{cases} 5c + 250b + 13{,}500a = 923 \\ 1000b + 100{,}000a = 6020 & (-50)\text{Eq. }1 + \text{Eq. }2 \\ 100{,}000b + 10{,}140{,}000a = 609{,}200 & (-2700)\text{Eq. }1 + (3)\text{Eq. }3 \end{cases}$

$\begin{cases} 5c + 250b + 13{,}500a = 923 \\ 1000b + 100{,}000a = 6020 \\ 140{,}000a = 7200 & (-100)\text{Eq. }2 + \text{Eq. }3 \end{cases}$

$$140{,}000a = 7200 \Rightarrow a \approx 0.0514$$
$$1000b + 100{,}000(0.0514) = 6020 \Rightarrow b \approx 0.8771$$
$$5c + 250(0.8771) + 13{,}500(0.0514) = 923 \Rightarrow c \approx 1.8857$$

Least-squares regression parabola: $y = 0.0514x^2 + 0.8771x + 1.8857$

(b)

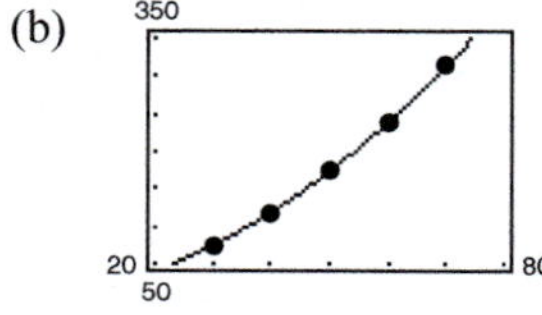

The model fits the data well.

(c) When $x = 75$, $y = 0.0514(75)^2 + 0.8771(75) + 1.8857 \approx 356$ feet.

71. $\begin{cases} 2x - 2x\lambda = 0 \Rightarrow 2x(1 - \lambda) = 0 \Rightarrow \lambda = 1 \text{ or } x = 0 \\ -2y + \lambda = 0 \\ y - x^2 = 0 \end{cases}$

If $\lambda = 1$: $2y = \lambda \Rightarrow y = \frac{1}{2}$

$x^2 = y \Rightarrow x = \pm\sqrt{\frac{1}{2}} = \pm\frac{\sqrt{2}}{2}$

If $x = 0$: $x^2 = y \Rightarrow y = 0$

$2y = \lambda \Rightarrow \lambda = 0$

Solution: $x = \pm\frac{\sqrt{2}}{2}$ or $x = 0$

$y = \frac{1}{2}$ $\quad$ $y = 0$

$\lambda = 1$ $\quad$ $\lambda = 0$

73. False. For example, refer to Example 6 on page 655,

$\begin{cases} x - 2y + z = 2 \\ 2x - y - z = 1 \end{cases}$

has the solution set of all ordered triples of the form $(a, a - 1, a)$ where a is a real number. Therefore, it is not an unique solution.

75. No, they are not equivalent. There are two arithmetic errors. The constant in the second equation should be -11 and the coefficient of z in the third equation should be 2.

77. Answers will vary. *Sample answer:*

$(2, 0, -1)$ is a solution to $\begin{cases} x + y + z = 1 \\ 3x + y + z = 5 \\ -x + 2y + 3z = -5 \end{cases}$

79. Answers will vary. *Sample answer:*

$\left(\frac{1}{2}, -3, 0\right)$ is a solution to $\begin{cases} 2x + y + z = -2 \\ 4x + y + z = -1 \\ -2x + 2y + 3z = -7 \end{cases}$

Section 7.4 Partial Fractions

1. partial fraction decomposition

3. partial fraction

5. $\frac{3x - 1}{x(x - 4)} = \frac{A}{x} + \frac{B}{x - 4}$

Matches (b).

6. $\frac{3x - 1}{x^2(x - 4)} = \frac{A}{x} + \frac{B}{x^2} + \frac{C}{x - 4}$

Matches (c).

7. $\frac{3x - 1}{x(x^2 + 4)} = \frac{A}{x} + \frac{Bx + C}{x^2 + 4}$

Matches (d).

8. $\frac{3x - 1}{x(x^2 - 4)} = \frac{3x - 1}{x(x - 2)(x + 2)}$

$= \frac{A}{x} + \frac{B}{x - 2} + \frac{C}{x + 2}$

Matches (a).

9. $\frac{3}{x^2 - 2x} = \frac{3}{x(x - 2)} = \frac{A}{x} + \frac{B}{x - 2}$

11. $\frac{6x + 5}{(x + 2)^4} = \frac{6x + 5}{(x + 2)(x + 2)(x + 2)(x + 2)}$

$= \frac{A}{x + 2} + \frac{B}{(x + 2)^2} + \frac{C}{(x + 2)^3} + \frac{D}{(x + 2)^4}$

13. $\frac{2x - 3}{x^3 + 10x} = \frac{2x - 3}{x(x^2 + 10)} = \frac{A}{x} + \frac{Bx + C}{x^2 + 10}$

15. $\frac{8x}{x^2(x^2 + 3)^2} = \frac{A}{x} + \frac{B}{x^2} + \frac{Cx + D}{x^2 + 3} + \frac{Ex + F}{(x^2 + 3)^2}$

17. $\frac{1}{x^2 + x} = \frac{A}{x} + \frac{B}{x + 1}$

$1 = A(x + 1) + Bx$

Let $x = 0$: $1 = A$

Let $x = -1$: $1 = -B \Rightarrow B = -1$

$\frac{1}{x^2 + x} = \frac{1}{x} - \frac{1}{x + 1}$

19. $\dfrac{3}{x^2 + x - 2} = \dfrac{A}{x - 1} + \dfrac{B}{x + 2}$

$3 = A(x + 2) + B(x - 1)$

Let $x = 1$: $3 = 3A \Rightarrow A = 1$

Let $x = -2$: $3 = -3B \Rightarrow B = -1$

$$\frac{3}{x^2 + x - 2} = \frac{1}{x - 1} - \frac{1}{x + 2}$$

21. $\dfrac{1}{x^2 - 1} = \dfrac{A}{x + 1} + \dfrac{B}{x - 1}$

$1 = A(x - 1) + B(x + 1)$

Let $x = -1$: $1 = -2A \Rightarrow A = -\dfrac{1}{2}$

Let $x = 1$: $1 = 2B \Rightarrow B = \dfrac{1}{2}$

$$\frac{1}{x^2 - 1} = \frac{1/2}{x - 1} - \frac{1/2}{x + 1} = \frac{1}{2}\left(\frac{1}{x - 1} - \frac{1}{x + 1}\right)$$

23. $\dfrac{x^2 + 12x + 12}{x^3 - 4x} = \dfrac{A}{x} + \dfrac{B}{x + 2} + \dfrac{C}{x - 2}$

$x^2 + 12x + 12 = A(x + 2)(x - 2) + Bx(x - 2) + Cx(x + 2)$

Let $x = 0$: $12 = -4A \Rightarrow A = -3$

Let $x = -2$: $-8 = 8B \Rightarrow B = -1$

Let $x = 2$: $40 = 8C \Rightarrow C = 5$

$$\frac{x^2 + 12x + 12}{x^3 - 4x} = -\frac{3}{x} - \frac{1}{x + 2} + \frac{5}{x - 2}$$

25. $\dfrac{3x}{(x - 3)^2} = \dfrac{A}{x - 3} + \dfrac{B}{(x - 3)^2}$

$3x = A(x - 3) + B$

Let $x = 3$: $9 = B$

Let $x = 0$: $0 = -3A + B$

$0 = -3A + 9$

$3 = A$

$$\frac{3x}{(x - 3)^2} = \frac{3}{x - 3} + \frac{9}{(x - 3)^2}$$

27. $\dfrac{4x^2 + 2x - 1}{x^2(x + 1)} = \dfrac{A}{x} + \dfrac{B}{x^2} + \dfrac{C}{x + 1}$

$4x^2 + 2x - 1 = Ax(x + 1) + B(x + 1) + Cx^2$

Let $x = 0$: $-1 = B$

Let $x = -1$: $1 = C$

Let $x = 1$: $5 = 2A + 2B + C$

$5 = 2A - 2 + 1$

$6 = 2A$

$3 = A$

$$\frac{4x^2 + 2x - 1}{x^2(x + 1)} = \frac{3}{x} - \frac{1}{x^2} + \frac{1}{x + 1}$$

29. $\dfrac{x^2 + 2x + 3}{x^3 + x} = \dfrac{A}{x} + \dfrac{Bx + C}{x^2 + 1}$

$x^2 + 2x + 3 = A(x^2 + 1) + (Bx + C)(x)$

$x^2 + 2x + 3 = x^2(A + B) + Cx + A$

Equating coefficients of like terms gives $A + B = 1$, $C = 2$, and $A = 3$.

So, $A = 3$, $B = -2$, and $C = 2$.

$$\frac{x^2 + 2x + 3}{x^3 + x} = \frac{3}{x} - \frac{2x - 2}{x^2 + 1}$$

31. $\dfrac{x}{x^3 - x^2 - 2x + 2} = \dfrac{x}{(x-1)(x^2-2)} = \dfrac{A}{x-1} + \dfrac{Bx+C}{x^2-2}$

$$\begin{aligned} x &= A(x^2 - 2) + (Bx + C)(x - 1) \\ &= Ax^2 - 2A + Bx^2 - Bx + Cx - C \\ &= (A + B)x^2 + (C - B)x - (2A + C) \end{aligned}$$

Equating coefficients of like terms gives $0 = A + B, 1 = C - B,$ and $0 = 2A + C$. So, $A = -1, B = 1,$ and $C = 2$.

$$\frac{x}{x^3 - x^2 - 2x + 2} = -\frac{1}{x-1} + \frac{x+2}{x^2-2}$$

33. $\dfrac{x}{16x^4 - 1} = \dfrac{x}{(4x^2-1)(4x^2+1)} = \dfrac{x}{(2x+1)(2x-1)(4x^2+1)} = \dfrac{A}{2x+1} + \dfrac{B}{2x-1} + \dfrac{Cx+D}{4x^2+1}$

$$\begin{aligned} x &= A(2x-1)(4x^2+1) + B(2x+1)(4x^2+1) + (Cx+D)(2x+1)(2x-1) \\ &= A(8x^3 - 4x^2 + 2x - 1) + B(8x^3 + 4x^2 + 2x + 1) + (Cx + D)(4x^2 - 1) \\ &= 8Ax^3 - 4Ax^2 + 2Ax - A + 8Bx^3 + 4Bx^2 + 2Bx + B + 4Cx^3 + 4Dx^2 - Cx - D \\ &= (8A + 8B + 4C)x^3 + (-4A + 4B + 4D)x^2 + (2A + 2B - C)x + (-A + B - D) \end{aligned}$$

Equating coefficients of like terms gives $0 = 8A + 8B + 4C, 0 = -4A + 4B + 4D, 1 = 2A + 2B - C,$ and $0 = -A + B - D$.

Using the first and third equations, $2A + 2B + C = 0$ and $2A + 2B - C = 1$; by subtraction, $2C = -1$, so $C = -\frac{1}{2}$.

Using the second and fourth equations, $-A + B + D = 0$ and $-A + B - D = 0$; by subtraction $2D = 0$, so $D = 0$.

Substituting $-\frac{1}{2}$ for C and 0 for D in the first and second equations, $8A + 8B = 2$ and $-4A + 4B = 0$, so $A = \frac{1}{8}$ and $B = \frac{1}{8}$.

$$\frac{x}{16x^4 - 1} = \frac{\frac{1}{8}}{2x+1} + \frac{\frac{1}{8}}{2x-1} + \frac{\left(-\frac{1}{2}\right)x}{4x^2+1} = \frac{1}{8(2x+1)} + \frac{1}{8(2x-1)} - \frac{x}{2(4x^2+1)} = \frac{1}{8}\left(\frac{1}{2x+1} + \frac{1}{2x-1} - \frac{4x}{4x^2+1}\right)$$

35. $\dfrac{x^2 + 5}{(x+1)(x^2 - 2x + 3)} = \dfrac{A}{x+1} + \dfrac{Bx+C}{x^2 - 2x + 3}$

$$\begin{aligned} x^2 + 5 &= A(x^2 - 2x + 3) + (Bx + C)(x + 1) = Ax^2 - 2Ax + 3A + Bx^2 + Bx + Cx + C \\ &= (A + B)x^2 + (-2A + B + C)x + (3A + C) \end{aligned}$$

Equating coefficients of like terms gives $1 = A + B, 0 = -2A + B + C,$ and $5 = 3A + C$.

Subtracting both sides of the second equation from the first gives $1 = 3A - C$; combining this with the third equation gives $A = 1$ and $C = 2$. Because $A + B = 1, B = 0$.

$$\frac{x^2 + 5}{(x+1)(x^2 - 2x + 3)} = \frac{1}{x+1} + \frac{2}{x^2 - 2x + 3}$$

37. $\dfrac{2x^2 + x + 8}{\left(x^2 + 4\right)^2} = \dfrac{Ax + B}{x^2 + 4} + \dfrac{Cx + D}{\left(x^2 + 4\right)^2}$

$2x^2 + x + 8 = (Ax + B)(x^2 + 4) + Cx + D$

$2x^2 + x + 8 = Ax^3 + Bx^2 + (4A + C)x + (4B + D)$

Equating coefficients of like terms gives

$0 = A$

$2 = B$

$1 = 4A + C \Rightarrow C = 1$

$8 = 4B + D \Rightarrow D = 0$

$\dfrac{2x^2 + x + 8}{\left(x^2 + 4\right)^2} = \dfrac{2}{x^2 + 4} + \dfrac{x}{\left(x^2 + 4\right)^2}$

39. $\dfrac{5x^2 - 2}{\left(x^2 + 3\right)^3} = \dfrac{Ax + B}{x^2 + 3} + \dfrac{Cx + D}{\left(x^2 + 3\right)^2} + \dfrac{Ex + F}{\left(x^2 + 3\right)^3}$

$$\begin{aligned} 5x^2 - 2 &= (Ax + B)(x^2 + 3)^2 + (Cx + D)(x^2 + 3) + Ex + F \\ &= Ax^5 + 6Ax^3 + 9Ax + Bx^4 + 6Bx^2 + 9B + Cx^3 + 3Cx + Dx^2 + 3D + F + Ex \\ &= Ax^5 + Bx^4 + (6A + C)x^3 + (6B + D)x^2 + (9A + 3C + E)x + (9B + 3D + F) \end{aligned}$$

Equating coefficients of like terms gives

$0 = A$

$0 = B$

$0 = 6A + C \Rightarrow C = 0$

$5 = 6B + D \Rightarrow D = 5$

$0 = 9A + 3C + E \Rightarrow E = 0$

$-2 = 9B + 3D + F \Rightarrow F = -17$

$\dfrac{5x^2 - 2}{\left(x^2 + 3\right)^3} = \dfrac{5}{\left(x^2 + 3\right)^2} - \dfrac{17}{\left(x^2 + 3\right)^3}$

41. $\dfrac{8x - 12}{x^2\left(x^2 + 2\right)^2} = \dfrac{A}{x} + \dfrac{B}{x^2} + \dfrac{Cx + D}{x^2 + 2} + \dfrac{Ey + F}{\left(x^2 + 2\right)^2}$

$$\begin{aligned} 8x - 12 &= Ax(x^2 + 2)^2 + B(x^2 + 2)^2 + (Cx + D)x^2(x^2 + 2) + (Ex + F)x^2 \\ &= Ax^5 + 4Ax^3 + 4Ax + Bx^4 + 4Bx^2 + 4B + Cx^5 + 2Cx^3 + Dx^4 + 2Dx^2 + Ex^3 + Fx^2 \\ &= (A + C)x^5 + (B + D)x^4 + (4A + 2C + E)x^3 + (4B + 2D + F)x^2 + 4Ax + 4B \end{aligned}$$

Equating coefficients of like terms gives

$$\begin{aligned} A + C &= 0, \\ B + D &= 0, \\ 4A + 2C + E &= 0, \\ 4B + 2D + F &= 0, \\ 4A &= 8, \text{ and} \\ 4B &= -12. \end{aligned}$$

So, $A = 2, B = -3, C = -2, D = 3, E = -4$, and $F = 6$.

$\dfrac{8x - 12}{x^2\left(x^2 + 2\right)^2} = \dfrac{2}{x} + \dfrac{-3}{x^2} + \dfrac{-2x + 3}{x^2 + 2} + \dfrac{-4x + 6}{\left(x^2 + 2\right)^2}$

43. $\dfrac{x^2 - x}{x^2 + x + 1} = 1 + \dfrac{-2x - 1}{x^2 + x + 1} = 1 - \dfrac{2x + 1}{x^2 + x + 1}$

45. $\dfrac{2x^3 - x^2 + x + 5}{x^2 + 3x + 2} = 2x - 7 + \dfrac{18x + 19}{(x + 1)(x + 2)}$

$$\frac{18x + 19}{(x + 1)(x + 2)} = \frac{A}{x + 1} + \frac{B}{x + 2}$$

$$18x + 19 = A(x + 2) + B(x + 1)$$

Let $x = -1$: $1 = A$

Let $x = -2$: $-17 = -B \Rightarrow B = 17$

$$\frac{2x^3 - x^2 + x + 5}{x^2 + 3x + 2} = 2x - 7 + \frac{1}{x + 1} + \frac{17}{x + 2}$$

47. $\dfrac{x^4}{(x - 1)^3} = \dfrac{x^4}{x^3 - 3x^2 + 3x - 1}$

$$= x + 3 + \frac{6x^2 - 8x + 3}{(x - 1)^3}$$

$$\frac{6x^2 - 8x + 3}{(x - 1)^3} = \frac{A}{x - 1} + \frac{B}{(x - 1)^2} + \frac{C}{(x - 1)^3}$$

$$6x^2 - 8x + 3 = A(x - 1)^2 + B(x - 1) + C$$

Let $x = 1$: $1 = C$

Let $x = 0$: $3 = A - B + 1$ ⎫ $A - B = 2$
Let $x = 2$: $11 = A + B + 1$ ⎭ $A + B = 10$

So, $A = 6$ and $B = 4$.

$$\frac{x^4}{(x - 1)^3} = x + 3 + \frac{6}{x - 1} + \frac{4}{(x - 1)^2} + \frac{1}{(x - 1)^3}$$

49. $\dfrac{x^4 + 2x^3 + 4x^2 + 8x + 2}{x^3 + 2x^2 + x} = x + \dfrac{3x^2 + 8x + 2}{x^3 + 2x^2 + x}$

$$= x + \frac{3x^2 + 8x + 2}{x(x + 1)^2}$$

$$\frac{3x^2 + 8x + 2}{x(x + 1)^2} = \frac{A}{x} + \frac{B}{x + 1} + \frac{C}{(x + 1)^2}$$

$$3x^2 + 8x + 2 = A(x + 1)^2 + B(x)(x + 1) + C(x)$$

$$3x^2 + 8x + 2 = Ax^2 + 2Ax + A + Bx^2 + Bx + Cx$$

$$3x^2 + 8x + 2 = (A + B)x^2 + (2A + B + C)x + A$$

Equating coefficients of like terms gives
$A + B = 3$, $2A + B + C = 8$, and $A = 2$.

So, $A = 2$, $B = 1$, and $C = 3$.

$$\frac{x^4 + 2x^3 + 4x^2 + 8x + 2}{x^3 + 2x^2 + x} = x + \frac{2}{x} + \frac{1}{x + 1} + \frac{3}{(x + 1)^2}$$

51. $\dfrac{5 - x}{2x^2 + x - 1} = \dfrac{A}{2x - 1} + \dfrac{B}{x + 1}$

$$-x + 5 = A(x + 1) + B(2x - 1)$$

Let $x = \frac{1}{2}$: $\frac{9}{2} = \frac{3}{2}A \Rightarrow A = 3$

Let $x = -1$: $6 = -3B \Rightarrow B = -2$

$$\frac{5 - x}{2x^2 + x - 1} = \frac{3}{2x - 1} - \frac{2}{x + 1}$$

2
−6 6
−6

53. $\dfrac{3x^2 - 7x - 2}{x^3 - x} = \dfrac{A}{x} + \dfrac{B}{x + 1} + \dfrac{C}{x - 1}$

$$3x^2 - 7x - 2 = A(x^2 - 1) + Bx(x - 1) + Cx(x + 1)$$

Let $x = 0$: $-2 = -A \Rightarrow A = 2$

Let $x = -1$: $8 = 2B \Rightarrow B = 4$

Let $x = 1$: $-6 = 2C \Rightarrow C = -3$

$$\frac{3x^2 - 7x - 2}{x^3 - x} = \frac{2}{x} + \frac{4}{x + 1} - \frac{3}{x - 1}$$

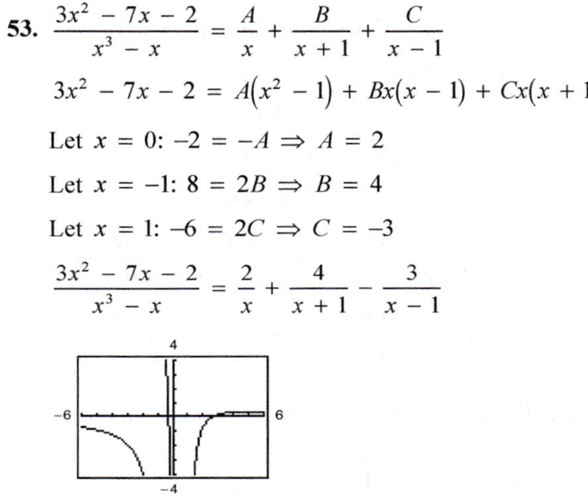

55. $\dfrac{x^2 + x + 2}{(x^2 + 2)^2} = \dfrac{Ax + B}{x^2 + 2} + \dfrac{Cx + D}{(x^2 + 2)^2}$

$$x^2 + x + 2 = (Ax + B)(x^2 + 2) + Cx + D$$

$$x^2 + x + 2 = Ax^3 + Bx^2 + (2A + C)x + (2B + D)$$

Equating coefficients of like terms gives

$0 = A$

$1 = B$

$1 = 2A + C \Rightarrow C = 1$

$2 = 2B + D \Rightarrow D = 0$

$$\frac{x^2 + x + 2}{(x^2 + 2)^2} = \frac{1}{x^2 + 2} + \frac{x}{(x^2 + 2)^2}$$

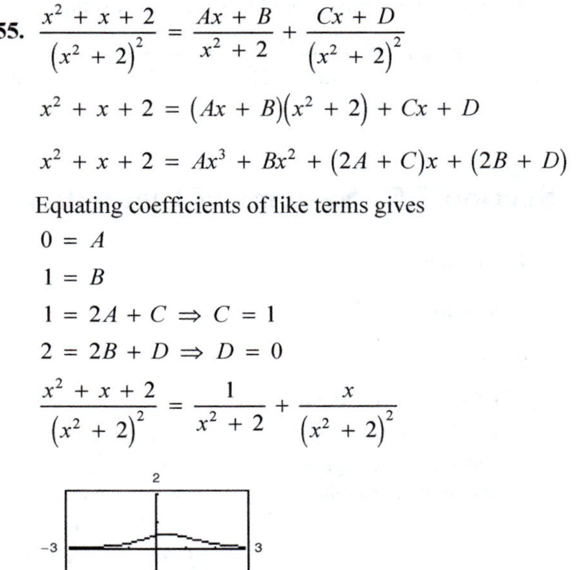

57. $\dfrac{2x^3 - 4x^2 - 15x + 5}{x^2 - 2x - 8} = 2x + \dfrac{x + 5}{(x + 2)(x - 4)}$

$$\frac{x + 5}{(x + 2)(x - 4)} = \frac{A}{x + 2} + \frac{B}{x - 4}$$

$$x + 5 = A(x - 4) + B(x + 2)$$

Let $x = -2$: $3 = -6A \Rightarrow A = -\dfrac{1}{2}$

Let $x = 4$: $9 = 6B \Rightarrow B = \dfrac{3}{2}$

$$\frac{2x^3 - 4x^2 - 15x + 5}{x^2 - 2x - 8} = 2x + \frac{1}{2}\left(\frac{3}{x - 4} - \frac{1}{x + 2}\right)$$

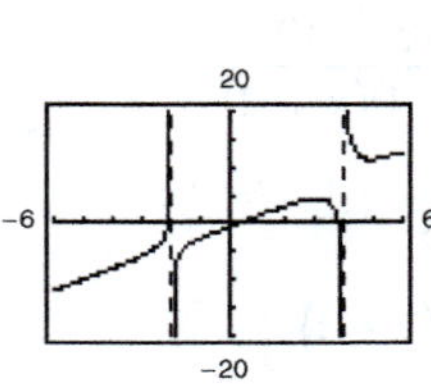

59. $C = \dfrac{120p}{10{,}000 - p^2} = \dfrac{120p}{(100 + p)(100 - p)} = \dfrac{A}{100 + p} + \dfrac{B}{100 - p}$

$$120p = A(100 - p) + B(100 + p)$$

Let $p = 100$: $200B = 12{,}000$

$B = 60$

Let $p = -100$: $200A = -12{,}000$

$A = -60$

$$C = \frac{120p}{10{,}000 - p^2} = -\frac{60}{100 + p} + \frac{60}{100 - p}$$

Let $y_1 = \dfrac{120p}{10{,}000 - p^2}$ and $y_2 = -\dfrac{60}{100 + p} + \dfrac{60}{100 - p}$.

X	Y1	Y2
0	0	0
10	.12121	.12121
20	.25	.25
30	.3956	.3956
40	.57143	.57143
50	.8	.8
60	1.125	1.125

X=0

61. False. The partial fraction decomposition is

$$\frac{A}{x + 10} + \frac{B}{x - 10} + \frac{C}{(x - 10)^2}.$$

63. False. The degrees could be equal. For example,

Exercises #55: $\dfrac{x^2 + x + 2}{(x^2 + 2)^2} = \dfrac{1}{x^2 + 2} + \dfrac{x}{(x^2 + 2)^2}$.

65. The expression is improper, $\dfrac{x^2 + 1}{x(x - 1)} = \dfrac{x^2 + 1}{x^2 - x}$ so first divide the denominator into the numerator to obtain

$$\frac{x^2 + 1}{x^2 - x} = 1 + \frac{x + 1}{x^2 - x} = 1 + \frac{x + 1}{x(x - 1)}.$$

Then find the partial fraction decomposition of

$$\frac{x + 1}{x(x - 1)} = \frac{A}{x} + \frac{B}{x - 1}$$

Section 7.5 Systems of Inequality

1. solution

3. solution

5. $y < 5 - x^2$

Using a dashed line, graph $y = 5 - x^2$, and shade the region inside the parabola.

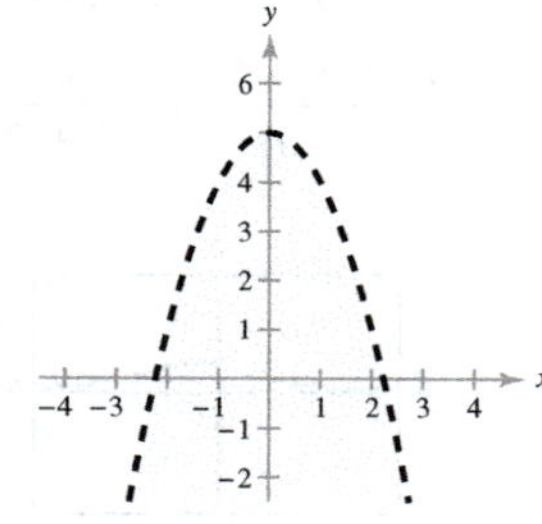

7. $x \geq 6$

Using a solid line, graph the vertical line $x = 6$, and shade to the right of this line.

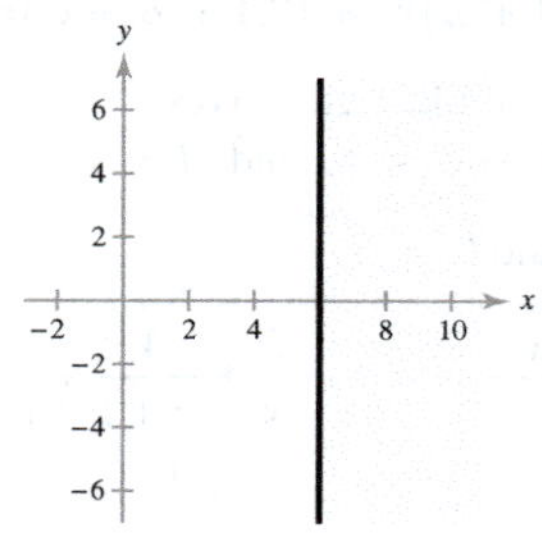

9. $y > -7$

Using a dashed line, graph the horizontal line $y = -7$, and shade above the line.

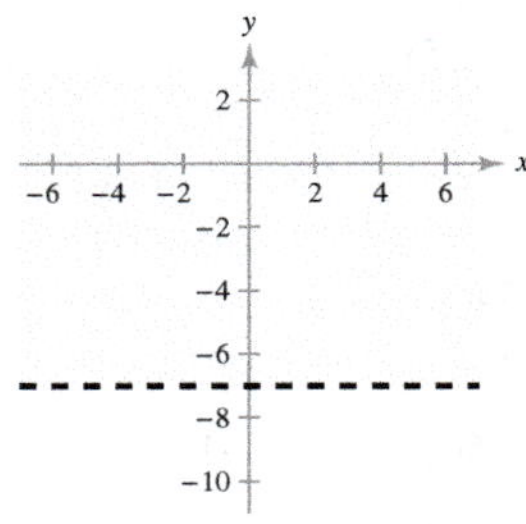

11. $y < 2 - x$

Using a dashed line, graph $y = 2 - x$, and then shade below the line. (Use $(0, 0)$ as a test point.)

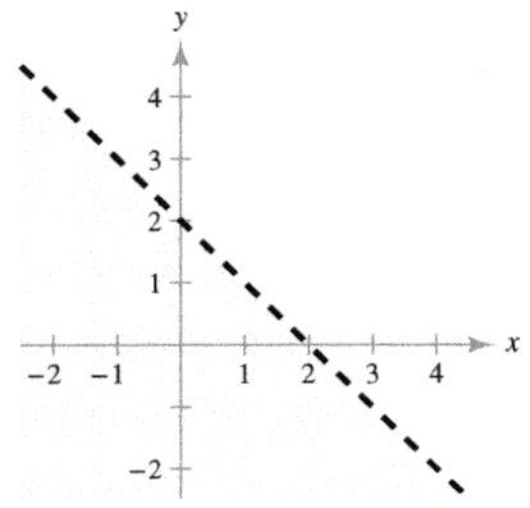

13. $2y - x \geq 4$

Using a solid line, graph $2y - x = 4$, and then shade above the line. (Use $(0, 0)$ as a test point.)

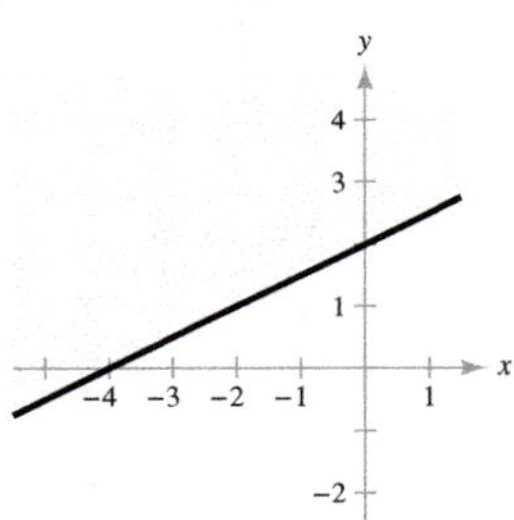

15. $x^2 + (y - 3)^2 < 4$

Using a dashed line, sketch the circle $x^2 + (y - 3)^2 = 4$.

Center: $(0, 3)$

Radius: 2

Test point: $(0, 0)$

Shade the inside of the circle.

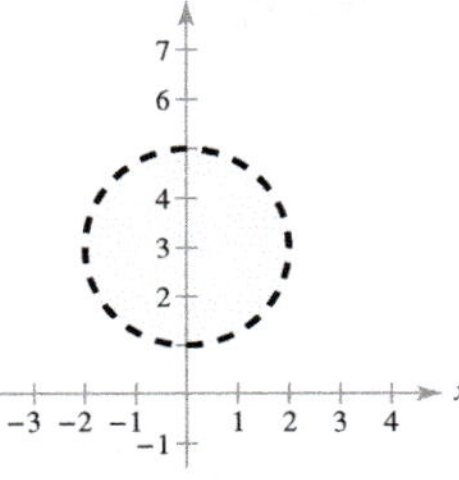

17. $y > -\dfrac{2}{x^2 + 1}$

Using a solid line, graph $y = -\dfrac{2}{x^2 + 1}$. Use $(0, 0)$ as a test point. Then shade above the curve.

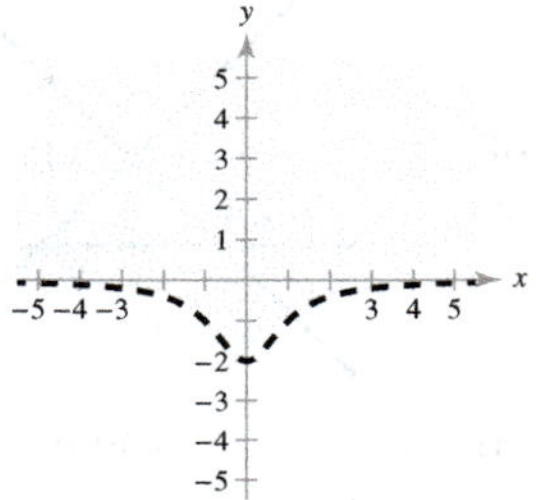

19. $y \geq -\ln(x - 1)$

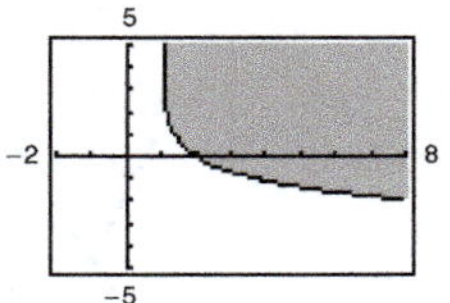

21. $y < 2^x$

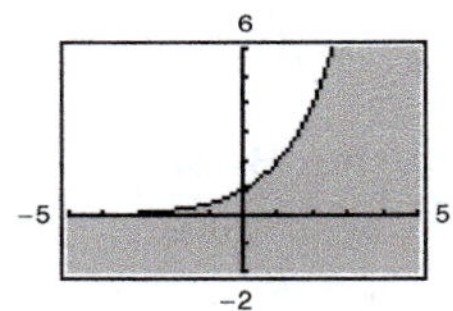

23. $y \leq 2 - \frac{1}{5}x$

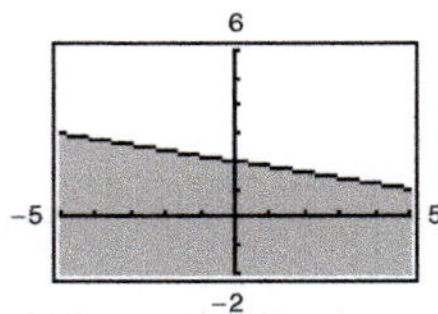

25. $\frac{2}{3}y + 2x^2 - 5 \geq 0$

$$\frac{2}{3}y \geq 5 - 2x^2$$

$$y \geq \frac{3}{2}(5 - 2x^2)$$

$$y \geq \frac{15}{2} - 3x^2$$

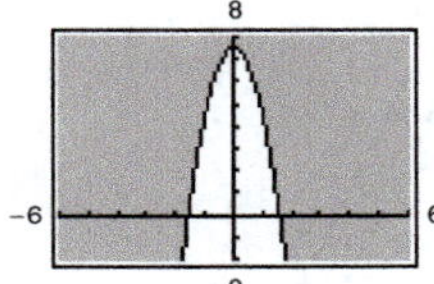

27. The line through $(-5, 0)$ and $(-1, 0)$ is $y = 5x + 5$. The shaded region below the line gives $y < 5x + 5$.

29. The line through $(0, 2)$ and $(3, 0)$ is $y = -\frac{2}{3}x + 2$.

The shaded region above the line gives $y \geq -\frac{2}{3}x + 2$.

31. 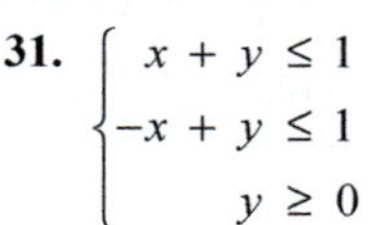

$$\begin{cases} x + y \leq 1 \\ -x + y \leq 1 \\ y \geq 0 \end{cases}$$

First, find the points of intersection of each pair of equations.

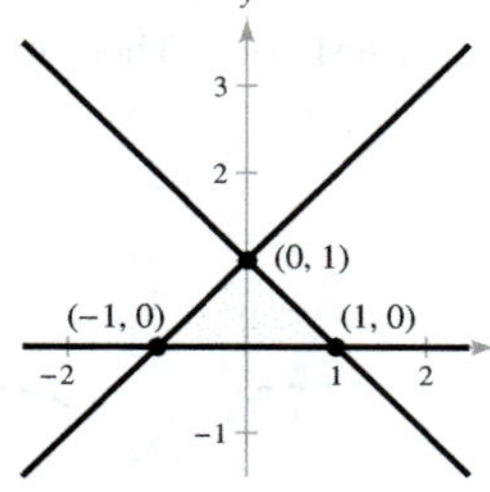

Vertex A	Vertex B	Vertex C
$x + y = 1$	$x + y = 1$	$-x + y = 1$
$-x + y = 1$	$y = 0$	$y = 0$
$(0, 1)$	$(1, 0)$	$(-1, 0)$

33. 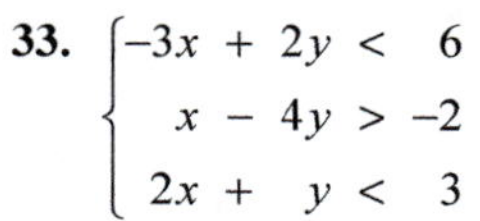

$$\begin{cases} -3x + 2y < 6 \\ x - 4y > -2 \\ 2x + y < 3 \end{cases}$$

First, find the points of intersection of each pair of equations.

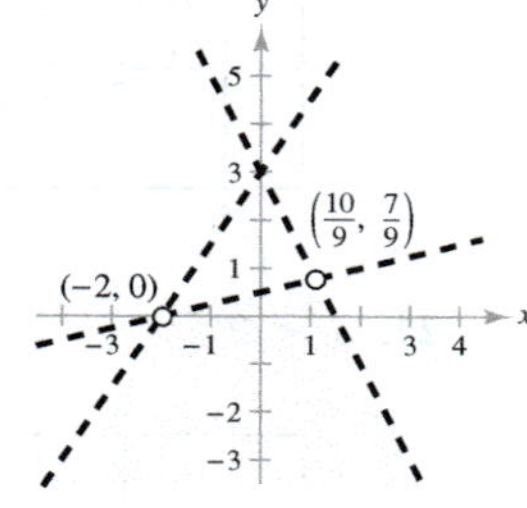

Vertex A	Vertex B
$-3x + 2y = 6$	$-3x + 2y = 6$
$x - 4y = -2$	$2x + y = 3$
$(-2, 0)$	$(0, 3)$

Vertex C

$x - 4y = -2$

$2x + y = 3$

$\left(\frac{10}{9}, \frac{7}{9}\right)$

Note that B is not a vertex of the solution region.

35. $\begin{cases} 2x + y > 2 \\ 6x + 3y < 2 \end{cases}$

The graphs of $2x + y = 2$ and $6x + 3y = 2$ are parallel lines. The first inequality has the region above the line shaded. The second inequality has the region below the line shaded. There are no points that satisfy both inequalities.

No solution

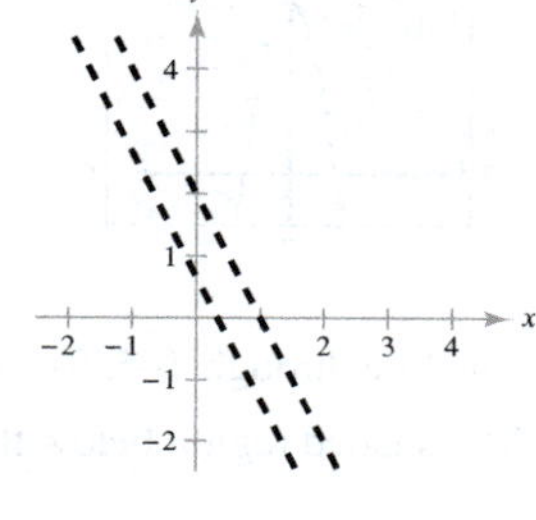

37. $\begin{cases} 2x - 3y > 7 \\ 5x + y < 9 \end{cases}$

$2x - 3y = 7$

$5x + y = 9 \Rightarrow y = -5x + 9$

$2x - 3(-5x + 9) = 7$

$2x + 15x - 27 = 7$

$17x = 34$

$x = 2$

$y = -5(2) + 9 = -1$

$(2, -1)$

Point of intersection: $(2, -1)$

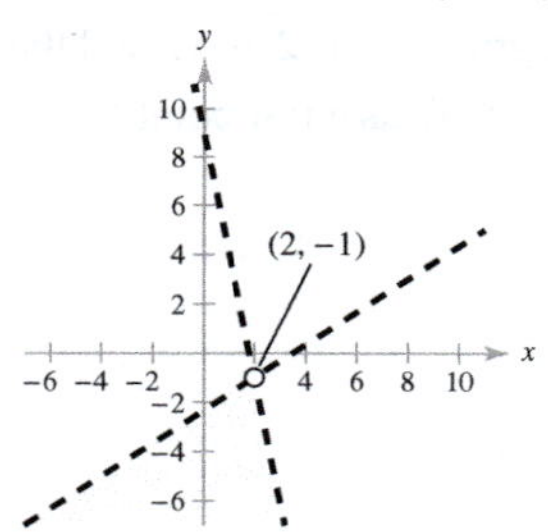

39. $\begin{cases} x^2 + y \leq 7 \\ x \geq -2 \\ y \geq 0 \end{cases}$

First, find the points of intersection of each pair of equations.

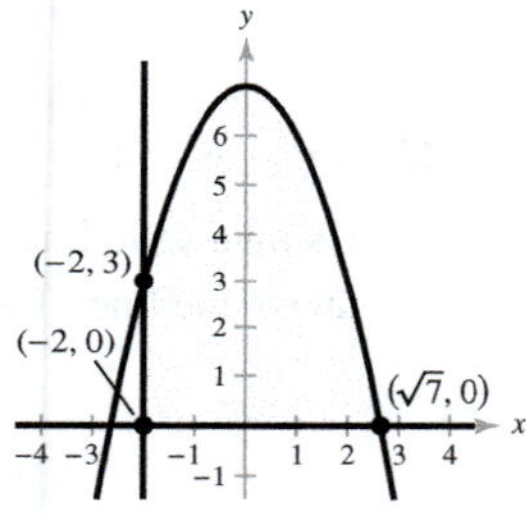

Vertex A	Vertex B
$x^2 + y = 7, x = -2$	$x^2 + y = 7, y = 0$
$4 + y = 7$	$x^2 = 7$
$y = 3$	$x = \sqrt{7}$
$(-2, 3)$	$(\sqrt{7}, 0)$

Vertex C

$x = -2, y = 0$

$(-2, 0)$

41. $\begin{cases} x - y^2 > 0 \\ x - y > 2 \end{cases}$

Points of intersection:

$$y^2 = y + 2$$
$$y^2 - y - 2 = 0$$
$$(y + 1)(y - 2) = 0$$
$$y = -1, 2$$

$(1, -1), (4, 2)$

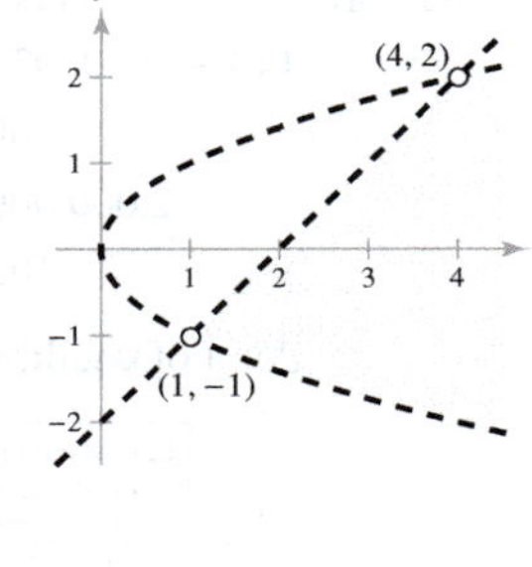

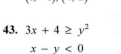

43. $3x + 4 \geq y^2$

$x - y < 0$

Points of intersection:

$x - y = 0 \Rightarrow y = x$

$3y + 4 = y^2$

$0 = y^2 - 3y - 4$

$0 = (y - 4)(y + 1)$

$y = 4 \text{ or } y = -1$

$x = 4 \quad x = -1$

$(4, 4)$ and $(-1, -1)$

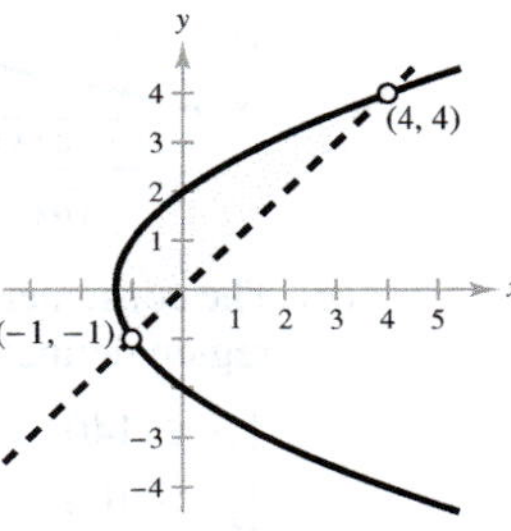

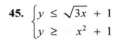

45. $\begin{cases} y \leq \sqrt{3x} + 1 \\ y \geq x^2 + 1 \end{cases}$

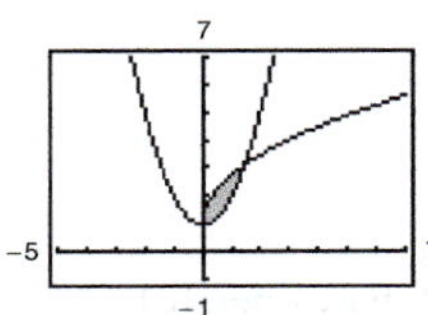

47. $\begin{cases} y < -x^2 + 2x + 3 \\ y > x^2 - 4x + 3 \end{cases}$

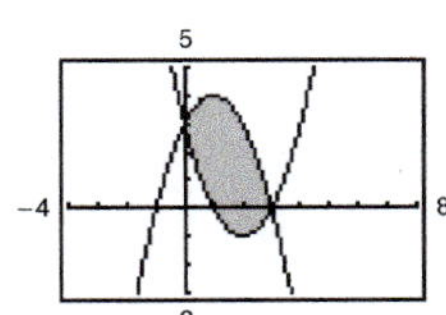

49. $\begin{cases} x^2y \geq 1 \Rightarrow y \geq \dfrac{1}{x^2} \\ 0 < x \leq 4 \\ y \leq 4 \end{cases}$

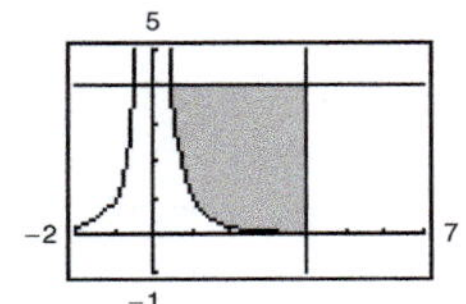

51. Line through points $(6, 0)$ and $(0, 6)$: $y = 6 - x$

$\begin{cases} x \geq 0 \\ y \geq 0 \\ y \leq 6 - x \end{cases}$

53. $(8, 0), (0, 8)$

$\begin{cases} x \geq 0 \\ y \geq 0 \\ x^2 + y^2 < 64 \end{cases}$

55. Rectangular region with vertices at $(4, 3), (9, 3), (9, 9), (4, 9)$

$\begin{cases} x \geq 4 \\ x \leq 9 \\ y \geq 3 \\ y \leq 9 \end{cases}$

This system may be written as:

$\begin{cases} 4 \leq x \leq 9 \\ 3 \leq y \leq 9 \end{cases}$

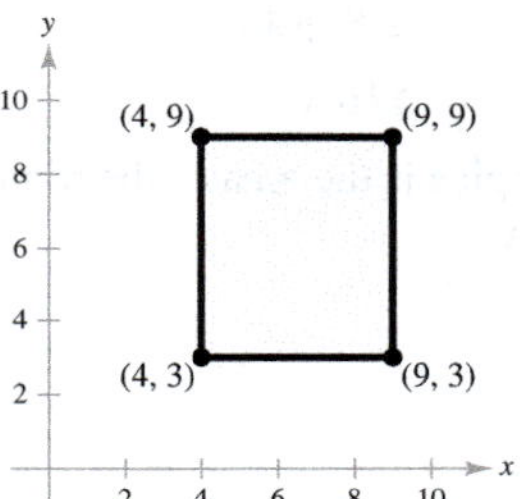

57. Triangle with vertices at $(0, 0), (6, 0), (1, 5)$

$(0, 0), (6, 0)$: $y = 0$

$(0, 0), (1, 5)$: $y = 5x$

$(6, 0), (1, 5)$: $y = -x + 6$

$\begin{cases} y \geq 0 \\ y \leq 5x \\ y \leq -x + 6 \end{cases}$

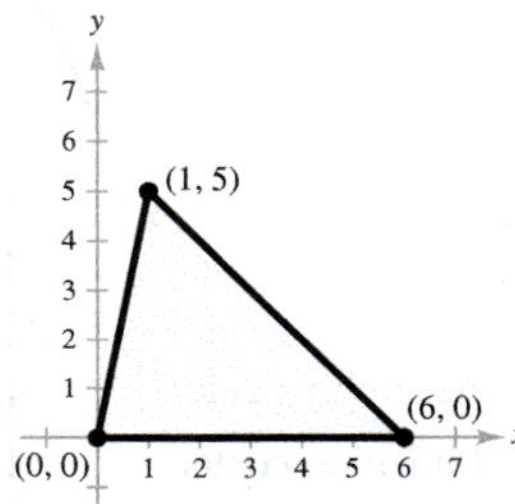

59. (a)

$$\begin{aligned} \text{Demand} &= \text{Supply} \\ 50 - 0.5x &= 0.125x \\ 50 &= 0.625x \\ 80 &= x \\ 10 &= p \end{aligned}$$

Point of equilibrium: $(80, 10)$

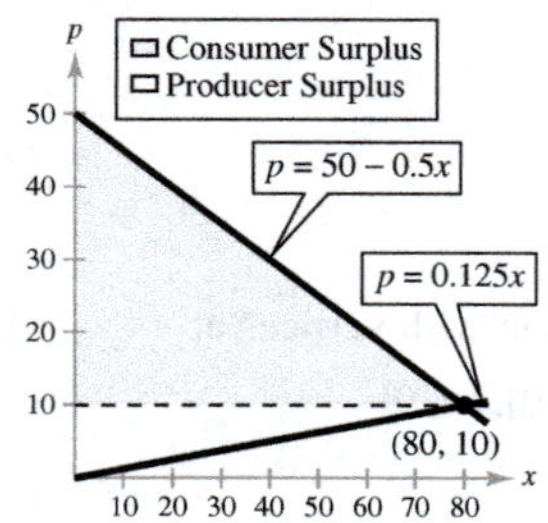

(b) The consumer surplus is the area of the triangular region defined by

$$\begin{cases} p \le 50 - 0.5x \\ p \ge 10 \\ x \ge 0. \end{cases}$$

$$\begin{aligned} \text{Consumer surplus} &= \tfrac{1}{2}(\text{base})(\text{height}) \\ &= \tfrac{1}{2}(80)(40) \\ &= \$1600 \end{aligned}$$

The producer surplus is the area of the triangular region defined by

$$\begin{cases} p \ge 0.125x \\ p \le 10 \\ x \ge 0. \end{cases}$$

$$\begin{aligned} \text{Producer surplus} &= \tfrac{1}{2}(\text{base})(\text{height}) \\ &= \tfrac{1}{2}(80)(10) \\ &= \$400 \end{aligned}$$

61. (a)

$$\begin{aligned} \text{Demand} &= \text{Supply} \\ 140 - 0.00002x &= 80 + 0.00001x \\ 60 &= 0.00003x \\ 2{,}000{,}000 &= x \\ 100 &= p \end{aligned}$$

Point of equilibrium: $(2{,}000{,}000, 100)$

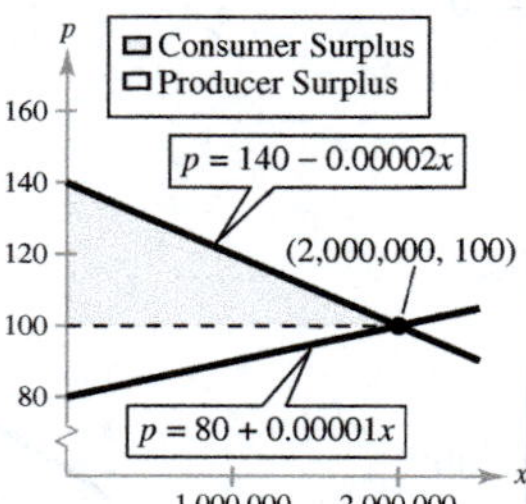

(b) The consumer surplus is the area of the triangular region defined by

$$\begin{cases} p \le 140 - 0.00002x \\ p \ge 100 \\ x \ge 0. \end{cases}$$

$$\begin{aligned} \text{Consumer surplus} &= \tfrac{1}{2}(\text{base})(\text{height}) \\ &= \tfrac{1}{2}(2{,}000{,}000)(40) \\ &= \$40{,}000{,}000 \end{aligned}$$

The producer surplus is the area of the triangular region defined by

$$\begin{cases} p \ge 80 + 0.00001x \\ p \le 100 \\ x \ge 0. \end{cases}$$

$$\begin{aligned} \text{Producer surplus} &= \tfrac{1}{2}(\text{base})(\text{height}) \\ &= \tfrac{1}{2}(2{,}000{,}000)(20) \\ &= \$20{,}000{,}000 \end{aligned}$$

63. x = amount in smaller account

y = amount in larger account

Account constraints:

$$\begin{cases} x + y \le 20{,}000 \\ y \ge 2x \\ x \ge 5{,}000 \\ y \ge 5{,}000 \end{cases}$$

65. x = number of tables

y = number of chairs

$$\begin{cases} x + \tfrac{3}{2}y \le 12 & \text{Assembly center} \\ \tfrac{4}{3}x + \tfrac{3}{2}y \le 15 & \text{Finishing center} \\ x \ge 0 \\ y \ge 0 \end{cases}$$

67. (a) x = number of ounces of food X

y = number of ounces of food Y

$$\begin{cases} 180x + 100y \geq 1000 \text{ (calcium)} \\ 6x + y \geq 18 \text{ (iron)} \\ 220x + 40y \geq 400 \text{ (magnesium)} \\ x \geq 0 \\ y \geq 0 \end{cases}$$

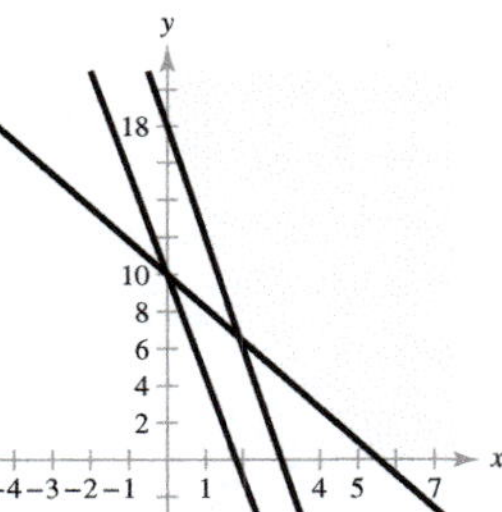

(b) Answers will vary. Some possible solutions which would satisfy the minimum daily requirements for calcium, iron, and magnesium:

$(5, 10) \Rightarrow$ 5 ounces of food X and 10 ounces of food Y

$(4, 12) \Rightarrow$ 4 ounces of food X and 12 ounces of food Y

Either of these will satisfy the minimum daily requirements of the dietician's special dietary diet plan.

69. (a) Let x = number of bags of gravel

Let y = number of bags of stone.

The delivery requirements are:

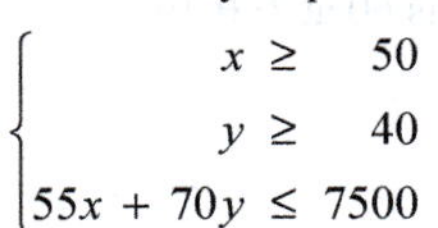

$$\begin{cases} x \geq 50 \\ y \geq 40 \\ 55x + 70y \leq 7500 \end{cases}$$

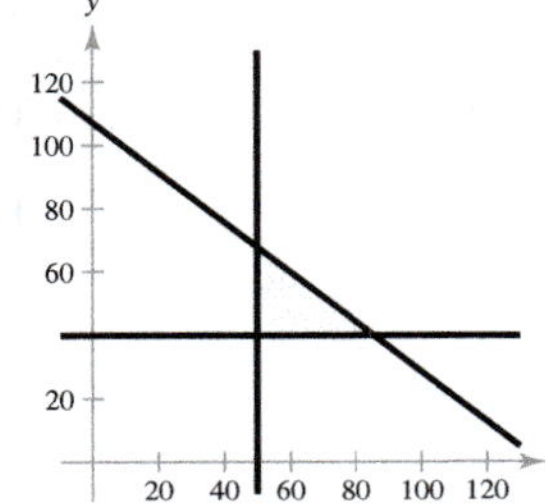

(b) The points $(60, 60)$ and $(70, 52)$ lie in the solution region. These values would represent the number of bags of each type of fill while maintaining the maximum weight capacity of the truck. The first $(60, 60)$ is to ship 60 bags of gravel and 60 bags of stone. The second $(70, 52)$ is to is to ship 70 bags of gravel and 52 bags of stone.

71. True. The figure is a rectangle with a length of 9 units and a width of 11 units.

73. Test a point on each side of the line $y = -x + 3$. Because the origin $(0, 0)$ satisfies the inequality, the solution set of the inequality lies below the dashed line.

75. (a) $\begin{cases} x^2 + y^2 \leq 16 & \Rightarrow \text{region inside the circle} \\ x + y \geq 4 & \Rightarrow \text{region above the line} \end{cases}$

Matches graph (iv).

(b) $\begin{cases} x^2 + y^2 \leq 16 & \Rightarrow \text{region inside the circle} \\ x + y \leq 4 & \Rightarrow \text{region below the line} \end{cases}$

Matches graph (ii).

(c) $\begin{cases} x^2 + y^2 \geq 16 & \Rightarrow \text{region outside the circle} \\ x + y \geq 4 & \Rightarrow \text{region above the line} \end{cases}$

Matches graph (iii).

(d) $\begin{cases} x^2 + y^2 \geq 16 & \Rightarrow \text{region outside the circle} \\ x + y \leq 4 & \Rightarrow \text{region below the line} \end{cases}$

Matches graph (i).

Section 7.6 Linear Programming

1. optimization

3. objective

5. inside; on

7. $z = 4x + 3y$

At $(0, 5)$: $z = 4(0) + 3(5) = 15$

At $(0, 0)$: $z = 4(0) + 3(0) = 0$

At $(5, 0)$: $z = 4(5) + 3(0) = 20$

The minimum value is 0 at $(0, 0)$.

The maximum value is 20 at $(5, 0)$.

9. $z = 2x + 5y$

At $(1, 0)$: $z = 2(1) + 5(0) = 2$

At $(4, 0)$: $z = 2(4) + 5(0) = 8$

At $(3, 4)$: $z = 2(3) + 5(4) = 26$

At $(0, 5)$: $z = 2(0) + 5(5) = 25$

The minimum value is 2 at $(1, 0)$.

The maximum value is 26 at $(3, 4)$.

11. $z = 10x + 7y$

At $(0, 20)$: $z = 10(0) + 7(20) = 140$

At $(30, 45)$: $z = 10(30) + 7(45) = 615$

At $(60, 20)$: $z = 10(60) + 7(20) = 740$

At $(60, 0)$: $z = 10(60) + 7(0) = 600$

At $(0, 45)$: $z = 10(0) + 7(45) = 315$

The minimum value is 140 at $(0, 20)$.

The maximum value is 740 at $(60, 20)$.

13. $z = 3x + 2y$

At $(3, 0)$: $z = 3(3) + 2(0) = 9$

The minimum value is 9 at $(3, 0)$.

The maximum value is 24 at any point on the line $3x + 2y = 24$, that is any point on the line segment between $(0, 12)$ and $(8, 0)$.

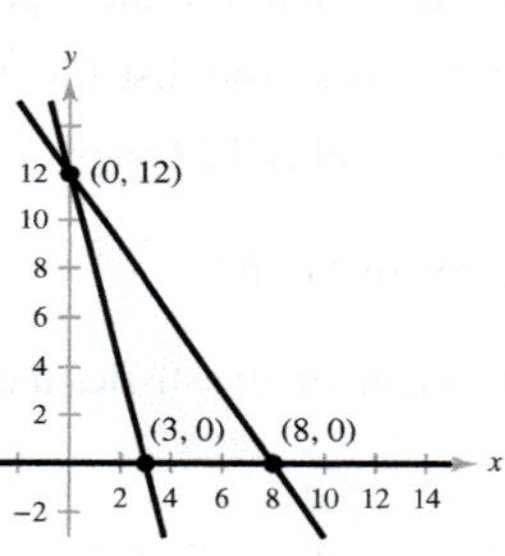

15. $z = 4x + 5y$

At $(10, 0)$: $z = 4(10) + 5(0) = 40$

At $(5, 3)$: $z = 4(5) + 5(3) = 35$

At $(0, 8)$: $z = 4(0) + 5(8) = 40$

The minimum value is 35 at $(5, 3)$.

The region is unbounded. There is no maximum.

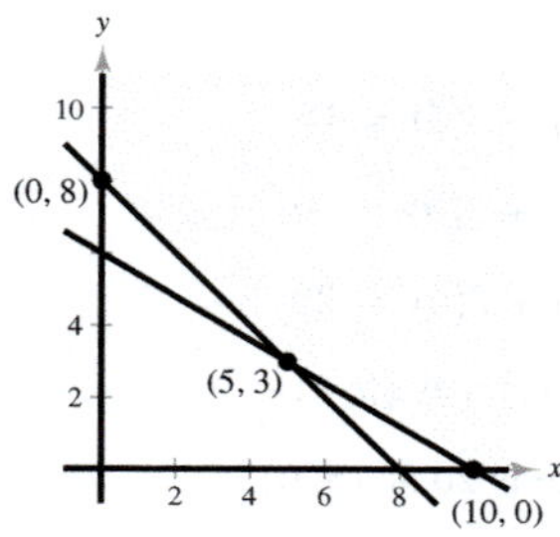

17. $z = 3x + y$

At $(16, 0)$: $z = 3(16) + 0 = 48$

At $(60, 0)$: $z = 3(60) + 0 = 180$

At $(7.2, 13.2)$: $z = 3(7.2) + 13.2 = 34.8$

The minimum value is 34.8 at $(7.2, 13.2)$.

The maximum value is 180 at $(60, 0)$.

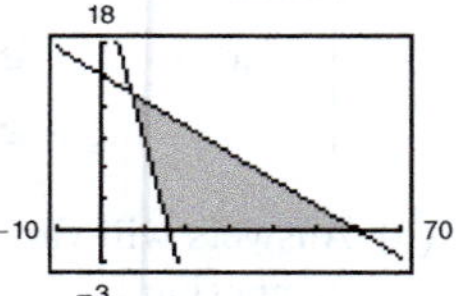

19. $z = x$

At $(60, 0)$: $z = 60$

At $(7.2, 13.2)$: $z = 7.2$

At $(16, 0)$: $z = 16$

The minimum value is 7.2 at $(7.2, 13.2)$.

The maximum value is 60 at $(60, 0)$.

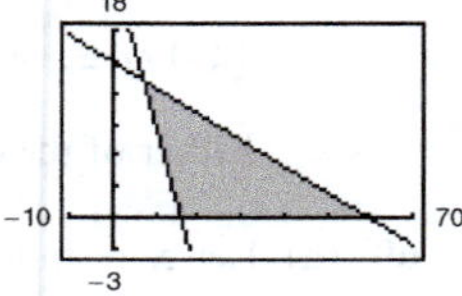

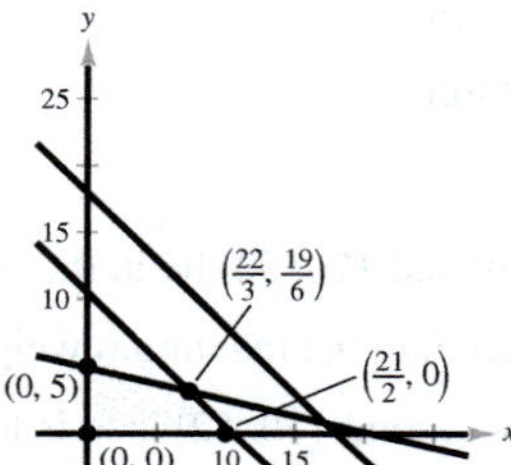

Figure for Exercises 21–23

21. $z = x + 5y$

At $(0, 5)$: $z = 0 + 5(5) = 25$

At $\left(\frac{22}{3}, \frac{19}{6}\right)$: $z = \frac{22}{3} + 5\left(\frac{19}{6}\right) = \frac{139}{6}$

At $\left(\frac{21}{2}, 0\right)$: $z = \frac{21}{2} + 5(0) = \frac{21}{2}$

At $(0, 0)$: $z = 0 + 5(0) = 0$

The minimum value is 0 at $(0, 0)$.

The maximum value is 25 at $(0, 5)$.

23. $z = 4x + 5y$

At $(0, 5)$: $z = 4(0) + 5(5) = 25$

At $\left(\frac{22}{3}, \frac{19}{6}\right)$: $z = 4\left(\frac{22}{3}\right) + 5\left(\frac{19}{6}\right) = \frac{271}{6}$

At $\left(\frac{21}{2}, 0\right)$: $z = 4\left(\frac{21}{2}\right) + 5(0) = 42$

At $(0, 0)$: $z = 4(0) + 5(0) = 0$

The minimum value is 0 at $(0, 0)$.

The maximum value is $\frac{271}{6}$ at $\left(\frac{22}{3}, \frac{19}{6}\right)$.

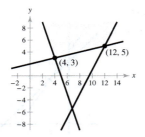

Figure for Exercises 25–27

25. $z = x + 2y$

At $(4, 3)$: $z = 4 + 2(3) = 10$

At $(12, 5)$: $z = 12 + 2(5) = 22$

The minimum value is 10 at $(4, 3)$.

There is no maximum value, the region is unbounded.

27. $z = x - y$

At $(4, 3)$: $z = 4 - 3 = 1$

At $(12, 5)$: $z = 12 - 5 = 7$

There is no minimum value.

The maximum value is 7 at $(12, 5)$.

29. Objective function: $z = 2.5x + y$

Constraints:
$x \geq 0, y \geq 0, 3x + 5y \leq 15, 5x + 2y \leq 10$

At $(0, 0)$: $z = 0$

At $(2, 0)$: $z = 5$

At $\left(\frac{20}{19}, \frac{45}{19}\right)$: $z = \frac{95}{19} = 5$

At $(0, 3)$: $z = 3$

The minimum value is 0 at $(0, 0)$.

The maximum value of 5 occurs at any point on the line segment connecting $(2, 0)$ and $\left(\frac{20}{19}, \frac{45}{19}\right)$.

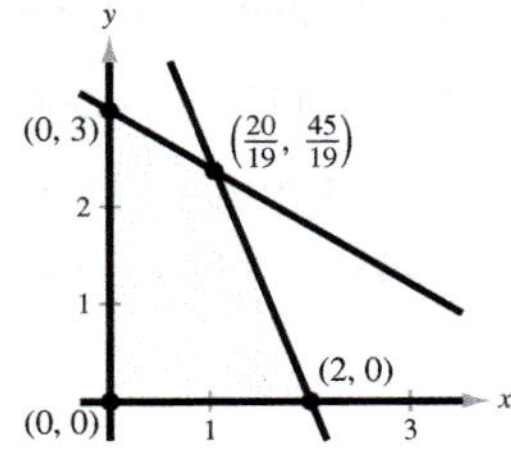

31. Objective function: $z = -x + 2y$

Constraints: $x \geq 0, y \geq 0, x \leq 10, x + y \leq 7$

At $(0, 0)$: $z = -0 + 2(0) = 0$

At $(0, 7)$: $z = -0 + 2(7) = 14$

At $(7, 0)$: $z = -7 + 2(0) = -7$

The constraint $x \leq 10$ is extraneous.

The minimum value is -7 at $(7, 0)$.

The maximum value is 14 at $(0, 7)$.

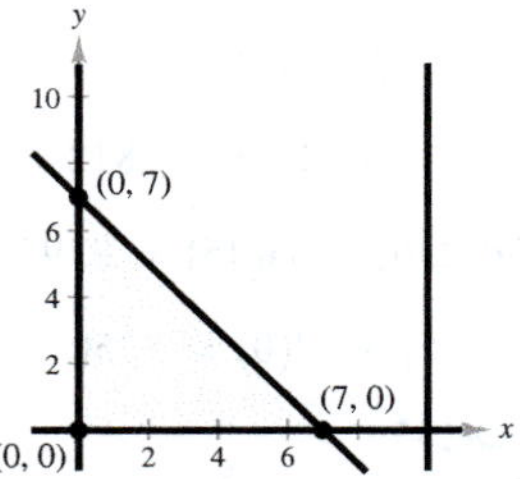

33. Objective function: $z = 3x + 4y$

Constraints: $x \geq 0, y \geq 0, x + y \leq 1, 2x + y \geq 4$

The feasible set is empty.

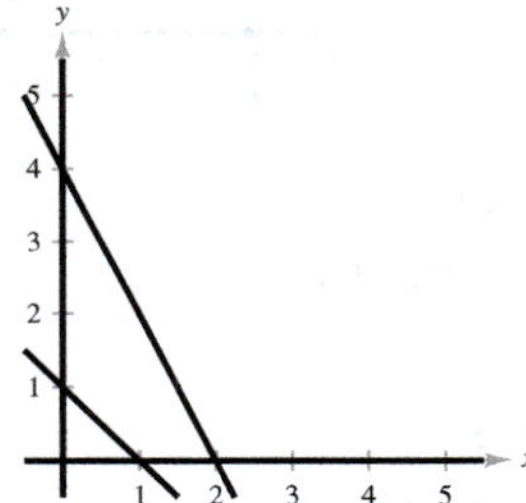

35. Objective function: $z = x + y$

Constraints: $x \geq 9, 0 \leq y \leq 7, -x + 3y \leq -6$

At $(9, 0)$: $z = 9 + 0 = 9$

At $(9, 1)$: $z = 9 + 1 = 10$

At $(27, 7)$: $z = 27 + 7 = 34$

The solution region is unbounded.

The minimum value is 9 at $(9, 0)$.

There is no maximum value.

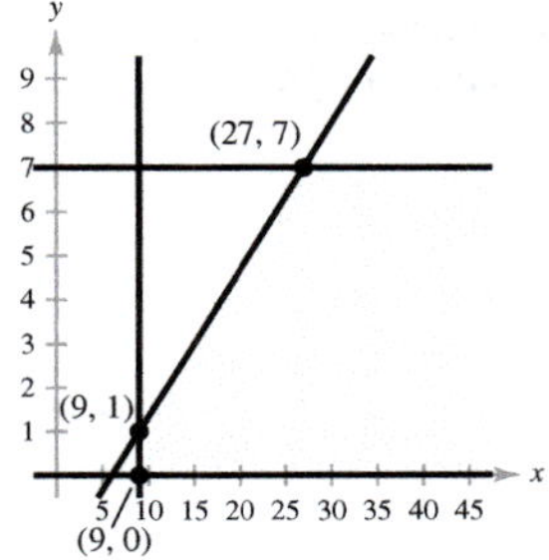

37. x = number of \$225 models

y = number of \$250 models

Constraints:

$$225x + 250y \le 63{,}000$$
$$x + y \le 275$$
$$x \ge 0$$
$$y \ge 0$$

Objective function: $P = 30x + 31y$

Vertices: $(0, 0)$, $(0, 252)$, $(230, 45)$ and $(275, 0)$

At $(0, 0)$: $P = 30(0) + 31(0) = 0$

At $(0, 252)$: $P = 30(0) + 31(252) = 7812$

At $(230, 45)$: $P = 30(230) + 31(45) = 8295$

At $(275, 0)$: $P = 30(275) + 31(0) = 8250$

An optimal profit of \$8295 occurs when 230 units of the \$225 model and 45 units of the \$250 model are stocked in inventory.

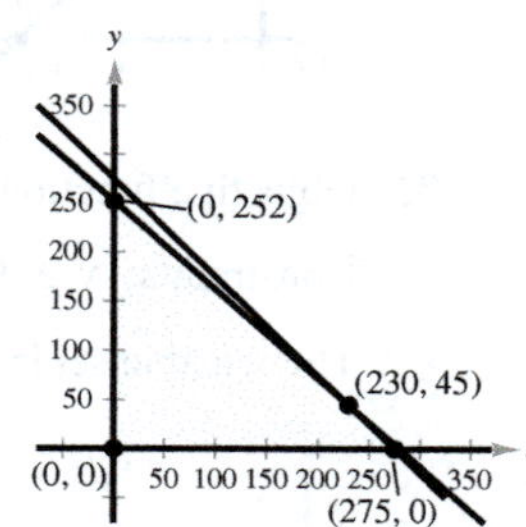

39. x = number of bags of Brand X

y = number of bags of Brand Y

Constraints: $3x + 9y \ge 30$

$$3x + 2y \ge 16$$
$$7x + 2y \ge 24$$
$$x \ge 0$$
$$y \ge 0$$

Objective function: $C = 25x + 15y$

Vertices: $(0, 12)$, $(4, 2)$, $(2, 5)$, $(10, 0)$

At $(0, 12)$: $C = 25(0) + 15(12) = 180$

At $(4, 2)$: $C = 25(4) + 15(2) = 130$

At $(2, 5)$: $C = 25(2) + 15(5) = 125$

At $(10, 0)$: $C = 25(10) + 15(0) = 250$

To minimize cost, use two bags of Brand X and five bags of Brand Y for a minimal cost of \$125.

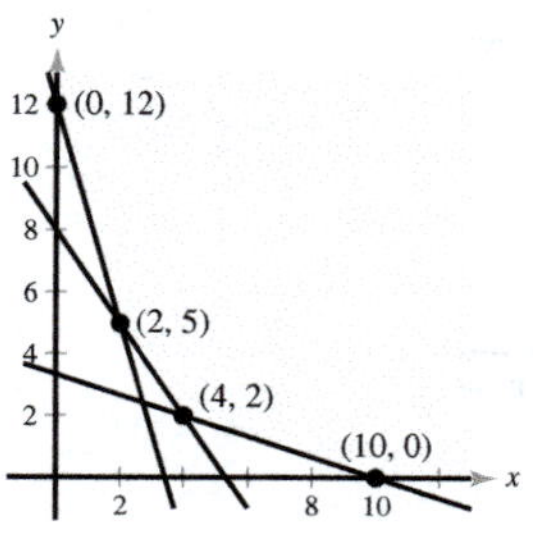

41. x = number of audits

y = number of tax returns

Constraints:

$$60x + 10y \le 780$$
$$16x + 4y \le 272$$
$$x \ge 0$$
$$y \ge 0$$

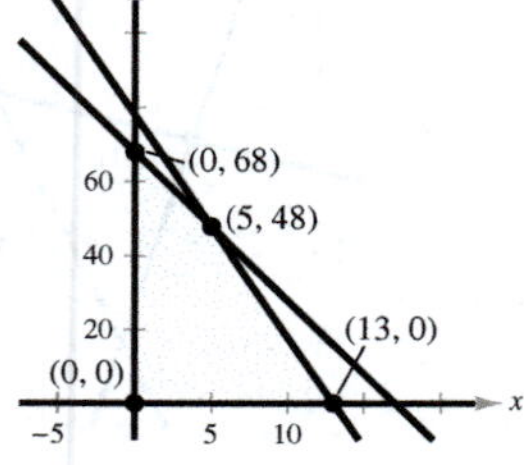

Objective function:

$R = 1600x + 250y$

Vertices: $(0, 0)$, $(13, 0)$, $(5, 48)$, $(0, 68)$

At $(0, 0)$: $R = 1600(0) + 250(0) = 0$

At $(13, 0)$: $R = 1600(13) + 250(0) = 20{,}800$

At $(5, 48)$: $R = 1600(5) + 250(48) = 20{,}000$

At $(0, 68)$: $R = 1600(0) + 250(68) = 17{,}000$

A maximum revenue of \$20,800 occurs when the firm conducts 13 audits and 0 tax returns.

43. x = acres of crop A

y = acres of crop B

Constraints: $x + y \le 150$, $x + 2y \le 240$,

$$0.3x + 0.1y \le 30$$

Objective function: $z = 300x + 500y$

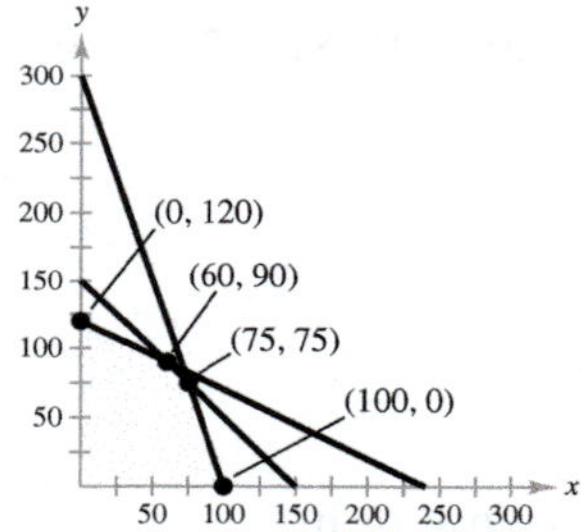

At $(0, 0)$: $z = 300(0) + 500(0) = 0$

At $(0, 20)$: $z = 300(0) + 500(120) = 60{,}000$

At $(60, 90)$: $z = 300(60) + 500(90) = 63{,}000$

At $(75, 75)$: $z = 300(75) + 500(75) = 60{,}000$

At $(100, 0)$: $z = 300(100) + 500(0) = 30{,}000$

So, 60 acres of crop A and 90 acres of crop B yield 63,000 bushels.

45. x = number of TV ads

y = number of newspaper ads

Constraints:
$$\begin{aligned} 100{,}000x + 20{,}000y &\le 1{,}000{,}000 \\ 100{,}000x &\le 800{,}000 \\ x &\ge 0 \\ y &\ge 0 \end{aligned}$$

Objective function: $A = 20x + 5y$ (A in millions)

Vertices: $(0, 0), (0, 50), (8, 10), (8, 0)$

At $(0, 0)$: $A = 20(0) + 5(0) = 0$

At $(0, 50)$: $A = 20(0) + 5(50) = 250$ million

At $(8, 10)$: $A = 20(8) + 5(10) = 210$ million

At $(8, 0)$: $A = 20(8) + 5(0) = 160$ million

The company should spend \$0 on television ads and \$1,000,000 on newspaper ads. The optimal total audience is 250 million people.

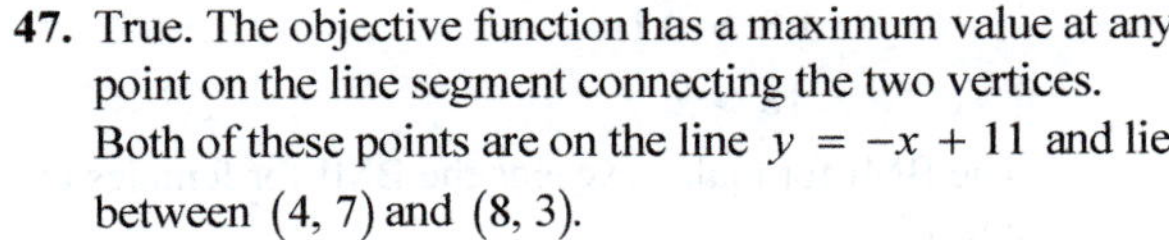

47. True. The objective function has a maximum value at any point on the line segment connecting the two vertices. Both of these points are on the line $y = -x + 11$ and lie between $(4, 7)$ and $(8, 3)$.

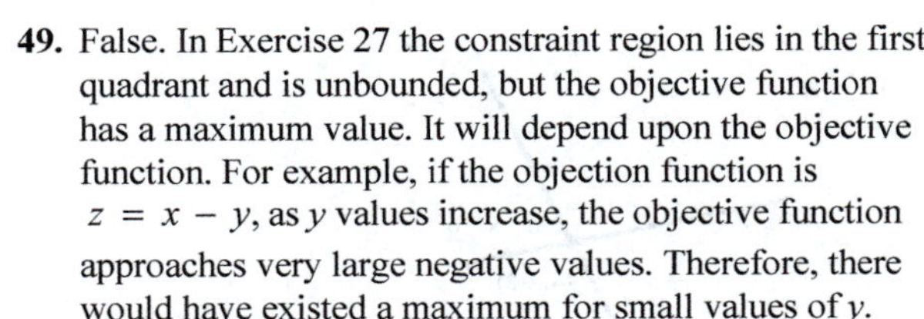

49. False. In Exercise 27 the constraint region lies in the first quadrant and is unbounded, but the objective function has a maximum value. It will depend upon the objective function. For example, if the objection function is $z = x - y$, as y values increase, the objective function approaches very large negative values. Therefore, there would have existed a maximum for small values of y.

51. If a linear programming problem has an objective function $z = 3x + 5y$ and an infinite number of optimal solutions then the slope of the line connecting two points is $m = -\frac{3}{5}$, that is $z = 3x + 5y \Rightarrow y = -\frac{3}{5}x - \frac{1}{5}z$.

Review Exercises for Chapter 7

1. $\begin{cases} x + y = 2 \\ x - y = 0 \Rightarrow x = y \end{cases}$

$$\begin{aligned} x + x &= 2 \\ 2x &= 2 \\ x &= 1 \\ y &= 1 \end{aligned}$$

Solution: $(1, 1)$

3. $\begin{cases} 4x - y - 1 = 0 \Rightarrow y = 4x - 1 \\ 8x + y - 17 = 0 \end{cases}$

$$\begin{aligned} 8x + (4x - 1) - 17 &= 0 \\ 12x &= 18 \\ x &= \tfrac{3}{2} \\ 4\left(\tfrac{3}{2}\right) - y - 1 &= 0 \\ -y + 5 &= 0 \\ y &= 5 \end{aligned}$$

Solution: $\left(\frac{3}{2}, 5\right)$

5. $\begin{cases} 0.5x + y = 0.75 \Rightarrow y = 0.75 - 0.5x \\ 1.25x - 4.5y = -2.5 \end{cases}$

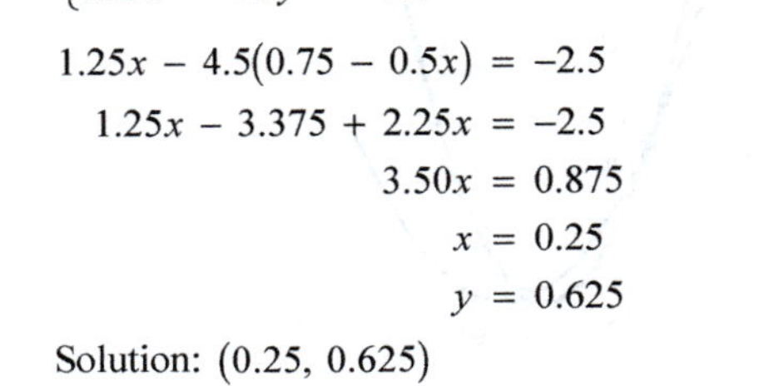

$$\begin{aligned} 1.25x - 4.5(0.75 - 0.5x) &= -2.5 \\ 1.25x - 3.375 + 2.25x &= -2.5 \\ 3.50x &= 0.875 \\ x &= 0.25 \\ y &= 0.625 \end{aligned}$$

Solution: $(0.25, 0.625)$

7. $\begin{cases} x^2 - y^2 = 9 \\ x - y = 1 \Rightarrow x = y + 1 \end{cases}$

$$\begin{aligned} (y + 1)^2 - y^2 &= 9 \\ 2y + 1 &= 9 \\ y &= 4 \\ x &= 5 \end{aligned}$$

Solution: $(5, 4)$

9. $\begin{cases} y = 2x^2 \\ y = x^4 - 2x^2 \end{cases} \Rightarrow 2x^2 = x^4 - 2x^2$

$0 = x^4 - 4x^2$

$0 = x^2(x^2 - 4)$

$0 = x^2(x + 2)(x - 2) \Rightarrow x = 0, -2, 2$

$x = 0: y = 2(0)^2 = 0$

$x = -2: y = 2(-2)^2 = 8$

$x = 2: y = 2(2)^2 = 8$

Solutions: $(0, 0), (-2, 8), (2, 8)$

11. $\begin{cases} 2x - y = 10 \\ x + 5y = -6 \end{cases}$

Point of intersection: $(4, -2)$

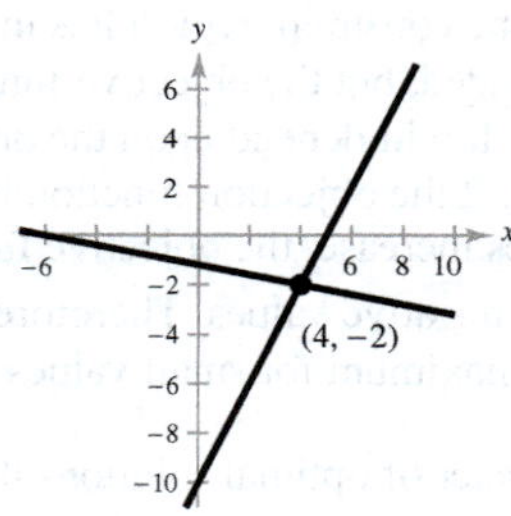

13. $\begin{cases} y = 2x^2 - 4x + 1 \\ y = x^2 - 4x + 3 \end{cases}$

Points of intersection: $(1.41, -0.66), (-1.41, 10.66)$

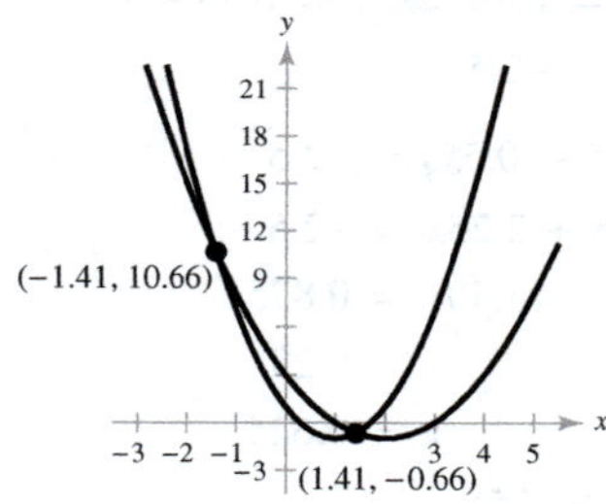

15. $\begin{cases} y = -2e^{-x} \\ 2e^x + y = 0 \Rightarrow y = -2e^x \end{cases}$

Point of intersection: $(0, -2)$

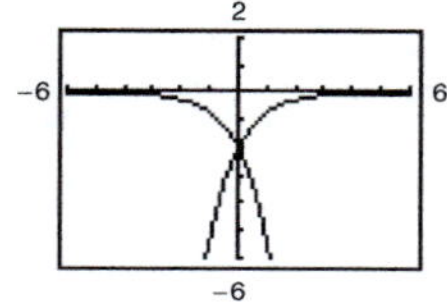

17. $\begin{cases} y = 2 + \log x \\ y = \frac{3}{4}x + 5 \end{cases}$

No Solution

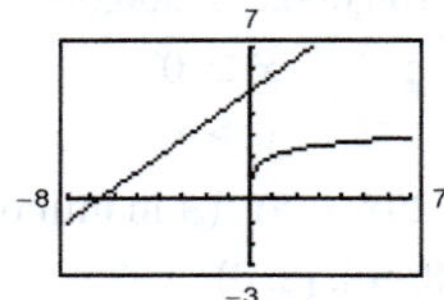

19.

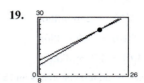

$0.68a + 13.5 > 0.78a + 11.7$

$1.8 > 0.1a$

$18 > a$

The BMI for males exceeds the BMI for females after age 18.

21. $\begin{cases} 2l + 2w = 68 \\ w = \frac{8}{9}l \end{cases}$

$2l + 2\left(\frac{8}{9}\right)l = 68$

$\frac{34}{9}l = 68$

$l = 18$

$w = \frac{8}{9}l = 16$

The width of the rectangle is 16 feet, and the length is 18 feet.

23. $\begin{cases} 2x - y = 2 \Rightarrow 16x - 8y = 16 \\ 6x + 8y = 39 \Rightarrow 6x + 8y = 39 \end{cases}$

$22x = 55$

$x = \frac{55}{22} = \frac{5}{2}$

Back-substitute $x = \frac{5}{2}$ into Equation 1.

$2\left(\frac{5}{2}\right) - y = 2$

$y = 3$

Solution: $\left(\frac{5}{2}, 3\right)$

25. $\begin{cases} 3x - 2y = 0 \\ 3x + 2y = 0 \end{cases}$

Add the equations $6x = 0 \Rightarrow x = 0$.

Back substitute into Equation 1.

$$\begin{aligned} 3(0) - 2y &= 0 \\ 2y &= 0 \\ y &= 0 \end{aligned}$$

Solution: $(0, 0)$

27. $\begin{cases} 1.25x - 2y = 3.5 \Rightarrow 5x - 8y = 14 \\ 5x - 8y = 14 \Rightarrow -5x + 8y = -14 \end{cases}$

$$0 = 0$$

There are infinitely many solutions.

Let $y = a$, then $5x - 8a = 14 \Rightarrow x = \frac{8}{5}a + \frac{14}{5}$.

Solution: $\left(\frac{8}{5}a + \frac{14}{5}, a\right)$ where a is any real number.

29. $\begin{cases} x + 5y = 4 \Rightarrow x + 5y = 4 \\ x - 3y = 6 \Rightarrow -x + 3y = -6 \end{cases}$

$$8y = -2 \Rightarrow y = -\frac{1}{4}$$

Matches graph (d). The system has one solution and is consistent.

31. $\begin{cases} 3x - y = 7 \Rightarrow 6x - 2y = 14 \\ -6x + 2y = 8 \Rightarrow -6x + 2y = 8 \end{cases}$

$$0 \neq 22$$

Matches graph (b). The system has no solution and is inconsistent.

33. $22 + 0.00001x = 43 - 0.0002x$

$$\begin{aligned} 0.00021x &= 21 \\ x &= 100{,}000, p = 2^3 \end{aligned}$$

Point of Equilibrium: $(100{,}000, 23)$

35. $\begin{cases} x - 4y + 3z = 3 \\ -y + z = -1 \\ z = -5 \end{cases}$

$$-y + (-5) = -1 \Rightarrow y = -4$$

$$x - 4(-4) + 3(-5) = 3 \Rightarrow x = 2$$

Solution: $(2, -4, -5)$

37. $\begin{cases} 4x - 3y - 2z = -65 \\ 8y - 7z = -14 \\ z = 10 \end{cases}$

$$8y - 7(10) = -14 \Rightarrow y = 7$$

$$4x - 3(7) - 2(10) = -65 \Rightarrow x = -6$$

Solution: $(-6, 7, 10)$

39. $\begin{cases} x + 2y + 6z = 4 & \text{Equation 1} \\ -3x + 2y - z = -4 & \text{Equation 2} \\ 4x + 2z = 16 & \text{Equation 3} \end{cases}$

$\begin{cases} x + 2y + 6z = 4 & \\ 8y + 17z = 8 & 3\text{Eq.1} + \text{Eq.2} \\ -8y - 22z = 0 & -4\text{Eq.1} + \text{Eq.3} \end{cases}$

$\begin{cases} x + 2y + 6z = 4 & \\ 8y + 17z = 8 & \\ -5z = 8 & \text{Eq.2} + \text{Eq.3} \end{cases}$

$\begin{cases} x + 2y + 6z = 4 & \\ 8y + 17z = 8 & \\ z = -\frac{8}{5} & -\frac{1}{5}\text{Eq.3} \end{cases}$

$$8y + 17\left(-\frac{8}{5}\right) = 8 \Rightarrow y = \frac{22}{5}$$

$$x + 2\left(\frac{22}{5}\right) + 6\left(-\frac{8}{5}\right) = 4 \Rightarrow x = \frac{24}{5}$$

Solution: $\left(\frac{24}{5}, \frac{22}{5}, -\frac{8}{5}\right)$

41. $\begin{cases} 2x + 6z = -9 & \text{Equation 1} \\ 3x - 2y + 11z = -16 & \text{Equation 2} \\ 3x - y + 7z = -11 & \text{Equation 3} \end{cases}$

$\begin{cases} -x + 2y - 5z = 7 & (-1)\text{Eq.2} + \text{Eq.1} \\ 3x - 2y + 11z = -16 & \\ 3x - y + 7z = -11 & \end{cases}$

$\begin{cases} -x + 2y - 5z = 7 & \\ 4y - 4z = 5 & 3\text{Eq.1} + \text{Eq.2} \\ 5y - 8z = 10 & 3\text{Eq.1} + \text{Eq.3} \end{cases}$

$\begin{cases} -x + 2y - 5z = 7 & \\ 4y - 4z = 5 & \\ -3y = 0 & (-2)\text{Eq.2} + \text{Eq.3} \end{cases}$

$\begin{cases} -x + 2y - 5z = 7 & \\ y - z = \frac{5}{4} & \left(\frac{1}{4}\right)\text{Eq.2} \\ y = 0 & \left(-\frac{1}{3}\right)\text{Eq.3} \end{cases}$

$$0 - z = \frac{5}{4} \Rightarrow z = -\frac{5}{4}$$

$$-x + 2(0) - 5\left(-\frac{5}{4}\right) = 7 \Rightarrow x = -\frac{3}{4}$$

Solution: $\left(-\frac{3}{4}, 0, -\frac{5}{4}\right)$

43. $\begin{cases} 5x - 12y + 7z = 16 \\ 3x - 7y + 4z = 9 \end{cases} \Rightarrow \begin{cases} 15x - 36y + 21z = 48 \\ -15x + 35y - 20z = -45 \end{cases}$

$$-y + z = 3$$

Let $z = a$. Then $y = a - 3$ and $5x - 12(a - 3) + 7a = 16 \Rightarrow x = a - 4$.

Solution: $(a - 4, a - 3, a)$ where a is any real number.

45. $y = ax^2 + bx + c$ through $(0, -5)$, $(1, -2)$, and $(2, 5)$.

$(0, -5)$: $-5 = c \Rightarrow c = -5$

$(1, -2)$: $-2 = a + b + c \Rightarrow \begin{cases} a + b = 3 \\ 2a + b = 5 \end{cases}$

$(2, 5)$: $5 = 4a + 2b + c \Rightarrow$

$\begin{cases} 2a + b = 5 \\ -a - b = -3 \end{cases}$

$a = 2$

$b = 1$

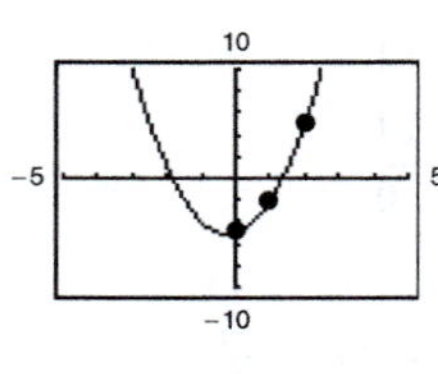

The equation of the parabola is $y = 2x^2 + x - 5$.

47. $x^2 + y^2 + Dx + Ey + F = 0$ through $(-1, -2)$, $(5, -2)$, and $(2, 1)$.

$(-1, -2)$: $5 - D - 2E + F = 0 \Rightarrow \begin{cases} D + 2E - F = 5 \\ 5D - 2E + F = -29 \\ 2D + E + F = -5 \end{cases}$

$(5, -2)$: $29 + 5D - 2E + F = 0 \Rightarrow$

$(2, 1)$: $5 + 2D + E + F = 0 \Rightarrow$

From the first two equations

$6D = -24$

$D = -4.$

Substituting $D = -4$ into the second and third equations yields:

$-20 - 2E + F = -29 \Rightarrow \begin{cases} -2E + F = -9 \\ -E - F = -3 \end{cases}$

$-8 + E + F = -5 \Rightarrow$

$-3E = -12$

$E = 4$

$F = -1$

The equation of the circle is $x^2 + y^2 - 4x + 4y - 1 = 0$.

To verify the result using a graphing utility, solve the equation for y.

$$(x^2 - 4x + 4) + (y^2 + 4y + 4) = 1 + 4 + 4$$

$$(x - 2)^2 + (y + 2)^2 = 9$$

$$(y + 2)^2 = 9 - (x - 2)^2$$

$$y = -2 \pm \sqrt{9 - (x - 2)^2}$$

Let $y_1 = -2 + \sqrt{9 - (x - 2)^2}$ and $y_2 = -2 - \sqrt{9 - (x - 2)^2}$.

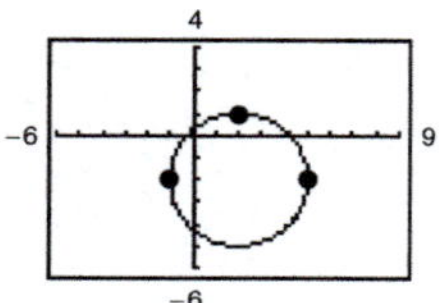

49. From the following chart we obtain our system of equations.

	A	B	C
Mixture X	$\frac{1}{5}$	$\frac{2}{5}$	$\frac{2}{5}$
Mixture Y	0	0	1
Mixture Z	$\frac{1}{3}$	$\frac{1}{3}$	$\frac{1}{3}$
Desired Mixture	$\frac{6}{27}$	$\frac{8}{27}$	$\frac{13}{27}$

$$\left.\begin{aligned}\tfrac{1}{5}x + \tfrac{1}{3}z &= \tfrac{6}{27}\\ \tfrac{2}{5}x + \tfrac{1}{3}z &= \tfrac{8}{27}\end{aligned}\right\} x = \tfrac{10}{27},\ z = \tfrac{12}{27}$$

$$\tfrac{2}{5}x + y + \tfrac{1}{3}z = \tfrac{13}{27} \Rightarrow y = \tfrac{5}{27}$$

To obtain the desired mixture, use 10 gallons of spray X, 5 gallons of spray Y, and 12 gallons of spray Z.

51. Let x = amount invested at 7%

y = amount invested at 9%

z = amount invested at 11%.

$y = x - 3000$ and

$z = x - 5000 \Rightarrow y + z = 2x - 8000$

$$\begin{cases} x + y + z = 40{,}000\\ 0.07x + 0.09y + 0.11z = 3500\\ y + z = 2x - 8000\end{cases}$$

$$x + (2x - 8000) = 40{,}000 \Rightarrow x = 16{,}000$$
$$y = 16{,}000 - 3000 \Rightarrow y = 13{,}000$$
$$z = 16{,}000 - 5000 \Rightarrow z = 11{,}000$$

So, \$16,000 was invested at 7%, \$13,000 at 9%, and \$11,000 at 11%.

53. $s = \frac{1}{2}at^2 + v_0t + s_0$

When $t = 1$: $s = 134$: $\frac{1}{2}a(1)^2 + v_0(1) + s_0 = 134 \Rightarrow a + 2v_0 + 2s_0 = 268$

When $t = 2$: $s = 86$: $\frac{1}{2}a(2)^2 + v_0(2) + s_0 = 86 \Rightarrow 2a + 2v_0 + s_0 = 86$

When $t = 3$: $s = 6$: $\frac{1}{2}a(3)^2 + v_0(3) + s_0 = 6 \Rightarrow 9a + 6v_0 + 2s_0 = 12$

$$\begin{cases} a + 2v_0 + 2s_0 = 268\\ 2a + 2v_0 + s_0 = 86\\ 9a + 6v_0 + 2s_0 = 12\end{cases}$$

$$\begin{cases} a + 2v_0 + 2s_0 = 268\\ -2v_0 - 3s_0 = -450 & (-2)\text{Eq.1} + \text{Eq.2}\\ -12v_0 - 16s_0 = -2400 & (-9)\text{Eq.1} + \text{Eq.3}\end{cases}$$

$$\begin{cases} a + 2v_0 + 2s_0 = 268\\ -2v_0 - 3s_0 = -450\\ 3v_0 + 4s_0 = 600 & \left(-\tfrac{1}{4}\right)\text{Eq.3}\end{cases}$$

$$\begin{cases} a + 2v_0 + 2s_0 = 268\\ -2v_0 - 3s_0 = -450\\ -s_0 = -150 & 3\text{Eq.2} + 2\text{Eq.3}\end{cases}$$

$$-s_0 = -150 \Rightarrow s_0 = 150$$
$$-2v_0 - 3(150) = -450 \Rightarrow v_0 = 0$$
$$a + 2(0) + 2(150) = 268 \Rightarrow a = -32$$

The position equation is $s = \frac{1}{2}(-32)t^2 + (0)t + 150$, or $s = -16t^2 + 150$.

55. $\dfrac{3}{x^2 + 20x} = \dfrac{3}{x(x + 20)} = \dfrac{A}{x} + \dfrac{B}{x + 20}$

57. $\dfrac{3x - 4}{x^3 - 5x^2} = \dfrac{3x - 4}{x^2(x - 5)} = \dfrac{A}{x} + \dfrac{B}{x^2} + \dfrac{C}{x - 5}$

59. $\dfrac{4 - x}{x^2 + 6x + 8} = \dfrac{A}{x + 2} + \dfrac{B}{x + 4}$

$4 - x = A(x + 4) + B(x + 2)$

Let $x = -2$: $6 = 2A \Rightarrow A = 3$

Let $x = -4$: $8 = -2B \Rightarrow B = -4$

$$\frac{4 - x}{x^2 + 6x + 8} = \frac{3}{x + 2} - \frac{4}{x + 4}$$

61. $\dfrac{x^2}{x^2 + 2x - 15} = 1 - \dfrac{2x - 15}{x^2 + 2x - 15}$

$\dfrac{-2x + 15}{(x + 5)(x - 3)} = \dfrac{A}{x + 5} + \dfrac{B}{x - 3}$

$-2x + 15 = A(x - 3) + B(x + 5)$

Let $x = -5$: $25 = -8A \Rightarrow A = -\dfrac{25}{8}$

Let $x = 3$: $9 = 8B \Rightarrow B = \dfrac{9}{8}$

$$\frac{x^2}{x^2 + 2x - 15} = 1 - \frac{25}{8(x + 5)} + \frac{9}{8(x - 3)}$$

63. $\dfrac{x^2 + 2x}{x^3 - x^2 + x - 1} = \dfrac{x^2 + 2x}{(x - 1)(x^2 + 1)}$

$= \dfrac{A}{x - 1} + \dfrac{Bx + C}{x^2 + 1}$

$x^2 + 2x = A(x^2 + 1) + (Bx + C)(x - 1)$

$= Ax^2 + A + Bx^2 - Bx + Cx - C$

$= (A + B)x^2 + (-B + C)x + (A - C)$

Equating coefficients of like terms gives $1 = A + B$, $2 = -B + C$, and $0 = A - C$. Adding both sides of all three equations gives $3 = 2A$. So, $A = \dfrac{3}{2}$, $B = -\dfrac{1}{2}$, and $C = \dfrac{3}{2}$.

$$\frac{x^2 + 2x}{x^3 - x^2 + x - 1} = \frac{\frac{3}{2}}{x - 1} + \frac{-\frac{1}{2}x + \frac{3}{2}}{x^2 + 1}$$

$$= \frac{1}{2}\left(\frac{3}{x - 1} - \frac{x - 3}{x^2 + 1}\right)$$

65. $\dfrac{3x^2 + 4x}{(x^2 + 1)^2} = \dfrac{Ax + B}{x^2 + 1} + \dfrac{Cx + D}{(x^2 + 1)^2}$

$3x^2 + 4x = (Ax + B)(x^2 + 1) + Cx + D$

$= Ax^3 + Bx^2 + (A + C)x + (B + D)$

Equating coefficients of like terms gives

$0 = A$

$3 = B$

$4 = 0 + C \Rightarrow C = 4$

$0 = B + D \Rightarrow D = -3$

$$\frac{3x^2 + 4x}{(x^2 + 1)^2} = \frac{3}{x^2 + 1} + \frac{4x - 3}{(x^2 + 1)^2}$$

67. $y \geq 5$

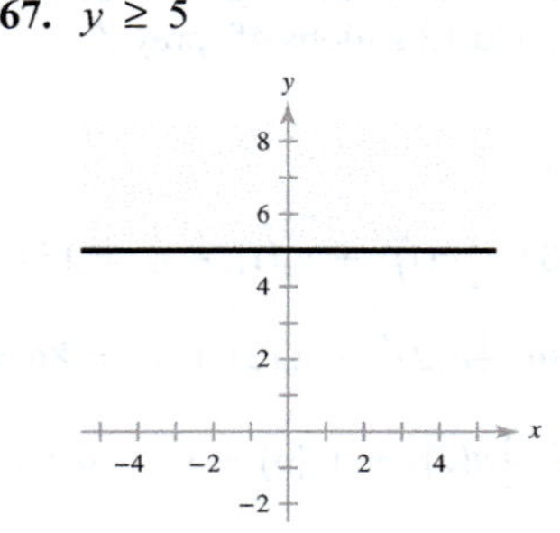

69. $y \leq 5 - 2x$

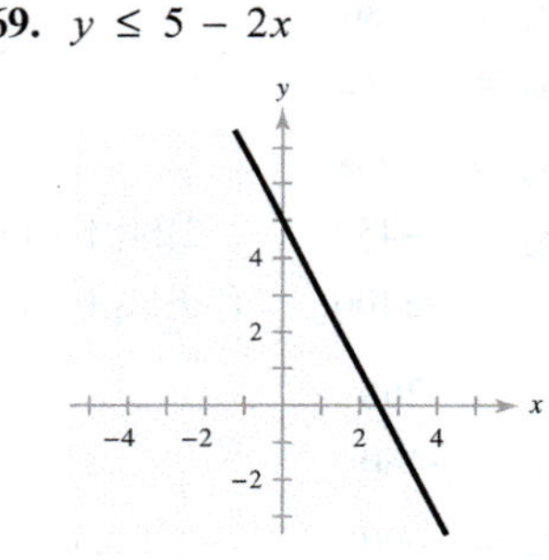

71. $(x - 1)^2 + (y - 3)^2 < 16$

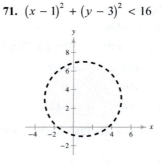

73. $\begin{cases} x + 2y \le 2 \\ -x + 2y \le 2 \\ y \ge 0 \end{cases}$

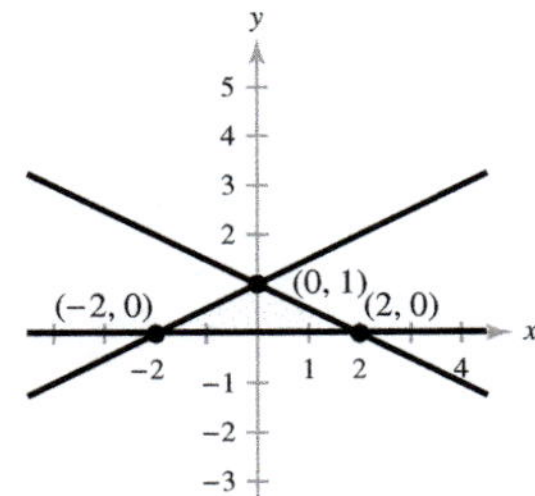

Vertex A

$\begin{cases} x + 2y = 2 \\ -x + 2y = 2 \end{cases}$

$4y = 4 \Rightarrow y = 1$

$x + 2(1) = 2 \Rightarrow x = 0$

$(0, 1)$

Vertex B

$\begin{cases} x + 2y = 2 \\ y = 0 \end{cases}$

$x + 2(0) = 2$

$x = 2$

$(2, 0)$

Vertex C

$\begin{cases} -x + 2y = 2 \\ y = 0 \end{cases}$

$-x + 2(0) = 2$

$x = -2$

$(-2, 0)$

75. $\begin{cases} 2x - y < -1 \\ -3x + 2y > 4 \\ y > 0 \end{cases}$

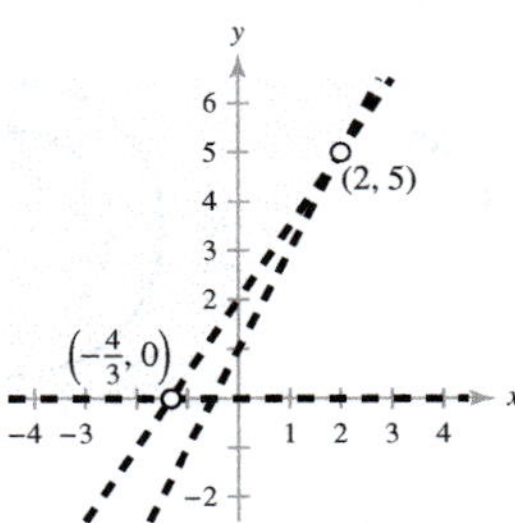

Vertex A

$\begin{cases} 2x - y = -1 \Rightarrow 4x - 2y = -2 \\ -3x + 2y = 4 \Rightarrow -3x + 2y = 4 \end{cases}$

$x = 2$

$2(2) - y = -1 \Rightarrow -y = -5 \Rightarrow y = 5$

$(2, 5)$

Vertex B

$\begin{cases} -3x + 2y = 4 \\ y = 0 \end{cases}$

$-3x + 2(0) = 4 \Rightarrow x = -\frac{4}{3}$

$\left(-\frac{4}{3}, 0\right)$

77. $\begin{cases} y < x + 1 \\ y > x^2 - 1 \end{cases}$

Vertices:

$x + 1 = x^2 - 1$

$0 = x^2 - x - 2 = (x + 1)(x - 2)$

$x = -1$ or $x = 2$

$y = 0 \qquad y = 3$

$(-1, 0) \qquad (2, 3)$

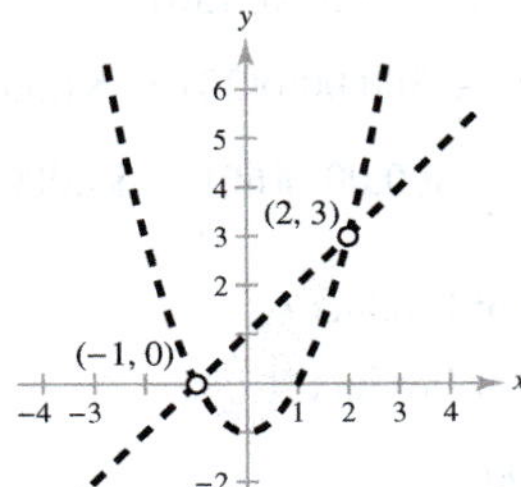

79. $\begin{cases} x^2 + y^2 > 4 \Rightarrow y^2 > 4 - x^2: \text{ The region outside the circle centered at } (0, 0) \text{ with radius of } 2. \\ x^2 + y^2 \le 9 \Rightarrow y^2 \le 9 - x^2: \text{ The region inside and on the circle centered at } (0, 0) \text{ with radius of } 3. \end{cases}$

Vertices: $4 - x^2 = 9 - x^2$

$$0 \ne 5$$

The circles do not intersect, so there are no vertices.

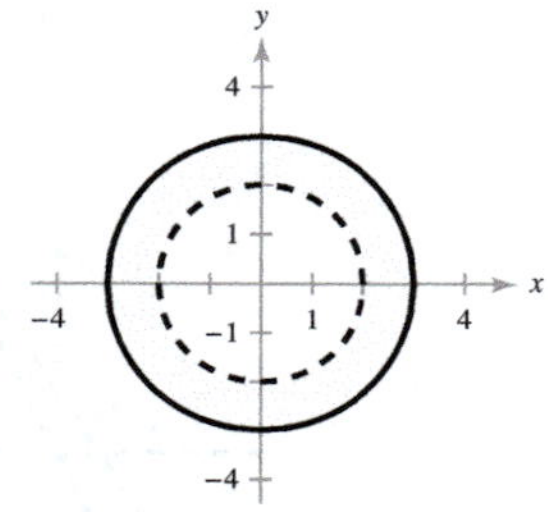

81. Rectangular region with vertices at:

$(3, 1)$, $(7, 1)$, $(7, 10)$, and $(3, 10)$

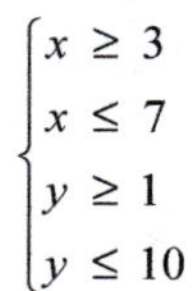

$$\begin{cases} x \ge 3 \\ x \le 7 \\ y \ge 1 \\ y \le 10 \end{cases}$$

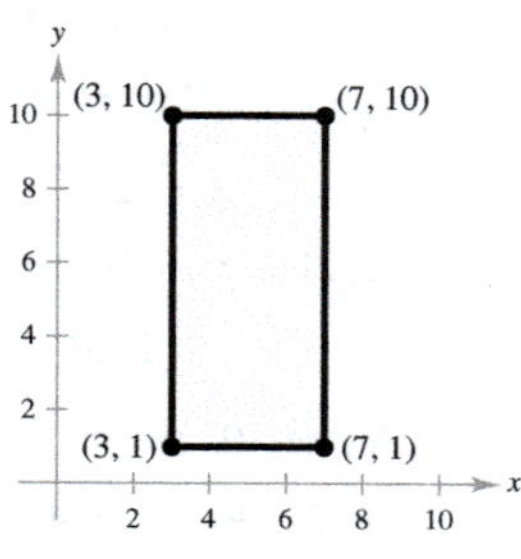

This system may be written as:

$$\begin{cases} 3 \le x \le 7 \\ 1 \le y \le 10 \end{cases}$$

83. (a)

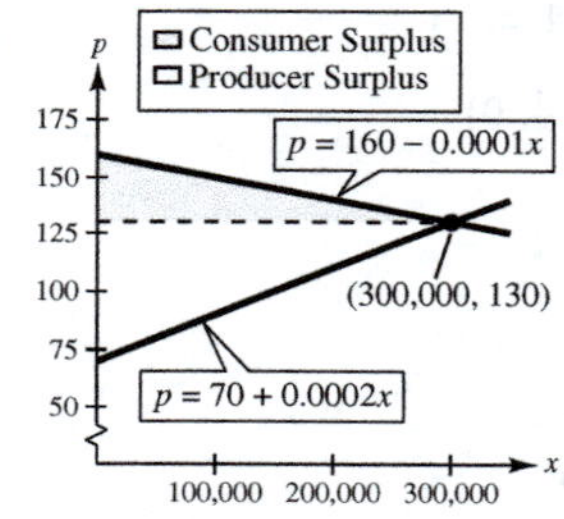

$$160 - 0.0001x = 70 + 0.0002x$$
$$90 = 0.0003x$$
$$x = 300{,}000 \text{ units}$$
$$p = \$130$$

Point of equilibrium: $(300{,}000, 130)$

(b) Consumer surplus: $\frac{1}{2}(300{,}000)(30) = \$4{,}500{,}000$

Producer surplus: $\frac{1}{2}(300{,}000)(60) = \$9{,}000{,}000$

85. x = number of units of Product I

y = number of units of Product II

$$\begin{cases} 20x + 30y \le 24{,}000 \\ 12x + 8y \le 12{,}400 \\ x \ge 0 \\ y \ge 0 \end{cases}$$

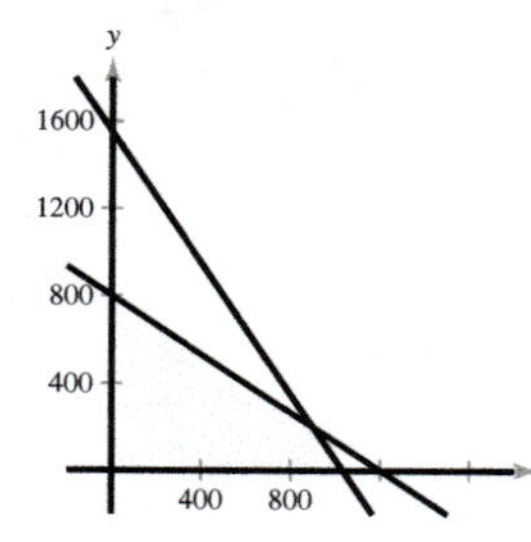

87. Objective function: $z = 3x + 4y$

Constraints: $\begin{cases} x \ge 0 \\ y \ge 0 \\ 2x + 5y \le 50 \\ 4x + y \le 28 \end{cases}$

At $(0, 0)$: $z = 0$

At $(0, 10)$: $z = 40$

At $(5, 8)$: $z = 47$

At $(7, 0)$: $z = 21$

The minimum value is 0 at $(0, 0)$.

The maximum value is 47 at $(5, 8)$.

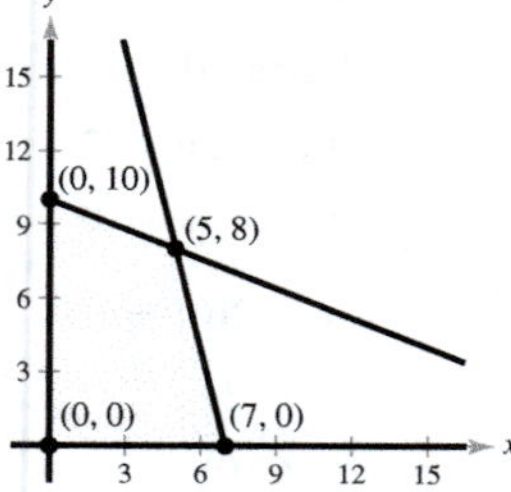

89. Objective function: $z = 1.75x + 2.25y$

Constraints: $\begin{cases} x \ge 0 \\ y \ge 0 \\ 2x + y \ge 25 \\ 3x + 2y \ge 45 \end{cases}$

At $(0, 25)$: $z = 56.25$

At $(5, 15)$: $z = 42.5$

At $(15, 0)$: $z = 26.25$

The minimum value is 26.25 at $(15, 0)$.

Because the region is unbounded, there is no maximum value.

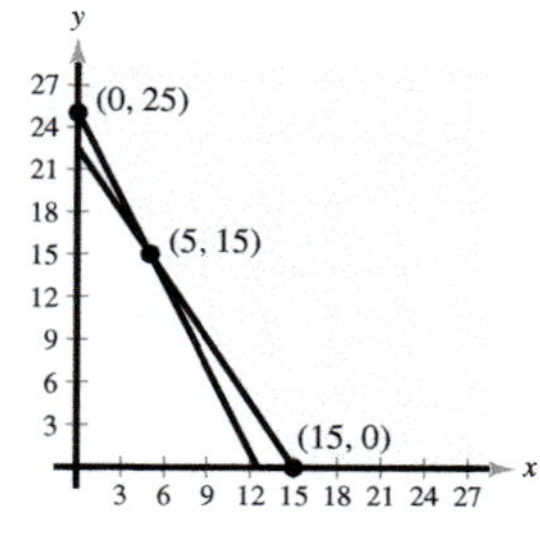

91. x = number of haircuts

y = number of permanents

Objective function: Optimize $R = 25x + 70y$ subject to the following constraints:

$$\begin{cases} x \geq 0 \\ y \geq 0 \\ \left(\frac{20}{60}\right)x + \left(\frac{70}{60}\right)y \leq 24 \Rightarrow 2x + 7y \leq 144 \end{cases}$$

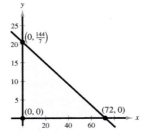

At $(0, 0)$: $R = 0$

At $(72, 0)$: $R = 1800$

At $\left(0, \frac{144}{7}\right)$: $R = 1440$

The revenue is optimal if the student does 72 haircuts and no permanents. The maximum revenue is \$1800.

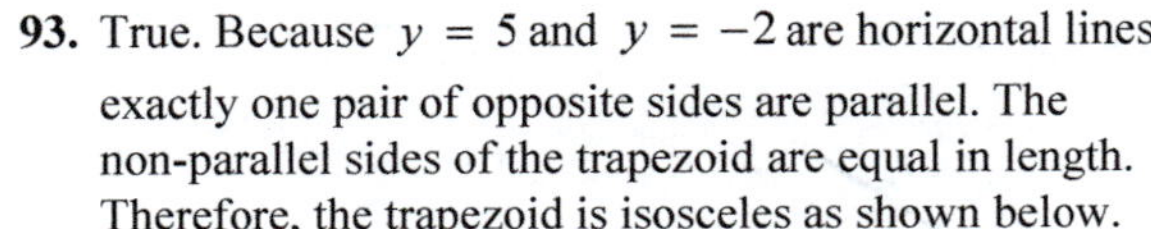

93. True. Because $y = 5$ and $y = -2$ are horizontal lines, exactly one pair of opposite sides are parallel. The non-parallel sides of the trapezoid are equal in length. Therefore, the trapezoid is isosceles as shown below.

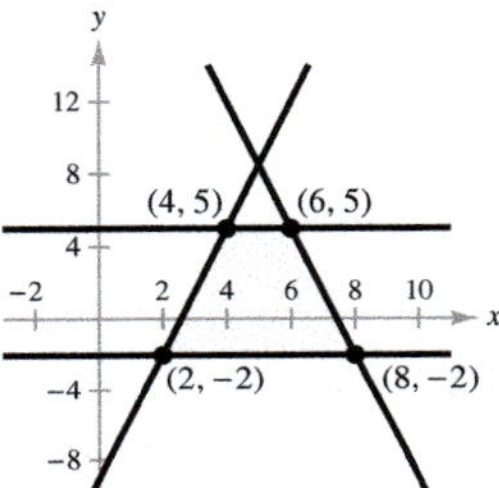

The distance from $(-4, 5)$ to $(2, -2)$ is equal to the distance from $(6, 5)$ to $(8, -2)$.

$$d_1 = \sqrt{(4-2)^2 + [5-(-2)]^2} = \sqrt{53}$$

$$d_2 = \sqrt{(8-6)^2 + (-2-5)^2} = \sqrt{53}$$

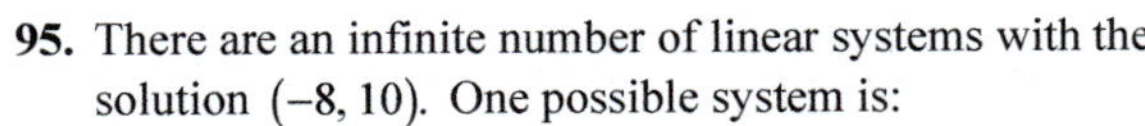

95. There are an infinite number of linear systems with the solution $(-8, 10)$. One possible system is:

$$\begin{cases} 4x + y = -22 \\ \frac{1}{2}x + y = 6 \end{cases}$$

97. There are infinite linear systems with the solution $\left(\frac{4}{3}, 3\right)$.

One possible system is:

$$\begin{cases} 3x + y = 7 \\ -6x + 3y = 1 \end{cases}$$

99. There are an infinite number of linear systems with the solution $(4, -1, 3)$. One possible system is as follows:

$$\begin{cases} x + y + z = 6 \\ x + y - z = 0 \\ x - y - z = 2 \end{cases}$$

101. There are an infinite number of linear systems with the solution $\left(5, \frac{3}{2}, 2\right)$. One possible system is:

$$\begin{cases} 2x + 2y - 3z = 7 \\ x - 2y + z = 4 \\ -x + 4y - z = -1 \end{cases}$$

103. A system of linear equations is inconsistent if it has no solution.

Problem Solving for Chapter 7

1. The longest side of the triangle is a diameter of the circle and has a length of 20.

The lines $y = \frac{1}{2}x + 5$ and $y = -2x + 20$ intersect at the point $(6, 8)$.

The distance between $(-10, 0)$ and $(6, 8)$ is: $d_1 = \sqrt{(6-(-10))^2 + (8-0)^2} = \sqrt{320} = 8\sqrt{5}$

The distance between $(6, 8)$ and $(10, 0)$ is: $d_2 = \sqrt{(10-6)^2 + (0-8)^2} = \sqrt{80} = 4\sqrt{5}$

Because $\left(\sqrt{320}\right)^2 + \left(\sqrt{80}\right)^2 = (20)^2$

$$400 = 400,$$

the sides of the triangle satisfy the Pythagorean Theorem. So, the triangle is a right triangle.

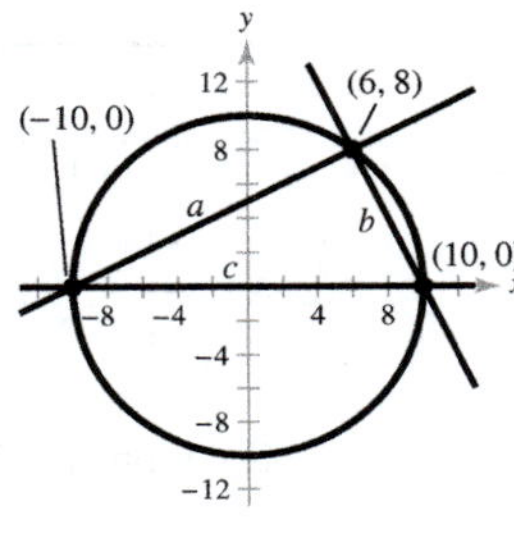

3. The system will have exactly one solution when the slopes of the line are *not* equal.

$$\begin{cases} ax + by = e \Rightarrow y = -\dfrac{a}{b}x + \dfrac{e}{b} \\ cx + dy = f \Rightarrow y = -\dfrac{c}{d}x + \dfrac{f}{d} \end{cases}$$

$$-\frac{a}{b} \neq -\frac{c}{d}$$

$$\frac{a}{b} \neq \frac{c}{d}$$

$$ad \neq bc$$

5. (a) $\begin{cases} x - 4y = -3 & \text{Eq. 1} \\ 5x - 6y = 13 & \text{Eq. 2} \end{cases}$

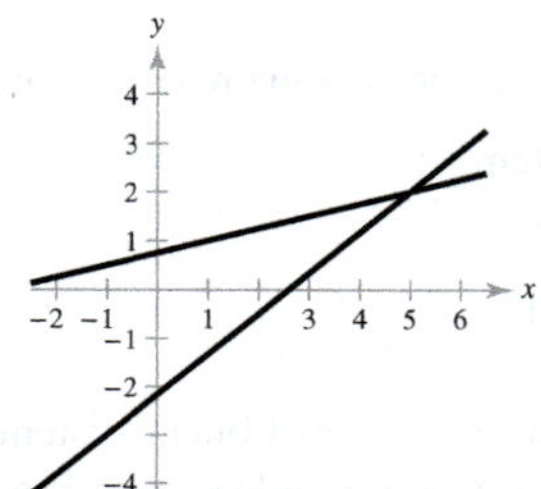

$\begin{cases} x - 4y = -3 \\ 14y = 28 \end{cases}$ $\quad -5\text{Eq.1} + \text{Eq.2}$

$\begin{cases} x - 4y = -3 \\ y = 2 \end{cases}$ $\quad \frac{1}{14}\text{Eq.2}$

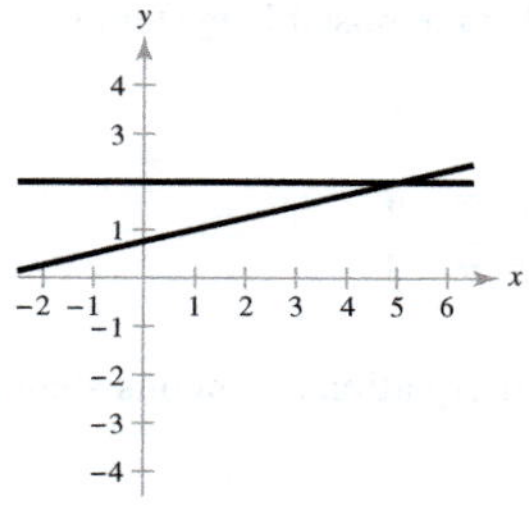

$\begin{cases} x = 5 \\ y = 2 \end{cases}$ $\quad 4\text{Eq.2} + \text{Eq.1}$

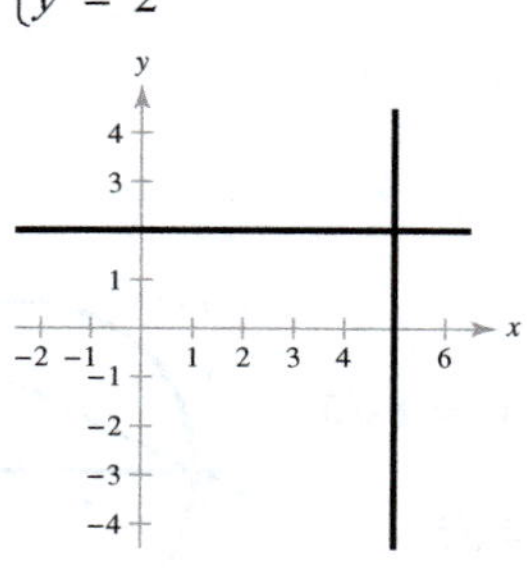

Solution: $(5, 2)$

(b) $\begin{cases} 2x - 3y = 7 & \text{Eq. 1} \\ -4x + 6y = -14 & \text{Eq. 2} \end{cases}$

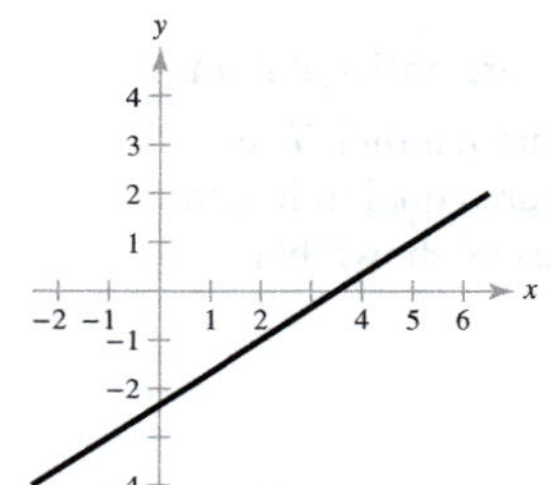

$\begin{cases} 2x - 3y = 7 \\ 0 = 0 \end{cases}$ $\quad 2\text{Eq.1} + \text{Eq.2}$

The lines coincide. Infinite solutions.

Let $y = a$, then $2x - 3a = 7 \Rightarrow x = \frac{3}{2}a + \frac{7}{2}$

Solution: $\left(\frac{3}{2}a + \frac{7}{2}, a\right)$

The solution(s) remain the same at each step of the process.

7. The point where the two sections meet is at a depth of 10.1 feet.
The distance between $(0, -10.1)$ and $(252.5, 0)$ is:

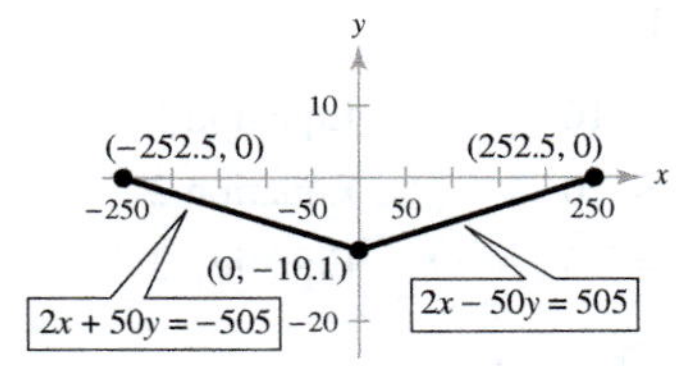

$$d = \sqrt{(252.5 - 0)^2 + (0 - (-10.1))^2} = \sqrt{63{,}858.26}$$

$$d \approx 252.7$$

Each section is approximately 252.7 feet long.

9. Let x = cost of the cable, per foot.
Let y = cost of a connector.

$$\begin{cases} 6x + 2y = 15.50 \Rightarrow & 6x + 2y = 15.50 \\ 3x + 2y = 10.25 \Rightarrow & -3x - 2y = -10.25 \end{cases}$$

$$\begin{aligned} 3x &= 5.25 \\ x &= 1.75 \\ y &= 2.50 \end{aligned}$$

For a four-foot cable with a connector on each end, the cost should be $4(1.75) + 2(2.50) = \$12.00$.

11. Let $X = \dfrac{1}{x}$, $Y = \dfrac{1}{y}$, and $Z = \dfrac{1}{z}$.

(a) $$\begin{cases} \dfrac{12}{x} - \dfrac{12}{y} = 7 \Rightarrow 12X - 12Y = 7 \Rightarrow 12X - 12Y = 7 \\ \dfrac{3}{x} - \dfrac{4}{y} = 0 \Rightarrow 3X + 4Y = 0 \Rightarrow 9X + 12Y = 0 \\ \qquad 21X = 7 \\ \qquad X = \dfrac{1}{3} \\ \qquad Y = -\dfrac{1}{4} \end{cases}$$

So, $\dfrac{1}{x} = \dfrac{1}{3} \Rightarrow x = 3$ and $\dfrac{1}{y} = -\dfrac{1}{4} \Rightarrow y = -4$.

Solution: $(3, -4)$

(b) $$\begin{cases} \dfrac{2}{x} + \dfrac{1}{y} - \dfrac{3}{z} = 4 \Rightarrow 2X + Y - 3Z = 4 & \text{Eq.1} \\ \dfrac{4}{x} + \dfrac{2}{z} = 10 \Rightarrow 4X + 2Z = 10 & \text{Eq.2} \\ -\dfrac{2}{x} + \dfrac{3}{y} - \dfrac{13}{z} = -8 \Rightarrow -2X + 3Y - 13Z = -8 & \text{Eq.3} \end{cases}$$

$$\begin{cases} 2X + Y - 3Z = 4 & \\ -2Y + 8Z = 2 & -2\text{Eq.1} + \text{Eq.2} \\ 4Y - 16Z = -4 & \text{Eq.1} + \text{Eq.3} \end{cases}$$

$$\begin{cases} 2X + Y - 3Z = 4 & \\ -2Y + 8Z = 2 & \\ 0 = 0 & 2\text{Eq.2} + \text{Eq.3} \end{cases}$$

The system has infinite solutions.

Let $Z = a$, then $Y = 4a - 1$ and $X = \dfrac{-a + 5}{2}$.

Then $\dfrac{1}{z} = a \Rightarrow z = \dfrac{1}{a}, \dfrac{1}{y} = 4a - 1 \Rightarrow y = \dfrac{1}{4a - 1}$, and $\dfrac{1}{x} = \dfrac{-a + 5}{2} \Rightarrow x = \dfrac{2}{-a + 5}$.

Solution: $\left(\dfrac{2}{-a + 5}, \dfrac{1}{4a - 1}, \dfrac{1}{a}\right), a \neq 5, \dfrac{1}{4}, 0$

13. Solution: $(1, -1, 2)$

$$\begin{cases} 4x - 2y + 5z = 16 & \text{Equation 1} \\ x + y = 0 & \text{Equation 2} \\ -x - 3y + 2z = 6 & \text{Equation 3} \end{cases}$$

(a) $$\begin{cases} 4x - 2y + 5z = 16 \\ x + y = 0 \end{cases}$$

$$\begin{cases} x + y = 0 \\ 4x - 2y + 5z = 16 \end{cases} \qquad \text{Interchange the equations.}$$

$$\begin{cases} x + y = 0 \\ -6y + 5z = 16 \end{cases} \qquad -4\text{Eq.}1 + \text{Eq.}2$$

Let $z = a$, then $y = \dfrac{5a - 16}{6}$ and $x = \dfrac{-5a + 16}{6}$.

Solution: $\left(\dfrac{-5a + 16}{6}, \dfrac{5a - 16}{6}, a\right)$

When $a = 2$, we have the original solution.

(b) $$\begin{cases} 4x - 2y + 5z = 16 \\ -x - 3y + 2z = 6 \end{cases}$$

$$\begin{cases} -x - 2y + 2z = 6 \\ 4x - 3y + 5z = 16 \end{cases} \qquad \text{Interchange the equations.}$$

$$\begin{cases} -x - 3y + 2z = 6 \\ -14y + 13z = 40 \end{cases} \qquad 4\text{Eq.}1 + \text{Eq.}2$$

Let $z = a$, then $y = \dfrac{13a - 40}{14}$ and $x = \dfrac{-11a + 36}{14}$.

Solution: $\left(\dfrac{-11a + 36}{14}, \dfrac{13a - 40}{14}, a\right)$

When $a = 2$, we have the original solution.

(c) $$\begin{cases} x + y = 0 \\ -x - 3y + 2z = 6 \end{cases}$$

$$\begin{cases} x + y = 0 \\ -2y + 2z = 6 \end{cases} \qquad \text{Eq.}1 + \text{Eq.}2$$

Let $z = a$, then $y = a - 3$ and $x = -a + 3$.

Solution: $(-a + 3, a - 3, a)$

When $a = 2$, we have the original solution.

(d) Each of these systems has infinite solutions.

15. t = amount of terrestrial vegetation in kilograms

a = amount of aquatic vegetation in kilograms

$$\begin{cases} a + t \le 32 \\ 0.15a \ge 1.9 \\ 193a + 772t \ge 11{,}000 \end{cases}$$

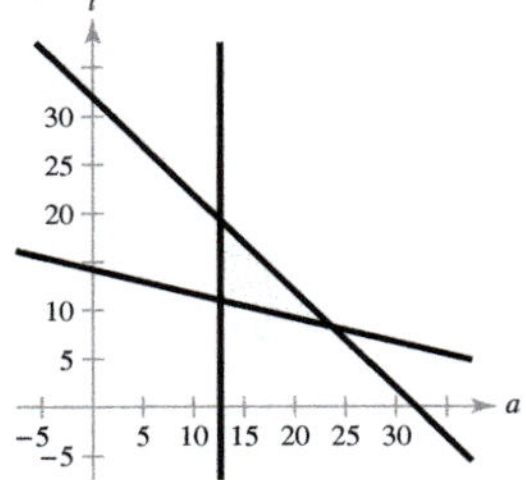

17. x = milligrams of HDL cholestrol

y = milligrams of LDL/VLDL cholestrol

(a)

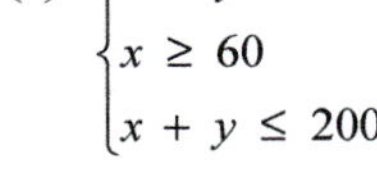

$$\begin{cases} 0 < y < 130 \\ x \ge 60 \\ x + y \le 200 \end{cases}$$

(b)

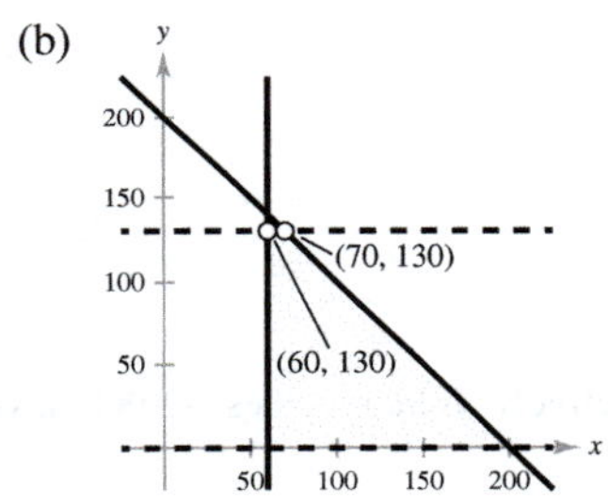

(c) $y = 120$ is in the region because $0 < y < 130$.

$x = 90$ is in the region because $60 \le x \le 200$.

$x + y = 210$ is not the region because $x + y \le 200$.

(d) *Sample answer*: If the LDL/VLDL reading is 135 and the HDL reading is 65, then $x \ge 60$ and $x + y \le 200$, but $y \not< 130$.

(e) $\dfrac{x + y}{x} < 4$

$x + y < 4x$

$y < 3x$

Sample answer: The point $(75, 90)$ is in the region, and $\dfrac{165}{75} = 2.2 < 4$.

Practice Test for Chapter 7

For Exercises 1–3, solve the given system by the method of substitution.

1. $\begin{cases} x + y = 1 \\ 3x - y = 15 \end{cases}$

2. $\begin{cases} x - 3y = -3 \\ x^2 + 6y = 5 \end{cases}$

3. $\begin{cases} x + y + z = 6 \\ 2x - y + 3z = 0 \\ 5x + 2y - z = -3 \end{cases}$

4. Find the two numbers whose sum is 110 and product is 2800.

5. Find the dimensions of a rectangle if its perimeter is 170 feet and its area is 1500 square feet.

For Exercises 6–8, solve the linear system by elimination.

6. $\begin{cases} 2x + 15y = 4 \\ x - 3y = 23 \end{cases}$

7. $\begin{cases} x + y = 2 \\ 38x - 19y = 7 \end{cases}$

8. $\begin{cases} 0.4x + 0.5y = 0.112 \\ 0.3x - 0.7y = -0.131 \end{cases}$

9. Herbert invests \$17,000 in two funds that pay 11% and 13% simple interest, respectively. If he receives \$2080 in yearly interest, how much is invested in each fund?

10. Find the least squares regression line for the points $(4, 3)$, $(1, 1)$, $(-1, -2)$, and $(-2, -1)$.

For Exercises 11–12, solve the system of equations.

11. $\begin{cases} x + y = -2 \\ 2x - y + z = 11 \\ 4y - 3z = -20 \end{cases}$

12. $\begin{cases} 3x + 2y - z = 5 \\ 6x - y + 5z = 2 \end{cases}$

13. Find the equation of the parabola $y = ax^2 + bx + c$ passing through the points $(0, -1)$, $(1, 4)$ and $(2, 13)$.

For Exercises 14–15, write the partial fraction decomposition of the rational functions.

14. $\dfrac{10x - 17}{x^2 - 7x - 8}$

15. $\dfrac{x^2 + 4}{x^4 + x^2}$

16. Graph $x^2 + y^2 \geq 9.$

17. Graph the solution of the system.

$$\begin{cases} x + y \leq 6 \\ x \geq 2 \\ y \geq 0 \end{cases}$$

18. Derive a set of inequalities to describe the triangle with vertices $(0, 0)$, $(0, 7)$, and $(2, 3)$.

19. Find the maximum value of the objective function, $z = 30x + 26y$, subject to the following constraints.

$$\begin{cases} x \geq 0 \\ y \geq 0 \\ 2x + 3y \leq 21 \\ 5x + 3y \leq 30 \end{cases}$$

20. Graph the system of inequalities.

$$\begin{cases} x^2 + y^2 \leq 4 \\ (x - 2)^2 + y^2 \geq 4 \end{cases}$$

For Exercises 21–22, write the partial fraction decomposition for the rational expression.

21. $\dfrac{1 - 2x}{x^2 + x}$

22. $\dfrac{6x - 17}{(x - 3)^2}$

CHAPTER 8
Matrices and Determinants

CHAPTER 8
Matrices and Determinants

Section 8.1 Matrices and Systems of Equations

1. square

3. augmented

5. row-equivalent

7. Because the matrix has one row and two columns, its dimension is 1×2.

9. Because the matrix has three rows and one column, its dimension is 3×1.

11. Because the matrix has two rows and two columns, its dimension is 2×2.

13. Because the matrix has three rows and three columns, its dimension is 3×3.

15. $\begin{cases} 2x - y = 7 \\ x + y = 2 \end{cases}$

$$\begin{bmatrix} 2 & -1 & \vdots & 7 \\ 1 & 1 & \vdots & 2 \end{bmatrix}$$

17. $\begin{cases} x - y + 2z = 2 \\ 4x - 3y + z = -1 \\ 2x + y = 0 \end{cases}$

$$\begin{bmatrix} 1 & -1 & 2 & \vdots & 2 \\ 4 & -3 & 1 & \vdots & -1 \\ 2 & 1 & 0 & \vdots & 0 \end{bmatrix}$$

19. $\begin{cases} 3x - 5y + 2z = 12 \\ 12x - 7z = 10 \end{cases}$

$$\begin{bmatrix} 3 & -5 & 2 & \vdots & 12 \\ 12 & 0 & -7 & \vdots & 10 \end{bmatrix}$$

21. $\begin{bmatrix} 1 & 1 & \vdots & 3 \\ 5 & -3 & \vdots & -1 \end{bmatrix}$

$$\begin{cases} x + y = 3 \\ 5x - 3y = -1 \end{cases}$$

23. $\begin{bmatrix} 2 & 0 & 5 & \vdots & -12 \\ 0 & 1 & -2 & \vdots & 7 \\ 6 & 3 & 0 & \vdots & 2 \end{bmatrix}$

$$\begin{cases} 2x + 5z = -12 \\ y - 2z = 7 \\ 6x + 3y = 2 \end{cases}$$

25. $\begin{bmatrix} 9 & 12 & 3 & 0 & \vdots & 0 \\ -2 & 18 & 5 & 2 & \vdots & 10 \\ 1 & 7 & -8 & 0 & \vdots & -4 \\ 3 & 0 & 2 & 0 & \vdots & -10 \end{bmatrix}$

$$\begin{cases} 9x + 12y + 3z = 0 \\ -2x + 18y + 5z + 2w = 10 \\ x + 7y - 8z = -4 \\ 3x + 2z = -10 \end{cases}$$

27. $\begin{bmatrix} -2 & 5 & 1 \\ 3 & -1 & -8 \end{bmatrix} \rightarrow \begin{bmatrix} 13 & 0 & -39 \\ 3 & -1 & -8 \end{bmatrix}$

Add 5 times Row 2 to Row 1.

29. $\begin{bmatrix} 0 & -1 & -5 & 5 \\ -1 & 3 & -7 & 6 \\ 4 & -5 & 1 & 3 \end{bmatrix} \rightarrow \begin{bmatrix} -1 & 3 & -7 & 6 \\ 0 & -1 & -5 & 5 \\ 0 & 7 & -27 & 27 \end{bmatrix}$

Interchange Row 1 and Row 2. Then add 4 times the new Row 1 to Row 3.

31. $\begin{bmatrix} 3 & 6 & 8 \\ 4 & -3 & 6 \end{bmatrix}$

$$\tfrac{1}{3}R_1 \rightarrow \begin{bmatrix} 1 & \boxed{2} & \frac{8}{3} \\ 4 & -3 & 6 \end{bmatrix}$$

33. $\begin{bmatrix} 1 & 1 & 1 \\ 5 & -2 & 4 \end{bmatrix}$

$$-5R_1 + R_2 \rightarrow \begin{bmatrix} 1 & 1 & 1 \\ 0 & \boxed{-7} & -1 \end{bmatrix}$$

35. $\begin{bmatrix} 1 & 5 & 4 & -1 \\ 0 & 1 & -2 & 2 \\ 0 & 0 & 1 & -7 \end{bmatrix}$

$-5R_2 + R_1 \rightarrow \begin{bmatrix} 1 & 0 & \boxed{14} & \boxed{-11} \\ 0 & 1 & -2 & 2 \\ 0 & 0 & 1 & -7 \end{bmatrix}$

37. $\begin{bmatrix} 1 & 1 & 4 & -1 \\ 3 & 8 & 10 & 3 \\ -2 & 1 & 12 & 6 \end{bmatrix}$

$\begin{matrix} \\ -3R_1 + R_2 \rightarrow \\ 2R_1 + R_3 \rightarrow \end{matrix} \begin{bmatrix} 1 & 1 & 4 & -1 \\ 0 & 5 & \boxed{-2} & \boxed{6} \\ 0 & 3 & \boxed{20} & \boxed{4} \end{bmatrix}$

$\frac{1}{5}R_2 \rightarrow \begin{bmatrix} 1 & 1 & 4 & -1 \\ 0 & 1 & -\frac{2}{5} & \frac{6}{5} \\ 0 & 3 & \boxed{20} & \boxed{4} \end{bmatrix}$

39. (a) $\left[\begin{array}{cc:c} -3 & 4 & 22 \\ 6 & -4 & -28 \end{array}\right]$

(i) $R_1 + R_2 \rightarrow \left[\begin{array}{cc:c} 3 & 0 & -6 \\ 6 & -4 & -28 \end{array}\right]$

(ii) $-2R_1 + R_2 \rightarrow \left[\begin{array}{cc:c} 3 & 0 & -6 \\ 0 & -4 & -16 \end{array}\right]$

(iii) $-\frac{1}{4}R_2 \rightarrow \left[\begin{array}{cc:c} 3 & 0 & -6 \\ 0 & 1 & 4 \end{array}\right]$

(iv) $\frac{1}{3}R_1 \rightarrow \left[\begin{array}{cc:c} 1 & 0 & -2 \\ 0 & 1 & 4 \end{array}\right]$

The solution is $x = -2$ and $y = 4$.

(b) $\begin{cases} -3x + 4y = 22 \\ 6x - 4y = -28 \end{cases}$

$$3x = -6$$
$$x = -2$$

Back-substitute $x = -2$ into $-3x + 4y = 22$.

$$-3(-2) + 4y = 22$$
$$4y = 16$$
$$y = 4$$

The solution is $x = -2$ and $y = 4$.

(c) Answers vary. *Sample answer:* In this case, solving the system of linear equations using the elimination method was more efficient.

41. $\begin{bmatrix} 1 & 0 & 0 & 0 \\ 0 & 1 & 1 & 5 \\ 0 & 0 & 0 & 0 \end{bmatrix}$

This matrix is in reduced row-echelon form.

43. $\begin{bmatrix} 1 & 0 & 0 & 1 \\ 0 & 1 & 0 & -1 \\ 0 & 0 & 0 & 2 \end{bmatrix}$

This matrix is not in row-echelon form.

45. $\begin{bmatrix} 1 & 1 & 0 & 5 \\ -2 & -1 & 2 & -10 \\ 3 & 6 & 7 & 14 \end{bmatrix}$

$\begin{matrix} \\ 2R_1 + R_2 \rightarrow \\ -3R_1 + R_3 \rightarrow \end{matrix} \begin{bmatrix} 1 & 1 & 0 & 5 \\ 0 & 1 & 2 & 0 \\ 0 & 3 & 7 & -1 \end{bmatrix}$

$\begin{matrix} \\ \\ -3R_2 + R_3 \rightarrow \end{matrix} \begin{bmatrix} 1 & 1 & 0 & 5 \\ 0 & 1 & 2 & 0 \\ 0 & 0 & 1 & -1 \end{bmatrix}$

47. $\begin{bmatrix} 1 & -1 & -1 & 1 \\ 5 & -4 & 1 & 8 \\ -6 & 8 & 18 & 0 \end{bmatrix}$

$\begin{matrix} \\ -5R_1 + R_2 \rightarrow \\ 6R_1 + R_3 \rightarrow \end{matrix} \begin{bmatrix} 1 & -1 & -1 & 1 \\ 0 & 1 & 6 & 3 \\ 0 & 2 & 12 & 6 \end{bmatrix}$

$\begin{matrix} \\ \\ -2R_2 + R_3 \rightarrow \end{matrix} \begin{bmatrix} 1 & -1 & -1 & 1 \\ 0 & 1 & 6 & 3 \\ 0 & 0 & 0 & 0 \end{bmatrix}$

49. Use the reduced row-echelon form feature of a graphing utility.

$\begin{bmatrix} 3 & 3 & 3 \\ -1 & 0 & -4 \\ 2 & 4 & -2 \end{bmatrix} \Rightarrow \begin{bmatrix} 1 & 0 & 0 \\ 0 & 1 & 0 \\ 0 & 0 & 1 \end{bmatrix}$

51. Use the reduced row-echelon form feature of a graphing utility.

$\begin{bmatrix} 1 & 2 & 3 & -5 \\ 1 & 2 & 4 & -9 \\ -2 & -4 & -4 & 3 \\ 4 & 8 & 11 & -14 \end{bmatrix} \Rightarrow \begin{bmatrix} 1 & 2 & 0 & 0 \\ 0 & 0 & 1 & 0 \\ 0 & 0 & 0 & 1 \\ 0 & 0 & 0 & 0 \end{bmatrix}$

53. Use the reduced row-echelon form feature of a graphing utility.

$\begin{bmatrix} -3 & 5 & 1 & 12 \\ 1 & -1 & 1 & 4 \end{bmatrix} \Rightarrow \begin{bmatrix} 1 & 0 & 3 & 16 \\ 0 & 1 & 2 & 12 \end{bmatrix}$

55. $\begin{cases} x - 2y = 4 \\ y = -1 \end{cases}$

$x - 2(-1) = 4$

$x = 2$

Solution: $(2, -1)$

57. $\begin{cases} x - y + 2z = 4 \\ y - z = 2 \\ z = -2 \end{cases}$

$y - (-2) = 2$

$y = 0$

$x - 0 + 2(-2) = 4$

$x = 8$

Solution: $(8, 0, -2)$

59. $\begin{cases} x + 2y = 7 \\ -x + y = 8 \end{cases}$

$$\left[\begin{array}{rr:r} 1 & 2 & 7 \\ -1 & 1 & 8 \end{array}\right]$$

$$R_1 + R_2 \rightarrow \left[\begin{array}{rr:r} 1 & 2 & 7 \\ 0 & 3 & 15 \end{array}\right]$$

$$\tfrac{1}{3}R_2 \rightarrow \left[\begin{array}{rr:r} 1 & 2 & 7 \\ 0 & 1 & 5 \end{array}\right]$$

$\begin{cases} x + 2y = 7 \\ y = 5 \end{cases}$

$x + 2(5) = 7 \Rightarrow x = -3$

Solution: $(-3, 5)$

61. $\begin{cases} 3x - 2y = -27 \\ x + 3y = 13 \end{cases}$

$$\left[\begin{array}{rr:r} 3 & -2 & -27 \\ 1 & 3 & 13 \end{array}\right]$$

$$\begin{matrix} R_1 \\ R_2 \end{matrix} \left[\begin{array}{rr:r} 1 & 3 & 13 \\ 3 & -2 & -27 \end{array}\right]$$

$$-3R_1 + R_2 \rightarrow \left[\begin{array}{rr:r} 1 & 3 & 13 \\ 0 & -11 & -66 \end{array}\right]$$

$$-\tfrac{1}{11}R_2 \rightarrow \left[\begin{array}{rr:r} 1 & 3 & 13 \\ 0 & 1 & 6 \end{array}\right]$$

$\begin{cases} x + 3y = 13 \\ y = 6 \end{cases}$

$y = 6$

$x + 3(6) = 13 \Rightarrow x = -5$

Solution: $(-5, 6)$

63. $\begin{cases} x + 2y - 3z = -28 \\ 4y + 2z = 0 \\ -x + y - z = -5 \end{cases}$

$$\left[\begin{array}{rrr:r} 1 & 2 & -3 & -28 \\ 0 & 4 & 2 & 0 \\ -1 & 1 & -1 & -5 \end{array}\right]$$

$$\begin{matrix} \\ \frac{1}{4}R_2 \rightarrow \\ R_1 + R_3 \rightarrow \end{matrix} \left[\begin{array}{rrr:r} 1 & 2 & -3 & -28 \\ 0 & 1 & \frac{1}{2} & 0 \\ 0 & 3 & -4 & -33 \end{array}\right]$$

$$\begin{matrix} \\ \\ -3R_2 + R_3 \rightarrow \end{matrix} \left[\begin{array}{rrr:r} 1 & 2 & -3 & -28 \\ 0 & 1 & \frac{1}{2} & 0 \\ 0 & 0 & -\frac{11}{2} & -33 \end{array}\right]$$

$$\begin{matrix} \\ \\ -\frac{2}{11}R_3 \rightarrow \end{matrix} \left[\begin{array}{rrr:r} 1 & 2 & -3 & -28 \\ 0 & 1 & \frac{1}{2} & 0 \\ 0 & 0 & 1 & 6 \end{array}\right]$$

$\begin{cases} x + 2y - 3z = -28 \\ y + \frac{1}{2}z = 0 \\ z = 6 \end{cases}$

$z = 6$

$y + \frac{1}{2}(6) = 0 \Rightarrow y = -3$

$x + 2(-3) - 3(6) = -28 \Rightarrow x = -4$

Solution: $(-4, -3, 6)$

65. $\begin{cases} -3x + 2y = -22 \\ 3x + 4y = 4 \\ 4x - 8y = 32 \end{cases}$

$$\left[\begin{array}{rr:r} -3 & 2 & -22 \\ 3 & 4 & 4 \\ 4 & -8 & 32 \end{array}\right]$$

$$\begin{matrix} R_1 + R_3 \rightarrow \\ \\ \\ \end{matrix} \left[\begin{array}{rr:r} 1 & -6 & 10 \\ 3 & 4 & 4 \\ 4 & -8 & 32 \end{array}\right]$$

$$\begin{matrix} \\ -3R_1 + R_2 \rightarrow \\ -4R_1 + R_3 \rightarrow \end{matrix} \left[\begin{array}{rr:r} 1 & -6 & 10 \\ 0 & 22 & -26 \\ 0 & 16 & -8 \end{array}\right]$$

$$\begin{matrix} \\ \frac{1}{22}R_2 \rightarrow \\ \frac{1}{16}R_3 \rightarrow \end{matrix} \left[\begin{array}{rr:r} 1 & -1 & 22 \\ 0 & 1 & -\frac{13}{10} \\ 0 & 1 & -\frac{1}{2} \end{array}\right]$$

$$\begin{matrix} \\ \\ -R_2 + R_3 \rightarrow \end{matrix} \left[\begin{array}{rr:r} 1 & -1 & 22 \\ 0 & 1 & -\frac{13}{10} \\ 0 & 0 & \frac{9}{5} \end{array}\right]$$

The system is inconsistent and there is no solution.

67. Use the reduced row-echelon form feature of a graphing utility.

$$\begin{cases} 3x + 2y - z + w = 0 \\ x - y + 4z + 2w = 25 \\ -2x + y + 2z - w = 2 \\ x + y + z + w = 6 \end{cases}$$

$$\left[\begin{array}{rrrr:r} 3 & 2 & -1 & 1 & 0 \\ 1 & -1 & 4 & 2 & 25 \\ -2 & 1 & 2 & -1 & 2 \\ 1 & 1 & 1 & 1 & 6 \end{array}\right] \Rightarrow \left[\begin{array}{rrrr:r} 1 & 0 & 0 & 0 & 3 \\ 0 & 1 & 0 & 0 & -2 \\ 0 & 0 & 1 & 0 & 5 \\ 0 & 0 & 0 & 1 & 0 \end{array}\right]$$

$x = 3$

$y = -2$

$z = 5$

$w = 0$

Solution: $(3, -2, 5, 0)$

69. $\left[\begin{array}{rr:r} 1 & 0 & 3 \\ 0 & 1 & -4 \end{array}\right]$

$x = 3$

$y = -4$

Solution: $(3, -4)$

71. $\begin{cases} -2x + 6y = -22 \\ x + 2y = -9 \end{cases}$

$$\left[\begin{array}{rr:r} -2 & 6 & -22 \\ 1 & 2 & -9 \end{array}\right]$$

$$\begin{matrix} R_1 \\ R_2 \end{matrix} \left[\begin{array}{rr:r} 1 & 2 & -9 \\ -2 & 6 & -22 \end{array}\right]$$

$$2R_1 + R_2 \rightarrow \left[\begin{array}{rr:r} 1 & 2 & -9 \\ 0 & 10 & -40 \end{array}\right]$$

$$\tfrac{1}{10}R_2 \rightarrow \left[\begin{array}{rr:r} 1 & 2 & -9 \\ 0 & 1 & -4 \end{array}\right]$$

$$\begin{cases} x + 2y = -9 \\ y = -4 \end{cases}$$

$y = -4$

$x + 2(-4) = -9 \Rightarrow x = -1$

Solution: $(-1, -4)$

73. $\begin{cases} x + 2y + z = 8 \\ 3x + 7y + 6z = 26 \end{cases}$

$$\left[\begin{array}{rrr:r} 1 & 2 & 1 & 8 \\ 3 & 7 & 6 & 26 \end{array}\right]$$

$$-3R_1 + R_2 \rightarrow \left[\begin{array}{rrr:r} 1 & 2 & 1 & 8 \\ 0 & 1 & 3 & 2 \end{array}\right]$$

$$\begin{cases} x + 2y + z = 8 \\ y + 3z = 2 \end{cases}$$

Let $z = a$.

$y + 3a = 2 \Rightarrow y = -3a + 2$

$x + 2(-3a + 2) + a = 8 \Rightarrow x = 5a + 4$

Solution: $(5a + 4, -3a + 2, a)$ where a is a real number

75. $\begin{cases} x - 3z = -2 \\ 3x + y - 2z = 5 \\ 2x + 2y + z = 4 \end{cases}$

$$\left[\begin{array}{rrr:r} 1 & 0 & -3 & -2 \\ 3 & 1 & -2 & 5 \\ 2 & 2 & 1 & 4 \end{array}\right]$$

$$\begin{matrix} \\ -3R_1 + R_2 \rightarrow \\ -2R_1 + R_3 \rightarrow \end{matrix} \left[\begin{array}{rrr:r} 1 & 0 & -3 & -2 \\ 0 & 1 & 7 & 11 \\ 0 & 2 & 7 & 8 \end{array}\right]$$

$$-2R_2 + R_3 \rightarrow \left[\begin{array}{rrr:r} 1 & 0 & -3 & -2 \\ 0 & 1 & 7 & 11 \\ 0 & 0 & -7 & -14 \end{array}\right]$$

$$-\tfrac{1}{7}R_3 \rightarrow \left[\begin{array}{rrr:r} 1 & 0 & -3 & -2 \\ 0 & 1 & 7 & 11 \\ 0 & 0 & 1 & 2 \end{array}\right]$$

$$\begin{cases} x - 3z = -2 \\ y + 7z = 11 \\ z = 2 \end{cases}$$

$z = 2$

$y + 7(2) = 11 \Rightarrow y = -3$

$x - 3(2) = -2 \Rightarrow x = 4$

Solution: $(4, -3, 2)$

77. $\begin{cases} -x + y - z = -14 \\ 2x - y + z = 21 \\ 3x + 2y + z = 19 \end{cases}$

$$\begin{bmatrix} -1 & 1 & -1 & \vdots & -14 \\ 2 & -1 & 1 & \vdots & 21 \\ 3 & 2 & 1 & \vdots & 19 \end{bmatrix}$$

$$-R_1 \rightarrow \begin{bmatrix} 1 & -1 & 1 & \vdots & 14 \\ 2 & -1 & 1 & \vdots & 21 \\ 3 & 2 & 1 & \vdots & 19 \end{bmatrix}$$

$$\begin{matrix} \\ -2R_1 + R_2 \rightarrow \\ -3R_1 + R_3 \rightarrow \end{matrix} \begin{bmatrix} 1 & -1 & 1 & \vdots & 14 \\ 0 & 1 & -1 & \vdots & -7 \\ 0 & 5 & -2 & \vdots & -23 \end{bmatrix}$$

$$\begin{matrix} \\ \\ -5R_2 + R_3 \rightarrow \end{matrix} \begin{bmatrix} 1 & -1 & 1 & \vdots & 14 \\ 0 & 1 & -1 & \vdots & -7 \\ 0 & 0 & 3 & \vdots & 12 \end{bmatrix}$$

$$\begin{matrix} \\ \\ \frac{1}{3}R_3 \rightarrow \end{matrix} \begin{bmatrix} 1 & -1 & 1 & \vdots & 14 \\ 0 & 1 & -1 & \vdots & -7 \\ 0 & 0 & 1 & \vdots & 4 \end{bmatrix}$$

$\begin{cases} x - y + z = 14 \\ y - z = -7 \\ z = 4 \end{cases}$

$z = 4$

$y - 4 = -7 \Rightarrow y = -3$

$x - (-3) + 4 = 14 \Rightarrow x = 7$

Solution: $(7, -3, 4)$

79. Use the reduced row-echelon form feature of a graphic utility.

$\begin{cases} 3x + 3y + 12z = 6 \\ x + y + 4z = 2 \\ 2x + 5y + 20z = 10 \\ -x + 2y + 8z = 4 \end{cases}$

$$\begin{bmatrix} 3 & 3 & 12 & \vdots & 6 \\ 1 & 1 & 4 & \vdots & 2 \\ 2 & 5 & 20 & \vdots & 10 \\ -1 & 2 & 8 & \vdots & 4 \end{bmatrix} \Rightarrow \begin{bmatrix} 1 & 0 & 0 & \vdots & 0 \\ 0 & 1 & 4 & \vdots & 2 \\ 0 & 0 & 0 & \vdots & 0 \\ 0 & 0 & 0 & \vdots & 0 \end{bmatrix} \Rightarrow \begin{cases} x = 0 \\ y + 4z = 2 \end{cases}$$

Let $z = a$.

$y = 2 - 4a$

$x = 0$

Solution: $(0, 2 - 4a, a)$ where a is any real number

81. Use the reduced row-echelon form feature of a graphing utility.

$\begin{cases} 2x + y - z + 2w = -6 \\ 3x + 4y + w = 1 \\ x + 5y + 2z + 6w = -3 \\ 5x + 2y - z - w = 3 \end{cases}$

$$\begin{bmatrix} 2 & 1 & -1 & 2 & \vdots & -6 \\ 3 & 4 & 0 & 1 & \vdots & 1 \\ 1 & 5 & 2 & 6 & \vdots & -3 \\ 5 & 2 & -1 & -1 & \vdots & 3 \end{bmatrix} \Rightarrow \begin{bmatrix} 1 & 0 & 0 & 0 & \vdots & 1 \\ 0 & 1 & 0 & 0 & \vdots & 0 \\ 0 & 0 & 1 & 0 & \vdots & 4 \\ 0 & 0 & 0 & 1 & \vdots & -2 \end{bmatrix}$$

$x = 1$

$y = 0$

$z = 4$

$w = -2$

Solution: $(1, 0, 4, -2)$

83. Use the reduced row-echelon form feature of a graphing utility.

$\begin{cases} x + y + z + w = 0 \\ 2x + 3y + z - 2w = 0 \\ 3x + 5y + z = 0 \end{cases}$

$$\begin{bmatrix} 1 & 1 & 1 & 1 & \vdots & 0 \\ 2 & 3 & 1 & -2 & \vdots & 0 \\ 3 & 5 & 1 & 0 & \vdots & 0 \end{bmatrix} \Rightarrow \begin{bmatrix} 1 & 0 & 2 & 0 & \vdots & 0 \\ 0 & 1 & -1 & 0 & \vdots & 0 \\ 0 & 0 & 0 & 1 & \vdots & 0 \end{bmatrix}$$

$\begin{cases} x + 2z = 0 \\ y - z = 0 \\ w = 0 \end{cases}$

Let $z = a$. Then $x = -2a$ and $y = a$.

Solution: $(-2a, a, a, 0)$ where a is a real number

85. The dimension of the matrix is 4×1.

87. $\begin{cases} a + b + c = 1 \\ 4a + 2b + c = -1 \\ 9a + 3b + c = -5 \end{cases}$

$$\begin{bmatrix} 1 & 1 & 1 & \vdots & 1 \\ 4 & 2 & 1 & \vdots & -1 \\ 9 & 3 & 1 & \vdots & -5 \end{bmatrix} \Rightarrow \begin{bmatrix} 1 & 0 & 0 & \vdots & -1 \\ 0 & 1 & 0 & \vdots & 1 \\ 0 & 0 & 1 & \vdots & 1 \end{bmatrix}$$

$a = 1$

$b = 1$

$c = 1$

So, $f(x) = -x^2 + x + 1$.

89. $\begin{cases} 4a - 2b + c = -15 \\ a - b + c = 7 \\ a + b + c = -3 \end{cases}$

$$\begin{bmatrix} 4 & -2 & 1 & \vdots & -15 \\ 4 & -1 & 1 & \vdots & 7 \\ 1 & 1 & 1 & \vdots & -3 \end{bmatrix} \Rightarrow \begin{bmatrix} 1 & 0 & 0 & \vdots & -9 \\ 0 & 1 & 0 & \vdots & -5 \\ 0 & 0 & 1 & \vdots & 11 \end{bmatrix}$$

$a = -9$

$b = -5$

$c = 11$

So, $f(x) = -9x^2 - 5x + 11$.

91. $\begin{cases} a + b + c = 8 \\ 4a + 2b + c = 13 \\ 9a + 3b + c = 20 \end{cases}$

$$\begin{bmatrix} 1 & 1 & 1 & \vdots & 8 \\ 4 & 2 & 1 & \vdots & 13 \\ 9 & 3 & 1 & \vdots & 20 \end{bmatrix} \Rightarrow \begin{bmatrix} 1 & 0 & 0 & \vdots & 1 \\ 0 & 1 & 0 & \vdots & 2 \\ 0 & 0 & 1 & \vdots & 5 \end{bmatrix}$$

$a = 1$

$b = 2$

$c = 5$

So, $f(x) = x^2 + 2x + 5$.

93. $\begin{cases} 12b + 66a = 831 \\ 66b + 506a = 5643 \end{cases}$

$$\begin{bmatrix} 12 & 66 & \vdots & 831 \\ 66 & 506 & \vdots & 5643 \end{bmatrix} \Rightarrow \begin{bmatrix} 1 & 0 & \vdots & 28 \\ 0 & 1 & \vdots & 7.5 \end{bmatrix}$$

$y = 7.5t + 28$

In 2020, the number of new cases of the waterborne disease will be $y = 7.5(15) + 28 = 140.5 \approx 141$ cases.

Because the data values increase in a linear pattern, this prediction is reasonable.

95. x = amount at 8%

y = amount at 9%

z = amount at 12%

$\begin{cases} x + y + z = 2{,}000{,}000 \\ 0.08x + 0.09y + 0.12z = 186{,}000 \\ x - 2z = 0 \end{cases}$

$$\begin{bmatrix} 1 & 1 & 1 & \vdots & 2{,}000{,}000 \\ 0.08 & 0.09 & 0.12 & \vdots & 186{,}000 \\ 1 & 0 & -2 & \vdots & 0 \end{bmatrix}$$

$$\begin{matrix} \\ -0.08R_1 + R_2 \\ -R_1 + R_3 \end{matrix} \begin{bmatrix} 1 & 1 & 1 & \vdots & 2{,}000{,}000 \\ 0 & 0.01 & 0.04 & \vdots & 26{,}000 \\ 0 & -1 & -3 & \vdots & -2{,}000{,}000 \end{bmatrix}$$

$$\begin{matrix} \\ 100R_2 \rightarrow \\ \\ \end{matrix} \begin{bmatrix} 1 & 1 & 1 & \vdots & 2{,}000{,}000 \\ 0 & 1 & 4 & \vdots & 2{,}600{,}000 \\ 0 & -1 & -3 & \vdots & -2{,}000{,}000 \end{bmatrix}$$

$$\begin{matrix} \\ \\ R_2 + R_3 \rightarrow \end{matrix} \begin{bmatrix} 1 & 1 & 1 & \vdots & 2{,}000{,}000 \\ 0 & 1 & 4 & \vdots & 2{,}600{,}000 \\ 0 & 0 & 1 & \vdots & 600{,}000 \end{bmatrix}$$

The matrix is now in row-echelon form, and the corresponding system is shown.

$\begin{cases} x + y + z = 2{,}000{,}000 \\ y + 4z = 2{,}600{,}000 \\ z = 600{,}000 \end{cases}$

Using back-substitution, you can determine the solution.

$$y + 4(600{,}000) = 2{,}600{,}000$$
$$y = 200{,}000$$
$$x + 200{,}000 + 600{,}000 = 2{,}000{,}000$$
$$x = 1{,}200{,}000$$

So, the zoo borrowed 1,200,000 at 8%, \$200,000 at 9%, and \$600,000 at 12%.

97. False. It is a 2×4 matrix.

99. They are the same.

Section 8.2 Operations with Matrices

1. equal

3. zero; O

5. $\begin{bmatrix} x & -2 \\ 7 & 23 \end{bmatrix} = \begin{bmatrix} -4 & -2 \\ 7 & y \end{bmatrix}$

$x = -4$

$y = 23$

7. $\begin{bmatrix} 16 & 4 & x & 4 \\ 0 & 2 & 4 & 0 \end{bmatrix} = \begin{bmatrix} 16 & 4 & 2x + 1 & 4 \\ 0 & 2 & 3y - 5 & 0 \end{bmatrix}$

$x = 2x + 1 \Rightarrow x = -1$

$4 = 3y - 5 \Rightarrow y = 3$

9. (a) $A + B = \begin{bmatrix} 1 & -1 \\ 2 & -1 \end{bmatrix} + \begin{bmatrix} 2 & -1 \\ -1 & 8 \end{bmatrix} = \begin{bmatrix} 1+2 & -1-1 \\ 2-1 & -1+8 \end{bmatrix} = \begin{bmatrix} 3 & -2 \\ 1 & 7 \end{bmatrix}$

(b) $A - B = \begin{bmatrix} 1 & -1 \\ 2 & -1 \end{bmatrix} - \begin{bmatrix} 2 & -1 \\ -1 & 8 \end{bmatrix} = \begin{bmatrix} 1-2 & -1+1 \\ 2+1 & -1-8 \end{bmatrix} = \begin{bmatrix} -1 & 0 \\ 3 & -9 \end{bmatrix}$

(c) $3A = 3\begin{bmatrix} 1 & -1 \\ 2 & -1 \end{bmatrix} = \begin{bmatrix} 3(1) & 3(-1) \\ 3(2) & 3(-1) \end{bmatrix} = \begin{bmatrix} 3 & -3 \\ 6 & -3 \end{bmatrix}$

(d) $3A - 2B = \begin{bmatrix} 3 & -3 \\ 6 & -3 \end{bmatrix} - 2\begin{bmatrix} 2 & -1 \\ -1 & 8 \end{bmatrix} = \begin{bmatrix} 3 & -3 \\ 6 & -3 \end{bmatrix} + \begin{bmatrix} -4 & 2 \\ 2 & -16 \end{bmatrix} = \begin{bmatrix} -1 & -1 \\ 8 & -19 \end{bmatrix}$

11. $A = \begin{bmatrix} 6 & 0 & 3 \\ -1 & -4 & 0 \end{bmatrix}, B = \begin{bmatrix} 8 & -1 \\ 4 & -3 \end{bmatrix}$

(a) $A + B$ is not possible. A and B do not have the same order.

(b) $A - B$ is not possible. A and B do not have the same order.

(c) $3A = \begin{bmatrix} 18 & 0 & 9 \\ -3 & -12 & 0 \end{bmatrix}$

(d) $3A - 2B$ is not possible. A and B do not have the same order.

13. $A = \begin{bmatrix} 8 & -1 \\ 2 & 3 \\ -4 & 5 \end{bmatrix}, B = \begin{bmatrix} 1 & 6 \\ -1 & -5 \\ 1 & 10 \end{bmatrix}$

(a) $A + B = \begin{bmatrix} 8 & -1 \\ 2 & 3 \\ -4 & 5 \end{bmatrix} + \begin{bmatrix} 1 & 6 \\ -1 & -5 \\ 1 & 10 \end{bmatrix} = \begin{bmatrix} 8+1 & -1+6 \\ 2-1 & 3-5 \\ -4+1 & 5+10 \end{bmatrix} = \begin{bmatrix} 9 & 5 \\ 1 & -2 \\ -3 & 15 \end{bmatrix}$

(b) $A - B = \begin{bmatrix} 8 & -1 \\ 2 & 3 \\ -4 & 5 \end{bmatrix} - \begin{bmatrix} 1 & 6 \\ -1 & -5 \\ 1 & 10 \end{bmatrix} = \begin{bmatrix} 8-1 & -1-6 \\ 2-(-1) & 3-(-5) \\ -4-1 & 5-10 \end{bmatrix} = \begin{bmatrix} 7 & -7 \\ 3 & 8 \\ -5 & -5 \end{bmatrix}$

(c) $3A = 3\begin{bmatrix} 8 & -1 \\ 2 & 3 \\ -4 & 5 \end{bmatrix} = \begin{bmatrix} 3(8) & 3(-1) \\ 3(2) & 3(3) \\ 3(-4) & 3(5) \end{bmatrix} = \begin{bmatrix} 24 & -3 \\ 6 & 9 \\ -12 & 15 \end{bmatrix}$

(d) $3A - 2B = \begin{bmatrix} 24 & -3 \\ 6 & 9 \\ -12 & 15 \end{bmatrix} - 2\begin{bmatrix} 1 & 6 \\ -1 & -5 \\ 1 & 10 \end{bmatrix} = \begin{bmatrix} 24-2 & -3-12 \\ 6+2 & 9+10 \\ -12-2 & 15-20 \end{bmatrix} = \begin{bmatrix} 22 & -15 \\ 8 & 19 \\ -14 & -5 \end{bmatrix}$

15. $A = \begin{bmatrix} 4 & 5 & -1 & 3 & 4 \\ 1 & 2 & -2 & -1 & 0 \end{bmatrix}, B = \begin{bmatrix} 1 & 0 & -1 & 1 & 0 \\ -6 & 8 & 2 & -3 & -7 \end{bmatrix}$

(a) $A + B = \begin{bmatrix} 4 & 5 & -1 & 3 & 4 \\ 1 & 2 & -2 & -1 & 0 \end{bmatrix} + \begin{bmatrix} 1 & 0 & -1 & 1 & 0 \\ -6 & 8 & 2 & -3 & -7 \end{bmatrix} = \begin{bmatrix} 4+1 & 5+0 & -1-1 & 3+1 & 4+0 \\ 1-6 & 2+8 & -2+2 & -1-3 & 0-7 \end{bmatrix}$

$= \begin{bmatrix} 5 & 5 & -2 & 4 & 4 \\ -5 & 10 & 0 & -4 & -7 \end{bmatrix}$

(b) $A - B = \begin{bmatrix} 4 & 5 & -1 & 3 & 4 \\ 1 & 2 & -2 & -1 & 0 \end{bmatrix} - \begin{bmatrix} 1 & 0 & -1 & 1 & 0 \\ -6 & 8 & 2 & -3 & -7 \end{bmatrix} = \begin{bmatrix} 4-1 & 5-0 & -1-(-1) & 3-1 & 4-0 \\ 1-(-6) & 2-8 & -2-2 & -1-(-3) & 0-(-7) \end{bmatrix}$

$= \begin{bmatrix} 3 & 5 & 0 & 2 & 4 \\ 7 & -6 & -4 & 2 & 7 \end{bmatrix}$

(c) $3A = 3\begin{bmatrix} 4 & 5 & -1 & 3 & 4 \\ 1 & 2 & -2 & -1 & 0 \end{bmatrix} = \begin{bmatrix} 3(4) & 3(5) & 3(-1) & 3(3) & 3(4) \\ 3(1) & 3(2) & 3(-2) & 3(-1) & 3(0) \end{bmatrix} = \begin{bmatrix} 12 & 15 & -3 & 9 & 12 \\ 3 & 6 & -6 & -3 & 0 \end{bmatrix}$

(d) $3A - 2B = \begin{bmatrix} 12 & 15 & -3 & 9 & 12 \\ 3 & 6 & -6 & -3 & 0 \end{bmatrix} - 2\begin{bmatrix} 1 & 0 & -1 & 1 & 0 \\ -6 & 8 & 2 & -3 & -7 \end{bmatrix} = \begin{bmatrix} 12-2 & 15+0 & -3+2 & 9-2 & 12-0 \\ 3+12 & 6-16 & -6-4 & -3+6 & 0+14 \end{bmatrix}$

$= \begin{bmatrix} 10 & 15 & -1 & 7 & 12 \\ 15 & -10 & -10 & 3 & 14 \end{bmatrix}$

17. $\begin{bmatrix} -5 & 0 \\ 3 & -6 \end{bmatrix} + \begin{bmatrix} 7 & 1 \\ -2 & -1 \end{bmatrix} + \begin{bmatrix} -10 & -8 \\ 14 & 6 \end{bmatrix} = \begin{bmatrix} -5+7+(-10) & 0+1+(-8) \\ 3+(-2)+14 & -6+(-1)+6 \end{bmatrix} = \begin{bmatrix} -8 & -7 \\ 15 & -1 \end{bmatrix}$

19. $4\left(\begin{bmatrix} -4 & 0 & 1 \\ 0 & 2 & 3 \end{bmatrix} - \begin{bmatrix} 2 & 1 & -2 \\ 3 & -6 & 0 \end{bmatrix}\right) = 4\begin{bmatrix} -6 & -1 & 3 \\ -3 & 8 & 3 \end{bmatrix} = \begin{bmatrix} -24 & -4 & 12 \\ -12 & 32 & 12 \end{bmatrix}$

21. $-3\left(\begin{bmatrix} 0 & -3 \\ 7 & 2 \end{bmatrix} + \begin{bmatrix} -6 & 3 \\ 8 & 1 \end{bmatrix}\right) - 2\begin{bmatrix} 4 & -4 \\ 7 & -9 \end{bmatrix} = -3\begin{bmatrix} -6 & 0 \\ 15 & 3 \end{bmatrix} - \begin{bmatrix} 8 & -8 \\ 14 & -18 \end{bmatrix} = \begin{bmatrix} 18 & 0 \\ -45 & -9 \end{bmatrix} - \begin{bmatrix} 8 & -8 \\ 14 & -18 \end{bmatrix} = \begin{bmatrix} 10 & 8 \\ -59 & 9 \end{bmatrix}$

23. $\frac{11}{25}\begin{bmatrix} 2 & 5 \\ -1 & -4 \end{bmatrix} + 6\begin{bmatrix} -3 & 0 \\ 2 & 2 \end{bmatrix} = \begin{bmatrix} -17.12 & 2.2 \\ 11.56 & 10.24 \end{bmatrix}$

25. $-2\begin{bmatrix} 1.23 & 4.19 & -3.85 \\ 7.21 & -2.60 & 6.54 \end{bmatrix} - \begin{bmatrix} 8.35 & -3.02 & 7.30 \\ -0.38 & -5.49 & 1.68 \end{bmatrix} = \begin{bmatrix} -10.81 & -5.39 & 0.4 \\ -14.04 & 10.69 & -14.76 \end{bmatrix}$

In Exercises 27-33, $A = \begin{bmatrix} -2 & 1 & 3 \\ -1 & 0 & 4 \end{bmatrix}$ and $B = \begin{bmatrix} 0 & 2 & -4 \\ 3 & 0 & 1 \end{bmatrix}$

27. $X = 2A + 2B = 2\begin{bmatrix} -2 & 1 & 3 \\ -1 & 0 & 4 \end{bmatrix} + 2\begin{bmatrix} 0 & 2 & -4 \\ 3 & 0 & 1 \end{bmatrix} = \begin{bmatrix} -4 & 2 & 6 \\ -2 & 0 & 8 \end{bmatrix} + \begin{bmatrix} 0 & 4 & -8 \\ 6 & 0 & 2 \end{bmatrix} = \begin{bmatrix} -4 & 6 & -2 \\ 4 & 0 & 10 \end{bmatrix}$

29. $2X = 2A - B$

$X = \frac{1}{2}(2A - B) = \frac{1}{2}\left(2\begin{bmatrix} -2 & 1 & 3 \\ -1 & 0 & 4 \end{bmatrix} - \begin{bmatrix} 0 & 2 & -4 \\ 3 & 0 & 1 \end{bmatrix}\right) = \frac{1}{2}\left(\begin{bmatrix} -4 & 2 & 6 \\ -2 & 0 & 8 \end{bmatrix} - \begin{bmatrix} 0 & 2 & -4 \\ 3 & 0 & 1 \end{bmatrix}\right) = \frac{1}{2}\begin{bmatrix} -4 & 0 & 10 \\ -5 & 0 & 7 \end{bmatrix} = \begin{bmatrix} -2 & 0 & 5 \\ -\frac{5}{2} & 0 & \frac{7}{2} \end{bmatrix}$

31. $2X + 3A = B$

$2X = B - 3A$

$X = \frac{1}{2}(B - 3A) = \frac{1}{2}\left(\begin{bmatrix} 0 & 2 & -4 \\ 3 & 0 & 1 \end{bmatrix} - 3\begin{bmatrix} -2 & 1 & 3 \\ -1 & 0 & 4 \end{bmatrix}\right) = \frac{1}{2}\left(\begin{bmatrix} 0 & 2 & -4 \\ 3 & 0 & 1 \end{bmatrix} - \begin{bmatrix} -6 & 3 & 9 \\ -3 & 0 & 12 \end{bmatrix}\right)$

$= \frac{1}{2}\begin{bmatrix} 6 & -1 & -13 \\ 6 & 0 & -11 \end{bmatrix} = \begin{bmatrix} 3 & -\frac{1}{2} & -\frac{13}{2} \\ 3 & 0 & -\frac{11}{2} \end{bmatrix}$

33. $4B = -2X - 2A$

$2X = -2A - 4B$

$X = \frac{1}{2}(-2A - 4B) = \frac{1}{2}\left(-2\begin{bmatrix} -2 & 1 & 3 \\ -1 & 0 & 4 \end{bmatrix} - 4\begin{bmatrix} 0 & 2 & -4 \\ 3 & 0 & 1 \end{bmatrix}\right) = \frac{1}{2}\left(\begin{bmatrix} 4 & -2 & -6 \\ 2 & 0 & -8 \end{bmatrix} - \begin{bmatrix} 0 & 8 & -16 \\ 12 & 0 & 4 \end{bmatrix}\right) = \frac{1}{2}\begin{bmatrix} 4 & -10 & 10 \\ -10 & 0 & -12 \end{bmatrix}$

$= \begin{bmatrix} 2 & -5 & 5 \\ -5 & 0 & -6 \end{bmatrix}$

35. A is 3×2, B is $2 \times 2 \Rightarrow AB$ is 3×2.

$$A = \begin{bmatrix} -1 & 6 \\ -4 & 5 \\ 0 & 3 \end{bmatrix}, B = \begin{bmatrix} 2 & 3 \\ 0 & 9 \end{bmatrix}$$

$$AB = \begin{bmatrix} -1 & 6 \\ -4 & 5 \\ 0 & 3 \end{bmatrix}\begin{bmatrix} 2 & 3 \\ 0 & 9 \end{bmatrix} = \begin{bmatrix} (-1)(2) + (6)(0) & (-1)(3) + (6)(9) \\ (-4)(2) + (5)(0) & (-4)(3) + (5)(9) \\ (0)(2) + (3)(0) & (0)(3) + (3)(9) \end{bmatrix} = \begin{bmatrix} -2 & 51 \\ -8 & 33 \\ 0 & 27 \end{bmatrix}$$

37. A is 3×2 and B is 3×3. AB is not possible.

39. A is 3×3, B is $3 \times 3 \Rightarrow AB$ is 3×3.

$$AB = \begin{bmatrix} 5 & 0 & 0 \\ 0 & -8 & 0 \\ 0 & 0 & 7 \end{bmatrix}\begin{bmatrix} \frac{1}{5} & 0 & 0 \\ 0 & -\frac{1}{8} & 0 \\ 0 & 0 & \frac{1}{2} \end{bmatrix} = \begin{bmatrix} 1 & 0 & 0 \\ 0 & 1 & 0 \\ 0 & 0 & \frac{7}{2} \end{bmatrix}$$

41. $A = \begin{bmatrix} 7 & 5 & -4 \\ -2 & 5 & 1 \\ 10 & -4 & -7 \end{bmatrix}, B = \begin{bmatrix} 2 & -2 & 3 \\ 8 & 1 & 4 \\ -4 & 2 & -8 \end{bmatrix}$

$$AB = \begin{bmatrix} 7 & 5 & -4 \\ -2 & 5 & 1 \\ 10 & -4 & -7 \end{bmatrix}\begin{bmatrix} 2 & -2 & 3 \\ 8 & 1 & 4 \\ -4 & 2 & -8 \end{bmatrix} = \begin{bmatrix} 70 & -17 & 73 \\ 32 & 11 & 6 \\ 16 & -38 & 70 \end{bmatrix}$$

43. $\begin{bmatrix} -3 & 8 & -6 & 8 \\ -12 & 15 & 9 & 6 \\ 5 & -1 & 1 & 5 \end{bmatrix}\begin{bmatrix} 3 & 1 & 6 \\ 24 & 15 & 14 \\ 16 & 10 & 21 \\ 8 & -4 & 10 \end{bmatrix} = \begin{bmatrix} 151 & 25 & 48 \\ 516 & 279 & 387 \\ 47 & -20 & 87 \end{bmatrix}$

45. (a) $AB = \begin{bmatrix} 1 & 2 \\ 4 & 2 \end{bmatrix}\begin{bmatrix} 2 & -1 \\ -1 & 8 \end{bmatrix} = \begin{bmatrix} (1)(2) + (2)(-1) & (1)(-1) + (2)(8) \\ (4)(2) + (2)(-1) & (4)(-1) + (2)(8) \end{bmatrix} = \begin{bmatrix} 0 & 15 \\ 6 & 12 \end{bmatrix}$

(b) $BA = \begin{bmatrix} 2 & -1 \\ -1 & 8 \end{bmatrix}\begin{bmatrix} 1 & 2 \\ 4 & 2 \end{bmatrix} = \begin{bmatrix} (2)(1) + (-1)(4) & (2)(2) + (-1)(2) \\ (-1)(1) + (8)(4) & (-1)(2) + (8)(2) \end{bmatrix} = \begin{bmatrix} -2 & 2 \\ 31 & 14 \end{bmatrix}$

(c) $A^2 = \begin{bmatrix} 1 & 2 \\ 4 & 2 \end{bmatrix}\begin{bmatrix} 1 & 2 \\ 4 & 2 \end{bmatrix} = \begin{bmatrix} (1)(1) + (2)(4) & (1)(2) + (2)(2) \\ (4)(1) + (2)(4) & (4)(2) + (2)(2) \end{bmatrix} = \begin{bmatrix} 9 & 6 \\ 12 & 12 \end{bmatrix}$

47. (a) $AB = \begin{bmatrix} 5 & -9 & 0 \\ 3 & 0 & -8 \\ -1 & 4 & 11 \end{bmatrix}\begin{bmatrix} 1 & 0 & 0 \\ 0 & 1 & 0 \\ 0 & 0 & 1 \end{bmatrix} = \begin{bmatrix} (5)(1) + (-9)(0) + (0)(0) & (5)(0) + (-9)(1) + (0)(0) & (5)(0) + (-9)(0) + (0)(1) \\ (3)(1) + (0)(0) + (-8)(0) & (3)(0) + (0)(1) + (-8)(0) & (3)(0) + (0)(0) + (-8)(1) \\ (-1)(1) + (4)(0) + (11)(0) & (-1)(0) + (4)(1) + (11)(0) & (-1)(0) + (4)(0) + (11)(1) \end{bmatrix}$

$= \begin{bmatrix} 5 & -9 & 0 \\ 3 & 0 & -8 \\ -1 & 4 & 11 \end{bmatrix}$

(b) $BA = \begin{bmatrix} 1 & 0 & 0 \\ 0 & 1 & 0 \\ 0 & 0 & 1 \end{bmatrix}\begin{bmatrix} 5 & -9 & 0 \\ 3 & 0 & -8 \\ -1 & 4 & 11 \end{bmatrix} = \begin{bmatrix} (1)(5) + (0)(3) + (0)(-1) & (1)(-9) + (0)(0) + (0)(4) & (1)(0) + (0)(-8) + (0)(11) \\ (0)(5) + (1)(3) + (0)(-1) & (0)(-9) + (1)(0) + (0)(4) & (0)(0) + (1)(-8) + (0)(11) \\ (0)(5) + (0)(3) + (1)(-1) & (0)(-9) + (0)(0) + (1)(4) & (0)(0) + (0)(-8) + (1)(11) \end{bmatrix}$

$= \begin{bmatrix} 5 & -9 & 0 \\ 3 & 0 & -8 \\ -1 & 4 & 11 \end{bmatrix}$

(c) $AA = \begin{bmatrix} 5 & -9 & 0 \\ 3 & 0 & -8 \\ -1 & 4 & 11 \end{bmatrix}\begin{bmatrix} 5 & -9 & 0 \\ 3 & 0 & -8 \\ -1 & 4 & 11 \end{bmatrix} = \begin{bmatrix} (5)(5) + (-9)(3) + (0)(-1) & (5)(-9) + (-9)(0) + (0)(4) & (5)(0) + (-9)(-8) + (0)(11) \\ (3)(5) + (0)(3) + (-8)(-1) & (3)(-9) + (0)(0) + (-8)(4) & (3)(0) + (0)(-8) + (-8)(11) \\ (-1)(5) + (4)(3) + (11)(-1) & (-1)(-9) + (4)(0) + (11)(4) & (-1)(0) + (4)(-8) + (11)(11) \end{bmatrix}$

$= \begin{bmatrix} -2 & -45 & 72 \\ 23 & -59 & -88 \\ -4 & 53 & 89 \end{bmatrix}$

49. (a) $AB = \begin{bmatrix} -4 & -1 \\ 2 & 12 \end{bmatrix}\begin{bmatrix} -6 \\ 5 \end{bmatrix} = \begin{bmatrix} (-4)(-6) + (-1)(5) \\ (2)(-6) + (12)(5) \end{bmatrix} = \begin{bmatrix} 19 \\ 48 \end{bmatrix}$

(b) BA is not possible, B is 2×1 and A is 2×2.

(c) $A^2 = \begin{bmatrix} -4 & -1 \\ 2 & 12 \end{bmatrix}\begin{bmatrix} -4 & -1 \\ 2 & 12 \end{bmatrix} = \begin{bmatrix} (-4)(-4) + (-1)(2) & (-4)(-1) + (-1)(12) \\ (2)(-4) + (12)(2) & (2)(-1) + (12)(12) \end{bmatrix} = \begin{bmatrix} 14 & -8 \\ 16 & 142 \end{bmatrix}$

51. (a) $AB = \begin{bmatrix} 7 \\ 8 \\ -1 \end{bmatrix}\begin{bmatrix} 1 & 1 & 2 \end{bmatrix} = \begin{bmatrix} 7(1) & 7(1) & 7(2) \\ 8(1) & 8(1) & 8(2) \\ -1(1) & -1(1) & -1(2) \end{bmatrix} = \begin{bmatrix} 7 & 7 & 14 \\ 8 & 8 & 16 \\ -1 & -1 & -2 \end{bmatrix}$

(b) $BA = \begin{bmatrix} 1 & 1 & 2 \end{bmatrix}\begin{bmatrix} 7 \\ 8 \\ -1 \end{bmatrix} = \begin{bmatrix} (1)(7) + (1)(8) + (2)(-1) \end{bmatrix} = \begin{bmatrix} 13 \end{bmatrix}$

(c) A^2 is not possible.

53. $\begin{bmatrix} 3 & 1 \\ 0 & -2 \end{bmatrix}\begin{bmatrix} 1 & 0 \\ -2 & 2 \end{bmatrix}\begin{bmatrix} 1 & 0 \\ 2 & 4 \end{bmatrix} = \begin{bmatrix} 1 & 2 \\ 4 & -4 \end{bmatrix}\begin{bmatrix} 1 & 0 \\ 2 & 4 \end{bmatrix} = \begin{bmatrix} 5 & 8 \\ -4 & -16 \end{bmatrix}$

55. $\begin{bmatrix} 0 & 2 & -2 \\ 4 & 1 & 2 \end{bmatrix}\left(\begin{bmatrix} 4 & 0 \\ 0 & -1 \\ -1 & 2 \end{bmatrix} + \begin{bmatrix} -2 & 3 \\ -3 & 5 \\ 0 & -3 \end{bmatrix}\right) = \begin{bmatrix} 0 & 2 & -2 \\ 4 & 1 & 2 \end{bmatrix}\begin{bmatrix} 2 & 3 \\ -3 & 4 \\ -1 & -1 \end{bmatrix} = \begin{bmatrix} -4 & 10 \\ 3 & 14 \end{bmatrix}$

57. $\mathbf{u} = \langle 1, 5 \rangle, \mathbf{v} = \langle 3, 2 \rangle$

(a) $\mathbf{u} + \mathbf{v} = \begin{bmatrix} 1 \\ 5 \end{bmatrix} + \begin{bmatrix} 3 \\ 2 \end{bmatrix} = \begin{bmatrix} 4 \\ 7 \end{bmatrix} = \langle 4, 7 \rangle$

(b) $\mathbf{u} - \mathbf{v} = \begin{bmatrix} 1 \\ 5 \end{bmatrix} - \begin{bmatrix} 3 \\ 2 \end{bmatrix} = \begin{bmatrix} -2 \\ 3 \end{bmatrix} = \langle -2, 3 \rangle$

(c) $3\mathbf{v} - \mathbf{u} = 3\begin{bmatrix} 3 \\ 2 \end{bmatrix} - \begin{bmatrix} 1 \\ 5 \end{bmatrix} = \begin{bmatrix} 8 \\ 1 \end{bmatrix} = \langle 8, 1 \rangle$

59. $\mathbf{u} = \langle -2, 2 \rangle, \mathbf{v} = \langle 5, 4 \rangle$

(a) $\mathbf{u} + \mathbf{v} = \begin{bmatrix} -2 \\ 2 \end{bmatrix} + \begin{bmatrix} 5 \\ 4 \end{bmatrix} = \begin{bmatrix} 3 \\ 6 \end{bmatrix} = \langle 3, 6 \rangle$

(b) $\mathbf{u} - \mathbf{v} = \begin{bmatrix} -2 \\ 2 \end{bmatrix} - \begin{bmatrix} 5 \\ 4 \end{bmatrix} = \begin{bmatrix} -7 \\ -2 \end{bmatrix} = \langle -7, -2 \rangle$

(c) $3\mathbf{v} - \mathbf{u} = 3\begin{bmatrix} 5 \\ 4 \end{bmatrix} - \begin{bmatrix} -2 \\ 2 \end{bmatrix} = \begin{bmatrix} 17 \\ 10 \end{bmatrix} = \langle 17, 10 \rangle$

In Exercises 61–65, $\mathbf{v} = \langle 4, 2 \rangle = \begin{bmatrix} 4 \\ 2 \end{bmatrix}$

61. $A = \begin{bmatrix} 1 & 0 \\ 0 & -1 \end{bmatrix}, A\mathbf{v} = \begin{bmatrix} 1 & 0 \\ 0 & -1 \end{bmatrix}\begin{bmatrix} 4 \\ 2 \end{bmatrix} = \begin{bmatrix} 4 \\ -2 \end{bmatrix} = \langle 4, -2 \rangle$ is a reflection in the x-axis.

63. $A = \begin{bmatrix} 0 & 1 \\ 1 & 0 \end{bmatrix}, A\mathbf{v} = \begin{bmatrix} 0 & 1 \\ 1 & 0 \end{bmatrix}\begin{bmatrix} 4 \\ 2 \end{bmatrix} = \begin{bmatrix} 2 \\ 4 \end{bmatrix} = \langle 2, 4 \rangle$ is a reflection in the line $y = x$.

65. $A = \begin{bmatrix} 2 & 0 \\ 0 & 1 \end{bmatrix}, A\mathbf{v} = \begin{bmatrix} 2 & 0 \\ 0 & 1 \end{bmatrix}\begin{bmatrix} 4 \\ 2 \end{bmatrix} = \begin{bmatrix} 8 \\ 2 \end{bmatrix} = \langle 8, 2 \rangle$ is a horizontal stretch.

67. (a) $\begin{bmatrix} 2 & 3 \\ 1 & 4 \end{bmatrix}\begin{bmatrix} x_1 \\ x_2 \end{bmatrix} = \begin{bmatrix} 5 \\ 10 \end{bmatrix}$

(b) $\begin{matrix} R_2 \\ R_1 \end{matrix}\begin{bmatrix} 1 & 4 & \vdots & 10 \\ 2 & 3 & \vdots & 5 \end{bmatrix}$

$-2R_1 + R_2 \rightarrow \begin{bmatrix} 1 & 4 & \vdots & 10 \\ 0 & -5 & \vdots & -15 \end{bmatrix}$

$-\frac{1}{5}R_2 \rightarrow \begin{bmatrix} 1 & 4 & \vdots & 10 \\ 0 & 1 & \vdots & 3 \end{bmatrix}$

$\begin{matrix} -4R_2 + R_1 \rightarrow \\ -\frac{1}{5}R_2 \rightarrow \end{matrix}\begin{bmatrix} 1 & 0 & \vdots & -2 \\ 0 & 1 & \vdots & 3 \end{bmatrix}$

$X = \begin{bmatrix} -2 \\ 3 \end{bmatrix}$

69. (a) $\begin{bmatrix} 1 & -2 & 3 \\ -1 & 3 & -1 \\ 2 & -5 & 5 \end{bmatrix}\begin{bmatrix} x_1 \\ x_2 \\ x_3 \end{bmatrix} = \begin{bmatrix} 9 \\ -6 \\ 17 \end{bmatrix}$

(b) $\begin{bmatrix} 1 & -2 & 3 & \vdots & 9 \\ -1 & 3 & -1 & \vdots & -6 \\ 2 & -5 & 5 & \vdots & 17 \end{bmatrix}$

$\begin{matrix} \\ R_1 + R_2 \rightarrow \\ -2R_2 + R_3 \rightarrow \end{matrix}\begin{bmatrix} 1 & -2 & 3 & \vdots & 9 \\ 0 & 1 & 2 & \vdots & 3 \\ 0 & -1 & -1 & \vdots & -1 \end{bmatrix}$

$\begin{matrix} 2R_2 + R_1 \rightarrow \\ \\ R_2 + R_3 \rightarrow \end{matrix}\begin{bmatrix} 1 & 0 & 7 & \vdots & 15 \\ 0 & 1 & 2 & \vdots & 3 \\ 0 & 0 & 1 & \vdots & 2 \end{bmatrix}$

$\begin{matrix} -7R_3 + R_1 \rightarrow \\ -2R_3 + R_2 \rightarrow \\ \end{matrix}\begin{bmatrix} 1 & 0 & 0 & \vdots & 1 \\ 0 & 1 & 0 & \vdots & -1 \\ 0 & 0 & 1 & \vdots & 2 \end{bmatrix}$

$X = \begin{bmatrix} 1 \\ -1 \\ 2 \end{bmatrix}$

71. (a) $\begin{bmatrix} 1 & -5 & 2 \\ -3 & 1 & -1 \\ 0 & -2 & 5 \end{bmatrix}\begin{bmatrix} x_1 \\ x_2 \\ x_3 \end{bmatrix} = \begin{bmatrix} -20 \\ 8 \\ -16 \end{bmatrix}$

(b) $\begin{bmatrix} 1 & -5 & 2 & \vdots & -20 \\ -3 & 1 & -1 & \vdots & 8 \\ 0 & -2 & 5 & \vdots & -16 \end{bmatrix}$

$3R_1 + R_2 \rightarrow \begin{bmatrix} 1 & -5 & 2 & \vdots & -20 \\ 0 & -14 & 5 & \vdots & -52 \\ 0 & -2 & 5 & \vdots & -16 \end{bmatrix}$

$-R_3 + R_2 \rightarrow \begin{bmatrix} 1 & -5 & 2 & \vdots & -20 \\ 0 & -12 & 0 & \vdots & -36 \\ 0 & -2 & 5 & \vdots & -16 \end{bmatrix}$

$-\frac{1}{12}R_2 \rightarrow \begin{bmatrix} 1 & -5 & 2 & \vdots & -20 \\ 0 & 1 & 0 & \vdots & 3 \\ 0 & -2 & 5 & \vdots & -16 \end{bmatrix}$

$\begin{matrix} 5R_2 + R_1 \rightarrow \\ \\ 2R_2 + R_3 \rightarrow \end{matrix}\begin{bmatrix} 1 & 0 & 2 & \vdots & -5 \\ 0 & 1 & 0 & \vdots & 3 \\ 0 & 0 & 5 & \vdots & -10 \end{bmatrix}$

$\begin{matrix} \\ \\ \frac{1}{5}R_3 \rightarrow \end{matrix}\begin{bmatrix} 1 & 0 & 2 & \vdots & -5 \\ 0 & 1 & 0 & \vdots & 3 \\ 0 & 0 & 1 & \vdots & -2 \end{bmatrix}$

$-2R_3 + R_1 \rightarrow \begin{bmatrix} 1 & 0 & 0 & \vdots & -1 \\ 0 & 1 & 0 & \vdots & 3 \\ 0 & 0 & 1 & \vdots & -2 \end{bmatrix}$

$X = \begin{bmatrix} -1 \\ 3 \\ -2 \end{bmatrix}$

73. $1.10\begin{bmatrix} 100 & 90 & 70 & 30 \\ 40 & 20 & 60 & 60 \end{bmatrix} = \begin{bmatrix} 110 & 99 & 77 & 33 \\ 44 & 22 & 66 & 66 \end{bmatrix}$

75. $BA = \begin{bmatrix} 3.50 & 6.00 \end{bmatrix}\begin{bmatrix} 125 & 100 & 75 \\ 100 & 175 & 125 \end{bmatrix} = \begin{bmatrix} \$1037.50 & \$1400 & \$1012.50 \end{bmatrix}$

The entries represent the profits from both crops at each of the three outlets.

77. $ST = \begin{bmatrix} 1.0 & 0.5 & 0.2 \\ 1.6 & 1.0 & 0.2 \\ 2.5 & 2.0 & 1.4 \end{bmatrix}\begin{bmatrix} 15 & 13 \\ 12 & 11 \\ 11 & 10 \end{bmatrix} = \begin{bmatrix} \$23.20 & \$20.50 \\ \$38.20 & \$33.80 \\ \$76.90 & \$68.50 \end{bmatrix}$ The entries represent the labor costs at each plant for each size of boat.

79. $P^2 = \begin{bmatrix} 0.6 & 0.1 & 0.1 \\ 0.2 & 0.7 & 0.1 \\ 0.2 & 0.2 & 0.8 \end{bmatrix}\begin{bmatrix} 0.6 & 0.1 & 0.1 \\ 0.2 & 0.7 & 0.1 \\ 0.2 & 0.2 & 0.8 \end{bmatrix} = \begin{bmatrix} 0.40 & 0.15 & 0.15 \\ 0.28 & 0.53 & 0.17 \\ 0.32 & 0.32 & 0.68 \end{bmatrix}$

The P^2 matrix gives the proportion of the voting population that changed parties or remained loyal to their parties from the first election to the third.

81. True.

The sum of two matrices of different orders is undefined.

83. Answers will vary. *Sample answer:*

$$(A + B)^2 = \left(\begin{bmatrix} 2 & -1 \\ 1 & 3 \end{bmatrix} + \begin{bmatrix} -1 & 1 \\ 0 & -2 \end{bmatrix}\right)^2 = \begin{bmatrix} 1 & 0 \\ 2 & 1 \end{bmatrix} \neq$$

$$A^2 + 2AB + B^2 = \left(\begin{bmatrix} 2 & -1 \\ 1 & 3 \end{bmatrix}\right)^2 + 2\begin{bmatrix} 2 & -1 \\ 1 & 3 \end{bmatrix}\begin{bmatrix} -1 & 1 \\ 0 & -2 \end{bmatrix} + \left(\begin{bmatrix} -1 & 1 \\ 0 & -2 \end{bmatrix}\right)^2 = \begin{bmatrix} 0 & 0 \\ 3 & 2 \end{bmatrix}$$

85. Answers will vary. *Sample answer:*

$$(A + B)(A - B) = \left(\begin{bmatrix} 2 & -1 \\ 1 & 3 \end{bmatrix} + \begin{bmatrix} -1 & 1 \\ 0 & -2 \end{bmatrix}\right)\left(\begin{bmatrix} 2 & -1 \\ 1 & 3 \end{bmatrix} - \begin{bmatrix} -1 & 1 \\ 0 & -2 \end{bmatrix}\right) = \begin{bmatrix} 3 & -2 \\ 4 & 3 \end{bmatrix} \neq$$

$$A^2 - B^2 = \left(\begin{bmatrix} 2 & -1 \\ 1 & 3 \end{bmatrix}\right)^2 - \left(\begin{bmatrix} -1 & 1 \\ 0 & -2 \end{bmatrix}\right)^2 = \begin{bmatrix} 2 & -2 \\ 5 & 4 \end{bmatrix}$$

87. $AC = \begin{bmatrix} 0 & 1 \\ 0 & 1 \end{bmatrix}\begin{bmatrix} 2 & 3 \\ 2 & 3 \end{bmatrix} = \begin{bmatrix} 2 & 3 \\ 2 & 3 \end{bmatrix}$

$BC = \begin{bmatrix} 1 & 0 \\ 1 & 0 \end{bmatrix}\begin{bmatrix} 2 & 3 \\ 2 & 3 \end{bmatrix} = \begin{bmatrix} 2 & 3 \\ 2 & 3 \end{bmatrix}$

So, $AC = BC$ even though $A \neq B$.

89. Answer will vary. *Sample answer*:

$A = \begin{bmatrix} 1 & 0 \\ 0 & 1 \end{bmatrix}, B = \begin{bmatrix} 0 & 1 \\ 1 & 0 \end{bmatrix}$

$$AB = \begin{bmatrix} 1 & 0 \\ 0 & 1 \end{bmatrix}\begin{bmatrix} 0 & 1 \\ 1 & 0 \end{bmatrix} = \begin{bmatrix} (1)(0) + (0)(1) & (1)(1) + (0)(0) \\ (0)(0) + (1)(1) & (0)(1) + (1)(0) \end{bmatrix} = \begin{bmatrix} 0 & 1 \\ 1 & 0 \end{bmatrix}$$

$$BA = \begin{bmatrix} 0 & 1 \\ 1 & 0 \end{bmatrix}\begin{bmatrix} 1 & 0 \\ 0 & 1 \end{bmatrix} = \begin{bmatrix} (0)(1) + (1)(0) & (0)(0) + (1)(1) \\ (1)(1) + (0)(0) & (1)(0) + (0)(1) \end{bmatrix} = \begin{bmatrix} 0 & 1 \\ 1 & 0 \end{bmatrix}$$

So, $AB = BA$.

91. The product of two diagonal matrices of the same order is a diagonal matrix whose entries are the products of the corresponding diagonal entries of A and B.

Section 8.3 The Inverse of a Square Matrix

1. inverse

3. determinant

5. $AB = \begin{bmatrix} 2 & 1 \\ 5 & 3 \end{bmatrix}\begin{bmatrix} 3 & -1 \\ -5 & 2 \end{bmatrix} = \begin{bmatrix} 6 - 5 & -2 + 2 \\ 15 - 15 & -5 + 6 \end{bmatrix} = \begin{bmatrix} 1 & 0 \\ 0 & 1 \end{bmatrix}$

$BA = \begin{bmatrix} 3 & -1 \\ -5 & 2 \end{bmatrix}\begin{bmatrix} 2 & 1 \\ 5 & 3 \end{bmatrix} = \begin{bmatrix} 6 - 5 & 3 - 3 \\ -10 + 10 & -5 + 6 \end{bmatrix} = \begin{bmatrix} 1 & 0 \\ 0 & 1 \end{bmatrix}$

7. $AB = \frac{1}{10}\begin{bmatrix} 3 & 2 \\ 1 & 4 \end{bmatrix}\begin{bmatrix} 4 & -2 \\ -1 & 3 \end{bmatrix} = \frac{1}{10}\begin{bmatrix} 12 - 2 & -6 + 6 \\ 4 - 4 & -2 + 12 \end{bmatrix} = \frac{1}{10}\begin{bmatrix} 10 & 0 \\ 0 & 10 \end{bmatrix} = \begin{bmatrix} 1 & 0 \\ 0 & 1 \end{bmatrix}$

$BA = \frac{1}{10}\begin{bmatrix} 4 & -2 \\ -1 & 3 \end{bmatrix}\begin{bmatrix} 3 & 2 \\ 1 & 4 \end{bmatrix} = \frac{1}{10}\begin{bmatrix} 12 - 2 & 8 - 8 \\ -3 + 3 & -2 + 12 \end{bmatrix} = \frac{1}{10}\begin{bmatrix} 10 & 0 \\ 0 & 10 \end{bmatrix} = \begin{bmatrix} 1 & 0 \\ 0 & 1 \end{bmatrix}$

9. $$AB = \begin{bmatrix} 2 & -17 & 11 \\ -1 & 11 & -7 \\ 0 & 3 & -2 \end{bmatrix}\begin{bmatrix} 1 & 1 & 2 \\ 2 & 4 & -3 \\ 3 & 6 & -5 \end{bmatrix} = \begin{bmatrix} 2 - 34 + 33 & 2 - 68 + 66 & 4 + 51 - 55 \\ -1 + 22 - 21 & -1 + 44 - 42 & -2 - 33 + 35 \\ 6 - 6 & 12 - 12 & -9 + 10 \end{bmatrix} = \begin{bmatrix} 1 & 0 & 0 \\ 0 & 1 & 0 \\ 0 & 0 & 1 \end{bmatrix}$$

$$BA = \begin{bmatrix} 1 & 1 & 2 \\ 2 & 4 & -3 \\ 3 & 6 & -5 \end{bmatrix}\begin{bmatrix} 2 & -17 & 11 \\ -1 & 11 & -7 \\ 0 & 3 & -2 \end{bmatrix} = \begin{bmatrix} 2 - 1 & -17 + 11 + 6 & 11 - 7 - 4 \\ 4 - 4 & -34 + 44 - 9 & 22 - 28 + 6 \\ 6 - 6 & -51 + 66 - 15 & 33 - 42 + 10 \end{bmatrix} = \begin{bmatrix} 1 & 0 & 0 \\ 0 & 1 & 0 \\ 0 & 0 & 1 \end{bmatrix}$$

11. $$AB = \frac{1}{3}\begin{bmatrix} 2 & 0 & 2 & 1 \\ 3 & 0 & 0 & 1 \\ -1 & 1 & -2 & 1 \\ 3 & -1 & 1 & 0 \end{bmatrix}\begin{bmatrix} -1 & 3 & -2 & -2 \\ -2 & 9 & -7 & -10 \\ 1 & 0 & -1 & -1 \\ 3 & -6 & 6 & 6 \end{bmatrix}$$

$$= \frac{1}{3}\begin{bmatrix} -2 + 0 + 2 + 3 & 6 + 0 + 0 - 6 & -4 + 0 - 2 + 6 & -4 + 0 - 2 + 6 \\ -3 + 0 + 0 + 3 & 9 + 0 + 0 - 6 & -6 + 0 + 0 + 6 & -6 + 0 + 0 + 6 \\ 1 - 2 - 2 + 3 & -3 + 9 + 0 - 6 & 2 - 7 + 2 + 6 & 2 - 10 + 2 + 6 \\ -3 + 2 + 1 + 0 & 9 - 9 + 0 + 0 & -6 + 7 - 1 + 0 & -6 + 10 - 1 + 0 \end{bmatrix} = \begin{bmatrix} 1 & 0 & 0 & 0 \\ 0 & 1 & 0 & 0 \\ 0 & 0 & 1 & 0 \\ 0 & 0 & 0 & 1 \end{bmatrix}$$

$$BA = \frac{1}{3}\begin{bmatrix} -1 & 3 & -2 & -2 \\ -2 & 9 & -7 & -10 \\ 1 & 0 & -1 & -1 \\ 3 & -6 & 6 & 6 \end{bmatrix}\begin{bmatrix} 2 & 0 & 2 & 1 \\ 3 & 0 & 0 & 1 \\ -1 & 1 & -2 & 1 \\ 3 & -1 & 1 & 0 \end{bmatrix}$$

$$= \frac{1}{3}\begin{bmatrix} -2 + 9 + 2 - 6 & 0 + 0 - 2 + 2 & -2 + 0 + 4 - 2 & -1 + 3 - 2 + 0 \\ -4 + 27 + 7 - 30 & 0 + 0 - 7 + 10 & -4 + 0 + 14 - 10 & -2 + 9 - 7 + 0 \\ 2 + 0 + 1 - 3 & 0 + 0 - 1 + 1 & 2 + 0 + 2 - 1 & 1 + 0 - 1 + 0 \\ 6 - 18 - 6 + 18 & 0 + 0 + 6 - 6 & 6 + 0 - 12 + 6 & 3 - 6 + 6 + 0 \end{bmatrix} = \begin{bmatrix} 1 & 0 & 0 & 0 \\ 0 & 1 & 0 & 0 \\ 0 & 0 & 1 & 0 \\ 0 & 0 & 0 & 1 \end{bmatrix}$$

13. $$[A \;\vdots\; I] = \left[\begin{array}{cc|cc} 2 & 1 & 1 & 0 \\ 5 & 3 & 0 & 1 \end{array}\right]$$

$$\begin{array}{r} \\ -5R_1 + 2R_2 \rightarrow \end{array} \left[\begin{array}{cc|cc} 2 & 1 & 1 & 0 \\ 0 & 1 & -5 & 2 \end{array}\right]$$

$$\begin{array}{r} -R_2 + R_1 \rightarrow \\ \\ \end{array} \left[\begin{array}{cc|cc} 2 & 0 & 6 & -2 \\ 0 & 1 & -5 & 2 \end{array}\right]$$

$$\begin{array}{r} \frac{1}{2}R_1 \rightarrow \\ \\ \end{array} \left[\begin{array}{cc|cc} 1 & 0 & 3 & -1 \\ 0 & 1 & -5 & 2 \end{array}\right] = [I \;\vdots\; A^{-1}]$$

$$A^{-1} = \begin{bmatrix} 3 & -1 \\ -5 & 2 \end{bmatrix}$$

15. $$[A \;\vdots\; I] = \left[\begin{array}{cc|cc} 1 & -2 & 1 & 0 \\ 2 & -3 & 0 & 1 \end{array}\right]$$

$$\begin{array}{r} \\ -2R_1 + R_2 \rightarrow \end{array} \left[\begin{array}{cc|cc} 1 & -2 & 1 & 0 \\ 0 & 1 & -2 & 1 \end{array}\right]$$

$$\begin{array}{r} 2R_2 + R_1 \rightarrow \\ \\ \end{array} \left[\begin{array}{cc|cc} 1 & 0 & -3 & 2 \\ 0 & 1 & -2 & 1 \end{array}\right] = [I \;\vdots\; A^{-1}]$$

$$A^{-1} = \begin{bmatrix} -3 & 2 \\ -2 & 1 \end{bmatrix}$$

17. $$[A \;\vdots\; I] = \left[\begin{array}{cc|cc} 3 & 1 & 1 & 0 \\ 4 & 2 & 0 & 1 \end{array}\right]$$

$$\begin{array}{r} \\ \frac{1}{2}R_2 \rightarrow \end{array} \left[\begin{array}{cc|cc} 3 & 1 & 1 & 0 \\ 2 & 1 & 0 & \frac{1}{2} \end{array}\right]$$

$$\begin{array}{r} -R_2 + R_1 \rightarrow \\ \\ \end{array} \left[\begin{array}{cc|cc} 1 & 0 & 1 & -\frac{1}{2} \\ 2 & 1 & 0 & \frac{1}{2} \end{array}\right]$$

$$\begin{array}{r} \\ -2R_1 + R_2 \rightarrow \end{array} \left[\begin{array}{cc|cc} 1 & 0 & 1 & -\frac{1}{2} \\ 0 & 1 & -2 & \frac{3}{2} \end{array}\right] = [I \;\vdots\; A^{-1}]$$

$$A^{-1} = \begin{bmatrix} 1 & -\frac{1}{2} \\ -2 & \frac{3}{2} \end{bmatrix}$$

19. $[A \;\vdots\; I] = \left[\begin{array}{ccc|ccc} 1 & 1 & 1 & 1 & 0 & 0 \\ 3 & 5 & 4 & 0 & 1 & 0 \\ 3 & 6 & 5 & 0 & 0 & 1 \end{array}\right]$

$$\begin{array}{r} \\ -3R_1 + R_2 \to \\ -3R_1 + R_3 \to \end{array} \left[\begin{array}{ccc|ccc} 1 & 1 & 1 & 1 & 0 & 0 \\ 0 & 2 & 1 & -3 & 1 & 0 \\ 0 & 3 & 2 & -3 & 0 & 1 \end{array}\right]$$

$$\begin{array}{r} \\ \frac{1}{2}R_2 \to \\ \\ \end{array} \left[\begin{array}{ccc|ccc} 1 & 1 & 1 & 1 & 0 & 0 \\ 0 & 1 & \frac{1}{2} & -\frac{3}{2} & \frac{1}{2} & 0 \\ 0 & 3 & 2 & -3 & 0 & 1 \end{array}\right]$$

$$\begin{array}{r} -R_2 + R_1 \to \\ \\ -3R_2 + R_3 \to \end{array} \left[\begin{array}{ccc|ccc} 1 & 0 & \frac{1}{2} & \frac{5}{2} & -\frac{1}{2} & 0 \\ 0 & 1 & \frac{1}{2} & -\frac{3}{2} & \frac{1}{2} & 0 \\ 0 & 0 & \frac{1}{2} & \frac{3}{2} & -\frac{3}{2} & 1 \end{array}\right]$$

$$\begin{array}{r} -R_3 + R_1 \to \\ -R_3 + R_2 \to \\ \\ \end{array} \left[\begin{array}{ccc|ccc} 1 & 0 & 0 & 1 & 1 & -1 \\ 0 & 1 & 0 & -3 & 2 & -1 \\ 0 & 0 & \frac{1}{2} & \frac{3}{2} & -\frac{3}{2} & 1 \end{array}\right]$$

$$\begin{array}{r} \\ \\ 2R_3 \to \end{array} \left[\begin{array}{ccc|ccc} 1 & 0 & 0 & 1 & 1 & -1 \\ 0 & 1 & 0 & -3 & 2 & -1 \\ 0 & 0 & 1 & 3 & -3 & 2 \end{array}\right] = [I \;\vdots\; A^{-1}]$$

$$A^{-1} = \begin{bmatrix} 1 & 1 & -1 \\ -3 & 2 & -1 \\ 3 & -3 & 2 \end{bmatrix}$$

21. $[A \;\vdots\; I] = \left[\begin{array}{ccc|ccc} -5 & 0 & 0 & 1 & 0 & 0 \\ 2 & 0 & 0 & 0 & 1 & 0 \\ -1 & 5 & 7 & 0 & 0 & 1 \end{array}\right] \begin{array}{r} \\ \\ R_2 + 2R_3 \to \end{array} \left[\begin{array}{ccc|ccc} -5 & 0 & 0 & 1 & 0 & 0 \\ 2 & 0 & 0 & 0 & 1 & 0 \\ 0 & 10 & 14 & 0 & 1 & 2 \end{array}\right] \begin{array}{r} \\ 2R_1 + 5R_2 \to \\ \\ \end{array} \left[\begin{array}{ccc|ccc} -5 & 0 & 0 & 1 & 0 & 0 \\ 0 & 0 & 0 & 2 & 5 & 0 \\ 0 & 10 & 14 & 0 & 1 & 2 \end{array}\right]$

Because the first three entries of row 2 are all zeros, the inverse of A does not exist.

23. $[A \;\vdots\; I] = \left[\begin{array}{cccc|cccc} -8 & 0 & 0 & 0 & 1 & 0 & 0 & 0 \\ 0 & 1 & 0 & 0 & 0 & 1 & 0 & 0 \\ 0 & 0 & 4 & 0 & 0 & 0 & 1 & 0 \\ 0 & 0 & 0 & -5 & 0 & 0 & 0 & 1 \end{array}\right] \begin{array}{r} -\frac{1}{8}R_1 \to \\ \\ \frac{1}{4}R_3 \to \\ -\frac{1}{5}R_4 \to \end{array} \left[\begin{array}{cccc|cccc} 1 & 0 & 0 & 0 & -\frac{1}{8} & 0 & 0 & 0 \\ 0 & 1 & 0 & 0 & 0 & 1 & 0 & 0 \\ 0 & 0 & 1 & 0 & 0 & 0 & \frac{1}{4} & 0 \\ 0 & 0 & 0 & 1 & 0 & 0 & 0 & -\frac{1}{5} \end{array}\right] = [I \;\vdots\; A^{-1}]$

$$A^{-1} = \begin{bmatrix} -\frac{1}{8} & 0 & 0 & 0 \\ 0 & 1 & 0 & 0 \\ 0 & 0 & \frac{1}{4} & 0 \\ 0 & 0 & 0 & -\frac{1}{5} \end{bmatrix}$$

25. $A = \begin{bmatrix} 1 & 2 & -1 \\ 3 & 7 & -10 \\ -5 & -7 & -15 \end{bmatrix}$

$A^{-1} = \begin{bmatrix} -175 & 37 & -13 \\ 95 & -20 & 7 \\ 14 & -3 & 1 \end{bmatrix}$

27. $A = \begin{bmatrix} -\frac{1}{2} & \frac{3}{4} & \frac{1}{4} \\ 1 & 0 & -\frac{3}{2} \\ 0 & -1 & \frac{1}{2} \end{bmatrix}$

$A^{-1} = \begin{bmatrix} -12 & -5 & -9 \\ -4 & -2 & -4 \\ -8 & -4 & -6 \end{bmatrix}$

29. $A = \begin{bmatrix} 0.1 & 0.2 & 0.3 \\ -0.3 & 0.2 & 0.2 \\ 0.5 & 0.4 & 0.4 \end{bmatrix}$

$A^{-1} = \begin{bmatrix} 0 & -1.\overline{81} & 0.\overline{90} \\ -10 & 5 & 5 \\ 10 & -2.\overline{72} & -3.\overline{63} \end{bmatrix}$

31. $A = \begin{bmatrix} -1 & 0 & 1 & 0 \\ 0 & 2 & 0 & -1 \\ 2 & 0 & -1 & 0 \\ 0 & -1 & 0 & 1 \end{bmatrix}$

$A^{-1} = \begin{bmatrix} 1 & 0 & 1 & 0 \\ 0 & 1 & 0 & 1 \\ 2 & 0 & 1 & 0 \\ 0 & 1 & 0 & 2 \end{bmatrix}$

33. $A = \begin{bmatrix} a & b \\ c & d \end{bmatrix}, A^{-1} = \dfrac{1}{ad - bc}\begin{bmatrix} d & -b \\ -c & a \end{bmatrix}$

$A = \begin{bmatrix} 2 & 3 \\ -1 & 5 \end{bmatrix}$

$ad - bc = (2)(5) - (3)(-1) = 13$

$A^{-1} = \frac{1}{13}\begin{bmatrix} 5 & -3 \\ 1 & 2 \end{bmatrix} = \begin{bmatrix} \frac{5}{13} & -\frac{3}{13} \\ \frac{1}{13} & \frac{2}{13} \end{bmatrix}$

35. $A = \begin{bmatrix} -4 & -6 \\ 2 & 3 \end{bmatrix}$

$ad - bc = (-4)(3) - (-2)(-6) = 0$

Because $ad - bc = 0$, A^{-1} does not exist.

37. $A = \begin{bmatrix} 0.5 & 0.3 \\ 1.5 & 0.6 \end{bmatrix}$

$ad - bc = (0.5)(0.6) - (0.3)(1.5) = 0.3 - 0.45 = -0.15$

$A^{-1} = -\frac{1}{0.15}\begin{bmatrix} 0.6 & -0.3 \\ -1.5 & 0.5 \end{bmatrix} = \begin{bmatrix} -4 & 2 \\ 10 & -\frac{10}{3} \end{bmatrix}$

39. $\begin{bmatrix} x \\ y \end{bmatrix} = \begin{bmatrix} -3 & 2 \\ -2 & 1 \end{bmatrix}\begin{bmatrix} 5 \\ 10 \end{bmatrix} = \begin{bmatrix} 5 \\ 0 \end{bmatrix}$

Solution: $(5, 0)$

41. $\begin{bmatrix} x \\ y \end{bmatrix} = \begin{bmatrix} -3 & 2 \\ -2 & 1 \end{bmatrix}\begin{bmatrix} 4 \\ 2 \end{bmatrix} = \begin{bmatrix} -8 \\ -6 \end{bmatrix}$

Solution: $(-8, -6)$

43. $\begin{bmatrix} x \\ y \\ z \end{bmatrix} = \begin{bmatrix} 1 & 1 & -1 \\ -3 & 2 & -1 \\ 3 & -3 & 2 \end{bmatrix}\begin{bmatrix} 0 \\ 5 \\ 2 \end{bmatrix} = \begin{bmatrix} 3 \\ 8 \\ -11 \end{bmatrix}$

Solution: $(3, 8, -11)$

45. $\begin{bmatrix} x_1 \\ x_2 \\ x_3 \\ x_4 \end{bmatrix} = \begin{bmatrix} -24 & 7 & 1 & -2 \\ -10 & 3 & 0 & -1 \\ -29 & 7 & 3 & -2 \\ 12 & -3 & -1 & 1 \end{bmatrix}\begin{bmatrix} 0 \\ 1 \\ -1 \\ 2 \end{bmatrix} = \begin{bmatrix} 2 \\ 1 \\ 0 \\ 0 \end{bmatrix}$

Solution: $(2, 1, 0, 0)$

47. $A = \begin{bmatrix} 5 & 4 \\ 2 & 5 \end{bmatrix}$

$A^{-1} = \dfrac{1}{25 - 8}\begin{bmatrix} 5 & -4 \\ -2 & 5 \end{bmatrix}$

$\begin{bmatrix} x \\ y \end{bmatrix} = -\dfrac{1}{17}\begin{bmatrix} 5 & -4 \\ -2 & 5 \end{bmatrix}\begin{bmatrix} -1 \\ 3 \end{bmatrix} = -\dfrac{1}{17}\begin{bmatrix} -17 \\ 17 \end{bmatrix} = \begin{bmatrix} -1 \\ 1 \end{bmatrix}$

Solution: $(-1, 1)$

49. $A = \begin{bmatrix} -0.4 & 0.8 \\ 2 & -4 \end{bmatrix}$

$A^{-1} = \dfrac{1}{1.6 - 1.6}\begin{bmatrix} -4 & -0.8 \\ -2 & -0.4 \end{bmatrix}$

A^{-1} does not exist.

This implies that there is no unique solution; that is, either the system is inconsistent *or* there are infinitely many solutions.

Find the reduced row-echelon form of the matrix corresponding to the system.

$$\begin{bmatrix} -0.4 & 0.8 & \vdots & 1.6 \\ 2 & -4 & \vdots & 5 \end{bmatrix}$$

$$-2.5R_1 \rightarrow \begin{bmatrix} 1 & -2 & \vdots & -4 \\ 2 & -4 & \vdots & 5 \end{bmatrix}$$

$$-2R_1 + R_2 \rightarrow \begin{bmatrix} 1 & -2 & \vdots & -4 \\ 0 & 0 & \vdots & 13 \end{bmatrix}$$

The given system is inconsistent and there is no solution.

51. $A = \begin{bmatrix} -\frac{1}{4} & \frac{3}{8} \\ \frac{3}{2} & \frac{3}{4} \end{bmatrix}$

$A^{-1} = \dfrac{1}{-\frac{3}{16} - \frac{9}{16}}\begin{bmatrix} \frac{3}{4} & -\frac{3}{8} \\ -\frac{3}{2} & -\frac{1}{4} \end{bmatrix} = -\dfrac{4}{3}\begin{bmatrix} \frac{3}{4} & -\frac{3}{8} \\ -\frac{3}{2} & -\frac{1}{4} \end{bmatrix} = \begin{bmatrix} -1 & \frac{1}{2} \\ 2 & \frac{1}{3} \end{bmatrix}$

$\begin{bmatrix} x \\ y \end{bmatrix} = \begin{bmatrix} -1 & \frac{1}{2} \\ 2 & \frac{1}{3} \end{bmatrix}\begin{bmatrix} -2 \\ -12 \end{bmatrix} = \begin{bmatrix} -4 \\ -8 \end{bmatrix}$

Solution: $(-4, -8)$

53. $A = \begin{bmatrix} 4 & -1 & 1 \\ 2 & 2 & 3 \\ 5 & -2 & 6 \end{bmatrix}$

Find A^{-1}.

$$[A \ \vdots \ I] = \left[\begin{array}{ccc:ccc} 4 & -1 & 1 & 1 & 0 & 0 \\ 2 & 2 & 3 & 0 & 1 & 0 \\ 5 & -2 & 6 & 0 & 0 & 1 \end{array}\right]$$

$$\begin{matrix} R_1 \\ \\ R_3 \end{matrix} \left[\begin{array}{ccc:ccc} 5 & -2 & 6 & 0 & 0 & 1 \\ 2 & 2 & 3 & 0 & 1 & 0 \\ 4 & -1 & 1 & 1 & 0 & 0 \end{array}\right]$$

$$\begin{matrix} -R_3 + R_1 \rightarrow \\ \\ \\ \end{matrix} \left[\begin{array}{ccc:ccc} 1 & -1 & 5 & -1 & 0 & 1 \\ 2 & 2 & 3 & 0 & 1 & 0 \\ 4 & -1 & 1 & 1 & 0 & 0 \end{array}\right]$$

$$\begin{matrix} \\ -2R_1 + R_2 \rightarrow \\ -4R_1 + R_3 \rightarrow \end{matrix} \left[\begin{array}{ccc:ccc} 1 & -1 & 5 & -1 & 0 & 1 \\ 0 & 4 & -7 & 2 & 1 & -2 \\ 0 & 3 & -19 & 5 & 0 & -4 \end{array}\right]$$

$$\begin{matrix} \\ -R_3 + R_2 \rightarrow \\ \\ \end{matrix} \left[\begin{array}{ccc:ccc} 1 & -1 & 5 & -1 & 0 & 1 \\ 0 & 1 & 12 & -3 & 1 & 2 \\ 0 & 3 & -19 & 5 & 0 & -4 \end{array}\right]$$

$$\begin{matrix} R_2 + R_1 \rightarrow \\ \\ -3R_2 + R_3 \rightarrow \end{matrix} \left[\begin{array}{ccc:ccc} 1 & 0 & 17 & -4 & 1 & 3 \\ 0 & 1 & 12 & -3 & 1 & 2 \\ 0 & 0 & -55 & 14 & -3 & -10 \end{array}\right]$$

$$\begin{matrix} \\ \\ -\frac{1}{55}R_3 \rightarrow \end{matrix} \left[\begin{array}{ccc:ccc} 1 & 0 & 17 & -4 & 1 & 3 \\ 0 & 1 & 12 & -3 & 1 & 2 \\ 0 & 0 & 1 & -\frac{14}{55} & \frac{3}{55} & \frac{2}{11} \end{array}\right]$$

$$\begin{matrix} -17R_3 + R_1 \rightarrow \\ -12R_3 + R_2 \rightarrow \\ \\ \end{matrix} \left[\begin{array}{ccc:ccc} 1 & 0 & 0 & \frac{18}{55} & \frac{4}{55} & -\frac{1}{11} \\ 0 & 1 & 0 & \frac{3}{55} & \frac{19}{55} & -\frac{2}{11} \\ 0 & 0 & 1 & -\frac{14}{55} & \frac{3}{55} & \frac{2}{11} \end{array}\right] = [I \ \vdots \ A^{-1}]$$

$$A^{-1} = \frac{1}{55}\begin{bmatrix} 18 & 4 & -5 \\ 3 & 19 & -10 \\ -14 & 3 & 10 \end{bmatrix}$$

$$\begin{bmatrix} x \\ y \\ z \end{bmatrix} = \frac{1}{55}\begin{bmatrix} 18 & 4 & -5 \\ 3 & 19 & -10 \\ -14 & 3 & 10 \end{bmatrix}\begin{bmatrix} -5 \\ 10 \\ 1 \end{bmatrix} = \frac{1}{55}\begin{bmatrix} -55 \\ 165 \\ 110 \end{bmatrix} = \begin{bmatrix} -1 \\ 3 \\ 2 \end{bmatrix}$$

Solution: $(-1, 3, 2)$

55. $\begin{cases} 5x - 3y + 2z = 2 \\ 2x + 2y - 3z = 3 \\ x - 7y + 7z = -4 \end{cases}$

Using a graphing utility $(0.8125, 0.6875, 0) = \left(\frac{13}{16}, \frac{11}{16}, 0\right)$

57. $A = \begin{bmatrix} 1 & 1 & 1 \\ 0.045 & 0.05 & 0.09 \\ 0 & 2 & -1 \end{bmatrix}$

$$[A \;\vdots\; I] = \left[\begin{array}{ccc|ccc} 1 & 1 & 1 & 1 & 0 & 0 \\ 0.045 & 0.05 & 0.09 & 0 & 1 & 0 \\ 0 & 2 & -1 & 0 & 0 & 1 \end{array}\right]$$

$$200R_2 \rightarrow \left[\begin{array}{ccc|ccc} 1 & 1 & 1 & 1 & 0 & 0 \\ 9 & 10 & 18 & 0 & 200 & 0 \\ 0 & 2 & -1 & 0 & 0 & 1 \end{array}\right]$$

$$-9R_1 + R_2 \rightarrow \left[\begin{array}{ccc|ccc} 1 & 1 & 1 & 1 & 0 & 0 \\ 0 & 1 & 9 & -9 & 200 & 0 \\ 0 & 2 & -1 & 0 & 0 & 1 \end{array}\right]$$

$$\begin{array}{r} -R_2 + R_1 \rightarrow \\ \\ -2R_2 + R_3 \rightarrow \end{array} \left[\begin{array}{ccc|ccc} 1 & 0 & -8 & 10 & -200 & 0 \\ 0 & 1 & 9 & -9 & 200 & 0 \\ 0 & 0 & -19 & 18 & -400 & 1 \end{array}\right]$$

$$-\tfrac{1}{19}R_3 \rightarrow \left[\begin{array}{ccc|ccc} 1 & 0 & -8 & 10 & -200 & 0 \\ 0 & 1 & 9 & -9 & 200 & 0 \\ 0 & 0 & 1 & -\frac{18}{19} & \frac{400}{19} & -\frac{1}{19} \end{array}\right]$$

$$\begin{array}{r} 8R_3 + R_1 \rightarrow \\ -9R_3 + R_2 \rightarrow \\ \\ \end{array} \left[\begin{array}{ccc|ccc} 1 & 0 & 0 & \frac{46}{19} & -\frac{600}{19} & -\frac{8}{19} \\ 0 & 1 & 0 & -\frac{9}{19} & \frac{200}{19} & \frac{9}{19} \\ 0 & 0 & 1 & -\frac{18}{19} & \frac{400}{19} & -\frac{1}{19} \end{array}\right] = [I \;\vdots\; A^{-1}]$$

$$X = A^{-1}B = \frac{1}{19}\begin{bmatrix} 46 & -600 & -8 \\ -9 & 200 & 9 \\ -18 & 400 & -1 \end{bmatrix}\begin{bmatrix} 10{,}000 \\ 650 \\ 0 \end{bmatrix} \approx \begin{bmatrix} 3684.21 \\ 2105.26 \\ 4210.53 \end{bmatrix}$$

Solution: \$3684.21 in AAA-rated bonds, \$2105.26 in A-rated bonds, \$4210.53 in B-rated bonds

59. $\begin{cases} 2I_1 + 4I_3 = 15 \\ I_2 + 4I_3 = 17 \\ I_1 + I_2 - I_3 = 0 \end{cases}$

$$\underset{A}{\begin{bmatrix} 2 & 0 & 4 \\ 0 & 1 & 4 \\ 1 & 1 & -1 \end{bmatrix}} \underset{X}{\begin{bmatrix} I_1 \\ I_2 \\ I_3 \end{bmatrix}} = \underset{B}{\begin{bmatrix} 15 \\ 17 \\ 0 \end{bmatrix}}$$

$$X = A^{-1}B = \begin{bmatrix} \frac{5}{14} & -\frac{2}{7} & \frac{2}{7} \\ -\frac{2}{7} & \frac{3}{7} & \frac{4}{7} \\ \frac{1}{14} & \frac{1}{7} & -\frac{1}{7} \end{bmatrix}\begin{bmatrix} 15 \\ 17 \\ 0 \end{bmatrix} = \begin{bmatrix} \frac{1}{2} \\ 3 \\ \frac{7}{2} \end{bmatrix}$$

So, $I_1 = 0.5$ ampere, $I_2 = 3.0$ ampere, and $I_3 = 3.5$ ampere.

61. $\begin{cases} 2I_1 + 4I_3 = 28 \\ I_2 + 4I_3 = 21 \\ I_1 + I_2 - I_3 = 0 \end{cases}$

$$\underset{A}{\begin{bmatrix} 2 & 0 & 4 \\ 0 & 1 & 4 \\ 1 & 1 & -1 \end{bmatrix}} \underset{X}{\begin{bmatrix} I_1 \\ I_2 \\ I_3 \end{bmatrix}} = \underset{B}{\begin{bmatrix} 28 \\ 21 \\ 0 \end{bmatrix}}$$

$$X = A^{-1}B = \begin{bmatrix} \frac{5}{14} & -\frac{2}{7} & \frac{2}{7} \\ -\frac{2}{7} & \frac{3}{7} & \frac{4}{7} \\ \frac{1}{14} & \frac{1}{7} & -\frac{1}{7} \end{bmatrix}\begin{bmatrix} 28 \\ 21 \\ 0 \end{bmatrix} = \begin{bmatrix} 4 \\ 1 \\ 5 \end{bmatrix}$$

So, $I_1 = 4$ ampere, $I_2 = 1$ ampere, and $I_3 = 5$ ampere.

In Exercise 63, use the following:

Let x = bags of potting soil for seedlings,

y = bags of potting soil for general potting, and

z = bags of potting soil for hardwood plants.

$$AX = B = \begin{bmatrix} 2 & 1 & 2 \\ 1 & 2 & 1 \\ 1 & 1 & 2 \end{bmatrix}\begin{bmatrix} x \\ y \\ z \end{bmatrix} = \begin{bmatrix} \text{Sand} \\ \text{Loam} \\ \text{Peat Moss} \end{bmatrix}$$

$$A^{-1} = \begin{bmatrix} 1 & 0 & -1 \\ 0 & 1 & -1 \\ -\frac{1}{2} & -\frac{1}{2} & \frac{3}{2} \end{bmatrix}$$

63. $A^{-1}\begin{bmatrix} 500 \\ 500 \\ 400 \end{bmatrix} = \begin{bmatrix} 100 \\ 100 \\ 100 \end{bmatrix}$

Solution:

x = 100 bags of potting soil for seedlings,

y = 100 bags of potting soil for general potting,

z = 100 bags of potting soil for hardwood plants.

65. Let r = number of roses, l = number of lilies, and i = number of irises.

(a) $\begin{cases} r + l + i = 120 \\ 2.5r + 4l + 2i = 300 \\ -r + 2l + 2i = 0 \end{cases}$

$$\begin{bmatrix} 1 & 1 & 1 \\ 2.5 & 4 & 2 \\ -1 & 2 & 2 \end{bmatrix}\begin{bmatrix} r \\ l \\ i \end{bmatrix} = \begin{bmatrix} 120 \\ 300 \\ 0 \end{bmatrix}$$

$$A \quad X \quad = \quad B$$

(b) $A^{-1} = \begin{bmatrix} \frac{2}{3} & 0 & -\frac{1}{3} \\ -\frac{7}{6} & \frac{1}{2} & \frac{1}{12} \\ \frac{3}{2} & -\frac{1}{2} & -\frac{1}{4} \end{bmatrix}$

$$X = A^{-1}B = \begin{bmatrix} \frac{2}{3} & 0 & -\frac{1}{3} \\ -\frac{7}{6} & \frac{1}{2} & \frac{1}{12} \\ \frac{3}{2} & -\frac{1}{2} & \frac{1}{4} \end{bmatrix}\begin{bmatrix} 120 \\ 300 \\ 0 \end{bmatrix} = \begin{bmatrix} 80 \\ 10 \\ 30 \end{bmatrix}$$

So, 80 roses, 10 lilies, and 30 irises will create 40 centerpieces.

67. True. If B is the inverse of A, then $AB = I = BA$.

69. If the determinant of a 2×2 matrix is not equal to 0, then the inverse exists.

To find the inverse, take 1 divided by the determinant and multiply it by the matrix which has a diagonal from top left to bottom right that has the terms from the original matrix flipped and the other diagonal is the negative of the terms from the original matrix.

71. If A^{-1} does not exist, it is singular.

$\begin{bmatrix} 4 & 3 \\ -2 & k \end{bmatrix}$, $(4)(k) - (-2)(3) = 0 \Rightarrow 4k + 6 = 0 \Rightarrow k = -\frac{3}{2}$.

When $k \neq -\frac{3}{2}$, A^{-1} exists because the determinant does not equal zero.

73. (a) Given $A = \begin{bmatrix} a_{11} & 0 \\ 0 & a_{22} \end{bmatrix}$, $A^{-1} = \begin{bmatrix} \dfrac{1}{a_{11}} & 0 \\ 0 & \dfrac{1}{a_{22}} \end{bmatrix}$.

Given $A = \begin{bmatrix} a_{11} & 0 & 0 \\ 0 & a_{22} & 0 \\ 0 & 0 & a_{33} \end{bmatrix}$, $A^{-1} = \begin{bmatrix} \dfrac{1}{a_{11}} & 0 & 0 \\ 0 & \dfrac{1}{a_{22}} & 0 \\ 0 & 0 & \dfrac{1}{a_{33}} \end{bmatrix}$.

(b) In general, the inverse of a matrix in the form of A is

$$\begin{bmatrix} \dfrac{1}{a_{11}} & 0 & 0 & \cdots & 0 \\ 0 & \dfrac{1}{a_{22}} & 0 & \cdots & 0 \\ 0 & 0 & \dfrac{1}{a_{33}} & \cdots & 0 \\ \vdots & \vdots & \vdots & \cdots & \vdots \\ 0 & 0 & 0 & \cdots & \dfrac{1}{a_{nn}} \end{bmatrix}.$$

75. Answers will vary. *Sample Answer.* $A = \begin{bmatrix} a & b \\ c & d \end{bmatrix}$, $A^{-1} = \dfrac{1}{ad - bc}\begin{bmatrix} d & -b \\ -c & a \end{bmatrix}$

$$A \cdot A^{-1} = \begin{bmatrix} a & b \\ c & d \end{bmatrix}\left(\frac{1}{ad - bc}\begin{bmatrix} d & -b \\ -c & a \end{bmatrix}\right) = \frac{1}{ad - bc}\begin{bmatrix} ad - bc & -ab + ab \\ cd - cd & -bc + ad \end{bmatrix} = \frac{1}{ad - bc}\begin{bmatrix} ad - bc & 0 \\ 0 & ad - bc \end{bmatrix} = \begin{bmatrix} 1 & 0 \\ 0 & 1 \end{bmatrix} = I$$

$$A^{-1} \cdot A = \left(\frac{1}{ad - bc}\begin{bmatrix} d & -b \\ -c & a \end{bmatrix}\right)\begin{bmatrix} a & b \\ c & d \end{bmatrix} = \frac{1}{ad - bc}\begin{bmatrix} ad - bc & -ab + ab \\ cd - cd & -bc + ad \end{bmatrix} = \frac{1}{ad - bc}\begin{bmatrix} ad - bc & 0 \\ 0 & ad - bc \end{bmatrix} = \begin{bmatrix} 1 & 0 \\ 0 & 1 \end{bmatrix} = I$$

Section 8.4 The Determinant of a Square Matrix

1. determinant

3. cofactor

5. 4

7. $\begin{vmatrix} 8 & 4 \\ 2 & 3 \end{vmatrix} = (8)(3) - (4)(2) = 16$

9. $\begin{vmatrix} 6 & -3 \\ -5 & 2 \end{vmatrix} = (6)(2) - (-3)(-5) = -3$

11. $\begin{vmatrix} -7 & 0 \\ 3 & 0 \end{vmatrix} = -7(0) - 0(3) = 0$

13. $\begin{vmatrix} 2 & 6 \\ 0 & 3 \end{vmatrix} = 2(3) - 6(0) = 6$

15. $\begin{vmatrix} -3 & -2 \\ -6 & -4 \end{vmatrix} = (-3)(-4) - (-2)(-6) = 12 - 12 = 0$

17. $\begin{vmatrix} -2 & -7 \\ -3 & 1 \end{vmatrix} = (-2)(1) - (-3)(-7) = -23$

19. $\begin{vmatrix} -7 & 6 \\ \frac{1}{2} & 3 \end{vmatrix} = (-7)(3) - (6)\left(\frac{1}{2}\right) = -24$

21. $\begin{vmatrix} -\frac{1}{2} & \frac{1}{3} \\ -6 & \frac{1}{3} \end{vmatrix} = -\frac{1}{2}\left(\frac{1}{3}\right) - \frac{1}{3}(-6) = -\frac{1}{2} + 2 = \frac{11}{6}$

23. $\begin{vmatrix} 3 & 4 \\ -2 & 1 \end{vmatrix} = 11$

25. $\begin{vmatrix} 19 & 20 \\ 43 & -56 \end{vmatrix} = -1924$

27. $\begin{vmatrix} \frac{1}{10} & \frac{1}{5} \\ -\frac{3}{10} & \frac{1}{5} \end{vmatrix} = 0.08$

29. $\begin{bmatrix} 4 & 5 \\ 3 & -6 \end{bmatrix}$

(a) $M_{11} = -6$
$M_{12} = 3$
$M_{21} = 5$
$M_{22} = 4$

(b) $C_{11} = M_{11} = -6$
$C_{12} = -M_{12} = -3$
$C_{21} = -M_{21} = -5$
$C_{22} = M_{22} = 4$

31. $\begin{bmatrix} 4 & 0 & 2 \\ -3 & 2 & 1 \\ 1 & -1 & 1 \end{bmatrix}$

(a) $M_{11} = \begin{vmatrix} 2 & 1 \\ -1 & 1 \end{vmatrix} = 2 - (-1) = 3$

$M_{12} = \begin{vmatrix} -3 & 1 \\ 1 & 1 \end{vmatrix} = -3 - 1 = -4$

$M_{13} = \begin{vmatrix} -3 & 2 \\ 1 & -1 \end{vmatrix} = 3 - 2 = 1$

$M_{21} = \begin{vmatrix} 0 & 2 \\ -1 & 1 \end{vmatrix} = 0 - (-2) = 2$

$M_{22} = \begin{vmatrix} 4 & 2 \\ 1 & 1 \end{vmatrix} = 4 - 2 = 2$

$M_{23} = \begin{vmatrix} 4 & 0 \\ 1 & -1 \end{vmatrix} = -4 - 0 = -4$

$M_{31} = \begin{vmatrix} 0 & 2 \\ 2 & 1 \end{vmatrix} = 0 - 4 = -4$

$M_{32} = \begin{vmatrix} 4 & 2 \\ -3 & 1 \end{vmatrix} = 4 - (-6) = 10$

$M_{33} = \begin{vmatrix} 4 & 0 \\ -3 & 2 \end{vmatrix} = 8 - 0 = 8$

(b) $C_{11} = (-1)^2 M_{11} = 3$
$C_{12} = (-1)^3 M_{12} = 4$
$C_{13} = (-1)^4 M_{13} = 1$
$C_{21} = (-1)^3 M_{21} = -2$
$C_{22} = (-1)^4 M_{22} = 2$
$C_{23} = (-1)^5 M_{23} = 4$
$C_{31} = (-1)^4 M_{31} = -4$
$C_{32} = (-1)^5 M_{32} = -10$
$C_{33} = (-1)^6 M_{33} = 8$

33. $\begin{bmatrix} -4 & 6 & 3 \\ 7 & -2 & 8 \\ 1 & 0 & -5 \end{bmatrix}$

(a) $M_{11} = \begin{vmatrix} -2 & 8 \\ 0 & -5 \end{vmatrix} = (-2)(-5) - (8)(0) = 10$

$M_{12} = \begin{vmatrix} 7 & 8 \\ 1 & -5 \end{vmatrix} = (7)(-5) - (8)(1) = -43$

$M_{13} = \begin{vmatrix} 7 & -2 \\ 1 & 0 \end{vmatrix} = (7)(0) - (-2)(1) = 2$

$M_{21} = \begin{vmatrix} 6 & 3 \\ 0 & -5 \end{vmatrix} = (6)(-5) - (3)(0) = -30$

$M_{22} = \begin{vmatrix} -4 & 3 \\ 1 & -5 \end{vmatrix} = (-4)(-5) - (3)(1) = 17$

$M_{23} = \begin{vmatrix} -4 & 6 \\ 1 & 0 \end{vmatrix} = (-4)(0) - (6)(1) = -6$

$M_{31} = \begin{vmatrix} 6 & 3 \\ -2 & 8 \end{vmatrix} = (6)(8) - (3)(-2) = 54$

$M_{32} = \begin{vmatrix} -4 & 3 \\ 7 & 8 \end{vmatrix} = (-4)(8) - (3)(7) = -53$

$M_{33} = \begin{vmatrix} -4 & 6 \\ 7 & -2 \end{vmatrix} = (-4)(-2) - (6)(7) = -34$

(b) $C_{11} = (-1)^2 M_{11} = 10$

$C_{12} = (-1)^3 M_{12} = 43$

$C_{13} = (-1)^4 M_{13} = 2$

$C_{21} = (-1)^3 M_{21} = 30$

$C_{22} = (-1)^4 M_{22} = 17$

$C_{23} = (-1)^5 M_{23} = 6$

$C_{31} = (-1)^4 M_{31} = 54$

$C_{32} = (-1)^5 M_{32} = 53$

$C_{33} = (-1)^6 M_{33} = -34$

35. (a) $\begin{vmatrix} 2 & 5 \\ 6 & -3 \end{vmatrix} = 2(-3) - 5(6) = -36$

(b) $\begin{vmatrix} 2 & 5 \\ 6 & -3 \end{vmatrix} = 2(-3) - 6(5) = -36$

37. (a) $\begin{vmatrix} 5 & 0 & -3 \\ 0 & 12 & 4 \\ 1 & 6 & 3 \end{vmatrix} = 0\begin{vmatrix} 0 & -3 \\ 6 & 3 \end{vmatrix} + 12\begin{vmatrix} 5 & -3 \\ 1 & 3 \end{vmatrix} - 4\begin{vmatrix} 5 & 0 \\ 1 & 6 \end{vmatrix}$

$= 0(18) + 12(18) - 4(30)$

$= 96$

(b) $\begin{vmatrix} 5 & 0 & -3 \\ 0 & 12 & 4 \\ 1 & 6 & 3 \end{vmatrix} = 0\begin{vmatrix} 0 & 4 \\ 1 & 3 \end{vmatrix} + 12\begin{vmatrix} 5 & -3 \\ 1 & 3 \end{vmatrix} - 6\begin{vmatrix} 5 & -3 \\ 0 & 4 \end{vmatrix}$

$= 0(-4) + 12(18) - 6(20)$

$= 96$

39. (a) $\begin{vmatrix} -3 & 2 & 1 \\ 4 & 5 & 6 \\ 2 & -3 & 1 \end{vmatrix} = -3\begin{vmatrix} 5 & 6 \\ -3 & 1 \end{vmatrix} - 2\begin{vmatrix} 4 & 6 \\ 2 & 1 \end{vmatrix} + \begin{vmatrix} 4 & 5 \\ 2 & -3 \end{vmatrix}$

$= -3(23) - 2(-8) - 22$

$= -75$

(b) $\begin{vmatrix} -3 & 2 & 1 \\ 4 & 5 & 6 \\ 2 & -3 & 1 \end{vmatrix} = -2\begin{vmatrix} 4 & 6 \\ 2 & 1 \end{vmatrix} + 5\begin{vmatrix} -3 & 1 \\ 2 & 1 \end{vmatrix} + 3\begin{vmatrix} -3 & 1 \\ 4 & 6 \end{vmatrix}$

$= -2(-8) + 5(-5) + 3(-22)$

$= -75$

41. (a) $\begin{vmatrix} 6 & 0 & -3 & 5 \\ 4 & 0 & 6 & -8 \\ -1 & 0 & 7 & 4 \\ 8 & 0 & 0 & 2 \end{vmatrix} = -8\begin{vmatrix} 0 & -3 & 5 \\ 0 & 6 & -8 \\ 0 & 7 & 4 \end{vmatrix} + 0\begin{vmatrix} 6 & -3 & 5 \\ 4 & 6 & -8 \\ -1 & 7 & 4 \end{vmatrix} - 0\begin{vmatrix} 6 & 0 & 5 \\ 4 & 0 & -8 \\ -1 & 0 & 4 \end{vmatrix} + 2\begin{vmatrix} 6 & 0 & -3 \\ 4 & 0 & 6 \\ -1 & 0 & 0 \end{vmatrix} = -8(0) + 2(0) = 0$

(b) $\begin{vmatrix} 6 & 0 & -3 & 5 \\ 4 & 0 & 6 & -8 \\ -1 & 0 & 7 & 4 \\ 8 & 0 & 0 & 2 \end{vmatrix} = -0\begin{vmatrix} 4 & 6 & -8 \\ -1 & 7 & 4 \\ 8 & 0 & 2 \end{vmatrix} + 0\begin{vmatrix} 6 & -3 & 5 \\ -1 & 7 & 4 \\ 8 & 0 & 2 \end{vmatrix} - 0\begin{vmatrix} 6 & -3 & 5 \\ 4 & 6 & -8 \\ 8 & 0 & 2 \end{vmatrix} + 0\begin{vmatrix} 6 & -3 & 5 \\ 4 & 6 & -8 \\ -1 & 7 & 4 \end{vmatrix} = 0$

43. (a) $\begin{vmatrix} -2 & 4 & 7 & 1 \\ 3 & 0 & 0 & 0 \\ 8 & 5 & 10 & 5 \\ 6 & 0 & 5 & 0 \end{vmatrix} = -3\begin{vmatrix} 4 & 7 & 1 \\ 5 & 10 & 5 \\ 0 & 5 & 0 \end{vmatrix} + 0\begin{vmatrix} -2 & 7 & 1 \\ 8 & 10 & 5 \\ 6 & 5 & 0 \end{vmatrix} - 0\begin{vmatrix} -2 & 4 & 1 \\ 8 & 5 & 5 \\ 6 & 0 & 0 \end{vmatrix} + 0\begin{vmatrix} -2 & 4 & 7 \\ 8 & 5 & 10 \\ 6 & 0 & 5 \end{vmatrix} = -3(-75) = 225$

(b) $\begin{vmatrix} -2 & 4 & 7 & 1 \\ 3 & 0 & 0 & 0 \\ 8 & 5 & 10 & 5 \\ 6 & 0 & 5 & 0 \end{vmatrix} = -1\begin{vmatrix} 3 & 0 & 0 \\ 8 & 5 & 10 \\ 6 & 0 & 5 \end{vmatrix} + 0\begin{vmatrix} -2 & 4 & 7 \\ 8 & 5 & 10 \\ 6 & 0 & 5 \end{vmatrix} - 5\begin{vmatrix} -2 & 4 & 7 \\ 3 & 0 & 0 \\ 6 & 0 & 5 \end{vmatrix} + 0\begin{vmatrix} -2 & 4 & 7 \\ 3 & 0 & 0 \\ 8 & 5 & 10 \end{vmatrix} = (-1)(75) - 5(-60) = 225$

45. Expand along Column 1.

$\begin{vmatrix} -1 & 2 & -5 \\ 0 & 3 & 4 \\ 0 & 0 & 3 \end{vmatrix} = -1\begin{vmatrix} 3 & 4 \\ 0 & 3 \end{vmatrix} - 0\begin{vmatrix} 2 & -5 \\ 0 & 3 \end{vmatrix} + 0\begin{vmatrix} 2 & -5 \\ 3 & 4 \end{vmatrix}$

$= -1(9) - 0(6) + 0(23) = -9$

47. Expand along Row 2.

$\begin{vmatrix} 6 & 3 & -7 \\ 0 & 0 & 0 \\ 4 & -6 & 3 \end{vmatrix} = 0\begin{vmatrix} 3 & -7 \\ -6 & 3 \end{vmatrix} - 0\begin{vmatrix} 6 & -7 \\ 4 & 3 \end{vmatrix} + 0\begin{vmatrix} 6 & 3 \\ 4 & -6 \end{vmatrix} = 0$

49. Expand along Column 1.

$\begin{vmatrix} 2 & -1 & 0 \\ 4 & 2 & 1 \\ 4 & 2 & 1 \end{vmatrix} = 2\begin{vmatrix} 2 & 1 \\ 2 & 1 \end{vmatrix} - 4\begin{vmatrix} -1 & 0 \\ 2 & 1 \end{vmatrix} + 4\begin{vmatrix} -1 & 0 \\ 2 & 1 \end{vmatrix}$

$= 2(0) - 4(-1) + 4(-1) = 0$

51. Expand along Column 3.

$\begin{vmatrix} 1 & 4 & -2 \\ 3 & 2 & 0 \\ -1 & 4 & 3 \end{vmatrix} = -2\begin{vmatrix} 3 & 2 \\ -1 & 4 \end{vmatrix} + 3\begin{vmatrix} 1 & 4 \\ 3 & 2 \end{vmatrix}$

$= -2(14) + 3(-10) = -58$

53. Expand along Column 3.

$\begin{vmatrix} 2 & 6 & 0 & 2 \\ 2 & 7 & 3 & 6 \\ 1 & 0 & 0 & 1 \\ 3 & 7 & 0 & 7 \end{vmatrix} = 0\begin{vmatrix} 2 & 7 & 6 \\ 1 & 0 & 1 \\ 3 & 7 & 7 \end{vmatrix} - 3\begin{vmatrix} 2 & 6 & 2 \\ 1 & 0 & 1 \\ 3 & 7 & 7 \end{vmatrix} + 0\begin{vmatrix} 2 & 6 & 2 \\ 2 & 7 & 6 \\ 3 & 7 & 7 \end{vmatrix} - 0\begin{vmatrix} 2 & 6 & 2 \\ 2 & 7 & 6 \\ 1 & 0 & 1 \end{vmatrix} = -3(-24) = 72$

55. Expand along Column 1.

$\begin{vmatrix} 5 & 3 & 0 & 6 \\ 4 & 6 & 4 & 12 \\ 0 & 2 & -3 & 4 \\ 0 & 1 & -2 & 2 \end{vmatrix} = 5\begin{vmatrix} 6 & 4 & 12 \\ 2 & -3 & 4 \\ 1 & -2 & 2 \end{vmatrix} - 4\begin{vmatrix} 3 & 0 & 6 \\ 2 & -3 & 4 \\ 1 & -2 & 2 \end{vmatrix} = 5(0) - 4(0) = 0$

57. Expand along Column 2, then along Column 4.

$\begin{vmatrix} 3 & 2 & 4 & -1 & 5 \\ -2 & 0 & 1 & 3 & 2 \\ 1 & 0 & 0 & 4 & 0 \\ 6 & 0 & 2 & -1 & 0 \\ 3 & 0 & 5 & 1 & 0 \end{vmatrix} = -2\begin{vmatrix} -2 & 1 & 3 & 2 \\ 1 & 0 & 4 & 0 \\ 6 & 2 & -1 & 0 \\ 3 & 5 & 1 & 0 \end{vmatrix} = (-2)(-2)\begin{vmatrix} 1 & 0 & 4 \\ 6 & 2 & -1 \\ 3 & 5 & 1 \end{vmatrix} = 4(103) = 412$

59. $\begin{vmatrix} 3 & 8 & -7 \\ 0 & -5 & 4 \\ 8 & 1 & 6 \end{vmatrix} = -126$

61. $\begin{vmatrix} 1 & -1 & 8 & 4 \\ 2 & 6 & 0 & -4 \\ 2 & 0 & 2 & 6 \\ 0 & 2 & 8 & 0 \end{vmatrix} = -336$

63. (a) $\begin{vmatrix} -1 & 0 \\ 0 & 3 \end{vmatrix} = -3$

(b) $\begin{vmatrix} 2 & 0 \\ 0 & -1 \end{vmatrix} = -2$

(c) $\begin{bmatrix} -1 & 0 \\ 0 & 3 \end{bmatrix}\begin{bmatrix} 2 & 0 \\ 0 & -1 \end{bmatrix} = \begin{bmatrix} -2 & 0 \\ 0 & -3 \end{bmatrix}$

(d) $\begin{vmatrix} -2 & 0 \\ 0 & -3 \end{vmatrix} = 6$

65. (a) $\begin{vmatrix} 4 & 0 \\ 3 & -2 \end{vmatrix} = -8$

(b) $\begin{vmatrix} -1 & 1 \\ -2 & 2 \end{vmatrix} = 0$

(c) $\begin{bmatrix} 4 & 0 \\ 3 & -2 \end{bmatrix}\begin{bmatrix} -1 & 1 \\ -2 & 2 \end{bmatrix} = \begin{bmatrix} -4 & 4 \\ 1 & -1 \end{bmatrix}$

(d) $\begin{vmatrix} -4 & 4 \\ 1 & -1 \end{vmatrix} = 0$

67. (a) $\begin{vmatrix} -1 & 2 & 1 \\ 1 & 0 & 1 \\ 0 & 1 & 0 \end{vmatrix} = 2$

(b) $\begin{vmatrix} -1 & 0 & 0 \\ 0 & 2 & 0 \\ 0 & 0 & 3 \end{vmatrix} = -6$

(c) $\begin{bmatrix} -1 & 2 & 1 \\ 1 & 0 & 1 \\ 0 & 1 & 0 \end{bmatrix}\begin{bmatrix} -1 & 0 & 0 \\ 0 & 2 & 0 \\ 0 & 0 & 3 \end{bmatrix} = \begin{bmatrix} 1 & 4 & 3 \\ -1 & 0 & 3 \\ 0 & 2 & 0 \end{bmatrix}$

(d) $\begin{vmatrix} 1 & 4 & 3 \\ -1 & 0 & 3 \\ 0 & 2 & 0 \end{vmatrix} = -12$

69. Answers will vary. *Sample answer*:

$|A| = \begin{vmatrix} 3 & 2 \\ 3 & 3 \end{vmatrix} = 9 - 6 = 3$

71. Answers will vary. *Sample Answer*: $|A| = \begin{vmatrix} 4 & 2 & -1 \\ 2 & 1 & 0 \\ 1 & 1 & 3 \end{vmatrix} = -2\begin{vmatrix} 2 & -1 \\ 1 & 3 \end{vmatrix} + 1\begin{vmatrix} 4 & -1 \\ 1 & 3 \end{vmatrix} = -2(7) + 13 = -1$

73. Answers will vary. *Sample Answer*: $|A| = \begin{vmatrix} 2 & 3 \\ 8 & 12 \end{vmatrix} = 24 - 24 = 0$

75. $\begin{vmatrix} w & x \\ y & z \end{vmatrix} = wz - xy$

$-\begin{vmatrix} y & z \\ w & x \end{vmatrix} = -(xy - wz) = wz - xy$

So, $\begin{vmatrix} w & x \\ y & z \end{vmatrix} = -\begin{vmatrix} y & z \\ w & x \end{vmatrix}$.

77. $\begin{vmatrix} w & x \\ y & z \end{vmatrix} = wz - xy$

$\begin{vmatrix} w & x + cw \\ y & z + cy \end{vmatrix} = w(z + cy) - y(x + cw) = wz - xy$

So, $\begin{vmatrix} w & x \\ y & z \end{vmatrix} = \begin{vmatrix} w & x + cw \\ y & z + cy \end{vmatrix}$.

79. $\begin{vmatrix} 1 & x & x^2 \\ 1 & y & y^2 \\ 1 & z & z^2 \end{vmatrix} = \begin{vmatrix} y & y^2 \\ z & z^2 \end{vmatrix} - \begin{vmatrix} x & x^2 \\ z & z^2 \end{vmatrix} + \begin{vmatrix} x & x^2 \\ y & y^2 \end{vmatrix}$

$= (yz^2 - y^2z) - (xz^2 - x^2z) + (xy^2 - x^2y)$

$= yz^2 - xz^2 - y^2z + x^2z + xy(y - x)$

$= z^2(y - x) - z(y^2 - x^2) + xy(y - x)$

$= z^2(y - x) - z(y - x)(y + x) + xy(y - x)$

$= (y - x)[z^2 - z(y + x) + xy]$

$= (y - x)[z^2 - zy - zx + xy]$

$= (y - x)[z^2 - zx - zy + xy]$

$= (y - x)[z(z - x) - y(z - x)]$

$= (y - x)(z - x)(z - y)$

81. $\begin{vmatrix} x & 2 \\ 1 & x \end{vmatrix} = 2$

$$\begin{aligned} x^2 - 2 &= 2 \\ x^2 &= 4 \\ x &= \pm 2 \end{aligned}$$

83. $\begin{vmatrix} x+1 & 2 \\ -1 & x \end{vmatrix} = 4$

$$\begin{aligned} (x+1)(x) - (2)(-1) &= 4 \\ x^2 + x - 2 &= 0 \\ (x+2)(x-1) &= 0 \\ x &= -2 \text{ or } x = 1 \end{aligned}$$

85. $\begin{vmatrix} x+3 & 2 \\ 1 & x+2 \end{vmatrix} = 0$

$$\begin{aligned} (x+3)(x+2) - 2 &= 0 \\ x^2 + 5x + 4 &= 0 \\ (x+1)(x+4) &= 0 \\ x &= -1 \text{ or } x = -4 \end{aligned}$$

87. $\begin{vmatrix} 4u & -1 \\ -1 & 2v \end{vmatrix} = 8uv - 1$

89. $\begin{vmatrix} e^{2x} & e^{3x} \\ 2e^{2x} & 3e^{3x} \end{vmatrix} = 3e^{5x} - 2e^{5x} = e^{5x}$

91. $\begin{vmatrix} x & \ln x \\ 1 & \dfrac{1}{x} \end{vmatrix} = 1 - \ln x$

93. True. If an entire row is zero, then each cofactor in the expansion is multiplied by zero.

95. *Sample answer:* Let $A = \begin{bmatrix} 1 & 3 \\ -2 & 4 \end{bmatrix}$ and $B = \begin{bmatrix} -4 & 0 \\ 3 & 5 \end{bmatrix}$.

$|A| = \begin{vmatrix} 1 & 3 \\ -2 & 4 \end{vmatrix} = 10,\ |B| = \begin{vmatrix} -4 & 0 \\ 3 & 5 \end{vmatrix} = -20,$

$|A| + |B| = -10$

$A + B = \begin{bmatrix} -3 & 3 \\ 1 & 9 \end{bmatrix},\ |A + B| = \begin{vmatrix} -3 & 3 \\ 1 & 9 \end{vmatrix} = -30$

So, $|A + B| \neq |A| + |B|$.

97. The signs of the cofactors should be $-, +, -$.

$$\begin{vmatrix} 1 & 1 & 4 \\ 3 & 2 & 0 \\ 2 & 1 & 3 \end{vmatrix} = 3(-1)\begin{vmatrix} 1 & 4 \\ 1 & 3 \end{vmatrix} + 2(1)\begin{vmatrix} 1 & 4 \\ 2 & 3 \end{vmatrix} + (0)(-1)\begin{vmatrix} 1 & 1 \\ 2 & 1 \end{vmatrix}$$

$$= 3(1) + 2(-5) + 0 = -7.$$

99. (a) $\begin{vmatrix} 1 & 3 & 4 \\ -7 & 2 & -5 \\ 6 & 1 & 2 \end{vmatrix} = -115$

$-\begin{vmatrix} 1 & 4 & 3 \\ -7 & -5 & 2 \\ 6 & 2 & 1 \end{vmatrix} = -115$

Column 2 and Column 3 were interchanged.

(b) $\begin{vmatrix} 1 & 3 & 4 \\ -2 & 2 & 0 \\ 1 & 6 & 2 \end{vmatrix} = -40$

$-\begin{vmatrix} 1 & 6 & 2 \\ -2 & 2 & 0 \\ 1 & 3 & 4 \end{vmatrix} = -40$

Row 1 and Row 3 were interchanged.

101. (a) $A = \begin{bmatrix} 1 & 2 \\ 2 & -3 \end{bmatrix},\ B = \begin{bmatrix} 5 & 10 \\ 2 & -3 \end{bmatrix}$

$|B| = \begin{vmatrix} 5 & 10 \\ 2 & -3 \end{vmatrix} = -35$

$5|A| = 5\begin{vmatrix} 1 & 2 \\ 2 & -3 \end{vmatrix} = -35$

Row 1 was multiplied by 5.

$|B| = 5|A|$

(b) $A = \begin{bmatrix} 1 & 2 & -1 \\ 3 & -3 & 2 \\ 7 & 1 & 3 \end{bmatrix},\ B = \begin{bmatrix} 1 & 8 & -3 \\ 3 & -12 & 6 \\ 7 & 4 & 9 \end{bmatrix}$

$|B| = \begin{bmatrix} 1 & 8 & -3 \\ 3 & -12 & 6 \\ 7 & 4 & 9 \end{bmatrix} = -300$

$12|A| = 12\begin{vmatrix} 1 & 2 & -1 \\ 3 & -3 & 2 \\ 7 & 1 & 3 \end{vmatrix} = -300$

Column 2 was multiplied by 4 and Column 3 was multiplied by 3.

$|B| = (4)(3)|A| = 12|A|$

103. (a) $\begin{vmatrix} 7 & 0 \\ 0 & 4 \end{vmatrix} = 7(4) - 0 = 28$

(b) $\begin{vmatrix} -1 & 0 & 0 \\ 0 & 5 & 0 \\ 0 & 0 & 2 \end{vmatrix} = (-1)\begin{vmatrix} 5 & 0 \\ 0 & 2 \end{vmatrix} - 0\begin{vmatrix} 0 & 0 \\ 0 & 2 \end{vmatrix} + 0\begin{vmatrix} 0 & 5 \\ 0 & 0 \end{vmatrix} = (-1)(10) = -10$

(c) $\begin{vmatrix} 2 & 0 & 0 & 0 \\ 0 & -2 & 0 & 0 \\ 0 & 0 & 1 & 0 \\ 0 & 0 & 0 & 3 \end{vmatrix} = (-2)\begin{vmatrix} -2 & 0 & 0 \\ 0 & 1 & 0 \\ 0 & 0 & 3 \end{vmatrix} - 0\begin{vmatrix} 0 & 0 & 0 \\ 0 & 1 & 0 \\ 0 & 0 & 3 \end{vmatrix} + 0\begin{vmatrix} 0 & -2 & 0 \\ 0 & 0 & 0 \\ 0 & 0 & 3 \end{vmatrix} - 0\begin{vmatrix} 0 & -2 & 0 \\ 0 & 0 & 1 \\ 0 & 0 & 0 \end{vmatrix}$

$= (-2)\left((2)\begin{vmatrix} 1 & 0 \\ 0 & 3 \end{vmatrix} - 0\begin{vmatrix} 0 & 0 \\ 0 & 3 \end{vmatrix} + 0\begin{vmatrix} 0 & 1 \\ 0 & 0 \end{vmatrix}\right)$

$= (-2)(2)(3) = -12$

The determinant of a diagonal matrix is the product of the entries on the main diagonal.

Section 8.5 Applications of Matrices and Determinants

1. Cramer's Rule

3. $A = \pm\frac{1}{2}\begin{vmatrix} x_1 & y_1 & 1 \\ x_2 & y_2 & 1 \\ x_3 & y_3 & 1 \end{vmatrix}$

5. uncoded; coded

7. $\begin{cases} -5x + 9y = -14 \\ 3x - 7y = 10 \end{cases}$

$x = \dfrac{\begin{vmatrix} -14 & 9 \\ 10 & -7 \end{vmatrix}}{\begin{vmatrix} -5 & 9 \\ 3 & -7 \end{vmatrix}} = \dfrac{8}{8} = 1$

$y = \dfrac{\begin{vmatrix} -5 & -14 \\ 3 & 10 \end{vmatrix}}{\begin{vmatrix} -5 & 9 \\ 3 & -7 \end{vmatrix}} = \dfrac{-8}{8} = -1$

Solution: $(1, -1)$

9. $\begin{cases} 3x + 2y = -2 \\ 6x + 4y = 4 \end{cases}$

Because $\begin{vmatrix} 3 & 2 \\ 6 & 4 \end{vmatrix} = 0$, Cramer's Rule does not apply.

The system is inconsistent in this case and has no solution.

11. $\begin{cases} 4x - y + z = -5 \\ 2x + 2y + 3z = 10, \\ 5x - 2y + 6z = 1 \end{cases}$ $D = \begin{vmatrix} 4 & -1 & 1 \\ 2 & 2 & 3 \\ 5 & -1 & 6 \end{vmatrix} = 55$

$x = \dfrac{\begin{vmatrix} -5 & -1 & 1 \\ 10 & 2 & 3 \\ 1 & -2 & 6 \end{vmatrix}}{55} = \dfrac{-55}{55} = -1,\ y = \dfrac{\begin{vmatrix} 4 & -5 & 1 \\ 2 & 10 & 3 \\ 5 & 1 & 6 \end{vmatrix}}{55} = \dfrac{165}{55} = 3,\ z = \dfrac{\begin{vmatrix} 4 & -1 & -5 \\ 2 & 2 & 10 \\ 5 & -2 & 1 \end{vmatrix}}{55} = \dfrac{110}{55} = 2$

Solution: $(-1, 3, 2)$

13. $\begin{cases} x + 2y + 3z = -3 \\ -2x + y - z = 6, \\ 3x - 3y + 2z = -11 \end{cases}$ $D = \begin{vmatrix} 1 & 2 & 3 \\ -2 & 1 & -1 \\ 3 & -3 & 2 \end{vmatrix} = 10$

$$x = \frac{\begin{vmatrix} -3 & 2 & 3 \\ 6 & 1 & -1 \\ -11 & -3 & 2 \end{vmatrix}}{10} = \frac{-20}{10} = -2$$

$$y = \frac{\begin{vmatrix} 1 & -3 & 3 \\ -2 & 6 & -1 \\ 3 & -11 & 2 \end{vmatrix}}{10} = \frac{10}{10} = 1$$

$$z = \frac{\begin{vmatrix} 1 & 2 & -3 \\ -2 & 1 & 6 \\ 3 & -3 & -11 \end{vmatrix}}{10} = \frac{-10}{10} = -1$$

Solution: $(-2, 1, -1)$

15. Vertices: $(0, 0)$ $(3, 1)$, $(1, 5)$

$$\text{Area} = \frac{1}{2}\begin{vmatrix} 0 & 0 & 1 \\ 3 & 1 & 1 \\ 1 & 5 & 1 \end{vmatrix} = \frac{1}{2}\begin{vmatrix} 3 & 1 \\ 1 & 5 \end{vmatrix} = 7 \text{ square units}$$

17. Vertices: $(-2, -3)$, $(2, -3)$, $(0, 4)$

$$\text{Area} = \frac{1}{2}\begin{vmatrix} -2 & -3 & 1 \\ 2 & -3 & 1 \\ 0 & 4 & 1 \end{vmatrix}$$

$$= \frac{1}{2}\left(-2\begin{vmatrix} -3 & 1 \\ 4 & 1 \end{vmatrix} - 2\begin{vmatrix} -3 & 1 \\ 4 & 1 \end{vmatrix}\right)$$

$$= \frac{1}{2}(14 + 14)$$

$$= 14 \text{ square units}$$

19. $4 = \pm\frac{1}{2}\begin{vmatrix} -5 & 1 & 1 \\ 0 & 2 & 1 \\ -2 & y & 1 \end{vmatrix}$

$$\pm 8 = -5\begin{vmatrix} 2 & 1 \\ y & 1 \end{vmatrix} - 2\begin{vmatrix} 1 & 1 \\ 2 & 1 \end{vmatrix}$$

$$\pm 8 = -5(2 - y) - 2(-1)$$

$$\pm 8 = 5y - 8$$

$$y = \frac{8 \pm 8}{5}$$

$$y = \frac{16}{5} \text{ or } y = 0$$

21. Vertices: $(0, 25)$, $(10, 0)$, $(28, 5)$

$$\text{Area} = \frac{1}{2}\begin{vmatrix} 0 & 25 & 1 \\ 10 & 0 & 1 \\ 28 & 5 & 1 \end{vmatrix} = 250 \text{ square miles}$$

23. Points: $(2, -6)$, $(0, -2)$, $(3, -8)$

$$\begin{vmatrix} 2 & -6 & 1 \\ 0 & -2 & 1 \\ 3 & -8 & 1 \end{vmatrix} = 2\begin{vmatrix} -2 & 1 \\ -8 & 1 \end{vmatrix} + 3\begin{vmatrix} -6 & 1 \\ -2 & 1 \end{vmatrix}$$

$$= 2(6) + 3(-4)$$

$$= 0$$

The points are collinear.

25. Points: $\left(2, -\frac{1}{2}\right)$, $(-4, 4)$, $(6, -3)$

$$\begin{vmatrix} 2 & -\frac{1}{2} & 1 \\ -4 & 4 & 1 \\ 6 & -3 & 1 \end{vmatrix} = \begin{vmatrix} -4 & 4 \\ 6 & -3 \end{vmatrix} - \begin{vmatrix} 2 & -\frac{1}{2} \\ 6 & -3 \end{vmatrix} + \begin{vmatrix} 2 & -\frac{1}{2} \\ -4 & 4 \end{vmatrix}$$

$$= -12 + 3 + 6$$

$$= -3 \neq 0$$

The points are not collinear.

27. Points: $(0, 2)$, $(1, 2.4)$, $(-1, 1.6)$

$$\begin{vmatrix} 0 & 2 & 1 \\ 1 & 2.4 & 1 \\ -1 & 1.6 & 1 \end{vmatrix} = -2\begin{vmatrix} 1 & 1 \\ -1 & 1 \end{vmatrix} + \begin{vmatrix} 1 & 2.4 \\ -1 & 1.6 \end{vmatrix} = -2(2) + 4 = 0$$

The points are collinear.

29. $\begin{vmatrix} 2 & -5 & 1 \\ 4 & y & 1 \\ 5 & -2 & 1 \end{vmatrix} = 0$

$$2\begin{vmatrix} y & 1 \\ -2 & 1 \end{vmatrix} + 5\begin{vmatrix} 4 & 1 \\ 5 & 1 \end{vmatrix} + \begin{vmatrix} 4 & y \\ 5 & -2 \end{vmatrix} = 0$$

$$2(y + 2) + 5(-1) + (-8 - 5y) = 0$$

$$-3y - 9 = 0$$

$$y = -3$$

31. Points: $(0, 0), (5, 3)$

Equation: $\begin{vmatrix} x & y & 1 \\ 0 & 0 & 1 \\ 5 & 3 & 1 \end{vmatrix} = -\begin{vmatrix} x & y \\ 5 & 3 \end{vmatrix} = 5y - 3x = 0 \Rightarrow 3x - 5y = 0$

33. Points: $(-4, 3), (2, 1)$

Equation: $\begin{vmatrix} x & y & 1 \\ -4 & 3 & 1 \\ 2 & 1 & 1 \end{vmatrix} = x\begin{vmatrix} 3 & 1 \\ 1 & 1 \end{vmatrix} - y\begin{vmatrix} -4 & 1 \\ 2 & 1 \end{vmatrix} + \begin{vmatrix} -4 & 3 \\ 2 & 1 \end{vmatrix} = 2x + 6y - 10 = 0 \Rightarrow x + 3y - 5 = 0$

35. Points: $\left(-\frac{1}{2}, 3\right), \left(\frac{5}{2}, 1\right)$

Equation: $\begin{vmatrix} x & y & 1 \\ -\frac{1}{2} & 3 & 1 \\ \frac{5}{2} & 1 & 1 \end{vmatrix} = x\begin{vmatrix} 3 & 1 \\ 1 & 1 \end{vmatrix} - y\begin{vmatrix} -\frac{1}{2} & 1 \\ \frac{5}{2} & 1 \end{vmatrix} + \begin{vmatrix} -\frac{1}{2} & 3 \\ \frac{5}{2} & 1 \end{vmatrix} = 2x + 3y - 8 = 0$

37. A horizontal stretch, $k = 2$, of the square with vertices $(0, 0), (0, 3), (3, 0)$ and $(3, 3)$.

$\begin{bmatrix} 2 & 0 \\ 0 & 1 \end{bmatrix}\begin{bmatrix} 0 \\ 0 \end{bmatrix} = \begin{bmatrix} 0 \\ 0 \end{bmatrix}, \begin{bmatrix} 2 & 0 \\ 0 & 1 \end{bmatrix}\begin{bmatrix} 0 \\ 3 \end{bmatrix} = \begin{bmatrix} 0 \\ 3 \end{bmatrix}, \begin{bmatrix} 2 & 0 \\ 0 & 1 \end{bmatrix}\begin{bmatrix} 3 \\ 0 \end{bmatrix} = \begin{bmatrix} 6 \\ 0 \end{bmatrix}$ and $\begin{bmatrix} 2 & 0 \\ 0 & 1 \end{bmatrix}\begin{bmatrix} 3 \\ 3 \end{bmatrix} = \begin{bmatrix} 6 \\ 3 \end{bmatrix}$.

New Vertices: $(0, 0), (0, 3), (6, 0)$ and $(6, 3)$

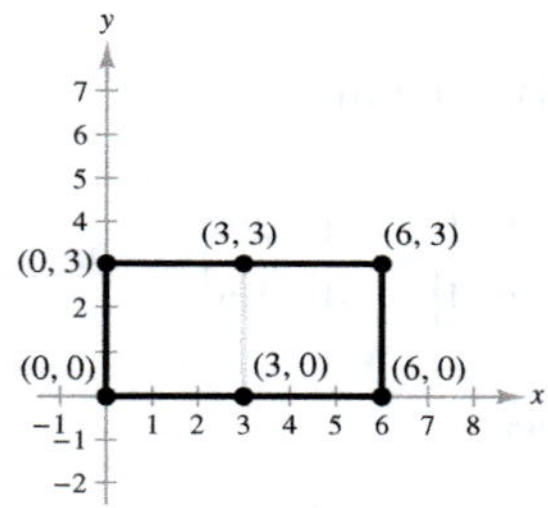

39. A reflection in the y-axis of the square with vertices $(4, 3), (5, 3), (4, 4)$ and $(5, 4)$.

$\begin{bmatrix} -1 & 0 \\ 0 & 1 \end{bmatrix}\begin{bmatrix} 4 \\ 3 \end{bmatrix} = \begin{bmatrix} -4 \\ 3 \end{bmatrix}, \begin{bmatrix} -1 & 0 \\ 0 & 1 \end{bmatrix}\begin{bmatrix} 5 \\ 3 \end{bmatrix} = \begin{bmatrix} -5 \\ 3 \end{bmatrix}, \begin{bmatrix} -1 & 0 \\ 0 & 1 \end{bmatrix}\begin{bmatrix} 4 \\ 4 \end{bmatrix} = \begin{bmatrix} -4 \\ 4 \end{bmatrix}$ and $\begin{bmatrix} -1 & 0 \\ 0 & 1 \end{bmatrix}\begin{bmatrix} 5 \\ 4 \end{bmatrix} = \begin{bmatrix} -5 \\ 4 \end{bmatrix}$.

New Vertices: $(-4, 3), (-5, 3), (-4, 4)$ and $(-5, 4)$

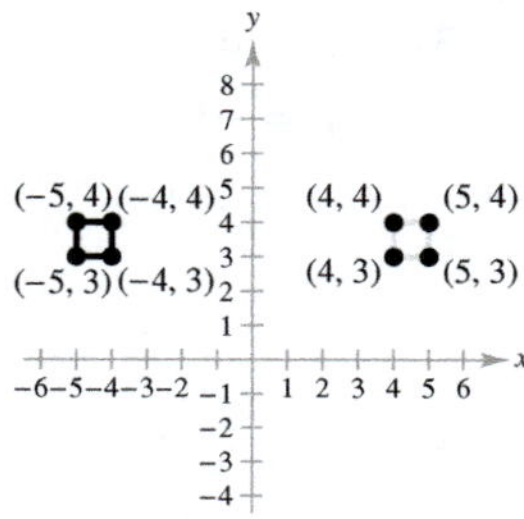

41. The area of the parallelogram with vertices: $(0, 0), (1, 0), (2, 2)$ and $(3, 2) \Rightarrow a = 1, b = 2, c = 2$ and $d = 2$.

$$A = \begin{bmatrix} 1 & 0 \\ 2 & 2 \end{bmatrix}$$

Area $= |\det(A)| = |2 - 0| = 2$ square units.

43. The area of the parallelogram with vertices: $(0, 0), (-2, 0), (3, 5)$ and $(1, 5) \Rightarrow a = -2, b = 0, c = 3$ and $d = 5$.

$$A = \begin{bmatrix} -2 & 0 \\ 3 & 5 \end{bmatrix}$$

Area $= |\det(A)| = |-10 - 0| = 10$ square units.

45. (a) Uncoded: C O M E _ H O M E _

$[3\ \ 15][13\ \ 5][0\ \ 8][15\ \ 13][5\ \ 0]$

S O O N

$[19\ \ 15][15\ \ 14]$

(b) $[3\ \ 15]\begin{bmatrix} 1 & 2 \\ 3 & 5 \end{bmatrix} = [48\ \ 81]$

$[13\ \ 5]\begin{bmatrix} 1 & 2 \\ 3 & 5 \end{bmatrix} = [28\ \ 51]$

$[0\ \ 8]\begin{bmatrix} 1 & 2 \\ 3 & 5 \end{bmatrix} = [24\ \ 40]$

$[15\ \ 13]\begin{bmatrix} 1 & 2 \\ 3 & 5 \end{bmatrix} = [54\ \ 95]$

$[5\ \ 0]\begin{bmatrix} 1 & 2 \\ 3 & 5 \end{bmatrix} = [5\ \ 10]$

$[19\ \ 15]\begin{bmatrix} 1 & 2 \\ 3 & 5 \end{bmatrix} = [64\ \ 113]$

$[15\ \ 14]\begin{bmatrix} 1 & 2 \\ 3 & 5 \end{bmatrix} = [57\ \ 100]$

Encoded: 48 81 28 51 24 40 54
95 5 10 64 113 57 100

47. (a) Uncoded:

C A L L _ M E _ T O M O

$[3\ \ 1\ \ 12][12\ \ 0\ \ 13][5\ \ 0\ \ 20][15\ \ 13\ \ 15]$

R R O W _ _

$[18\ \ 18\ \ 15][23\ \ 0\ \ 0]$

(b) $[3\ \ 1\ \ 12]\begin{bmatrix} 1 & -1 & 0 \\ 1 & 0 & -1 \\ -6 & 2 & 3 \end{bmatrix} = [-68\ \ 21\ \ 35]$

$[12\ \ 0\ \ 13]\begin{bmatrix} 1 & -1 & 0 \\ 1 & 0 & -1 \\ -6 & 2 & 3 \end{bmatrix} = [-66\ \ 14\ \ 39]$

$[5\ \ 0\ \ 20]\begin{bmatrix} 1 & -1 & 0 \\ 1 & 0 & -1 \\ -6 & 2 & 3 \end{bmatrix} = [-115\ \ 35\ \ 60]$

$[15\ \ 13\ \ 15]\begin{bmatrix} 1 & -1 & 0 \\ 1 & 0 & -1 \\ -6 & 2 & 3 \end{bmatrix} = [-62\ \ 15\ \ 32]$

$[18\ \ 18\ \ 15]\begin{bmatrix} 1 & -1 & 0 \\ 1 & 0 & -1 \\ -6 & 2 & 3 \end{bmatrix} = [-54\ \ 12\ \ 27]$

$[23\ \ 0\ \ 0]\begin{bmatrix} 1 & -1 & 0 \\ 1 & 0 & -1 \\ -6 & 2 & 3 \end{bmatrix} = [23\ \ -23\ \ 0]$

Encoded:
–68 21 35 –66 14 39 –115 35 60
–62 15 32 –54 12 27 23 –23 0

In Exercises 49–51, use the matrix $A = \begin{bmatrix} 1 & 2 & 2 \\ 3 & 7 & 9 \\ -1 & -4 & -7 \end{bmatrix}$.

49. L A N D I N G _ S U C C E S S F U L

$[12 \quad 1 \quad 14][4 \quad 9 \quad 14][7 \quad 0 \quad 19][21 \quad 3 \quad 3][5 \quad 19 \quad 19][6 \quad 21 \quad 12]$

$$[12 \quad 1 \quad 14]\begin{bmatrix} 1 & 2 & 2 \\ 3 & 7 & 9 \\ -1 & -4 & -7 \end{bmatrix} = [1 \quad -25 \quad -65]$$

$$[4 \quad 9 \quad 14]\begin{bmatrix} 1 & 2 & 2 \\ 3 & 7 & 9 \\ -1 & -4 & -7 \end{bmatrix} = [17 \quad 15 \quad -9]$$

$$[7 \quad 0 \quad 19]\begin{bmatrix} 1 & 2 & 2 \\ 3 & 7 & 9 \\ -1 & -4 & -7 \end{bmatrix} = [-12 \quad -62 \quad -119]$$

$$[21 \quad 3 \quad 3]\begin{bmatrix} 1 & 2 & 2 \\ 3 & 7 & 9 \\ -1 & -4 & -7 \end{bmatrix} = [27 \quad 51 \quad 48]$$

$$[5 \quad 19 \quad 19]\begin{bmatrix} 1 & 2 & 2 \\ 3 & 7 & 9 \\ -1 & -4 & -7 \end{bmatrix} = [43 \quad 67 \quad 48]$$

$$[6 \quad 21 \quad 12]\begin{bmatrix} 1 & 2 & 2 \\ 3 & 7 & 9 \\ -1 & -4 & -7 \end{bmatrix} = [57 \quad 111 \quad 117]$$

Cryptogram: 1 −25 −65 17 15 −9 −12 −62 −119 27 51 48 43 67 48 57 111 117

51. H A P P Y _ B I R T H D A Y _

$[8 \quad 1 \quad 16][16 \quad 25 \quad 0][2 \quad 9 \quad 18][20 \quad 8 \quad 4][1 \quad 25 \quad 0]$

$$[8 \quad 1 \quad 16]\begin{bmatrix} 1 & 2 & 2 \\ 3 & 7 & 9 \\ -1 & -4 & -7 \end{bmatrix} = [-5 \quad -41 \quad -87]$$

$$[16 \quad 25 \quad 0]\begin{bmatrix} 1 & 2 & 2 \\ 3 & 7 & 9 \\ -1 & -4 & -7 \end{bmatrix} = [91 \quad 207 \quad 257]$$

$$[2 \quad 9 \quad 18]\begin{bmatrix} 1 & 2 & 2 \\ 3 & 7 & 9 \\ -1 & -4 & -7 \end{bmatrix} = [11 \quad -5 \quad -41]$$

$$[20 \quad 8 \quad 4]\begin{bmatrix} 1 & 2 & 2 \\ 3 & 7 & 9 \\ -1 & -4 & -7 \end{bmatrix} = [40 \quad 80 \quad 84]$$

$$[1 \quad 25 \quad 0]\begin{bmatrix} 1 & 2 & 2 \\ 3 & 7 & 9 \\ -1 & -4 & -7 \end{bmatrix} = [76 \quad 177 \quad 227]$$

Cryptogram: −5 −41 −87 91 207 257 11 −5 −41 40 80 84 76 177 227

53. $A^{-1} = \begin{bmatrix} 1 & 2 \\ 3 & 5 \end{bmatrix}^{-1} = \begin{bmatrix} -5 & 2 \\ 3 & -1 \end{bmatrix}$

$$\begin{bmatrix} 11 & 21 \\ 64 & 112 \\ 25 & 50 \\ 29 & 53 \\ 23 & 46 \\ 40 & 75 \\ 55 & 92 \end{bmatrix} \begin{bmatrix} -5 & 2 \\ 3 & -1 \end{bmatrix} = \begin{bmatrix} 8 & 1 \\ 16 & 16 \\ 25 & 0 \\ 14 & 5 \\ 23 & 0 \\ 25 & 5 \\ 1 & 18 \end{bmatrix} \begin{matrix} H & A \\ P & P \\ Y & \_ \\ N & E \\ W & \_ \\ Y & E \\ A & R \end{matrix}$$

Message: HAPPY NEW YEAR

55. $A^{-1} = \begin{bmatrix} 1 & -1 & 0 \\ 1 & 0 & -1 \\ -6 & 2 & 3 \end{bmatrix}^{-1} = \begin{bmatrix} -2 & -3 & -1 \\ -3 & -3 & -1 \\ -2 & -4 & -1 \end{bmatrix}$

$$\begin{bmatrix} 9 & -1 & -9 \\ 38 & -19 & -19 \\ 28 & -9 & -19 \\ -80 & 25 & 41 \\ -64 & 21 & 31 \\ 9 & -5 & -4 \end{bmatrix} \begin{bmatrix} -2 & -3 & -1 \\ -3 & -3 & -1 \\ -2 & -4 & -1 \end{bmatrix} = \begin{bmatrix} 3 & 12 & 1 \\ 19 & 19 & 0 \\ 9 & 19 & 0 \\ 3 & 1 & 14 \\ 3 & 5 & 12 \\ 5 & 4 & 0 \end{bmatrix} \begin{matrix} C & L & A \\ S & S & \_ \\ I & S & \_ \\ C & A & N \\ C & E & L \\ E & D & \_ \end{matrix}$$

Message: CLASS IS CANCELED

57. $A^{-1} = \begin{bmatrix} 1 & 2 & 2 \\ 3 & 7 & 9 \\ -1 & -4 & -7 \end{bmatrix}^{-1} = \begin{bmatrix} -13 & 6 & 4 \\ 12 & -5 & -3 \\ -5 & 2 & 1 \end{bmatrix}$

$$\begin{bmatrix} 20 & 17 & -15 \\ -12 & -56 & -104 \\ 1 & -25 & -65 \\ 62 & 143 & 181 \end{bmatrix} \begin{bmatrix} -13 & 6 & 4 \\ 12 & -5 & -3 \\ -5 & 2 & 1 \end{bmatrix} = \begin{bmatrix} 19 & 5 & 14 \\ 4 & 0 & 16 \\ 12 & 1 & 14 \\ 5 & 19 & 0 \end{bmatrix} \begin{matrix} S & E & N \\ D & \_ & P \\ L & A & N \\ E & S & \_ \end{matrix}$$

Message: SEND PLANES

59. Let A be the 2×2 matrix needed to decode the message.

$$\begin{bmatrix} -18 & -18 \\ 1 & 16 \end{bmatrix} A = \begin{bmatrix} 0 & 18 \\ 15 & 14 \end{bmatrix} \begin{matrix} \_ & R \\ O & N \end{matrix}$$

$$A = \begin{bmatrix} -18 & -18 \\ 1 & 16 \end{bmatrix}^{-1} \begin{bmatrix} 0 & 18 \\ 15 & 14 \end{bmatrix} = \begin{bmatrix} -\frac{8}{135} & -\frac{1}{15} \\ \frac{1}{270} & \frac{1}{15} \end{bmatrix} \begin{bmatrix} 0 & 18 \\ 15 & 14 \end{bmatrix} = \begin{bmatrix} -1 & -2 \\ 1 & 1 \end{bmatrix}$$

$$\begin{bmatrix} 8 & 21 \\ -15 & -10 \\ -13 & -13 \\ 5 & 10 \\ 5 & 25 \\ 5 & 19 \\ -1 & 6 \\ 20 & 40 \\ -18 & -18 \\ 1 & 16 \end{bmatrix} \begin{bmatrix} -1 & -2 \\ 1 & 1 \end{bmatrix} = \begin{bmatrix} 13 & 5 \\ 5 & 20 \\ 0 & 13 \\ 5 & 0 \\ 20 & 15 \\ 14 & 9 \\ 7 & 8 \\ 20 & 0 \\ 0 & 18 \\ 15 & 14 \end{bmatrix} \begin{matrix} M & E \\ E & T \\ \_ & M \\ E & \_ \\ T & O \\ N & I \\ G & H \\ T & \_ \\ \_ & R \\ O & N \end{matrix}$$

Message: MEET ME TONIGHT RON

61. $D = \begin{vmatrix} 4 & 0 & 8 \\ 0 & 2 & 8 \\ 1 & 1 & -1 \end{vmatrix} = -56$

$I_1 = \dfrac{\begin{vmatrix} 2 & 0 & 8 \\ 6 & 2 & 8 \\ 0 & 1 & -1 \end{vmatrix}}{-56} = -\dfrac{28}{56} = -\dfrac{1}{2}$ $\quad I_2 = \dfrac{\begin{vmatrix} 4 & 2 & 8 \\ 0 & 6 & 8 \\ 1 & 0 & -1 \end{vmatrix}}{-56} = \dfrac{-56}{-56} = 1$ $\quad I_3 = \dfrac{\begin{vmatrix} 4 & 0 & 2 \\ 0 & 2 & 6 \\ 1 & 1 & 0 \end{vmatrix}}{-56} = \dfrac{-28}{-56} = \dfrac{1}{2}$

So, the solution is $I_1 = -0.5$ ampere, $I_2 = 1$ ampere, and $I_3 = 0.5$ ampere.

63. False. In Cramer's Rule, the denominator is the determinant of the coefficient matrix.

65. If the determinant of the coefficient matrix is zero, the system has either no solution or infinitely many solutions.

67. $\text{Area} = \frac{1}{2}\begin{vmatrix} 3 & -1 & 1 \\ 7 & -1 & 1 \\ 7 & 5 & 1 \end{vmatrix}$

$= \frac{1}{2}\left(3\begin{vmatrix} -1 & 1 \\ 5 & 1 \end{vmatrix} + 1\begin{vmatrix} 7 & 1 \\ 7 & 1 \end{vmatrix} + 1\begin{vmatrix} 7 & -1 \\ 7 & 5 \end{vmatrix}\right)$

$= \frac{1}{2}(-18 + 0 + 42)$

$= 12$ square units

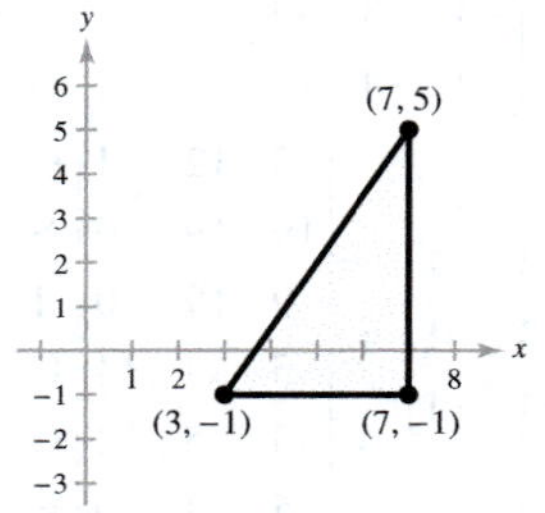

$\text{Area} = \frac{1}{2}(\text{base})(\text{height}) = \frac{1}{2}(7-3)(5-(-1)) = \frac{1}{2}(4)(6) = 12$ square units

Review Exercises for Chapter 8

1. $\begin{bmatrix} -1 & 3 \end{bmatrix}$

Order: 1×2

3. $\begin{bmatrix} 2 & 1 & 0 & 4 & -1 \\ 6 & 2 & 1 & 8 & 0 \end{bmatrix}$

Order: 2×5

5. $\begin{cases} 3x - 10y = 15 \\ 5x + 4y = 22 \end{cases}$

$\left[\begin{array}{cc|c} 3 & -10 & 15 \\ 5 & 4 & 22 \end{array}\right]$

7. $\left[\begin{array}{ccc|c} 1 & 0 & 2 & -8 \\ 2 & -2 & 3 & 12 \\ 4 & 7 & 1 & 3 \end{array}\right]$

$\begin{cases} x + 2z = -8 \\ 2x - 2y + 3z = 12 \\ 4x + 7y + z = 3 \end{cases}$

9. $\begin{bmatrix} 0 & 1 & 1 \\ 1 & 2 & 3 \\ 2 & 2 & 2 \end{bmatrix}$

$\begin{matrix} R_1 \\ R_2 \\ \end{matrix}\begin{bmatrix} 1 & 2 & 3 \\ 0 & 1 & 1 \\ 2 & 2 & 2 \end{bmatrix}$

$-2R_1 + R_3 \rightarrow \begin{bmatrix} 1 & 2 & 3 \\ 0 & 1 & 1 \\ 0 & -2 & -4 \end{bmatrix}$

$2R_2 + R_3 \rightarrow \begin{bmatrix} 1 & 2 & 3 \\ 0 & 1 & 1 \\ 0 & 0 & -2 \end{bmatrix}$

$-\frac{1}{2}R_3 \rightarrow \begin{bmatrix} 1 & 2 & 3 \\ 0 & 1 & 1 \\ 0 & 0 & 1 \end{bmatrix}$

11. $\begin{bmatrix} 1 & 2 & 3 & \vdots & 9 \\ 0 & 1 & -2 & \vdots & 2 \\ 0 & 0 & 1 & \vdots & -1 \end{bmatrix} \Rightarrow \begin{cases} x + 2y + 3z = 9 \\ y - 2z = 2 \\ z = -1 \end{cases}$

$y - 2(-1) = 2 \Rightarrow y = 0'$

$x + 2(0) + 3(-1) = 9 \Rightarrow x = 12$

Solution: $(12, 0, -1)$

13. $\begin{bmatrix} 1 & 3 & 4 & \vdots & 1 \\ 0 & 1 & 2 & \vdots & 3 \\ 0 & 0 & 1 & \vdots & 4 \end{bmatrix} \Rightarrow \begin{cases} x + 3y + 4z = 1 \\ y + 2z = 3 \\ z = 4 \end{cases}$

$y + 2(4) = 3 \Rightarrow y = -5$

$x + 3(-5) + 4(4) = 1 \Rightarrow x = 0$

Solution: $(0, -5, 4)$

15. $\begin{bmatrix} 5 & 4 & \vdots & 2 \\ -1 & 1 & \vdots & -22 \end{bmatrix}$

$4R_2 + R_1 \to \begin{bmatrix} 1 & 8 & \vdots & -86 \\ -1 & 1 & \vdots & -22 \end{bmatrix}$

$R_1 + R_2 \to \begin{bmatrix} 1 & 8 & \vdots & -86 \\ 0 & 9 & \vdots & -108 \end{bmatrix}$

$\frac{1}{9}R_2 \to \begin{bmatrix} 1 & 8 & \vdots & -86 \\ 0 & 1 & \vdots & -12 \end{bmatrix}$

$\begin{cases} x + 8y = -86 \\ y = -12 \end{cases}$

$y = -12$

$x + 8(-12) = -86 \Rightarrow x = 10$

Solution: $(10, -12)$

17. $\begin{bmatrix} 0.3 & -0.1 & \vdots & -0.13 \\ 0.2 & -0.3 & \vdots & -0.25 \end{bmatrix}$

$\begin{matrix} 10R_1 \to \\ 10R_2 \to \end{matrix} \begin{bmatrix} 3 & -1 & \vdots & -1.3 \\ 2 & -3 & \vdots & -2.5 \end{bmatrix}$

$-R_2 + R_1 \to \begin{bmatrix} 1 & 2 & \vdots & 1.2 \\ 2 & -3 & \vdots & -2.5 \end{bmatrix}$

$-2R_1 + R_2 \to \begin{bmatrix} 1 & 2 & \vdots & 1.2 \\ 0 & -7 & \vdots & -4.9 \end{bmatrix}$

$-\frac{1}{7}R_2 \to \begin{bmatrix} 1 & 2 & \vdots & 1.2 \\ 0 & 1 & \vdots & 0.7 \end{bmatrix}$

$\begin{cases} x + 2y = 1.2 \\ y = 0.7 \end{cases}$

$y = 0.7$

$x + 2(0.7) = 1.2 \Rightarrow x = -0.2$

Solution: $(-0.2, 0.7) = \left(-\frac{1}{5}, \frac{7}{10}\right)$

19. $\begin{cases} -x + 2y = 3 \\ 2x - 4y = 6 \end{cases}$

$\begin{bmatrix} -1 & 2 & \vdots & 3 \\ 2 & -4 & \vdots & 6 \end{bmatrix}$

$2R_1 + R_2 \to \begin{bmatrix} -1 & 2 & \vdots & 3 \\ 0 & 0 & \vdots & 12 \end{bmatrix}$

Because the last row consists of all zeros except for the last entry, the system is inconsistent and there is no solution.

21. $\begin{cases} x - 2y + z = 7 \\ 2x + y - 2z = -4 \\ -x + 3y + 2z = -3 \end{cases}$

$\begin{bmatrix} 1 & -2 & 1 & \vdots & 7 \\ 2 & 1 & -2 & \vdots & -4 \\ -1 & 3 & 2 & \vdots & -3 \end{bmatrix}$

$\begin{matrix} \\ -2R_1 + R_2 \to \\ R_1 + R_3 \to \end{matrix} \begin{bmatrix} 1 & -2 & 1 & \vdots & 7 \\ 0 & 5 & -4 & \vdots & -18 \\ 0 & 1 & 3 & \vdots & 4 \end{bmatrix}$

$R_2 + (-5)R_3 \to \begin{bmatrix} 1 & -2 & 1 & \vdots & 7 \\ 0 & 0 & -19 & \vdots & -38 \\ 0 & 1 & 3 & \vdots & 4 \end{bmatrix}$

$-19z = -38$

$z = 2$

$y + 3(2) = 4 \Rightarrow y = -2$

$x - 2(-2) + 2 = 7 \Rightarrow x = 1$

Solution: $(1, -2, 2)$

23. $\begin{bmatrix} 2 & 1 & 2 & \vdots & 4 \\ 2 & 2 & 0 & \vdots & 5 \\ 2 & -1 & 6 & \vdots & 2 \end{bmatrix}$

$\begin{matrix} \\ -R_1 + R_2 \to \\ -R_1 + R_3 \to \end{matrix} \begin{bmatrix} 2 & 1 & 2 & \vdots & 4 \\ 0 & 1 & -2 & \vdots & 1 \\ 0 & -2 & 4 & \vdots & -2 \end{bmatrix}$

$\begin{matrix} -R_2 + R_1 \to \\ \\ 2R_2 + R_3 \to \end{matrix} \begin{bmatrix} 2 & 0 & 4 & \vdots & 3 \\ 0 & 1 & -2 & \vdots & 1 \\ 0 & 0 & 0 & \vdots & 0 \end{bmatrix}$

$\frac{1}{2}R_1 \to \begin{bmatrix} 1 & 0 & 2 & \vdots & \frac{3}{2} \\ 0 & 1 & -2 & \vdots & 1 \\ 0 & 0 & 0 & \vdots & 0 \end{bmatrix}$

Let $z = a$, then:

$y - 2a = 1 \Rightarrow y = 2a + 1$

$x + 2a = \frac{3}{2} \Rightarrow x = -2a + \frac{3}{2}$

Solution: $\left(-2a + \frac{3}{2}, 2a + 1, a\right)$ where a is any real number

25. $\begin{bmatrix} 2 & 3 & 1 & \vdots & 10 \\ 2 & -3 & -3 & \vdots & 22 \\ 4 & -2 & 3 & \vdots & -2 \end{bmatrix}$

$$\begin{matrix} \\ -R_1 + R_2 \to \\ -2R_1 + R_3 \to \end{matrix} \begin{bmatrix} 2 & 3 & 1 & \vdots & 10 \\ 0 & -6 & -4 & \vdots & 12 \\ 0 & -8 & 1 & \vdots & -22 \end{bmatrix}$$

$$\begin{matrix} \frac{1}{2}R_1 \to \\ -\frac{1}{6}R_2 \to \\ \\ \end{matrix} \begin{bmatrix} 1 & \frac{3}{2} & \frac{1}{2} & \vdots & 5 \\ 0 & 1 & \frac{2}{3} & \vdots & -2 \\ 0 & -8 & 1 & \vdots & -22 \end{bmatrix}$$

$$\begin{matrix} \\ \\ 8R_2 + R_3 \to \end{matrix} \begin{bmatrix} 1 & \frac{3}{2} & \frac{1}{2} & \vdots & 5 \\ 0 & 1 & \frac{2}{3} & \vdots & -2 \\ 0 & 0 & \frac{19}{3} & \vdots & -38 \end{bmatrix}$$

$$\begin{matrix} \\ \\ \frac{3}{19}R_3 \to \end{matrix} \begin{bmatrix} 1 & \frac{3}{2} & \frac{1}{2} & \vdots & 5 \\ 0 & 1 & \frac{2}{3} & \vdots & -2 \\ 0 & 0 & 1 & \vdots & -6 \end{bmatrix}$$

$z = -6$

$y + \frac{2}{3}(-6) = -2 \Rightarrow y = 2$

$x + \frac{3}{2}(2) + \frac{1}{2}(-6) = 5 \Rightarrow x = 5$

Solution: $(5, 2, -6)$

27. $\begin{cases} x + 2y - z = 3 \\ x - y - z = -3 \\ 2x + y + 3z = 10 \end{cases}$

$$\begin{bmatrix} 1 & 2 & -1 & \vdots & 3 \\ 1 & -1 & -1 & \vdots & -3 \\ 2 & 1 & 3 & \vdots & 10 \end{bmatrix}$$

$$\begin{matrix} \\ -R_1 + R_2 \to \\ -2R_2 + R_3 \to \end{matrix} \begin{bmatrix} 1 & 2 & -1 & \vdots & 3 \\ 0 & -3 & 0 & \vdots & -6 \\ 0 & 3 & 5 & \vdots & 16 \end{bmatrix}$$

$$\begin{matrix} \\ \\ R_2 + R_3 \to \end{matrix} \begin{bmatrix} 1 & 2 & -1 & \vdots & 3 \\ 0 & -3 & 0 & \vdots & -6 \\ 0 & 0 & 5 & \vdots & 10 \end{bmatrix}$$

$$\begin{matrix} 3R_1 + 2R_2 \to \\ \\ \\ \end{matrix} \begin{bmatrix} 3 & 0 & -3 & \vdots & -3 \\ 0 & -3 & 0 & \vdots & -6 \\ 0 & 0 & 5 & \vdots & 10 \end{bmatrix}$$

$$\begin{matrix} 5R_1 + 3R_3 \to \\ \\ \\ \end{matrix} \begin{bmatrix} 15 & 0 & 0 & \vdots & 15 \\ 0 & -3 & 0 & \vdots & -6 \\ 0 & 0 & 5 & \vdots & 10 \end{bmatrix}$$

$$\begin{matrix} \frac{1}{15}R_1 \to \\ \frac{1}{3}R_2 \to \\ \frac{1}{5}R_3 \to \end{matrix} \begin{bmatrix} 1 & 0 & 0 & \vdots & 1 \\ 0 & 1 & 0 & \vdots & 2 \\ 0 & 0 & 1 & \vdots & 2 \end{bmatrix}$$

$x = 1$

$y = 2$

$z = 2$

Solution: $(1, 2, 2)$

29. $\begin{bmatrix} -1 & 1 & 2 & \vdots & 1 \\ 2 & 3 & 1 & \vdots & -2 \\ 5 & 4 & 2 & \vdots & 4 \end{bmatrix}$

$$\begin{matrix} -R_1 \to \\ \\ \\ \end{matrix} \begin{bmatrix} 1 & -1 & -2 & \vdots & -1 \\ 2 & 3 & 1 & \vdots & -2 \\ 5 & 4 & 2 & \vdots & 4 \end{bmatrix}$$

$$\begin{matrix} \\ -2R_1 + R_2 \to \\ -5R_1 + R_3 \to \end{matrix} \begin{bmatrix} 1 & -1 & -2 & \vdots & -1 \\ 0 & 5 & 5 & \vdots & 0 \\ 0 & 9 & 12 & \vdots & 9 \end{bmatrix}$$

$$\begin{matrix} \\ \frac{1}{5}R_2 \to \\ \\ \end{matrix} \begin{bmatrix} 1 & -1 & -2 & \vdots & -1 \\ 0 & 1 & 1 & \vdots & 0 \\ 0 & 9 & 12 & \vdots & 9 \end{bmatrix}$$

$$\begin{matrix} R_2 + R_1 \to \\ \\ -9R_2 + R_3 \to \end{matrix} \begin{bmatrix} 1 & 0 & -1 & \vdots & -1 \\ 0 & 1 & 1 & \vdots & 0 \\ 0 & 0 & 3 & \vdots & 9 \end{bmatrix}$$

$$\begin{matrix} \\ \\ \frac{1}{3}R_3 \to \end{matrix} \begin{bmatrix} 1 & 0 & -1 & \vdots & -1 \\ 0 & 1 & 1 & \vdots & 0 \\ 0 & 0 & 1 & \vdots & 3 \end{bmatrix}$$

$$\begin{matrix} R_3 + R_1 \to \\ -R_3 + R_2 \to \\ \\ \end{matrix} \begin{bmatrix} 1 & 0 & 0 & \vdots & 2 \\ 0 & 1 & 0 & \vdots & -3 \\ 0 & 0 & 1 & \vdots & 3 \end{bmatrix}$$

$x = 2, y = -3, z = 3$

Solution: $(2, -3, 3)$

31. Use the reduced row-echelon form feature of a graphing utility.

$$\begin{bmatrix} 3 & -1 & 5 & -2 & \vdots & -44 \\ 1 & 6 & 4 & -1 & \vdots & 1 \\ 5 & -1 & 1 & 3 & \vdots & -15 \\ 0 & 4 & -1 & -8 & \vdots & 58 \end{bmatrix} \Rightarrow \begin{bmatrix} 1 & 0 & 0 & 0 & \vdots & 2 \\ 0 & 1 & 0 & 0 & \vdots & 6 \\ 0 & 0 & 1 & 0 & \vdots & -10 \\ 0 & 0 & 0 & 1 & \vdots & -3 \end{bmatrix}$$

$x = 2, y = 6, z = -10, w = -3$

Solution: $(2, 6, -10, -3)$

33. $\begin{bmatrix} -1 & x \\ y & 9 \end{bmatrix} = \begin{bmatrix} -1 & 12 \\ 11 & 9 \end{bmatrix} \Rightarrow x = 12$ and $y = 11$

35. $\begin{bmatrix} x + 3 & -4 & 44 \\ 0 & -3 & 2 \\ -2 & y + 5 & 6 \end{bmatrix} = \begin{bmatrix} 5x - 1 & -4 & 44 \\ 0 & -3 & 2 \\ -2 & 16 & 6 \end{bmatrix}$

$$\left.\begin{aligned} x + 3 &= 5x - 1 \\ 4 &= 4x \\ y + 5 &= 16 \\ y &= 11 \end{aligned}\right\} x = 1 \text{ and } y = 11$$

37. (a) $A + B = \begin{bmatrix} 2 & -2 \\ 3 & 5 \end{bmatrix} + \begin{bmatrix} -3 & 10 \\ 12 & 8 \end{bmatrix} = \begin{bmatrix} -1 & 8 \\ 15 & 13 \end{bmatrix}$

(b) $A - B = \begin{bmatrix} 2 & -2 \\ 3 & 5 \end{bmatrix} - \begin{bmatrix} -3 & 10 \\ 12 & 8 \end{bmatrix} = \begin{bmatrix} 5 & -12 \\ -9 & -3 \end{bmatrix}$

(c) $4A = 4\begin{bmatrix} 2 & -2 \\ 3 & 5 \end{bmatrix} = \begin{bmatrix} 8 & -8 \\ 12 & 20 \end{bmatrix}$

(d) $$\begin{aligned} 2A + 2B &= 2\begin{bmatrix} 2 & -2 \\ 3 & 5 \end{bmatrix} + 2\begin{bmatrix} -3 & 10 \\ 12 & 8 \end{bmatrix} \\ &= \begin{bmatrix} 4 & -4 \\ 6 & 10 \end{bmatrix} + \begin{bmatrix} -6 & 20 \\ 24 & 16 \end{bmatrix} \\ &= \begin{bmatrix} -2 & 16 \\ 30 & 26 \end{bmatrix} \end{aligned}$$

39. (a) $A + B = \begin{bmatrix} 5 & 4 \\ -7 & 2 \\ 11 & 2 \end{bmatrix} + \begin{bmatrix} 0 & 3 \\ 4 & 12 \\ 20 & 40 \end{bmatrix} = \begin{bmatrix} 5 & 7 \\ -3 & 14 \\ 31 & 42 \end{bmatrix}$

(b) $A - B = \begin{bmatrix} 5 & 4 \\ -7 & 2 \\ 11 & 2 \end{bmatrix} - \begin{bmatrix} 0 & 3 \\ 4 & 12 \\ 20 & 40 \end{bmatrix} = \begin{bmatrix} 5 & 1 \\ -11 & -10 \\ -9 & -38 \end{bmatrix}$

(c) $4A = 4\begin{bmatrix} 5 & 4 \\ -7 & 2 \\ 11 & 2 \end{bmatrix} = \begin{bmatrix} 20 & 16 \\ -28 & 8 \\ 44 & 8 \end{bmatrix}$

(d) $$\begin{aligned} 2A + 2B &= 2\begin{bmatrix} 5 & 4 \\ -7 & 2 \\ 11 & 2 \end{bmatrix} + 2\begin{bmatrix} 0 & 3 \\ 4 & 12 \\ 20 & 40 \end{bmatrix} \\ &= \begin{bmatrix} 10 & 8 \\ -14 & 4 \\ 22 & 4 \end{bmatrix} + \begin{bmatrix} 0 & 6 \\ 8 & 24 \\ 40 & 80 \end{bmatrix} \\ &= \begin{bmatrix} 10 & 14 \\ -6 & 28 \\ 62 & 84 \end{bmatrix} \end{aligned}$$

41. $\begin{bmatrix} 7 & 3 \\ -1 & 5 \end{bmatrix} + \begin{bmatrix} 10 & -20 \\ 14 & -3 \end{bmatrix} + \begin{bmatrix} 5 & 0 \\ 1 & 9 \end{bmatrix} = \begin{bmatrix} 7 + 10 + 5 & 3 - 20 + 0 \\ -1 + 14 + 1 & 5 - 3 + 9 \end{bmatrix} = \begin{bmatrix} 22 & -17 \\ 14 & 11 \end{bmatrix}$

43. $-2\left(\begin{bmatrix} 1 & 2 \\ 5 & -4 \\ 6 & 0 \end{bmatrix} + \begin{bmatrix} 7 & 1 \\ 1 & 2 \\ 1 & 4 \end{bmatrix}\right) = -2\begin{bmatrix} 8 & 3 \\ 6 & -2 \\ 7 & 4 \end{bmatrix} = \begin{bmatrix} -16 & -6 \\ -12 & 4 \\ -14 & -8 \end{bmatrix}$

45. $X = 2A - 3B = 2\begin{bmatrix} -4 & 0 \\ 1 & -5 \\ -3 & 2 \end{bmatrix} - 3\begin{bmatrix} 1 & 2 \\ -2 & 1 \\ 4 & 4 \end{bmatrix} = \begin{bmatrix} -8 & 0 \\ 2 & -10 \\ -6 & 4 \end{bmatrix} + \begin{bmatrix} -3 & -6 \\ 6 & -3 \\ -12 & -12 \end{bmatrix} = \begin{bmatrix} -11 & -6 \\ 8 & -13 \\ -18 & -8 \end{bmatrix}$

47. $X = \frac{1}{3}[B - 2A] = \frac{1}{3}\left(\begin{bmatrix} 1 & 2 \\ -2 & 1 \\ 4 & 4 \end{bmatrix} - 2\begin{bmatrix} -4 & 0 \\ 1 & -5 \\ -3 & 2 \end{bmatrix}\right) = \frac{1}{3}\begin{bmatrix} 9 & 2 \\ -4 & 11 \\ 10 & 0 \end{bmatrix} = \begin{bmatrix} 3 & \frac{2}{3} \\ -\frac{4}{3} & \frac{11}{3} \\ \frac{10}{3} & 0 \end{bmatrix}$

49. A and B are both 2×2, so AB exists and has dimensions 2×2.

$$AB = \begin{bmatrix} 2 & -2 \\ 3 & 5 \end{bmatrix}\begin{bmatrix} -3 & 10 \\ 12 & 8 \end{bmatrix} = \begin{bmatrix} 2(-3) + (-2)(12) & 2(10) + (-2)(8) \\ 3(-3) + 5(12) & 3(10) + 5(8) \end{bmatrix} = \begin{bmatrix} -30 & 4 \\ 51 & 70 \end{bmatrix}$$

51. Because A is 3×2 and B is 2×2, AB exists and has dimensions 3×2.

$$AB = \begin{bmatrix} 5 & 4 \\ -7 & 2 \\ 11 & 2 \end{bmatrix}\begin{bmatrix} 4 & 12 \\ 20 & 40 \end{bmatrix} = \begin{bmatrix} 5(4) + 4(20) & 5(12) + 4(40) \\ -7(4) + 2(20) & -7(12) + 2(40) \\ 11(4) + 2(20) & 11(12) + 2(40) \end{bmatrix} = \begin{bmatrix} 100 & 220 \\ 12 & -4 \\ 84 & 212 \end{bmatrix}$$

53. $\begin{bmatrix} 4 & 1 \\ 11 & -7 \\ 12 & 3 \end{bmatrix}\begin{bmatrix} 3 & -5 & 6 \\ 2 & -2 & -2 \end{bmatrix} = \begin{bmatrix} 14 & -22 & 22 \\ 19 & -41 & 80 \\ 42 & -66 & 66 \end{bmatrix}$

55. Not possible. The number of columns of the first matrix does not equal the number of rows of the second matrix.

57. (a) $AB = \begin{bmatrix} 1 & 3 \\ 4 & 1 \end{bmatrix}\begin{bmatrix} 5 & -1 \\ -2 & 0 \end{bmatrix} = \begin{bmatrix} (1)(5) + (3)(-2) & (1)(-1) + (3)(0) \\ (4)(5) + (1)(-2) & (4)(-1) + (1)(0) \end{bmatrix} = \begin{bmatrix} -1 & -1 \\ 18 & -4 \end{bmatrix}$

(b) $BA = \begin{bmatrix} 5 & -1 \\ -2 & 0 \end{bmatrix}\begin{bmatrix} 1 & 3 \\ 4 & 1 \end{bmatrix} = \begin{bmatrix} (5)(1) + (-1)(4) & (5)(3) + (-1)(1) \\ (-2)(1) + (0)(4) & (-2)(3) + (0)(1) \end{bmatrix} = \begin{bmatrix} 1 & 14 \\ -2 & -6 \end{bmatrix}$

(c) $A^2 = \begin{bmatrix} 1 & 3 \\ 4 & 1 \end{bmatrix}\begin{bmatrix} 1 & 3 \\ 4 & 1 \end{bmatrix} = \begin{bmatrix} (1)(1) + (3)(4) & (1)(3) + (3)(1) \\ (4)(1) + (1)(4) & (4)(3) + (1)(1) \end{bmatrix} = \begin{bmatrix} 13 & 6 \\ 8 & 13 \end{bmatrix}$

In Exercises 59-61, $\mathbf{v} = \langle 2, 5 \rangle = \begin{bmatrix} 2 \\ 5 \end{bmatrix}$

59. $A\mathbf{v} = \begin{bmatrix} 1 & 0 \\ 0 & -1 \end{bmatrix}\begin{bmatrix} 2 \\ 5 \end{bmatrix} = \begin{bmatrix} 2 \\ -5 \end{bmatrix} = \langle 2, -5 \rangle$ is a reflection in the x-axis.

61. $A\mathbf{v} = \begin{bmatrix} \frac{1}{2} & 0 \\ 0 & 1 \end{bmatrix}\begin{bmatrix} 2 \\ 5 \end{bmatrix} = \begin{bmatrix} 1 \\ 5 \end{bmatrix} = \langle 1, 5 \rangle$ is a horizontal shrink.

63. $0.95A = 0.95\begin{bmatrix} 80 & 120 & 140 \\ 40 & 100 & 80 \end{bmatrix} = \begin{bmatrix} 76 & 114 & 133 \\ 38 & 95 & 76 \end{bmatrix}$

65. $AB = \begin{bmatrix} -4 & -1 \\ 7 & 2 \end{bmatrix}\begin{bmatrix} -2 & -1 \\ 7 & 4 \end{bmatrix} = \begin{bmatrix} -4(-2) + (-1)(7) & -4(-1) + (-1)(4) \\ 7(-2) + 2(7) & 7(-1) + 2(4) \end{bmatrix} = \begin{bmatrix} 1 & 0 \\ 0 & 1 \end{bmatrix} = I$

$BA = \begin{bmatrix} -2 & -1 \\ 7 & 4 \end{bmatrix}\begin{bmatrix} -4 & -1 \\ 7 & 2 \end{bmatrix} = \begin{bmatrix} -2(-4) + (-1)(7) & -2(-1) + (-1)(2) \\ 7(-4) + 4(7) & 7(-1) + 4(2) \end{bmatrix} = \begin{bmatrix} 1 & 0 \\ 0 & 1 \end{bmatrix} = I$

67. $AB = \begin{bmatrix} 1 & 1 & 0 \\ 1 & 0 & 1 \\ 6 & 2 & 3 \end{bmatrix}\begin{bmatrix} -2 & -3 & 1 \\ 3 & 3 & -1 \\ 2 & 4 & -1 \end{bmatrix} = \begin{bmatrix} 1(-2) + 1(3) + 0(2) & 1(-3) + 1(3) + 0(4) & 1(1) + 1(-1) + 0(-1) \\ 1(-2) + 0(3) + 1(2) & 1(-3) + 0(3) + 1(4) & 1(1) + 0(-1) + 1(-1) \\ 6(-2) + 2(3) + 3(2) & 6(-3) + 2(3) + 3(4) & 6(1) + 2(-1) + 3(-1) \end{bmatrix} = \begin{bmatrix} 1 & 0 & 0 \\ 0 & 1 & 0 \\ 0 & 0 & 1 \end{bmatrix} = I$

$BA = \begin{bmatrix} -2 & -3 & 1 \\ 3 & 3 & -1 \\ 2 & 4 & -1 \end{bmatrix}\begin{bmatrix} 1 & 1 & 0 \\ 1 & 0 & 1 \\ 6 & 2 & 3 \end{bmatrix} = \begin{bmatrix} -2(1) + (-3)(1) + 1(6) & -2(1) + (-3)(0) + 1(2) & -2(0) + (-3)(1) + 1(3) \\ 3(1) + 3(1) + (-1)(6) & 3(1) + 3(0) + (-1)(2) & 3(0) + 3(1) + (-1)(3) \\ 2(1) + 4(1) + (-1)(6) & 2(1) + 4(0) + (-1)(2) & 2(0) + 4(1) + (-1)(3) \end{bmatrix}$

$= \begin{bmatrix} 1 & 0 & 0 \\ 0 & 1 & 0 \\ 0 & 0 & 1 \end{bmatrix} = I$

69. $[A \;\vdots\; I] = \left[\begin{array}{cc|cc} -6 & 5 & 1 & 0 \\ -5 & 4 & 0 & 1 \end{array}\right]$

$-\frac{1}{6}R_1 \rightarrow \left[\begin{array}{cc|cc} 1 & -\frac{5}{6} & -\frac{1}{6} & 0 \\ -5 & 4 & 0 & 1 \end{array}\right]$

$5R_1 + R_2 \rightarrow \left[\begin{array}{cc|cc} 1 & -\frac{5}{6} & -\frac{1}{6} & 0 \\ 0 & -\frac{1}{6} & -\frac{5}{6} & 1 \end{array}\right]$

$-6R_2 \rightarrow \left[\begin{array}{cc|cc} 1 & -\frac{5}{6} & -\frac{1}{6} & 0 \\ 0 & 1 & 5 & -6 \end{array}\right]$

$\frac{5}{6}R_2 + R_1 \rightarrow \left[\begin{array}{cc|cc} 1 & 0 & 4 & -5 \\ 0 & 1 & 5 & -6 \end{array}\right] = [I \;\vdots\; A^{-1}]$

$A^{-1} = \begin{bmatrix} 4 & -5 \\ 5 & -6 \end{bmatrix}$

71. $[A \vdots I] = \begin{bmatrix} 2 & 0 & 3 & \vdots & 1 & 0 & 0 \\ -1 & 1 & 1 & \vdots & 0 & 1 & 0 \\ 2 & -2 & 1 & \vdots & 0 & 0 & 1 \end{bmatrix}$

$$2R_2 + R_3 \rightarrow \begin{bmatrix} 2 & 0 & 3 & \vdots & 1 & 0 & 0 \\ -1 & 1 & 1 & \vdots & 0 & 1 & 0 \\ 0 & 0 & 3 & \vdots & 0 & 2 & 1 \end{bmatrix}$$

$$-R_3 + R_1 \rightarrow \begin{bmatrix} 2 & 0 & 0 & \vdots & 1 & -2 & -1 \\ -1 & 1 & 1 & \vdots & 0 & 1 & 0 \\ 0 & 0 & 3 & \vdots & 0 & 2 & 1 \end{bmatrix}$$

$$\begin{matrix} \frac{1}{2}R_1 \rightarrow \\ \\ \frac{1}{3}R_3 \rightarrow \end{matrix} \begin{bmatrix} 1 & 0 & 0 & \vdots & \frac{1}{2} & -1 & -\frac{1}{2} \\ -1 & 1 & 1 & \vdots & 0 & 1 & 0 \\ 0 & 0 & 1 & \vdots & 0 & \frac{2}{3} & \frac{1}{3} \end{bmatrix}$$

$$R_1 + R_2 \rightarrow \begin{bmatrix} 1 & 0 & 0 & \vdots & \frac{1}{2} & -1 & -\frac{1}{2} \\ 0 & 1 & 1 & \vdots & \frac{1}{2} & 0 & -\frac{1}{2} \\ 0 & 0 & 1 & \vdots & 0 & \frac{2}{3} & \frac{1}{3} \end{bmatrix}$$

$$-R_3 + R_2 \rightarrow \begin{bmatrix} 1 & 0 & 0 & \vdots & \frac{1}{2} & -1 & -\frac{1}{2} \\ 0 & 1 & 0 & \vdots & \frac{1}{2} & -\frac{2}{3} & -\frac{5}{6} \\ 0 & 0 & 1 & \vdots & 0 & \frac{2}{3} & \frac{1}{3} \end{bmatrix} = [I \vdots A^{-1}]$$

$$A^{-1} = \begin{bmatrix} \frac{1}{2} & -1 & -\frac{1}{2} \\ \frac{1}{2} & -\frac{2}{3} & -\frac{5}{6} \\ 0 & \frac{2}{3} & \frac{1}{3} \end{bmatrix}$$

73. $\begin{bmatrix} -1 & -2 & -2 \\ 3 & 7 & 9 \\ 1 & 4 & 7 \end{bmatrix}^{-1} = \begin{bmatrix} 13 & 6 & -4 \\ -12 & -5 & 3 \\ 5 & 2 & -1 \end{bmatrix}$

75. $A = \begin{bmatrix} -7 & 2 \\ -8 & 2 \end{bmatrix}$

$$A^{-1} = \frac{1}{-7(2) - 2(-8)}\begin{bmatrix} 2 & -2 \\ 8 & -7 \end{bmatrix} = \frac{1}{2}\begin{bmatrix} 2 & -2 \\ 8 & -7 \end{bmatrix} = \begin{bmatrix} 1 & -1 \\ 4 & -\frac{7}{2} \end{bmatrix}$$

77. $A = \begin{bmatrix} -12 & 6 \\ 10 & -5 \end{bmatrix}$

$ad - bc = (-12)(-5) - (6)(10) = 0$

A^{-1} does not exist.

79. $\begin{cases} -x + 4y = 8 \\ 2x - 7y = -5 \end{cases}$

$$\begin{bmatrix} x \\ y \end{bmatrix} = \begin{bmatrix} -1 & 4 \\ 2 & -7 \end{bmatrix}^{-1}\begin{bmatrix} 8 \\ -5 \end{bmatrix} = \begin{bmatrix} 7 & 4 \\ 2 & 1 \end{bmatrix}\begin{bmatrix} 8 \\ -5 \end{bmatrix} = \begin{bmatrix} 7(8) + 4(-5) \\ 2(8) + 1(-5) \end{bmatrix} = \begin{bmatrix} 36 \\ 11 \end{bmatrix}$$

Solution: $(36, 11)$

81. $\begin{cases} -3x + 10y = 8 \\ 5x - 17y = -13 \end{cases}$

$$\begin{bmatrix} x \\ y \end{bmatrix} = \begin{bmatrix} -3 & 10 \\ 5 & -17 \end{bmatrix}^{-1}\begin{bmatrix} 8 \\ -13 \end{bmatrix} = \begin{bmatrix} -17 & -10 \\ -5 & -3 \end{bmatrix}\begin{bmatrix} 8 \\ -13 \end{bmatrix} = \begin{bmatrix} -17(8) + (-10)(-13) \\ -5(8) + (-3)(-13) \end{bmatrix} = \begin{bmatrix} -6 \\ -1 \end{bmatrix}$$

Solution: $(-6, -1)$

83. $\begin{cases} \frac{1}{2}x + \frac{1}{3}y = 2 \\ -3x + 2y = 0 \end{cases}$

$$\begin{bmatrix} x \\ y \end{bmatrix} = \begin{bmatrix} \frac{1}{2} & \frac{1}{3} \\ -3 & 2 \end{bmatrix}^{-1}\begin{bmatrix} 2 \\ 0 \end{bmatrix} = \begin{bmatrix} 1 & -\frac{1}{6} \\ \frac{3}{2} & \frac{1}{4} \end{bmatrix}\begin{bmatrix} 2 \\ 0 \end{bmatrix} = \begin{bmatrix} 2 \\ 3 \end{bmatrix}$$

Solution: $(2, 3)$

85. $\begin{cases} 0.3x + 0.7y = 10.2 \\ 0.4x + 0.6y = 7.6 \end{cases}$

$$\begin{bmatrix} x \\ y \end{bmatrix} = \begin{bmatrix} 0.3 & 0.7 \\ 0.4 & 0.6 \end{bmatrix}^{-1}\begin{bmatrix} 10.2 \\ 7.6 \end{bmatrix} = \begin{bmatrix} -6 & 7 \\ 4 & -3 \end{bmatrix}\begin{bmatrix} 10.2 \\ 7.6 \end{bmatrix} = \begin{bmatrix} -8 \\ 18 \end{bmatrix}$$

Solution: $(-8, 18)$

87. $\begin{cases} 3x + 2y - z = 6 \\ x - y + 2z = -1 \\ 5x + y + z = 7 \end{cases}$

$$\begin{bmatrix} x \\ y \\ z \end{bmatrix} = \begin{bmatrix} 3 & 2 & -1 \\ 1 & -1 & 2 \\ 5 & 1 & 1 \end{bmatrix}^{-1}\begin{bmatrix} 6 \\ -1 \\ 7 \end{bmatrix} = \begin{bmatrix} -1 & -1 & 1 \\ 3 & \frac{8}{3} & -\frac{7}{3} \\ 2 & \frac{7}{3} & -\frac{5}{3} \end{bmatrix}\begin{bmatrix} 6 \\ -1 \\ 7 \end{bmatrix}$$

$$= \begin{bmatrix} -1(6) - 1(-1) + 1(7) \\ 3(6) + \frac{8}{3}(-1) - \frac{7}{3}(7) \\ 2(6) + \frac{7}{3}(-1) - \frac{5}{3}(7) \end{bmatrix} = \begin{bmatrix} 2 \\ -1 \\ -2 \end{bmatrix}$$

Solution: $(2, -1, -2)$

89. $\begin{cases} x + 2y = -1 \\ 3x + 4y = -5 \end{cases}$

$$\begin{bmatrix} x \\ y \end{bmatrix} = \begin{bmatrix} 1 & 2 \\ 3 & 4 \end{bmatrix}^{-1}\begin{bmatrix} -1 \\ -5 \end{bmatrix} = \begin{bmatrix} -2 & 1 \\ \frac{3}{2} & -\frac{1}{2} \end{bmatrix}\begin{bmatrix} -1 \\ -5 \end{bmatrix} = \begin{bmatrix} -3 \\ 1 \end{bmatrix}$$

Solution: $(-3, 1)$

91. $\begin{cases} \frac{6}{5}x - \frac{4}{7}y = \frac{6}{5} \\ -\frac{12}{5}x + \frac{12}{7}y = -\frac{17}{5} \end{cases}$

$$\begin{bmatrix} x \\ y \end{bmatrix} = \begin{bmatrix} \frac{6}{5} & -\frac{4}{7} \\ -\frac{12}{5} & \frac{12}{7} \end{bmatrix}^{-1}\begin{bmatrix} \frac{6}{5} \\ -\frac{17}{5} \end{bmatrix} = \begin{bmatrix} \frac{5}{2} & \frac{5}{6} \\ \frac{7}{2} & \frac{7}{4} \end{bmatrix}\begin{bmatrix} \frac{6}{5} \\ -\frac{17}{5} \end{bmatrix} = \begin{bmatrix} \frac{1}{6} \\ -\frac{7}{4} \end{bmatrix}$$

Solution: $\left(\frac{1}{6}, -\frac{7}{4}\right)$

93. $A = \begin{bmatrix} 2 & 5 \\ -4 & 3 \end{bmatrix} : \begin{vmatrix} 2 & 5 \\ -4 & 3 \end{vmatrix} = (2)(3) - (-4)(5) = 26$

95. $A = \begin{bmatrix} 10 & -2 \\ 18 & 8 \end{bmatrix} : \begin{vmatrix} 10 & -2 \\ 18 & 8 \end{vmatrix} = (10)(8) - (18)(-2) = 116$

97. $\begin{bmatrix} 2 & -1 \\ 7 & 4 \end{bmatrix}$

(a) $M_{11} = 4$
$M_{12} = 7$
$M_{21} = -1$
$M_{22} = 2$

(b) $C_{11} = M_{11} = 4$
$C_{12} = -M_{12} = -7$
$C_{21} = -M_{21} = 1$
$C_{22} = M_{22} = 2$

99. $\begin{bmatrix} 3 & 2 & -1 \\ -2 & 5 & 0 \\ 1 & 8 & 6 \end{bmatrix}$

(a) $M_{11} = \begin{vmatrix} 5 & 0 \\ 8 & 6 \end{vmatrix} = 30$

$M_{12} = \begin{vmatrix} -2 & 0 \\ 1 & 6 \end{vmatrix} = -12$

$M_{13} = \begin{vmatrix} -2 & 5 \\ 1 & 8 \end{vmatrix} = -21$

$M_{21} = \begin{vmatrix} 2 & -1 \\ 8 & 6 \end{vmatrix} = 20$

$M_{22} = \begin{vmatrix} 3 & -1 \\ 1 & 6 \end{vmatrix} = 19$

$M_{23} = \begin{vmatrix} 3 & 2 \\ 1 & 8 \end{vmatrix} = 22$

$M_{31} = \begin{vmatrix} 2 & -1 \\ 5 & 0 \end{vmatrix} = 5$

$M_{32} = \begin{vmatrix} 3 & -1 \\ -2 & 0 \end{vmatrix} = -2$

$M_{33} = \begin{vmatrix} 3 & 2 \\ -2 & 5 \end{vmatrix} = 19$

(b) $C_{11} = M_{11} = 30$
$C_{12} = -M_{12} = 12$
$C_{13} = M_{13} = -21$
$C_{21} = -M_{21} = -20$
$C_{22} = M_{22} = 19$
$C_{23} = -M_{23} = -22$
$C_{31} = M_{31} = 5$
$C_{32} = -M_{32} = 2$
$C_{33} = M_{33} = 19$

101. Expand using Row 1.

$$\begin{vmatrix} -2 & 0 & 0 \\ 2 & -1 & 0 \\ -1 & 1 & -3 \end{vmatrix} = -2\begin{vmatrix} -1 & 0 \\ 1 & -3 \end{vmatrix} - 0\begin{vmatrix} 2 & 0 \\ -1 & 3 \end{vmatrix} + 0\begin{vmatrix} 2 & -1 \\ -1 & 1 \end{vmatrix}$$
$$= -2(3) - 0(6) + 0(1)$$
$$= -6$$

103. Expand using Row 3.

$$\begin{vmatrix} 4 & 1 & -1 \\ 2 & 3 & 2 \\ 1 & -1 & 0 \end{vmatrix} = 1\begin{vmatrix} 1 & -1 \\ 3 & 2 \end{vmatrix} + 1\begin{vmatrix} 4 & -1 \\ 2 & 2 \end{vmatrix} + 0\begin{vmatrix} 4 & 1 \\ 2 & 3 \end{vmatrix}$$
$$= 1(5) + 1(10) + 0(10)$$
$$= 15$$

105. Expand using Column 2.

$$\begin{vmatrix} -2 & 4 & 1 \\ -6 & 0 & 2 \\ 5 & 3 & 4 \end{vmatrix} = -4\begin{vmatrix} -6 & 2 \\ 5 & 4 \end{vmatrix} - 3\begin{vmatrix} -2 & 1 \\ -6 & 2 \end{vmatrix}$$
$$= -4(-34) - 3(2)$$
$$= 130$$

107. $\begin{cases} 5x - 2y = 6 \\ -11x + 3y = -23 \end{cases}$

$$x = \frac{\begin{vmatrix} 6 & -2 \\ -23 & 3 \end{vmatrix}}{\begin{vmatrix} 5 & -2 \\ -11 & 3 \end{vmatrix}} = \frac{-28}{-7} = 4$$

$$y = \frac{\begin{vmatrix} 5 & 6 \\ -11 & -23 \end{vmatrix}}{\begin{vmatrix} 5 & -2 \\ -11 & 3 \end{vmatrix}} = \frac{-49}{-7} = 7$$

Solution: $(4, 7)$

109. $\begin{cases} -2x + 3y - 5z = -11 \\ 4x - y + z = -3 \\ -x - 4y + 6z = 15 \end{cases}$

$$D = \begin{vmatrix} -2 & 3 & -5 \\ 4 & -1 & 1 \\ -1 & -4 & 6 \end{vmatrix} = -2(-1)^2\begin{vmatrix} -1 & 1 \\ -4 & 6 \end{vmatrix} + 4(-1)^3\begin{vmatrix} 3 & -5 \\ -4 & 6 \end{vmatrix} - 1(-1)^4\begin{vmatrix} 3 & -5 \\ -1 & 1 \end{vmatrix} = -2(-2) - 4(-2) - (-2) = 14$$

$$x = \frac{\begin{vmatrix} -11 & 3 & -5 \\ -3 & -1 & 1 \\ 15 & -4 & 6 \end{vmatrix}}{14} = \frac{-11(-1)^2\begin{vmatrix} -1 & 1 \\ -4 & 6 \end{vmatrix} - 3(-1)^3\begin{vmatrix} 3 & -5 \\ -4 & 6 \end{vmatrix} + 15(-1)^4\begin{vmatrix} 3 & -5 \\ -1 & 1 \end{vmatrix}}{14} = \frac{-11(-2) + 3(-2) + 15(-2)}{14} = \frac{-14}{14} = -1$$

$$y = \frac{\begin{vmatrix} -2 & -11 & -5 \\ 4 & -3 & 1 \\ -1 & 15 & 6 \end{vmatrix}}{14} = \frac{-2(-1)^2\begin{vmatrix} -3 & 1 \\ 15 & 6 \end{vmatrix} + 4(-1)^3\begin{vmatrix} -11 & -5 \\ 15 & 6 \end{vmatrix} - 1(-1)^4\begin{vmatrix} -11 & -5 \\ -3 & 1 \end{vmatrix}}{14} = \frac{-2(-33) - 4(9) - 1(-26)}{14} = \frac{56}{14} = 4$$

$$z = \frac{\begin{vmatrix} -2 & 3 & -11 \\ 4 & -1 & -3 \\ -1 & -4 & 15 \end{vmatrix}}{14} = \frac{-2(-1)^2\begin{vmatrix} -1 & -3 \\ -4 & 15 \end{vmatrix} + 4(-1)^3\begin{vmatrix} 3 & -11 \\ -4 & 15 \end{vmatrix} - 1(-1)^4\begin{vmatrix} 3 & -11 \\ -1 & -3 \end{vmatrix}}{14} = \frac{-2(-27) - 4(1) - 1(-20)}{14} = \frac{70}{14} = 5$$

Solution: $(-1, 4, 5)$

111. $(1, 0), (5, 0), (5, 8)$

$$\text{Area} = \tfrac{1}{2}\begin{vmatrix} 1 & 0 & 1 \\ 5 & 0 & 1 \\ 5 & 8 & 1 \end{vmatrix} = \tfrac{1}{2}\left(1\begin{vmatrix} 0 & 1 \\ 8 & 1 \end{vmatrix} + 1\begin{vmatrix} 5 & 0 \\ 5 & 8 \end{vmatrix}\right) = \tfrac{1}{2}(-8 + 40) = \tfrac{1}{2}(32) = 16 \text{ square units}$$

113. $(-1, 7), (3, -9), (-3, 15)$

$$\begin{vmatrix} -1 & 7 & 1 \\ 3 & -9 & 1 \\ -3 & 15 & 1 \end{vmatrix} = \begin{vmatrix} 3 & -9 \\ -3 & 15 \end{vmatrix} - \begin{vmatrix} -1 & 7 \\ -3 & 15 \end{vmatrix} + \begin{vmatrix} -1 & 7 \\ 3 & -9 \end{vmatrix}$$

$$= 18 - 6 - 12 = 0$$

The points are collinear.

115. $(-4, 0), (4, 4)$

$$\begin{vmatrix} x & y & 1 \\ -4 & 0 & 1 \\ 4 & 4 & 1 \end{vmatrix} = 0$$

$$1\begin{vmatrix} -4 & 0 \\ 4 & 4 \end{vmatrix} - 1\begin{vmatrix} x & y \\ 4 & 4 \end{vmatrix} + 1\begin{vmatrix} x & y \\ -4 & 0 \end{vmatrix} = 0$$

$$-16 - (4x - 4y) + 4y = 0$$

$$-4x + 8y - 16 = 0$$

$$x - 2y + 4 = 0$$

117. $\left(-\frac{5}{2}, 3\right), \left(\frac{7}{2}, 1\right)$

$$\begin{vmatrix} x & y & 1 \\ -\frac{5}{2} & 3 & 1 \\ \frac{7}{2} & 1 & 1 \end{vmatrix} = 0$$

$$1\begin{vmatrix} -\frac{5}{2} & 3 \\ \frac{7}{2} & 1 \end{vmatrix} - 1\begin{vmatrix} x & y \\ \frac{7}{2} & 1 \end{vmatrix} + 1\begin{vmatrix} x & y \\ -\frac{5}{2} & 3 \end{vmatrix} = 0$$

$$-13 - \left(x - \tfrac{7}{2}y\right) + \left(3x + \tfrac{5}{2}y\right) = 0$$

$$2x + 6y - 13 = 0$$

119. The area of the parallelogram with vertices: $(0, 0)$, $(2, 0)$, $(1, 4)$ and $(3, 4) \Rightarrow a = 2, b = 0, c = 1$ and $d = 4$.

$$A = \left\|\begin{matrix} 2 & 0 \\ 1 & 4 \end{matrix}\right\| = |8 - 0| = 8 \text{ square units.}$$

121. $A^{-1} = \begin{bmatrix} -1 & 2 & -3 \\ 2 & 1 & 0 \\ 4 & -2 & 5 \end{bmatrix}$

$[-5 \ 11 \ -2]\begin{bmatrix} -1 & 2 & -3 \\ 2 & 1 & 0 \\ 4 & -2 & 5 \end{bmatrix} = [19 \ 5 \ 5]$ S E E

$[370 \ -265 \ 225]\begin{bmatrix} -1 & 2 & -3 \\ 2 & 1 & 0 \\ 4 & -2 & 5 \end{bmatrix} = [0 \ 25 \ 15]$ _ Y O

$[-57 \ 48 \ -33]\begin{bmatrix} -1 & 2 & -3 \\ 2 & 1 & 0 \\ 4 & -2 & 5 \end{bmatrix} = [21 \ 0 \ 6]$ U _ F

$[32 \ -15 \ 20]\begin{bmatrix} -1 & 2 & -3 \\ 2 & 1 & 0 \\ 4 & -2 & 5 \end{bmatrix} = [18 \ 9 \ 4]$ R I D

$[245 \ -171 \ 147]\begin{bmatrix} -1 & 2 & -3 \\ 2 & 1 & 0 \\ 4 & -2 & 5 \end{bmatrix} = [1 \ 25 \ 0]$ A Y _

Message: SEE YOU FRIDAY

123. False. The matrix must be square.

125. If A is a square matrix, the cofactor C_{ij} of the entry a_{ij} is $(-1)^{i+j}M_{ij}$, where M_{ij} is the determinant obtained by deleting the ith row and jth column of A. The determinant of A is the sum of the entries of any row or column of A multiplied by their respective cofactors.

Problem Solving for Chapter 8

1. $A = \begin{bmatrix} 0 & -1 \\ 1 & 0 \end{bmatrix}$ $T = \begin{bmatrix} 1 & 2 & 3 \\ 1 & 4 & 2 \end{bmatrix}$

(a) $AT = \begin{bmatrix} -1 & -4 & -2 \\ 1 & 2 & 3 \end{bmatrix}$ $AAT = \begin{bmatrix} -1 & -2 & -3 \\ -1 & -4 & -2 \end{bmatrix}$

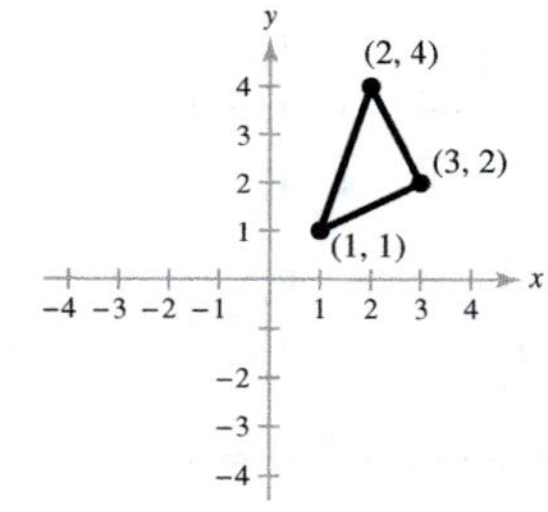

Original Triangle

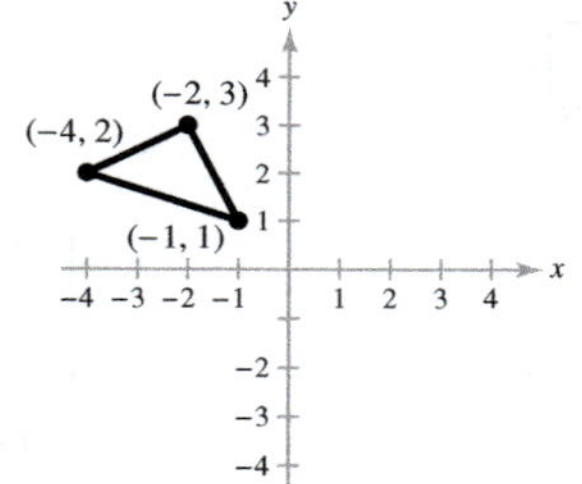

AT Triangle

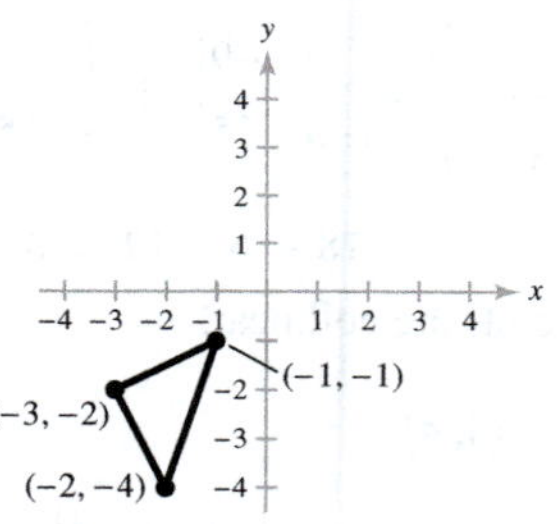

AAT Triangle

The transformation A interchanges the x and y coordinates and then takes the negative of the x coordinate. A represents a counterclockwise rotation by $90°$.

(b) AAT is rotated clockwise $90°$ to obtain AT. AT is then rotated clockwise $90°$ to obtain T.

3. (a) $A^2 = \begin{bmatrix} 1 & 0 \\ 0 & 0 \end{bmatrix}\begin{bmatrix} 1 & 0 \\ 0 & 0 \end{bmatrix} = \begin{bmatrix} 1 & 0 \\ 0 & 0 \end{bmatrix} = A$

A *is* idempotent.

(b) $A^2 = \begin{bmatrix} 0 & 1 \\ 1 & 0 \end{bmatrix}\begin{bmatrix} 0 & 1 \\ 1 & 0 \end{bmatrix} = \begin{bmatrix} 1 & 0 \\ 0 & 1 \end{bmatrix} \neq A$

A is *not* idempotent.

(c) $A^2 = \begin{bmatrix} 2 & 3 \\ -1 & -2 \end{bmatrix}\begin{bmatrix} 2 & 3 \\ -1 & -2 \end{bmatrix} = \begin{bmatrix} 1 & 0 \\ 0 & 1 \end{bmatrix} \neq A$

A is *not* idempotent.

(d) $A^2 = \begin{bmatrix} 2 & 3 \\ 1 & 2 \end{bmatrix}\begin{bmatrix} 2 & 3 \\ 1 & 2 \end{bmatrix} = \begin{bmatrix} 7 & 12 \\ 4 & 7 \end{bmatrix} \neq A$

A is *not* idempotent.

(e) $A^2 = \begin{bmatrix} 0 & 0 & 1 \\ 0 & 1 & 0 \\ 1 & 0 & 0 \end{bmatrix}\begin{bmatrix} 0 & 0 & 1 \\ 0 & 1 & 0 \\ 1 & 0 & 0 \end{bmatrix} = \begin{bmatrix} 1 & 0 & 0 \\ 0 & 1 & 0 \\ 0 & 0 & 1 \end{bmatrix} \neq A$

A is *not* idempotent.

(f) $A^2 = \begin{bmatrix} 0 & 1 & 0 \\ 1 & 0 & 0 \\ 0 & 0 & 1 \end{bmatrix}\begin{bmatrix} 0 & 1 & 0 \\ 1 & 0 & 0 \\ 0 & 0 & 1 \end{bmatrix} = \begin{bmatrix} 1 & 0 & 0 \\ 0 & 1 & 0 \\ 0 & 0 & 1 \end{bmatrix} \neq A$

A is *not* idempotent.

5. $A = \begin{bmatrix} 1 & 2 \\ -2 & 1 \end{bmatrix}$

(a)
$$\begin{aligned} A^2 - 2A + 5I &= \begin{bmatrix} 1 & 2 \\ -2 & 1 \end{bmatrix}\begin{bmatrix} 1 & 2 \\ -2 & 1 \end{bmatrix} - 2\begin{bmatrix} 1 & 2 \\ -2 & 1 \end{bmatrix} + 5\begin{bmatrix} 1 & 0 \\ 0 & 1 \end{bmatrix} \\ &= \begin{bmatrix} -3 & 4 \\ -4 & -3 \end{bmatrix} + \begin{bmatrix} -2 & -4 \\ 4 & -2 \end{bmatrix} + \begin{bmatrix} 5 & 0 \\ 0 & 5 \end{bmatrix} \\ &= \begin{bmatrix} 0 & 0 \\ 0 & 0 \end{bmatrix} = 0 \end{aligned}$$

(b)
$$A^{-1} = \frac{1}{(1) - (-4)}\begin{bmatrix} 1 & -2 \\ 2 & 1 \end{bmatrix} = \frac{1}{5}\begin{bmatrix} 1 & -2 \\ 2 & 1 \end{bmatrix}$$

$$\frac{1}{5}(2I - A) = \frac{1}{5}\left[\begin{bmatrix} 2 & 0 \\ 0 & 2 \end{bmatrix} - \begin{bmatrix} 1 & 2 \\ -2 & 1 \end{bmatrix}\right] = \frac{1}{5}\begin{bmatrix} 1 & -2 \\ 2 & 1 \end{bmatrix}$$

So, $A^{-1} = \frac{1}{5}(2I - A)$.

(c)
$$\begin{aligned} A^2 - 2A + 5I &= 0 \\ A^2 - 2A &= -5I \\ (A - 2I)A &= -5I \\ -\frac{1}{5}(A - 2I)A &= I \\ \frac{1}{5}(2I - A)A &= I \end{aligned}$$

So, $A^{-1} = \frac{1}{5}(2I - A)$.

7. $A = \begin{bmatrix} -1 & 1 & -2 \\ 2 & 0 & 1 \end{bmatrix}$, $\quad B = \begin{bmatrix} -3 & 0 \\ 1 & 2 \\ 1 & -1 \end{bmatrix}$

$A^T = \begin{bmatrix} -1 & 2 \\ 1 & 0 \\ -2 & 1 \end{bmatrix}$, $\quad B^T = \begin{bmatrix} -3 & 1 & 1 \\ 0 & 2 & -1 \end{bmatrix}$

$AB = \begin{bmatrix} 2 & 4 \\ -5 & -1 \end{bmatrix}$, $\quad (AB)^T = \begin{bmatrix} 2 & -5 \\ 4 & -1 \end{bmatrix}$

$$B^T A^T = \begin{bmatrix} -3 & 1 & 1 \\ 0 & 2 & -1 \end{bmatrix}\begin{bmatrix} -1 & 2 \\ 1 & 0 \\ -2 & 1 \end{bmatrix} = \begin{bmatrix} 2 & -5 \\ 4 & -1 \end{bmatrix}$$

So, $(AB)^T = B^T A^T$.

9. If $A = \begin{bmatrix} 4 & x \\ -2 & -3 \end{bmatrix}$ is singular then

$ad - bc = -12 + 2x = 0.$

So, $x = 6$.

11. $(a-b)(b-c)(c-a)(a+b+c) = -a^3b + a^3c + ab^3 - ac^3 - b^3c + bc^3$

$$\begin{vmatrix} 1 & 1 & 1 \\ a & b & c \\ a^3 & b^3 & c^3 \end{vmatrix} = \begin{vmatrix} b & c \\ b^3 & c^3 \end{vmatrix} - \begin{vmatrix} a & c \\ a^3 & c^3 \end{vmatrix} + \begin{vmatrix} a & b \\ a^3 & b^3 \end{vmatrix} = bc^3 - b^3c - ac^3 + a^3c + ab^3 - a^3b$$

So, $\begin{vmatrix} 1 & 1 & 1 \\ a & b & c \\ a^3 & b^3 & c^3 \end{vmatrix} = (a-b)(b-c)(c-a)(a+b+c).$

13. $$\begin{vmatrix} x & 0 & 0 & d \\ -1 & x & 0 & c \\ 0 & -1 & x & b \\ 0 & 0 & -1 & a \end{vmatrix} = x\begin{vmatrix} x & 0 & c \\ -1 & x & b \\ 0 & -1 & a \end{vmatrix} - d\begin{vmatrix} -1 & x & 0 \\ 0 & -1 & x \\ 0 & 0 & -1 \end{vmatrix} = x\underbrace{\left(ax^2 + bx + c\right)}_{\text{From Exercise 12}} - d\left(-\begin{vmatrix} -1 & x \\ 0 & -1 \end{vmatrix}\right) = ax^3 + bx^2 + cx + d$$

15. $$\begin{aligned} 4S + 4N \qquad &= 184 \\ S \qquad + 6F &= 146 \\ 2N + 4F &= 104 \end{aligned}$$

$$D = \begin{vmatrix} 4 & 4 & 0 \\ 1 & 0 & 6 \\ 0 & 2 & 4 \end{vmatrix} = -64$$

$$S = \frac{\begin{vmatrix} 184 & 4 & 0 \\ 146 & 0 & 6 \\ 104 & 2 & 4 \end{vmatrix}}{-64} = \frac{-2048}{-64} = 32$$

$$N = \frac{\begin{vmatrix} 4 & 184 & 0 \\ 1 & 146 & 6 \\ 0 & 104 & 4 \end{vmatrix}}{-64} = \frac{-896}{-64} = 14$$

$$F = \frac{\begin{vmatrix} 4 & 4 & 184 \\ 1 & 0 & 146 \\ 0 & 2 & 104 \end{vmatrix}}{-64} = \frac{-1216}{-64} = 19$$

Element	*Atomic mass*
Sulfur	32
Nitrogen	14
Fluoride	19

17. $A = \begin{bmatrix} 1 & -2 & 2 \\ 1 & 1 & -3 \\ 1 & -1 & 4 \end{bmatrix} \Rightarrow A^{-1} = \begin{bmatrix} \frac{1}{11} & \frac{6}{11} & \frac{4}{11} \\ -\frac{7}{11} & \frac{2}{11} & \frac{5}{11} \\ -\frac{2}{11} & -\frac{1}{11} & \frac{3}{11} \end{bmatrix}$

$$\begin{bmatrix} 23 & 13 & -34 \\ 31 & -34 & 63 \\ 25 & -17 & 61 \\ 24 & 14 & -37 \\ 41 & -17 & -8 \\ 20 & -29 & 40 \\ 38 & -56 & 116 \\ 13 & -11 & 1 \\ 22 & -3 & -6 \\ 41 & -53 & 85 \\ 28 & -32 & 16 \end{bmatrix} \begin{bmatrix} \frac{1}{11} & \frac{6}{11} & \frac{4}{11} \\ -\frac{7}{11} & \frac{2}{11} & \frac{5}{11} \\ -\frac{2}{11} & -\frac{1}{11} & \frac{3}{11} \end{bmatrix} \begin{bmatrix} 0 & 18 & 5 \\ 13 & 5 & 13 \\ 2 & 5 & 18 \\ 0 & 19 & 5 \\ 16 & 20 & 5 \\ 13 & 2 & 5 \\ 18 & 0 & 20 \\ 8 & 5 & 0 \\ 5 & 12 & 5 \\ 22 & 5 & 14 \\ 20 & 8 & 0 \end{bmatrix}$$

0 18 5 13 5 13 2 5 18 0
_ R E M E M B E R _

19 5 16 20 5 13 2 5 18 0
S E P T E M B E R _

20 8 5 0 5 12 5 22 5 14 20 8 0
T H E _ E L E V E N T H _

Message: REMEMBER SEPTEMBER THE ELEVENTH

19. $A = \begin{bmatrix} 6 & 4 & 1 \\ 0 & 2 & 3 \\ 1 & 1 & 2 \end{bmatrix} \qquad A^{-1} = \begin{bmatrix} \frac{1}{16} & -\frac{7}{16} & \frac{5}{8} \\ \frac{3}{16} & \frac{11}{16} & -\frac{9}{8} \\ -\frac{1}{8} & -\frac{1}{8} & \frac{3}{4} \end{bmatrix}$

$|A| = 16$ and $|A^{-1}| = \dfrac{1}{16}$

Conjecture: $|A^{-1}| = \dfrac{1}{|A|}$

Practice Test for Chapter 8

1. Put the matrix in reduced row-echelon form.

$$\begin{bmatrix} 1 & -2 & 4 \\ 3 & -5 & 9 \end{bmatrix}$$

For Exercises 2–4, use matrices to solve the system of equations.

2. $\begin{cases} 3x + 5y = 3 \\ 2x - y = -11 \end{cases}$

3. $\begin{cases} 2x + 3y = -3 \\ 3x + 2y = 8 \\ x + y = 1 \end{cases}$

4. $\begin{cases} x + 3z = -5 \\ 2x + y = 0 \\ 3x + y - z = 3 \end{cases}$

5. Multiply $\begin{bmatrix} 1 & 4 & 5 \\ 2 & 0 & -3 \end{bmatrix}\begin{bmatrix} 1 & 6 \\ 0 & -7 \\ -1 & 2 \end{bmatrix}$.

6. Given $A = \begin{bmatrix} 9 & 1 \\ -4 & 8 \end{bmatrix}$ and $B = \begin{bmatrix} 6 & -2 \\ 3 & 5 \end{bmatrix}$, find $3A - 5B$.

7. Find $f(A)$.

$f(x) = x^2 - 7x + 8,\ A = \begin{bmatrix} 3 & 0 \\ 7 & 1 \end{bmatrix}$

8. True or false:

$(A + B)(A + 3B) = A^2 + 4AB + 3B^2$ where A and B are matrices.

(Assume that A^2, AB, and B^2 exist.)

For Exercises 9–10, find the inverse of the matrix, if it exists.

9. $\begin{bmatrix} 1 & 2 \\ 3 & 5 \end{bmatrix}$

10. $\begin{bmatrix} 1 & 1 & 1 \\ 3 & 6 & 5 \\ 6 & 10 & 8 \end{bmatrix}$

11. Use an inverse matrix to solve the systems.

(a) $\begin{cases} x + 2y = 4 \\ 3x + 5y = 1 \end{cases}$

(b) $\begin{cases} x + 2y = 3 \\ 3x + 5y = -2 \end{cases}$

For Exercises 12–14, find the determinant of the matrix.

12. $\begin{bmatrix} 6 & -1 \\ 3 & 4 \end{bmatrix}$

13. $\begin{bmatrix} 1 & 3 & -1 \\ 5 & 9 & 0 \\ 6 & 2 & -5 \end{bmatrix}$

14. $\begin{bmatrix} 1 & 4 & 2 & 3 \\ 0 & 1 & -2 & 0 \\ 3 & 5 & -1 & 1 \\ 2 & 0 & 6 & 1 \end{bmatrix}$

15. Evaluate $\begin{vmatrix} 6 & 4 & 3 & 0 & 6 \\ 0 & 5 & 1 & 4 & 8 \\ 0 & 0 & 2 & 7 & 3 \\ 0 & 0 & 0 & 9 & 2 \\ 0 & 0 & 0 & 0 & 1 \end{vmatrix}$.

16. Use a determinant to find the area of the triangle with vertices $(0, 7)$, $(5, 0)$, and $(3, 9)$.

17. Use a determinant to find the equation of the line passing through $(2, 7)$ and $(-1, 4)$.

For Exercises 18–20, use Cramer's Rule to find the indicated value.

18. Find x.

$$\begin{cases} 6x - 7y = 4 \\ 2x + 5y = 11 \end{cases}$$

19. Find z.

$$\begin{cases} 3x + z = 1 \\ y + 4z = 3 \\ x - y = 2 \end{cases}$$

20. Find y.

$$\begin{cases} 721.4x - 29.1y = 33.77 \\ 45.9x + 105.6y = 19.85 \end{cases}$$

CHAPTER 9
Sequences, Series, and Probability

CHAPTER 9
Sequences, Series, and Probability

Section 9.1 Sequences and Series

1. infinite sequence

3. recursively

5. index; upper; lower

7. $a_n = 4n - 7$
$a_1 = 4(1) - 7 = -3$
$a_2 = 4(2) - 7 = 1$
$a_3 = 4(3) - 7 = 5$
$a_4 = 4(4) - 7 = 9$
$a_5 = 4(5) - 7 = 13$

9. $a_n = (-1)^{n+1} + 4$
$a_1 = (-1)^{1+1} + 4 = 5$
$a_2 = (-1)^{2+1} + 4 = 3$
$a_3 = (-1)^{3+1} + 4 = 5$
$a_4 = (-1)^{4+1} + 4 = 3$
$a_5 = (-1)^{5+1} + 4 = 5$

11. $a_n = (-2)^n$
$a_1 = (-2)^1 = -2$
$a_2 = (-2)^2 = 4$
$a_3 = (-2)^3 = -8$
$a_4 = (-2)^4 = 16$
$a_5 = (-2)^5 = -32$

13. $a_n = \frac{2}{3}$
$a_1 = \frac{2}{3}$
$a_2 = \frac{2}{3}$
$a_3 = \frac{2}{3}$
$a_4 = \frac{2}{3}$
$a_5 = \frac{2}{3}$

15. $a_n = \frac{1}{3}n^3$
$a_1 = \frac{1}{3}(1)^3 = \frac{1}{3}$
$a_2 = \frac{1}{3}(2)^3 = \frac{8}{3}$
$a_3 = \frac{1}{3}(3)^3 = 9$
$a_4 = \frac{1}{3}(4)^3 = \frac{64}{3}$
$a_5 = \frac{1}{3}(5)^3 = \frac{125}{3}$

17. $a_n = \frac{n}{n + 2}$
$a_1 = \frac{1}{1 + 2} = \frac{1}{3}$
$a_2 = \frac{2}{2 + 2} = \frac{1}{2}$
$a_3 = \frac{3}{3 + 2} = \frac{3}{5}$
$a_4 = \frac{4}{4 + 2} = \frac{2}{3}$
$a_5 = \frac{5}{5 + 2} = \frac{5}{7}$

19. $a_n = n(n - 1)(n - 2)$
$a_1 = (1)(0)(-1) = 0$
$a_2 = (2)(1)(0) = 0$
$a_3 = (3)(2)(1) = 6$
$a_4 = (4)(3)(2) = 24$
$a_5 = (5)(4)(3) = 60$

21. $a_n = (-1)^n\left(\frac{n}{n+1}\right)$

$a_1 = (-1)^1\frac{1}{1+1} = -\frac{1}{2}$

$a_2 = (-1)^2\frac{2}{2+1} = \frac{2}{3}$

$a_3 = (-1)^3\frac{3}{3+1} = -\frac{3}{4}$

$a_4 = (-1)^4\frac{4}{4+1} = \frac{4}{5}$

$a_5 = (-1)^5\frac{5}{5+1} = -\frac{5}{6}$

23. $a_{25} = (-1)^{25}(3(25) - 2) = -73$

25. $a_{11} = \frac{4(11)}{2(11)^2 - 3} = \frac{44}{239}$

27. $a_n = \frac{2}{3}n$

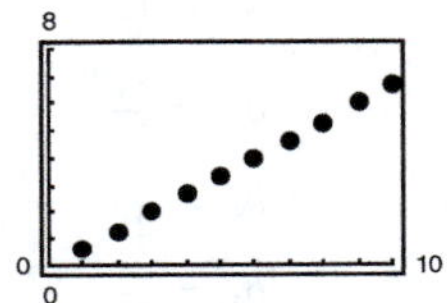

29. $a_n = 16(-0.5)^{n-1}$

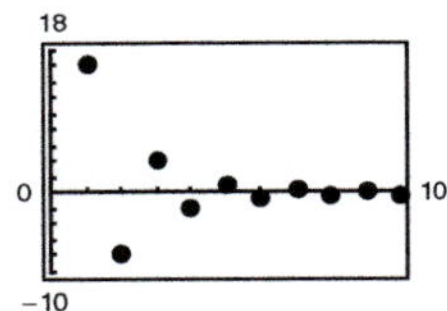

31. $a_n = \frac{2n}{n+1}$

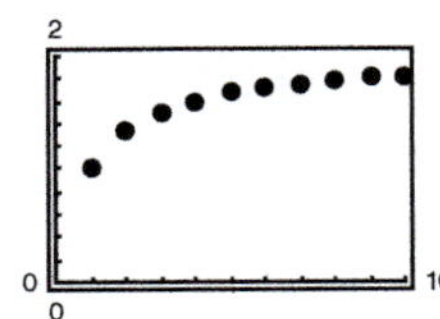

33. $a_n = \frac{8}{n+1}$

$a_1 = 4, a_{10} = \frac{8}{11}$

The sequence decreases.

Matches graph (c).

34. $a_n = \frac{8n}{n+1}$

$a_1 = 4, a_3 = \frac{24}{4} = 6$

The sequence increases.

Matches graph (b).

35. $a_n = 4(0.5)^{n-1}$

$a_1 = 4, a_{10} = \frac{1}{128}$

The sequence decreases.

Matches graph (d).

36. $a_n = n\left(2 - \frac{n}{10}\right)$

$a_1 = \frac{9}{10}, a_{10} = 10$

The sequence increases.

Matches graph (a).

37. 3, 7, 11, 15, 19, ...

n:	1	2	3	4	5	...	n
Terms:	3	7	11	15	19	...	a_n

Apparent pattern:

Each term is one less than four times n, which implies that $a_n = 4n - 1$.

39. 3, 10, 29, 66, 127, ...

n:	1	2	3	4	5	...	n
Terms:	3	10	29	66	127	...	a_n

Apparent pattern:

Each term is more than n cubed, which implies that $a_n = n^3 + 2$.

41. 1, −1, 1, −1, 1, ...

n:	1	2	3	4	5	...	n
Terms:	1	−1	1	−1	1	...	a_n

Apparent pattern:

Each term is either 1 or −1 which implies that $a_n = (-1)^{n+1}$.

43. $-\frac{2}{3}, \frac{3}{4}, -\frac{4}{5}, \frac{5}{6}, -\frac{6}{7}, \dots$

$a_n = (-1)^n\left(\frac{n+1}{n+2}\right)$

45. $\frac{2}{1}, \frac{3}{3}, \frac{4}{5}, \frac{5}{7}, \frac{6}{9}, \ldots$

$a_n = \frac{n+1}{2n-1}$

47. $1, \frac{1}{2}, \frac{1}{6}, \frac{1}{24}, \frac{1}{120}, \ldots$

n:	1	2	3	4	5	…	n
Terms:	1	$\frac{1}{2}$	$\frac{1}{6}$	$\frac{1}{24}$	$\frac{1}{120}$	…	a_n

Apparent pattern:

Each term is the reciprocal of $n!$, which implies that $a_n = \frac{1}{n!}$.

49. $\frac{1}{1}, \frac{3}{1}, \frac{9}{2}, \frac{27}{6}, \frac{81}{24}, \ldots$

n:	1	2	3	4	…	n
Terms:	$\frac{1}{1}$	$\frac{3}{1}$	$\frac{9}{2}$	$\frac{27}{6}$	…	a_n

Apparent pattern:

Each term is a power of three, 3^{n-1} divided by a factorial, $(n-1)!$. After trial and error, $a_n = \frac{3^{n-1}}{(n-1)!}$.

51. $a_1 = 28$ and $a_{k+1} = a_k - 4$

$a_1 = 28$

$a_2 = a_1 - 4 = 28 - 4 = 24$

$a_3 = a_2 - 4 = 24 - 4 = 20$

$a_4 = a_3 - 4 = 20 - 4 = 16$

$a_5 = a_4 - 4 = 16 - 4 = 12$

53. $a_1 = 81$ and $a_{k+1} = \frac{1}{3}a_k$

$a_1 = 81$

$a_2 = \frac{1}{3}a_1 = \frac{1}{3}(81) = 27$

$a_3 = \frac{1}{3}a_2 = \frac{1}{3}(27) = 9$

$a_4 = \frac{1}{3}a_3 = \frac{1}{3}(9) = 3$

$a_5 = \frac{1}{3}a_4 = \frac{1}{3}(3) = 1$

55. $a_0 = 1, a_1 = 2, a_k = a_{k-2} + \frac{1}{2}a_{k-1}$

$a_0 = 1$

$a_1 = 2$

$a_2 = a_0 + \frac{1}{2}a_1 = 1 + \frac{1}{2}(1) = 2$

$a_3 = a_1 + \frac{1}{2}a_2 = 2 + \frac{1}{2}(2) = 3$

$a_4 = a_2 + \frac{1}{2}a_3 = 2 + \frac{1}{2}(3) = \frac{7}{2}$

57. $a_1 = 1, a_2 = 1, a_k = a_{k-1} + a_{k-2}, k \ge 1$

$a_1 = 1$	$b_1 = \frac{1}{1} = 1$
$a_2 = 1$	$b_2 = \frac{2}{1} = 2$
$a_3 = 1 + 1 = 2$	$b_3 = \frac{3}{2}$
$a_4 = 2 + 1 = 3$	$b_4 = \frac{5}{3}$
$a_5 = 3 + 2 = 5$	$b_5 = \frac{8}{5}$
$a_6 = 5 + 3 = 8$	$b_6 = \frac{13}{8}$
$a_7 = 8 + 5 = 13$	$b_7 = \frac{21}{13}$
$a_8 = 13 + 8 = 21$	$b_8 = \frac{34}{21}$
$a_9 = 21 + 13 = 34$	$b_9 = \frac{55}{34}$
$a_{10} = 34 + 21 = 55$	$b_{10} = \frac{89}{55}$
$a_{11} = 55 + 34 = 89$	
$a_{12} = 89 + 55 = 144$	

59. $a_n = \frac{5}{n!}$

$a_0 = \frac{5}{0!} = \frac{5}{1} = 5$

$a_1 = \frac{5}{1!} = \frac{5}{1} = 5$

$a_2 = \frac{5}{2!} = \frac{5}{2}$

$a_3 = \frac{5}{3!} = \frac{5}{6}$

$a_4 = \frac{5}{4!} = \frac{5}{24}$

61. $a_n = \dfrac{(-1)^n(n+3)!}{n!}$

$a_0 = \dfrac{(-1)^0(0+3)!}{0!} = \dfrac{3!}{0!} = 6$

$a_1 = \dfrac{(-1)^1(1+3)!}{1!} = \dfrac{4!}{1!} = -24$

$a_2 = \dfrac{(-1)^2(2+3)!}{2!} = \dfrac{5!}{2!} = 60$

$a_3 = \dfrac{(-1)^3(3+3)!}{3!} = -\dfrac{6!}{3!} = -120$

$a_4 = \dfrac{(-1)^4(4+3)!}{4!} = \dfrac{7!}{4!} = 210$

63. $\dfrac{4!}{6!} = \dfrac{\cancel{1\cdot 2\cdot 3\cdot 4}}{\cancel{1\cdot 2\cdot 3\cdot 4}\cdot 5\cdot 6} = \dfrac{1}{5\cdot 6} = \dfrac{1}{30}$

65. $\dfrac{(n+1)!}{n!} = \dfrac{\cancel{1\cdot 2\cdot 3\cdots n}\cdot(n+1)}{\cancel{1\cdot 2\cdot 3\cdots n}} = \dfrac{n+1}{1}$

$= n + 1$

67. $\displaystyle\sum_{i=0}^{4} 3i^2 = 3\cdot 0^2 + 3\cdot 1^2 + 3\cdot 2^2 + 3\cdot 3^2 + 3\cdot 4^2 = 90$

69. $\displaystyle\sum_{j=3}^{5} \frac{1}{j^2-3} = \frac{1}{3^2-3} + \frac{1}{4^2-3} + \frac{1}{5^2-3} = \frac{124}{429}$

71. $\displaystyle\sum_{k=2}^{5} (k+1)^2(k-3) = (3)^2(-1) + (4)^2(0) + (5)^2(1) + (6)^2(2) = 88$

73. $\displaystyle\sum_{i=1}^{4} \frac{i!}{2^i} = \frac{1!}{2^1} + \frac{2!}{2^2} + \frac{3!}{2^3} + \frac{4!}{2^4} = \frac{1}{2} + \frac{2}{4} + \frac{6}{8} + \frac{24}{16} = \frac{1}{2} + \frac{1}{2} + \frac{3}{4} + \frac{3}{2} = \frac{13}{4}$

75. $\displaystyle\sum_{k=0}^{4} \frac{(-1)^k}{k!} = \frac{3}{8}$

77. $\displaystyle\sum_{n=0}^{25} \frac{1}{4^n} \approx 1.33$

79. $\displaystyle\frac{1}{3(1)} + \frac{1}{3(2)} + \frac{1}{3(3)} + \cdots + \frac{1}{3(9)} = \sum_{i=1}^{9} \frac{1}{3i}$

81. $\displaystyle\left[2\left(\frac{1}{8}\right)+3\right] + \left[2\left(\frac{2}{8}\right)+3\right] + \left[2\left(\frac{3}{8}\right)+3\right] + \cdots + \left[2\left(\frac{8}{8}\right)+3\right] = \sum_{i=1}^{8}\left[2\left(\frac{i}{8}\right)+3\right]$

83. $\displaystyle 3 - 9 + 27 - 81 + 243 - 729 = \sum_{i=1}^{6} (-1)^{i+1}3i$

85. $\displaystyle\frac{1^2}{2} + \frac{2^2}{6} + \frac{3^2}{24} + \frac{4^2}{120} + \frac{5^2}{720} + \frac{6^2}{5040} + \frac{7^2}{40{,}320} = \sum_{i=1}^{7} \frac{i^2}{(i+1)!}$

87. $\displaystyle\frac{1}{4} + \frac{3}{8} + \frac{7}{16} + \frac{15}{32} + \frac{31}{64} = \sum_{i=1}^{5} \frac{2^i-1}{2^{i+1}}$

89. (a) $\displaystyle\sum_{i=1}^{3}\left(\frac{1}{2}\right)^i = \left(\frac{1}{2}\right) + \left(\frac{1}{2}\right)^2 + \left(\frac{1}{2}\right)^3 = \frac{7}{8}$

(b) $\displaystyle\sum_{i=1}^{4}\left(\frac{1}{2}\right)^i = \left(\frac{1}{2}\right) + \left(\frac{1}{2}\right)^2 + \left(\frac{1}{2}\right)^3 + \left(\frac{1}{2}\right)^4 = \frac{15}{16}$

(c) $\displaystyle\sum_{i=1}^{5}\left(\frac{1}{2}\right)^i = \left(\frac{1}{2}\right) + \left(\frac{1}{2}\right)^2 + \left(\frac{1}{2}\right)^3 + \left(\frac{1}{2}\right)^4 + \left(\frac{1}{2}\right)^5 = \frac{21}{32}$

91. (a) $\sum_{n=1}^{3} 4\left(-\frac{1}{2}\right)^n = 4\left(-\frac{1}{2}\right) + 4\left(-\frac{1}{2}\right)^2 + 4\left(-\frac{1}{2}\right)^3 = -\frac{3}{2}$

(b) $\sum_{n=1}^{4} 4\left(-\frac{1}{2}\right)^n = 4\left(-\frac{1}{2}\right) + 4\left(-\frac{1}{2}\right)^2 + 4\left(-\frac{1}{2}\right)^3 + 4\left(-\frac{1}{2}\right)^4 = -\frac{5}{4}$

(c) $\sum_{n=1}^{5} 4\left(-\frac{1}{2}\right)^n = 4\left(-\frac{1}{2}\right) + 4\left(-\frac{1}{2}\right)^2 + 4\left(-\frac{1}{2}\right)^3 + 4\left(-\frac{1}{2}\right)^4 + 4\left(-\frac{1}{2}\right)^5 = -\frac{11}{8}$

93. $\sum_{i=1}^{\infty} 6\left(\frac{1}{10}\right)^i = 0.6 + 0.06 + 0.006 + 0.0006 + \cdots = \frac{2}{3}$

95. By using a calculator,

$\sum_{k=1}^{10} 7\left(\frac{1}{10}\right)^k \approx 0.7777777777$

$\sum_{k=1}^{50} 7\left(\frac{1}{10}\right)^k \approx 0.7777777778$

$\sum_{k=1}^{100} 7\left(\frac{1}{10}\right)^k \approx \frac{7}{9}.$

The terms approach zero as $n \to \infty$.

So, $\sum_{k=1}^{\infty} 7\left(\frac{1}{10}\right)^k = \frac{7}{9}.$

97. (a) $A_1 = \$10{,}087.50$
$A_2 \approx \$10{,}175.77$
$A_3 \approx \$10{,}264.80$
$A_4 \approx \$10{,}354.62$
$A_5 \approx \$10{,}445.22$
$A_6 \approx \$10{,}536.62$
$A_7 \approx \$10{,}628.81$
$A_8 \approx \$10{,}721.82$

(b) $A_{40} \approx \$14{,}169.09$

(c) No; The balance after 20 years, $A_{80} = \$20{,}076.31$ is not twice the balance after 10 years, $A_{40} \approx \$14{,}169.09$.

99. True, $\sum_{i=1}^{4} \left(i^2 + 2i\right) = \sum_{i=1}^{4} i^2 + 2\sum_{i=1}^{4} i$ by the Properties of Sums.

101. $\dfrac{327.15 + 785.69 + 433.04 + 265.38 + 604.12 + 590.30}{6} \approx \$500.95$

103.
$$\begin{aligned}
\sum_{i=1}^{n} (x_i - \bar{x})^2 &= \sum_{i=1}^{n} \left(x_i^2 - 2x_i\bar{x} + \bar{x}^2\right) \\
&= \sum_{i=1}^{n} x_i^2 - 2\bar{x}\sum_{i=1}^{n} x_i + n\bar{x}^2 \\
&= \sum_{i=1}^{n} x_i^2 - 2 \cdot \frac{1}{n}\sum_{i=1}^{n} x_i \sum_{i=1}^{n} x_i + n \cdot \frac{1}{n}\sum_{i=1}^{n} x_i \cdot \frac{1}{n}\sum_{i=1}^{n} x_i \\
&= \sum_{i=1}^{n} x_i^2 + \sum_{i=1}^{n} x_i \sum_{i=1}^{n} x_i\left(-\frac{2}{n} + \frac{1}{n}\right) \\
&= \sum_{i=1}^{n} x_i^2 - \frac{1}{n}\left(\sum_{i=1}^{n} x_i\right)^2
\end{aligned}$$

105. The error was not summing the constant, $\sum_{k=1}^{4} 3 = 12$.

$$\begin{aligned}
\sum_{k=1}^{4} \left(3 + 2k^2\right) &= \sum_{k=1}^{4} 3 + \sum_{k=1}^{4} 2k^2 = (3 + 3 + 3 + 3) + \left(2(1)^2 + 2(2)^2 + 2(2)^3 + 2(4)^2\right) \\
&= (12) + (2 + 8 + 18 + 32) \\
&= 75
\end{aligned}$$

107. (a) and (b)

Number of blue cube faces	0	1	2	3
$3 \times 3 \times 3$	1	6	12	8
$4 \times 4 \times 4$	8	24	24	8
$5 \times 5 \times 5$	27	54	36	8
$6 \times 6 \times 6$	64	96	48	8

(c)

Number of blue cube faces	0	1	2	3
$n \times n \times n$	$(n-2)^3$	$6(n-2)^2$	$12(n-2)$	8

Section 9.2 Arithmetic Sequences and Partial Sums

1. arithmetic; common

3. recursion

5. $1, 2, 4, 8, 16, \ldots$

Not an arithmetic sequence

7. $10, 8, 6, 4, 2, \ldots$

Arithmetic sequence, $d = -2$

9. $\frac{5}{4}, \frac{3}{2}, \frac{7}{4}, 2, \frac{9}{4}, \ldots$

Arithmetic sequence, $d = \frac{1}{4}$

11. $1^2, 2^2, 3^2, 4^2, 5^2, \ldots$

Not an arithmetic sequence

13. $a_n = 5 + 3n$

$8, 11, 14, 17, 20$

Arithmetic sequence, $d = 3$

15. $a_n = 3 - 4(n-2)$

$7, 3, -1, -5, -9$

Arithmetic sequence, $d = -4$

17. $a_n = (-1)^n$

$-1, 1, -1, 1, -1$

Not an arithmetic sequence

19. $a_n = (2^n)n$

$2, 8, 24, 64, 160$

Not an arithmetic sequence

21. $a_1 = 1, d = 3$

$a_n = a_1 + (n-1)d = 1 + (n-1)(3) = 3n - 2$

23. $a_1 = 100, d = -8$

$$a_n = a_1 + (n-1)d = 100 + (n-1)(-8)$$
$$= -8n + 108$$

25. $4, \frac{3}{2}, -1, -\frac{7}{2}, \ldots$

$d = -\frac{5}{2}$

$a_n = a_1 + (n-1)d = 4 + (n-1)\left(-\frac{5}{2}\right) = -\frac{5}{2}n + \frac{13}{2}$

27. $a_1 = 5, a_4 = 15$

$a_4 = a_1 + 3d \Rightarrow 15 = 5 + 3d \Rightarrow d = \frac{10}{3}$

$a_n = a_1 + (n-1)d = 5 + (n-1)\left(\frac{10}{3}\right) = \frac{10}{3}n + \frac{5}{3}$

29. $a_3 = 94, a_6 = 103$

$a_6 = a_3 + 3d \Rightarrow 103 = 94 + 3d \Rightarrow d = 3$

$a_1 = a_3 - 2d \Rightarrow a_1 = 94 - 2(3) = 88$

$$a_n = a_1 + (n-1)d = 88 + (n-1)(3)$$
$$= 3n + 85$$

31. $a_1 = 5, d = 6$

$a_1 = 5$

$a_2 = 5 + 6 = 11$

$a_3 = 11 + 6 = 17$

$a_4 = 17 + 6 = 23$

$a_5 = 23 + 6 = 29$

33. $a_1 = 2, a_{12} = -64$

$-64 = 2 + (12 + 1)d$

$-66 = 11d$

$-6 = d$

$a_1 = 2$

$a_2 = 2 - 6 = -4$

$a_3 = -4 - 6 = -10$

$a_4 = -10 - 6 = -16$

$a_5 = -16 - 6 = -22$

35. $a_8 = 26, a_{12} = 42$

$a_{12} = a_8 + 4d$

$42 = 26 + 4d \Rightarrow d = 4$

$a_8 = a_1 + 7d$

$26 = a_1 + 28 \Rightarrow a_1 = -2$

$a_1 = -2$

$a_2 = -2 + 4 = 2$

$a_3 = 2 + 4 = 6$

$a_4 = 6 + 4 = 10$

$a_5 = 10 + 4 = 14$

37. $a_1 = 15, a_{n+1} = a_n + 4$

$a_2 = 15 + 4 = 19$

$a_3 = 19 + 4 = 23$

$a_4 = 23 + 4 = 27$

$a_5 = 27 + 4 = 31$

39. $a_{n+1} = a_n - 2 \Rightarrow a_n = a_{n+1} + 2$

$a_5 = 7$

$a_4 = 7 + 2 = 9$

$a_3 = 9 + 2 = 11$

$a_2 = 11 + 2 = 13$

$a_1 = 13 + 2 = 15$

41. $a_1 = 5, a_2 = -1 \Rightarrow d = -1 - 5 = -6$

$a_n = a_1 + (n - 1)d \Rightarrow a_{10} = 5 + 9(-6) = -49$

43. $a_1 = \frac{1}{8}, a_2 = \frac{3}{8} \Rightarrow d = \frac{3}{4} - \frac{1}{8} = \frac{5}{8}$

$a_n = a_1 + (n - 1)d \Rightarrow a_7 = \frac{1}{8} + 6\left(\frac{5}{8}\right) = \frac{31}{8}$

45. $S_{10} = \frac{10}{2}(2 + 20) = 110$

47. $S_5 = \frac{5}{2}(-1 + (-9)) = -25$

49. $a_n = 2n - 1$

$a_1 = 1, a_{100} = 199$

$S_{100} = \frac{100}{2}(1 + 199) = 10{,}000$

51. $8, 20, 32, 44, \ldots$

$a_1 = 8, d = 12, n = 50$

$a_{50} = 8 + 49(12) = 596$

$S_{10} = \frac{50}{2}(8 + 596) = 15{,}100$

53. $0, -9, -18, -27, \ldots, n = 40$

$a_1 = 0, d = -9$

$a_{40} = 0 + (39)(-9) = -351$

$S_{100} = \frac{40}{2}(0 + (-351)) = -7020$

55. $a_1 = 1, a_{50} = 50, n = 50$

$\sum_{n=1}^{50} n = \frac{50}{2}(1 + 50) = 1275$

57. $a_1 = 9, a_{500} = 508, n = 500$

$\sum_{n=1}^{500} (n + 8) = \frac{500}{2}(9 + 508) = 129{,}250$

59. $a_1 = 14, a_{100} = -580, n = 100$

$\sum_{n=1}^{100} (-6n + 20) = \frac{100}{2}(14 + (-580)) = -28{,}300$

61. $a_n = -\frac{3}{4}n + 8$

$d = -\frac{3}{4}$ so the sequence is decreasing and $a_1 = 7\frac{1}{4}$.

Matches (b).

62. $a_n = 3n - 5$

$d = 3$ so the sequence is increasing and $a_1 = -2$.

Matches (d).

63. $a_n = 2 + \frac{3}{4}n$

$d = \frac{3}{4}$ so the sequence is increasing and $a_1 = 2\frac{3}{4}$.

Matches (c).

64. $a_n = 25 - 3n$

$d = -3$ so the sequence is decreasing and $a_1 = 22$.

Matches (a).

65. $a_n = 15 - \frac{3}{2}n$

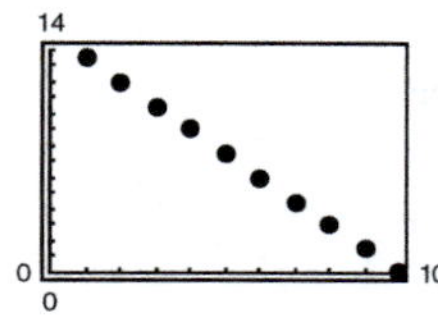

67. 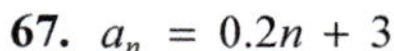
$a_n = 0.2n + 3$

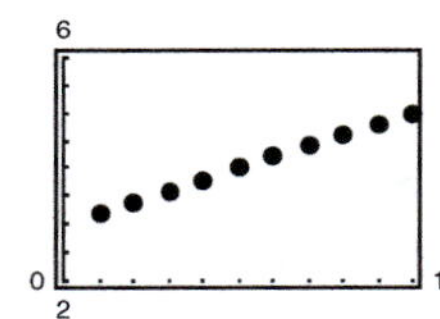

69. (a) $a_1 = 32{,}500,\ d = 1500$

$a_6 = a_1 + 5d = 32{,}500 + 5(1500) = \$40{,}000$

(b) $S_6 = \frac{6}{2}[32{,}500 + 40{,}000] = \$217{,}500$

71. $a_1 = 15,\ d = 3,\ n = 36$

$a_{36} = 15 + 35(3) = 120$

$S_{36} = \frac{36}{2}(15 + 120) = 2430$ seats

73. $a_1 = 16,\ a_2 = 48,\ a_3 = 80,\ a_4 = 112$

$d = 32$

$a_n = dn + c = 32n + c$

$c = a_n - dn$

$c = a_1 - d = 16 - 32 = -16$

$a_n = 32n - 16$

$$\text{Distance} = \sum_{n=1}^{7} (32n - 16) = 784 \text{ ft}$$

75. $a_1 = 15{,}000$

$d = 5{,}000$

$n = 1, \ldots, 10$

$a_n = dn + c = 5000n + c$

$c = a_1 - d = 15{,}000 - 5000 = 10{,}000$

$a_n = 5000n + 10{,}000$

$$\text{Total sales} = \sum_{n=1}^{10} (5000n + 10{,}000) = \frac{10}{2}(15{,}000 + 60{,}000) = \$375{,}000$$

Answers will vary.

77. (a)

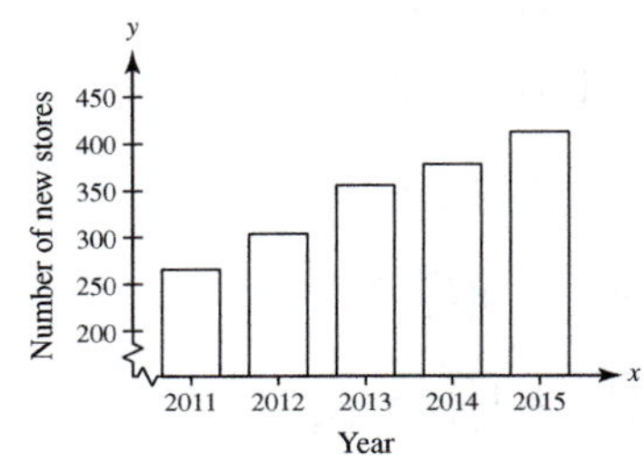

(b) The increase in net number of new stores

$\frac{413 - 266}{5 - 1} = \frac{147}{4} = 36.75$ each year from 2011 to 2015. Because the common difference is $d = 36.75$ and the first term is $a_1 = 266$, the arithmetic sequence is

$a_n = 266 + 36.75(n - 1) = 229.25 + 36.75n$

(c)

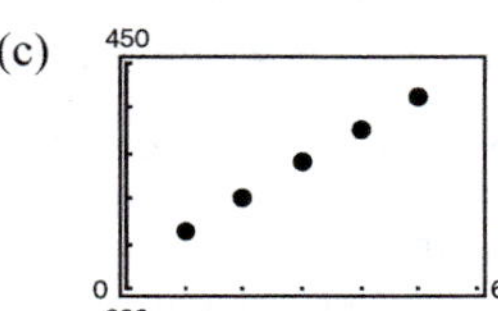

(d) $\displaystyle\sum_{n=1}^{5} (229.25 + 36.75n) = \frac{5}{2}(266 + 413)$

$= 1697.5 \approx 1698$ stores

79. True; given a_1 and a_2 then $d = a_2 - a_1$ and $a_n = a_1 + (n - 1)d$.

81. (a) 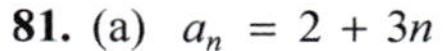
$a_n = 2 + 3n$

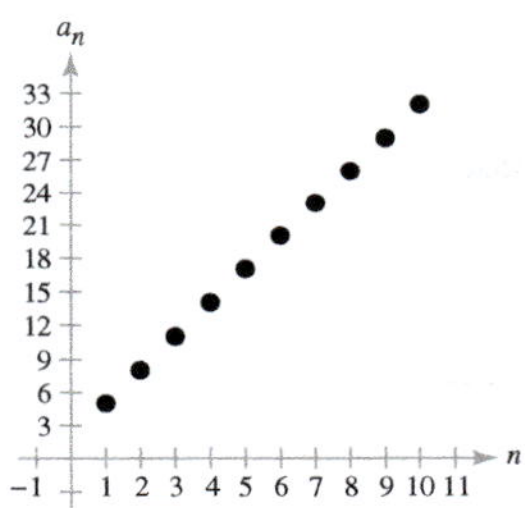

(b) 
$y = 3x + 2$

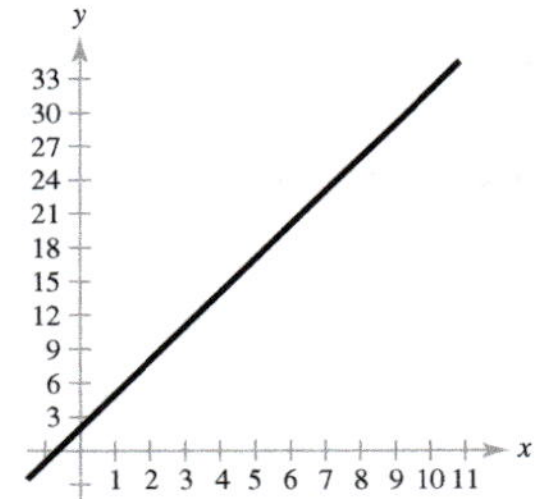

(c) The graph of $a_n = 2 + 3n$ contains only points at the positive integers. The graph of $y = 3x + 2$ is a solid line which contains these points.

(d) The slope $m = 3$ is equal to the common difference $d = 3$. In general, these should be equal.

83. $a_1 = x, d = 2x$

$a_n = x + (n - 1)2x$

$a_n = 2xn - x$

$a_1 = 2x(1) - x = x$ $\quad a_6 = 2x(6) - x = 11x$

$a_2 = 2x(2) - x = 3x$ $\quad a_7 = 2x(7) - x = 13x$

$a_3 = 2x(3) - x = 5x$ $\quad a_8 = 2x(8) - x = 15x$

$a_4 = 2x(4) - x = 7x$ $\quad a_9 = 2x(9) - x = 17x$

$a_5 = 2x(5) - x = 9x$ $\quad a_{10} = 2x(10) - x = 19x$

85. The error was the value of

$n = 50, a_{50} = 1 + (50 - 1)(2) = 99$. So

$S_{50} = \frac{50}{2}(1 + 99) = 2500$.

87. (a) $1 + 3 = 4$

$1 + 3 + 5 = 9$

$1 + 3 + 5 + 7 = 16$

$1 + 3 + 5 + 7 + 9 = 36$

$1 + 3 + 5 + 7 + 9 + 11 = 36$

(b) $S_n = n^2$

$S_7 = 1 + 3 + 5 + 7 + 9 + 11 + 13 = 49 = 7^2$

(c) $S_n = \frac{n}{2}[1 + (2n - 1)] = \frac{n}{2}(2n) = n^2$

Section 9.3 Geometric Sequences and Series

1. geometric; common

3. $S_n = a_1\left(\frac{1 - r^n}{1 - r}\right)$

5. 3, 6, 12, 24, …

Geometric sequence, $r = 2$

7. $\frac{1}{27}, \frac{1}{9}, \frac{1}{3}, 1, \ldots$

Geometric sequence, $r = 3$

9. $1, \frac{1}{2}, \frac{1}{3}, \frac{1}{4}, \ldots$

Not a geometric sequence

11. $1, -\sqrt{7}, 7, -7\sqrt{7}, \ldots$

Geometric sequence, $r = -\sqrt{7}$

13. $a_1 = 4, r = 3$

$a_1 = 4$

$a_2 = 4(3) = 12$

$a_3 = 12(3) = 36$

$a_4 = 36(3) = 108$

$a_5 = 108(3) = 324$

15. $a_1 = 1, r = \frac{1}{2}$

$a_1 = 1$

$a_2 = 1\left(\frac{1}{2}\right) = \frac{1}{2}$

$a_3 = \frac{1}{2}\left(\frac{1}{2}\right) = \frac{1}{4}$

$a_4 = \frac{1}{4}\left(\frac{1}{2}\right) = \frac{1}{8}$

$a_5 = \frac{1}{8}\left(\frac{1}{2}\right) = \frac{1}{16}$

17. $a_1 = 1, r = e$

$a_1 = 1$

$a_2 = 1(e) = e$

$a_3 = (e)(e) = e^2$

$a_4 = (e^2)(e) = e^3$

$a_5 = (e^3)(e) = e^4$

19. $a_1 = 3, r = \sqrt{5}$

$a_1 = 3$

$a_2 = 3(\sqrt{5})^1 = 3\sqrt{5}$

$a_3 = 3(\sqrt{5})^2 = 15$

$a_4 = 3(\sqrt{5})^3 = 15\sqrt{5}$

$a_5 = 3(\sqrt{5})^4 = 75$

21. $a_1 = 2, r = 3x$

$a_1 = 2$

$a_2 = 2(3x) = 6x$

$a_3 = 6x(3x) = 18x^2$

$a_4 = 18x^2(3x) = 54x^3$

$a_5 = 54x^3(3x) = 162x^4$

23. $a_1 = 4, r = \frac{1}{2}, n = 10$

$a_n = a_1r^{n-1} = 4\left(\frac{1}{2}\right)^{n-1}$

$a_{10} = 4\left(\frac{1}{2}\right)^9 = \left(\frac{1}{2}\right)^7 = \frac{1}{128}$

25. $a_1 = 6, r = -\dfrac{1}{3}, n = 12$

$a_n = a_1 r^{n-1} = 6\left(-\dfrac{1}{3}\right)^{n-1}$

$a_{12} = 6\left(-\dfrac{1}{3}\right)^{11} = -\dfrac{2}{3^{10}} = -\dfrac{2}{59{,}049}$

27. $a_1 = 100, r = e^x, n = 9$

$a_n = a_1 r^{n-1} = 100\left(e^x\right)^{n-1} = 100e^{x(n-1)}$

$a_9 = 100\left(e^x\right)^8 = 100e^{8x}$

29. $a_1 = 1, r = \sqrt{2}, n = 12$

$a_n = 1\left(\sqrt{2}\right)^{n-1} = \left(\sqrt{2}\right)^{n-1}$

$a_{12} = \left(\sqrt{2}\right)^{12-1} = 32\sqrt{2}$

31. $a_1 = 500, r = 1.02, n = 40$

$a_n = a_1 r^{n-1} = 500(1.02)^{n-1}$

$a_{40} = 500(1.02)^{39} \approx 1082.372$

33. $64, 32, 16, \ldots$

$r = \dfrac{32}{64} = \dfrac{1}{2}$

$a_n = 64\left(\frac{1}{2}\right)^{n-1}$

35. $9, 18, 36, \ldots$

$r = \dfrac{18}{9} = 2$

$a_n = 9(2)^{n-1}$

37. $6, -9, \frac{27}{2}, \ldots$

$r = \dfrac{-9}{6} = -\dfrac{3}{2}$

$a_n = 6\left(-\frac{3}{2}\right)^{n-1}$

39. $6, 18, 54, \ldots$

$r = \dfrac{18}{6} = 3$

$a_n = 6(3)^{n-1}$

$a_8 = 6(3)^{8-1}$

$= 6(3)^7$

$= 13{,}122$

41. $\frac{1}{3}, \frac{-1}{6}, \frac{1}{12}, \ldots$

$r = \dfrac{-\frac{1}{6}}{\frac{1}{3}} = -\frac{1}{2}$

$a_n = \dfrac{1}{3}\left(-\frac{1}{2}\right)^{n-1}$

$a_9 = \dfrac{1}{3}\left(-\frac{1}{2}\right)^{9-1}$

$= \dfrac{1}{3}\left(-\frac{1}{2}\right)^{8}$

$= \dfrac{1}{768}$

43. $a_1 = 16, a_4 = \frac{27}{4}$

$a_4 = a_1 r^3$

$\frac{27}{4} = 16r^3$

$\frac{27}{64} = r^3$

$\frac{3}{4} = r$

$a_n = 16\left(\frac{3}{4}\right)^{n-1}$

$a_3 = 16\left(\frac{3}{4}\right)^2 = 9$

45. $a_4 = -18, a_7 = \dfrac{2}{3}$

$a_7 = a_4 r^3$

$\dfrac{2}{3} = -18r^3$

$-\dfrac{1}{27} = r^3$

$-\dfrac{1}{3} = r$

$a_6 = \dfrac{a_7}{r} = \dfrac{2/3}{-1/3} = -2$

47. $a_n = 18\left(\frac{2}{3}\right)^{n-1}$

$a_1 = 18$ and $r = \frac{2}{3}$

Because $0 < r < 1$, the sequence is decreasing.

Matches (a).

48. $a_n = 18\left(-\frac{2}{3}\right)^{n-1}$

Because $r = \left(-\frac{2}{3}\right) > -1$, the sequence alternates as it approaches 0.

Matches (c).

49. $a_n = 18\left(\frac{3}{2}\right)^{n-1}$

Because $a_1 = 18$ and $r = \frac{3}{2} > 1$, the sequence is increasing.
Matches (b).

50. $a_n = 18\left(-\frac{3}{2}\right)^{n-1}$

Because $r = \left(-\frac{3}{2}\right) < -1$, the sequence alternates as it approaches ∞.
Matches (d).

51. $a_n = 14(1.4)^{n-1}$

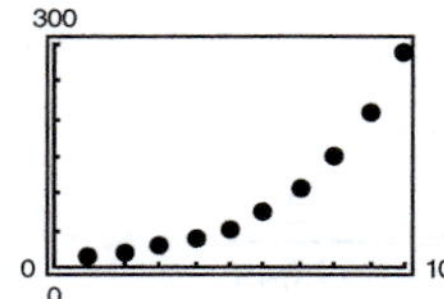

53. $a_n = 8(-0.3)^{n-1}$

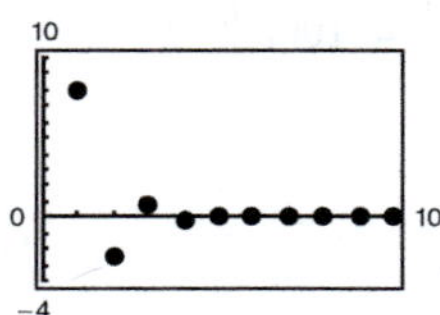

55. $\sum_{n=1}^{7} 4^{n-1} = 1 + 4^1 + 4^2 + 4^3 + 4^4 + 4^5 + 4^6 \Rightarrow a_1 = 1, r = 4$

$$S_7 = \frac{1(1 - 4^7)}{1 - 4} = 5461$$

57. $\sum_{n=1}^{6} (-7)^{n-1} = 1 + (-7) + (-7)^2 + \cdots + (-7)^5 \Rightarrow a_1 = 1, r = -7$

$$S_6 = \frac{1\left(1 - (-7)^6\right)}{1 - (-7)} = -14{,}706$$

59. $\sum_{n=0}^{20} 3\left(\frac{3}{2}\right)^n = \sum_{n=1}^{21} 3\left(\frac{3}{2}\right)^{n-1} = 3 + 3\left(\frac{3}{2}\right)^1 + 3\left(\frac{3}{2}\right)^2 + \cdots + 3\left(\frac{3}{2}\right)^{20} \Rightarrow a_1 = 3, r = \frac{3}{2}$

$$S_{21} = 3\left[\frac{1 - \left(\frac{3}{2}\right)^{21}}{1 - \frac{3}{2}}\right] = -6\left[1 - \left(\frac{3}{2}\right)^{21}\right] \approx 29{,}921.311$$

61. $\sum_{n=0}^{5} 200(1.05)^n = 200 + \sum_{n=1}^{5} 200(1.05)^5 = 200 + \left[200(1.05)^1 + 200(1.05)^2 + \cdots + 200(1.05)^5\right]$

$a_1 = 210, r = 1.05$

$$S_5 = 200 + 210\left[\frac{1 - (1.05)^5}{1 + (1.05)}\right] \approx 1360.383$$

63. $\sum_{n=0}^{40} 2\left(-\frac{1}{4}\right)^n = 2 + 2\left(-\frac{1}{4}\right) + 2\left(-\frac{1}{4}\right)^2 + \cdots + 2\left(-\frac{1}{4}\right)^{40} \Rightarrow a_1 = 2, r = -\frac{1}{4}, n = 41$

$$S_{41} = 2\left[\frac{1 - \left(-\frac{1}{4}\right)^{41}}{1 - \left(-\frac{1}{4}\right)}\right] = \frac{8}{5}\left[1 - \left(-\frac{1}{4}\right)^{41}\right] \approx 1.6 = \frac{8}{5}$$

65. $10 + 30 + 90 + \cdots + 7290$

$r = 3$ and $7290 = 10(3)^{n-1}$

$729 = 3^{n-1}$

$6 = n - 1 \Rightarrow n = 7$

So, the sum can be written as $\sum_{n=1}^{7} 10(3)^{n-1}$.

67. $0.1 + 0.4 + 1.6 + \cdots + 102.4$

$r = 4$ and $102.4 = 0.1(4)^{n-1}$

$1024 = 4^{n-1} \Rightarrow 5 = n - 1 \Rightarrow n = 6$

So, the sum can be written as $\sum_{n=1}^{6} 0.1(4)^{n-1}$.

69. $\sum_{n=0}^{\infty} \left(\frac{1}{2}\right)^n = 1 + \left(\frac{1}{2}\right)^1 + \left(\frac{1}{2}\right)^2 + \cdots$

$a_1 = 1, r = \frac{1}{2}$

$\sum_{n=0}^{\infty} \left(\frac{1}{2}\right)^n = \frac{a_1}{1-r} = \frac{1}{1 - \left(\frac{1}{2}\right)} = 2$

71. $\sum_{n=0}^{\infty} \left(-\frac{1}{2}\right)^n = 1 + \left(-\frac{1}{2}\right)^1 + \left(-\frac{1}{2}\right)^2 + \cdots$

$a_1 = 1, r = -\frac{1}{2}$

$\sum_{n=0}^{\infty} \left(-\frac{1}{2}\right)^n = \frac{a_1}{1-r} = \frac{1}{1 - \left(-\frac{1}{2}\right)} = \frac{2}{3}$

73. $\sum_{n=0}^{\infty} (0.8)^n = 1 + (0.8)^1 + (0.8)^2 + \cdots$

$a_1 = 1, r = 0.8$

$\sum_{n=0}^{\infty} (0.8)^n = \frac{1}{1 - 0.8} = 5$

75. $8 + 6 + \frac{9}{2} + \frac{27}{8} + \cdots = \sum_{n=0}^{\infty} 8\left(\frac{3}{4}\right)^n = \frac{8}{1 - \frac{3}{4}} = 32$

77. $\frac{1}{9} - \frac{1}{3} + 1 - 3 + \cdots = \sum_{n=0}^{\infty} \frac{1}{9}(-3)^n$

The sum is undefined because $|r| = |-3| = 3 > 1$.

79. $0.\overline{36} = \sum_{n=0}^{\infty} 0.36(0.01)^n = \frac{0.36}{1 - 0.01} = \frac{0.36}{0.99} = \frac{36}{99} = \frac{4}{11}$

81. $f(x) = 6\left[\frac{1 - (0.5)^x}{1 - (0.5)}\right], \sum_{n=0}^{\infty} 6\left(\frac{1}{2}\right)^n = \frac{6}{1 - \frac{1}{2}} = 12$

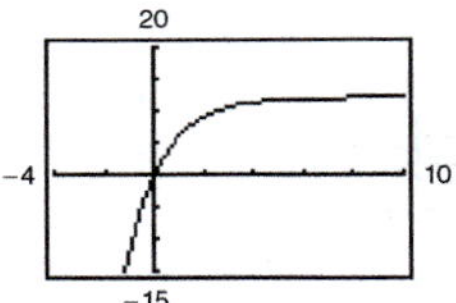

The horizontal asymptote of $f(x)$ is $y = 12$.

This corresponds to the sum of the series.

83. $V_5 = 175{,}000(0.70)^5 = \$29{,}412.25$

85. Let $N = 12t$ be the total number of deposits.

$$A = P\left(1 + \frac{r}{12}\right) + P\left(1 + \frac{r}{12}\right)^2 + \cdots + P\left(1 + \frac{r}{12}\right)^N$$

$$= \left(1 + \frac{r}{12}\right)\left[P + P\left(1 + \frac{r}{12}\right) + \cdots + P\left(1 + \frac{r}{12}\right)^{N-1}\right]$$

$$= P\left(1 + \frac{r}{12}\right)\sum_{n=1}^{N}\left(1 + \frac{r}{12}\right)^{n-1}$$

$$= P\left(1 + \frac{r}{12}\right)\left[\frac{1 - \left(1 + \frac{r}{12}\right)^N}{1 - \left(1 + \frac{r}{12}\right)}\right]$$

$$= P\left(1 + \frac{r}{12}\right)\left(-\frac{12}{r}\right)\left[1 - \left(1 + \frac{r}{12}\right)^N\right]$$

$$= P\left(\frac{12}{r} + 1\right)\left[-1 + \left(1 + \frac{r}{12}\right)^N\right]$$

$$= P\left[\left(1 + \frac{r}{12}\right)^N - 1\right]\left(1 + \frac{12}{r}\right)$$

$$= P\left[\left(1 + \frac{r}{12}\right)^{12t} - 1\right]\left(1 + \frac{12}{r}\right)$$

87. $\sum_{n=0}^{\infty} 400(0.75)^n = \frac{400}{1 - 0.75} = \$1600$

89. $27^2\left(\frac{1}{9}\right) + 27^2\left(\frac{1}{9}\right)\left(\frac{8}{9}\right) + 27^2\left(\frac{1}{9}\right)\left(\frac{8}{9}\right)^2 + 27^2\left(\frac{1}{9}\right)\left(\frac{8}{9}\right)^3 = \sum_{n=0}^{3} 27^2\left(\frac{1}{9}\right)\left(\frac{8}{9}\right)^n = \frac{2465}{9} = 273\frac{8}{9}$ square inches

91. $a_n = (45{,}000)(1.05)^{n-1}$

$$S_{40} = (45{,}000)\frac{1 - (1.05)^{40}}{1 - (1.05)} = \$5{,}435{,}989.84$$

93. False. A sequence is geometric if the ratios of consecutive terms are the same.

95. $y = \left(\dfrac{1 - r^x}{1 - r}\right)$

(a)

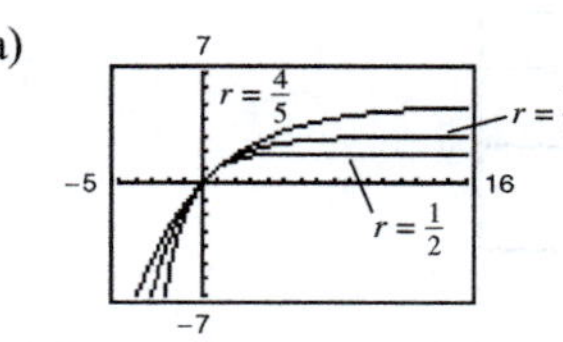

As $x \to \infty$, $y \to \dfrac{1}{1 - r}$.

(b)

As $x \to \infty$, $y \to \infty$.

Section 9.4 Mathematical Induction

1. mathematical induction

3. arithmetic

5. $P_k = \dfrac{5}{k(k + 1)}$

$$P_{k+1} = \frac{5}{(k + 1)[(k + 1) + 1]} = \frac{5}{(k + 1)(k + 2)}$$

7. $P_k = k^2(k + 3)^2$

$$P_{k+1} = (k + 1)^2[(k + 1) + 3]^2 = (k + 1)^2(k + 4)^2$$

9. $P_k = \dfrac{3}{(k + 2)(k + 3)}$

$$P_{k+1} = \frac{3}{[(k + 1) + 2][(k + 1) + 3]} = \frac{3}{(k + 3)(k + 4)}$$

11. 1. When $n = 1$, $S_1 = 2 = 1(1 + 1)$.

2. Assume that

$$S_k = 2 + 4 + 6 + 8 + \cdots + 2k = k(k + 1).$$

Then,

$$\begin{aligned} S_{k+1} &= 2 + 4 + 6 + 8 + \cdots + 2k + 2(k + 1) \\ &= S_k + 2(k + 1) \\ &= k(k + 1) + 2(k + 1) \\ &= (k + 1)(k + 2). \end{aligned}$$

So, we conclude that the formula is valid for all positive integer values of n.

13. 1. When $n = 1$, $S_1 = 2 = \frac{1}{2}(5(1) - 1)$.

2. Assume that

$$S_k = 2 + 7 + 12 + 17 + \cdots + (5k - 3) = \frac{k}{2}(5k - 1).$$

Then,

$$\begin{aligned} S_{k+1} &= 2 + 7 + 12 + 17 + \cdots + (5k - 3) + [5(k + 1) - 3] \\ &= S_k + (5k + 5 - 3) = \frac{k}{2}(5k - 1) + 5k + 2 \\ &= \frac{5k^2 - k + 10k + 4}{2} = \frac{5k^2 + 9k + 4}{2} \\ &= \frac{(k + 1)(5k + 4)}{2} = \frac{(k + 1)}{2}[5(k + 1) - 1]. \end{aligned}$$

So, we conclude that this formula is valid for all positive integer values of n.

15. 1. When $n = 1, S_1 = 1 = 2^1 - 1$.

2. Assume that

$S_k = 1 + 2 + 2^2 + 2^3 + \cdots + 2^{k-1} = 2^k - 1.$

Then,

$S_{k+1} = 1 + 2 + 2^2 + 2^3 + \cdots + 2^{k-1} + 2^k = S_k + 2^k = 2^k - 1 + 2^k = 2\left(2^k\right) - 1 = 2^{k+1} - 1.$

So, we conclude that this formula is valid for all positive integer values of n.

17. 1. When $n = 1, S_1 = 1 = \dfrac{1(1+1)}{2}$.

2. Assume that

$S_k = 1 + 2 + 3 + 4 + \cdots + k = \dfrac{k(k+1)}{2}.$

Then,

$S_{k+1} = 1 + 2 + 3 + 4 + \cdots + k + (k+1) = S_k + (k+1) = \dfrac{k(k+1)}{2} + \dfrac{2(k+1)}{2} = \dfrac{(k+1)(k+2)}{2}.$

So, we conclude that this formula is valid for all positive integer values of n.

19. 1. When $n = 1, S_1 = 1^2 = \dfrac{1(2(1)-1)(2(1)+1)}{3}$

2. Assume that

$S_k = 1^2 + 3^2 + \cdots + (2k-1)^2 = \dfrac{k(2k-1)(2k+1)}{3}$

Then,

$$
\begin{aligned}
S_{k+1} &= 1^2 + 3^2 + \cdots + (2k-1)^2 + (2k+1)^2 \\
&= S_k + (2k+1)^2 = \frac{k(2k-1)(2k+1)}{3} + (2k+1)^2 \\
&= (2k+1)\left[\frac{k(2k-1)}{3} + (2k+1)\right] = \frac{2k+1}{3}\left[2k^2 - k + 6k + 3\right] \\
&= \frac{2k+1}{3}(2k+3)(k+1) = \frac{(k+1)(2(k+1)-1)(2(k+1)+1)}{3}
\end{aligned}
$$

So, we conclude that this formula is valid for all positive integer values of n.

21. 1. When $n = 1, S_1 = 1 = \dfrac{(1)^2(1+1)^2\left(2(1)^2 + 2(1) - 1\right)}{12}.$

2. Assume that

$S_k = \displaystyle\sum_{i=1}^{k} i^5 = \frac{k^2(k+1)^2\left(2k^2 + 2k - 1\right)}{12}.$

Then,

$$
\begin{aligned}
S_{k+1} &= \sum_{i=1}^{k+1} i^5 = \left(\sum_{i=1}^{k} i^5\right) + (k+1)^5 = \frac{k^2(k+1)^2\left(2k^2+2k-1\right)}{12} + \frac{12(k+1)^5}{12} = \frac{(k+1)^2\left[k^2\left(2k^2+2k-1\right) + 12(k+1)^3\right]}{12} \\
&= \frac{(k+1)^2\left[2k^4 + 2k^3 - k^2 + 12\left(k^3 + 3k^2 + 3k + 1\right)\right]}{12} = \frac{(k+1)^2\left[2k^4 + 14k^3 + 35k^2 + 36k + 12\right]}{12} \\
&= \frac{(k+1)^2\left(k^2 + 4k + 4\right)\left(2k^2 + 6k + 3\right)}{12} = \frac{(k+1)^2(k+2)^2\left[2(k+1)^2 + 2(k+1) - 1\right]}{12}.
\end{aligned}
$$

So, we conclude that this formula is valid for all positive integer values of n.

Note: The easiest way to complete the last two steps is to "work backwards." Start with the desired expression for S_{k+1} and multiply out to show that it is equal to the expression you found for $S_k + (k+1)^5$.

23. 1. When $n = 1$, $S_1 = 2 = \dfrac{1(2)(3)}{3}$.

2. Assume that

$$S_k = 1(2) + 2(3) + 3(4) + \cdots + k(k + 1) = \frac{k(k + 1)(k + 2)}{3}.$$

Then,

$$S_{k+1} = 1(2) + 2(3) + 3(4) + \cdots + k(k + 1) + (k + 1)(k + 2)$$

$$= S_k + (k + 1)(k + 2) = \frac{k(k + 1)(k + 2)}{3} + \frac{3(k + 1)(k + 2)}{3} = \frac{(k + 1)(k + 2)(k + 3)}{3}.$$

So, we conclude that this formula is valid for all positive integer values of n.

25. 1. When $n = 4$, $4! = 24$ and $2^4 = 16$, thus $4! > 2^4$.

2. Assume

$k! > 2^k, k > 4.$

Then,

$(k + 1)! = k!(k + 1) > 2^k(2)$ since $k! > 2^k$ and $k + 1 > 2$.

Thus, $(k + 1)! > 2^{k+1}$.

So, by extended mathematical induction, the inequality is valid for all integers n such that $n \geq 4$.

27. 1. When $n = 2$, $\dfrac{1}{\sqrt{1}} + \dfrac{1}{\sqrt{2}} \approx 1.707$ and $\sqrt{2} \approx 1.414$, thus $\dfrac{1}{\sqrt{1}} + \dfrac{1}{\sqrt{2}} > \sqrt{2}$.

2. Assume that

$$\frac{1}{\sqrt{1}} + \frac{1}{\sqrt{2}} + \frac{1}{\sqrt{3}} + \cdots + \frac{1}{\sqrt{k}} > \sqrt{k}, k > 2.$$

Then,

$$\frac{1}{\sqrt{1}} + \frac{1}{\sqrt{2}} + \frac{1}{\sqrt{3}} + \cdots + \frac{1}{\sqrt{k}} + \frac{1}{\sqrt{k + 1}} > \sqrt{k} + \frac{1}{\sqrt{k + 1}}.$$

Now it is sufficient to show that

$$\sqrt{k} + \frac{1}{\sqrt{k + 1}} > \sqrt{k + 1}, k > 2,$$

or equivalently $\left(\text{multiplying by } \sqrt{k + 1}\right)$,

$$\sqrt{k}\sqrt{k + 1} + 1 > k + 1.$$

This is true because

$$\sqrt{k}\sqrt{k + 1} + 1 > \sqrt{k}\sqrt{k} + 1 = k + 1.$$

Therefore,

$$\frac{1}{\sqrt{1}} + \frac{1}{\sqrt{2}} + \frac{1}{\sqrt{3}} + \cdots + \frac{1}{\sqrt{k}} + \frac{1}{\sqrt{k + 1}} > \sqrt{k + 1}.$$

So, by extended mathematical induction, the inequality is valid for all integers n such that $n \geq 2$.

29. 1. When $n = 1, \left(\frac{x}{y}\right)^2 < \left(\frac{x}{y}\right)$ and $(0 < x < y)$.

2. Assume that

$$\left(\frac{x}{y}\right)^{k+1} < \left(\frac{x}{y}\right)^k$$

$$\left(\frac{x}{y}\right)^{k+1} < \left(\frac{x}{y}\right)^k \Rightarrow \left(\frac{x}{y}\right)\left(\frac{x}{y}\right)^{k+1} < \left(\frac{x}{y}\right)\left(\frac{x}{y}\right)^k \Rightarrow \left(\frac{x}{y}\right)^{k+2} < \left(\frac{x}{y}\right)^{k+1}.$$

So, $\left(\frac{x}{y}\right)^{n+1} < \left(\frac{x}{y}\right)^n$ for all integers $n \geq 1$.

31. 1. When $n = 1, \left[1^3 + 3(1)^2 + 2(1)\right] = 6$ and 3 is a factor.

2. Assume that 3 is a factor of $k^3 + 3k^2 + 2k$.

Then, $(k + 1)^3 + 3(k + 1)^2 + 2(k + 1) = k^3 + 3k^2 + 3k + 1 + 3k^2 + 6k + 3 + 2k + 2$

$$= \left(k^3 + 3k^2 + 2k\right) + \left(3k^2 + 9k + 6\right)$$

$$= \left(k^3 + 3k^2 + 2k\right) + 3\left(k^2 + 3k + 2\right).$$

Because 3 is a factor of each term, 3 is a factor of the sum.

So, 3 is a factor of $\left(n^3 + 3n^2 + 2n\right)$ for every positive integer n.

33. Prove 3 is a factor of $2^{2n+1} + 1$ for all positive integers n.

1. When $n = 1, 2^{2\cdot 1+1} + 1 = 2^3 + 1 = 8 + 1 = 9$ and 3 is a factor.

2. Assume 3 is a factor of $2^{2k+1} + 1$.

Then,

$$2^{2(k+1)+1} + 1 = 2^{2k+2+1} + 1 = 2^{(2k+1)+2} + 1 = 2^{2k+1} \cdot 2^2 + 1 = 4 \cdot 2^{2k+1} + 1 = 4\left(2^{2k+1} + 1\right) - 3.$$

Because 3 is a factor of each term, 3 is a factor of the sum.

So, 3 is a factor of $2^{2n+1} + 1$ for all positive integers n.

35. 1. When $n = 1, (ab)^1 = a^1b^1 = ab$.

2. Assume that $(ab)^k = a^kb^k$.

Then, $(ab)^{k+1} = (ab)^k(ab)$

$$= a^kb^kab$$

$$= a^{k+1}b^{k+1}.$$

So, $(ab)^n = a^nb^n$.

37. 1. When $n = 2, (x_1x_2)^{-1} = \frac{1}{x_1x_2} = \frac{1}{x_1} \cdot \frac{1}{x_2} = x_1^{-1}x_2^{-1}$.

2. Assume that

$$(x_1x_2x_3 \cdots x_k)^{-1} = x_1^{-1}x_2^{-1}x_3^{-1} \cdots x_k^{-1}.$$

Then,

$$(x_1x_2x_3 \cdots x_kx_{k+1})^{-1} = \left[(x_1x_2x_3 \cdots x_k)x_{k+1}\right]^{-1}$$

$$= (x_1x_2x_3 \cdots x_k)^{-1}x_{k+1}^{-1}$$

$$= x_1^{-1}x_2^{-1}x_3^{-1} \cdots x_k^{-1}x_{k+1}^{-1}.$$

So, the formula is valid.

39. 1. When $n = 1$, $x(y_1) = xy_1$.

2. Assume that

$x(y_1 + y_2 + \cdots + y_k) = xy_1 + xy_2 + \cdots + xy_k$.

Then,

$$xy_1 + xy_2 + \cdots + xy_k + xy_{k+1} = x(y_1 + y_2 + \cdots + y_k) + xy_{k+1} = x\left[(y_1 + y_2 + \cdots + y_k) + y_{k+1}\right]$$
$$= x(y_1 + y_2 + \cdots + y_k + y_{k+1}).$$

So, the formula holds.

41. $S_n = 1 + 5 + 9 + 13 + \cdots + (4n - 3)$

$S_1 = 1 = 1 \cdot 1$

$S_2 = 1 + 5 = 6 = 2 \cdot 3$

$S_3 = 1 + 5 + 9 = 15 = 3 \cdot 5$

$S_4 = 1 + 5 + 9 + 13 = 28 = 4 \cdot 7$

From this sequence, it appears that $S_n = n(2n - 1)$.

This can be verified by mathematical induction.

1. The formula has already been verified for $n = 1$.
2. Assume that the formula is valid for $n = k$: $1 + 5 + 9 + 13 + \cdots + (4n - 3) = n(2n - 1)$.

   Then,

$$S_{k+1} = \left[1 + 5 + 9 + 13 + \cdots + (4k - 3)\right] + \left[4(k + 1) - 3\right] = k(2k - 1) + (4k + 1)$$
$$= 2k^2 + 3k + 1 = (k + 1)(2k + 1) = (k + 1)\left[2(k + 1) - 1\right]$$

So, the formula is valid.

43. $S_n = \dfrac{1}{4} + \dfrac{1}{12} + \dfrac{1}{24} + \dfrac{1}{40} + \cdots + \dfrac{1}{2n(n + 1)}$

$S_1 = \dfrac{1}{4} = \dfrac{1}{2(2)}$

$S_2 = \dfrac{1}{4} + \dfrac{1}{12} = \dfrac{4}{12} = \dfrac{2}{6} = \dfrac{2}{2(3)}$

$S_3 = \dfrac{1}{4} + \dfrac{1}{12} + \dfrac{1}{24} = \dfrac{9}{24} = \dfrac{3}{8} = \dfrac{3}{2(4)}$

$S_4 = \dfrac{1}{4} + \dfrac{1}{12} + \dfrac{1}{24} + \dfrac{1}{40} = \dfrac{16}{40} = \dfrac{4}{10} = \dfrac{4}{2(5)}$

From the sequence, it appears that $S_n = \dfrac{n}{2(n + 1)}$.

This can be verified by mathematical induction.

1. The formula has already by verified for $n = 1$.
2. Assume that the formula is valid for $n = k$: $\dfrac{1}{4} + \dfrac{1}{12} + \dfrac{1}{24} + \dfrac{1}{40} + \cdots + \dfrac{1}{2n(n + 1)} = \dfrac{n}{2(n + 1)}$

   Then, $S_{k+1} = \left[\dfrac{1}{4} + \dfrac{1}{12} + \dfrac{1}{40} + \cdots + \dfrac{1}{2k(k + 1)}\right] + \dfrac{1}{2(k + 1)(k + 2)} = \dfrac{k}{2(k + 1)} + \dfrac{1}{2(k + 1)(k + 2)}$

$$= \frac{k(k + 2) + 1}{2(k + 1)(k + 2)} = \frac{k^2 + 2k + 1}{2(k + 1)(k + 2)} = \frac{(k + 1)^2}{2(k + 1)(k + 2)} = \frac{k + 1}{2(k + 2)}.$$

So, the formula is valid.

45. $\sum_{n=1}^{15} n = \frac{15(15+1)}{2} = 120$

47. $\sum_{n=1}^{6} n^2 = \frac{6(6+1)[2(6)+1]}{6} = 91$

49. $\sum_{n=1}^{5} n^4 = \frac{5(5+1)[2(5)+1][3(5)^2+3(5)-1]}{30} = 979$

51. $\sum_{n=1}^{6}(n^2 - n) = \sum_{n=1}^{6} n^2 - \sum_{n=1}^{6} n$

$= \frac{6(6+1)[2(6)+1]}{6} - \frac{6(6+1)}{2}$

$= 91 - 21 = 70$

53. $\sum_{i=1}^{6}(6i - 8i^3) = 6\sum_{i=1}^{6} i - 8\sum_{i=1}^{6} i^3$

$= 6\left[\frac{6(6+1)}{2}\right] - 8\left[\frac{(6)^2(6+1)^2}{4}\right]$

$= 6(21) - 8(441)$

$= -3402$

55. 5, 14, 23, 32, 41, 50, …

Linear

Note: This is an arithmetic sequence.

$a_1 = 5, d = 9$

$a_n = 5 + (n-1)(9)$

$a_n = 9n - 4$

57. 4, 10, 20, 34, 52, 74, …

Quadratic

$$\begin{cases} a + b + c = 4 \\ 4a + 2b + c = 10 \\ 9a + 3b + c = 20 \end{cases}$$

Solving this system yields $a = 2, b = 0$, and $c = 2$.

So, $a_n = 2n^2 + 2$.

59. −1, 11, 31, 59, 95, 139, …

Quadratic

$$\begin{cases} a + b + c = -1 \\ 4a + 2b + c = 11 \\ 9a + 3b + c = 31 \end{cases}$$

Solving this system yields $a = 4, b = 0$, and $c = -5$.

So, $a_n = 4n^2 - 5$.

61. $a_1 = 0, a_n = a_{n-1} + 3$

$a_1 = a_1 = 0$

$a_2 = a_1 + 3 = 0 + 3 = 3$

$a_3 = a_2 + 3 = 3 + 3 = 6$

$a_4 = a_3 + 3 = 6 + 3 = 9$

$a_5 = a_4 + 3 = 9 + 3 = 12$

$a_6 = a_5 + 3 = 12 + 3 = 15$

a_n: 0 3 6 9 12 15

First differences: 3 3 3 3 3

Second differences: 0 0 0 0

Because the first differences are equal, the sequence has a linear model.

63. $a_1 = 4, a_n = a_{n-1} + 3n$

$a_1 = a_1 = 4$

$a_2 = a_1 + 3n = 4 + 3(2) = 10$

$a_3 = a_2 + 3n = 10 + 3(3) = 19$

$a_4 = a_3 + 3n = 19 + 3(4) = 31$

$a_5 = a_4 + 3n = 31 + 3(5) = 46$

$a_6 = a_5 + 3n = 46 + 3(6) = 64$

a_n: 4 10 19 31 46 64

First differences: 6 9 12 15 18

Second differences: 3 3 3 3

Because the second differences are all the same, the sequence has a quadratic model.

65. $a_1 = 3, a_n = a_{n-1} + n^2$

$a_1 = a_1 = 3$

$a_2 = a_1 + 2^2 = 3 + 4 = 7$

$a_3 = a_2 + 3^2 = 7 + 9 = 16$

$a_4 = a_3 + 4^2 = 16 + 16 = 32$

$a_5 = a_4 + 5^2 = 32 + 25 = 57$

$a_6 = a_5 + 6^2 = 57 + 36 = 93$

a_n: 3 7 16 32 57 93

First differences: 4 9 16 25 36

Second differences: 5 7 9 11

Neither linear nor quadratic.

67. $a_1 = 5, a_n = 4n - a_{n-1}$

$a_1 = a_1 = 5$

$a_2 = 4(2) - a_1 = 8 - 5 = 3$

$a_3 = 4(3) - a_2 = 12 - 3 = 9$

$a_4 = 4(4) - a_3 = 16 - 9 = 7$

$a_5 = 4(5) - a_4 = 20 - 7 = 13$

$a_6 = 4(6) - a_5 = 24 - 13 = 11$

a_n:	5	3	9	7	13	11
First differences:	−2	6	−2	6	−2	
Second differences:	8	−8	8	−8		

Neither linear nor quadratic.

69. $a_0 = 3, a_1 = 3, a_4 = 15$

Let $a_n = an^2 + bn + c$.

Then:

$$a_0 = a(0)^2 + b(0) + c = 3 \Rightarrow \quad c = 3$$

$$a_1 = a(1)^2 + b(1) + c = 3 \Rightarrow a + b + c = 3$$

$$a + b = 0$$

$$a_4 = a(4)^2 + b(4) + c = 15 \Rightarrow 16a + 4b + c = 15$$

$$16a + 4b = 12$$

$$4a + b = 3$$

By elimination: $-a - b = 0$

$$\frac{4a + b = 3}{3a = 3}$$

$$a = 1 \Rightarrow b = -1$$

So, $a_n = n^2 - n + 3$.

71. $a_0 = -1, a_2 = 5, a_4 = 15$

Let $a_n = an^2 + bn + c$.

Then: $a_0 = a(0)^2 + b(0) + c = -1 \Rightarrow c = -1$

$$a_2 = a(2)^2 + b(2) + c = 5 \Rightarrow 4a + 2b + c = 5$$

$$4a + 2b = 6$$

$$2a + b = 3$$

$$a_4 = a(4)^2 + b(4) + c = 15 \Rightarrow 16a + 4b + c = 15$$

$$16a + 4b = 16$$

$$4a + b = 4$$

By elimination: $-2a - b = -3$

$$\frac{4a + b = 4}{2a = 1}$$

$$a = \tfrac{1}{2}$$

$$4\left(\tfrac{1}{2}\right) + b = 4$$

$$b = 2$$

So, $a_n = \frac{1}{2}n^2 + 2n - 1$.

73. $a_1 = 0, a_2 = 7, a_4 = 27$

Let $a_n = an^2 + bn + c.$

Then: $a_1 = a(1)^2 + b(1) + c = 0 \Rightarrow a + b + c = 0$

$a_2 = a(2)^2 + b(2) + c = 7 \Rightarrow 4a + 2b + c = 7$

$a_4 = a(4)^2 + b(4) + c = 27 \Rightarrow 16a + 4b + c = 27$

$$\begin{cases} a + b + c = 0 \\ 4a + 2b + c = 7 \\ 16a + 4b + c = 27 \end{cases}$$

$$\begin{cases} a + b + c = 0 \\ -2b - 3c = 7 \\ -12b - 15c = 27 \end{cases}$$

$$\begin{cases} a + b + c = 0 \\ -2b - 3c = 7 \\ 3c = -15 \end{cases}$$

$3c = -15 \rightarrow c = -5$

$-2b - 3(-5) = 7 \rightarrow b = 4$

$a + 4 + (-5) = 0 \rightarrow a = 1$

So, $a_n = n^2 + 4n - 5.$

75. (a) n: 10 11 12 13 14 15

Terms: 4785 4801 4816 4831 4846 4859

First differences: 16 15 15 15 13

Sample Answer: Using common difference $d = 15, a_n = 15n + 4636$

(b) Using a graphing utility, a linear model for the data is $a_n \approx 14.9n + 4637$. So, the models are comparable.

(c) Using $a_n = 15n + 4636$, the number of residents in 2021 is $a_{16} = 15(21) + 4636 = 4{,}951{,}000$.

Using $a_n = 14.9n + 4637$, the number of residents in 2021 is $a_{16} = 14.9(21) + 4637 = 4{,}949{,}900$.

So, the predictions are similar.

77. False. P_1 must be proven to be true.

Section 9.5 The Binomial Theorem

1. expanding

3. Binomial Theorem; Pascal's Triangle

5. ${}_5C_3 = \dfrac{5!}{3! \cdot 2!} = \dfrac{5 \cdot 4}{2 \cdot 1} = 10$

7. ${}_{12}C_0 = \dfrac{12!}{0! \cdot 12!} = 1$

9. $\dbinom{10}{4} = \dfrac{10!}{6! \cdot 4!} = \dfrac{10 \cdot 9 \cdot 8 \cdot 7 \cdot 6!}{6!(24)} = 210$

11. $\dbinom{100}{98} = \dfrac{100!}{2! \cdot 98!} = \dfrac{100 \cdot 99}{2 \cdot 1} = 4950$

13.

1
1 1
1 2 1
1 3 3 1
1 4 6 4 1
1 5 10 10 5 1
1 6 15 (20) 15 6 1

$\dbinom{6}{3} = 20$, the 4th entry in the 6th row.

15.

$$
\begin{array}{ccccccccccc}
 & & & & & 1 & & & & & \\
 & & & & 1 & & 1 & & & & \\
 & & & 1 & & 2 & & 1 & & & \\
 & & 1 & & 3 & & 3 & & 1 & & \\
 & 1 & & 4 & & 6 & & 4 & & 1 & \\
1 & & \textcircled{5} & & 10 & & 10 & & 5 & & 1
\end{array}
$$

$\binom{5}{1} = 5$, the 2nd entry in the 5th row.

17. $(x + 1)^6 = {}_6C_0x^6 + {}_6C_1x^5(1) + {}_6C_2x^4(1)^2 + {}_6C_3x^3(1)^3 + {}_6C_4x^2(1)^4 + {}_6C_5x(1)^5 + {}_6C_6(1)^6$
$= x^6 + 6x^5 + 15x^4 + 20x^3 + 15x^2 + 6x + 1$

19. $(y - 3)^3 = {}_3C_0y^3 - {}_3C_1y^2(3) + {}_3C_2y(3)^2 - {}_3C_3(3)^3$
$= 1y^3 - 3y^2(3) + 3y(3)^2 - 1(3)^3$
$= y^3 - 9y^2 + 27y - 27$

21. $(r + 3s)^3 = {}_3C_0r^3 + {}_3C_1r^2(3s) + {}_3C_2r(3s)^2 + {}_3C_3(3s)^3$
$= 1r^3 + 3r^2(3s) + 3r(3s)^2 + 1(3s)^3$
$= r^3 + 9r^2s + 27rs^2 + 27s^3$

23. $(3a - 4b)^5 = {}_5C_0(3a)^5 - {}_5C_1(3a)^4(4b) + {}_5C_2(3a)^3(4b)^2 - {}_5C_3(3a)^2(4b)^3 + {}_5C_4(3a)(4b)^4 - {}_5C_5(4b)^5$
$= (1)(243a^5) - 5(81a^4)(4b) + 10(27a^3)(16b^2) - 10(9a^2)(64b^3) + 5(3a)(256b^4) - (1)(1024b^5)$
$= 243a^5 - 1620a^4b + 4320a^3b^2 - 5760a^2b^3 + 3840ab^4 - 1024b^5$

25. $(a + 6)^4 = {}_4C_0a^4 + {}_4C_1a^3(6) + {}_4C_2a^2(6)^2 + {}_4C_3a(6)^3 + {}_4C_4(6)^4$
$= 1a^4 + 4a^3(6) + 6a^2(6)^2 + 4a(6)^3 + 1(6)^4$
$= a^4 + 24a^3 + 216a^2 + 864a + 1296$

27. $(y - 1)^6 = {}_6C_0y^6 - {}_6C_1y^5(1) + {}_6C_2y^4(1)^2 - {}_6C_3y^3(1)^3 + {}_6C_4y^2(1)^4 - {}_6C_5y(1)^5 + {}_6C_6(1)^6$
$= 1y^6 - 6y^5(1) + 15y^4(1)^2 - 20y^3(1)^3 + 15y^2(1)^4 - 6y(1)^5 + 1(1)^6$
$= y^6 - 6y^5 + 15y^4 - 20y^3 + 15y^2 - 6y + 1$

29. 4th Row of Pascal's Triangle: 1 4 6 4 1

$(3 - 2z)^4 = 3^4 - 4(3)^3(2z) + 6(3)^2(2z)^2 - 4(3)(2z)^3 + (2z)^4$
$= 81 - 216z + 216z^2 - 96z^3 + 16z^4$

31. 5th Row of Pascal's Triangle: 1 5 10 10 5 1

$(x + 2y)^5 = 1x^5 + 5x^4(2y) + 10x^3(2y)^2 + 10x^2(2y)^3 + 5x(2y)^4 + 1(2y)^5$
$= x^5 + 10x^4y + 40x^3y^2 + 80x^2y^3 + 80xy^4 + 32y^5$

33. $(x^2 + y^2)^4 = {}_4C_0(x^2)^4 + {}_4C_1(x^2)^3(y^2) + {}_4C_2(x^2)^2(y^2)^2 + {}_4C_3(x^2)(y^2)^3 + {}_4C_4(y^2)^4$
$= (1)(x^8) + (4)(x^6y^2) + (6)(x^4y^4) + (4)(x^2y^6) + (1)(y^8)$
$= x^8 + 4x^6y^2 + 6x^4y^4 + 4x^2y^6 + y^8$

35. $\left(\frac{1}{x} + y\right)^5 = {}_5C_0\left(\frac{1}{x}\right)^5 + {}_5C_1\left(\frac{1}{x}\right)^4 y + {}_5C_2\left(\frac{1}{x}\right)^3 y^2 + {}_5C_3\left(\frac{1}{x}\right)^2 y^3 + {}_5C_4\left(\frac{1}{x}\right)y^4 + {}_5C_5 y^5$

$= 1\left(\frac{1}{x}\right)^5 + 5\left(\frac{1}{x}\right)^4 y + 10\left(\frac{1}{x}\right)^3 y^2 + 10\left(\frac{1}{x}\right)^2 y^3 + 5\left(\frac{1}{x}\right)y^4 + 1y^5$

$= \frac{1}{x^5} + \frac{5y}{x^4} + \frac{10y^2}{x^3} + \frac{10y^3}{x^2} + \frac{5y^4}{x} + y^5$

37. $2(x-3)^4 + 5(x-3)^2 = 2\left[{}_4C_0x^4 - {}_4C_1x^3(3) + {}_4C_2x^2(3)^2 - {}_4C_3x(3)^3 + {}_4C_4(3)^4\right] + 5\left[{}_2C_0x^2 + {}_2C_1x(3) + {}_2C_2(3)^2\right]$

$= 2\left[x^4 - 4(x^3)(3) + 6(x^2)(3)^2 - 4(x)(3^3) + 3^4\right] + 5\left[x^2 - 2(x)(3) + 3^2\right]$

$= 2(x^4 - 12x^3 + 54x^2 - 108x + 81) + 5(x^2 - 6x + 9)$

$= 2x^4 - 24x^3 + 113x^2 - 246x + 207$

39. The 4th term in the expansion of $(x + y)^{10}$ is

${}_{10}C_3x^{10-3}y^3 = 120x^7y^3.$

41. The 3rd term in the expansion of $(x - 6y)^5$ is

${}_5C_2x^{5-2}(-6y)^2 = 10x^3(36y^2) = 360x^3y^2.$

43. The 8th term in the expansion of $(4x + 3y)^9$ is

${}_9C_7(4x)^{9-7}(3y)^7 = 36(16x^2)(2187y^7) = 1{,}259{,}712x^2y^7.$

45. The 10th term in the expansion of $(10x - 3y)^{12}$ is

${}_{12}C_9(10x)^{12-9}(-3y)^9 = 220(1000x^3)(-19{,}683y^9)$

$= -4{,}330{,}260{,}000x^3y^9.$

47. The new term involving ax^3 in the expansion of

$(x + 2)^6$ is ${}_6C_3x^3(2)^3 = \frac{6!}{3! \cdot 3!} \cdot 2^3x^3 = 160x^3.$

The coefficient is 160.

49. The term involving x^2y^8 in the expansion of

$(4x - y)^{10}$ is

${}_{10}C_8(4x)^2(-y)^8 = \frac{10!}{(10-8)!8!} \cdot 16x^2y^8 = 720x^2y^8.$

The coefficient is 720.

51. The term involving x^4y^5 in the expansion of $(2x - 5x)^9$

is ${}_9C_5(2x)^4(-5y)^5 = 126(16x^4)(-3125y^5)$

$= -6{,}300{,}000x^4y^5.$

The coefficient is $-6{,}300{,}000$.

53. The term involving $x^8y^6 = (x^2)^4y^6$ in the expansion of

$(x^2 + y)^{10}$ is ${}_{10}C_6(x^2)^4y^6 = \frac{10!}{4!6!}(x^2)^4y^6 = 210x^8y^6.$

The coefficient is 210.

55. $(\sqrt{x} + 5)^3 = (\sqrt{x})^3 + 3(\sqrt{x})^2(5) + 3(\sqrt{x})(5^2) + 5^3$

$= x^{3/2} + 15x + 75x^{1/2} + 125$

57. $(x^{2/3} - y^{1/3})^3 = (x^{2/3})^3 - 3(x^{2/3})^2(y^{1/3}) + 3(x^{2/3})(y^{1/3})^2 - (y^{1/3})^3 = x^2 - 3x^{4/3}y^{1/3} + 3x^{2/3}y^{2/3} - y$

59. $(3\sqrt{t} + \sqrt[4]{t})^4 = (3\sqrt{t})^4 + 4(3\sqrt{t})^3(\sqrt[4]{t}) + 6(3\sqrt{t})^2(\sqrt[4]{t})^2 + 4(3\sqrt{t})(\sqrt[4]{t})^3 + (\sqrt[4]{t})^4 = 81t^2 + 108t^{7/4} + 54t^{3/2} + 12t^{5/4} + t$

61. $\frac{f(x+h) - f(x)}{h} = \frac{(x+h)^3 - x^3}{h} = \frac{x^3 + 3x^2h + 3xh^2 + h^3 - x^3}{h} = \frac{h(3x^2 + 3xh + h^2)}{h} = 3x^2 + 3xh + h^2, h \neq 0$

63. $\frac{f(x+h) - f(x)}{h} = \frac{(x+h)^6 - x^6}{h}$

$= \frac{x^6 + 6x^5h + 15x^4h^2 + 20x^3h^3 + 15x^2h^4 + 6xh^5 + h^6 - x^6}{h}$

$= \frac{h(6x^5 + 15x^4h + 20x^3h^2 + 15x^2h^3 + 6xh^4 + h^5)}{h}$

$= 6x^5 + 15x^4h + 20x^3h^2 + 15x^2h^3 + 6xh^4 + h^5, h \neq 0$

65. $\dfrac{f(x+h)-f(x)}{h} = \dfrac{\sqrt{x+h}-\sqrt{x}}{h} = \dfrac{\sqrt{x+h}-\sqrt{x}}{h}\cdot\dfrac{\sqrt{x+h}+\sqrt{x}}{\sqrt{x+h}+\sqrt{x}} = \dfrac{(x+h)-x}{h\left(\sqrt{x+h}+\sqrt{x}\right)} = \dfrac{1}{\sqrt{x+h}+\sqrt{x}},\ h \neq 0$

67. $(1+i)^4 = {}_4C_0(1)^4 + {}_4C_1(1)^3 i + {}_4C_2(1)^2 i^2 + {}_4C_3(1)i^3 + {}_4C_4 i^4 = 1 + 4i - 6 - 4i + 1 = -4$

69. $(2-3i)^6 = {}_6C_0 2^6 - {}_6C_1 2^5(3i) + {}_6C_2 2^4(3i)^2 - {}_6C_3 2^3(3i)^3 + {}_6C_4 2^2(3i)^4 - {}_6C_5 2(3i)^5 + {}_6C_6(3i)^6$

$= (1)(64) - (6)(32)(3i) + 15(16)(-9) - 20(8)(-27i) + 15(4)(81) - 6(2)(243i) + (1)(-729)$

$= 64 - 576i - 2160 + 4320i + 4860 - 2916i - 729$

$= 2035 + 828i$

71. $\left(-\dfrac{1}{2}+\dfrac{\sqrt{3}}{2}i\right)^3 = \dfrac{1}{8}\left[(-1)^3 + 3(-1)^2\left(\sqrt{3}i\right) + 3(-1)\left(\sqrt{3}i\right)^2 + \left(\sqrt{3}i\right)^3\right] = \dfrac{1}{8}\left[-1 + 3\sqrt{3}i + 9 - 3\sqrt{3}i\right] = 1$

73. $(1.02)^8 = (1+0.02)^8$

$= 1 + 8(0.02) + 28(0.02)^2 + 56(0.02)^3 + 70(0.02)^4 + 56(0.02)^5 + 28(0.02)^6 + 8(0.02)^7 + (0.02)^8$

$= 1 + 0.16 + 0.0112 + 0.000448 + \cdots$

≈ 1.172

75. $(2.99)^{12} = (3-0.01)^{12}$

$= 3^{12} - 12(3)^{11}(0.01) + 66(3)^{10}(0.01)^2 - 220(3)^9(0.01)^3 + 495(3)^8(0.01)^4$

$-792(3)^7(0.01)^5 + 924(3)^6(0.01)^6 - 792(3)^5(0.01)^7 + 495(3)^4(0.01)^8$

$-220(3)^3(0.01)^9 + 66(3)^2(0.01)^{10} - 12(3)(0.01)^{11} + (0.01)^{12}$

$\approx 531{,}441 - 21{,}257.64 + 389.7234 - 4.3303 + 0.0325 - 0.0002 + \cdots \approx 510{,}568.785$

77. ${}_7C_4\left(\dfrac{1}{2}\right)^4\left(\dfrac{1}{2}\right)^3 = \dfrac{7!}{3!4!}\left(\dfrac{1}{16}\right)\left(\dfrac{1}{8}\right) = 35\left(\dfrac{1}{16}\right)\left(\dfrac{1}{8}\right) \approx 0.273$

79. ${}_8C_4\left(\dfrac{1}{3}\right)^4\left(\dfrac{2}{3}\right)^4 = \dfrac{8!}{4!4!}\left(\dfrac{1}{81}\right)\left(\dfrac{16}{81}\right) = 70\left(\dfrac{1}{81}\right)\left(\dfrac{16}{81}\right) \approx 0.171$

81. $f(x) = x^2 - 4x$

$g(x) = f(x+4)$

$= (x+4)^3 - 4(x+4)$

$= x^3 + 3x^2(4) + 3x(4)^2 + (4)^3 - 4x - 16$

$= x^3 + 12x^2 + 48x + 64 - 4x - 16$

$= x^3 + 12x^2 + 44x + 48$

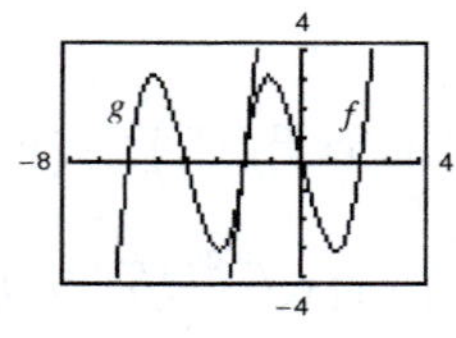

The graph of g is the same as the graph of f shifted four units to the left.

83.

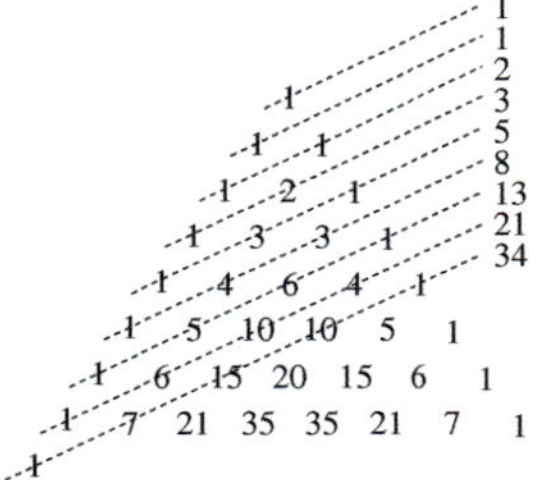

The first nine terms of the sequence are 1, 1, 2, 3, 5, 8, 13, 21, 34, …

After the first two terms, the next terms are formed by adding the previous two terms.

$$\begin{aligned} a_1 &= 1, a_2 = 1 \\ a_3 &= a_1 + a_2 = 1 + 1 = 2 \\ a_4 &= a_2 + a_3 = 1 + 2 = 3 \\ a_5 &= a_3 + a_4 = 2 + 3 = 5 \\ a_6 &= a_4 + a_5 = 3 + 5 = 8 \\ a_7 &= a_5 + a_6 = 5 + 8 = 13 \end{aligned}$$

This is called the Fibonacci sequence.

85. (a)
$$\begin{aligned} f(t) &= -0.056t^2 + 1.62t + 16.4 \\ g(t) &= f(t+5) \\ &= -0.056(t+5)^2 + 1.62(t+5) + 16.4 \\ &= -0.056(t^2 + 10t + 25) + 1.62(t+5) + 16.4 \\ &= -0.056t^2 - 0.56t - 1.4 + 1.62t + 8.1 + 16.4 \\ &= -0.056t^2 + 1.06t + 23.1 \end{aligned}$$

(b) and (c)

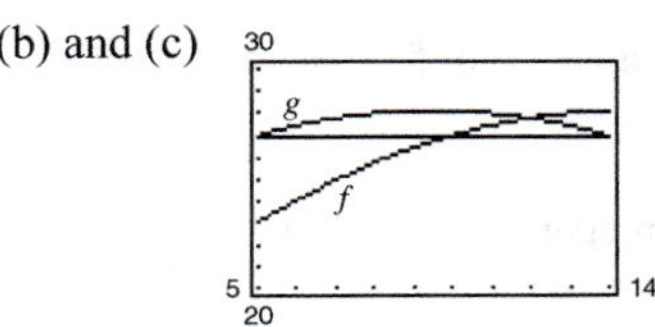

Graphically, child support collections exceeded \$27 billion in about 2010.

87. True. The coefficients from the Binomial Theorem can be used to find the numbers in Pascal's Triangle.

89. The first and last numbers in each row are 1. Every other number in each row is formed by adding the two numbers immediately above the number.

91. The functions $f(x) = (1 - x)^3$ and
$k(x) = 1 - 3x + 3x^2 + x^3$

f and k have identical graphs, because $k(x)$ is the expansion of $f(x)$.

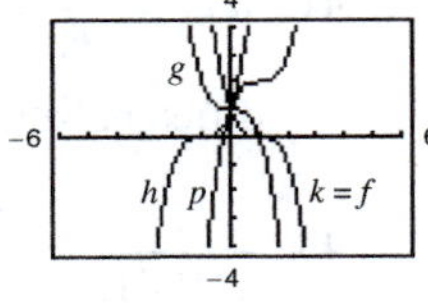

93.
$$\begin{aligned} {}_nC_{n-r} &= \frac{n!}{(n-(n-r))!(n-r)!} \\ &= \frac{n!}{r!(n-r)!} \\ &= \frac{n!}{(n-r)!r!} \\ &= {}_nC_r \end{aligned}$$

95.
$$\begin{aligned} {}_nC_r + {}_nC_{r-1} &= \frac{n!}{(n-r)!r!} + \frac{n!}{(n-r+1)!(r-1)!} \\ &= \frac{n!(n-r+1)!(r-1)! + n!(n-r)!r!}{(n-r)!r!(n-r+1)!(r-1)!} \\ &= \frac{n!\left[(n-r+1)!(r-1)! + r!(n-r)!\right]}{(n-r)!r!(n-r+1)!(r-1)!} \\ &= \frac{n!\cancel{(r-1)!}\left[(n-r+1)! + r(n-r)!\right]}{(n-r)!r!(n-r+1)!\cancel{(r-1)!}} \\ &= \frac{n!\cancel{(n-r)!}\left[(n-r+1) + r\right]}{\cancel{(n-r)!}r!(n-r+1)!} \\ &= \frac{n![n+1]}{r!(n-r+1)!} \\ &= \frac{(n+1)!}{\left[(n+1)-r\right]!r!} \\ &= {}_{n+1}C_r \end{aligned}$$

97.

n	r	$_nC_r$	$_nC_{n-r}$
9	5	126	126
7	1	7	7
12	4	495	495
6	0	1	1
10	7	120	120

$_nC_r = {_nC_{n-r}}$

The table illustrates the symmetry of Pascal's Triangle.

Section 9.6 Counting Principles

1. Fundamental Counting Principle

3. $_nP_r = \dfrac{n!}{(n-r)!}$

5. combinations

7. Odd integers: 1, 3, 5, 7, 9, 11
6 ways

9. Prime integers: 2, 3, 5, 7, 11
5 ways

11. Divisible by 4: 4, 8, 12
3 ways

13. Sum is 9: $1 + 8, 2 + 7, 3 + 6, 4 + 5, 5 + 4,$
$6 + 3, 7 + 2, 8 + 1$

8 ways

15. Amplifiers: 3 choices
Compact disc players: 2 choices
Speakers: 5 choices
Total: $3 \cdot 2 \cdot 5 = 30$ ways

17. Math courses: 2
Science courses: 3
Social sciences and humanities courses: 5
Total: $2 \cdot 3 \cdot 5 = 30$ schedules

19. $2^6 = 64$

21. $26 \cdot 26 \cdot 26 \cdot 10 \cdot 10 \cdot 10 \cdot 10 = 175{,}760{,}000$

distinct license plate numbers

23. (a) $9 \cdot 10 \cdot 10 = 900$

(b) $9 \cdot 9 \cdot 8 = 648$

(c) $9 \cdot 10 \cdot 2 = 180$

(d) $6 \cdot 10 \cdot 10 = 600$

25. $40^3 = 64{,}000$

27. (a) $8 \cdot 7 \cdot 6 \cdot 5 \cdot 4 \cdot 3 \cdot 2 \cdot 1 = 40{,}320$

(b) $8 \cdot 1 \cdot 6 \cdot 1 \cdot 4 \cdot 1 \cdot 2 \cdot 1 = 384$

29. $5! = 120$ ways

31. $_5P_2 = \dfrac{5!}{(5-2)!} = \dfrac{5!}{3!} = 5 \cdot 4 = 20$

33. $_{12}P_2 = \dfrac{12!}{(12-2)!} = \dfrac{12!}{10!} = 12 \cdot 11 = 132$

35. $_{15}P_3 = 2730$

37. $_{50}P_4 = 5{,}527{,}200$

39. The number of permutations of 9 possible donors taken 3 at a time is

$_9P_3 = \dfrac{9!}{(9-3)!} = \dfrac{9!}{6!} = 9 \cdot 8 \cdot 7 = 504$ possible orders.

41. $_{15}P_9 = \dfrac{15!}{6!} = 1{,}816{,}214{,}400$

different batting orders

43. $\dfrac{7!}{2!1!3!1!} = \dfrac{7!}{2!3!} = 420$

45. $\dfrac{7!}{2!1!1!1!1!1!} = \dfrac{7!}{2!} = 7 \cdot 6 \cdot 5 \cdot 4 \cdot 3 = 2520$

47.

ABCD	BACD	CABD	DABC
ABDC	BADC	CADB	DACB
ACBD	BCAD	CBAD	DBAC
ACDB	BCDA	CBDA	DBCA
ADBC	BDAC	CDAB	DCAB
ADCB	BDCA	CDBA	DCBA

49. ${}_6C_4 = \dfrac{6!}{4!(6-4)!} = \dfrac{6!}{4!2!} = 15$

51. ${}_9C_9 = \dfrac{9!}{9!(9-9)!} = \dfrac{9!}{9!0!} = 1$

53. ${}_{16}C_2 = 120$

55. ${}_{20}C_6 = 38{,}760$

57. There are ${}_6C_2 = 15$ different combinations: AB, AC, AD, AE, AF, BC, BD, BE, BF, CD, CE, CF, DE, DF, EF

59. ${}_{40}C_{12} = \dfrac{40!}{28!12!} = 5{,}586{,}853{,}480$ ways

61. ${}_{35}C_5 = \dfrac{35!}{30!5!} = 324{,}632$ ways

63. There are 22 good units and 3 defective units.

(a) ${}_{22}C_4 = \dfrac{22!}{4!8!} = 7315$ ways

(b) ${}_{22}C_2 \cdot {}_3C_2 = \dfrac{22!}{2!20!} \cdot \dfrac{3!}{2!1!} = 231 \cdot 3 = 693$ ways

(c) ${}_{22}C_4 + {}_{22}C_3 \cdot {}_3C_1 + {}_{22}C_2 \cdot {}_3C_2 = \dfrac{22!}{4!18!} + \dfrac{22!}{3!19!} \cdot \dfrac{3!}{1!2!} + \dfrac{22!}{2!20!} \cdot \dfrac{3!}{2!1!} = 7315 + 1540 \cdot 3 + 231 \cdot 3 = 12{,}628$ ways

65. (a) Select type of card for three of a kind: ${}_{13}C_1$

Select three of four cards for three of a kind: ${}_4C_3$

Select type of card for pair: ${}_{12}C_1$

Select two of four cards for pair: ${}_4C_2$

$${}_{13}C_1 \cdot {}_4C_3 \cdot {}_{12}C_1 \cdot {}_4C_2 = \frac{13!}{(13-1)!1!} \cdot \frac{4!}{(4-3)!3!} \cdot \frac{12!}{(12-1)!1!} \cdot \frac{4!}{(4-2)!2!} = 3744$$

(b) Select two jacks: ${}_4C_2$

Select three aces: ${}_4C_3$

$${}_4C_2 \cdot {}_4C_3 = \frac{4!}{(4-2)!2!} \cdot \frac{4!}{(4-3)!3!} = 24$$

67. ${}_7C_1 \cdot {}_{12}C_3 \cdot {}_{20}C_2 = \dfrac{7!}{(7-1)!1!} \cdot \dfrac{12!}{(12-3)!3!} \cdot \dfrac{20!}{(20-2)!2!} = 292{,}600$

69. ${}_5C_2 - 5 = 10 - 5 = 5$ diagonals

71. ${}_8C_2 - 8 = 28 - 8 = 20$ diagonals

73. ${}_9C_2 = \dfrac{9!}{2!7!} = 36$ lines

75. $4 \cdot {}_{n+1}P_2 = {}_{n+2}P_3$ **Note:** $n \geq 1$ for this to be defined.

$$4 \cdot \frac{(n+1)!}{(n-1)!} = \frac{(n+2)!}{(n-1)!}$$

$4(n+1)(n) = (n+2)(n+1)n$ (We can divide by $(n+1)n$ because $n \neq -1$ and $n \neq 0$.)

$4 = n + 2$

$2 = n$

77. ${}_{n+1}P_3 = 4 \cdot {}_nP_2$ **Note:** $n \geq 2$ for this to be defined.

$$\frac{(n+1)!}{(n-2)!} = 4 \cdot \frac{n!}{(n-2)!}$$

$(n+1)(n)(n-1) = 4(n)(n-1)$ (We can divide by $n(n-1)$ because $n \neq 0$, and $n \neq 1$.)

$n + 1 = 4$

$n = 3$

79. $14 \cdot {}_nP_3 = {}_{n+2}P_4$ **Note:** $n \geq 3$ for this to be defined.

$$14\left(\frac{n!}{(n-3)!}\right) = \frac{(n+2)!}{(n-2)!}$$
$$14n(n-1)(n-2) = (n+2)(n+1)n(n-1)$$ (We can divide here by $n(n-1)$ because $n \neq 0$, $n \neq 1$.)
$$14(n-2) = (n+2)(n+1)$$
$$14n - 28 = n^2 + 3n + 2$$
$$0 = n^2 - 11n + 30$$
$$0 = (n-5)(n-6)$$
$$n = 5 \text{ or } n = 6$$

81. ${}_nP_4 = 10 \cdot {}_{n-1}P_3$ **Note:** $n \geq 4$ for this to be defined.

$$\frac{n!}{(n-4)!} = 10 \cdot \frac{(n-1)!}{(n-4)!}$$
$$n(n-1)(n-2)(n-3) = 10(n-1)(n-2)(n-3)$$ (We can divide by $(n-1)(n-2)(n-3)$ because $n \neq 1$, $n \neq 2$, and $n \neq 3$.)
$$n = 10$$

83. False.

It is an example of a combination.

85. ${}_{10}P_6 > {}_{10}C_6$

Changing the order of any of the six elements selected results in a different permutation but the same combination.

87. $${}_nP_{n-1} = \frac{n!}{(n-(n-1))!} = \frac{n!}{1!} = \frac{n!}{0!} = {}_nP_n$$

89. $${}_nC_{n-1} = \frac{n!}{(n-(n-1))!(n-1)!} = \frac{n!}{(1)!(n-1)!}$$
$$= \frac{n!}{(n-1)!1!} = {}_nC_1$$

91. ${}_{100}P_{80} \approx 3.836 \times 10^{139}$

This number is too large for some calculators to evaluate.

Section 9.7 Probability

1. experiment; outcomes

3. probability

5. mutually exclusive

7. complement

9. $\{(H, 1), (H, 2), (H, 3), (H, 4), (H, 5), (H, 6), (T, 1), (T, 2), (T, 3), (T, 4), (T, 5), (T, 6)\}$

11. {ABC, ACB, BAC, BCA, CAB, CBA}

13. {AB, AC, AD, AE, BC, BD, BE, CD, CE, DE}

15. $E = \{HHT, HTH, THH\}$

$$P(E) = \frac{n(E)}{n(S)} = \frac{3}{8}$$

17. $E = \{HHH, HHT, HTH, HTT\}$

$$P(E) = \frac{n(E)}{n(S)} = \frac{4}{8} = \frac{1}{2}$$

19. $E = \{HHH, HHT, HTH, HTT, THH, THT, TTH\}$

$$P(E) = \frac{n(E)}{n(S)} = \frac{7}{8}$$

21. $E = \{K\clubsuit, K\diamondsuit, K\heartsuit, K\spadesuit, Q\clubsuit, Q\diamondsuit, Q\heartsuit, Q\spadesuit, J\clubsuit, J\diamondsuit, J\heartsuit, J\spadesuit\}$

$$P(E) = \frac{n(E)}{n(S)} = \frac{12}{52} = \frac{3}{13}$$

23. $E = \{K\diamondsuit, K\heartsuit, Q\diamondsuit, Q\heartsuit, J\diamondsuit, J\heartsuit\}$

$$P(E) = \frac{n(E)}{n(S)} = \frac{6}{52} = \frac{3}{26}$$

25. $E = \{(1, 5), (2, 4), (3, 3), (4, 2), (5, 1)\}$

$$P(E) = \frac{n(E)}{n(S)} = \frac{5}{36}$$

27. Use the complement.

$$E' = \{(5, 6), (6, 5), (6, 6)\}$$

$$P(E') = \frac{n(E')}{n(S)} = \frac{3}{36} = \frac{1}{12}$$

$$P(E) = 1 - P(E') = 1 - \frac{1}{12} = \frac{11}{12}$$

29. $E_3 = \{(1, 2), (2, 1)\}, n(E_3) = 2$

$E_5 = \{(1, 4), (2, 3), (3, 2), (4, 1)\}, n(E_5) = 4$

$E_7 = \{(1, 6), (2, 5), (3, 4), (4, 3), (5, 2), (6, 1)\}, n(E_7) = 6$

$E = E_3 \cup E_5 \cup E_7$

$n(E) = 2 + 4 + 6 = 12$

$$P(E) = \frac{n(E)}{n(S)} = \frac{12}{36} = \frac{1}{3}$$

31. $P(E) = \dfrac{{}_3C_2}{{}_6C_2} = \dfrac{3}{15} = \dfrac{1}{5}$

33. $P(E) = \dfrac{{}_4C_2}{{}_6C_2} = \dfrac{6}{15} = \dfrac{2}{5}$

35. (a) $0.12(8.3) \approx 0.996$ million $= 996{,}000$

(b) $18\% = \frac{9}{50}$

(c) $54\% = \frac{27}{50}$

(d) $12\% + 4\% = 16\% = \frac{4}{25}$

37. (a) $\dfrac{104}{128} = \dfrac{13}{16}$

(b) $\dfrac{24}{128} = \dfrac{3}{16}$

(c) $\dfrac{52 - 48}{128} = \dfrac{1}{32}$

39. $1 - 0.37 - 0.44 = 0.19 = 19\%$

41. (a) $\dfrac{{}_{15}C_{10}}{{}_{20}C_{10}} = \dfrac{3003}{184{,}756} = \dfrac{21}{1292} \approx 0.016$

(b) $\dfrac{{}_{15}C_8 \cdot {}_5C_2}{{}_{20}C_{10}} = \dfrac{64{,}350}{184{,}756} = \dfrac{225}{646} \approx 0.348$

(c) $\dfrac{{}_{15}C_9 \cdot {}_5C_1}{{}_{20}C_{10}} + \dfrac{{}_{15}C_{10}}{{}_{20}C_{10}} + \dfrac{25{,}025 + 3003}{184{,}756} = \dfrac{28{,}028}{184{,}756}$

$= \dfrac{49}{323}$

≈ 0.152

43. (a) $\dfrac{1}{{}_5P_5} = \dfrac{1}{120}$

(b) $\dfrac{1}{{}_4P_4} = \dfrac{1}{24}$

45. (a) $\frac{20}{52} = \frac{5}{13}$

(b) $\frac{26}{52} = \frac{1}{2}$

(c) $\frac{16}{52} = \frac{4}{13}$

47. (a) $\dfrac{{}_9C_4}{{}_{12}C_4} = \dfrac{126}{495} = \dfrac{14}{55}$ (4 good units)

(b) $\dfrac{{}_9C_2 \cdot {}_3C_2}{{}_{12}C_4} = \dfrac{108}{495} = \dfrac{12}{55}$ (2 good units)

(c) $\dfrac{{}_9C_3 \cdot {}_3C_1}{{}_{12}C_4} = \dfrac{252}{495} = \dfrac{28}{55}$ (3 good units)

At least 2 good units: $\dfrac{12}{55} + \dfrac{28}{55} + \dfrac{14}{55} = \dfrac{54}{55}$

49. (a) $P(EE) = \frac{20}{40} \cdot \frac{20}{40} = \frac{1}{4}$

(b) $P(EO \text{ or } OE) = 2\left(\frac{20}{40}\right)\left(\frac{20}{40}\right) = \frac{1}{2}$

(c) $P(N_1 < 30, N_2 < 30) = \frac{29}{40} \cdot \frac{29}{40} = \frac{841}{1600}$

(d) $P(N_1N_1) = \frac{30}{40} \cdot \frac{1}{40} = \frac{1}{40}$

51. $P(E') = 1 - P(E) = 1 - 0.73 = 0.27$

53. $P(E') = 1 - P(E) = 1 - \frac{1}{5} = \frac{4}{5}$

55. $P(E) = 1 - P(E') = 1 - 0.29 = 0.71$

57. $P(E) = 1 - P(E') = 1 - \frac{14}{25} = \frac{11}{25}$

59. (a) $P(SS) = (0.985)^2 \approx 0.9702$

(b) $P(FF) = (0.015)^2 \approx 0.0002$

(c) $P(S) = 1 - P(FF) = 1 - (0.015)^2 \approx 0.9998$

61. (a) $\dfrac{1}{38}$

(b) $\dfrac{18}{38} = \dfrac{9}{19}$

(c) $\dfrac{2}{38} + \dfrac{18}{38} = \dfrac{20}{38} = \dfrac{10}{19}$

(d) $\dfrac{1}{38} \cdot \dfrac{1}{38} = \dfrac{1}{1444}$

(e) $\dfrac{18}{38} \cdot \dfrac{18}{38} \cdot \dfrac{18}{38} = \dfrac{5832}{54{,}872} = \dfrac{729}{6859}$

63. $1 - \frac{(45)^2}{(60)^2} = 1 - \left(\frac{45}{60}\right)^2 = 1 - \left(\frac{3}{4}\right)^2 = 1 - \frac{9}{16} = \frac{7}{16}$

65. True. Two events are independent if the occurrence of one has no effect on the occurrence of the other.

67. (a) As you consider successive people with distinct birthdays, the probabilities must decrease to take into account the birth dates already used. Because the birth dates of people are independent events, multiply the respective probabilities of distinct birthdays.

(b) $\frac{365}{365} \cdot \frac{364}{365} \cdot \frac{363}{365} \cdot \frac{362}{365}$

(c) $P_1 = \frac{365}{365} = 1$

$P_2 = \frac{365}{365} \cdot \frac{364}{365} = \frac{364}{365}P_1 = \frac{365 - (2 - 1)}{365}P_1$

$P_3 = \frac{365}{365} \cdot \frac{364}{365} \cdot \frac{363}{365} = \frac{363}{365}P_2 = \frac{365 - (3 - 1)}{365}P_2$

$P_n = \frac{365}{365} \cdot \frac{364}{365} \cdot \frac{363}{365} \cdot \dots \cdot \frac{365 - (n - 1)}{365} = \frac{365 - (n - 1)}{365}P_{n-1}$

(d) Q_n is the probability that the birthdays are not distinct which is equivalent to at least two people having the same birthday.

(e)

n	10	15	20	23	30	40	50
P_n	0.88	0.75	0.59	0.49	0.29	0.11	0.03
Q_n	0.12	0.25	0.41	0.51	0.71	0.89	0.97

(f) 23; $Q_n > 0.5$ for $n \ge 23$.

Review Exercises for Chapter 9

1. $a_n = 3 + \frac{12}{n}$

$a_1 = 3 + \frac{12}{1} = 15$

$a_2 = 3 + \frac{12}{2} = 9$

$a_3 = 3 + \frac{12}{3} = 7$

$a_4 = 3 + \frac{12}{4} = 6$

$a_5 = 3 + \frac{12}{5} = \frac{27}{5}$

3. $a_n = \frac{120}{n!}$

$a_1 = \frac{120}{1!} = 120$

$a_2 = \frac{120}{2!} = 60$

$a_3 = \frac{120}{3!} = 20$

$a_4 = \frac{120}{4!} = 5$

$a_5 = \frac{120}{5!} = 1$

5. $-2, 2, -2, 2, -2, \dots$

$a_n = 2(-1)^n$

7. $4, 2, \frac{4}{3}, 1, \frac{4}{5}, \dots$

$a_n = \frac{4}{n}$

9. $\frac{3!}{5!} = \frac{3 \cdot 2 \cdot 1}{5 \cdot 4 \cdot 3 \cdot 2 \cdot 1} = \frac{1}{20}$

11. $\frac{(n-1)!}{(n+1)!} = \frac{(n-1)(n-2) \cdot \dots \cdot 3 \cdot 2 \cdot 1}{(n+1)(n)(n-1)(n-2) \cdot \dots \cdot 3 \cdot 2 \cdot 1}$

$= \frac{1}{n(n+1)}$

13. $\sum_{j=1}^{4} \frac{6}{j^2} = \frac{6}{1^2} + \frac{6}{2^2} + \frac{6}{3^2} + \frac{6}{4^2} = 6 + \frac{3}{2} + \frac{2}{3} + \frac{3}{8} = \frac{205}{24}$

15. $\frac{1}{2(1)} + \frac{1}{2(2)} + \frac{1}{2(3)} + \dots + \frac{1}{2(20)} = \sum_{k=1}^{20} \frac{1}{2k}$

17. $\sum_{i=1}^{\infty} \frac{4}{10^i} = \sum_{i=1}^{\infty} 4\left(\frac{1}{10^i}\right) = \frac{\frac{4}{10}}{1 - \frac{1}{10}} = \frac{4}{9}$

19. (a) $A_1 = \$10,018.75$ $\quad A_6 \approx \$10,113.03$

$A_2 \approx \$10,037.54$ $\quad A_7 \approx \$10,131.99$

$A_3 \approx \$10,056.36$ $\quad A_8 \approx \$10,150.99$

$A_4 \approx \$10,075.21$ $\quad A_9 \approx \$10,170.02$

$A_5 \approx \$10,094.10$ $\quad A_{10} \approx \$10,189.09$

(b) The balance in the account after 10 years is

$$A_{120}\ 10,000\left(1 + \frac{0.0225}{12}\right)^{120} \approx \$12,520.59$$

21. $5, -1, -7, -13, -19, \ldots$

Arithmetic sequence, $d = -6$

23. $\frac{1}{8}, \frac{1}{4}, \frac{1}{2}, 1, 2, \ldots$

Not an arithmetic sequence.

25. $a_1 = 7, d = 12$

$$\begin{aligned} a_n &= 7 + (n - 1)12 \\ &= 7 + 12n - 12 \\ &= 12n - 5 \end{aligned}$$

27. $a_3 = 96, a_7 = 24$

$$\begin{aligned} a_7 &= a_3 + 4d \Rightarrow 24 = 96 + 4d \Rightarrow -72 = 4d \Rightarrow d = -18 \\ a_1 &= a_3 - 2d \Rightarrow a_1 = 96 - 2(-18) = 132 \\ a_n &= 132 + (n - 1)(-18) \\ &= -18n + 150 \end{aligned}$$

29. $a_1 = 4, d = 17$

$$\begin{aligned} a_1 &= 4 \\ a_2 &= 4 + 17 = 21 \\ a_3 &= 21 + 17 = 38 \\ a_4 &= 38 + 17 = 55 \\ a_5 &= 45 + 17 = 72 \end{aligned}$$

31. $\sum_{k=1}^{100} 9k$ is arithmetic. Therefore, $a_1 = 9, a_{100} = 900,$

$S_{700} = \frac{100}{2}(9 + 900) = 45,450.$

33. $\sum_{j=1}^{10} (2j - 3)$ is arithmetic. Therefore,

$a_1 = -1, a_{10} = 17, S_{10} = \frac{10}{2}[-1 + 17] = 80.$

35. $\sum_{k=1}^{11} \left(\frac{2}{3}k + 4\right)$ is arithmetic. Therefore, $a_1 = \frac{14}{3}, a_{11} = \frac{34}{3}, S_{11} = \frac{11}{2}\left[\frac{14}{3} + \frac{34}{3}\right] = 88.$

37. $a_n = 43,800 + (n - 1)(1950)$

(a) $a_5 = 43,800 + 4(1950) = \$51,600$

(b) $S_5 = \frac{5}{2}(43,800 + 51,600) = \$238,500$

39. $2, 6, 18, 54, 162, \ldots$

$$r = \frac{6}{2} = 3$$

Geometric sequence, $r = 3$

41. $\frac{1}{5}, -\frac{3}{5}, \frac{9}{5}, -\frac{27}{5}, \ldots$

Geometric sequence, $r = -3$

43. $a_1 = 2, r = 15$

$$\begin{aligned} a_1 &= 2 \\ a_2 &= 2(15) = 30 \\ a_3 &= 30(15) = 450 \\ a_4 &= 450(15) = 6750 \\ a_5 &= 6750(15) = 101,250 \end{aligned}$$

45. $a_1 = 9, a_3 = 4$

$$\begin{aligned} a_3 &= a_1 r^2 \\ 4 &= 9r^2 \\ \tfrac{4}{9} &= r^2 \Rightarrow r = \pm\tfrac{2}{3} \end{aligned}$$

$a_1 = 9$ $\qquad\qquad a_1 = 9$

$a_2 = 9\left(\frac{2}{3}\right) = 6$ $\qquad a_2 = 9\left(-\frac{2}{3}\right) = -6$

$a_3 = 6\left(\frac{2}{3}\right) = 4$ $\quad$ or $\quad a_3 = -6\left(-\frac{2}{3}\right) = 4$

$a_4 = 4\left(\frac{2}{3}\right) = \frac{8}{3}$ $\qquad a_4 = 4\left(-\frac{2}{3}\right) = -\frac{8}{3}$

$a_5 = \frac{8}{3}\left(\frac{2}{3}\right) = \frac{16}{9}$ $\qquad a_5 = -\frac{8}{3}\left(-\frac{2}{3}\right) = \frac{16}{9}$

47. $a_1 = 100, r = 1.05$

$$\begin{aligned} a_n &= 100(1.05)^{n-1} \\ a_{10} &= 100(1.05)^9 \approx 155.133 \end{aligned}$$

49. $a_1 = 18, a_2 = -9$

$a_2 = a_1 r$

$-9 = 18r$

$-\frac{1}{2} = r$

$a_n = 18\left(-\frac{1}{2}\right)^{n-1}$

$a_{10} = 18\left(-\frac{1}{2}\right)^{9} = \frac{-9}{256}$

51. $\sum_{i=1}^{7} 2^{i-1} = \frac{1-2^7}{1-2} = 127$

53. $\sum_{i=1}^{4}\left(\frac{1}{2}\right)^{i} = \frac{1}{2} + \frac{1}{4} + \frac{1}{8} + \frac{1}{16} = \frac{15}{16}$

55. $\sum_{i=1}^{5} (2)^{i-1} = 1 + 2 + 4 + 8 + 16 = 31$

57. $\sum_{i=1}^{5} 10(0.6)^{i-1} = 23.056$

59. $\sum_{i=1}^{\infty}\left(\frac{7}{8}\right)^{i-1} = \frac{1}{1-\frac{7}{8}} = 8$

61. $\sum_{k=1}^{\infty} 4\left(\frac{2}{3}\right)^{k-1} = \frac{4}{1-\frac{2}{3}} = 12$

63. (a) $a_n = 120{,}000(0.7)^n$

(b) $a_5 = 120{,}000(0.7)^5$

$= \$20{,}168.40$

65. 1. When $n = 1, 3 = 1(1 + 2)$.

2. Assume that

$S_k = 3 + 5 + 7 + \cdots + (2k+1) = k(k+2).$

Then,

$S_{k+1} = 3 + 5 + 7 + \cdots + (2k+1) + [2(k+1)+1]$

$= S_k + (2k+3)$

$= k(k+2) + 2k + 3$

$= k^2 + 4k + 3$

$= (k+1)(k+3)$

$= (k+1)[(k+1)+2].$

So, by mathematical induction, the formula is valid for all positive integer values of n.

67. 1. When $n = 1, a = a\left(\frac{1-r}{1-r}\right)$.

2. Assume that $S_k = \sum_{i=0}^{k-1} ar^i = \frac{a(1-r^k)}{1-r}$.

Then $S_{k+1} = \sum_{i=0}^{k} ar^i = \left(\sum_{i=0}^{k-1} ar^i\right) + ar^k$

$= \frac{a(1-r^k)}{1-r} + ar^k$

$= \frac{a(1-r^k+r^k-r^{k+1})}{1-r}$

$= \frac{a(1-r^{k+1})}{1-r}.$

So, by mathematical induction, the formula is valid for all positive integer values of n.

69. $S_1 = 9 = 1(9) = 1[2(1)+7]$

$S_2 = 9 + 13 = 22 = 2(11) = 2[2(2)+7]$

$S_3 = 9 + 13 + 17 = 39 = 3(13) = 3[2(3)+7]$

$S_4 = 9 + 13 + 17 + 21 = 60 = 4(15) = 4[2(4)+7]$

$S_n = n(2n+7)$

1. When $n = 1, S_1 = 9 = 1(2+7)$

2. Assume $S_k = 9 + 13 + \ldots + [4k+5] = k(2k+7)$

$S_{k+1} = S_k + a_{k+1}$

$= k(2k+7) + [4(k+1)+5]$

$= 2k^2 + 7k + 4k + 9$

$= 2k^2 + 11k + 9$

$= (k+1)(2k+9)$

$= (k+1)[2(k+1)+7]$

So, the formula holds for all positive integers n.

71. $S_1 = 1$

$S_2 = 1 + \frac{3}{5} = \frac{8}{5}$

$S_3 = 1 + \frac{3}{5} + \frac{9}{25} = \frac{49}{25}$

$S_4 = 1 + \frac{3}{5} + \frac{9}{25} + \frac{27}{125} = \frac{272}{125}$

From these sums, there is no apparent pattern. Because the series is geometric, the formula for the sum is

$$S_n = \frac{5}{2}\left[1 - \left(\frac{3}{5}\right)^n\right].$$

1. When $n = 1$, $S_1 = 1 = \frac{5}{2}\left[1 - \left(\frac{3}{5}\right)^1\right]$

2. Assume

$$S_k = 1 + \frac{3}{5} + \frac{9}{25} + \cdots + \left(\frac{3}{5}\right)^{k-1}$$
$$= \frac{5}{2}\left[1 - \left(\frac{3}{5}\right)^k\right]$$
$$= \frac{1 - \left(\frac{3}{5}\right)^k}{1 - \frac{3}{5}}$$

$$S_{k+1} = S_k + a_{k+1} = \frac{1 - \left(\frac{3}{5}\right)^k}{1 - \frac{3}{5}} + \left(\frac{3}{5}\right)^{(k+1)-1}$$
$$= \frac{1 - \left(\frac{3}{5}\right)^k}{1 - \frac{3}{5}} + \left(\frac{3}{5}\right)^k$$
$$= \frac{1 - \left(\frac{3}{5}\right)^k + \left(1 - \frac{3}{5}\right)\left(\frac{3}{5}\right)^k}{1 - \frac{3}{5}}$$
$$= \frac{1 - \left(\frac{3}{5}\right)^k + \left(\frac{3}{5}\right)^k - \left(\frac{3}{5}\right)^{k+1}}{1 - \frac{3}{5}}$$
$$= \frac{1 - \left(\frac{3}{5}\right)^{k+1}}{1 - \frac{3}{5}}$$
$$= \frac{5}{2}\left(1 - \left(\frac{3}{5}\right)^{k+1}\right)$$

So, the formula holds for all positive integers n.

73. $\sum_{n=1}^{75} n = \frac{75(76)}{2} = 2850$

75. $a_1 = f(1) = 5$, $a_n = a_{n-1} + 5$

$a_1 = 5$

$a_2 = 5 + 5 = 10$

$a_3 = 10 + 5 = 15$

$a_4 = 15 + 5 = 20$

$a_5 = 20 + 5 = 25$

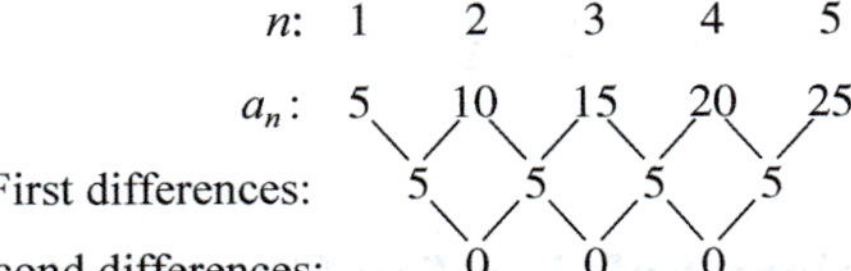

Because the first differences are all the same, the sequence has a linear model.

77. ${}_6C_4 = \frac{6!}{2!4!} = 15$

79.

```
                1
              1   1
            1   2   1
          1   3   3   1
        1   4   6   4   1
      1   5  10  10   5   1
    1   6  15  20  15   6   1
  1   7 (21) 35  35  21   7   1
```

$\binom{7}{2} = 21$, the 3rd entry in the 7th row.

81. $(x + 4)^4 = x^4 + 4x^3(4) + 6x^2(4)^2 + 4x(4)^3 + 4^4$

$= x^4 + 16x^3 + 96x^2 + 256x + 256$

83. $(4 - 5x)^3 = {}_3C_0(4^3) + {}_3C_1(4^2)(-5x) + {}_3C_2(4)(-5x)^2 + {}_3C_3(-5x)^3$

$= 4^3 - 3(4)^2(5x) + 3(4)(5x)^2 - (5x)^3$

$= 64 - 240x + 300x^2 + 125x^3$

85. First number: 1 2 3 4 5 6

Second number: 6 5 4 3 2 1

From this list, you can see that a sum of 7 occurs 6 different ways.

87. $(10)(10)(10)(10) = 10{,}000$ different telephone numbers

89. $5 \cdot 4 \cdot 3 \cdot 2 \cdot 1 = 120$

91. ${}_{32}C_{12} = \frac{32!}{20!12!} = 225{,}792{,}840$

93. (a) $P(E) = \dfrac{n(E)}{n(S)} = \dfrac{2}{10} = \dfrac{1}{5} = 0.2$

(b) $P(E) = \dfrac{n(E)}{n(S)} = \dfrac{6}{10} = \dfrac{3}{5} = 0.6$

95. (a) $25\% + 18\% = 43\%$

(b) $100\% - 18\% = 82\%$

97. $\left(\frac{1}{6}\right)\left(\frac{1}{6}\right)\left(\frac{1}{6}\right)\left(\frac{1}{6}\right) = \frac{1}{1296}$

99. $1 - \frac{13}{52} = 1 - \frac{1}{4} = \frac{3}{4}$

101. False. $\dfrac{(n+2)!}{n!} = \dfrac{(n+2)(n+1)\cancel{n!}}{\cancel{n!}}$

$$= (n+2)(n+1)$$

$$\neq \frac{n+2}{n}$$

103. True. $\sum_{k=1}^{8} 3k = 3\sum_{k=1}^{8} k$ by the Properties of Sums.

105. The domain of an infinite sequence is the set of natural numbers.

107. Each term of the sequence is defined in terms of preceding terms.

Problem Solving for Chapter 9

1. $a_n = \dfrac{n+1}{n^2+1}$

(a)

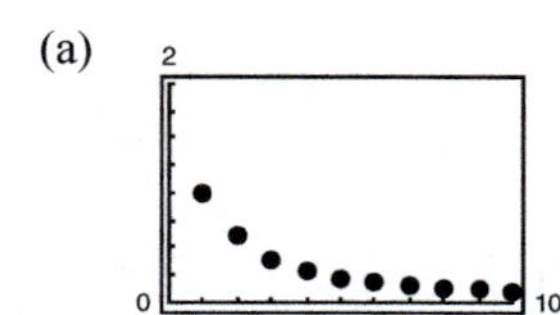

(b) $a_n \to 0$ as $n \to \infty$

(c)

n	1	10	100	1000	10,000
a_n	1	0.1089	0.0101	0.0010	0.0001

(d) $a_n \to 0$ as $n \to \infty$

3. Distance: $\displaystyle\sum_{n=1}^{\infty} 20\left(\frac{1}{2}\right)^{n-1} = \frac{20}{1 - \frac{1}{2}} = 40$

Time: $\displaystyle\sum_{n=1}^{\infty} \left(\frac{1}{2}\right)^{n-1} = \frac{1}{1 - \frac{1}{2}} = 2$

In two seconds, both Achilles and the tortoise will be 40 feet away from Achilles' starting point.

5. Let $a_n = dn + c$, an arithmetic sequence with a common difference of d.

(a) If C is added to each term, then the resulting sequence, $b_n = a_n + C = dn + c + C$, is still arithmetic with a common difference of d.

(b) If each term is multiplied by a nonzero constant C, then the resulting sequence,

$b_n = C(dn + c) = Cdn + Cc$, is still arithmetic.

The common difference is Cd.

(c) If each term is squared, the resulting sequence,

$b_n = a_n^2 = (dn + c)^2$, is not arithmetic.

7. $a_n = \begin{cases} \dfrac{a_{n-1}}{2}, & \text{if } a_{n-1} \text{ is even} \\ 3a_{n-1} + 1 & \text{if } a_{n-1} \text{ is odd} \end{cases}$

(a)

$a_1 = 7$	$a_{11} = \frac{20}{2} = 10$
$a_2 = 3(7) + 1 = 22$	$a_{12} = \frac{10}{2} = 5$
$a_3 = \frac{22}{3} = 11$	$a_{13} = 3(5) + 1 = 16$
$a_4 = 3(11) + 1 = 34$	$a_{14} = \frac{16}{2} = 8$
$a_5 = \frac{34}{2} = 17$	$a_{15} = \frac{8}{2} = 4$
$a_6 = 3(17) + 1 = 52$	$a_{16} = \frac{4}{2} = 2$
$a_7 = \frac{52}{2} = 26$	$a_{17} = \frac{2}{2} = 1$
$a_8 = \frac{26}{2} = 13$	$a_{18} = 3(1) + 1 = 4$
$a_9 = 3(13) + 1 = 40$	$a_{19} = \frac{4}{2} = 2$
$a_{10} = \frac{40}{2} = 20$	$a_{20} = \frac{2}{2} = 1$

(b)

$a_1 = 4$	$a_1 = 5$	$a_1 = 12$
$a_2 = 2$	$a_2 = 16$	$a_2 = 6$
$a_3 = 1$	$a_3 = 8$	$a_3 = 3$
$a_4 = 4$	$a_4 = 4$	$a_4 = 10$
$a_5 = 2$	$a_5 = 2$	$a_5 = 5$
$a_6 = 1$	$a_6 = 1$	$a_6 = 16$
$a_7 = 4$	$a_7 = 4$	$a_7 = 8$
$a_8 = 2$	$a_8 = 2$	$a_8 = 4$
$a_9 = 1$	$a_9 = 1$	$a_9 = 2$
$a_{10} = 4$	$a_{10} = 4$	$a_{10} = 1$

Eventually the terms repeat: 4, 2, 1

9. The numbers 1, 5, 12, 22, 35, 51, … can be written recursively as $P_n = P_{n-1} + (3n - 2)$. Show that $P_n = n(3n - 1)/2$.

1. For $n = 1$: $1 = \dfrac{1(3 - 1)}{2}$

2. Assume $P_k = \dfrac{k(3k - 1)}{2}$.

Then, $P_{k+1} = P_k + \left[3(k + 1) - 2\right] = \dfrac{k(3k - 1)}{2} + (3k + 1)$

$$= \frac{k(3k - 1) + 2(3k + 1)}{2} = \frac{3k^2 + 5k + 2}{2}$$

$$= \frac{(k + 1)(3k + 2)}{2} = \frac{(k + 1)\left[3(k + 1) - 1\right]}{2}.$$

So, by mathematical induction, the formula is valid for all integers $n \geq 1$.

11. Side lengths: $1, \frac{1}{2}, , \frac{1}{8}, \ldots$

$$S_n = \left(\frac{1}{2}\right)^{n-1} \text{ for } n \geq 1$$

Areas: $\dfrac{\sqrt{3}}{4}, \dfrac{\sqrt{3}}{4}\left(\dfrac{1}{2}\right)^2, \dfrac{\sqrt{3}}{4}\left(\dfrac{1}{4}\right)^2, \dfrac{\sqrt{3}}{4}\left(\dfrac{1}{8}\right)^2, \ldots$

$$A_n = \frac{\sqrt{3}}{4}\left[\left(\frac{1}{2}\right)^{n-1}\right]^2 = \frac{\sqrt{3}}{4}\left(\frac{1}{2}\right)^{2n-2} = \frac{\sqrt{3}}{4}S_n^2$$

13. $\frac{1}{3}$

15. (a) Odds in favor of choosing a blue marble $= \dfrac{\text{number of blue marbles}}{\text{number of yellow marbles}} = \dfrac{3}{7}$

Odds against choosing a blue marble $= \dfrac{\text{number of yellow marbles}}{\text{number of blue marbles}} = \dfrac{7}{3}$

(b) Odds against choosing a red marble $= \dfrac{\text{number of non-red marbles}}{\text{number of red marbles}}$

$$\frac{4}{1} = \frac{x}{6}$$

$$24 = x \quad \text{(number of non-red marbles)}$$

$$\text{Total marbles} = 6 + 24 = 30$$

(c) $P(E) = \dfrac{n(E)}{n(S)} = \dfrac{n(E)}{n(E) + n(E')} = \dfrac{n(E)/n(E')}{n(E)/n(E') + n(E')/n(E')}$

$$P(E) = \frac{\text{odds in favor of } E}{\text{odds in favor of } E + 1}$$

(d) $P(E) = \dfrac{n(E)}{n(S)}$ $\qquad P(E') = \dfrac{n(E')}{n(S)}$

$n(S)P(E) = n(E)$ $\qquad n(S)P(E') = n(E')$

Odds in favor of event $E = \dfrac{n(E)}{n(E')} = \dfrac{n(S)P(E)}{n(S)P(E')} = \dfrac{P(E)}{P(E')}$

Practice Test for Chapter 9

1. Write out the first five terms of the sequence $a_n = \dfrac{2n}{(n+2)!}$.

2. Write an expression for the nth term of the sequence $\frac{4}{3}, \frac{5}{9}, \frac{6}{27}, \frac{7}{81}, \frac{8}{243}, \ldots$.

3. Find the sum $\sum_{i=1}^{6}(2i - 1)$.

4. Write out the first five terms of the arithmetic sequence where $a_1 = 23$ and $d = -2$.

5. Find a_n for the arithmetic sequence with $a_1 = 12$, $d = 3$, and $n = 50$.

6. Find the sum of the first 200 positive integers.

7. Write out the first five terms of the geometric sequence with $a_1 = 7$ and $r = 2$.

8. Evaluate $\sum_{n=1}^{10} 6\left(\frac{2}{3}\right)^{n-1}$.

9. Evaluate $\sum_{n=0}^{\infty}(0.03)^n$.

10. Use mathematical induction to prove that $1 + 2 + 3 + 4 + \cdots + n = \dfrac{n(n+1)}{2}$.

11. Use mathematical induction to prove that $n! > 2^n$, $n \ge 4$.

12. Evaluate ${}_{13}C_4$.

13. Expand $(x + 3)^5$.

14. Find the term involving x^7 in $(x - 2)^{12}$.

15. Evaluate ${}_{30}P_4$.

16. How many ways can six people sit at a table with six chairs?

17. Twelve cars run in a race. How many different ways can they come in first, second, and third place? (Assume that there are no ties.)

18. Two six-sided dice are tossed. Find the probability that the total of the two dice is less than 5.

19. Two cards are selected at random from a deck of 52 playing cards without replacement. Find the probability that the first card is a King and the second card is a black ten.

20. A manufacturer has determined that for every 1000 units it produces, 3 will be faulty. What is the probability that an order of 50 units will have one or more faulty units?

CHAPTER 10
Topics in Analytic Geometry

CHAPTER 10
Topics in Analytic Geometry

Section 10.1 Lines

1. inclination

3. $\left|\dfrac{m_2 - m_1}{1 + m_1 m_2}\right|$

5. $m = \tan\dfrac{\pi}{6} = \dfrac{\sqrt{3}}{3}$

7. $m = \tan\dfrac{3\pi}{4} = -1$

9. $m = \tan\dfrac{\pi}{3} = \sqrt{3}$

11. $m = \tan 0.39 \approx 0.4111$

13. $m = \tan 1.27 \approx 3.2236$

15. $m = \tan 1.81 \approx -4.1005$

17. $m = 1$
$1 = \tan\theta$
$\theta = \dfrac{\pi}{4}$ radian $= 45°$

19. $m = \dfrac{2}{3}$
$\dfrac{2}{3} = \tan\theta$
$\theta = \arctan\left(\dfrac{2}{3}\right) \approx 0.5880$ radian $\approx 33.7°$

21. $m = -1$
$-1 = \tan\theta$
$\theta = 180° + \arctan(-1) = \dfrac{3\pi}{4}$ radians $= 135°$

23. $m = -\dfrac{3}{2}$
$-\dfrac{3}{2} = \tan\theta$
$\theta = \tan^{-1}\left(-\dfrac{3}{2}\right) + \pi \approx 2.1588$ radians $\approx 123.7°$

25. $\left(\sqrt{3}, 2\right), (0, 1)$
$m = \dfrac{1 - 2}{0 - \sqrt{3}} = \dfrac{-1}{-\sqrt{3}} = \dfrac{1}{\sqrt{3}}$
$\dfrac{1}{\sqrt{3}} = \tan\theta$
$\theta = \arctan\dfrac{1}{\sqrt{3}} = \dfrac{\pi}{6}$ radian $= 30°$

27. $\left(-\sqrt{3}, -1\right), (0, -2)$
$m = \dfrac{-2 - (-1)}{0 - \left(-\sqrt{3}\right)} = \dfrac{-1}{\sqrt{3}}$
$-\dfrac{1}{\sqrt{3}} = \tan\theta$
$\theta = \arctan\left(-\dfrac{1}{\sqrt{3}}\right) = \dfrac{5\pi}{6}$ radians $= 150°$

29. $(6, 1), (10, 8)$
$m = \dfrac{8 - 1}{10 - 6} = \dfrac{7}{4}$
$\dfrac{7}{4} = \tan\theta$
$\theta = \arctan\dfrac{7}{4} \approx 1.0517$ radians $\approx 60.3°$

31. $(-2, 20), (10, 0)$
$m = \dfrac{0 - 20}{10 - (-2)} = -\dfrac{20}{12} = -\dfrac{5}{3}$
$-\dfrac{5}{3} = \tan\theta$
$\theta = \pi + \arctan\left(-\dfrac{5}{3}\right) \approx 2.1112$ radians $\approx 121.0°$

33. $\left(\dfrac{1}{4}, \dfrac{3}{2}\right), \left(\dfrac{1}{3}, \dfrac{1}{2}\right)$
$m = \dfrac{\frac{1}{2} - \frac{3}{2}}{\frac{1}{3} - \frac{1}{4}} = -\dfrac{1}{\frac{1}{12}} = -12$
$-12 = \tan\theta$
$\theta = \arctan(-12) + \pi \approx 1.6539$ radians $\approx 94.8°$

35. $2x + 2y - 5 = 0$
$y = -x + \dfrac{5}{2} \Rightarrow m = -1$
$-1 = \tan\theta$
$\theta = \arctan(-1) = \dfrac{3\pi}{4}$ radians $= 135°$

37. $3x - 3y + 1 = 0$
$y = x + \dfrac{1}{3} \Rightarrow m = 1$
$1 = \tan\theta$
$\theta = \arctan 1 = \dfrac{\pi}{4}$ radian $= 45°$

39. $x + \sqrt{3y} + 2 = 0$

$$y = -\frac{1}{\sqrt{3}}x - \frac{2}{\sqrt{3}} \Rightarrow m = -\frac{1}{\sqrt{3}}$$

$$-\frac{1}{\sqrt{3}} = \tan\theta$$

$$\theta = \arctan\left(-\frac{1}{\sqrt{3}}\right) = \frac{5\pi}{6} \text{ radians} = 150°$$

41. $6x - 2y + 8 = 0$

$$y = 3x + 4 \Rightarrow m = 3$$

$$3 = \tan\theta$$

$$\theta = \arctan 3 \approx 1.2490 \text{ radians} \approx 71.6°$$

43. $4x + 5y - 9 = 0$

$$y = -\tfrac{4}{5}x + \tfrac{9}{5} \Rightarrow m = -\tfrac{4}{5}$$

$$-\tfrac{4}{5} = \tan\theta$$

$$\theta = \tan^{-1}\left(-\tfrac{4}{5}\right) + \pi$$

$$\approx 2.4669 \text{ radians} \approx 141.3°$$

45. $3x + y = 3 \Rightarrow y = -3x + 3 \Rightarrow m_1 = -3$

$x - y = 2 \Rightarrow y = x - 2 \Rightarrow m_2 = 1$

$$\tan\theta = \left|\frac{1 - (-3)}{1 + (-3)(1)}\right| = 2$$

$$\theta = \arctan 2 \approx 1.1071 \text{ radians} \approx 63.4°$$

47. $x - y = 0 \Rightarrow y = x \Rightarrow m_1 = 1$

$3x - 2y = -1 \Rightarrow y = \frac{3}{2}x + \frac{1}{2} \Rightarrow m_2 = \frac{3}{2}$

$$\tan\theta = \left|\frac{\frac{3}{2} - 1}{1 + \left(\frac{3}{2}\right)(1)}\right| = \frac{1}{5}$$

$$\theta = \arctan\frac{1}{5} \approx 0.1974 \text{ radian} \approx 11.3°$$

49. $x - 2y = 7 \Rightarrow y = \frac{1}{2}x - \frac{7}{2} \Rightarrow m_1 = \frac{1}{2}$

$6x + 2y = 5 \Rightarrow y = -3x + \frac{5}{2} \Rightarrow m_2 = -3$

$$\tan\theta = \left|\frac{-3 - \frac{1}{2}}{1 + \left(\frac{1}{2}\right)(-3)}\right| = 7$$

$$\theta = \arctan 7 \approx 1.4289 \text{ radians} \approx 81.9°$$

51. $x + 2y = 8 \Rightarrow y = -\frac{1}{2}x + 4 \Rightarrow m_1 = -\frac{1}{2}$

$x - 2y = 2 \Rightarrow y = \frac{1}{2}x - 1 \Rightarrow m_2 = \frac{1}{2}$

$$\tan\theta = \left|\frac{\frac{1}{2} - \left(-\frac{1}{2}\right)}{1 + \left(-\frac{1}{2}\right)\left(\frac{1}{2}\right)}\right| = \frac{4}{3}$$

$$\theta = \arctan\left(\frac{4}{3}\right) \approx 0.9273 \text{ radian} \approx 53.1°$$

53. $0.05x - 0.03y = 0.21 \Rightarrow y = \frac{5}{3}x - 7 \Rightarrow m_1 = \frac{5}{3}$

$0.07x + 0.02y = 0.16 \Rightarrow y = -\frac{7}{2}x + 8 \Rightarrow m_2 = -\frac{7}{2}$

$$\tan\theta = \left|\frac{\left(-\frac{7}{2}\right) - \left(\frac{5}{3}\right)}{1 + \left(\frac{5}{3}\right)\left(-\frac{7}{2}\right)}\right| = \frac{31}{29}$$

$$\theta = \arctan\left(\frac{31}{29}\right) \approx 0.8187 \text{ radian} \approx 46.9°$$

55. Let $A = (1, 5)$, $B = (3, 8)$, and $C = (4, 5)$.

Slope of AB: $m_1 = \dfrac{8 - 5}{3 - 1} = \dfrac{3}{2}$

Slope of BC: $m_2 = \dfrac{5 - 8}{4 - 3} = \dfrac{-3}{1} = -3$

Slope of AC: $m_3 = \dfrac{5 - 5}{4 - 1} = \dfrac{0}{3} = 0$

$$\tan A = \left|\frac{0 - \frac{3}{2}}{1 + \left(\frac{3}{2}\right)(0)}\right| = \frac{\frac{3}{2}}{1} = \frac{3}{2}$$

$$A = \arctan\left(\frac{3}{2}\right) \approx 56.3°$$

$$\tan B = \left|\frac{\frac{3}{2} - (-3)}{1 + (-3)\left(\frac{3}{2}\right)}\right| = \frac{\frac{9}{2}}{\frac{7}{2}} = \frac{9}{7}$$

$$B = \arctan\frac{9}{7} \approx 52.1°$$

$$\tan C = \left|\frac{-3 - 0}{1 + (0)(-3)}\right| = \frac{3}{1} = 3$$

$$C = \arctan 3 \approx 71.6°$$

57. Let $A = (-4, -1)$, $B = (3, 2)$, and $C = (1, 0)$.

Slope of AB: $m_1 = \dfrac{-1 - 2}{-4 - 3} = \dfrac{3}{7}$

Slope of BC: $m_2 = \dfrac{2 - 0}{3 - 1} = 1$

Slope of AC: $m_3 = \dfrac{-1 - 0}{-4 - 1} = \dfrac{1}{5}$

$$\tan A = \left|\frac{\frac{1}{5} - \frac{3}{7}}{1 + \left(\frac{3}{7}\right)\left(\frac{1}{5}\right)}\right| = \frac{\frac{8}{35}}{\frac{38}{35}} = \frac{4}{9}$$

$$A = \arctan\left(\frac{4}{19}\right) \approx 11.9°$$

$$\tan B = \left|\frac{1 - \frac{3}{7}}{1 + \left(\frac{3}{7}\right)(1)}\right| = \frac{\frac{4}{7}}{\frac{10}{7}} = \frac{2}{5}$$

$$B = \arctan\left(\frac{2}{5}\right) \approx 21.8°$$

$$C = 180° - A - B$$

$$\approx 180° - 11.9° - 21.8° = 146.3°$$

59. $(x_1, y_1) = (1, 2)$

$y = x + 2 \Rightarrow x - y + 2 = 0$

$$d = \frac{|(1)(1) + (-1)(2) + 2|}{\sqrt{1^2 + (-1)^2}} = \frac{1}{\sqrt{2}} = \frac{\sqrt{2}}{2} \approx 0.7071$$

61. $(x_1, y_1) = (2, 3)$

$y = 2x - 3 \Rightarrow 2x - y - 3 = 0$

$$d = \frac{|2(2) + (-1)(3) + (-3)|}{\sqrt{2^2 + (-1)^2}} = \frac{2}{\sqrt{5}} = \frac{2\sqrt{5}}{5} \approx 0.8944$$

63. $(x_1, y_1) = (-2, 4)$

$y = -x + 6 \Rightarrow x + y - 6 = 0$

$$d = \frac{|(1)(-2) + (1)(4) + (-6)|}{\sqrt{1^2 + 1^2}} = \frac{4}{\sqrt{2}} = 2\sqrt{2}$$

$$\approx 2.8284$$

65. $(x_1, y_1) = (1, -2)$

$y = 3x - 6 \Rightarrow 3x - y - 6 = 0$

$$d = \frac{|(3)(1) + (-1)(-2) + (-6)|}{\sqrt{3^2 + (-1)^2}} = \frac{1}{\sqrt{10}} = \frac{\sqrt{10}}{10}$$

$$\approx 0.3162$$

67. $(x_1, y_1) = (2, 3)$

$3x + y = 1 \Rightarrow 3x + y - 1 = 0$

$$d = \frac{|3(2) + (1)(3) + (-1)|}{\sqrt{3^2 + 1^2}} = \frac{8}{\sqrt{10}} = \frac{8\sqrt{10}}{10} = \frac{4\sqrt{10}}{5}$$

$$\approx 2.5298$$

69. $(x_1, y_1) = (6, 2)$

$-3x + 4y = -5 \Rightarrow -3x + 4y + 5 = 0$

$$d = \frac{|(-3)(6) + (4)(2) + (5)|}{\sqrt{(-3)^2 + (4)^2}} = \frac{5}{\sqrt{25}} = 1$$

71. $(x_1, y_1) = (-2, 4)$

$4x + 3y = 5 \Rightarrow 4x + 3y - 5 = 0$

$$d = \frac{|(4)(-2) + (3)(4) + (-5)|}{\sqrt{4^2 + 3^2}} = \frac{1}{\sqrt{25}} = \frac{1}{5}$$

73. (a)

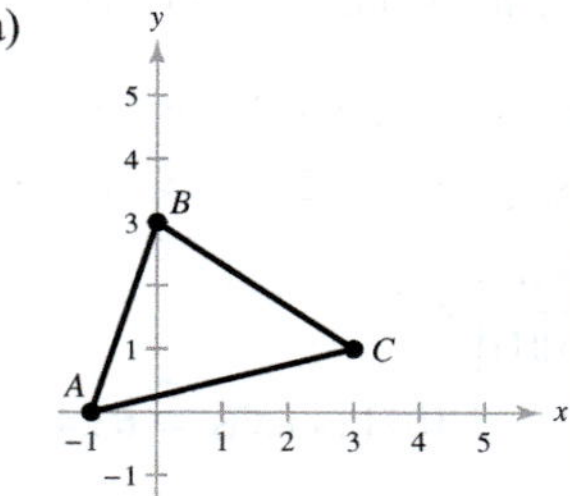

(b) Slope of the line AC: $m = \dfrac{1 - 0}{3 - (-1)} = \dfrac{1}{4}$

Equation of the line AC: $y - 0 = \dfrac{1}{4}(x + 1)$

$$x - 4y + 1 = 0$$

Altitude from $B = (0, 3)$:

$$h = \frac{|(1)(0) + (-4)(3) + (1)|}{\sqrt{1^2 + (-4)^2}} = \frac{11}{\sqrt{17}} = \frac{11\sqrt{17}}{17}$$

(c) Length of the base AC:

$$b = \sqrt{(3 + 1)^2 + (1 - 0)^2} = \sqrt{17}$$

Area of the triangle:

$$A = \frac{1}{2}bh = \frac{1}{2}\left(\sqrt{17}\right)\left(\frac{11}{\sqrt{17}}\right) = \frac{11}{2} \text{ units}^2$$

75. (a)

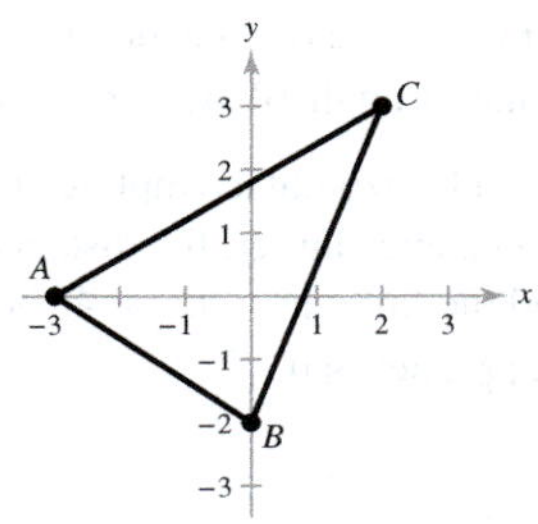

(b) Slope of the line AC: $m = \dfrac{3-0}{2+3} = \dfrac{3}{5}$

Equation of the line AC: $y - 0 = \dfrac{3}{5}(x+3)$

$$3x - 5y + 9 = 0$$

Altitude from $B = (0, -2)$:

$$h = \frac{|3(0) + (-5)(-2) + (9)|}{\sqrt{3^2 + (-5)^2}} = \frac{19}{\sqrt{34}} = \frac{19\sqrt{34}}{34}$$

(c) Length of the base AC:

$$b = \sqrt{(2+3)^2 + (3-0)^2} = \sqrt{34}$$

Area of the triangle:

$$A = \frac{1}{2}bh = \frac{1}{2}\left(\sqrt{34}\right)\left(\frac{19}{\sqrt{34}}\right) = \frac{19}{2} \text{ units}^2$$

77. (a)

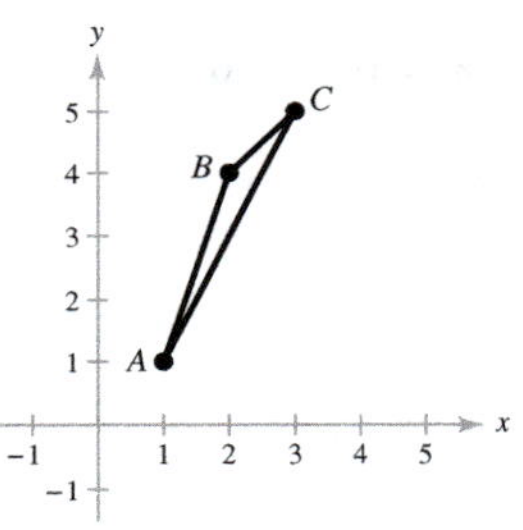

(b) Slope of the line AC:

$$b = \sqrt{(5+2)^2 + (1-0)^2} = \sqrt{50} = 5\sqrt{2}$$

Equation of the line AC: $y - 1 = 2(x - 1)$

$$2x - y - 1 = 0$$

Altitude from $B = (2, 4)$:

$$h = \frac{|(2)(2) + (-1)(4) + (-1)|}{\sqrt{2^2 + (-1)^2}} = \frac{1}{\sqrt{5}} = \frac{\sqrt{5}}{5}$$

(c) Length of the base AC:

$$b = \sqrt{(3-1)^2 + (5-1)^2} = \sqrt{20} = 2\sqrt{5}$$

Area of the triangle:

$$A = \frac{1}{2}bh = \frac{1}{2}\left(2\sqrt{5}\right)\left(\frac{\sqrt{5}}{5}\right) = 1 \text{ unit}^2$$

79. $x + y = 1 \Rightarrow (0, 1)$ is a point on the line $\Rightarrow x_1 = 0$ and $y_1 = 1$

$x + y = 5 \Rightarrow A = 1, B = 1$, and $C = -5$

$$d = \frac{|1(0) + 1(1) + (-5)|}{\sqrt{1^2 + 1^2}} = \frac{4}{\sqrt{2}} = 2\sqrt{2}$$

81. Slope: $m = \tan 0.1 \approx 0.1003$

Change in elevation: $\sin 0.1 = \dfrac{x}{2(5280)}$

$$x \approx 1054 \text{ feet}$$

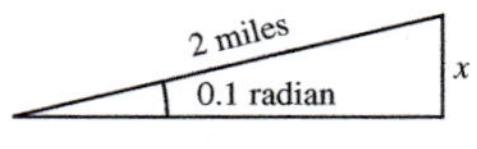

Not drawn to scale

83. Slope $= \frac{3}{5}$

Inclination $= \tan^{-1}\frac{3}{5} \approx 31.0°$

85. $\tan\gamma = \frac{6}{9}$

$\gamma = \arctan\left(\frac{2}{3}\right) \approx 33.69°$

$\beta = 90 - \gamma \approx 56.31°$

Also, because the right triangles containing α and β are equal, $\alpha = \gamma \approx 33.69°$

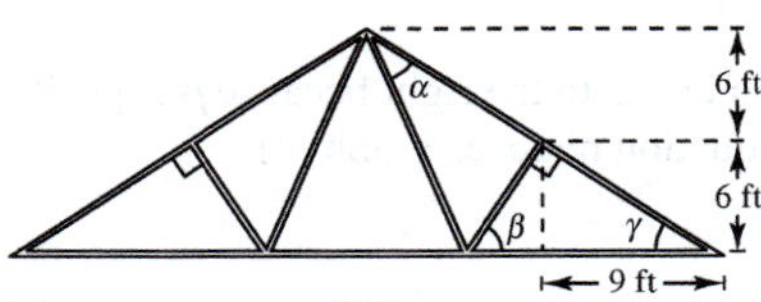

87. True. The inclination of a line is related to its slope by $m = \tan\theta$. If the line has an inclination of 0 radians, then the slope is 0 radians.

89. False. Substitute $m_1 = \tan\theta_1$ and $m_2 = \tan\theta_2$ into the formula for the angle between two lines.

91. False. By definition, the inclination of a nonhorizontal line is the positive angle θ measured counter clockwise from the x-axis to the line. So, the angle θ can be acute, right or obtuse. The angle θ between two lines is less than $\pi/2$ because, if $\theta > \dfrac{\pi}{2}$, then $\tan\theta < 0$.

Because the formula for the angle between two lines involves absolute value, then $\tan\theta$ will always be positive. So, θ cannot be larger than $\pi/2$.

93. (a) $(0, 0) \Rightarrow x_1 = 0$ and $y_1 = 0$

$y = mx + 4 \Rightarrow 0 = mx - y + 4$

$$d = \frac{|m(0) + (-1)(0) + 4|}{\sqrt{m^2 + (-1)^2}} = \frac{4}{\sqrt{m^2 + 1}}$$

(b)

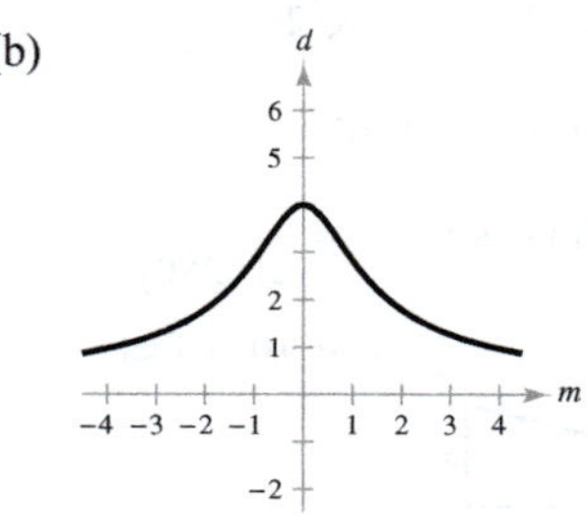

(c) The maximum distance of 4 occurs when the slope m is 0 and the line through $(0, 4)$ is horizontal.

(d) The graph has a horizontal asymptote at $d = 0$. As the slope becomes larger, the distance between the origin and the line, $y = mx + 4$, becomes smaller and approaches 0.

Section 10.2 Introduction to Conics: Parabolas

1. conic

3. locus

5. axis

7. focal chord

9. $y^2 = 4x$

Vertex: $(0, 0)$

$p = 1 > 0$

The graph opens to the right because p is positive. So, the equation matches graph (c).

10. $x^2 = 2y$

Vertex: $(0, 0)$

$p = \frac{1}{2} > 0$

The graph opens upward because p is positive. So, the equation matches graph (a).

11. $x^2 = -8y$

Vertex: $(0, 0)$

$p = -2 < 0$

The graph opens downward because p is negative. So, the equation matches graph (b).

12. $y^2 = -12x$

Vertex: $(0, 0)$

$p = -3 < 0$

The graph opens to the left because p is negative. So, the equation matches graph (d).

13. Vertex: $(0, 0) \Rightarrow h = 0, k = 0$

Graph opens upward.

$x^2 = 4py$

Focus: $(0, 1)$

$p = 1$

$x^2 = 4(1)y$

$x^2 = 4y$

15. Vertex: $(0, 0) \Rightarrow h = 0, k = 0$

Focus: $\left(0, \frac{1}{2}\right) \Rightarrow p = \frac{1}{2}$

$x^2 = 4py$

$x^2 = 4\left(\frac{1}{2}\right)y$

$x^2 = 2y$

17. Focus: $(-2, 0) \Rightarrow p = -2$

$y^2 = 4px$

$y^2 = 4(-2)x$

$y^2 = -8x$

19. Vertex: $(0, 0) \Rightarrow h = 0, k = 0$

Directrix: $y = 2 \Rightarrow p = -2$

$x^2 = 4py$

$x^2 = 4(-2)y$

$x^2 = -8y$

21. Vertex: $(0, 0) \Rightarrow h = 0, k = 0$

Directrix: $x = -1 \Rightarrow p = 1$

$y^2 = 4px$

$y^2 = 4(1)x$

$y^2 = 4x$

23. Vertex: $(0, 0) \Rightarrow h = 0, k = 0$

Vertical axis

Passes through: $(4, 6)$

$$x^2 = 4py$$
$$4^2 = 4p(6)$$
$$16 = 24p$$
$$p = \tfrac{2}{3}$$
$$x^2 = 4\left(\tfrac{2}{3}\right)y$$
$$x^2 = \tfrac{8}{3}y$$

25. Vertex: $(0, 0) \Rightarrow h = 0, k = 0$

Horizontal axis

Passes through: $(-2, 5)$

$$y^2 = 4px$$
$$5^2 = 4p(-2)$$
$$25 = -8p$$
$$p = -\tfrac{25}{8}$$
$$y^2 = 4\left(-\tfrac{25}{8}\right)x$$
$$y^2 = -\tfrac{25}{2}x$$

27. Vertex: $(2, 6) \Rightarrow h = 2, k = 6$

Focus: $(2, 4) \Rightarrow p = -2$

$$(x - h)^2 = 4p(y - k)$$
$$(x - 2)^2 = 4(-2)(y - 6)$$
$$(x - 2)^2 = -8(y - 6)$$

29. Vertex: $(6, 3) \Rightarrow h = 6, k = 3$

Focus: $(4, 3) \Rightarrow p = -2$

$$(y - k)^2 = 4p(x - h)$$
$$(y - 3)^2 = 4(-2)(x - 6)$$
$$(y - 3)^2 = -8(x - 6)$$

31. Vertex: $(0, 2)$

Directrix: $y = 4$

Vertical axis

$$p = 2 - 4 = -2$$
$$(x - 0)^2 = 4(-2)(y - 2)$$
$$x^2 = -8(y - 2)$$

33. Focus: $(2, 2)$

Directrix: $x = -2$

Horizontal axis

Vertex: $(0, 2)$

$$p = 2 - 0 = 2$$
$$(y - 2)^2 = 4(2)(x - 0)$$
$$(y - 2)^2 = 8x$$

35. Vertex: $(3, -3) \Rightarrow h = 3, k = -3$

Vertical Axis; Passes through $(0, 0)$

$$(x - h)^2 = 4p(y - k)$$
$$(x - 3)^2 = 4p(y + 3)$$
$$(0 - 3)^2 = 4p(0 + 3)$$
$$9 = 12p$$
$$p = \frac{3}{4}$$
$$(x - 3)^2 = 3(y + 3)$$

37. $y = \tfrac{1}{2}x^2$

$$x^2 = 2y$$
$$x^2 = 4\left(\tfrac{1}{2}\right)y \Rightarrow h = 0, k = 0, p = \tfrac{1}{2}$$

Vertex: $(0, 0)$

Focus: $\left(0, \tfrac{1}{2}\right)$

Directrix: $y = -\tfrac{1}{2}$

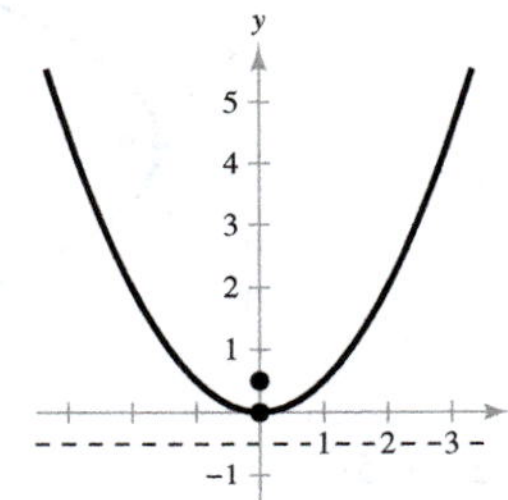

39. $y^2 = -6x$

$$y^2 = 4\left(-\tfrac{3}{2}\right)x \Rightarrow h = 0, k = 0, p = -\tfrac{3}{2}$$

Vertex: $(0, 0)$

Focus: $\left(-\tfrac{3}{2}, 0\right)$

Directrix: $x = \tfrac{3}{2}$

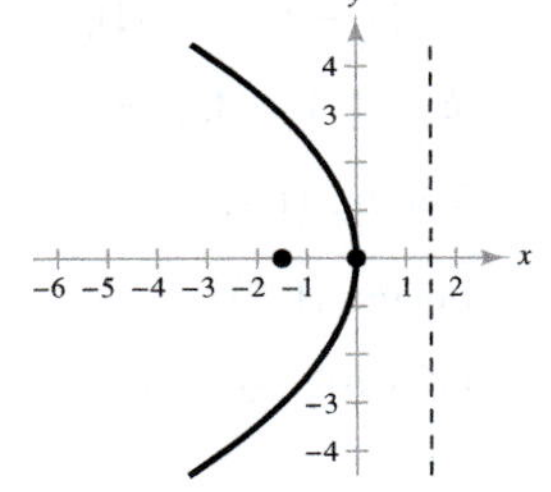

41. $x^2 + 12y = 0$

$$x^2 = -12y = 4(-3)y \Rightarrow h = 0,\ k = 0,\ p = -3$$

Vertex: $(0, 0)$

Focus: $(0, -3)$

Directrix: $y = 3$

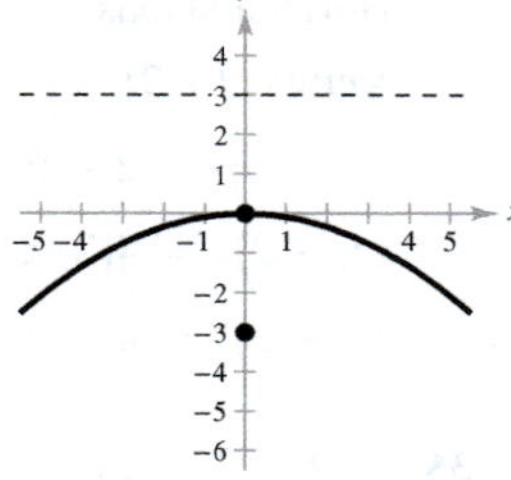

43. $(x - 1)^2 + 8(y + 2) = 0$

$$(x - 1)^2 = 4(-2)(y + 2)$$

$h = 1,\ k = -2,\ p = -2$

Vertex: $(1, -2)$

Focus: $(1, -4)$

Directrix: $y = 0$

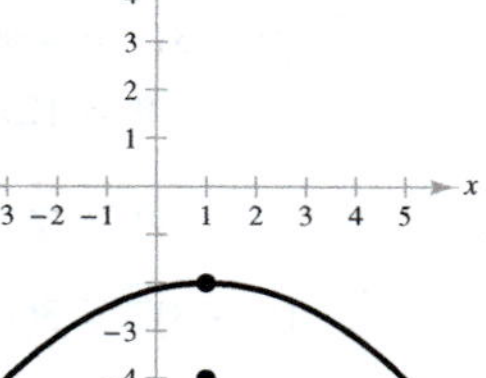

45. $(y + 7)^2 = 4\left(x - \dfrac{3}{2}\right)$

$$(y + 7)^2 = 4(1)\left(x - \frac{3}{2}\right)$$

$h = \dfrac{3}{2},\ k = -7,\ p = 1$

Vertex: $\left(\dfrac{3}{2}, -7\right)$

Focus: $\left(\dfrac{5}{2}, -7\right)$

Directrix: $x = \dfrac{1}{2}$

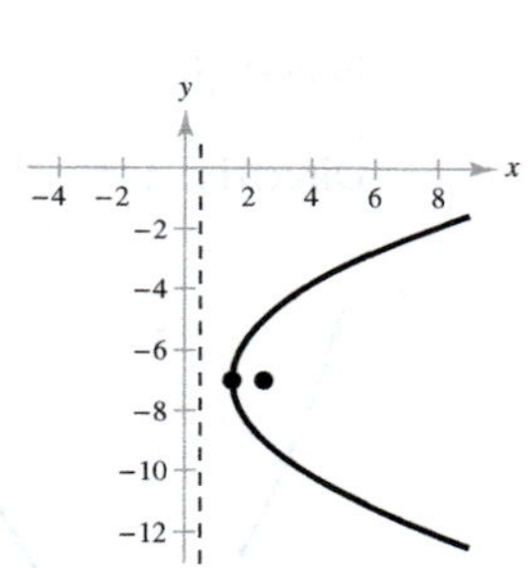

47.

$$y = \tfrac{1}{4}(x^2 - 2x + 5)$$
$$4y = x^2 - 2x + 5$$
$$4y - 5 + 1 = x^2 - 2x + 1$$
$$4y - 4 = (x - 1)^2$$
$$(x - 1)^2 = 4(1)(y - 1)$$

$h = 1,\ k = 1,\ p = 1$

Vertex: $(1, 1)$

Focus: $(1, 2)$

Directrix: $y = 0$

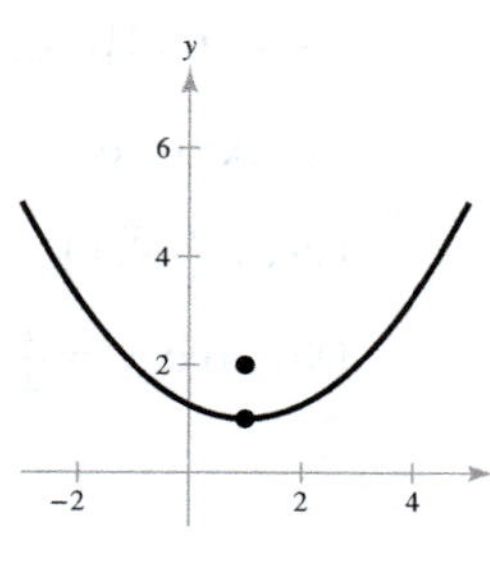

49. $y^2 + 6y + 8x + 25 = 0$

$$y^2 + 6y + 9 = -8x - 25 + 9$$
$$(y + 3)^2 = 4(-2)(x + 2)$$

$h = -2,\ k = -3,\ p = -2$

Vertex: $(-2, -3)$

Focus: $(-4, -3)$

Directrix: $x = 0$

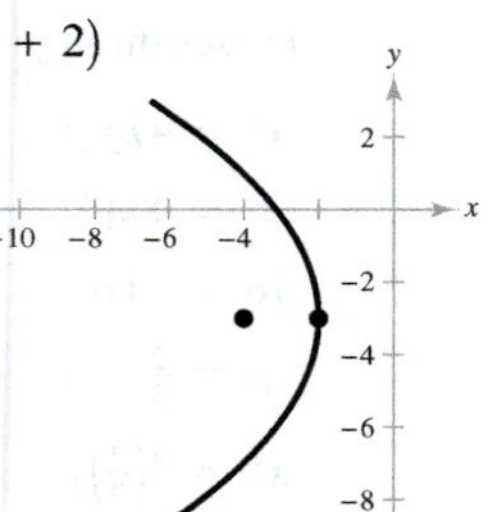

51. $x^2 + 4x - 6y = -10$

$$x^2 + 4x + 4 = 6y - 10 + 4$$
$$x^2 + 4x + 4 = 6y - 6$$
$$(x + 2)^2 = 6(y - 1)$$
$$(x + 2)^2 = 4\left(\frac{3}{2}\right)(y - 1)$$

$h = -2,\ k = 1,\ p = \dfrac{3}{2}$

Vertex: $(-2, 1)$

Focus: $\left(-2, \dfrac{5}{2}\right)$

Directrix: $y = -\dfrac{1}{2}$

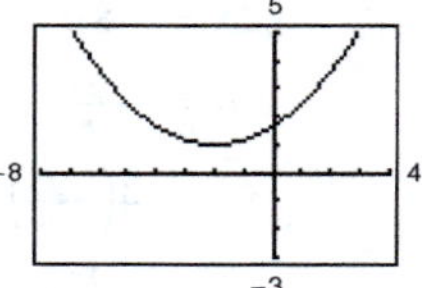

53. $y^2 + x + y = 0$

$$y^2 + y + \tfrac{1}{4} = -x + \tfrac{1}{4}$$
$$\left(y + \tfrac{1}{2}\right)^2 = 4\left(-\tfrac{1}{4}\right)\left(x - \tfrac{1}{4}\right)$$

$h = \tfrac{1}{4},\ k = -\tfrac{1}{2},\ p = -\tfrac{1}{4}$

Vertex: $\left(\tfrac{1}{4}, -\tfrac{1}{2}\right)$

Focus: $\left(0, -\tfrac{1}{2}\right)$

Directrix: $x = \tfrac{1}{2}$

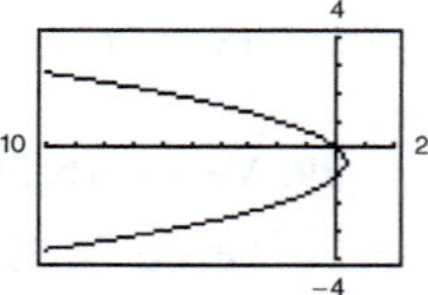

55. $x^2 = 8y$

$x^2 = 4(2)y \Rightarrow p = 2$

Focus: $(0, 2)$

$$d_1 = 2 - b$$
$$d_2 = \sqrt{(6 - 0)^2 + \left(\frac{9}{2} - 2\right)^2} = \sqrt{36 + \frac{25}{4}}$$
$$2 - b = \frac{13}{2}$$
$$b = -\frac{9}{2}$$
$$m = \frac{-(9/2) - (9/2)}{0 - 6} = \frac{3}{2}$$

Tangent line: $y = \dfrac{3}{2}x - \dfrac{9}{2}$

57. $x^2 = 2y \Rightarrow p = \frac{1}{2}$

Point: $(4, 8)$

Focus: $\left(0, \frac{1}{2}\right)$

$$d_1 = \frac{1}{2} - b$$

$$d_2 = \sqrt{(4-0)^2 + \left(8 - \frac{1}{2}\right)^2} = \frac{17}{2}$$

$$d_1 = d_2 \Rightarrow b = -8$$

$$m = \frac{8 - (-8)}{4 - 0} = 4$$

Tangent line: $y = 4x - 8$

59. $y = -2x^2$

$$x^2 = -\frac{1}{2}y \Rightarrow p = -\frac{1}{8}$$

Point: $(-1, -2)$

Focus: $\left(0, -\frac{1}{8}\right)$

$$d_1 = b - \left(-\frac{1}{8}\right) = b + \frac{1}{8}$$

$$d_2 = \sqrt{(-1-0)^2 + \left(-2 - \left(-\frac{1}{8}\right)\right)^2} = \frac{17}{8}$$

$$d_1 = d_2 \Rightarrow b = 2$$

$$m = \frac{-2 - 2}{-1 - 0} = 4$$

$$y = 4x + 2$$

61. $y^2 = 4px,\ p = 1.5$

$$y^2 = 4(1.5)x$$

$$y^2 = 6x$$

63. Vertex: $(0, 0)$

$$(y - 0)^2 = 4p(x - 0)$$

$$y^2 = 4px$$

At $(1000, 800)$: $800^2 = 4p(1000) \Rightarrow p = 160$

$$y^2 = 4(160)x$$

$$y^2 = 640x$$

65. (a)

$$x^2 = 4py$$

$$32^2 = 4p\left(\frac{1}{12}\right)$$

$$1024 = \frac{1}{3}p$$

$$3072 = p$$

$$x^2 = 4(3072)y$$

$$x^2 = 12{,}288y \text{ (in feet)}$$

(b)

$$\frac{1}{24} = \frac{x^2}{12{,}288}$$

$$\frac{12{,}288}{24} = x^2$$

$$512 = x^2$$

$$x \approx 22.6 \text{ feet}$$

67. Vertex: $(0, 48) \Rightarrow h = 0, k = 48$

Passes through $\left(10\sqrt{3}, 0\right)$

Vertical axis

$$(x - 0)^2 = 4p(y - 48)$$

$$\left(10\sqrt{3} - 0\right)^2 = 4p(0 - 48)$$

$$300 = -192p$$

$$-\tfrac{25}{16} = p$$

$$x^2 = 4\left(-\tfrac{25}{16}\right)(y - 48)$$

$$x^2 = -\tfrac{25}{4}(y - 48)$$

69. $x^2 = 4p(y - 12)$

$(4, 10)$ on curve:

$$16 = 4p(10 - 12) = -8p \Rightarrow p = -2$$

$$x^2 = 4(-2)(y - 12) = -8y + 96$$

$$y = \frac{-x^2 + 96}{8}$$

$y = 0$ if $x^2 = 96 \Rightarrow x = 4\sqrt{6}$

So, the width is about $2\left(4\sqrt{6}\right) \approx 19.6$ meters.

71. (a) $x^2 = 4py$

$$60^2 = 4p(20) \Rightarrow p = 45$$

Focus: $(0, 45)$

(b) $x^2 = 4(45)y$ or $y = \frac{1}{180}x^2$

73. (a) $V = 17{,}500\sqrt{2}$ mi/h
$\approx 24{,}750$ mi/h

(b) $p = -4100,\ (h, k) = (0, 4100)$

$$(x - 0)^2 = 4(-4100)(y - 4100)$$
$$x^2 = -16{,}400(y - 4100)$$

75. (a) $x^2 = -\dfrac{v^2}{16}(y - s)$

$$x^2 = -\frac{(28)^2}{16}(y - 100)$$
$$x^2 = -49(y - 100)$$

(b) The ball bits the ground when $y = 0$.

$$x^2 = -49(0 - 100)$$
$$x^2 = 4900$$
$$x = 70$$

The ball travels 70 feet.

77. False. It is not possible for a parabola to intersect its directrix. If the graph crossed the directrix there would exist points closer to the directrix than the focus.

79. True. If the axis (line connecting the vertex and focus) is horizontal, then the directrix must be vertical.

81. Both (a) and (b) are parabolas with vertical axes, while (c) is a parabola with a horizontal axis.

So, equations (a) and (b) are equivalent when $p = \dfrac{1}{4a}$.

(a) $y = a(x - h)^2 + k$

(b)
$$(x - h)^2 = 4p(y - k)$$
$$(x - h)^2 = 4py - 4pk$$
$$(1/4p)\left((x - h)^2 + 4pk\right) = 4py(1/4p)$$
$$(1/4p)(x - h)^2 + k = y = a(x - h)^2 + k$$
$$a = \left(\frac{1}{4p}\right)$$
$$4a = (1/p)$$
$$p = \frac{1}{4a}$$

83. The graph of $x^2 + y^2 = 0$ is a single point, $(0, 0)$.

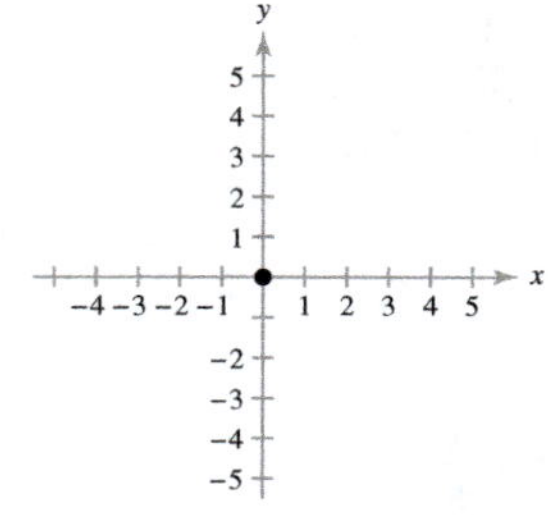

The plane intersects the double-napped cone at the vertices of the cones.

85. (a) $A = \dfrac{8}{3}(2)^{1/2}(4)^{3/2} = \dfrac{8}{3}(\sqrt{2})(8) = \dfrac{64\sqrt{2}}{3}$ square units

(b) As p approaches zero, the parabola becomes narrower and narrower, so the area becomes smaller and smaller.

Section 10.3 Ellipses

1. ellipse; foci

3. minor axis

5. $\dfrac{x^2}{4} + \dfrac{y^2}{9} = 1$

Center: $(0, 0)$

$a = 3, b = 2$

Vertical major axis

Matches graph (b).

6. $\dfrac{x^2}{9} + \dfrac{y^2}{4} = 1$

Center: $(0, 0)$

$a = 3, b = 2$

Horizontal major axis

Matches graph (c).

7. $\dfrac{(x - 2)^2}{16} + (y + 1)^2 = 1$

Center: $(2, -1)$

$a = 4, b = 1$

Horizontal major axis

Matches graph (a).

8. $\dfrac{(x + 2)^2}{9} + \dfrac{(y + 2)^2}{4} = 1$

Center: $(-2, -2)$

$a = 3, b = 2$

Horizontal major axis

Matches graph (d).

9. Center: $(0, 0)$

$a = 4, b = 2$

Vertical major axis

$$\frac{(x-h)^2}{b^2} + \frac{(y-k)^2}{a^2} = 1$$

$$\frac{x^2}{4} + \frac{y^2}{16} = 1$$

11. Center: $(0, 0)$

Vertices: $(\pm 7, 0) \Rightarrow a = 7$

Foci: $(\pm 2, 0) \Rightarrow c = 2$

$b^2 = a^2 - c^2 = 49 - 4 = 45$

$$\frac{x^2}{a^2} + \frac{y^2}{b^2} = 1$$

$$\frac{x^2}{49} + \frac{y^2}{45} = 1$$

13. Center: $(0, 0)$

Foci: $(\pm 4, 0) \Rightarrow c = 4$

Length of horizontal major axis: $10 \Rightarrow a = 5$

$b^2 = a^2 - c^2 = 25 - 16 = 9$

$$\frac{x^2}{a^2} + \frac{y^2}{b^2} = 1$$

$$\frac{x^2}{25} + \frac{y^2}{9} = 1$$

15. Major axis vertical

Passes through: $(0, 6)$ and $(3, 0)$

$a = 6, b = 3$

$$\frac{x^2}{b^2} + \frac{y^2}{a^2} = 1$$

$$\frac{x^2}{9} + \frac{y^2}{36} = 1$$

17. Vertices: $(\pm 6, 0) \Rightarrow a = 6$

Major axis horizontal

Passes through: $(4, 1)$

$$\frac{x^2}{36} + \frac{y^2}{b^2} = 1$$

$$\frac{4^2}{36} + \frac{1^2}{b^2} = 1$$

$$16b^2 + 36 = 36b^2$$

$$36 = 20b^2$$

$$\frac{9}{5} = b^2$$

$$\frac{x^2}{36} + \frac{y^2}{\frac{9}{5}} = 1 \text{ or } \frac{x^2}{36} + \frac{5y^2}{9} = 1$$

19. Center: $(2, 3)$

$a = 3, b = 1$

Vertical major axis

$$\frac{(x-h)^2}{b^2} + \frac{(y-k)^2}{a^2} = 1$$

$$\frac{(x-2)^2}{1} + \frac{(y-3)^2}{9} = 1$$

21. Vertices: $(2, 0), (10, 0) \Rightarrow a = 4$

Horizontal major axis

Length of minor axis: $4 \Rightarrow b = 2$

Center: $(6, 0) \Rightarrow h = 6, k = 0$

$$\frac{(x-h)^2}{a^2} + \frac{(y-k)^2}{b^2} = 1$$

$$\frac{(x-6)^2}{16} + \frac{(y-0)^2}{4} = 1$$

$$\frac{(x-6)^2}{16} + \frac{y^2}{4} = 1$$

23. Foci: $(0, 0), (4, 0) \Rightarrow c = 2$

Length of major axis: $6 \Rightarrow a = 3$

Center: $(2, 0) = (h, k)$

$b^2 = a^2 - c^2 = 9 - 4 = 5$

$$\frac{(x-h)^2}{a^2} + \frac{(y-k)^2}{b^2} = 1$$

$$\frac{(x-2)^2}{9} + \frac{y^5}{5} = 1$$

25. Center: $(1, 3)$

Vertex: $(-2, 3) \Rightarrow a = 3$

Major axis horizontal

Length of minor axis: $4 \Rightarrow b = 2$

$$\frac{(x-h)^2}{a^2} + \frac{(y-k)^2}{b^2} = 1$$

$$\frac{(x-1)^2}{9} + \frac{(y-3)^2}{4} = 1$$

27. Center: $(1, 4)$

Vertices: $(1, 0)$ and $(1, 8) \Rightarrow a = 4$

Major axis vertical

$$a = 2c$$
$$4 = 2c$$
$$c = 2$$
$$b^2 = a^2 - c^2$$
$$b^2 = 16 - 4 = 12$$

$$\frac{(x-h)^2}{b^2} + \frac{(y-k)^2}{a^2} = 1$$

$$\frac{(x-1)^2}{12} + \frac{(y-4)^2}{16} = 1$$

29. Vertices: $(0, 2), (4, 2) \Rightarrow a = 2$

Center: $(2, 2)$

Endpoints of the minor axis: $(2, 3), (2, 1) \Rightarrow b = 1$

Horizontal major axis:

$$\frac{(x-h)^2}{a^2} + \frac{(y-k)^2}{b^2} = 1$$

$$\frac{(x-2)^2}{4} + \frac{(y-2)^2}{1} = 1$$

31. $\dfrac{x^2}{25} + \dfrac{y^2}{16} = 1$

$a = 5, b = 4, c = 3$

Center: $(0, 0)$

Vertices: $(\pm 5, 0)$

Foci: $(\pm 3, 0)$

Eccentricity: $e = \dfrac{3}{5}$

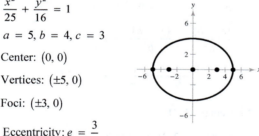

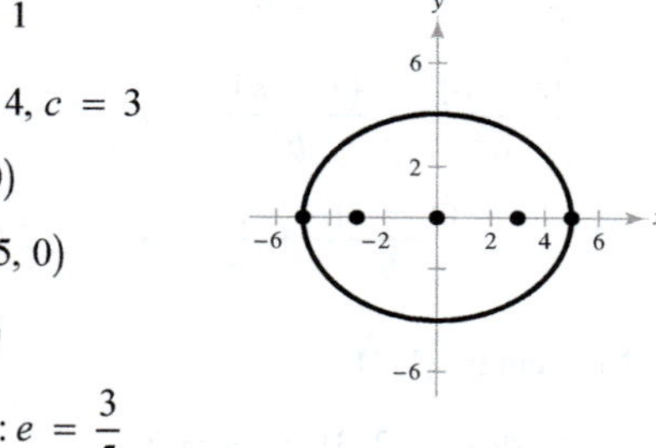

33. $9x^2 + y^2 = 36$

$\dfrac{x^2}{4} + \dfrac{y^2}{36} = 1$

$a = 6, b = 2$

$c^2 = a^2 - b^2$

$\quad = 36 - 4 = 32$

Center: $(0, 0)$

Vertices: $(0, \pm 6)$

Foci: $(0, \pm 4\sqrt{2})$

Eccentricity: $e = \dfrac{4\sqrt{2}}{6} = \dfrac{2\sqrt{2}}{3}$

35. $\dfrac{(x-4)^2}{16} + \dfrac{(y+1)^2}{25} = 1$

$a = 5, b = 4$

$c^2 = a^2 - b^2 = 25 - 16 = 9 \Rightarrow c = 3$

Center: $(4, -1)$

Vertices: $(4, 4), (4, -6)$

Foci: $(4, 2), (4, -4)$

Eccentricity: $e = \dfrac{3}{5}$

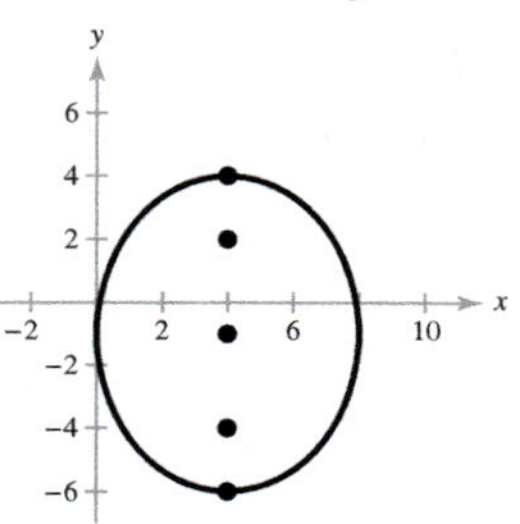

37. $\dfrac{(x+5)^2}{9/4} + (y-1)^2 = 1$

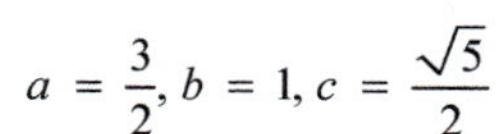

$a = \dfrac{3}{2}, b = 1, c = \dfrac{\sqrt{5}}{2}$

Center: $(-5, 1)$

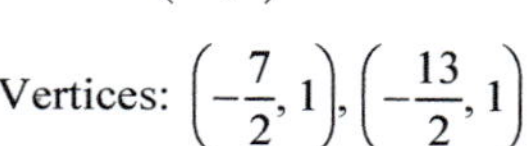

Vertices: $\left(-\dfrac{7}{2}, 1\right), \left(-\dfrac{13}{2}, 1\right)$

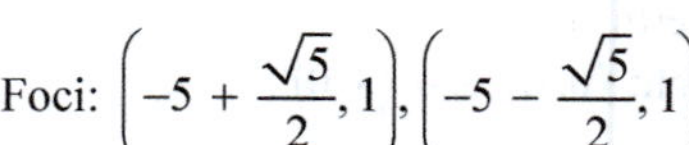

Foci: $\left(-5 + \dfrac{\sqrt{5}}{2}, 1\right), \left(-5 - \dfrac{\sqrt{5}}{2}, 1\right)$

Eccentricity: $e = \dfrac{\sqrt{5}}{3}$

39.

$$9x^2 + 4y^2 + 36x - 24y + 36 = 0–$$
$$9(x^2 + 4x + 4) + 4(y^2 - 6y + 9) = -36 + 36 + 36$$
$$9(x+2)^2 + 4(y-3)^2 = 36$$
$$\frac{(x+2)^2}{4} + \frac{(y-3)^2}{9} = 1$$

$a = 3, b = 2, c = \sqrt{5}$

Center: $(-2, 3)$

Vertices: $(-2, 6), (-2, 0)$

Foci: $(-2, 3 \pm \sqrt{5})$

Eccentricity: $e = \dfrac{\sqrt{5}}{3}$

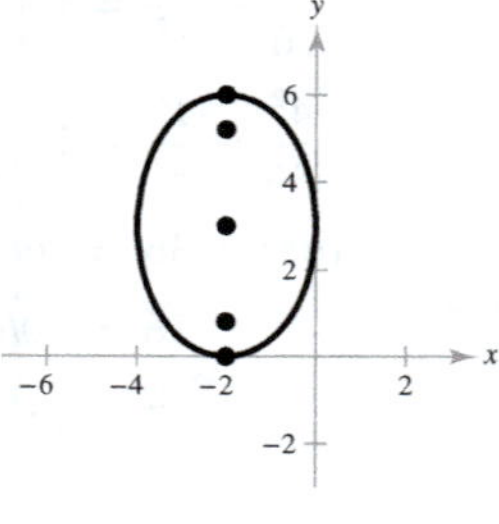

41. $$x^2 + 5y^2 - 8x - 30y - 39 = 0$$
$$(x^2 - 8x + 16) + 5(y^2 - 6y + 9) = 39 + 16 + 45$$
$$(x - 4)^2 + 5(y - 3)^2 = 100$$
$$\frac{(x - 4)^2}{100} + \frac{(y - 3)^2}{20} = 1$$

$a = 10, b = \sqrt{20} = 2\sqrt{5},$

$c = \sqrt{80} = 4\sqrt{5}$

Center: $(4, 3)$

Foci: $\left(4 \pm 4\sqrt{5}, 3\right)$

Vertices: $(14, 3), (-6, 3)$

Eccentricity: $e = \dfrac{4\sqrt{5}}{10} = \dfrac{2\sqrt{5}}{5}$

43. $$6x^2 + 2y^2 + 18x - 10y + 2 = 0$$
$$6\left(x^2 + 3x + \frac{9}{4}\right) + 2\left(y^2 - 5y + \frac{25}{4}\right) = -2 + \frac{27}{2} + \frac{25}{2}$$
$$6\left(x + \frac{3}{2}\right)^2 + 2\left(y - \frac{5}{2}\right)^2 = 24$$
$$\frac{\left(x + \frac{3}{2}\right)^2}{4} + \frac{\left(y - \frac{5}{2}\right)^2}{12} = 1$$

$a = \sqrt{12} = 2\sqrt{3}, b = 2, c = \sqrt{8} = 2\sqrt{2}$

Center: $\left(-\dfrac{3}{2}, \dfrac{5}{2}\right)$

Vertices: $\left(-\dfrac{3}{2}, \dfrac{5}{2} \pm 2\sqrt{3}\right)$

Foci: $\left(-\dfrac{3}{2}, \dfrac{5}{2} \pm 2\sqrt{2}\right)$

Eccentricity: $e = \dfrac{2\sqrt{2}}{2\sqrt{3}} = \dfrac{\sqrt{6}}{3}$

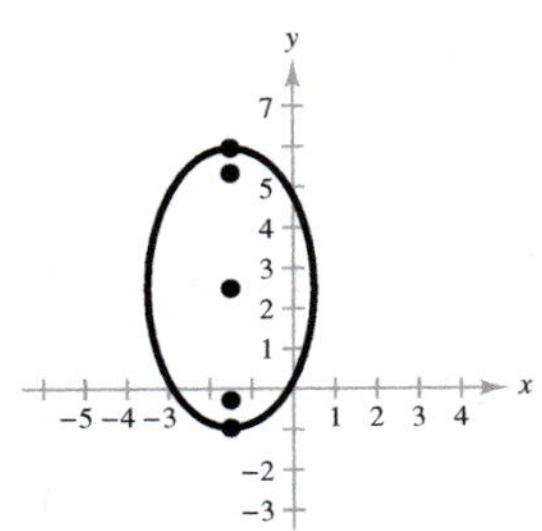

45. $$12x^2 + 20y^2 - 12x + 40y - 37 = 0$$
$$12\left(x^2 - x + \frac{1}{4}\right) + 20(y^2 + 2y + 1) = 37 + 3 + 20$$
$$12\left(x - \frac{1}{2}\right)^2 + 20(y + 1)^2 = 60$$
$$\frac{\left(x - \frac{1}{2}\right)^2}{5} + \frac{(y + 1)^2}{3} = 1$$

$a^2 = 5, b^2 = 3$

$c^2 = 5 - 3 = 2$

Center: $\left(\dfrac{1}{2}, -1\right)$

Vertices: $\left(\dfrac{1}{2} \pm \sqrt{5}, -1\right)$

Foci: $\left(\dfrac{1}{2} \pm \sqrt{2}, -1\right)$

Eccentricity: $e = \dfrac{\sqrt{2}}{\sqrt{5}} = \dfrac{\sqrt{10}}{5}$

47. $5x^2 + 3y^2 = 15$

$$\frac{x^2}{3} + \frac{y^2}{5} = 1$$

$a = \sqrt{5}, b = \sqrt{3}, c = \sqrt{2}$

Center: $(0, 0)$

Vertices: $\left(0, \pm\sqrt{5}\right)$

Foci: $\left(0, \pm\sqrt{2}\right)$

Eccentricity: $e = \dfrac{\sqrt{10}}{5}$

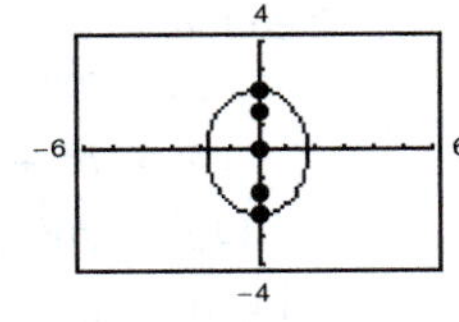

49. $$x^2 + 9y^2 - 10x + 36y + 52 = 0$$
$$(x^2 - 10x + 25) + 9(y^2 + 4y + 4) = -52 + 25 + 36$$
$$(x - 5)^2 + 9(y + 2)^2 = 9$$
$$\frac{(x - 5)^2}{9} + \frac{(y + 2)}{1} = 1$$

$a = 3, b = 1, c = 2\sqrt{2}$

Center: $(5, -2)$

Vertices: $(8, -2), (2, -2)$

Foci: $\left(5 \pm 2\sqrt{2}, -2\right)$

Eccentricity: $e = \dfrac{2\sqrt{2}}{3}$

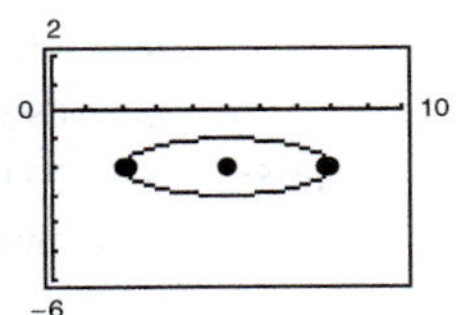

51. Vertices: $(\pm 5, 0) \Rightarrow a = 5$

$e = \frac{3}{5} \Rightarrow c = \frac{3}{5}a = 3$

$b^2 = a^2 - c^2 = 25 - 9 = 16$

Center: $(0, 0) = (h, k)$

$$\frac{(x-h)^2}{a^2} + \frac{(y-k)^2}{b^2} = 1$$

$$\frac{x^2}{25} + \frac{y^2}{16} = 1$$

53. (a)

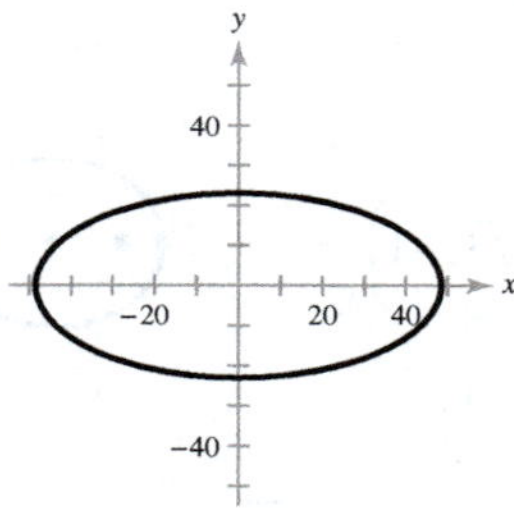

$$\frac{x^2}{2352.25} + \frac{y^2}{23^2} = 1 \text{ or } \frac{x^2}{529} + \frac{y^2}{2352.25} = 1$$

$$a = \frac{97}{2}, b = 23, c = \sqrt{\left(\frac{97}{2}\right)^2 - (23)^2} \approx 4.7$$

(b) Distance between foci: $2(4.7) \approx 85.4$ feet

55. The length of the major axis and minor axis are 280 millimeters and 160 millimeters, respectively. Therefore,

$2a = 280 \Rightarrow a = 140$ and $2b = 160 \Rightarrow b = 80$.

$$a^2 = b^2 + c^2$$
$$140^2 = 80^2 + c^2$$
$$13{,}200 = c^2$$
$$\sqrt{13{,}200} = c$$
$$20\sqrt{33} = c$$

The kidney stone and spark plug are each located at a focus, therefore they are 2c millimeters apart, or

$2(20\sqrt{33}) = 40\sqrt{33} \approx 229.8$ millimeters apart.

57. $a + c = 6378 + 939 = 7317$

$a - c = 6378 + 215 = 6593$

Solving this system for a and c yields

$a + c = 7317$

$a - c = 6593$

$2a = 13{,}910$

$a = 6955$

$6955 + c = 7317$

$c = 362$

Eccentricity: $e = \frac{c}{a} = \frac{362}{6955} \approx 0.0520$

59. $\frac{x^2}{9} + \frac{y^2}{16} = 1$

$a = 4, b = 3, c = \sqrt{7}$

Points on the ellipse: $(\pm 3, 0), (0, \pm 4)$

Length of latus recta: $\frac{2b^2}{a} = \frac{2(3)^2}{4} = \frac{9}{2}$

Additional points: $\left(\pm\frac{9}{4}, -\sqrt{7}\right), \left(\pm\frac{9}{4}, \sqrt{7}\right)$

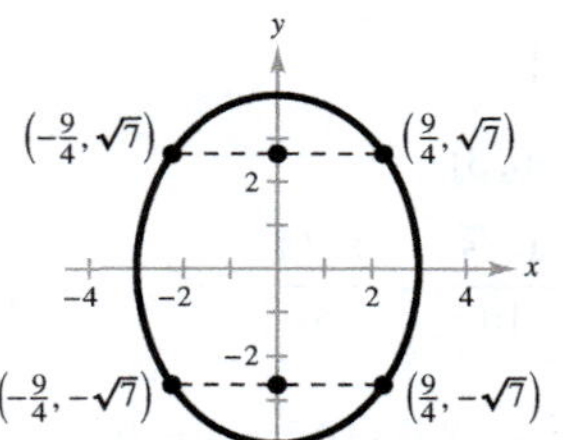

61. $5x^2 + 3y^2 = 15$

$\frac{x^2}{3} + \frac{y^2}{5} = 1$

$a = \sqrt{5}, b = \sqrt{3}, c = \sqrt{2}$

Points on the ellipse: $(\pm\sqrt{3}, 0), (0, \pm\sqrt{5})$

Length of latus recta: $\frac{2b^2}{a} = \frac{2 \cdot 3}{\sqrt{5}} = \frac{6\sqrt{5}}{5}$

Additional points: $\left(\pm\frac{3\sqrt{5}}{5}, \pm\sqrt{2}\right)$

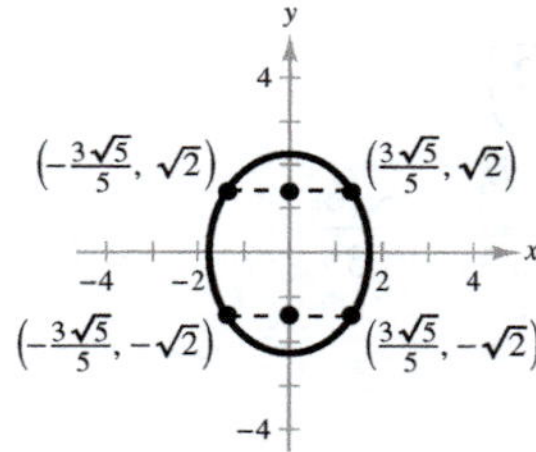

63. False. The graph of $\frac{x^2}{4} + y^4 = 1$ is not an ellipse.

The degree of y is 4, not 2.

65. *Sample answer:* Foci: $(2, 2), (10, 2) \Rightarrow c = 4$

Center: $(6, 2)$

Let $a^2 = 324$ and $b^2 = 308$

So that $c^2 = a^2 - b^2$.

$$\frac{(x-6)^2}{324} + \frac{(y-2)^2}{308} = 1$$

67. $\dfrac{x^2}{a^2} + \dfrac{y^2}{b^2} = 1$

(a) $a + b = 20 \Rightarrow b = 20 - a$

$A = \pi ab = \pi a(20 - a)$

(b) $264 = \pi a(20 - a)$

$0 = -\pi a^2 + 20\pi a - 264$

$0 = \pi a^2 - 20\pi a + 264$

By the Quadratic Formula: $a \approx 14$ or $a \approx 6$. Choosing the larger value of a, you have $a \approx 14$ and $b \approx 6$.

The equation of an ellipse with an area of 264 is

$\dfrac{x^2}{196} + \dfrac{y^2}{36} = 1.$

69.

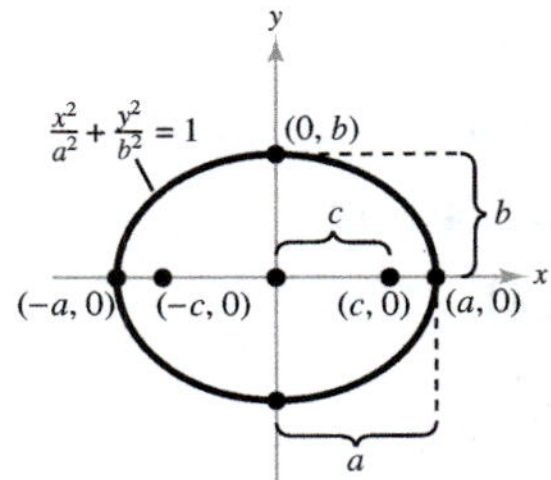

The length of half the major axis is a and the length of half the minor axis is b.

Find the distance between $(0, b)$ and $(c, 0)$ and $(0, b)$ and $(-c, 0)$.

$$d_1 = \sqrt{(0 - c)^2 + (b - 0)^2} = \sqrt{c^2 + b^2}$$

$$d_2 = \sqrt{(0 - (-c))^2 + (b - 0)^2} = \sqrt{c^2 + b^2}$$

The sum of the distances from any point on the ellipse to the two foci is constant. Using the vertex $(a, 0)$, the constant sum is $(a + c) + (a - c) = 2a$.

So, the sum of the distances from $(0, b)$ to the two foci is

$$\sqrt{c^2 + b^2} + \sqrt{c^2 + b^2} = 2a$$
$$2\sqrt{c^2 + b^2} = 2a$$
$$\sqrt{c^2 + b^2} = a$$
$$c^2 + b^2 = a^2$$

So, $a^2 = b^2 + c^2$ for the ellipse $\dfrac{x^2}{a^2} + \dfrac{y^2}{b^2} = 1$, where $a > 0, b > 0$.

Section 10.4 Hyperbolas

1. hyperbola; foci

3. transverse axis; center

5. $\dfrac{y^2}{9} - \dfrac{x^2}{25} = 1$

Center: $(0, 0)$

$a = 3, b = 5$

Vertical transverse axis

Matches graph (b).

6. $\dfrac{x^2}{9} - \dfrac{y^2}{25} = 1$

Center: $(0, 0)$

$a = 3, b = 5$

Horizontal transverse axis

Matches graph (d).

7. $\dfrac{x^2}{25} - \dfrac{(y + 2)^2}{9} = 1$

Center: $(0, -2)$

$a = 5, b = 3$

Horizontal transverse axis

Matches graph (c).

8. $\dfrac{(y + 4)^2}{25} - \dfrac{(x - 2)^2}{9} = 1$

Center: $(2, -4)$

$a = 5, b = 3$

Vertical transverse axis

Matches graph (a).

9. Vertices: $(0, \pm 2) \Rightarrow a = 2$

Foci: $(0, \pm 4) \Rightarrow c = 4$

$b^2 = c^2 - a^2 = 16 - 4 = 12$

Center: $(0, 0) = (h, k)$

$$\frac{(y-k)^2}{a^2} - \frac{(x-h)^2}{b^2} = 1$$

$$\frac{y^2}{4} - \frac{x^2}{12} = 1$$

11. Vertices: $(2, 0), (6, 0) \Rightarrow a = 2$

Foci: $(0, 0), (8, 0) \Rightarrow c = 4$

$b^2 = c^2 - a^2 = 16 - 4 = 12$

Center: $(4, 0) = (h, k)$

$$\frac{(x-h)^2}{a^2} - \frac{(y-k)^2}{b^2} = 1$$

$$\frac{(x-4)^2}{4} - \frac{y^2}{12} = 1$$

13. Vertices: $(4, 1), (4, 9) \Rightarrow a = 4$

Foci: $(4, 0), (4, 10) \Rightarrow c = 5$

$b^2 = c^2 - a^2 = 25 - 16 = 9$

Center: $(4, 5) = (h, k)$

$$\frac{(y-k)^2}{a^2} - \frac{(x-h)^2}{b^2} = 1$$

$$\frac{(y-5)^2}{16} - \frac{(x-4)^2}{9} = 1$$

15. Vertices: $(2, 3), (2, -3) \Rightarrow a = 3$

Passes through the point: $(0, 5)$

Center: $(2, 0) = (h, k)$

$$\frac{(y-k)^2}{a^2} - \frac{(x-h)^2}{b^2} = 1$$

$$\frac{y^2}{9} - \frac{(x-2)^2}{b^2} = 1$$

$$\frac{(x-2)^2}{b^2} = \frac{y^2}{9} - 1 = \frac{y^2-9}{9}$$

$$b^2 = \frac{9(x-2)^2}{y^2-9} = \frac{9(-2)^2}{25-9} = \frac{36}{16} = \frac{9}{4}$$

$$\frac{y^2}{9} - \frac{(x-2)^2}{9/4} = 1$$

$$\frac{y^2}{9} - \frac{4(x-2)^2}{9} = 1$$

17. Vertices: $(0, -3), (4, -3) \Rightarrow a = 2$

Center: $(2, -3)$

Passes through: $(-4, 5)$

$$\frac{(x-h)^2}{a^2} - \frac{(y-k)^2}{b^2} = 1$$

$$\frac{(x-2)^2}{4} - \frac{(y+3)^2}{b^2} = 1$$

$$\frac{(-4-2)^2}{4} - \frac{(5+3)^2}{b^2} = 1$$

$$9 - \frac{64}{b^2} = 1 \Rightarrow b^2 = 8$$

$$\frac{(x-2)^2}{4} - \frac{(y+3)^2}{8} = 1$$

19. $x^2 - y^2 = 1$

$a = 1, b = 1, c = \sqrt{2}$

Center: $(0, 0)$

Vertices: $(\pm 1, 0)$

Foci: $(\pm\sqrt{2}, 0)$

Asymptotes: $y = \pm x$

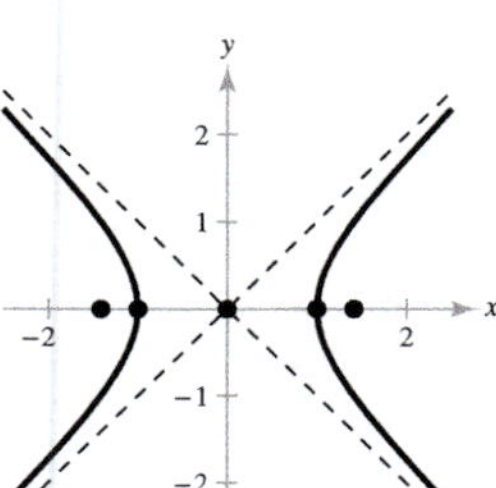

21. $\dfrac{y^2}{36} - \dfrac{x^2}{100} = 1$

$a = 6, b = 10$

$c^2 = a^2 + b^2 = 136 \Rightarrow c = 2\sqrt{34}$

Center: $(0, 0)$

Vertices: $(0, \pm 6)$

Foci: $(0, \pm 2\sqrt{34})$

Asymptotes: $y = \pm\dfrac{3}{5}x$

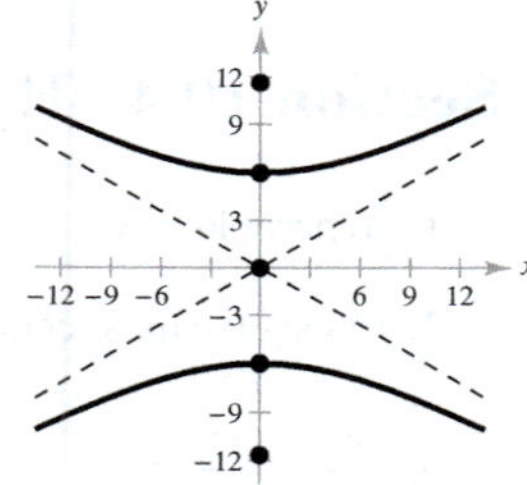

23. $2y^2 - \dfrac{x^2}{2} = 2$

$y^2 - \dfrac{x^2}{4} = 1$

$a = 1, b = 2,$

$c^2 = a^2 + b^2 = 5 \Rightarrow c = \sqrt{5}$

Center: $(0, 0)$

Vertices: $(0, \pm 1)$

Foci: $(0, \pm\sqrt{5})$

Asymptotes: $y = \pm\dfrac{1}{2}x$

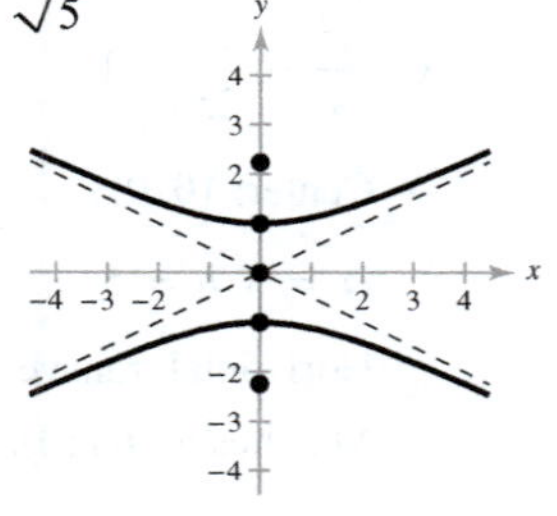

25. $\dfrac{(x-1)^2}{4} - \dfrac{(y+2)^2}{1} = 1$

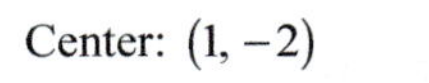

$a = 2, b = 1, c = \sqrt{5}$

Center: $(1, -2)$

Vertices: $(-1, -2), (3, -2)$

Foci: $\left(1 \pm \sqrt{5}, -2\right)$

Asymptotes: $y = -2 \pm \dfrac{1}{2}(x - 1)$

27. $\dfrac{(y+6)^2}{1/9} - \dfrac{(x-2)^2}{1/4} = 1$

$a = \dfrac{1}{3}, b = \dfrac{1}{2},$

$c = \dfrac{\sqrt{13}}{6}$

Center: $(2, -6)$

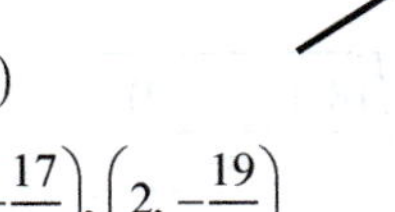

Vertices: $\left(2, -\dfrac{17}{3}\right), \left(2, -\dfrac{19}{3}\right)$

Foci: $\left(2, -6 \pm \dfrac{\sqrt{13}}{6}\right)$

Asymptotes: $y = -6 \pm \dfrac{2}{3}(x - 2)$

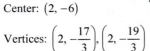

29.

$$9x^2 - y^2 - 36x - 6y + 18 = 0$$
$$9\left(x^2 - 4x + 4\right) - \left(y^2 + 6y + 9\right) = -18 + 36 - 9$$
$$9(x-2)^2 - (y+3)^2 = 9$$
$$\frac{(x-2)^2}{1} - \frac{(y+3)^2}{9} = 1$$

$a = 1, b = 3, c = \sqrt{10}$

Center: $(2, -3)$

Vertices: $(1, -3), (3, -3)$

Foci: $\left(2 \pm \sqrt{10}, -3\right)$

Asymptotes: $y = -3 \pm 3(x - 2)$

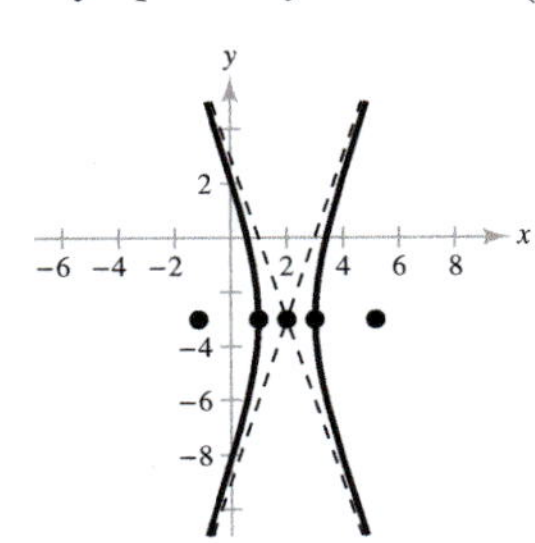

31.

$$4x^2 - y^2 + 8x + 2y - 1 = 0$$
$$4\left(x^2 + 2x + 1\right) - \left(y^2 - 2y + 1\right) = 1 + 4 - 1$$
$$4(x+1)^2 - (y-1)^2 = 4$$
$$\frac{(x+1)^2}{1} - \frac{(y-1)^2}{4} = 1$$

$a = 1, b = 2, c = \sqrt{5}$

Center: $(-1, 1)$

Vertices: $(-2, 1), (0, 1)$

Foci: $\left(-1 \pm \sqrt{5}, 1\right)$

Asymptotes: $y = 1 \pm 2(x + 1)$

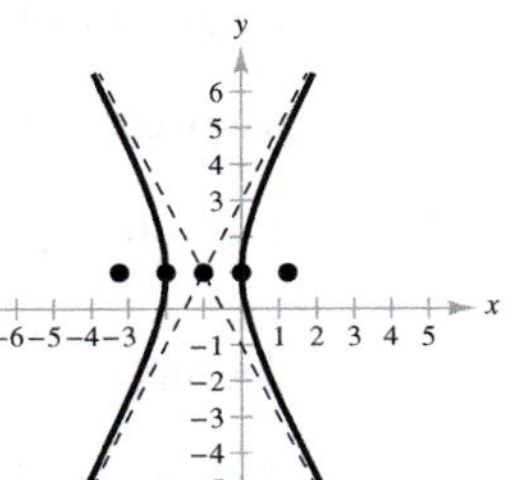

33. $2x^2 - 3y^2 = 6$

$\dfrac{x^2}{3} - \dfrac{y^2}{2} = 1$

$a = \sqrt{3}, b = \sqrt{2}, c = \sqrt{5}$

Center: $(0, 0)$

Vertices: $\left(\pm\sqrt{3}, 0\right)$

Foci: $\left(\pm\sqrt{5}, 0\right)$

Asymptotes: $y = \pm\sqrt{\dfrac{2}{3}}x = \pm\dfrac{\sqrt{6}}{3}x$

To use a graphing utility, solve for y first.

$y^2 = \dfrac{2x^2 - 6}{3}$

$\left.\begin{aligned} y_1 &= \sqrt{\dfrac{2x^2 - 6}{3}} \\ y_2 &= -\sqrt{\dfrac{2x^2 - 6}{3}} \end{aligned}\right\}$ Hyperbola

$\left.\begin{aligned} y_3 &= \dfrac{\sqrt{6}}{3}x \\ y_4 &= -\dfrac{\sqrt{6}}{3}x \end{aligned}\right\}$ Asymptotes

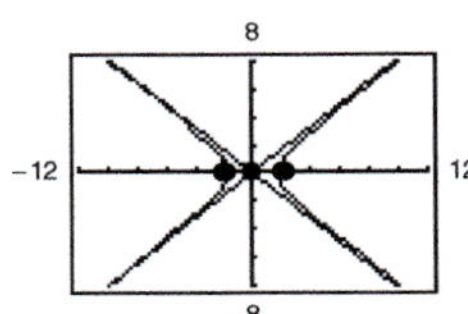

35. $25y^2 - 9x^2 = 225$

$$\frac{y^2}{9} - \frac{x^2}{25} = 1$$

$a = 3, b = 5,$

$c^2 = a^2 + b^2 = 9 + 25 = 34 \Rightarrow c = \sqrt{34}$

Center: $(0, 0)$

Vertices: $(0, \pm 3)$

Foci: $\left(0, \pm\sqrt{34}\right)$

Asymptotes: $y = \pm\frac{3}{5}x$

To use a graphing utility, solve for y first.

$$y^2 = \frac{225 + 9x^2}{25}$$

$$\left.\begin{aligned} y_1 &= \sqrt{\frac{9x^2 + 225}{25}} \\ y_2 &= -\sqrt{\frac{9x^2 + 225}{25}} \end{aligned}\right\} \text{Hyperbola}$$

$$\left.\begin{aligned} y_3 &= \frac{3}{5}x \\ y_4 &= -\frac{3}{5}x \end{aligned}\right\} \text{Asymptotes}$$

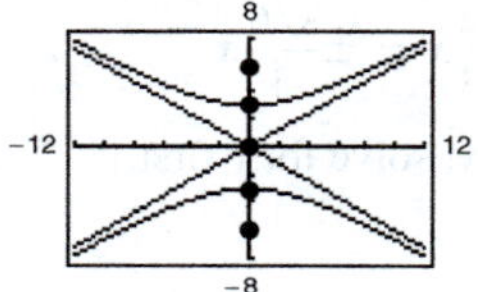

37. $9y^2 - x^2 + 2x + 54y + 62 = 0$

$$9\left(y^2 + 6y + 9\right) - \left(x^2 - 2x + 1\right) = -62 - 1 + 81$$

$$9(y + 3)^2 - (x - 1)^2 = 18$$

$$\frac{(y + 3)^2}{2} - \frac{(x - 1)^2}{18} = 1$$

$a = \sqrt{2}, b = 3\sqrt{2}, c = 2\sqrt{5}$

Center: $(1, -3)$

Vertices: $\left(1, -3 \pm \sqrt{2}\right)$

Foci: $\left(1, -3 \pm 2\sqrt{5}\right)$

Asymptotes: $y = -3 \pm \frac{1}{3}(x - 1)$

To use a graphing utility, solve for y first.

$$9(y + 3)^2 = 18 + (x - 1)^2$$

$$y = -3 \pm \sqrt{\frac{18 + (x - 1)^2}{9}}$$

$$\left.\begin{aligned} y_1 &= -3 + \frac{1}{3}\sqrt{18 + (x - 1)^2} \\ y_2 &= -3 - \frac{1}{3}\sqrt{18 + (x - 1)^2} \end{aligned}\right\} \text{Hyperbola}$$

$$\left.\begin{aligned} y_3 &= -3 + \frac{1}{3}(x - 1) \\ y_4 &= -3 - \frac{1}{3}(x - 1) \end{aligned}\right\} \text{Asymptotes}$$

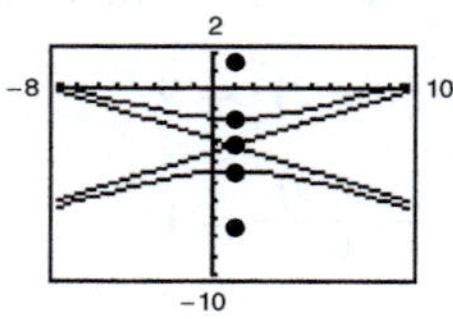

39. Vertices: $(\pm 1, 0) \Rightarrow a = 1$

Asymptotes: $y = \pm 5x \Rightarrow \frac{b}{a} = 5, b = 5$

Center: $(0, 0) = (h, k)$

$$\frac{(x - h)^2}{a^2} - \frac{(y - k)^2}{b^2} = 1$$

$$\frac{x^2}{1} - \frac{y^2}{25} = 1$$

41. Foci: $(0, \pm 8) \Rightarrow c = 8$

Asymptotes: $y = \pm 4x \Rightarrow \dfrac{a}{b} = 4 \Rightarrow a = 4b$

Center: $(0, 0) = (h, k)$

$$c^2 = a^2 + b^2 \Rightarrow 64 = 16b^2 + b^2$$

$$\frac{64}{17} = b^2 \Rightarrow a^2 = \frac{1024}{17}$$

$$\frac{(y-k)^2}{a^2} - \frac{(x-h)^2}{b^2} = 1$$

$$\frac{y^2}{1024/17} - \frac{x^2}{64/17} = 1$$

$$\frac{17y^2}{1024} - \frac{17x^2}{64} = 1$$

43. Vertices: $(1, 2), (3, 2) \Rightarrow a = 1$

Asymptotes: $y = x, y = 4 - x$

$$\frac{b}{a} = 1 \Rightarrow \frac{b}{1} = 1 \Rightarrow b = 1$$

Center: $(2, 2) = (h, k)$

$$\frac{(x-h)^2}{a^2} - \frac{(y-k)^2}{b^2} = 1$$

$$\frac{(x-2)^2}{1} - \frac{(y-2)^2}{1} = 1$$

45. Vertices: $(3, 0), (3, 4) \Rightarrow a = 2$

Asymptotes: $y = \dfrac{2}{3}x, y = 4 - \dfrac{2}{3}x$

$$\frac{a}{b} = \frac{2}{3} \Rightarrow b = 3$$

Center: $(3, 2) = (h, k)$

$$\frac{(y-k)^2}{a^2} - \frac{(x-h)^2}{b^2} = 1$$

$$\frac{(y-2)^2}{4} - \frac{(x-3)^2}{9} = 1$$

47. Foci: $(-1, -1), (9, -1) \Rightarrow c = 5$

Asymptotes: $y = \dfrac{3}{4}x - 4, y = -\dfrac{3}{4}x + 2$

$$\frac{b}{a} = \frac{3}{4} \Rightarrow b = 3, a = 4$$

Center: $(4, -1) = (h, k)$

$$\frac{(x-h)^2}{a^2} - \frac{(y-k)^2}{b^2} = 1$$

$$\frac{(x-4)^2}{16} - \frac{(y+1)^2}{9} = 1$$

49. (a) Vertices: $(\pm 1, 0) \Rightarrow a = 1$

Horizontal transverse axis

Center: $(0, 0)$

$$\frac{x^2}{a^2} - \frac{y^2}{b^2} = 1$$

Point on the graph: $(2, 13)$

$$\frac{2^2}{1^2} - \frac{13^2}{b^2} = 1$$

$$4 - \frac{169}{b^2} = 1$$

$$3b^2 = 169$$

$$b^2 = \frac{169}{3}$$

So, $\dfrac{x^2}{1} - \dfrac{y^2}{169/3} = 1.$

(b) When $y = 5$: $x^2 = 1 + \dfrac{5^2}{56.33}$

$$x = \sqrt{1 + \frac{25}{56.33}} \approx 1.2016$$

So, the width is about $2x \approx 2.403$ feet.

51.

$$2c = 4 \text{ mi} = 21{,}120 \text{ ft}$$

$$c = 10{,}560 \text{ ft}$$

$$(1100 \text{ ft/s})(18 \text{ s}) = 19{,}800 \text{ ft}$$

The lightning occurred 19,800 feet further from B than from A:

$$d_2 - d_1 = 2a = 19{,}800 \text{ ft}$$

$$a = 9900 \text{ ft}$$

$$b^2 = c^2 - a^2 = (10{,}560)^2 - (9900)^2$$

$$b^2 = 13{,}503{,}600$$

$$\frac{x^2}{(9900)^2} - \frac{y^2}{13{,}503{,}600} = 1$$

$$\frac{x^2}{98{,}010{,}000} - \frac{y^2}{13{,}503{,}600} = 1$$

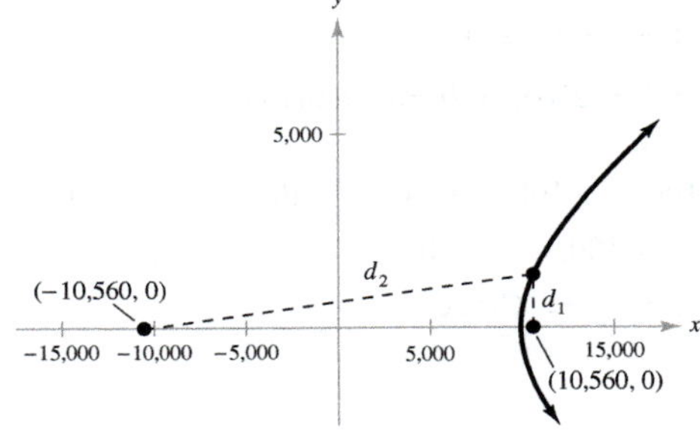

53. (a) Foci: $(\pm 150, 0) \Rightarrow c = 150$

Center: $(0, 0) = (h, k)$

$$\frac{d_2}{186{,}000} - \frac{d_1}{186{,}000} = 0.001 \Rightarrow 2a = 186, a = 93$$

$$b^2 = c^2 - a^2 = 150^2 - 93^2 = 13{,}851$$

$$\frac{x^2}{93^2} - \frac{y^2}{13{,}851} = 1$$

$$x^2 = 93^2\left(1 + \frac{75^2}{13{,}851}\right) \approx 12{,}161$$

$$x \approx 110.3 \text{ miles}$$

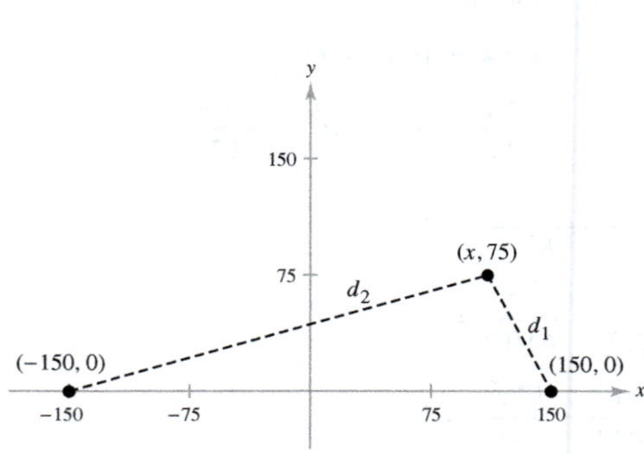

(b) $c - a = 150 - 93 = 57$ miles

(c) Using the asymptote with positive slope,

$$y = k \pm \frac{b}{a}(x - h)$$

$$y = \frac{\sqrt{13{,}851}}{\sqrt{8694}}x$$

$$y = \frac{27\sqrt{19}}{93}x$$

55. $9x^2 + 4y^2 - 18x + 16y - 119 = 0$

$A = 9, C = 4$

$AC = (9)(4) = 36 > 0 \Rightarrow$ Ellipse

57. $4x^2 - y^2 - 4x - 3 = 0$

$A = 4, C = -1$

$AC = (4)(-1) = -4 < 0 \Rightarrow$ Hyperbola

59. $y^2 - 4x^2 + 4x - 2y - 4 = 0$

$A = -4, C = 1$

$AC = (-4)(1) = -4 < 0 \Rightarrow$ Hyperbola

61. $4x^2 + 25y^2 + 16x + 250y + 541 = 0$

$A = 4, C = 25$

$AC = (4)(25) = 100 > 0 \Rightarrow$ Ellipse

63. $25x^2 - 10x - 200y - 119 = 0$

$A = 25, C = 0$

$AC = 25(0) = 0 \Rightarrow$ Parabola

65. $100x^2 + 100y^2 - 100x + 400y + 409 = 0$

$A = 100, C = 100$

$A = C \Rightarrow$ Circle

67. True. For a hyperbola, $c^2 = a^2 + b^2$ or

$$e^2 = \frac{c^2}{a^2} = 1 + \frac{b^2}{a^2}.$$

The larger the ratio of b to a, the larger the eccentricity $e = c/a$ of the hyperbola.

69. False. The graph is two intersecting lines.

$$x^2 - y^2 + 4x - 4y = 0$$

$$(x^2 + 4x + 4) - (y^2 + 4y + 4) = 4 - 4$$

$$(x - 2)^2 - (y + 2)^2 = 0$$

$$(x - 2)^2 = (y + 2)^2$$

$$x - 2 = \pm(y - 2)$$

$$y = x \text{ and } y = -x + 4$$

71. Draw a rectangle through the vertices and the endpoints of the conjugate axis. Sketch the asymptotes by drawing lines through the opposite corners of the rectangle.

73. Because the transverse axis is vertical,

$\dfrac{(y + 5)^2}{9} - \dfrac{(x - 3)^2}{4} = 1$, where $a = 3, b = 2, h = 3,$ and $k = -5$ the equations of the asymptotes should be

$$y = k \pm \frac{a}{b}(x - h)$$

$$y = -5 \pm \frac{3}{2}(x - 3)$$

$$y = \frac{3}{2}x - \frac{19}{2} \text{ and } y = -\frac{3}{2}x - \frac{1}{2}.$$

75.

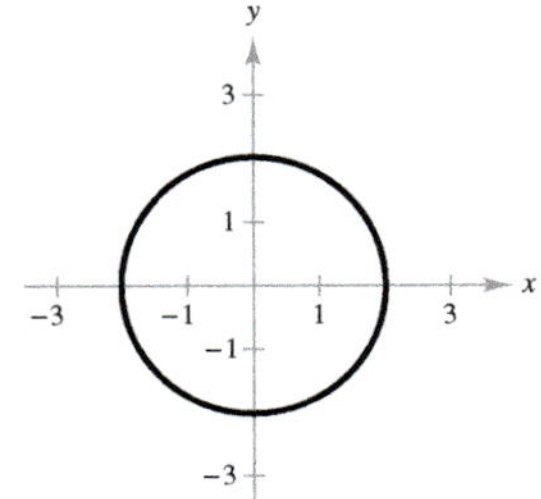

Value of C	Possible number of points of intersection
$C > 2$	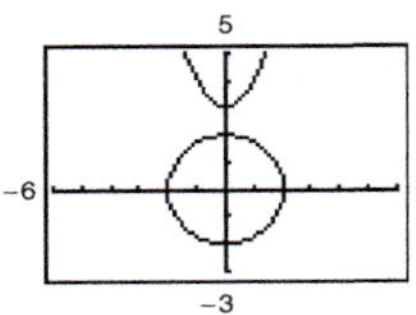
$C = 2$	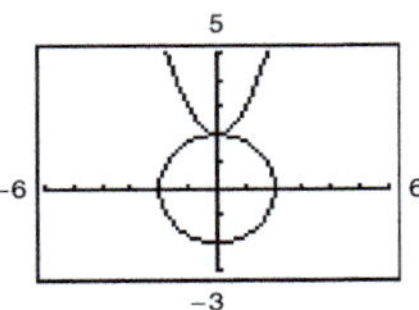
$-2 < C < 2$	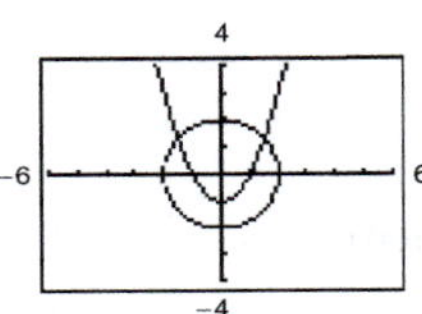
$C = -2$	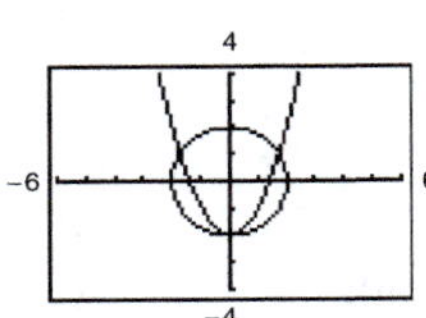
$C < -2$	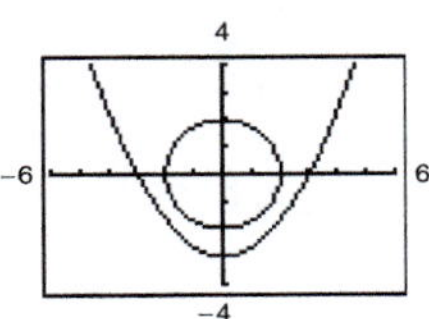
or	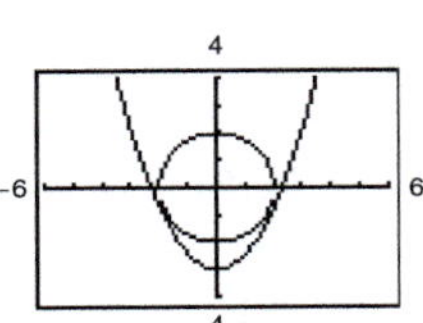
or	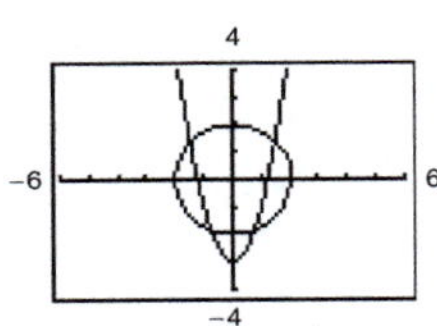

For $C \le -2$, analyze the two curves to determine the number of points of intersection.

$C = -2$: $x^2 + y^2 = 4$ and $y = x^2 - 2$

$$x^2 = y + 2$$

Substitute: $(y+2)+y^2=4$

$$y^2+y-2=0$$
$$(y+2)(y-1)=0$$
$$y=-2,1$$

$x^2=y+2$	$x^2=y+2$
$x^2=-2+2$	$x^2=1+2$
$x^2=0$	$x^2=3$
$x=0$	$x=\pm\sqrt{3}$
$(0,-2)$	$\left(-\sqrt{3},1\right),\left(\sqrt{3},1\right)$

There are three points of intersection when $C=-2$.

$C<-2$: $x^2+y^2=4$ and $y=x^2+C$

$$x^2=y-C$$

Substitute: $(y-C)+y^2=4$

$$y^2+y-4-C=0$$
$$y=\frac{-1\pm\sqrt{(1)^2-(4)(1)(-C-4)}}{2}$$
$$y=\frac{-1\pm\sqrt{1+4(C+4)}}{2}$$

If $1+4(C+4)<0$, there are no real solutions (no points of intersection):

$$1+4C+16<0$$
$$4C<-17$$
$$C<\frac{-17}{4}\text{, no points of intersection}$$

If $1+4(C+4)=0$, there is one real solution (two points of intersection):

$$1+4C+16=0$$
$$4C=-17$$
$$C=\frac{-17}{4}\text{, two points of intersection}$$

If $1+4(C+4)>0$, there are two real solutions (four points of intersection):

$$1+4C+16>0$$
$$4C>-17$$
$$C>\frac{-17}{4}\text{, (but } C<-2\text{), four points of intersection}$$

Summary:

a. no points of intersection: $C>2$ or $C<\frac{-17}{4}$

b. one point of intersection: $C=2$

c. two points of intersection: $-2<C<2$ or $C=\frac{-17}{4}$

d. three points of intersection: $C=-2$

e. four points of intersection: $\frac{-17}{4}<C<-2$

Section 10.5 Rotation of Conics

1. rotation; axes

3. invariant under rotation

5. $\theta = 90°$; Point: $(2, 0)$

$$\begin{aligned} x &= x' \cos\theta - y' \sin\theta \\ 2 &= x' \cos 90° - y' \sin 90° \\ 2 &= -y' \\ y' &= -2 \end{aligned} \qquad \begin{aligned} y &= x' \sin\theta + y' \cos\theta \\ 0 &= x' \sin 90° + y' \cos 90° \\ 0 &= x' \end{aligned}$$

So, $(x', y') = (0, -2)$.

7. $\theta = 30°$; Point: $(1, 3)$

$$\begin{aligned} x &= x' \cos\theta - y' \sin\theta \\ y &= x' \sin\theta + y' \cos\theta \end{aligned} \Rightarrow \begin{cases} 1 = x' \cos 30° - y' \sin 30° \\ 3 = x' \sin 30° + y' \cos 30° \end{cases}$$

Solving the system yields $(x', y') = \left(\dfrac{3 + \sqrt{3}}{2}, \dfrac{3\sqrt{3} - 1}{2}\right)$.

9. $\theta = 45°$; Point: $(2, 1)$

$$\begin{aligned} x &= x' \cos\theta - y' \sin\theta \\ y &= x' \sin\theta + y' \cos\theta \end{aligned} \Rightarrow \begin{cases} 2 = x' \cos 45° - y' \sin 45° \\ 1 = x' \sin 45° + y' \cos 45° \end{cases}$$

Solving the system yields $(x', y') = \left(\dfrac{3\sqrt{2}}{2}, -\dfrac{\sqrt{2}}{2}\right)$.

11. $\theta = 60°$; Point: $(1, 2)$

$$\begin{aligned} x &= x' \cos\theta - y' \sin\theta \\ y &= x' \sin\theta + y' \cos\theta \end{aligned} \Rightarrow \begin{cases} 1 = x' \cos 60° - y' \sin 60° \\ 2 = x' \sin 60° + y' \cos 60° \end{cases}$$

Solving the system yields $(x', y') = \left(\dfrac{1}{2} + \sqrt{3}, 1 - \dfrac{\sqrt{3}}{2}\right) = \left(\dfrac{1 + 2\sqrt{3}}{2}, \dfrac{2 - \sqrt{3}}{2}\right)$

13. $xy + 3 = 0$, $A = 0$, $B = 1$, $C = 0$

$$\cot 2\theta = \frac{A - C}{B} = 0 \Rightarrow 2\theta = \frac{\pi}{2} \Rightarrow \theta = \frac{\pi}{4}$$

$$\begin{aligned} x &= x' \cos\frac{\pi}{4} - y' \sin\frac{\pi}{4} \\ &= x'\left(\frac{\sqrt{2}}{2}\right) - y'\left(\frac{\sqrt{2}}{2}\right) \\ &= \frac{x' - y'}{\sqrt{2}} \end{aligned} \qquad \begin{aligned} y &= x' \sin\frac{\pi}{4} + y' \cos\frac{\pi}{4} \\ &= x'\left(\frac{\sqrt{2}}{2}\right) + y'\left(\frac{\sqrt{2}}{2}\right) \\ &= \frac{x' + y'}{\sqrt{2}} \end{aligned}$$

$$\begin{aligned} xy + 3 &= 0 \\ \left(\frac{x' - y'}{\sqrt{2}}\right)\left(\frac{x' + y'}{\sqrt{2}}\right) + 3 &= 0 \\ \frac{(x')^2}{2} - \frac{(y')^2}{2} &= -3 \\ \frac{(y')^2}{6} - \frac{(x')^2}{6} &= 1 \end{aligned}$$

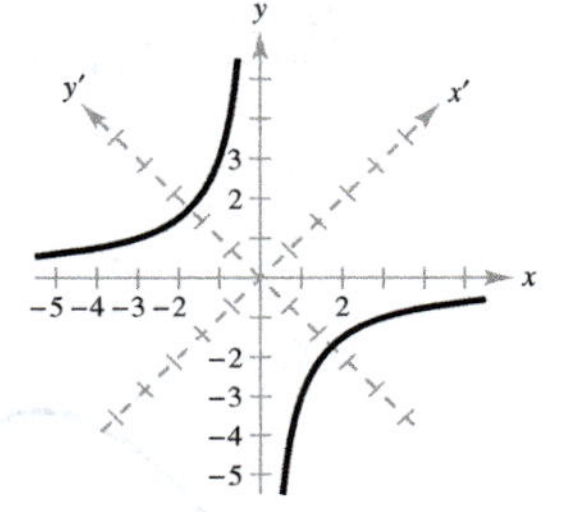

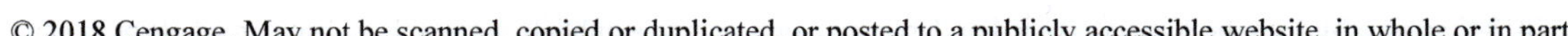

15. $xy + 2x - y + 4 = 0$

$A = 0, B = 1, C = 0$

$\cot 2\theta = \dfrac{A - C}{B} = 0 \Rightarrow 2\theta = \dfrac{\pi}{2} \Rightarrow \theta = \dfrac{\pi}{4}$

$x = x'\cos\dfrac{\pi}{4} - y'\sin\dfrac{\pi}{4} = \dfrac{x' - y'}{\sqrt{2}}$ $\qquad y = x'\sin\dfrac{\pi}{4} + y'\cos\dfrac{\pi}{4} = \dfrac{x' + y'}{\sqrt{2}}$

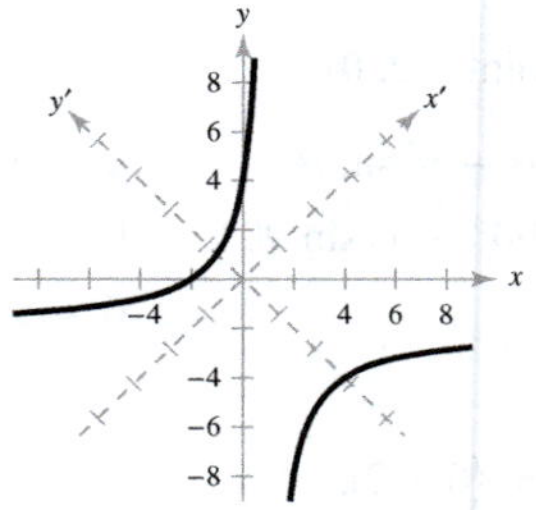

$$xy + 2x - y + 4 = 0$$

$$\left(\frac{x' - y'}{\sqrt{2}}\right)\left(\frac{x' + y'}{\sqrt{2}}\right) + 2\left(\frac{x' - y'}{\sqrt{2}}\right) - \left(\frac{x' + y'}{\sqrt{2}}\right) + 4 = 0$$

$$\frac{(x')^2}{2} - \frac{(y')^2}{2} + \frac{2x'}{\sqrt{2}} - \frac{2y'}{\sqrt{2}} - \frac{x'}{\sqrt{2}} - \frac{y'}{\sqrt{2}} + 4 = 0$$

$$\left[(x')^2 + \sqrt{2}x' + \left(\frac{\sqrt{2}}{2}\right)^2\right] - \left[(y')^2 + 3\sqrt{2}y' + \left(\frac{3\sqrt{2}}{2}\right)^2\right] = -8 + \left(\frac{\sqrt{2}}{2}\right)^2 - \left(\frac{3\sqrt{2}}{2}\right)^2$$

$$\left(x' + \frac{\sqrt{2}}{2}\right)^2 - \left(y' + \frac{3\sqrt{2}}{2}\right)^2 = -12$$

$$\frac{\left(y' + \frac{3\sqrt{2}}{2}\right)^2}{12} - \frac{\left(x' + \frac{\sqrt{2}}{2}\right)^2}{12} = 1$$

17. $5x^2 - 6xy + 5y^2 - 12 = 0$

$A = 5, B = -6, C = 5$

$\cot 2\theta = \dfrac{A - C}{B} = 0 \Rightarrow 2\theta = \dfrac{\pi}{2} \Rightarrow \theta = \dfrac{\pi}{4}$

$x = x'\cos\dfrac{\pi}{4} - y'\sin\dfrac{\pi}{4} = x'\left(\dfrac{\sqrt{2}}{2}\right) - y'\left(\dfrac{\sqrt{2}}{2}\right) = \dfrac{x' - y'}{\sqrt{2}}$

$y = x'\sin\dfrac{\pi}{4} + y'\cos\dfrac{\pi}{4} = x'\left(\dfrac{\sqrt{2}}{2}\right) + y'\left(\dfrac{\sqrt{2}}{2}\right) = \dfrac{x' + y'}{\sqrt{2}}$

$$5x^2 - 6xy + 5y^2 - 12 = 0$$

$$5\left(\frac{x' - y'}{\sqrt{2}}\right)^2 - 6\left(\frac{x' - y'}{\sqrt{2}}\right)\left(\frac{x' + y'}{\sqrt{2}}\right) + 5\left(\frac{x' + y'}{\sqrt{2}}\right)^2 - 12 = 0$$

$$\frac{5(x')^2}{2} - 5x'y' + \frac{5(y')^2}{2} - 3(x')^2 + 3(y')^2 + \frac{5(x')^2}{2} + 5x'y' + \frac{5(y')^2}{2} - 12 = 0$$

$$2(x')^2 + 8(y')^2 = 12$$

$$\frac{(x')^2}{6} + \frac{(y')^2}{3/2} = 1$$

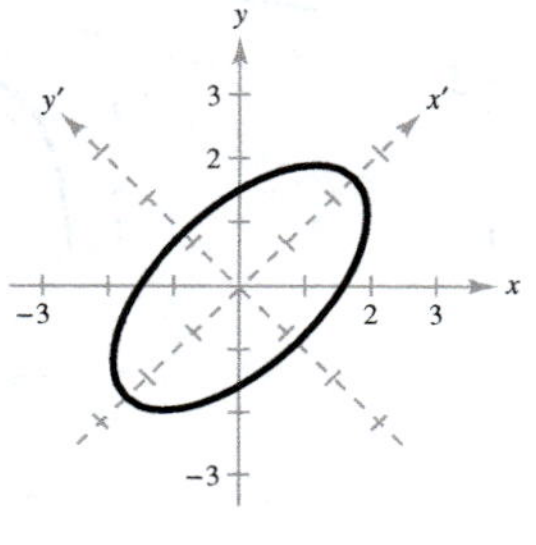

19. $13x^2 + 6\sqrt{3}xy + 7y^2 - 16 = 0$

$A = 13, B = 6\sqrt{3}, C = 7$

$\cot 2\theta = \dfrac{A - C}{B} = \dfrac{1}{\sqrt{3}} \Rightarrow 2\theta = \dfrac{\pi}{3} \Rightarrow \theta = \dfrac{\pi}{6}$

$$x = x'\cos\frac{\pi}{6} - y'\sin\frac{\pi}{6} = x'\left(\frac{\sqrt{3}}{2}\right) - y'\left(\frac{1}{2}\right) = \frac{\sqrt{3}x' - y'}{2}$$

$$y = x'\sin\frac{\pi}{6} + y'\cos\frac{\pi}{6} = x'\left(\frac{1}{2}\right) + y'\left(\frac{\sqrt{3}}{2}\right) = \frac{x' + \sqrt{3}y'}{2}$$

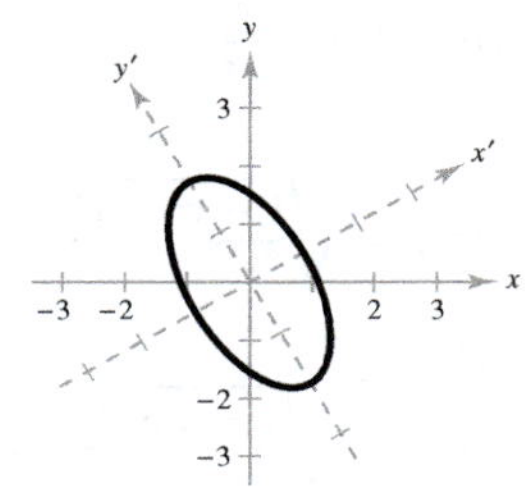

$$13x^2 + 6\sqrt{3}xy + 7y^2 - 16 = 0$$

$$13\left(\frac{\sqrt{3}x' - y'}{2}\right)^2 + 6\sqrt{3}\left(\frac{\sqrt{3}x' - y'}{2}\right)\left(\frac{x' + \sqrt{3}y'}{2}\right) + 7\left(\frac{x' + \sqrt{3}y'}{2}\right)^2 - 16 = 0$$

$$\frac{39(x')^2}{4} - \frac{13\sqrt{3}x'y'}{2} + \frac{13(y')^2}{4} + \frac{18(x')^2}{4} + \frac{18\sqrt{3}x'y'}{4} - \frac{6\sqrt{3}x'y'}{4}$$

$$-\frac{18(y')^2}{4} + \frac{7(x')^2}{4} + \frac{7\sqrt{3}x'y'}{2} + \frac{21(y')^2}{4} - 16 = 0$$

$$16(x')^2 + 4(y')^2 = 16$$

$$\frac{(x')^2}{1} + \frac{(y')^2}{4} = 1$$

21. $x^2 + 2xy + y^2 + \sqrt{2}x - \sqrt{2}y = 0, A = 1, B = 2, C = 1$

$\cot 2\theta = \dfrac{A - C}{B} = \dfrac{1 - 1}{2} = 0 \Rightarrow 2\theta = \dfrac{\pi}{2} \Rightarrow \theta = \dfrac{\pi}{4}$

$$x = x'\cos\frac{\pi}{4} - y'\sin\frac{\pi}{4} = \frac{x' - y'}{\sqrt{2}}$$

$$y = x'\sin\frac{\pi}{4} + y'\cos\frac{\pi}{4} = \frac{x' + y'}{\sqrt{2}}$$

$$x^2 + 2xy + y^2 + \sqrt{2}x - \sqrt{2}y = 0$$

$$\left(\frac{x' - y'}{\sqrt{2}}\right)^2 + 2\left(\frac{x' - y'}{\sqrt{2}}\right)\left(\frac{x' + y'}{\sqrt{2}}\right) + \left(\frac{x' + y'}{\sqrt{2}}\right)^2 + \sqrt{2}\left(\frac{x' - y'}{\sqrt{2}}\right) - \sqrt{2}\left(\frac{x' + y'}{\sqrt{2}}\right) = 0$$

$$\frac{(x')^2}{2} - x'y' + \frac{(y')^2}{2} + (x')^2 - (y')^2 + \frac{(x')^2}{2} + x'y' + \frac{(y')^2}{2} + x' - y' - x' - y' = 0$$

$$2(x')^2 - 2y' = 0$$

$$2(x')^2 = 2y'$$

$$(x')^2 = y'$$

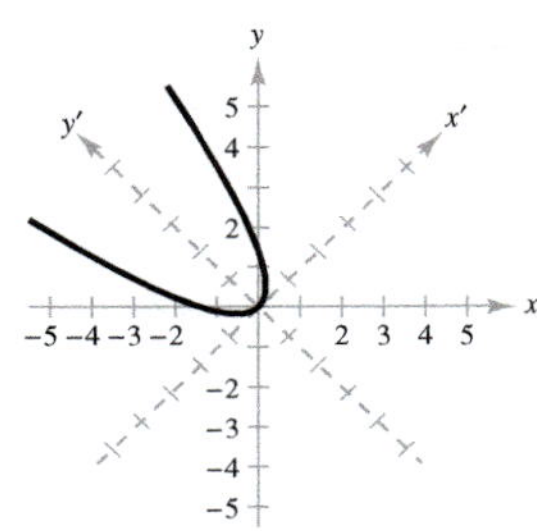

23. $9x^2 + 24xy + 16y^2 + 19x - 130y = 0$

$A = 9, B = 24, C = 16$

$\cot 2\theta = \dfrac{A - C}{B} = -\dfrac{7}{24} \Rightarrow \theta \approx 53.13°$

$\cos 2\theta = -\dfrac{7}{25}$

$\sin\theta = \sqrt{\dfrac{1 - \cos 2\theta}{2}} = \sqrt{\dfrac{1 - \left(-\frac{7}{25}\right)}{2}} = \dfrac{4}{5}$

$\cos\theta = \sqrt{\dfrac{1 + \cos 2\theta}{2}} = \sqrt{\dfrac{1 + \left(-\frac{7}{25}\right)}{2}} = \dfrac{3}{5}$

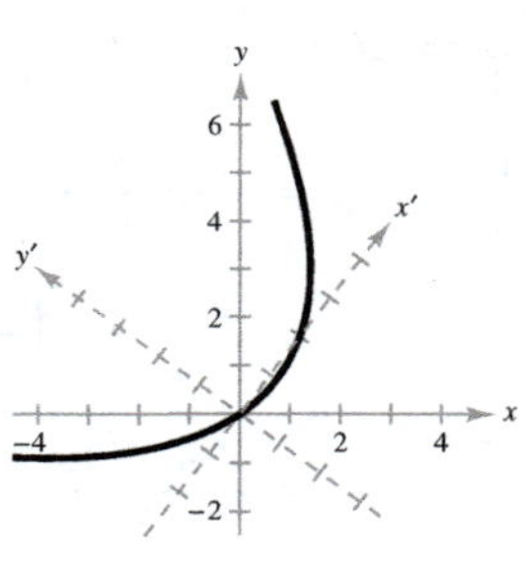

$x = x'\cos\theta - y'\sin\theta = x'\left(\dfrac{3}{5}\right) - y'\left(\dfrac{4}{5}\right) = \dfrac{3x' - 4y'}{5}$

$y = x'\sin\theta + y'\cos\theta = x'\left(\dfrac{4}{5}\right) + y'\left(\dfrac{3}{5}\right) = \dfrac{4x' + 3y'}{5}$

$$9x^2 + 24xy + 16y^2 + 90x - 130y = 0$$

$$9\left(\frac{3x' - 4y'}{5}\right)^2 + 24\left(\frac{3x' - 4y'}{5}\right)\left(\frac{4x' + 3y'}{5}\right) + 16\left(\frac{4x' + 3y'}{5}\right)^2 + 90\left(\frac{3x' - 4y'}{5}\right) - 130\left(\frac{4x' + 3y'}{5}\right) = 0$$

$$\frac{81(x')^2}{25} - \frac{216x'y'}{25} + \frac{144(y')^2}{25} + \frac{288(x')^2}{25} - \frac{168x'y'}{25} - \frac{288(y')^2}{25} + \frac{256(x')^2}{25} + \frac{384x'y'}{25} + \frac{144(y')^2}{25}$$

$$+ 54x' - 72y' - 104x' - 78y' = 0$$

$$25(x')^2 - 50x' - 150y' = 0$$

$$(x')^2 - 2x' = 6y'$$

$$(x')^2 - 2x' + 1 = 6y' + 1$$

$$(x' - 1)^2 = 6\left(y' + \frac{1}{6}\right)$$

25. $x^2 - 4xy + 2y^2 = 6$

$A = 1, B = -4, C = 2$

$\cot 2\theta = \dfrac{A - C}{B} = \dfrac{1 - 2}{-4} = \dfrac{1}{4}$

$\dfrac{1}{\tan 2\theta} = \dfrac{1}{4}$

$\tan 2\theta = 4$

$2\theta \approx 75.96$

$\theta \approx 37.98°$

To graph conic with a graphing calculator, solve for y in terms of x.

$x^2 - 4xy + 2y^2 = 6$

$y^2 - 2xy + x^2 = 3 - \dfrac{x^2}{2} + x^2$

$(y - x)^2 = 3 + \dfrac{x^2}{2}$

$y - x = \pm\sqrt{3 + \dfrac{x^2}{2}}$

$y = x \pm \sqrt{3 + \dfrac{x^2}{2}}$

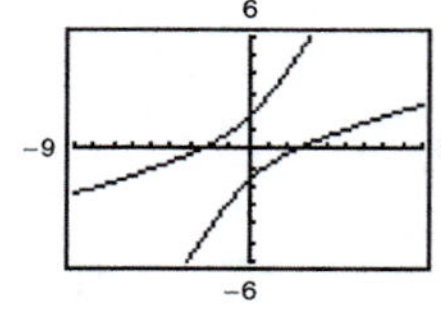

Enter $y_1 = x + \sqrt{3 + \dfrac{x^2}{2}}$ and $y_2 = x - \sqrt{3 + \dfrac{x^2}{2}}$.

27. $14x^2 + 16xy + 9y^2 = 44$

$A = 14, B = 16, C = 9$

$$\cot 2\theta = \frac{A - C}{B} = \frac{14 - 9}{16} = \frac{5}{16}$$

$$\tan 2\theta = \frac{16}{5}$$

$$2\theta \approx 72.65°$$

$$\theta \approx 36.32°$$

Solve for y in terms of x using the Quadratic Formula.

$$(9)y^2 + (16x)y + (14x^2 - 44) = 0$$

$$y = \frac{-b \pm \sqrt{b^2 - 4ac}}{2a}$$

$$y = \frac{-(16x) \pm \sqrt{(16x)^2 - 4(9)(14x^2 - 44)}}{2(9)}$$

$$y = \frac{-16x \pm \sqrt{-248x^2 + 1548}}{18}$$

Use $y_1 = \dfrac{-16x + \sqrt{-248x^2 + 1548}}{18}$

and $y_2 = \dfrac{-16x - \sqrt{-248x^2 + 1548}}{18}$

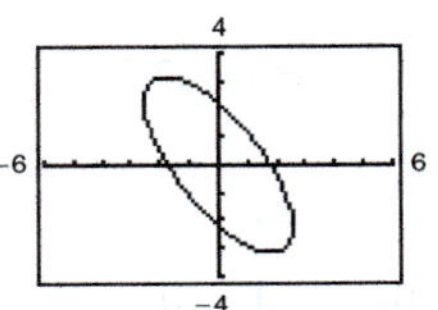

29. $2x^2 + 4xy + 2y^2 + \sqrt{26}x + 3y = -15$

$A = 2, B = 4, C = 2$

$$\cot 2\theta = \frac{A - C}{B} = 0 \Rightarrow 2\theta = \frac{\pi}{2} \Rightarrow \theta = \frac{\pi}{4} \text{ or } 45°$$

Solve for y in terms of x using the Quadratic Formula.

$$2y^2 + (4x + 3)y + (2x^2 + \sqrt{26}x + 15) = 0$$

$$y = \frac{-(4x + 3) \pm \sqrt{(4x + 3)^2 - 4(2)(2x^2 + \sqrt{26}x + 15)}}{2(2)}$$

$$= \frac{-(4x + 3) \pm \sqrt{(4x + 3)^2 - 8(2x^2 + \sqrt{26}x + 15)}}{4}$$

Enter $y_1 = \dfrac{-(4x + 3) + \sqrt{(4x + 3)^2 - 8(2x^2 + \sqrt{26}x + 15)}}{4}$ and

$y_2 = \dfrac{-(4x + 3) - \sqrt{(4x + 3)^2 - 8(2x^2 + \sqrt{26}x + 15)}}{4}.$

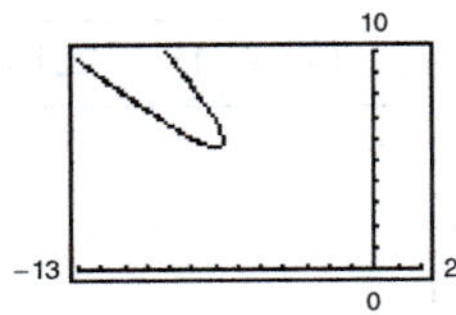

31. $xy + 2 = 0$

$B^2 - 4AC = 1 \Rightarrow$ The graph is a hyperbola.

$\cot 2\theta = \dfrac{A - C}{B} = 0 \Rightarrow \theta = 45^\circ$

Matches graph (e).

32. $x^2 - xy + 3y^2 - 5 = 0$

$A = 1, B = -1, C = 3$

$B^2 - 4AC = (-1)^2 - 4(1)(3) = -11$

The graph is an ellipse.

$\cot 2\theta = \dfrac{A - C}{B} = \dfrac{1 - 3}{-1} = 2 \Rightarrow \theta \approx 13.28^\circ$

Matches graph (a).

33. $3x^2 + 2xy + y^2 - 10 = 0$

$B^2 - 4AC = (2)^2 - 4(3)(1) = -8 \Rightarrow$

The graph is an ellipse or circle.

$\cot 2\theta = \dfrac{A - C}{B} = 1 \Rightarrow \theta = 22.5^\circ$

Matches graph (d).

34. $x^2 - 4xy + 4y^2 + 10x - 30 = 0$

$A = 1, B = -4, C = 4$

$B^2 - 4AC = (-4)^2 - 4(1)(4) = 0$

The graph is a parabola.

$\cot 2\theta = \dfrac{A - C}{B} = \dfrac{1 - 4}{-4} = \dfrac{3}{4} \Rightarrow \theta \approx 26.57^\circ$

Matches graph (c).

35. $x^2 + 2xy + y^2 = 0$

$$(x + y)^2 = 0$$
$$x + y = 0$$
$$y = -x$$

The graph is a line. Matches graph (f).

36. $-2x^2 + 3xy + 2y^2 + 3 = 0$

$B^2 - 4AC = (3)^2 - 4(-2)(2) = 25 \Rightarrow$

The graph is a hyperbola.

$\cot 2\theta = \dfrac{A - C}{B} = -\dfrac{4}{3} \Rightarrow \theta \approx -18.43^\circ$

Matches graph (b).

37. (a) $16x^2 - 8xy + y^2 - 10x + 5y = 0$

$B^2 - 4AC = (-8)^2 - 4(16)(1) = 0$

The graph is a parabola.

(b) $y^2 + (-8x + 5)y + (16x^2 - 10x) = 0$

$$y = \frac{-(-8x + 5) \pm \sqrt{(-8x + 5)^2 - 4(1)(16x^2 - 10x)}}{2(1)}$$
$$= \frac{(8x - 5) \pm \sqrt{(8x - 5)^2 - 4(16x^2 - 10x)}}{2}$$

(c)

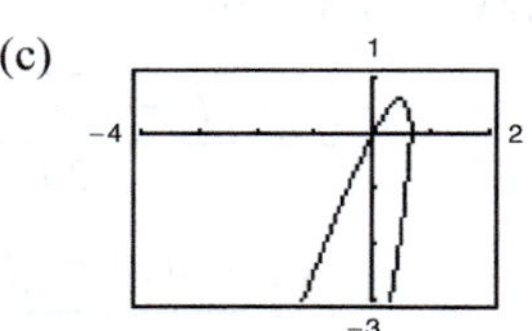

39. (a) $12x^2 - 6xy + 7y^2 - 45 = 0$

$B^2 - 4AC = (-6)^2 - 4(12)(7) = -300 < 0$

The graph is an ellipse.

(b) $7y^2 + (-6x)y + (12x^2 - 45) = 0$

$$y = \frac{-(-6x) \pm \sqrt{(-6x)^2 - 4(7)(12x^2 - 45)}}{2(7)}$$
$$= \frac{6x \pm \sqrt{36x^2 - 28(12x^2 - 45)}}{14}$$

(c)

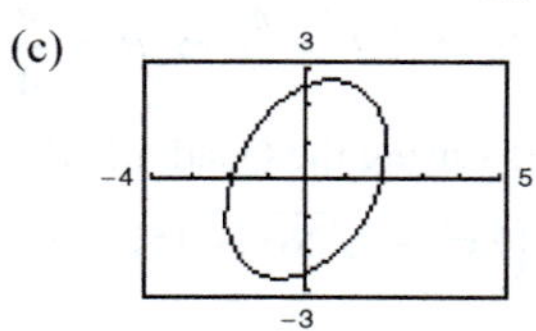

41. (a) $x^2 - 6xy - 5y^2 + 4x - 22 = 0$

$B^2 - 4AC = (-6)^2 - 4(1)(-5) = 56 > 0$

The graph is a hyperbola.

(b) $-5y^2 + (-6x)y + (x^2 + 4x - 22) = 0$

$$y = \frac{-(-6x) \pm \sqrt{(-6x)^2 - 4(-5)(x^2 + 4x - 22)}}{2(-5)}$$
$$= \frac{6x \pm \sqrt{36x^2 + 20(x^2 + 4x - 22)}}{-10}$$
$$= \frac{-6x \pm \sqrt{36x^2 + 20(x^2 + 4x - 22)}}{10}$$

(c)

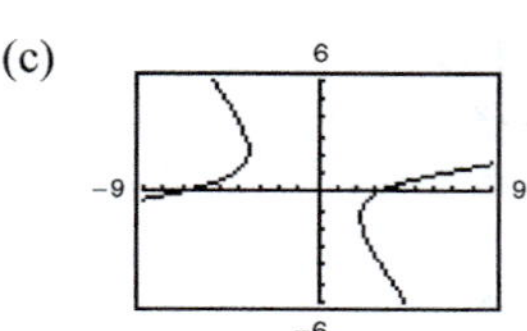

43. (a) $x^2 + 4xy + 4y^2 - 5x - y - 3 = 0$

$$B^2 - 4AC = (4)^2 - 4(1)(4) = 0$$

The graph is a parabola.

(b) $4y^2 + (4x - 1)y + (x^2 - 5x - 3) = 0$

$$y = \frac{-(4x-1) \pm \sqrt{(4x-1)^2 - 4(4)(x^2 - 5x - 3)}}{2(4)}$$

$$= \frac{-(4x-1) \pm \sqrt{(4x-1)^2 - 16(x^2 - 5x - 3)}}{8}$$

(c)

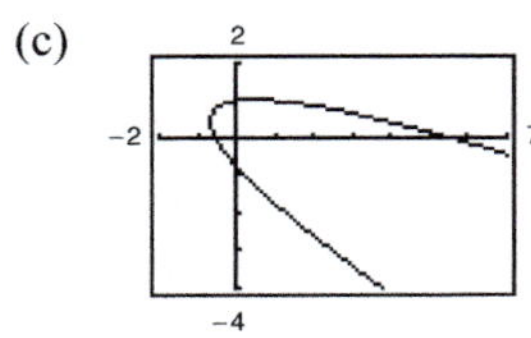

45. $y^2 - 16x^2 = 0$

$$y^2 = 16x^2$$

$$y = \pm 4x$$

Two intersecting lines

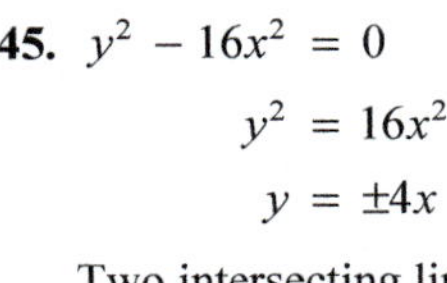

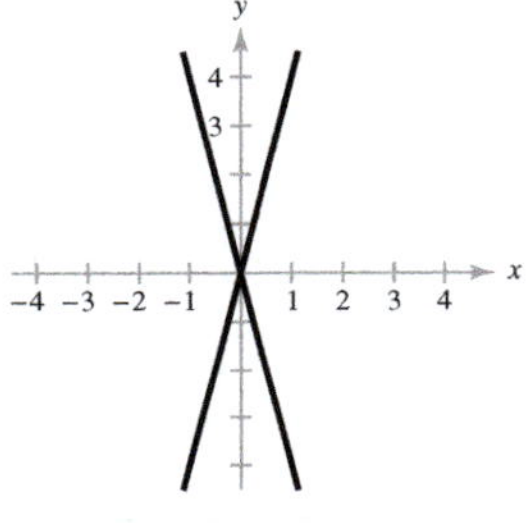

47. $15x^2 - 2xy - y^2 = 0$

$$(5x - y)(3x + y) = 0$$

$5x - y = 0 \qquad 3x + y = 0$

$y = 5x \qquad y = -3x$

Two intersecting lines

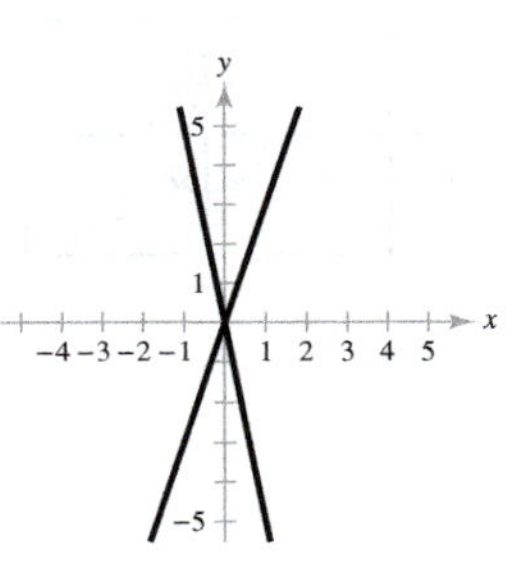

49. $x^2 - 2xy + y^2 = 0$

$$y^2 - 2xy + x^2 = x^2 - x^2$$

$$(y - x)^2 = 0$$

$$y - x = 0$$

$$y = x$$

Line

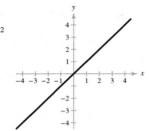

51. $x^2 + y^2 + 2x - 4y + 5 = 0$

$$x^2 + 2x + 1 + y^2 - 4y + 4 = -5 + 1 + 4$$

$$(x + 1)^2 + (y - 2)^2 = 0$$

Point $(-1, 2)$

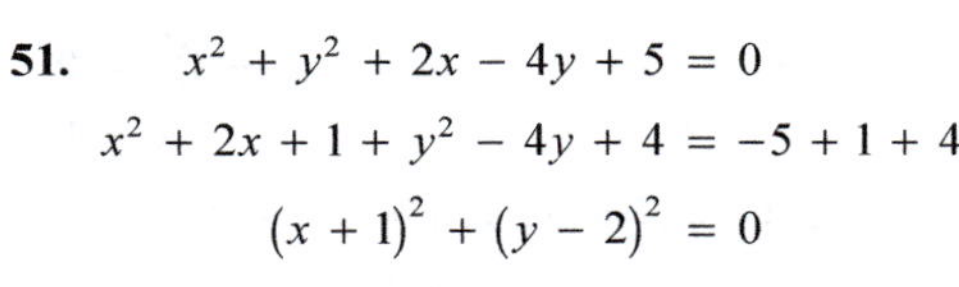

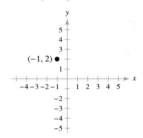

53. $x^2 + 2xy + y^2 - 1 = 0$

$$(x + y)^2 - 1 = 0$$

$$(x + y)^2 = 1$$

$$x + y = \pm 1$$

$$y = -x \pm 1$$

Two parallel lines

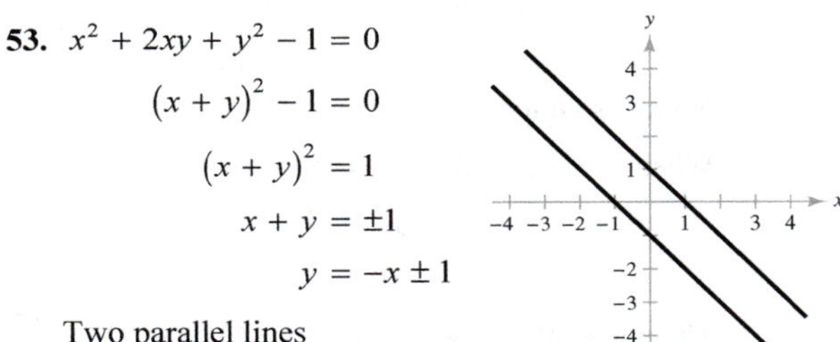

55.

$$\begin{aligned} x^2 - 4y^2 - 20x - 64y - 172 &= 0 \Rightarrow (x - 10)^2 - 4(y + 8)^2 = 16 \\ \underline{16x^2 + 4y^2 - 320x + 64y + 1600} &= 0 \Rightarrow 16(x - 10)^2 + 4(y + 8)^2 = 256 \\ 17x^2 \quad - 340x \quad 1428 &= 0 \\ (17x - 238)(x - 6) &= 0 \\ x = 6 \text{ or } x &= 14 \end{aligned}$$

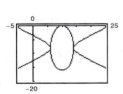

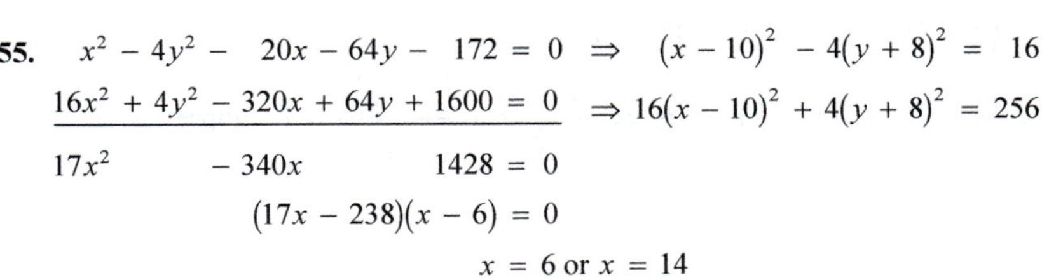

When $x = 6$:

$$\begin{aligned} 6^2 - 4y^2 - 20(6) - 64y - 172 &= 0 \\ -4y^2 - 64y - 256 &= 0 \\ y^2 + 16y + 64 &= 0 \\ (y + 8)^2 &= 0 \\ y &= -8 \end{aligned}$$

When $x = 14$:

$$\begin{aligned} 14^2 - 4y^2 - 20(14) - 64y - 172 &= 0 \\ -4y^2 - 64y - 256 &= 0 \\ y^2 + 16y + 64 &= 0 \\ (y + 8)^2 &= 0 \\ y &= -8 \end{aligned}$$

Points of intersection: $(6, -8), (14, -8)$

57. $\begin{array}{r} x^2 + 4y^2 - 2x - 8y + 1 = 0 \\ -x^2 \qquad + 2x - 4y - 1 = 0 \\ \hline 4y^2 \qquad - 12y \quad = 0 \end{array}$ $\begin{array}{l} \Rightarrow (x-1)^2 + 4(y-1)^2 = 4 \\ \Rightarrow y = -\frac{1}{4}(x-1)^2 \end{array}$

$$4y(y - 3) = 0$$

$$y = 0 \text{ or } y = 3$$

When $y = 0$: $x^2 + 4(0)^2 - 2x - 8(0) + 1 = 0$

$$x^2 - 2x + 1 = 0$$

$$(x - 1)^2 = 0$$

$$x = 1$$

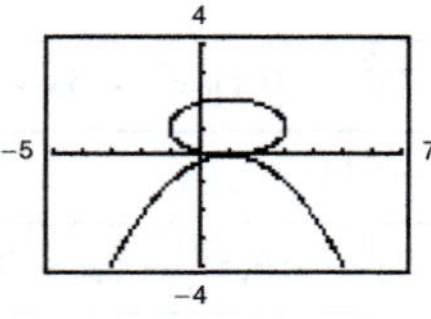

When $y = 3$: $-x^2 + 2x - 4(3) - 1 = 0$

$$x^2 - 2x + 13 = 0$$

No real solution

Point of intersection: $(1, 0)$

59. $\begin{array}{r} x^2 \qquad + y^2 - 4 = 0 \\ 3x - y^2 \qquad = 0 \\ \hline x^2 + 3x \qquad - 4 = 0 \end{array}$

$$(x + 4)(x - 1) = 0$$

$$x = -4 \text{ or } x = 1$$

When $x = -4$: $3(-4) - y^2 = 0$

$$y^2 = -12$$

No real solution

When $x = 1$: $3(1) - y^2 = 0$

$$y^2 = 3$$

$$y = \pm\sqrt{3}$$

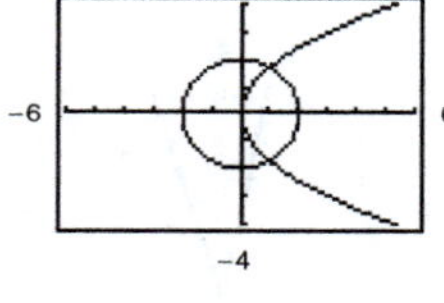

The points of intersection are $(1, \sqrt{3})$ and $(1, -\sqrt{3})$.

The standard forms of the equations are: $x^2 + y^2 = 4$

$$y^2 = 3x$$

61. $\begin{array}{r} -x^2 - \ y^2 - 8x + 20y - 7 = 0 \\ x^2 + 9y^2 + 8x + \ 4y + 7 = 0 \\ \hline \end{array}$ $\begin{array}{l} \Rightarrow (x+4)^2 + (y-10)^2 = 123\text{: Circle} \\ \Rightarrow (x+4)^2 + 9\left(y + \frac{2}{9}\right)^2 = \frac{85}{9} \Rightarrow \dfrac{(x+4)}{\frac{85}{9}} + \dfrac{\left(y + \frac{2}{9}\right)^2}{\frac{85}{81}} = 1\text{: Ellipse} \end{array}$

$$8y^2 + 24y = 0$$

$$8y(y + 3) = 0$$

$$y = 0 \text{ or } y = -3$$

When $y = 0$: $x^2 + 9(0)^2 + 8x + 4(0) + 7 = 0$

$$x^2 + 8x + 7 = 0$$

$$(x + 7)(x + 1) = 0$$

$$x = -7 \text{ or } x = -1$$

When $y = -3$: $x^2 + 9(-3)^2 + 8x + 4(-3) + 7 = 0$

$$x^2 + 8x + 76 = 0$$

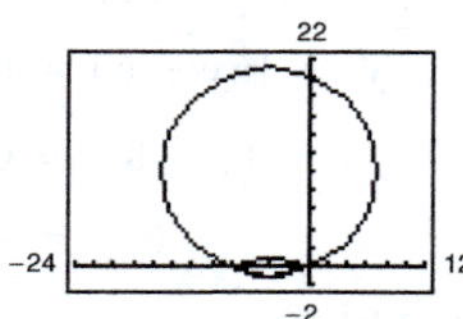

No real solutions

The points of intersection are $(-7, 0)$ and $(-1, 0)$.

63. (a) Because $A = 1$, $B = -2$ and $c = 1$ you have $\cot 2\theta = \dfrac{A - C}{B} = \dfrac{1 - 1}{-2} = 0 \Rightarrow 2\theta = \dfrac{\pi}{2} \Rightarrow \theta = \dfrac{\pi}{4}$

which implies that $x = x'\cos\dfrac{\pi}{4} - y'\sin\dfrac{\pi}{4} = x'\left(\dfrac{1}{\sqrt{2}}\right) - y'\left(\dfrac{1}{\sqrt{2}}\right) = \dfrac{x' - y'}{\sqrt{2}}$

and $y = x'\sin\dfrac{\pi}{4} + y'\cos\dfrac{\pi}{4} = x'\left(\dfrac{1}{\sqrt{2}}\right) + y'\left(\dfrac{1}{\sqrt{2}}\right) = \dfrac{x' + y'}{\sqrt{2}}.$

The equation in the $x'y'$-system is obtained by $x^2 - 2xy - 27\sqrt{2}x + y^2 + 9\sqrt{2}y + 378 = 0.$

$$\left(\frac{x' - y'}{\sqrt{2}}\right)^2 - 2\left(\frac{x' - y'}{\sqrt{2}}\right)\left(\frac{x' + y'}{\sqrt{2}}\right) - 27\sqrt{2}\left(\frac{x' - y'}{\sqrt{2}}\right) + \left(\frac{x' + y'}{\sqrt{2}}\right)^2 + 9\sqrt{2}\left(\frac{x' + y'}{\sqrt{2}}\right) + 378 = 0$$

$$\frac{(x')^2}{2} - \frac{2x'y'}{\sqrt{2}} + \frac{(y')^2}{2} - (x')^2 + (y')^2 - 27x' + 27y' + \frac{(x')^2}{2} + \frac{2x'y'}{\sqrt{2}} + \frac{(y')^2}{2} + 9x' + 9y' + 378 = 0$$

$$2(y')^2 + 36y' - 18x' + 378 = 0$$

$$(y')^2 + 18y' + 81 = 9x' - 189 + 81$$

$$(y' + 9)^2 = 9(x' - 12)$$

$$(y' + 9)^2 = 4(9/4)(x' - 12)$$

(b) Since $p = 9/4 = 2.25$, the distance from the vertex to the receiver is 2.25 feet.

65. $x^2 + xy + ky^2 + 6x + 10 = 0$

$B^2 - 4AC = 1^2 - 4(1)(k) = 1 - 4k > 0 \Rightarrow -4k > -1 \Rightarrow k < \frac{1}{4}$

True. For the graph to be a hyperbola, the discriminant must be greater than zero.

67. $r^2 = x^2 + y^2 = (x'\cos\theta - y'\sin\theta)^2 + (y'\cos\theta + x'\sin\theta)^2$

$= (x')^2\cos^2\theta - 2x'y'\cos\theta\sin\theta + (y')^2\sin^2\theta + (y')^2\cos^2\theta + 2x'y'\cos\theta\sin\theta + (x')^2\sin^2\theta$

$= (x')^2(\cos^2\theta + \sin^2\theta) + (y')^2(\sin^2\theta + \cos^2\theta) = (x')^2 + (y')^2$

So, $(x')^2 + (y')^2 = r^2$.

Section 10.6 Parametric Equations

1. plane curve

3. eliminating; parameter

5. (a) $x = \sqrt{t}$, $y = 3 - t$

t	0	1	2	3	4
x	0	1	$\sqrt{2}$	$\sqrt{3}$	2
y	3	2	1	0	−1

(b)

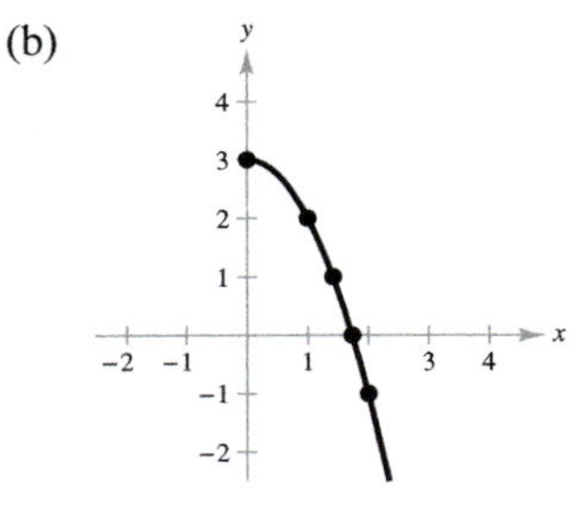

(c) $x = \sqrt{t} \Rightarrow x^2 = t$

$y = 3 - t \Rightarrow y = 3 - x^2$

The graph of the parametric equations only shows the right half of the parabola, whereas the rectangular equation yields the entire parabola.

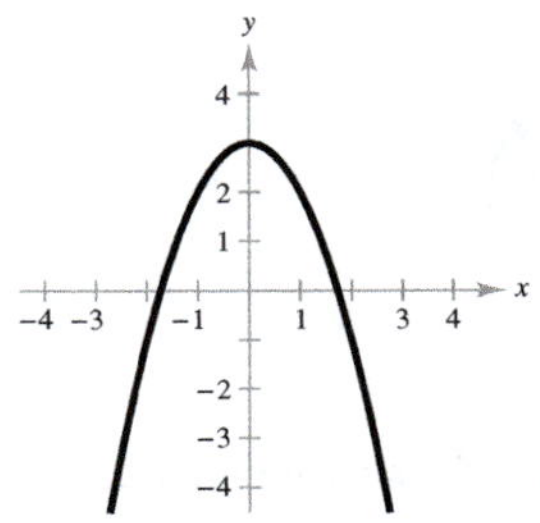

7. $x = t,\ y = -5t$

t	-2	-1	0	1	2
x	-2	-1	0	1	2
y	10	5	0	-5	-10

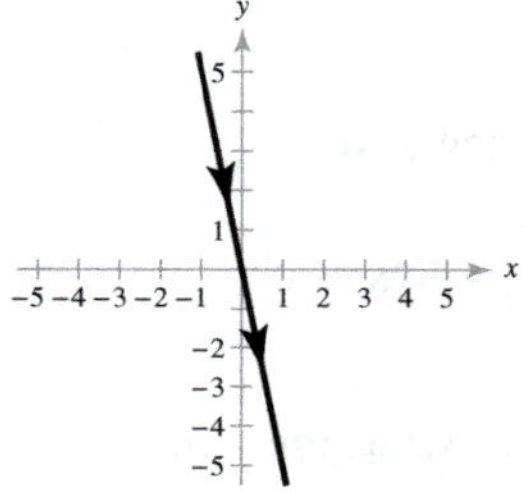

The curve is traced from left to right.

9. $x = t^2,\ y = 3t$

t	-2	-1	0	1	2
x	4	1	0	1	4
y	-6	-3	0	3	6

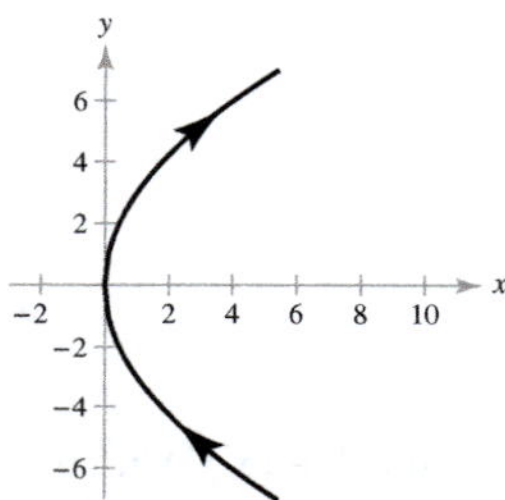

The curve is traced clockwise.

11. $x = 3\cos\theta,\ y = 2\sin^2\theta,\ 0 \le \theta \le \pi$

θ	0	$\frac{\pi}{4}$	$\frac{\pi}{2}$	$\frac{3\pi}{4}$	π
x	3	$\frac{3\sqrt{2}}{2}$	0	$-\frac{3\sqrt{2}}{2}$	-3
y	0	1	2	1	0

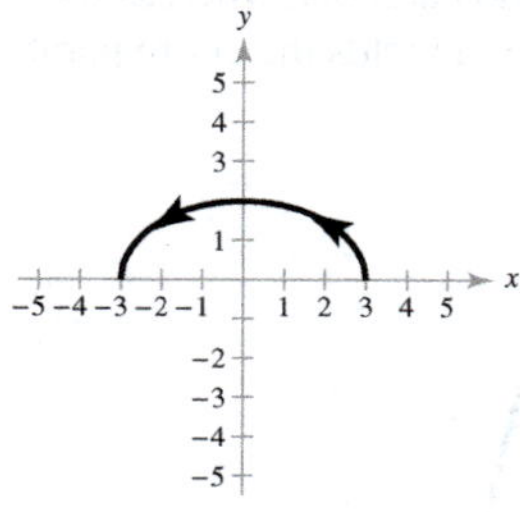

The curve is traced from right to left.

13. (a) $x = t,\ y = 4t$

t	-2	-1	0	1	2
x	-2	-1	0	1	2
y	-8	-4	0	4	8

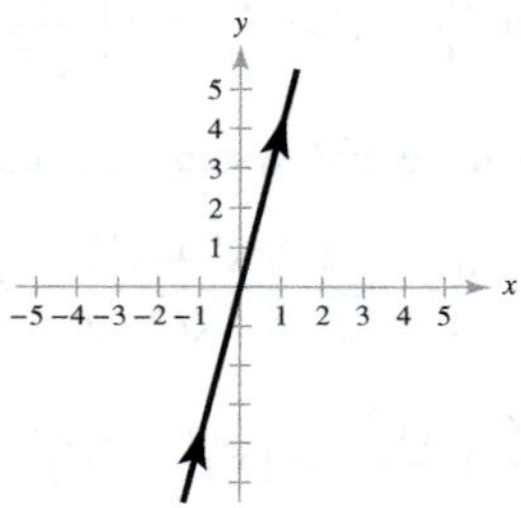

(b) $x = t \Rightarrow t = x$

$y = 4t \Rightarrow y = 4x$

15. (a) $x = -t + 1,\ y = -3t$

t	-2	-1	0	1	2
x	3	2	1	0	-1
y	6	3	0	-3	-6

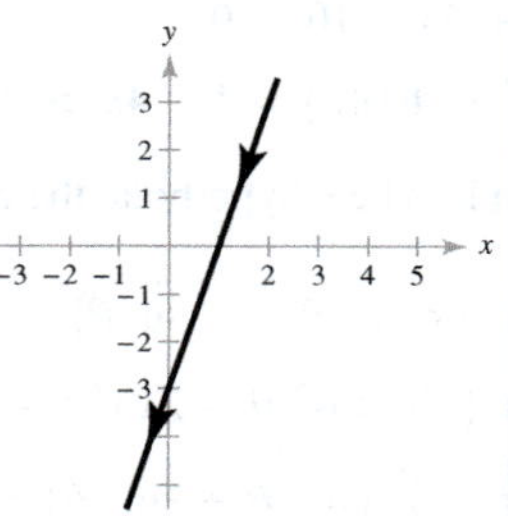

(b) $x = -t + 1 \Rightarrow t = 1 - x$

$y = -3t$

$y = -3(1 - x)$

$y = -3 + 3x$

$y = 3x - 3$

17. (a) $x = \frac{1}{4}t,\ y = t^2$

t	-2	-1	0	1	2
x	$-\frac{1}{2}$	$-\frac{1}{4}$	0	$\frac{1}{4}$	$\frac{1}{2}$
y	4	1	0	1	4

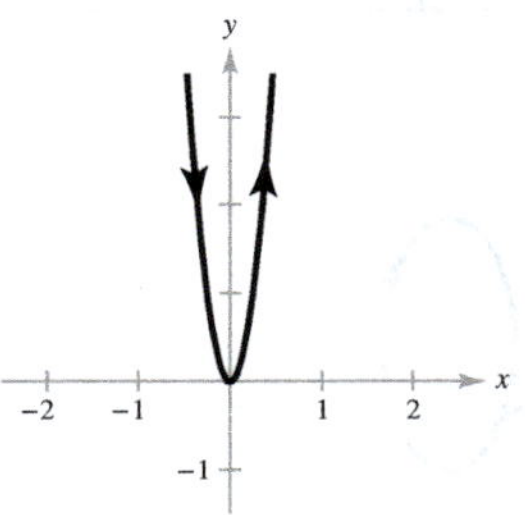

(b) $x = \frac{1}{4}t \Rightarrow t = 4x$

$y = t^2 \Rightarrow y = 16x^2$

19. (a) $x = t^2,\ y = -2t$

t	-2	-1	0	1	2
x	4	1	0	1	4
y	4	2	0	-2	-4

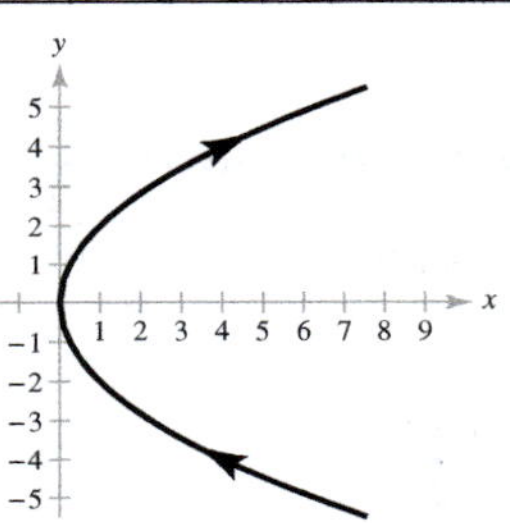

(b) $x = t^2 \Rightarrow t = \pm\sqrt{x}$

$y = -2t = \pm 2\sqrt{x}$

21. (a) $x = \sqrt{t},\ y = 1 - t$

t	0	1	2	3
x	0	1	$\sqrt{2}$	$\sqrt{3}$
y	1	0	-1	-2

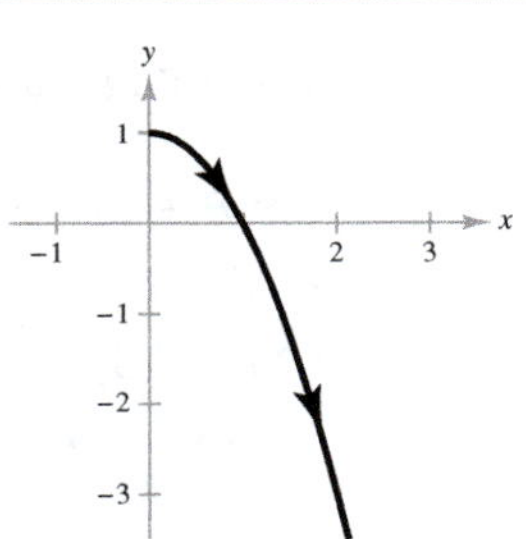

(b) $x = \sqrt{t} \Rightarrow x^2 = t,\ t \geq 0$

$y = 1 - t = 1 - x^2,\ x \geq 0$

23. (a) $x = \sqrt{t} - 3,\ y = t^3$

t	0	1	2	3	4
x	-3	-2	$\sqrt{2} - 3$	$\sqrt{3} - 3$	-1
y	0	1	8	27	64

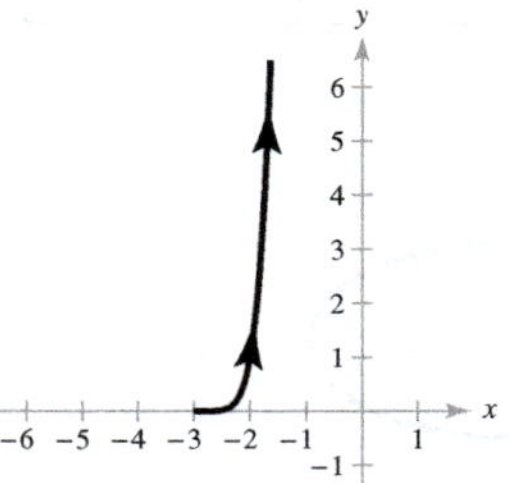

(b) $x = \sqrt{t} - 3 \Rightarrow t = (x + 3)^2$

$y = t^3 \Rightarrow y = \left[(x + 3)^2\right]^3 = (x + 3)^6,\ x \geq -3$

25. (a) $x = t + 1,\ y = \dfrac{t}{t + 1}$

t	-3	-2	0	1	2
x	-2	-1	1	2	3
y	$\frac{3}{2}$	2	0	$\frac{1}{2}$	$\frac{2}{3}$

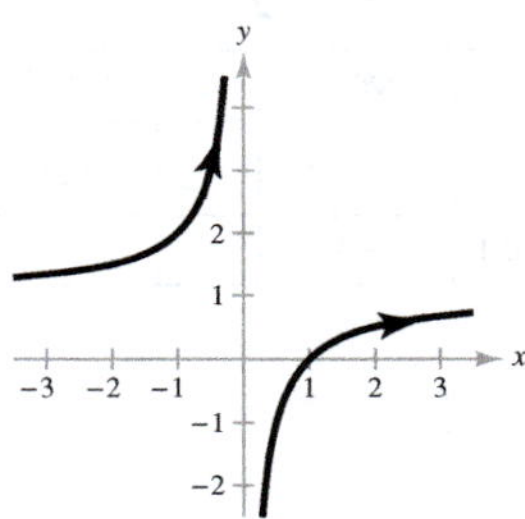

(b) $x = t + 1 \Rightarrow t = x - 1$

$y = \dfrac{t}{t + 1} \Rightarrow y = \dfrac{x - 1}{x}$

27. (a) $x = 4\cos\theta,\ y = 2\sin\theta$

θ	0	$\frac{\pi}{2}$	π	$\frac{3\pi}{2}$	2π
x	4	0	−4	0	4
y	0	2	0	−2	0

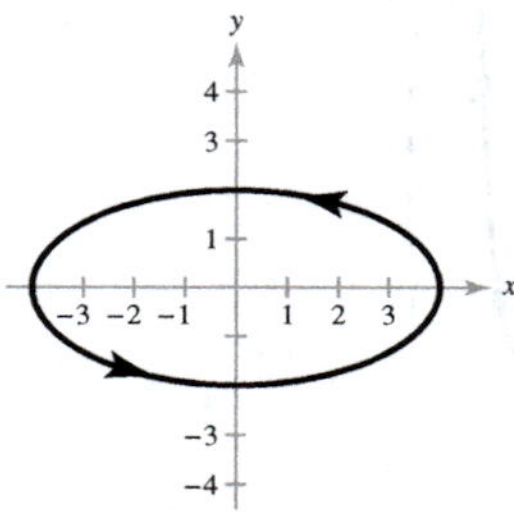

(b) $x = 4\cos\theta \Rightarrow \left(\frac{x}{4}\right)^2 = \cos^2\theta$

$y = 2\sin\theta \Rightarrow \left(\frac{y}{2}\right)^2 = \sin^2\theta$

$\left(\frac{x}{4}\right)^2 + \left(\frac{y}{2}\right)^2 = 1$

$\frac{x^2}{16} + \frac{y^2}{4} = 1$

29. (a) $x = 1 + \cos\theta,\ y = 1 + 2\sin\theta$

θ	0	$\frac{\pi}{2}$	π	$\frac{3\pi}{2}$	2π
x	2	1	0	1	2
y	1	3	1	−1	1

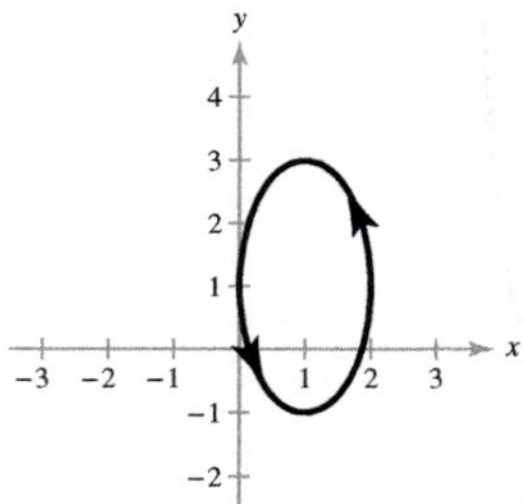

(b) $x = 1 + \cos\theta \Rightarrow (x-1)^2 = \cos^2\theta$

$y = 1 + 2\sin\theta \Rightarrow \left(\frac{y-1}{2}\right)^2 = \sin^2\theta$

$\frac{(x-1)^2}{1} + \frac{(y-1)^2}{4} = 1$

31. (a) $x = 2\sec\theta,\ y = \tan\theta,\ \frac{\pi}{2} \le \theta \le \frac{3\pi}{2}$

θ	$\pi/2$	$3\pi/4$	π	$5\pi/4$	$3\pi/2$
x	undefined	$-2\sqrt{2}$	−2	$-2\sqrt{2}$	undefined
y	undefined	−1	0	1	undefined

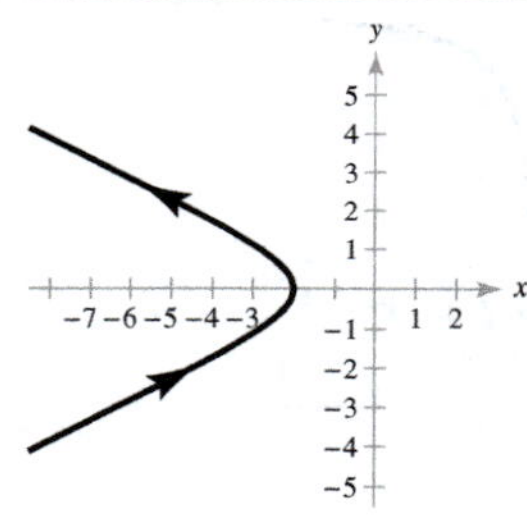

(b) $x = 2\sec\theta \Rightarrow \sec\theta = \frac{x}{2}$

$y = \tan\theta$

$1 + \tan^2\theta = \sec^2\theta$

$1 + y^2 = \left(\frac{x}{2}\right)^2$

$1 + y^2 = \frac{x^2}{4}$

$\frac{x^2}{4} - y^2 = 1,\ x \le -2$

33. (a) $x = 3\cos\theta,\ y = 3\sin\theta$

θ	0	$\pi/4$	$\pi/2$	$3\pi/4$	π	$5\pi/4$	$3\pi/2$	$7\pi/4$	2π
x	3	$3\sqrt{2}/2$	0	$-3\sqrt{2}/2$	−3	$-3\sqrt{2}/2$	0	$3\sqrt{2}/2$	3
y	0	$3\sqrt{2}/2$	3	$3\sqrt{2}/2$	0	$-3\sqrt{2}/2$	−3	$-3\sqrt{2}/2$	0

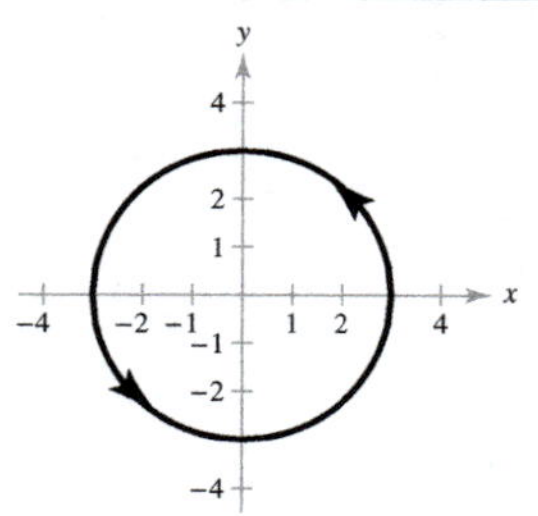

(b) $x = 3\cos\theta \Rightarrow \cos\theta = \frac{x}{3}$

$y = 3\sin\theta \Rightarrow \sin\theta = \frac{x}{3}$

$\sin^2\theta + \cos^2\theta = 1$

$\left(\frac{x}{3}\right)^2 + \left(\frac{x}{3}\right)^2 = 1$

$x^2 + y^2 = 9$

35. (a) $x = e^t, y = e^{3t}$

t	−3	−2	−1	0	1
x	0.0498	0.1353	0.3679	1	2.7183
y	0.0001	0.0024	0.0498	1	20.0855

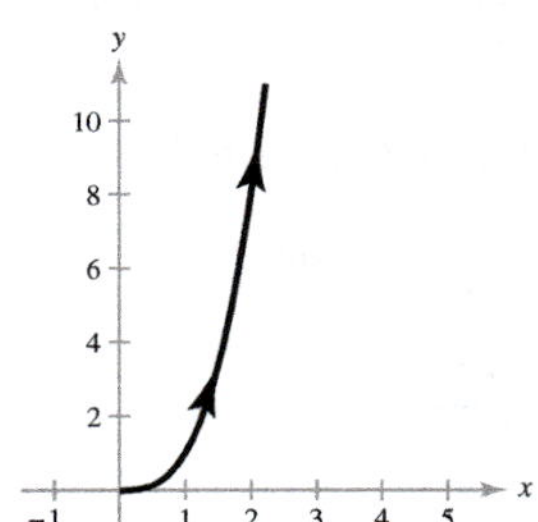

(b) $x = e^t$

$y = e^{3t} = \left(e^t\right)^3 \Rightarrow y = x^3, x > 0$

37. (a) $x = t^3, y = 3 \ln t$

t	$\frac{1}{2}$	1	2	3	4
x	$\frac{1}{8}$	1	8	27	64
y	−2.0794	0	2.0794	3.2958	4.1589

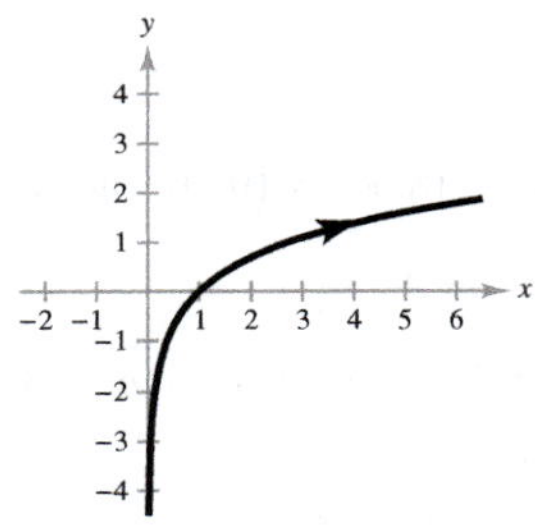

(b) $x = t^3 \quad \Rightarrow x^{1/3} = t$

$y = 3 \ln t \Rightarrow y = \ln t^3$

$y = \ln\left(x^{1/3}\right)^3$

$y = \ln x$

39. $x = t, y = \sqrt{t}$

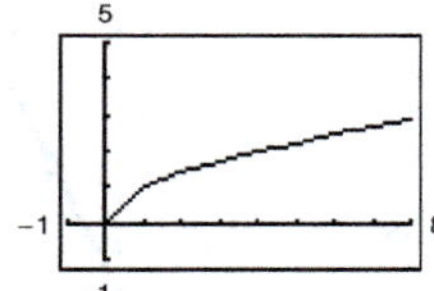

41. $x = 2t, y = |t + 1|$

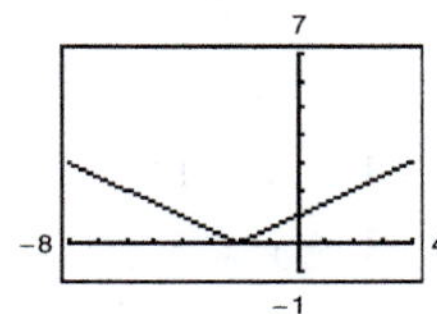

43. $x = 4 + 3 \cos \theta, y = -2 + \sin \theta$

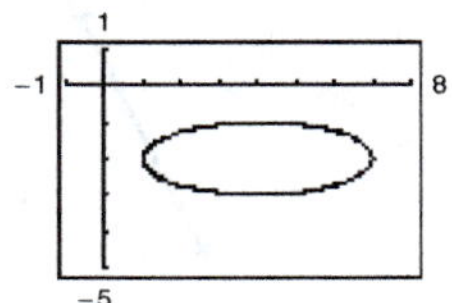

45. $x = 2 \csc \theta, y = 4 \cot \theta$

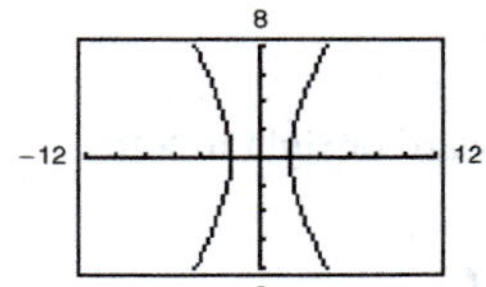

47. $x = \dfrac{t}{2}, y = \ln\left(t^2 + 1\right)$

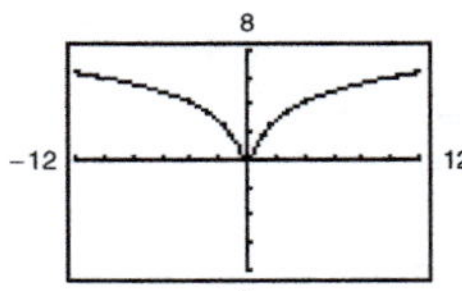

49. By eliminating the parameter, each curve becomes $y = 2x + 1$.

(a) $x = t$

$y = 2t + 1$

There are no restrictions on x and y.

Domain: $(-\infty, \infty)$

Orientation: Left to right

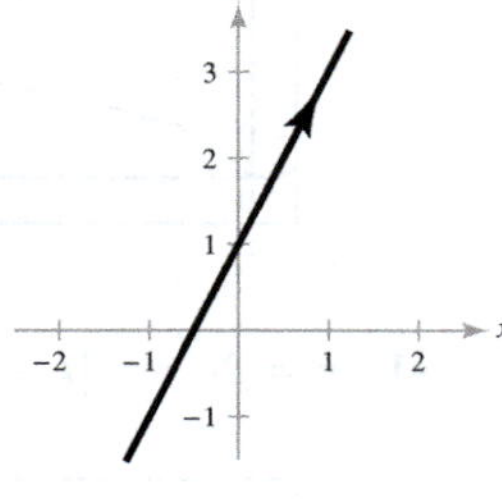

(b) $x = \cos\theta \Rightarrow -1 \le x \le 1$

$y = 2\cos\theta + 1 \Rightarrow -1 \le y \le 3$

The graph oscillates.

Domain: $[-1, 1]$

Orientation: Depends on θ

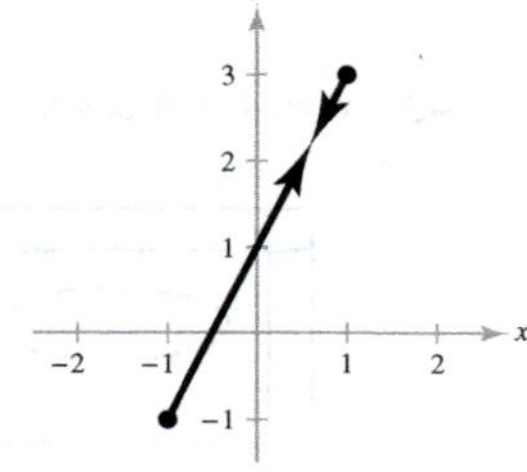

(c) $x = e^{-t} \Rightarrow x > 0$

$y = 2e^{-t} + 1 \Rightarrow y > 1$

Domain: $(0, \infty)$

Orientation: Downward or right to left

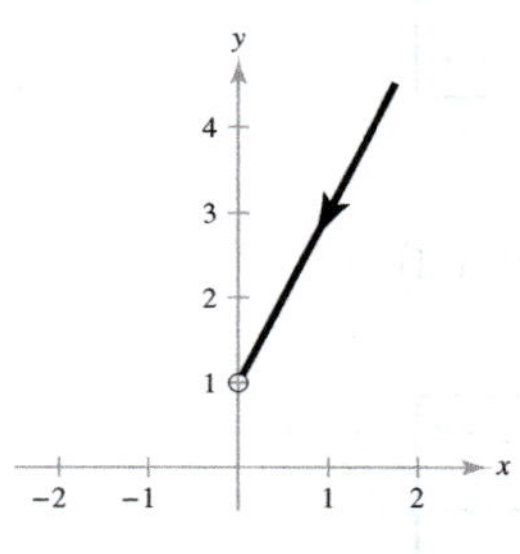

(d) $x = e^{t} \Rightarrow x > 0$

$y = 2e^{t} + 1 \Rightarrow y > 1$

Domain: $(0, \infty)$

Orientation: Upward or left to right

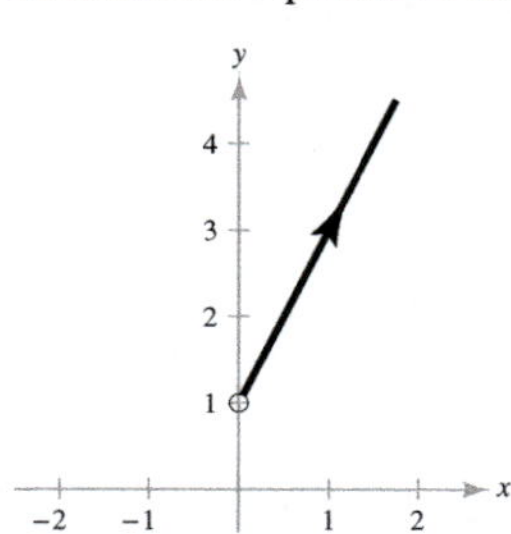

51. $x = x_1 + t(x_2 - x_1),\ y = y_1 + t(y_2 - y_1)$

$$\frac{x - x_1}{x_2 - x_1} = t$$

$$y = y_1 + \left(\frac{x - x_1}{x_2 - x_1}\right)(y_2 - y_1)$$

$$y - y_1 = \frac{y_2 - y_1}{x_2 - x_1}(x - x_1) = m(x - x_1)$$

53. $x = h + a\cos\theta,\ y = k + b\sin\theta$

$$\frac{x - h}{a} = \cos\theta,\ \frac{y - k}{b} = \sin\theta$$

$$\frac{(x - h)^2}{a^2} + \frac{(y - k)^2}{b^2} = 1$$

55. Line through $(0, 0)$ and $(3, 6)$

From Exercise 51:

$x = x_1 + t(x_2 - x_1)$ $y = y_1 + t(y_2 - y_1)$

$= 0 + t(3 - 0)$ $= 0 + t(6 - 0)$

$= 3t$ $= 6t$

57. Circle with center $(3, 2)$; radius 4

From Exercise 52:

$x = 3 + 4\cos\theta$

$y = 2 + 4\sin\theta$

59. Ellipse

Vertices: $(\pm 5, 0) \Rightarrow (h, k) = (0, 0)$ and $a = 5$

Foci: $(\pm 4, 0) \Rightarrow c = 4$

$b^2 = a^2 - c^2 \Rightarrow 25 - 16 = 9 \Rightarrow b = 3$

From Exercise 53:

$x = h + a\cos\theta = 5\cos\theta$

$y = k + b\sin\theta = 3\sin\theta$

61. Hyperbola

Vertices: $(1, 0), (9, 0) \Rightarrow (h, k) = (5, 0)$ and $a = 4$

Foci: $(0, 0), (10, 0) \Rightarrow c = 5$

$c^2 = a^2 + b^2 \Rightarrow 25 = 16 + b^2 \Rightarrow b = 3$

From Exercise 54:

$x = 5 + 4\sec\theta$

$y = 3\tan\theta$

63. Line segment between $(0, 0)$ and $(-5, 2)$.

$x = x_1 + t(x_2 - x_1)$ and $y = y_1 + t(y_2 - y_1)$

$x = 0 + t(-5 - 0)$ and $y = 0_1 + t(2 - 0)$

$x = -5t$

$y = 2t$

$0 \le t \le 1$

65. Left branch of the hyperbola with vertices $(\pm 3, 0)$ and foci $(\pm 5, 0)$.

$a = 3, c = 5$

Center: $(0, 0), h = 0, k = 0.$

$b^2 = c^2 - a^2 = 25 - 9 = 16$

$x = h + a \sec\theta \Rightarrow x = 0 + 3\sec\theta$

$y = k + b\tan\theta \Rightarrow y = 0 + 4\tan\theta$

$x = 3\sec\theta$

$y = 4\tan\theta$

$\dfrac{\pi}{2} < \theta < \dfrac{3\pi}{2}$

67. $y = 3x - 2$

(a) $t = x \Rightarrow x = t$ and $y = 3t - 2$

(b) $t = 2 - x \Rightarrow x = -t + 2$ and

$y = 3(-t + 2) - 2 = -3t + 4$

69. $x = 2y + 1$

(a) $t = x \Rightarrow x = t$ and $t = 2y + 1 \Rightarrow y = \dfrac{1}{2}t - \dfrac{1}{2}$

(b) $t = 2 - x \Rightarrow x = -t + 2$ and

$-t + 2 = 2y + 1 \Rightarrow y = -\dfrac{1}{2}t + \dfrac{1}{2}$

71. $y = x^2 + 1$

(a) $t = x \Rightarrow x = t$ and $y = t^2 + 1$

(b) $t = 2 - x \Rightarrow x = -t + 2$ and

$y = (-t + 2)^2 + 1 = t^2 - 4t + 5$

73. $y = 1 - 2x^2$

(a) $t = x \Rightarrow x = t$

$\Rightarrow y = 1 - 2t^2$

(b) $t = 2 - x \Rightarrow x = -t + 2$

$\Rightarrow y = 1 - 2(-t + 2)^2 = -2t^2 + 8t - 7$

75. $y = \dfrac{1}{x}$

(a) $t = x \Rightarrow x = t$ and $y = \dfrac{1}{t}$

(b) $t = 2 - x \Rightarrow x = -t + 2$ and

$y = \dfrac{1}{-t + 2} = \dfrac{-1}{t - 2}$

77. $y = e^x$

(a) $t = x \Rightarrow x = t$ and $y = e^t$

(b) $t = 2 - x \Rightarrow x = -t + 2$ and $y = e^{-t+2}$

79. $x = 4(\theta - \sin\theta)$

$y = 4(1 - \cos\theta)$

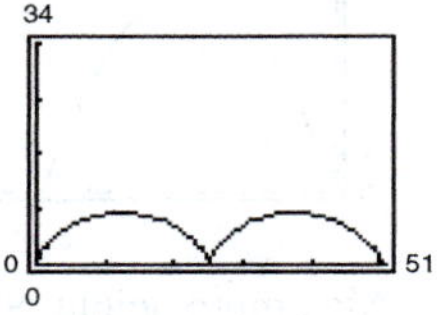

81. $x = 2\theta - 4\sin\theta$

$y = 2 - 4\cos\theta$

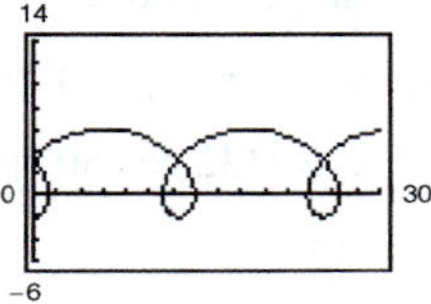

83. $x = 3\cos^3\theta$

$y = 3\sin^3\theta$

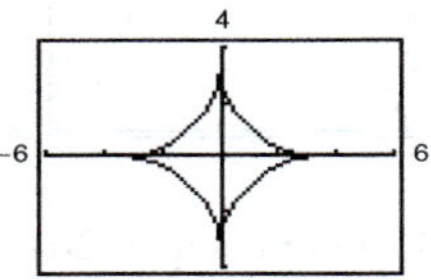

85. $x = 2\cot\theta$

$y = 2\sin^2\theta$

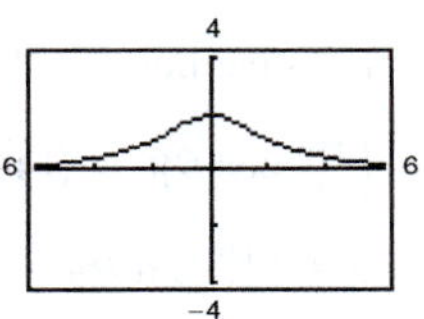

87. $x = 2\cos\theta \Rightarrow -2 \le x \le 2$

$y = \sin 2\theta \Rightarrow -1 \le y \le 1$

Matches graph (b).

Domain: $[-2, 2]$

Range: $[-1, 1]$

88. $x = 4\cos^3\theta \Rightarrow -4 \le x \le 4$

$y = 6\sin^3\theta \Rightarrow -6 \le y \le 6$

Matches graph (c).

Domain: $[-4, 4]$

Range: $[-6, 6]$

89. $x = \frac{1}{2}(\cos\theta + \theta\sin\theta)$

$y = \frac{1}{2}(\sin\theta - \theta\cos\theta)$

Matches graph (d).

Domain: $(-\infty, \infty)$

Range: $(-\infty, \infty)$

90. $x = \frac{1}{2}\cot\theta \Rightarrow -\infty < x < \infty$

$y = 4\sin\theta\cos\theta \Rightarrow -2 \le y \le 2$

Matches graph (a).

Domain: $(-\infty, \infty)$

Range: $[-2, 2]$

91. $x = (v_0\cos\theta)t$ and $y = h + (v_0\sin\theta)t - 16t^2$

(a) $\theta = 60°$, $v_0 = 88$ ft/sec

$x = (88\cos 60°)t$ and $y = (88\sin 60°)t - 16t^2$

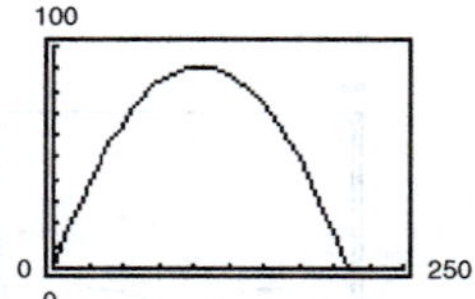

Maximum height: 90.7 feet

Range: 209.6 feet

(b) $\theta = 60°$, $v_0 = 132$ ft/sec

$x = (132\cos 60°)t$ and $y = (132\sin 60°)t - 16t^2$

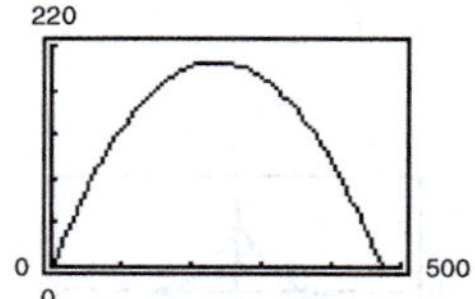

Maximum height: 204.2 feet

Range: 471.6 feet

(c) $\theta = 45°$, $v_0 = 88$ ft/sec

$x = (88\cos 45°)t$ and $y = (88\sin 45°)t - 16t^2$

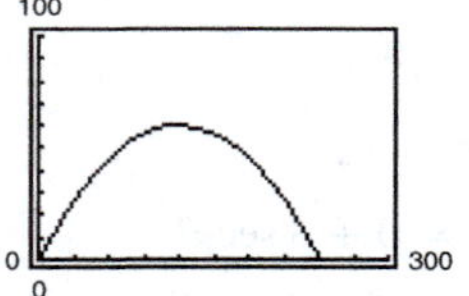

Maximum height: 60.5 ft

Range: 242.0 ft

(d) $\theta = 45°$, $v_0 = 132$ ft/sec

$x = (132\cos 45°)t$ and $y = (132\sin 45°)t - 16t^2$

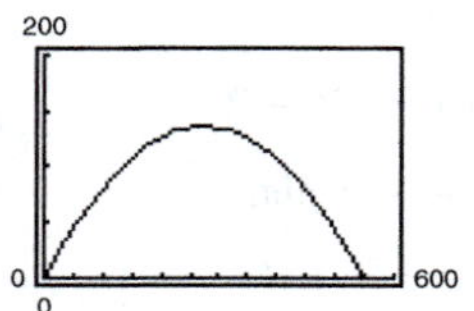

Maximum height: 136.1 ft

Range: 544.5 ft

93. (a) 100 miles per hour $= 100\left(\frac{5280}{3600}\right)$ ft/sec $= \frac{440}{3}$ ft/sec

$x = \left(\frac{440}{3}\cos\theta\right)t \approx (146.67\cos\theta)t$

$y = 3 + \left(\frac{440}{3}\sin\theta\right)t - 16t^2 \approx 3 + (146.67\sin\theta)t - 16t^2$

(b) For $\theta = 15°$:

$x = \left(\frac{440}{3}\cos 15°\right)t \approx 141.7t$

$y = 3 + \left(\frac{440}{3}\sin 15°\right)t - 16t^2 \approx 3 + 38.0t - 16t^2$

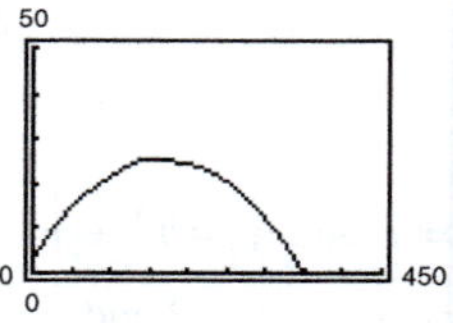

The ball hits the ground inside the ballpark, so it is not a home run.

(c) For $\theta = 23°$:

$x = \left(\frac{440}{3}\cos 23°\right)t \approx 135.0t$

$y = 3 + \left(\frac{440}{3}\sin 23°\right)t - 16t^2 \approx 3 + 57.3t - 16t^2$

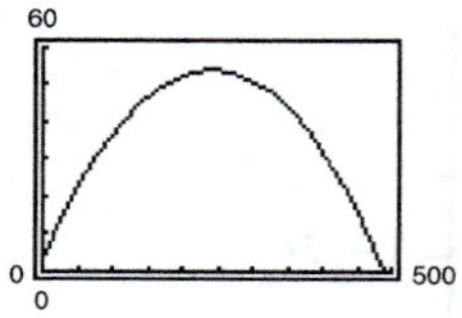

The ball easily clears the 7-foot fence at 408 feet so it is a home run.

(d) Find θ so that $y = 7$ when $x = 408$ by graphing the parametric equations for θ values between 15° and 23°. This occurs when $\theta \approx 19.3°$.

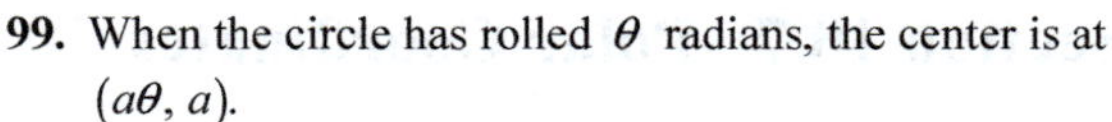

95. (a) $x = (\cos 35°)v_0 t$

$y = 7 + (\sin 35°)v_0 t - 16t^2$

(b) If the ball is caught at time t_1, then:

$90 = (\cos 35°)v_0 t_1$

$4 = 7 + (\sin 35°)v_0 t_1 - 16t_1^2$

$$v_0 t_1 = \frac{90}{\cos 35°} \Rightarrow -3 = (\sin 35°)\frac{90}{\cos 35°} - 16t_1^2$$

$$\Rightarrow 16t_1^2 = 90 \tan 35° + 3$$

$$\Rightarrow t_1 \approx 2.03 \text{ seconds}$$

$$\Rightarrow v_0 = \frac{90}{t_1 \cos 35°} \approx 54.09 \text{ ft/sec}$$

(c) 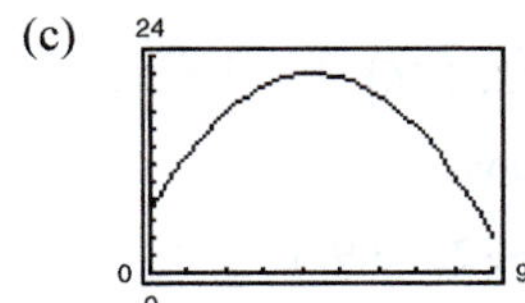

Maximum height $\approx$ 22 feet

(d) From part (b), $t_1 \approx 2.03$ seconds.

97. $y = 7 + x - 0.02x^2$

(a) Exercise 98 result:

$$y = -\frac{16 \sec^2 \theta}{v_0^2}x^2 + (\tan \theta)x + h$$

$h = 7$

$\tan \theta = 1 \Rightarrow \theta = 45°$

$$\frac{16 \sec^2 45°}{v_0^2} = 0.02 \Rightarrow v_0 = 40$$

$x = (v_0 \cos \theta)t = (40 \cos 45°)t$

$y = h + (v_0 \sin \theta)t - 16t^2$

$\approx 7 + (40 \sin 45°)t - 16t^2$

(b) 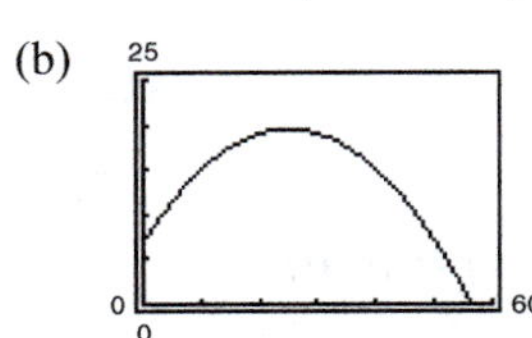

(c) Maximum height: 19.5 feet

Range: 56.2 feet

99. When the circle has rolled θ radians, the center is at $(a\theta, a)$.

$$\sin \theta = \sin(180° - \theta) = \frac{|AC|}{b} = \frac{|BD|}{b} \Rightarrow |BD| = b \sin \theta$$

$$\cos \theta = -\cos(180° - \theta) = \frac{|AP|}{-b} \Rightarrow |AP| = -b \cos \theta$$

So, $x = a\theta - b \sin \theta$ and $y = a - b \cos \theta$.

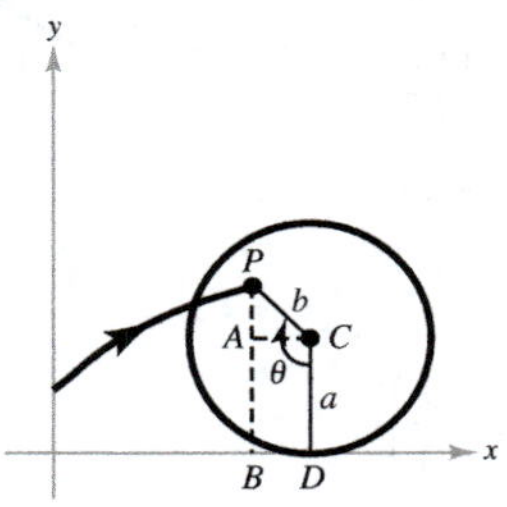

101. True

$x = t$

$y = t^2 + 1 \Rightarrow y = x^2 + 1$

$x = 3t$

$y = 9t^2 + 1 \Rightarrow y = x^2 + 1$

103. False. It is possible for both x and y to be functions of t, but y cannot be a function of x. For example, consider the parametric equations $x = 3 \cos t$ and $y = 3 \sin t$. Both x and y are functions of t. However, after eliminating the parameter and finding the rectangular equation $\frac{x^2}{9} + \frac{y^2}{9} = 1$, you can see that y is not a function of x.

105. The use of parametric equations is useful when graphing two functions simultaneously on the same coordinate system. For example, this is useful when tracking the path of an object so the position and the time associated with that position can be determined.

107. $x = \sqrt{t - 1} \Rightarrow x^2 = t - 1$

$x^2 + 1 = t$

$y = 2t$

$y = 2(x^2 + 1)$

$y = 2x^2 + 2, x \ge 0$

The parametric equation for x is defined only when $t \ge 1$, so the domain of the rectangular equation is $x \ge 0$.

109. The graph is the same, but the orientation is reversed.

Section 10.7 Polar Coordinates

1. pole

3. polar

5. Polar coordinates: $\left(2, \dfrac{5\pi}{6}\right)$

Additional representations:

$$\left(-2, \frac{\pi}{6} - \pi\right) = \left(-2, -\frac{5\pi}{6}\right)$$
$$\left(2, \frac{\pi}{6} - 2\pi\right) = \left(2, -\frac{11\pi}{6}\right)$$
$$\left(-2, \frac{\pi}{6} + \pi\right) = \left(-2, \frac{7\pi}{6}\right)$$

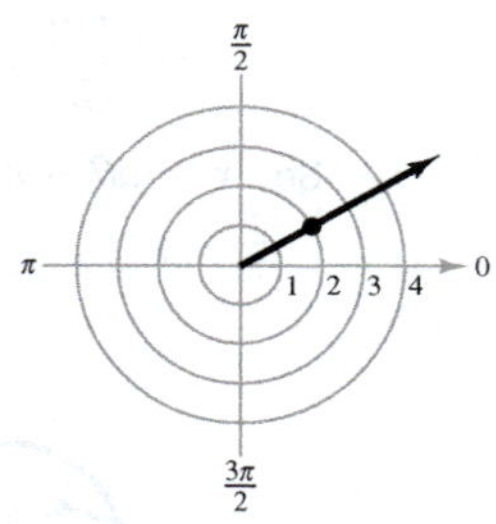

7. Polar coordinates: $\left(4, -\dfrac{\pi}{3}\right)$

Additional representations:

$$\left(4, -\frac{\pi}{3} + 2\pi\right) = \left(4, \frac{5\pi}{3}\right)$$
$$\left(-4, -\frac{\pi}{3} - \pi\right) = \left(-4, -\frac{4\pi}{3}\right)$$
$$\left(-4, -\frac{\pi}{3} + \pi\right) = \left(-4, \frac{2\pi}{3}\right)$$

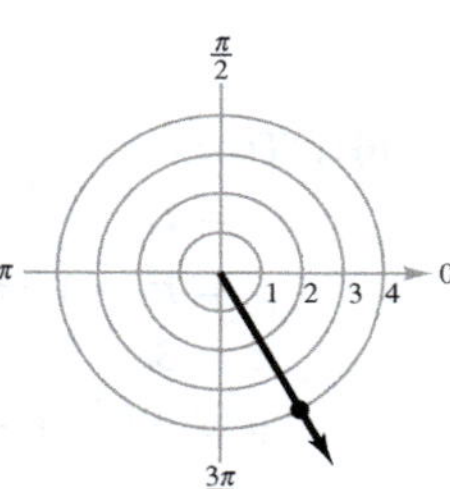

9. Polar coordinates: $(2, 3\pi)$

Additional representations:

$$(2, 3\pi - 2\pi) = (2, \pi)$$
$$(-2, 3\pi - 2\pi) = (-2, \pi)$$
$$(-2, 3\pi - 3\pi) = (-2, 0)$$

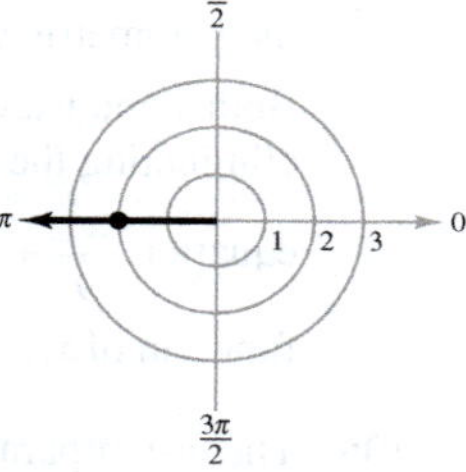

11. Polar coordinates: $\left(-2, \dfrac{2\pi}{3}\right)$

Additional representations:

$$\left(-2, \frac{2\pi}{3} - 2\pi\right) = \left(-2 -\frac{4\pi}{3}\right)$$
$$\left(2, \frac{2\pi}{3} + \pi\right) = \left(2, \frac{5\pi}{3}\right)$$
$$\left(-2, \frac{2\pi}{3} - \pi\right) = \left(-2, -\frac{\pi}{3}\right)$$

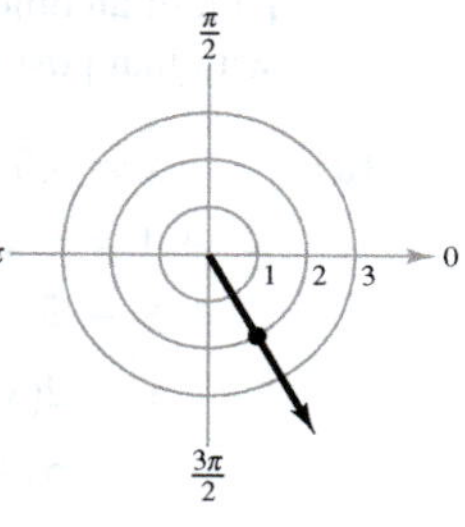

13. Polar coordinates: $\left(0, \dfrac{7\pi}{6}\right)$

Additional representations:

$$\left(0, \frac{7\pi}{6} - \pi\right) = \left(0, \frac{\pi}{6}\right)$$
$$\left(0, \frac{7\pi}{6} - 2\pi\right) = \left(0, -\frac{5\pi}{6}\right)$$
$$\left(0, \frac{7\pi}{6} - 3\pi\right) = \left(0, -\frac{11\pi}{6}\right)$$

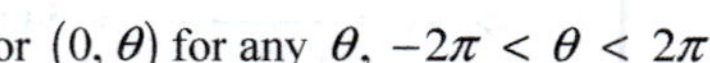

or $(0, \theta)$ for any θ, $-2\pi < \theta < 2\pi$

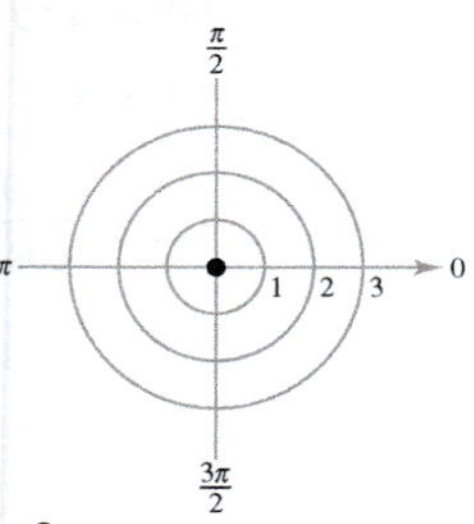

15. Polar coordinates: $\left(\sqrt{2}, 2.36\right)$

Additional representations:

$$\left(\sqrt{2}, 2.36 - 2\pi\right) \approx \left(\sqrt{2}, -3.92\right)$$
$$\left(-\sqrt{2}, 2.36 - \pi\right) \approx \left(-\sqrt{2}, -0.78\right)$$
$$\left(-\sqrt{2}, 2.36 + \pi\right) \approx \left(-\sqrt{2}, 5.50\right)$$

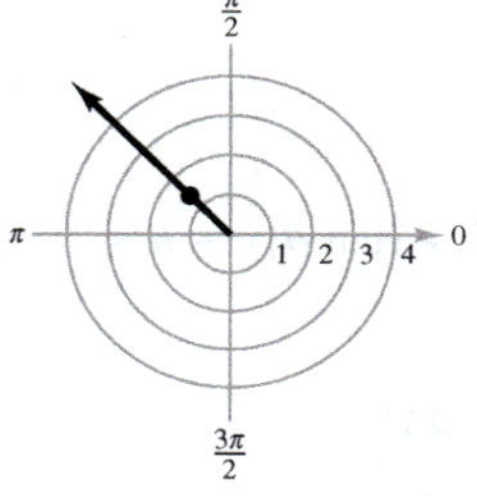

17. Polar coordinates: $(-3, -1.57)$

Additional representations:

$(3, 1.57)$

$(-3, 4.71)$

$(3, -4.71)$

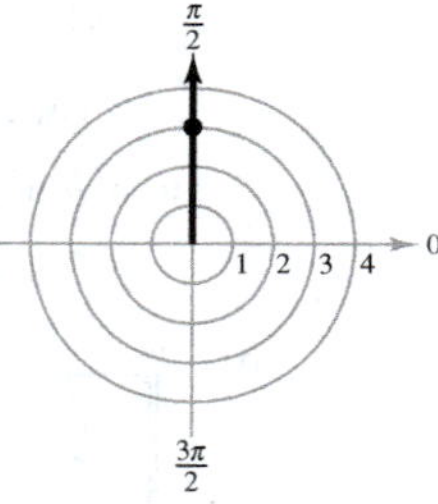

19. Polar coordinates: $(0, \pi) = (r, \theta)$

$$x = 0 \cos \pi = (0)(-1) = 0$$
$$y = 0 \sin \pi = (0)(0) = 0$$

Rectangular coordinates: $(0, 0)$

21. Polar coordinates: $\left(3, \frac{\pi}{2}\right)$

$x = 3 \cos \frac{\pi}{2} = 0$

$y = 3 \sin \frac{\pi}{2} = 3$

Rectangular coordinates: $(0, 3)$

23. Polar coordinates: $\left(2, \frac{3\pi}{4}\right)$

$x = 2 \cos \frac{3\pi}{4} = -\sqrt{2}$

$y = 2 \sin \frac{3\pi}{4} = \sqrt{2}$

Rectangular coordinates: $\left(-\sqrt{2}, \sqrt{2}\right)$

25. Polar coordinates: $\left(-2, \frac{7\pi}{6}\right) = (r, \theta)$

$x = r \cos \theta = -2 \cos \frac{7\pi}{6} = \sqrt{3}$

$y = r \sin \theta = -2 \sin \frac{7\pi}{6} = 1$

Rectangular coordinates: $\left(\sqrt{3}, 1\right)$

27. Polar coordinates: $\left(-3, -\frac{\pi}{3}\right)$

$x = r \cos \theta = (-3) \cos\left(-\frac{\pi}{3}\right) = (-3)\left(\frac{1}{2}\right) = -\frac{3}{2}$

$y = r \sin \theta = (-3) \sin\left(-\frac{\pi}{3}\right) = (-3)\left(-\frac{\sqrt{3}}{2}\right) = \frac{3\sqrt{3}}{2}$

Rectangular coordinates: $\left(-\frac{3}{2}, \frac{3\sqrt{3}}{2}\right)$

29. Polar coordinates: $\left(2, \frac{7\pi}{8}\right) = (r, \theta)$

$x = r \cos \theta = 2 \cos \frac{7\pi}{8} \approx -1.85$

$y = r \sin \theta = 2 \sin \frac{7\pi}{8} \approx 0.77$

Regular coordinates: $(-1.85, 0.77)$

31. Polar coordinates: $\left(1, \frac{5\pi}{12}\right)$

$x = \cos \frac{5\pi}{12} \approx 0.26$

$y = \sin \frac{5\pi}{12} \approx 0.97$

Rectangular coordinates: $(0.26, 0.97)$

33. Polar coordinates: $(-2.5, 1.1) = (r, \theta)$

$x = r \cos \theta = -2.5 \cos 1.1 \approx -1.13$

$y = r \sin \theta = -2.5 \sin 1.1 \approx -2.23$

Rectangular coordinates: $(-1.13, -2.23)$

35. Polar coordinates: $(2.5, -2.9) = (r, \theta)$

$x = r \cos \theta = 2.5 \cos(-2.9) \approx -2.43$

$y = r \sin \theta = 2.5 \sin(-2.9) \approx -0.60$

Rectangular coordinates: $(-2.43, -0.60)$

37. Polar coordinates: $(-3.1, 7.92) = (r, \theta)$

$x = r \cos \theta = -3.1 \cos 7.92 \approx 0.20$

$y = r \sin \theta = -3.1 \sin 7.92 \approx -3.09$

Rectangular coordinates: $(0.20, -3.09)$

39. Rectangular coordinates: $(1, 1)$

$r = \pm\sqrt{2}, \tan \theta = 1, \theta = \frac{\pi}{4}$

Polar coordinates: $\left(\sqrt{2}, \frac{\pi}{4}\right)$

41. Rectangular coordinates: $(-3, -3)$

$r = 3\sqrt{2}, \tan \theta = 1, \theta = \frac{5\pi}{4}$

Polar coordinates: $\left(3\sqrt{2}, \frac{5\pi}{4}\right)$

43. Rectangular coordinates: $(3, 0)$

$r = \sqrt{9 + 0} = 3, \tan \theta = 0, \theta = 0$

Polar coordinates: $(3, 0)$

45. Rectangular coordinates: $(0, -5)$

$r = 5, \tan \theta$ undefined, $\theta = \frac{\pi}{2}$

Polar coordinates: $\left(5, \frac{3\pi}{2}\right)$

47. Rectangular coordinates: $\left(-\sqrt{3}, -\sqrt{3}\right)$

$r = \pm\sqrt{3 + 3} = \pm\sqrt{6}$, $\tan \theta = 1$, $\theta = \dfrac{5\pi}{4}$

Polar coordinates: $\left(\sqrt{6}, \dfrac{5\pi}{4}\right)$

49. Rectangular coordinates: $\left(\sqrt{3}, -1\right)$

$r = \sqrt{3 + 1} = 2$, $\tan \theta = -\dfrac{1}{\sqrt{3}}$, $\theta = \dfrac{11\pi}{6}$

Polar coordinates: $\left(2, \dfrac{11\pi}{6}\right)$

51. Rectangular coordinates: $(3, -2)$

$R \blacktriangleright Pr(3, -2) \approx 3.61 = r$

$R \blacktriangleright P\theta(3, -2) \approx 5.70 = \theta$

Polar coordinates: $(3.61, 5.70)$

53. Rectangular coordinates: $(-5, 2)$

$R \blacktriangleright Pr(-5, -2) \approx 5.39 = r$

$R \blacktriangleright P\theta(-5, -2) \approx 2.76 = \theta$

Polar coordinates: $(5.39, 2.76)$

55. Rectangular coordinates: $\left(-\sqrt{3}, -4\right)$

$R \blacktriangleright \text{Pr}\left(-\sqrt{3}, -4\right) \approx 4.36 = r$

$R \blacktriangleright P\theta\left(-\sqrt{3}, -4\right) \approx 4.30 = \theta$

Polar coordinates: $(4.36, 4.30)$

57. Rectangular coordinates: $\left(\frac{5}{2}, \frac{4}{3}\right)$

$R \blacktriangleright Pr\left(\frac{5}{2}, \frac{4}{3}\right) \approx 2.83 = r$

$R \blacktriangleright P\theta\left(\frac{5}{2}, \frac{4}{3}\right) \approx 0.49 = \theta$

Polar coordinates: $(2.83, 0.49)$

59.
$$\begin{aligned} x^2 + y^2 &= 9 \\ r &= 3 \end{aligned}$$

61.
$$\begin{aligned} y &= x \\ r\cos\theta &= r\sin\theta \\ 1 &= \tan\theta \\ \theta &= \frac{\pi}{4} \end{aligned}$$

63.
$$\begin{aligned} x &= 10 \\ r\cos\theta &= 10 \\ r &= 10\sec\theta \end{aligned}$$

65.
$$\begin{aligned} 3x - y + 2 &= 0 \\ 3r\cos\theta - r\sin\theta + 2 &= 0 \\ r(3\cos\theta - \sin\theta) &= -2 \\ r &= \frac{-2}{3\cos\theta - \sin\theta} \end{aligned}$$

67.
$$\begin{aligned} xy &= 16 \\ (r\cos\theta)(r\sin\theta) &= 16 \\ r^2 &= 16\sec\theta\csc\theta = 32\csc 2\theta \end{aligned}$$

69.
$$\begin{aligned} x &= a \\ r\cos\theta &= a \\ r &= a\sec\theta \end{aligned}$$

71.
$$\begin{aligned} x^2 + y^2 &= a^2 \\ r^2 &= a^2 \\ r &= a \end{aligned}$$

73.
$$\begin{aligned} x^2 + y^2 - 2ax &= 0 \\ r^2 - 2a\,r\cos\theta &= 0 \\ r(r - 2a\cos\theta) &= 0 \\ r - 2a\cos\theta &= 0 \\ r &= 2a\cos\theta \end{aligned}$$

75.
$$\begin{aligned} \left(x^2 + y^2\right)^2 &= x^2 - y^2 \\ \left(r^2\right)^2 &= x^2 - y^2 \\ r^4 &= x^2 - y^2 \\ r^2 &= \frac{x^2}{r^2} - \frac{y^2}{r^2} \\ r^2 &= \left(\frac{x}{r}\right)^2 - \left(\frac{y}{r}\right)^2 \\ r^2 &= \cos^2\theta - \sin^2\theta \\ r^2 &= \cos 2\theta \end{aligned}$$

77.
$$\begin{aligned} y^3 &= x^2 \\ (r\sin\theta)^3 &= (r\cos\theta)^2 \\ r^3\sin^3\theta &= r^2\cos^2\theta \\ \frac{r\sin^3\theta}{\cos^2\theta} &= 1 \\ r\sin\theta\tan^2\theta &= 1 \\ r &= \csc\theta\cot^2\theta \end{aligned}$$

79.
$$\begin{aligned} r &= 5 \\ r^2 &= 25 \\ x^2 + y^2 &= 25 \end{aligned}$$

81.
$$\theta = \frac{2\pi}{3}$$
$$\tan\theta = \tan\frac{2\pi}{3}$$
$$\frac{y}{x} = -\sqrt{3}$$
$$y = -\sqrt{3}x$$
$$\sqrt{3}x + y = 0$$

83. $\tan\left(\frac{\pi}{2}\right)$ is undefined, therefore it is the vertical line $x = 0$.

85.
$$r = 4\csc\theta$$
$$r\sin\theta = 4$$
$$y = 4$$

87.
$$r = -3\sec\theta$$
$$\frac{r}{\sec\theta} = -3$$
$$r\cos\theta = -3$$
$$x = -3$$

89.
$$r = -2\cos\theta$$
$$r^2 = -2r\cos\theta$$
$$x^2 + y^2 = -2x$$
$$x^2 + y^2 + 2x = 0$$

91.
$$r^2 = \cos\theta$$
$$r^3 = r\cos\theta$$
$$\left(\pm\sqrt{x^2 + y^2}\right)^3 = x$$
$$\pm\left(x^2 + y^2\right)^{3/2} = x$$
$$\left(x^2 + y^2\right)^3 = x^2$$
$$x^2 + y^2 = x^{2/3}$$
$$x^2 + y^2 - x^{2/3} = 0$$

93.
$$r^2 = \sin 2\theta = 2\sin\theta\cos\theta$$
$$r^2 = 2\left(\frac{y}{r}\right)\left(\frac{x}{r}\right) = \frac{2xy}{r^2}$$
$$r^4 = 2xy$$
$$\left(x^2 + y^2\right)^2 = 2xy$$

95.
$$r = 2\sin 3\theta$$
$$r = 2\sin(\theta + 2\theta)$$
$$r = 2[\sin\theta\cos 2\theta + \cos\theta\sin 2\theta]$$
$$r = 2\left[\sin\theta\left(1 - 2\sin^2\theta\right) + \cos\theta(2\sin\theta\cos\theta)\right]$$
$$r = 2\left[\sin\theta - 2\sin^3\theta + 2\sin\theta\cos^2\theta\right]$$
$$r = 2\left[\sin\theta - 2\sin^3\theta + 2\sin\theta\left(1 - \sin^2\theta\right)\right]$$
$$r = 2\left(3\sin\theta - 4\sin^3\theta\right)$$
$$r^4 = 6r^3\sin\theta - 8r^3\sin^3\theta$$
$$\left(x^2 + y^2\right)^2 = 6\left(x^2 + y^2\right)y - 8y^3$$
$$\left(x^2 + y^2\right)^2 = 6x^2y - 2y^3$$

97.
$$r = \frac{2}{1 + \sin\theta}$$
$$r(1 + \sin\theta) = 2$$
$$r + r\sin\theta = 2$$
$$r = 2 - r\sin\theta$$
$$\pm\sqrt{x^2 + y^2} = 2 - y$$
$$x^2 + y^2 = (2 - y)^2$$
$$x^2 + y^2 = 4 - 4y + y^2$$
$$x^2 + 4y - 4 = 0$$

99.
$$r = \frac{6}{2 - 3\sin\theta}$$
$$r(2 - 3\sin\theta) = 6$$
$$2r = 6 + 3r\sin\theta$$
$$2\left(\pm\sqrt{x^2 + y^2}\right) = 6 + 3y$$
$$4\left(x^2 + y^2\right) = (6 + 3y)^2$$
$$4x^2 + 4y^2 = 36 + 36y + 9y^2$$
$$4x^2 - 5y^2 - 36y - 36 = 0$$

101. The graph of the polar equation consists of all points that are six units from the pole.

$$r = 6$$
$$r^2 = 36$$
$$x^2 + y^2 = 36$$

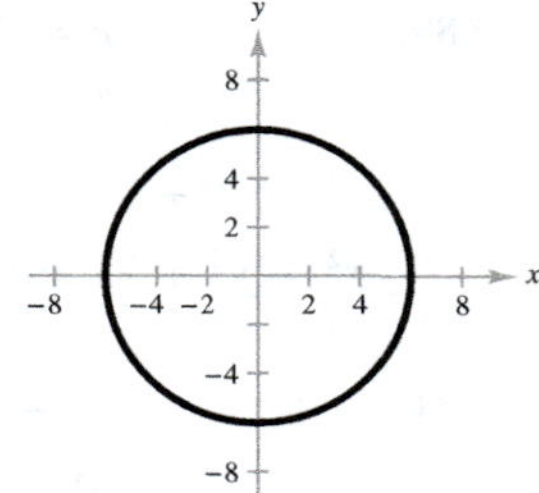

103. The graph of the polar equation consists of all points that make an angle of $\pi/6$ with the polar axis.

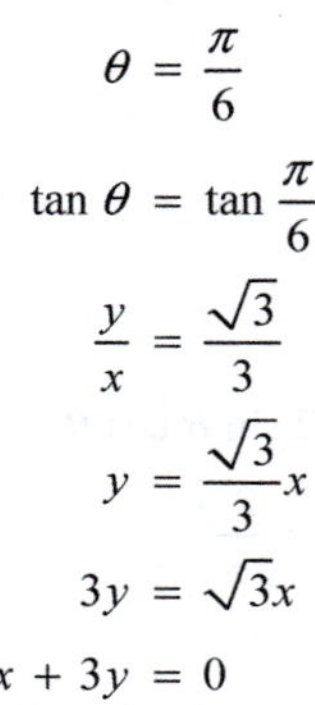

$$\theta = \frac{\pi}{6}$$
$$\tan\theta = \tan\frac{\pi}{6}$$
$$\frac{y}{x} = \frac{\sqrt{3}}{3}$$
$$y = \frac{\sqrt{3}}{3}x$$
$$3y = \sqrt{3}x$$
$$-\sqrt{3}x + 3y = 0$$

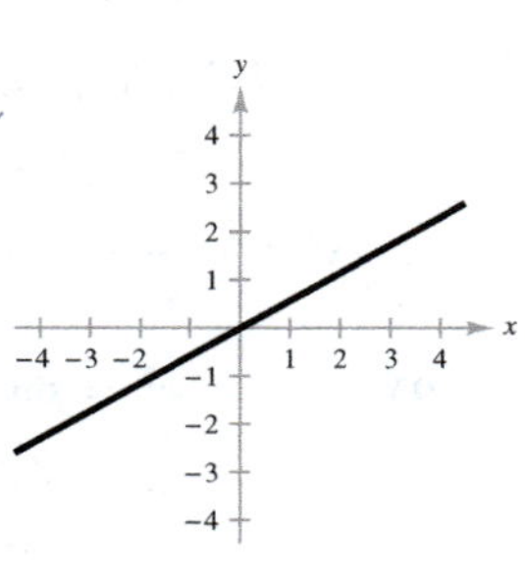

105. The graph of the polar equation is not evident by simple inspection. Convert to rectangular form first.

$$r = 3\sec\theta$$
$$r\cos\theta = 3$$
$$x = 3$$
$$x - 3 = 0$$

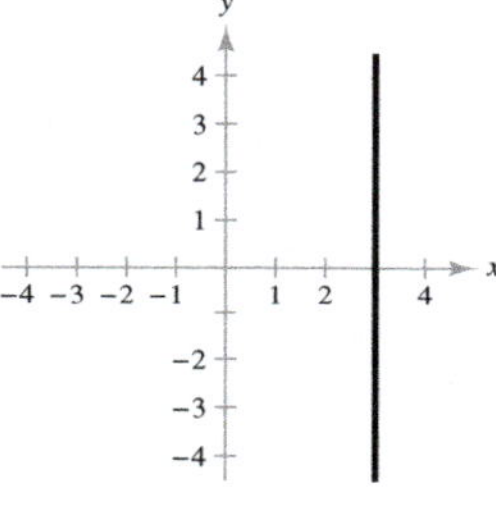

107. The graph of the polar equation consists of all points on the circle with radius 1 and center $(0, 1)$.

$$r = 2\sin\theta$$
$$r^2 = 2r\sin\theta$$
$$x^2 + y^2 = 2y$$
$$x^2 + y^2 - 2y = 0$$
$$x^2 + y^2 - 2y + 1 = 1$$
$$x^2 + (y - 1)^2 = 1$$

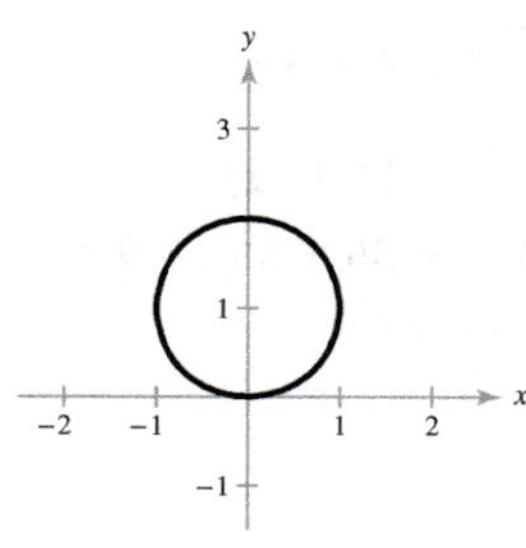

109. (a)

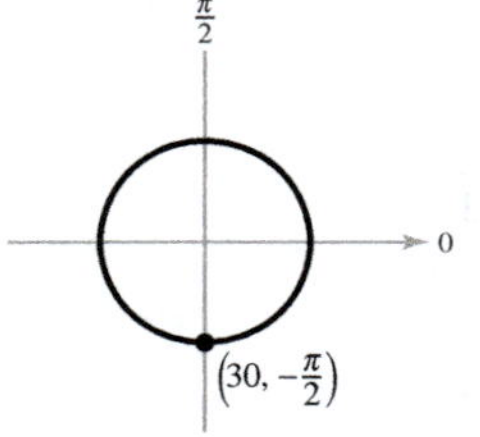

Since the passengers enter a car at the point $(r, \theta) = \left(30, -\frac{\pi}{2}\right)$, $r = 30$ is the polar equation for the model.

(b) Since it takes 45 seconds for the Ferris wheel to complete one revolution clockwise, after 15 seconds a passenger car makes one-third of one revolution or an angle of $\frac{2\pi}{3}$ radians.

Because $\theta = -\frac{\pi}{2} - \frac{2\pi}{3} = -\frac{7\pi}{6}$, the passenger car is at $\left(30, -\frac{7\pi}{6}\right) = \left(30, \frac{5\pi}{6}\right)$.

(c) Polar coordinates: $\left(30, \frac{5\pi}{6}\right)$

$$x = r\cos\theta = 30\cos\frac{5\pi}{6} = 15\sqrt{3} \approx 25.98$$
$$y = r\sin\theta = 30\sin\frac{5\pi}{6} = 15$$

Rectangular coordinates: (25.98, 15)

The car is about 25.98 feet to the left of the center and 15 feet above the center.

111. True. Because r is a directed distance, then the point (r, θ) can be represented as $(r, \theta \pm 2n\pi)$.

113. Rectangular coordinates: $\left(1, -\sqrt{3}\right)$

$r = \sqrt{1 + 3} = \sqrt{4} = 2$, $\tan\theta = \frac{-\sqrt{3}}{1}$, $\theta = -\frac{\pi}{3}$,

the point lies in Quadrant IV.

Polar coordinates: $(r, \theta)\left(2, -\frac{\pi}{3}\right)$ or $\left(2, \frac{5\pi}{3}\right)$

115. (a)

$$r = 2(h\cos\theta + k\sin\theta)$$
$$r = 2\left(h\left(\frac{x}{r}\right) + k\left(\frac{y}{r}\right)\right)$$
$$r = \frac{2hx + 2ky}{r}$$
$$r^2 = 2hx + 2ky$$
$$x^2 + y^2 = 2hx + 2ky$$
$$x^2 - 2hx + y^2 - 2ky = 0$$
$$\left(x^2 - 2hx + h^2\right) + \left(y^2 - 2ky + k^2\right) = h^2 + k^2$$
$$(x - h)^2 + (y - k)^2 = h^2 + k^2$$

Center: (h, k)

Radius: $\sqrt{h^2 + k^2}$

(b) $r = \cos\theta + 3\sin\theta = 2\left(\frac{1}{2}\cos\theta + \frac{3}{2}\sin\theta\right)$

$$h = \frac{1}{2}, k = \frac{3}{2}$$

$$\left(x - \frac{1}{2}\right)^2 + \left(y - \frac{3}{2}\right)^2 = \left(\frac{1}{2}\right)^2 + \left(\frac{3}{2}\right)^2$$
$$\left(x - \frac{1}{2}\right)^2 + \left(y - \frac{3}{2}\right)^2 = \frac{5}{2}$$

Center: $\left(\frac{1}{2}, \frac{3}{2}\right)$

Radius: $r = \sqrt{\frac{5}{2}} = \frac{\sqrt{10}}{2}$

Section 10.8 Graphs of Polar Equations

1. $\theta = \frac{\pi}{2}$

3. convex limaçon

5. lemniscate

7. $r = 3\cos\theta$

Circle

9. $r = 3(1 - 2\cos\theta)$

Limaçon with inner loop

11. $r = 6\cos 2\theta$

Rose curve with 4 petals

13. $r = 6 + 3\cos\theta$

$\theta = \frac{\pi}{2}$: $-r = 6 + 3\cos(-\theta)$

$-r = 6 + 3\cos\theta$

Not an equivalent equation

Polar axis: $r = 6 + 3\cos(-\theta)$

$r = 6 + 3\cos\theta$

Equivalent equation

Pole: $-r = 6 + 3\cos\theta$

Not an equivalent equation

Answer: Symmetric with respect to polar axis.

15. $r = \dfrac{2}{1 + \sin\theta}$

$\theta = \dfrac{\pi}{2}$: $\quad r = \dfrac{2}{1 + \sin(\pi - \theta)}$

$r = \dfrac{2}{1 + \sin\pi\cos\theta - \cos\pi\sin\theta}$

$r = \dfrac{2}{1 + \sin\theta}$

Equivalent equation

Polar axis: $\quad r = \dfrac{2}{1 + \sin(-\theta)}$

$r = \dfrac{2}{1 - \sin\theta}$

Not an equivalent equation

Pole: $\quad -r = \dfrac{2}{1 + \sin\theta}$

Answer: Symmetric with respect to $\theta = \pi/2$

17. $r^2 = 36\cos 2\theta$

$\theta = \dfrac{\pi}{2}$: $\quad (-r)^2 = 36\cos 2(-\theta)$

$r^2 = 36\cos 2\theta$

Equivalent equation

Polar axis: $\quad r^2 = 36\cos 2(-\theta)$

$r^2 = 36\cos 2\theta$

Equivalent equation

Pole: $\quad (-r)^2 = 36\cos 2\theta$

$r^2 = 36\cos 2\theta$

Equivalent equation

Answer: Symmetric with respect to $\theta = \dfrac{\pi}{2}$, the polar axis, and the pole

19. $|r| = |10 - 10\sin\theta| = 10|1 - \sin\theta| \le 10(2) = 20$

$|1 - \sin\theta| = 2$

$1 - \sin\theta = 2$ or $1 - \sin\theta = -2$

$\sin\theta = -1$ $\qquad \sin\theta = 3$

$\theta = \dfrac{3\pi}{2}$ $\qquad$ Not possible

Maximum: $|r| = 20$ when $\theta = \dfrac{3\pi}{2}$

$0 = 10(1 - \sin\theta)$

$\sin\theta = 1$

$\theta = \dfrac{\pi}{2}$

Zero: $r = 0$ when $\theta = \dfrac{\pi}{2}$

21. $|r| = |4\cos 3\theta| = 4|\cos 3\theta| \le 4$

$|\cos 3\theta| = 1$

$\cos 3\theta = \pm 1$

$\theta = 0, \dfrac{\pi}{3}, \dfrac{2\pi}{3}$

Maximum: $|r| = 4$ when $\theta = 0, \dfrac{\pi}{3}, \dfrac{2\pi}{3}$

$0 = 4\cos 3\theta$

$\cos 3\theta = 0$

$\theta = \dfrac{\pi}{6}, \dfrac{\pi}{2}, \dfrac{5\pi}{6}$

Zero: $r = 0$ when $\theta = \dfrac{\pi}{6}, \dfrac{\pi}{2}, \dfrac{5\pi}{6}$

23. $r = 5$

Symmetric with respect to $\theta = \dfrac{\pi}{2}$, polar axis, pole

Circle with radius 5

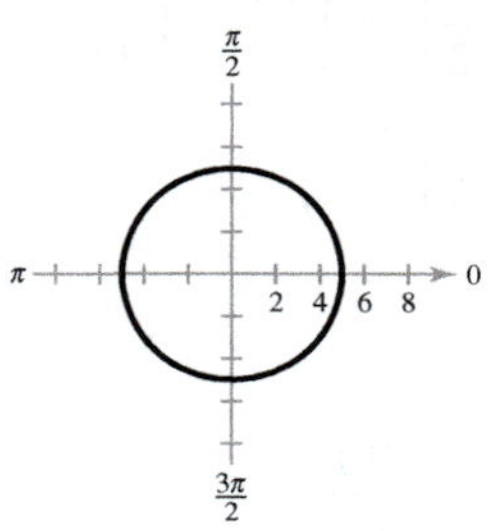

25. $r = \dfrac{\pi}{4}$

Symmetric with respect to $\theta = \dfrac{\pi}{2}$, polar axis, pole

Circle with radius $\dfrac{\pi}{4}$

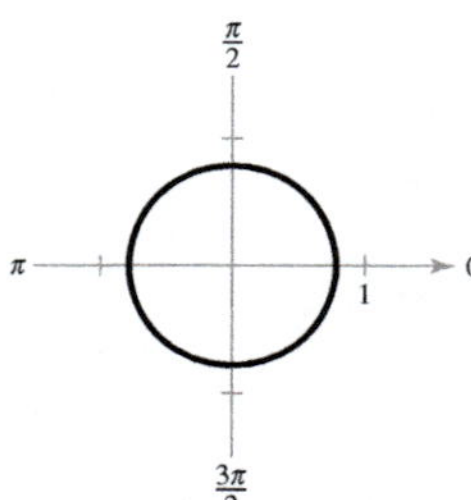

27. $r = 3\sin\theta$

Symmetric with respect to $\theta = \dfrac{\pi}{2}$

Circle with radius $\dfrac{3}{2}$

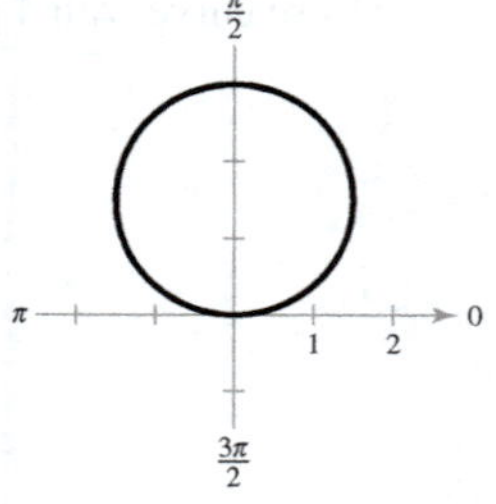

29. $r = 3(1 - \cos\theta)$

Symmetric with respect to the polar axis

$\frac{a}{b} = \frac{3}{3} = 1 \Rightarrow$ Cardioid

$|r| = 6$ when $\theta = \pi$

$r = 0$ when $\theta = 0$

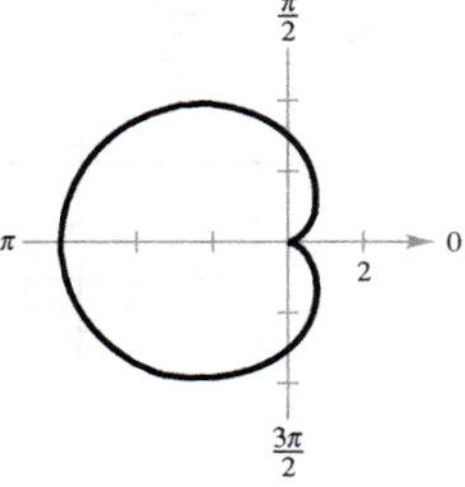

31. $r = 4(1 + \sin\theta)$

Symmetric with respect to $\theta = \frac{\pi}{2}$

$\frac{a}{b} = \frac{4}{4} = 1 \Rightarrow$ Cardioid

$|r| = 8$ when $\theta = \frac{\pi}{2}$

$r = 0$ when $\theta = \frac{3\pi}{2}$

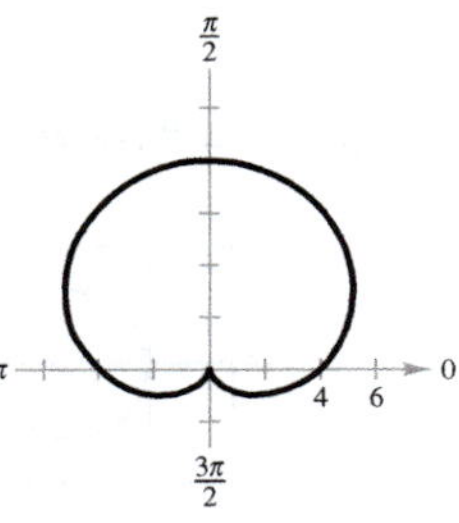

33. $r = 5 + 2\cos\theta$

Symmetric with respect to the polar axis

$\frac{a}{b} = \frac{5}{2} > 1 \Rightarrow$ Dimpled limaçon

$|r| = 7$ when $\theta = 0$

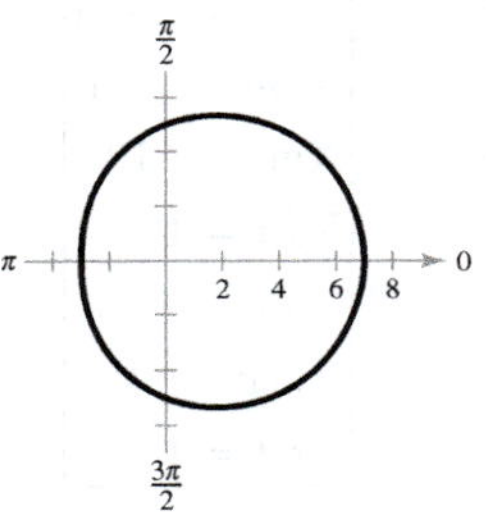

35. $r = 1 - 3\sin\theta$

Symmetric with respect to $\theta = \frac{\pi}{2}$

$\frac{a}{b} = \frac{1}{3} < 1 \Rightarrow$ Limaçon with inner loop

$|r| = 4$ when $\theta = \frac{3\pi}{2}$

$r = 0$ when $\theta = 0.3398, 2.802$

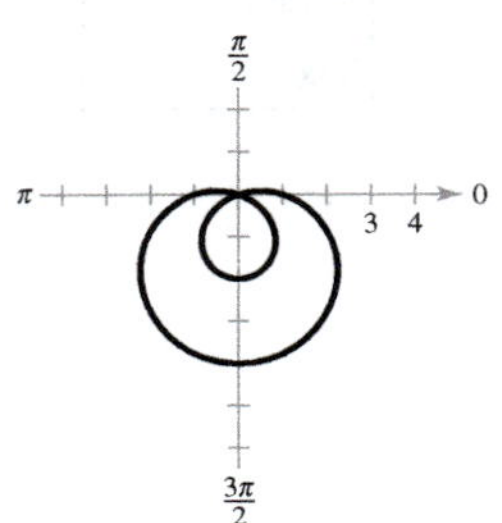

37. $r = 3 - 6\cos\theta$

Symmetric with respect to the polar axis

$\frac{a}{b} = \frac{1}{2} < 1 \Rightarrow$ Limaçon with inner loop

$|r| = 9$ when $\theta = \pi$

$r = 0$ when $\theta = \frac{\pi}{3}, \frac{5\pi}{3}$

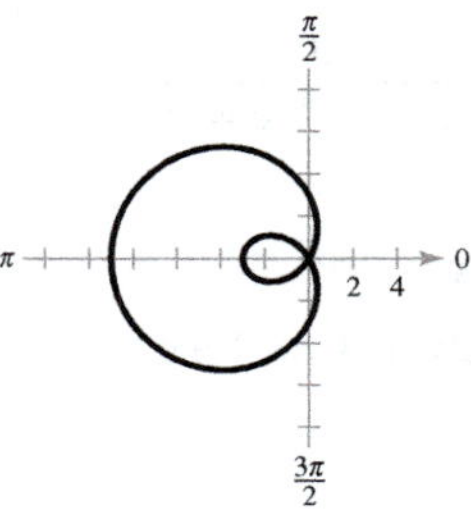

39. $r = 5\sin 2\theta$

Symmetric with respect to $\theta = \pi/2$, the polar axis, and the pole

Rose curve $(n = 2)$ with 4 petals

$|r| = 5$ when $\theta = \frac{\pi}{4}, \frac{3\pi}{4}, \frac{5\pi}{4}, \frac{7\pi}{4}$

$r = 0$ when $\theta = 0, \frac{\pi}{2}, \pi$

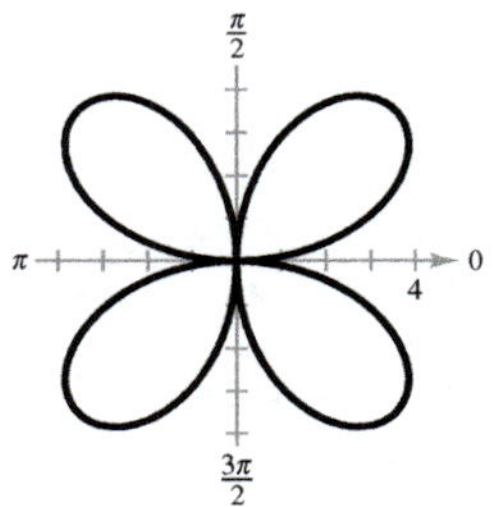

41. $r = 6\cos 3\theta$

Symmetric with respect to polar axis

Rose curve $(n = 3)$ with three petals

$|r| = 6$ when $\theta = 0, \frac{\pi}{3}, \frac{2\pi}{3}, \pi$

$r = 0$ when $\theta = \frac{\pi}{6}, \frac{\pi}{2}, \frac{5\pi}{6}$

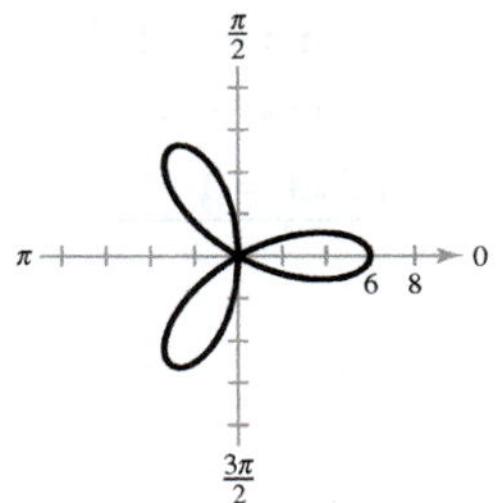

43. $r = 2\sec\theta$

$r = \dfrac{2}{\cos\theta}$

$r\cos\theta = 2$

$x = 2 \Rightarrow$ Line

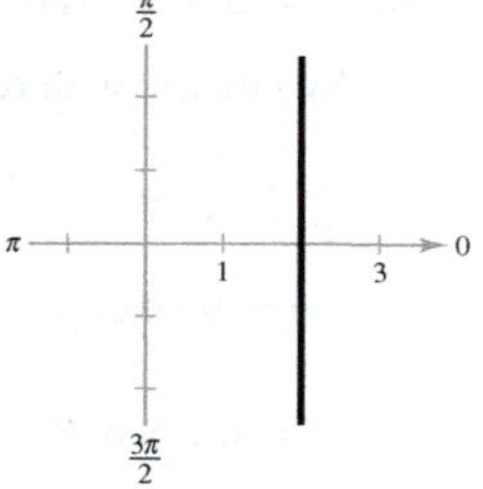

45. $r = \dfrac{3}{\sin\theta - 2\cos\theta}$

$r(\sin\theta - 2\cos\theta) = 3$

$y - 2x = 3$

$y = 2x + 3 \Rightarrow$ Line

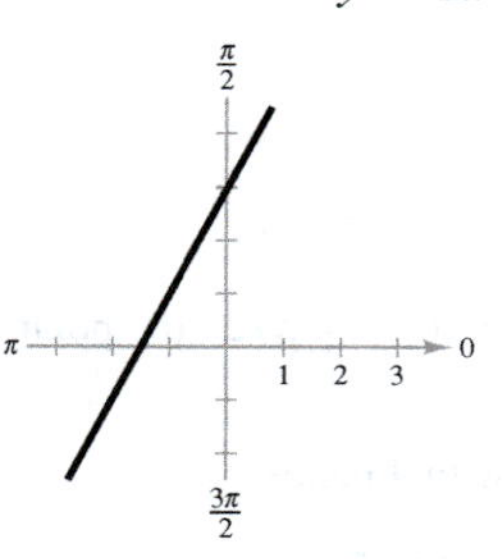

47. $r^2 = 9\cos 2\theta$

Symmetric with respect to the polar axis, $\theta = \pi/2$, and the pole

Lemniscate

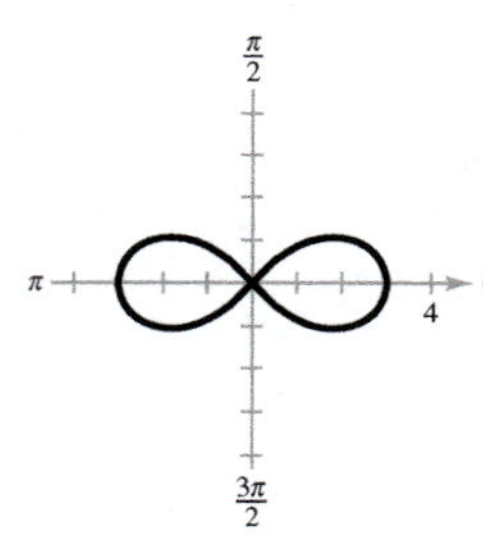

49. $r = \dfrac{9}{4}$

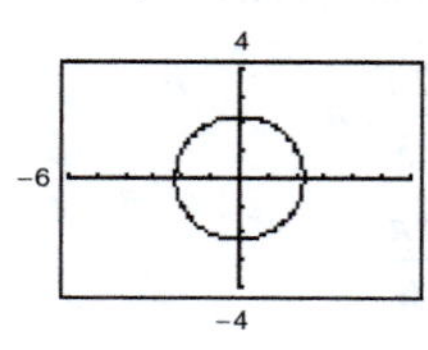

$0 \le \theta \le 2\pi$

θmin = 0
θmax = 2π
θstep = $\pi/24$
Xmin = –6
Xmax = 6
Xscl = 1
Ymin = –4
Ymax = 4
Yscl = 1

51. $r = \dfrac{5\pi}{8}$

$0 \le \theta \le 2\pi$

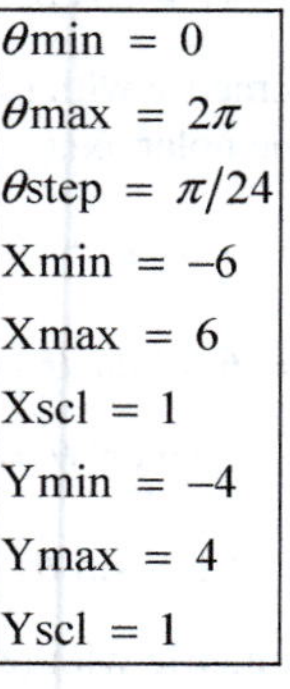

θmin = 0
θmax = 2π
θstep = $\pi/24$
Xmin = –6
Xmax = 6
Xscl = 1
Ymin = –4
Ymax = 4
Yscl = 1

53. $r = 8\cos\theta$

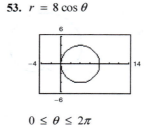

$0 \le \theta \le 2\pi$

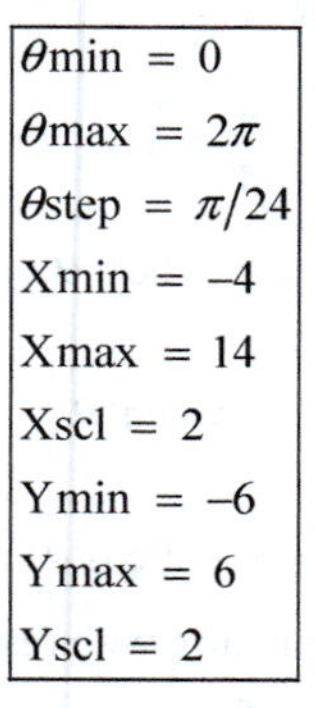

θmin = 0
θmax = 2π
θstep = $\pi/24$
Xmin = –4
Xmax = 14
Xscl = 2
Ymin = –6
Ymax = 6
Yscl = 2

55. $r = 3(2 - \sin\theta)$

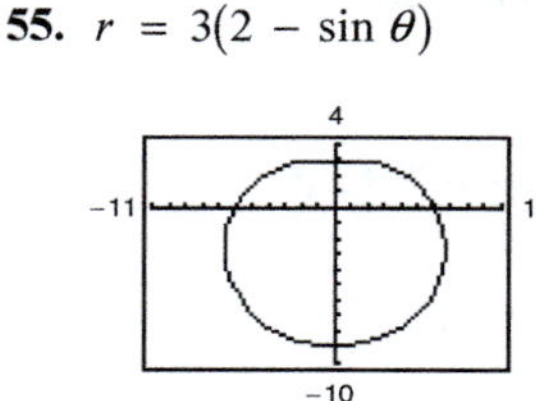

$0 \le \theta \le 2\pi$

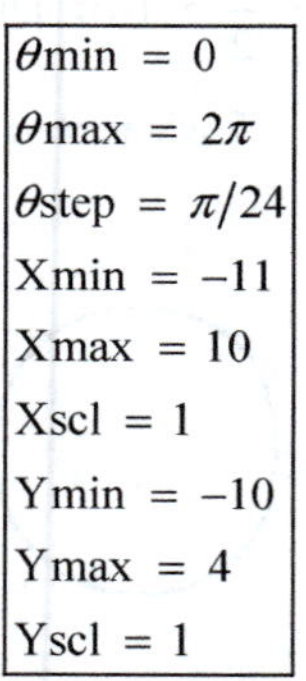

θmin = 0
θmax = 2π
θstep = $\pi/24$
Xmin = –11
Xmax = 10
Xscl = 1
Ymin = –10
Ymax = 4
Yscl = 1

57. $r = 8\sin\theta\cos^2\theta$

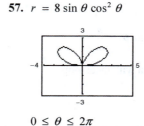

$0 \le \theta \le 2\pi$

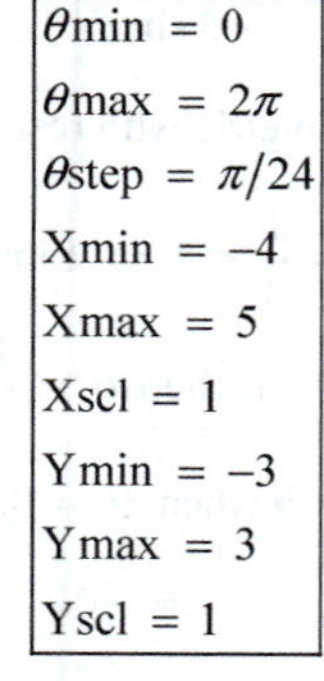

θmin = 0
θmax = 2π
θstep = $\pi/24$
Xmin = –4
Xmax = 5
Xscl = 1
Ymin = –3
Ymax = 3
Yscl = 1

59. $r = 3 - 8\cos\theta$

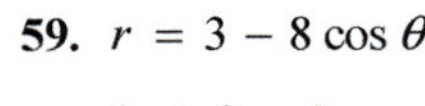

$0 \le \theta < 2\pi$

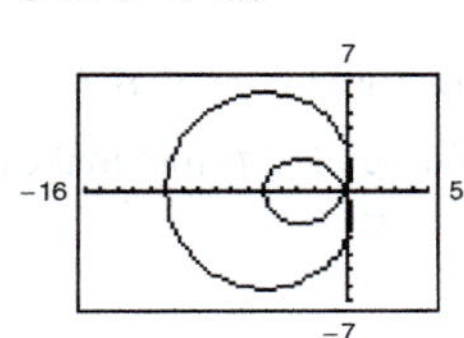

61. $r = 2\cos\left(\dfrac{3\theta}{2}\right)$

$0 \le \theta < 4\pi$

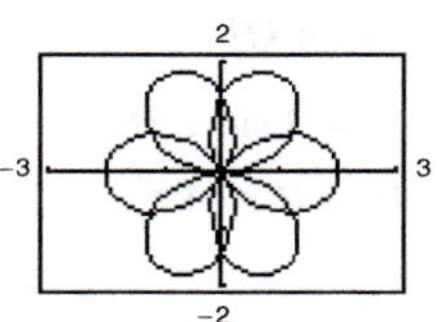

63. $r^2 = 9\sin 2\theta$

$0 \le \theta < \pi$

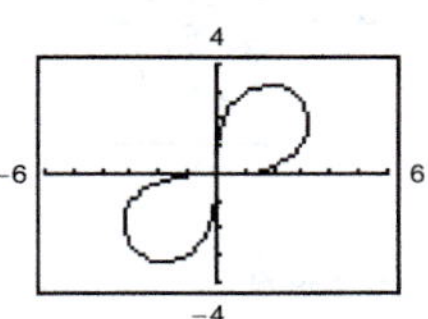

65.

$$\begin{aligned}
r &= 2 - \sec\theta = 2 - \frac{1}{\cos\theta}\\
r\cos\theta &= 2\cos\theta - 1\\
r(r\cos\theta) &= 2r\cos\theta - r\\
\left(\pm\sqrt{x^2 + y^2}\right)x &= 2x - \left(\pm\sqrt{x^2 + y^2}\right)\\
\left(\pm\sqrt{x^2 + y^2}\right)(x + 1) &= 2x\\
\left(\pm\sqrt{x^2 + y^2}\right) &= \frac{2x}{x + 1}\\
x^2 + y^2 &= \frac{4x^2}{(x + 1)^2}\\
y^2 &= \frac{4x^2}{(x + 1)^2} - x^2 = \frac{4x^2 - x^2(x + 1)^2}{(x + 1)^2} = \frac{4x^2 - x^2(x^2 + 2x + 1)}{(x + 1)^2}\\
&= \frac{-x^4 - 2x^3 + 3x^2}{(x + 1)^2} = \frac{-x^2(x^2 + 2x - 3)}{(x + 1)^2}\\
y &= \pm\sqrt{\frac{x^2(3 - 2x - x^2)}{(x + 1)^2}} = \pm\left|\frac{x}{x + 1}\right|\sqrt{3 - 2x - x^2}
\end{aligned}$$

The graph has an asymptote at $x = -1$.

67. $r = \dfrac{3}{\theta}$

$\theta = \dfrac{3}{r} = \dfrac{3\sin\theta}{r\sin\theta} = \dfrac{3\sin\theta}{y}$

$y = \dfrac{3\sin\theta}{\theta}$

As $\theta \to 0$, $y \to 3$

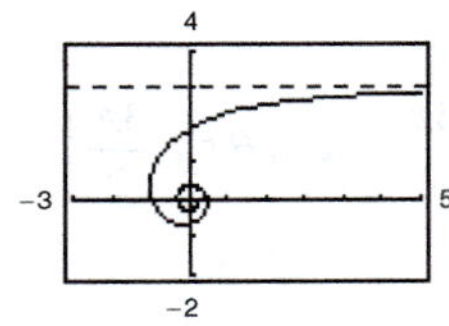

69. (a)

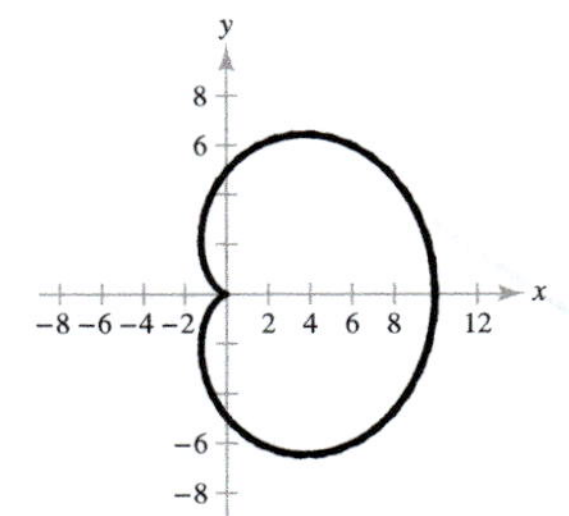

The graph is a cardioid.

(b) Since $|r|$ is at a maximum when $r = 10$ at $\theta = 0$ radians, the microphone is most sensitive to sound when $\theta = 0$.

71. True. The equation is of the form $r = a\sin n\theta$, where n is odd, so it has five petals.

73. $r = 3 \sin k\theta$

(a) $r = 3 \sin 1.5\theta$

$0 \le \theta < 4\pi$

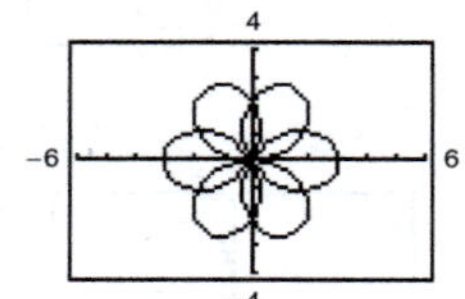

(b) $r = 3 \sin 2.5\theta$

$0 \le \theta < 4\pi$

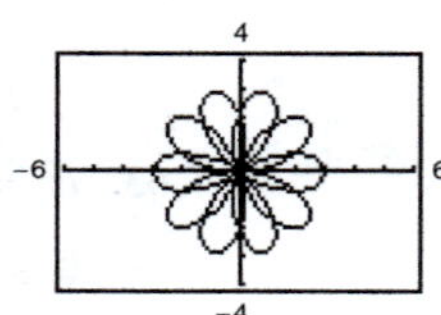

(c) Yes. $r = 3 \sin(k\theta)$.

Find the minimum value of θ, $(\theta > 0)$, that is a multiple of 2π that makes $k\theta$ a multiple of 2π.

75. $r = 10 \cos \theta$

(a) $0 \le \theta \le \dfrac{\pi}{2}$

Upper half of circle

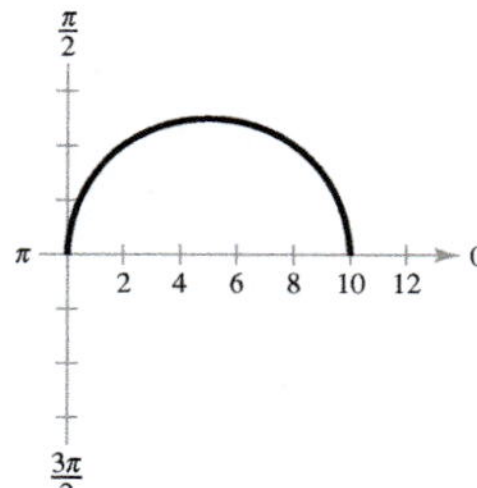

(b) $\dfrac{\pi}{2} \le \theta \le \pi$

Lower half of circle

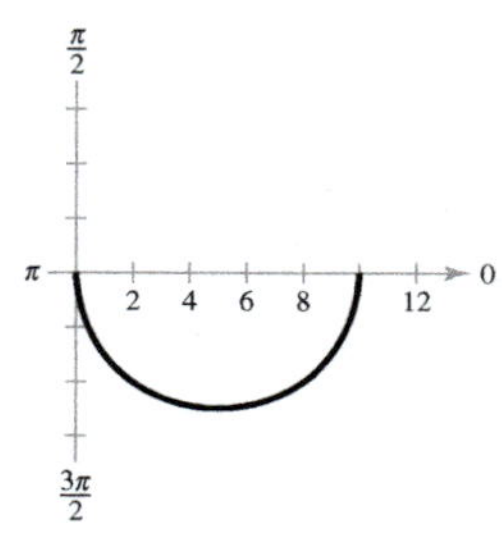

(c) $-\dfrac{\pi}{2} \le \theta \le \dfrac{\pi}{2}$

Entire circle

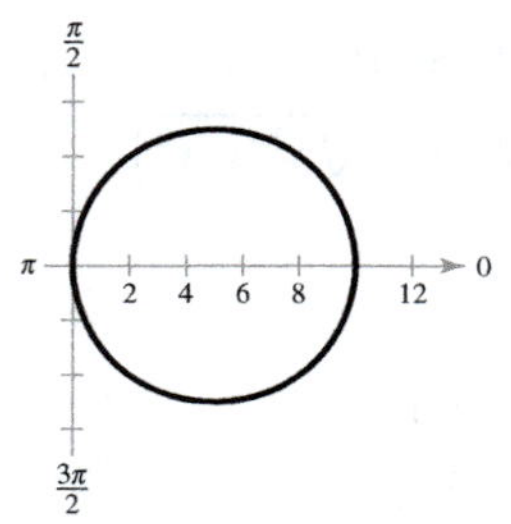

(d) $\dfrac{\pi}{4} \le \theta \le \dfrac{3\pi}{4}$

Left half of circle

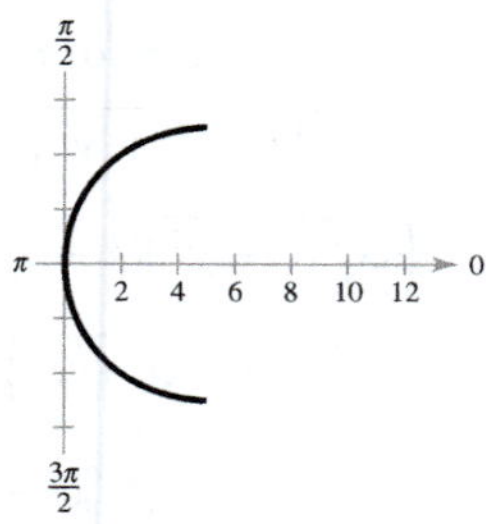

77. Let the curve $r = f(\theta)$ be rotated by ϕ to form the curve $r = g(\theta)$.

If (r_1, θ_1) is a point on $r = f(\theta)$, then $(r_1, \theta_1 + \phi)$ is on $r = g(\theta)$.

That is, $g(\theta_1 + \phi) = r_1 = f(\theta_1)$. Letting $\theta = \theta_1 + \phi$, or $\theta_1 = \theta - \phi$, you see that

$g(\theta) = g(\theta_1 + \phi) = f(\theta_1) = f(\theta - \phi)$.

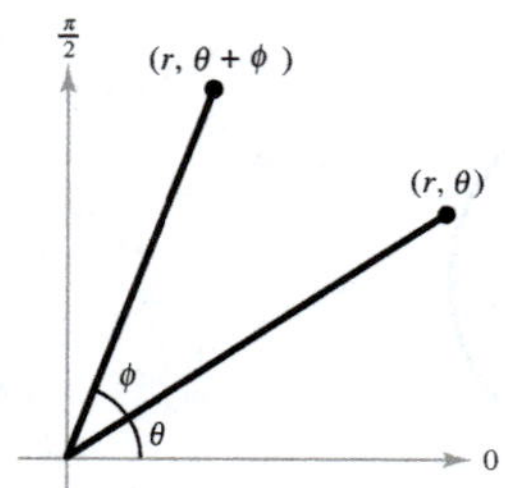

79. (a) $r = 2 - \sin\left(\theta - \dfrac{\pi}{4}\right)$

$= 2 - \left[\sin\theta\cos\dfrac{\pi}{4} - \cos\theta\sin\dfrac{\pi}{4}\right]$

$= 2 - \dfrac{\sqrt{2}}{2}(\sin\theta - \cos\theta)$

(b) $r = 2 - \sin\left(\theta - \dfrac{\pi}{2}\right)$

$= 2 - \left[\sin\theta\cos\dfrac{\pi}{2} - \cos\theta\sin\dfrac{\pi}{2}\right]$

$= 2 + \cos\theta$

(c) $r = 2 - \sin(\theta - \pi)$

$= 2 - [\sin\theta\cos\pi - \cos\theta\sin\pi]$

$= 2 + \sin\theta$

(d) $r = 2 - \sin\left(\theta - \dfrac{3\pi}{2}\right)$

$= 2 - \left[\sin\theta\cos\dfrac{3\pi}{2} - \cos\theta\sin\dfrac{3\pi}{2}\right]$

$= 2 - \cos\theta$

Section 10.9 Polar Equations of Conics

1. conic

3. vertical; left

5. $r = \dfrac{2e}{1 + e\cos\theta}$

$e = 1$: $r = \dfrac{2}{1 + \cos\theta} \Rightarrow$ parabola

$e = 0.5$: $r = \dfrac{1}{1 + 0.5\cos\theta} \Rightarrow$ ellipse

$e = 1.5$: $r = \dfrac{3}{1 + 1.5\cos\theta} \Rightarrow$ hyperbola

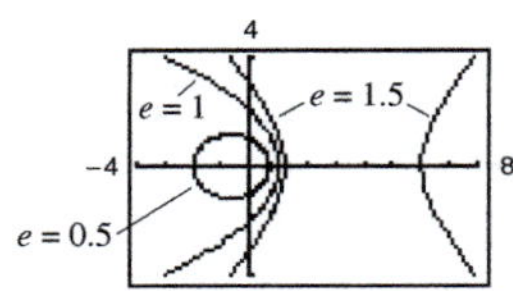

7. $r = \dfrac{2e}{1 - e\sin\theta}$

$e = 1$: $r = \dfrac{2}{1 - \sin\theta} \Rightarrow$ parabola

$e = 0.5$: $r = \dfrac{1}{1 - 0.5\sin\theta} \Rightarrow$ ellipse

$e = 1.5$: $r = \dfrac{3}{1 - 1.5\sin\theta} \Rightarrow$ hyperbola

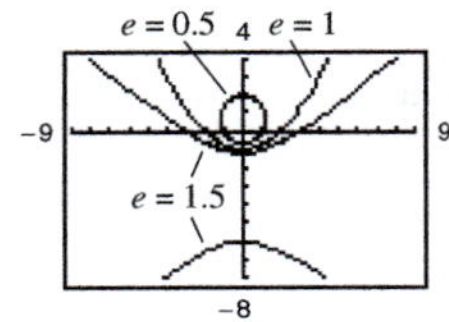

9. $r = \dfrac{4}{1 - \cos\theta}$

$e = 1 \Rightarrow$ Parabola

Vertical directrix to the left of the pole

Matches graph (c).

10. $r = \dfrac{3}{2 + \cos\theta}$

$e = \dfrac{1}{2} \Rightarrow$ Ellipse

Vertical directrix to the right of the pole

Matches graph (d).

11. $r = \dfrac{4}{1 + \sin\theta}$

$e = 1 \Rightarrow$ Parabola

Horizontal directrix above the pole

Matches graph (a).

12. $r = \dfrac{4}{1 - 3\sin\theta}$

$e = 3 \Rightarrow$ Hyperbola

Horizontal directrix below pole

Matches graph (b).

13. $r = \dfrac{3}{1 - \cos\theta}$

$e = 1 \Rightarrow$ Parabola

Vertex: $\left(\dfrac{3}{2}, \pi\right)$

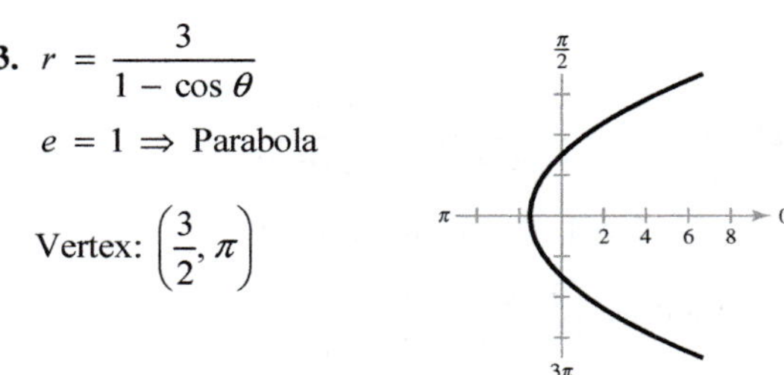

15. $r = \dfrac{5}{1 - \sin\theta}$

$e = 1$, the graph is a parabola.

Vertex: $\left(\dfrac{5}{2}, -\dfrac{\pi}{2}\right)$

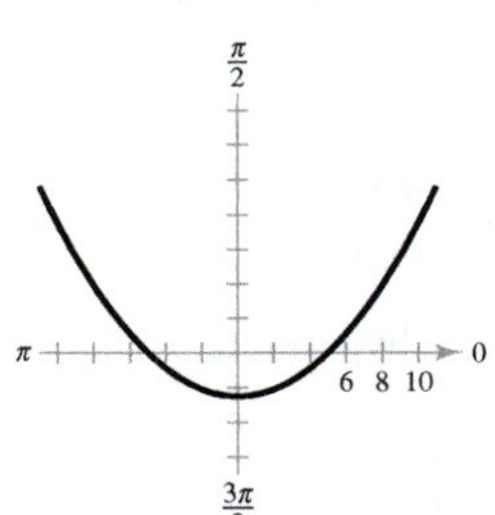

17. $r = \dfrac{2}{2 - \cos\theta} = \dfrac{1}{1 - (1/2)\cos\theta}$

$e = \dfrac{1}{2} < 1$, the graph is an ellipse.

Vertices: $(2, 0), \left(\dfrac{2}{3}, \pi\right)$

19. $r = \dfrac{6}{2 + \sin\theta} = \dfrac{3}{1 + (1/2)\sin\theta}$

$e = \dfrac{1}{2} < 1$, the graph is an ellipse.

Vertices: $\left(2, \dfrac{\pi}{2}\right), \left(6, \dfrac{3\pi}{2}\right)$

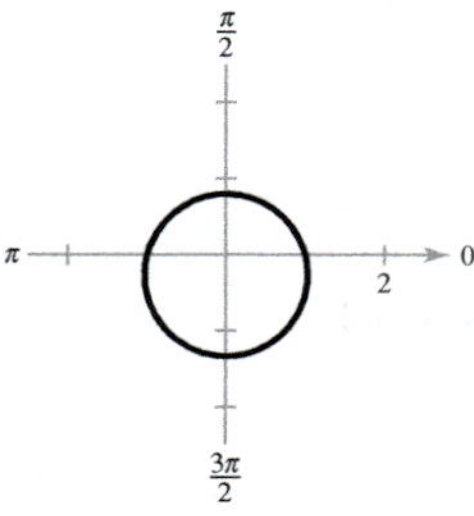

21. $r = \dfrac{3}{2 + 4\sin\theta} = \dfrac{3/2}{1 + 2\sin\theta}$

$e = 2 > 1$, the graph is a hyperbola.

Vertices: $\left(\dfrac{1}{2}, \dfrac{\pi}{2}\right), \left(-\dfrac{3}{2}, \dfrac{3\pi}{2}\right)$

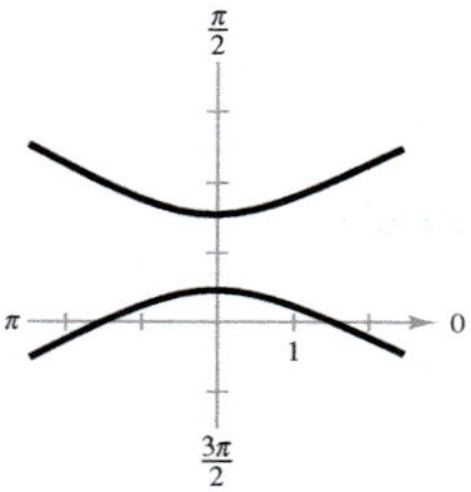

23. $r = \dfrac{3}{2 - 6\cos\theta} = \dfrac{3/2}{1 - 3\cos\theta}$

$e = 3 > 1$, the graph is a hyperbola.

Vertices: $\left(-\dfrac{3}{4}, 0\right), \left(\dfrac{3}{8}, \pi\right)$

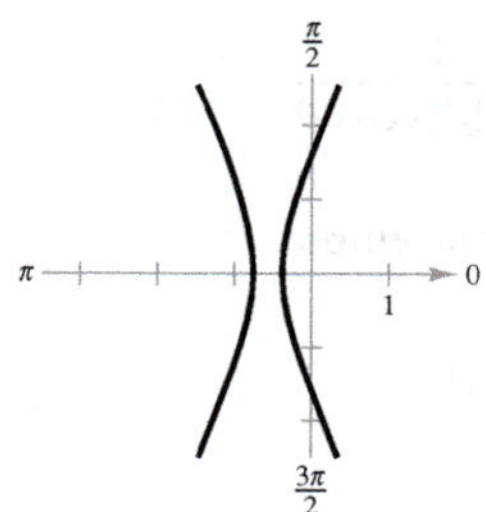

25. $r = \dfrac{-1}{1 - \sin\theta}$

$e = 1 \Rightarrow$ Parabola

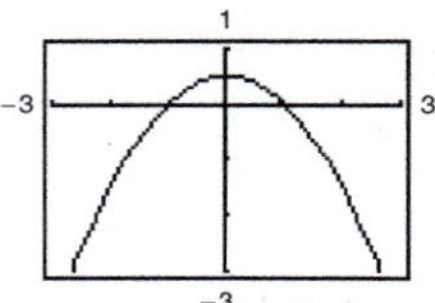

27. $r = \dfrac{3}{-4 + 2\cos\theta}$

$e = \dfrac{1}{2} \Rightarrow$ Ellipse

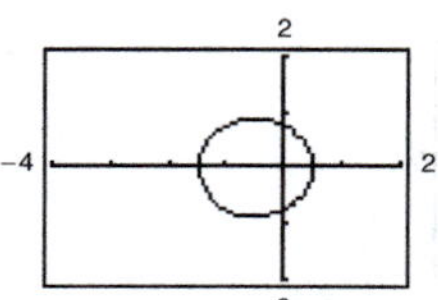

29.

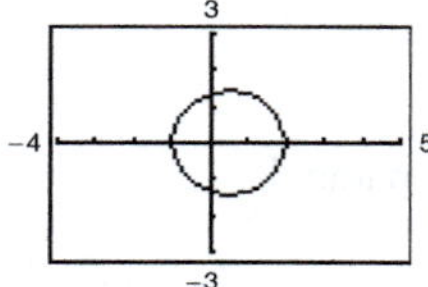

Ellipse

31. $r = \dfrac{14}{14 + 17\sin\theta} = \dfrac{1}{1 + (17/14)\sin\theta}$

$e = \dfrac{17}{14} > 1 \Rightarrow$ Hyperbola

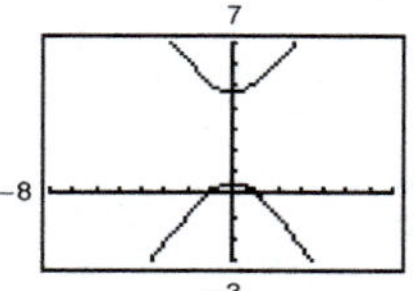

33. $r = \dfrac{3}{1 - \cos(\theta - \pi/4)}$

Rotate the graph in Exercise 13 through the angle $\pi/4$.

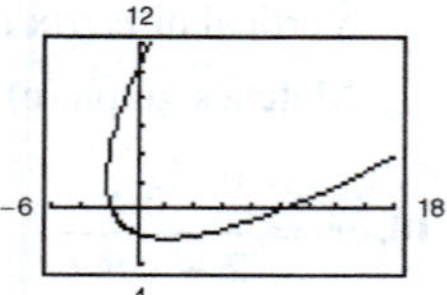

35. $r = \dfrac{6}{2 + \sin\left(\theta + \dfrac{\pi}{6}\right)}$

Rotate the graph in Exercise 19 through the angle $-\pi/6$.

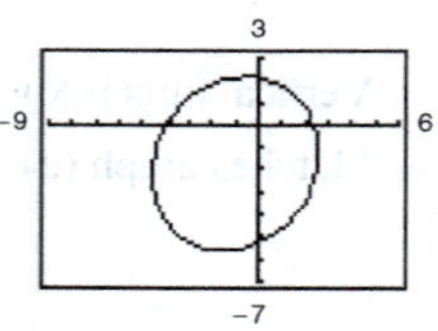

37. Parabola: $e = 1$

Directrix: $x = -1$

Vertical directrix to the left of the pole

$$r = \frac{1(1)}{1 - 1\cos\theta} = \frac{1}{1 - \cos\theta}$$

39. Ellipse: $e = \frac{1}{2}$

Directrix: $x = 3$

$p = 3$

Vertical directrix to the right of the pole

$$r = \frac{(1/2)(3)}{1 + (1/2)\cos\theta} = \frac{\frac{3}{2}}{1 + \frac{1}{2}\cos\theta} = \frac{3}{2 + \cos\theta}$$

41. Hyperbola: $e = 2$

Directrix: $x = 1$

$p = 1$

Vertical directrix to the right of the pole

$$r = \frac{2(1)}{1 + 2\cos\theta} = \frac{2}{1 + 2\cos\theta}$$

43. Parabola

Vertex: $(2, 0) \Rightarrow e = 1, p = 4$

Vertical directrix to the right of the pole

$$r = \frac{1(4)}{1 + 1\cos\theta} = \frac{4}{1 + \cos\theta}$$

45. Parabola

Vertex: $(5, \pi) \Rightarrow e = 1, p = 10$

Vertical directrix to the left of the pole

$$r = \frac{1(10)}{1 - 1\cos\theta} = \frac{10}{1 - \cos\theta}$$

47. Ellipse: Vertices $(2, 0), (10, \pi)$

Center: $(4, \pi)$; $c = 4, a = 6, e = \frac{2}{3}$

Vertical directrix to the right of the pole

$$r = \frac{(2/3)p}{1 + (2/3)\cos\theta} = \frac{2p}{3 + 2\cos\theta}$$

$$2 = \frac{2p}{3 + 2\cos 0}$$

$$p = 5$$

$$r = \frac{2(5)}{3 + 2\cos\theta} = \frac{10}{3 + 2\cos\theta}$$

49. Ellipse: Vertices $(20, 0), (4, \pi)$

Center: $(8, 0)$; $c = 8, a = 12, e = \frac{2}{3}$

Vertical directrix to the left of the pole

$$r = \frac{(2/3)p}{1 - (2/3)\cos\theta} = \frac{2p}{3 - 2\cos\theta}$$

$$20 = \frac{2p}{3 - 2\cos 0}$$

$$p = 10$$

$$r = \frac{2(10)}{3 - 2\cos\theta} = \frac{20}{3 - 2\cos\theta}$$

51. Hyperbola: Vertices $\left(1, \frac{3\pi}{2}\right), \left(9, \frac{3\pi}{2}\right)$

Center: $\left(5, \frac{3\pi}{2}\right)$; $c = 5, a = 4, e = \frac{5}{4}$

Horizontal directrix below the pole

$$r = \frac{(5/4)p}{1 - (5/4)\sin\theta} = \frac{5p}{4 - 5\sin\theta}$$

$$1 = \frac{5p}{4 - 5\sin(3\pi/2)}$$

$$p = \frac{9}{5}$$

$$r = \frac{5(9/5)}{4 - 5\sin\theta} = \frac{9}{4 - 5\sin\theta}$$

53. When $\theta = 0, r = c + a = ea + a = a(1 + e)$.

Therefore,

$$a(1 + e) = \frac{ep}{1 - e\cos 0}$$

$$a(1 + e)(1 - e) = ep$$

$$a(1 - e^2) = ep.$$

So, $r = \dfrac{ep}{1 - e\cos\theta} = \dfrac{(1 - e^2)a}{1 - e\cos\theta}$

55. Earth:

(a) $r = \dfrac{\left[1 - (0.0167)^2(9.2957 \times 10^7)\right]}{1 - 0.0167 \cos\theta} \approx \dfrac{9.2931 \times 10^7}{1 - 0.0167 \cos\theta}$

(b) Perihelion distance: $r = 9.2957 \times 10^7(1 - 0.0167) \approx 9.1405 \times 10^7$ miles

Aphelion distance: $r = 9.2957 \times 10^7(1 + 0.0167) \approx 9.4509 \times 10^7$ miles

57. Venus:

(a) $r = \dfrac{\left[1 - (0.0067)^2(1.0821 \times 10^8)\right]}{1 - 0.0067 \cos\theta} \approx \dfrac{1.0821 \times 10^8}{1 - 0.0067 \cos\theta}$

(b) Perihelion distance: $r = 1.0821 \times 10^8(1 - 0.0067) \approx 1.0748 \times 10^8$ kilometers

Aphelion distance: $r = 1.0821 \times 10^8(1 + 0.0067) \approx 1.0894 \times 10^8$ kilometers

59. Mars:

(a) $r = \dfrac{\left[1 - (0.0935)^2(1.4162 \times 10^8)\right]}{1 - 0.0935 \cos\theta} \approx \dfrac{1.4038 \times 10^8}{1 - 0.0935 \cos\theta}$

(b) Perihelion distance: $r = 1.4162 \times 10^8(1 - 0.0935) \approx 1.2838 \times 10^8$ miles

Aphelion distance: $r = 1.4162 \times 10^8(1 + 0.0935) \approx 1.5486 \times 10^8$ miles

61. $r = \dfrac{3}{2 + \sin\theta} = \dfrac{3/2}{1 + 1/2 \sin\theta}$

Because $e = \frac{1}{2}$, the equation represents an ellipse.

63. True. The graphs represent the same hyperbola, although the graphs are not traced out in the same order as θ goes from 0 to 2π.

65. True. The conic is an ellipse because the eccentricity is less than 1.

$e = \frac{2}{3} < 1$

67.

$$\frac{x^2}{a^2} + \frac{y^2}{b^2} = 1$$

$$\frac{r^2 \cos^2\theta}{a^2} + \frac{r^2 \sin^2\theta}{b^2} = 1$$

$$\frac{r^2 \cos^2\theta}{a^2} + \frac{r^2(1 - \cos^2\theta)}{b^2} = 1$$

$$r^2b^2 \cos^2\theta + r^2a^2 - r^2a^2 \cos^2\theta = a^2b^2$$

$$r^2(b^2 - a^2)\cos^2\theta + r^2a^2 = a^2b^2$$

Since $b^2 - a^2 = -c^2$, we have:

$$-r^2c^2 \cos^2\theta + r^2a^2 = a^2b^2$$

$$-r^2\left(\frac{c}{a}\right)^2 \cos^2\theta + r^2 = b^2, e = \frac{c}{a}$$

$$-r^2e^2 \cos^2\theta + r^2 = b^2$$

$$r^2(1 - e^2 \cos^2\theta) = b^2$$

$$r^2 = \frac{b^2}{1 - e^2 \cos^2\theta}$$

69. $\dfrac{x^2}{169} + \dfrac{y^2}{144} = 1$

$a = 13, b = 12, c = 5, e = \dfrac{5}{13}$

$r^2 = \dfrac{144}{1 - (25/169)\cos^2\theta} = \dfrac{24{,}336}{169 - 25\cos^2\theta}$

71. $\dfrac{x^2}{9} - \dfrac{y^2}{16} = 1$

$a = 3, b = 4, c = 5, e = \dfrac{5}{3}$

$r^2 = \dfrac{-16}{1 - (25/9)\cos^2\theta} = \dfrac{144}{25\cos^2\theta - 9}$

73. One focus: $(5, 0)$

Vertices: $(4, 0), (4, \pi)$

$a = 4, c = 5 \Rightarrow b = 3$ and $e = \dfrac{5}{4}$

$$\frac{x^2}{16} - \frac{y^2}{9} = 1$$

$r^2 = \dfrac{-9}{1 - (25/16)\cos^2\theta} = \dfrac{-144}{16 - 25\cos^2\theta} = \dfrac{144}{25\cos^2\theta}$

75. The graph of $r = \dfrac{5}{1 - \sin\theta}$ is a parabola with a horizontal directrix below the pole and opens upward.

(a) The graph of $r = \dfrac{5}{1 - \cos\theta}$ is a parabola with a vertical directrix to the left of the pole and opens to the right.

(b) The graph of $r = \dfrac{5}{1 + \sin\theta}$ is a parabola with a horizontal directrix above the pole and opens downward.

(c) The graph of $r = \dfrac{5}{1 + \cos\theta}$ is a parabola with a vertical directrix to the right of the pole and opens to the left.

(d) The graph of $r = \dfrac{5}{1 - \sin\left[\theta = (\pi/4)\right]}$ is the graph of $r = \dfrac{5}{1 - \sin\theta}$ rotated through the angle $\pi/4$.

77. $r = \dfrac{4}{1 - 0.4\cos\theta}$

(a) Because $e < 1$, the conic is an ellipse.

(b) $r = \dfrac{4}{1 + 0.4\cos\theta}$ has a vertical directrix to the right of the pole and $r = \dfrac{4}{1 - 0.4\sin\theta}$ has a horizontal directrix below the pole. The given polar equation, $r = \dfrac{4}{1 - 0.4\cos\theta}$, has a vertical directrix to the left of the pole.

(c)

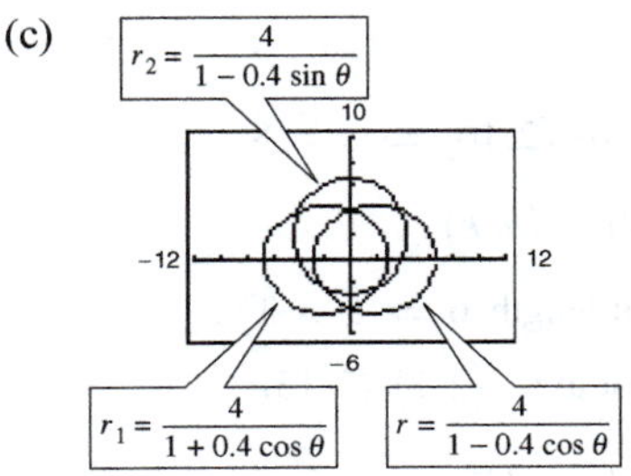

Review Exercises for Chapter 10

1. Points: (−1, 2) and (2, 5)

$$m = \frac{5 - 2}{2 - (-1)} = \frac{3}{3} = 1$$

$$\tan\theta = 1 \Rightarrow \theta = \frac{\pi}{4} \text{ radian} = 45^\circ$$

3. $5x + 2y + 4 = 0$

$$2y = -5x - 4$$

$$y = -\frac{5}{2}x - 2$$

$$m = -\frac{5}{2}$$

$$\tan\theta = -\frac{5}{2}$$

$$\theta = \arctan\left(-\frac{5}{2}\right)$$

$$\approx \pi - 1.1902$$

$$= 1.9513 \text{ radians or } 111.8^\circ$$

5. $4x + y = 2 \;\Rightarrow\; y = -4x + 2 \;\Rightarrow\; m_1 = -4$

$-5x + y = -1 \;\Rightarrow\; y = 5x - 1 \;\Rightarrow\; m_2 = 5$

$$\tan\theta = \left|\frac{5 - (-4)}{1 + (-4)(5)}\right| = \frac{9}{19}$$

$$\theta = \arctan\frac{9}{19} \approx 0.4424 \text{ radian} \approx 25.35^\circ$$

7. $2x - 7y = 8 \;\Rightarrow\; y = \frac{2}{7}x - \frac{8}{7} \;\Rightarrow\; m_1 = \frac{2}{7}$

$0.4x + y = 0 \;\Rightarrow\; y = -0.4x \;\Rightarrow\; m_2 = -0.4$

$$\tan\theta = \left|\frac{-0.4 - (2/7)}{1 + (2/7)(-0.4)}\right| = \frac{24}{31}$$

$$\theta = \arctan\left(\frac{24}{31}\right) \approx 0.6588 \text{ radian} \approx 37.7^\circ$$

9. $(4, 3) \Rightarrow x_1 = 4, y_1 = 3$

$2x - y - 1 = 0 \Rightarrow A = 2, B = -1, C = -1$

$$d = \frac{|(2)(4) + (-1)(3) + (-1)|}{\sqrt{(2)^2 + (-1)^2}} = \frac{4}{\sqrt{5}} = \frac{4\sqrt{5}}{5}$$

11. Hyperbola

13. Vertex: $(0, 0) = (h, k)$

Focus: $(0, 3) \Rightarrow p = 3$

$(x - h)^2 = 4p(y - k)$

$(x - 0)^2 = 4(3)(y - 0)$

$x^2 = 12y$

15. Vertex: $(0, 2) = (h, k)$

Focus: $x = -3 \Rightarrow p = 3$

$(y - k)^2 = 4p(x - h)$

$(y - 2)^2 = 12x$

17. $y = 2x^2 \Rightarrow x^2 = \frac{1}{2}y \Rightarrow p = \frac{1}{8}$

Focus: $\left(0, \frac{1}{8}\right)$

$d_1 = b + \frac{1}{8}$

$d_2 = \sqrt{(-1 - 0)^2 + \left(2 - \frac{1}{8}\right)^2}$

$= \sqrt{1 + \frac{225}{64}} = \frac{17}{8}$

$d_1 = d_2$

$b + \frac{1}{8} = \frac{17}{8}$

$b = 2$

slope $m = \frac{-2 - 2}{0 + 1} = -4.$

Point-slope: $y - 2 = -4(x + 1)$

Tangent line: $y = -4x - 2$

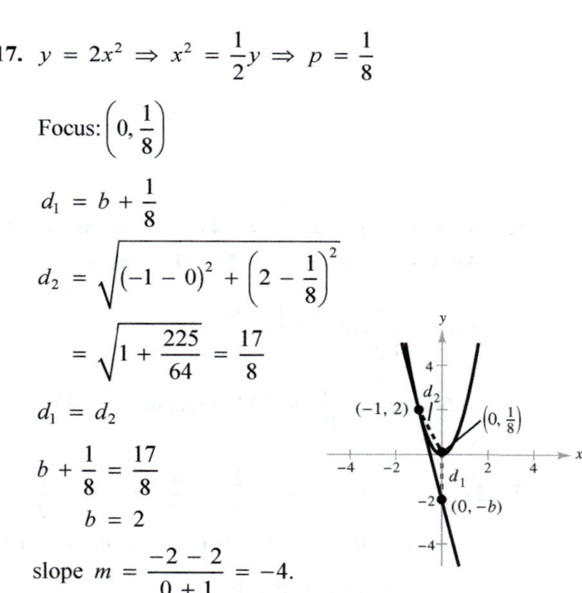

19. Parabola

Opens downward

Vertex: $(0, 10)$

$(x - h)^2 = 4p(y - k)$

$x^2 = 4p(y - 10)$

Solution points: $(\pm 3, 8)$

$9 = 4p(8 - 10)$

$9 = -8p$

$-\frac{9}{8} = p$

$x^2 = 4\left(-\frac{9}{8}\right)(y - 10)$

$x^2 = -\frac{9}{2}(y - 10)$

To find the x-intercepts, let $y = 0$.

$x^2 = 45$

$x = \pm\sqrt{45} = \pm 3\sqrt{5}$

At the base, the archway is $2\left(3\sqrt{5}\right) = 6\sqrt{5} \approx 13.4$ meters wide.

21. Vertices: $(2, 0), (2, 16) \Rightarrow a = 8$

Center: $(2, 8) = (h, k)$

Minor axis of length $6 \Rightarrow b = 3$

Ends of minor axis: $(2, 3), (2, 13)$

$$\frac{(x - h)^2}{b^2} + \frac{(y - k)^2}{a^2} = 1$$

$$\frac{(x - 2)^2}{9} + \frac{(y - 8)^2}{64} = 1$$

23. Vertices: $(0, 1), (4, 1) \Rightarrow a = 2, (h, k) = (2, 1)$

Endpoints of minor axis: $(2, 0), (2, 2) \Rightarrow b = 1$

$$\frac{(x - h)^2}{a^2} + \frac{(y - k)^2}{b^2} = 1$$

$$\frac{(x - 2)^2}{4} + (y - 1)^2 = 1$$

25. $2a = 10 \Rightarrow a = 5$

$b = 4$

$c^2 = a^2 - b^2 = 25 - 16 = 9 \Rightarrow c = 3$

The foci occur 3 feet from the center of the arch on a line connecting the tops of the pillars.

27. $\dfrac{(x+2)^2}{64} + \dfrac{(y-5)^2}{36} = 1$

$a = 8, b = 6$

$c^2 = a^2 - b^2 = 64 - 36 = 28 \Rightarrow c = 2\sqrt{7}$

Center: $(-2, 5)$

Vertices: $(-10, 5), (6, 5)$

Foci: $\left(-2 \pm 2\sqrt{7}, 5\right)$

Eccentricity: $e = \dfrac{2\sqrt{7}}{8} = \dfrac{\sqrt{7}}{4}$

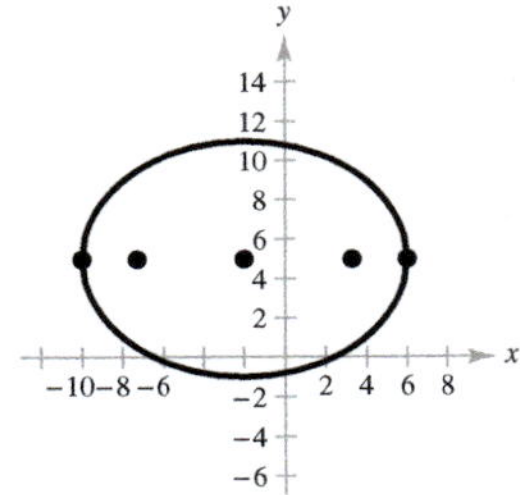

29. $16x^2 + 9y^2 - 32x + 72y + 16 = 0$

$16\left(x^2 - 2x + 1\right) + 9\left(y^2 + 8y + 16\right) = -16 + 16 + 144$

$16(x-1)^2 + 9(y+4)^2 = 144$

$\dfrac{(x-1)^2}{9} + \dfrac{(y+4)^2}{16} = 1$

$a = 4, b = 3, c = \sqrt{7}$

Center: $(1, -4)$

Vertices: $(1, 0)$, and $(1, -8)$

Foci: $\left(1, -4 \pm \sqrt{7}\right)$

Eccentricity: $e = \dfrac{\sqrt{7}}{4}$

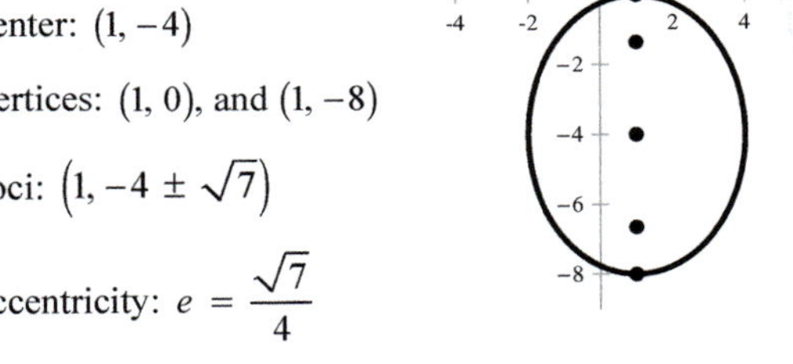

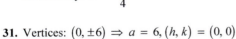

31. Vertices: $(0, \pm 6) \Rightarrow a = 6, (h, k) = (0, 0)$

Foci: $(0, \pm 8) \Rightarrow c = 8$

$b^2 = c^2 - a^2 = 64 - 36 = 28$

$\dfrac{(y-k)^2}{a^2} - \dfrac{(x-h)^2}{b^2} = 1$

$\dfrac{y^2}{36} - \dfrac{x^2}{28} = 1$

33. Foci: $(\pm 5, 0) \Rightarrow c = 5, (h, k) = (0, 0)$

Asymptotes:

$y = \pm\dfrac{3}{4}x \Rightarrow y = \pm\dfrac{b}{a}x \Rightarrow b = 3, a = 4$

$\dfrac{(x-h)^2}{a^2} - \dfrac{(y-k)^2}{b^2} = 1$

$\dfrac{x^2}{4^2} - \dfrac{y^2}{3^2} = 1 \Rightarrow \dfrac{x^2}{16} - \dfrac{y^2}{9} = 1$

35. $\dfrac{(x-4)^2}{49} - \dfrac{(y+2)^2}{25} = 1$

$a = 7, b = 5$

$c^2 = a^2 + b^2 = 49 + 25 = 74 \Rightarrow c = \sqrt{74}$

Center: $(4, -2)$

Vertices: $(11, -2), (-3, -2)$

Foci: $\left(4 \pm \sqrt{74}, -2\right)$

Asymptotes: $y = -2 \pm \dfrac{5}{7}(x - 4)$

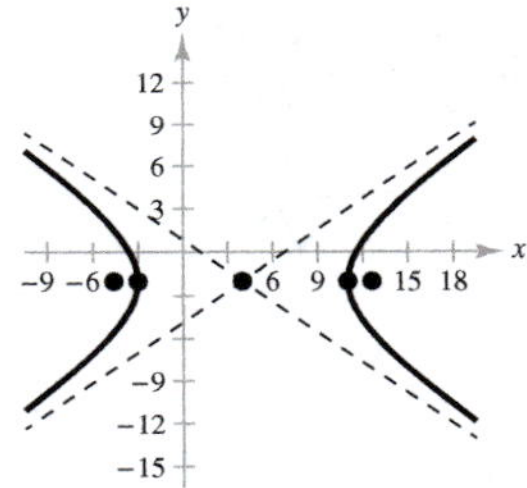

37. $9x^2 - 16y^2 - 18x - 32y - 151 = 0$

$9\left(x^2 - 2x + 1\right) - 16\left(y^2 + 2y + 1\right) = 151 + 9 - 16$

$9(x-1)^2 - 16(y+1)^2 = 144$

$\dfrac{(x-1)^2}{16} - \dfrac{(y+1)^2}{9} = 1$

$a = 4, b = 3, c = 5$

Center: $(1, -1)$

Vertices: $(5, -1)$ and $(-3, -1)$

Foci: $(6, -1)$ and $(-4, -1)$

Asymptotes: $y = -1 \pm \dfrac{3}{4}(x - 1)$

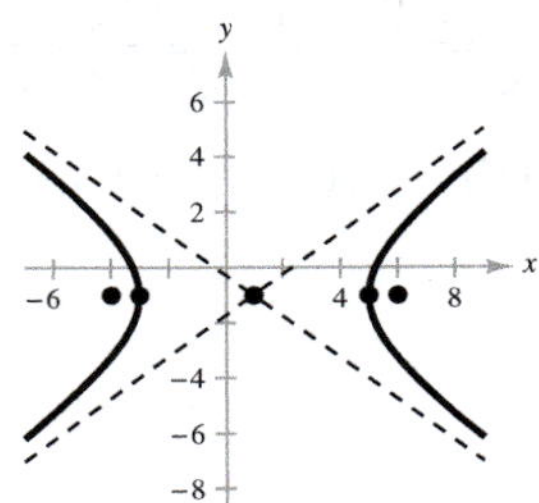

39. Since the microphones were two miles apart, 2 miles $\cdot$ 5280 $\Rightarrow c = 5280$. Since sound travels at 1100 feet per second, $|d_1 - d_2| = 2a = 6600 \Rightarrow a = 3300$.

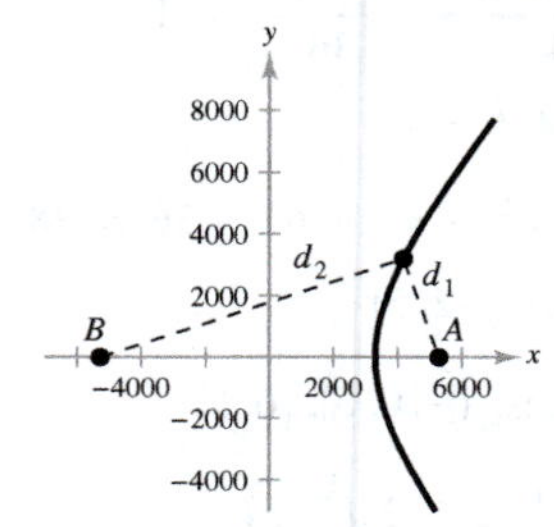

$b^2 = c^2 - a^2 = 5280^2 - 3300^2 = 27{,}878{,}400 - 10{,}890{,}000 = 16{,}988{,}400$

Since the explosion took place 6600 feet farther from B than from A. The locus of all points that are 6600 feet closer to A than to B is one branch of the hyperbola of the form $\dfrac{x^2}{a^2} - \dfrac{y^2}{b^2} = 1$, so it follows that

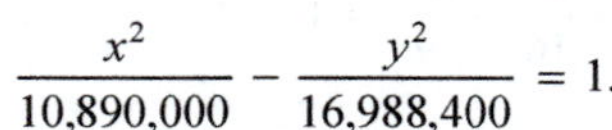

$$\frac{x^2}{10{,}890{,}000} - \frac{y^2}{16{,}988{,}400} = 1.$$

41. $5x^2 - 2y^2 + 10x - 4y + 17 = 0$

$AC = 5(-2) = -10 < 0$

Hyperbola

43. $3x^2 + 2y^2 - 12x + 12y + 29 = 0$

$A = 3, C = 2$

$AC = 3(2) = 6 > 0$

Ellipse

45. $xy + 5 = 0$

$A = C = 0, B = 1$

$B^2 - 4AC = 1^2 - 4(0)(0) = 1 > 0 \Rightarrow$ Hyperbola

$\cot 2\theta = \dfrac{A - C}{B} = \dfrac{0 - 0}{1} = 0 \Rightarrow 2\theta = \dfrac{\pi}{2} \Rightarrow \theta = \dfrac{\pi}{4}$

$x = x' \cos\dfrac{\pi}{4} - y' \sin\dfrac{\pi}{4} = \dfrac{x' - y'}{\sqrt{2}}$

$y = x' \sin\dfrac{\pi}{4} + y' \cos\dfrac{\pi}{4} = \dfrac{x' + y'}{\sqrt{2}}$

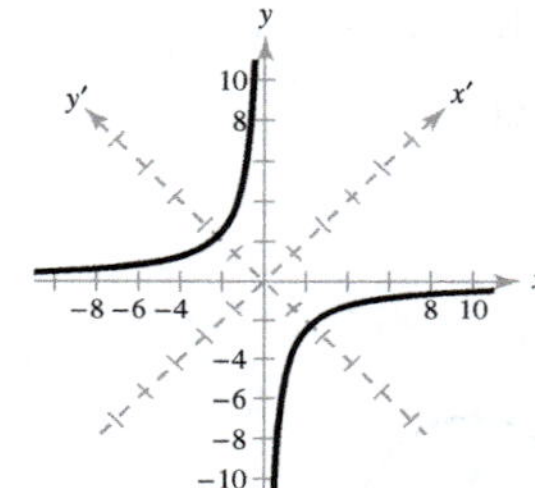

$$\left(\frac{x' - y'}{\sqrt{2}}\right)\left(\frac{x' + y'}{\sqrt{2}}\right) + 5 = 0$$

$$\frac{(x')^2 - (y')^2}{2} = -5$$

$$\frac{(y')^2}{10} - \frac{(x')^2}{10} = 1$$

47. $5x^2 - 2xy + 5y^2 - 12 = 0$

$A = C = 5, B = -2$

$B^2 - 4AC = (-2)^2 - 4(5)(5) = -96 < 0$

The graph is an ellipse.

$\cot 2\theta = 0 \Rightarrow 2\theta = \dfrac{\pi}{2} \Rightarrow \theta = \dfrac{\pi}{4}$

$x = x' \cos\dfrac{\pi}{4} - y' \sin\dfrac{\pi}{4} = \dfrac{x' - y'}{\sqrt{2}}$

$y = x' \sin\dfrac{\pi}{4} + y' \cos\dfrac{\pi}{4} = \dfrac{x' + y'}{\sqrt{2}}$

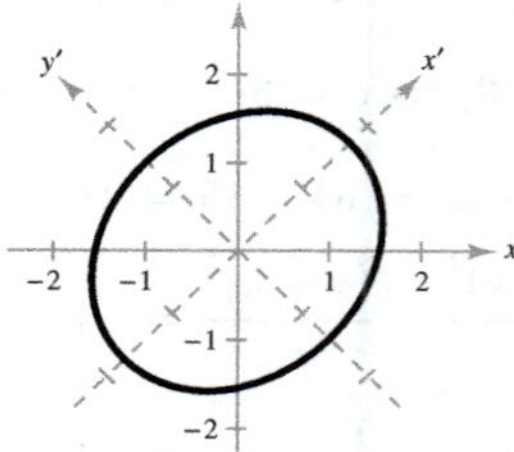

$$5\left(\frac{x' - y'}{\sqrt{2}}\right)^2 - 2\left(\frac{x' - y'}{\sqrt{2}}\right)\left(\frac{x' + y'}{\sqrt{2}}\right) + 5\left(\frac{x' + y'}{\sqrt{2}}\right)^2 - 12 = 0$$

$$\frac{5}{2}\left[(x')^2 - 2(x'y') + (y')^2\right] - \left[(x')^2 - (y')^2\right] + \frac{5}{2}\left[(x')^2 + 2(x'y') + (y')^2\right] = 12$$

$$4(x')^2 + 6(y')^2 = 12$$

$$\frac{(x')^2}{3} + \frac{(y')^2}{2} = 1$$

49. (a) $16x^2 - 24xy + 9y^2 - 30x - 40y = 0$

$B^2 - 4AC = (-24)^2 - 4(16)(9) = 0$

The graph is a parabola.

(b) To use a graphing utility, we need to solve for y in terms of x.

$9y^2 + (-24x - 40)y + (16x^2 - 30x) = 0$

$$y = \frac{-(-24x - 40) \pm \sqrt{(-24x - 40)^2 - 4(9)(16x^2 - 30x)}}{2(9)}$$

$$= \frac{(24x + 40) \pm \sqrt{(24x + 40)^2 - 36(16x^2 - 30x)}}{18}$$

(c)

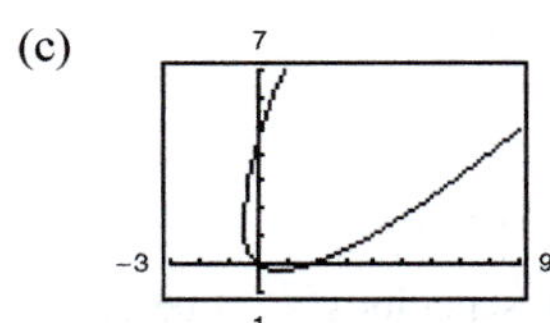

51. (a) $x^2 - 10xy + y^2 + 1 = 0$

Since

$B^2 - 4AC = (-10)^2 - 4(1)(1) > 0 \Rightarrow$ Hyperbola

(b) Use the Quadratic Formula to solve for y in terms of x:

$y^2 - 10xy + x^2 + 1 = 0$

$$y = \frac{10x \pm \sqrt{100x^2 - 4(x^2 + 1)}}{2}$$

(c)

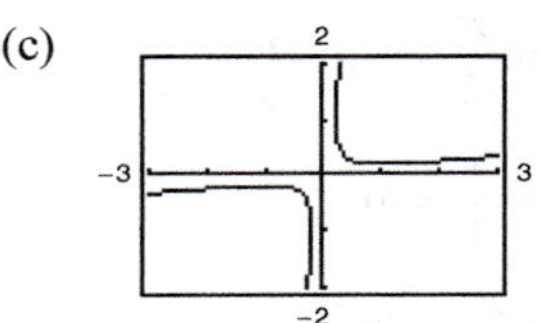

53. $x = 3t - 2,\ y = 7 - 4t$

(a)

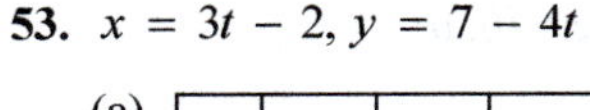

t	−2	−1	0	1	2
x	−8	−5	−2	1	4
y	15	11	7	3	−1

(b)

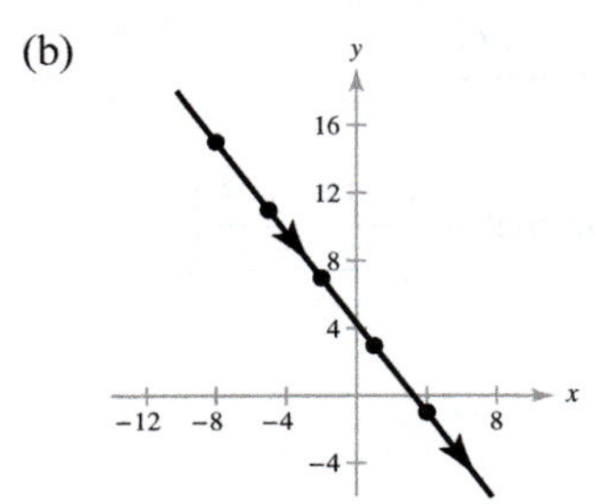

55. (a)

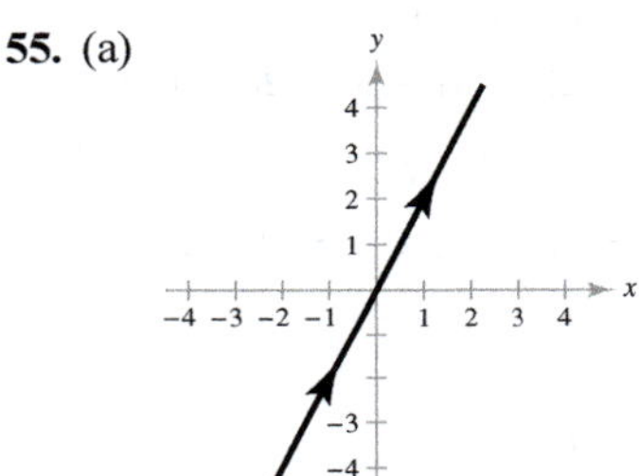

(b) $x = 2t \Rightarrow \dfrac{x}{2} = t$

$y = 4t \Rightarrow y = 4\left(\dfrac{x}{2}\right) = 2x$

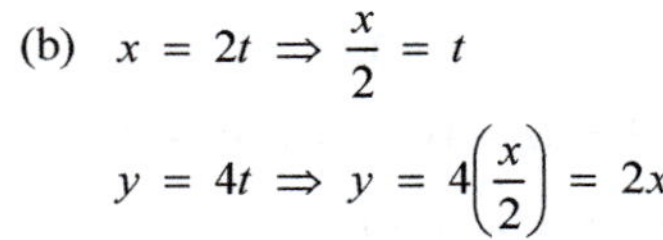

57. (a)

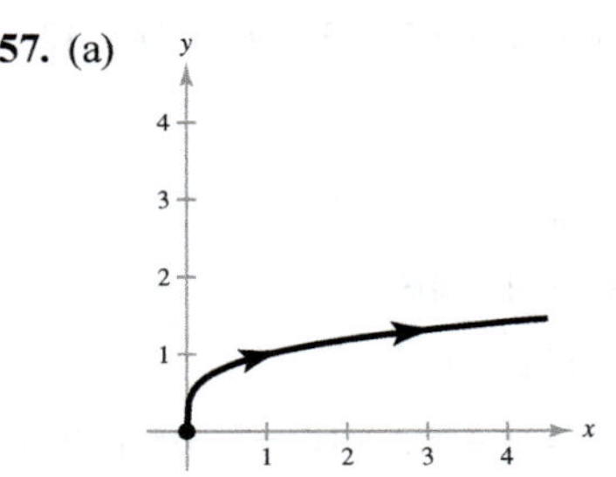

(b) $x = t^2,\ x \ge 0$

$y = \sqrt{t} \Rightarrow y^2 = t$

$x = (y^2)^2 \Rightarrow x$

$= y^4 \Rightarrow y = \sqrt[4]{x}$

59. (a)

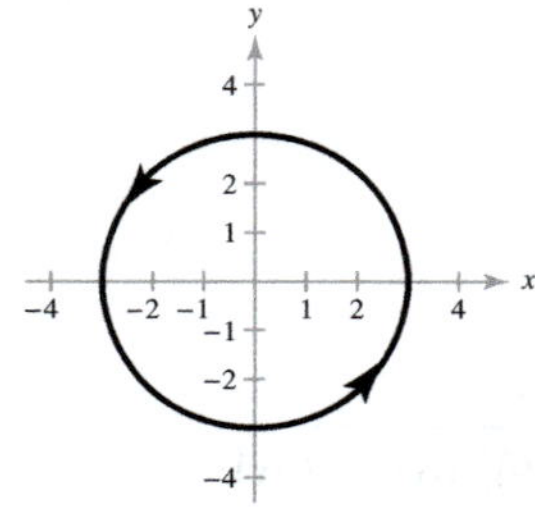

(b) $x = 3\cos\theta,\ y = 3\sin\theta$

$\left(\frac{x}{3}\right)^2 = \cos^2\theta, \left(\frac{y}{3}\right)^2 = \sin^2\theta$

$x^2 + y^2 = 9$

61. $y = 2x + 3$

(a) $t = x \Rightarrow x = t$

$y = 2x + 3 = 2t + 3$

(b) $t = x + 1 \Rightarrow x = t - 1$

$y = 2x + 3 = 2(t - 1) + 3 = 2t + 1$

(c) $t = 3 - x \Rightarrow x = 3 - t$

$y = 2x + 3 = 2(3 - t) + 3 = 9 - 2t$

63. $y = x^2 + 3$

(a) $t = x \Rightarrow x = t$

$y = x^2 + 3 = t^2 + 3$

(b) $t = x + 1 \Rightarrow x = t - 1$

$y = x^2 + 3 = (t - 1)^2 + 3 = t^2 - 2t + 4$

(c) $t = 3 - x \Rightarrow x = 3 - t$

$y = x^2 + 3 = (3 - t)^2 + 3 = t^2 - 6t + 12$

65. $y = 1 - 4x^2$

(a) $t = x \Rightarrow x = t$

$y = 1 - 4x^2 = 1 - 4t^2$

(b) $t = x + 1 \Rightarrow x = t - 1$

$y = 1 - 4x^2 = 1 - 4(t - 1)^2 = -4t^2 + 8t - 3$

(c) $t = 3 - x \Rightarrow x = 3 - t$

$y = 1 - 4x^2 = 1 - 4(3 - t)^2 = -4t^2 + 24t - 35$

67. Polar coordinates: $\left(4, \frac{5\pi}{6}\right)$

Additional polar representations:

$\left(4, -\frac{7\pi}{6}\right), \left(-4, -\frac{\pi}{6}\right), \left(-4, \frac{11\pi}{6}\right)$

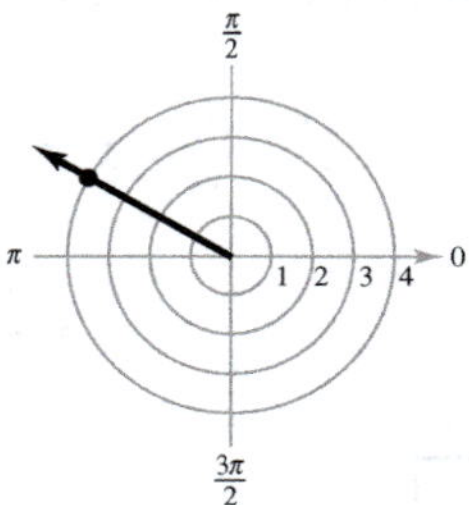

69. Polar coordinates: $(-7, 4.19)$

Additional polar representations: $(7, 1.05), (-7, -2.09)$

$(7, -5.23)$

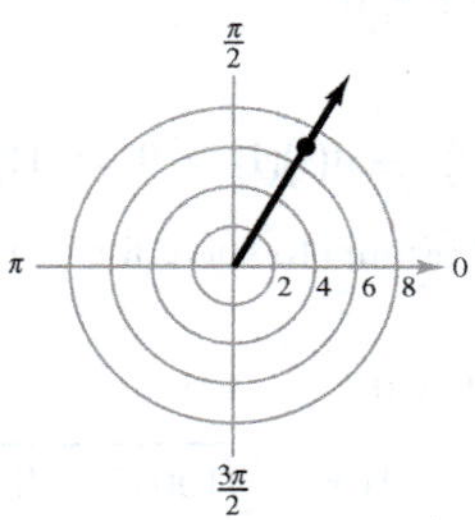

71. Polar coordinates: $\left(0, \frac{\pi}{2}\right) = (r, \theta)$

$x = r\cos\theta = 0\cos\frac{\pi}{2} = 0$

$y = r\sin\theta = 0\sin\frac{\pi}{2} = 0$

Rectangular coordinates: $(0, 0)$

73. Polar coordinates: $\left(-1, \frac{\pi}{3}\right)$

$x = -1\cos\frac{\pi}{3} = -\frac{1}{2}$

$y = -1\sin\frac{\pi}{3} = -\frac{\sqrt{3}}{2}$

Rectangular coordinates: $\left(-\frac{1}{2}, -\frac{\sqrt{3}}{2}\right)$

75. Rectangular coordinates: $(3, 3)$

$$r = \sqrt{(3)^2 + (3)^2} = \sqrt{18} = 3\sqrt{2}$$

$$\tan \theta = 1, \theta = \frac{\pi}{4}$$

Polar coordinates: $\left(3\sqrt{2}, \dfrac{\pi}{4}\right)$

77. Rectangular coordinates: $\left(-\sqrt{5}, \sqrt{5}\right)$

$$r = \sqrt{\left(-\sqrt{5}\right)^2 + \left(\sqrt{5}\right)^2} = \sqrt{10}$$

$$\tan \theta = -1, \theta = \frac{3\pi}{4}$$

Polar coordinates: $\left(\sqrt{10}, \dfrac{3\pi}{4}\right)$

79.
$$x^2 + y^2 = 81$$
$$r^2 = 81$$
$$r = 9$$

81.
$$x = 5$$
$$r \cos \theta = 5$$
$$r = \frac{5}{\cos \theta}$$
$$r = 5 \sec \theta$$

83.
$$xy = 5$$
$$(r \cos \theta)(r \sin \theta) = 5$$
$$r^2 = \frac{5}{\sin \theta \cos \theta}$$
$$= \frac{10}{\sin 2\theta} = 10 \csc 2\theta$$

85.
$$r = 4$$
$$r^2 = 16$$
$$x^2 + y^2 = 16$$

87.
$$r = 3 \cos \theta$$
$$r^2 = 3r \cos \theta$$
$$x^2 + y^2 = 3x$$

89.
$$r^2 = \sin \theta$$
$$r^3 = r \sin \theta$$
$$\left(\pm\sqrt{x^2 + y^2}\right)^3 = y$$
$$\left(x^2 + y^2\right)^3 = y^2$$
$$x^2 + y^2 = y^{2/3}$$

91. $r = 6$

Circle of radius 6 centered at the pole

Symmetric with respect to $\theta = \pi/2$, the polar axis and the pole

Maximum value of $|r| = 6$, for all values of θ

Zeros: None

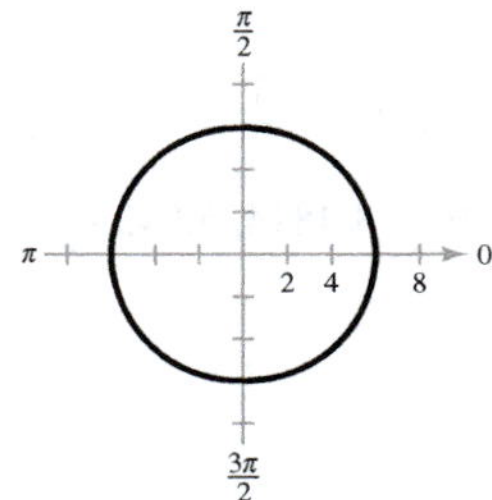

93. $r = -2(1 + \cos \theta)$

Symmetric with respect to the polar axis

Maximum value of $|r| = 4$ when $\theta = 0$

Zeros: $r = 0$ when $\theta = \pi$

$\dfrac{a}{b} = \dfrac{2}{2} = 1 \Rightarrow$ Cardioid

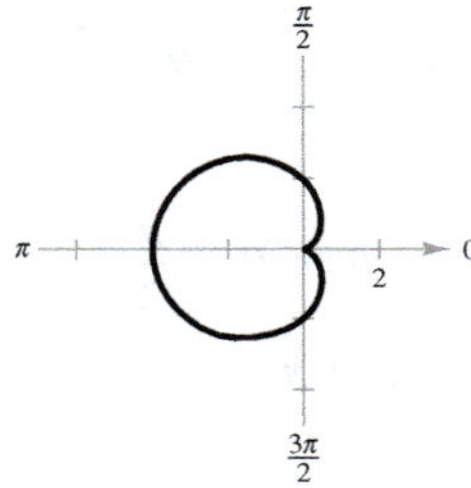

95. $r = 4 \sin 2\theta$

Rose curve $(n = 2)$ with 4 petals

Symmetric with respect to $\theta = \pi/2$, the polar axis, and the pole

Maximum value of $|r| = 4$ when $\theta = \dfrac{\pi}{4}, \dfrac{3\pi}{4}, \dfrac{5\pi}{4}, \dfrac{7\pi}{4}$

Zeros: $r = 0$ when $\theta = 0, \dfrac{\pi}{2}, \pi, \dfrac{3\pi}{2}$

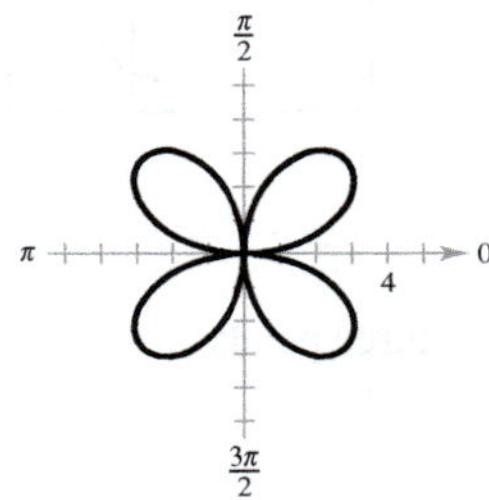

97. $r = 2 + 6 \sin \theta$

Limaçon with inner loop

$r = f(\sin \theta) \Rightarrow \theta = \dfrac{\pi}{2}$ symmetry

Maximum value: $|r| = 8$ when $\theta = \dfrac{\pi}{2}$

Zeros:

$2 + 6 \sin \theta = 0 \Rightarrow \sin \theta = -\dfrac{1}{3}$

$\Rightarrow \theta \approx 3.4814, 5.9433$

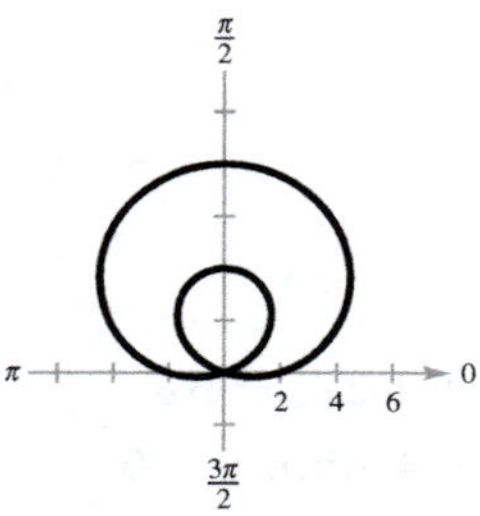

99. $r^2 = 9 \sin \theta$

$r = \pm 3\sqrt{\sin \theta}$

Symmetric with respect to polar axis, $\theta = \dfrac{\pi}{2}$, and the pole.

Maximum value of $|r| = 3$ when $\theta = \dfrac{\pi}{2}$

Zeros: $r = 0$ when $\theta = 0$ and π

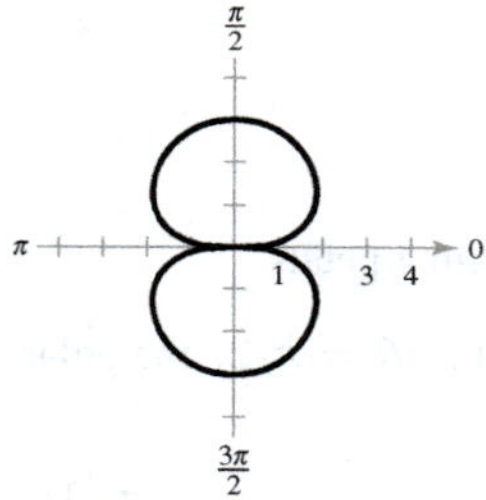

101. $r = 3(2 - \cos \theta)$

$= 6 - 3 \cos \theta$

$\dfrac{a}{b} = \dfrac{6}{3} = 2$

Convex limaçon

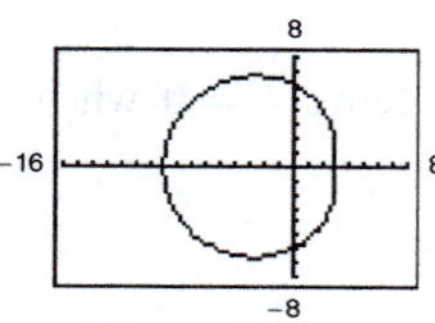

103. $r = 8 \cos 3\theta$

Rose curve $(n = 3)$ with three petals

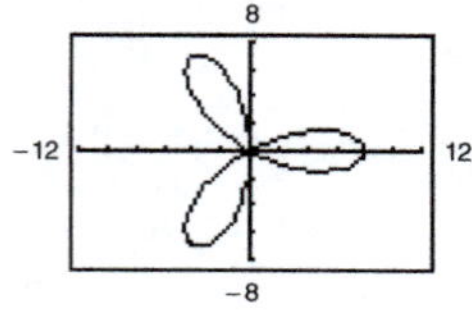

105. $r = \dfrac{1}{1 + 2 \sin \theta}, e = 2$

Hyperbola symmetric with respect to $\theta = \dfrac{\pi}{2}$ and having vertices at $\left(\dfrac{1}{3}, \dfrac{\pi}{2}\right)$ and $\left(-1, \dfrac{3\pi}{2}\right)$

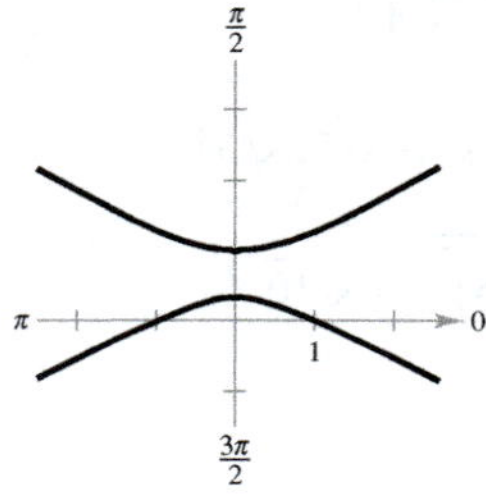

107. $r = \dfrac{4}{5 - 3 \cos \theta}$

$r = \dfrac{4/5}{1 - (3/5) \cos \theta}, e = \dfrac{3}{5}$

Ellipse symmetric with respect to the polar axis and having vertices at $(2, 0)$ and $(1/2, \pi)$.

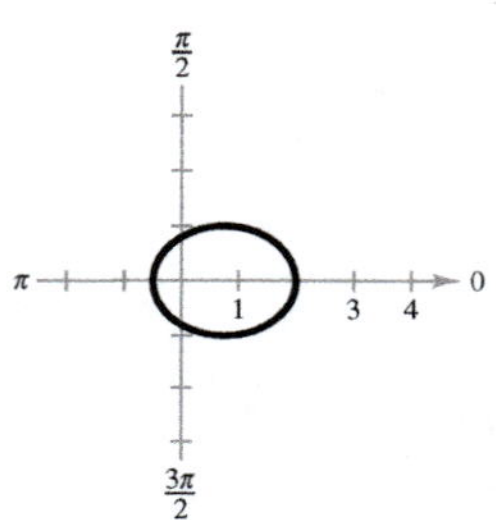

109. Parabola: $r = \dfrac{ep}{1 - e \cos \theta}, e = 1$

Vertex: $(2, \pi)$

Focus: $(0, 0) \Rightarrow p = 4$

$r = \dfrac{4}{1 - \cos \theta}$

111. Ellipse: $r = \dfrac{ep}{1 - e \cos \theta}$

Vertices: $(5, 0), (1, \pi) \Rightarrow a = 3$

One focus: $(0, 0) \Rightarrow c = 2$

$e = \dfrac{c}{a} = \dfrac{2}{3}, p = \dfrac{5}{2}$

$r = \dfrac{(2/3)(5/2)}{1 - (2/3) \cos \theta} = \dfrac{5/3}{1 - (2/3) \cos \theta} = \dfrac{5}{3 - 2 \cos \theta}$

113. $a + c = 122{,}800 + 4000 \Rightarrow a + c = 126{,}800$

$a - c = 119 + 4000 \Rightarrow a - c = 4{,}119$

$2a = 130{,}919$

$a = 65{,}459.5$

$c = 61{,}340.5$

$$e = \frac{c}{a} = \frac{61{,}340.5}{65{,}459.5} \approx 0.937$$

$$r = \frac{ep}{1 - e\cos\theta} \approx \frac{0.937p}{1 - 0.937\cos\theta}$$

$r = 126{,}800$ when $\theta = 0$

$$126{,}800 = \frac{ep}{1 - e\cos 0}$$

$$ep = 126{,}800\left(1 - \frac{61{,}340.5}{65{,}459.5}\right) \approx 7978.81$$

So, $r \approx \dfrac{7978.81}{1 - 0.937\cos\theta}$.

When

$\theta = \dfrac{\pi}{3}, r \approx \dfrac{7978.81}{1 - 0.937\cos(\pi/3)} \approx 15{,}011.87$ miles.

The distance from the surface of Earth and the satellite is $15{,}011.87 - 4000 \approx 11{,}011.87$ miles.

115. False.

$\dfrac{x^2}{4} - y^4 = 1$ is a fourth-degree equation.

The equation of a hyperbola is a second degree equation.

117. False.

$(r, \theta), (r, \theta + 2\pi), (-r, \theta + \pi)$, etc.

All represent the same point.

119. (a) $x^2 + y^2 = 25$

$r = 5$

The graphs are the same. They are both circles centered at $(0, 0)$ with a radius of 5.

(b) $x - y = 0 \Rightarrow y = x$

$$\theta = \frac{\pi}{4}$$

The graphs are the same. They are both lines with slope 1 and intercept $(0, 0)$.

Problem Solving for Chapter 10

1. (a) $\theta = \pi - 1.10 - 0.84 \approx 1.2016$ radians

(b) $\sin 0.84 = \dfrac{x}{3250} \Rightarrow x = 3250 \sin 0.84 \approx 2420$ feet

$\sin 1.10 = \dfrac{y}{6700} \Rightarrow y = 6700 \sin 1.10 \approx 5971$ feet

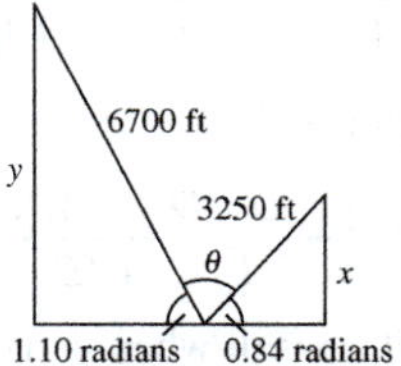

3. Let (x, x) be the corner of the square in Quadrant I.

$A = 4x^2$

$$\frac{x^2}{a^2} + \frac{x^2}{b^2} = 1 \Rightarrow x^2 = \frac{a^2b^2}{a^2 + b^2}$$

So, $A = \dfrac{4a^2b^2}{a^2 + b^2}$.

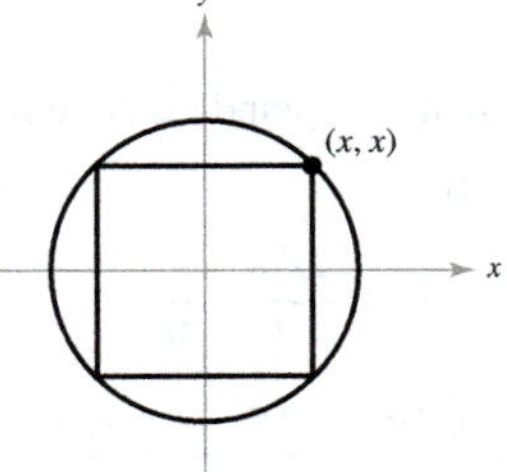

5. (a)

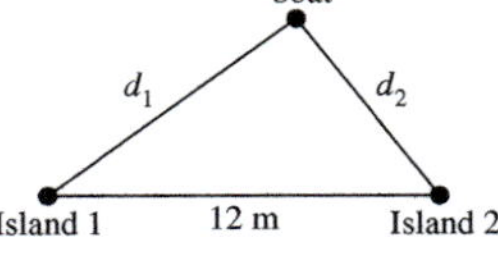

Because $d_1 + d_2 \le 20$, by definition, the outer bound that the boat can travel is an ellipse. The islands are the foci.

(b)

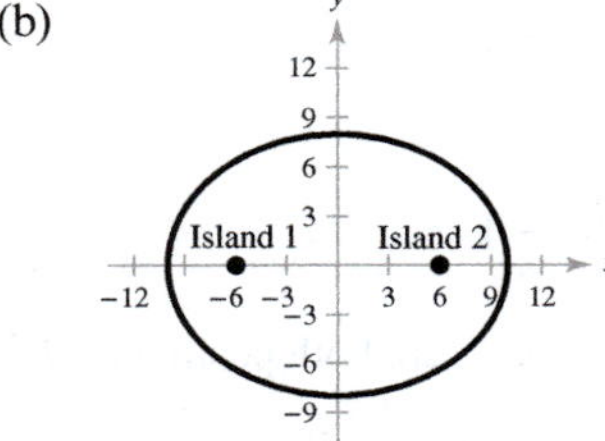

Island 1 is located at $(-6, 0)$ and Island 2 is located at $(6, 0)$.

(c) $d_1 + d_2 = 2a = 20 \Rightarrow a = 10$

The boat traveled 20 miles.

The vertex is $(10, 0)$.

(d) $c = 6, a = 10 \Rightarrow b^2 = a^2 - c^2 = 64$

$$\frac{x^2}{100} + \frac{y^2}{64} = 1$$

7. $Ax^2 + Cy^2 + Dx + Ey + F = 0$

Assume that the conic is *not* degenerate.

(a) $A = C, A \neq 0$. Complete the square with respect to x and y, to write the standard equation of a circle.

$$Ax^2 + Ay^2 + Dx + Ey + F = 0$$

$$x^2 + y^2 + \frac{D}{A}x + \frac{E}{A}y + \frac{F}{A} = 0$$

$$\left(x^2 + \frac{D}{A}x + \frac{D^2}{4A^2}\right) + \left(y^2 + \frac{E}{A}y + \frac{E^2}{4A^2}\right) = -\frac{F}{A} + \frac{D^2}{4A^2} + \frac{E^2}{4A^2}$$

$$\left(x + \frac{D}{2A}\right)^2 + \left(y + \frac{E}{2A}\right)^2 = \frac{D^2 + E^2 - 4AF}{4A^2}$$

$$(x - h)^2 + (y - k)^2 = r^2$$

This is a circle with center $\left(-\frac{D}{2A}, -\frac{E}{2A}\right)$ and radius $\frac{\sqrt{D^2 + E^2 - 4AF}}{2|A|}$.

(b) $A = 0$ or $C = 0$ (but not both).

Case 1: Let $C = 0$. Complete the square with respect to x to write the standard equation of a parabola with horizontal axis.

$$Ax^2 + Dx + Ey + F = 0$$

$$x^2 + \frac{D}{A}x = -\frac{E}{A}y - \frac{F}{A}$$

$$x^2 + \frac{D}{A}x + \frac{D^2}{4A^2} = -\frac{E}{A}y - \frac{F}{A} + \frac{D^2}{4A^2}$$

$$\left(x + \frac{D}{2A}\right)^2 = -\frac{E}{A}\left(y + \frac{F}{E} - \frac{D^2}{4AE}\right)$$

$$\left(x + \frac{D}{2A}\right)^2 = -\frac{E}{A}\left(y + \left(\frac{4AF}{E} - \frac{D^2}{4AE}\right)\right)$$

$$(x - h)^2 = 4p(y - k)$$

This is a parabola with vertex $\left(-\frac{D}{2A}, \frac{D^2 - 4AF}{4AE}\right)$.

Case 2: $A = 0$ yields a similar result when you complete the square with respect to y to have a parabola with vertical axis.

(c) $AC > 0 \Rightarrow A$ and C are either both positive or are both negative, if that is the case, move the terms to the other side of the equation so that they are both positive.

Complete the square with respect to x and to y to write the standard equation of an ellipse.

$$Ax^2 + Cy^2 + Dx + Ey + F = 0$$

$$A\left(x^2 + \frac{D}{A}x + \frac{D^2}{4A^2}\right) + C\left(y^2 + \frac{E}{C}y + \frac{E^2}{4C^2}\right) = -F + \frac{D^2}{4A} + \frac{E^2}{4C}$$

$$A\left(x + \frac{D}{2A}\right)^2 + C\left(y + \frac{E}{2C}\right)^2 = \frac{CD^2 + AE^2 - 4ACF}{4AC}$$

$$\frac{\left(x + \frac{D}{2A}\right)^2}{\left(\frac{CD^2 + AE^2 - 4ACF}{4A^2C}\right)} + \frac{\left(y + \frac{E}{2C}\right)^2}{\left(\frac{CD^2 + AE^2 - 4ACF}{4AC^2}\right)} = 1$$

$$\frac{(x - h)^2}{a^2} + \frac{(y - k)^2}{b^2} = 1$$

Because A and C are both positive, $4A^2C$ and $4AC^2$ are both positive. $CD^2 + AE^2 - 4ACF$ must be positive or the conic is degenerate. So, we have an ellipse with center $\left(-\frac{D}{2A}, -\frac{E}{2C}\right)$. The values of $\frac{CD^2 + AE^2 - 4ACF}{4A^2C}$ and $\frac{CD^2 + AE^2 - 4ACF}{4AC^2}$ will determine if the major axis is vertical or horizontal.

(d) $AC < 0 \Rightarrow A$ and C have opposite signs. Let's assume that A is positive and C is negative.

If A is negative and C is positive, move the terms to the other side of the equation. From part (c) above completing the square with respect to x and to y yields the standard equation of the hyperbola.

$$\frac{\left(x + \frac{D}{2A}\right)^2}{\left(\frac{CD^2 + AE^2 - 4ACF}{4A^2C}\right)} + \frac{\left(y + \frac{E}{2C}\right)^2}{\left(\frac{CD^2 + AE^2 - 4ACF}{4AC^2}\right)} = 1$$

$$\frac{(x-h)^2}{a^2} + \frac{(y-k)^2}{b^2} = 1.$$

Because $A > 0$ and $C < 0$, the first denominator is positive if $CD^2 + AE^2 - 4ACF < 0$ and is negative if $CD^2 + AE^2 - 4ACF > 0$, since $4A^2C$ is negative. Recall in the first sentence we assumed A is positive and C is negative. The second denominator would have the *opposite* sign because $4AC^2 > 0$. So, we have a hyperbola with center $\left(-\frac{D}{2A}, -\frac{E}{2C}\right)$.

9. At the point $(a, 0)$, the difference of the distances to the foci $(\pm c, 0)$ is $(c + a) - (c - a) = 2a$. Let (x, y) be a point on the hyperbola.

$$2a = \sqrt{(x+c)^2 + y^2} - \sqrt{(x-c)^2 + y^2}$$

$$2a + \sqrt{(x-c)^2 + y^2} = \sqrt{(x+c)^2 + y^2}$$

$$4a^2 + 4a\sqrt{(x-c)^2 + y^2} + (x-c)^2 + y^2 = (x+c)^2 + y^2$$

$$4a\sqrt{(x-c)^2 + y^2} = 4cx - 4a^2$$

$$a\sqrt{(x-c)^2 + y^2} = cx - a^2$$

$$a^2\left(x^2 - 2cx + c^2 + y^2\right) = c^2x^2 - 2a^2cx + a^4$$

$$a^2\left(c^2 - a^2\right) = \left(c^2 - a^2\right)x^2 - a^2y^2$$

$$1 = \frac{x^2}{a^2} - \frac{y^2}{c^2 - a^2}$$

Thus, $c^2 = a^2 + b^2$.

11. To change the orientation, replace t with $-t$.

$x = \cos(-t) = \cos t$

$y = 2\sin(-t) = -2\sin t$

13. (a)

$$r = 2\cos 2\theta \sec\theta$$

$$r = \frac{2(\cos^2\theta - \sin^2\theta)}{\cos\theta}$$

$$\sqrt{x^2+y^2} = \frac{2\left[\left(\frac{x}{\sqrt{x^2+y^2}}\right)^2 - \left(\frac{y}{\sqrt{x^2+y^2}}\right)^2\right]}{\frac{x}{\sqrt{x^2+y^2}}}$$

$$\frac{x}{\sqrt{x^2+y^2}}\cdot\sqrt{x^2+y^2} = 2\left[\frac{x^2}{x^2+y^2} - \frac{y^2}{x^2+y^2}\right]$$

$$x = 2\left[\frac{x^2-y^2}{x^2+y^2}\right]$$

$$x(x^2+y^2) = 2(x^2-y^2)$$

$$x^3 + xy^2 = 2x^2 - 2y^2$$

$$2y^2 + xy^2 = 2x^2 - x^3$$

$$y^2(2+x) = x^2(2-x)$$

$$y^2 = x^2\left(\frac{2-x}{2+x}\right)$$

(b) $x = \dfrac{2-2t^2}{1+t^2}$ and $y = \dfrac{t(2-2t^2)}{1+t^2}$

(c)

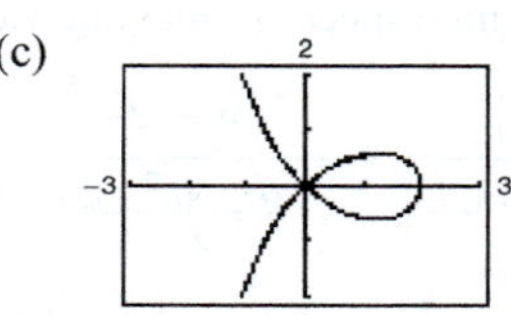

15. $x = (a-b)\cos t + b\cos\left(\frac{a-b}{b}t\right)$

$y = (a-b)\sin t - b\sin\left(\frac{a-b}{b}t\right)$

(a) $a = 2, b = 1$

$x = \cos t + \cos t = 2\cos t$

$y = \sin t - \sin t = 0$

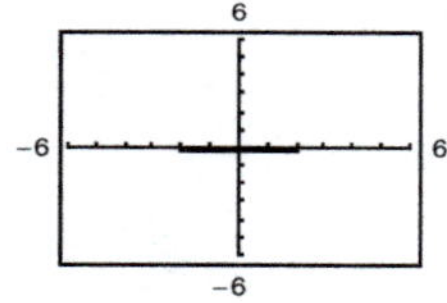

The graph oscillates between -2 and 2 on the x-axis.

(b) $a = 3, b = 1$

$x = 2\cos t + \cos 2t$

$y = 2\sin t - \sin 2t$

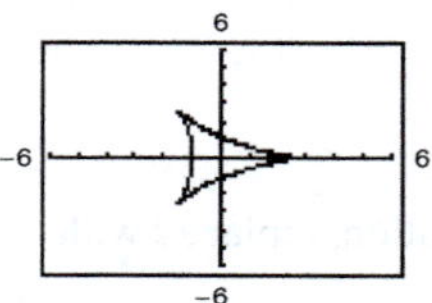

The graph is a three-sided figure with counterclockwise orientation.

(c) $a = 4, b = 1$

$x = 3\cos t + \cos 3t$

$y = 3\sin t - \sin 3t$

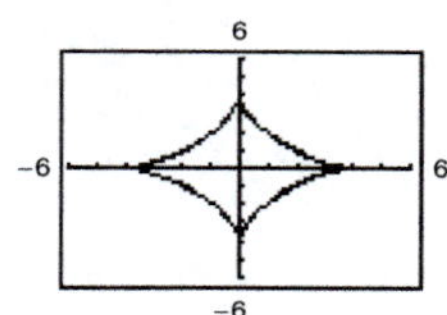

The graph is a four-sided figure with counterclockwise orientation.

(d) $a = 10, b = 1$

$x = 9\cos t + \cos 9t$

$y = 9\sin t - \sin 9t$

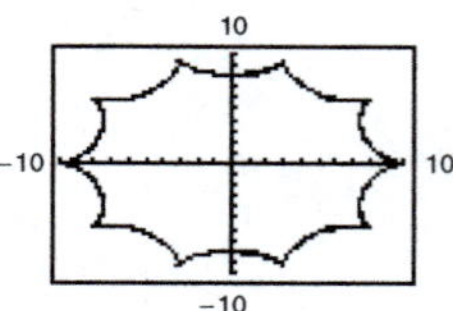

The graph is a ten-sided figure with counterclockwise orientation.

(e) $a = 3, b = 2$

$$x = \cos t + 2\cos\frac{t}{2}$$

$$y = \sin t - 2\sin\frac{t}{2}$$

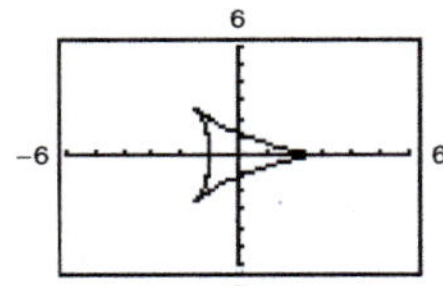

The graph looks the same as the graph in part (b), but is oriented clockwise instead of counterclockwise.

(f) $a = 4, b = 3$

$$x = \cos t + 3\cos\frac{t}{3}$$

$$y = \sin t - 3\sin\frac{t}{3}$$

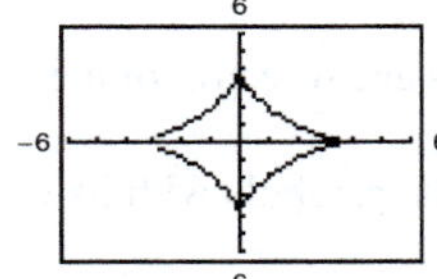

The graph is the same as the graph in part (c), but is oriented clockwise instead of counterclockwise.

17. *Sample answer:*

$r = 2\cos\left(\frac{1}{2}\theta\right)$

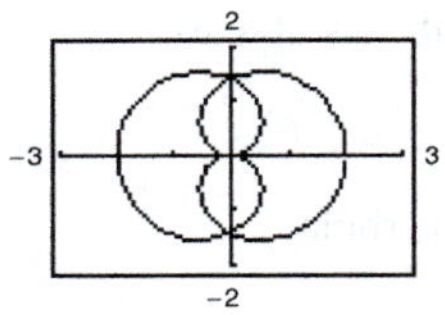

$r = 3\sin\left(\frac{5\theta}{2}\right)$

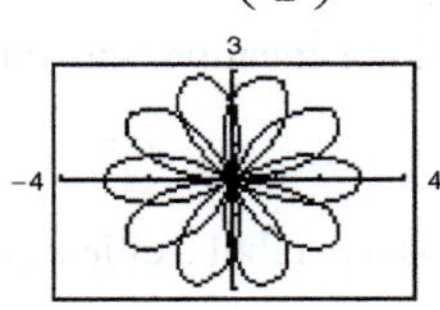

$r = -\cos\left(\sqrt{2}\theta\right), -2\pi \le \theta \le 2\pi$

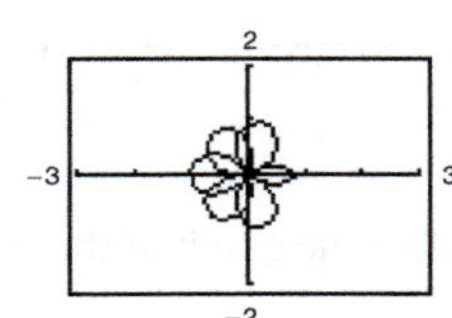

$r = -2\sin\left(\frac{4\theta}{7}\right)$

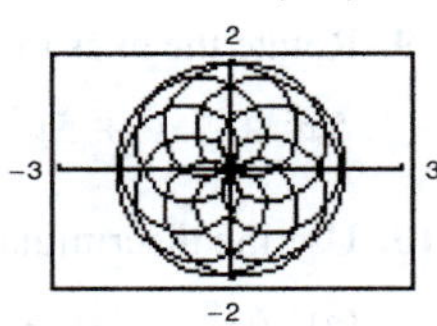

If n is a rational number, then the curve has a finite number of petals. If n is an irrational number, then the curve has an infinite number of petals.

Chapter 10 Practice Test

1. Find the angle, θ, between the lines $3x + 4y = 12$ and $4x - 3y = 12$.

2. Find the distance between the point $(5, -9)$ and the line $3x - 7y = 21$.

3. Find the vertex, focus and directrix of the parabola $x^2 - 6x - 4y + 1 = 0$.

4. Find an equation of the parabola with its vertex at $(2, -5)$ and focus at $(2, -6)$.

5. Find the center, foci, vertices, and eccentricity of the ellipse $x^2 + 4y^2 - 2x + 32y + 61 = 0$.

6. Find an equation of the ellipse with vertices $(0, \pm 6)$ and eccentricity $e = \frac{1}{2}$.

7. Find the center, vertices, foci, and asymptotes of the hyperbola $16y^2 - x^2 - 6x - 128y + 231 = 0$.

8. Find an equation of the hyperbola with vertices at $(\pm 3, 2)$ and foci at $(\pm 5, 2)$.

9. Rotate the axes to eliminate the xy-term. Sketch the graph of the resulting equation, showing both sets of axes.

$5x^2 + 2xy + 5y^2 - 10 = 0$

10. Use the discriminant to determine whether the graph of the equation is a parabola, ellipse, or hyperbola.

(a) $6x^2 - 2xy + y^2 = 0$

(b) $x^2 + 4xy + 4y^2 - x - y + 17 = 0$

11. Convert the polar point $\left(\sqrt{2}, \dfrac{3\pi}{4}\right)$ to rectangular coordinates.

12. Convert the rectangular point $\left(\sqrt{3}, -1\right)$ to polar coordinates.

13. Convert the rectangular equation $4x - 3y = 12$ to polar form.

14. Convert the polar equation $r = 5 \cos \theta$ to rectangular form.

15. Sketch the graph of $r = 1 - \cos \theta$.

16. Sketch the graph of $r = 5 \sin 2\theta$.

17. Sketch the graph of $r = \dfrac{3}{6 - \cos \theta}$.

18. Find a polar equation of the parabola with its vertex at $\left(6, \dfrac{\pi}{2}\right)$ and focus at $(0, 0)$.

For Exercises 19 and 20, eliminate the parameter and write the corresponding rectangular equation.

19. $x = 3 - 2 \sin \theta$, $y = 1 + 5 \cos \theta$

20. $x = e^{2t}$, $y = e^{4t}$

APPENDIX A
Review of Fundamental Concepts of Algebra

Appendix A.1 Real Numbers and Their Properties

1. irrational

3. absolute value

5. terms

7. $-9, -\frac{7}{2}, 5, \frac{2}{3}, \sqrt{2}, 0, 1, -4, 2, -11$

(a) Natural numbers: 5, 1, 2

(b) Whole numbers: 0, 5, 1, 2

(c) Integers: $-9, 5, 0, 1, -4, 2, -11$

(d) Rational numbers: $-9, -\frac{7}{2}, 5, \frac{2}{3}, 0, 1, -4, 2, -11$

(e) Irrational numbers: $\sqrt{2}$

9. $2.01, 0.\overline{6}, -13, 0.010110111\ldots, 1, -6$

(a) Natural numbers: 1

(b) Whole numbers: 1

(c) Integers: $-13, 1, -6$

(d) Rational numbers: $2.01, 0.\overline{6}, -13, 1, -6$

(e) Irrational numbers: $0.010110111\ldots$

11. (a)

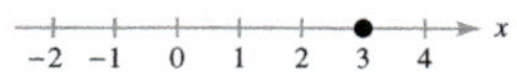

(b)

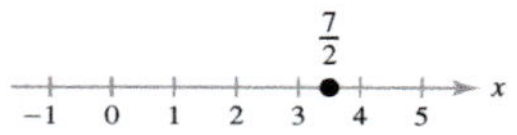

(c)

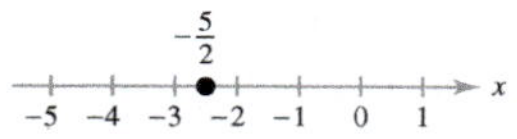

(d)

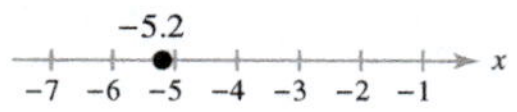

13. $-4 > -8$

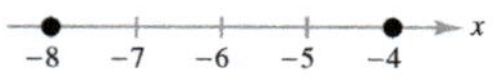

15. $\frac{5}{6} > \frac{2}{3}$

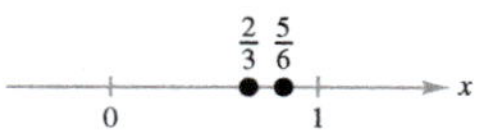

17. (a) The inequality $x \le 5$ denotes the set of all real numbers less than or equal to 5.

(b)

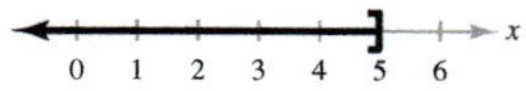

(c) The interval is unbounded.

19. (a) The inequality $-2 < x < 2$ denotes the set of all real numbers greater than -2 and less than 2.

(b)

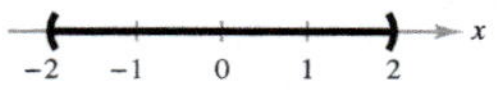

(c) The interval is bounded.

21. (a) The interval $[4, \infty)$ denotes the set of all real numbers greater than or equal to 4.

(b)

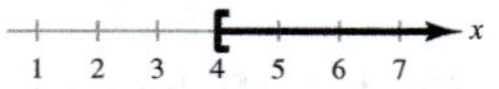

(c) The interval is unbounded.

23. (a) The interval $[-5, 2)$ denotes the set of all real numbers greater than or equal to -5 and less than 2.

(b)

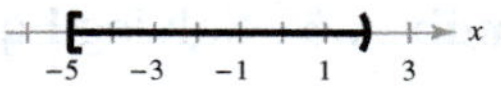

(c) The interval is bounded.

25. $y \ge 0; [0, \infty)$

27. $10 \le t \le 22; [10, 22]$

29. $|-10| = -(-10) = 10$

31. $|3 - 8| = |-5| = -(-5) = 5$

33. $|-1| - |-2| = 1 - 2 = -1$

35. $5|-5| = 5(5) = 25$

37. If $x < -2$, then $x + 2$ is negative.

So, $\dfrac{|x + 2|}{x + 2} = \dfrac{-(x + 2)}{x + 2} = -1.$

39. $|-4| = |4|$ because $|-4| = 4$ and $|4| = 4$.

41. $-|-6| < |-6|$ because $|-6| = 6$ and $-|-6| = -(6) = -6$.

43. $d(126, 75) = |75 - 126| = 51$

45. $d\left(-\frac{5}{2}, 0\right) = \left|0 - \left(-\frac{5}{2}\right)\right| = \frac{5}{2}$

47. $d(x, 5) = |x - 5|$ and $d(x, 5) \le 3$, so $|x - 5| \le 3$.

	Receipts, R	Expenditures, E	$\lvert R - E \rvert$
49.	\$2524.0 billion	\$2982.5 billion	$\lvert 2524.0 - 2982.5 \rvert = \$458.5$ billion
51.	\$2450.0 billion	\$3537.0 billion	$\lvert 2450.0 - 3537.0 \rvert = \$1087.0$ billion

53. $7x + 4$

Terms: $7x$, 4

Coefficient: 7

55. $6x^3 - 5x$

Terms: $6x^3, -5x$

Coefficients: $6, -5$

57. $3\sqrt{3}x^2 + 1$

Terms: $3\sqrt{3}x^2, 1$

Coefficient: $3\sqrt{3}$

59. $4x - 6$

(a) $4(-1) - 6 = -4 - 6 = -10$

(b) $4(0) - 6 = 0 - 6 = -6$

61. $x^2 - 3x + 2$

(a) $(0)^2 - 3(0) + 2 = 2$

(b) $(-1)^2 - 3(-1) + 2 = 1 + 3 + 2 = 6$

63. $\dfrac{x+1}{x-1}$

(a) $\dfrac{1+1}{1-1} = \dfrac{2}{0}$

Division by zero is undefined.

(b) $\dfrac{-1+1}{-1-1} = \dfrac{0}{-2} = 0$

65. $\dfrac{1}{(h+6)}(h + 6) = 1, h \neq -6$

Multiplicative Inverse Property

67. $x(3y) = (x \cdot 3)y$ Associative Property of Multiplication

$= (3x)y$ Commutative Property of Multiplication

69. $\dfrac{2x}{3} - \dfrac{x}{4} = \dfrac{8x}{12} - \dfrac{3x}{12} = \dfrac{5x}{12}$

71. $\dfrac{3x}{10} \cdot \dfrac{5}{6} = \dfrac{x}{2} \cdot \dfrac{1}{2} = \dfrac{x}{4}$

73. False. Because zero is nonnegative but not positive, not every nonnegative number is positive.

75. The product of two negative numbers is positive.

77. (a)

n	0.0001	0.01	1	100	10,000
$5/n$	50,000	500	5	0.05	0.0005

(b) (i) As n approaches 0, the value of $5/n$ increases without bound (approaches infinity).

(ii) As n increases without bound (approaches infinity), the value of $5/n$ approaches 0.

Appendix A.2 Exponents and Radicals

1. exponent; base

3. square root

5. like radicals

7. rationalizing

9. (a) $5 \cdot 5^3 = 5^4 = 625$

(b) $\dfrac{5^2}{5^4} = 5^{-2} = \dfrac{1}{5^2} = \dfrac{1}{25}$

11. (a) $(2^3 \cdot 3^2)^2 = 2^{3 \cdot 2} \cdot 3^{2 \cdot 2}$

$= 2^6 \cdot 3^4 = 64 \cdot 81 = 5184$

(b) $\left(-\dfrac{3}{5}\right)^3\left(\dfrac{5}{3}\right)^2 = (-1)^3\dfrac{3^3}{5^3} \cdot \dfrac{5^2}{3^2} = -1 \cdot 3^{3-2} \cdot 5^{2-3}$

$= -3 \cdot 5^{-1} = -\dfrac{3}{5}$

13. (a) $\dfrac{4 \cdot 3^{-2}}{2^{-2} \cdot 3^{-1}} = 4 \cdot 2^2 \cdot 3^{-2-(-1)} = 4 \cdot 4 \cdot 3^{-1} = \dfrac{16}{3}$

(b) $(-2)^0 = 1$

15. When $x = 2$,

$-3x^3 = -3(2)^3 = -24.$

17. When $x = 10$,

$6x^0 = 6(10)^0 = 6(1) = 6.$

19. When $x = -2$,

$-3x^4 = -3(-2)^4$

$= -3(16) = -48.$

21. (a) $(5z)^3 = 5^3z^3 = 125z^3$

(b) $5x^4(x^2) = 5x^{4+2} = 5x^6$

23. (a) $6y^2(2y^0)^2 = 6y^2(2 \cdot 1)^2 = 6y^2(4) = 24y^2$

(b) $(-z)^3(3z^4) = (-1)^3(z^3)3z^4$

$= -1 \cdot 3 \cdot z^{3+4} = -3z^7$

25. (a) $\left(\frac{4}{y}\right)^3\left(\frac{3}{y}\right)^4 = \frac{4^3}{y^3} \cdot \frac{3^4}{y^4} = \frac{64 \cdot 81}{y^{3+4}} = \frac{5184}{y^7}$

(b) $\left(\frac{b^{-2}}{a^{-2}}\right)\left(\frac{b}{a}\right)^2 = \left(\frac{a^2}{b^2}\right)\left(\frac{b^2}{a^2}\right) = 1,\ a \neq 0,\ b \neq 0$

27. (a) $(x + 5)^0 = 1,\ x \neq -5$

(b) $(2x^2)^{-2} = \frac{1}{(2x^2)^2} = \frac{1}{4x^4}$

29. (a) $\left(\frac{x^{-3}y^4}{5}\right)^{-3} = \left(\frac{5x^3}{y^4}\right)^3 = \frac{125x^9}{y^{12}}$

(b) $\left(\frac{a^{-2}}{b^{-2}}\right)\left(\frac{b}{a}\right)^3 = \left(\frac{b^2}{a^2}\right)\left(\frac{b^3}{a^3}\right) = \frac{b^5}{a^5}$

31. $10{,}250.4 = 1.02504 \times 10^4$

33. $3.14 \times 10^{-4} = 0.000314$

35. $9.46 \times 10^{12} = 9{,}460{,}000{,}000{,}000$ kilometers

37. (a) $(2.0 \times 10^9)(3.4 \times 10^{-4}) = 6.8 \times 10^5$

(b) $(1.2 \times 10^7)(5.0 \times 10^{-3}) = 6.0 \times 10^4$

39. (a) $\sqrt{9} = 3$

(b) $\sqrt[3]{\frac{27}{8}} = \frac{\sqrt[3]{27}}{\sqrt[3]{8}} = \frac{3}{2}$

41. (a) $(\sqrt[5]{2})^5 = 2^{5/5} = 2^1 = 2$

(b) $\sqrt[5]{32x^5} = \sqrt[5]{(2x)^5} = 2x$

43. (a) $\sqrt{20} = \sqrt{4 \cdot 5} = \sqrt{4}\sqrt{5} = 2\sqrt{5}$

(b) $\sqrt[3]{128} = \sqrt[3]{64 \cdot 2} = \sqrt[3]{64}\sqrt[3]{2} = 4\sqrt[3]{2}$

45. (a) $\sqrt{72x^3} = \sqrt{36x^2 \cdot 2x} = 6x\sqrt{2x}$

(b) $\sqrt{54xy^4} = \sqrt{6 \cdot 3^2 \cdot x \cdot (y^2)^2} = 3y^2\sqrt{6x}$

47. (a) $\sqrt[3]{16x^5} = \sqrt[3]{(2x)^3 \cdot 2x^2} = 2x\sqrt[3]{2x^2}$

(b) $\sqrt{75x^2y^{-4}} = \sqrt{\frac{75x^2}{y^4}}$

$= \frac{\sqrt{25x^2 \cdot 3}}{\sqrt{y^4}}$

$= \frac{\sqrt{(5x)^2 \cdot 3}}{\sqrt{(y^2)^2}}$

$= \frac{5|x|\sqrt{3}}{y^2}$

49. (a) $2\sqrt{20x^2} + 5\sqrt{125x^2} = 2\sqrt{4x^2 \cdot 5} + 5\sqrt{25x^2 \cdot 5}$

$= 2\sqrt{(2x)^2 \cdot 5} + 5\sqrt{(5x)^2 \cdot 5}$

$= 4|x|\sqrt{5} + 25|x|\sqrt{5}$

$= 29|x|\sqrt{5}$

(b) $8\sqrt{147x} - 3\sqrt{48x} = 8\sqrt{49 \cdot 3x} - 3\sqrt{16 \cdot 3x}$

$= 8\sqrt{7^2 \cdot 3x} - 3\sqrt{4^2 \cdot 3x}$

$= 56\sqrt{3x} - 12\sqrt{3x}$

$= 44\sqrt{3x}$

51. $\frac{1}{\sqrt{3}} = \frac{1}{\sqrt{3}} \cdot \frac{\sqrt{3}}{\sqrt{3}} = \frac{\sqrt{3}}{3}$

53. $\frac{5}{\sqrt{14} - 2} = \frac{5}{\sqrt{14} - 2} \cdot \frac{\sqrt{14} + 2}{\sqrt{14} + 2} = \frac{5(\sqrt{14} + 2)}{(\sqrt{14})^2 - (2)^2} = \frac{5(\sqrt{14} + 2)}{14 - 4} = \frac{5(\sqrt{14} + 2)}{10} = \frac{\sqrt{14} + 2}{2}$

55. $\dfrac{\sqrt{5}+\sqrt{3}}{3} = \dfrac{\sqrt{5}+\sqrt{3}}{3} \cdot \dfrac{\sqrt{5}-\sqrt{3}}{\sqrt{5}-\sqrt{3}} = \dfrac{5-3}{3\left(\sqrt{5}-\sqrt{3}\right)} = \dfrac{2}{3\left(\sqrt{5}-\sqrt{3}\right)}$

57. $\sqrt[3]{64} = 64^{1/3}$

59. $3x^{-2/3} = \dfrac{3}{x^{2/3}} = \dfrac{3}{\sqrt[3]{x^2}}, x \neq 0$

61. (a) $32^{-3/5} = \dfrac{1}{32^{3/5}} = \dfrac{1}{\left(\sqrt[5]{32}\right)^3} = \dfrac{1}{(2)^3} = \dfrac{1}{8}$

(b) $\left(\dfrac{16}{81}\right)^{-3/4} = \left(\dfrac{81}{16}\right)^{3/4} = \left(\sqrt[4]{\dfrac{81}{16}}\right)^3 = \left(\dfrac{3}{2}\right)^3 = \dfrac{27}{8}$

63. (a) $\sqrt[4]{3^2} = 3^{2/4} = 3^{1/2} = \sqrt{3}$

(b) $\sqrt[6]{(x+1)^4} = (x+1)^{4/6} = (x+1)^{2/3} = \sqrt[3]{(x+1)^2}$

65. (a) $\sqrt{\sqrt{32}} = \left(32^{1/2}\right)^{1/2}$

$= 32^{1/4} = \sqrt[4]{32} = \sqrt[4]{16 \cdot 2} = 2\sqrt[4]{2}$

(b) $\sqrt{\sqrt[4]{2x}} = \left((2x)^{1/4}\right)^{1/2} = (2x)^{1/8} = \sqrt[8]{2x}$

67. (a) $(x-1)^{1/3}(x-1)^{2/3} = (x-1)^{3/3} = x-1$

(b) $(x-1)^{1/3}(x-1)^{-4/3} = (x-1)^{-3/3}$

$= (x-1)^{-1}$

$= \dfrac{1}{x-1}$

69. $t = 0.03\left[12^{5/2} - (12-h)^{5/2}\right], 0 \leq h \leq 12$

h (in centimeters)	t (in seconds)
0	0
1	2.93
2	5.48
3	7.67
4	9.53
5	11.08
6	12.32
7	13.29
8	14.00
9	14.50
10	14.80
11	14.93
12	14.96

71. False. When $x = 0$, the expressions are not equal.

73. False. When a sum is raised to a power, you multiply the sum by itself using the Distributive Property.

$(a+b)^2 = a^2 + 2ab + b^2 \neq a^2 + b^2$

Appendix A.3 Polynomials and Factoring

1. n; a_n; a_0

3. like terms

5. factoring

7. perfect square binomial

9. (a) Standard form: $7x$

(b) Degree: 1
Leading coefficient: 7

(c) Monomial

11. (a) Standard form: $-\frac{1}{2}x^5 + 14x$

(b) Degree: 5
Leading coefficient: $-\frac{1}{2}$

(c) Binomial

13. (a) Standard form: $-4x^5 + 6x^4 + 1$

(b) Degree: 5
Leading coefficient: -4

(c) Trinomial

15. $(6x+5) - (8x+15) = 6x + 5 - 8x - 15$

$= (6x - 8x) + (5 - 15)$

$= -2x - 10$

17. $(15x^2 - 6) + (-8.3x^3 - 14.7x^2 - 17) = 15x^2 - 6 - 8.3x^3 - 14.7x^2 - 17$
$= -8.3x^3 + (15x^2 - 14.7x^2) + (-6 - 17)$
$= -8.3x^3 + 0.3x^2 - 23$

19. $3x(x^2 - 2x + 1) = 3x(x^2) + 3x(-2x) + 3x(1)$
$= 3x^3 - 6x^2 + 3x$

21. $-5z(3z - 1) = -5z(3z) + (-5z)(-1)$
$= -15z^2 + 5z$

23. $(3x - 5)(2x + 1) = 6x^2 + 3x - 10x - 5$
$= 6x^2 - 7x - 5$

25. $(x^2 - x + 2)(x^2 + x + 1)$

$$\begin{array}{r} x^2 - x + 2 \\ \times\, x^2 + x + 1 \\ \hline x^4 - x^3 + 2x^2 \quad\quad\quad \\ x^3 - x^2 + 2x \quad \\ x^2 - x + 2 \\ \hline x^4 + 0x^3 + 2x^2 + x + 2 \end{array} = x^4 + 2x^2 + x + 2$$

27. $(x + 10)(x - 10) = x^2 - 10^2 = x^2 - 100$

29. $(2x + 3)^2 = (2x)^2 + 2(2x)(3) + 3^2$
$= 4x^2 + 12x + 9$

31. $(x + 3)^3 = x^3 + 3(x)^2(3) + 3(x)(3)^2 + 3^3$
$= x^3 + 9x^2 + 27x + 27$

33. $[(x - 3) + y]^2 = (x - 3)^2 + 2y(x - 3) + y^2$
$= x^2 - 6x + 9 + 2xy - 6y + y^2$
$= x^2 + 2xy + y^2 - 6x - 6y + 9$

35. $[(m - 3) + n][(m - 3) - n] = (m - 3)^2 - (n)^2$
$= m^2 - 6m + 9 - n^2$
$= m^2 - n^2 - 6m + 9$

37. $2x^3 - 6x = 2x(x^2 - 3)$

39. $3x(x - 5) + 8(x - 5) = (x - 5)(3x + 8)$

41. $25y^2 - 4 = (5y)^2 - 2^2 = (5y + 2)(5y - 2)$

43. $(x - 1)^2 - 4 = (x - 1)^2 - (2)^2$
$= [(x - 1) + 2][(x - 1) - 2]$
$= (x + 1)(x - 3)$

45. $x^2 - 4x + 4 = x^2 - 2(2)x + 2^2 = (x - 2)^2$

47. $25z^2 - 30z + 9 = (5z)^2 - 2(5z)(3) + 3^2 = (5z - 3)^2$

49. $4y^2 - 12y + 9 = (2y)^2 - 2(2y)(3) + (3)^2$
$= (2y - 3)^2$

51. $x^3 + 125 = x^3 + 5^3 = (x + 5)(x^2 - 5x + 25)$

53. $8t^3 - 1 = (2t)^3 - 1^3 = (2t - 1)(4t^2 + 2t + 1)$

55. $x^2 + x - 2 = (x + 2)(x - 1)$

57. $3x^2 + 10x - 8 = (3x - 2)(x + 4)$

59. $5x^2 + 31x + 6 = (5x + 1)(x + 6)$

61. $-5y^2 - 8y + 4 = -(5y^2 + 8y - 4)$
$= -(5y - 2)(y + 2)$

63. $x^3 - x^2 + 2x - 2 = x^2(x - 1) + 2(x - 1)$
$= (x - 1)(x^2 + 2)$

65. $2x^3 - x^2 - 6x + 3 = x^2(2x - 1) - 3(2x - 1)$
$= (2x - 1)(x^2 - 3)$

67. $3x^5 + 6x^3 - 2x^2 - 4 = 3x^3(x^2 + 2) - 2(x^2 + 2)$
$= (3x^3 - 2)(x^2 + 2)$

69. $a \cdot c = (2)(9) = 18$. Rewrite the middle term,
$9x = 6x + 3x$, because $(6)(3) = 18$ and $6 + 3 = 9$.
$2x^2 + 9x + 9 = 2x^2 + 6x + 3x + 9$
$= 2x(x + 3) + 3(x + 3)$
$= (x + 3)(2x + 3)$

71. $a \cdot c = (6)(-15) = -90$. Rewrite the middle term,

$-x = -10x + 9x$, because $(-10)(9) = -90$ and $-10 + 9 = -1$.

$$\begin{aligned} 6x^2 - x - 15 &= 6x^2 - 10x + 9x - 15 \\ &= 2x(3x - 5) + 3(3x - 5) \\ &= (2x + 3)(3x - 5) \end{aligned}$$

73. $6x^2 - 54 = 6(x^2 - 9) = 6(x + 3)(x - 3)$

75. $x^3 - x^2 = x^2(x - 1)$

77. $$\begin{aligned} 2x^2 + 4x - 2x^3 &= -2x(-x - 2 + x^2) \\ &= -2x(x^2 - x - 2) \\ &= -2x(x + 1)(x - 2) \end{aligned}$$

79. $$\begin{aligned} 5 - x + 5x^2 - x^3 &= 1(5 - x) + x^2(5 - x) \\ &= (5 - x)(1 + x^2) \end{aligned}$$

81. $$\begin{aligned} 2(x - 2)(x + 1)^2 - 3(x - 2)^2(x + 1) &= (x - 2)(x + 1)[2(x + 1) - 3(x - 2)] \\ &= (x - 2)(x + 1)[2x + 2 - 3x + 6] \\ &= (x - 2)(x + 1)(-x + 8) \\ &= -(x - 2)(x + 1)(x - 8) \end{aligned}$$

83. (a) $$\begin{aligned} V &= \pi R^2 h - \pi r^2 h \\ &= \pi h(R^2 - r^2) \\ &= \pi h(R + r)(R - r) \end{aligned}$$

(b) Let w = thickness of the shell and let p = average radius of the shell.

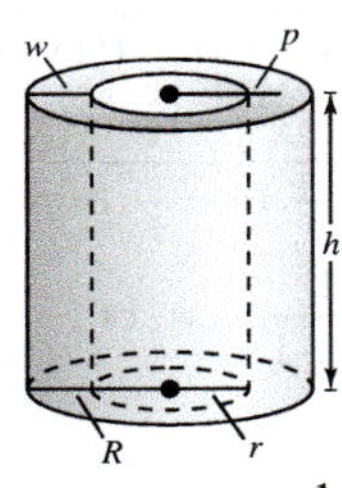

So, $R = p + \frac{1}{2}w$ and $r = p - \frac{1}{2}w$

$$\begin{aligned} V &= \pi h(R + r)(R - r) \\ &= \pi h\left[\left(p + \frac{1}{2}w\right) + \left(p - \frac{1}{2}w\right)\right]\left[\left(p + \frac{1}{2}w\right) - \left(p - \frac{1}{2}w\right)\right] \\ &= \pi h(2p)(w) \\ &= 2\pi pwh \\ &= 2\pi(\text{average radius})(\text{thickness of shell})\, h \end{aligned}$$

85. False. $(4x^2 + 1)(3x + 1) = 12x^3 + 4x^2 + 3x + 1$

87. True. $a^2 - b^2 = (a + b)(a - b)$

89. Because $x^m x^n = x^{m+n}$, the degree of the product is $m + n$.

91. The unknown polynomial may be found by adding $-x^3 + 3x^2 + 2x - 1$ and $5x^2 + 8$:

$$\begin{aligned} (-x^3 + 3x^2 + 2x - 1) + (5x^2 + 8) &= -x^3 + (3x^2 + 5x^2) + 2x + (-1 + 8) \\ &= -x^3 + 8x^2 + 2x + 7 \end{aligned}$$

93. Answers will vary. *Sample answer:* $x^2 - 3$

95. $x^{2n} - y^{2n} = (x^n)^2 - (y^n)^2 = (x^n + y^n)(x^n - y^n)$

This is not completely factored unless $n = 1$.

For $n = 2$: $(x^2 + y^2)(x^2 - y^2) = (x^2 + y^2)(x + y)(x - y)$

For $n = 3$: $(x^3 + y^3)(x^3 - y^3) = (x + y)(x^2 - xy + y^2)(x - y)(x^2 + xy + y^2)$

For $n = 4$: $(x^4 + y^4)(x^4 - y^4) = (x^4 + y^4)(x^2 + y^2)(x + y)(x - y)$

Appendix A.4 Rational Expressions

1. domain

3. complex

5. The domain of $3x^2 - 4x + 7$ is the set of all real numbers.

7. The domain of $\dfrac{1}{3 - x}$ is the set of all real numbers x such that $x \neq 3$.

9. The domain of $\dfrac{x + 6}{3x + 2}$ is the set of all real numbers x such that $x \neq -\dfrac{2}{3}$.

11. The domain of $\dfrac{x^2 - 5x + 6}{x^2 + 6x + 8} = \dfrac{(x - 2)(x - 3)}{(x + 4)(x + 2)}$ is the set of all real numbers x such that $x \neq -4, -2$.

13. The domain of $\sqrt{x - 7}$ is the set of all real numbers x such that $x \geq 7$.

15. The domain of $\dfrac{1}{\sqrt{x - 3}}$ is the set of all real numbers x such that $x > 3$.

17. $\dfrac{15x^2}{10x} = \dfrac{5x(3x)}{5x(2)} = \dfrac{3x}{2}, x \neq 0$

19. $\dfrac{x - 5}{10 - 2x} = \dfrac{x - 5}{-2(x - 5)} = -\dfrac{1}{2}, x \neq 5$

21. $\dfrac{y^2 - 16}{y + 4} = \dfrac{(y + 4)(y - 4)}{y + 4} = y - 4, y \neq -4$

23. $\dfrac{6y + 9y^2}{12y + 8} = \dfrac{3y(3y + 2)}{4(3y + 2)} = \dfrac{3y}{4}, y \neq -\dfrac{2}{3}$

25. $\dfrac{x^2 + 4x - 5}{x^2 + 8x + 15} = \dfrac{(x + 5)(x - 1)}{(x + 5)(x + 3)} = \dfrac{x - 1}{x + 3}, x \neq -5$

27. $\dfrac{x^2 - x - 2}{10 - 3x - x^2} = \dfrac{x^2 - x - 2}{-(x^2 + 3x - 10)}$

$= \dfrac{(x + 1)(x - 2)}{-(x + 5)(x - 2)} = -\dfrac{x + 1}{x + 5}, x \neq 2$

29. $\dfrac{x^2 - 16}{x^3 + x^2 - 16x - 16} = \dfrac{x^2 - 16}{x^2(x + 1) - 16(x + 1)}$

$= \dfrac{x^2 - 16}{(x + 1)(x^2 - 16)}$

$= \dfrac{1}{x + 1}, x \neq \pm 4$

31. $\dfrac{5x^3}{2x^3 + 4} = \dfrac{5x^3}{2(x^3 + 2)}$

When simplifying fractions, only common factors can be divided out, not terms.

33. $\dfrac{5}{x - 1} \cdot \dfrac{x - 1}{25(x - 2)} = \dfrac{1}{5(x - 2)}, x \neq 1$

35. $\dfrac{x^2 - 4}{12} \div \dfrac{2 - x}{2x + 4} = \dfrac{x^2 - 4}{12} \cdot \dfrac{2x + 4}{2 - x}$

$= \dfrac{(x + 2)(x - 2)}{12} \cdot \dfrac{2(x + 2)}{-(x - 2)}$

$= -\dfrac{(x + 2)^2}{6}, x \neq \pm 2$

37. $\dfrac{x^2 + xy - 2y^2}{x^3 + x^2y} \cdot \dfrac{x}{x^2 + 3xy + 2y^2} = \dfrac{(x + 2y)(x - y)}{x^2(x + y)} \cdot \dfrac{x}{(x + 2y)(x + y)} = \dfrac{x - y}{x(x + y)^2}, x \neq -2y$

39. $\dfrac{x - 1}{x + 2} - \dfrac{x - 4}{x + 2} = \dfrac{x - 1 - (x - 4)}{x + 2} = \dfrac{x - 1 - x + 4}{x + 2} = \dfrac{3}{x + 2}$

41. $\frac{1}{3x+2}+\frac{x}{x+1}=\frac{(1)(x+1)}{(3x+2)(x+1)}+\frac{x(3x+2)}{(3x+2)(x+1)}$

$=\frac{x+1+3x^2+2x}{(3x+2)(x+1)}$

$=\frac{3x^2+3x+1}{(3x+2)(x+1)}$

43. $\frac{3}{2x+4}-\frac{x}{x+2}=\frac{3}{2(x+2)}-\frac{x}{x+2}$

$=\frac{3}{2(x+2)}-\frac{2x}{2(x+2)}$

$=\frac{3-2x}{2(x+2)}$

45. $-\frac{1}{x}+\frac{2}{x^2+1}+\frac{1}{x^3+x}=\frac{-(x^2+1)}{x(x^2+1)}+\frac{2x}{x(x^2+1)}+\frac{1}{x(x^2+1)}$

$=\frac{-x^2-1+2x+1}{x(x^2+1)}=\frac{-x^2+2x}{x(x^2+1)}=\frac{-x(x-2)}{x(x^2+1)}$

$=-\frac{x-2}{x^2+1}=\frac{2-x}{x^2+1}, x\neq 0$

47. The minus sign should be distributed to each term in the numerator of the second fraction.

$\frac{x+4}{x+2}-\frac{3x-8}{x+2}=\frac{(x+4)-(3x-8)}{x+2}$

$=\frac{x+4-3x+8}{x+2}$

$=\frac{-2x+12}{x+2}$

$=\frac{-2(x-6)}{x+2}$

49. $\frac{\left(\frac{x}{2}-1\right)}{(x-2)}=\frac{\left(\frac{x}{2}-\frac{2}{2}\right)}{\left(\frac{x-2}{1}\right)}=\frac{x-2}{2}\cdot\frac{1}{x-2}=\frac{1}{2}, x\neq 2$

51. $\frac{\left[\frac{x^2}{(x+1)^2}\right]}{\left[\frac{x}{(x+1)^3}\right]}=\frac{x^2}{(x+1)^2}\cdot\frac{(x+1)^3}{x}$

$=x(x+1), x\neq -1, 0$

53. $\frac{\left(\sqrt{x}-\frac{1}{2\sqrt{x}}\right)}{\sqrt{x}}=\frac{\left(\sqrt{x}-\frac{1}{2\sqrt{x}}\right)}{\sqrt{x}}\cdot\frac{2\sqrt{x}}{2\sqrt{x}}$

$=\frac{2x-1}{2x}, x>0$

55. $x^2(x^2+3)^{-4}+(x^2+3)^3=(x^2+3)^{-4}\left[x^2+(x^2+3)^7\right]$

$=\frac{x^2+(x^2+3)^7}{(x^2+3)^4}$

57. $2x^2(x-1)^{1/2}-5(x-1)^{-1/2}=(x-1)^{-1/2}\left[2x^2(x-1)^1-5\right]=\frac{2x^3-2x^2-5}{(x-1)^{1/2}}$

59. $\frac{3x^{1/3}-x^{-2/3}}{3x^{-2/3}}=\frac{3x^{1/3}-x^{-2/3}}{3x^{-2/3}}\cdot\frac{x^{2/3}}{x^{2/3}}=\frac{3x^1-x^0}{3x^0}=\frac{3x-1}{3}, x\neq 0$

61. $\frac{\left(\frac{1}{x+h}-\frac{1}{x}\right)}{h}=\frac{\left(\frac{1}{x+h}-\frac{1}{x}\right)}{h}\cdot\frac{x(x+h)}{x(x+h)}=\frac{x-(x+h)}{hx(x+h)}=\frac{-h}{hx(x+h)}=-\frac{1}{x(x+h)}, h\neq 0$

63. $$\frac{\left(\frac{1}{x+h-4}-\frac{1}{x-4}\right)}{h}=\frac{\left(\frac{1}{x+h-4}-\frac{1}{x-4}\right)}{h}\cdot\frac{(x-4)(x+h-4)}{(x-4)(x+h-4)}$$
$$=\frac{(x-4)-(x+h-4)}{h(x-4)(x+h-4)}$$
$$=\frac{-h}{h(x-4)(x+h-4)}$$
$$=-\frac{1}{(x-4)(x+h-4)},\ h\neq 0$$

65. $$\frac{\sqrt{x+2}-\sqrt{x}}{2}=\frac{\sqrt{x+2}-\sqrt{x}}{2}\cdot\frac{\sqrt{x+2}+\sqrt{x}}{\sqrt{x+2}+\sqrt{x}}$$

67. $$\frac{\sqrt{t+3}-\sqrt{3}}{t}=\frac{\sqrt{t+3}-\sqrt{3}}{t}\cdot\frac{\sqrt{t+3}+\sqrt{3}}{\sqrt{t+3}+\sqrt{3}}=\frac{(t+3)-3}{t\left(\sqrt{t+3}+\sqrt{3}\right)}=\frac{t}{t\left(\sqrt{t+3}+\sqrt{3}\right)}=\frac{1}{\sqrt{t+3}+\sqrt{3}},\ t\neq 0$$

69. $$\frac{\sqrt{x+h+1}-\sqrt{x+1}}{h}=\frac{\sqrt{x+h+1}-\sqrt{x+1}}{h}\cdot\frac{\sqrt{x+h+1}+\sqrt{x+1}}{\sqrt{x+h+1}+\sqrt{x+1}}$$
$$=\frac{(x+h+1)-(x+1)}{h\left(\sqrt{x+h+1}+\sqrt{x+1}\right)}$$
$$=\frac{h}{h\left(\sqrt{x+h+1}+\sqrt{x+1}\right)}$$
$$=\frac{1}{\sqrt{x+h+1}+\sqrt{x+1}},\ h\neq 0$$

71. $T=10\left(\dfrac{4t^2+16t+75}{t^2+4t+10}\right)$

(a)

t	0	2	4	6	8	10	12	14	16	18	20	22
T	75°	55.9°	48.3°	45°	43.3°	42.3°	41.7°	41.3°	41.1°	40.9°	40.7°	40.6°

(b) T is approaching 40°.

73. $$\text{Probability}=\frac{\text{Shaded area}}{\text{Total area}}=\frac{x(x/2)}{x(2x+1)}=\frac{x/2}{2x+1}\cdot\frac{2}{2}=\frac{x}{2(2x+1)},\ x\neq 0$$

75. (a)

Year, t	Online Banking	Mobile Banking
11	79.1	17.9
12	80.9	24.0
13	83.1	29.6
14	86.0	34.8

(b) The values from the models are close to the actual data.

(c) $\dfrac{\text{Number of households using mobile banking}}{\text{Number of households using online banking}}$

$$= \frac{(0.661t^2 - 47)/(0.007t^2 + 1)}{(-2.9709t + 70.517)/(-0.0474t + 1)}$$

$$= \frac{0.661t^2 - 47}{0.007t^2 + 1} \cdot \frac{-0.0474t + 1}{-2.9709t + 70.517}$$

$$= \frac{(0.661t^2 - 47)(-0.0474t + 1)}{(0.007t^2 + 1)(-2.9709t + 70.517)}$$

$$= \frac{-0.313t^2 + 0.661t^2 + 2.23t - 47}{-0.0208t^3 + 0.494t^2 - 2.97t + 70.5}$$

$$= \frac{0.0313t^3 - 0.661t^2 - 2.23t + 47}{0.0208t^3 - 0.494t^2 + 2.97t - 70.5}$$

$$= \frac{0.0313t^3 - 0.661t^2 - 2.23t + 47}{0.0208t^3 - 0.494t^2 + 2.97t - 70.5}$$

(d) When $t = 11$, $\dfrac{0.0313(11)^3 - 0.661(11)^2 - 2.23(11) + 47}{0.0208(11)^3 - 0.494(11)^2 + 2.97(11) - 70.5} \approx 0.2267.$

When $t = 12$, $\dfrac{0.0313(12)^3 - 0.661(12)^2 - 2.23(12) + 47}{0.0208(12)^3 - 0.494(12)^2 + 2.97(12) - 70.5} \approx 0.2977.$

When $t = 13$, $\dfrac{0.0313(13)^3 - 0.661(13)^2 - 2.23(13) + 47}{0.0208(13)^3 - 0.494(13)^2 - 2.97(13) - 70.5} \approx 0.3578.$

When $t = 14$, $\dfrac{0.0313(14)^3 - 0.661(14)^2 - 2.23(14) + 47}{0.0208(14)^3 - 0.494(14)^2 + 2.97(14) - 70.5} \approx 0.4061.$

Answers will vary.

77. $R_T = \dfrac{1}{\dfrac{1}{R_1} + \dfrac{1}{R_2}} = \dfrac{1}{\dfrac{R_2 + R_1}{R_1R_2}} = \dfrac{R_1R_2}{R_1 + R_2}$

79. False. In order for the simplified expression to be equivalent to the original expression, the domain of the simplified expression needs to be restricted. If n is even, $x \neq \pm 1$. If n is odd, $x \neq 1$.

Appendix A.5 Solving Equations

1. equation

3. extraneous

5.
$$\begin{aligned} x + 11 &= 15 \\ x + 11 - 11 &= 15 - 11 \\ x &= 4 \end{aligned}$$

7.
$$\begin{aligned} 7 - 2x &= 25 \\ 7 - 7 - 2x &= 25 - 7 \\ -2x &= 18 \\ \frac{-2x}{-2} &= \frac{18}{-2} \\ x &= -9 \end{aligned}$$

9.
$$\begin{aligned} 3x - 5 &= 2x + 7 \\ 3x - 2x - 5 &= 2x - 2x + 7 \\ x - 5 &= 7 \\ x - 5 + 5 &= 7 + 5 \\ x &= 12 \end{aligned}$$

11.
$$\begin{aligned} x - 3(2x + 3) &= 8 - 5x \\ x - 6x - 9 &= 8 - 5x \\ -5x - 9 &= 8 - 5x \\ -5x + 5x - 9 &= 8 - 5x + 5x \\ -9 &\neq 8 \end{aligned}$$

Because $-9 = 8$ is a contradiction, the equation has no solution.

13.
$$\frac{3x}{8} - \frac{4x}{3} = 4$$
$$(24)\frac{3x}{8} - (24)\frac{4x}{3} = (24)4$$
$$9x - 32x = 96$$
$$-23x = 96$$
$$x = -\frac{96}{23}$$

15.
$$\frac{5x - 4}{5x + 4} = \frac{2}{3}$$
$$3(5x - 4) = 2(5x + 4)$$
$$15x - 12 = 10x + 8$$
$$5x = 20$$
$$x = 4$$

17.
$$10 - \frac{13}{x} = 4 + \frac{5}{x}$$
$$\frac{10x - 13}{x} = \frac{4x + 5}{x}$$
$$10x - 13 = 4x + 5$$
$$6x = 18$$
$$x = 3$$

19.
$$\frac{x}{x + 4} + \frac{4}{x + 4} + 2 = 0$$
$$\frac{x + 4}{x + 4} + 2 = 0$$
$$1 + 2 = 0$$
$$3 \neq 0$$

Because $3 = 0$ is a contradiction, the equation has no solution.

21.
$$\frac{2}{(x - 4)(x - 2)} = \frac{1}{x - 4} + \frac{2}{x - 2}$$ Multiply each term by $(x - 4)(x - 2)$.
$$2 = 1(x - 2) + 2(x - 4)$$
$$2 = x - 2 + 2x - 8$$
$$2 = 3x - 10$$
$$12 = 3x$$
$$4 = x$$

A check reveals that $x = 4$ yields a denominator of zero. So, $x = 4$ is an extraneous solution, and the original equation has no real solution.

23.
$$\frac{1}{x - 3} + \frac{1}{x + 3} = \frac{10}{x^2 - 9}$$ Multiply each term by $(x + 3)(x - 3)$.
$$\frac{1}{x - 3} + \frac{1}{x + 3} = \frac{10}{(x + 3)(x - 3)}$$
$$1(x + 3) + 1(x - 3) = 10$$
$$2x = 10$$
$$x = 5$$

25.
$$6x^2 + 3x = 0$$
$$3x(2x + 1) = 0$$
$$3x = 0 \quad \text{or} \quad 2x + 1 = 0$$
$$x = 0 \quad \text{or} \quad x = -\tfrac{1}{2}$$

27.
$$x^2 + 10x + 25 = 0$$
$$(x + 5)(x + 5) = 0$$
$$x + 5 = 0$$
$$x = -5$$

29.
$$3 + 5x - 2x^2 = 0$$
$$(3 - x)(1 + 2x) = 0$$
$$3 - x = 0 \quad \text{or} \quad 1 + 2x = 0$$
$$x = 3 \quad \text{or} \quad x = -\tfrac{1}{2}$$

31.
$$16x^2 - 9 = 0$$
$$(4x + 3)(4x + 3) = 0$$
$$4x + 3 = 0 \Rightarrow x = -\tfrac{3}{4}$$
$$4x - 3 = 0 \Rightarrow x = \tfrac{3}{4}$$

33.
$$\tfrac{3}{4}x^2 + 8x + 20 = 0$$
$$4\left(\tfrac{3}{4}x^2 + 8x + 20\right) = 4(0)$$
$$3x^2 + 32x + 80 = 0$$
$$(3x + 20)(x + 4) = 0$$
$$3x + 20 = 0 \quad \text{or} \quad x + 4 = 0$$
$$x = -\tfrac{20}{3} \quad \text{or} \quad x = -4$$

35.
$$x^2 = 49$$
$$x = \pm 7$$

37. $3x^2 = 81$

$$x^2 = 27$$

$$x = \pm 3\sqrt{3}$$

$$\approx \pm 5.20$$

39. $(x - 4)^2 = 49$

$$x - 4 = \pm 7$$

$$x = 4 \pm 7$$

$$x = 11 \text{ or } x = -3$$

41. $(2x - 1)^2 = 18$

$$2x - 1 = \pm\sqrt{18}$$

$$2x = 1 \pm 3\sqrt{2}$$

$$x = \frac{1 \pm 3\sqrt{2}}{2}$$

$$\approx 2.62, \ -1.62$$

43. $x^2 + 4x - 32 = 0$

$$x^2 + 4x = 32$$

$$x^2 + 4x + 2^2 = 32 + 2^2$$

$$(x + 2)^2 = 36$$

$$x + 2 = \pm 6$$

$$x = -2 \pm 6$$

$$x = 4 \quad \text{or} \quad x = -8$$

45. $x^2 + 4x + 2 = 0$

$$x^2 + 4x = -2$$

$$x^2 + 4x + 2^2 = -2 + 2^2$$

$$(x + 2)^2 = 2$$

$$x + 2 = \pm\sqrt{2}$$

$$x = -2 \pm \sqrt{2}$$

47. $6x^2 - 12x = -3$

$$x^2 - 2x = -\frac{1}{2}$$

$$x^2 - 2x + 1^2 = -\frac{1}{2} + 1^2$$

$$(x - 1)^2 = \frac{1}{2}$$

$$x - 1 = \pm\sqrt{\frac{1}{2}}$$

$$x = 1 \pm \sqrt{\frac{1}{2}}$$

$$x = 1 \pm \frac{\sqrt{2}}{2}$$

49. $2x^2 + 5x - 8 = 0$

$$2x^2 + 5x = 8$$

$$x^2 + \frac{5}{2}x = 4$$

$$x^2 + \frac{5}{2}x + \left(\frac{5}{4}\right)^2 = 4 + \left(\frac{5}{4}\right)^2$$

$$\left(x + \frac{5}{4}\right)^2 = \frac{89}{16}$$

$$x + \frac{5}{4} = \pm\frac{\sqrt{89}}{4}$$

$$x = -\frac{5}{4} \pm \frac{\sqrt{89}}{4}$$

$$x = \frac{-5 \pm \sqrt{89}}{4}$$

51. $2x^2 + x - 1 = 0$

$$x = \frac{-b \pm \sqrt{b^2 - 4ac}}{2a}$$

$$= \frac{-1 \pm \sqrt{1^2 - 4(2)(-1)}}{2(2)}$$

$$= \frac{-1 \pm 3}{4} = \frac{1}{2}, -1$$

53. $9x^2 + 30x + 25 = 0$

$$x = \frac{-b \pm \sqrt{b^2 - 4ac}}{2a}$$

$$= \frac{-30 \pm \sqrt{30^2 - 4(9)(25)}}{2(9)}$$

$$= \frac{-30 \pm 0}{18} = -\frac{5}{3}$$

55. $2x^2 - 7x + 1 = 0$

$$x = \frac{-b \pm \sqrt{b^2 - 4ac}}{2a}$$

$$= \frac{-(-7) \pm \sqrt{(-7)^2 - 4(2)(1)}}{2(2)}$$

$$= \frac{7 \pm \sqrt{49 - 8}}{2(2)}$$

$$= \frac{7 \pm \sqrt{41}}{4}$$

$$= \frac{7}{4} \pm \frac{\sqrt{41}}{4}$$

57. $12x - 9x^2 = -3$

$-9x^2 + 12x + 3 = 0$

$$x = \frac{-b \pm \sqrt{b^2 - 4ac}}{2a}$$

$$= \frac{-12 \pm \sqrt{12^2 - 4(-9)(3)}}{2(-9)}$$

$$= \frac{-12 \pm 6\sqrt{7}}{-18} = \frac{2}{3} \pm \frac{\sqrt{7}}{3}$$

59. $2 + 2x - x^2 = 0$

$-x^2 + 2x + 2 = 0$

$$x = \frac{-b \pm \sqrt{b^2 - 4ac}}{2a}$$

$$= \frac{-2 \pm \sqrt{2^2 - 4(-1)(2)}}{2(-1)}$$

$$= \frac{-2 \pm 2\sqrt{3}}{-2}$$

$$= 1 \pm \sqrt{3}$$

61. $8t = 5 + 2t^2$

$-2t^2 + 8t - 5 = 0$

$$t = \frac{-b \pm \sqrt{b^2 - 4ac}}{2a}$$

$$= \frac{-8 \pm \sqrt{8^2 - 4(-2)(-5)}}{2(-2)}$$

$$= \frac{-8 \pm 2\sqrt{6}}{-4} = 2 \pm \frac{\sqrt{6}}{2}$$

63. $(y - 5)^2 = 2y$

$y^2 - 12y + 25 = 0$

$$y = \frac{-b \pm \sqrt{b^2 - 4ac}}{2a}$$

$$= \frac{-(-12) \pm \sqrt{(-12)^2 - 4(1)(25)}}{2(1)}$$

$$= \frac{12 \pm 2\sqrt{11}}{2} = 6 \pm \sqrt{11}$$

65. $x^2 - 2x - 1 = 0$ Complete the square.

$x^2 - 2x = 1$

$x^2 - 2x + 1^2 = 1 + 1^2$

$(x - 1)^2 = 2$

$x - 1 = \pm\sqrt{2}$

$x = 1 \pm \sqrt{2}$

67. $(x + 2)^2 = 64$ Extract square roots.

$x + 2 = \pm 8$

$x + 2 = 8$ or $x + 2 = -8$

$x = 6$ or $x = -10$

69. $x^2 - x - \frac{11}{4} = 0$ Complete the square.

$x^2 - x = \frac{11}{4}$

$x^2 - x + \left(\frac{1}{2}\right)^2 = \frac{11}{4} + \left(\frac{1}{2}\right)^2$

$\left(x - \frac{1}{2}\right)^2 = \frac{12}{4}$

$x - \frac{1}{2} = \pm\sqrt{\frac{12}{4}}$

$x = \frac{1}{2} \pm \sqrt{3}$

71. $3x + 4 = 2x^2 - 7$ Quadratic Formula

$0 = 2x^2 - 3x - 11$

$$x = \frac{-(-3) \pm \sqrt{(-3)^2 - 4(2)(-11)}}{2(2)}$$

$$= \frac{3 \pm \sqrt{97}}{4}$$

$$= \frac{3}{4} \pm \frac{\sqrt{97}}{4}$$

73. $6x^4 - 54x^2 = 0$

$6x^2(x^2 - 9) = 0$

$6x^2 = 0 \Rightarrow x = 0$

$x^2 - 9 = 0 \Rightarrow x = \pm 3$

75. $x^3 + 2x^2 - 8x = 16$

$x^3 + 2x^2 - 8x - 16 = 0$

$x^2(x + 2) - 8(x + 2) = 0$

$(x + 2)(x^2 - 8) = 0$

$x + 2 = 0 \Rightarrow x = -2$

$x^2 - 8 = 0 \Rightarrow \pm\sqrt{8} = \pm 2\sqrt{2}$

77. $\sqrt{5x} - 10 = 0$

$\sqrt{5x} = 10$

$\left(\sqrt{5x}\right)^2 = (10)^2$

$5x = 100$

$x = 20$

79. $4 + \sqrt[3]{2x - 9} = 0$

$$\begin{aligned} \sqrt[3]{2x - 9} &= -4 \\ \left(\sqrt[3]{2x - 9}\right)^3 &= (-4)^3 \\ 2x - 9 &= -64 \\ 2x &= -55 \\ x &= -\frac{55}{2} \end{aligned}$$

81.

$$\begin{aligned} \sqrt{x + 8} &= 2 + x \\ \left(\sqrt{x + 8}\right)^2 &= (2 + x)^2 \\ x + 8 &= x^2 + 4x + 4 \\ 0 &= x^2 + 3x - 4 \\ x^2 + 3x - 4 &= 0 \\ (x + 4)(x - 1) &= 0 \end{aligned}$$

$x + 4 = 0 \Rightarrow x = -4$, extraneous

$x - 1 = 0 \Rightarrow x = 1$

83. $\sqrt{x - 3} + 1 = \sqrt{x}$

$$\begin{aligned} \sqrt{x - 3} &= \sqrt{x} - 1 \\ \left(\sqrt{x - 3}\right)^2 &= \left(\sqrt{x} - 1\right)^2 \\ x - 3 &= x - 2\sqrt{x} + 1 \\ -4 &= -2\sqrt{x} \\ 2 &= \sqrt{x} \\ (2)^2 &= \left(\sqrt{x}\right)^2 \\ 4 &= x \end{aligned}$$

85. $(x - 5)^{3/2} = 8$

$$\begin{aligned} (x - 5)^3 &= 8^2 \\ x - 5 &= \sqrt[3]{64} \\ x &= 5 + 4 = 9 \end{aligned}$$

87. $3x(x - 1)^{1/2} + 2(x - 1)^{3/2} = 0$

$$\begin{aligned} (x - 1)^{1/2}\left[3x + 2(x - 1)\right] &= 0 \\ (x - 1)^{1/2}(5x - 2) &= 0 \end{aligned}$$

$(x - 1)^{1/2} = 0 \Rightarrow x - 1 = 0 \Rightarrow x = 1$

$5x - 2 = 0 \Rightarrow x = \frac{2}{5}$, extraneous

89. $|2x - 5| = 11$

$2x - 5 = 11 \Rightarrow x = 8$

$-(2x - 5) = 11 \Rightarrow x = -3$

91. $|x + 1| = x^2 - 5$

First equation:

$$\begin{aligned} x + 1 &= x^2 - 5 \\ x^2 - x - 6 &= 0 \\ (x - 3)(x + 2) &= 0 \end{aligned}$$

$x - 3 = 0 \Rightarrow x = 3$

$x + 2 = 0 \Rightarrow x = -2$

Second equation:

$$\begin{aligned} -(x + 1) &= x^2 - 5 \\ -x - 1 &= x^2 - 5 \\ x^2 + x - 4 &= 0 \\ x &= \frac{-1 + \sqrt{17}}{2} \end{aligned}$$

Only $x = 3$ and $x = \dfrac{-1 - \sqrt{17}}{2}$ are solutions of the original equation. $x = -2$ and $x = \dfrac{-1 + \sqrt{17}}{2}$ are extraneous.

93.

$$\begin{aligned} V &= \frac{4}{3}\pi r^3 \\ 5.96 &= \frac{4}{3}\pi r^3 \\ 17.88 &= 4\pi r^3 \\ \frac{17.88}{4\pi} &= r^3 \\ r &= \sqrt[3]{\frac{4.47}{\pi}} \approx 1.12 \text{ inches} \end{aligned}$$

95. Let $y = 18$.

$$\begin{aligned} y &= 0.514x - 14.75 \\ 18 &= 0.514x - 14.75 \\ 32.75 &= 0.514x \\ \frac{32.75}{0.514} &= x \\ 63.7 &= x \end{aligned}$$

So, the height of the female is about 63.7 inches.

97. False.

$$\begin{aligned}\sqrt{2x+1} &= -2 + \sqrt{x+1}\\ 2x + 1 &= 4 - 4\sqrt{x+1} + (x+1)\\ x - 4 &= -4\sqrt{x+1}\\ x^2 - 8x + 16 &= 16(x+1)\\ x^2 - 24x &= 0\\ x(x-24) &= 0\end{aligned}$$

$x = 0 \quad 1 \neq -2 + 1, \quad x = 24 \quad 5 \neq -2 + 5$

99. $\sqrt{x-10} - \sqrt{x-10} = 0$

$\sqrt{x-10} = \sqrt{x-10}$

False. The equation is an identity, so every real number is a solution.

Appendix A.6 Linear Inequalities in One Variable

1. solution set

3. double

5. Interval: $[-2, 6)$

Inequality: $-2 \leq x < 6$; The interval is bounded.

7. Interval: $[-1, 5]$

Inequality: $-1 \leq x \leq 5$; The interval is bounded.

9. Interval: $(11, \infty)$

Inequality: $x > 11$; The interval is unbounded.

11. Interval: $(-\infty, -2)$

Inequality: $x < -2$; The interval is unbounded.

13. $4x < 12$

$\frac{1}{4}(4x) < \frac{1}{4}(12)$

$x < 3$

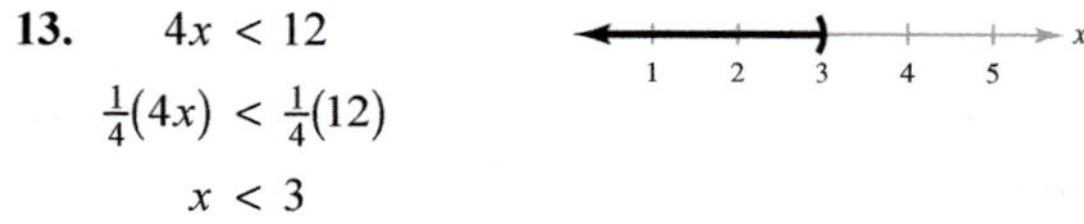

15. $-2x > -3$

$-\frac{1}{2}(-2x) < \left(-\frac{1}{2}\right)(-3)$

$x < \frac{3}{2}$

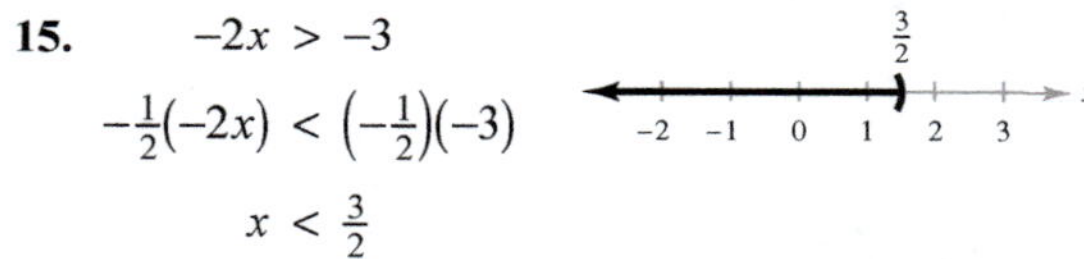

17. $x - 5 \geq 7$

$x \geq 12$

19. $2x + 7 < 3 + 4x$

$-2x < -4$

$x > 2$

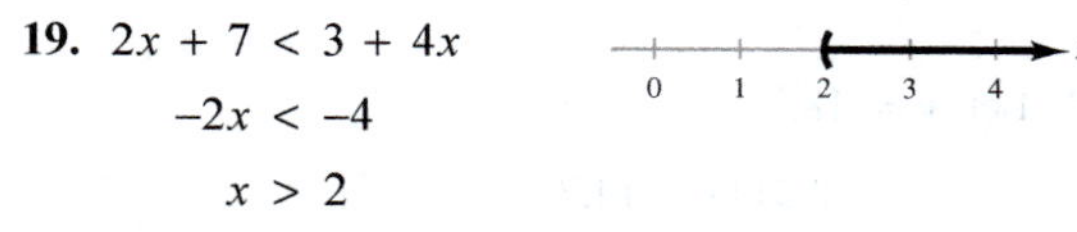

21. $3x - 4 \geq 4 - 5x$

$8x \geq 8$

$x \geq 1$

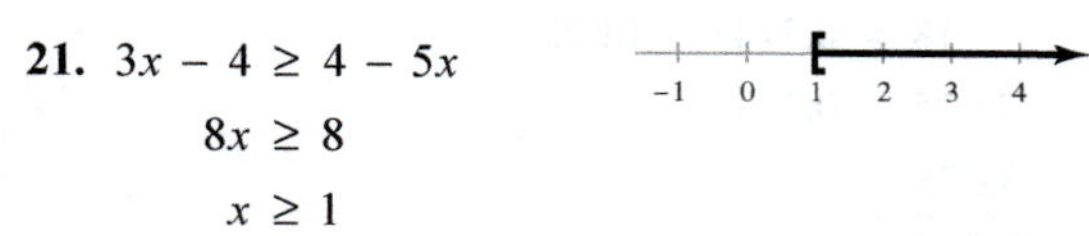

23. $4 - 2x < 3(3 - x)$

$4 - 2x < 9 - 3x$

$x < 5$

25. $\frac{3}{4}x - 6 \leq x - 7$

$-\frac{1}{4}x \leq -1$

$x \geq 4$

27. $\frac{1}{2}(8x + 1) \geq 3x + \frac{5}{2}$

$4x + \frac{1}{2} \geq 3x + \frac{5}{2}$

$x \geq 2$

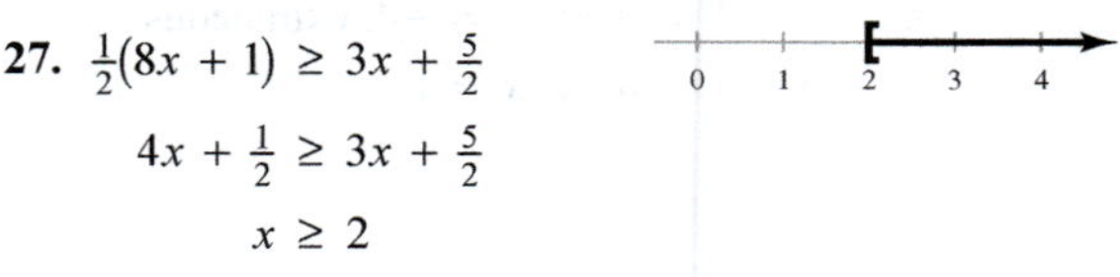

29. $3.6x + 11 \geq -3.4$

$3.6x \geq 14.4$

$x \geq -4$

31. $1 < 2x + 3 < 9$

$-2 < 2x < 6$

$-1 < x < 3$

33. $0 < 3(x + 7) \leq 20$

$0 < x + 7 \leq \frac{20}{3}$

$-7 < x \leq -\frac{1}{3}$

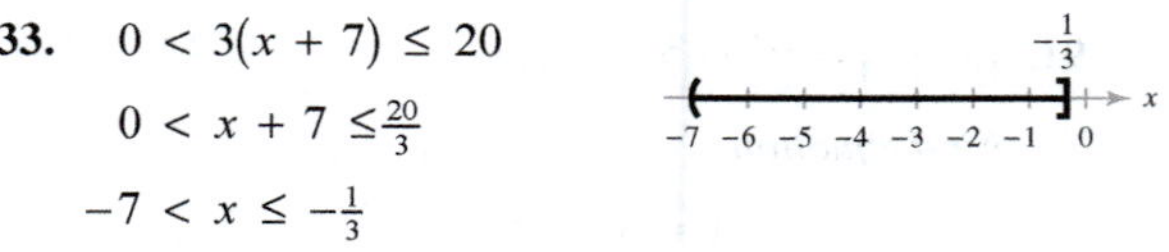

35. $-4 < \dfrac{2x - 3}{3} < 4$

$-12 < 2x - 3 < 12$

$-9 < 2x < 15$

$-\dfrac{9}{2} < x < \dfrac{15}{2}$

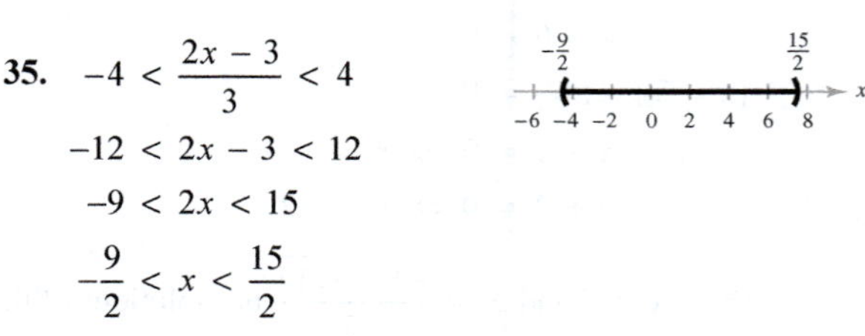

37. $-1 < \dfrac{-x - 2}{3} \leq 1$

$-3 < -x - 2 \leq 3$

$-1 < -x \leq 5$

$1 > x \geq -5$

$-5 \leq x < 1$

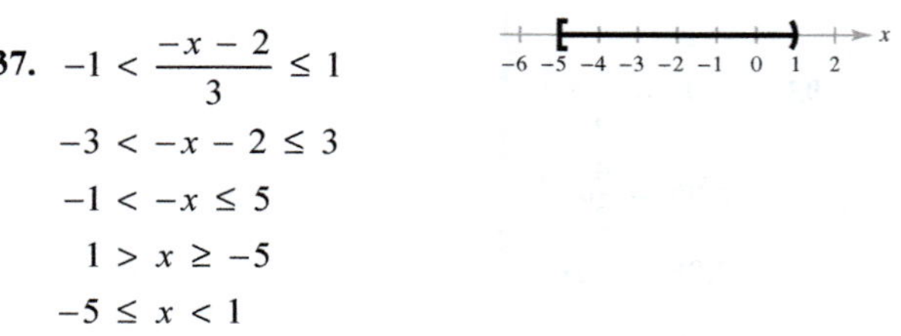

39. $\frac{3}{4} > x + 1 > \frac{1}{4}$

$-\frac{1}{4} > x > -\frac{3}{4}$

$-\frac{3}{4} < x < -\frac{1}{4}$

41. $3.2 \le 0.4x - 1 \le 4.4$

$4.2 \le 0.4x \le 5.4$

$10.5 \le x \le 13.5$

43. $|x| < 5$

$-5 < x < 5$

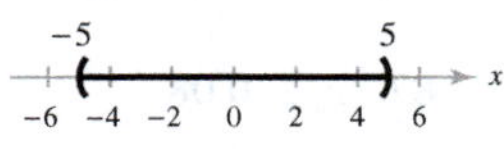

45. $\left|\frac{x}{2}\right| > 1$

$\frac{x}{2} < -1$ or $\frac{x}{2} > 1$

$x < -2 \qquad x > 2$

47. $|x - 5| < -1$

No solution. The absolute value of a number cannot be less than a negative number.

49. $|x - 20| \le 6$

$-6 \le x - 20 \le 6$

$14 \le x \le 26$

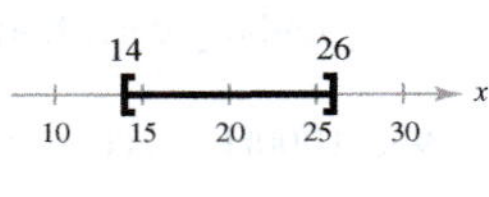

51. $|7 - 2x| \ge 9$

$7 - 2x \le -9$ or $7 - 2x \ge 9$

$-2x \le -16 \qquad -2x \ge 2$

$x \ge 8 \qquad x \le -1$

53. $\left|\frac{x - 3}{2}\right| \ge 4$

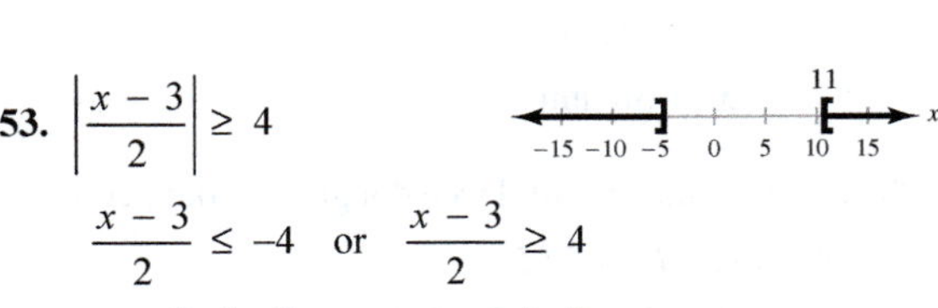

$\frac{x - 3}{2} \le -4$ or $\frac{x - 3}{2} \ge 4$

$x - 3 \le -8 \qquad x - 3 \ge 8$

$x \le -5 \qquad x \ge 11$

55. $|9 - 2x| - 2 < -1$

$|9 - 2x| < 1$

$-1 < 9 - 2x < 1$

$-10 < -2x < -8$

$5 > x > 4$

$4 < x < 5$

57. $2|x + 10| \ge 9$

$|x + 10| \ge \frac{9}{2}$

$x + 10 \le -\frac{9}{2}$ or $x + 10 \ge \frac{9}{2}$

$x \le -\frac{29}{2} \qquad x \ge -\frac{11}{2}$

59. $7x > 21$

$x > 3$

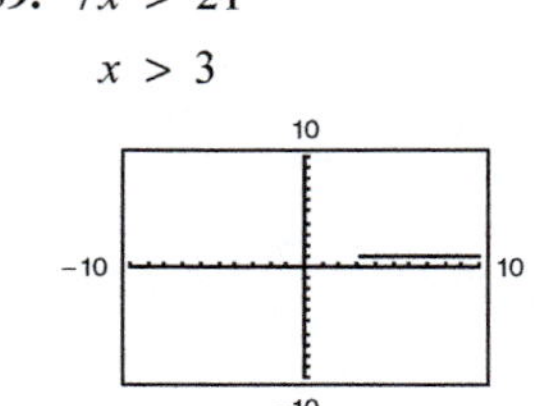

61. $8 - 3x \ge 2$

$-3x \ge -6$

$x \le 2$

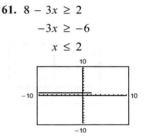

63. $4(x - 3) \le 8 - x$

$4x - 12 \le 8 - x$

$5x \le 20$

$x \le 4$

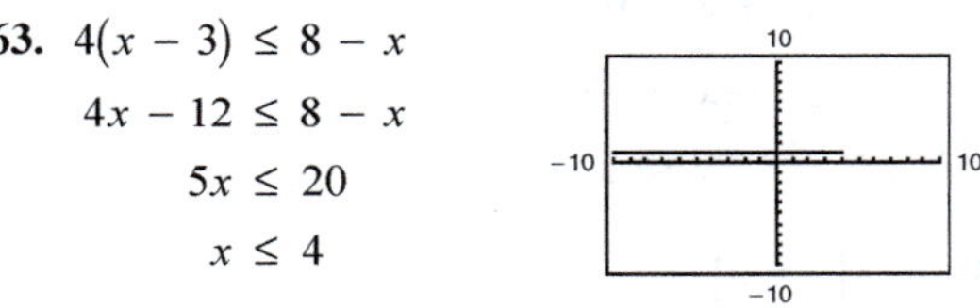

65. $|x - 8| \le 14$

$-14 \le x - 8 \le 14$

$-6 \le x \le 22$

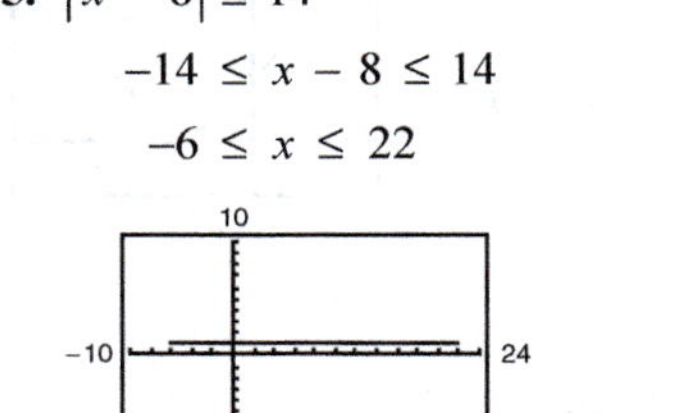

67. $2|x + 7| \ge 13$

$|x + 7| \ge \frac{13}{2}$

$x + 7 \le -\frac{13}{2}$ or $x + 7 \ge \frac{13}{2}$

$x \le -\frac{27}{2} \qquad x \ge -\frac{1}{2}$

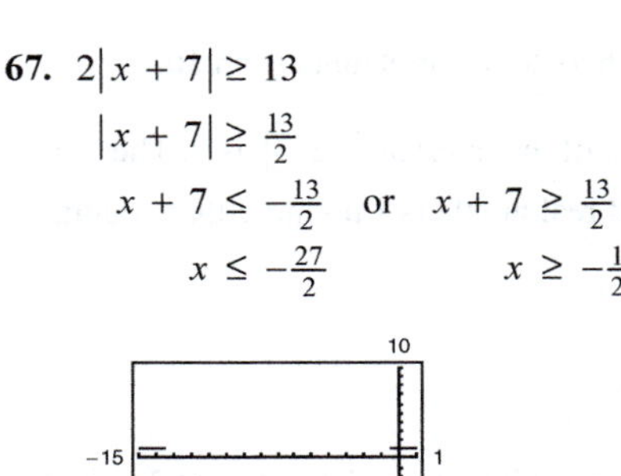

69. $y = 3x - 1$

(a) $y \geq 2$

$3x - 1 \geq 2$

$3x \geq 3$

$x \geq 1$

(b) $y \leq 0$

$3x - 1 \leq 0$

$3x \leq 1$

$x \leq \frac{1}{3}$

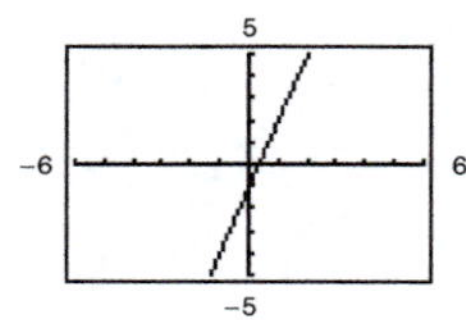

71. $y = -\frac{1}{2}x + 2$

(a) $0 \leq y \leq 3$

$0 \leq -\frac{1}{2}x + 2 \leq 3$

$-2 \leq -\frac{1}{2}x \leq 1$

$4 \geq x \geq -2$

(b) $y \geq 0$

$-\frac{1}{2}x + 2 \geq 0$

$-\frac{1}{2}x \geq -2$

$x \leq 4$

73. $y = |x - 3|$

(a) $y \leq 2$

$|x - 3| \leq 2$

$-2 \leq x - 3 \leq 2$

$1 \leq x \leq 5$

(b) $y \geq 4$

$|x - 3| \geq 4$

$x - 3 \leq -4$ or $x - 3 \geq 4$

$x \leq -1$ or $x \geq 7$

75. All real numbers less than 8 units from 10.

77. The midpoint of the interval $[-3, 3]$ is 0. The interval represents all real numbers x no more than 3 units from 0.

$|x - 0| \leq 3$

$|x| \leq 3$

79. The graph shows all real numbers at least 3 units from 7.

$|x - 7| \geq 3$

81. All real numbers less than 3 units from 7

$|x - 7| \geq 3$

83. All real numbers less than 4 units from −3

$|x - (-3)| < 4$

$|x + 3| < 4$

85. $\$7.25 \leq P \leq \$7.75$

87. $r \leq 0.08$

89. $r = 220 - A = 220 - 20 = 200$ beats per minute

$0.50(200) \leq r \leq 0.85(200)$

$100 \leq r \leq 170$

The target heart rate is at least 100 beats per minute and at most 170 beats per minute.

91. $9.00 + 0.75x > 13.50$

$0.75x > 4.50$

$x > 6$

You must produce at least 6 units each hour in order to yield a greater hourly wage at the second job.

93. $1000(1 + r(10)) > 2000.00$

$1 + 10r > 2$

$10r > 1$

$r > 0.1$

The rate must be greater than 10%.

95. $R > C$

$115.95x > 95x + 750$

$20.95x > 750$

$x \geq 35.7995$

$x \geq 36$ units

97. Let x = number of dozen doughnuts sold per day.

Revenue: $R = 7.95x$

Cost: $C = 1.45x + 165$

$P = R - C$

$= 7.95x - (1.45x + 165)$

$= 6.50x - 165$

$400 \leq P \leq 1200$

$400 \leq 6.50x - 165 \leq 1200$

$565 \leq 6.50x \leq 1365$

$86.9 \leq x \leq 210$

The daily sales vary between 87 and 210 dozen doughnuts per day.

99. (a)

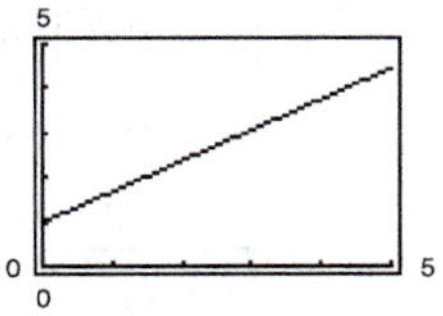

(b) From the graph you see that $y \geq 3$ when $x \geq 2.9$.

(c) Algebraically:

$$3 \leq 0.692x + 0.988$$
$$2.012 \leq 0.692x$$
$$2.91 \leq x$$
$$x \geq 2.91$$

(d) IQ scores are not a good predictor of GPAs. Other factors include study habits, class attendance, and attitudes.

101. (a) $W = 0.903t + 26.08$

$$30 \leq 0.903t + 26.08 \leq 32$$
$$3.92 \leq 0.903t \leq 6.56$$
$$4.34 \leq t \leq 5.92$$

Between the years 2004 and 2006, the mean hourly wage was between \$30 and \$32.

(b) $0.903t + 26.08 \geq 45$

$$0.903t \geq 18.92$$
$$t \geq 20.95$$

The mean hourly wage will exceed \$45 sometime during the year 2020.

103. $$\left|\frac{t - 15.6}{1.9}\right| < 1$$

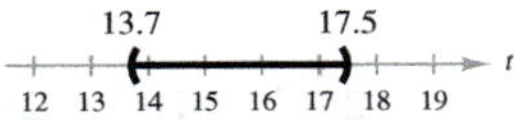

$$-1 < \frac{t - 15.6}{1.9} < 1$$
$$-1.9 < t - 15.6 < 1.9$$
$$13.7 < t < 17.5$$

Two-thirds of the workers could perform the task in the time interval between 13.7 minutes and 17.5 minutes.

105. $1 \text{ oz} = \frac{1}{16} \text{ lb}$, so $\frac{1}{2} \text{ oz} = \frac{1}{32} \text{ lb}$.

Because $8.99 \cdot \frac{1}{32} = 0.2809375$, you may be undercharged or overcharged by \$0.28.

107. $|s - 10.4| \leq \frac{1}{16}$

$$-\tfrac{1}{16} \leq s - 10.4 \leq \tfrac{1}{16}$$
$$-0.0625 \leq s - 10.4 \leq 0.0625$$
$$10.3375 \leq s \leq 10.4625$$

Because $A = s^2$,

$$(10.3375)^2 \leq \text{area} \leq (10.4625)^2$$
$$106.864 \text{ in.}^2 \leq \text{area} \leq 109.464 \text{ in.}^2.$$

109. True. This is the Addition of a Constant Property of Inequalities.

111. False. If $-10 \leq x \leq 8$, then $10 \geq -x$ and $-x \geq -8$.

113. Answer not unique. Sample answer: $x < x + 1$

115. Answer not unique. *Sample answer:* $|ax - b| \leq c$, if $a = 1$, $b = 5$, and $c = 5$, then

$$|x - 5| \leq 5$$
$$-5 \leq x - 5 \leq 5$$
$$0 \leq x \leq 10.$$

Appendix A.7 Errors and the Algebra of Calculus

1. numerator

3. The middle term needs to be included.

$$(x + 3)^2 = x^2 + 6x + 9$$

5. $\sqrt{x + 9} \neq \sqrt{x} + 3$

Do not apply the radical to the terms.

$\sqrt{x + 9}$ does not simplify.

7. $$\frac{2x^2 + 1}{5x} \neq \frac{2x + 1}{5}$$

Divide out common factors not common terms.

$\frac{2x^2 + 1}{5x}$ cannot be simplified.

9. $(4x)^2 \neq 4x^2$

The exponent applies to the coefficient also.

$$(4x)^2 = 16x^2$$

11. $\dfrac{3}{x} + \dfrac{4}{y} = \dfrac{3}{x}\cdot\dfrac{y}{y} + \dfrac{4}{y}\cdot\dfrac{x}{x} = \dfrac{3y + 4x}{xy}$

To add fractions, they must have a common denominator.

13. $$2x(x + 2)^{-1/2} + (x + 2)^{1/2} = (x + 2)^{-1/2}\left[2x + (x + 2)\right]$$
$$= (x + 2)^{-1/2}(3x + 2)$$
$$= \frac{3x + 2}{(x + 2)^{1/2}}$$

15. $$4x^3(2x - 1)^{3/2} - 2x(2x - 1)^{-1/2} = 2x(2x - 1)^{-1/2}\left[2x^2(2x - 1)^2 - 1\right]$$
$$= 2x(2x - 1)^{-1/2}\left[2x^2(4x^2 - 4x + 1) - 1\right]$$
$$= 2x(2x - 1)^{-1/2}(8x^4 - 8x^3 + 2x^2 - 1)$$
$$= \frac{2x(8x^4 - 8x^3 + 2x^2 - 1)}{(2x - 1)^{1/2}}$$

17. $\dfrac{5x + 3}{4} = \dfrac{1}{4}(5x + 3)$

The required factor is $5x + 3$.

19. $\frac{2}{3}x^2 + \frac{1}{3}x + 5 = \frac{2}{3}x^2 + \frac{1}{3}x + \frac{15}{3} = \frac{1}{3}(2x^2 + x + 15)$

The required factor is $2x^2 + x + 15$.

21. $x^{1/3} - 5x^{4/3} = x^{1/3}(1 - 5x^{3/3}) = x^{1/3}(1 - 5x)$

The required factor is $1 - 5x$.

23. $$\tfrac{1}{10}(2x + 1)^{5/2} - \tfrac{1}{6}(2x + 1)^{3/2} = \tfrac{3}{30}(2x + 1)^{3/2}(2x + 1)^1 - \tfrac{5}{30}(2x + 1)^{3/2}$$
$$= \tfrac{1}{30}(2x + 1)^{3/2}\left[3(2x + 1) - 5\right]$$
$$= \tfrac{1}{30}(2x + 1)^{3/2}(6x - 2)$$
$$= \tfrac{1}{30}(2x + 1)^{3/2}2(3x - 1)$$
$$= \tfrac{1}{15}(2x + 1)^{3/2}(3x - 1)$$

The required factor is $3x - 1$.

25. $x^2(x^3 - 1)^4 = \frac{1}{3}(x^3 - 1)^4(3x^2)$

The required factor is $\frac{1}{3}$.

27. $$\frac{4x + 6}{(x^2 + 3x + 7)^3} = \frac{2(2x + 3)}{(x^2 + 3x + 7)^3}$$
$$= \frac{2}{1}\cdot\frac{(2x + 3)}{1}\cdot\frac{1}{(x^2 + 3x + 7)^3}$$
$$= (2)\frac{1}{(x^2 + 3x + 7)^3}(2x + 3)$$

The required factor is 2.

29. $$4x^2 + \frac{6y^2}{10} = 4x^2\frac{(1/4)}{(1/4)} + \frac{6y^2}{10}\frac{(1/2)}{(1/2)}$$
$$= \frac{x^2}{1/4} + \frac{3y^2}{5}$$

31. $$\frac{25x^2}{36} + \frac{4y^2}{9} = \frac{25x^2}{36}\frac{(1/25)}{(1/25)} + \frac{4y^2}{9}\frac{(1/4)}{(1/4)}$$
$$= \frac{x^2}{36/25} + \frac{y^2}{9/4}$$

33. $\dfrac{x^2}{\frac{3}{10}}\dfrac{(10)}{(10)} - \dfrac{y^2}{\frac{4}{5}}\dfrac{(5)}{(5)} = \dfrac{10x^2}{3} - \dfrac{5y^2}{4}$

35. $\dfrac{7}{(x + 3)^5} = 7(x + 3)^{-5}$

37. $\dfrac{2x^5}{(3x + 5)^4} = 2x^5(3x + 5)^{-4}$

39. $\dfrac{4}{3x} + \dfrac{4}{x^4} - \dfrac{7x}{\sqrt[3]{2x}} = \dfrac{4}{3}x^{-1} + 4x^{-4} - 7x(2x)^{-1/3}$

41. $$\frac{x^2 + 6x + 12}{3x} = \frac{x^2}{3x} + \frac{6x}{3x} + \frac{12}{3x}$$
$$= \frac{x}{3} + 2 + \frac{4}{x}$$

43. $$\frac{4x^3 - 7x^2 + 1}{x^{1/3}} = \frac{4x^3}{x^{1/3}} - \frac{7x^2}{x^{1/3}} + \frac{1}{x^{1/3}}$$
$$= 4x^{3-1/3} - 7x^{2-1/3} + \frac{1}{x^{1/3}}$$
$$= 4x^{8/3} - 7x^{5/3} + \frac{1}{x^{1/3}}$$

45. $$\frac{3 - 5x^2 - x^4}{\sqrt{x}} = \frac{3}{\sqrt{x}} - \frac{5x^2}{\sqrt{x}} - \frac{x^4}{\sqrt{x}}$$
$$= \frac{3}{\sqrt{x}} - 5x^{2-1/2} - x^{4-1/2}$$
$$= \frac{3}{x^{1/2}} - 5x^{3/2} - x^{7/2}$$

47. $$\frac{-2(x^2 - 3)^{-3}(2x)(x + 1)^3 - 3(x + 1)^2(x^2 - 3)^{-2}}{\left[(x + 1)^3\right]^2} = \frac{(x^2 - 3)^{-3}(x + 1)^2\left[-4x(x + 1) - 3(x^2 - 3)\right]}{(x + 1)^6}$$
$$= \frac{-4x^2 - 4x - 3x^2 + 9}{(x^2 - 3)^3(x + 1)^4}$$
$$= \frac{-7x^2 - 4x + 9}{(x^2 - 3)^3(x + 1)^4}$$

49. $$\frac{(6x + 1)^3(27x^2 + 2) - (9x^3 + 2x)(3)(6x + 1)^2(6)}{\left[(6x + 1)^3\right]^2} = \frac{(6x + 1)^2\left[(6x + 1)(27x^2 + 2) - 18(9x^3 + 2x)\right]}{(6x + 1)^6}$$
$$= \frac{162x^3 + 12x + 27x^2 + 2 - 162x^3 - 36x}{(6x + 1)^4}$$
$$= \frac{27x^2 - 24x + 2}{(6x + 1)^4}$$

51. $$\frac{(x + 2)^{3/4}(x + 3)^{-2/3} - (x + 3)^{1/3}(x + 2)^{-1/4}}{\left[(x + 2)^{3/4}\right]^2} = \frac{(x + 2)^{-1/4}(x + 3)^{-2/3}\left[(x + 2) - (x + 3)\right]}{(x + 2)^{6/4}}$$
$$= \frac{x + 2 - x - 3}{(x + 2)^{1/4}(x + 3)^{2/3}(x + 2)^{6/4}}$$
$$= -\frac{1}{(x + 3)^{2/3}(x + 2)^{7/4}}$$

53. $$\frac{2(3x - 1)^{1/3} - (2x + 1)(1/3)(3x - 1)^{-2/3}(3)}{(3x - 1)^{2/3}} = \frac{(3x - 1)^{-2/3}\left[2(3x - 1) - (2x + 1)\right]}{(3x - 1)^{2/3}}$$
$$= \frac{6x - 2 - 2x - 1}{(3x - 1)^{2/3}(3x - 1)^{2/3}}$$
$$= \frac{4x - 3}{(3x - 1)^{4/3}}$$

55. $$\frac{1}{(x^2 + 4)^{1/2}} \cdot \frac{1}{2}(x^2 + 4)^{-1/2}(2x) = \frac{1}{(x^2 + 4)^{1/2}} \cdot \frac{1}{(x^2 + 4)^{1/2}} \cdot \frac{1}{2}(2x)$$
$$= \frac{1}{(x^2 + 4)^1}(x)$$
$$= \frac{x}{x^2 + 4}$$

57. $\left(x^2+5\right)^{1/2}\left(\dfrac{3}{2}\right)(3x-2)^{1/2}(3)+(3x-2)^{3/2}\left(\dfrac{1}{2}\right)\left(x^2+5\right)^{-1/2}(2x)=\dfrac{9}{2}\left(x^2+5\right)^{1/2}(3x-2)^{1/2}+x\left(x^2+5\right)^{-1/2}(3x-2)^{3/2}$

$$=\frac{9}{2}\left(x^2+5\right)^{1/2}(3x-2)^{1/2}+\frac{2}{2}x\left(x^2+5\right)^{-1/2}(3x-2)^{3/2}$$

$$=\frac{1}{2}\left(x^2+5\right)^{-1/2}(3x-2)^{1/2}\left[9\left(x^2+5\right)^1+2x(3x-2)^1\right]$$

$$=\frac{1}{2}\left(x^2+5\right)^{-1/2}(3x-2)^{1/2}\left(9x^2+45+6x^2-4x\right)$$

$$=\frac{(3x-2)^{1/2}\left(15x^2-4x+45\right)}{2\left(x^2+5\right)^{1/2}}$$

59. (a) $y_1=x^2\left(\dfrac{1}{3}\right)\left(x^2+1\right)^{-2/3}(2x)+\left(x^2+1\right)^{1/3}(2x)$

$$=2x\left(x^2+1\right)^{-2/3}\left[\frac{x^2}{3}+\left(x^2+1\right)\right]$$

$$=2x\left(x^2+1\right)^{-2/3}\left[\frac{x^2}{3}+\frac{3\left(x^2+1\right)}{3}\right]$$

$$=\frac{2x}{\left(x^2+1\right)^{2/3}}\cdot\frac{4x^2+3}{3}$$

$$=\frac{2x\left(4x^2+3\right)}{3\left(x^2+1\right)^{2/3}}$$

$$=y_2$$

(b)

x	-2	-1	$-\frac{1}{2}$	0	1	2	$\frac{5}{2}$
y_1	-8.7	-2.9	-1.1	0	2.9	8.7	12.5
y_2	-8.7	-2.9	-1.1	0	2.9	8.7	12.5

(c) The graphs appear to coincide, so they are equal.

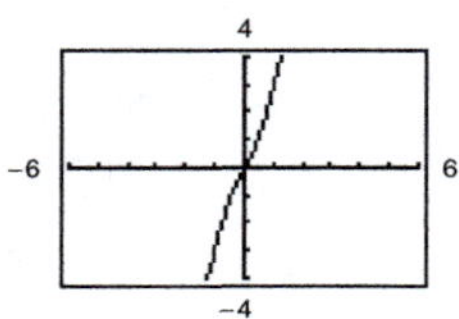

61. You cannot move term-by-term from the denominator to the numerator.

C H E C K P O I N T S

CHECKPOINTS
Chapter 1

Checkpoints for Section 1.1

1.

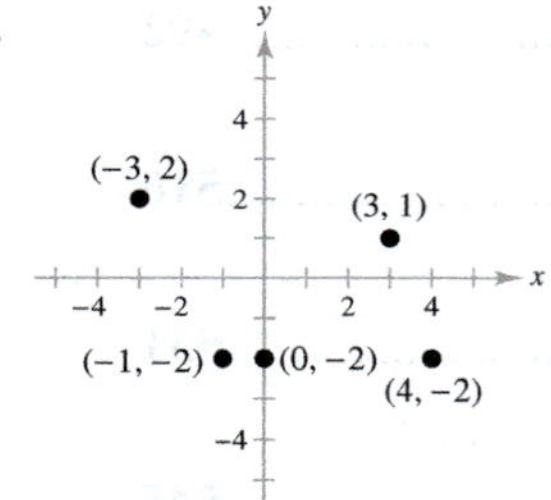

2. To sketch a scatter plot of the data shown in the table, first draw a vertical axis to represent the number of employees E (in thousands) and a horizontal axis to represent the year. Then plot the resulting points. Note that the break in the *t*-axis indicates that the numbers through 2003 have been omitted.

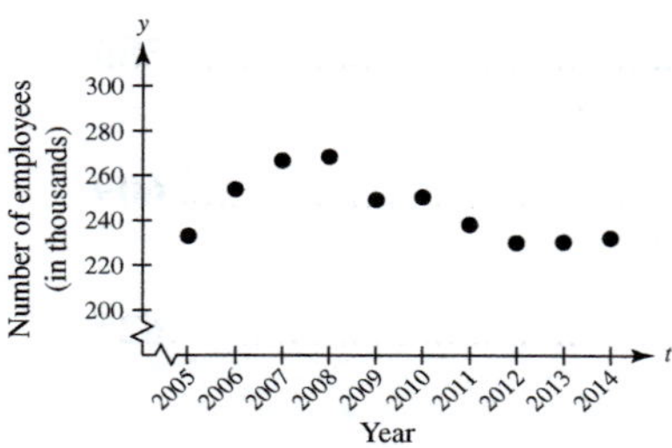

3. Let $(x_1, y_1) = (3, 1)$ and $(x_2, y_2) = (-3, 0)$.

Then apply the Distance Formula.

$$\begin{aligned} d &= \sqrt{(x_2 - x_1)^2 + (y_2 - y_1)^2} \\ &= \sqrt{(-3 - 3)^2 + (0 - 1)^2} \\ &= \sqrt{(-6)^2 + (-1)^2} \\ &= \sqrt{36 + 1} \\ &= \sqrt{37} \\ &\approx 6.08 \end{aligned}$$

So, the distance between the points is about 6.08 units.

4. The three points are plotted in the figure.

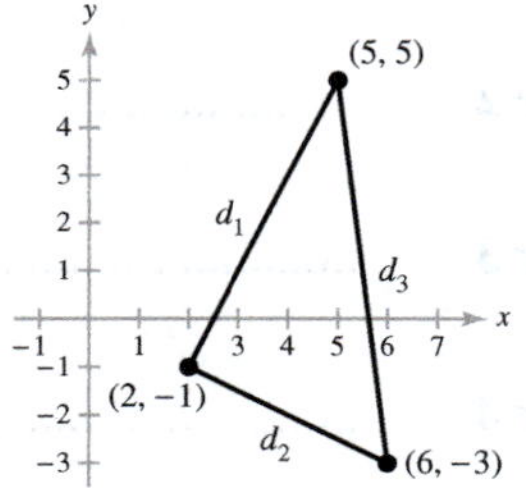

Using the Distance Formula, the lengths of the three sides are as follows.

$$\begin{aligned} d_1 &= \sqrt{(5 - 2)^2 + (5 - (-1))^2} \\ &= \sqrt{3^2 + 6^2} \\ &= \sqrt{9 + 36} \\ &= \sqrt{45} \end{aligned}$$

$$\begin{aligned} d_2 &= \sqrt{(6 - 2)^2 + (-3 - (-1))^2} \\ &= \sqrt{4^2 + (-2)^2} \\ &= \sqrt{16 + 4} \\ &= \sqrt{20} \end{aligned}$$

$$\begin{aligned} d_3 &= \sqrt{(6 - 5)^2 + (-3 - 5)^2} \\ &= \sqrt{(1)^2 + (-8)^2} \\ &= \sqrt{1 + 64} \\ &= \sqrt{65} \end{aligned}$$

Because $(d_1)^2 + (d_2)^2 = 45 + 20 = \left(\sqrt{65}\right)^2 = (d_3)^2$

you can conclude by the Pythagorean Theorem that the triangle must be a right triangle.

5. Let $(x_1, y_1) = (-2, 8)$ and $(x_2, y_2) = (-4, -10)$.

$$\begin{aligned} \text{Midpoint} &= \left(\frac{x_1 + x_2}{2}, \frac{y_1 + y_2}{2}\right) \\ &= \left(\frac{-2 + 4}{2}, \frac{8 + (-10)}{2}\right) \\ &= \left(\frac{2}{2}, -\frac{2}{2}\right) \\ &= (1, -1) \end{aligned}$$

The midpoint of the line segment is $(1, -1)$.

6. You can find the length of the pass by finding the distance between the points $(10, 10)$ and $(25, 32)$.

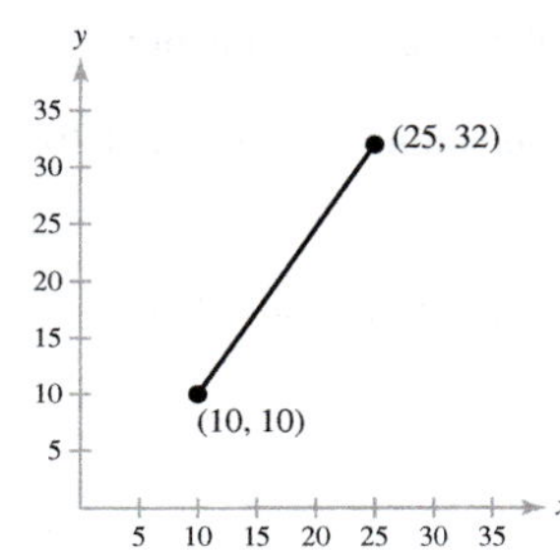

$$\begin{aligned} d &= \sqrt{(x_2 - x_1)^2 + (y_2 - y_1)^2} \\ &= \sqrt{(25 - 10)^2 + (32 - 10)^2} \\ &= \sqrt{15^2 + 22^2} \\ &= \sqrt{225 + 484} \\ &= \sqrt{709} \\ &\approx 26.6 \text{ yards} \end{aligned}$$

So, the pass is about 26.6 yards long.

7. Assuming that the annual revenue from Yahoo! Inc. followed a linear pattern, you can estimate the 2013 annual revenue by finding the midpoint of the line segment connecting the points $(2012, 5.0)$ and $(2014, 4.6)$.

$$\begin{aligned} \text{Midpoint} &= \left(\frac{x_1 + x_2}{2}, \frac{y_1 + y_2}{2}\right) \\ &= \left(\frac{2012 + 2014}{2}, \frac{5.0 + 4.6}{2}\right) \\ &= (2013, 4.8) \end{aligned}$$

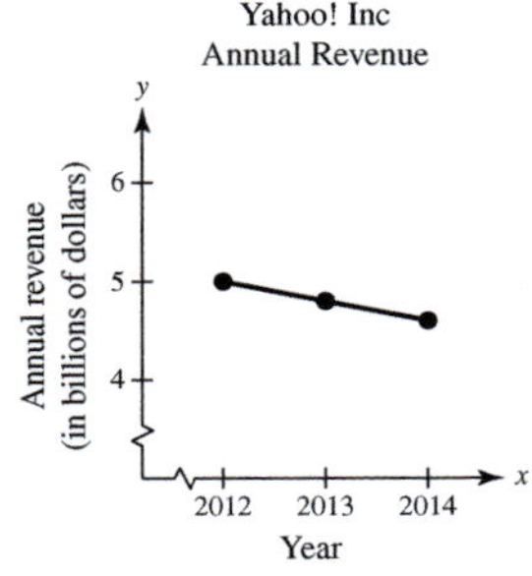

So, you can estimate the annual revenue for Yahoo! Inc. was $4.8 billion in 2013.

8. To shift the vertices two units to the left, subtract 2 from each of the x-coordinates. To shift the vertices four units down, subtract 4 from each of the y-coordinates.

Original point	**Translated Point**
$(1, 4)$	$(1 - 2, 4 - 4) = (-1, 0)$
$(1, 0)$	$(1 - 2, 0 - 4) = (-1, -4)$
$(3, 2)$	$(3 - 2, 2 - 4) = (1, -2)$
$(3, 6)$	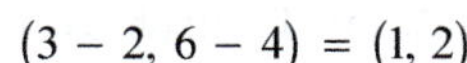 $(3 - 2, 6 - 4) = (1, 2)$

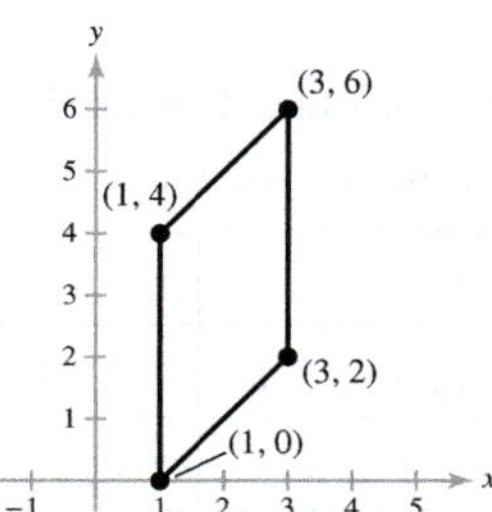

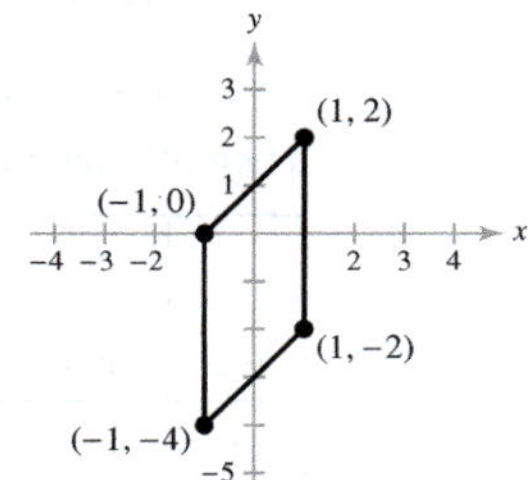

Checkpoints for Section 1.2

1. (a) $y = 14 - 6x$ Write original equation.

$-5 \stackrel{?}{=} 14 - 6(3)$ Substitute 3 for x and -5 for y.

$-5 \stackrel{?}{=} 14 - 18$

$-5 \neq -4$ $(3, -5)$ is not a solution.

(b) $y = 14 - 6x$ Write original equation.

$26 \stackrel{?}{=} 14 - 6(-2)$ Substitute -2 for x and 26 for y.

$26 \stackrel{?}{=} 14 + 12$

$26 = 26$ $(-2, 26)$ is a solution. ✓

2. (a) To graph $3x + y = 2$, first isolate the variable y.

$y = -3x + 2$

Next construct a table of values that consists of several solution points.
Then plot the points and connect them.

x	$y = -3x + 2$	(x, y)
-2	$y = -3(-2) + 2 = 8$	$(-2, 8)$
-1	$y = -3(-1) + 2 = 5$	$(-1, 5)$
0	$y = -3(0) + 2 = 2$	$(0, 2)$
1	$y = -3(1) + 2 = -1$	$(1, -1)$
2	$y = -3(2) + 2 = -4$	$(2, -4)$

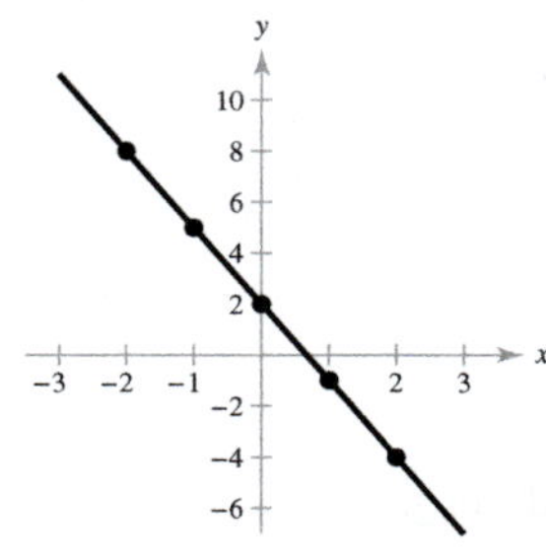

(b) To graph $-2x + y = 1$, first isolate the variable y.

$y = 2x + 1$

Next construct a table of values that consists of several solution points.
Then plot the points and connect them.

x	$y = x^2 + 3$	(x, y)
-2	$y = 2(-2) + 1 = -3$	$(-2, -3)$
-1	$y = 2(-1) + 1 = -1$	$(-1, -1)$
0	$y = 2(0) + 1 = 1$	$(0, 1)$
1	$y = 2(1) + 1 = 3$	$(1, 3)$
2	$y = 2(2) + 1 = 5$	$(2, 5)$

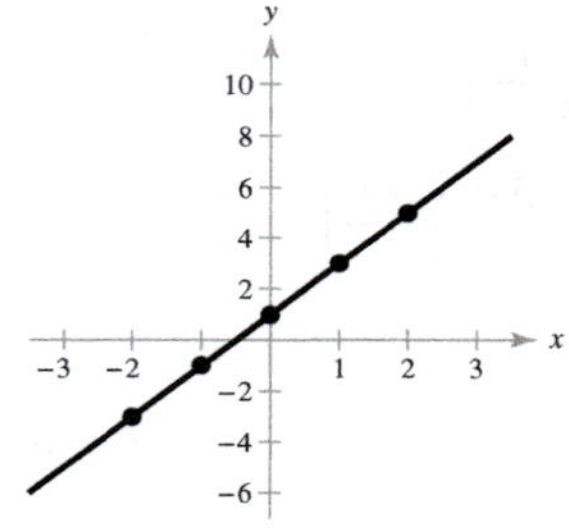

3. (a) To graph $y = x^2 + 3$, construct a table of values that consists of several solution points.
Then plot the points and connect them with a smooth curve.

x	$y = x^2 + 3$	(x, y)
-2	$y = (-2)^2 + 3 = 7$	$(-2, 7)$
-1	$y = (-1)^2 + 3 = 4$	$(-1, 4)$
0	$y = (0)^2 + 3 = 3$	$(0, 3)$
1	$y = (1)^2 + 3 = 4$	$(1, 4)$
2	$y = 2(2) + 3 = 7$	$(2, 7)$

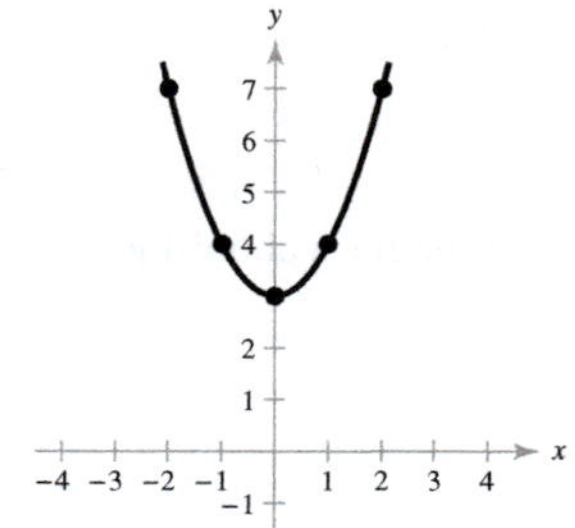

(b) To graph $y = 1 - x^2$, construct a table of values that consists of several solution points. Then plot the points and connect them with a smooth curve.

x	$y = 1 - x^2$	(x, y)
-2	$y = 1 - (-2)^2 = -3$	$(-2, -3)$
-1	$y = 1 - (-1)^2 = 0$	$(-1, 0)$
0	$y = 1 - (0)^2 = 1$	$(0, 1)$
1	$y = 1 - (1)^2 = 0$	$(1, 0)$
2	$y = 1 - (2)^2 = -3$	$(2, -3)$

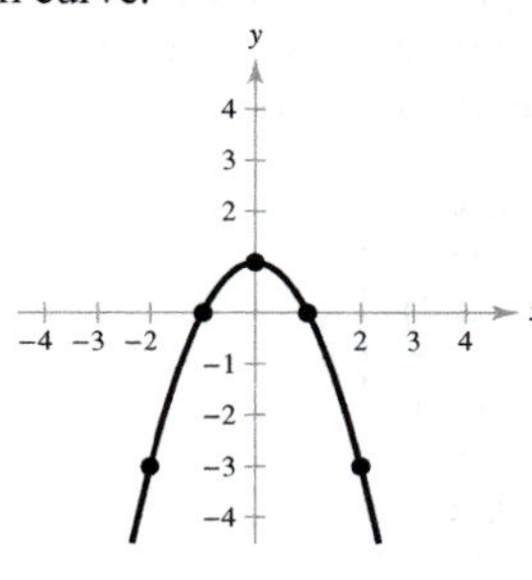

4. To find the x-intercepts, let $y = 0$ and solve for x.

$y = -x^2 - 5x$

$0 = -x^2 - 5x$

$0 = -x(x + 5)$

$x = 0$ and $x = -5$

x-intercepts: $(0, 0), (-5, 0)$

To find the y-intercept, let $x = 0$ and solve for y.

$y = -x^2 - 5x$

$y = -(0)^2 - 5(0)$

$y = 0$

y-intercept: $(0, 0)$

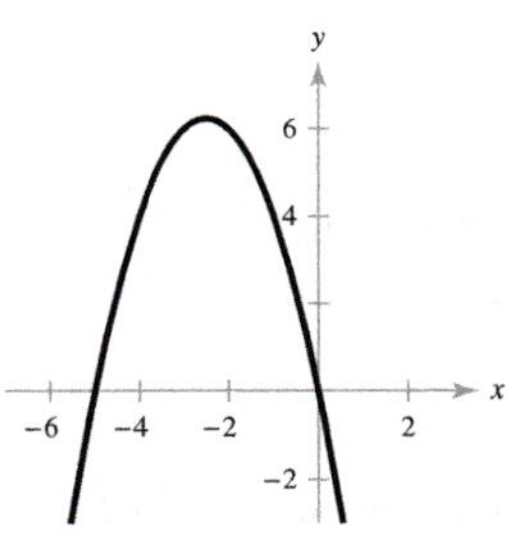

5. x-Axis:

$y^2 = 6 - x$ Write original equation.

$(-y)^2 = 6 - x$ Replace y with $-y$.

$y^2 = 6 - x$ Result is the original equation.

y-Axis:

$y^2 = 6 - x$ Write original equation.

$y^2 = 6 - (-x)$ Replace x with $-x$.

$y^2 = 6 + x$ Result is *not* an equivalent equation.

Origin:

$y^2 = 6 - x$ Write original equation.

$(-y)^2 = 6 - (-x)$ Replace y with $-y$ and x with $-x$.

$y^2 = 6 + x$ Result is *not* an equivalent equation.

Of the three tests for symmetry, the only one that is satisfied is the test for x-axis symmetry.

6. Of the three test of symmetry, the only one that is satisfied is the test for y-axis symmetry because $y = (-x)^2 - 4$ is equivalent to $y = x^2 - 4$. Using symmetry, you only need to find solution points to the right of the y-axis and then reflect them about the y-axis to obtain the graph.

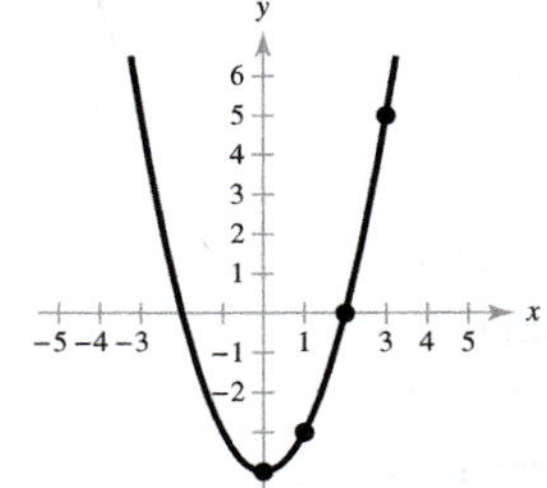

7. The equation $y = |x - 2|$ fails all three tests for symmetry and consequently its graph is not symmetric with respect to either axis or to the origin. So, construct a table of values. Then plot and connect the points.

x	$y = \|x - 2\|$	(x, y)
-2	$y = \|(-2) - 2\| = 4$	$(-2, 4)$
-1	$y = \|(-1) - 2\| = 3$	$(-1, 3)$
0	$y = \|(0) - 2\| = 2$	$(0, 2)$
1	$y = \|(1) - 2\| = 1$	$(1, 1)$
2	$y = \|(2) - 2\| = 0$	$(3, 0)$
3	$y = \|(2) - 2\| = 0$	$(3, 1)$
4	$y = \|(2) - 2\| = 0$	$(4, 2)$

From the table, you can see that the x-intercept is $(2, 0)$ and the y-intercept is $(0, 2)$.

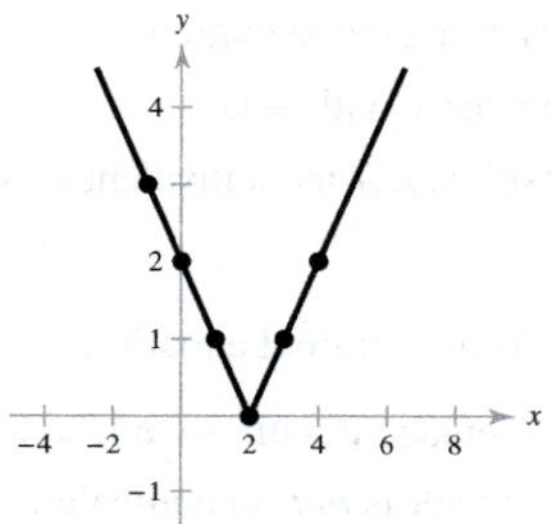

8. The radius of the circle is the distance between $(1, -2)$ and $(-3, -5)$.

$$\begin{aligned} r &= \sqrt{(x-h)^2 + (y-k)^2} \\ &= \sqrt{[1-(-3)]^2 + [-2-(-5)]^2} \\ &= \sqrt{4^2 + 3^2} \\ &= \sqrt{16 + 9} \\ &= \sqrt{25} \\ &= 5 \end{aligned}$$

Using $(h, k) = (-3, -5)$ and $r = 5$, the equation of the circle is

$$(x-h)^2 + (y-k)^2 = r^2$$
$$[x-(-3)]^2 + [y-(-5)]^2 = (5)^2$$
$$(x+3)^2 + (y+5)^2 = 25.$$

9. From the graph, you can estimate that a height of 75 inches corresponds to a maximum weight of about 221 pounds.

$$\begin{aligned} \text{Let } x = 75{:}\ y &= 0.040x^2 - 0.11x + 3.9 \\ &= 0.040(75)^2 - 0.11(75) + 3.9 \\ &= 220.65 \end{aligned}$$

Algebraically, you can conclude that a height of 75 inches corresponds to a maximum weight of 220.65 pounds. So, the graphical estimate of 221 is fairly good.

Checkpoints for Section 1.3

1. (a) $y = 3x + 2$: Because $b = 2$, the y-intercept is $(0, 2)$. Because the slope is $m = 3$, the line rises three units for each unit the line moves to the right.

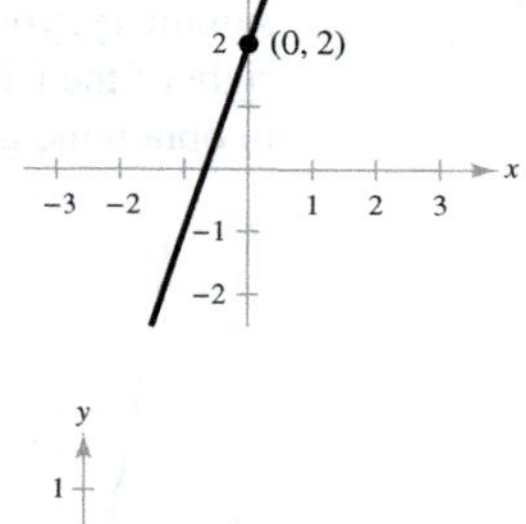

(b) $y = -3$: By writing this equation in the form $y = (0)x - 3$, you can see that the y-intercept is $(0, -3)$ and the slope is $m = 0$. A zero slope implies that the line is horizontal.

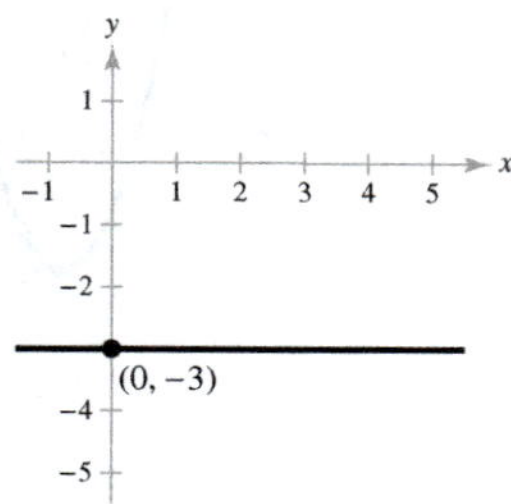

(c) $4x + y = 5$: By writing this equation in slope-intercept form

$$4x + y = 5$$
$$y = -4x + 5$$

you can see that the y-intercept is $(0, 5)$. Because the slope is $m = -4$, the line falls four units for each unit the line moves to the right.

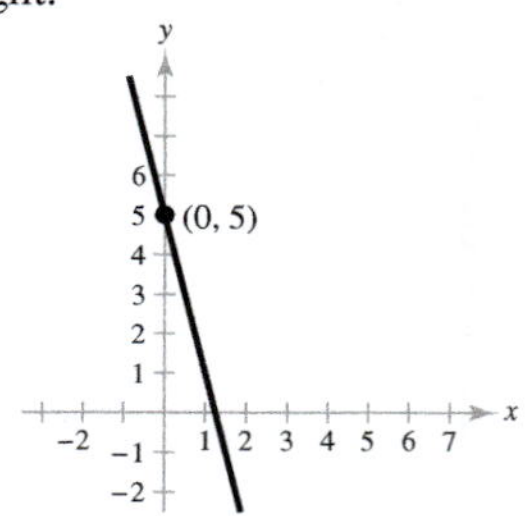

2. (a) The slope of the line passing through $(-5, -6)$ and $(2, 8)$ is $m = \dfrac{8 - (-6)}{2 - (-5)} = \dfrac{14}{7} = 2.$

(b) The slope of the line passing through $(4, 2)$ and $(2, 5)$ is $m = \dfrac{5 - 2}{2 - 4} = \dfrac{3}{-2} = -\dfrac{3}{2}.$

(c) The slope of the line passing through $(0, 0)$ and $(0, -6)$ is $m = \dfrac{-6 - 0}{0 - 0} = \dfrac{-6}{0}.$ Because division by 0 is undefined, the slope is undefined and the line is vertical.

(d) The slope of the line passing through $(0, -1)$ and $(3, -1)$ is $m = \dfrac{-1 - (-1)}{3 - 0} = \dfrac{0}{3} = 0.$

3. (a) Use the point-slope form with $m = 2$ and $(x_1, y_1) = (3, -7).$

$$\begin{aligned} y - y_1 &= m(x - x_1) \\ y - (-7) &= 2(x - 3) \\ y + 7 &= 2x - 6 \\ y &= 2x - 13 \end{aligned}$$

The slope-intercept form of this equation is $y = 2x - 13.$

(b) Use the point-slope form with $m = -\dfrac{2}{3}$ and

$(x_1, y_1) = (1, 1)$

$$\begin{aligned} y - y_1 &= m(x - x_1) \\ y - 1 &= -\frac{2}{3}(x - 1) \\ y - 1 &= \frac{-2}{3}x + \frac{2}{3} \\ y &= \frac{-2}{3}x + \frac{5}{3} \end{aligned}$$

The slope-intercept form of this equation is $y = -\dfrac{2}{3}x + \dfrac{5}{3}.$

(c) Use the point-slope form with $m = 0$ and $(x_1, y_1) = (1, 1).$

$$\begin{aligned} y - y_1 &= m(x - x_1) \\ y - 1 &= 0(x - 1) \\ y - 1 &= 0 \\ y &= 1 \end{aligned}$$

The slope-intercept of the equation is the line $y = 1.$

4. By writing the equation of the given line in slope-intercept form

$$\begin{aligned} 5x - 3y &= 8 \\ -3y &= -5x + 8 \\ y &= \frac{5}{3}x - \frac{8}{3} \end{aligned}$$

You can see that it has a slope of $m = \dfrac{5}{3}.$

(a) Any line parallel to the given line must also have a slope of $m = \dfrac{5}{3}.$ So, the line through $(-4, 1)$ that is parallel to the given line has the following equation.

$$\begin{aligned} y - y_1 &= m(x - x_1) \\ y - 1 &= \frac{5}{3}(x - (-4)) \\ y - 1 &= \frac{5}{3}(x + 4) \\ y - 1 &= \frac{5}{3}x + \frac{20}{3} \\ y &= \frac{5}{3}x + \frac{23}{3} \end{aligned}$$

(b) Any line perpendicular to the given line must also have a slope of $m = -\dfrac{3}{5}$ because $-\dfrac{3}{5}$ is the negative reciprocal of $\dfrac{5}{3}.$ So, the line through $(-4, 1)$ that is perpendicular to the given line has the following equation.

$$\begin{aligned} y - y_1 &= m(x - x_1) \\ y - 1 &= -\frac{3}{5}(x - (-4)) \\ y - 1 &= -\frac{3}{5}(x + 4) \\ y - 1 &= -\frac{3}{5}x - \frac{12}{5} \\ y &= -\frac{3}{5}x - \frac{7}{5} \end{aligned}$$

5. The horizontal length of the ramp is 32 feet or $12(32) = 384$ inches.

So, the slope of the ramp is

$$\text{Slope} = \frac{\text{vertical change}}{\text{horizontal change}} = \frac{36 \text{ in.}}{384 \text{ in.}} \approx 0.094.$$

Because $\dfrac{1}{12} \approx 0.083,$ the slope of the ramp is steeper than recommended.

6. The y-intercept $(0, 1500)$ tells you that the value of the copier when it was purchased $(t = 0)$ was \$1500. The slope of $m = -300$ tells you that the value of the copier decreases \$300 each year after it was purchased.

7. Let V represent the value of the machine at the end of year t. The initial value of the machine can be represented by the data point $(0, 24{,}750)$ and the salvage value of the machine can be represented by the data point $(6, 0)$. The slope of the line is

$$m = \frac{0 - 24{,}750}{6 - 0}$$

$$m = -\$4125$$

The slope represents the annual depreciation in dollars per year. Using the point-slope form, you can write the equation of the line as follows

$$V - 24{,}750 = -4125(t - 0)$$
$$V - 24{,}750 = -4125t$$
$$V = -4125t + 24{,}750$$

The equation $V = -4125t + 24{,}750$ represents the book value of the machine each year.

8. Let $t = 3$ represent 2013. Then the two given values are represented by the data points $(3, 6.5)$ and $(4, 7.2)$.

The slope of the line through these points is

$$m = \frac{7.2 - 6.5}{4 - 3} = -0.7.$$

You can find the equation that relates the sales y and the year t to be,

$$y - 6.5 = -0.7(t - 3)$$
$$y - 6.5 = -0.7t + 2.1$$
$$y = -0.7t + 4.4.$$

According to this equation, the sales for Foot Locker in 2017 will be $y = -0.7(7) + 4.4$

$$= 4.9 + 4.4$$
$$= \$9.3 \text{ billion}.$$

Checkpoints for Section 1.4

1. (a) This mapping *does not* describe y as a function of x. The input value of -1 is assigned or matched to two different y-values.

(b) The table *does* describe y as a function of x. Each input value is matched with exactly one output value.

2. (a) Solving for y yields

$x^2 + y^2 = 8$ Write original equation.

$y^2 = 8 - x^2$ Subtract x^2 from each side.

$y = \pm\sqrt{8 - x^2}$. Solve for y.

The $\pm$ indicates that to a given value of x there corresponds two values of y. So y is *not* a function of x.

(b) Solving for y yields,

$y - 4x^2 = 36$ Write original equation.

$y = 36 + 4x^2$ Add $4x^2$ to each side.

To each value of x there corresponds exactly one value of y. So, y is a function of x.

3. (a) Replacing x with 2 in $f(x) = 10 - 3x^2$ yields the following.

$$f(2) = 10 - 3(2)^2$$
$$= 10 - 12$$
$$= -2$$

(b) Replacing x with -4 yields the following.

$$f(-4) = 10 - 3(-4)^2$$
$$= 10 - 48$$
$$= -38$$

(c) Replacing x with $x - 1$ yields the following.

$$f(x - 1) = 10 - 3(x - 1)^2$$
$$= 10 - 3(x^2 - 2x + 1)$$
$$= 10 - 3x^2 + 6x - 3$$
$$= -3x^2 + 6x + 7$$

4. Because $x = -2$ is less than 0, use $f(x) = x^2 + 1$ to obtain $f(-2) = (-2)^2 + 1 = 4 + 1 = 5$.

Because $x = 2$ is greater than or equal to 0, use $f(x) = x - 1$ to obtain $f(2) = 2 - 1 = 1$.

For $x = 3$, use $f(x) = x - 1$ to obtain $f(3) = 3 - 1 = 2$.

5. Set $f(x) = 0$ and solve for x.

$$\begin{aligned} f(x) &= 0 \\ x^2 - 16 &= 0 \\ (x+4)(x-4) &= 0 \\ x + 4 = 0 &\Rightarrow x = -4 \\ x - 4 = 0 &\Rightarrow x = 4 \end{aligned}$$

So, $f(x) = 0$ when $x = -4$ or $x = 4$.

6.

$x^2 + 6x - 24 = 4x - x^2$	Set $f(x)$ equal to $g(x)$.
$2x^2 + 2x - 24 = 0$	Write in general form.
$2(x^2 + x - 12) = 0$	Factor out common factor.
$x^2 + x - 12 = 0$	Divide each side by 2.
$(x+4)(x-3) = 0$	Factor.
$x + 4 = 0 \Rightarrow x = -4$	Set 1st factor equal to 0.
$x - 3 = 0 \Rightarrow x = 3$	Set 2nd factor equal to 0.

So, $f(x) = g(x)$, when $x = -4$ or $x = 3$.

7. (a) The domain of f consists of all first coordinates in the set of ordered pairs.

Domain $= \{-2, -1, 0, 1, 2\}$

(b) Excluding x-values that yield zero in the denominator, the domain of g is the set of all real numbers x except $x = 3$.

(c) Because the function represents the circumference of a circle, the values of the radius r must be positive. So, the domain is the set of real numbers r such that $r > 0$.

(d) This function is defined only for x-values for which $x - 16 \geq 0$. You can conclude that $x \geq 16$. So, the domain is the interval $[16, \infty)$.

8. Use the formula for surface area of a cylinder, $s = 2\pi r^2 + 2\pi rh$.

(a) $$\begin{aligned} s(r) &= 2\pi r^2 + 2\pi r(4r) \\ &= 2\pi r^2 + 8\pi r^2 \\ &= 10\pi r^2 \end{aligned}$$

(b) $$\begin{aligned} s(h) &= 2\pi\left(\frac{h}{4}\right)^2 + 2\pi\left(\frac{h}{4}\right)h \\ &= 2\pi\left(\frac{h^2}{16}\right) + \frac{\pi h^2}{2} \\ &= \frac{1}{8}\pi h^2 + \frac{1}{2}\pi h^2 \\ &= \frac{5}{8}\pi h^2 \end{aligned}$$

9. When $x = 60$, you can find the height of the baseball as follows

$f(x) = -0.004x^2 + 0.3x + 6$	Write original function.
$f(60) = -0.004(60)^2 + 0.3(60) + 6$	Substitute 60 for x.
$= 9.6$	Simplify.

When $x = 60$, the height of the ball thrown from the second baseman is 9.6 feet. So, the first baseman cannot catch the baseball without jumping.

10. From 2009 through 2011, use $S(t) = 69t + 151$.

2009: $S(t) = 69(9) + 151 = 772$ fuel stations.

2010: $S(t) = 69(10) + 151 = 841$ fuel stations.

2011: $S(t) = 69(11) + 151 = 910$ fuel stations.

From 2012 through 2015, use $S(t) = 160t - 803$.

2012: $S(12) = 160(12) - 803 = 1117$ fuel stations.

2013: $S(13) = 160(13) - 803 = 1277$ fuel stations.

2014: $S(14) = 160(14) - 803 = 1437$ fuel stations.

2015: $S(15) = 160(15) - 803 = 1597$ fuel stations.

11.
$$\begin{aligned}\frac{f(x+h)-f(x)}{h} &= \frac{\left[(x+h)^2 + 2(x+h) - 3\right] - (x^2 + 2x - 3)}{h}\\ &= \frac{x^2 + 2xh + h^2 + 2x + 2h - 3 - x^2 - 2x + 3}{h}\\ &= \frac{2xh + h^2 + 2h}{h}\\ &= \frac{2(2x + h + 2)}{h}\\ &= 2x + h + 2, \quad h \neq 0\end{aligned}$$

Checkpoints for Section 1.5

1. (a) The open dot at $(-3, -6)$ indicates that $x = -3$ is not in the domain of f. So, the domain of f is all real numbers, except $x \neq -3$, or $(-\infty, -3) \cup (-3, \infty)$.

(b) Because $(0, 3)$ is a point on the graph of f, it follows that $f(0) = 3$. Similarly, because the point $(3, -6)$ is a point on the graph of f, it follows that $f(3) = -6$.

(c) Because the graph of f does not extend above $f(0) = 3$, the range of f is the interval $(-\infty, 3]$.

2.

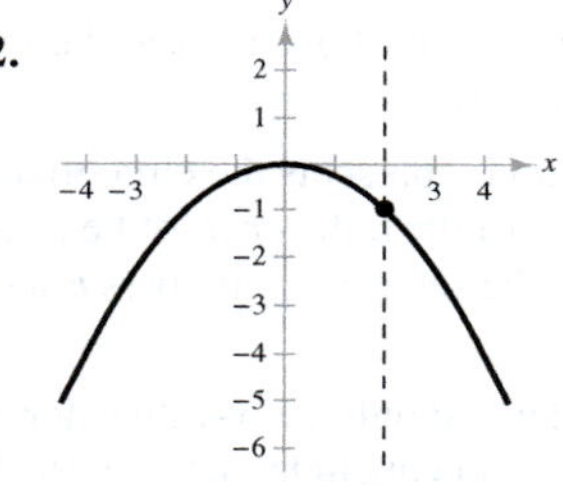

This *is* a graph of y as a function of x, because every vertical line intersects the graph at most once. That is, for a particular input x, there is at most one output y.

3. To find the zeros of a function, set the function equal to zero, and solve for the independent variable.

(a)

$2x^2 + 13x - 24 = 0$	Set $f(x)$ equal to 0.
$(2x - 3)(x + 8) = 0$	Factor.
$2x - 3 = 0 \Rightarrow x = \dfrac{3}{2}$	Set 1st factor equal to 0.
$x + 8 = 0 \Rightarrow x = -8$	Set 2nd factor equal to 0.

The zeros of f are $x = \dfrac{3}{2}$ and $x = -8$. The graph of f has $\left(\dfrac{3}{2}, 0\right)$ and $(-8, 0)$ as its x-intercepts.

(b)

$\sqrt{t - 25} = 0$	Set $g(t)$ equal to 0.
$\left(\sqrt{t - 25}\right)^2 = (0)^2$	Square each side.
$t - 25 = 0$	Simplify.
$t = 25$	Add 25 to each side.

The zero of g is $t = 25$. The graph of g has $(25, 0)$ as its t-intercept.

(c) $\dfrac{x^2 - 2}{x - 1} = 0$ Set $h(x)$ equal to zero.

$(x - 1)\left(\dfrac{x^2 - 2}{x - 1}\right) = (x - 1)(0)$ Multiply each side by $x - 1$.

$x^2 - 2 = 0$ Simplify.

$x^2 = 2$ Add 2 to each side.

$x = \pm\sqrt{2}$ Extract square roots.

The zeros of h are $x = \pm\sqrt{2}$. The graph of h has $(\sqrt{2}, 0)$ and $(-\sqrt{2}, 0)$ as its x-intercepts.

4.

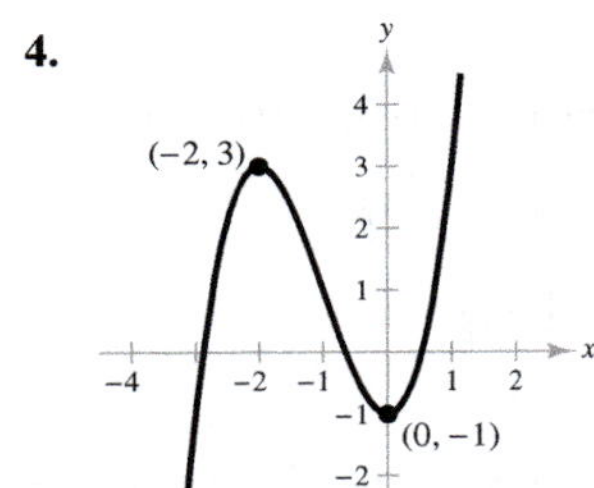

This function is increasing on the interval $(-\infty, -2)$, decreasing on the interval $(-2, 0)$, and increasing on the interval $(0, \infty)$.

5.

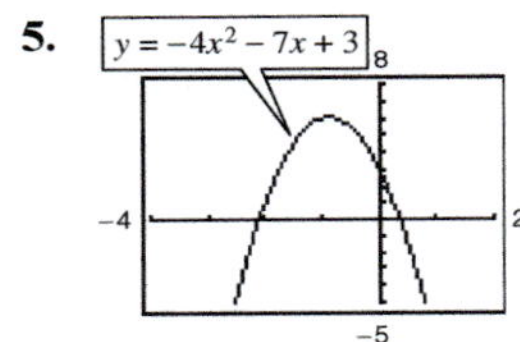

By using the zoom and the trace features or the maximum feature of a graphing utility, you can determine that the function has a relative maximum at the point $\left(-\dfrac{7}{8}, \dfrac{97}{16}\right)$ or $(-0.875, 6.0625)$.

6. (a) The average rate of change of f from $x_1 = -3$ to $x_2 = -2$ is

$$\frac{f(x_2) - f(x_1)}{x_2 - x_1} = \frac{f(-2) - f(-3)}{-2 - (-3)} = \frac{0 - 3}{1} = -3.$$

(b) The average rate of change of f from $x_1 = -2$ to $x_2 = 0$ is

$$\frac{f(x_2) - f(x_1)}{x_2 - x_1} = \frac{f(0) - f(-2)}{0 - (-2)} = \frac{0 - 0}{2} = 0.$$

7. (a) The average speed of the car from $t_1 = 0$ to $t_2 = 1$ second is

$$\frac{s(t_2) - s(t_1)}{t_2 - t_1} = \frac{20 - 0}{1 - 0} = 20 \text{ feet per second.}$$

(b) The average speed of the car from $t_1 = 1$ to $t_2 = 4$ seconds is

$$\frac{s(t_2) - s(t_1)}{t_2 - t_1} = \frac{160 - 20}{4 - 1} = \frac{140}{3} \approx 46.7 \text{ feet per second.}$$

8. (a) The function $f(x) = 5 - 3x$ is neither odd nor even because $f(-x) \neq -f(x)$ and $f(-x) \neq f(x)$ as follows.

$f(-x) = 5 - 3(-x)$

$= 5 + 3x \neq -f(x)$ not odd

$\neq f(x)$ not even

So, the graph of f is not symmetric to the origin nor the y-axis.

(b) The function $g(x) = x^4 - x^2 - 1$ is even because $g(-x) = g(x)$ as follows.

$$\begin{aligned} g(-x) &= (-x)^4 - (-x)^2 - 1 \\ &= x^4 - x^2 - 1 \\ &= g(x) \end{aligned}$$

So, the graph of g is symmetric to the y-axis.

(c) The function $h(x) = 2x^3 + 3x$ is odd because $h(-x) = -h(x)$,

$$\begin{aligned} h(-x) &= 2(-x)^3 + 3(-x) \\ &= -2x^3 - 3x \\ &= -(2x^3 + 3x) \\ &= -h(x) \end{aligned}$$

So, the graph of h is symmetric to the origin.

Checkpoints for Section 1.6

1. To find the equation of the line that passes through the points $(x_1, y_1) = (-2, 6)$ and $(x_2, y_2) = (4, -4)$, first find the slope of the line.

$$m = \frac{y_2 - y_1}{x_2 - x_1} = \frac{-9 - 6}{4 - (-2)} = \frac{-15}{6} = \frac{-5}{2}$$

Next, use the point-slope form of the equation of the line.

$y - y_1 = m(x - x_1)$ Point-slope form

$y - 6 = -\frac{5}{2}[x - (-2)]$ Substitute x_1, y_1 and m.

$y - 6 = -\frac{5}{2}(x + 2)$ Simplify.

$y - 6 = -\frac{5}{2}x - 5$ Simplify.

$y = -\frac{5}{2}x + 1$ Simplify.

$f(x) = -\frac{5}{2}x + 1$ Function notation

2. For $x = -\frac{3}{2}$, $f\left(-\frac{3}{2}\right) = \left[\!\left[-\frac{3}{2} + 2\right]\!\right]$

$= \left[\!\left[\frac{1}{2}\right]\!\right]$

$= 0$

Since the greatest integer $\leq \frac{1}{2}$ is 0, $f\left(-\frac{3}{2}\right) = 0$.

For $x = 1$, $f(1) = [\![1 + 2]\!]$

$= [\![3]\!]$

$= 3$

Since the greatest integer ≤ 3 is 3, $f(1) = 3$.

For $x = -\frac{5}{2}$, $f\left(-\frac{5}{2}\right) = \left[\!\left[-\frac{5}{2} + 2\right]\!\right]$

$= \left[\!\left[-\frac{1}{2}\right]\!\right]$

$= -1$

Since the greatest integer $\leq -\frac{1}{2}$ is -1, $f\left(-\frac{5}{2}\right) = -1$.

3. This piecewise-defined function consists of two linear functions. At $x = -4$ and to the left of $x = -4$, the graph is the line $y = -\frac{1}{2}x - 6$, and to the right of $x = -4$ the graph is the line $y = x + 5$. Notice that the point $(-4, -2)$ is a solid dot and $(-4, 1)$ is an open dot. This is because $f(-4) = -2$.

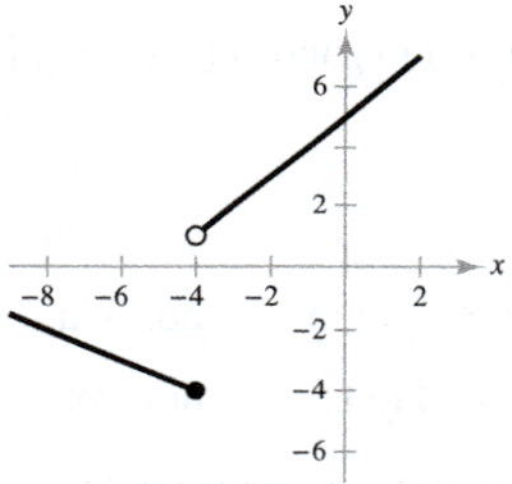

Checkpoints for Section 1.7

1. (a) Relative to the graph of $f(x) = x^3$, the graph of $h(x) = x^3 + 5$ is an upward shift of five units.

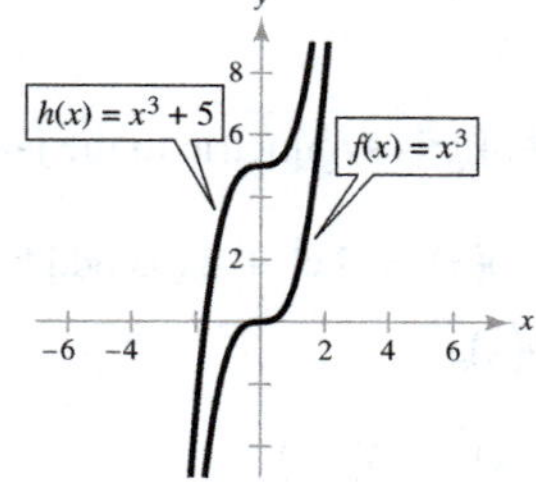

(b) Relative to the graph of $f(x) = x^3$, the graph of $g(x) = (x - 3)^3 + 2$ involves a right shift of three units and an upward shift of two units.

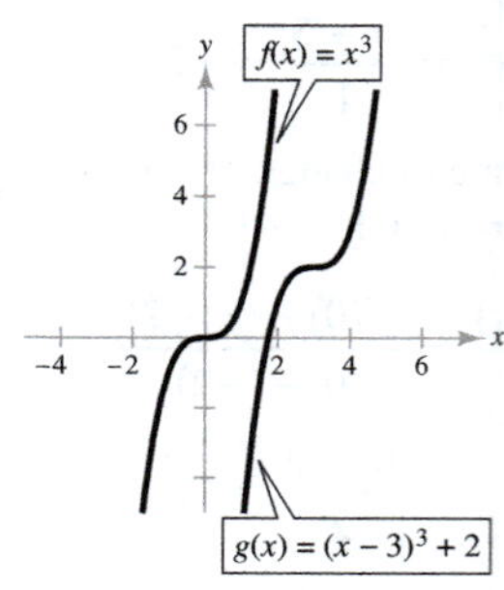

2. The graph of j is a horizontal shift of three units to the left followed by a reflection in the x-axis of the graph of $f(x) = x^4$. So, the equation for j is $j(x) = -(x + 3)^4$.

3. (a) **Algebraic Solution:**

The graph of g is a reflection of the graph of f in the x-axis because

$$\begin{aligned} g(x) &= -\sqrt{x-1} \\ &= -f(x). \end{aligned}$$

Graphical Solution:

Graph f and g on the same set of coordinate axes. From the graph, you can see that the graph of g is a reflection of the graph of f in the x-axis.

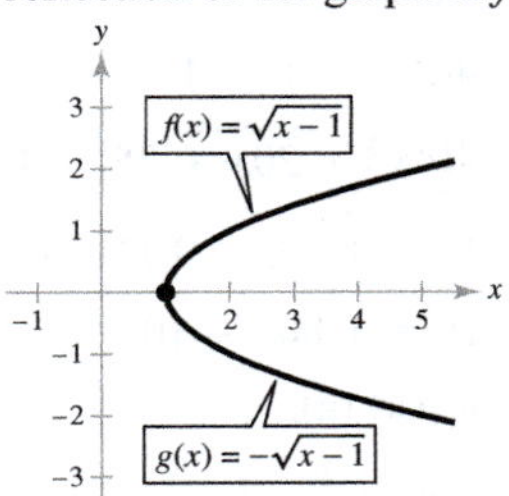

(b) **Algebraic Solution:**

The graph of h is a reflection of the graph of f in the y-axis because

$$\begin{aligned} h(x) &= \sqrt{-x-1} \\ &= f(-x). \end{aligned}$$

Graphical Solution:

Graph f and h on the same set of coordinate axes. From the graph, you can see that the graph h is a reflection of the graph of f, in the y-axis.

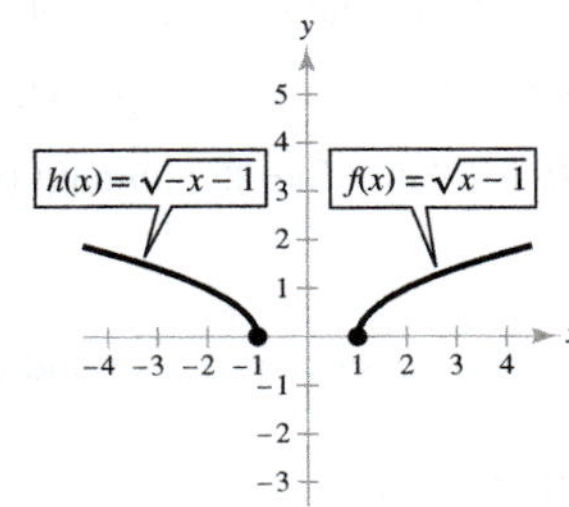

4. (a) Relative to the graph of $f(x) = x^2$, the graph of $g(x) = 4x^2 = 4f(x)$ is a vertical stretch (each y-value is multiplied by 4) of the graph of f.

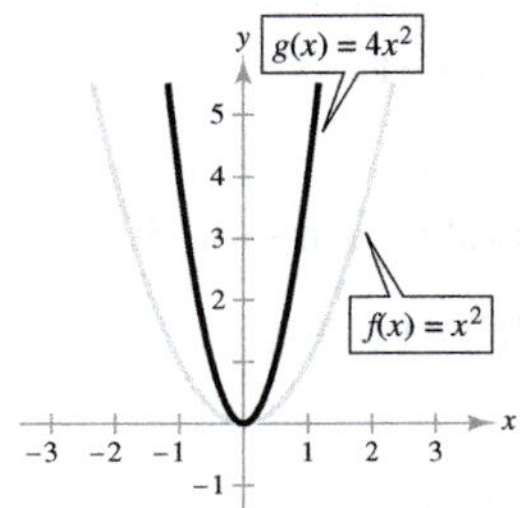

(b) Relative to the graph of $f(x) = x^2$, the graph of $h(x) = \frac{1}{4}x^2 = \frac{1}{4}f(x)$ is a vertical shrink $\left(\text{each } y\text{-value is multiplied by } \frac{1}{4}\right)$ of the graph of f.

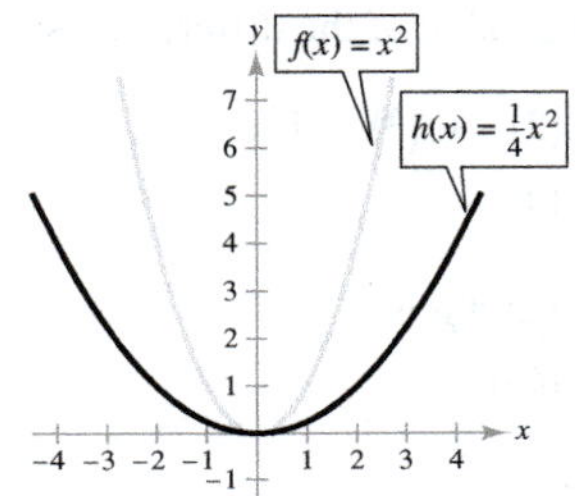

5. (a) Relative to the graph of $f(x) = x^2 + 3$, the graph of $g(x) = f(2x) = (2x)^2 + 3 = 4x^2 + 3$ is a horizontal shrink $(c > 1)$ of the graph of f.

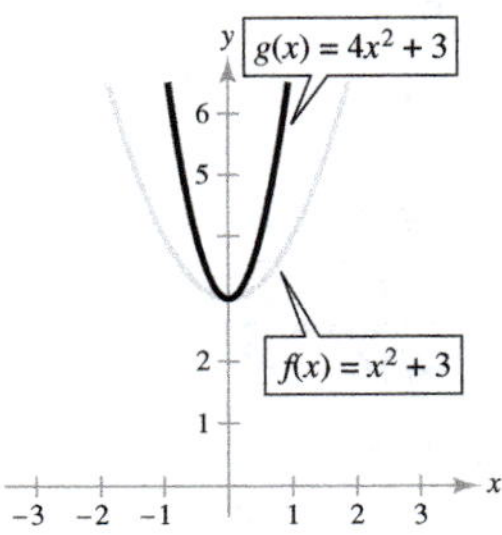

(b) Relative to the graph of $f(x) = x^2 + 3$, the graph of $h(x) = f\left(\frac{1}{2}x\right) = \left(\frac{1}{2}x\right)^2 + 3 = \frac{1}{4}x^2 + 3$ is a horizontal stretch $(0 < c < 1)$ of the graph of f.

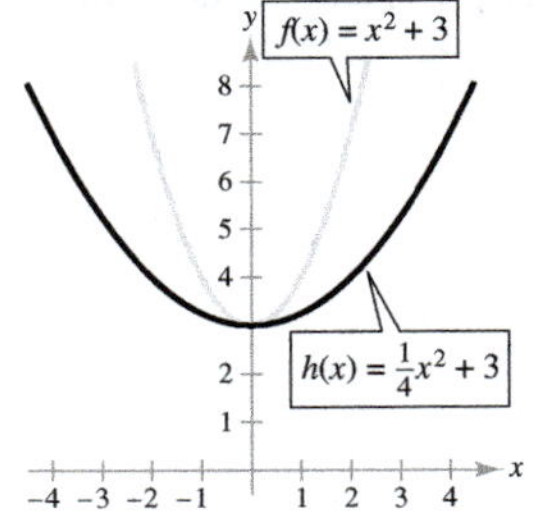

Checkpoints for Section 1.8

1. The sum of f and g is

$$\begin{aligned}(f+g)(x) &= f(x)+g(x)\\ &= (x^2)+(1-x)\\ &= x^2-x+1.\end{aligned}$$

When $x = 2$, the value of this sum is

$$\begin{aligned}(f+g)(2) &= (2)^2-(2)+1\\ &= 3.\end{aligned}$$

2. The difference of f and g is

$$\begin{aligned}(f-g)(x) &= f(x)-g(x)\\ &= (x^2)-(1-x)\\ &= x^2+x-1.\end{aligned}$$

When $x = 3$, the value of the difference is

$$\begin{aligned}(f-g)(3) &= (3)^2+(3)-1\\ &= 11.\end{aligned}$$

3. The product of f and g is

$$\begin{aligned}(fg) &= f(x)g(x)\\ &= (x^2)(1-x)\\ &= x^2-x^3\\ &= -x^3+x^2.\end{aligned}$$

When $x = 3$, the value of the product is

$$\begin{aligned}(fg)(3) &= -(3)^3-(3)^2\\ &= -27+9\\ &= -18.\end{aligned}$$

4. The quotient of f and g is

$$\left(\frac{f}{g}\right)(x) = \frac{f(x)}{g(x)} = \frac{\sqrt{x-3}}{\sqrt{16-x^2}}.$$

The quotient of g and f is

$$\left(\frac{g}{f}\right)(x) = \frac{g(x)}{f(x)} = \frac{\sqrt{16-x^2}}{\sqrt{x-3}}.$$

The domain of f is $[3, \infty)$ and the domain of g is $[-4, 4]$. The intersection of these two domains is $[3, 4]$. So, the domain of f/g is $[3, 4)$ and the domain of g/f is $(3, 4]$.

5. (a) The composition of f with g is as follows.

$$\begin{aligned}(f\circ g)(x) &= f(g(x))\\ &= f(4x^2+1)\\ &= 2(4x^2+1)+5\\ &= 8x^2+2+5\\ &= 8x^2+7\end{aligned}$$

(b) The composition of g with f is as follows.

$$\begin{aligned}(g\circ f)(x) &= g(f(x))\\ &= g(2x+5)\\ &= 4(2x+5)^2+1\\ &= 4(4x^2+20x+25)+1\\ &= 16x^2+80x+100+1\\ &= 16x^2+80x+101\end{aligned}$$

(c) Use the result of part (a).

$$\begin{aligned}(f\circ g)\left(-\tfrac{1}{2}\right) &= 8\left(-\tfrac{1}{2}\right)^2+7\\ &= 8\left(\tfrac{1}{4}\right)+7\\ &= 2+7\\ &= 9\end{aligned}$$

6. The composition of f with g is as follows.

$$\begin{aligned}(f\circ g)(x) &= f(g(x))\\ &= f(x^2+4)\\ &= \sqrt{x^2+4}\end{aligned}$$

The domain of f is $[0, \infty)$ and the domain of g is the set of all real numbers. The range of g is $[4, \infty)$, which is in the range of f, $[0, \infty)$. Therefore the domain of $f\circ g$ is all real numbers.

7. Let the inner function be $g(x) = 8 - x$ and the outer function be $f(x) = \dfrac{\sqrt[3]{x}}{5}$.

$$\begin{aligned}h(x) &= \frac{\sqrt[3]{8-x}}{5}\\ &= f(8-x)\\ &= f(g(x))\end{aligned}$$

8. (a) $(N \circ T)(t) = N(T(t))$

$= 8(2t + 2)^2 - 14(2t + 2) + 200$

$= 8(4t^2 + 8t + 4) - 28t - 28 + 200$

$= 32t^2 + 64t + 32 - 28t - 28 + 200$

$= 32t^2 + 36t + 204$

The composite function $(N \circ T)(t)$ represents the number of bacteria in the food as a function of the amount of time the food has been out of refrigeration.

(b) Let $(N \circ T)(t) = 1000$ and solve for t.

$32t^2 + 36t + 204 = 1000$

$32t^2 + 36t - 796 = 0$

$4(8t^2 + 9t - 199) = 0$

$8t^2 + 9t - 199 = 0$

Use the quadratic formula:

$$t = \frac{-9 \pm \sqrt{(9)^2 - 4(8)(-199)}}{2(8)}$$

$$= \frac{-9 \pm \sqrt{6449}}{16}$$

$t \approx 4.5$ and $t \approx -5.6$.

Using $t \approx 4.5$ hours, the bacteria count reaches approximately 1000 about 4.5 hours after the food is removed from the refrigerator.

Checkpoints for Section 1.9

1. The function f multiplies each input by $\frac{1}{5}$. To "undo" this function, you need to multiply each input by 5. So, the inverse function of $f(x) = \frac{1}{5}x$ is $f^{-1}(x) = 5x$.

To verify this, show that

$f(f^{-1}(x)) = x$ and $f^{-1}(f(x)) = x$.

$f(f^{-1}(x)) = f(5x) = \frac{1}{5}(5x) = x$

$f^{-1}(f(x)) = f^{-1}\left(\frac{1}{5}x\right) = 5\left(\frac{1}{5}x\right) = x$

So, the inverse function of $f(x) = \frac{1}{5}x$ is $f^{-1}(x) = 5x$.

2. By forming the composition of f and g, you have

$$f(g(x)) = f(7x + 4) = \frac{(7x + 4) - 4}{7} = \frac{7x}{7} = x.$$

So, it appears that g is the inverse function of f. To confirm this, form the composition of g and f.

$$g(f(x)) = g\left(\frac{x - 4}{7}\right) = 7\left(\frac{x - 4}{7}\right) + 4 = x - 4 + 4 = x$$

By forming the composition of f and h, you can see that h is *not* the inverse function of f, since the result is not the identity function x.

$$f(h(x)) = f\left(\frac{7}{x - 4}\right) = \frac{\left(\frac{7}{x - 4}\right) - 4}{7} = \frac{23 - 4x}{7(x - 4)} \neq x$$

So, g is the inverse function of f.

3. First, sketch the graphs of $f(x) = 4x - 1$ and $g(x) = \frac{1}{4}(x + 1)$ as shown. You can see that they appear to be reflections of each other in the line $y = x$. This reflective property can be tested using a few points and the fact that if the point (a, b) is on the graph of f then the point (b, a) is on the graph of g.

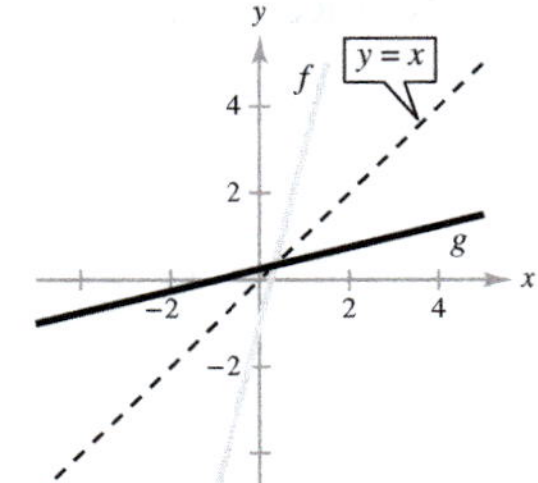

Graph of $f(x) = 4x - 1$	Graph of $g(x) = \frac{1}{4}(x + 1)$
$(-1, -5)$	$(-5, -1)$
$(0, -1)$	$(-1, 0)$
$(1, 3)$	$(3, 1)$
$(2, 7)$	$(7, 2)$

4. First, sketch the graphs of $f(x) = x^2 + 1, x \geq 0$ and $g(x) = \sqrt{x - 1}$ as shown. You can see that they appear to be reflections of each other in the line $y = x$. This reflective property can be tested using a few points and the fact that if the point (a, b) is on the graph of f then the point (b, a) is on the graph of g.

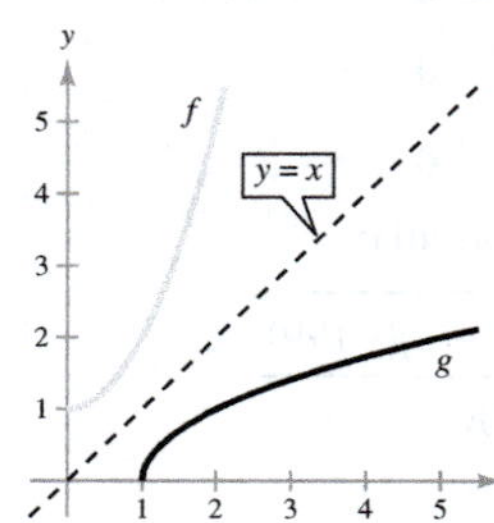

Graph of $f(x) = x^2 + 1, x \geq 0$	Graph of $g(x) = \sqrt{x - 1}$
(0, 1)	(1, 0)
(1, 2)	(2, 1)
(2, 5)	(5, 2)
(3, 10)	(10, 3)

5. (a) The graph of $f(x) = \frac{1}{2}(3 - x)$ is shown.

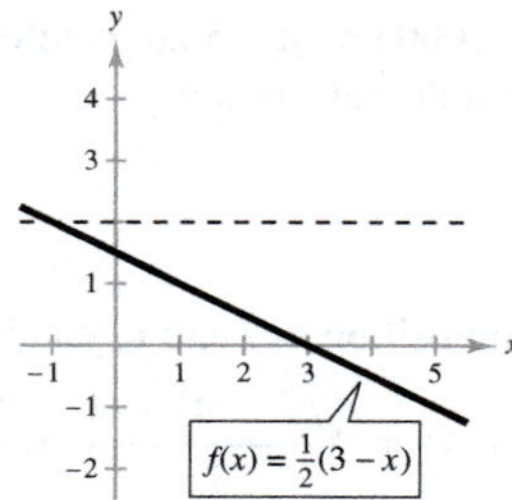

Because no horizontal line intersects the graph of f at more than one point, f is a one-to-one function and *does* have an inverse function.

(b) The graph of $f(x) = |x|$ is shown.

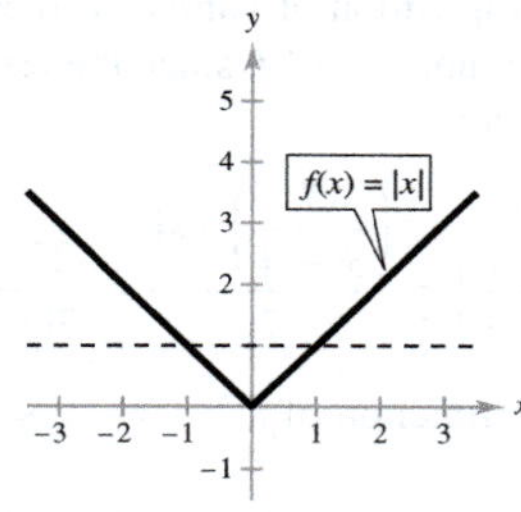

Because it is possible to find a horizontal line that intersects the graph of f at more than one point, f *is not* a one-to-one function and *does not* have an inverse function.

6. The graph of $f(x) = \dfrac{5 - 3x}{x + 2}$ is shown.

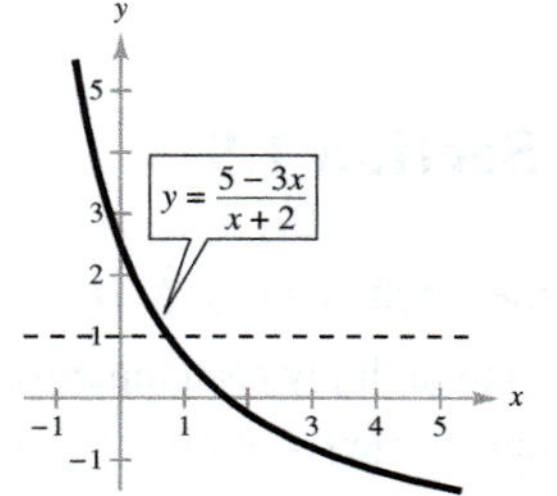

This graph passes the Horizontal Line Test. So, you know f is one-to-one and has an inverse function.

$f(x) = \dfrac{5 - 3x}{x + 2}$ Write original function.

$y = \dfrac{5 - 3x}{x + 2}$ Replace $f(x)$ with y.

$x = \dfrac{5 - 3y}{y + 2}$ Interchange x and y.

$x(y + 2) = 5 - 3y$ Multiply each side by $y + 2$.

$xy + 2x = 5 - 3y$ Distribute Property

$xy + 3y = 5 - 2x$ Collect like terms with y.

$y(x + 3) = 5 - 2x$ Factor.

$y = \dfrac{5 - 2x}{x + 3}$ Solve for y.

$f^{-1}(x) = \dfrac{5 - 2x}{x + 3}$ Replace y with $f^{-1}(x)$.

7. The graph of $f(x) = \sqrt[3]{10 + x}$ is shown.

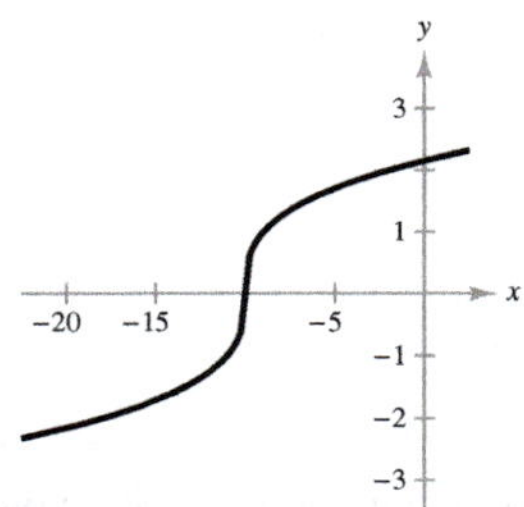

Because this graph passes the Horizontal Line Test, you know that f is one-to-one and has an inverse function.

$$\begin{aligned} f(x) &= \sqrt[3]{10 + x} \\ y &= \sqrt[3]{10 + x} \\ x &= \sqrt[3]{10 + y} \\ x^3 &= 10 + y \\ y &= x^3 - 10 \\ f^{-1}(x) &= x^3 - 10 \end{aligned}$$

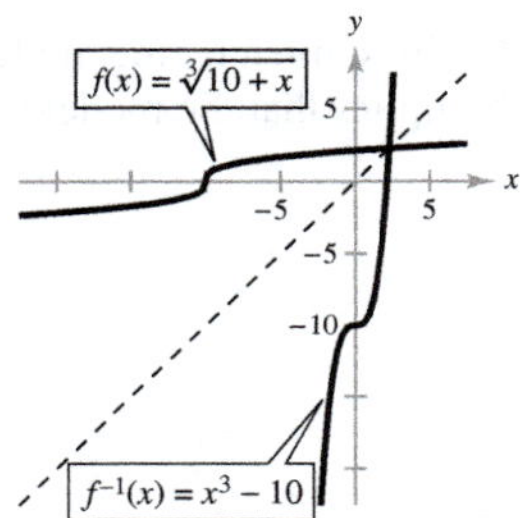

The graphs of f and f^{-1} are reflections of each other in the line $y = x$. So, the inverse of $f(x) = \sqrt[3]{10 + x}$ is $f^{-1}(x) = x^3 - 10$.

To verify, check that $f(f^{-1}(x)) = x$ and $f^{-1}(f(x)) = x$.

$$\begin{aligned} f(f^{-1}(x)) &= f(x^3 - 10) \\ &= \sqrt[3]{10 + (x^3 - 10)} \\ &= \sqrt[3]{x^3} \\ &= x \end{aligned} \qquad \begin{aligned} f^{-1}(f(x)) &= f^{-1}\left(\sqrt[3]{10 + x}\right) \\ &= \left(\sqrt[3]{10 + x}\right)^3 - 10 \\ &= 10 + x - 10 \\ &= x \end{aligned}$$

Checkpoints for Section 1.10

1.

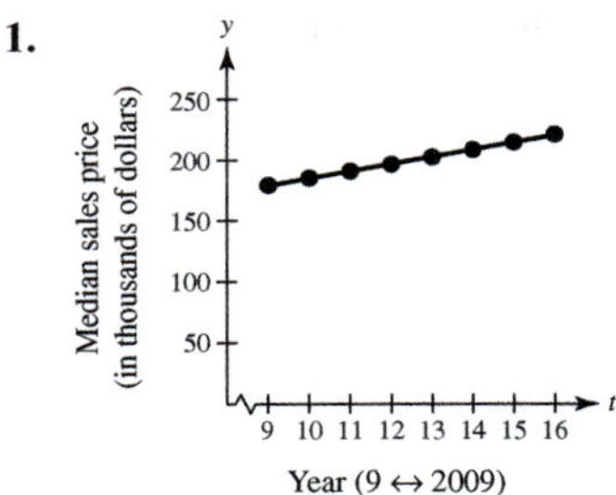

The actual data are plotted, along with the graph of the linear model. From the graph, it appears that the model is a "good fit" for the actual data. You can see how well the model fits by comparing the actual values of y with the values of y given by the model. The values given by the model are labeled y^* in the table below.

t	9	10	11	12	13	14	15	16
y	179.4	185.4	191.0	196.7	202.6	208.7	214.9	221.4
y^*	179.1	185.1	191.1	197.0	203.0	208.9	214.9	220.9

2.

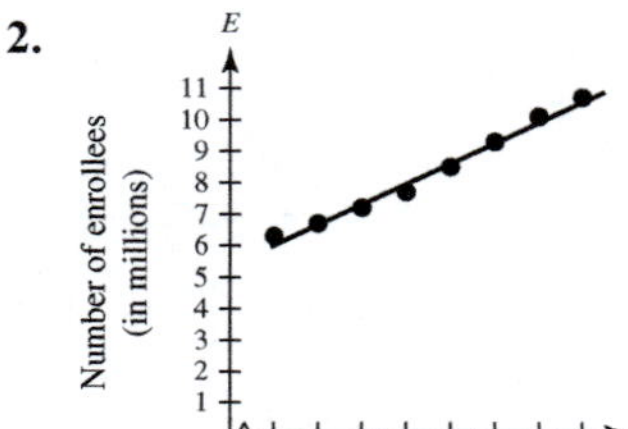

Let $t = 8$ represent 2008. The scatter plot for the data is shown. Using the *regression* feature of a graphing utility, you can determine that the equation of the least squares regression line is

$E = 0.65t + 0.8$

To check this model, compare the actual E values with the E^* values given by the model, which are labeled E^* in the table. The correlation coefficient for this model is $r \approx 0.991$, which implies that the model is a good fit.

t	8	9	10	11	12	13	14	15
E	6.3	6.7	7.2	7.7	8.5	9.3	10.1	10.7
E^*	6.0	6.7	7.3	8.0	8.6	9.3	9.9	10.6

3. *Verbal Model*: $\boxed{\text{Simple interest}} = \boxed{r}\ \boxed{\text{Amount of investment}}$

Labels: Simple interest $= I$ (dollars)

Amount of investment $= P$ (dollars)

Interest rate $= r$ (percent in decimal form)

Equation: $I = rP$

To solve for r, substitute the given information into the equation $I = rP$, and then solve for r.

$I = rP$ Write direct variation model.

$187.50 = r(2500)$ Substitute 187.50 for t and 2500 for P.

$\dfrac{187.50}{2500} = r$ Divide each side by 2500.

$0.075 = r$ Simplify.

So, the mathematical model is $I = 0.075P$.

4. Letting s be the distance (in feet) the object falls and letting t be the time (in seconds) that the object falls, you have $s = Kt^2$.

Because $s = 144$ feet when $t = 3$ seconds, you can see that $K = \frac{144}{9}$ as follows.

$s = Kt^2$ Write direct variation model.

$144 = K(3)^2$ Substitute 144 for s and 3 for t.

$144 = 9K$ Simplify.

$\frac{144}{9} = K$ Divide each side by 9.

$16 = K$

So, the equation relating distance to time is $s = 16t^2$.

To find the distance the object falls in 6 seconds, let $t = 6$.

$s = 16t^2$ Write direct variation model.

$s = 16(6)^2$ Substitute 6 for t.

$s = 16(36)$ Simplify.

$s = 576$ Simplify.

So, the object falls 576 feet in 6 seconds.

5. Let p be the price and let x be the demand. Because x varies inversely as p, you have

$$x = \frac{k}{p}$$

Now because $x = 600$ then $p = 2.75$ you have

$$x = \frac{k}{p} \quad \text{Write inverse variation model.}$$

$$600 = \frac{k}{2.75} \quad \text{Substitute 600 for } x \text{ and 2.75 for } p.$$

$$(600)(2.75) = k \quad \text{Multiply each side by 2.75.}$$

$$1650 = k. \quad \text{Simplify.}$$

So, the equation relating price and demand is

$$x = \frac{1650}{p}.$$

When $p = 3.25$ the demand is

$$x = \frac{1650}{p} \quad \text{Write inverse variation model.}$$

$$= \frac{1650}{3.25} \quad \text{Substitute 3.25 for } p.$$

$$\approx 508 \text{ units.} \quad \text{Simplify.}$$

So, the demand for the product is 508 units when the price of the product is \$3.25.

6. Let R be the resistance (in ohms), let L be the length (in inches), and let A be the cross-sectional area (in square inches).

Because R varies directly as L and inversely as A, you have

$$R = \frac{kL}{A}.$$

Now, because $R = 64.9$ ohms when

$L = 1000$ feet $= 12{,}000$ inches and

$$A = \pi\left(\frac{0.0126}{2}\right)^2 \approx 1.2469 \times 10^{-4} \text{ square inches,}$$

you have

$$64.9 = \frac{k(12,000)}{1.2469 \times 10^{-4}}$$

$$6.7437 \times 10^{-7} \approx k.$$

So, the equation relating resistance, length, and the cross-sectional area is $R = 6.7437 \times 10^{-7}\,\frac{L}{A}$.

To find the length of copper wire that will produce a resistance of 33.5 ohms, let $R = 33.5$ and

$$A = \pi\left(\frac{0.0201}{2}\right)^2 \approx 3.1731 \times 10^{-4} \text{ square inches, and}$$

solve for L.

$$R = \left(6.7437 \times 10^{-7}\right)\frac{L}{A}$$

$$33.5 = \left(6.7437 \times 10^{-7}\right)\frac{L}{\left(3.1731 \times 10^{-4}\right)}$$

$$(33.5)\left(3.1731 \times 10^{-4}\right) = \left(6.7437 \times 10^{-7}\right)L$$

$$\frac{(33.5)\left(3.1731 \times 10^{-4}\right)}{\left(6.7437 \times 10^{-7}\right)} = L$$

$$15{,}762.7 \text{ inches} \approx L$$

So, the length of the wire is approximately 15,762.7 inches, or about 1314 feet.

7. $E =$ kinetic energy, $m =$ mass, and $V =$ velocity.

Because E varies jointly with the object's mass, m and the square of the object's velocity, V you have

$$E = kmV^2$$

For $E = 6400$ joules, $m = 50$ kg, and $V = 16$m/sec, you have

$$E = kmV^2$$

$$6400 = k(50)(16)^2$$

$$6400 = k(12{,}800)$$

$$\tfrac{1}{2} = k$$

So, the equation relating kinetic energy, mass, and velocity is $E = \frac{1}{2}mV$.

When $m = 70$ and $V = 20$, the kinetic energy is

$$E = \tfrac{1}{2}mV^2 = \tfrac{1}{2}(70)(20)^2 = \tfrac{1}{2}(70)(400)$$

$$= 14{,}000 \text{ joules.}$$

Chapter 2

Checkpoints for Section 2.1

1. (a) Compared with the graph of $y = x^2$, each output of $f(x) = \frac{1}{4}x^2$ shrinks by a factor of $\frac{1}{4}$, creating a broader parabola.

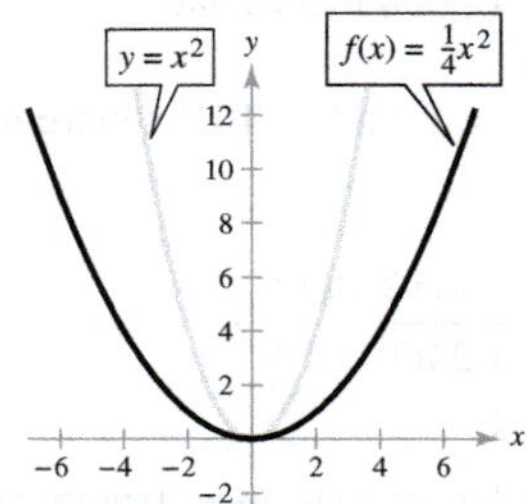

(b) Compared with the graph of $y = x^2$, each output of $f(x) = -\frac{1}{6}x^2$ is reflected in the x-axis and shrinks by a factor of $\frac{1}{6}$, creating a broader parabola.

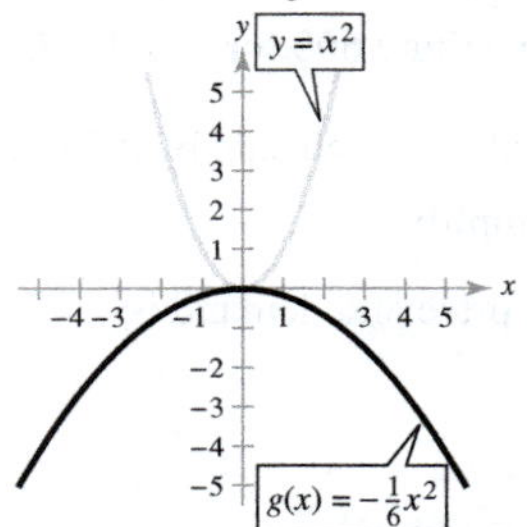

(c) Compared with $y = x^2$, each output of of $h(x) = \frac{5}{2}x^2$ "stretches" by a factor of $\frac{5}{2}$, creating a narrower parabola.

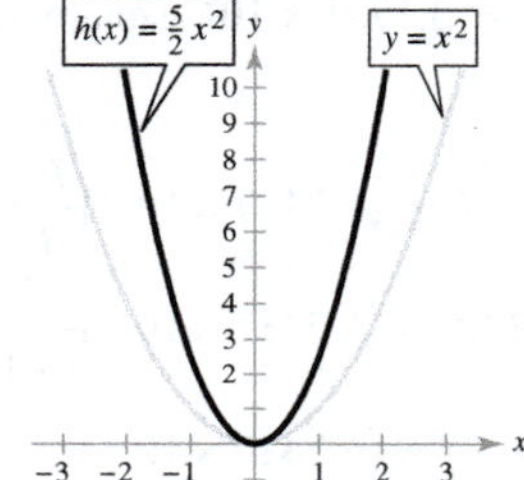

(d) Compared with $y = -4x^2$, each output of $k(x)$ is reflected in the x-axis and "stretches" by a factor of 4.

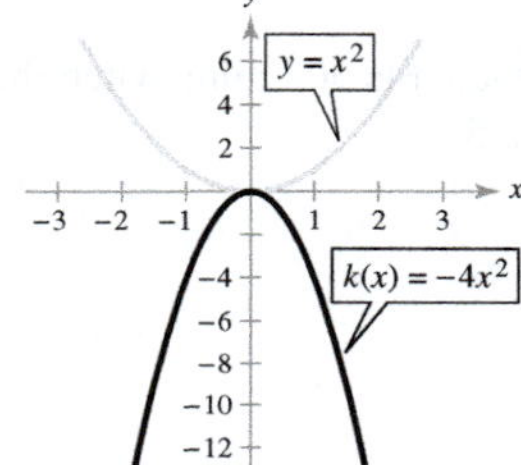

2.

$f(x) = 3x^2 - 6x + 4$	Write original function.
$= 3(x^2 - 2x) + 4$	Factor 3 out of x-terms.
$= 3(x^2 - 2x + 1 - 1) + 4$	Add and subtract 1 within parenthesis.
$= 3(x^2 - 2x + 1) - 3(1) + 4$	Regroup terms.
$= 3(x^2 - 2x + 1) - 3 + 4$	Simplify.
$= 3(x - 1)^2 + 1$	Write in standard form.

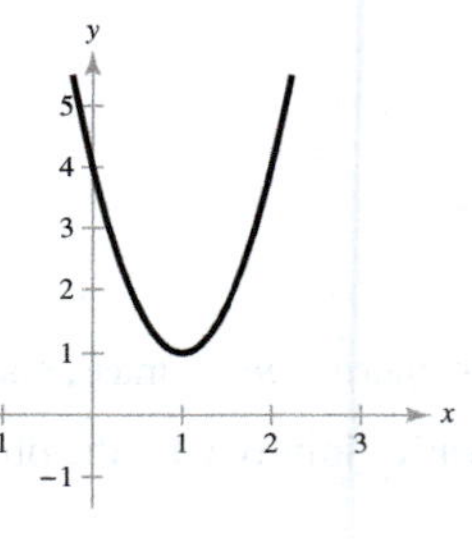

You can see that the graph of f is a parabola that opens upward and has its vertex at $(1, 1)$. This corresponds to a right shift of one unit and an upward shift of one unit relative to the graph of $y = 3x^2$, which is a "stretch" of $y = x^2$.

The axis of the parabola is the vertical line through the vertex, $x = 1$.

3. $f(x) = x^2 - 4x + 3$ Write original function.

$= \left(x^2 - 4x + 4 - 4\right) + 3$ Add and subtract 4 within parenthesis.

$= \left(x^2 - 4x + 4\right) - 4 + 3$ Regroup terms.

$= \left(x^2 - 4x + 4\right) - 1$ Simplify.

$= \left(x^2 - 2\right)^2 - 1$ Write in standard form.

In standard form, you can see that f is a parabola that opens upward with vertex $(2, -1)$.

The x-intercepts of the graph are determined as follows.

$$x^2 - 4x + 3 = 0$$

$$(x - 3)(x - 1) = 0$$

$$x - 3 = 0 \Rightarrow x = 3$$

$$x - 1 = 0 \Rightarrow x = 1$$

So, the x-intercepts are $(3, 0)$ and $(1, 0)$.

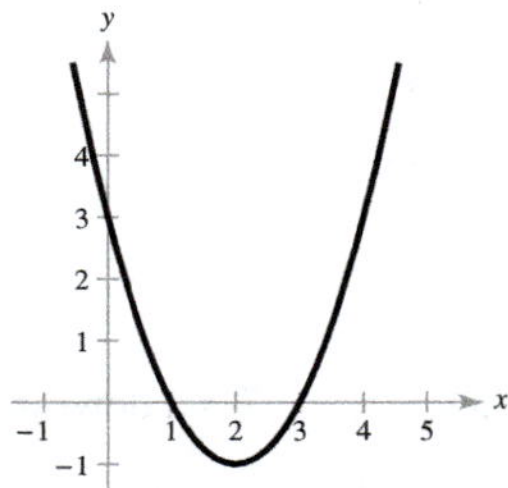

4. The vertex is $(h, k) = (-4, 11)$ so the equation has the form

$f(x) = a(x + 4)^2 + 11$.

The parabola passes through the point $(-6, 15)$ so it follows that $f(-6) = 15$.

$f(x) = a(x + 4)^2 + 11$ Write standard form.

$15 = a(-6 + 4)^2 + 11$ Substitute -6 for x and 15 for $f(x)$.

$15 = a(-2)^2 + 11$ Simplify.

$4 = 4a$ Subtract 11 from each side.

$1 = a$ Divide each side by 4.

The equation in standard form is $f(x) = (x + 4)^2 + 11$.

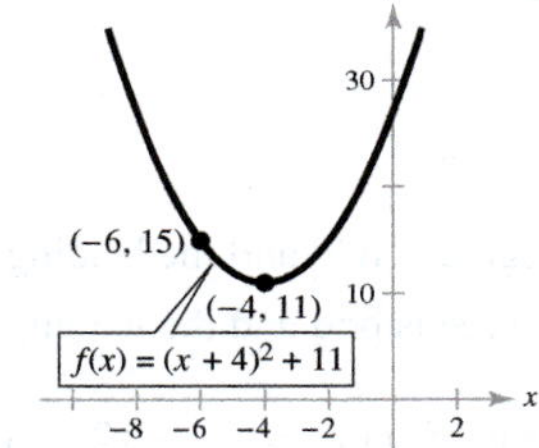

5. For this quadratic function,

$f(x) = ax^2 + bx + c = -0.007x^2 + x + 4$

which implies that $a = -0.007$ and $b = 1$.

Because $a < 0$, the function has a maximum at $x = -\dfrac{b}{2a}$. So, the baseball reaches its maximum height when it is

$$x = -\frac{b}{2a} = -\frac{1}{2(-0.007)} = \frac{1}{0.014} \approx 71.4 \text{ feet from home plate.}$$

At this distance, the maximum height is $f(71.4) = -0.007(71.4)^2 + (71.4) + 4 \approx 39.7$ feet.

Checkpoints for Section 2.2

1. (a) The graph of $f(x) = (x + 5)^4$ is a left shift by five units of the graph of $y = x^4$.

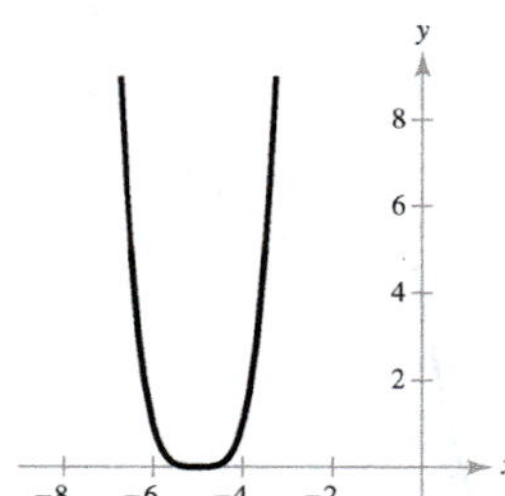

(b) The graph of $g(x) = x^4 - 7$ is a downward shift of seven units of the graph of $y = x^4$.

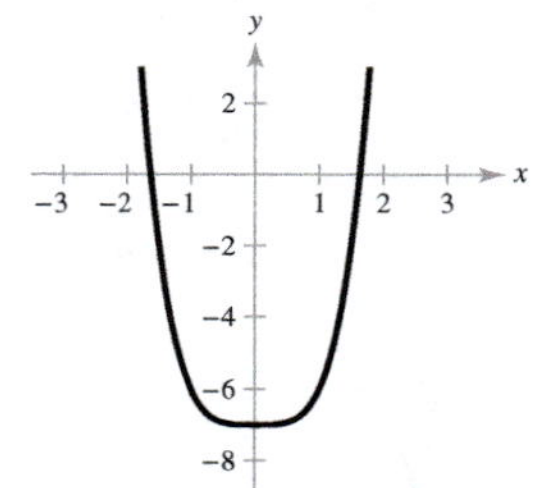

(c) The graph of $h(x) = 7 - x^4 = -x^4 + 7$ is a reflection in the x-axis then an upward shift of seven units of the graph of $y = x^4$.

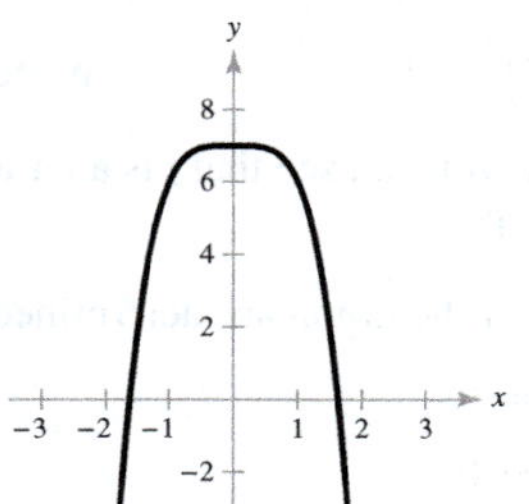

(d) The graph of $k(x) = \frac{1}{4}(x - 3)^4$ is a right shift by three units and a vertical "shrink" by a factor of $\frac{1}{4}$ of the graph of $y = x^4$.

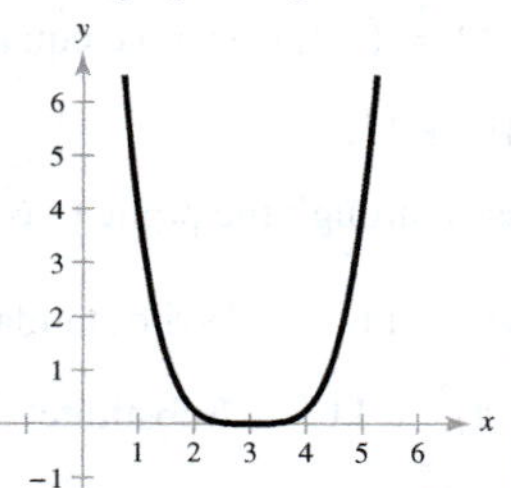

2. (a) Because the degree is odd and the leading coefficient is positive, the graph falls to the left and rises to the right.

(b) Because the degree is odd and the leading coefficient is negative, the graph rises to the left and falls to the right.

3. To find the real zeros of $f(x) = x^3 - 12x^2 + 36x$, set $f(x)$ equal to zero, and solve for x.

$$x^3 - 12x^2 + 36x = 0$$

$$x(x^2 - 12x + 36) = 0$$

$$x(x - 6)^2 = 0$$

$$x = 0$$

$$x - 6 = 0 \Rightarrow x = 6$$

So, the real zeros are $x = 0$ and $x = 6$. Because the function is a third-degree polynomial, the graph of f can have at most $3 - 1 = 2$ turning points. In this case, the graph of f has two turning points.

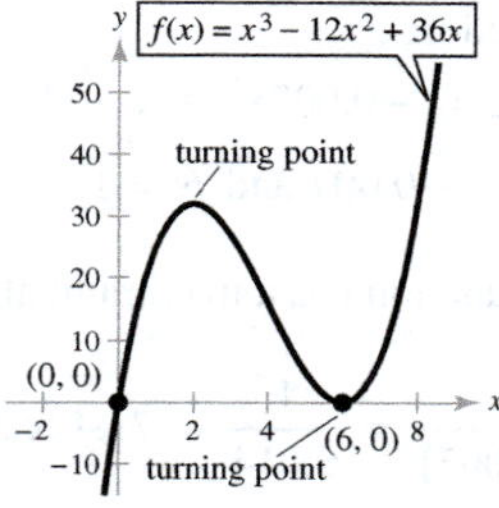

4. 1. *Apply the Leading Coefficient Test.*

Because the leading coefficient is positive and the degree is odd, you know that the graph eventually falls to the left and rises to the right.

2. *Find the Real Zeros of the Polynomial.*

By factoring $f(x) = 2x^3 - 6x^2$

$= 2x^2(x - 3)$

you can see that the real zeros of f are $x = 0$ (even multiplicity) and $x = 3$ (odd multiplicity). So, the x-intercepts occur at $(0, 0)$ and $(3, 0)$

3. *Plot a Few Additional Points.*

x	-1	1	2	4
$f(x)$	-8	-4	-8	32

4. *Draw the Graph.*

Draw a continuous curve through all of the points. Because $x = 0$ is of even multiplicity, you know that the graph touches the x-axis but does not cross it at $(0, 0)$. Because $x = 3$ is of odd multiplicity, you know that the graph should cross the x-axis at $(3, 0)$.

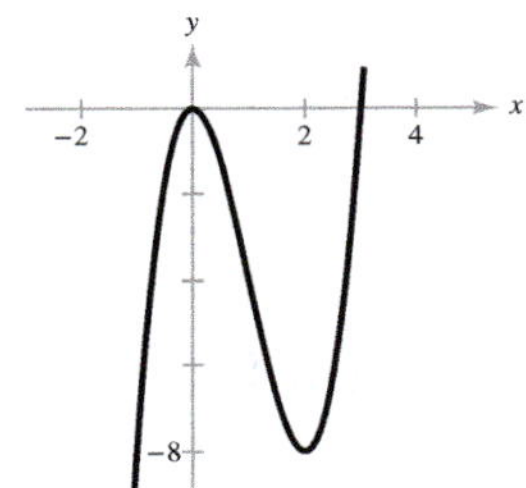

5. 1. *Apply the Leading Coefficient Test.*

Because the leading coefficient is negative and the degree is even, you know that the graph eventually falls to the left and falls to the right.

2. *Find the Real Zeros of the Polynomial.*

By factoring $f(x) = -\frac{1}{4}x^4 + \frac{3}{2}x^3 - \frac{9}{4}x^2$

$= -\frac{1}{4}x^2(x^2 - 6x + 9)$

$= -\frac{1}{4}x^2(x - 3)^2$

you can see that the real zeros of f are $x = 0$ (even multiplicity) and $x = 3$ (even multiplicity). So, the x-intercepts occur at $(0, 0)$ and $(3, 0)$.

3. *Plot a Few Additional Points.*

x	-1	1	2	4
$f(x)$	-4	-1	-1	-4

4. *Draw the graph.*

Draw a continuous curve through the points. As indicated by the multiplicities of the zeros, the graph touches but does not cross the x-axis at $(0, 0)$ and $(3, 0)$.

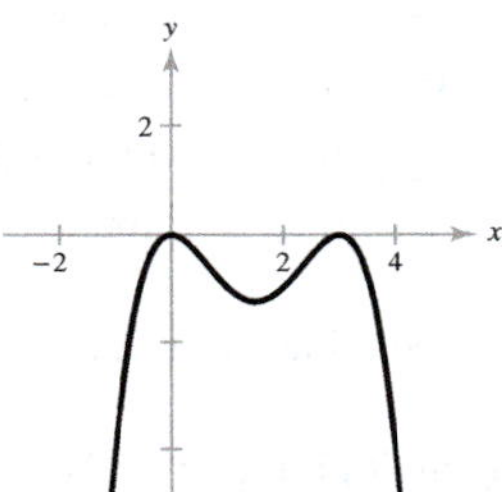

6. Begin by computing a few function values of $f(x) = x^3 - 3x^2 - 2$.

x	-1	0	1	2	3	4
$f(x)$	-6	-2	-4	-6	-2	14

Because $f(3)$ is negative and $f(4)$ is positive, you can apply the Intermediate Value Theorem to conclude that the function has a real zero between $x = 3$ and $x = 4$. To find this real zero more closely, divide the interval $[3, 4]$ into tenths and evaluate the function at each point.

x	3.1	3.2	3.3	3.4	3.5	3.6	3.7	3.8	3.9
$f(x)$	-1.039	0.048	1.267	2.624	4.125	5.776	7.583	9.552	11.689

So, f must have a real zero between 3.1 and 3.2.
To find a more accurate approximation, you can compute the function value between $f(3.1)$ and $f(3.2)$ and apply the Intermediate Value Theorem again to verify that $x \approx 3.196$.

Checkpoints for Section 2.3

1. To divide $9x^3 + 36x^2 - 49x - 196$ by $x + 4$ using long division, you can set up the operation as shown.

$$\begin{array}{rll}
 & 9x^2 \qquad\quad - 49 & \\
x + 4 & \overline{)9x^3 + 36x^2 - 49x - 196} & \\
 & \underline{9x^3 + 36x^2 \qquad\qquad\quad} & \text{Multiply by: } 9x^2(x + 4) \\
 & - 49x - 196 & \text{Subtract.} \\
 & \underline{- 49x - 196} & \text{Multiply by: } -49(x + 4). \\
 & 0 & \text{Subtract.}
\end{array}$$

By factoring, you have

$$\begin{aligned}
9x^3 + 36x^2 - 49x - 196 &= (x + 4)(9x^2 - 49) \\
&= (x + 4)(3x + 7)(3x - 7).
\end{aligned}$$

2. To divide $x^3 - 2x^2 - 9$ by $x - 3$ using long division, you can set up the operation as shown. Because there is no x-term in the dividend, rewrite the dividend as $x^3 - 2x^2 + 0x - 9$ before you apply the Division Algorithm.

$$\begin{array}{rll}
 & x^2 + x + 3 & \\
x - 3 & \overline{)x^3 - 2x^2 + 0x - 9} & \\
 & \underline{x^3 - 3x^2} & \text{Multiply } x^2 \text{ by } x - 3. \\
 & x^2 + 0x - 9 & \text{Subtract.} \\
 & \underline{x^2 - 3x} & \text{Multiply } x \text{ by } x - 3. \\
 & 3x - 9 & \text{Subtract.} \\
 & \underline{3x - 9} & \text{Multiply 3 by } x - 3. \\
 & 0 & \text{Subtract.}
\end{array}$$

So, $x - 3$ divides evenly into $x^3 - 2x^2 - 9$, and you can write $\dfrac{x^3 - 2x^2 - 9}{x - 3} = x^2 + x + 3, x \neq 3.$

Check the result by multiplying $(x + 3)(x^2 + x + 3) = x^3 - 3x^2 + x^2 - 3x + 3x - 9$

$$= x^3 - 2x^2 - 9.$$

3. To divide $-x^3 + 9x + 6x^4 - x^2 - 3$ by $1 + 3x$ using long division, begin by rewriting the dividend and divisor in descending powers of x.

$$\begin{array}{rll}
 & 2x^3 - x^2 \qquad\quad + 3 & \\
3x + 1 & \overline{)6x^4 - x^3 - x^2 + 9x - 3} & \\
 & \underline{6x^4 + 2x^3} & \text{Multiply } 2x^3 \text{ by } 3x + 1. \\
 & - 3x^3 - x^2 + 9x - 3 & \text{Subtract.} \\
 & \underline{- 3x^3 - x^2} & \text{Multiply } -x^2 \text{ by } 3x + 1. \\
 & 9x - 3 & \text{Subtract.} \\
 & \underline{9x + 3} & \text{Multiply 3 by } 3x + 1. \\
 & -6 & \text{Subtract.}
\end{array}$$

So, you have $\dfrac{6x^4 - x^3 - x^2 + 9x - 3}{3x + 1} = 2x^3 - x^2 + 3 - \dfrac{6}{3x + 1}.$

Check the result by multiplying $(3x + 1)\left(2x^3 - x^2 + 3 - \dfrac{6}{3x + 1}\right) = 6x^4 - 3x^3 + 9x + 2x^3 - x^2 + 3 - 6$

$$= 6x^4 - x^3 - x^2 + 9x - 3$$

4. To divide $5x^3 + 8x^2 - x + 6$ by $x + 2$ using synthetic division, you can set up the array as shown.

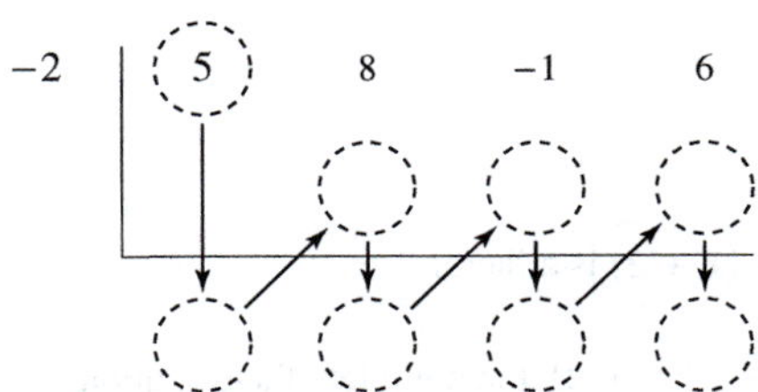

Then, use the synthetic division pattern by adding terms in columns and multiplying the results by -2.

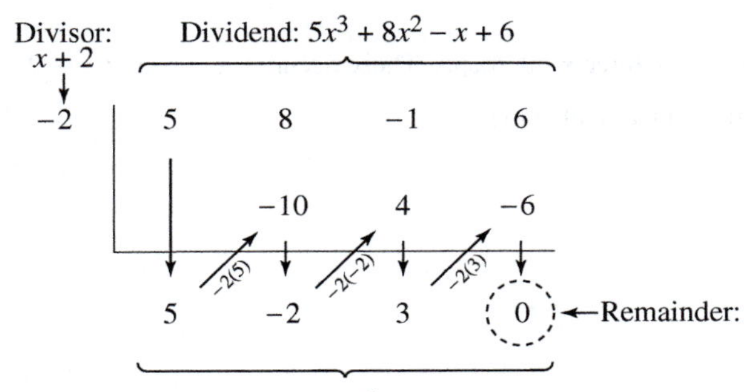

So, you have, $\dfrac{5x^3 + 8x^2 - x + 6}{x + 2} = 5x^2 - 2x + 3.$

5. (a) $f(-1)$

$$\begin{array}{r|rrrr} -1 & 4 & 10 & -3 & -8 \\ & & -4 & -6 & 9 \\ \hline & 4 & 6 & -9 & 1 \end{array}$$

Because the remainder is $r = 1$, $f(-1) = 1$.

Check: $f(-1) = 4(-1)^3 + 10(-1)^2 - 3(-1) - 8$

$= 4(-1) + 10(1) + 3 - 8$

$= 1$

(b) $f(4)$

$$\begin{array}{r|rrrr} 4 & 4 & 10 & -3 & -8 \\ & & 16 & 104 & 404 \\ \hline & 4 & 26 & 101 & 396 \end{array}$$

Because the remainder is $r = 396$, $f(4) = 396$.

Check: $f(4) = 4(4)^3 + 10(4)^2 - 3(4) - 8$

$= 4(64) + 10(16) - 12 - 8$

$= 396$

(c) $f\left(\frac{1}{2}\right)$

$$\begin{array}{r|rrrr} \frac{1}{2} & 4 & 10 & -3 & -8 \\ & & 2 & 6 & \frac{3}{2} \\ \hline & 4 & 12 & 3 & -\frac{13}{2} \end{array}$$

Because the remainder is $r = -\frac{13}{2}$, $f\left(\frac{1}{2}\right) = -\frac{13}{2}$.

Check: $f\left(\frac{1}{2}\right) = 4\left(\frac{1}{2}\right)^3 + 10\left(\frac{1}{2}\right)^2 - 3\left(\frac{1}{2}\right) - 8$

$= 4\left(\frac{1}{8}\right) + 10\left(\frac{1}{4}\right) - \frac{3}{2} - 8$

$= \frac{1}{2} + \frac{5}{2} - \frac{3}{2} - 8$

$= -\frac{13}{2}$

(d) $f(-3)$

$$\begin{array}{r|rrrr} -3 & 4 & 10 & -3 & -8 \\ & & -12 & 6 & -9 \\ \hline & 4 & -2 & 3 & -17 \end{array}$$

Because the remainder is $r = -17$, $f(-3) = -17$.

Check: $f(-3) = 4(-3)^3 + 10(-3)^2 - 3(-3) - 8$

$= 4(-27) + 10(9) + 9 - 8$

$= -17$

6. Algebraic Solution:

Using synthetic division with the factor $(x + 3)$, you obtain the following.

$$\begin{array}{r|rrrr} -3 & 1 & 0 & -19 & -30 \\ & & -3 & 9 & 30 \\ \hline & 1 & -3 & -10 & 0 \end{array}$$

$\rightarrow$ 0 remainder, so $f(-3) = 0$ and $(x + 3)$ is a factor.

Because the resulting quadratic expression factors as $x^2 - 3x - 10 = (x - 5)(x + 2)$, the complete factorization of $f(x)$ is $f(x) = x^3 - 19x - 30 = (x + 3)(x - 5)(x + 2)$.

Graphical Solution:

From the graph of $f(x) = x^3 - 19x - 30$, you can see there are three x-intercepts. These occur at $x = -3$, $x = -2$, and $x = 5$. This implies that $(x + 3)$, $x + 2$, and $(x - 5)$ are factors of $f(x)$.

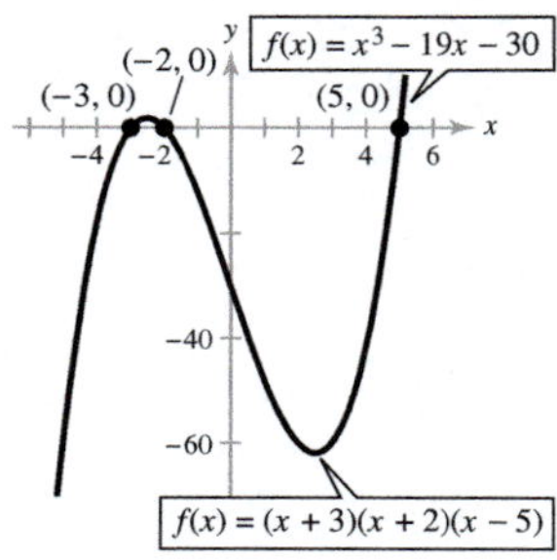

Checkpoints for Section 2.4

1. (a) $(7 + 3i) + (5 - 4i) = 7 + 3i + 5 - 4i$ Remove parentheses.

$= (7 + 5) + (3 - 4)i$ Group like terms.

$= 12 - i$ Write in standard form.

(b) $(3 + 4i) - (5 - 3i) = 3 + 4i - 5 + 3i$ Remove parentheses.

$= (3 - 5) + (4 + 3)i$ Group like terms.

$= -2 + 7i$ Write in standard form.

(c) $2i + (-3 - 4i) - (-3 - 3i) = 2i - 3 - 4i + 3 + 3i$ Remove parentheses.

$= (-3 + 3) + (2 - 4 + 3)i$ Group like terms.

$= i$ Write in standard form.

(d) $(5 - 3i) + (3 + 5i) - (8 + 2i) = 5 - 3i + 3 + 5i - 8 - 2i$ Remove parentheses.

$= (5 + 3 - 8) + (-3 + 5 - 2)i$ Group like terms.

$= 0 + 0i$ Simplify.

$= 0$ Write in standard form.

2. (a) $-5(3 - 2i) = -5(3) - (-5)(2i)$ Distributive Property

$= -15 + 10i$ Simplify

(b) $(2 - 4i)(3 + 3i) = 6 + 6i - 12i - 12i^2$ FOIL Method

$= 6 + 6i - 12i - 12(-1)$ $i^2 = -1$

$= 6 - 6i + 12$ Simplify.

$= 18 - 6i$ Write in standard form.

(c) $(4 + 5i)(4 - 5i) = 4(4 - 5i) + 5i(4 - 5i)$ Distributive Property

$= 16 - 20i + 20i - 25i^2$ Distributive Property

$= 16 - 20i + 20i - 25(-1)$ $i^2 = -1$

$= 16 + 25$ Simplify.

$= 41$ Write in standard form.

(d) $(4 + 2i)^2 = (4 + 2i)(4 + 2i)$ Square of a binomial

$= 4(4 + 2i) + 2i(4 + 2i)$ Distributive Property

$= 16 + 8i + 8i + 4i^2$ Distributive Property

$= 16 + 8i + 8i + 4(-1)$ $i^2 = -1$

$= (16 - 4) + (8i + 8i)$ Group like terms.

$= 12 + 16i$ Write in standard form.

3. (a) The complex conjugate of $3 + 6i$ is $3 - 6i$.

$$(3 + 6i)(3 - 6i) = (3)^2 - (6i)^2$$
$$= 9 - 36i^2$$
$$= 9 - 36(-1)$$
$$= 45$$

(b) The complex conjugate of $2 - 5i$ is $2 + 5i$.

$$(2 - 5i)(2 + 5i) = (2)^2 - (5i)^2$$
$$= 4 - 25i^2$$
$$= 4 - 25(-1)$$
$$= 29$$

4. $\dfrac{2 + i}{2 - i} = \dfrac{2 + i}{2 - i} \cdot \dfrac{2 + i}{2 + i}$ Multiply numerator and denominator by complex conjugate of the denominator.

$= \dfrac{4 + 2i + 2i + i^2}{4 - i^2}$ Expand.

$= \dfrac{4 - 1 + 4i}{4 - (-1)}$ $i^2 = -1$

$= \dfrac{3 + 4i}{5}$ Simplify.

$= \dfrac{3}{5} + \dfrac{4}{5}i$ Write in standard form.

5. $\sqrt{-14}\sqrt{-2} = \sqrt{14}i\sqrt{2}i = \sqrt{28}i^2 = 2\sqrt{7}(-1) = -2\sqrt{7}$

6. To solve $8x^2 + 14x + 9 = 0$, use the Quadratic formula

$$x = \frac{-b \pm \sqrt{b^2 - 4ac}}{2a}.$$

$$x = \frac{-14 \pm \sqrt{14^2 - 4(8)(9)}}{2(8)}$$ Substitute $a = 8$, $b = 14$, and $c = 9$.

$$= \frac{-14 \pm \sqrt{-92}}{16}$$ Simplify.

$$= \frac{-14 \pm 2\sqrt{23}i}{16}$$ Write $\sqrt{-92}$ in standard form.

$$= \frac{-14}{16} + \frac{2\sqrt{23}i}{16}$$ Write in standard form.

$$= -\frac{7}{8} \pm \frac{\sqrt{23}i}{8}$$ Simplify.

Checkpoints for Section 2.5

1. $f(x) = x^4 - 1 = (x^2 + 1)(x^2 - 1)$

$= (x^2 + 1)(x + 1)(x - 1)$

$x^2 + 1 \Rightarrow x = \pm i$

$x + 1 \Rightarrow x = -1$

$x - 1 \Rightarrow x = 1$

The fourth-degree polynomial function $f(x) = x^4 - 1$ has exactly four zeros, $x = \pm i$ and $x = \pm 1$

2. $f(x) = x^3 + 2x^2 + 6x - 4$

Possible rational zeros: ± 1, ± 2, and ± 4

By testing these possible zeros,

$f(-1) = (-1)^3 + 2(-1)^2 + 6(-1) - 4 = -9$

$f(1) = (1)^3 + 2(1)^2 + 6(1) - 4 = 5$

$f(-2) = (-2)^3 + 2(-2)^2 + 6(-2) - 4 = -16$

$f(2) = (2)^3 + 2(2)^2 + 6(2) - 4 = 24$

$f(-4) = (-4)^3 + 2(-4)^2 + 6(-4) - 4 = -60$

$f(4) = (4)^3 + 2(4)^2 + 6(4) - 4 = 116$

you can conclude that the polynomial $f(x) = x^3 + 2x^2 + 6x - 4$ has no rational zeros.

3. Because the leading coefficient is 1, the possible rational zeros are the factors of the constant term.

Possible rational zeros: $\pm1, \pm5, \pm25, \pm125$

$$\begin{array}{r|rrrr} 5 & 1 & -15 & 75 & -125 \\ & & 5 & -50 & 125 \\ \hline & 1 & -10 & 25 & 0 \end{array}$$

$\rightarrow$ 0 remainder, so $x = 5$ is a factor.

$$\begin{array}{r|rrr} 5 & 1 & -10 & 25 \\ & & 5 & -25 \\ \hline & 1 & -5 & 0 \end{array}$$

$\rightarrow$ 0 remainder, so $x = 5$ is a factor.

By applying synthetic division successively, you can determine that $x = 5$ is the only rational zero.

So, $f(x) = x^3 - 15x^2 + 75x - 125$ factors as $f(x) = (x - 5)(x - 5)(x - 5) = (x - 5)^3$.

Because the rational zero $x = 5$ has multiplicity of three, which is odd, the graph of f crosses the x-axis at the x-intercept, $(5, 0)$.

4. The leading coefficient is 2 and the constant term is 6.

Possible rational zeros: $\dfrac{\text{Factors of } 6}{\text{Factors of } 2} = \dfrac{\pm1, \pm2, \pm3, \pm6}{\pm1, \pm2} = \pm1, \pm2, \pm3, \pm6, \pm\dfrac{1}{2}, \pm\dfrac{3}{2}$

Choose a value of x and use synthetic division.

$x = -3$

$$\begin{array}{r|rrrr} -3 & 2 & 1 & -13 & 6 \\ & & -6 & 15 & -6 \\ \hline & 2 & -5 & 2 & 0 \end{array}$$

$\rightarrow$ 0 remainder, so $x + 3$ is a factor.

So, $f(x) = 2x^3 + x^2 - 13x + 6$ factors as $f(x) = (x + 3)(2x^2 - 5x + 2) = (x + 3)(2x - 1)(x - 2)$ and you can conclude that the rational zeros of f are $x = -3$, $x = \dfrac{1}{2}$, and $x = 2$.

5. The leading coefficient is -2 and the constant term is 18.

Possible rational zeros: $\dfrac{\text{Factor of } 18}{\text{Factors of } -2} = \dfrac{\pm1, \pm2, \pm3, \pm6, \pm9, \pm18}{\pm1, \pm2} = \pm1, \pm2, \pm3, \pm6, \pm9, \pm18, \pm\dfrac{1}{2}, \pm\dfrac{3}{2}, \pm\dfrac{9}{2}$

A graph can assist you to narrow the list to reasonable possibilities.

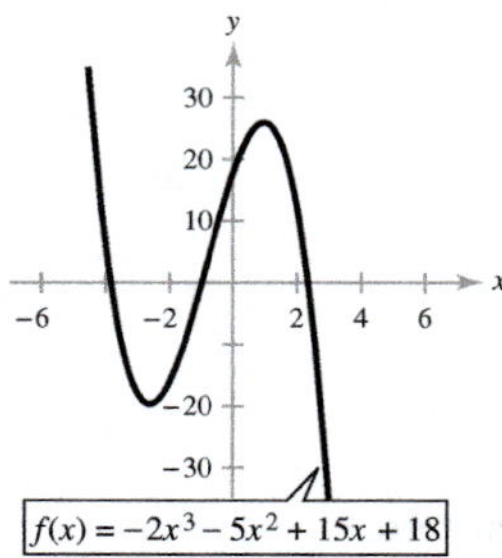

Start by testing $x = -1$.

$$\begin{array}{r|rrrr} -1 & -2 & -5 & 15 & 18 \\ & & 2 & 3 & -18 \\ \hline & -2 & -3 & 18 & 0 \end{array} \rightarrow \text{ 0 remainder, so } x + 1 \text{ is a factor.}$$

So, $f(x) = -2x^3 - 5x^2 + 15x + 18$ factors as $f(x) = (x + 1)(-2x^2 - 3x + 18)$

$$-2x^2 - 3x + 18 = 0$$

$$x = \frac{-b \pm \sqrt{b^2 - 4ac}}{2a}$$

$$x = \frac{-(-3) \pm \sqrt{(-3)^2 - 4(-2)(18)}}{2(-2)}$$

$$x = \frac{3 \pm \sqrt{153}}{-4}$$

$$x = \frac{-3 \pm 3\sqrt{17}}{4} \approx -3.8423, 2.3423$$

And you can conclude that the rational zeros of f are $x = -1$ and $x = \dfrac{-3(1 \pm 3\sqrt{37})}{4} \approx -3.8423, 2.3423.$

6. Because $-7i$ is a zero, you know that the conjugate $7i$ must also be a zero.

So, the four zeros are $2, -2, 7i,$ and $-7i$.

Then, using the Linear Factorization Theorem, $f(x)$ can be written as

$f(x) = a(x - 2)(x + 2)(x - 7i)(x + 7i)$.

For simplicity, let $a = 1$. Then multiply the factors with real coefficients to obtain $(x + 2)(x - 2) = x^2 - 4$ and multiply the complex conjugates to obtain $(x - 7i)(x + 7i) = x^2 + 49$.

So, you obtain the following fourth-degree polynomial function.

$$\begin{aligned} f(x) = (x^2 - 4)(x^2 + 49) &= x^4 + 49x^2 - 4x^2 - 196 \\ &= x^4 + 45x^2 - 196. \end{aligned}$$

7. Because $2i$ is a zero, you know that the conjugate $-2i$ must also be a zero.

This means that both $(x - 2i)$ and $(x + 2i)$ are factors of f.

$(x - 2i)(x + 2i) = x^2 - 4i^2 = x^2 + 4$. The other zeros are $x = -2$ and $x = 1$.

$$\begin{aligned} f(x) &= a(x + 2)(x - 1)(x^2 + 4) \\ &= a(x^2 + x - 2)(x^2 + 4) \\ &= a(x^4 + x^3 + 2x^2 + 4x - 8) \end{aligned}$$

Because $f(-1) = 10$

$$\begin{aligned} 10 &= a\left[(-1)^4 + (-1)^3 + 2(-1)^2 + 4(-1) - 8\right] \\ 10 &= a(-10) \\ -1 &= a \end{aligned}$$

So, $f(x) = (-1)(x^4 + x^3 + 2x^2 + 4x - 8)$

$$= -x^4 - x^3 - 2x^2 - 4x + 8.$$

8. Because complex zeros occur in conjugate pairs you know that if $4i$ is a zero of f, so is $-4i$.

This means that both $(x - 4i)$ and $(x + 4i)$ are factors of f.

$(x - 4i)(x + 4i) = x^2 - 16i^2 = x^2 + 16$

Using long division, you can divide $x^2 + 16$ into $f(x)$ to obtain the following.

$$\begin{array}{r} 3x - 12 \\ x^2 + 16 \overline{)\,3x^3 - 2x^2 + 48x - 32} \\ \underline{3x^3 \qquad\quad + 48x \qquad} \\ -2x^2 \qquad\quad - 32 \\ \underline{-2x^2 \qquad\quad - 32} \\ 0 \end{array}$$

So, you have $f(x) = (x^2 + 16)(3x - 2)$ and you can conclude that the real zeros of f are $x = -4i$, $x = 4i$, and $x = \frac{2}{3}$.

9. $f(x) = x^4 + 8x^2 - 9$

Because the leading coefficient is 1, the possible rational zeros are the factors of the constant term.
Possible rational zeros: $\pm 1, \pm 3$, and ± 9

Synthetic division produces the following.

$$\begin{array}{r|rrrrr} 1 & 1 & 0 & 8 & 0 & -9 \\ & & 1 & 1 & 9 & 9 \\ \hline & 1 & 1 & 9 & 9 & 0 \end{array} \quad \rightarrow \quad 1 \text{ is a zero, so } x - 1 \text{ is a factor.}$$

$$\begin{array}{r|rrrr} -1 & 1 & 1 & 9 & 9 \\ & & -1 & 0 & -9 \\ \hline & 1 & 0 & 9 & 0 \end{array} \quad \rightarrow \quad -1 \text{ is a zero, so } x + 1 \text{ is a factor.}$$

So, you have $f(x) = x^4 + 8x^2 - 9 = (x - 1)(x + 1)(x^2 + 9)$.

You can factor $x^2 + 9$ as $x^2 - (-9) = (x + \sqrt{-9})(x - \sqrt{-9}) = (x + 3i)(x - 3i)$.

So, you have $f(x) = (x - 1)(x + 1)(x + 3i)(x - 3i)$ and you can conclude that the zeros of f are $x = 1$, $x = -1$, $x = 3i$, and $x = -3i$.

10. Determine the possible numbers of positive and negative real zeros of $f(x) = 2x^3 + 5x^2 + x + 8$.

The original polynomial has *zero* variations in sign.

$f(x) = 2x^3 + 5x^2 + x + 8$

The polynomial $f(-x) = 2(-x)^3 + 5(-x)^2 + (-x) + 8 = -2x^3 + 5x^2 - x + 8$ has *three* variations in sign.

$$\begin{array}{cccccc} & - \text{ to } + & & - \text{ to } + & \\ & \downarrow\ \downarrow & & \downarrow\ \ \ \downarrow & \\ f(x) = & -2x^3 + 5x^2 & & - x + 8 & \\ & \uparrow & \uparrow & & \\ & + \text{ to } - & & & \end{array}$$

So, from Descarte's Rule of Signs, the polynomial $f(x) = 2x^3 + 5x^2 + x + 8$ has either three negative real zeros or one negative real zero and no positive real zeros.

From the graph, you can see that the function has only one negative real zero.

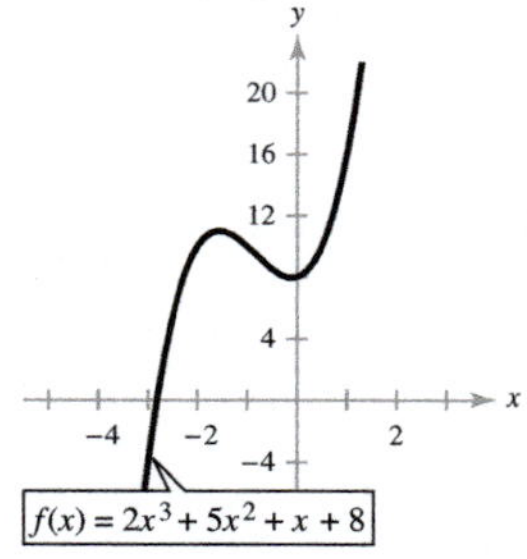

11. The possible real zeros are as follows.

$$\frac{\text{Factors of } -3}{\text{Factors of } 8} = \frac{\pm1 \pm3}{\pm1 \pm2 \pm4 \pm8} = \pm\frac{1}{8}, \pm\frac{1}{4}, \pm\frac{3}{8}, \pm\frac{1}{2}, \pm\frac{3}{4}, \pm1, \pm\frac{3}{2}, \pm3$$

The original polynomial $f(x)$ has three variations in sign. The polynomial

$f(-x) = 8(-x)^3 - 4(-x)^2 + 6(-x) - 3 = -8x^3 - 4x^2 - 6x - 3$

has no variations in sign. So, you can apply Descarte's Rule of Signs to conclude that there are either three positive real zeros or one positive real zero, and no negative real zeros.

Using $x = 1$, synthetic division produces the following.

$$\begin{array}{r|rrrr} 1 & 8 & -4 & 6 & -3 \\ & & 8 & 4 & 10 \\ \hline & 8 & 4 & 10 & 7 \end{array} \quad \rightarrow \quad 1 \text{ is not a zero.}$$

So, $x = 1$ is not a zero, but because the last row has all positive entries, you know that $x = 1$ is an upper bound for the real zeros. So, you can restrict the search to real zeros between 0 and 1. Using $x = \frac{1}{2}$, synthetic division produces the following.

$$\begin{array}{r|rrrr} \frac{1}{2} & 8 & -4 & 6 & -3 \\ & & 4 & 0 & 3 \\ \hline & 8 & 0 & 6 & 0 \end{array} \quad \rightarrow \quad \tfrac{1}{2} \text{ is a real zero.}$$

$$\begin{aligned} f(x) &= 8x^3 - 4x^2 + 6x - 3 \\ &= \left(x - \frac{1}{2}\right)\left(8x^2 + 6\right) \end{aligned}$$

Because $8x^2 + 6$ has no real zeros, it follows that $x = \dfrac{1}{2}$ is the only real zero.

12. The volume of a pyramid is $V = \frac{1}{3}Bh$, where B is the area of the base and h is the height. The area at the base is x^2 and the height is $x + 2$. So, the volume of the pyramid is $V = \frac{1}{3}x^2(x + 2)$. Substituting 147 for the volume yields the following.

$$147 = \frac{1}{3}x^2(x + 2)$$
$$441 = x^3 + 2x^2$$
$$0 = x^3 + 2x^2 - 441$$

The possible rational zeros are $x = \pm1, \pm3, \pm7, \pm9, \pm21, \pm49, \pm63, \pm147$, and ±441.

Use synthetic division to test some of the possible solutions. So, you can determine that $x = 7$ is a solution.

$$\begin{array}{r|rrrr} 7 & 1 & 2 & 0 & -441 \\ & & 7 & 63 & 441 \\ \hline & 1 & 9 & 63 & 0 \end{array}$$

The other two solutions that satisfy $x^2 + 9x + 63 = 0$ are imaginary and can be discarded. You can conclude that the base of the candle mold should be 7 inches by 7 inches and the height should be $7 + 2 = 9$ inches.

Checkpoints for Section 2.6

1. Because the denominator is zero when $x = 1$, the domain of f is all real numbers except $x = 1$.
To determine the behavior of f, near this excluded value, evaluate $f(x)$ to the left and to the right of $x = 1$, as shown in the table.

x	0	0.5	0.9	0.99	0.999	$\to 1$
$f(x)$	0	−3	−27	−297	−2997	$\to -\infty$

x	$1 \leftarrow$	1.001	1.01	1.1	1.5	2
$f(x)$	$\infty \leftarrow$	3003	303	33	9	6

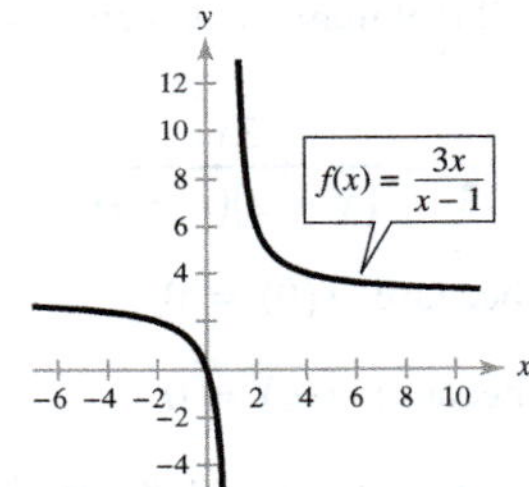

As x approaches 1 from the left, $f(x)$ decreases without bound.
As x approaches 1 from the right, $f(x)$ increases without bound.

2. For this rational function, the degree of the numerator is equal to the degree of the denominator. The leading coefficient of the numerator is 3 and the leading coefficient of the denominator is 1, so the graph of the function has the line $y = \frac{3}{1} = 3$ as a horizontal asymptote. To find any vertical asymptotes, first factor the numerator and denominator as follows.

$$f(x) = \frac{3x^2 + 7x - 6}{x^2 + 4x + 3} = \frac{(3x - 2)\cancel{(x + 3)}}{(x + 1)\cancel{(x + 3)}}$$
$$= \frac{3x - 2}{x + 1}, \ x \neq -3$$

By setting the denominator $x + 1$ (of the simplified function) equal to zero, you can determine that the graph has the line $x = -1$ as a vertical asymptote.

3. $f(x) = \dfrac{1}{x + 3}$

y-intercept: $\left(0, \frac{1}{3}\right)$, because $f(0) = \frac{1}{3}$

x-intercept: none, because there are no zeros in the numerator

Vertical asymptote: $x = -3$, zero of denominator

Horizontal asymptote: $y = 0$, because degree of $N(x) <$ degree of $D(x)$

Additional points:

x	-5	-4	-2	-1	1	2
$f(x)$	$-\frac{1}{2}$	-1	1	$\frac{1}{2}$	$\frac{1}{4}$	$\frac{1}{5}$

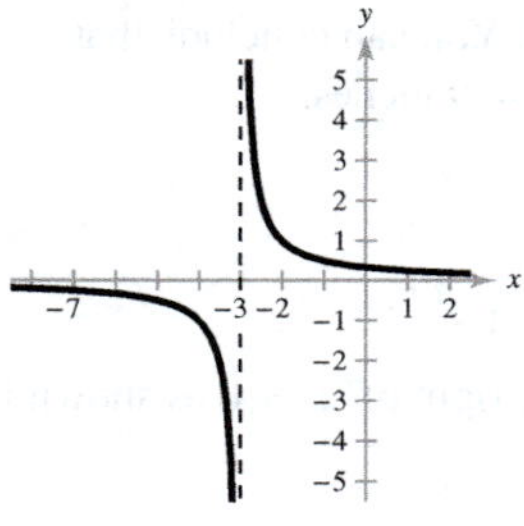

The domain of f is all real numbers except $x = -3$.

4. $g(x) = \dfrac{3 + 2x}{1 + x}$

y-intercept: $(0, 3)$, because $g(0) = 3$

x-intercept: $\left(-\frac{3}{2}, 0\right)$, because $g\left(-\frac{3}{2}\right) = 0$

Vertical asymptote: $x = -1$, zero of denominator

Horizontal asymptote: $y = 2$, because degree of $N(x) =$ degree of $D(x)$

Additional Points:

x	-3	-2	1	3
$g(x)$	$\frac{3}{2}$	1	$\frac{5}{2}$	$\frac{9}{4}$

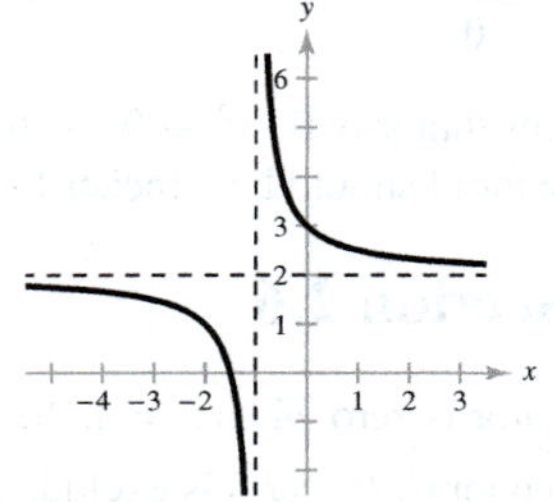

The domain of f is all real numbers except $x = -1$.

5. $f(x) = \dfrac{3x}{x^2 + x - 2} = \dfrac{3x}{(x + 2)(x - 1)}$

y-intercept: $(0, 0)$, because $f(0) = 0$

x-intercept: $(0, 0)$, because $f(0) = 0$

Vertical asymptotes: $x = -2$, $x = 1$, zeros of denominator

Horizontal asymptote: $y = 0$, because degree of $N(x) <$ degree of $D(x)$

Additional points:

x	-3	-1	2	3
$f(x)$	$-\frac{9}{4}$	$\frac{3}{2}$	$\frac{3}{2}$	$\frac{9}{10}$

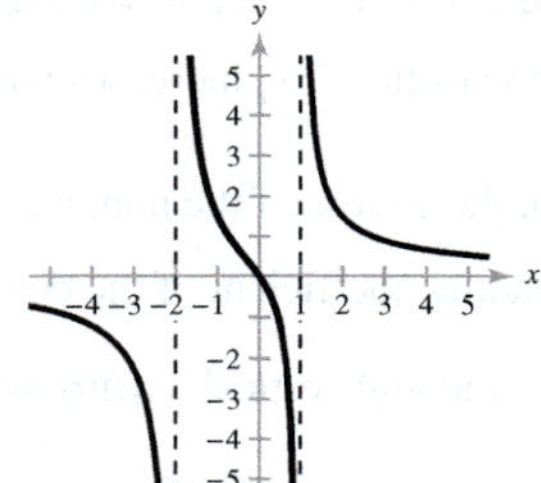

The domain of f is all real numbers except $x = -2$ and $x = 1$.

6. $f(x) = \dfrac{x^2 - 4}{x^2 - x - 6}$

$= \dfrac{(x + 2)(x - 2)}{(x - 3)(x + 2)}$

$= \dfrac{x - 2}{x - 3}, x \neq -2$

y-intercept: $\left(0, \dfrac{2}{3}\right)$, because $f(0) = \dfrac{2}{3}$

x-intercept: $(2, 0)$, because $f(2) = 0$

Vertical asymptote: $x = 3$, zero of (simplified) denominator

Horizontal asymptote: $y = 1$, because degree of $N(x)$ = degree of $D(x)$

Additional points:

x	-7	-5	-1	1	4	5
$f(x)$	$\frac{9}{10}$	$\frac{7}{8}$	$\frac{3}{4}$	$\frac{1}{2}$	2	$\frac{3}{2}$

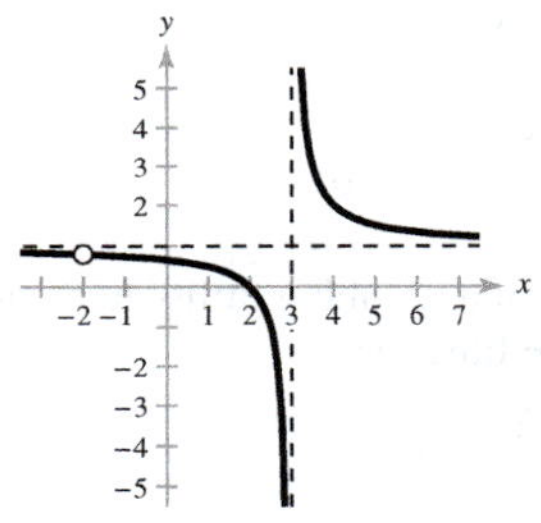

Notice that there is a hole in the graph at $x = -2$ because the numerator and denominator have a common factor of $x + 2$. The domain is all real number except $x = -2$ and $x = 3$.

7. $f(x) = \dfrac{3x^2 + 1}{x} = 3x + \dfrac{1}{x}$

So, the start asymptote is $y = 3x$.

y-intercept: none, since $f(0)$ is undefined.

x-intercept: none, since $3x^2 + 1 \neq 0$ for real numbers.

Vertical asymptote: $x = 0$, zero of denominator

Start asymptote: $y = 3x$

Additional points:

x	-2	-1	-0.5	0.5	1	2
$f(x)$	$-\frac{13}{2}$	-4	$-\frac{7}{2}$	$\frac{7}{2}$	4	$\frac{13}{2}$

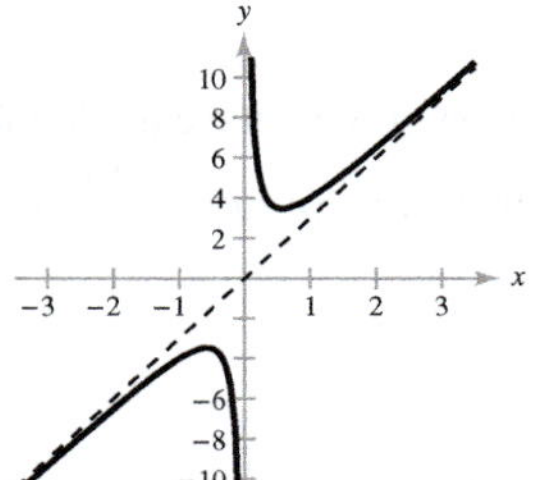

The domain of f is all real numbers except $x = 0$.

8. (a) The cost to remove 20% of the pollutants is

$C = \dfrac{255(20)}{100 - (20)} = \$63.75$ million.

The cost to remove 45% of the pollutants is

$C = \dfrac{255(45)}{100 - 45} \approx \$208.64$ million.

The cost to remove 80% of the pollutants is

$C = \dfrac{255(80)}{100 - 80} = \$1020$ million.

(b) The cost to remove 100% of the pollutants is

$C = \dfrac{255(100)}{100 - (100)}$ which is undefined.

So, it would not be possible to remove 100% of the pollutants.

9. Graphical Solution

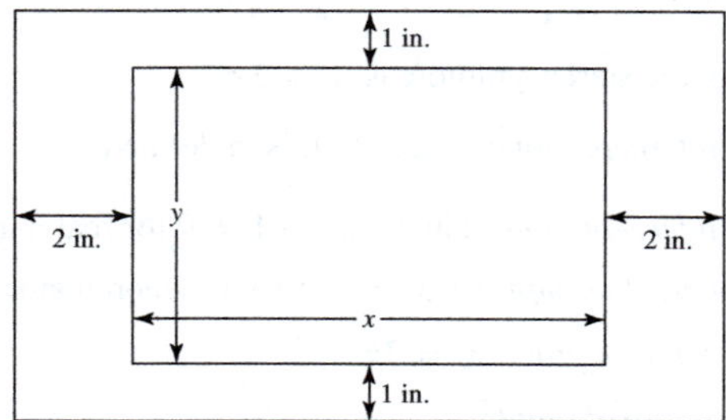

Let A be the area to be minimized.

$A = (x + 4)(y + 2)$

The printed area inside the margins is modeled by

$40 = xy$ or $y = \frac{40}{x}$.

To find the minimum area, rewrite the equation for A in terms of just one variable by substituting $\frac{40}{x}$ for y.

$$\begin{aligned} A &= (x + 4)\left(\frac{40}{x} + 2\right) \\ &= (x + 4)\left(\frac{40 + 2x}{x}\right) \\ &= \frac{(x + 4)(40 + 2x)}{x}, \ x > 0 \end{aligned}$$

The graph of this rational function is shown below. Because x represents the width of the printed area, you need to consider only the portion of the graph for which x is positive. Using a graphing utility, you can approximate the minimum value of A to occur when $x \approx 8.9$ inches. The corresponding value of y is $\frac{40}{8.9} \approx 4.5$ inches.

So, the dimensions should be $8.9 + 4 = 12.9$ inches by $4.5 + 2 = 6.5$ inches.

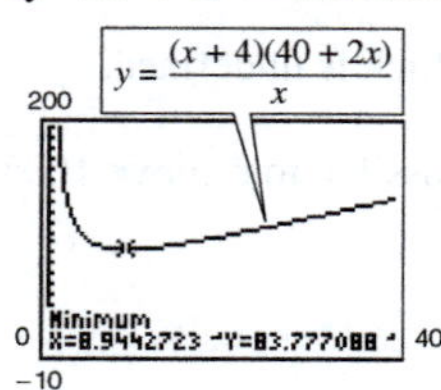

Numerical Solution

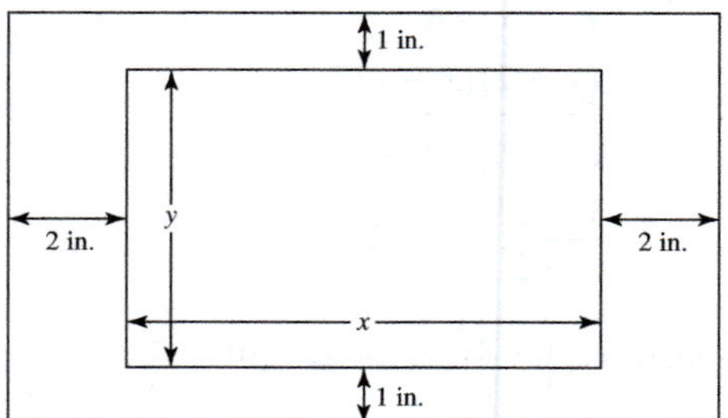

Let A be the area to be minimized.

$A = (x + 4)(y + z)$

The printed area inside the margins is modeled by

$40 = xy$ or $y = \frac{40}{x}$.

To find the minimum area, rewrite the equation for A in terms of just one variable by substituting $\frac{40}{x}$ for y.

$$\begin{aligned} A &= (x + 4)\left(\frac{40}{x} + 2\right) \\ &= (x + 4)\left(\frac{40 + 2x}{x}\right) \\ &= \frac{(x + 4)(40 + 2x)}{x}, \ x > 0 \end{aligned}$$

Use the *table* feature of a graphing utility to create a table of values for the function

$$y_1 = \frac{(x + 4)(40 + 2x)}{x}, \ x > 0$$

beginning at $x = 6$. From the table, you can see that the minimum value of y_1 occurs when x is somewhere between 8 and 9, as shown.

To approximate the minimum value of y_1 to one decimal place, change the table so that it starts at $x = 8$ and increases by 0.1. The minimum value of y_1 occurs when $x \approx 8.9$ as shown.

The corresponding value of y is $\frac{40}{8.9} \approx 4.5$ inches.

So, the dimensions should be $8.9 + 4 = 12.9$ inches by $4.5 + 2 = 6.5$ inches.

x	y_1
6	86.667
7	84.857
8	84
9	83.778
10	84
11	84.545

x	y_1
8.8	83.782
8.9	83.778
9.0	83.778
9.1	83.782

Checkpoints for Section 2.7

1. By factoring the polynomial $x^2 - x - 20 < 0$ as $x^2 - x - 20 = (x + 4)(x - 5)$ you can see that the key numbers are $x = -4$ and $x = 5$. So, the polynomial's test intervals are $(-\infty, -4)$, $(-4, 5)$, and $(5, \infty)$.

In each test interval, choose a representative x-value and evaluate the polynomial.

Test interval	x-value	Polynomial value	Conclusion
$(-\infty, -4)$	$x = -5$	$(-5)^2 - (-5) - 20 = 10$	Positive
$(-4, 5)$	$x = 0$	$(0)^2 - (0) - 20 = -20$	Negative
$(5, \infty)$	$x = 6$	$(6)^2 - (6) - 20 = 10$	Positive

From this, you can conclude that the inequality is satisfied for all x-values in $(-4, 5)$.

This implies that the solution of the inequality $x^2 - x - 20 < 0$ is the interval $(-4, 5)$.

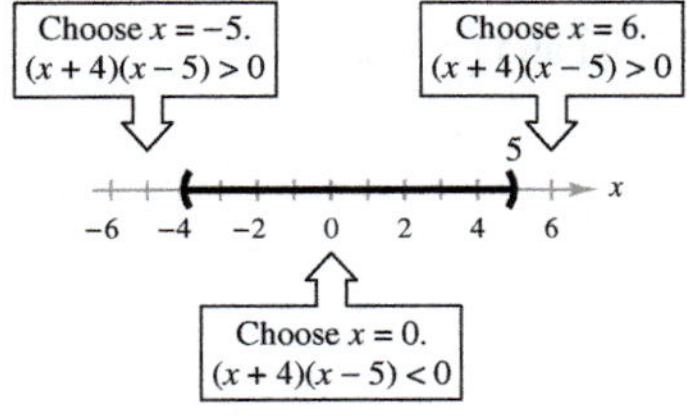

2. (a) **Algebraic solution**

$2x^2 + 3x < 5$ Write original inequality.

$2x^2 + 3x - 5 < 0$ Write in general form.

$(2x + 5)(x - 1) < 0$ Factor.

Key numbers: $x = -\frac{5}{2}$ and $x = 1$

Test intervals: $\left(-\infty, -\frac{5}{2}\right), \left(-\frac{5}{2}, 1\right), (1, \infty)$

Test: Is $(2x + 5)(x - 1) < 0$?

After testing the intervals, you can see that the polynomial $2x^2 + 3x - 5$ is negative on the open interval $\left(-\frac{5}{2}, 1\right)$.

So, the solution set of the inequality is $\left(-\frac{5}{2}, 1\right)$.

(b) **Graphical solution**

First write the polynomial inequality $2x^2 + 3x < 5$ as $2x^2 + 3x - 5 > 0$. Then use a graphing utility to graph $y = 2x^2 + 3x - 5$. You can see that the graph is below the x-axis when x is greater than $-\frac{5}{2}$ and when x is less than 1. So, the solution set is $\left(-\frac{5}{2}, 1\right)$.

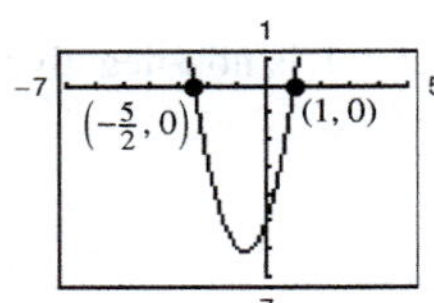

3.

$3x^3 - x^2 - 12x > -4$	Write original inequality.
$3x^2 - x^2 - 12x + 4 > 0$	Write in general form.
$x^2(3x - 1) - 4(3x - 1) > 0$	Factor.
$(3x - 1)(x^2 - 4) > 0$	Factor.
$(3x - 1)(x + 2)(x - 2) > 0$	Factor.

The key numbers are $x = -2, x = \frac{1}{3}$, and $x = 2$, and the test intervals are $(-\infty, -2)$, $\left(-2, \frac{1}{3}\right)$, $\left(\frac{1}{3}, 2\right)$, and $(2, \infty)$.

Test interval	***x*-value**	**Polynomial value**	**Conclusion**
$(-\infty, -2)$	$x = -3$	$3(-3)^3 - (-3)^2 - 12(-3) + 4 = -50$	Negative
$\left(-2, \frac{1}{3}\right)$	$x = 0$	$3(0)^3 - (0)^2 - 12(0) + 4 = 4$	Positive
$\left(\frac{1}{3}, 2\right)$	$x = 1$	$3(1)^3 - (1)^2 - 12(1) + 4 = -6$	Negative
$(2, \infty)$	$x = 3$	$3(3)^3 - (3)^2 - 12(3) + 4 = 40$	Positive

From this, you can conclude that the inequality is satisfied on the open intervals $\left(-2, \frac{1}{3}\right)$ and $(2, \infty)$.

So, the solution set is $\left(-2, \frac{1}{3}\right) \cup (2, \infty)$.

4. (a) The solution set of $x^2 + 6x + 9 < 0$ is empty. In other words, the quadratic $x^2 + 6x + 9$ is not less than 0 for any value of x.

(b) The solution set of $x^2 + 4x + 4 \le 0$ consists of the single real number $\{-2\}$, because the quadratic $x^2 + 4x + 4$ has only one key number, $x = -2$, and it is the only value that satisfies the inequality.

(c) The solution set of $x^2 - 6x + 970$ consists of all real numbers except $x = 3$. In interval notation, the solution set can be written as $(-\infty, 3) \cup (3, \infty)$.

(d) The solution set of $x^2 - 2x + 1 \ge 0$ consists of the entire set of real numbers $(-\infty, \infty)$. In other words, the value of the quadratic $x^2 - 2x + 1$ is non-negative for every real value of x.

5. (a)

$$\frac{x-2}{x-3} \geq -3 \quad \text{Write original inequality.}$$

$$\frac{x-2}{x-3} + 3 \geq 0 \quad \text{Write in general form.}$$

$$\frac{x-2}{x-3} + \frac{3(x-3)}{x-3} \geq 0 \quad \text{Rewrite fraction using LCD.}$$

$$\frac{x-2+3x-9}{x-3} \geq 0 \quad \text{Add fractions.}$$

$$\frac{4x-11}{x-3} \geq 0 \quad \text{Simplify.}$$

Key numbers: $x = \frac{11}{4}, x = 3$

Test intervals: $\left(-\infty, \frac{11}{4}\right), \left(\frac{11}{4}, 3\right), (3, \infty)$

Test: Is $\frac{4x-11}{x-3} \geq 0$?

Test interval	x-value	Polynomial value	Conclusion
$\left(-\infty, \frac{11}{4}\right)$	$x = 0$	$\frac{4(0)-11}{0-3} = \frac{11}{3}$	Positive
$\left(\frac{11}{4}, 3\right)$	$x = 2.9$	$\frac{4(2.9)-11}{2.9-3} = \frac{0.6}{-0.1} = -6$	Negative
$(3, \infty)$	$x = 4$	$\frac{4(4)-11}{4-3} = \frac{5}{1} = 5$	Positive

After testing these intervals, you can see that the inequality is satisfied on the open intervals $\left(-\infty, \frac{11}{4}\right)$ and $(3, \infty)$. Moreover, because $\frac{4x-11}{x-3} = 0$ when $x = \frac{11}{4}$, you can conclude that the solution set consists of all real numbers in the intervals $\left(-\infty, \frac{11}{4}\right] \cup (3, \infty)$.

$\frac{11}{4}$

1 2 3 4 5 x

(b)

$$\frac{4x - 1}{x - 6} > 3 \quad \text{Write original inequality.}$$

$$\frac{4x - 1}{x - 6} - 3 > 0 \quad \text{Write in general form.}$$

$$\frac{4x - 1 - 3(x - 6)}{x - 6} > 0 \quad \text{Combine fractions with LCD.}$$

$$\frac{4x - 1 - 3x + 18}{x - 6} > 0 \quad \text{Simplify.}$$

$$\frac{x + 17}{x - 6} > 0 \quad \text{Simplify.}$$

Key numbers: $x = -17, x = 6$

Test intervals: $(-\infty, -17)$, $(-17, 6)$, and $(6, \infty)$

Test: Is $\frac{x + 17}{x - 6} > 0$?

Test interval	x-value	Polynomial value	Conclusion
$(-\infty, -17)$	-20	$\frac{(-20) + 17}{(-20) - 6} = \frac{-3}{-26} = \frac{3}{26}$	Positive
$(-17, 6)$	0	$\frac{(0) + 17}{(0) - 6} = -\frac{17}{6}$	Negative
$(6, \infty)$	8	$\frac{(8) + 17}{(8) - 6} = \frac{25}{2}$	Positive

After testing these intervals, you can see that the inequality is satisfied on the open intervals $(-\infty, -17)$ and $(6, \infty)$.

So, you can conclude that the solution set consists of all real numbers in the intervals $(-\infty, -17) \cup (6, \infty)$.

−17 6
−24 −12 0 12 24 x

6. *Verbal Model:* [Profit] = [Review] − [Cost]

Equation:

$$P = R - C$$

$$P = x(60 - 0.0001x) - (12x + 1{,}800{,}000)$$

$$P = -0.0001x^2 + 48x = 1{,}800{,}000$$

To answer the question, solve the inequality as follows.

$$P \ge 3{,}600{,}000$$

$$-0.0001x^2 + 48x - 1{,}800{,}000 \ge 3{,}600{,}000$$

$$-0.0001x^2 + 48x - 5{,}400{,}000 \ge 0$$

$$0.0001x^2 - 48x + 5{,}400{,}000 \le 0$$

$$x^2 - 480{,}000x + 54{,}000{,}000{,}000 \le 0$$

$$(x - 180{,}000)(x - 300{,}000) \le 0$$

After finding the key point and testing the intervals, you can find the solution set is $[180{,}000, 300{,}000]$.

So, by selling at least 180,000 units but not more than 300,000 units, the profit is at least \$3,600,000.

7. Algebraic solution

Recall that the domain of an expression is the set of all x-values for which the expression is defined. Because $\sqrt{x^2 - 7x + 10}$ is defined only of $x^2 - 7x + 10$ is non-negative, the domain is given by $x^2 - 7x + 10 \geq 0$.

$x^2 - 7x + 10 \geq 0$ Write in general form.

$(x - 2)(x - 5) \geq 0$ Factor.

So, the inequality has two key numbers: $x = 2$ and $x = 5$.

Key numbers: $x = 2, x = 5$

Test intervals: $(-\infty, 2), (2, 5), (5, \infty)$

Test: Is $(x - 2)(x - 5) \geq 0$?

A test shows that the inequality is satisfied in the unbounded half-closed intervals $(-\infty, 2]$ or $[5, \infty)$. So, the domain of the expression $\sqrt{x^2 - 7x + 10}$ is $(-\infty, 2] \cup [5, \infty)$.

Graphical solution

Begin by sketching the graph of the equation $y = \sqrt{x^2 - 7x + 10}$. From the graph, you can determine that the x-values extend up to 2 (including 2) and from 5 and beyond (including 5). So, the domain of the expression $\sqrt{x^2 - 7x + 10}$ is $(-\infty, 2] \cup [5, \infty)$.

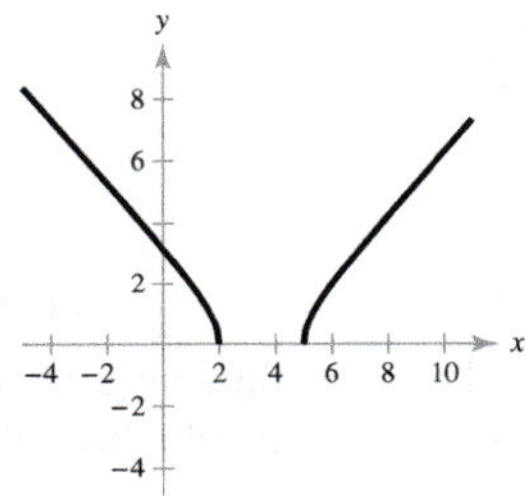

Chapter 3

Checkpoints for Section 3.1

1. Function Value

$f(\sqrt{2}) = 8^{-\sqrt{2}}$

Graphing Calculator Keystrokes

8 [∧] [(] [(−)] [√] 2 [)] [Enter]

Display

0.052824803759

2. The table lists some values for each function and the graph shows a sketch of the two functions. Note that both graphs are increasing and the graph of $g(x) = 9^x$ is increasing more rapidly than the graph of $f(x) = 3^x$.

x	-3	-2	-1	0	1	2
3^x	$\frac{1}{27}$	$\frac{1}{9}$	$\frac{1}{3}$	1	3	9
9^x	$\frac{1}{729}$	$\frac{1}{81}$	$\frac{1}{9}$	1	9	81

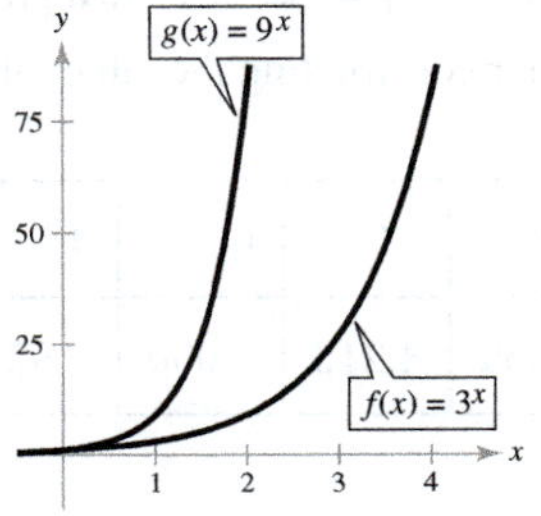

3. The table lists some values for each function and the graph shows a sketch for each function. Note that both graphs are decreasing and the graph of $g(x) = 9^{-x}$ is decreasing more rapidly than the graph of $f(x) = 3^{-x}$.

x	-2	-1	0	1	2	3
9^{-x}	64	8	1	$\frac{1}{8}$	$\frac{1}{64}$	$\frac{1}{512}$
3^{-x} $g(x)$	9	3	1	$\frac{1}{3}$	$\frac{1}{9}$	$\frac{1}{27}$

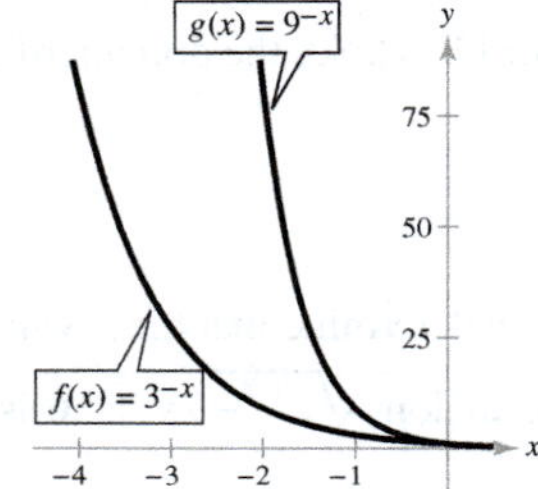

4. (a)

$8 = 2^{2x-1}$ Write Original equation.

$2^3 = 2^{2x-1}$ $8 = 2^3$

$3 = 2x - 1$ One-to-One Property

$4 = 2x$

$2 = x$ Solve for x.

(b)

$\left(\frac{1}{3}\right)^{-x} = 27$ Write Original equation.

$3^x = 27$ $\left(\frac{1}{3}\right)^{-x} = 3^x$

$3^x = 3^3$ $27 = 3^3$

$x = 3$ One-to-One Property

5. (a) Because $g(x) = 4^{x-2} = f(x - 2)$, the graph of g can be obtained by shifting the graph of f two units to the right.

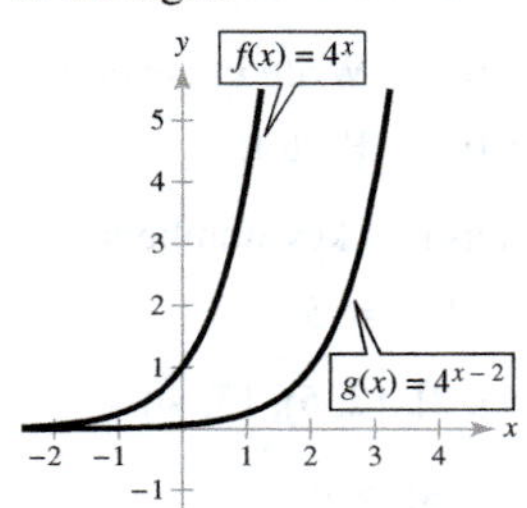

(b) Because $h(x) = 4^x + 3 = f(x) + 3$ the graph of h can be obtained by shifting the graph of f up three units.

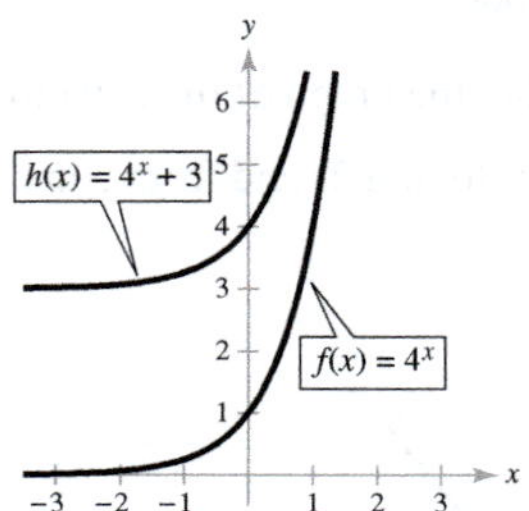

(c) Because $k(x) = 4^{-x} - 3 = f(-x) - 3$, the graph of k can be obtained by reflecting the graph of f in the y-axis and shifting the graph of f down three units.

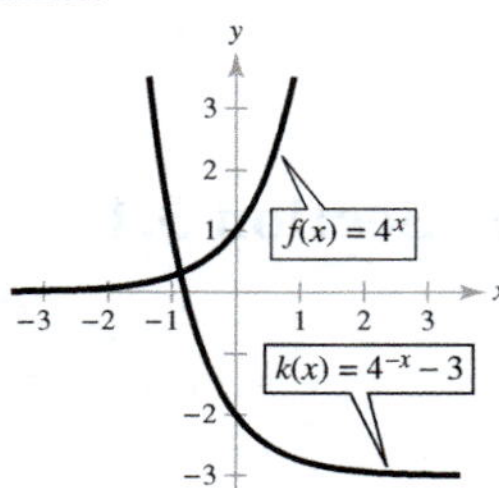

6.

Function Value	**Graphing Calculator Keystrokes**	**Display**
(a) $f(0.3) = e^{0.3}$	[e^x] 0.3 [Enter]	1.3498588
(b) $f(-1.2) = e^{-1.2}$	[e^x] [(−)] 1.2 [Enter]	0.3011942
(c) $f(6, 2) = e^{6.2}$	[e^x] 6.2 [Enter]	492.7490411

7. To sketch the graph of $f(x) = 5e^{0.17x}$, use a graphing utility to construct a table of values. After constructing the table, plot the points and draw a smooth curve.

x	-3	-2	-1	0	1	2	3
$f(x)$	3.002	3.559	4.218	5.000	5.927	7.025	8.326

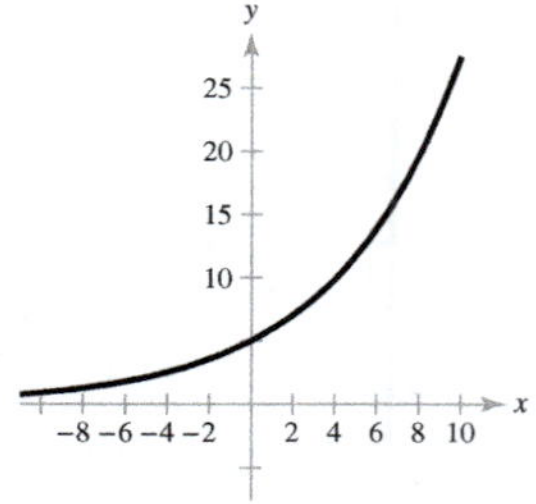

8. (a) For quarterly compounding, you have $n = 4$. So, in 7 years at 4%, the balance is as follows.

$A = P\left(1 + \frac{r}{n}\right)^{nt}$ Formula for compound interest.

$= 6000\left(1 + \frac{0.04}{4}\right)^{4(7)}$ Substitute P, r, n, and t.

$\approx \$7927.75$ Use a calculator.

(b) For monthly compounding, you have $n = 12$. So in 7 years at 4%, the balance is as follows.

$A = P\left(1 + \frac{r}{n}\right)^{nt}$ Formula for compound interest.

$= 6000\left(1 + \frac{0.04}{12}\right)^{12(7)}$ Substitute P, r, n, and t.

$\approx \$7935.08$ Use a calculator.

(c) For continuous compounding, the balance is as follows.

$A = Pe^{rt}$ Formula for continuous compounding.

$= 6000e^{0.04(7)}$ Substitute P, r, and t.

$\approx \$7938.78$ Use a calculator.

9. Use the model for the amount of Plutonium that remains from an initial amount of 10 pounds after t years, where $t = 0$ represents the year 1986.

$P = 10\left(\frac{1}{2}\right)^{t/24,100}$

To find the amount that remains in the year 2089, let $t = 103$.

$P = 10\left(\frac{1}{2}\right)^{t/24,100}$ Write original model.

$P = 10\left(\frac{1}{2}\right)^{103/24,100}$ Substitute 103 for t.

$P \approx 9.970$ Use a calculator.

In the year 2089, 9.970 pounds of plutonium will remain.

To find the amount that remains after 125,000 years, let $t = 125,000$.

$P = 10\left(\frac{1}{2}\right)^{t/24,100}$ Write original model.

$P = 10\left(\frac{1}{2}\right)^{125,000/24,100}$ Substitute 125,000 for t.

$P \approx 0.275$ Use a calculator.

After 125,000 years 0.275 pound of plutonium will remain.

Checkpoints for Section 3.2

1. (a) $f(1) = \log_6 1 = 0$ because $6^0 = 1$.

(b) $f\left(\frac{1}{125}\right) = \log_5 \frac{1}{125} = -3$ because $5^{-3} = \frac{1}{125}$.

(c) $f(343) = \log_7 343 = 3$ because $7^3 = 343$.

2.

Function Value	Graphing Calculator Keystrokes	Display
(a) $f(275) = \log 275$	[LOG] 275 [ENTER]	2.4393327
(b) $f\left(-\frac{1}{2}\right) = \log -\frac{1}{2}$	[LOG] [(] [(−)] [(] 1 [÷] 2 [)] [)] [ENTER]	ERROR
(c) $f\left(\frac{1}{2}\right) = \log \frac{1}{2}$	[LOG] [(] 1 [÷] 2 [)] [ENTER]	− 0.3010300

3. (a) Using Property 2, $\log_9 9 = 1$.

(b) Using Property 3, $20^{\log_{20} 3} = 3$.

(c) Using Property 1, $\log_{\sqrt{3}} 1 = 0$.

4. $\log_5 (x^2 + 3) = \log_5 12$

$x^2 + 3 = 12$

$x^2 = 9$

$x = \pm 3$

5. (a) For $f(x) = 8^x$, construct a table of values. Then plot the points and draw a smooth curve.

x	-2	-1	0	1	2
$f(x) = 8^x$	$\frac{1}{64}$	$\frac{1}{8}$	1	8	64

(b) Because $g(x) = \log_8 x$ is the inverse function of $f(x) = 8^x$, the graph of g is obtained by plotting the points $(f(x), x)$ and connecting them with a smooth curve. The graph of g is a reflection of the graph of f in the line $y = x$.

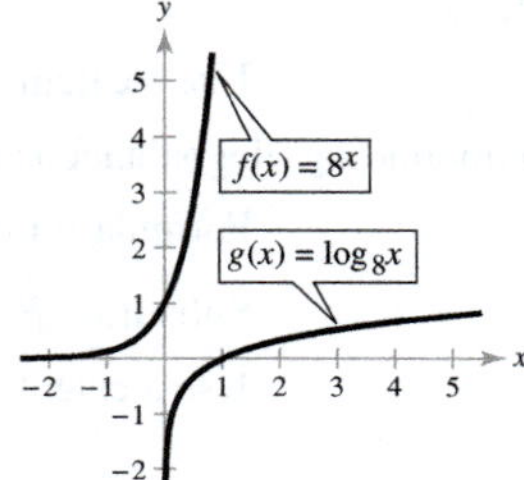

x	$\frac{1}{64}$	$\frac{1}{8}$	1	8	64
$g(x) = \log_8 x$	-2	-1	0	1	2

6. Begin by constructing a table of values. Note that some of the values can be obtained without a calculator by using the properties of logarithms. Then plot the points and draw a smooth curve.

x	$\frac{1}{9}$	$\frac{1}{3}$	1	3	9
$f(x) = \log_3 x$	-2	-1	0	1	2

Vertical asymptote: $x = 0$

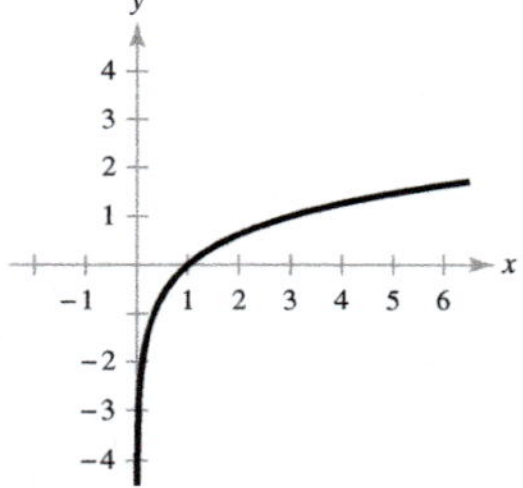

7. (a) Because $g(x) = -1 + \log_3 x = f(x) - 1$, the graph of g can be obtained by shifting the graph of f one unit down.

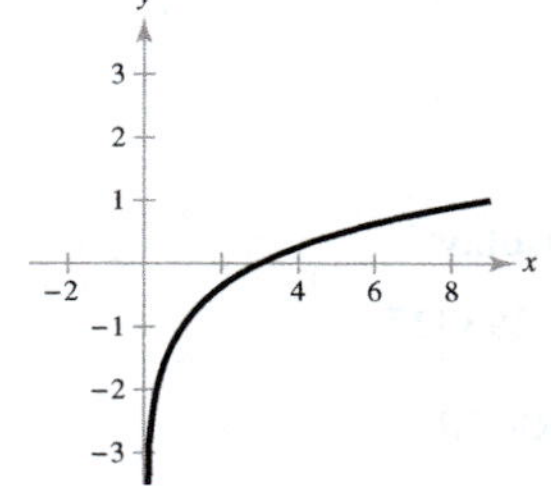

(b) Because $h(x) = \log_3 (x + 3) = f(x + 3)$, the graph of h can be obtained by shifting the graph of f three units to the left.

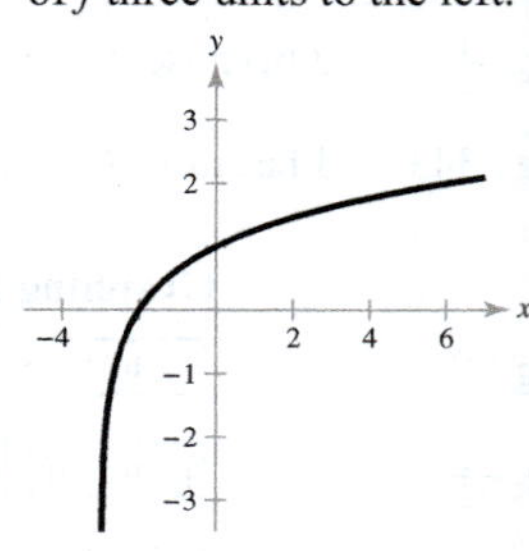

8.

Function Value	Graphing Calculator Keystrokes	Display
$f(0.01) = \ln 0.01$	[LN] 0.01 [ENTER]	-4.6051702
$f(4) = \ln 4$	[LN] 4 [ENTER]	1.3862944
$f(\sqrt{3} + 2) = \ln(\sqrt{3} + 2)$	[LN] [(] [(] [√] 3 [)] [+] 2 [)] [ENTER]	1.3169579
$f(\sqrt{3} - 2) = \ln(\sqrt{3} - 2)$	[LN] [(] [(] [√] 3 [)] [−] 2 [)] [ENTER]	ERROR

9. (a) $\ln e^{1/3} = \frac{1}{3}$ Inverse Property

(b) $5 \ln 1 = 5(0) = 0$ Property 1

(c) $\frac{3}{4} \ln e = \frac{3}{4}(1) = \frac{3}{4}$ Property 2

(d) $e^{\ln 7} = 7$ Inverse Property

10. Because $\ln(x + 3)$ is defined only when $x + 3 > 0$, it follows that the domain of f is $(-3, \infty)$. The graph of f is shown.

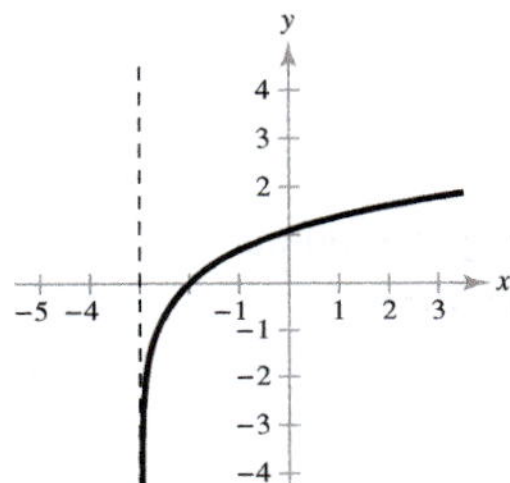

11. (a) After 1 month, the average score was the following.

$f(1) = 75 - 6 \ln(1 + 1)$ Substitute 1 for t.

$= 75 - 6 \ln 2$ Simplify.

$\approx 75 - 6(0.6931)$ Use a calculator.

≈ 70.84 Solution

(b) After 9 months, the average score was the following.

$f(9) = 75 - 6 \ln(9 + 1)$ Substitute 9 for t.

$= 75 - 6 \ln 10$ Simplify.

$\approx 75 - 6(2.3026)$ Use a calculator.

≈ 61.18 Solution

(c) After 12 months, the average score was the following.

$f(12) = 75 - 6 \ln(12 + 1)$ Substitute 9 for t.

$= 75 - 6 \ln 13$ Simplify.

$\approx 75 - 6(2.5649)$ Use a calculator.

≈ 59.61 Solution

Checkpoints for Section 3.3

1. $\log_2 12 = \dfrac{\log 12}{\log 2}$ $\quad \log_a x = \dfrac{\log x}{\log a}$

$\approx \dfrac{1.07918}{0.30103}$ Use a calculator.

≈ 3.5850 Simplify.

2. $\log_2 12 = \dfrac{\ln 12}{\ln 2}$ $\quad \log_a x = \dfrac{\ln x}{\ln a}$

$\approx \dfrac{2.48491}{0.69315}$ Use a calculator.

≈ 3.5850 Simplify.

3. (a) $\log 75 = \log(3.25)$ Rewrite 75 as $3 \cdot 25$.

$= \log 3 + \log 25$ Product Property

$= \log 3 + \log 5^2$ Rewrite 25 as 5^2.

$= \log 3 + 2 \log 5$ Power Property

(b) $\log \frac{9}{125} = \log 9 - \log 125$ Quotient Property

$= \log 3^2 - \log 5^3$ Rewrite 9 as 3^2 and 125 as 5^2.

$= 2\log 3 - 3 \log 5$ Power Property

4. $\ln e^6 - \ln e^2 = 6 \ln e - 2\ln e$

$= 6(1) - 2(1)$

$= 4$

5. $\log_3 \dfrac{4x^2}{\sqrt{y}} = \log_3 \dfrac{4x^2}{y^{1/2}}$ Rewrite using rational exponent.

$= \log_3 4x^2 - \log_3 y^{1/2}$ Quotient Property

$= \log_3 4 + \log_3 x^2 - \log_3 y^{1/2}$ Product Property

$= \log_3 4 + 2 \log_3 x - \dfrac{1}{2} \log_3 y$ Power Property

6. $2\left[\log(x+3) - 2\log(x-2)\right] = 2\left[\log(x+3) - \log(x-2)^2\right]$ Power Property

$= 2\left[\log\left(\dfrac{x+3}{(x-2)^2}\right)\right]$ Quotient Property

$= \log\left(\dfrac{x+3}{(x-2)^2}\right)^2$ Power Property

$= \log\dfrac{(x+3)^2}{(x-2)^4}$ Simplify.

7. To solve this problem, take the natural logarithm of each of the x- and y-values of the ordered pairs.

$(\ln x, \ln y)$: $(-0.994, -0.673)$, $(0.000, 0.000)$, $(1.001, 0.668)$, $(2.000, 1.332)$, $(3.000, 2.000)$

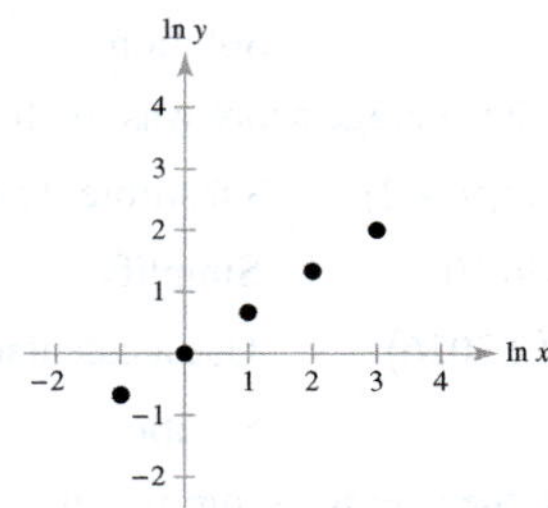

By plotting the ordered pairs, you can see that all five points appear to lie in a line. Choose any two points to determine the slope of the line. Using the points $(0, 0)$ and $(1.001, 0.668)$, the slope of the line is

$$m = \frac{0.668 - 0}{1 - 0} = 0.668 \approx \frac{2}{3}.$$

By the point-slope form, the equation of the line is $y = \frac{2}{3}x$, where $y = \ln y$ and $x = \ln x$. So, the logarithmic equation is $\ln y = \frac{2}{3}\ln x$.

Checkpoints for Section 3.4

1.

Original Equation	Rewritten Equation	Solution	Property
(a) $2^x = 512$	$2^x = x^9$	$x = 9$	One-to-One
(b) $\log_6 x = 3$	$6^{\log_6 x} = 6^3$	$x = 216$	Inverse
(c) $5 - e^x = 0$ $5 = e^x$	$\ln 5 = \ln e^x$	$\ln 5 = x$	Inverse
(d) $9^x = \frac{1}{3}$	$3^{2x} = 3^{-1}$	$2x = -1$ $x = -\frac{1}{2}$	One-to-One

2. (a)

$$\begin{aligned} e^{2x} &= e^{x^2-8} && \text{Write original equation.} \\ 2x &= x^2 - 8 && \text{One-to-One Property} \\ 0 &= x^2 - 2x - 8 && \text{Write in general form.} \\ 0 &= (x-4)(x+2) && \text{Factor.} \\ x - 4 = 0 &\Rightarrow x = 4 && \text{Set 1st factor equal to 0.} \\ x + 2 = 0 &\Rightarrow x = -2 && \text{Set 2nd factor equal to 0.} \end{aligned}$$

The solutions are $x = 4$ and $x = -2$

Check $x = -2$: $x = 4$:

$$\begin{aligned} e^{2x} &= e^{x^2-8} \\ e^{2(-2)} &\stackrel{?}{=} e^{(-2)^2-8} \\ e^{-4} &\stackrel{?}{=} e^{4-8} \\ e^{-4} &= e^{-4}\ \checkmark \end{aligned} \qquad \begin{aligned} e^{2(4)} &\stackrel{?}{=} e^{(4)^2-8} \\ e^{8} &\stackrel{?}{=} e^{16-8} \\ e^{8} &= e^{8}\ \checkmark \end{aligned}$$

(b)

$$\begin{aligned} 2(5^x) &= 32 && \text{Write original equation.} \\ 5^x &= 16 && \text{Divide each side by 2.} \\ \log_5 5^x &= \log_5 16 && \text{Take log(base 5) of each side.} \\ x &= \log_5 16 && \text{Inverse Property} \\ x &= \frac{\ln 16}{\ln 5} \approx 1.723 && \text{Change of base formula} \end{aligned}$$

The solution is $x = \log_5 16 \approx 1.723$.

Check $x = \log_5 16$:

$$\begin{aligned} 2(5^x) &= 32 \\ 2\left[5^{(\log_5 16)}\right] &\stackrel{?}{=} 32 \\ 2(16) &\stackrel{?}{=} 32 \\ 32 &= 32\ \checkmark \end{aligned}$$

3.

$$\begin{aligned} e^x - 7 &= 23 && \text{Write original equation.} \\ e^x &= 30 && \text{Add 7 to each side.} \\ \ln e^x &= \ln 30 && \text{Take natural log of each side.} \\ x &= \ln 30 \approx 3.401 && \text{Inverse Property} \end{aligned}$$

Check $x = \ln 30$:

$$\begin{aligned} e^x - 7 &= 23 \\ e^{(\ln 30)-7} &\stackrel{?}{=} 23 \\ 30 - 7 &\stackrel{?}{=} 23 \\ 23 &= 23\ \checkmark \end{aligned}$$

4. $6\left(2^{t+5}\right) + 4 = 11$ Write original equation.

$6\left(2^{t+5}\right) = 7$ Subtract 4 from each side.

$2^{t+5} = \dfrac{7}{6}$ Divide each side by 6.

$\log_2 2^{t+5} = \log_2\left(\dfrac{7}{6}\right)$ Take log (base 2) of each side.

$t + 5 = \log_2\left(\dfrac{7}{6}\right)$ Inverse Property

$t = \log_2\left(\dfrac{7}{6}\right) - 5$ Subtract 5 from each side.

$t = \dfrac{\ln\left(\dfrac{1}{6}\right)}{\ln 2} - 5$ Change of base formula.

$t \approx -4.778$ Use a calculator.

The solution is $t = \log_2\left(\dfrac{7}{6}\right) - 5 \approx -4.778$.

Check $t \approx -4.778$:

$$6\left(2^{t+5}\right) + 4 = 11$$

$$6\left[2^{(-4.778+5)}\right] + 4 \stackrel{?}{=} 11$$

$$6(1.166) + 4 \stackrel{?}{=} 11$$

$$10.998 \approx 11 \checkmark$$

5. Algebraic Solution

$e^{2x} - 7e^x + 12 = 0$	Write original equation.
$(e^x)^2 - 7e^x + 12 = 0$	Write in quadratic form.
$(e^x - 3)(e^x - 4) = 0$	Factor.
$e^x - 3 = 0 \Rightarrow e^x = 3$	Set 1st factor equal to 0.
$x = \ln 3$	Solution
$e^x - 4 = 0 \Rightarrow e^x = 4$	Set 2nd factor equal to 0.
$x = \ln 4$	Solution

The solutions are $x = \ln 3 \approx 1.099$ and $x = \ln 4 \approx 1.386$.

Check $x = \ln 3$:

$$e^{2x} - 7e^x + 12 = 0$$
$$e^{2(\ln 3)} - 7e^{(\ln 3)} + 12 \stackrel{?}{=} 0$$
$$e^{\ln(3^2)} - 7e^{\ln 3} + 12 \stackrel{?}{=} 0$$
$$3^2 - 7(3) + 12 \stackrel{?}{=} 0$$
$$0 = 0 \checkmark$$

$x = \ln 4$:

$$e^{2(\ln 4)} - 7e^{(\ln 4)} + 12 = 0$$
$$e^{\ln 4^2} - 7e^{\ln 4} + 12 \stackrel{?}{=} 0$$
$$4^2 - 7(4) + 12 \stackrel{?}{=} 0$$
$$0 = 0 \checkmark$$

Graphical Solution

Use a graphing utility to graph $y = e^{2x} - 7e^x + 12$ and then find the zeros.

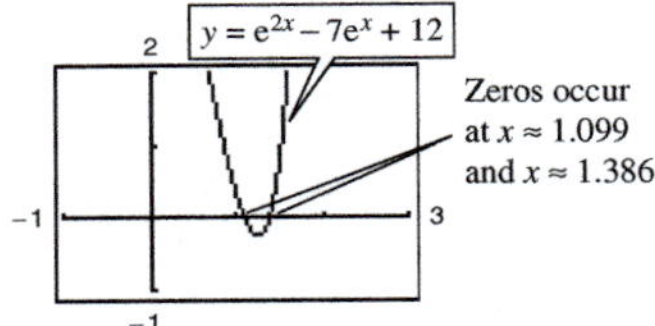

So, you can conclude that the solutions are $x \approx 1.099$ and $x \approx 1.386$.

6. (a)

$\ln x = \frac{2}{3}$	Write original equation.
$e^{\ln x} = e^{2/3}$	Exponentiate each side.
$x = e^{2/3}$	Inverse Property

(b)

$\log_2(2x - 3) = \log_2(x + 4)$	Write original equation.
$2x - 3 = x + 4$	One-to-One Property
$x = 7$	Solution

(c)

$\log 4x - \log(12 + x) = \log 2$	Write Original equation.
$\log\left(\frac{4x}{12 + x}\right) = \log 2$	Quotient Property of Logarithms
$\frac{4x}{12 + x} = 2$	One-to-One Property
$4x = 2(12 + x)$	Multiply each side by $(12 + x)$.
$4x = 24 + 2x$	Distribute.
$2x = 24$	Subtract $2x$ from each side.
$x = 12$	Solution

7. Algebraic Solution

$7 + 3 \ln x = 5$	Write original equation.
$3 \ln x = -2$	Subtract 7 from each side.
$\ln x = -\frac{2}{3}$	Divide each side by 3.
$e^{\ln x} = e^{-2/3}$	Exponentiate each side.
$x = e^{-2/3}$	Inverse Property
$x \approx 0.513$	Use a calculator.

Graphical Solution

Use a graphing utility to graph $y_1 = 7 + 3 \ln x$ and $y_2 = 5$. Then find the intersection point.

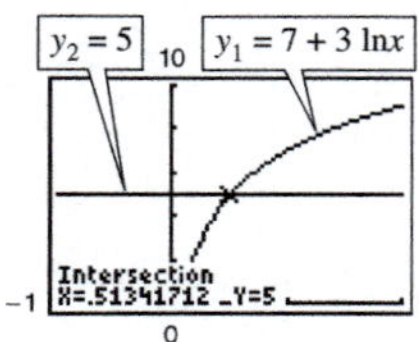

The point of intersection is about $(0.513, 5)$. So, the solution is $x \approx 0.513$.

8.

$3 \log_4 6x = 9$	Write original equation.
$\log_4 6x = 3$	Divide each side by 3.
$4^{\log_4 6x} = 4^3$	Exponentiate each side (base 4).
$6x = 64$	Inverse Property
$x = \frac{32}{3}$	Divide each side by 6 and simplify.

Check $x = \frac{32}{3}$:

$$3 \log_4 6x = 9$$

$$3 \log_4 6\left(\tfrac{32}{3}\right) \stackrel{?}{=} 9$$

$$3 \log_4 64 \stackrel{?}{=} 9$$

$$3 \log_4 4^3 \stackrel{?}{=} 9$$

$$3 \cdot 3 \stackrel{?}{=} 9$$

$$9 = 9 \checkmark$$

9. Algebraic Solution

$\log x + \log(x-9) = 1$	Write original equation.
$\log[x(x-9)] = 1$	Product Property of Logarithms
$10^{\log[x(x-9)]} = 10^1$	Exponentiate each side (base 10).
$x(x-9) = 10$	Inverse Property
$x^2 - 9x - 10 = 0$	Write in general form.
$(x-10)(x+1) = 0$	Factor.
$x - 10 = 0 \Rightarrow x = 10$	Set 1st factor equal to 0.
$x + 1 = 0 \Rightarrow x = -1$	Set 2nd factor equal to 0.

Check $x = 10$:

$$\log x + \log(x-9) = 1$$
$$\log(10) + \log(10-9) \stackrel{?}{=} 1$$
$$\log 10 + \log 1 \stackrel{?}{=} 1$$
$$1 + 0 \stackrel{?}{=} 1$$
$$1 = 1 \checkmark$$

$x = -1$:

$$\log x + \log(x-9) = 1$$
$$\log(-1) + \log(-1-9) \stackrel{?}{=} 1$$
$$\log(-1) + \log(-10) \stackrel{?}{=} 1$$

-1 and -10 are not in the domain of $\log x$. So, it does not check.

The solutions appear to be $x = 10$ and $x = -1$. But when you check these in the original equation, you can see that $x = 10$ is the only solution.

Graphical Solution

First, rewrite the original solution as $\log x + \log(x-9) - 1 = 0$.

Then use a graphing utility to graph the equation $y = \log x + \log(x-9) - 1$ and find the zeros.

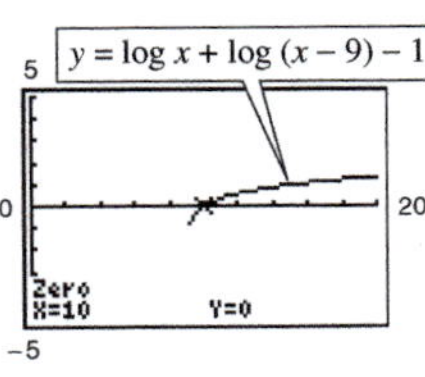

10. Using the formula for continuous compounding, the balance is

$A = Pe^{rt}$

$A = 500e^{0.0525t}$.

To find the time required for the balance to double, let $A = 1000$ and solve the resulting equation for t

$500e^{0.0525t} = 1000$	Let $A = 1000$.
$e^{0.0525t} = 2$	Divide each side by 500.
$\ln e^{0.0525t} = \ln 2$	Take natural log of each side.
$0.0525t = \ln 2$	Inverse Property
$t = \dfrac{\ln 2}{0.0525}$	Divide each side by 0.0525.
$t \approx 13.20$	Use a calculator.

The balance in the account will double after approximately 13.20 years.

Because the interest rate is lower than the interest rate in Example 2, it will take more time for the account balance to double.

11. To find when sales reached \$180 billion, let $y = 80$ and solve for t.

$$\begin{aligned}
-614 + 342.2 \ln t &= y && \text{Write original equation} \\
-614 + 342.2 \ln t &= 180 && \text{Substitute 180 for } y. \\
342.2 \ln t &= 794 && \text{Add 614 to each side.} \\
\ln t &= \frac{794}{342.2} && \text{Divide each side by 342.2.} \\
e^{\ln t} &= e^{794/342.2} && \text{Exponentiate each side (base } e\text{).} \\
t &= e^{794/342.2} && \text{Inverse Property} \\
t &\approx 10.2 && \text{Use a calculator.}
\end{aligned}$$

The solution is $t \approx 10.2$. Because $t = 9$ represents 2009, it follows that $t = 10$ represents 2010. So, sales reached \$180 billion in 2010.

Checkpoints for Section 3.5

1. Algebraic Solution

To find when the amount of U.S. online advertising spending will reach \$100 billion, let $s = 100$ and solve for t.

$$\begin{aligned}
0.00036e^{0.7563t} &= S && \text{Write original model.} \\
0.00036e^{0.7563t} &= 300 && \text{Substitute 300 for } s. \\
e^{0.7563t} &\approx 833{,}333.33 && \text{Divide each side by 0.0036.} \\
\ln e^{0.7563t} &\approx \ln 833{,}333.33 && \text{Take natural log of each side.} \\
0.7563t &\approx 13.6332 && \text{Inverse Property} \\
t &\approx 18.03 && \text{Divide each side by 0.7563.}
\end{aligned}$$

According to the model, the amount of U.S. online advertising spending will reach \$300 million in 2018.

Graphical Solution

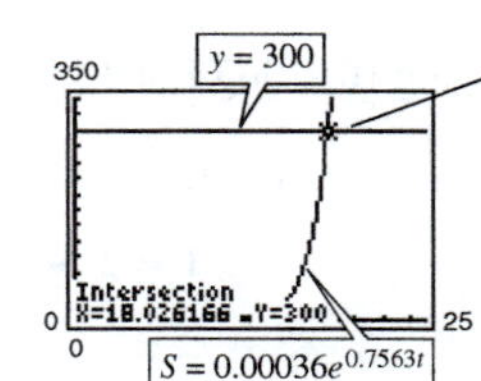

The intersection point of the model and the line $y = 300$ is about (18.03, 300). So, according to the model, the amount of U.S. online advertising spending will reach \$300 billion in 2018.

2. Let y be the number of bacteria at time t. From the given information you know that $y = 100$ when $t = 1$ and $y = 200$ when $t = 2$. Substituting this information into the model $y = ae^{bt}$ produces $100 = ae^{(1)b}$ and $200 = ae^{(2)b}$. To solve for b, solve for a in the first equation.

$$\begin{aligned}
100 &= ae^{b} && \text{Write first equation.} \\
\frac{100}{e^b} &= a && \text{Solve for } a.
\end{aligned}$$

Then substitute the result into the second equation.

$$\begin{aligned}
200 &= ae^{2b} && \text{Write second equation} \\
200 &= \left(\frac{100}{e^b}\right)e^{2b} && \text{Substitute } \frac{100}{e^b} \text{ for } a. \\
\frac{200}{100} &= e^b && \text{Simplify and divide each side by 100.} \\
2 &= e^b && \text{Simplify.} \\
\ln 2 &= \ln e^b && \text{Take natural log of each side} \\
\ln 2 &= b && \text{Inverse Property}
\end{aligned}$$

Use $b = \ln 2$ and the equation you found for a.

$$\begin{aligned}
a &= \frac{100}{e^{\ln 2}} && \text{Substitute } \ln 2 \text{ for } b. \\
&= \frac{100}{2} && \text{Inverse Property} \\
&= 50 && \text{Simplify.}
\end{aligned}$$

So, with $a = 50$ and $b = \ln 2$, the exponential growth model is $y = 50e^{(\ln 2)t}$.

After 3 hours, the number of bacteria will be $y = 50e^{\ln 2(3)} = 400$ bacteria.

3. Algebraic Solution

In the carbon dating model, substitute the given value of R to obtain the following.

$$\frac{1}{10^{12}}e^{-t/8223} = R \qquad \text{Write original model.}$$

$$\frac{e^{-t/8223}}{10^{12}} = \frac{1}{10^{14}} \qquad \text{Substitute } \frac{1}{10^{14}} \text{ for } R.$$

$$e^{-t/8223} = \frac{1}{10^{2}} \qquad \text{Multiply each side by } 10^{12}.$$

$$e^{-t/8223} = \frac{1}{100} \qquad \text{Simplify.}$$

$$\ln e^{-t/8223} = \ln \frac{1}{100} \qquad \text{Take natural log of each side.}$$

$$-\frac{t}{8223} \approx -4.6052 \qquad \text{Inverse Property}$$

$$t \approx 37{,}869 \qquad \text{Multiply each side by } -8223.$$

So, to the nearest thousand years, the age of the fossil is about 38,000 years.

Graphical Solution

Use a graphing utility to graph the formula for the ratio of carbon 14 to carbon 12 at any time t as

$$y_1 = \frac{1}{10^{12}}e^{-x/8223}.$$

In the same viewing window, graph $y_2 = \dfrac{1}{10^{14}}$

Use the *intersect* feature to estimate that $x \approx 18{,}934$ when $y = 1/10^{13}$.

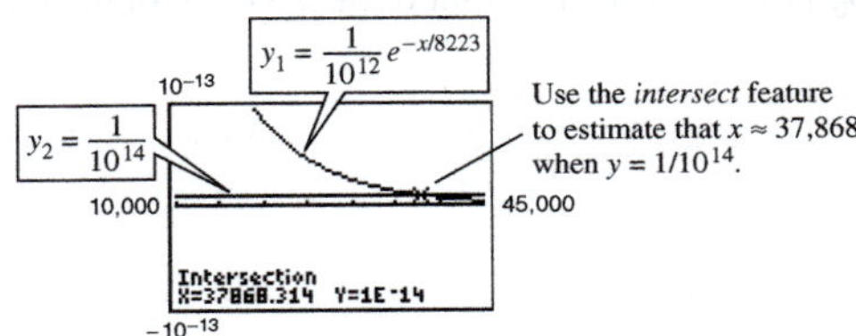

So, to the nearest thousand years, the age of fossil is about 38,000 years.

4. The graph of the function is shown below. On this bell-shaped curve, the maximum value of the curve represents the average score. From the graph, you can estimate that the average reading score for college-bound seniors in the United States in 2015 was 495.

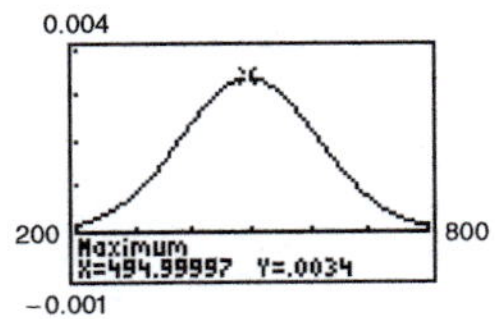

5. To find the number of days that 250 students are infected, let $y = 250$ and solve for t.

$\dfrac{5000}{1 + 4999e^{-0.8t}} = y$	Write original model.
$\dfrac{5000}{1 + 4999e^{-0.8t}} = 250$	Substitute 250 for y.
$\dfrac{5000}{250} = 1 + 4999e^{-0.8t}$	Divide each side by 250 and multiply each side by $1 + 4999e^{-0.8t}$.
$20 = 1 + 4999e^{-0.8t}$	Simplify.
$19 = 4999e^{-0.8t}$	Subtract 1 from each side.
$\dfrac{19}{4999} = e^{-0.8t}$	Divide each side by 4999.
$\ln\left(\dfrac{19}{4999}\right) = \ln e^{-0.8t}$	Take natural log of each side.
$\ln\left(\dfrac{19}{4999}\right) = -0.8t$	Inverse Property
$-5.5726 \approx -0.8t$	Use a calculator.
$t \approx 6.97$	Divide each side by -0.8.

So, after about 7 days, 250 students will be infected.

Graphical Solution

To find the number of days that 250 students are infected, use a graphing utility to graph.

$$y_1 = \frac{5000}{1 + 4999e^{-0.8x}} \text{ and } y_2 = 250$$

in the same viewing window. Use the *intersect* feature of the graphing utility to find the point of intersection of the graphs.

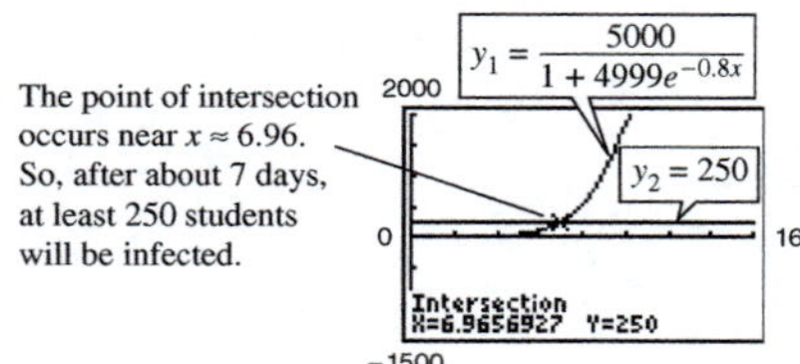

6. (a) Because $I_0 = 1$ and $R = 6.0$, you have the following.

$R = \log \dfrac{I}{I_0}$	
$6.0 = \log \dfrac{I}{1}$	Substitute 1 for I_0 and 6.0 for R.
$10^{6.0} = 10^{\log I}$	Exponentiate each side (base 10).
$10^{6.0} = I$	Inverse Property
$1{,}000{,}000 = I$	Simplify.

(b) Because $I_0 = 1$ and $R = 7.9$, you have the following.

$7.9 = \log \dfrac{I}{1}$	Substitute 1 for I_0 and 7.9 for R.
$10^{7.9} = 10^{\log I}$	Exponentiate each side (base 10).
$10^{7.9} = I$	Inverse Property
$79{,}432{,}823 \approx I$	Simplify.

Chapter 4

Checkpoints for Section 4.1

1. (a) *Sample answers:* $\frac{9\pi}{4} - 4\pi = -\frac{7\pi}{4}$

$\frac{9\pi}{4} - 2\pi = \frac{\pi}{4}$

(b) *Sample answers:* $\frac{-\pi}{3} + 2\pi = \frac{5\pi}{3}$

$\frac{-\pi}{3} - 2\pi = -\frac{7\pi}{3}$

2. (a) $\frac{\pi}{2} - \frac{\pi}{6} = \frac{3\pi}{6} - \frac{\pi}{6} = \frac{2\pi}{6} = \frac{\pi}{3}$

The complement of $\frac{\pi}{6}$ is $\frac{\pi}{3}$.

$\pi - \frac{\pi}{6} = \frac{6\pi}{6} - \frac{\pi}{6} = \frac{5\pi}{6}$

The supplement of $\frac{\pi}{6}$ is $\frac{5\pi}{6}$.

(b) Because $\frac{5\pi}{6}$ is greater than $\frac{\pi}{2}$, it has no complement.

$\pi - \frac{5\pi}{6} = \frac{6\pi}{6} - \frac{5\pi}{6} = \frac{\pi}{6}$

The supplement of $\frac{5\pi}{6}$ is $\frac{\pi}{6}$.

3. (a) $60° = (60 \cancel{\text{deg}})\left(\frac{\pi \text{ rad}}{180 \cancel{\text{deg}}}\right) = \frac{\pi}{3}$ radians

(b) $320° = (320 \cancel{\text{deg}})\left(\frac{\pi \text{ rad}}{180 \cancel{\text{deg}}}\right) = \frac{16\pi}{9}$ radians

4. (a) $\frac{\pi}{6} = \left(\frac{\cancel{\pi}}{6} \cancel{\text{rad}}\right)\left(\frac{180 \text{ deg}}{\cancel{\pi} \cancel{\text{rad}}}\right) = 30°$

(b) $\frac{5\pi}{3} = \left(\frac{5\cancel{\pi}}{3} \cancel{\text{rad}}\right)\left(\frac{180 \text{ deg}}{\cancel{\pi} \cancel{\text{rad}}}\right) = 300°$

5. To use the formula $s = r\theta$ first convert 160° to radian measure.

$160° = (160 \cancel{\text{deg}})\left(\frac{\pi \text{ rad}}{180 \cancel{\text{deg}}}\right) = \frac{8\pi}{9}$ radians

Then, using a radius of $r = 27$ inches, you can find the arc length to be

$$\begin{aligned} s &= r\theta \\ &= (27)\left(\frac{8\pi}{9}\right) \\ &= 24\pi \\ &\approx 75.40 \text{ inches.} \end{aligned}$$

6. In one revolution, the arc length traveled is

$$\begin{aligned} s &= 2\pi r \\ &= 2\pi(8) \\ &= 16\pi \text{ centimeters.} \end{aligned}$$

The time required for the second hand to travel this distance is

$t = 1 \text{ minute} = 60 \text{ seconds}.$

So, the linear speed of the tip of the second hand is

$$\begin{aligned} \text{Linear speed} &= \frac{s}{t} \\ &= \frac{16\pi \text{ centimeters}}{60 \text{ seconds}} \\ &\approx 0.838 \text{ centimeters per second.} \end{aligned}$$

7. (a) Because each revolution generates 2π radians, it follows that the saw blade turns $(2400)(2\pi) = 4800\pi$ radians per minute. In other words, the angular speed is

$$\text{Angular speed} = \frac{\theta}{t} = \frac{4800\pi \text{ radians}}{1 \text{ minute}} = 4800\pi \text{ radians per minute.}$$

(b) The radius is $r = 4$. The linear speed is

$$\text{Linear speed} = \frac{s}{t} = \frac{r\theta}{t} = \frac{(4)(4800\pi) \text{ inches}}{60 \text{ seconds}} = 60{,}319 \text{ inches per minute.}$$

8. First convert 80° to radian measure as follows.

$$\theta = 80° = (80\ \cancel{\text{deg}})\left(\frac{\pi \text{ rad}}{180\ \cancel{\text{deg}}}\right) = \frac{4\pi}{9} \text{ radians}$$

Then, using $\theta = \frac{4\pi}{9}$ and $r = 40$ feet, the area is

$A = \frac{1}{2}r^2\theta$	Formula for area of a sector of a circle
$= \frac{1}{2}(40)^2\left(\frac{4\pi}{9}\right)$	Substitute for r and θ
$= \frac{3200\pi}{9}$	Multiply
≈ 1117 square feet.	Simplify.

Checkpoints for Section 4.2

1. (a) $t = \frac{\pi}{2}$ corresponds to the point $(x, y) = (0, 1)$

$\sin\frac{\pi}{2} = 1$ $\csc\frac{\pi}{2} = 1$

$\cos\frac{\pi}{2} = 0$ $\sec\frac{\pi}{2}$ is undefined.

$\tan\frac{\pi}{2}$ is undefined. $\cot\frac{\pi}{2} = 0$

(b) $t = 0$ corresponds to the point $(x, y) = (1, 0)$

$\sin 0 = 0$ $\csc 0$ is undefined.

$\cos 0 = 1$ $\sec 0 = 1$

$\tan 0 = 0$ $\cot 0$ is undefined.

(c) $t = -\frac{5\pi}{6}$ corresponds to the point $(x, y) = \left(-\frac{\sqrt{3}}{2}, -\frac{1}{2}\right)$

$\sin\left(-\frac{5\pi}{6}\right) = -\frac{1}{2}$ $\csc\left(-\frac{5\pi}{6}\right) = -2$

$\cos\left(-\frac{5\pi}{6}\right) = -\frac{\sqrt{3}}{2}$ $\sec\left(-\frac{5\pi}{6}\right) = -\frac{2\sqrt{3}}{3}$

$\tan\left(-\frac{5\pi}{6}\right) = \frac{\sqrt{3}}{3}$ $\cot\left(-\frac{5\pi}{6}\right) = \sqrt{3}$

(d) $t = -\frac{3\pi}{4}$ corresponds to the point $(x, y) = \left(-\frac{\sqrt{2}}{2}, -\frac{\sqrt{2}}{2}\right)$

$\sin\left(-\frac{3\pi}{4}\right) = -\frac{\sqrt{2}}{2}$ $\csc\left(-\frac{3\pi}{4}\right) = -\sqrt{2}$

$\cos\left(-\frac{3\pi}{4}\right) = -\frac{\sqrt{2}}{2}$ $\sec\left(-\frac{3\pi}{4}\right) = -\sqrt{2}$

$\tan\left(-\frac{3\pi}{4}\right) = 1$ $\cot\left(-\frac{3\pi}{4}\right) = 1$

2. (a) Because $\frac{9\pi}{2} = 4\pi + \frac{\pi}{2}$, you have $\cos\frac{9\pi}{2} = \cos\left(4\pi + \frac{\pi}{2}\right) = \cos\frac{\pi}{2} = 0.$

(b) Because $-\frac{7\pi}{3} = -2\pi - \frac{\pi}{3}$, you have $\sin -\frac{7\pi}{3} = \sin\left(-2\pi - \frac{\pi}{3}\right) = \sin -\frac{\pi}{3} = -\frac{\sqrt{3}}{2}$

(c) For $\cos(-t) = 0.3$, $\cos t = 0.3$ because the cosine function is even.

3. (a) 0.78183148

(b) 1.0997502

Checkpoints for Section 4.3

1.

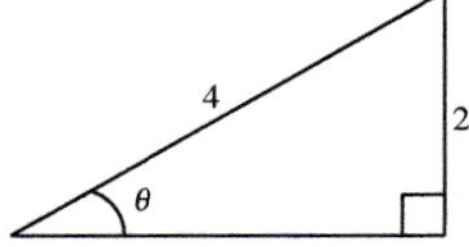

By the Pythagorean Theorem, $(\text{hyp})^2 = (\text{opp})^2 + (\text{adj})^2$, it follows that

$\text{adj} = \sqrt{4^2 - 2^2} = \sqrt{12} = 2\sqrt{3}.$

So, the six trigonometric functions of θ are

$$\sin\theta = \frac{\text{opp}}{\text{hyp}} = \frac{2}{4} = \frac{1}{2} \qquad \csc\theta = \frac{\text{hyp}}{\text{opp}} = \frac{4}{2} = 2$$

$$\cos\theta = \frac{\text{adj}}{\text{hyp}} = \frac{2\sqrt{3}}{4} = \frac{\sqrt{3}}{2} \qquad \sec\theta = \frac{\text{hyp}}{\text{adj}} = \frac{4}{2\sqrt{3}} = \frac{2}{\sqrt{3}} = \frac{2\sqrt{3}}{3}$$

$$\tan\theta = \frac{\text{opp}}{\text{adj}} = \frac{2}{2\sqrt{3}} = \frac{1}{\sqrt{3}} = \frac{\sqrt{3}}{3} \qquad \cot\theta = \frac{\text{adj}}{\text{opp}} = \frac{2\sqrt{3}}{2} = \sqrt{3}$$

2.

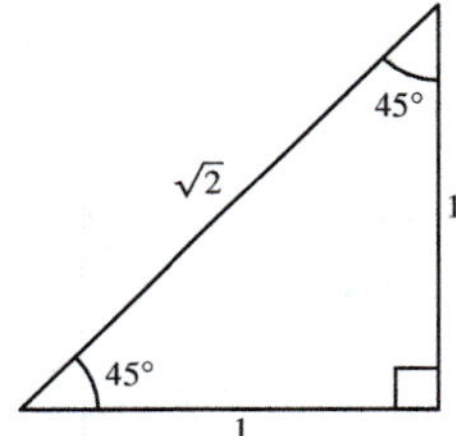

$$\cot 45° = \frac{\text{adj}}{\text{opp}} = \frac{1}{1} = 1$$

$$\sec 45° = \frac{\text{hyp}}{\text{adj}} = \frac{\sqrt{2}}{1} = \sqrt{2}$$

$$\csc 45° = \frac{\text{hyp}}{\text{opp}} = \frac{\sqrt{2}}{1} = \sqrt{2}$$

3. 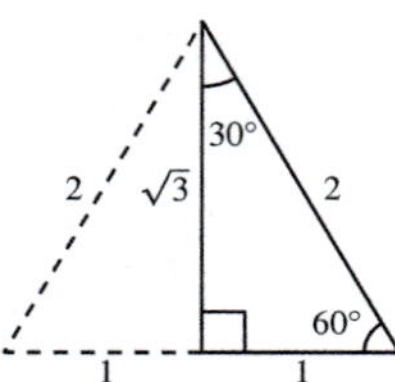

For $\theta = 60°$, you have adj $= 1$, opp $= \sqrt{3}$ and hyp $= 2$.

So, $\tan 60° = \dfrac{\text{opp}}{\text{adj}} = \dfrac{\sqrt{3}}{1} = \sqrt{3}$.

For $\theta = 30°$, you have adj $= \sqrt{3}$, opp $= 1$ and hyp $= 2$.

So, $\tan 30° = \dfrac{\text{opp}}{\text{adj}} = \dfrac{1}{\sqrt{3}} = \dfrac{\sqrt{3}}{3}$.

4. $34°\,30'\,36'' = 34° + \left(\dfrac{30}{60}\right)^{\circ} + \left(\dfrac{36}{3600}\right)^{\circ} = 34.51°$

$\csc(34°\,30'\,36'') = \csc 34.51° = \dfrac{1}{\sin 34.51°} \approx 1.765069$

5. (a) To find the value of $\sin\theta$, use the Pythagorean Identity $\sin^2\theta + \cos^2\theta = 1$.

So, you have

$$\sin^2\theta + (0.96)^2 = 1$$
$$\sin^2\theta = 1 - (0.96)^2$$
$$\sin^2\theta = 1 - 0.9216$$
$$\sin^2\theta = 0.0784$$
$$\sin\theta = \sqrt{0.0784}$$
$$= 0.28$$

(b) Now, knowing the sine and cosine of θ, you can find the tangent of θ to be

$$\tan\theta = \frac{\sin\theta}{\cos\theta} = \frac{0.28}{0.96} \approx 0.2917.$$

6. Given $\tan\theta = 2$

$\cot\theta = \dfrac{1}{\tan\theta}$ Reciprocal Identity.

$= \dfrac{1}{2}$

$\sec^2\theta = 1 + \tan^2\theta$ Pythagorean Identity

$\sec^2\theta = 1 + (2)^2$

$\sec^2\theta = 5$

$\sec\theta = \sqrt{5}$

Use the definitions of $\cot\theta$ and $\sec\theta$ and the triangle to check these results.

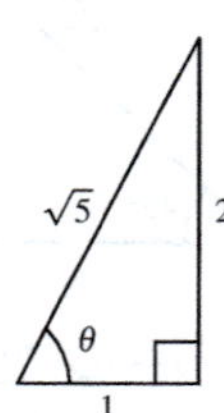

7. (a) $\tan\theta\csc\theta$

$= \left(\dfrac{\sin\theta}{\cos\theta}\right)\left(\dfrac{1}{\sin\theta}\right)$ Use a Quotient Identity and a Reciprocal Identity.

$= \dfrac{1}{\cos\theta}$ Simplify.

$= \sec\theta$ Use a Reciprocal Identity

(b) $(\csc\theta + 1)(\csc\theta - 1) = \csc^2\theta - \csc\theta + \csc\theta - 1$ FOIL Method.

$= \csc^2\theta - 1$ Simplify.

$= \cot^2\theta$ Pythagorean identity

8. Drawing a sketch of the situation can assist you in solving the problem.

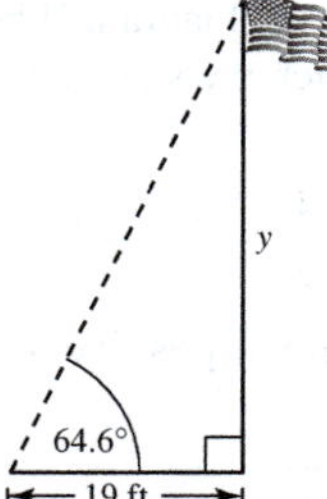

$$\tan 64.6° = \frac{\text{opp}}{\text{adj}} = \frac{y}{x}$$

Where $x = 19$ and y is the height of the flagpole. So, the height of the flagpole is

$$\begin{aligned} y &= 19 \tan 64.6° \\ &\approx 19(2.106) \\ &\approx 40 \text{ feet.} \end{aligned}$$

9. From the figure, you can see that the cosine of the angle θ is

$$\cos\theta = \frac{\text{adj}}{\text{hyp}} = \frac{3}{6} = \frac{1}{2}.$$

You should recognize that $\theta = 60°$.

10. From the figure, you can see that

$$\sin 11.5° = \frac{\text{opp}}{\text{hyp}} = \frac{3.5}{c}.$$

$$\begin{aligned} \sin 11.5° &= \frac{3.5}{c} \\ c \sin 11.5° &= 3.5 \\ c &= \frac{3.5}{\sin 11.5°} \end{aligned}$$

So, the length c of the loading ramp is

$$c = \frac{3.5}{\sin 11.5} \approx \frac{3.5}{0.1994} \approx 17.6 \text{ feet.}$$

Also from the figure, you can see that

$$\tan 11.5° = \frac{\text{opp}}{\text{adj}} = \frac{3.5}{a}.$$

So, the length a of the ramp is

$$a = \frac{3.5}{\tan 11.5°} \approx \frac{3.5}{0.2034} \approx 17.2 \text{ feet.}$$

Checkpoints for Section 4.4

1. Referring to the figure shown, you can see that $x = -2$, $y = 3$ and

$$r = \sqrt{x^2 + y^2} = \sqrt{(-2)^2 + (3)^2} = \sqrt{13}.$$

So, you have the following.

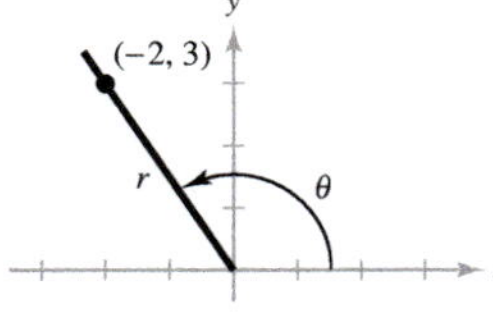

$$\sin\theta = \frac{y}{r} = \frac{3}{\sqrt{13}} = \frac{3\sqrt{13}}{13}$$

$$\cos\theta = \frac{x}{r} = -\frac{2}{\sqrt{13}} = -\frac{2\sqrt{13}}{13}$$

$$\tan\theta = \frac{y}{x} = -\frac{3}{2}$$

2. Note that θ lies in Quadrant II because that is the only quadrant in which the sine is positive and the tangent is negative.

Using $\sin\theta = \frac{4}{5} = \frac{y}{r}$

and the fact that y is positive in Quadrant II, let $y = 4$ and $r = 5$.

$$r = \sqrt{x^2 + y^2}$$
$$5 = \sqrt{x^2 + 4^2}$$
$$25 = x^2 + 16$$
$$9 = x^2$$
$$\pm 3 = x$$

Since x is negative in Quadrant II, $x = -3$.

$$\cos\theta = \frac{x}{r} = -\frac{3}{5} \text{ and } \tan\theta = \frac{y}{x} = \frac{4}{-3} = -\frac{4}{3}$$

3. To begin, choose a point on the terminal side of the angle $\frac{3\pi}{2}$.

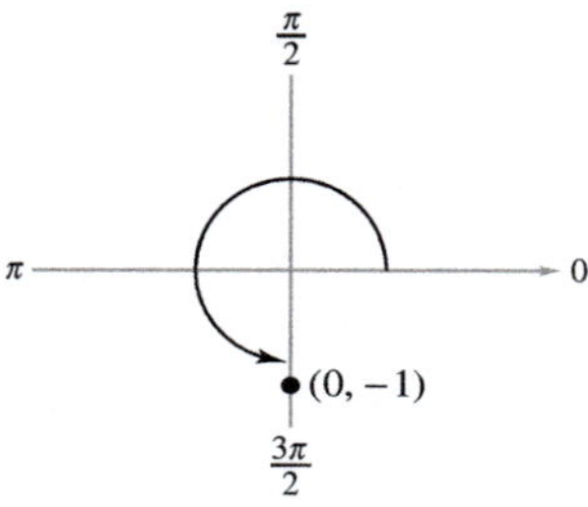

For the point $(0, -1)$, $r = 1$ and you have the following.

$$\sin\frac{3\pi}{2} = \frac{y}{r} = \frac{-1}{1} = -1$$
$$\cot\frac{3\pi}{2} = \frac{x}{y} = \frac{0}{-1} = 0$$

4. (a) Because 213° lies in Quadrant III, the angle it makes with the x-axis is $\theta' = 213° - 180° = 33°$.

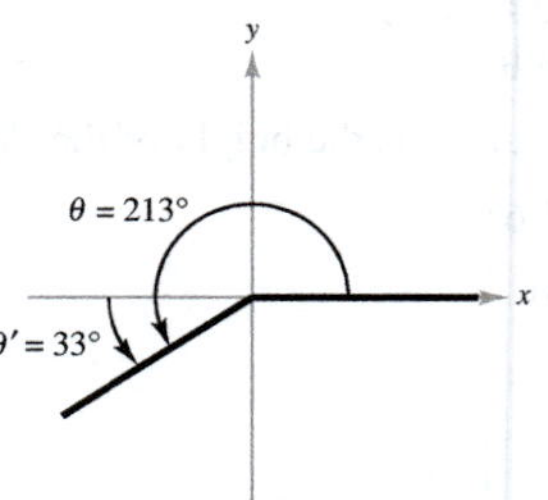

(b) Because $\frac{14\pi}{9}$ lies in Quadrant IV, the angle it makes with the x-axis is

$$\theta' = 2\pi - \frac{14\pi}{9}$$
$$= \frac{18\pi}{9} - \frac{14\pi}{9}$$
$$= \frac{4\pi}{9}.$$

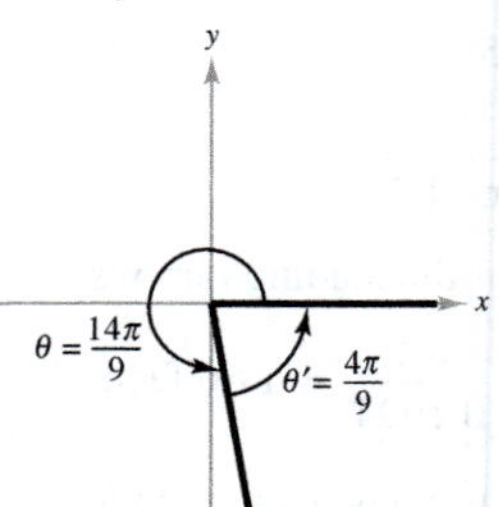

(c) Because $\frac{4\pi}{5}$ lies in Quadrant II, the angle it makes with the x-axis is

$$\theta' = \pi - \frac{4\pi}{5}$$
$$= \frac{\pi}{5}.$$

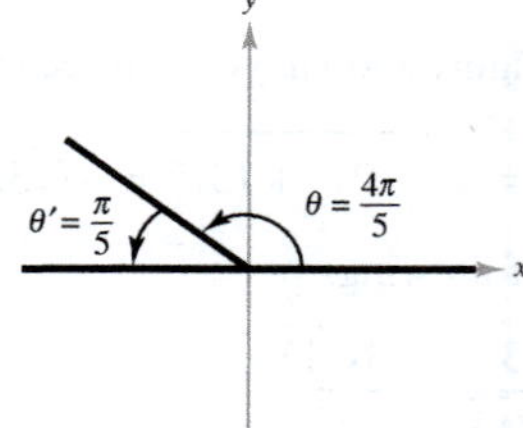

5. (a) Because $\theta = \dfrac{7\pi}{4}$ lies in Quadrant IV, the reference angle is $\theta' = 2\pi - \dfrac{7\pi}{4} = \dfrac{\pi}{4}$ as shown.

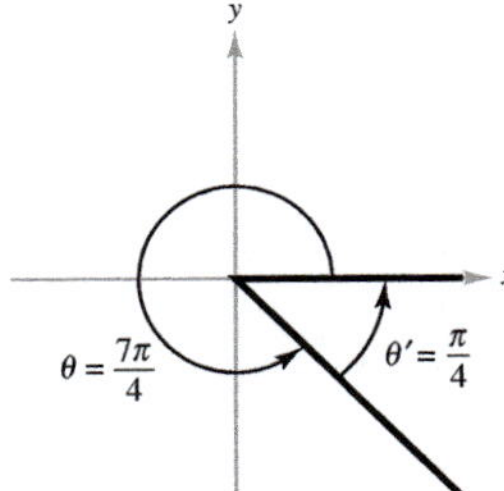

Because the sine is negative in Quadrant IV, you have $\sin \dfrac{7\pi}{4} = (-)\sin \dfrac{\pi}{7}$

$= -\dfrac{\sqrt{2}}{2}.$

(b) Because $-120° + 360° = 240°$, it follows that $-120°$ is coterminal with the third-quadrant angle $240°$. So, the reference angle is $\theta' = 240° - 180° = 60°$ as shown.

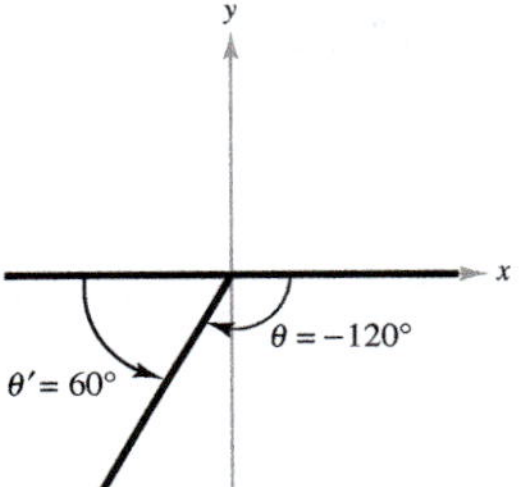

Because the cosine is negative in Quadrant III, you have $\cos(-120°) = (-)\cos 60° = -\dfrac{1}{2}.$

(c) Because $\theta = \dfrac{11\pi}{6}$ lies in Quadrant IV, the reference angle is $\theta' = 2\pi - \dfrac{11\pi}{6} = \dfrac{\pi}{6}$ as shown.

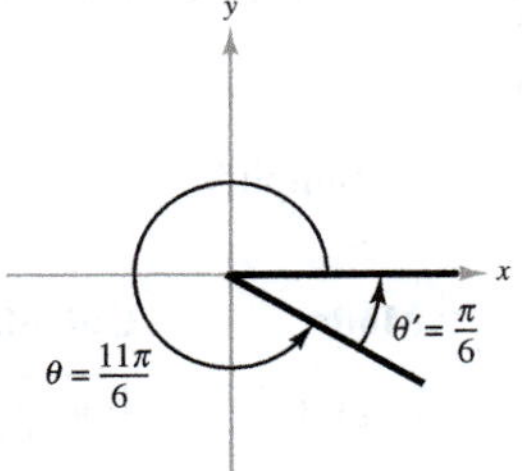

Because the tangent is negative in Quadrant IV, you have

$$\tan \frac{11\pi}{6} = (-)\tan \frac{\pi}{6} = -\frac{\sqrt{3}}{3}.$$

6. (a) Using the Pythagorean Identity $\sin^2 \theta + \cos^2 \theta = 1$, you obtain the following.

$\sin^2 \theta + \cos^2 \theta = 1$ — Write Identity

$\left(-\dfrac{4}{5}\right)^2 + \cos^2 \theta = 1$ — Substitute $-\dfrac{4}{5}$ for $\sin \theta$.

$\dfrac{16}{25} + \cos^2 \theta = 1$ — Simplify.

$\cos^2 \theta = 1 - \dfrac{16}{25}$ — Subtract $\dfrac{16}{25}$ from each side.

$\cos^2 \theta = \dfrac{9}{25}$ — Simplify.

Because $\cos \theta < 0$ in Quadrant III, you can use the negative root to obtain

$$\cos \theta = -\sqrt{\frac{9}{25}} = -\frac{3}{5}.$$

(b) Using the trigonometric identity $\tan\theta = \dfrac{\sin\theta}{\cos\theta}$, you obtain

$$\tan\theta = \frac{-\frac{4}{5}}{-\frac{3}{5}} \quad \text{Substitute for } \sin\theta \text{ and } \cos\theta.$$

$$= \frac{4}{3}. \quad \text{Simplify.}$$

7.

	Function	Mode	Calculator Keystrokes	Display
(a)	$\tan 119°$	Degree	[tan] [(] 119 [)] [ENTER]	-1.8040478
(b)	$\csc 5$	Radian	[(] [sin] [(] 5 [)] [)] [x^{-1}] [ENTER]	-1.0428352
(c)	$\cos\dfrac{\pi}{5}$	Radian	[cos] [(] π [÷] 5 [)] [ENTER]	0.8090170

Checkpoints for Section 4.5

1. Note that $y = 2\cos x = 2(\cos x)$ indicates that the y-values for the key points will have twice the magnitude of those on the graph of $y = \cos x$. Divide the period 2π into four equal parts to get the key points.

Maximum	Intercept	Minimum	Intercept	Maximum
$(0, 2)$	$\left(\frac{\pi}{2}, 0\right)$	$(\pi, -2)$	$\left(\frac{3\pi}{2}, 0\right)$	$(2\pi, 2)$

By connecting these key points with a smooth curve and extending the curve in both directions over the interval $\left[-\frac{\pi}{2}, \frac{9\pi}{2}\right]$, you obtain the graph shown.

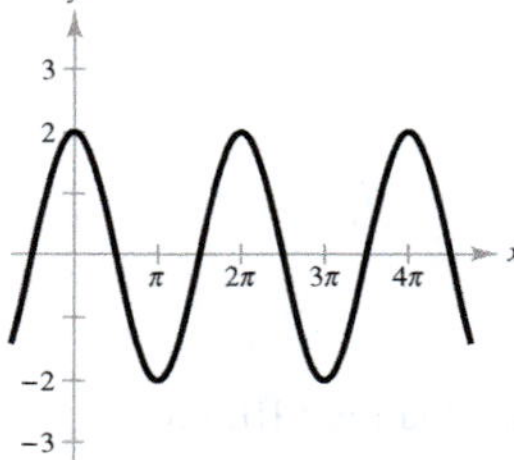

2. (a) Because the amplitude of $y = \frac{1}{3}\sin x$ is $\frac{1}{3}$, the maximum value is $\frac{1}{3}$ and the minimum value is $-\frac{1}{3}$.

Divide one cycle, $0 \le x \le 2\pi$, into for equal parts to get the key points.

Intercept	Maximum	Intercept	Minimum	Intercept
$(0, 0)$	$\left(\frac{\pi}{2}, \frac{1}{3}\right)$	$(\pi, 0)$	$\left(\frac{3\pi}{2}, -\frac{1}{3}\right)$	$(2\pi, 0)$

(b) A similar analysis shows that the amplitude of $y = 3\sin x$ is 3, and the key points are as follows.

Intercept	Maximum	Intercept	Minimum	Intercept
$(0, 0)$	$\left(\frac{\pi}{2}, 3\right)$	$(\pi, 0)$	$\left(\frac{3\pi}{2}, -3\right)$	$(2\pi, 0)$

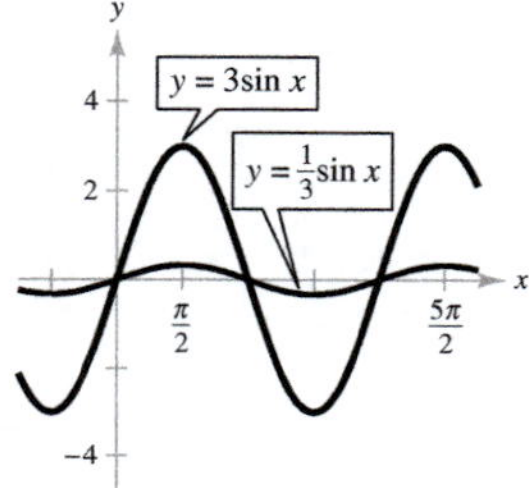

3. The amplitude is 1. Moreover, because $b = \frac{1}{3}$, the period is

$$\frac{2\pi}{b} = \frac{2\pi}{\frac{1}{3}} = 6\pi. \quad \text{Substitute } \frac{1}{3} \text{ for } b.$$

Now, divide the period-interval $[0, 6\pi]$ into four equal parts using the values $\frac{3\pi}{2}$, 3π, and $\frac{9\pi}{2}$ to obtain the key points.

Maximum	Intercept	Minimum	Intercept	Maximum
$(0, 1)$	$\left(\frac{3\pi}{2}, 0\right)$	$(3\pi, -1)$	$\left(\frac{9\pi}{2}, 0\right)$	$(6\pi, 1)$

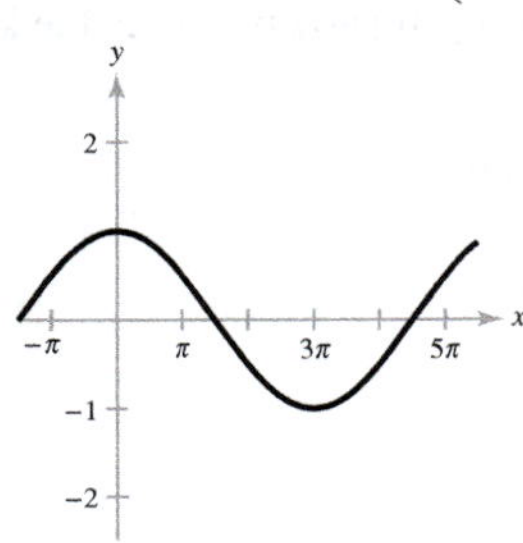

4. Algebraic Solution

The amplitude is 2 and the period is 2π.

By solving the equations

$$x - \frac{\pi}{2} = 0 \Rightarrow x = \frac{\pi}{2}$$

and

$$x - \frac{\pi}{2} = 2\pi \Rightarrow x = \frac{5\pi}{2}$$

you see that the interval $\left[\frac{\pi}{2}, \frac{5\pi}{2}\right]$ corresponds to one cycle of the graph. Dividing this interval into four equal parts produces the key points.

Maximum	Intercept	Minimum	Intercept	Maximum
$\left(\frac{\pi}{2}, 2\right)$	$(\pi, 0)$	$\left(\frac{3\pi}{2}, -2\right)$	$(2\pi, 0)$	$\left(\frac{5\pi}{2}, 2\right)$

Graphical Solution

Use a graphing utility set in *radian* mode to graph $y = 2\cos\left(x - \frac{\pi}{2}\right)$ as shown.

Use the *minimum*, *maximum*, and *zero* or *root* features of the graphing utility to approximate the key points $(1.57, 2)$, $(3.14, 0)$, $(4.71, -2)$, $(6.28, 0)$ and $(7.85, 2)$.

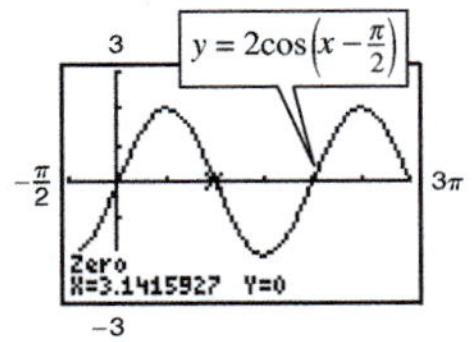

5. The amplitude is $\frac{1}{2}$ and the period is $\frac{2\pi}{b} = \frac{2\pi}{\pi} = 2$.

By solving the equations

$$\begin{aligned} \pi x + \pi &= 0 \\ \pi x &= -\pi \\ x &= -1 \end{aligned}$$

and

$$\begin{aligned} \pi x + \pi &= 2\pi \\ \pi x &= \pi \\ x &= 1 \end{aligned}$$

you see that the interval $[-1, 1]$ corresponds to one cycle of the graph. Dividing this into four equal parts produces the key points.

Intercept	Minimum	Intercept	Maximum	Intercept
$(-1, 0)$	$\left(-\frac{1}{2}, -\frac{1}{2}\right)$	$(0, 0)$	$\left(\frac{1}{2}, \frac{1}{2}\right)$	$(1, 0)$

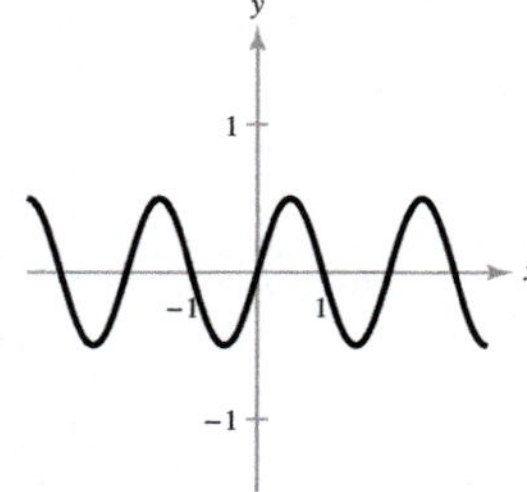

6. The amplitude is 2 and the period is 2π. The key points over the interval $[0, 2\pi]$ are

$(0, -3)$, $\left(\frac{\pi}{2}, -5\right)$, $(\pi, -7)$, $\left(\frac{3\pi}{2} - 5\right)$, and $(2\pi, -3)$.

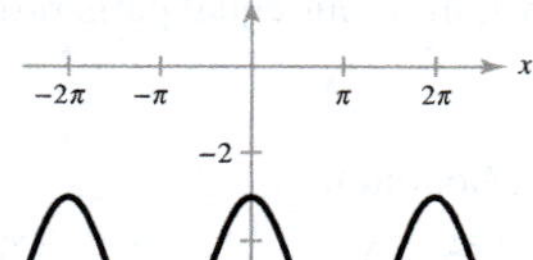

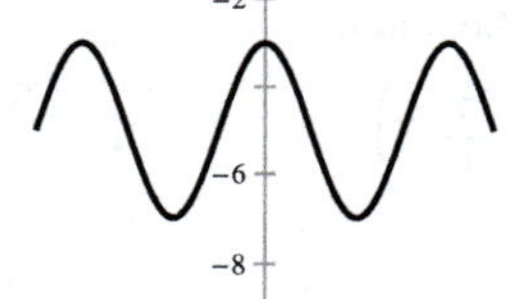

7. Use a sine model of the form $y = a\sin(bt - c) + d$.

The difference between the maximum value and minimum value is twice the amplitude of the function. So, the amplitude is

$$\begin{aligned} a &= \frac{1}{2}\left[(\text{maximum depth}) - (\text{minimum depth})\right] \\ &= \frac{1}{2}(11.3 - 0.1) = 5.6. \end{aligned}$$

The sine function completes one half cycle between the times at which the maximum and minimum depths occur. So, the period p is

$$\begin{aligned} p &= 2\left[(\text{time of min. depth}) - (\text{time of max. depth})\right] \\ &= 2(10 - 4) = 12 \end{aligned}$$

which implies that $b = \frac{2\pi}{p} \approx 0.524$. Because high tide occurs 4 hours after midnight, consider the maximum to be

$$bt - c = \frac{\pi}{2} \approx 1.571.$$

So, $(0.524)(4) - c \approx 1.571$

$$c \approx 0.525.$$

Because the average depth is $\frac{1}{2}(11.3 + 0.1) = 5.7$, it follows that $d = 5.7$. So, you can model the depth with the function

$$\begin{aligned} y &= a\sin(bt - c) + d \\ &= 5.6\sin(0.524t - 0.525) + 5.7. \end{aligned}$$

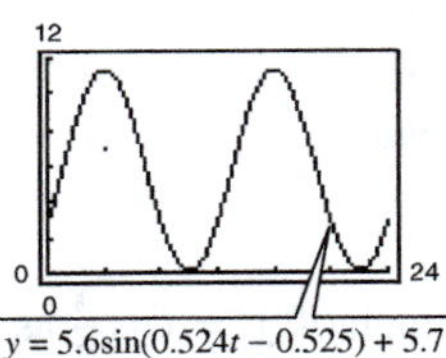

Checkpoints for Section 4.6

1. By solving the equations

$$\frac{x}{4} = -\frac{\pi}{2} \quad \text{and} \quad \frac{x}{4} = \frac{\pi}{2}$$
$$x = -2\pi \qquad\qquad x = 2\pi$$

you can see that two consecutive vertical asymptotes occur at $x = -2\pi$ and $x = 2\pi$. Between these two asymptotes, plot a few points including the x-intercept.

x	-2π	$-\pi$	0	π	2π
$f(x)$	Undef.	-1	0	1	Undef.

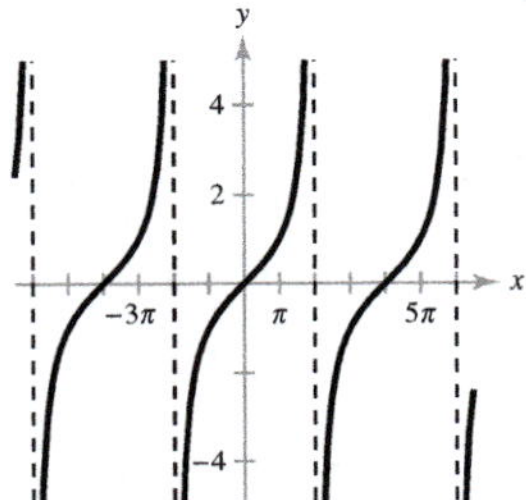

2. By solving the equations

$$2x = -\frac{\pi}{2} \quad \text{and} \quad 2x = \frac{\pi}{2}$$

$$x = -\frac{\pi}{4} \qquad\quad x = \frac{\pi}{4}$$

you can see that two consecutive vertical asymptotes occur at $x = -\frac{\pi}{4}$ and $x = \frac{\pi}{4}$. Between these two asymptotes, plot a few points including the x-intercept.

x	$-\frac{\pi}{4}$	$-\frac{\pi}{8}$	0	$\frac{\pi}{8}$	$\frac{\pi}{4}$
$\tan 2x$	Undef.	-1	0	1	Undef.

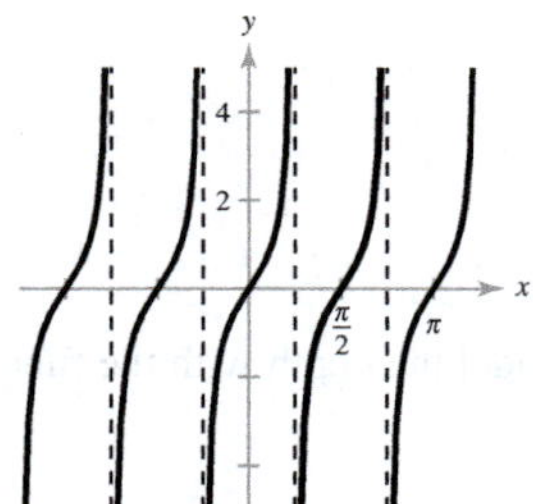

3. By solving the equations

$$\frac{x}{4} = 0 \quad \text{and} \quad \frac{x}{4} = \pi$$

$$x = 0 \qquad\quad x = 4\pi$$

you can see that two consecutive vertical asymptotes occur at $x = 0$ and $x = 4\pi$. Between these two asymptotes, plot a few points, including the x-intercept.

x	0	π	2π	3π	4π
$\cot \frac{x}{4}$	Undef.	1	0	-1	Undef.

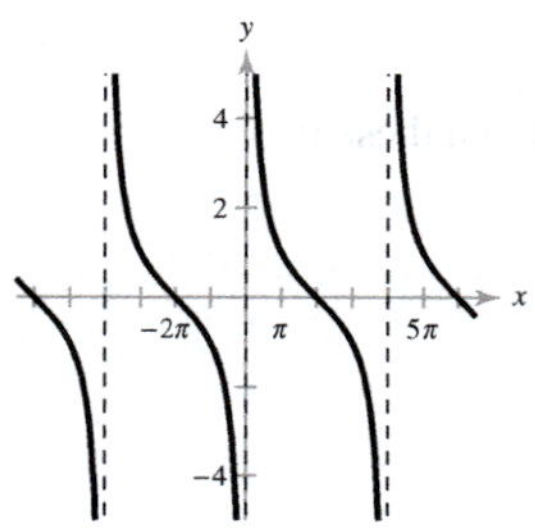

4. Begin by sketching the graph of $y = 2 \sin\left(x + \frac{\pi}{2}\right)$. For this function, the amplitude is 2 and the period is 2π. By solving the equations

$$x + \frac{\pi}{2} = 0 \quad \text{and} \quad x + \frac{\pi}{2} = 2\pi$$

$$x = -\frac{\pi}{2} \qquad\quad x = \frac{3\pi}{2}$$

you can see that one cycle of the sine function corresponds to the interval from $x = -\frac{\pi}{2}$ to $x = \frac{3\pi}{2}$. The graph of this sine function is represented by the gray curve. Because the sine function is zero at the midpoint and endpoints of this interval, the corresponding cosecant function

$$y = 2 \csc\left(x + \frac{\pi}{2}\right)$$

$$= 2\left(\frac{1}{\sin\left(x + \frac{\pi}{2}\right)}\right)$$

has vertical asymptotes at $x = -\frac{\pi}{2}, x = \frac{\pi}{2}, x = \frac{3\pi}{2}$, and so on. The graph of the cosecant curve is represented by the black curve.

5. Begin by sketching the graph of $y = \cos \frac{x}{2}$ as indicated by the gray curve. Then, form the graph of $y = \sec \frac{x}{2}$ as the black curve. Note that the x-intercepts of $y = \cos \frac{x}{2}$, $(\pi, 0), (3\pi, 0), (5\pi, 0), \ldots$ correspond to the vertical asymptotes $x = \pi, x = 3\pi, x = 5\pi, \ldots$ of the graph of $y = \sec \frac{x}{2}$. Moreover, notice that the period of $y = \cos \frac{x}{2}$ and $y = \sec \frac{x}{2}$ is $\frac{2\pi}{\frac{1}{2}} = 4\pi$.

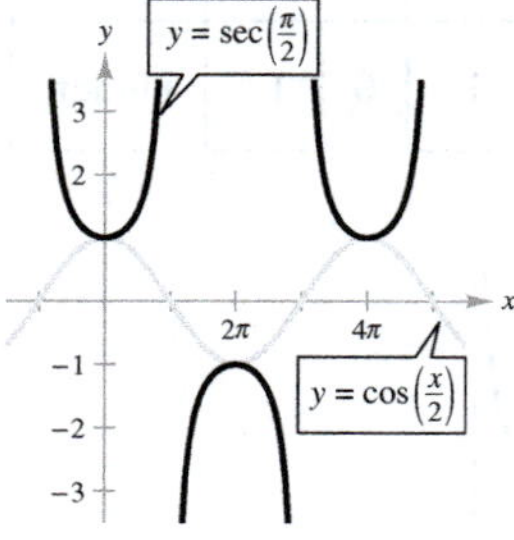

6. Consider $f(x)$ as the product of these two functions

$y = e^x$ and $y = \sin 4x$

each of which has a set of real numbers as its domain. For any real number x, you know that $e^x|\sin 4x| \le e^x$

which means that $-e^x \le e^x \sin 4x \le e^x$.

Furthermore, because

$f(x) = e^x \sin 4x = \pm e^x$ at $x = \dfrac{\pi}{8} \pm \dfrac{n\pi}{4}$ since

$\sin 4x = \pm 1$ at $4x = \dfrac{\pi}{2} + n\pi$

and

$f(x) = e^x \sin 4x = 0$ at $x = \dfrac{n\pi}{4}$ since

$\sin 4x = 0$ at $4x = n\pi$

the graph of f touches the curve $y = -e^x$ or $y = e^x$

at $x = \dfrac{\pi}{8} + \dfrac{n\pi}{4}$ and has x-intercepts at $x = \dfrac{n\pi}{4}$.

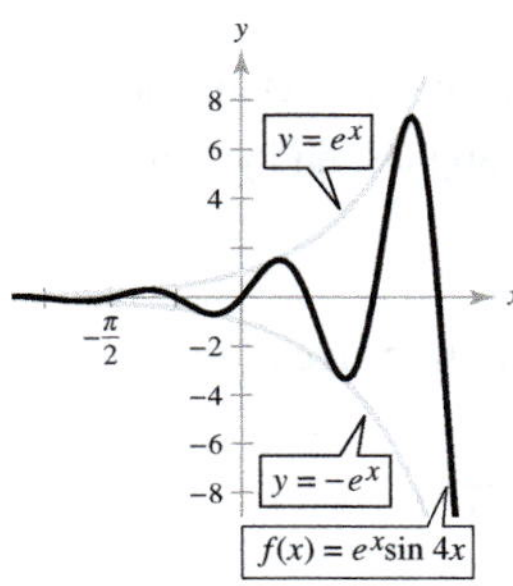

Checkpoints for Section 4.7

1. (a) Because $\sin \dfrac{\pi}{2} = 1$, and $\dfrac{\pi}{2}$ lies in $\left[-\dfrac{\pi}{2}, \dfrac{\pi}{2}\right]$, it follows that $\arcsin 1 = \dfrac{\pi}{2}$.

(b) It is not possible to evaluate $y = \sin^{-1} x$ when $x = -2$ because there is no angle whose sine is -2. Remember that the domain of the inverse sine function is $[-1, 1]$.

2. Using a graphing utility you can graph the three functions with the following keystrokes

Function	Keystroke	Display
$y = \sin x$	[y =] [SIN] [(] [x] [)]	$y_1 = \sin(x)$
$y = \arcsin x$	[y =] [2ND] [SIN] [(] [x] [)]	$y_2 = \sin^{-1}(x)$
$y = x$	[y =] [x]	$y_3 = x$

Remember to check the mode to make sure the angle measure is set to radian mode. Although the graphing utility will graph the sine function for all real values of x, restrict the viewing window to values of x to be the interval $\left[-\dfrac{\pi}{2}, \dfrac{\pi}{2}\right]$.

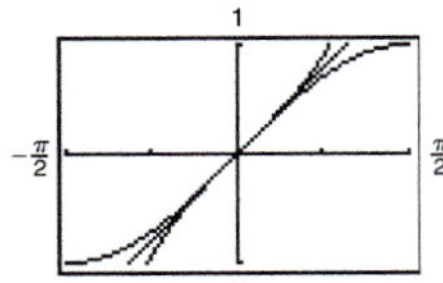

Notice that the graphs of $y_1 = \sin x$, $\left(-\dfrac{\pi}{2}, \dfrac{\pi}{2}\right)$ and $y_2 = \sin^{-1} x$ are reflections of each other in the line $y_3 = x$. So, g is the inverse of f.

3. Because $\cos \pi = -1$ and π lies in $[0, \pi]$, it follows that $\arccos(-1) = \cos^{-1}(-1) = \pi$.

4.

	Function	Mode	Calculator Keystrokes
(a)	arctan 4.84	Radian	TAN⁻¹ (4.84) ENTER

From the display, it follows that arctan 4.84 $\approx$ 1.3670516.

(b)	arcsin (-1.1)	Radian	SIN⁻¹ ((−) 1.1) ENTER

In radian mode the calculator should display an *error* message because the domain of the inverse sine function is $[-1, 1]$.

(c)	arccos (-0.349)	Radian	COS⁻¹ ((−) 0.349) ENTER

From the display, it follows that arccos $(-0.349) \approx 1.9273001$.

5. (a) Because -14 lies in the domain of the arctangent function, the inverse property applies, and you have $\tan\left[\tan^{-1}(-14)\right] = -14$.

(b) In this case, $\frac{7\pi}{4}$ does not lie in the range of the arcsine function, $-\frac{\pi}{2} \le y \le \frac{\pi}{2}$.

However, $\frac{7\pi}{4}$ is coterminal with $\frac{7\pi}{4} - 2\pi = -\frac{\pi}{4}$ which does lie in the range of the arcsine function, and you have

$$\sin^{-1}\left(\sin\frac{7\pi}{4}\right) = \sin^{-1}\left[\sin\left(-\frac{\pi}{4}\right)\right]$$
$$= -\frac{\pi}{4}.$$

(c) Because 0.54 lies in the domain of the arccosine function, the inverse property applies and you have $\cos(\arccos 0.54) = 0.54$.

6. If you let $u = \arctan\left(-\frac{3}{4}\right)$, then $\tan u = -\frac{3}{4}$. Because the range of the inverse tangent function is the first and fourth quadrants and tan u is negative, u is a fourth-quadrant angle. You can sketch and label angle u.

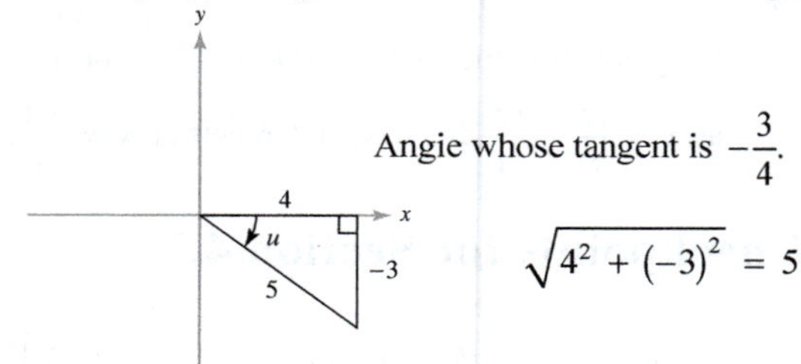

So, $\cos\left[\arctan\left(-\frac{3}{4}\right)\right] = \cos u = \frac{4}{5}$.

7. If you let $u = \arctan x$, then $\tan u = x$, where x is any real number. Because $\tan u = \frac{\text{opp}}{\text{adj}} = \frac{x}{1}$ you can sketch a right triangle with acute angle u as shown. From this triangle, you can convert to algebraic form.

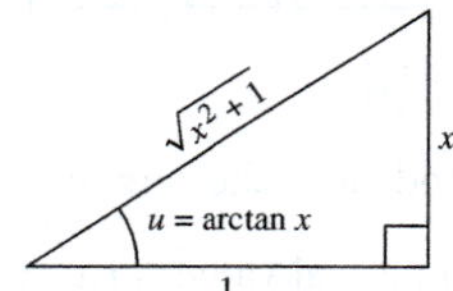

$$\sec(\arctan x) = \sec u = \frac{\sqrt{x^2+1}}{1} = \sqrt{x^2+1}$$

Checkpoints for Section 4.8

1. Because $c = 90°$, it follows that $A + B = 90°$ and $B° = 90° - 20° = 70°$.
To solve for a, use the fact that

$$\tan A = \frac{\text{opp}}{\text{adj}} = \frac{a}{b} \Rightarrow a = b \tan A.$$

So, $a = 15 \tan 20° \approx 5.46$. Similarly, to solve for c, use the fact that $\cos A = \frac{\text{adj}}{\text{hyp}} = \frac{b}{c} \Rightarrow c = \frac{b}{\cos A}$

So, $c = \frac{15}{\cos 20°} \approx 15.96$.

2.

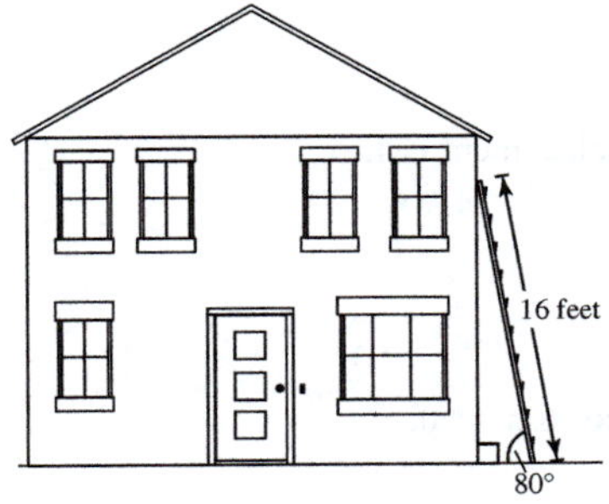

From the equation $\sin A = \frac{a}{c}$, it follows that

$$\begin{aligned} a &= c \sin A \\ &= 16 \sin 80° \\ &\approx 15.8. \end{aligned}$$

So, the height from the top of the ladder to the ground is about 15.8 feet.

3.

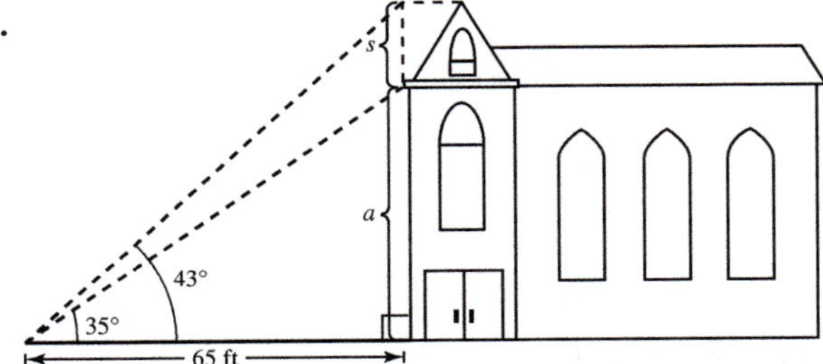

Note that this problem involves two right triangles. For the smaller right triangle, use the fact that

$\tan 35° = \frac{a}{65}$ to conclude that the height of the church is $a = 65 \tan 35°$.

For the larger right triangle use the equation

$\tan 43° = \frac{a + s}{65}$ to conclude that $a + s = 65 \tan 43°$.

So, the height of the steeple is

$$\begin{aligned} s &= 65 \tan 43° - a \\ &= 65 \tan 43° - (65 \tan 35°) \\ &\approx 15.1 \text{ feet.} \end{aligned}$$

4.

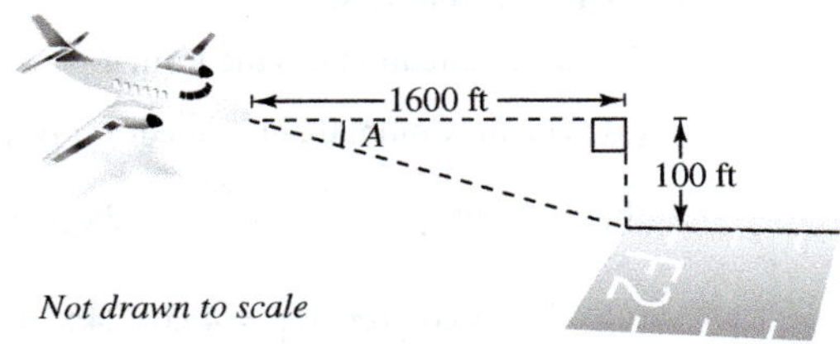

Not drawn to scale

Using the tangent function, you can see that

$$\tan A = \frac{\text{opp}}{\text{adj}} = \frac{100}{1600} = 0.0625$$

So, the angle of depression is

$$\begin{aligned} A &= \arctan(0.0625) \text{ radian} \\ &\approx 0.06242 \text{ radian} \\ &\approx 3.58°. \end{aligned}$$

5.

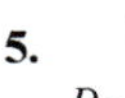

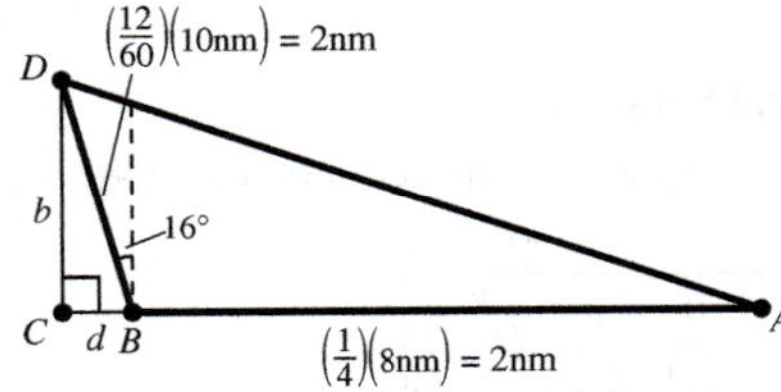

For triangle BCD, you have $B = 90° - 16° = 74°$. The two sides of this triangle can be determined to be $b = 2 \sin 74°$ and $d = 2 \cos 74°$.

For triangle ACD, you can find angle A as follows.

$$\tan A = \frac{b}{d + 2} = \frac{2 \sin 74°}{2 \cos 74° + 2} \approx 0.7535541$$

$$A = \arctan A \approx \arctan 0.7535541 \text{ radian} \approx 37°$$

The angle with the north south line is $90° - 37° = 53°$.

So, the bearing of the ship is N 53° W.

Finally, from triangle ACD you have

$\sin A = \frac{b}{c}$, which yields

$$c = \frac{b}{\sin A} = \frac{2 \sin 74°}{\sin 37°} \approx 3.2 \text{ nautical miles.}$$

6. Because the spring is at equilibrium $(d = 0)$ when $t = 0$, use the equation $d = a \sin \omega t$.

Because the maximum displacement from zero is 6 and the period is 3, you have the following.

$$\text{Amplitude} = |a| = 6$$

$$\text{Period} = \frac{2\pi}{\omega} = 3 \Rightarrow \omega = \frac{2\pi}{3}.$$

So, an equation of motion is $d = 6 \sin \frac{2\pi}{3}t$.

7. Algebraic Solution

The given equation has the form $d = 4 \cos 6\pi t$, with $a \approx 4$ and $w = 6\pi$.

(a) The maximum displacement is given by the amplitude. So, the maximum displacement is 4.

(b) Frequency $= \dfrac{w}{2\pi} = \dfrac{6\pi}{2\pi} = 3$ cycles per unit of time

(c) $d = 4 \cos\left[6\pi(4)\right] = 4 \cos 24\pi = 4(1) = 4$

(d) To find the least positive value of t, for which $d = 0$, solve the equation $4 \cos 6\pi t = 0$.

First divide each side by 4 to obtain $\cos 6\pi t = 0$.

This equation is satisfied when $6\pi t = \dfrac{\pi}{2}, \dfrac{3\pi}{2}, \dfrac{5\pi}{2}, \ldots$.

Divide each of these values by 6π to obtain $t = \dfrac{1}{12}, \dfrac{1}{4}, \dfrac{5}{12}, \ldots$.

So, the least positive value of t is $t = \dfrac{1}{12}$.

Graphical Solution

(a) Use a graphing utility set in radian mode.

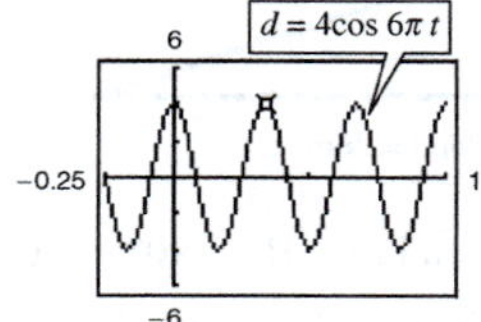

The maximum displacement is from the point of equilibrium $(d = 0)$ is 4.

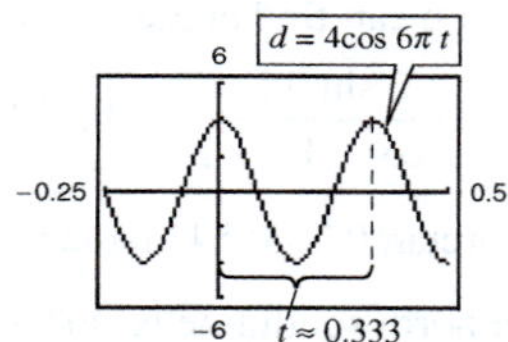

(b) The period is the time for the graph to complete one cycle, which is $t \approx 0.333$. So, the frequency is about $\dfrac{1}{0.333} \approx 3$ per unit of time.

(c)

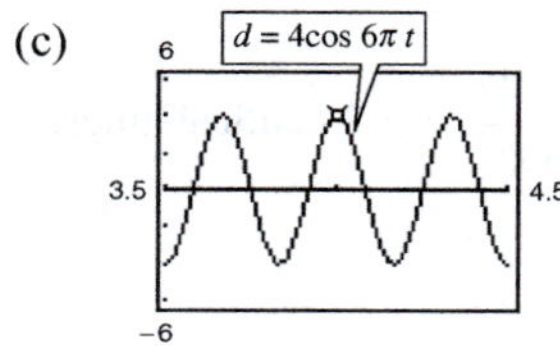

The value of d when $t = 4$ is $d = 4$

(d)

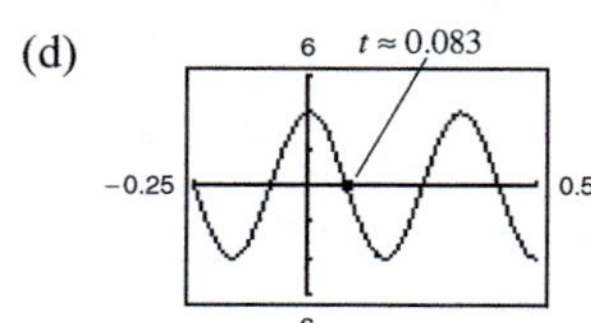

The least positive value of t for which $d = 0$ is $t \approx 0.083$.

Chapter 5

Checkpoints for Section 5.1

1. Using a reciprocal identity, you have $\cot x = \dfrac{1}{\tan x} = \dfrac{1}{\frac{1}{3}} = 3.$

Using a Pythagorean identity, you have

$$\sec^2 x = 1 + \tan^2 x = 1 + \left(\frac{1}{3}\right)^2 = 1 + \frac{1}{9} = \frac{10}{9}.$$

Because $\tan x > 0$ and $\cos x < 0$, you know that the angle x lies in Quadrant III.

Moreover, because $\sec x$ is negative when x is in Quadrant III, choose the negative root and obtain

$$\sec x = -\sqrt{\frac{10}{9}} = -\frac{\sqrt{10}}{3}.$$

Using a reciprocal identity, you have

$$\cos x = \frac{1}{\sec x} = -\frac{1}{\sqrt{10}/3} = -\frac{3}{\sqrt{10}} = -\frac{3\sqrt{10}}{10}.$$

Using a quotient identity, you have

$$\tan x = \frac{\sin x}{\cos x} \Rightarrow \sin x = \cos x \tan x = \left(-\frac{3\sqrt{10}}{10}\right)\left(\frac{1}{3}\right) = -\frac{\sqrt{10}}{10}.$$

Using a reciprocal identity, you have

$$\csc x = \frac{1}{\sin x} = -\frac{1}{\sqrt{10}/10} = -\frac{10}{\sqrt{10}} = -\sqrt{10}.$$

$\sin x = -\dfrac{\sqrt{10}}{10}$ $\csc x = -\sqrt{10}$

$\cos x = -\dfrac{3\sqrt{10}}{10}$ $\sec x = -\dfrac{\sqrt{10}}{3}$

$\tan x = \dfrac{1}{3}$ $\cot x = 3$

2. First factor out a common monomial factor then use a fundamental identity.

$\cos^2 x \csc x - \csc x = \csc x(\cos^2 x - 1)$	Factor out a common monomial factor.
$= -\csc x(1 - \cos^2 x)$	Factor out -1.
$= -\csc x \sin^2 x$	Pythagorean identity
$= -\left(\dfrac{1}{\sin x}\right)\sin^2 x$	Reciprocal identity
$= -\sin x$	Multiply.

3. (a) This expression has the form $u^2 - v^2$, which is the difference of two squares. It factors as

$$1 - \cos^2 \theta = (1 - \cos\theta)(1 + \cos\theta).$$

(b) This expression has the polynomial form $ax^2 + bx + c$, and it factors as

$$2\csc^2\theta - 7\csc\theta + 6 = (2\csc\theta - 3)(\csc\theta - 2).$$

4. Use the identity $\sec^2 x = 1 + \tan^2 x$ to rewrite the expression.

$$\begin{aligned} \sec^2 x + 3\tan x + 1 &= (1 + \tan^2 x) + 3\tan x + 1 && \text{Pythagorean identity} \\ &= \tan^2 x + 3\tan x + 2 && \text{Combine like terms.} \\ &= (\tan x + 2)(\tan x + 1) && \text{Factor.} \end{aligned}$$

5. $$\begin{aligned} \csc x - \cos x \cot x &= \frac{1}{\sin x} - \cos x\left(\frac{\cos x}{\sin x}\right) && \text{Quotient and reciprocal identities} \\ &= \frac{1}{\sin x} - \frac{\cos^2 x}{\sin x} && \text{Multiply.} \\ &= \frac{1 - \cos^2 x}{\sin x} && \text{Add fractions.} \\ &= \frac{\sin^2 x}{\sin x} && \text{Pythagorean identity.} \\ &= \sin x && \text{Simplify.} \end{aligned}$$

6. $$\begin{aligned} \frac{1}{1 + \sin\theta} + \frac{1}{1 - \sin\theta} &= \frac{1 - \sin\theta + 1 + \sin\theta}{(1 + \sin\theta)(1 - \sin\theta)} && \text{Add fractions.} \\ &= \frac{2}{1 - \sin^2\theta} && \text{Combine like terms in numerator and multiply factors in denominator.} \\ &= \frac{2}{\cos^2\theta} && \text{Pythagorean identity} \\ &= 2\sec^2\theta && \text{Reciprocal identity} \end{aligned}$$

7. $$\begin{aligned} \frac{\cos^2\theta}{1 - \sin\theta} &= \frac{1 - \sin^2\theta}{1 - \sin\theta} && \text{Pythagorean identity} \\ &= \frac{(1 + \sin\theta)(1 - \sin\theta)}{1 - \sin\theta} && \text{Factor the numerator as the difference of squares.} \\ &= 1 + \sin\theta && \text{Simplify.} \end{aligned}$$

8. Begin by letting $x = 3\sin x$, then you obtain the following

$$\begin{aligned} \sqrt{9 - x^2} &= \sqrt{9 - (3\sin\theta)^2} && \text{Substitute } 3\sin\theta \text{ for } x. \\ &= \sqrt{9 - 9\sin^2\theta} && \text{Rule of exponents.} \\ &= \sqrt{9(1 - \sin^2\theta)} && \text{Factor.} \\ &= \sqrt{9\cos^2\theta} && \text{Pythagorean identity} \\ &= 3\cos\theta && \cos\theta > 0 \text{ for } 0 < \theta = \frac{\pi}{2} \end{aligned}$$

9. $$\begin{aligned} \ln|\sec x| + \ln|\sin x| &= \ln|\sec x \sin x| && \text{Product Property of Logarithms} \\ &= \ln\left|\frac{1}{\cos s} \cdot \sin x\right| && \text{Reciprocal identity} \\ &= \ln\left|\frac{\sin x}{\cos x}\right| && \text{Simplify.} \\ &= \ln|\tan x| && \text{Quotient identity} \end{aligned}$$

Checkpoints for Section 5.2

1. Start with the left side because it is more complicated.

$$\frac{\sin^2\theta + \cos^2\theta}{\cos^2\theta\sec^2\theta} = \frac{1}{\cos^2\theta\sec^2\theta} \qquad \text{Pythagorean identity}$$

$$= \frac{1}{\cos^2\theta\left(\dfrac{1}{\cos^2\theta}\right)} \qquad \text{Reciprocal identity}$$

$$= 1 \qquad \text{Simplify.}$$

2. Algebraic Solution

Start with the right side because it is more complicated.

$$\frac{1}{1-\cos\beta} + \frac{1}{1+\cos\beta} = \frac{1+\cos\beta+1-\cos\beta}{(1-\cos\beta)(1+\cos\beta)} \qquad \text{Add fractions.}$$

$$= \frac{2}{1-\cos^2\beta} \qquad \text{Simplify.}$$

$$= \frac{2}{\sin^2\beta} \qquad \text{Pythagorean identity}$$

$$= 2\csc^2\beta \qquad \text{Reciprocal identity}$$

Numerical Solution

Use a graphing utility to create a table that shows the values of

$y_1 = 2\csc^2 x$ and $y_2 = \dfrac{1}{1-\cos x} + \dfrac{1}{1+\cos x}$ for different values of x.

X	Y1	Y2
-3	100.43	100.43
-2	2.4189	2.4189
-1	2.8246	2.8246
0	ERROR	ERROR
1	2.8246	2.8246
2	2.4189	2.4189
3	100.43	100.43

X=-3

The values for y_1 and y_2 appear to be identical, so the equation appears to be an identity.

3. Algebraic Solution

By applying identities before multiplying, you obtain the following.

$$(\sec^2 x - 1)(\sin^2 x - 1) = (\tan^2 x)(-\cos^2 x) \qquad \text{Pythagorean identities}$$

$$= \left(\frac{\sin x}{\cos x}\right)^2(-\cos^2 x) \qquad \text{Quotient identity}$$

$$= \left(\frac{\sin^2 x}{\cos^2 x}\right)(-\cos^2 x) \qquad \text{Property of exponents}$$

$$= -\sin^2 x \qquad \text{Multiply.}$$

Graphical Solution

Using a graphing utility, let $y_1 = (\sec^2 x - 1)(\sin^2 x - 1)$ and $y_2 = -\sin^2 x$.

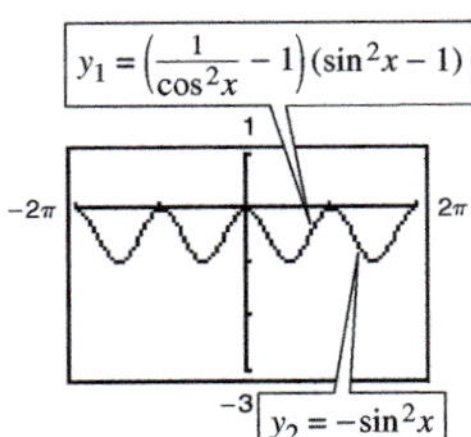

Because the graphs appear to coincide the given equation, $(\sec^2 x - 1)(\sin^2 x - 1) = -\sin^2 x$ appears to be an identity.

4. (a) $\cot x \sec x = \left(\dfrac{\cos x}{\sin x}\right)\left(\dfrac{1}{\cos x}\right)$ Convert the left into sines and cosines.

$= \dfrac{1}{\sin x}$ Cancel like factors of cosines.

$= \csc x$ Rewrite using reciprocal identities.

(b) Convert the left into sines and cosines.

$\csc x - \sin x = \dfrac{1}{\sin x} - \sin x$

$= \dfrac{1 - \sin^2 x}{\sin x}$ Add fractions.

$= \dfrac{\cos^2 x}{\sin x}$ Pythagorean identity

$= \left(\dfrac{\cos x}{1}\right)\left(\dfrac{\cos x}{\sin x}\right)$ Product of fractions

$= \cos x \cot x$ Quotient identity

5. Algebraic Solution

Begin with the right side and create a monomial denominator by multiplying the numerator and denominator by $1 + \cos x$.

$\dfrac{\sin x}{1 - \cos x} = \dfrac{\sin x}{1 - \cos x}\left(\dfrac{1 + \cos x}{1 + \cos x}\right)$ Multiply numerator and denomintor by $1 + \cos x$.

$= \dfrac{\sin x + \sin x \cos x}{1 - \cos^2 x}$ Multiply.

$= \dfrac{\sin x + \sin x \cos x}{\sin^2 x}$ Pythagorean identity

$= \dfrac{\sin x}{\sin^2 x} + \dfrac{\sin x \cos x}{\sin^2 x}$ Write as separate functions.

$= \dfrac{1}{\sin x} + \dfrac{\cos x}{\sin x}$ Simplify.

$= \csc x + \cot x$ Identities

Graphical Solution

Using a graphing utility, let $y_1 = \csc x + \cot x$ and $y_2 = \dfrac{\sin x}{1 - \cos x}$.

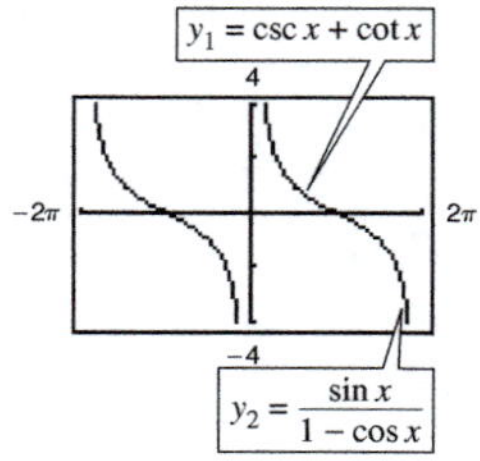

Because the graphs appear to coincide, the given equation appears to be an identity.

6. Algebraic Solution

Working with the left side, you have the following.

$$\frac{\tan^2 \theta}{1 + \sec \theta} = \frac{\sec^2 \theta - 1}{\sec \theta + 1} \qquad \text{Pythagorean identity}$$

$$= \frac{(\sec \theta + 1)(\sec \theta - 1)}{\sec \theta + 1} \qquad \text{Factor.}$$

$$= \sec \theta - 1 \qquad \text{Simplify.}$$

Now, working with the right side, you have the following.

$$\frac{1 - \cos \theta}{\cos \theta} = \frac{1}{\cos \theta} - \frac{\cos \theta}{\cos \theta} \qquad \text{Write as separate fractions.}$$

$$= \sec \theta - 1 \qquad \text{Identity and simplify.}$$

This verifies the identity because both sides are equal to $\sec \theta - 1$.

Numerical Solution

Use a graphing utility to create a table that shows the values of

$y_1 = \dfrac{\tan^2 x}{1 + \sec x}$ and $y_2 = \dfrac{1 - \cos x}{\cos x}$ for different values of x.

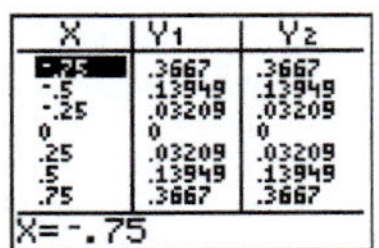

X	Y1	Y2
-.75	.3667	.3667
-.5	.13949	.13949
-.25	.03209	.03209
0	0	0
.25	.03209	.03209
.5	.13949	.13949
.75	.3667	.3667

X=-.75

The values of y_1 and y_2 appear to be identical, so the equation appears to be an identity.

7. (a) $\tan x \sec^2 x - \tan x = \tan x(\sec^2 x - 1)$ Factor.

$= \tan x \tan^2 x$ Pythagorean identity

$= \tan^3 x$ Multiply.

(b) $(\cos^4 x - \cos^6 x)\sin x = \cos^4 x(1 - \cos^2 x)\sin x$ Factor.

$= \cos^4 x\,(\sin^2 x)\sin x$ Pythagorean identity

$= \sin^3 x \cos^4 x$ Multiply.

Checkpoints for Section 5.3

1. Begin by isolating $\sin x$ on one side of the equation.

$\sin x - \sqrt{2} = -\sin x$ Write original equation.

$\sin x + \sin x - \sqrt{2} = 0$ Add $\sin x$ to each side.

$\sin x + \sin x = \sqrt{2}$ Add $\sqrt{2}$ to each side.

$2\sin x = \sqrt{2}$ Combine like terms.

$\sin x = \dfrac{\sqrt{2}}{2}$ Divide each side by 2.

Because $\sin x$ has a period of 2π, first find all solutions in the interval $[0, 2\pi)$. These solutions are $x = \dfrac{\pi}{4}$ and $x = \dfrac{3\pi}{4}$.

Finally, add multiples of 2π to each of these solutions to obtain the general form

$x = \dfrac{\pi}{4} + 2n\pi$ and $x = \dfrac{3\pi}{4} + 2n\pi$ where n is an integer.

2. Begin by isolating sin x on one side of the equation.

$4\sin^2 x - 3 = 0$	Write original equation.
$4\sin^2 x = 3$	Add 3 to each side.
$\sin^2 x = \dfrac{3}{4}$	Divide each side by 4.
$\sin x = \pm\sqrt{\dfrac{3}{4}}$	Extract square roots.
$\sin x = \pm\dfrac{\sqrt{3}}{2}$	Simplify.

Because sin x has a period of 2π, first find all solutions in the interval $[0, 2\pi)$. These solutions are $x = \frac{\pi}{3}$, $x = \frac{2\pi}{3}$, $x = \frac{4\pi}{3}$, and $x = \frac{5\pi}{3}$.

Finally, add multiples of 2π to each of these solutions to obtain the general form.

$x = \frac{\pi}{3} + 2n\pi$, $x = \frac{2\pi}{3} + 2n\pi$, $x = \frac{4\pi}{3} + 2n\pi$, and $x = \frac{5\pi}{3} + 2n\pi$ where n is an integer.

3. Begin by collecting all terms on one side of the equation and factoring.

$\sin^2 x = 2\sin x$	Write original equation.
$\sin^2 x - 2\sin x = 0$	Subtract $2\sin x$ from each side.
$\sin x(\sin x - 2) = 0$	Factor.

By setting each of these factors equal to zero, you obtain

$\sin x = 0$ and $\sin x - 2 = 0$

$\sin x = 2.$

In the interval $[0, 2\pi)$, the equation $\sin x = 0$ has solutions $x = 0$ and $x = \pi$. Because sin x has a period of 2π, you would obtain the general forms $x = 0 + 2n\pi$ and $x = \pi + 2n\pi$ where n is an integer by adding multiples of 2π.

No solution exists for $\sin x = 2$ because 2 is outside the range of the sine function, $[-1, 1]$, so the solutions are of the form $x = n\pi$, where n is an integer. Confirm this graphically by graphing $y = \sin^2 x - 2\sin x$.

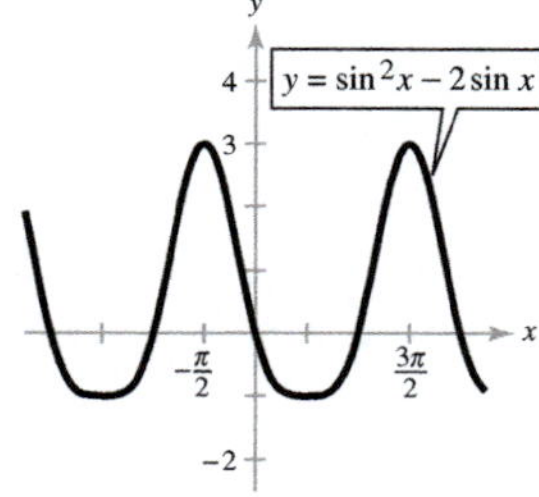

Notice that the x-intercepts occur at $-2\pi, -\pi, 0, \pi, 2\pi$ and so on. These x-intercepts correspond to the solutions of $\sin^2 x - 2\sin x = 0$.

4. Algebraic Solution:

Treat the equation as a quadratic in sin x and factor.

$2\sin^2 x - 3\sin x + 1 = 0$	Write original equation.
$(2\sin x - 1)(\sin x - 1) = 0$	Factor.

Setting each factor equal to zero, you obtain the following solutions in the interval $[0, 2\pi)$.

$2\sin x - 1 = 0$	and	$\sin x - 1 = 0$
$\sin x = \dfrac{1}{2}$		$\sin x = 1$
$x = \dfrac{\pi}{6}, \dfrac{5\pi}{6}$		$x = \dfrac{\pi}{2}$

Graphical Solution:

The x-intercepts are $x \approx 0.524$, $x = 2.618$, and $x = 1.571$.

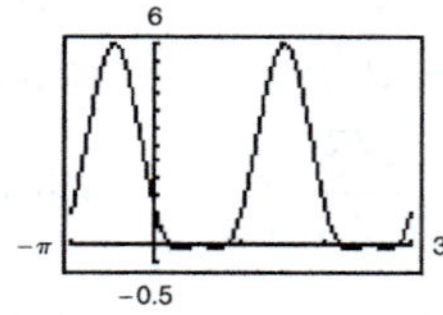

From the graph, you can conclude that the approximate solutions of $2\sin^2 x - 3\sin x + 1 = 0$ in the interval $[0, 2\pi)$ are $x \approx 0.524 = \frac{\pi}{6}$, $x \approx 2.618 = \frac{5\pi}{6}$, and $x \approx 1.571 = \frac{\pi}{2}$.

5. This equation contains both tangent and secant functions. You can rewrite the equation so that it has only tangent functions by using the identity $\sec^2 x = \tan^2 x + 1$.

$3\sec^2 x - 2\tan^2 x - 4 = 0$	Write original equation.
$3(\tan^2 x + 1) - 2\tan^2 x - 4 = 0$	Pythagorean identity
$3\tan^2 x + 3 - 2\tan^2 x - 4 = 0$	Distributive property
$\tan^2 x - 1 = 0$	Simplify.
$\tan^2 x = 1$	Add 1 to each side.
$\tan x = \pm 1$	Extract square roots.

Because $\tan x$ has a period of π, you can find the solutions in the interval $[0, \pi)$ to be $x = \frac{\pi}{4}$ and $x = \frac{3\pi}{4}$.

The general solution is $x = \frac{\pi}{4} + n\pi$ and $x = \frac{3\pi}{4} + n\pi$ where n is an integer.

6. Solution It is not clear how to rewrite this equation in terms of a single trigonometric function. Notice what happens when you square each side of the equation.

$\sin x + 1 = \cos x$	Write original equation.
$\sin^2 x + 2\sin x + 1 = \cos^2 x$	Square each side.
$\sin^2 x + 2\sin x + 1 = 1 - \sin^2 x$	Pythagorean identity
$\sin^2 x + \sin^2 x + 2\sin x + 1 - 1 = 0$	Rewrite equation.
$2\sin^2 x + 2\sin x = 0$	Combine like terms.
$2\sin x(\sin x + 1) = 0$	Factor.

Setting each factor equal to zero produces the following.

$2\sin x = 0$	and	$\sin x + 1 = 0$
$\sin x = 0$		$\sin x = -1$
$x = 0, \pi$		$x = \frac{3\pi}{2}$

Because you squared the original equation, check for extraneous solutions.

Check $x = 0$:	$\sin 0 + 1 \stackrel{?}{=} \cos 0$	Substitute 0 for x.
	$0 + 1 = 1$	Solution checks. ✓
Check $x = \pi$:	$\sin \pi + 1 \stackrel{?}{=} \cos \pi$	Substitute π for x.
	$0 + 1 \neq -1$	Solution does not check.
Check $x = \frac{3\pi}{2}$:	$\sin \frac{3\pi}{2} + 1 \stackrel{?}{=} \cos \frac{3\pi}{2}$	Substitute $\frac{3\pi}{2}$ for x.
	$-1 + 1 = 0$	Solution checks. ✓

Of the three possible solutions, $x = \pi$ is extraneous. So, in the interval $[0, 2\pi)$, the two solutions are $x = 0$ and $x = \frac{3\pi}{2}$.

7. $2\sin 2t - \sqrt{3} = 0$ Write original equation.

$2\sin 2t = \sqrt{3}$ Add $\sqrt{3}$ to each side.

$\sin 2t = \dfrac{\sqrt{3}}{2}$ Divide each side by 2.

In the interval $[0, 2\pi)$, you know that

$2t = \dfrac{\pi}{3}$ and $2t = \dfrac{2\pi}{3}$ are the only solutions.

So, in general you have

$2t = \dfrac{\pi}{3} + 2n\pi$ and $2t = \dfrac{2\pi}{3} + 2n\pi$.

Dividing these results by 2, you obtain the general solution

$t = \dfrac{\pi}{6} + n\pi$ and $t = \dfrac{\pi}{3} + n\pi$.

8. $2\tan\dfrac{x}{2} - 2 = 0$ Write original equation.

$2\tan\dfrac{x}{2} = 2$ Add 2 to each side.

$\tan\dfrac{x}{2} = 1$ Divide each side by 2.

In the interval $[0, \pi)$, you know that $\dfrac{x}{2} = \dfrac{\pi}{4}$ is the only solution. So, in general, you have

$\dfrac{x}{2} = \dfrac{\pi}{4} + n\pi.$

Multiplying this result by 2, you obtain the general solution

$x = \dfrac{\pi}{2} + 2n\pi$

Where n is an integer.

9. $4\tan^2 x + 5\tan x - 6 = 0$ Write original equation.

$(4\tan x - 3)(\tan x + 2) = 0$ Factor.

$4\tan x - 3 = 0$ and $\tan x + 2 = 0$ Set each factor equal to zero.

$\tan x = \dfrac{3}{4}$ $\tan x = -2$

$x = \arctan\dfrac{3}{4}$ $x = \arctan(-2)$ Use inverse tangent function to solve for x.

These two solutions are in the interval $\left(-\dfrac{\pi}{2}, \dfrac{\pi}{2}\right)$. Recall that the range of the inverse tangent function is $\left(-\dfrac{\pi}{2}, \dfrac{\pi}{2}\right)$.

Finally, because $\tan x$ has a period of π, you add multiples of π to obtain

$x = \arctan\dfrac{3}{4} + n\pi$ and $x = \arctan(-2) + n\pi$

where n is an integer.

You can use a calculator to approximate the values of $x = \arctan\dfrac{3}{4} \approx 0.6435$ and $x = \arctan(-2) \approx -1.1071$.

10. $\sin^2 x + 2\sin x - 1 = 0$ Write original equation.

$\sin x = \dfrac{-2 \pm \sqrt{2^2 - 4(1)(-1)}}{2(1)}$ Use the Quadratic Formula to solve for $\sin x$.

$\sin x = \dfrac{-2 \pm \sqrt{8}}{2}$ Simplify.

$\sin x = -1 \pm \sqrt{2}$

$x = \arcsin\left(-1 + \sqrt{2}\right)$ and $x = \arcsin\left(-1 - \sqrt{2}\right)$ Use inverse sine function to solve for x.

Using the solution, $x = \arcsin\left(-1 + \sqrt{2}\right)$: $x \approx 0.4271$ and $x \approx \pi - 0.4271 = 2.7145$. These solutions lie in Quadrant I and Quadrant II.

The solution $x = \arcsin\left(-1 - \sqrt{2}\right)$ is not in the domain of the arcsine function.

11. Start with $S = 6hs + 1.5s^2\left[\left(\sqrt{3} - \cos\theta\right)/\sin\theta\right]$ and let $h = 3.2$ inches and $s = 0.75$ inch.

Next graph the function $S = 14.4 + 0.84375\left[\left(\sqrt{3} - \cos\theta\right)/\sin\theta\right]$ with a graphing utility.

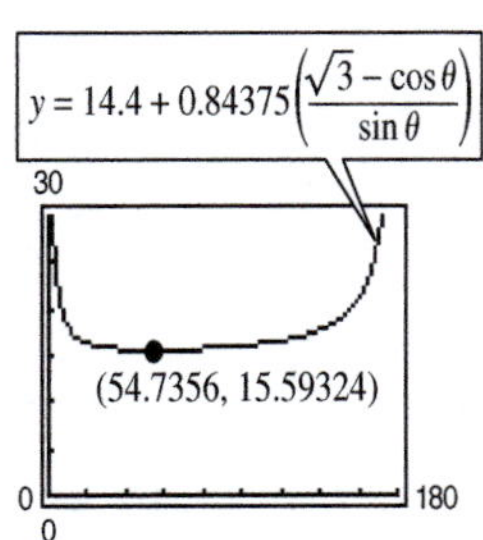

Use the minimum feature to approximate the minimum value on the graph. So, the minimum surface area of 15.6 square inches occurs when $\theta \approx 54.7°$.

Checkpoints for Section 5.4

1. To find the exact value of $\cos\dfrac{\pi}{12}$, use the fact that

$$\frac{\pi}{12} = \frac{\pi}{3} - \frac{\pi}{4}.$$

The formula for $\cos(u - v)$ yields the following.

$$\begin{aligned}\cos\frac{\pi}{12} &= \cos\left(\frac{\pi}{3} - \frac{\pi}{4}\right)\\ &= \cos\frac{\pi}{3}\cos\frac{\pi}{4} + \sin\frac{\pi}{3}\sin\frac{\pi}{4}\\ &= \left(\frac{1}{2}\right)\left(\frac{\sqrt{2}}{2}\right) + \left(\frac{\sqrt{3}}{2}\right)\left(\frac{\sqrt{2}}{2}\right)\\ &= \frac{\sqrt{2}}{4} + \frac{\sqrt{6}}{4}\\ &= \frac{\sqrt{2} + \sqrt{6}}{4}\end{aligned}$$

2. Using the fact that $75° = 30° + 45°$, together with the formula for $\sin(u + v)$, you obtain the following.

$$\begin{aligned}\sin 75° &= \sin(30° + 45°)\\ &= \sin 30°\cos 45° + \cos 30°\sin 45°\\ &= \left(\frac{1}{2}\right)\left(\frac{\sqrt{2}}{2}\right) + \left(\frac{\sqrt{3}}{2}\right)\left(\frac{\sqrt{2}}{2}\right)\\ &= \frac{\sqrt{2}}{4} + \frac{\sqrt{6}}{4}\\ &= \frac{\sqrt{2} + \sqrt{6}}{4}\end{aligned}$$

3. Because $\sin u = \dfrac{12}{13}$ and u is in Quadrant I,

$\cos u = \dfrac{5}{13}$ as shown.

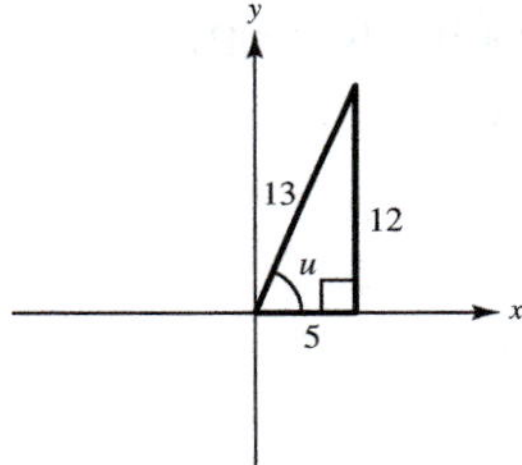

Because $\cos v = -\dfrac{3}{5}$ and v is in Quadrant II, $\sin v = \dfrac{4}{5}$ as shown.

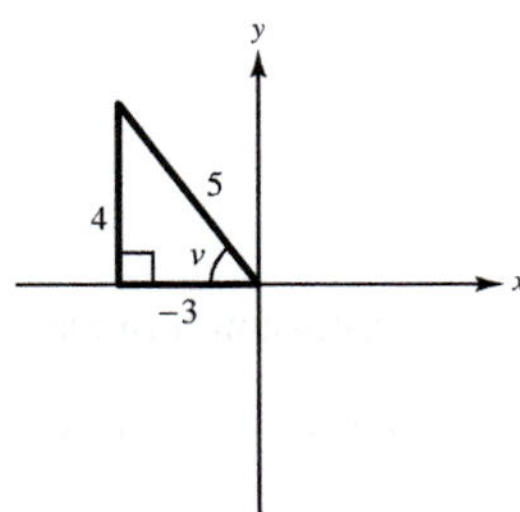

You can find $\cos(u + v)$ as follows.

$$\begin{aligned}\cos(u + v) &= \cos u\cos v - \sin u\sin v\\ &= \left(\frac{5}{13}\right)\left(-\frac{3}{5}\right) - \left(\frac{12}{13}\right)\left(\frac{4}{5}\right)\\ &= -\frac{63}{65}\end{aligned}$$

4. This expression fits the formula for $\sin(u + v)$. The figures show the angles $u = \arctan 1$ and $v = \arccos x$.

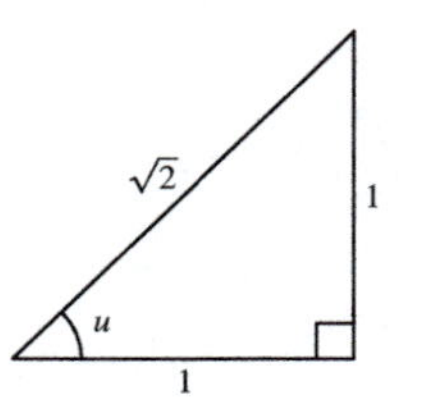

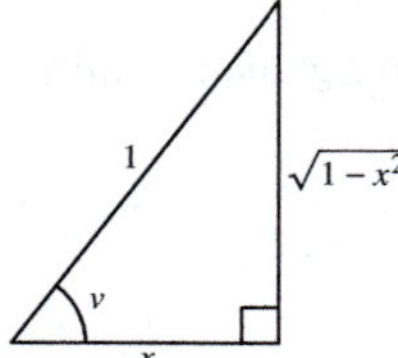

$$\begin{aligned}\sin(u + v) &= \sin u \cos v + \cos u \sin v\\ &= \sin(\arctan 1)\cos(\arccos x) + \cos(\arctan 1)\sin(\arccos x)\\ &= \left(\frac{1}{\sqrt{2}}\right)(x) + \left(\frac{1}{\sqrt{2}}\right)\left(\sqrt{1 - x^2}\right)\\ &= \frac{x}{\sqrt{2}} + \frac{\sqrt{1 - x^2}}{\sqrt{2}}\\ &= \frac{x + \sqrt{1 - x^2}}{\sqrt{2}}\end{aligned}$$

5. Using the formula for $\sin(u - v)$, you have

$$\begin{aligned}\sin\left(x - \frac{\pi}{2}\right) &= \sin x \cos\frac{\pi}{2} - \cos x \sin\frac{\pi}{2}\\ &= (\sin x)(0) - (\cos x)(1)\\ &= -\cos x.\end{aligned}$$

6. (a) Using the formula for $\sin(u - v) = \sin u \cos v - \cos u \sin v$, you have

$$\begin{aligned}\sin\left(\frac{3\pi}{2} - \theta\right) &= \sin\frac{3\pi}{2}\cos\theta - \cos\frac{3\pi}{2}\sin\theta\\ &= (-1)(\cos\theta) - (0)(\sin\theta)\\ &= -\cos\theta.\end{aligned}$$

(b) Using the formula for $\tan(u - v) = \dfrac{\tan u - \tan v}{1 + \tan u \tan v}$, you have

$$\begin{aligned}\tan\left(\theta - \frac{\pi}{4}\right) &= \frac{\tan\theta - \tan\frac{\pi}{4}}{1 + \tan\theta\tan\frac{\pi}{4}}\\ &= \frac{\tan\theta - 1}{1 + (\tan\theta)(1)}\\ &= \frac{\tan\theta - 1}{1 + \tan\theta}.\end{aligned}$$

7. Using the appropriate product-to-sum formula

$\sin u \cos v = \frac{1}{2}[\sin(u + v) + \sin(u - v)]$, you obtain the following.

$$\begin{aligned}\sin 5x \cos 3x &= \frac{1}{2}[\sin(5x + 3x) + \sin(5x - 3x)]\\ &= \frac{1}{2}(\sin 8x + \sin 2x)\\ &= \frac{1}{2}\sin 8x + \frac{1}{2}\sin 2x\end{aligned}$$

8. Using the appropriate sum-to-product formula,

$\sin u + \sin v = 2\sin\left(\dfrac{u+v}{2}\right)\cos\left(\dfrac{u-v}{2}\right)$, you obtain the following.

$$\begin{aligned}\sin 195° + \sin 105° &= 2\sin\left(\frac{195° + 105°}{2}\right)\cos\left(\frac{195° - 105°}{2}\right)\\ &= 2\sin 150° \cos 45°\\ &= 2\left(\frac{1}{2}\right)\left(\frac{\sqrt{2}}{2}\right)\\ &= \frac{\sqrt{2}}{2}\end{aligned}$$

9.

$\sin 4x - \sin 2x = 0$		Write orignal equation.
$2\cos\left(\dfrac{4x+2x}{2}\right)\sin\left(\dfrac{4x-2x}{2}\right) = 0$		Sum-to-product formula
$2\cos 3x \sin x = 0$		Simplify.
$\cos 3x \sin x = 0$		Divide each side by 2.
$\cos 3x = 0$	$\sin x = 0$	Set each factor equal to zero.

The solutions in the interval $[0, 2\pi)$ are $3x = \dfrac{\pi}{2}, \dfrac{3\pi}{2}$ and $x = 0, \pi$.

The general solutions for the equation $\cos 3x = 0$ are $3x = \dfrac{\pi}{2} + 2n\pi$ and $3x = \dfrac{3\pi}{2} + 2n\pi$.

So, by solving these equations for x, you have $x = \dfrac{\pi}{6} + \dfrac{2n\pi}{3}$ and $x = \dfrac{\pi}{2} + \dfrac{2n\pi}{3}$.

The general solution for the equation $\sin x = 0$ is $x = 0 + 2n\pi$ and $x = \pi + 2n\pi$.

These can be combined as $x = n\pi$.

So, the general solutions to the equation, $\sin 4x - \sin 2x = 0$ are

$x = \dfrac{\pi}{6} + \dfrac{2n\pi}{3}$, $x = \dfrac{\pi}{2} + \dfrac{2n\pi}{3}$, and $x = n\pi$ where n is an integer.

To verify these solutions you can graph $y = \sin 4x - \sin 2x$ and approximate the x-intercepts.

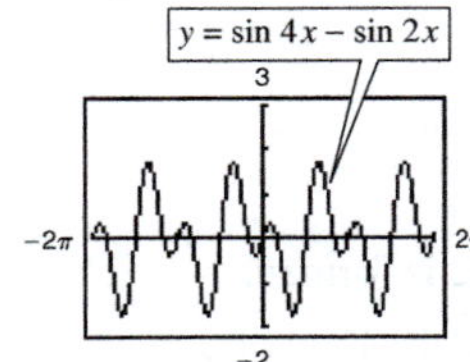

The x-intercepts occur at $0, \dfrac{\pi}{6}, \dfrac{\pi}{2}, \dfrac{5\pi}{6}, \pi, \dfrac{7\pi}{6}, \ldots$

10. Given that a football player can kick a football from ground level with an initial velocity of 80 feet per second, you have the following

$r = \frac{1}{32}v_0^2 \sin 2\theta$	Write projectile motion model.
$r = \frac{1}{32}(80)^2 \sin 2\theta$	Substitute 80 for v_0.
$r = 200 \sin 2\theta$	Simplify.

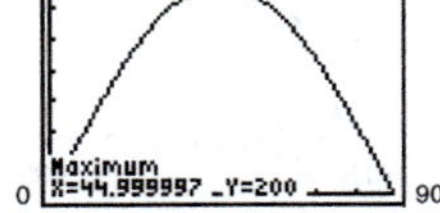

Use a graphing utility to graph the model, $r = 200 \sin 2\theta$.

The maximum point on the graph over the interval $(0°, 90°)$ occurs at $\theta = 45°$.

So, the player must kick the football at an angle of 45° to yield the maximum horizontal distance of 200 feet.

Chapter 6

Checkpoints for Section 6.1

1. The third angle of the triangle is

$C = 180° - A - B = 180° - 30° - 45° = 105°$.

By the Law of Sines, you have

$$\frac{a}{\sin A} = \frac{b}{\sin B} = \frac{c}{\sin C}.$$

Using $a = 32$ centimeters produces

$$b = \frac{a}{\sin A}(\sin B) = \frac{32}{\sin 30°}(\sin 45°) \approx 45.25 \text{ centimeters}$$

and

$$c = \frac{a}{\sin A}(\sin C) = \frac{32}{\sin 30°}(\sin 105°) \approx 61.82 \text{ centimeters}$$

2. From the figure, note that $A = 23°$ and $C = 96°$.

So, the third angle is $B = 180° - A - C = 180° - 23° - 96° = 61°$.

By the Law of Sines, you have

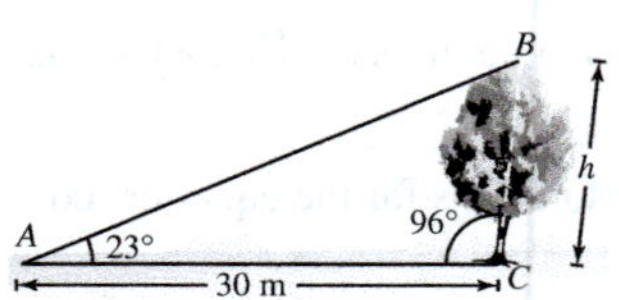

$$\frac{h}{\sin A} = \frac{b}{\sin B}$$

$$h = \frac{b}{\sin B}(\sin A) = \frac{30}{\sin 61°}(\sin 23°)$$

$$\approx 13.40$$

So, the height of the tree h is approximately 13.40 meters.

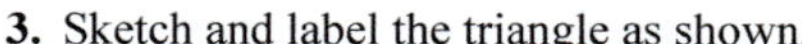

3. Sketch and label the triangle as shown.

$A = 31°$, $a = 12$ inches, and $b = 5$ inches.

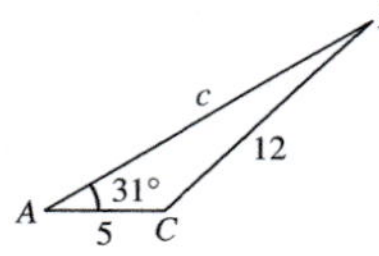

By the Law of Sines, you have

$\dfrac{\sin B}{b} = \dfrac{\sin A}{a}$ Reciprocal form

$\sin B = b\left(\dfrac{\sin A}{a}\right)$ Multiply each side by b.

$\sin B = 5\left(\dfrac{\sin 31°}{12}\right)$ Substitute for A, a, and b.

$B \approx 12.39°$

Now, you can determine that

$C = 180° - A - B \approx 180° - 31° - 12.39° \approx 136.61°$.

Then, the remaining side is

$$\frac{c}{\sin C} = \frac{a}{\sin A}$$

$$c = \frac{a}{\sin A}(\sin C) \approx \frac{12}{\sin 31°}(\sin 136.61°)$$

$$\approx 16.01 \text{ inches.}$$

4. Sketch and label the triangle.

$a = 4$ feet, $b = 5$ feet and $A = 58°$

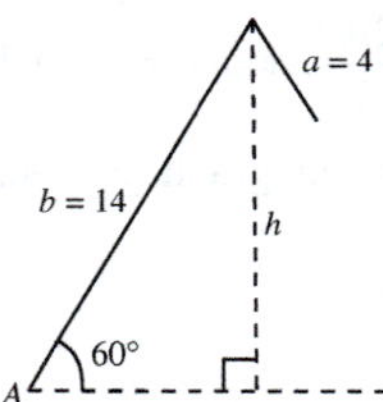

It appears that no triangle is formed.

You can verify this using the Law of Sines.

$$\frac{\sin B}{b} = \frac{\sin A}{a}$$

$$\sin B = b\left(\frac{\sin A}{a}\right)$$

$$\sin B = 14\left(\frac{\sin 60°}{4}\right) \approx 3.0311 > 1$$

This contradicts the fact that $|\sin B| \le 1$.

So, no triangle can be formed having sides $a = 4$ feet and $b = 5$ feet and angle $A = 58°$.

5. By the Law of Sines, you have

$$\frac{\sin B}{b} = \frac{\sin A}{a}$$

$$\sin B = b\left(\frac{\sin A}{a}\right) = 5\left(\frac{\sin 58^\circ}{4.5}\right) \approx 0.9423.$$

There are two angles $B_1 \approx 70.4^\circ$ and $B_2 \approx 180^\circ - 70.4^\circ = 109.6^\circ$ between 0° and 180° whose sine is approximately 0.9423.

For $B_1 \approx 70.4^\circ$, you obtain the following.

$$C = 180^\circ - A - B_1 = 180^\circ - 58^\circ - 70.4^\circ = 51.6^\circ$$

$$c = \frac{a}{\sin A}(\sin C) = \frac{4.5}{\sin 58^\circ}(\sin 51.6^\circ) \approx 4.16 \text{ feet}$$

For $B_2 = 109.6^\circ$, you obtain the following.

$$C = 180^\circ - A - B_2 = 180^\circ - 58^\circ - 109.6^\circ = 12.4^\circ$$

$$c = \frac{a}{\sin A}(\sin C) = \frac{4.5}{\sin 58^\circ}(\sin 12.4^\circ) \approx 1.14 \text{ feet}$$

The resulting triangles are shown.

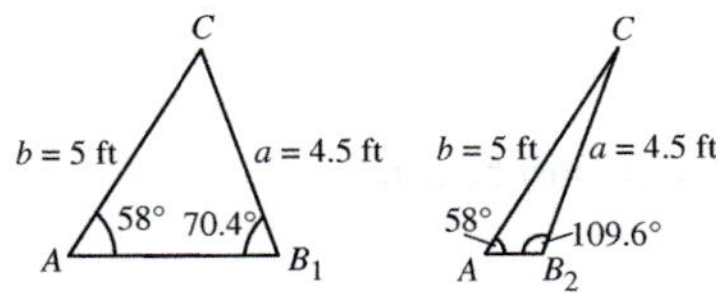

6. Consider $a = 24$ inches, $b = 18$ inches, and angle $C = 80^\circ$ as shown. Then, the area of the triangle is

$$A = \frac{1}{2}ab \sin C = \frac{1}{2}(24)(18)\sin 80^\circ \approx 213 \text{ square yards.}$$

7. Because lines AC and BD are parallel, it follows that $\angle ACB \cong \angle CBD$.

So, triangle ABC has the following measures as shown.

The measure of angle B is $180^\circ - A - C = 180^\circ - 28^\circ - 58^\circ = 94^\circ$.

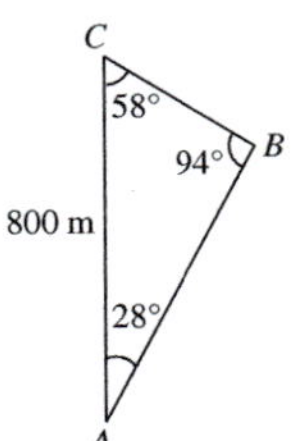

Using the Law of Sines, $\dfrac{a}{\sin 28^\circ} = \dfrac{b}{\sin 94^\circ} = \dfrac{c}{\sin 58^\circ}$.

Because $b = 800$, $C = \dfrac{800}{\sin 94^\circ}(\sin 58^\circ) \approx 680.1$ meters and $a = \dfrac{800}{\sin 94^\circ}(\sin 28^\circ) \approx 376.5$ meters

The total distance that you swim is approximately

Distance $= 680.1 + 376.5 + 800 = 1856.6$ meters.

Checkpoints for Section 6.2

1. 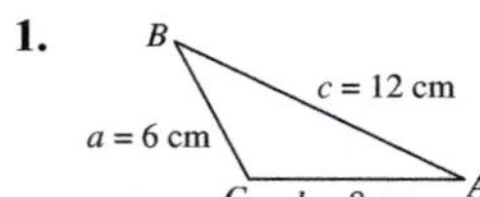

First, find the angle opposite the longest side – side c in this case. Using the alternative form of the Law of Cosines, you find that

$$\cos C = \frac{a^2 + b^2 - c^2}{2ab} = \frac{6^2 + 8^2 - 12^2}{2(6)(8)} \approx -0.4583.$$

Because $\cos C$ is negative, C is an *obtuse* angle given by

$C \approx \cos^{-1}(-0.4583) \approx 117.28°.$

At this point, it is simpler to use the Law of Sines to determine angle B.

$$\sin B = b\left(\frac{\sin C}{c}\right)$$

$$\sin B = 8\left(\frac{\sin 117.28°}{12}\right) \approx 0.5925$$

Because C is obtuse and a triangle can have at most one obtuse angle, you know that B must be acute.

So, $B \approx \sin^{-1}(0.5925) \approx 36.34°$

So, $A = 180° - B - C \approx 180° - 36.34° - 117.28° \approx 26.38°.$

2. $A = 80°, b = 16$ meters and $c = 12$ meters.

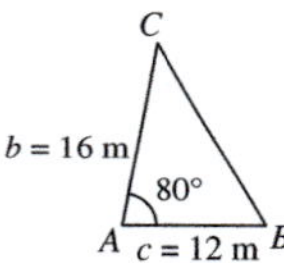

Use the Law of Cosines to find the unknown side a in the figure.

$a^2 = b^2 + c^2 - 2bc \cos A$

$a^2 = 16^2 + 12^2 - 2(16)(12) \cos 80°$

$a^2 \approx 333.3191$

$a \approx 18.26$

Use the Law of Sines to find angle B.

$$\frac{\sin B}{b} = \frac{\sin A}{a}$$

$$\sin B = b\left(\frac{\sin A}{a}\right)$$

$$\sin B = 16\left(\frac{\sin 80°}{18.26}\right)$$

$\sin B \approx 0.863$

There are two angles between 0° and 180° whose sine is 0.863. The two angles are $B_1 \approx 59.66°$ and $B_2 \approx 180° - 59.66° \approx 120.34°$.

Because side a is the longest side of the triangle, angle A must be the largest angle, therefore B must be less than 80°. So, $B \approx 59.66°$.

Therefore, $C = 180° - A - B \approx 180° - 80° - 59.66° \approx 40.34°$.

3.

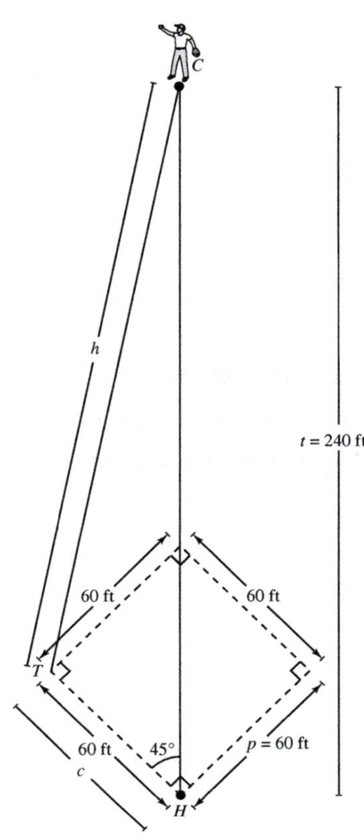

In triangle HCT, $H = 45°$ (line HC bisects the right angle at H), $t = 240$, and $c = 60$.

Using the Law of Cosines for this SAS case, you have

$$h^2 = c^2 + t^2 - 2\,ct\cos H$$

$$h^2 = 60^2 + 240^2 - 2\,(60)(240)\cos 45°$$

$$h^2 \approx 40835.3$$

$$h \approx 202.1$$

So, the center fielder is approximately 202.1 feet from the third base.

4. You have $a = 30$, $b = 56$, and $c = 40$.

So, using the alternative form of the Law of Cosines, you have

$$\cos B = \frac{a^2 + c^2 - b^2}{2ac} = \frac{30^2 + 40^2 - 56^2}{2(30)(40)} = -0.265.$$

So, $B = \cos^{-1}(-0.265) \approx 105.37°$, and thus the bearing from due north from point B to point C is $105.37° - 90° = 15.37°$, or N 15.37° E.

5. $a = 5$ inches, $b = 9$ inches and $c = 8$ inches.

Because $s = \dfrac{a + b + c}{2} = \dfrac{5 + 9 + 8}{2} = \dfrac{22}{2} = 11$,

Heron's Area Formula yields

$$\begin{aligned}\text{Area} &= \sqrt{s(s-a)(s-b)(s-c)}\\ &= \sqrt{11(11-5)(11-9)(11-8)}\\ &= \sqrt{(11)(6)(2)(3)}\\ &= \sqrt{396}\\ &\approx 19.90 \text{ square units.}\end{aligned}$$

Checkpoints for Section 6.3

1. From the Distance Formula, it follows that $\overline{PQ}$ and $\overline{RS}$ have the *same magnitude*.

$$\|\overline{PQ}\| = \sqrt{(3-0)^2 + (1-0)^2} = \sqrt{10}$$

$$\|\overline{RS}\| = \sqrt{(5-2)^2 + (3-2)^2} = \sqrt{10}$$

Moreover, both line segments have the *same direction* because they are both directed toward the upper right on lines having a slope of

$$\frac{1-0}{3-0} = \frac{3-2}{5-2} = \frac{1}{3}.$$

Because, $\overline{PQ}$ and $\overline{RS}$ have the same magnitude and direction, **u** and **v** are equivalent.

2. Algebraic Solution

Let $P(-2, 3) = (p_1, p_2)$ and $Q(-7, 9) = (q_1, q_2)$.

Then, the components of $\mathbf{v} = \langle v_1, v_2 \rangle$ are

$$v_1 = q_1 - p_1 = -7 - (-2) = -5$$
$$v_2 = q_2 - p_2 = 9 - 3 = 6.$$

So, $\mathbf{v} = \langle -5, 6 \rangle$ and the magnitude of $\mathbf{v}$ is $\|\mathbf{v}\| = \sqrt{(-5)^2 + (6)^2} = \sqrt{61}$.

Graphical Solution

Use centimeter graph paper to plot the points $P(-2, 3)$ and $Q(-7, 9)$. Carefully sketch the vector **v**.

Use the sketch to find the components of $\mathbf{v} = \langle v_1, v_2 \rangle$. Then use a centimeter ruler to find the magnitude of **v**.

The figure shows that the components of **v** are $v_1 = -5$ and $v_2 = 6$, so $\mathbf{v} = \langle -5, 6 \rangle$. The figure also shows that the magnitude of **v** is $\|\mathbf{v}\| = \sqrt{61}$.

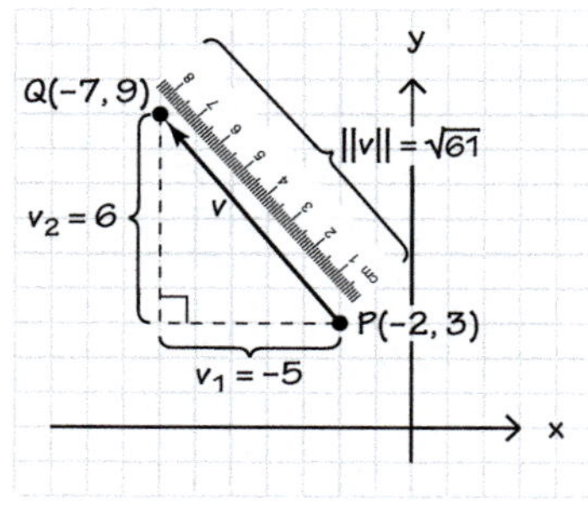

3. (a) The sum of **u** and **v** is

$$\mathbf{u} + \mathbf{v} = \langle 1, 4 \rangle + \langle 3, 2 \rangle$$
$$= \langle 1 + 3, 4 + 2 \rangle$$
$$= \langle 4, 6 \rangle.$$

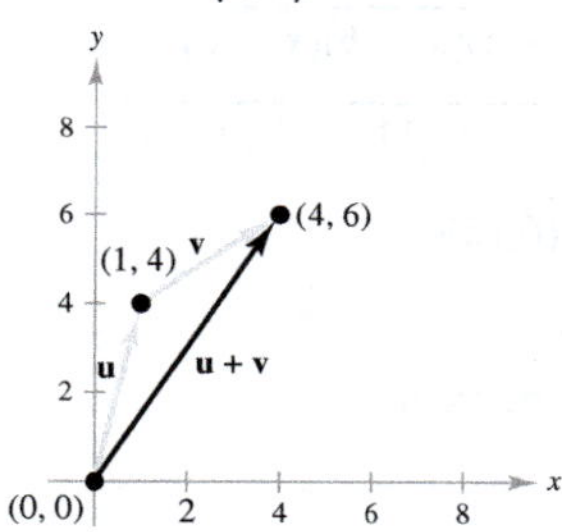

(b) The difference of **u** and **v** is

$$\mathbf{u} + \mathbf{v} = \langle 1, 4 \rangle - \langle 3, 2 \rangle$$
$$= \langle 1 - 3, 4 - 2 \rangle$$
$$= \langle -2, 2 \rangle.$$

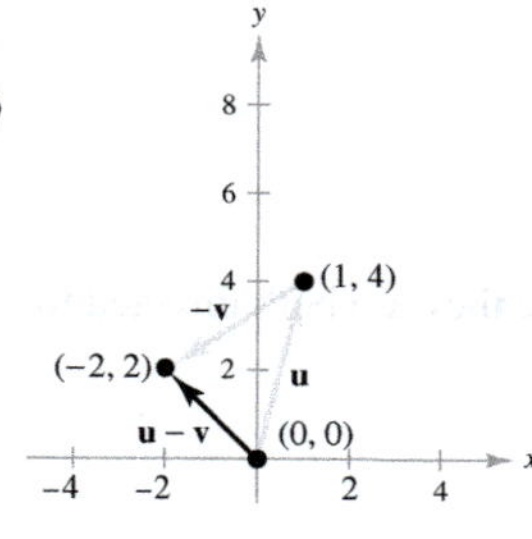

(c) The difference of 2**u** and 3**v** is

$$2\mathbf{u} - 3\mathbf{v} = 2\langle 1, 4 \rangle - 3\langle 3, 2 \rangle$$
$$= \langle 2, 8 \rangle - \langle 9, 6 \rangle$$
$$= \langle 2 - 9, 8 - 6 \rangle$$
$$= \langle -7, 2 \rangle.$$

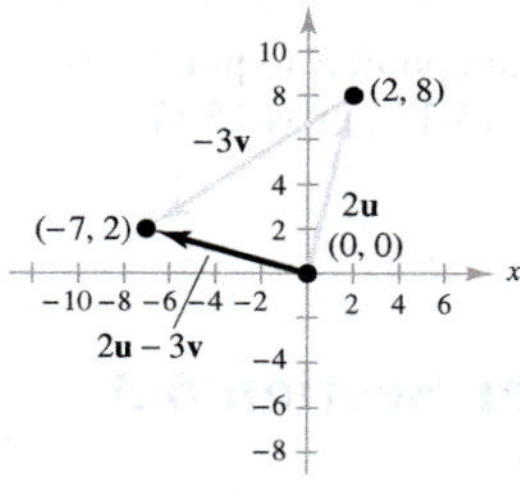

4. $\mathbf{u} = \langle 4, -1 \rangle$ **and** $\mathbf{v} = \langle 3, 2 \rangle$

(a) $\|3\mathbf{u}\| = |3|\|\mathbf{u}\| = |3|\|\langle 4, -1\rangle\| = |3|\sqrt{4^2 + (-1)^2} = |3|\sqrt{17} = 3\sqrt{17}$

(b) $\|-2\mathbf{v}\| = |-2|\|\mathbf{v}\| = |-2|\|\langle 3, 2\rangle\| = |-2|\sqrt{3^2 + 2^2} = |-2|\sqrt{13} = 2\sqrt{13}$

(c) $\|5\mathbf{v}\| = |5|\|\mathbf{v}\| = |5|\|\langle 3, 2\rangle\| = |5|\sqrt{3^2 + 2^2} = |5|\sqrt{13} = 5\sqrt{13}$

5. The unit vector in the direction of $\mathbf{v}$ is

$$\frac{\mathbf{v}}{\|\mathbf{v}\|} = \frac{\langle 6, -1\rangle}{\sqrt{(6)^2 + (-1)^2}} = \frac{1}{\sqrt{37}}\langle 6, -1\rangle = \left\langle \frac{6}{\sqrt{37}}, -\frac{1}{\sqrt{37}} \right\rangle.$$

This vector has a magnitude of 1 because

$$\sqrt{\left(\frac{6}{\sqrt{37}}\right)^2 + \left(-\frac{1}{\sqrt{37}}\right)^2} = \sqrt{\frac{36}{37} + \frac{1}{37}} = \sqrt{\frac{37}{37}} = 1.$$

6. Begin by writing the component form of vector $\mathbf{u}$.

$$\mathbf{u} = \langle -8 - (-2), 3 - 6\rangle = \langle -6, -3\rangle = -6\mathbf{i} - 3\mathbf{j}$$

The result is shown graphically.

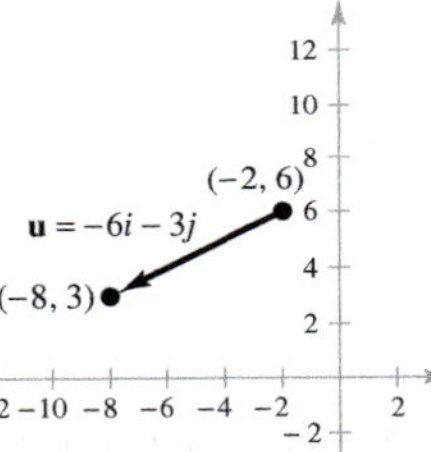

7. Perform the operations in unit vector form.

$$5\mathbf{u} - 2\mathbf{v} = 5(\mathbf{i} - 2\mathbf{j}) - 2(-3\mathbf{i} + 2\mathbf{j}) = 5\mathbf{i} - 10\mathbf{j} + 6\mathbf{i} - 4\mathbf{j} = 11\mathbf{i} - 14\mathbf{j}$$

8. (a) The direction angle is determined from

$$\tan\theta = \frac{b}{a} = \frac{6}{-6} = -1.$$

Because $\mathbf{v} = -6\mathbf{i} + 6\mathbf{j}$ lies in Quadrant II, θ lies in Quadrant II and its reference angle is

$$\theta' = |\arctan(-1)| = \left|-\frac{\pi}{4}\right| = 45°.$$

So, it follows that the direction angle is $\theta = 180° - 45° = 135°$.

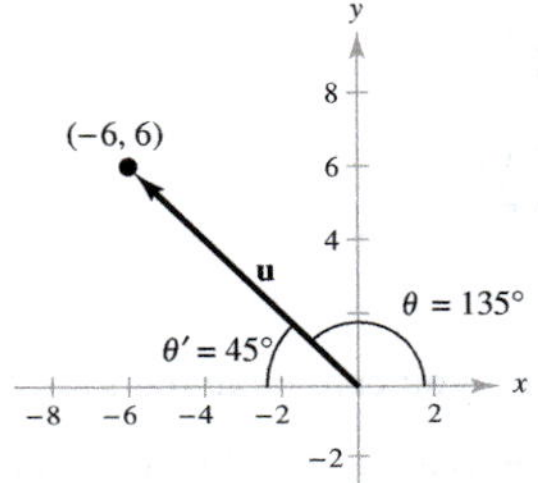

(b) The direction angle is determined from

$$\tan\theta = \frac{b}{a} = \frac{-4}{-7} = \frac{4}{7}.$$

Because $\mathbf{v} = -7\mathbf{i} - 4\mathbf{j}$ lies in Quadrant III, θ lies in Quadrant III and its reference angle is

$$\theta' = \left|\arctan\left(\frac{4}{7}\right)\right| \approx |0.51915 \text{ radian}| \approx 29.74°.$$

So, it follows that the direction angle is $\theta = 180° + 29.74° = 209.74°$.

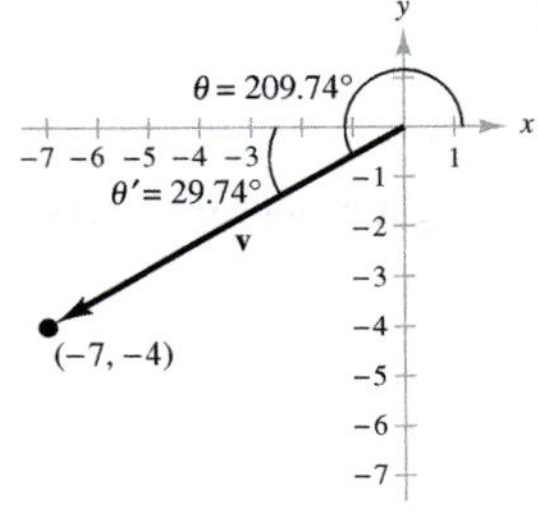

9. The velocity vector v has a magnitude of 100 and a direction angle of $\theta = 195°$.

$$\begin{aligned}\mathbf{v} &= \|\mathbf{v}\|(\cos\theta)\mathbf{i} + \|\mathbf{v}\|(\sin\theta)\mathbf{j}\\ &= 100(\cos 195°)\mathbf{i} + 100(\sin 195°)\mathbf{j}\\ &\approx 100(-0.9659)\mathbf{i} + 100(-0.2588)\mathbf{j}\\ &= -96.59\mathbf{i} - 25.88\mathbf{j}\\ &\approx \langle -96.59, -25.88\rangle\end{aligned}$$

You can check that **v** has a magnitude of 100, as follows.

$$\begin{aligned}\|\mathbf{v}\| &= \sqrt{(-96.59)^2 + (-25.88)^2}\\ &\approx \sqrt{9999.4025}\\ &\approx 100\end{aligned}$$

10.

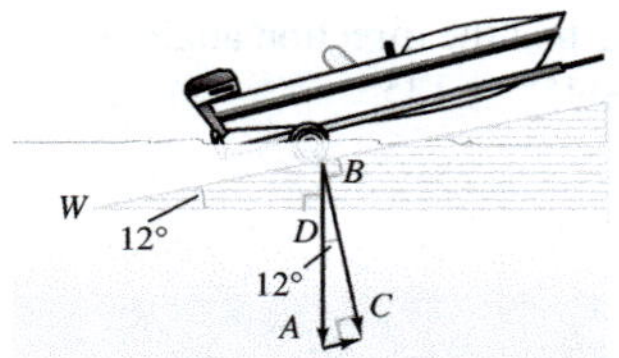

Solution

Based in the figure, you can make the following observations.

$\|\overline{BA}\|$ = force of gravity = combined weight of boat and trailer

$\|\overline{BC}\|$ = force against ramp

$\|\overline{AC}\|$ = force required to move boat up ramp = 500 pounds

By construction, triangles *BWD* and *ABC* are similar. So, angle *ABC* is 12°. In triangle *ABC*, you have

$$\sin 12° = \frac{\|\overline{AC}\|}{\|\overline{BA}\|}$$

$$\sin 12° = \frac{500}{\|\overline{BA}\|}$$

$$\|\overline{BA}\| = \frac{500}{\sin 12°}$$

$$\|\overline{BA}\| \approx 2405.$$

So, the combined weight is approximately 2405 pounds. (In the figure, note that $\overline{AC}$ is parallel to the ramp).

11. 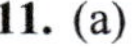 (a)

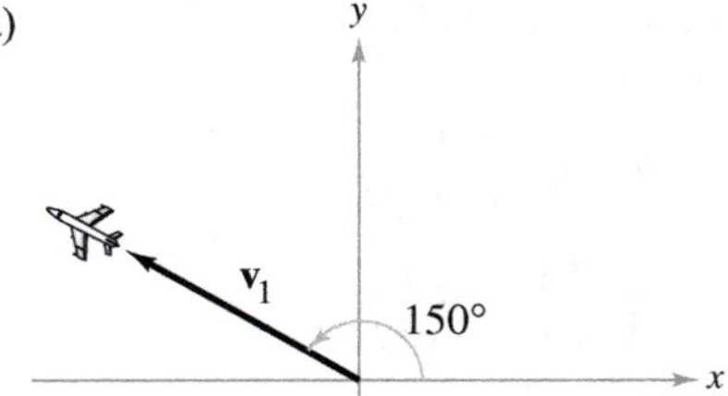

(b)

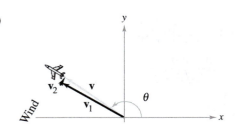

Solution

Using the figure, the velocity of the airplane (alone) is $\mathbf{v}_1 = 450\langle\cos 150°, \sin 150°\rangle = \langle -225\sqrt{3}, 225\rangle$

and the velocity of the wind is $\mathbf{v}_2 = 40\langle\cos 60°, \sin 60°\rangle = \langle 20, 20\sqrt{3}\rangle$.

So, the velocity of the airplane (in the wind) is $\mathbf{v} = \mathbf{v}_1 + \mathbf{v}_2 = \langle -225\sqrt{3} + 20, 225 + 20\sqrt{3}\rangle \approx \langle -369.7, 259.6\rangle$

and the resultant speed of the airplane is $\|\mathbf{v}\| \approx \sqrt{(-369.7)^2 + (259.6)^2} \approx 451.8$ miles per hour.

Finally, given that θ is the direction angle of the flight path, you have $\tan\theta \approx \dfrac{259.6}{-369.7} \approx 0.7022$

which implies that $\theta \approx 180° - 35.1° = 144.9°$.

So, the true direction of the airplane is approximately $270° + (180° - 144.9°) = 305.1°$.

Checkpoints for Section 6.4

1. (a) $\langle 3, 4\rangle \cdot \langle 2, -3\rangle = 3(2) + 4(-3)$
$= 6 - 12$
$= -6$

(b) $\langle -3, -5\rangle \cdot \langle 1, -8\rangle = (-3)(1) + (-5)(-8)$
$= -3 + 40$
$= 37$

(c) $\langle -6, 5\rangle \cdot \langle 5, 6\rangle = (-6)(5) + (5)(6)$
$= -30 + 30$
$= 0$

2. (a) Begin by finding the dot product of **u** and **v**.

$$\begin{aligned}\mathbf{u}\cdot\mathbf{v} &= \langle 3, 4\rangle \cdot \langle -2, 6\rangle \\ &= 3(-2) + 4(6) \\ &= -6 + 24 \\ &= 18\end{aligned}$$

$$\begin{aligned}(\mathbf{u}\cdot\mathbf{v})\mathbf{v} &= 18\langle -2, 6\rangle \\ &= \langle -36, 108\rangle\end{aligned}$$

(b) Begin by finding **u** + **v**.

$$\begin{aligned}\mathbf{u}+\mathbf{v} &= \langle 3, 4\rangle + \langle -2, 6\rangle \\ &= \langle 3 + (-2), 4 + 6\rangle \\ &= \langle 1, 10\rangle\end{aligned}$$

$$\begin{aligned}\mathbf{u}\cdot(\mathbf{u}+\mathbf{v}) &= \langle 3, 4\rangle \cdot \langle 1, 10\rangle \\ &= 3(1) + 4(10) \\ &= 3 + 40 \\ &= 43\end{aligned}$$

(c) Begin by finding the dot product of **v** and **v**.

$$\begin{aligned}\mathbf{v}\cdot\mathbf{v} &= \langle -2, 6\rangle \cdot \langle -2, 6\rangle \\ &= -2(-2) + 6(6) \\ &= 4 + 36 \\ &= 40\end{aligned}$$

Because $\|\mathbf{v}\|^2 = \mathbf{v}\cdot\mathbf{v} = 40$, it follows that $\|\mathbf{v}\| = \sqrt{\mathbf{v}\cdot\mathbf{v}}$
$= \sqrt{40}$
$= 2\sqrt{10}$.

3. $\cos\theta = \dfrac{\mathbf{u}\cdot\mathbf{v}}{\|\mathbf{u}\|\,\|\mathbf{v}\|} = \dfrac{\langle 2, 1\rangle \cdot \langle 1, 3\rangle}{\|\langle 2, 1\rangle\|\,\|\langle 1, 3\rangle\|}$

$$= \frac{2(1) + 1(3)}{\sqrt{2^2 + 1^2}\sqrt{1^2 + 3^2}}$$

$$= \frac{5}{\sqrt{5}\sqrt{10}}$$

$$= \frac{5}{\sqrt{50}}$$

$$= \frac{5}{5\sqrt{2}}$$

$$= \frac{1}{\sqrt{2}}$$

$$= \frac{\sqrt{2}}{2}$$

This implies that the angle between the two vectors is

$$\theta = \cos^{-1}\left(\frac{\sqrt{2}}{2}\right) = \frac{\pi}{4} = 45°.$$

4. Find the dot product of the two vectors.

$$\mathbf{u}\cdot\mathbf{v} = \langle 6, 10\rangle \cdot \left\langle -\frac{1}{3}, \frac{1}{5}\right\rangle$$

$$= 6\left(-\frac{1}{3}\right) + 10\left(\frac{1}{5}\right)$$

$$= -2 + 2$$

$$= 0$$

Because the dot product is 0, the two vectors are orthogonal.

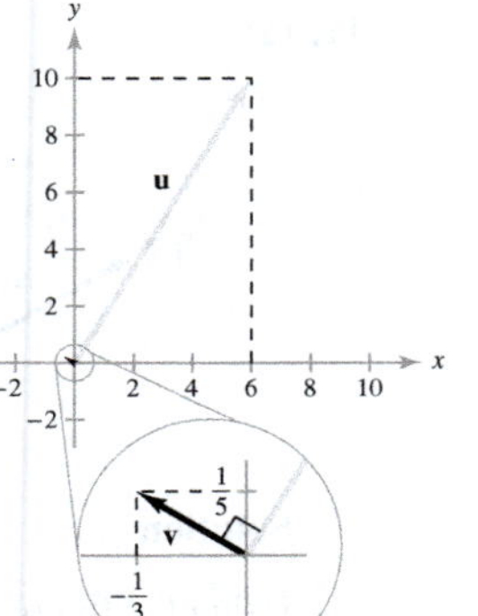

5. The projection of **u** onto **v** is

$$\mathbf{w}_1 = \text{proj}_{\mathbf{v}}\,\mathbf{u} = \left(\frac{\mathbf{u}\cdot\mathbf{v}}{\|\mathbf{v}\|^2}\right)\mathbf{v}$$

$$= \left(\frac{\langle 3, 4\rangle \cdot \langle 8, 2\rangle}{\langle 8, 2\rangle \cdot \langle 8, 2\rangle}\right)\langle 8, 2\rangle$$

$$= \left(\frac{3(8) + 4(2)}{8(8) + 2(2)}\right)\langle 8, 2\rangle$$

$$= \left(\frac{32}{68}\right)\langle 8, 2\rangle$$

$$= \left(\frac{8}{17}\right)\langle 8, 2\rangle$$

$$= \left\langle \frac{64}{17}, \frac{16}{17}\right\rangle$$

$$= \frac{1}{17}\langle 64, 16\rangle.$$

The other component, $\mathbf{w}_2$ is

$$\mathbf{w}_2 = \mathbf{u} - \mathbf{w}_1 = \langle 3, 4\rangle - \left\langle \frac{64}{17}, \frac{16}{17}\right\rangle = \left\langle -\frac{13}{17}, \frac{52}{17}\right\rangle = \frac{1}{17}\langle -13, 52\rangle.$$

So, $\mathbf{u} = \mathbf{w}_1 + \mathbf{w}_2 = \left\langle \frac{64}{17}, \frac{16}{17}\right\rangle + \left\langle -\frac{13}{17}, \frac{52}{17}\right\rangle = \langle 3, 4\rangle.$

6. Solution

Because the force due to gravity is vertical and downward, you can represent the gravitational force by the vector

$\mathbf{F} = -150\mathbf{j}$. Force due to gravity

To find the force required to keep the cart from rolling down the ramp, project $\mathbf{F}$ onto a unit vector $\mathbf{v}$ in the direction of the ramp, as follows.

$$\begin{aligned}\mathbf{v} &= (\cos 15°)\mathbf{i} + (\sin 15°)\mathbf{j}\\ &= 0.966\mathbf{i} + 0.259\mathbf{j} \qquad \text{Unit vector along ramp}\end{aligned}$$

So, the projection of $\mathbf{F}$ onto $\mathbf{v}$ is

$$\begin{aligned}\mathbf{w}_1 &= \text{proj}_{\mathbf{v}}\mathbf{F}\\ &= \left(\frac{\mathbf{F}\cdot\mathbf{v}}{\|\mathbf{v}\|^2}\right)\mathbf{v}\\ &= (\mathbf{F}\cdot\mathbf{v})\mathbf{v} \approx (\langle 0, -150\rangle \cdot \langle 0.966, 0.259\rangle)\mathbf{v}\\ &\approx (-38.85)\mathbf{v}\\ &\approx -37.5\mathbf{i} - 10.1\mathbf{j}.\end{aligned}$$

The magnitude of this force is approximately 38.8. So, a force of approximately 38.8 pounds is required to keep the cart from rolling down the ramp.

7.

Using a projection, you can calculate the work as follows.

$$\begin{aligned}W &= \left\|\text{proj}_{\overrightarrow{PQ}}\mathbf{F}\right\| \left\|\overrightarrow{PQ}\right\|\\ &= (\cos 30°)\|\mathbf{F}\| \left\|\overrightarrow{PQ}\right\|\\ &= \frac{\sqrt{3}}{2}(35)(40)\\ &= 700\sqrt{3}\\ &\approx 1212.436 \qquad \text{foot-pounds}\end{aligned}$$

So, the work done is about 1212 foot-pounds.

Checkpoints for Section 6.5

1. The number $z = 3 - 4i$ is plotted in the complex plane.

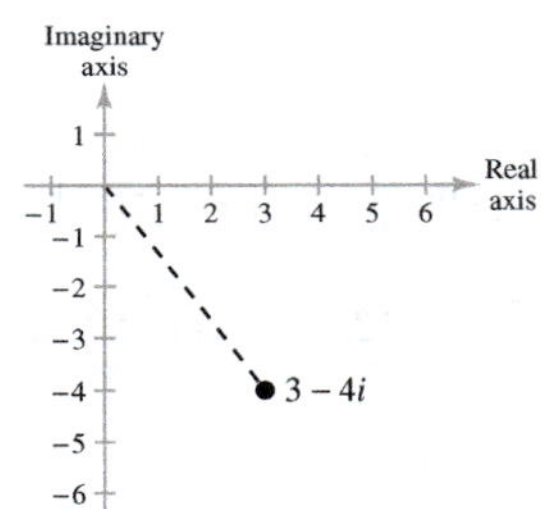

It has an absolute value of

$$\begin{aligned}|z| &= \sqrt{3^2 + (-4)^2}\\ &= \sqrt{9 + 16}\\ &= \sqrt{25}\\ &= 5.\end{aligned}$$

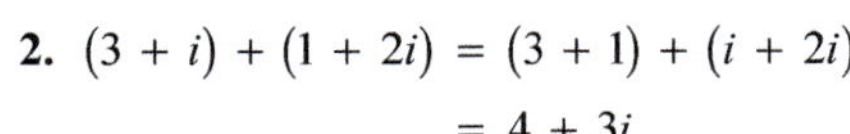

2. $(3 + i) + (1 + 2i) = (3 + 1) + (i + 2i)$
$= 4 + 3i$

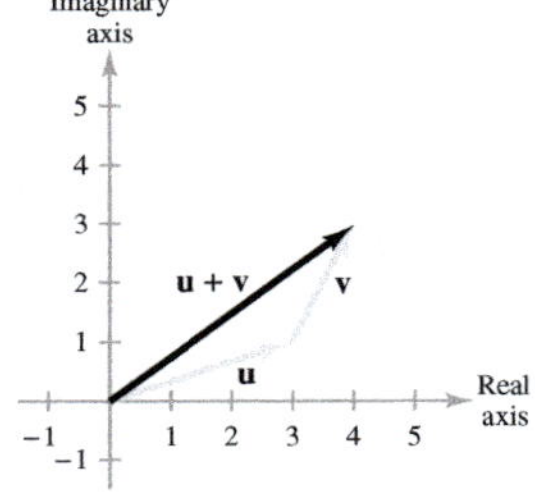

3. $(2 - 4i) - (1 + i) = (2 - 1) + (-4i - i)$
$= 1 - 5i$

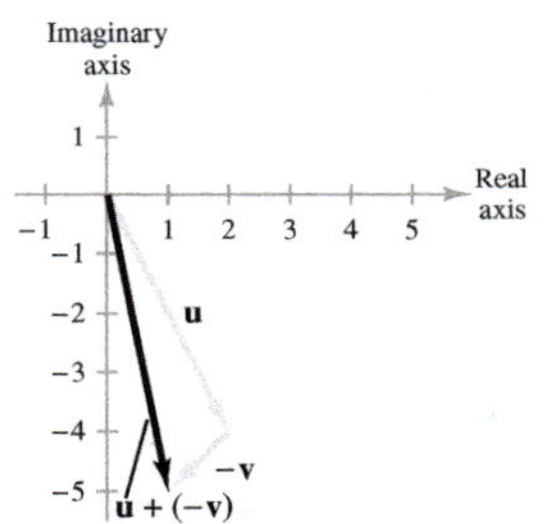

4.

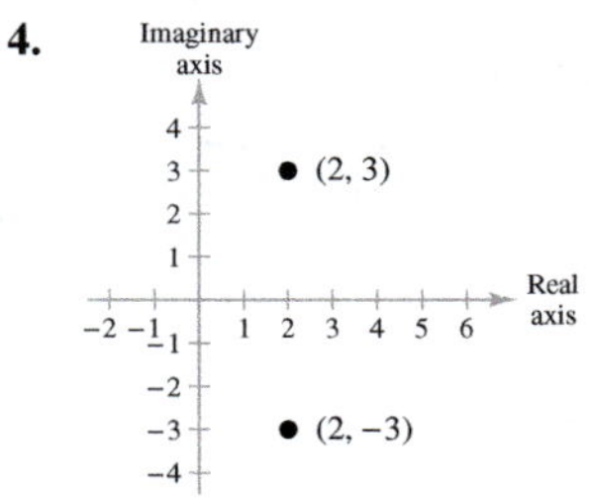

The complex conjugate of $z = 2 - 3i$ is $z = 2 + 3i$.

5. The distance between $5 - 4i$ and $6 + 5i$ is

$$d = \sqrt{(5-6)^2 + (-4-5)^2}$$
$$= \sqrt{(-1)^2 + (-9)^2}$$
$$= \sqrt{82} \approx 9.06 \text{ units}$$

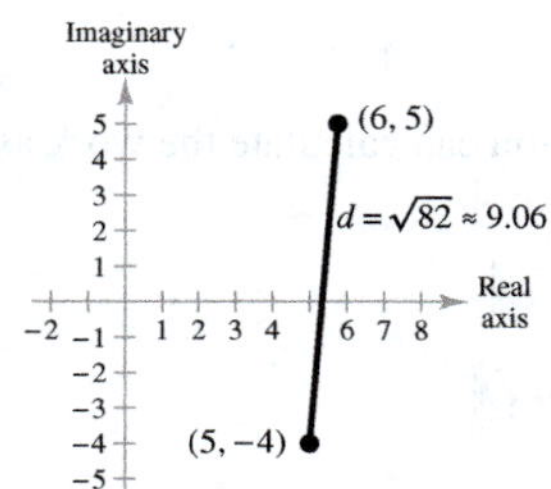

6. The midpoint of the line segment joining the points $2 + i$ and $5 - 5i$ is

$$\text{Midpoint} = \left(\frac{2+5}{2}, \frac{1+(-5)}{2}\right) = \left(\frac{7}{2}, -2\right)$$

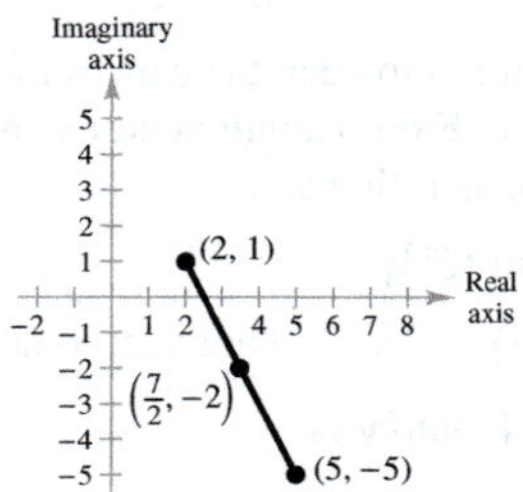

Checkpoints for Section 6.6

1. $z = 6 - 6i$

The modulus of $z = 6 - 6i$ is

$$r = \sqrt{6^2 + (-6)^2}$$
$$= \sqrt{36 + 36} = \sqrt{72} = 6\sqrt{2}$$

and the argument θ is determined from

$$\tan\theta = \frac{b}{a} = \frac{-6}{6} = -1.$$

Because $z = 6 - 6i$ lies in Quadrant IV.

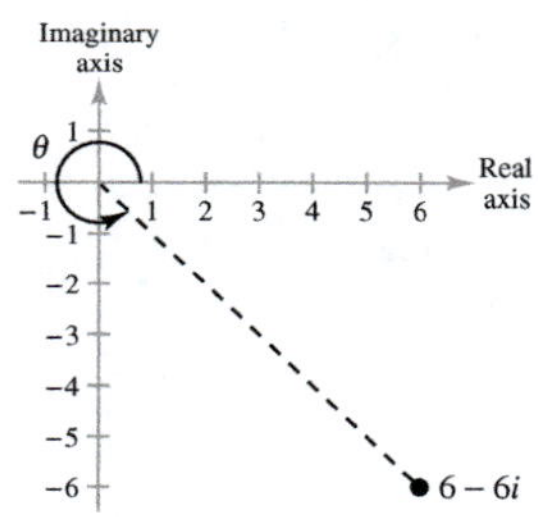

$$\theta = 2\pi - \left|\arctan(-1)\right| = 2\pi - \frac{\pi}{4} = \frac{7\pi}{4}.$$

So, the trigonometric form is

$$z = r(\cos\theta + i\sin\theta) = 6\sqrt{2}\left(\cos\frac{7\pi}{4} + i\sin\frac{7\pi}{4}\right).$$

2. To write $z = 8\left[\cos\left(\frac{2\pi}{3}\right) + i\sin\left(\frac{2\pi}{3}\right)\right]$ in standard form, first find the trigonometric ratios. Because $\cos\left(\frac{2\pi}{3}\right) = \frac{-1}{2}$ and $\sin\left(\frac{2\pi}{3}\right) = \frac{\sqrt{3}}{2}$, you can write

$$z = 8\left[\cos\left(\frac{2\pi}{3}\right) + i\sin\left(\frac{2\pi}{3}\right)\right]$$
$$= 8\left(-\frac{1}{2} + \frac{\sqrt{3}}{2}i\right)$$
$$= -4 + 4\sqrt{3}\,i.$$

3. $$z_1z_2 = 2\left(\cos\frac{5\pi}{6} + i\sin\frac{5\pi}{6}\right)\cdot 5\left(\cos\frac{7\pi}{6} + i\sin\frac{7\pi}{6}\right)$$
$$= (2)(5)\left[\cos\left(\frac{5\pi}{6} + \frac{7\pi}{6}\right) + i\sin\left(\frac{5\pi}{6} + \frac{7\pi}{6}\right)\right]$$
$$= 10(\cos 2\pi + i\sin 2\pi)$$
$$= 10[1 + i(0)]$$
$$= 10$$

4. $z_1z_2 = 3\left(\cos\frac{\pi}{3} + i\sin\frac{\pi}{3}\right)\cdot 4\left(\cos\frac{\pi}{6} + i\sin\frac{\pi}{6}\right)$

$= (3)(4)\left[\cos\left(\frac{\pi}{3} + \frac{\pi}{6}\right) + i\sin\left(\frac{\pi}{3} + \frac{\pi}{6}\right)\right]$

$= 12\left(\cos\frac{\pi}{2} + i\sin\frac{\pi}{2}\right)$

$= 12\left[0 + i(1)\right]$

$= 12i$

You can check this by first converting the complex numbers to their standard forms and then multiplying algebraically.

$$z_1 = 3\left(\cos\frac{\pi}{3} + i\sin\frac{\pi}{3}\right) = 3\left(\frac{1}{2} + \frac{\sqrt{3}}{2}i\right) = \frac{3}{2} + \frac{3\sqrt{3}}{2}i$$

$$z_2 = 4\left(\cos\frac{\pi}{6} + i\sin\frac{\pi}{6}\right) = 4\left(\frac{\sqrt{3}}{2} + \frac{1}{2}i\right) = 2\sqrt{3} + 2i$$

So, $z_1\,z_2 = \left(\frac{3}{2} + \frac{3\sqrt{3}}{2}i\right)\left(2\sqrt{3} + 2i\right)$

$= 3\sqrt{3} + 3i + 9i + 3\sqrt{3}i^2$

$= 3\sqrt{3} + 12i + 3\sqrt{3}(-1)$

$= 3\sqrt{3} + 12i - 3\sqrt{3}$

$= 12i.$

5. $\frac{z_1}{z_2} = \frac{\cos 40° + i\sin 40°}{\cos 10° + i\sin 10°}$

$= \left[\cos(40° - 10°) + i\sin(40° - 10°)\right]$

$= \cos 30° + i\sin 30°$

$= \frac{\sqrt{3}}{2} + \frac{1}{2}i$

6. $z_1 = 2\left(\cos\frac{\pi}{4} + i\sin\frac{\pi}{4}\right)$ and $z_2 = 4\left(\cos\frac{3\pi}{4} + i\sin\frac{3\pi}{4}\right)$

To find z_1z_2 in the complex plane, let

$$\mathbf{u} = 2\left(\cos\frac{\pi}{4} + i\sin\frac{\pi}{4}\right) = \left\langle\sqrt{2}, \sqrt{2}\right\rangle \text{ and } \mathbf{v} = 4\left(\cos\frac{3\pi}{4} + i\sin\frac{3\pi}{4}\right) = \left\langle -2\sqrt{2}, 2\sqrt{2}\right\rangle$$

$$\|\mathbf{u}\| = \sqrt{\left(\sqrt{2}\right)^2 + \left(\sqrt{2}\right)^2} = \sqrt{4} = 2 \text{ and } \|\mathbf{v}\| = \sqrt{\left(-2\sqrt{2}\right)^2 + \left(2\sqrt{2}\right)^2} = \sqrt{16} = 4$$

So, the magnitude of the product vector is $\|\mathbf{u}\|\|\mathbf{v}\| = (2)(4) = 8$. The sum of the direction angles is $\frac{\pi}{4} + \frac{3\pi}{4} = \pi$. The product vector lies on the negative real axis and is represented in vector form as $\langle 0, 8\rangle$. This means that $z_1z_2 = \langle -8, 0\rangle = -8$.

7. The modulus of $z = -1 - i$ is $r = \sqrt{(-1)^2 + (-1)^2} = \sqrt{1 + 1} = \sqrt{2}$ and the argument θ given by $\tan\theta = \dfrac{b}{a} = \dfrac{-1}{-1} = 1.$

Because $z = -1 - i$ lies in Quadrant III, $\theta = \pi + \arctan 1 = \pi + \dfrac{\pi}{4} = \dfrac{5\pi}{4}.$

So, the trigonometric form is $z = -1 - i = \sqrt{2}\left(\cos\dfrac{5\pi}{4} + i\sin\dfrac{5\pi}{4}\right).$

Then, by DeMoivre's Theorem, you have

$$\begin{aligned}(-1-i)^4 &= \left[\sqrt{2}\left(\cos\frac{5\pi}{4} + i\sin\frac{5\pi}{4}\right)\right]^4\\ &= \left(\sqrt{2}\right)^4\left(\cos\left[\frac{4(5\pi)}{4}\right] + i\sin\left[\frac{4(5\pi)}{4}\right]\right)\\ &= 4(\cos 5\pi + i\sin 5\pi)\\ &= 4\left[-1 + i(0)\right]\\ &= -4.\end{aligned}$$

8. First, write 1 in trigonometric form $z = 1(\cos 0 + i\sin 0)$. Then by the nth root formula, with $n = 4$ and $r = 1$, the roots are of the form

$$z_k = \sqrt[4]{1}\left(\cos\frac{0 + 2\pi k}{4} + i\sin\frac{0 + 2\pi k}{4}\right) = (1)\left(\cos\frac{\pi k}{2} + i\sin\frac{\pi k}{2}\right) = \cos\frac{\pi k}{2} + i\sin\frac{\pi k}{2}.$$

So, for $k = 0, 1, 2,$ and 3, the fourth roots are as follows.

$z_0 = \cos 0 + i\sin 0 = 1 + i(0) = 1$

$z_1 = \cos\dfrac{\pi}{2} + i\sin\dfrac{\pi}{2} = 0 + i(1) = i$

$z_2 = \cos\pi + i\sin\pi = -1 + i(0) = -1$

$z_3 = \cos\dfrac{3\pi}{2} + i\sin\dfrac{3\pi}{2} = 0 + i(-1) = -i$

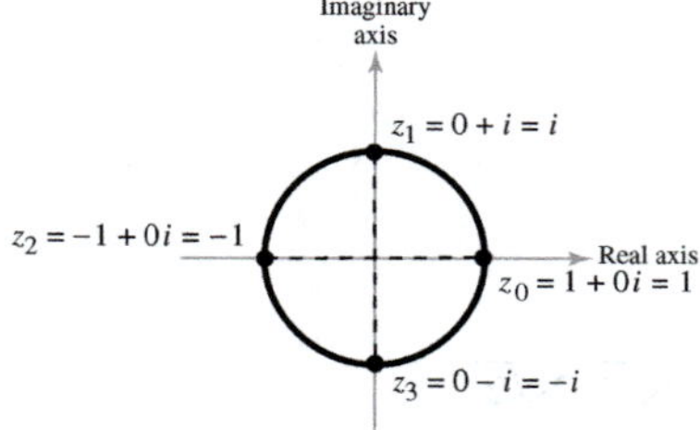

9. The modulus of $z = -6 + 6i$ is $r = \sqrt{(-6)^2 + 6^2} = \sqrt{36 + 36} = \sqrt{72} = 6\sqrt{2}$

and the argument θ is given by $\tan\theta = \dfrac{b}{a} = \dfrac{6}{-6} = -1.$

Because $z = -6 + 6i$ lies in Quadrant II, the trigonometric form of z is $z = -6 + 6i = 6\sqrt{2}(\cos 135° + i\sin 135°)$.

By the formula for nth roots, the cube roots have the form $z_k = \sqrt[3]{6\sqrt{2}}\left[\cos\left(\dfrac{135° + 360°(0)}{3}\right) + i\sin\left(\dfrac{135° + 360°k}{3}\right)\right].$

Finally, for $k = 0, 1,$ and 2, you obtain the roots

$$\begin{aligned}
z_0 &= \sqrt[3]{6\sqrt{2}}\left[\cos\left(\frac{135° + 360°(0)}{3}\right) + i\sin\left(\frac{135° + 360°(0)}{3}\right)\right]. \\
&= \sqrt[3]{6\sqrt{2}}\,(\cos 45° + i\sin 45°) \\
&= \sqrt[3]{6\sqrt{2}}\left(\frac{\sqrt{2}}{2} + \frac{\sqrt{2}}{2}i\right) \\
&= \sqrt[3]{3} + \sqrt[3]{3}\,i \approx 1.4422 + 1.4422i \\
z_1 &= \sqrt[3]{6\sqrt{2}}\left[\cos\left(\frac{135° + 360°(1)}{3}\right) + i\sin\left(\frac{135° + 360°(1)}{3}\right)\right]. \\
&= \sqrt[3]{6\sqrt{2}}(\cos 165° + i\sin 165°) \\
&\approx -1.9701 + 0.5279i \\
z_2 &= \sqrt[3]{6\sqrt{2}}\left[\cos\left(\frac{135° + 360°(2)}{3}\right) + i\sin\left(\frac{135° + 360°(2)}{3}\right)\right]. \\
&= \sqrt[3]{6\sqrt{2}}(\cos 285° + i\sin 285°) \\
&\approx 0.5279 - 1.9701i
\end{aligned}$$

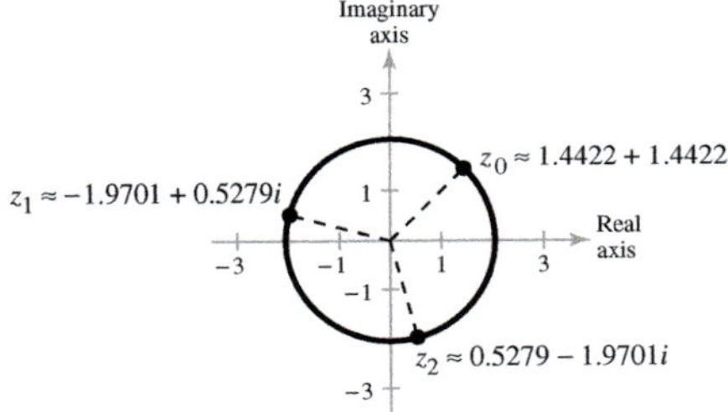

Chapter 7

Checkpoints for Section 7.1

1. $\begin{cases} x - y = 0 & \text{Equation 1} \\ 5x - 3y = 6 & \text{Equation 2} \end{cases}$

Begin by solving for y in Equation 1.

$x - y = 0$

$y = x$

Next substitute this expression for y into Equation 2 and solve the resulting single-variable equation for x.

$5x - 3y = 6$ Write Equation 2.

$5x - 3(x) = 6$ Substitute x for y.

$2x = 6$ Collect like terms.

$x = 3$ Divide each side by 2.

Finally, solve for y by back-substituting $x = 3$ into equation $y = x$, to obtain the corresponding value for y.

$y = x$ Write revised Equation 1.

$y = 3$ Substitute 3 for x.

The solution is the ordered pair $(3, 3)$.

Check

Substitute $(3, 3)$ into Equation 1:

$x - y = 0$ Write Equation 1.

$3 - 3 \stackrel{?}{=} 0$ Substitute for x and y

$0 = 0$ Solution checks in Equation 1.

Substitute $(3, 3)$ into Equation 2:

$5x - 3y = 6$ Write Equation 2.

$5(3) - 3(3) \stackrel{?}{=} 6$ Substitute for x and y.

$15 - 9 \stackrel{?}{=} 6$

$6 = 6$ Solution checks in Equation 2.

Because $(3, 3)$ satisfies both equations in the system, it is a solution of the system of equations.

2. *Verbal Model:* $\boxed{\text{Amount in 6.5\% fund}} + \boxed{\text{Amount in 8.5\% fund}} = \boxed{\text{Total investment}}$

$\boxed{\text{Interest for 6.5\% fund}} + \boxed{\text{Interest for 8.5\% fund}} = \boxed{\text{Total interest}}$

Labels: Amount in 6.5% fund $= x$ (dollars)

Interest for 6.5% fund $= 0.065x$ (dollars)

Amount in 8.5% fund $= y$ (dollars)

Interest for 8.5% fund $= 0.085y$ (dollars)

Total investment $= 25{,}000$ (dollars)

Total interest $= 2600$ (dollars)

System:
$$\begin{cases} x + \quad y = 25{,}000 & \text{Equation 1} \\ 0.065x + 0.085y = 2000 & \text{Equation 2} \end{cases}$$

To begin, it is convenient to multiply each side of Equation 2 by 1000. This eliminates the need to work with decimals.

$1000(.065x + 0.085y) = 1000(2000)$ Multiply each side of Equation 2 by 1000.

$65x + 85y = 2{,}000{,}000$ Revised Equation 2

To solve this system, you can solve for x in Equation 1.

$x = 25{,}000 - y$ Revised Equation 1

Then, substitute this expression for x into revised Equation 2 and solve the resulting equation for y.

$65x + 85y = 2{,}000{,}000$ Write revised Equation 2.

$65(25{,}000 - y) + 85y = 2{,}000{,}000$ Substitute 1 $25000 - y$ for x.

$1{,}625{,}000 - 65y + 85y = 2{,}000{,}000$ Distributive Property

$20y = 375{,}000$ Combine like terms.

$y = 18{,}750$ Divide each side by 20.

Next, back-substitute $y = 18{,}750$ to solve for x.

$x = 25000 - y$ Write revised Equation 1.

$x = 25000 - (18750)$ Substitute 18750 for y.

$x = 6250$ Subtract.

The solution is $(6250, 18{,}750)$. So, \$6250 is invested at 6.5% and \$18,750 is invested at 8.5%.

3.
$$\begin{cases} -2x + y = 5 & \text{Equation 1} \\ x^2 - y + 3x = 1 & \text{Equation 2} \end{cases}$$

Begin by solving for y in Equation 1 to obtain $y = 2x + 5$. Next, substitute this expression for y into Equation 2 and solve for x.

$x^2 - y + 3x = 1$ Write Equation 2.

$x^2 - (2x + 5) + 3x = 1$ Substitute $2x + 5$ for y into Equation 2.

$x^2 - 2x - 5 + 3x = 1$ Simplify.

$x^2 + x - 6 = 0$ Write in standard form.

$(x + 3)(x - 2) = 0$ Factor.

$x + 3 = 0 \Rightarrow x = -3$ Solve for x.

$x - 2 = 0 \Rightarrow x = 2$

Back-substituting these values of x to solve for the corresponding values of y produces the following solutions.

$y = 2x + 5$

$y = 2(-3) + 5 = -1$

$y = 2(2) + 5 = 9$

So, the solutions of the system are $(-3, -1)$ and $(2, 9)$.

4. $\begin{cases} 2x - y = -3 & \text{Equation 1} \\ 2x^2 + 4x - y^2 = 0 & \text{Equation 2} \end{cases}$

Begin by solving for y in Equation 1 to obtain $y = 2x + 3$. Next, substitute this expression for y into Equation 2 and solve for x.

$2x^2 + 4x - y^2 = 0$ Write Equation 2.

$2x^2 + 4x - (2x + 3)^2 = 0$ Substitute $2x + 3$ for y into Equation 2.

$2x^2 + 4x - (4x^2 + 12x + 9) = 0$ Simplify.

$-2x^2 - 8x - 9 = 0$ Combine like terms.

$2x^2 + 8x + 9 = 0$ Write in standard form

$x = \dfrac{-(8) \pm \sqrt{(8)^2 - 4(2)(9)}}{2(2)}$ Use the Quadratic Formula.

$x = \dfrac{-8 \pm \sqrt{-8}}{4}$ Simplify.

Because the discriminant is negative, the equation $2x^2 + 8x + 9 = 0$ has no (real) solution. So, the original system of equations has no (real) solution.

5.

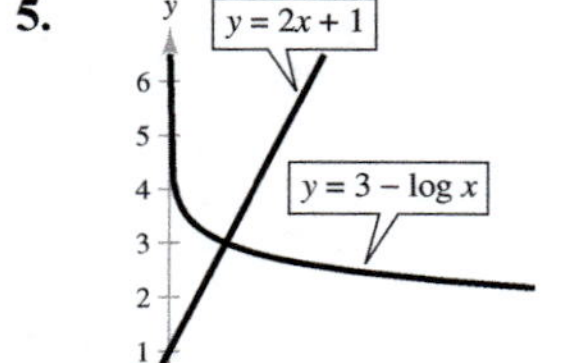

There is only one point of intersection of the graphs of the two equations, and (1, 3) is the solution point.

Check (1, 3) in Equation 1:

$y = 3 - \log x$ Write Equation 1.

$3 \stackrel{?}{=} 3 - \log 1$ Substitute for x and y.

$3 \stackrel{?}{=} 3 - 0$

$3 = 3$ Solution checks in Equation 1.

Check (1, 3) in Equation 2:

$-2x + y = 1$ Write Equation 2.

$-2(1) + 3 \stackrel{?}{=} 1$ Substitute for x and y.

$-2 + 3 \stackrel{?}{=} 1$

$1 = 1$ Solution checks in Equation 2.

6. Algebraic Solution

The total cost of producing x units is

$\boxed{\text{Total cost}} = \boxed{\text{Cost per unit}} \cdot \boxed{\text{Number of units}} + \boxed{\text{Initial cost}}$

$C = 12x + 300{,}000.$ Equation 1

The revenue obtained by selling x units is

$\boxed{\text{Total revenue}} = \boxed{\text{Price per unit}} \cdot \boxed{\text{Number of units}}$

$R = 70x.$ Equation 2

Because the break-even point occurs when $R = C$, you have $C = 70x$, and the system of equations to solve is

$$\begin{cases} C = 12x + 300{,}000 \\ C = 70x \end{cases}.$$

Solve by substitution.

$70x = 12x + 300{,}000$ Substitute $70x$ for C in Equation 1.

$58x = 300{,}000$ Subtract $12x$ from each side.

$x \approx 5172$ Divide each side by 58

So, the company must sell 5172 pairs of shoes to break even.

Graphical Solution

The system of equations to solve is

$$\begin{cases} C = 12x + 300{,}000 \\ C = 70x \end{cases}.$$

Use a graphing utility to graph $y_1 = 12x + 300{,}000$ and $y_2 = 70x$ in the same viewing window.

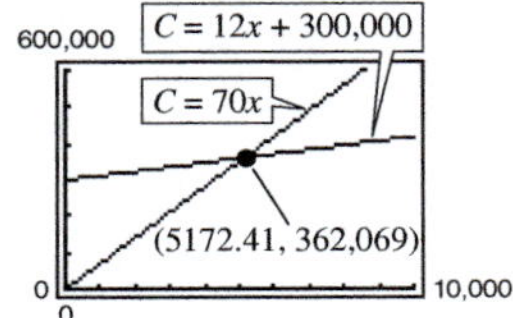

So, the company must sell about 5172 pairs of shoes to break even.

7. Algebraic Solution

Because both equations are already solved for S in terms of x, substitute either expression for S into the other equation and solve for x.

$$\begin{cases} S = 108 - 9.4x & \text{Animated} \\ S = 16 + 9x & \text{Horror} \end{cases}$$

$16 + 9x = 108 - 9.4x$ Substitute for S in Equation 1.

$9.4x + 9x = 108 - 16$ Add $9.4x$ and -16 to each side.

$18.4x = 92$

$x = 5$ Divide each side by 18.4.

So, since $x = 1$ corresponds to week one, the weekly ticket sales in millions of dollars for the two movies will be equal after 5 weeks.

Numerical Solution

You can create a table of values for each model to determine when ticket sales for the two movies will be equal.

Number of weeks x	1	2	3	4	5	6	7
Sales S (Animated)	98.6	89.2	79.8	70.4	61	51.6	70
Sales S (Horror)	25	34	43	52	61	42.2	79

So, from the table, the weekly ticket sales in millions of dollars for the two movies will be equal after 5 weeks.

Checkpoints for Section 7.2

1. Because the coefficients of y differ only in sign, eliminate the y-terms by adding the two equations.

$2x + y = 4$ Write Equation 1.

$2x - y = -1$ Write Equation 2.

$4x = 3$ Add equations.

$x = \frac{3}{4}$ Solve for x.

Solve for y by back-substituting $x = \frac{3}{4}$ into Equation 1.

$$2\left(\tfrac{3}{4}\right) + y = 4$$
$$\tfrac{3}{2} + y = 4$$
$$y = \tfrac{5}{2}$$

The solution is $\left(\frac{3}{4}, \frac{5}{2}\right)$.

Check this in the original system.

$2\left(\frac{3}{4}\right) + \left(\frac{5}{2}\right) \stackrel{?}{=} 4$ Write Equation 1.

$\frac{3}{2} + \frac{5}{2} = 4$ Solution checks in Equation 1. ✓

$2\left(\frac{3}{4}\right) - \left(\frac{5}{2}\right) \stackrel{?}{=} -1$ Write Equation 2.

$\frac{3}{2} - \frac{5}{2} = -1$ Solution checks in Equation 2. ✓

2. To obtain coefficients that differ only in sign, multiply Equation 2 by 3.

$2x + 3y = 17 \Rightarrow 2x + 3y = 17$ Write Equation 1.

$5x - y = 17 \Rightarrow 15x - 3y = 51$ Multiply Equation 2 by 3.

$17x = 68$ Add Equations.

$x = 4$ Solve for x.

Solve for y by back-substituting $x = 4$ into Equation 2.

$5x - y = 17$ Write Equation 2.

$5(4) - y = 17$ Substitute 4 for x.

$20 - y = 17$ Simplify.

$y = 3$ Solve for y.

The solution is $(4, 3)$.

Check this in the original system.

$2(4) + 3(3) \stackrel{?}{=} 17$ Write Equation 1.

$8 + 9 = 17$ Solution Checks in Equation 1. ✓

$5(4) - (3) \stackrel{?}{=} 17$ Write Equation 2.

$20 - 3 = 17$ Solution Checks in Equation 2. ✓

3. Algebraic Solution

You can obtain coefficients that differ only in sign by multiplying Equation 1 by 2 and multiplying Equation 2 by -3.

$$\begin{aligned} 3x + 2y &= 7 \Rightarrow \quad 6x + \ \ 4y = 14 && \text{Multiply Equation 1 by 2.} \\ 2x + 5y &= 1 \Rightarrow \underline{-6x - 15y = -3} && \text{Multiply Equation 2 by } -3. \\ & \qquad\qquad\quad -11y = 11 && \text{Add Equations.} \\ & \qquad\qquad\qquad\quad y = -1 && \text{Solve for } y. \end{aligned}$$

Solve for x by back-substituting $y = 1$ into Equation 1.

$$\begin{aligned} 3x + 2y &= 7 && \text{Write Equation 1.} \\ 3x + 2(-1) &= 7 && \text{Substitute } -1 \text{ for } y \\ 3x - 2 &= 7 \\ 3x &= 9 \\ x &= 3 \end{aligned}$$

The solution is $(3, -1)$.

Graphical Solution

Solve each equation for y and use a graphing utility to graph the equations in the same viewing window.

From the graph, the solution is $(3, -1)$.

Check this in the original system.

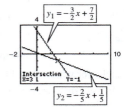

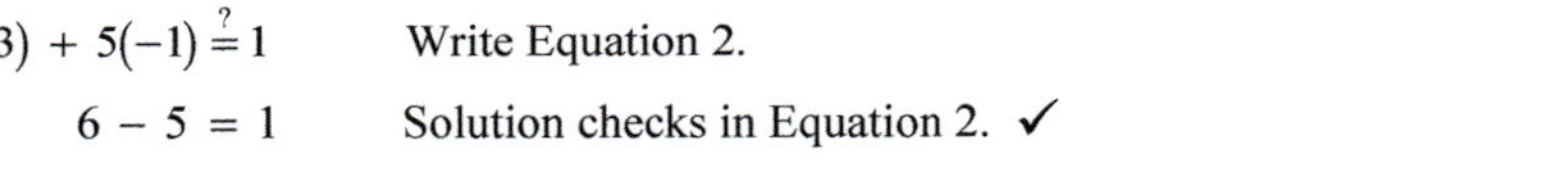

$$\begin{aligned} 3(3) + 2(-1) &\stackrel{?}{=} 7 && \text{Write Equation 1.} \\ 9 - 2 &= 7 && \text{Solution checks in Equation 1. } \checkmark \\ 2(3) + 5(-1) &\stackrel{?}{=} 1 && \text{Write Equation 2.} \\ 6 - 5 &= 1 && \text{Solution checks in Equation 2. } \checkmark \end{aligned}$$

4. Because the coefficients in this system have two decimal places, you can begin by multiplying each equation by 100. This produces a system in which the coefficients are all integers.

$$\begin{aligned} 0.03x + 0.04y &= 0.75 \Rightarrow 3x + 4y = 75 \\ 0.02x + 0.06y &= 0.90 \Rightarrow 2x + 6y = 90 \end{aligned}$$

Now, to obtain coefficients that differ only in sign, multiply Equation 1 by 2 and Equation 2 by –3.

$$\begin{aligned} 3x + 4y &= 75 \Rightarrow \quad 6x + \ \ 8y = \ \ 150 && \text{Multiply Equation 1 by 2.} \\ 2x + 6y &= 90 \Rightarrow \underline{-6x - 18y = -270} && \text{Multiply Equation 2 by } -3. \\ & \qquad\qquad\quad -10y = -120 && \text{Add Equations.} \\ & \qquad\qquad\qquad\quad y = \ \ 12 && \text{Solve for } y. \end{aligned}$$

Back-substitute $y = 12$ into revised Equation 1 to solve for x.

$$\begin{aligned} 3x + 4y &= 75 \\ 3x + 4(12) &= 75 \\ 3x + 48 &= 75 \\ 3x &= 27 \\ x &= 9 \end{aligned}$$

The solution is $(9, 12)$.

Check this in the original system, as follows.

$$\begin{aligned} 0.03(9) + 0.04(12) &\stackrel{?}{=} 0.75 && \text{Write Equation 1.} \\ 0.27 + 0.48 &= 0.75 && \text{Solution Checks in Equation 1. } \checkmark \\ 0.02(9) + 0.06(12) &\stackrel{?}{=} 0.90 && \text{Write Equation 2.} \\ 0.18 + 0.72 &= 0.90 && \text{Solution Checks in Equation 2. } \checkmark \end{aligned}$$

5. First, write each equation in slope-intercept form.

$$\begin{cases} 2x + 3y = 6 & \Rightarrow \; y = \frac{2}{3}x + 2 \\ 4x - 6y = -9 & \Rightarrow \; y = \frac{2}{3}x + \frac{3}{2} \end{cases}$$

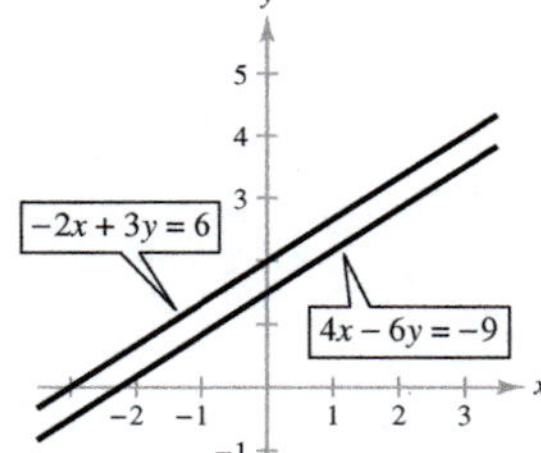

The graph of the system is a pair of parallel lines. The lines have no point of intersection, so the system has no solution. The system is inconsistent.

6. To obtain coefficients that differ only in sign, multiply Equation 1 by 2.

$$\begin{array}{rcll} 6x - 5y = 3 & \Rightarrow & 12x - 10y = 6 & \text{Multiply Equation by 2.} \\ -12x + 10y = 5 & \Rightarrow & \underline{-12x + 10y = 5} & \text{Write Equation 2.} \\ & & 0 = 11 & \text{Add equations.} \end{array}$$

Because there are no values of x and y for which $0 = 11$, you can conclude that the system is inconsistent and has no solution. The graph shows the lines corresponding to the two equations in this system. Note that the two lines are parallel, so they have no point of intersection.

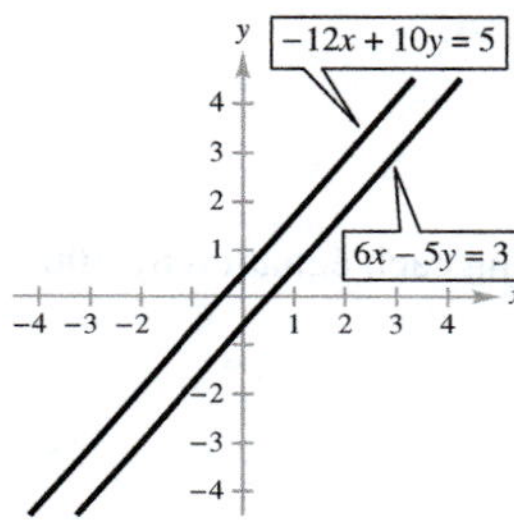

7. To obtain coefficients that differ only in sign, multiply Equation 1 by 8.

$$\begin{array}{rcll} \frac{1}{2}x - \frac{1}{8}y = -\frac{3}{8} & \Rightarrow & 4x - y = -3 & \text{Multiply Equation by 8} \\ -4x + y = 3 & \Rightarrow & \underline{-4x + y = 3} & \text{Write Equation 2.} \\ & & 0 = 0 & \text{Add equations.} \end{array}$$

Because the two equations are equivalent (have the same solution set), the system has infinitely many solutions. The solution set consists of all points (x, y) lying on the line $-4x + y = 3$ as shown. Letting $x = a$, where a is any real number, the solutions of the system are $(a, 4a + 3)$

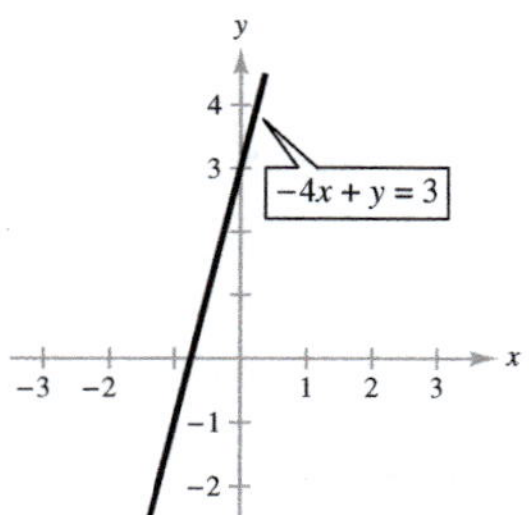

8. The two unknown quantities are the speeds of the wind and of the plane. If r_1 is the speed of the plane and r_2 is the speed of the wind, then

$$r_1 - r_2 = \text{speed of the plane against the wind}$$
$$r_1 + r_2 = \text{speed of the plane with the wind.}$$

Using the formula

$$\text{distance} = (\text{rate})(\text{time})$$

for these two speeds, you obtain the following equations.

$$2000 = (r_1 - r_2)\left(4 + \frac{24}{60}\right)$$

$$2000 = (r_1 - r_2)\left(4 + \frac{6}{60}\right)$$

These two equations simplify as follows.

$$\begin{cases} 5000 = 11r_1 - 11r_2 & \text{Equation 1} \\ 20{,}000 = 41r_1 + 41r_2 & \text{Equation 2} \end{cases}$$

To solve this system by elimination, multiply Equation 1 by 41 and Equation 2 by 11.

$$205{,}000 = 451r_1 - 451r_2 \qquad \text{Multiply Equation 1 by 41.}$$
$$\underline{220{,}000 = 451r_1 + 451r_2} \qquad \text{Multiply Equation 2 by 11.}$$
$$425{,}000 = 902r_1 \qquad \text{Add equations.}$$

So, $r_1 = \dfrac{425{,}000}{902} \approx 471.18$ miles per hour

and $r_2 = \frac{1}{11}(11r_1 - 5000)$

$$r_2 = \frac{1}{11}\left(11 \cdot \frac{425{,}000}{902} - 5000\right) \approx 16.63 \text{ miles per hour.}$$

Check this solution in the original system of equations.

$$2000 \approx (471.18 - 16.63)\left(4 + \frac{24}{60}\right) \checkmark$$

$$2000 \approx (471.18 + 16.63)\left(4 + \frac{6}{60}\right) \checkmark$$

9. Because p is written in terms of x, begin by substituting the value of p given in the supply equation into the demand equation.

$$p = 567 - 0.00002x \qquad \text{Write demand equation.}$$
$$492 + 0.00003x = 567 - 0.00002x \qquad \text{Substitute } 492 + 0.00003x \text{ for } p.$$
$$0.00005x = 75 \qquad \text{Combine like terms.}$$
$$x = 1{,}500{,}000 \qquad \text{Solve for } x.$$

So, the equilibrium point occurs when the demand and supply are each 1.5 million units. Obtain the price that corresponds to this x-value by back-substituting $x = 1{,}500{,}000$ into either of the original equations. For instance, back-substituting into the demand equation produces

$$p = 567 - 0.00002(1{,}500{,}000) = 567 - 30 = \$537.$$

The solution is $(1{,}500{,}000,\ 537)$. Check this by substituting into the demand and supply equations.

$$p = 567 - 0.00002x$$
$$537 = 567 - 0.00002(1{,}500{,}000) \checkmark$$
$$p = 492 + 0.00003x$$
$$537 = 492 + 0.00003(1{,}500{,}000) \checkmark$$

Checkpoints for Section 7.3

1. $\begin{cases} x - y + 5z = 22 \\ \quad y + 3z = 6 \\ \qquad z = 3 \end{cases}$

From Equation 3, you know the value of z. To solve for y, back-substitute $z = 3$ into Equation 2 to obtain the following.

$y + 3z = 6$ Write Equation 2.

$y + 3(3) = 6$ Substitute 3 for z.

$y = -3$ Solve for y.

Then back-substitute $y = -3$ and $z = 3$ into Equation 1 to obtain the following.

$x - y + 5z = 22$ Write Equation 1.

$x - (-3) + 5(3) = 22$ Substitute -3 for y and 3 for z.

$x + 18 = 4$ Combine like terms.

$x = 4$ Solve for x.

The solution is $x = 4$, $y = -3$, and $z = 3$, which can be written as the ordered triple $(4, -3, 3)$. Check this in the original system of equations.

Check

Equation 1: $x - y + 5z = 22$

$4 - (-3) + 5(3) = 22$

$4 + 3 + 15 = 22$ ✓

Equation 2: $y + 3z = 6$

$(-3) + 3(3) \stackrel{?}{=} 6$

$-3 + 9 = 6$ ✓

Equation 3: $z = 3$

$(3) = 3$ ✓

2. $\begin{cases} 2x + y = 3 \\ x + 2y = 3 \end{cases}$ Write Equation 1. Write Equation 2.

$\begin{cases} x + 2y = 3 \\ 2x + y = 3 \end{cases}$ Interchange the two equations in the system.

$\begin{cases} -2x - 4y = -6 \\ 2x + y = 3 \end{cases}$ Multiply the first equation by -2.

$$\begin{array}{r} -2x - 4y = -6 \\ \underline{2x + y = 3} \\ -3y = -3 \\ y = 1 \end{array}$$

Add the multiple of the first equation to the second equation to obtain a new second equation.

$\begin{cases} x + 2y = 3 \\ \quad y = 1 \end{cases}$ New system in row-echelon form.

Now back-substitute $y = 1$ into the first equation in row-echelon form and solve for x.

$x + 2(1) = 3$ Substitute 1 for y.

$x = 1$ Solve for x.

The solution is $x = 1$ and $y = 1$, which can be written as the ordered pair $(1, 1)$.

3. Because the leading coefficient of the first equation is 1, begin by keeping the x in the upper left position and eliminating the other x-terms from the first column.

$$\begin{aligned} -2x - 2y - 2z &= -12 \\ \underline{2x - y + z} &= \underline{3} \\ -3y - z &= -9 \end{aligned}$$

Multiply Equation 1 by -2.
Write Equation 2.
Add revised Equation 1 to Equation 2.

$$\begin{cases} x + y + z = 6 \\ -3y - z = -9 \\ 3x + y - z = 2 \end{cases}$$

Adding -2 times the first equation to the second equation produces a new second equation.

$$\begin{aligned} -3x - 3y - 3z &= -18 \\ \underline{3x + y - z} &= \underline{2} \\ -2y - 4z &= -16 \end{aligned}$$

Multiply Equation 1 by -3.
Write Equation 3.
Add revised Equation 1 to Equation 3.

$$\begin{cases} x + y + z = 6 \\ -3y - z = -9 \\ -2y - 4z = -16 \end{cases}$$

Adding -3 times the first equation to the third equation produces a new third equation.

Now that you have eliminated all but the x in the upper position of the first column, work on the second column.

$$\begin{cases} x + y + z = 6 \\ -3y - z = -9 \\ -y - 2z = -8 \end{cases}$$

Multiplying the third equation by 2, produces a new third equation.

$$\begin{aligned} -3y - z &= -9 \\ \underline{3y + 6z} &= \underline{24} \\ 5z &= 15 \end{aligned}$$

Write Equation 2.
Multiply Equation 3 by -3.
Add equations.

$$\begin{cases} x + y + z = 6 \\ -3y - z = -9 \\ 5z = 15 \end{cases}$$

Adding the second equation to -3 times the third equation produces a new third equation.

$$\begin{cases} x + y + z = 6 \\ y + \frac{1}{3}z = 3 \\ 5z = 15 \end{cases}$$

Multiplying the second equation by $-\frac{1}{3}$ produces a new second equation.

$$\begin{cases} x + y + z = 6 \\ y + \frac{1}{3}z = 3 \\ z = 3 \end{cases}$$

Multiplying the third equation by $\frac{1}{5}$ produces a new third equation.

To solve for y, back-substitute $z = 3$ into Equation 2 to obtain the following.

$$\begin{aligned} y + \tfrac{1}{3}(3) &= 3 \\ y &= 2. \end{aligned}$$

Then back-substitute $y = 2$ and $z = 3$ into Equation 1 to obtain the following.

$$\begin{aligned} x + (2) + (3) &= 6 \\ x &= 1 \end{aligned}$$

The solution is $x = 1$, $y = 2$, and $z = 3$, which can be written as $(1, 2, 3)$.

4. $\begin{cases} x + y - 2z = 3 \\ 3x - 2y + 4z = 1 \\ 2x - 3y + 6z = 8 \end{cases}$

$\begin{cases} x + y - 2z = 3 \\ -5y + 10z = -8 \\ 2x - 3y + 6z = 8 \end{cases}$ Adding -3 times the first equation to the second equation produces a new second equation.

$\begin{cases} x + y - 2z = 3 \\ -5y + 10z = -8 \\ -5y + 10z = 2 \end{cases}$ Adding -2 times the first equation to the third equation produces a new third equation.

$\begin{cases} x + y - 2z = 3 \\ -5y + 10z = -8 \\ 0 = 10 \end{cases}$ Adding -1 times the second equation to the third equation produces a new third equation.

Because $0 = 10$ is a false statement, this is an inconsistent system and has no solution. Moreover, because this system is equivalent to the original system, the original system has no solution.

5. $\begin{cases} x + 2y - 7z = -4 \\ 2x + 3y + z = 5 \\ 3x + 7y - 36z = -25 \end{cases}$

$\begin{cases} x + 2y - 7z = -4 \\ -y + 15z = 13 \\ 3x + 7y - 36z = -25 \end{cases}$ Adding -2 times the first equation to the second equation produces a new second equation.

$\begin{cases} x + 2y - 7z = -4 \\ -y + 15z = 13 \\ y - 15z = -13 \end{cases}$ Adding -3 times the first equation to the third equation produces a new third equation.

$\begin{cases} x + 2y - 7z = -4 \\ -y + 15z = 13 \\ 0 = 0 \end{cases}$ Adding the second equation to the third equation to produces a new third equation.

This result means that Equation 3 depends on Equations 1 and 2 in the sense that it gives no additional information about the variables. Because $0 = 0$ is a true statement, this system has infinitely many solutions. However, it is incorrect to say that the solution is "infinite." You must also specify the correct form of the solution. So, the original system is equivalent to the system.

$\begin{cases} x + 2y - 7z = -4 \\ -y + 15z = 13. \end{cases}$

In the second equation, solve for y in terms of z to obtain the following.

$$-y + 15z = 13$$
$$-y = -15z + 13$$
$$y = 15z - 13$$

Back-substituting in the first equation produces the following.

$$x + 2y - 7z = -4$$
$$x + 2(15z - 13) - 7z = -4$$
$$x + 30z - 26 - 7z = -4$$
$$x = -23z + 22$$

Finally, letting $z = a$ where a is a real number, the solutions of the given system are all of the form $x = -23a + 22$, $y = 15a - 13$, and $z = a$. So, every ordered triple of the form $(-23a + 22, 15a - 13, a)$ is a solution of the system.

6. $\begin{cases} x - y + 4z = 3 \\ 4x \qquad -z = 0 \end{cases}$

$\begin{cases} x - y + \quad z = \quad 3 \\ \quad 4y - 17z = -12 \end{cases}$ Adding -4 times the first equation to the second equation produces a new second equation.

$\begin{cases} x - y + \quad z = \quad 3 \\ \quad y - \frac{17}{4}z = -3 \end{cases}$ Multiplying the second equation by $\frac{1}{4}$ produces a new second equation.

Solve for y in terms of z to obtain the following.

$$y - \tfrac{17}{4}z = -3$$

$$y = \tfrac{17}{4}z - 3$$

Solve for x by back-substituting $y = \frac{17}{4}z - 3$ into Equation 1.

$$x - y + 4z = 3$$

$$x - \left(\tfrac{17}{4}z - 3\right) + 4z = 3$$

$$x - \tfrac{17}{4}z + 3 + 4z = 3$$

$$x = \tfrac{1}{4}z$$

Finally, by letting $z = a$, where a is a real number, you have the solution

$x = \frac{1}{4}a$, $y = \frac{17}{4}a - 3$, and $z = a$.

So, every ordered triple of the form $\left(\frac{1}{4}a, \frac{17}{4}a - 3, a\right)$ is a solution of the system. Because the original system had three variables and only two equations, the system cannot have a unique solution and has infinitely many solutions.

7. By substituting the three values of t and s into the position equation, you can obtain three linear equations in a, v_0 and s_0.

When $t = 1$: $\frac{1}{2}a(1)^2 + v_0(1) + s_0 = 104 \Rightarrow a + 2v_0 + 2s_0 = 208$

When $t = 2$: $\frac{1}{2}a(2)^2 + v_0(2) + s_0 = 76 \Rightarrow 2a + 2v_0 + s_0 - 76$

When $t = 3$: $\frac{1}{2}a(3)^2 + v_0(3) + s_0 = 16 \Rightarrow 9a + 6v_0 + 2s_0 = 32$

This produces the following system of linear equation.

$$\begin{cases} a + 2v_0 + 2s_0 = 208 \\ 2a + 2v_0 + s_0 = 76 \\ 9a + 6v_0 + 2s_0 = 32 \end{cases}$$

Now solve the system using Gaussian Elimination.

$$\begin{cases} a + 2v_0 + 2s_0 = 208 \\ -2v_0 - 3s_0 = -340 \\ 9a + 6v_0 + 2s_0 = 32 \end{cases}$$

Adding -2 times the first equation to the second equation produces a new second equation.

$$\begin{cases} a + 2v_0 + 2s_0 = 208 \\ -2v_0 - 3s_0 = -340 \\ -12v_0 - 16s_0 = -1840 \end{cases}$$

Adding -9 times the first equation to the third equation produces a new third equation.

$$\begin{cases} a + 2v_0 + 2s_0 = 208 \\ -2v_0 - 3s_0 = -340 \\ 2s_0 = 200 \end{cases}$$

Adding -6 times the second equation to the third equation produces a new third equation.

$$\begin{cases} a + 2v_0 + 2s_0 = 208 \\ v_0 + \frac{3}{2}s_0 = 170 \\ s_0 = 100 \end{cases}$$

Multiplying the second equation by $-\frac{1}{2}$ produces a new second equation and multiplying the third equation by $\frac{1}{2}$ produces a new third equation.

So, $s_0 = 100$.

Find v_0 by back-substituting $s_0 = 100$ into Equation 2.

$$v_0 + \tfrac{3}{2}(100) = 170$$
$$v_0 = 20$$

Find a by back-substituting $s_0 = 100$ and $v_0 = 20$ into Equation 1.

$$a + 2(20) + 2(100) = 208$$
$$a = -32$$

So, the solution of this system is $a = -32$, $v_0 = 20$, and $s_0 = 100$, which can be written as $(-32, 20, 100)$.

This results in a position equation of $s = \frac{1}{2}(-32)t^2 + 20t + 100$

$$= -16t^2 + 20t + 100$$

and implies that the object was thrown upward at a velocity of 20 feet per second from a height of 100 feet.

8. Because the graph of $y = ax^2 + bx + c$ passes through the points $(0, 0)$, $(3, -3)$, and $(6, 0)$, you can write the following.

When $x = 0$, $y = 0$: $a(0)^2 + b(0) + c = 0$

When $x = 3$, $y = -3$: $a(3)^2 + b(3) + c = -3$

When $x = 6$, $y = 0$: $a(6)^2 + b(6) + c = 0$

This produces the following system of linear equations.

$$\begin{cases} \quad\quad\quad\; c = 0 & \text{Equation 1} \\ 9a + 3b + c = -3 & \text{Equation 2} \\ 36a + 6b + c = 0 & \text{Equation 3} \end{cases}$$

You can reorder these equations as shown.

$$\begin{cases} 36a + 6b + c = 0 \\ 9a + 3b + c = -3 \\ c = 0 \end{cases}$$

$$\begin{cases} 36a + 6b + c = 0 \\ -6b - 3c = 12 \\ c = 0 \end{cases}$$

Adding -4 times the second equation to the first equation produces a new second equation.

$$\begin{cases} a + \frac{1}{6}b + \frac{1}{36}c = 0 \\ b + \frac{1}{2}c = -2 \\ c = 0 \end{cases}$$

Multiplying the first equation by $\frac{1}{36}$ produces a new first equation and multiplying the second equation by $-\frac{1}{6}$ produces a new second equation.

So, $c = 0$,

$$b + \tfrac{1}{2}(0) = -2$$
$$b = -2,$$

and $a + \frac{1}{6}(-2) + \frac{1}{36}(0) = 0$

$$a = \tfrac{1}{3}.$$

The solution of this system is $a = \frac{1}{3}$, $b = -2$, and $c = 0$.

So, the equation of the parabola is $y = \frac{1}{3}x^2 - 2x$.

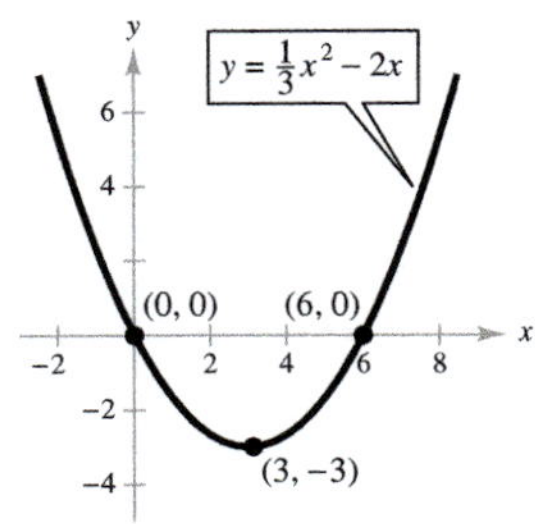

Checkpoints for Section 7.4

1. The expression is proper, so you should begin by factoring the denominator. Because

$$2x^2 - x - 1 = (2x + 1)(x - 1)$$

you should include one partial fraction with a constant numerator for each linear factor of the denominator. Write the form of the decomposition as follows.

$$\frac{x + 5}{2x^2 - x - 1} = \frac{A}{2x + 1} + \frac{B}{x - 1}$$

Multiplying each side of this equation by the least common denominator, $(2x + 1)(x - 1)$, leads to the **basic equation**

$$x + 5 = A(x - 1) + B(2x + 1).$$

Because this equation is true for all x, substitute any *convenient* values of x that will help determine the constants A and B. Values of x that are especially convenient are those that make the factors $x - 1$ and $2x + 1$ equal to zero. For instance, to solve for B, let $x = 1$. Then

$$1 + 5 = A(1 - 1) + B\left[2(1) + 1\right] \quad \text{Substitute 1 for } x$$

$$6 = A(0) + B(3)$$

$$6 = 3B$$

$$2 = B.$$

To solve for A, let $x = -\frac{1}{2}$ and then

$$-\frac{1}{2} + 5 = A\left(-\frac{1}{2} - 1\right) + B\left[2\left(-\frac{1}{2}\right) + 1\right] \quad \text{Substitute } -\frac{1}{2} \text{ for } x$$

$$\frac{9}{2} = A\left(-\frac{3}{2}\right) + B(0)$$

$$\frac{9}{2} = -\frac{3}{2}A$$

$$-3 = A.$$

So, the partial fraction decomposition is

$$\frac{x + 5}{2x^2 - x - 1} = -\frac{3}{2x + 1} + \frac{2}{x - 1}.$$

Check this result by combining the two partial fractions on the right side of the equation, or by using your graphing utility.

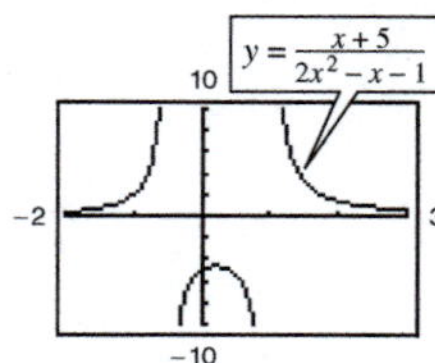

2. This rational expression is improper, so you should begin by dividing the numerator by the denominator.

$$\frac{x^4 + x^3 + x + 4}{x^3 + x^2} \Rightarrow \begin{array}{r} x \\ x^3 + x^2 \overline{\big) x^4 + x^3 + 0x^2 + x + 4} \\ \underline{x^4 + x^3 } \\ x + 4 \end{array}$$

So, $\dfrac{x^4 + x^3 + x + 4}{x^3 + x^2} = x + \dfrac{x + 4}{x^3 + x^2}$.

Because the denominator of the remainder factors as $x^3 + x^2 = x^2(x + 1)$, you should include one partial fraction with a constant numerator for each power of x and $x + 1$, and write the form of the decomposition as follows.

$$\frac{x + 4}{x^3 + x^2} = \frac{A}{x} + \frac{B}{x^2} + \frac{C}{x + 1}$$

Multiplying each side by the LCD, $x^2(x + 1)$, leads to the basic equation

$$x + 4 - Ax(x + 1) + B(x + 1) + Cx^2.$$

Letting $x = -1$ eliminates the A- and B-terms and yields the following.

$$\begin{aligned} -1 + 4 &= A(-1)(-1 + 1) + B(-1 + 1) + C(-1)^2 \\ 3 &= 0 + 0 + C \\ 3 &= C \end{aligned}$$

Letting $x = 0$, eliminates the A- and C-terms.

$$\begin{aligned} 0 + 4 &= A(0)(0 + 1) + B(0 + 1) + C(0)^2 \\ 4 &= 0 + B + 0 \\ 4 &= B \end{aligned}$$

At this point, you have exhausted the most convenient values of x, so to find the value of A, use any other value of x along with the known values of B and C.

So, using $x = 1$, $B = 4$ and $C = 3$,

$$\begin{aligned} 1 + 4 &= A(1)(1 + 1) + 4(1 + 1) + 3(1)^2 \\ 5 &= 2A + 8 + 3 \\ -6 &= 2A \\ -3 &= A. \end{aligned}$$

So, the partial fraction decomposition is

$$\frac{x^4 + x^3 + x + 4}{x^3 + x^2} = x - \frac{3}{x} + \frac{4}{x^2} + \frac{3}{x + 1}.$$

3. This expression is proper, so begin by factoring the denominator. Because the denominator factors as

$x^3 + x = x(x^2 + 1)$

you should include one partial fraction with a constant numerator and one partial fraction with a linear numerator, and write the form of the decomposition as follows.

$$\frac{2x^2 - 5}{x^3 + x} = \frac{A}{x} + \frac{Bx + C}{x^2 + 1}$$

Multiplying each side by the LCD, $x(x^2 + 1)$ yields the basic equation

$$2x^2 - 5 = A(x^2 + 1) + (Bx + C)x$$

Expanding this basic equation and collecting like terms produces

$$\begin{aligned} 2x^2 - 5 &= Ax^2 + A + Bx^2 + Cx \\ &= (A + B)x^2 + Cx + A. \qquad \text{Polynomial form} \end{aligned}$$

Finally, because two polynomials are equal if and only if the coefficients of like terms are equal, equate the coefficients of like terms on opposite sides of the equation.

$$2x^2 + 0x - 5 = (A + B)x^2 + Cx + A$$

Now write the following system of linear equations.

$$\begin{cases} A + B = 2 & \text{Equation 1} \\ C = 0 & \text{Equation 2} \\ A = -5 & \text{Equation 3} \end{cases}$$

From this system, you can see that $A = -5$ and $C = 0$.

Moreover, back-substituting $A = -5$ into Equation 1 yields $-5 + B = 2 \Rightarrow B = 7$.

So, the partial fraction decomposition is $\dfrac{2x^2 - 5}{x^3 + x} = -\dfrac{5}{x} + \dfrac{7x}{x^2 + 1}$.

4. Include one partial fraction with a linear numerator for each power of $(x^2 + 4)$.

$$\frac{x^3 + 3x^2 - 2x + 7}{(x^2 + 4)^2} = \frac{Ax + B}{x^2 + 4} + \frac{Cx + D}{(x^2 + 4)^2} \qquad \text{Write form of decomposition.}$$

Multiplying each side by the LCD, $(x^2 + 4)^2$, yields the basic equation

$$\begin{aligned} x^3 + 3x^2 - 2x + 7 &= (Ax + B)(x^2 + 4) + Cx + D \qquad \text{Basic equation} \\ &= Ax^3 + 4Ax + Bx^2 + 4B + Cx + D \\ &= Ax^3 + Bx^2 + (4A + C)x + (4B + D). \qquad \text{Polynomial form} \end{aligned}$$

Equating coefficients of like terms on opposite sides of the equation

$$x^3 + 3x^2 - 2x + 7 = Ax^3 + Bx^2 + (4A + C)x + (4B + D)$$

produces the following system of linear equations.

$$\begin{cases} A = 1 & \text{Equation 1} \\ B = 3 & \text{Equation 2} \\ 4A + C = -2 & \text{Equation 3} \\ 4B + D = 7 & \text{Equation 4} \end{cases}$$

Use the values $A = 1$ and $B = 3$ to obtain the following.

$4(1) + C = -2$ Substitute 1 for A in Equation 3.

$C = -6$

$4(3) + D = 7$ Substitute 3 for B in Equation 4.

$D = -5$

So, using $A = 1$, $B = 3$, $C = -6$, and $D = -5$

The partial fraction decomposition is $\dfrac{x^3 + 3x^2 - 2x + 7}{(x^2 + 4)^2} = \dfrac{x + 3}{x^2 + 4} - \dfrac{6x + 5}{(x^2 + 4)^2}$.

Check this result by combining the two partial fractions on the right side of the equation, or by using your graphing utility.

5. Include one partial fraction with a constant numerator for each power of x and one partial fraction with a linear numerator for each power of $(x^2 + 2)$.

$$\frac{4x - 8}{x^2(x^2 + 2)^2} = \frac{A}{x} + \frac{B}{x^2} + \frac{Cx + D}{x^2 + 2} + \frac{Ex + F}{(x^2 + 2)^2} \qquad \text{Write form of decomposition.}$$

Multiplying each side by the LCD, $x^2(x^2 + 2)^2$, yields the basic equation

$$\begin{aligned}4x - 8 &= Ax(x^2 + 2)^2 B(x^2 + 2)^2 + (Cx + D)x^2(x^2 + 2) + (Ex + F)x^2\\ &= Ax(x^4 + 4x^2 + 4) + B(x^4 + 4x^2 + 4) + (Cx + D)(x^4 + 2x^2) + (Ex + F)x^2\\ &= Ax^5 + 4Ax^3 + 4Ax + Bx^4 + 4Bx^2 + 4B + Cx^5 + 2Cx^3 + Dx^4 + 2Dx^2 + Ex^3 + Fx^2\\ &= (A + C)x^5 + (B + D)x^4 + (4A + 2C + E)x^3 + (4B + 2D + F)x^2 + (4A)x + 4B\end{aligned}$$

Equating coefficients yields this system of linear equations.

$$\begin{cases} A + C = 0 & \text{Equation 1}\\ B + D = 0 & \text{Equation 2}\\ 4A + 2C + E = 0 & \text{Equation 3}\\ 4B + 2D + F = 0 & \text{Equation 4}\\ 4A = 4 & \text{Equation 5}\\ 4B = -8 & \text{Equation 6}\end{cases}$$

So, from Equations 5 and 6, $A = 1$ and $B = -2$.

Then back-substituting into Equations 1 and 2, $1 + C = 0 \Rightarrow C = -1$ and $-2 + D = 0 \Rightarrow D = 2$.

Using these values and Equations 3 and 4, you have

$4(1) + 2(-1) + E = 0 \Rightarrow E = -2$ and $4(-2) + 2(2) + F = 0 \Rightarrow F = 4$.

So, $A = 1$, $B = -2$, $C = -1$, $D = 2$, $E = -2$, and $F = 4$.

The partial fraction decomposition is

$$\frac{4x - 8}{x^2(x^2 + 2)^2} = \frac{1}{x} - \frac{2}{x^2} - \frac{x - 2}{x^2 + 2} - \frac{2x - 4}{(x^2 + 2)^2}.$$

Checkpoints for Section 7.5

1. Begin by graphing the corresponding equation $(x + 2)^2 + (y - 2)^2 = 16$, which is a circle, with center $(-2, 2)$ and a radius of 4 units as shown.

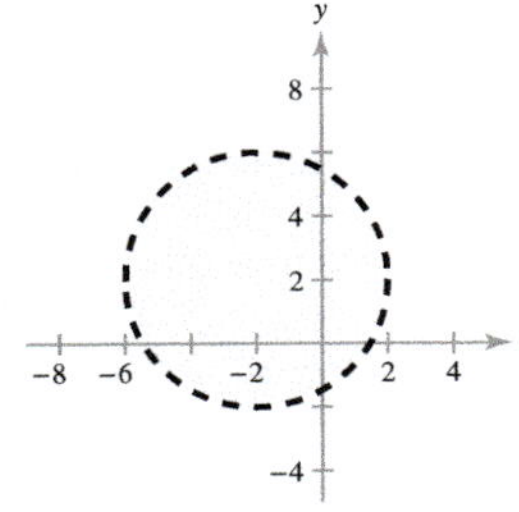

Test a point inside the circle such as $(-2, 2)$ and a point outside the circle such as $(4, 2)$.

The points that satisfy the inequality are those lying inside the circle but not on the circle.

$(-2, 2)$: $(x + 2)^2 + (y - 2)^2 \overset{?}{<} 16$

$(-2 + 2)^2 + (2 - 2)^2 \overset{?}{<} 16$

$0 < 16$

$(4, 2)$: $(x + 2)^2 + (y - 2)^2 < 16$

$(4 + 2)^2 + (2 - 2)^2 \overset{?}{<} 16$

$36 \not< 16$

$(-2, 2)$ is a solution. $(4, 2)$ is not a solution.

2. The graph of the corresponding equation $x = 3$ is a vertical line. The points that satisfy the inequality $x \geq 3$ are those lying to the right of (or on) this line.

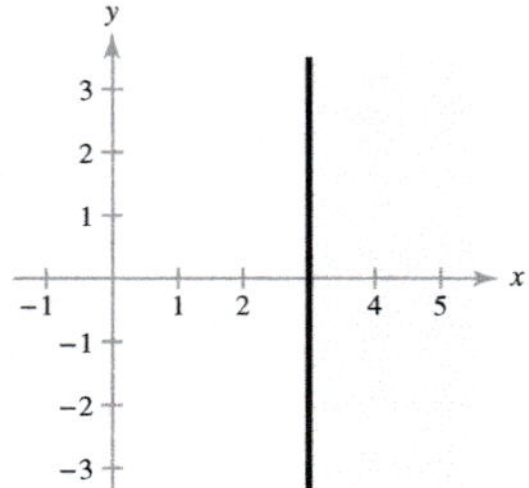

3. The graph of the corresponding equation $x + y = -2$ is a line as shown. Because the origin $(0, 0)$ satisfies the inequality, the graph consists of the half-plane lying above the line.

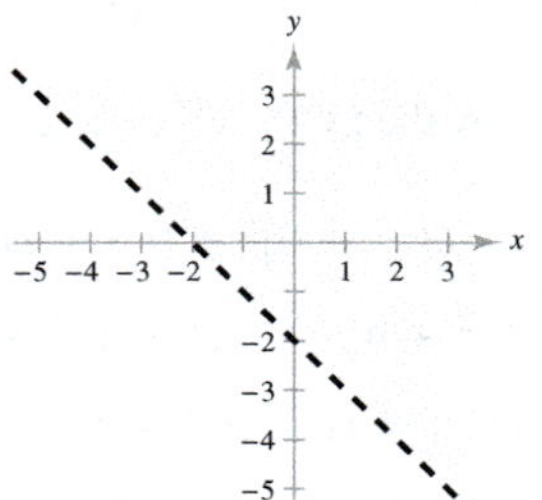

4. The graphs of each of these inequalities are shown independently.

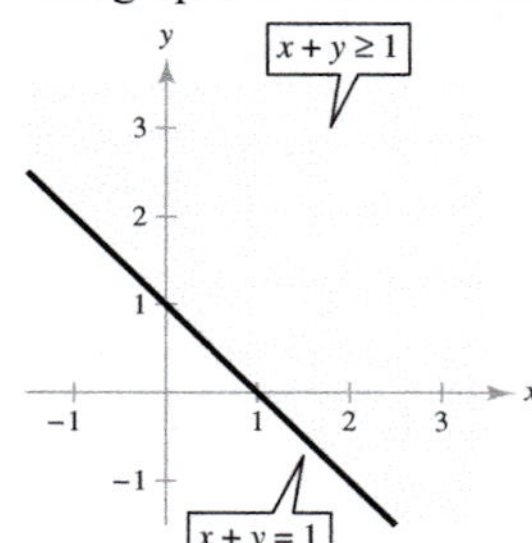

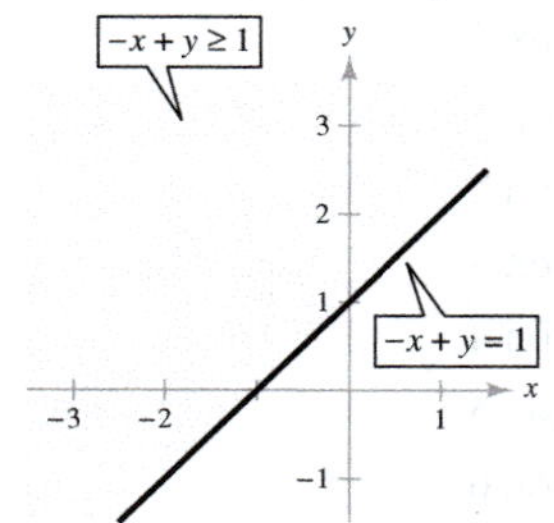

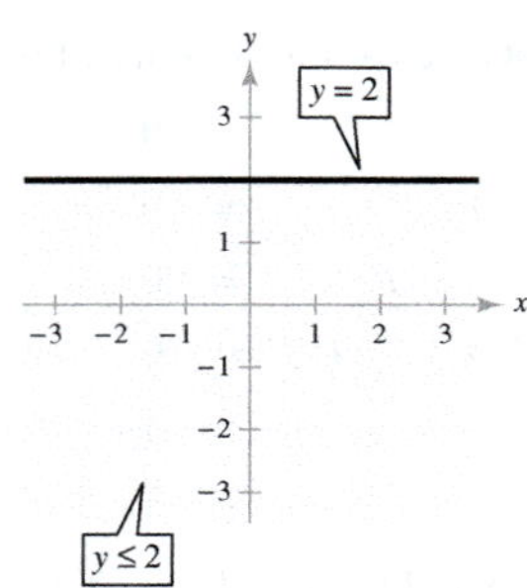

By superimposing the graphs on the same coordinate system, the region common to all three graphs can be found. To find the vertices of the region, solve the three systems of corresponding equations by taking pairs of equations representing the boundaries of the individual regions.

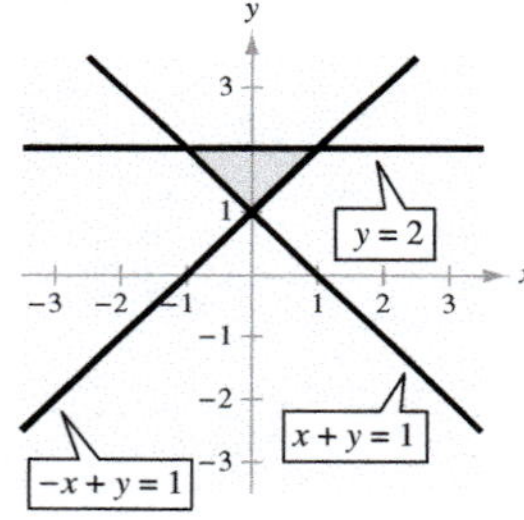

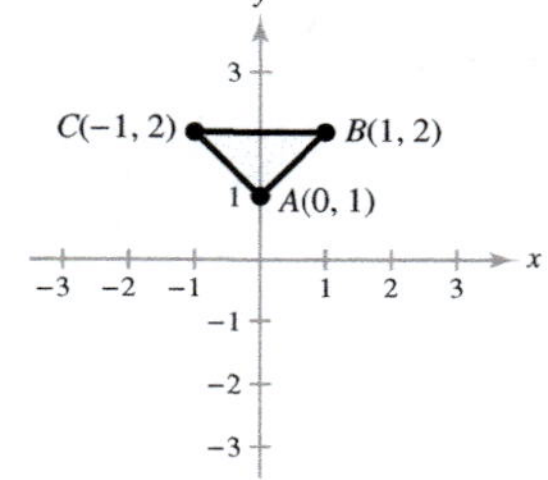

Vertex A: $(0, 1)$

$$\begin{cases} x + y = 1 \\ -x + y = 1 \end{cases}$$

Vertex B: $(1, 2)$

$$\begin{cases} -x + y = 1 \\ y = 2 \end{cases}$$

Vertex C: $(-1, 2)$

$$\begin{cases} x + y = 1 \\ y = 2 \end{cases}$$

Note that the vertices of the region are represented by solid dots. This means that the vertices *are* solutions of the system of equations as well as all of the points that lie on the lines.

5. The points that satisfy the inequality $x - y^2 > 0$ are the points inside the parabola (but not on) the parabola $x = y^2$.

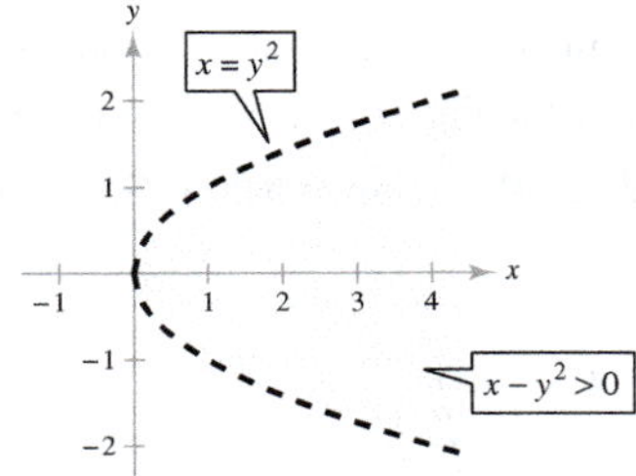

The points satisfying the inequality $x + y < 2$ are the points lying below (but not on) the line $x + y = 2$.

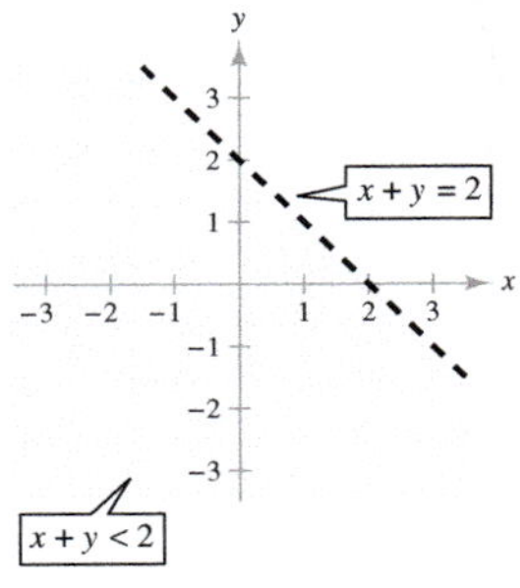

To find the points of the intersection of the parabola and the line, solve the system of corresponding equations.

$$\begin{cases} x - y^2 = 0 \\ x + y = 2 \end{cases}$$

$$x + y = 2 \Rightarrow y = 2 - x$$

$$x - y^2 = 0$$

$$x - (2 - x)^2 = 0$$

$$x - (4 - 4x + x^2) = 0$$

$$x - 4 + 4x - x^2 = 0$$

$$-x^2 + 5x - 4 = 0$$

$$-(x^2 - 5x + 4) = 0$$

$$(x - 4)(x - 1) = 0$$

$$x - 4 = 0 \qquad x - 1 = 0$$

$$x = 4 \qquad x = 1$$

When $x = 4$, $y = 2 - 4 = -2$.

When $x = 1$, $y = 2 - 1 = 1$.

Using the method of substitution, you can find the solutions to be $(4, -2)$ and $(1, 1)$.

So, the region containing all points that satisfy the system is indicated by the shaded region.

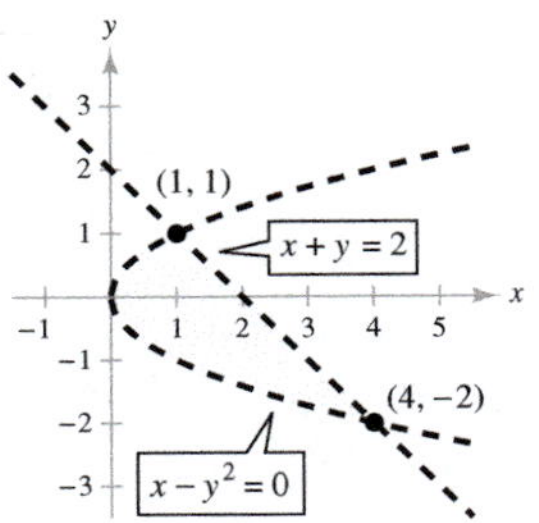

6. From the way the system is written, it should be clear that the system has no solution because the quantity $2x - y$ cannot be both less than -3 and greater than 1. The graph of the inequality $2x - y < -3$ is the half-plane lying above the line $2x - y = -3$ and the graph of the inequality $2x - y > 1$ is the half-plane lying below the line $2x - y = 1$ as shown. These two half-planes have no points in common. So, the system of inequalities has no solution.

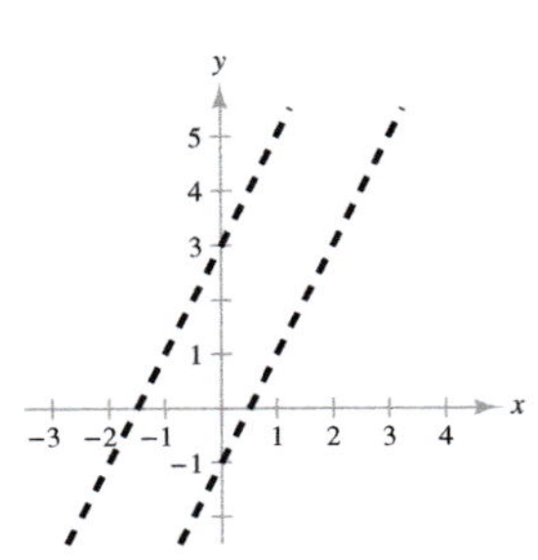

7. The graph of the inequality $x^2 - y < 0$ is the region inside the parabola $x^2 - y = 0$. The graph of the inequality $x - y < -2$ is the half-plane that lies above the line $x - y = -2$. The intersection of these regions is an infinite region having points of intersection at $(-1, 1)$ and $(2, 4)$, as shown below. So, the solution set of the system of inequalities is unbounded.

Points of intersection:

$$\begin{cases} x^2 - y = 0 \Rightarrow x^2 = y \\ x - y = -2 \end{cases}$$

$$x - \left(x^2\right) = -2$$

$$x^2 - x - 2 = 0$$

$$(x - 2)(x + 1) = 0$$

$$x - 2 = 0 \qquad x + 1 = 0$$

$$x = 2 \qquad x = -1$$

When $x = 2, y = (2)^2 = 4.$

When $x = -1 \;\; y = (-1)^2 = 1.$

8. Begin by finding the equilibrium point (when supply and demand are equal) by solving the equation $492 + 0.00003x = 567 - 0.00002x.$

In checkpoint 9 in Section 9.2, you saw that the solution is $x = 1{,}500{,}000$ units, which corresponds to an equilibrium price of $p = 537$. So, the consumer surplus and producer surplus are the areas of the following triangular regions.

Consumer Surplus

$$\begin{cases} p \le 567 - 0.00002x \\ p \ge 537 \\ x \ge 0 \end{cases}$$

Producer Surplus

$$\begin{cases} p \ge 492 + 0.00003x \\ p \le 537 \\ x \ge 0 \end{cases}$$

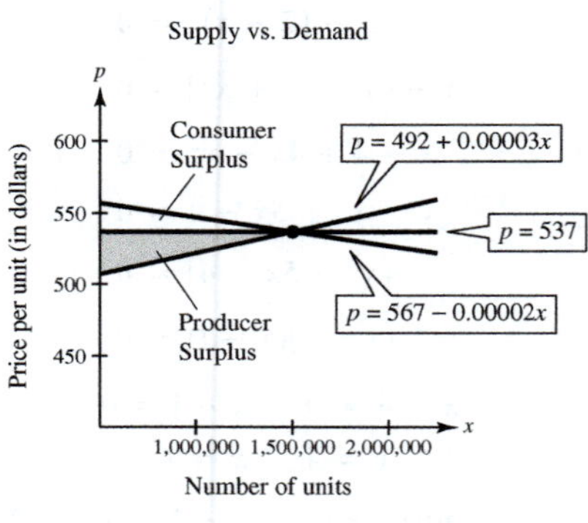

The consumer and producer surpluses are the areas of the shaded triangles shown.

$$\boxed{\text{Consumer surplus}} = \tfrac{1}{2}(\text{base})(\text{height})$$
$$= \tfrac{1}{2}(1{,}500{,}000)(30)$$
$$= \$22{,}500{,}000$$

$$\boxed{\text{Producer surplus}} = \tfrac{1}{2}(\text{base})(\text{height})$$
$$= \tfrac{1}{2}(1{,}500{,}000)(45)$$
$$= \$33{,}750{,}000$$

9. Begin by letting x represent the number of bottles of brand X coral nutrients and y represent the number of bottles of brand Y coral nutrients.

To meet the minimum required amounts of nutrients, the following inequalities must be satisfied.

$$\begin{cases} 8x + 2y \geq 16 & \text{Nutrient A} \\ x + y \geq 5 & \text{Nutrient B} \\ 2x + 7y \geq 20 & \text{Nutrient C} \\ x \geq 0 \\ y \geq 0 \end{cases}$$

The graph of this system of inequalities is shown.

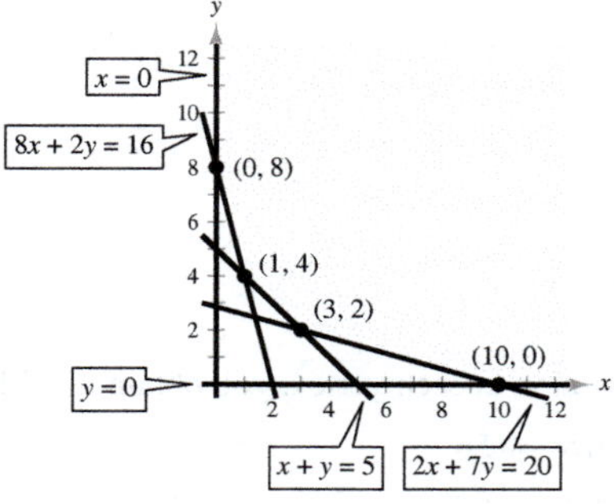

Nutrients A and B

$$\begin{cases} 8x + 2y = 16 \\ x + y = 5 \end{cases}$$

$$\begin{cases} 8x + 2y = 16 \\ -2x - 2y = -10 \end{cases}$$

$$\begin{cases} 6x = 6 \rightarrow x = 1 \\ 1 + y = 5 \rightarrow y = 4 \end{cases}$$

Nutrients B and C

$$\begin{cases} x + y = 5 \\ 2x + 7y = 20 \end{cases}$$

$$\begin{cases} 2x + 2y = 10 \\ -2x - 7y = -20 \end{cases}$$

$$\begin{cases} -5y = -10 \rightarrow y = 2 \\ x + 2 = 5 \rightarrow x = 3 \end{cases}$$

Checkpoints for Section 7.6

1. The constraints form the region shown.

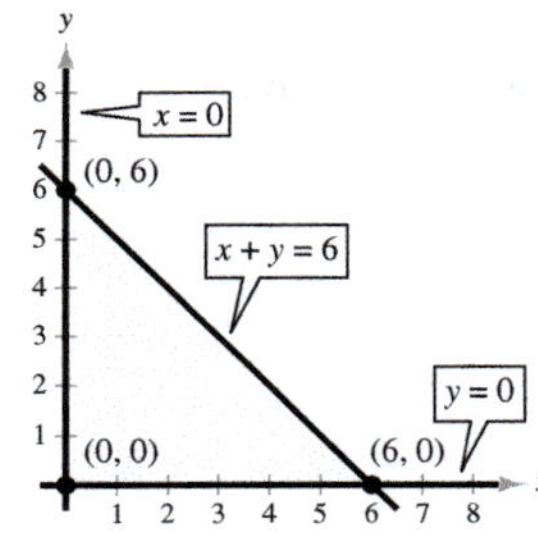

At the three vertices of the region, the objective function has the following values,

At $(0, 0)$: $z = 4(0) + 5(0) = 0$

At $(0, 6)$: $z = 4(0) + 5(6) = 30$

At $(6, 0)$: $z = 4(6) + 5(0) = 24$

So the maximum value of z is 30, and this occurs when $x = 0$ and $y = 6$.

2. The constraints form the region shown.

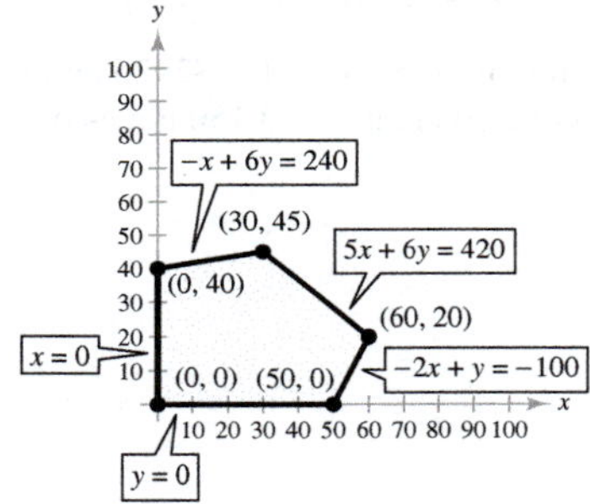

By testing the objective function at each vertex, you obtain the following.

At $(0, 0)$: $z = 12(0) + 8(0) = 0$

At $(0, 40)$: $z = 12(0) + 8(40) = 320$

At $(30, 45)$: $z = 12(30) + 8(45) = 720$

At $(60, 20)$: $z = 12(60) + 8(20) = 880$

At $(50, 0)$: $z = 12(50) + 8(0) = 600$

So, the minimum value of z is 0, which occurs when $x = 0$ and $y = 0$.

3. Using the values of z at the vertices shown in Checkpoint Example 2, the maximum value of z is

$$z = 12(60) + 8(20)$$
$$= 880$$

and occurs when $x = 60$ and $y = 20$.

4. The constraints form the region shown.

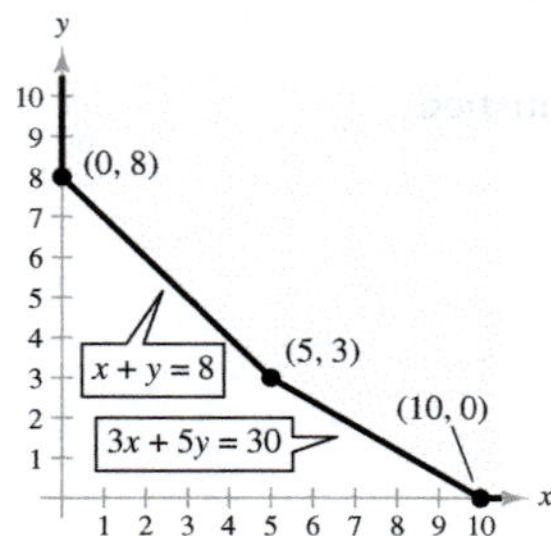

By testing the objective function at each vertex, you obtain the following.

At $(0, 8)$: $z = 3(0) + 7(8) = 56$

At $(5, 3)$: $z = 3(5) + 7(3) = 36$

At $(10, 0)$: $z = 3(10) + 7(0) = 30$

So, the minimum of z is 30, which occurs when $x = 10$ and $y = 0$.

5. Let x be the number of boxes of chocolate-covered creams and let y be the number of boxes of chocolate-covered nuts. So, the objective function (for the combined profit) is

$P = 2.5x + 2y$ Objective function

To find the maximum monthly profit, test the values of P at the vertices of the region.

At $(0, 0)$: $P = 2.5(0) + 2(0) = 0$

At $(800, 400)$: $P = 2.5(800) + 2(400) = 2800$

At $(1050, 150)$: $P = 2.5(1050) + 2(150) = 2925$ Maximum Profit

At $(600, 0)$: $P = 2.5(600) + 2(0) = 1500$

So, the maximum monthly profit is \$2925 and it occurs when the monthly production consists of 1050 boxes of chocolate-covered creams and 150 boxes of chocolate-covered nuts.

6. Begin by letting x represent the number of bottles of brand X coral nutrients and y represent the number of bottles of brand Y coral nutrients.

To meet the minimum required amounts of nutrients, the following inequalities must be satisfied.

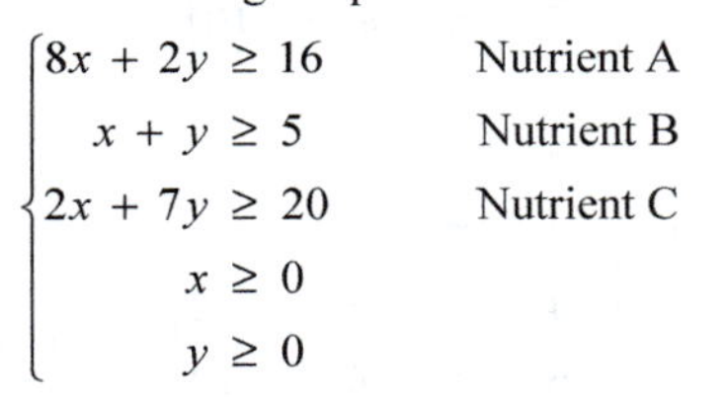

$$\begin{cases} 8x + 2y \geq 16 & \text{Nutrient A} \\ x + y \geq 5 & \text{Nutrient B} \\ 2x + 7y \geq 20 & \text{Nutrient C} \\ x \geq 0 & \\ y \geq 0 & \end{cases}$$

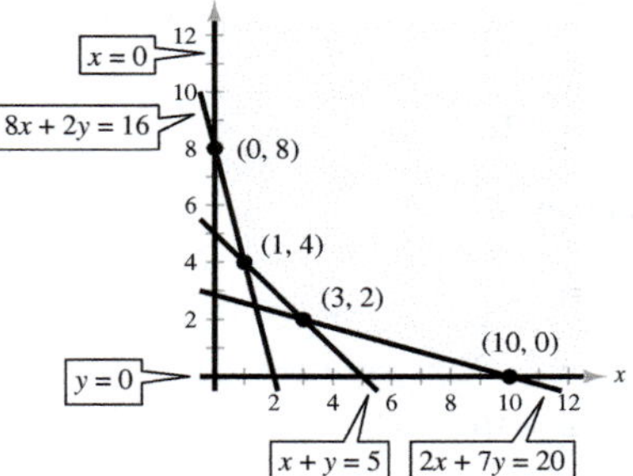

The graph of this system of inequalities is shown.

Nutrients A and B

$$\begin{cases} 8x + 2y = 16 \\ x + y = 5 \end{cases}$$

$$\begin{cases} 8x + 2y = 16 \\ -2x - 2y = -10 \end{cases}$$

$$\begin{cases} 6x = 6 \rightarrow x = 1 \\ 1 + y = 5 \rightarrow y = 4 \end{cases}$$

Nutrients B and C

$$\begin{cases} x + y = 5 \\ 2x + 7y = 20 \end{cases}$$

$$\begin{cases} 2x + 2y = 10 \\ -2x - 7y = -20 \end{cases}$$

$$\begin{cases} -5y = -10 \rightarrow y = 2 \\ x + 2 = 5 \rightarrow x = 3 \end{cases}$$

The figure shows the graph of the region corresponding to the constraints.

The cost function is given by $C = 15x + 30y$.

Because you want to incur as little cost as possible, you want to determine the minimum cost.

At $(0, 8)$: $C = 15(0) + 30(8) = 240$

At $(1, 4)$: $C = 15(1) + 30(4) = 135$

At $(3, 2)$: $C = 15(3) + 30(2) = 105$ Minimum Cost

At $(10, 0)$: $C = 15(10) + 30(0) = 150$

So, the minimum cost is \$7.20 and occurs when 3 bottles of Brand X and 2 bottles of Brand Y are added.

Chapter 8

Checkpoints for Section 8.1

1. The matrix has *two* rows and *three* columns. The order of the matrix is 2×3.

2. $$\begin{cases} x + y + z = 2 \\ 2x - y + 3z = -1 \\ -x + 2y - z = 4 \end{cases}$$

All of the variables are aligned in the system. Next, use the coefficients and constant terms as the matrix entries.

$$\begin{matrix} R_1 \\ R_2 \\ R_3 \end{matrix} \begin{bmatrix} 1 & 1 & 1 & \vdots & 2 \\ 2 & -1 & 3 & \vdots & -1 \\ -1 & 2 & -1 & \vdots & 4 \end{bmatrix}$$

The augmented matrix has three rows and four columns, so it is a 3×4 matrix.

3. Add -3 times the first row of the original matrix to the second row.

Original Matrix **New Row-Equivalent Matrix**

$$\begin{bmatrix} 1 & 0 & 2 \\ 3 & 1 & 7 \\ 2 & -6 & 14 \end{bmatrix} \quad -3R_1 + R_2 \rightarrow \begin{bmatrix} 1 & 0 & 2 \\ 0 & 1 & 1 \\ 2 & -6 & 14 \end{bmatrix}$$

4. Linear System **Associated Augmented Matrix**

$$\begin{cases} 2x + y - z = -3 \\ 4x - 2y + 2z = -2 \\ -6x + 5y + 4z = 10 \end{cases} \qquad \begin{bmatrix} 2 & 1 & -1 & : & -3 \\ 4 & -2 & 2 & : & -2 \\ -6 & 5 & 4 & : & 10 \end{bmatrix}$$

Multiply the first equation by $\frac{1}{2}$.

$$\begin{cases} x + \frac{1}{2}y - \frac{1}{2}z = -\frac{3}{2} \\ 4x - 2y + 2z = -2 \\ -6x + 5y + 4z = 10 \end{cases} \qquad \frac{1}{2}R_1 \rightarrow \begin{bmatrix} 1 & \frac{1}{2} & -\frac{1}{2} & : & -\frac{3}{2} \\ 4 & -2 & 2 & : & -2 \\ -6 & 5 & 4 & : & 10 \end{bmatrix}$$

Add -4 times the first equation to the second equation.

$$\begin{cases} x + \frac{1}{2}y - \frac{1}{2}z = -\frac{3}{2} \\ -4y + 4z = 4 \\ -6x + 5y + 4z = 10 \end{cases} \qquad -4R_1 + R_2 \rightarrow \begin{bmatrix} 1 & \frac{1}{2} & -\frac{1}{2} & : & -\frac{3}{2} \\ 0 & -4 & 4 & : & 4 \\ -6 & 5 & 4 & : & 10 \end{bmatrix}$$

Multiply the second equation by $-\frac{1}{4}$.

$$\begin{cases} x + \frac{1}{2}y - \frac{1}{2}z = -\frac{3}{2} \\ y - z = -1 \\ -6x + 5y + 4z = 10 \end{cases} \qquad -\frac{1}{4}R_2 \rightarrow \begin{bmatrix} 1 & \frac{1}{2} & -\frac{1}{2} & : & -\frac{3}{2} \\ 0 & 1 & -1 & : & -1 \\ -6 & 5 & 4 & : & 10 \end{bmatrix}$$

Add 6 times the first equation to the third equation.

$$\begin{cases} x + \frac{1}{2}y - \frac{1}{2}z = -\frac{3}{2} \\ y - z = -1 \\ 8y + z = 1 \end{cases} \qquad 6R_1 + R_3 \rightarrow \begin{bmatrix} 1 & \frac{1}{2} & -\frac{1}{2} & : & -\frac{3}{2} \\ 0 & 1 & -1 & : & -1 \\ 0 & 8 & 1 & : & 1 \end{bmatrix}$$

Add -8 times the second equation to the third equation.

$$\begin{cases} x + \frac{1}{2}y - \frac{1}{2}z = -\frac{3}{2} \\ y - z = -1 \\ 9z = 9 \end{cases} \qquad -8R_2 + R_3 \rightarrow \begin{bmatrix} 1 & \frac{1}{2} & -\frac{1}{2} & : & -\frac{3}{2} \\ 0 & 1 & -1 & : & -1 \\ 0 & 0 & 9 & : & 9 \end{bmatrix}$$

Multiply the third equation by $\frac{1}{9}$.

$$\begin{cases} x + \frac{1}{2}y - \frac{1}{2}z = -\frac{3}{2} \\ y - z = -1 \\ z = 1 \end{cases} \qquad \frac{1}{9}R_3 \rightarrow \begin{bmatrix} 1 & \frac{1}{2} & -\frac{1}{2} & : & -\frac{3}{2} \\ 0 & 1 & -1 & : & -1 \\ 0 & 0 & 1 & : & 1 \end{bmatrix}$$

At this point, you can use back-substitution to find x and y.

$$\begin{aligned} y - z &= -1 \\ y - (1) &= -1 \\ y &= 0 \\ x + \tfrac{1}{2}y - \tfrac{1}{2}z &= -\tfrac{3}{2} \\ x + \tfrac{1}{2}(0) - \tfrac{1}{2}(1) &= -\tfrac{3}{2} \\ x &= -1 \end{aligned}$$

The solution is $x = -1$, $y = 0$, and $z = 1$.

5. The Matrix is in row-echelon form, because the row consisting entirely of zeros occurs at the bottom of the matrix, and for each row that does not consist entirely of zeros, the first nonzero entry is 1. Furthermore the matrix is in reduced row-echelon form, since every column that has a leading 1 has zeros in every position above and below its leading 1.

6.

$$\begin{bmatrix} -3 & 5 & 3 & \vdots & -19 \\ 3 & 4 & 4 & \vdots & 8 \\ 4 & -8 & -6 & \vdots & 26 \end{bmatrix}$$

Write augmented matrix.

$$R_3 + R_1 \rightarrow \begin{bmatrix} 1 & -3 & -3 & \vdots & 7 \\ 3 & 4 & 4 & \vdots & 8 \\ 4 & -8 & -6 & \vdots & 26 \end{bmatrix}$$

Add R_3 to R_1 so first column has leading 1 in upper left corner.

$$\begin{matrix} \\ -3R_1 + R_2 \rightarrow \\ -4R_1 + R_3 \rightarrow \end{matrix} \begin{bmatrix} 1 & -3 & -3 & \vdots & 7 \\ 0 & 13 & 13 & \vdots & -13 \\ 0 & 4 & 6 & \vdots & -2 \end{bmatrix}$$

Perform operations on R_2 and R_3 so first column has zeros below its leading 1.

$$\begin{matrix} \\ R_2 + (-3)R_3 \rightarrow \\ \\ \end{matrix} \begin{bmatrix} 1 & -3 & -3 & \vdots & 5 \\ 0 & 1 & -5 & \vdots & -7 \\ 0 & 4 & 6 & \vdots & -2 \end{bmatrix}$$

Perform operations on R_2 so second column has a leading 1.

$$\begin{matrix} \\ \\ -4R_2 + R_3 \rightarrow \end{matrix} \begin{bmatrix} 1 & -3 & -3 & \vdots & 7 \\ 0 & 1 & -5 & \vdots & -7 \\ 0 & 0 & 26 & \vdots & 26 \end{bmatrix}$$

Perform operations on R_3 so second column has a zero below its leading 1.

$$\begin{matrix} \\ \\ \frac{1}{26}R_3 \rightarrow \end{matrix} \begin{bmatrix} 1 & -3 & -3 & \vdots & 7 \\ 0 & 1 & -5 & \vdots & -7 \\ 0 & 0 & 1 & \vdots & 1 \end{bmatrix}$$

Perform operations on R_3 so third column has a leading 1.

The matrix is now in row-echelon form, and the corresponding system is

$$\begin{cases} x - 3y - 3z = 7 \\ \quad\;\; y - 5z = -7 \\ \qquad\quad\;\; z = 1 \end{cases}$$

Using back-substitution, you can determine that the solution is $x = 4$, $y = -2$ and $z = 1$.

7.

$$\begin{bmatrix} 1 & 1 & 1 & \vdots & 1 \\ 1 & 2 & 2 & \vdots & 2 \\ 1 & -1 & -1 & \vdots & 1 \end{bmatrix}$$

Write augmented matrix.

$$\begin{matrix} \\ -R_1 + R_2 \rightarrow \\ -R_1 + R_3 \rightarrow \end{matrix} \begin{bmatrix} 1 & 1 & 1 & \vdots & 1 \\ 0 & 1 & 1 & \vdots & 1 \\ 0 & -2 & -2 & \vdots & 0 \end{bmatrix}$$

Perform row operations.

$$\begin{matrix} \\ \\ 2R_2 + R_3 \rightarrow \end{matrix} \begin{bmatrix} 1 & 1 & 1 & \vdots & 1 \\ 0 & 1 & 1 & \vdots & 1 \\ 0 & 0 & 0 & \vdots & 2 \end{bmatrix}$$

Perform row operations.

Note that the third row of this matrix consists entirely of zeros except for the last entry. This means that the original system of linear equations is inconsistent. You can see why this is true by converting back to a system of linear equations.

$$\begin{cases} x + y + z = 1 \\ \quad\;\; y + z = 1 \\ \qquad\;\; 0 = 2 \end{cases}$$

Because the third equation is not possible, the system has no solution.

8.

$$\begin{bmatrix} -3 & 7 & 2 & \vdots & 1 \\ -5 & 3 & -5 & \vdots & -8 \\ 2 & -2 & -3 & \vdots & 15 \end{bmatrix}$$

$$R_2 + R_1 \rightarrow \begin{bmatrix} -1 & 5 & -1 & \vdots & 16 \\ -5 & 3 & -5 & \vdots & -8 \\ 2 & -2 & -3 & \vdots & 15 \end{bmatrix}$$

$$-R_1 \rightarrow \begin{bmatrix} 1 & -5 & 1 & \vdots & -16 \\ -5 & 3 & -5 & \vdots & -8 \\ 2 & -2 & -3 & \vdots & 15 \end{bmatrix}$$

$$5R_1 + R_2 \rightarrow \begin{bmatrix} 1 & -5 & 1 & \vdots & -16 \\ 0 & -22 & 0 & \vdots & -88 \\ 2 & -2 & -3 & \vdots & 15 \end{bmatrix}$$

$$-2R_1 + R_3 \rightarrow \begin{bmatrix} 1 & -5 & 1 & \vdots & -16 \\ 0 & -22 & 0 & \vdots & -88 \\ 0 & 8 & -5 & \vdots & 47 \end{bmatrix}$$

$$-\tfrac{1}{22}R_2 \rightarrow \begin{bmatrix} 1 & -5 & 1 & \vdots & -16 \\ 0 & 1 & 0 & \vdots & 4 \\ 0 & 8 & -5 & \vdots & 47 \end{bmatrix}$$

$$-8R_2 + R_3 \rightarrow \begin{bmatrix} 1 & -5 & 1 & \vdots & -16 \\ 0 & 1 & 0 & \vdots & 4 \\ 0 & 0 & -5 & \vdots & 15 \end{bmatrix}$$

$$-\tfrac{1}{5}R_3 \rightarrow \begin{bmatrix} 1 & -5 & 1 & \vdots & -16 \\ 0 & 1 & 0 & \vdots & 4 \\ 0 & 0 & 1 & \vdots & -3 \end{bmatrix}$$

At this point, the matrix is in row-echelon form. Now, apply elementary row operations until you obtain zeros above each of the leading is, as follows.

$$5R_2 + R_1 \rightarrow \begin{bmatrix} 1 & 0 & 1 & \vdots & 4 \\ 0 & 1 & 0 & \vdots & 4 \\ 0 & 0 & 1 & \vdots & -3 \end{bmatrix}$$

$$-R_3 + R_1 \rightarrow \begin{bmatrix} 1 & 0 & 0 & \vdots & 7 \\ 0 & 1 & 0 & \vdots & 4 \\ 0 & 0 & 1 & \vdots & -3 \end{bmatrix}$$

The matrix is now in reduced row-echelon form. Converting back to a system of linear equations, you have

$$\begin{cases} x = 7 \\ y = 4. \\ z = -3 \end{cases}$$

So, the solution is $x = 7$, $y = 4$, and $z = -3$, which can be written as the ordered triple $(7, 4, -3)$.

9.

$$\begin{bmatrix} 2 & -6 & 6 & \vdots & 46 \\ 2 & -3 & 0 & \vdots & 31 \end{bmatrix}$$

$$\begin{bmatrix} 1 & -3 & 3 & \vdots & 23 \\ 2 & -3 & 0 & \vdots & 31 \end{bmatrix}$$

$$-2R_1 + R_2 \rightarrow \begin{bmatrix} 1 & -3 & 3 & \vdots & 23 \\ 0 & 3 & -6 & \vdots & -15 \end{bmatrix}$$

$$\tfrac{1}{3}R_2 \rightarrow \begin{bmatrix} 1 & -3 & 3 & \vdots & 23 \\ 0 & 1 & -2 & \vdots & -5 \end{bmatrix}$$

$$R_1 + 3R_2 \rightarrow \begin{bmatrix} 1 & 0 & -3 & \vdots & 8 \\ 0 & 1 & -2 & \vdots & -5 \end{bmatrix}$$

The corresponding system of equations is

$$\begin{cases} x - 3z = 8 \\ y - 2z = -5. \end{cases}$$

Solving for x and y in terms of z, you have $x = 3z + 8$ and $y = 2z - 5$.

To write a solution of the system that does not use any of the three variables of the system, let a represent any real number and let $z = a$. Substituting a for z in the equations for x and y, you have

$x = 3z + 8 = 3a + 8$ and $y = 2z - 5 = 2a - 5$.

So, the solution set can be written as an ordered triple of the form

$(3a + 8, 2a - 5, a)$

where a is any real number. Remember that a solution set of this form represents an infinite number of solutions. Try substituting values for a to obtain a few solutions. Then check each solution in the original system of equations.

Checkpoints for Section 8.2

1. $\begin{bmatrix} a_{11} & a_{12} \\ a_{21} & a_{22} \end{bmatrix} = \begin{bmatrix} 6 & 3 \\ -2 & 4 \end{bmatrix}$

Because two matrices are equal when their corresponding entries are equal you can conclude that $a_{11} = 6, a_{12} = 3, a_{21} = -2$, and $a_{22} = 4$.

2. (a) $\begin{bmatrix} 4 & -1 \\ 2 & -3 \end{bmatrix} + \begin{bmatrix} 2 & -1 \\ 0 & 6 \end{bmatrix} = \begin{bmatrix} 4+2 & -1+(-1) \\ 2+0 & -3+6 \end{bmatrix} = \begin{bmatrix} 6 & -2 \\ 2 & 3 \end{bmatrix}$

(b) $\begin{bmatrix} 2 & -1 \\ 3 & 4 \\ 0 & -2 \end{bmatrix} + \begin{bmatrix} -2 & 1 \\ -3 & -4 \\ 0 & 2 \end{bmatrix} = \begin{bmatrix} 0 & 0 \\ 0 & 0 \\ 0 & 0 \end{bmatrix}$

(c) $\begin{bmatrix} 3 & 9 & 6 \\ 0 & 4 & -2 \\ 1 & -1 & 0 \end{bmatrix} + \begin{bmatrix} 3 & 9 & 6 \\ 0 & 2 & -4 \end{bmatrix}$, not possible. The matrices are not of the same dimension.

(d) $\begin{bmatrix} 1 \\ -1 \\ 1 \end{bmatrix} + \begin{bmatrix} -1 \\ 1 \\ 1 \end{bmatrix} = \begin{bmatrix} 0 \\ 0 \\ 2 \end{bmatrix}$

3. $A = \begin{bmatrix} 4 & -1 \\ 0 & 4 \\ -3 & 8 \end{bmatrix}$ and $B = \begin{bmatrix} 0 & 4 \\ -1 & 3 \\ 1 & 7 \end{bmatrix}$

(a) $A - B = \begin{bmatrix} 4 & -1 \\ 0 & 4 \\ -3 & 8 \end{bmatrix} - \begin{bmatrix} 0 & 4 \\ -1 & 3 \\ 1 & 7 \end{bmatrix} = \begin{bmatrix} 4 & -5 \\ 1 & 1 \\ -4 & 1 \end{bmatrix}$

(b) $3A = 3\begin{bmatrix} 4 & -1 \\ 0 & 4 \\ -3 & 8 \end{bmatrix} = \begin{bmatrix} 12 & -3 \\ 0 & 12 \\ -9 & 24 \end{bmatrix}$

(c) $3A - 2B = 3\begin{bmatrix} 4 & -1 \\ 0 & 4 \\ -3 & 8 \end{bmatrix} - 2\begin{bmatrix} 0 & 4 \\ -1 & 3 \\ 1 & 7 \end{bmatrix} = \begin{bmatrix} 12 & -3 \\ 0 & 12 \\ -9 & 24 \end{bmatrix} - \begin{bmatrix} 0 & 8 \\ -2 & 6 \\ 2 & 14 \end{bmatrix} = \begin{bmatrix} 12 & -11 \\ 2 & 6 \\ -11 & 10 \end{bmatrix}$

4. $\begin{bmatrix} 3 & -8 \\ 0 & 2 \end{bmatrix} + \begin{bmatrix} -2 & 3 \\ 6 & -5 \end{bmatrix} + \begin{bmatrix} 0 & 7 \\ 4 & -1 \end{bmatrix} = \begin{bmatrix} 3+(-2)+0 & -8+3+7 \\ 0+6+4 & 2+(-5)+(-1) \end{bmatrix} = \begin{bmatrix} 1 & 2 \\ 10 & -4 \end{bmatrix}$

5. $$\begin{aligned} 2\left(\begin{bmatrix} 1 & 3 \\ -2 & 2 \end{bmatrix} + \begin{bmatrix} -4 & 0 \\ -3 & 1 \end{bmatrix}\right) &= 2\begin{bmatrix} 1 & 3 \\ -2 & 2 \end{bmatrix} + 2\begin{bmatrix} -4 & 0 \\ -3 & 1 \end{bmatrix} \\ &= \begin{bmatrix} 2 & 6 \\ -4 & 4 \end{bmatrix} + \begin{bmatrix} -8 & 0 \\ -6 & 2 \end{bmatrix} \\ &= \begin{bmatrix} -6 & 6 \\ -10 & 6 \end{bmatrix} \end{aligned}$$

6. Begin by solving the matrix equation for X to obtain

$$2X - A = B$$
$$2X = B + A$$
$$X = \tfrac{1}{2}(B + A).$$

Now, using the matrices A and B you have the following

$$X = \frac{1}{2}\left(\begin{bmatrix} 4 & -1 \\ -2 & 5 \end{bmatrix} + \begin{bmatrix} 6 & 1 \\ 0 & 3 \end{bmatrix}\right) = \frac{1}{2}\begin{bmatrix} 10 & 0 \\ -2 & 8 \end{bmatrix} = \begin{bmatrix} 5 & 0 \\ -1 & 4 \end{bmatrix}$$

7. $AB = \begin{bmatrix} -1 & 4 \\ 2 & 0 \\ 1 & 2 \end{bmatrix}\begin{bmatrix} 1 & -2 \\ 0 & 7 \end{bmatrix}$

$= \begin{bmatrix} (-1)(1) + (4)(0) & (-1)(-2) + (4)(7) \\ (2)(1) + (0)(0) & (2)(-2) + (0)(7) \\ (1)(1) + (2)(0) & (1)(-2) + (2)(7) \end{bmatrix}$

$= \begin{bmatrix} -1 & 30 \\ 2 & -4 \\ 1 & 12 \end{bmatrix}$

8. $AB = \begin{bmatrix} 0 & 4 & -3 \\ 2 & 1 & 7 \\ 3 & -2 & 1 \end{bmatrix}\begin{bmatrix} -2 & 0 \\ 0 & -4 \\ 1 & 2 \end{bmatrix} = \begin{bmatrix} (0)(-2) + (4)(0) + (-3)(1) & (0)(0) + (4)(-4) + (-3)(2) \\ (2)(-2) + (1)(0) + (7)(1) & (2)(0) + (1)(-4) + (7)(2) \\ (3)(-2) + (-2)(0) + (1)(1) & (3)(0) + (-2)(-4) + (1)(2) \end{bmatrix} = \begin{bmatrix} -3 & -22 \\ 3 & 10 \\ -5 & 10 \end{bmatrix}$

9. (a) $BA = \begin{bmatrix} 1 \\ -3 \end{bmatrix}[3 \ -1] = \begin{bmatrix} (1)(3) & (1)(-1) \\ (-3)(3) & (-3)(-1) \end{bmatrix} = \begin{bmatrix} 3 & -1 \\ -9 & 3 \end{bmatrix}$

(b) $[3 \ -1]\begin{bmatrix} 1 \\ -3 \end{bmatrix} = [(3)(1) + (-1)(-3)] = [6]$

(c) $\begin{bmatrix} 3 & 1 & 2 \\ 7 & 0 & -2 \end{bmatrix}\begin{bmatrix} 6 & 4 \\ 2 & -1 \end{bmatrix}$ is not defined, since the first matrix has dimensions 2×3 and the second matrix has dimensions 2×2. The number of columns of the first matrix is not equal to the number of rows of the second matrix.

10. $A^2 = AA = \begin{bmatrix} 2 & 1 \\ 3 & -2 \end{bmatrix}\begin{bmatrix} 2 & 1 \\ 3 & -2 \end{bmatrix} = \begin{bmatrix} (2)(2) + (1)(3) & (2)(1) + (1)(-2) \\ (3)(2) + (-2)(3) & (3)(1) + (-2)(-2) \end{bmatrix} = \begin{bmatrix} 7 & 0 \\ 0 & 7 \end{bmatrix}$

11. $\mathbf{v} = \langle 3, 6 \rangle$ and $\mathbf{w} = \langle 8, 5 \rangle$

(a) $\mathbf{v} - \mathbf{w} = \langle 3, 6 \rangle - \langle 8, 5 \rangle = \langle -5, 1 \rangle$

(b) $3\mathbf{v} + \mathbf{w} = 3\langle 3, 6 \rangle + \langle 8, 5 \rangle$

$= \langle 9, 18 \rangle + \langle 8, 5 \rangle$

$= \langle 17, 23 \rangle$

12. $A = \begin{bmatrix} -1 & 0 \\ 0 & 1 \end{bmatrix}$ and $\mathbf{v} = \langle 3, 1 \rangle$

$A\mathbf{v} = \begin{bmatrix} -1 & 0 \\ 0 & 1 \end{bmatrix}\begin{bmatrix} 3 \\ 1 \end{bmatrix} = [(-1)(3) + (0)(1) \quad (0)(3) + (1)(1)]$

$= [-3 \quad 1]$

$A\mathbf{v} = \langle -3, 1 \rangle$ is a reflection in the y-axis.

13. $\begin{cases} -2x_1 - 3x_2 = -4 \\ 6x_1 + x_2 = -36 \end{cases}$

(a) In matrix form $Ax = B$, the system can be written as follows.

$$\begin{bmatrix} -2 & -3 \\ 6 & 1 \end{bmatrix} \begin{bmatrix} x_1 \\ x_2 \end{bmatrix} = \begin{bmatrix} -4 \\ -36 \end{bmatrix}$$

(b) The augmented matrix is formed by adjoining matrix B to matrix A.

$$[A \vdots B] = \begin{bmatrix} -2 & -3 & \vdots & -4 \\ 6 & 1 & \vdots & -36 \end{bmatrix}$$

Use Gauss-Jordan elimination to rewrite the matrix.

$$-\tfrac{1}{2}R_1 \to \begin{bmatrix} 1 & \frac{3}{2} & \vdots & 2 \\ 6 & 1 & \vdots & -36 \end{bmatrix}$$

$$\begin{matrix} \\ -6R_1 + R_2 \end{matrix} \begin{matrix} = \\ \to \end{matrix} \begin{bmatrix} 1 & \frac{3}{2} & \vdots & 2 \\ 0 & -8 & \vdots & -48 \end{bmatrix}$$

$$\begin{matrix} \\ -\tfrac{1}{8}R_2 \end{matrix} \begin{matrix} = \\ \to \end{matrix} \begin{bmatrix} 1 & \frac{3}{2} & \vdots & -2 \\ 0 & 1 & \vdots & 6 \end{bmatrix}$$

$$-\tfrac{3}{2}R_2 + R_1 \to = \begin{bmatrix} 1 & 0 & \vdots & 7 \\ 0 & 1 & \vdots & 6 \end{bmatrix}$$

$$= [I \vdots X]$$

So, the solution of the matrix equation is $X = \begin{bmatrix} x_1 \\ x_2 \end{bmatrix} = \begin{bmatrix} -7 \\ 6 \end{bmatrix}$.

14. The equipment lists E and the costs per item C can be written in matrix form as

$$E = \begin{bmatrix} 12 & 15 \\ 45 & 38 \\ 15 & 17 \end{bmatrix} \text{ and } C = [100 \quad 3 \quad 65].$$

The total cost of equipment for each team is given by the following product.

$$CE = [100 \quad 3 \quad 65] \begin{bmatrix} 12 & 15 \\ 45 & 38 \\ 15 & 17 \end{bmatrix} = [(100)(12) + (3)(45) + (65)(15) \quad (100)(15) + (3)(38) + (65)(17)] = [2310 \quad 2719]$$

So, the total cost of equipment for the women's team is \$2310 and the total cost of equipment for the men's team is \$2719.

Checkpoints for Section 8.3

1. To show that B is the inverse of A, show that $AB = I = BA$, as follows

$$AB = \begin{bmatrix} 2 & -1 \\ -3 & 1 \end{bmatrix} \begin{bmatrix} -1 & -1 \\ -3 & -2 \end{bmatrix} = \begin{bmatrix} -2+3 & -2+2 \\ 3-3 & 3-2 \end{bmatrix} = \begin{bmatrix} 1 & 0 \\ 0 & 1 \end{bmatrix}$$

$$BA = \begin{bmatrix} -1 & -1 \\ -3 & -2 \end{bmatrix} \begin{bmatrix} 2 & -1 \\ -3 & 1 \end{bmatrix} = \begin{bmatrix} -2+3 & 1-1 \\ -6+6 & 3-2 \end{bmatrix} = \begin{bmatrix} 1 & 0 \\ 0 & 1 \end{bmatrix}$$

Because $AB = I = BA$, B is the inverse of A.

2. To find the inverse of A, solve the matrix equation $AX = I$ for X.

$$\begin{matrix} A & X & = & I \end{matrix}$$
$$\begin{bmatrix} 1 & -2 \\ -1 & 3 \end{bmatrix}\begin{bmatrix} X_{11} & X_{12} \\ X_{21} & X_{22} \end{bmatrix} = \begin{bmatrix} 1 & 0 \\ 0 & 1 \end{bmatrix}$$
$$\begin{bmatrix} X_{11} - 2X_{21} & X_{12} - 2X_{22} \\ -X_{11} + 3X_{21} & -X_{12} + 3X_{22} \end{bmatrix} = \begin{bmatrix} 1 & 0 \\ 0 & 1 \end{bmatrix}$$

Equating corresponding entries, you obtain two system of linear equations.

$$\begin{cases} X_{11} - 2X_{21} = 1 \\ -X_{11} + 3X_{21} = 0 \end{cases} \qquad \begin{cases} X_{12} - 2X_{22} = 0 \\ -X_{12} + 3X_{22} = 1 \end{cases}$$

Solving the first system yields

$X_{11} = 3$ and $X_{21} = 1$.

Solving the second system yields.

$X_{12} = 2$ and $X_{22} = 1$.

So, the inverse of A is $X = A^{-1} = \begin{bmatrix} 3 & 2 \\ 1 & 1 \end{bmatrix}$.

You can check this by finding AA^{-1} and $A^{-1}A$.

Check:

$$AA^{-1} = \begin{bmatrix} 1 & -2 \\ -1 & 3 \end{bmatrix}\begin{bmatrix} 3 & 2 \\ 1 & 1 \end{bmatrix} = \begin{bmatrix} 1 & 0 \\ 0 & 1 \end{bmatrix} = I \checkmark$$
$$A^{-1}A = \begin{bmatrix} 3 & 2 \\ 1 & 1 \end{bmatrix}\begin{bmatrix} 1 & -2 \\ -1 & 3 \end{bmatrix} = \begin{bmatrix} 1 & 0 \\ 0 & 1 \end{bmatrix} = I \checkmark$$

3. Begin by adjoining the identity matrix to A to form the matrix

$$[A \,\vdots\, I] = \begin{bmatrix} 1 & -2 & -1 & \vdots & 1 & 0 & 0 \\ 0 & -1 & 2 & \vdots & 0 & 1 & 0 \\ 1 & -2 & 0 & \vdots & 0 & 0 & 1 \end{bmatrix}.$$

Use elementary row operations to obtain the form $[I \,\vdots\, A^{-1}]$

$$\begin{matrix} \\ -R_2 \rightarrow \\ -R_1 + R_3 \rightarrow \end{matrix}\begin{bmatrix} 1 & -2 & -1 & \vdots & 1 & 0 & 0 \\ 0 & 1 & -2 & \vdots & 0 & -1 & 0 \\ 0 & 0 & 1 & \vdots & -1 & 0 & 1 \end{bmatrix}$$
$$\begin{matrix} 2R_2 + R_1 \rightarrow \\ 2R_3 - R_2 \rightarrow \\ \\ \end{matrix}\begin{bmatrix} 1 & 0 & -5 & \vdots & 1 & -2 & 0 \\ 0 & 1 & 0 & \vdots & -2 & -1 & 2 \\ 0 & 0 & 1 & \vdots & -1 & 0 & 1 \end{bmatrix}$$
$$\begin{matrix} 5R_3 + R_1 \rightarrow \\ \\ \\ \end{matrix}\begin{bmatrix} 1 & 0 & 0 & \vdots & -4 & -2 & 5 \\ 0 & 1 & 0 & \vdots & -2 & -1 & 2 \\ 0 & 0 & 1 & \vdots & -1 & 0 & 1 \end{bmatrix} = [I \,\vdots\, A^{-1}]$$

So, the matrix A is invertable and its inverse is

$$A^{-1} = \begin{bmatrix} -4 & -2 & 5 \\ -2 & -1 & 2 \\ -1 & 0 & 1 \end{bmatrix}.$$

Confirm this result by multiplying AA^{-1} to obtain I.

Check:

$$AA^{-1} = \begin{bmatrix} 1 & -2 & -1 \\ 0 & -1 & 2 \\ 1 & -2 & 0 \end{bmatrix}\begin{bmatrix} -4 & -2 & 5 \\ -2 & -1 & 2 \\ -1 & 0 & 1 \end{bmatrix} = \begin{bmatrix} 1 & 0 & 0 \\ 0 & 1 & 0 \\ 0 & 0 & 1 \end{bmatrix} = I$$

4. For the matrix A, apply the formula for the inverse of a 2×2 matrix to obtain $ad - bc = (5)(4) - (-1)(3) = 23$

Because this quantity is not zero, the matrix is invertible. The inverse is formed by interchanging the entries on the main diagonal, changing the signs of the other two entries, and multiplying by the scalar $\frac{1}{23}$, as follows.

$$A^{-1} = \frac{1}{ad - bc}\begin{bmatrix} d & -b \\ -c & a \end{bmatrix} \qquad \text{Formula for the inverse of a } 2 \times 2 \text{ matrix}$$
$$= \frac{1}{23}\begin{bmatrix} 4 & 1 \\ -3 & 5 \end{bmatrix} \qquad \text{Substitute for } a, b, c, d, \text{ and the determinant}$$
$$= \begin{bmatrix} \frac{4}{23} & \frac{1}{23} \\ \frac{-3}{23} & \frac{5}{23} \end{bmatrix} \qquad \text{Multiply by the scalar } \frac{1}{23}.$$

5. Begin by writing the system in the matrix form $AX = B$.

$$\begin{bmatrix} 2 & 3 & 1 \\ 3 & 3 & 1 \\ 2 & 4 & 1 \end{bmatrix}\begin{bmatrix} x \\ y \\ z \end{bmatrix} = \begin{bmatrix} -1 \\ 1 \\ -2 \end{bmatrix}$$

Then, use Gauss-Jordan elimination to find A^{-1}.

$$[A \vdots I] = \left[\begin{array}{ccc:ccc} 2 & 3 & 1 & 1 & 0 & 0 \\ 3 & 3 & 1 & 0 & 1 & 0 \\ 2 & 4 & 1 & 0 & 0 & 1 \end{array}\right] \qquad \tfrac{1}{2}R_1 \rightarrow \left[\begin{array}{ccc:ccc} 1 & \frac{3}{2} & \frac{1}{2} & \frac{1}{2} & 0 & 0 \\ 3 & 3 & 1 & 0 & 1 & 0 \\ 2 & 4 & 1 & 0 & 0 & 1 \end{array}\right]$$

$$\begin{array}{r} \\ -3R_1 + R_2 \rightarrow \\ -2R_1 + R_3 \rightarrow \end{array}\left[\begin{array}{ccc:ccc} 1 & \frac{3}{2} & \frac{1}{2} & \frac{1}{2} & 0 & 0 \\ 0 & -\frac{3}{2} & -\frac{1}{2} & -\frac{3}{2} & 1 & 0 \\ 0 & 1 & 0 & -1 & 0 & 1 \end{array}\right]$$

$$\begin{array}{r} \\ R_2 \rightarrow \\ R_3 \rightarrow \end{array}\left[\begin{array}{ccc:ccc} 1 & \frac{3}{2} & \frac{1}{2} & \frac{1}{2} & 0 & 0 \\ 0 & 1 & 0 & -1 & 0 & 1 \\ 0 & -\frac{3}{2} & -\frac{1}{2} & -\frac{3}{2} & 1 & 0 \end{array}\right]$$

$$\begin{array}{r} \\ \\ \frac{3}{2}R_2 + R_3 \rightarrow \end{array}\left[\begin{array}{ccc:ccc} 1 & \frac{3}{2} & \frac{1}{2} & \frac{1}{2} & 0 & 0 \\ 0 & 1 & 0 & -1 & 0 & 1 \\ 0 & 0 & -\frac{1}{2} & -3 & 1 & \frac{3}{2} \end{array}\right]$$

$$\begin{array}{r} \\ \\ -2R_3 \rightarrow \end{array}\left[\begin{array}{ccc:ccc} 1 & \frac{3}{2} & \frac{1}{2} & \frac{1}{2} & 0 & 0 \\ 0 & 1 & 0 & -1 & 0 & 1 \\ 0 & 0 & 1 & 6 & -2 & -3 \end{array}\right]$$

$$\begin{array}{r} -\frac{3}{2}R_2 + R_1 \rightarrow \\ \\ \\ \end{array}\left[\begin{array}{ccc:ccc} 1 & 0 & \frac{1}{2} & 2 & 0 & \frac{3}{2} \\ 0 & 1 & 0 & -1 & 0 & 1 \\ 0 & 0 & 1 & 6 & -2 & -3 \end{array}\right]$$

$$\begin{array}{r} -\frac{1}{2}R_3 + R_1 \rightarrow \\ \\ \\ \end{array}\left[\begin{array}{ccc:ccc} 1 & 0 & 0 & -1 & 1 & 0 \\ 0 & 1 & 0 & -1 & 0 & 1 \\ 0 & 0 & 1 & 6 & -2 & -3 \end{array}\right] = [I \ \vdots \ A^{-1}]$$

$$A^{-1} = \begin{bmatrix} -1 & 1 & 0 \\ -1 & 0 & 1 \\ 6 & -2 & -3 \end{bmatrix}$$

Finally, multiply B by A^{-1} on the left to obtain the solution.

$$X = A^{-1}B = \begin{bmatrix} -1 & 1 & 0 \\ -1 & 0 & 1 \\ 6 & -2 & -3 \end{bmatrix}\begin{bmatrix} -1 \\ 1 \\ -2 \end{bmatrix} = \begin{bmatrix} 2 \\ -1 \\ -2 \end{bmatrix}$$

The solution of the system is $x = 2$, $y = -1$, and $z = -2$

Checkpoints for Section 8.4

1. (a) $\det(A) = \begin{vmatrix} 1 & 2 \\ 3 & -1 \end{vmatrix}$
$= 1(-1) - 3(2)$
$= -1 - 6$
$= -7$

(b) $\det(B) = \begin{vmatrix} 5 & 0 \\ -4 & 2 \end{vmatrix}$
$= 5(2) - (-4)(0)$
$= 10 + 0$
$= 10$

(c) $\det(C) = \begin{vmatrix} 3 & 6 \\ 2 & 4 \end{vmatrix}$
$= 3(4) - (2)(6)$
$= 12 - 12$
$= 0$

2. To find the minor M_{11}, delete the first row and first column of A and evaluate the determinant of the resulting matrix.

$$\begin{bmatrix} 1 & 2 & 3 \\ 0 & -1 & 5 \\ 2 & 1 & 4 \end{bmatrix}, \quad M_{11} = \begin{vmatrix} -1 & 5 \\ 1 & 4 \end{vmatrix} = -1(4) - 1(5) = -9$$

Continuing this pattern, you obtain the minors.

$$M_{12} = \begin{vmatrix} 0 & 5 \\ 2 & 4 \end{vmatrix} = 0(4) - 2(5) = -10$$

$$M_{13} = \begin{vmatrix} 0 & -1 \\ 2 & 1 \end{vmatrix} = 0(1) - 2(-1) = 2$$

$$M_{21} = \begin{vmatrix} 2 & 3 \\ 1 & 4 \end{vmatrix} = 2(4) - 1(3) = 5$$

$$M_{22} = \begin{vmatrix} 1 & 3 \\ 2 & 4 \end{vmatrix} = 1(4) - 2(3) = -2$$

$$M_{23} = \begin{vmatrix} 1 & 2 \\ 2 & 1 \end{vmatrix} = 1(1) - 2(2) = -3$$

$$M_{31} = \begin{vmatrix} 2 & 3 \\ -1 & 5 \end{vmatrix} = 2(5) - (-1)(3) = 13$$

$$M_{32} = \begin{vmatrix} 1 & 3 \\ 0 & 5 \end{vmatrix} = 1(5) - 0(3) = 5$$

$$M_{33} = \begin{vmatrix} 1 & 2 \\ 0 & -1 \end{vmatrix} = 1(-1) - 0(2) = -1$$

Now, to find the cofactors, combine these minors with the checker board pattern of signs for a 3×3 matrix,

$$\begin{bmatrix} + & - & + \\ - & + & - \\ + & - & + \end{bmatrix}.$$

$C_{11} = -9 \qquad C_{12} = 10 \qquad C_{13} = 2$
$C_{21} = -5 \qquad C_{22} = -2 \qquad C_{23} = 3$
$C_{31} = 13 \qquad C_{32} = -5 \qquad C_{33} = -1$

3. The cofactors of the entries in the first row are as follows.

$$C_{11} = +M_{11} = \begin{vmatrix} 5 & 0 \\ 4 & 1 \end{vmatrix} = 5(1) - 4(0) = 5$$

$$C_{12} = -M_{12} = \begin{vmatrix} 3 & 0 \\ -1 & 1 \end{vmatrix} = -(3(1) - (-1)(0)) = -3$$

$$C_{13} = +M_{13} = \begin{vmatrix} 3 & 5 \\ -1 & 4 \end{vmatrix} = 3(4) - (-1)(5) = 17$$

$= 3(5) + 4(-3) + (-2)(17)$
$= -31$

So, by the definition of a determinant, you have the following.

$|A| = a_{11}C_{11} + a_{12}C_{12} + a_{13}C_{13}$

4. Notice that these are two zeros in the third column. So, you can eliminate some of the work in the expansion by using the third column.

$$|A| = a_{13}C_{13} + a_{23}C_{23} + a_{33}C_{33} + a_{43}C_{43}$$
$$= -4C_{13} + 3C_{23} + 0C_{33} + 0C_{43}$$

Because C_{33} and C_{43} have zero coefficients, you need only to find the cofactors of C_{13} and C_{23}.

$$C_{13} = (-1)^{1+3}\begin{vmatrix} 2 & -2 & 6 \\ 1 & 5 & 1 \\ 3 & 1 & -5 \end{vmatrix} = \begin{vmatrix} 2 & -2 & 6 \\ 1 & 5 & 1 \\ 3 & 1 & -5 \end{vmatrix}$$

Expanding by cofactors along the first row yields the following.

$$C_{13} = (2)(-1)^{1+1}\begin{vmatrix} 5 & 1 \\ 1 & -5 \end{vmatrix} + (-2)(-1)^{1+2}\begin{vmatrix} 1 & 1 \\ 3 & -5 \end{vmatrix} + (6)(-1)^{1+3}\begin{vmatrix} 1 & 5 \\ 3 & 1 \end{vmatrix}$$
$$= (2)(1)(-26) + (-2)(-1)(-8) + (6)(1)(-14)$$
$$= -152$$

$$C_{23} = (-1)^{2+3}\begin{vmatrix} 2 & 6 & 2 \\ 1 & 5 & 1 \\ 3 & 1 & -5 \end{vmatrix} = -\begin{vmatrix} 2 & 6 & 2 \\ 1 & 5 & 1 \\ 3 & 1 & -5 \end{vmatrix}$$

Expanding by cofactors along the first row yields the following.

$$C_{23} = -\left((2)(-1)^{1+1}\begin{vmatrix} 5 & 1 \\ 1 & -5 \end{vmatrix} + (6)(-1)^{1+2}\begin{vmatrix} 1 & 1 \\ 3 & -5 \end{vmatrix} + (2)(-1)^{1+3}\begin{vmatrix} 1 & 5 \\ 3 & 1 \end{vmatrix}\right)$$
$$= -((2)(1)(-26) + (6)(-1)(-8) + (2)(1)(-14))$$
$$= 32$$

So, $|A| = -4C_{13} + 3C_{23} + 0C_{33} + 0C_{43}$

$$= -4(-152) + 3(32) + 0 + 0$$
$$= 704.$$

Checkpoints for Section 8.5

1. To begin, find the determinant of the coefficient matrix.

$$D = \begin{vmatrix} 3 & 4 \\ 5 & 3 \end{vmatrix} = 9 - 20 = -11$$

Because this determinant is not zero, you can apply Cramer's Rule.

$$x = \frac{D_x}{D} = \frac{\begin{vmatrix} 1 & 4 \\ 9 & 3 \end{vmatrix}}{-11} = \frac{3 - 36}{-11} = \frac{-33}{-11} = 3$$

$$y = \frac{D_y}{D} = \frac{\begin{vmatrix} 3 & 1 \\ 5 & 9 \end{vmatrix}}{-11} = \frac{27 - 5}{-11} = \frac{22}{-11} = -2$$

So, the solution is $x = 3$ and $y = -2$.

2. To find the determinant of the coefficient matrix, expand along the first row, as follows.

$$\begin{bmatrix} 4 & -1 & 1 \\ 2 & 2 & 3 \\ 5 & -2 & 6 \end{bmatrix}$$

$$D = 4(-1)^2\begin{vmatrix} 2 & 3 \\ -2 & 6 \end{vmatrix} + (-1)(-1)^3\begin{vmatrix} 2 & 3 \\ 5 & 6 \end{vmatrix} + (1)(-1)^4\begin{vmatrix} 2 & 2 \\ 5 & -2 \end{vmatrix} = 4(18) + (1)(-3) + (1)(-14) = 55$$

Because this determinant is not zero, you can apply Cramer's Rule. Next, find D_x, D_y, and D_z.

$$\begin{aligned} D_x &= \begin{vmatrix} 12 & -1 & 1 \\ 1 & 2 & 3 \\ 22 & -2 & 6 \end{vmatrix} \\ &= (12)(-1)^2\begin{vmatrix} 2 & 3 \\ -2 & 6 \end{vmatrix} + (-1)(-1)^3\begin{vmatrix} 1 & 3 \\ 22 & 6 \end{vmatrix} + (1)(-1)^4\begin{vmatrix} 1 & 2 \\ 22 & -2 \end{vmatrix} \\ &= (12)(18) + (1)(-60) + (1)(-46) \\ &= 110 \end{aligned}$$

$$\begin{aligned} D_y &= \begin{vmatrix} 4 & 12 & 1 \\ 2 & 1 & 3 \\ 5 & 22 & 6 \end{vmatrix} \\ &= (4)(-1)^2\begin{vmatrix} 1 & 3 \\ 22 & 6 \end{vmatrix} + (12)(-1)^3\begin{vmatrix} 2 & 3 \\ 5 & 6 \end{vmatrix} + (1)(-1)^4\begin{vmatrix} 2 & 1 \\ 5 & 22 \end{vmatrix} \\ &= (4)(-60) + (-12)(-3) + (1)(39) \\ &= -165 \end{aligned}$$

$$\begin{aligned} D_z &= \begin{vmatrix} 4 & -1 & 12 \\ 2 & 2 & 1 \\ 5 & -2 & 22 \end{vmatrix} \\ &= (4)(-1)^2\begin{vmatrix} 2 & 1 \\ -2 & 22 \end{vmatrix} + (-1)(-1)^3\begin{vmatrix} 2 & 1 \\ 5 & 22 \end{vmatrix} + (12)(-1)^4\begin{vmatrix} 2 & 2 \\ 5 & -2 \end{vmatrix} \\ &= (4)(46) + (1)(39) + (12)(-14) \\ &= 55 \end{aligned}$$

Finally, you can determine the values of x, y, and z as follows.

$$x = \frac{D_x}{D} = \frac{110}{55} = 2$$

$$y = \frac{D_y}{D} = \frac{-165}{55} = -3$$

$$z = \frac{D_z}{D} = \frac{55}{55} = 1$$

So, the solution is $x = 2$, $y = -3$, and $z = 1$.

3.

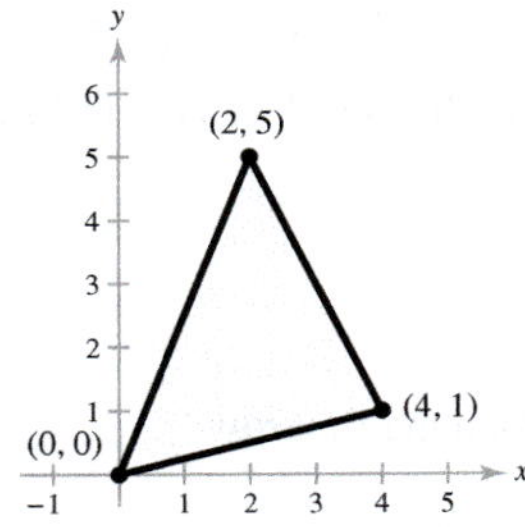

Let $(x_1, y_1) = (0, 0), (x_2, y_2) = (4, 1)$, and $(x_3, y_3) = (2, 5)$. Then, to find the area of the triangle, evaluate the determinant.

$$\begin{vmatrix} x_1 & y_1 & 1 \\ x_2 & y_2 & 1 \\ x_3 & y_3 & 1 \end{vmatrix} = \begin{vmatrix} 0 & 0 & 1 \\ 4 & 1 & 1 \\ 2 & 5 & 1 \end{vmatrix}$$

$$= (0)(-1)^2\begin{vmatrix} 1 & 1 \\ 5 & 1 \end{vmatrix} + (0)(-1)^3\begin{vmatrix} 4 & 1 \\ 2 & 1 \end{vmatrix} + (1)(-1)^4\begin{vmatrix} 4 & 1 \\ 2 & 5 \end{vmatrix}$$

$$= 0 + 0 + (1)(18)$$

$$= 18$$

Using this value, you can conclude that the area of the triangle is

$$\text{Area} = \frac{1}{2}\begin{vmatrix} 0 & 0 & 1 \\ 4 & 1 & 1 \\ 2 & 5 & 1 \end{vmatrix} = \frac{1}{2}(18) = 9 \text{ square units.}$$

4.

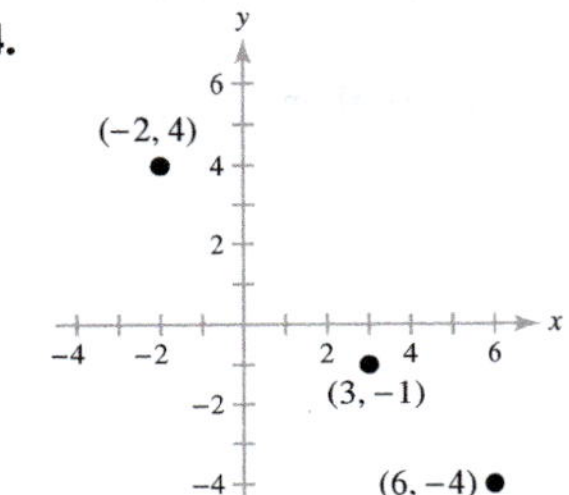

To determine if the points are collinear, let $(x_1, y_1) = (-2, 4), (x_2, y_2) = (3, -1)$, and $(x_3, y_3) = (6, -4)$. Then, evaluate the determinant as follows.

$$\begin{vmatrix} x_1 & y_1 & 1 \\ x_2 & y_2 & 1 \\ x_3 & y_3 & 1 \end{vmatrix} = \begin{vmatrix} -2 & 4 & 1 \\ 3 & -1 & 1 \\ 6 & -4 & 1 \end{vmatrix}$$

$$= (-2)(-1)^2\begin{vmatrix} -1 & 1 \\ -4 & 1 \end{vmatrix} + (4)(-1)^3\begin{vmatrix} 3 & 1 \\ 6 & 1 \end{vmatrix} + (1)(-1)^4\begin{vmatrix} 3 & -1 \\ 6 & -4 \end{vmatrix}$$

$$= (-2)(3) + (-4)(-3) + (1)(-6)$$

$$= 0$$

Because the value of this determinant is equal to zero, you can conclude that the three points are collinear.

5.

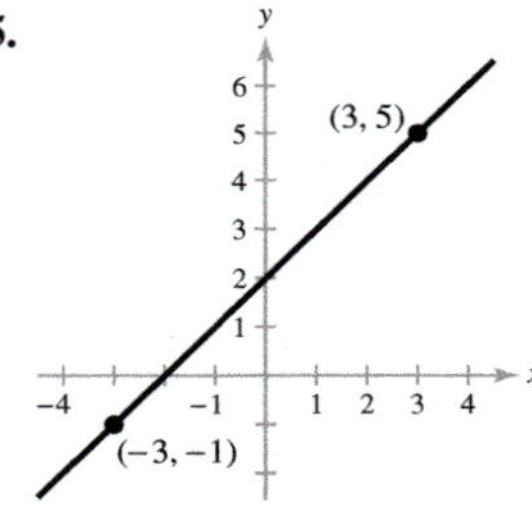

To find an equation of the line, let $(x_1, y_1) = (-3, -1)$ and $(x_2, y_2) = (3, 5)$.

Applying the determinant formula for the equation of a line produces the following.

$$\begin{vmatrix} x & y & 1 \\ -3 & -1 & 1 \\ 3 & 5 & 1 \end{vmatrix} = 0$$

To evaluate this determinant, expand by cofactors along the first row.

$$x(-1)^2\begin{vmatrix} -1 & 1 \\ 5 & 1 \end{vmatrix} + y(-1)^3\begin{vmatrix} -3 & 1 \\ 3 & 1 \end{vmatrix} + (1)(-1)^4\begin{vmatrix} -3 & -1 \\ 3 & 5 \end{vmatrix} = 0$$

$$(x)(-6) + (-y)(-6) + (1)(-12) = 0$$

$$-6x + 6y - 12 = 0$$

$$x - y + 2 = 0$$

So, an equation passing through the two points is $x - y + 2 = 0$.

6. To find the image of the square with vertices $(0, 0)$, $(2, 0)$, $(0, 2)$ and $(2, 2)$ after a vertical stretch by a factor of $k = 2$, multiply the vertices matrix by $\begin{bmatrix} 1 & 0 \\ 0 & 2 \end{bmatrix}$

$\begin{bmatrix} 1 & 0 \\ 0 & 2 \end{bmatrix}\begin{bmatrix} 0 \\ 0 \end{bmatrix} = \begin{bmatrix} 0 \\ 0 \end{bmatrix}$, $\begin{bmatrix} 1 & 0 \\ 0 & 2 \end{bmatrix}\begin{bmatrix} 2 \\ 0 \end{bmatrix} = \begin{bmatrix} 2 \\ 0 \end{bmatrix}$, $\begin{bmatrix} 1 & 0 \\ 0 & 2 \end{bmatrix}\begin{bmatrix} 0 \\ 2 \end{bmatrix} = \begin{bmatrix} 0 \\ 4 \end{bmatrix}$ and $\begin{bmatrix} 1 & 0 \\ 0 & 2 \end{bmatrix}\begin{bmatrix} 2 \\ 2 \end{bmatrix} = \begin{bmatrix} 2 \\ 4 \end{bmatrix}$. The vertices of the image are $(0, 0)$, $(2, 0)$, $(0, 4)$ and $(2, 4)$.

7. To find the area of the parallelogram with vertices $(0, 0)$, (a, b), (c, d) and $(a + c, b + d)$, you can use the formula

$Area = |\det(A)|$ where $A = \begin{bmatrix} a & b \\ c & d \end{bmatrix}$. If the parallelogram has vertices $(0, 0)$, $(5, 5)$, $(2, 4)$ and $(7, 9)$ then $a = 2$, $b = 4$, $c = 5$, and $d = 5$.

$$Area = |\det(A)| = \begin{vmatrix} 2 & 4 \\ 5 & 5 \end{vmatrix}$$

$$= |(2)(5) - (4)(5)|$$

$$= |-10| = 10 \text{ square units.}$$

8. Partitioning the message (including blank spaces, but ignoring any punctuation) into groups of three produces the following uncoded 1×3 row matrices.

[15	23	12]	[19	0	1]	[18	5	0]	[14	15	3]	[20	21	18]	[14	1	12]
O	W	L	S		A	R	E		N	O	C	T	U	R	N	A	L

9. The coded row matrices are obtained by multiplying each of the uncoded row matrices found in Checkpoint Example 6 by the matrix A, as follows.

Uncoded Matrix	**Encoding Matrix A**		**Coded Matrix**
$\begin{bmatrix} 15 & 23 & 12 \end{bmatrix}$	$\begin{bmatrix} 1 & -1 & 0 \\ 1 & 0 & -1 \\ 6 & -2 & -3 \end{bmatrix}$	$=$	$\begin{bmatrix} 110 & -39 & -59 \end{bmatrix}$
$\begin{bmatrix} 19 & 0 & 1 \end{bmatrix}$	$\begin{bmatrix} 1 & -1 & 0 \\ 1 & 0 & -1 \\ 6 & -2 & -3 \end{bmatrix}$	$=$	$\begin{bmatrix} 25 & -21 & -3 \end{bmatrix}$
$\begin{bmatrix} 18 & 5 & 0 \end{bmatrix}$	$\begin{bmatrix} 1 & -1 & 0 \\ 1 & 0 & -1 \\ 6 & -2 & -3 \end{bmatrix}$	$=$	$\begin{bmatrix} 23 & -18 & -5 \end{bmatrix}$
$\begin{bmatrix} 14 & 15 & 3 \end{bmatrix}$	$\begin{bmatrix} 1 & -1 & 0 \\ 1 & 0 & -1 \\ 6 & -2 & -3 \end{bmatrix}$	$=$	$\begin{bmatrix} 47 & -20 & -24 \end{bmatrix}$
$\begin{bmatrix} 20 & 21 & 18 \end{bmatrix}$	$\begin{bmatrix} 1 & -1 & 0 \\ 1 & 0 & -1 \\ 6 & -2 & -3 \end{bmatrix}$	$=$	$\begin{bmatrix} 149 & -56 & -75 \end{bmatrix}$
$\begin{bmatrix} 14 & 1 & 12 \end{bmatrix}$	$\begin{bmatrix} 1 & -1 & 0 \\ 1 & 0 & -1 \\ 6 & -2 & -3 \end{bmatrix}$	$=$	$\begin{bmatrix} 87 & -38 & -37 \end{bmatrix}$

So, the crytogram is $110\ -39\ -59\ 25\ -21\ -3\ 23\ -18\ -5\ 47\ -20\ -24\ 149\ -56\ -75\ 87\ -38\ -37$.

10. First find the decoding matrix A^{-1} using matrix A from the Checkpoint Example 7.

$$[A \;\vdots\; I] = \begin{bmatrix} 1 & -1 & 0 & \vdots & 1 & 0 & 0 \\ 1 & 0 & -1 & \vdots & 0 & 1 & 0 \\ 6 & -2 & -3 & \vdots & 0 & 0 & 1 \end{bmatrix}$$

$$\begin{matrix} R_2 \rightarrow \\ R_1 \rightarrow \\ -6R_1 + R_3 \rightarrow \end{matrix} \begin{bmatrix} 1 & 0 & -1 & \vdots & 0 & 1 & 0 \\ 1 & -1 & 0 & \vdots & 1 & 0 & 0 \\ 0 & 4 & -3 & \vdots & -6 & 0 & 1 \end{bmatrix}$$

$$\begin{matrix} \\ -R_1 + R_2 \rightarrow \\ \\ \end{matrix} \begin{bmatrix} 1 & 0 & -1 & \vdots & 0 & 1 & 0 \\ 0 & -1 & 1 & \vdots & 1 & -1 & 0 \\ 0 & 4 & -3 & \vdots & -6 & 0 & 1 \end{bmatrix}$$

$$\begin{matrix} \\ \\ 4R_2 + R_3 \rightarrow \end{matrix} \begin{bmatrix} 1 & 0 & -1 & \vdots & 0 & 1 & 0 \\ 0 & -1 & 1 & \vdots & 1 & -1 & 0 \\ 0 & 0 & 1 & \vdots & -2 & -4 & 1 \end{bmatrix}$$

$$\begin{matrix} \\ -R_2 \rightarrow \\ \\ \end{matrix} \begin{bmatrix} 1 & 0 & -1 & \vdots & 0 & 1 & 0 \\ 0 & 1 & -1 & \vdots & -1 & 1 & 0 \\ 0 & 0 & 1 & \vdots & -2 & -4 & 1 \end{bmatrix}$$

$$\begin{matrix} R_1 + R_3 \rightarrow \\ R_2 + R_3 \rightarrow \\ \\ \end{matrix} \begin{bmatrix} 1 & 0 & 0 & \vdots & -2 & -3 & 1 \\ 0 & 1 & 0 & \vdots & -3 & -3 & 1 \\ 0 & 0 & 1 & \vdots & -2 & -4 & 1 \end{bmatrix} = \left[I \,\vdots\, A^{-1}\right]$$

Partition the message into groups of three to form the coded row matrices. Finally, multiply each coded row matrix by A^{-1} (on the right).

Coded Matrix	**Decoding Matrix** A^{-1}		**Decoded Matrix**
$[110 \quad -39 \quad -59]$	$\begin{bmatrix} -2 & -3 & 1 \\ -3 & -3 & 1 \\ -2 & -4 & 1 \end{bmatrix}$	=	$[15 \quad 23 \quad 12]$
$[25 \quad -21 \quad -3]$	$\begin{bmatrix} -2 & -3 & 1 \\ -3 & -3 & 1 \\ -2 & -4 & 1 \end{bmatrix}$	=	$[19 \quad 0 \quad 1]$
$[23 \quad -18 \quad -5]$	$\begin{bmatrix} -2 & -3 & 1 \\ -3 & -3 & 1 \\ -2 & -4 & 1 \end{bmatrix}$	=	$[18 \quad 5 \quad 0]$
$[47 \quad -20 \quad -24]$	$\begin{bmatrix} -2 & -3 & 1 \\ -3 & -3 & 1 \\ -2 & -4 & 1 \end{bmatrix}$	=	$[14 \quad 15 \quad 3]$
$[149 \quad -56 \quad -75]$	$\begin{bmatrix} -2 & -3 & 1 \\ -3 & -3 & 1 \\ -2 & -4 & 1 \end{bmatrix}$	=	$[20 \quad 21 \quad 18]$
$[87 \quad -38 \quad -37]$	$\begin{bmatrix} -2 & -3 & 1 \\ -3 & -3 & 1 \\ -2 & -4 & 1 \end{bmatrix}$	=	$[14 \quad 1 \quad 12]$

So, the message is as follows.

[15	23	12]	[19	0	1]	[18	5	0]	[14	15	3]	[20	21	18]	[14	1	12]
O	W	L	S		A	R	E		N	O	C	T	U	R	N	A	L

Chapter 9

Checkpoints for Section 9.1

1. The first four terms of the sequence given by $a_n = 2n + 1$ are as follows.

$a_1 = 2(1) + 1 = 3$ 1st term

$a_2 = 2(2) + 1 = 5$ 2nd term

$a_3 = 2(3) + 1 = 7$ 3rd term

$a_4 = 2(4) + 1 = 9$ 4th term

2. The first four terms of the sequence given by $a_n = \dfrac{2 + (-1)^n}{n}$ are as follows.

$$a_1 = \frac{2 + (-1)^1}{1} = \frac{2 - 1}{1} = 1$$

$$a_2 = \frac{2 + (-1)^2}{2} = \frac{2 + 1}{2} = \frac{3}{2}$$

$$a_3 = \frac{2 + (-1)^3}{3} = \frac{2 - 1}{3} = \frac{1}{3}$$

$$a_4 = \frac{2 + (-1)^4}{4} = \frac{2 + 1}{4} = \frac{3}{4}$$

3. (a) n: 1 2 3 4 ... n

Terms: 1 5 9 13 ... a_n

Apparent pattern: Each term is 3 less than 4 times n, which implies that $a_n = 4n - 3$.

(b) n: 1 2 3 4 ... n

Terms: 2 −4 6 −8 ... a_n

Apparent pattern: The absolute value of each term is 2 times n, and the terms have alternating signs, with those in the even positions being negative. This implies that $a_n = (-1)^{n+1}2n$.

4. The first five terms of the sequence are as follows.

$a_1 = 6$ and $a_{k+1} = a_k + 1$

$a_1 = 6$ 1st term is given.

$a_2 = a_{1+1} = a_1 + 1 = 6 + 1 = 7$ Use recursion formula.

$a_3 = a_{2+1} = a_2 + 1 = 7 + 1 = 8$ Use recursion formula.

$a_4 = a_{3+1} = a_3 + 1 = 8 + 1 = 9$ Use recursion formula.

$a_5 = a_{4+1} = a_4 + 1 = 9 + 1 = 10$ Use recursion formula.

5. The first five terms of the sequence are as follows,

$a_0 = 1$ 0th term is given.

$a_1 = 3$ 1st term is given.

$a_2 = a_{2-2} + a_{2-1} = a_0 + a_1 = 1 + 3 = 4$ Use recursion formula.

$a_3 = a_{3-2} + a_{3-1} = a_1 + a_2 = 3 + 4 = 7$ Use recursion formula.

$a_4 = a_{4-2} + a_{4-1} = a_2 + a_3 = 4 + 7 = 11$ Use recursion formula.

6. Algebraic Solution

$$a_0 = \frac{3^0+1}{0!} = \frac{1+1}{1} = 2$$

$$a_1 = \frac{3^1+1}{1!} = \frac{3+1}{1} = 4$$

$$a_2 = \frac{3^2+1}{2!} = \frac{9+1}{2} = \frac{10}{2} = 5$$

$$a_3 = \frac{3^3+1}{3!} = \frac{27+1}{6} = \frac{28}{6} = \frac{14}{3}$$

$$a_4 = \frac{3^4+1}{4!} = \frac{81+1}{24} = \frac{82}{24} = \frac{41}{12}$$

Graphical Solution

Using a graphing utility set to *dot* and *sequence* modes, enter the sequence. Next, graph the sequence.

You can estimate the first five terms of the sequence as follows.

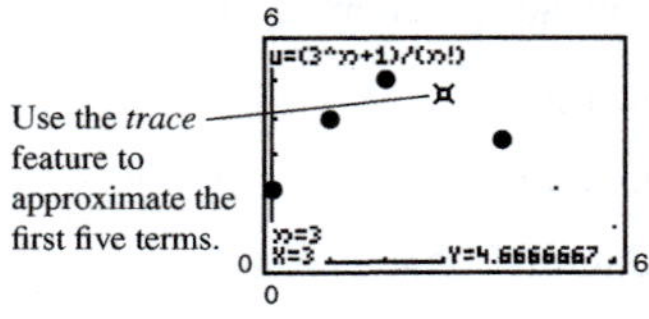
Use the *trace* feature to approximate the first five terms.
6
0
6
0

$$u_0 = 2$$

$$u_1 = 4$$

$$u_2 = 5$$

$$u_3 \approx 4.667 = \frac{14}{3}$$

$$u_4 \approx 3.417 = \frac{41}{12}$$

7. $$\frac{4!(n+1)!}{3!n!} = \frac{(1\cdot 2\cdot 3\cdot 4)[1\cdot 2\cdot 3\dots n\cdot(n+1)]}{(1\cdot 2\cdot 3)(1\cdot 2\cdot 3\dots n)}$$
$$= 4(n+1)$$

8. $$\sum_{i=1}^{4}(4i+1) = [4(1)+1]+[4(2)+1]+[4(3)+1]+[4(4)+1]$$
$$= 5 + 9 + 13 + 17$$
$$= 44$$

9. (a) The fourth partial sum is as follows.

$$\sum_{i=1}^{4}\frac{5}{10^i} = \frac{5}{10^1}+\frac{5}{10^2}+\frac{5}{10^3}+\frac{5}{10^4}$$
$$= 0.5 + 0.05 + 0.005 + 0.0005$$
$$= 0.5555$$

(b) The sum of the series is as follows.

$$\sum_{i=1}^{\infty}\frac{5}{10^i} = \frac{5}{10^1}+\frac{5}{10^2}+\frac{5}{10^3}+\frac{5}{10^4}+\frac{5}{10^5}+\cdots$$
$$= 0.5 + 0.05 + 0.005 + 0.0005 + 0.00005 + \cdots$$
$$= 0.55555\dots$$
$$= \frac{5}{9}$$

10. (a) The first three terms of the sequence are as follows.

$$A_0 = 1000\left(1+\frac{0.03}{12}\right)^0 = \$1000 \quad \text{Original deposit}$$

$$A_1 = 1000\left(1+\frac{0.03}{12}\right)^1 = \$1002.50 \quad \text{First-month balance}$$

$$A_2 = 1000\left(1+\frac{0.03}{12}\right)^2 \approx \$1005.01 \quad \text{Second-month balance.}$$

(b) The 48th term of the sequence is

$$A_{48} = 1000\left(1+\frac{0.03}{12}\right)^{48} \approx \$1127.33 \quad \text{Four-year balance}$$

Checkpoints for Section 9.2

1. The sequence whose nth term is $3n - 1$ is arithmetic. The first four terms are as follows.

$3(1) - 1 = 2$

$3(2) - 1 = 5$

$3(3) - 1 = 8$

$3(4) - 1 = 11$

For this sequence, the common difference between consecutive terms is 3.

$\underbrace{2, 5}_{5-2=3}, 8, 11, \ldots$

2. You know that the formula for the nth term is of the form $a_n = a_1 + (n-1)d$. Because the common difference is $d = 5$ and the first term is $a_1 = -1$, the formula must have the form

$a_n = a_1 + (n-1)d = -1 + 5(n-1)$.

So, the formula for the nth term is $a_n = 5n - 6$.

The sequence therefore has the following form.

$-1, 4, 9, 14, \ldots, 5n - 6, \ldots$

The figure below shows a graph of the first 15 terms of the sequence. Notice that the points lie on a line.

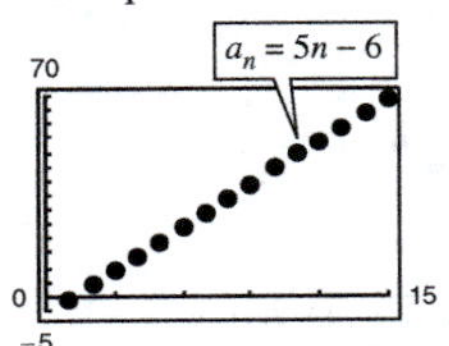

3. You know that $a_8 = 25$ and $a_{12} = 41$. So, you must add the common difference d four times to the eighth term to obtain the 12th term. Therefore, the eighth term and the 12th terms of the sequence are related by

$a_{12} = a_8 + 4d.$

Using $a_8 = 25$ and $a_{12} = 41$, solve for d.

$$\begin{aligned} a_{12} &= a_8 + 4d \\ 41 &= 25 + 4d \\ 16 &= 4d \\ 4 &= d \end{aligned}$$

Use the formula for the nth term of an arithmetic sequence to find a_1.

$$\begin{aligned} a_n &= a_1 + (n-1)d \\ a_8 &= a_1 + (8-1)(4) \\ 25 &= a_1 + (7)(4) \\ -3 &= a_1 \end{aligned}$$

So, the formula for the nth term of the sequence is

$a_n = -3 + (n-1)(4) = -3 + 4n - 4 = 4n - 7.$

The sequence is as follows.

a_1	a_2	a_3	a_4	a_5	a_6	a_7	a_8	a_9	a_{10}	a_{11}	$\ldots$
-3	1	5	9	13	17	21	25	29	33	37	$\ldots$

4. For this arithmetic sequence, the common difference is $d = 15 - 7 = 8$.

There are two ways to find the tenth term. One way is to write the first ten terms (by repeatedly adding 8).

7, 15, 23, 31, 39, 47, 55, 63, 71, 79

So, the tenth term is 79.

Another way to find the tenth term is to first find a formula for the nth term. Because the common difference is $d = 8$ and the first term is $a_1 = 7$, the formula must have the form

$a_n = a_1 + (n-1)d = 7 + (n-1)(8).$

Therefore, a formula for the nth term is $a_n = 8n - 1$ which implies that the tenth term is

$a_{10} = 8(10) - 1 = 79.$

5. To begin, notice that the sequence is arithmetic (with a common difference of $d = 37 - 40 = -3$).

Moreover, the sequence has 7 terms. So, the sum of the sequence is

$S_n = \frac{n}{2}(a_1 + a_n)$ Sum of a finite arithmetic sequence

$= \frac{7}{2}(40 + 22)$ Substitute 7 for n, 40 for a_1, and 22 for a_n.

$= 217.$ Simplify.

6. (a) The integers from 1 to 35 form an arithmetic sequence that has 35 terms. So, you can use the formula for the sum of a finite arithmetic sequence, as follows.

$S_n = 1 + 2 + 3 + \ldots + 34 + 35$

$= \frac{n}{2}(a_1 + a_n)$ Sum of a finite arithmetic sequence

$= \frac{35}{2}(1 + 35)$ Substitute 35 for n, 1 for a_1, and 35 for a_n.

$= 630$ Simplify.

(b) The sum of the integers from 1 to $2N$ form an arithmetic sequence that has $2N$ terms.

$S_n = 1 + 2 + 3 + \ldots + (2N - 1) + 2N$

$= \frac{n}{2}(a_1 + a_n)$ Sum of a finite arithmetic sequence.

$= \frac{2N}{2}(1 + 2N)$ Substitute $2N$ for n, 1 for a_1 and $2N$ for a_n.

$= N(1 + 2N)$ Simplify.

7. For this arithmetic sequence, $a_1 = 6$ and $d = 12 - 6 = 6$.

So, $a_n = a_1 + (n - 1)d = 6 + 6(n - 1)$ and the nth term is $a_n = 6n$.

Therefore, $a_{120} = 6(120) = 720$, and the sum of the first 120 terms is

$$S_{120} = \frac{n}{2}(a_1 + a_{120})$$
$$= \frac{120}{2}(6 + 720)$$
$$= 60(726)$$
$$= 43{,}560.$$

8. For this arithmetic sequence, $a_1 = 78$ and $d = 76 - 78 = -2$.

So, $a_n = 78 + (-2)(n - 1)$ and the nth term is $a_n = -2n + 80$.

Therefore, $a_{30} = -2(30) + 80 = 20$, and the sum of the first 30 terms is

$$S_{30} = \frac{n}{2}(a_1 + a_{30})$$
$$= \frac{30}{2}(78 + 20)$$
$$= 15(98)$$
$$= 1470.$$

9. The annual sales form an arithmetic sequence in which $a_1 = 160{,}000$ and $d = 20{,}000$.

So, $a_n = 160{,}000 + 20{,}000(n - 1)$ and the nth term of the sequence is $a_n = 20{,}000n + 140{,}000$.

Therefore, the 10th term of the sequence is

$$a_{10} = 20{,}000(10) + 140{,}000$$
$$= 340{,}000.$$

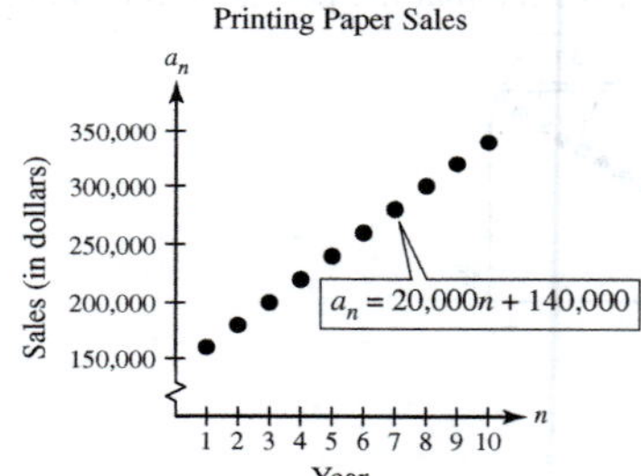

The sum of the first 10 terms of the sequence is

$$S_{10} = \frac{n}{2}(a_1 + a_{10})$$
$$= \frac{10}{2}(160{,}000 + 340{,}000)$$
$$= 5(500{,}000)$$
$$= 2{,}500{,}000.$$

So, the total sales for the first 10 years will be \$2,500,000.

Checkpoints for Section 9.3

1. The sequence whose nth term is $6(-2)^n$ is geometric.

For this sequence, the common ratio of consecutive terms is -2.

The first four terms, beginning with $n = 1$ are as follows.

$a_1 = 6(-2)^1 = 6(-2) = -12$

$a_2 = 6(-2)^2 = 6(4) = 24$

$a_3 = 6(-2)^3 = 6(-8) = -48$

$a_4 = 6(-2)^4 = 6(16) = 96$

So the sequence of terms are

$\underbrace{-12, 24,}_{\frac{24}{-12}=-2} -48, 96, \ldots, 6(-2)^n, \ldots$

2. Starting with $a_1 = 2$, repeatedly multiply by 4 to obtain the following.

$a_1 = 2$	1st term
$a_2 = 2(4^1) = 8$	2nd term
$a_3 = 2(4^2) = 32$	3rd term
$a_4 = 2(4^3) = 128$	4th term
$a_5 = 2(4)^4 = 512$	5th term

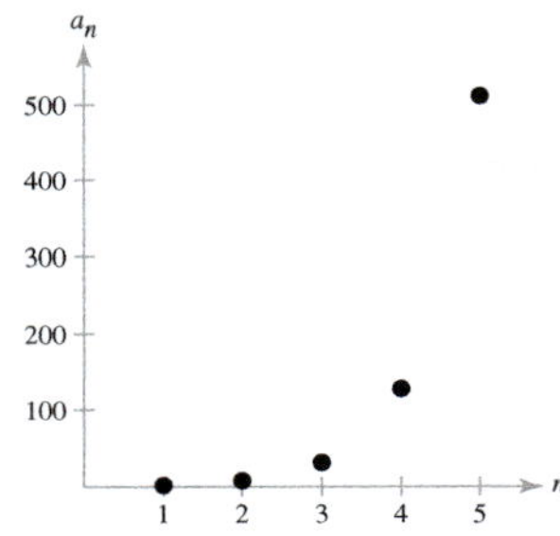

3. Algebraic Solution

Use the formula for the nth term of a geometric sequence.

$a_n = a_1 r^{n-1}$

$a_{12} = 14(1.2)^{12-1}$

$= 14(1.2)^{11}$

≈ 104.02

Numerical Solution

For this sequence, $r = 1.2$ and $a_1 = 14$. So $a_n = 14(1.2)^{n-1}$. Use a graphing utility to create a table that shows the terms of the sequence.

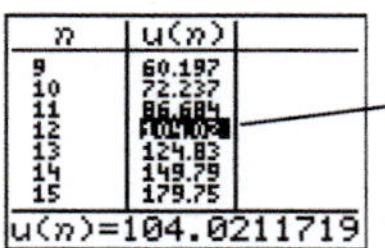

4. The first few terms of the geometric sequence are 4, 20, 100, ... You can find the common ratio of this geometric sequence by dividing any term by the previous term so $r = \frac{20}{4} = 5$. Because the first term is $a_1 = 4$, the formula for the nth term of a geometric sequence is as follows.

$a_n = a_1 r^{n-1}$

$a_n = 4(5)^{n-1}$

The 12th term of the sequence is as follows.

$a_{12} = 4(5)^{11} = 195{,}312{,}500$

5. The fifth term is related to the second term by the equation $a_5 = a_2 r^3$.

Because $a_5 = \frac{81}{4}$ and $a_2 = 6$, you can solve for r as follows.

$a_5 = a_2 r^3$	Multiply the second term by r^{5-3}.
$\frac{81}{4} = 6r^3$	Substitute $\frac{81}{4}$ for a_5 and 6 for a_2.
$\frac{27}{8} = r^3$	Divide each side by 6.
$\frac{3}{2} = r$	Take the cube root of each side.

You can obtain the eighth term by multiplying the fifth term by r^3.

$a_8 = a_5 r^3$	Multiply the fifth term by r^{8-5}.
$= \frac{81}{4}\left(\frac{3}{2}\right)^3$	Substitute $\frac{81}{3}$ for a_5 and $\frac{3}{2}$ for r
$= \frac{81}{4}\left(\frac{27}{8}\right)$	Evaluate power.
$= \frac{2187}{32}$	Multiply fractions.

6. You have $\sum_{i=1}^{10} 2(0.25)^{i-1} = 2(0.25)^0 + 2(0.25)^1 + 2(0.25)^2 + \ldots + 2(0.25)^9$.

Now, $a_1 = 2$, $r = 0.25$, and $n = 10$, so applying the formula for the sum of a finite geometric sequence, you obtain the following.

$$S_n = a_1\left(\frac{1 - r^n}{1 - r}\right) \qquad \text{Sum of a finite geometric series}$$

$$\sum_{i=1}^{10} 2(0.25)^{i-1} = 2\left[\frac{1 - (0.25)^{10}}{1 - 0.25}\right] \qquad \text{Substitute 2 for } a_1\text{, 0.25 for } r\text{, and 10 for } n.$$

$$\approx 2.667 \qquad \text{Use a calculator.}$$

7. (a) $\sum_{n=0}^{\infty} 5(0.5)^n = 5 + 5(0.5) + 5(0.5)^2 + \ldots + 5(0.5)^n + \ldots$

$$= \frac{5}{1 - 0.5} \qquad \text{Use } \frac{a_1}{1 - r} \text{ and Subsitute 5 for } a_1 \text{ and 0.5 for } r.$$

$$= \frac{5}{0.5}$$

$$= 10$$

(b) To find the common ratio, divide any term by the preceding term.

So, $r = \frac{1}{5} = 0.2$.

The sum of the infinite geometric series is as follows.

$$5 + 1 + 0.2 + 0.04 + \ldots = 5(0.2)^0 + 5(0.2)^1 + 5(0.2)^2 + 5(0.2)^3 + \ldots$$

$$= \frac{a_1}{1 - r}$$

$$= \frac{5}{1 - 0.2}$$

$$= 6.25$$

8. To find the balance in the account after 48 months, consider each of the 48 deposits separately. The first deposit will gain interest for 48 months, and its balance will be

$$A_{48} = 70\left(1 + \frac{0.02}{12}\right)^{48}.$$

The second deposit will gain interest for 47 months, and its balance will be

$$A_{47} = 70\left(1 + \frac{0.02}{12}\right)^{47}.$$

The last deposit will gain interest for only 1 month, and its balance will be

$$A_1 = 70\left(1 + \frac{0.02}{12}\right)^{1}.$$

The total balance in the annuity will be the sum of the balances of the 48 deposits.

Using the formula for the sum of a finite geometric sequence, with $A_1 = 70(1.0017)$ $r = 1.0017$, and $n = 48$ you have

$$S_n = A_1\left(\frac{1 - r^n}{1 - r}\right)$$

$$S_{48} = 70\left(1 + \frac{0.02}{12}\right)\left[\frac{1 - \left(1 + \frac{0.02}{12}\right)^{48}}{1 - \left(1 + \frac{0.02}{12}\right)}\right] \approx \$3500.85.$$

Checkpoints for Section 9.4

1. (a) $P_{k+1} : S_{k+1} = \dfrac{6}{(k+1)(k+1+3)}$

$= \dfrac{6}{(k+1)(k+4)}$

(b) $P_{k+1} : k + 1 + 2 \le 3(k+1-1)^2$

$k + 3 \le 3k^2$

(c) $P_{k+1} : 2^{4(k+1)-2} + 1 > 5(k+1)$

$2^{4k+4-2} + 1 > 5k + 5$

$2^{4k+2} + 1 > 5k + 5$

2. Mathematical induction consists of two distinct parts.

1. First, you must show that the formula is true when $n = 1$. When $n = 1$, the formula is valid, because $S_1 = 1(1+4) = 5$.
2. The second part of mathematical induction has two steps. The first step is to *assume* that the formula is valid for some integer k. The second step is to use this assumption to prove that the formula is valid for the *next* integer, $k+1$. Assuming that the formula

   $S_k = 5 + 7 + 9 + 11 + \ldots + (2k+3) = k(k+4)$

   is true, you must show that the formula

   $S_{k+1} = (k+1)(k+1+4)$

   $= (k+1)(k+5)$ is true.

   $S_{k+1} = 5 + 7 + 9 + 11 + \ldots + (2k+3) + [2(k+1)+3]$

   $= [5 + 7 + 9 + 11 + \ldots + (2k+3)] + (2k+2+3)$

   $= S_k + 2(k+5)$

   $= k(k+4) + 2k + 5$

   $= k^2 + 4k + 2k + 5$

   $= k^2 + 6k + 5$

   $= (k+1)(k+5)$

   Combining the results of parts (1) and (2), you can conclude by mathematical induction that the formula is valid for all integers $n \ge 1$.

3. 1. When $n = 1$, the formula is valid, because $S_1 = (1)(1-1) = \dfrac{(1)(1-1)(1+1)}{3} = 0.$

2. Assuming that $S_k = 1(1-1) + 2(2-1) + 3(3-1) + \ldots + k(k-1) = \dfrac{k(k-1)(k+1)}{3}$

   You must show that $S_{k+1} = \dfrac{(k+1)(k+1-1)(k+1+1)}{3} = \dfrac{k(k+1)(k+2)}{3}.$

   To do this, write the following.

   $S_{k+1} = S_k + a_{k+1}$

   $= [1(1-1) + 2(2-1) + 3(3-1) + \ldots + k(k-1)] + (k+1)(k+1-1)$ Substitution

   $= \dfrac{k(k-1)(k+1)}{3} + k(k+1)$ By assumption

   $= \dfrac{k(k-1)(k+1) + 3k(k+1)}{3}$ Combine fractions.

   $= \dfrac{k(k+1)[(k-1)+3]}{3}$ Factor.

   $= \dfrac{k(k+1)(k+2)}{3}$ Simplify.

   So, S_k implies S_{k+1}.

   Combining the results of parts (1) and (2), you can conclude by mathematical induction that the formula is valid for all positive integers n.

4. 1. For $n = 1$ and $n = 2$, the statement is true because

$1! \geq 1$ and $2! \geq 2$.

2. Assuming that

$k! \geq k$

You need to show that $(k + 1)! \geq k + 1$. For $n = k$, you have $(k + 1)! = (k + 1)k! \geq (k + 1)k$.

Because $(k + 1)! > k + 1$ for all $k > 1$, it follows that $(k + 1)! \geq k + 1$.

Combining the results of parts (1) and (2), you can conclude by mathematical induction that $n! \geq n$ for all integers $n \geq 1$.

5. 1. For $n = 1$, the statement is true because

$3^1 + 1 = 4$.

So, 2 is a factor.

2. Assuming that 2 is a factor of $3^k + 1$, you must show that 2 is a factor of $3^{k+1} + 1$.

To do this, write the following.

$3^{k+1} + 1 = 3^{k+1} - 3^k + 3^k + 1$	Subtract and add 3^k.
$= 3^k(3 - 1) + 3^k + 1$	Regroup terms.
$= 3^k \cdot 2 + 3^k + 1$	Simplify.

Because 2 is a factor of $3^k \cdot 2$ and 2 is also a factor of $3^k + 1$, it follows that 2 is a factor of $3^{k+1} + 1$. Combining the results of parts (1) and (2), you can conclude by mathematical induction that 2 is a factor of $3^n + 1$ for all positive integers n.

6. Begin by writing the first few sums.

$S_1 = 3 = 1(3)$

$S_2 = 3 + 7 = 10 = 2(5)$

$S_3 = 3 + 7 + 11 = 21 = 3(7)$

$S_4 = 3 + 7 + 11 + 15 = 36 = 4(9)$

$S_5 = 3 + 7 + 11 + 15 + 19 = 55 = 5(11)$

From this sequence, it appears that the formula for the kth sum is

$S_k = 3 + 7 + 11 + 15 + 19 + \ldots + 4k - 1 = k(2k + 1)$.

To prove the validity of this hypothesis, use mathematical induction. Note that you have already verified the formula for $n = 1$, so begin by assuming that the formula is valid for $n = k$ and trying to show that it is valid for $n = k + 1$.

$$\begin{aligned}
S_{k+1} &= [3 + 7 + 11 + 15 + \ldots + 4k - 1] + 4(k + 1) - 1 \\
&= k(2k + 1) + 4k + 3 \\
&= 2k^2 + k + 4k + 3 \\
&= 2k^2 + 5k + 3 \\
&= (k + 1)(2k + 3) \\
&= (k + 1)[2(k + 1) + 1]
\end{aligned}$$

So, by mathematical induction, the hypothesis is valid.

7. (a) Using the formula for the sum of the first n positive integers, you obtain

$$\sum_{i=1}^{20} i = 1 + 2 + 3 + \ldots + 20 = \frac{20(20+1)}{2} = \frac{20(21)}{2} = 210.$$

(b)
$$\begin{aligned}\sum_{i=1}^{5} 2i^2 + 3i^3 &= \sum_{i=1}^{5} 2i^2 + \sum_{i=1}^{5} 3i^3\\ &= 2\sum_{i=1}^{5} i^2 + 3\sum_{i=1}^{5} i^3\\ &= 2\left[\frac{5(5+1)[2(5)+1]}{6}\right] + 3\left[\frac{(5)^2(5+1)^2}{4}\right]\\ &= 2\left[\frac{(5)(6)(11)}{6}\right] + 3\left[\frac{(25)(36)}{4}\right]\\ &= 2(55) + 3(225)\\ &= 785\end{aligned}$$

8. Begin by finding the first and second differences.

n:	1	2	3	4	5	6
a_n:	−2	0	4	10	18	28
First differences:		2	4	6	8	10
Second differences:			2	2	2	2

You know from the second differences that the model is quadratic and has the form $a_n = an^2 + bn + c$.

By substituting 1, 2, and 3 for n, you can obtain a system of three linear equations in three variables.

$$a_1 = a(1)^2 + b(1) + c = -2$$

$$a_2 = a(2)^2 + b(2) + c = 0$$

$$a_3 = a(3)^2 + b(3) + c = 4$$

$$\begin{cases} a + b + c = -2\\ 4a + 2b + c = 0\\ 9a + 3b + c = 4\end{cases}$$

Solving this system using techniques from Chapter 9, you can find the solution to be $a = 1$, $b = -1$, and $c = -2$.

So, the quadratic model is $a_n = n^2 - n - 2$.

Checkpoints for Section 9.5

1. (a) $\binom{11}{5} = \dfrac{11!}{6!5!} = \dfrac{(11\cdot10\cdot9\cdot8\cdot7)\cdot \cancel{6!}}{\cancel{6!}\,5!} = \dfrac{11\cdot10\cdot9\cdot8\cdot7}{5\cdot4\cdot3\cdot2\cdot1} = 462$

(b) ${}_9C_2 = \dfrac{9!}{7!2!} = \dfrac{(9\cdot8)\cdot\cancel{7!}}{\cancel{7!}\,2!} = \dfrac{9\cdot8}{2\cdot1} = 36$

(c) $\binom{5}{0} = \dfrac{\cancel{5!}}{\cancel{5!}\,0!} = 1$

(d) ${}_{15}C_{15} = \dfrac{\cancel{15!}}{0!\,\cancel{15!}} = 1$

2. (a) $_7C_5 = \dfrac{7!}{2!5!} = \dfrac{7 \cdot 6 \cdot \cancel{5} \cdot \cancel{4} \cdot \cancel{3}}{\cancel{5} \cdot \cancel{4} \cdot \cancel{3} \cdot \cancel{2} \cdot 1} = 21$

(b) $\dbinom{7}{2} = \dfrac{7!}{5!2!} = \dfrac{7.6}{2.1} = 21$

(c) $_{14}C_{13} = \dfrac{14!}{1!13!} = \dfrac{14}{1} = 14$

(d) $\dbinom{14}{1} = \dfrac{14!}{13!1!} = \dfrac{14}{1} = 14$

3.

1
1 1
1 2 1
1 3 3 1
1 4 6 4 1
1 5 10 10 5 1
1 6 15 20 15 6 1
1 7 21 35 35 21 7 1
1 8 28 56 70 56 28 8 1 8th row
1 9 36 84 126 126 84 36 9 1

$\downarrow$ $\downarrow$ $\downarrow$ $\downarrow$ $\downarrow$ $\downarrow$ $\downarrow$ $\downarrow$ $\downarrow$ $\downarrow$

C_{9-0} C_{9-1} C_{9-2} C_{9-3} C_{9-4} C_{9-5} C_{9-6} C_{9-7} C_{9-8} C_{9-9}

4. The binomial coefficients from the fourth row of Pascal's Triangle are 1, 4, 6, 4, 1.

So, the expansion is as follows.

$$(x+2)^4 = (1)x^4 + (4)x^3(2) + (6)x^2(2^2) + (4)x(2^3) + (1)(2^4)$$
$$= x^4 + 8x^3 + 24x^2 + 32x + 16$$

5. (a) $(y-2)^4 = (1)y^4 - (4)y^3(2) + (6)y^2(2^2) - (4)y(2^3) + (1)(2)^4$

$= y^4 - 8y^3 + 24y^2 - 32y + 16$

(b) $(2x-y)^5 = (1)(2x)^5 - (5)(2x)^4y + (10)(2x)^3y^2 - (10)(2x)^2y^3 + (5)(2x)y^4 - (1)y^5$

$= 32x^5 - 80x^4y + 80x^3y^2 - 40x^2y^3 + 10xy^4 - y^5$

6. Use the third row of Pascal's Triangle to write the expansion of $(5+y^2)^3 = (y^2+5)^3$, as follows.

$$(y^2+5)^3 = (1)(y^2)^3 + (3)(y^2)^2(5) + (3)(y^2)(5^2) + (1)(5^3)$$
$$= y^6 + 15y^4 + 75y^2 + 125$$

7. (a) Remember that the formula is for the $(r+1)$th term, so r is one less than the number of the term you need. So, to find the fifth term in this binomial expansion, use $r = 4$, $n = 8$, $x = a$, and $y = 2b$, as shown.

$$_nC_rx^{n-r}y^r = {}_8C_4a^{8-4}(2b)^4 = (70)(a^4)(2b)^4$$
$$= 70(2^4)a^4b^4$$
$$= 1120a^4b^4$$

(b) In this case, $n = 11$, $r = 7$, $x = 3a$, and $y = -2b$. Substitute these values to obtain the following.

$$_nC_rx^{n-r}y^r = {}_{11}C_7(3a)^4(-2b)^7$$
$$= (330)(81a^4)(-128b^7)$$
$$= -3{,}421{,}440a^4b^7$$

So, the coefficient is $-3{,}421{,}440$.

Checkpoints for Section 9.6

1. To solve this problem, count the different ways to obtain a sum of 14 using two numbers from 1 to 8.

First number:	6	7	8
Second number:	8	7	6

So, a sum of 14 can occur in three different ways.

2. To solve this problem, count the different ways to obtain a sum of 14 *using two different numbers* from 1 to 8.

First number:	6	8
Second number:	8	6

So, a sum of 14 can occur in 2 ways.

3. There are three events in this situation. The first event is the choice of the first number, the second event is the choice of the second number, and the third event is the choice of the third number. Because there is a choice of 30 numbers for each event, it follows that the number of different lock combinations is

$30 \cdot 30 \cdot 30 = 27{,}000.$

4. Because the product's catalog number is made up of one letter from the English alphabet followed by a five-digit number, there are 26 choices for the first digit and 10 choices for each of the other 5 digits.

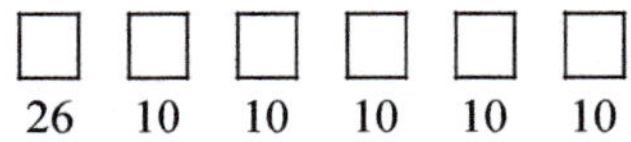

26 10 10 10 10 10

So, the number of possible catalog numbers is $26 \cdot 10 \cdot 10 \cdot 10 \cdot 10 \cdot 10 = 2{,}600{,}000.$

5. *First position*: Any of the *four* letters

Second position: Any of the remaining *three* letters

Third position: Either of the remaining *two* letters

Fourth position: The *one* remaining letter

So, the numbers of choices for the four positions are as follows.

Permutations of four letters

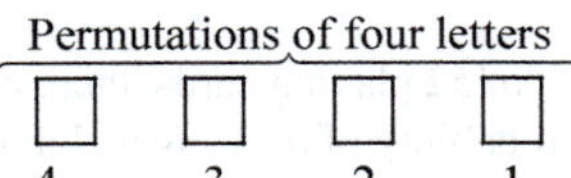

4 3 2 1

The total number of permutations of the four letters is

$$\begin{aligned} 4! &= 4 \cdot 3 \cdot 2 \cdot 1 \\ &= 24. \end{aligned}$$

6. Here are the different possibilities.

President (first position): *Five* choices

Vice-President (second position): *Four* choices

Using the Fundamental Counting Principle, multiply these two numbers to obtain the following.

Different orders of offices

☐ ☐

President Vice-President

5 4

So, there are $5 \cdot 4 = 20$ different ways there can be a President and Vice-President.

7. The word M I T O S I S has seven letters, of which there are two I's, two S's, and one M, T, and O. So, the number of distinguishable ways the letters can be written is

$$\frac{n!}{n_1!\,n_2!} = \frac{7!}{2!\,2!} = \frac{7 \cdot 6 \cdot 5 \cdot 4 \cdot 3 \cdot \cancel{2!}}{2!\,\cancel{2!}} = 1260.$$

8. The following subsets represent the different combinations of two letters that can be chosen from the seven letters.

{A, B} {B, C} {C, D} {D, E} {E, F} {F, G}
{A, C} {B, D} {C, E} {D, F} {E, G}
{A, D} {B, E} {C, F} {D, G}
{A, E} {B, F} {C, G}
{A, F} {B, G}
{A, G}

From this list, you can conclude that there are 21 different ways that two letters can be chosen from seven letters.

9. To find the number of three card poker hands, use the formula for the number of combinations of 52 elements taken three at a time, as follows.

$$\begin{aligned} {}_{52}C_3 &= \frac{52!}{(52-3)!\,3!} \\ &= \frac{52!}{49!\,3!} \\ &= \frac{52 \cdot 51 \cdot 50 \cdot \cancel{49!}}{\cancel{49!}\,3!} \\ &= \frac{52 \cdot 51 \cdot 50}{3 \cdot 2 \cdot 1} \\ &= 22{,}100 \end{aligned}$$

10. There are ${}_{10}C_6$ ways of choosing six girls from a group of ten girls and ${}_{15}C_6$ ways of choosing boys from a group of fifteen boys. By the Fundamental Counting Principle, there are ${}_{10}C_6 \cdot {}_{15}C_6$ ways of choosing six girls and six boys.

$$\begin{aligned} {}_{10}C_6 \cdot {}_{15}C_6 &= \frac{10!}{4! \cdot 6!} \cdot \frac{15!}{9! \cdot 6!} \\ &= 210 \cdot 5005 \\ &= 1{,}051{,}050 \end{aligned}$$

So, there are 1,051,050 12-member swim teams possible.

Checkpoints for Section 9.7

1. Because either coin can land heads up or tails up, and the six-sided die can land with a 1 through 6 up. So, the sample space is

$$\begin{aligned} S = \{&HH1, HH2, HH3, HH4, HH5, HH6 \\ &HT1, HT2, HT3, HT4, HT5, HT6, \\ &TH1, TH2, TH3, TH4, TH5, TH6, \\ &TT1, TT2, TT3, TT4, TT5, TT6\} \end{aligned}$$

2. (a) Let $E = \{TTT\}$ and $S = \{HHH, HHT, HTH, HTT, TTT, THT, TTH, THH\}$.

The probability of getting three tails is

$$P(E) = \frac{n(E)}{n(S)} = \frac{1}{8}.$$

(b) Because there are 52 cards in a standard deck of playing cards and there are 13 diamonds, the probability of drawing a diamond is

$$\begin{aligned} P(E) &= \frac{n(E)}{n(S)} \\ &= \frac{13}{52} \\ &= \frac{1}{4}. \end{aligned}$$

3. Because there are six possible outcomes on each die, use the Fundamental Counting Principle to conclude that there are $6 \cdot 6$ or 36 different outcomes when you toss two dice. To find the probability of rolling a total of 5, you must first count the number of ways in which this can occur.

First Die	Second Die
1	4
2	3
3	2
4	1

So, a total of 5 can be rolled in four ways, which means that the probability of rolling a 5 is

$$P(E) = \frac{n(E)}{n(S)} = \frac{4}{36} = \frac{1}{9}.$$

4. For a standard deck of 52 playing cards, there are 13 clubs. So, the probability of drawing a club is

$$P(E) = \frac{n(E)}{n(S)} = \frac{13}{52} = \frac{1}{4}.$$

For a set consisting of the aces, the sample space is 4 cards. So, the probability of drawing the ace at hearts is

$$P(E) = \frac{n(E)}{n(S)} = \frac{1}{4}.$$

So, the probability of drawing a club from a standard deck of cards is the same as drawing the ace of hearts from the set of aces.

5. The total number of colleges and universities is 4622. Because there are 640 colleges and universities in the Pacific region, the probability that the institution is in that region is

$$P(E) = \frac{n(E)}{n(S)} = \frac{640}{4622} \approx 0.138.$$

6. To find the number of elements in the sample space, use the formula for the number of combinations of 43 elements taken five at a time.

$$n(S) = {}_{43}C_5 = \frac{43 \cdot 42 \cdot 41 \cdot 40 \cdot 39}{5 \cdot 4 \cdot 3 \cdot 2 \cdot 1} = 962{,}598$$

When a player buys one ticket, the probability of winning is

$$P(E) = \frac{1}{962{,}598} \approx 0.000001.$$

7. Because the deck has 4 aces, the probability of drawing an ace (event A) is

$$P(A) = \frac{4}{52}.$$

Similarly, because the deck has 13 spades, the probability of drawing a spade (event B) is

$$P(B) = \frac{13}{52}.$$

Because one of the cards is an ace *and* a spade, the ace of spades, it follows that

$$P(A \cap B) = \frac{1}{52}.$$

Spades
2♠ 3♠ 4♠ 5♠ 6♠ 7♠ 8♠ 9♠ 10♠ J♠ Q♠ K♠
$P(A \cap B)$
A♠
Aces
A♦ A♣ A♥

Finally, applying the formula for the probability of the union of two events, the probability of drawing an ace or spade is as follows.

$$P(A \cup B) = P(A) + P(B) - P(A \cap B) = \frac{4}{52} + \frac{13}{52} - \frac{1}{52} = \frac{16}{52} = \frac{4}{13} \approx 0.308$$

8. To begin, add the number of employees to find that the total is 529. Next, let event A represent choosing an employee with 30-34 years of service, event B with 35-39 years of service, event C with 40-44 years of service, and event D with 45 or more years of service.

Because events A, B, C, and D have no outcomes in common, these four events are mutually exclusive and

$$P(A \cup B \cup C \cup D) = P(A) + P(B) + P(C) + +P(D) = \frac{35}{529} + \frac{21}{529} + \frac{8}{529} + \frac{2}{529} = \frac{66}{529} \approx 0.125.$$

So, the probability of choosing an employee who has 30 or more years of service is about 0.125.

9. The probability of selecting a number from 1 to 11 from a set of numbers from 1 to 30 is

$$P(A) = \frac{11}{30}.$$

So, the probability that both numbers are less than 12 is

$$P(A) \cdot P(A) = \frac{11}{30} \cdot \frac{11}{30} = \frac{121}{900} \approx 0.134.$$

10. Let event A represent selecting a person who expected much of the workforce to be automated within 50 years. The probability of event A is 0.65. Each of the 5 occurrence of event A is an independent event, so the probability that all 5 people expected much of the workforce to be automated within 50 years is

$$\left[P(A)\right]^5 = (0.65)^5 \approx 0.116.$$

11. To solve this problem as stated, you would need to find the probabilities of having exactly one faulty unit, exactly two faulty units, exactly three faulty units, and so on. However, using complements, you can find the probability that all units are perfect and then subtract this value from 1. Because the probability that any given unit is perfect is 499/500, the probability that all 300 units are perfect is

$$P(A) = \left(\frac{499}{500}\right)^{300} \approx 0.548.$$

So, the probability that at least one unit is faulty is

$$P(A') = 1 - P(A) \approx 1 - 0.548 = 0.452.$$

Chapter 10

Checkpoints for Section 10.1

1. (a) The slope of this line is $m = \frac{4}{5}$.

So, its inclination is determined from $\tan \theta = \frac{4}{5}$.

Note that $m \geq 0$. This means that

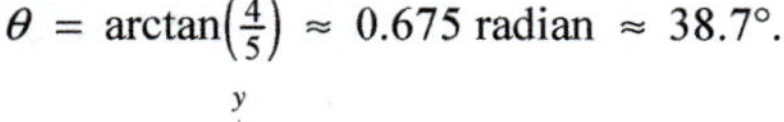

$$\theta = \arctan\left(\frac{4}{5}\right) \approx 0.675 \text{ radian} \approx 38.7°.$$

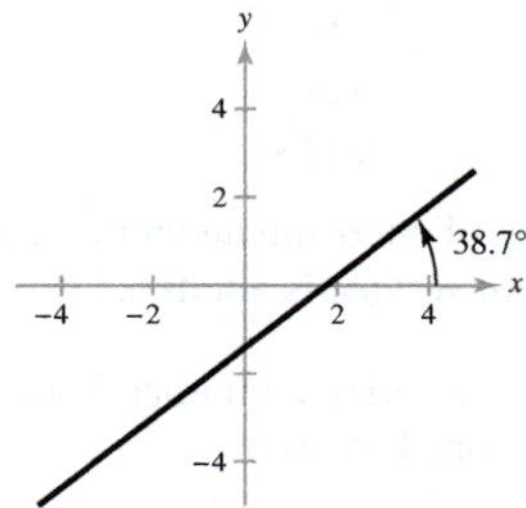

(b) The slope of the line $x + y = -1$ is $m = -1$. So, use $\tan \theta = -1$ to determine its inclination.

Note that $m \leq 0$.

This means that

$$\theta = \pi + \arctan(-1)$$

$$= \pi + \left(-\frac{\pi}{4}\right)$$

$$= \frac{3\pi}{4} \text{ radians} = 135°.$$

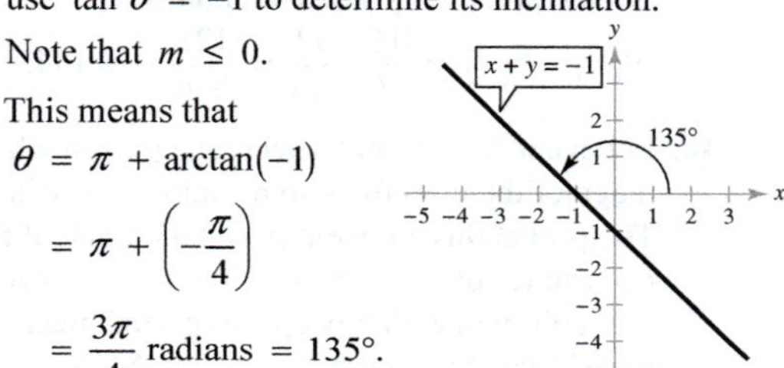

2. The two lines, $4x - 5y + 10 = 0$ and $3x + 2y + 5 = 0$, have slopes of $m_1 = \dfrac{4}{5}$ and $m_2 = -\dfrac{3}{2}$, respectively.

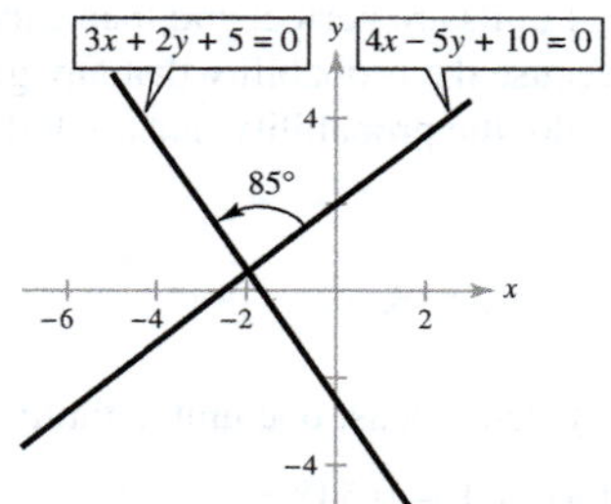

So, the tangent of the angle between the two lines is

$$\tan \theta = \left|\frac{m_2 - m_1}{1 + m_1 m_2}\right| = \left|\frac{-\frac{3}{2} - \frac{4}{5}}{1 + \left(-\frac{3}{2}\right)\left(\frac{4}{5}\right)}\right| = \left|\frac{-\frac{23}{10}}{-\frac{1}{5}}\right| = \frac{23}{2}$$

Finally, you can conclude that the angle is

$$\theta = \arctan\left(\frac{23}{2}\right) \approx 1.4841 \text{ radians} \approx 85.0°.$$

3. The general form of $y = -3x + 2$ is $3x + y - 2 = 0$. So, the distance between the point and a line is

$$d = \frac{|Ax_1 + By_1 + C|}{\sqrt{A^2 + B^2}}$$

$$= \frac{|3(5) + 1(-1) + (-2)|}{\sqrt{(3)^2 + (1)^2}}$$

$$= \frac{12}{\sqrt{10}}$$

$$\approx 3.79 \text{ units.}$$

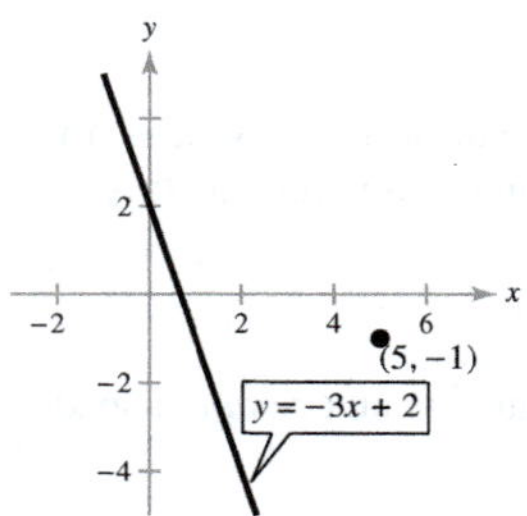

4. The general form is $3x - 5y - 2 = 0$. So, the distance between the point and a line is

$$d = \frac{|Ax_1 + By_1 + C|}{\sqrt{A^2 + B^2}}$$

$$= \frac{|3(3) + (-5)(2) + (-2)|}{\sqrt{(3)^2 + (-5)^2}}$$

$$= \frac{3}{\sqrt{34}} \approx 0.51 \text{ unit.}$$

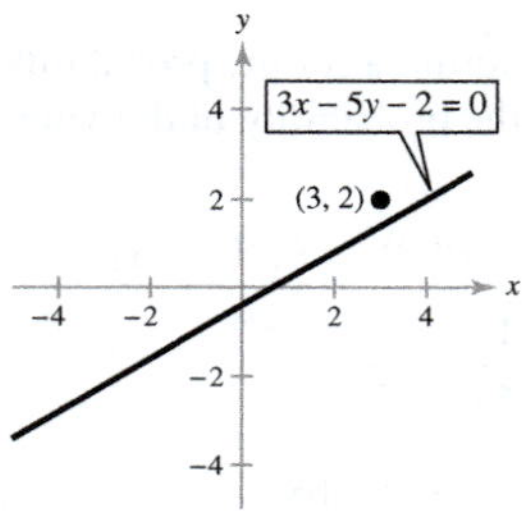

5.

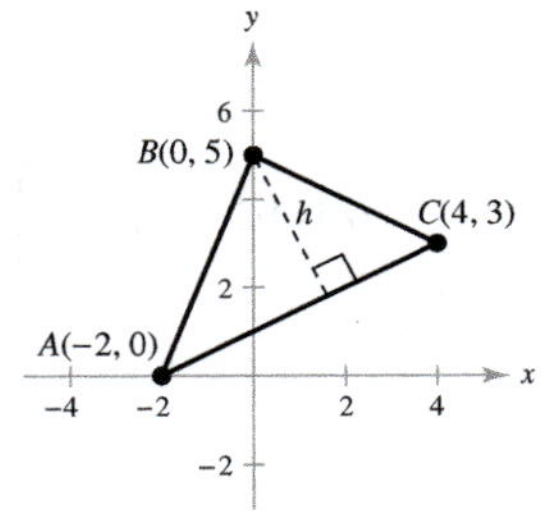

(a) To find the altitude, use the formula for the distance between line AC and the point $(0, 5)$.

The equation of line AC is obtained as follows.

Slope: $m = \dfrac{y_2 - y_1}{x_2 - x_1} = \dfrac{3 - 0}{4 - (-2)} = \dfrac{3}{6} = \dfrac{1}{2}$

Equation:

$$y - y_1 = m(x - x_1) \qquad \text{Point-slope form}$$

$$y - (0) = \frac{1}{2}[x - (-2)] \qquad \text{Substitute.}$$

$$y = \frac{1}{2}x + 1 \qquad \text{Slope-intercept form}$$

$$2y = x + 2 \qquad \text{Multiply each side by 2.}$$

$$x - 2y + 2 = 0 \qquad \text{General form}$$

So, the distance between this line and the point $(0, 5)$ is

$$\text{Altitude} = h = \frac{|(1)(0) + (-2)(5) + (2)|}{\sqrt{(1)^2 + (-2)^2}}$$

$$= \frac{8}{\sqrt{5}} \text{ units.}$$

(b) Using the formula for the distance between two points, you can find the length of the base AC to be

$$b = \sqrt{(x_2 - x_1)^2 + (y_2 - y_1)^2} \qquad \text{Distance Formula}$$

$$= \sqrt{[4 - (-2)]^2 + (3 - 0)^2} \qquad \text{Substitute.}$$

$$= \sqrt{6^2 + 3^2} \qquad \text{Simplify.}$$

$$= \sqrt{45} \qquad \text{Simplify.}$$

$$= 3\sqrt{5} \text{ units.} \qquad \text{Simplify.}$$

Finally, the area of the triangle is

$$A = \frac{1}{2}bh$$

$$= \frac{1}{2}(3\sqrt{5})\left(\frac{8}{\sqrt{5}}\right)$$

$$= 12 \text{ square units.}$$

Checkpoints for Section 10.2

1. The axis of the parabola is vertical, passing through $(0, 0)$ and $\left(0, \frac{3}{8}\right)$. The standard form is $x^2 = 4py$, where $p = \frac{3}{8}$.

So, the equation is
$$x^2 = 4\left(\frac{3}{8}\right)y$$
$$x^2 = \frac{3}{2}y.$$

You can use a graphing utility to confirm this equation. To do this, graph $y_1 = \frac{3}{2}x^2$.

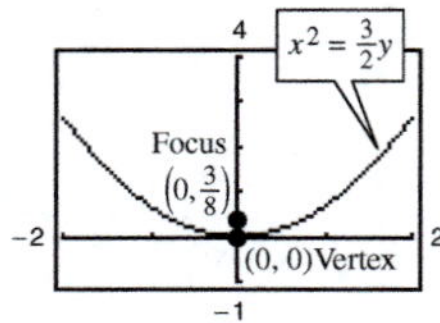

2. Because the axis of the parabola is horizontal, passing through $(2, -3)$ and $(4, -3)$, consider the equation

$$(y - k)^2 = 4p(x - h)$$

where $h = 2$, $k = -3$, and $p = 4 - 2 = 2$. So, the standard form is

$$[y - (-3)]^2 = 4(2)(x - 2)$$
$$(y + 3)^2 = 8(x - 2).$$

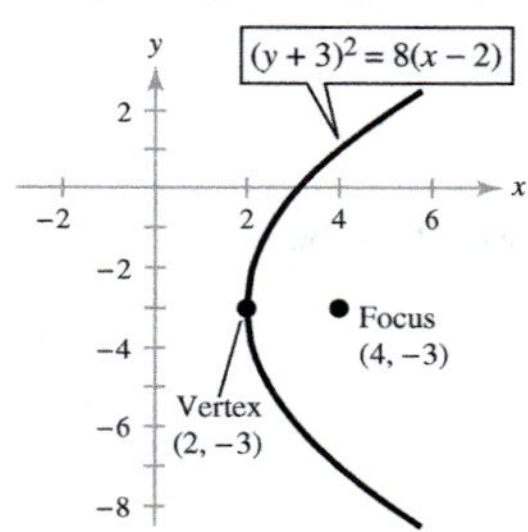

3. Convert to standard form by completing the square.

$x = \frac{1}{4}y^2 + \frac{3}{2}y + \frac{13}{4}$	Write orginal equation.
$4x = y^2 + 6y + 13$	Multiply each side by 4.
$4x - 13 = y^2 + 6y$	Subtract 13 from each side.
$4x - 13 + 9 = y^2 + 6y + 9$	Add 9 to each side.
$4x - 4 = y^2 + 6y + 9$	Combine like terms.
$4(x - 1) = (y + 3)^2$	Standard form

Comparing this equation with $(y - k)^2 = 4p(x - h)$, you can conclude that $h = 1$, $k = -3$, and $p = 1$. Because p is positive, the parabola opens to the right. So, the focus is $(h + p, k) = (1 + 1, -3) = (2, -3)$.

4. For this parabola, $p = \dfrac{1}{12}$ and the focus is $\left(0, \dfrac{1}{12}\right)$ as shown in the figure below.

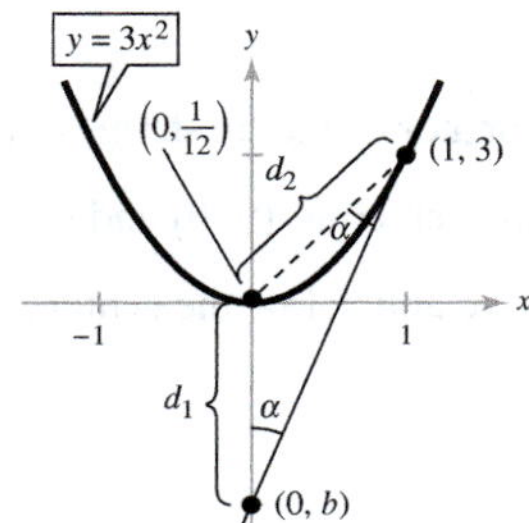

You can find the y-intercept $(0, b)$ of the tangent line by equating the lengths of the two sides of the isosceles triangle shown in the figure:

$$d_1 = \frac{1}{12} - b$$

and

$$d_2 = \sqrt{(1-0)^2 + \left(3 - \frac{1}{12}\right)^2} = \sqrt{1^2 + \left(\frac{35}{12}\right)^2}$$
$$= \frac{37}{12}.$$

Note that $d_1 = \dfrac{1}{12} - b$ rather than $b - \dfrac{1}{12}$. The order of subtraction for the distance is important because the distance must be positive. Setting $d_1 = d_2$ produces

$$\frac{1}{12} - b = \frac{37}{12}$$
$$b = -3.$$

So, the slope of the tangent line through $(0, -3)$ and $(1, 3)$ is

$$m = \frac{3 - (-3)}{1 - 0} = 6$$

and the equation of the tangent line in slope-intercept form is $y = 6x - 3$.

Checkpoints for Section 10.3

1. Because the foci occur at $(2, 0)$ and $(2, 6)$, the center of the ellipse is $(2, 3)$ and the distance from the center to one of the foci is $C = 3$. Because $2a = 8$, you know that $a = 4$. Now, from $c^2 = a^2 - b^2$, you have

$$b = \sqrt{a^2 - c^2} = \sqrt{4^2 - 3^2} = \sqrt{7}.$$

Because the major axis is vertical, the standard equation is as follows.

$$\frac{(y-k)^2}{a^2} + \frac{(x-h)^2}{b^2} = 1$$

$$\frac{(y-3)^2}{4^2} + \frac{(x-2)^2}{\left(\sqrt{7}\right)^2} = 1$$

$$\frac{(y-3)^2}{16} + \frac{(x-2)^2}{7} = 1$$

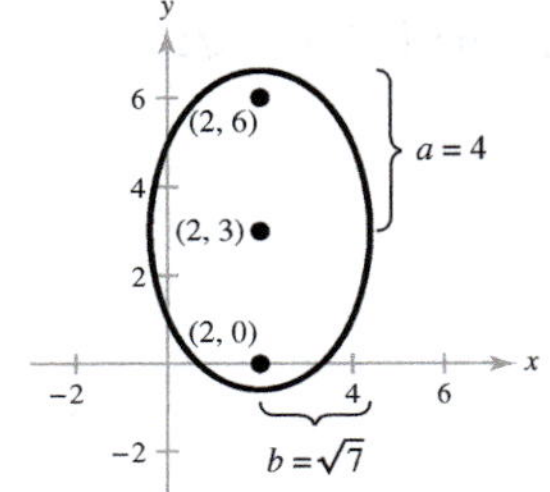

2. $x^2 + 9y^2 = 81$

$$\frac{x^2}{81} + \frac{y^2}{9} = 1$$

The center of the ellipse is $(0, 0)$. The denominator of the x^2-term is greater than the denominator of the y^2-term, so the major axis is horizontal. Because $a^2 = 81$, the endpoints of the major axis lie 9 units from the center at $(9, 0)$ and $(-9, 0)$. Because $b^2 = 9$, the minor axis is vertical and the endpoints of the minor axis lie 3 units above and below the center at $(0, 3)$ and $(0, -3)$.

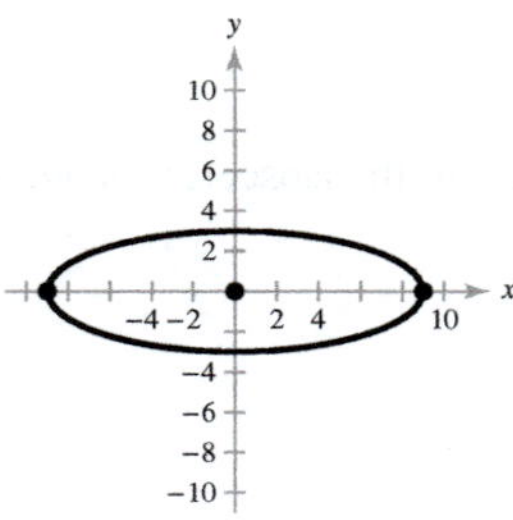

Center: $(0, 0)$

Vertices: $(-9, 0)$, $(9, 0)$

3. Begin by writing the original equation in standard form.

$$9x^2 + 4y^2 + 36x - 8y + 4 = 0 \qquad \text{Write original equation.}$$

$$9x^2 + 36x + 4y^2 - 8y = -4 \qquad \text{Group terms.}$$

$$9(x^2 + 4x + \square) + 4(y^2 - 2y + \square) = -4 \qquad \text{Factor out leading coefficients.}$$

$$9(x^2 + 4x + 4) + 4(y^2 - 2y + 1) = -4 + 36 + 4 \qquad \text{Complete the square.}$$

$$9(x + 2)^2 + 4(y - 1)^2 = 36 \qquad \text{Write in completed square form.}$$

$$\frac{(x + 2)^2}{4} + \frac{(y - 1)^2}{9} = 1 \qquad \text{Divide each side by 36.}$$

$$\frac{(x + 2)^2}{2^2} + \frac{(y - 1)^2}{3^2} = 1 \qquad \text{Write in standard form.}$$

The center is $(h, k) = (-2, 1)$. Because the denominator of the y-term is $a^2 = 3^2$, the endpoints of the major axis lie 3 units above and below the center. So, the vertices are $(-2, 4)$ and $(-2, -2)$.

Similarly, because the denominator of the x-term is $b^2 = 2^2$, the endpoints of the minor axis lie 2 units to the right and left of the center at $(-4, 1)$ and $(0, 1)$.

Now, from $c^2 = a^2 - b^2$, you have $c = \sqrt{3^2 - 2^2} = \sqrt{5}$.

So, the foci of the ellipse are $(-2, 1 + \sqrt{5})$ and $(-2, 1 - \sqrt{5})$.

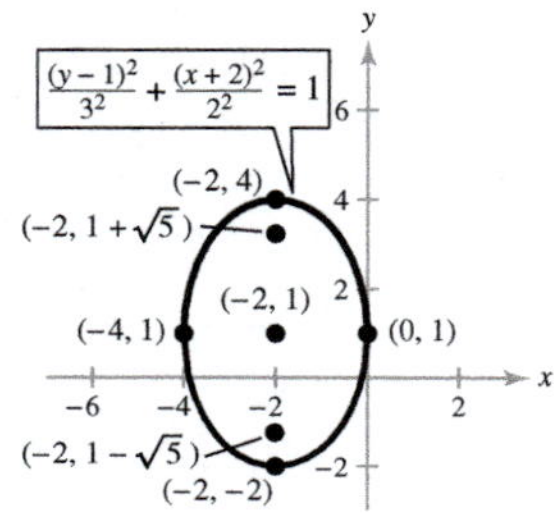

4. By completing the square, you can write the original equation in standard form.

$$5x^2 + 9y^2 + 10x - 54y + 41 = 0 \quad \text{Write original equation.}$$

$$5x^2 + 10x + 9y^2 - 54y = -41 \quad \text{Group terms.}$$

$$5(x^2 + 2x + \square) + 9(y^2 - 6y + \square) = -41 \quad \text{Factor out leading coefficients.}$$

$$5(x^2 + 2x + 1) + 9(y^2 - 6y + 9) = -41 + 5 + 81$$

$$5(x + 1)^2 + 9(y - 3)^2 = 45 \quad \text{Write in completed square form.}$$

$$\frac{(x + 1)^2}{9} + \frac{(y - 3)^2}{5} = 1 \quad \text{Divide each side by 45.}$$

$$\frac{(x + 1)^2}{3^2} + \frac{(y - 3)^2}{(\sqrt{5})^2} = 1 \quad \text{Write in standard form.}$$

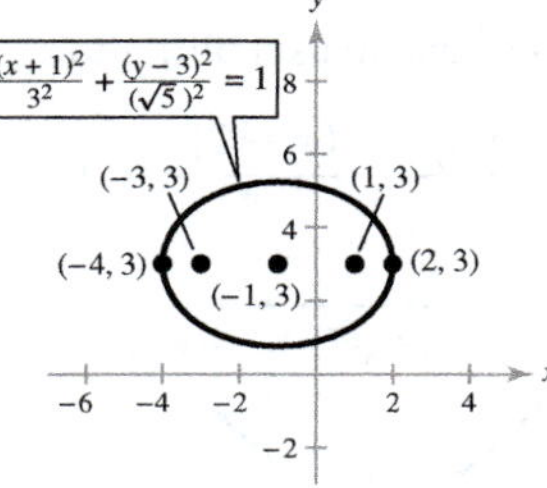

The major axis is horizontal,
where $h = -1$, $k = 3$, $a = 3$, $b = \sqrt{5}$ and $c = \sqrt{a^2 - b^2}$

$$= \sqrt{3^2 - (\sqrt{5})^2} = \sqrt{4} = 2.$$

So, you have the following.

Center: $(-1, 3)$ Vertices: $(-4, 3)$, $(2, 3)$ Foci: $(-3, 3)$, $(1, 3)$

5.

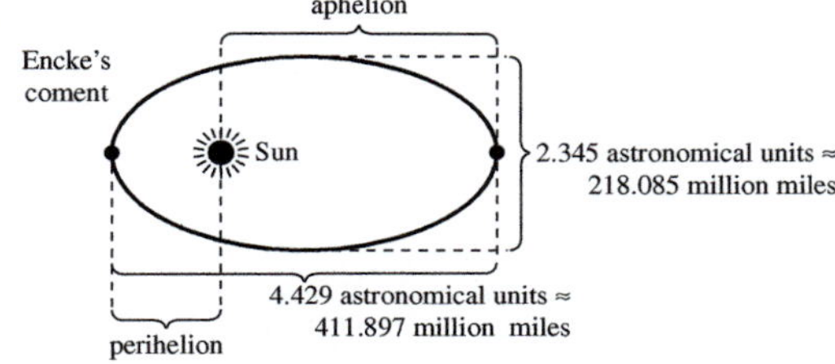

Because $2a = 411.897$ and $2b = 218.085$, you have $a \approx 205.95$ and $b \approx 109.04$

which implies that $c = \sqrt{a^2 - b^2}$

$$= \sqrt{205.95^2 - 109.04^2}$$

$$\approx 174.72.$$

So, the greatest distance, the aphelion, from the sun's center to the comet's center is

$a + c \approx 205.95 + 174.72 = 380.67$ million miles

and the least distance, the perihelion, is

$a - c = 205.95 - 174.72 = 31.23$ million miles.

Checkpoints for Section 10.4

1. By the Midpoint Formula, the center of the hyperbola occurs at the point of

$$(h, k) = \left(\frac{2+2}{2}, -\frac{4+2}{2}\right) = (2, -1).$$

Furthermore, $a = 2 - (-1) = 3$ and $c = 3 - (-1) = 4$, and it follows that $b = \sqrt{c^2 - a^2} = \sqrt{4^2 - 3^2} = \sqrt{7}$.

So, the hyperbola has a vertical transverse axis and the standard form of the equation is

$$\frac{(y-k)^2}{a^2} - \frac{(x-h)^2}{b^2} = 1$$

$$\frac{(y+1)^2}{3^2} - \frac{(x-2)^2}{\left(\sqrt{7}\right)^2} = 1.$$

This equation simplifies to

$$\frac{(y+1)^2}{9} - \frac{(x-2)^2}{7} = 1$$

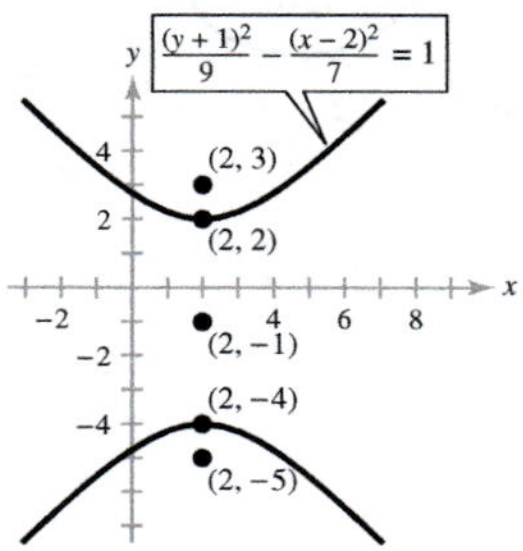

2. Algebraic Solution

Divide each side of the equation by 36, and write the equation in standard form.

$$4y^2 - 9x^2 = 36$$
$$\frac{y^2}{9} - \frac{x^2}{4} = 1$$
$$\frac{y^2}{3^2} - \frac{x^2}{2^2} = 1$$

From this, you can conclude that $a = 3$, $b = 2$, and the transverse axis is vertical. Because the center is $(0, 0)$, the vertices occur at $(0, 3)$ and $(0, -3)$. The endpoints of the conjugate axis occur at $(2, 0)$ and $(-2, 0)$. Using these four points, sketch a rectangle that is $2a = 6$ units tall and $2b = 4$ units wide.

Now from $c^2 = a^2 + b^2$, you have $c = \sqrt{a^2 + b^2} = \sqrt{3^2 + 2^2} = \sqrt{13}$. So, the foci of the hyperbola are $(0, \sqrt{13})$ and $(0, -\sqrt{13})$.

Finally, by drawing the asymptotes through the corners of this rectangle, you can complete the sketch. Note that the asymptotes are $y = \frac{3}{2}x$ and $y = -\frac{3}{2}x$.

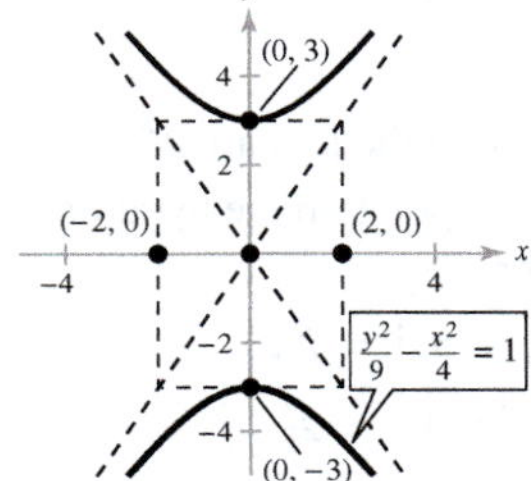

Graphical Solution

Solve the equation of the hyperbola for y as follows.

$$4y^2 - 9x^2 = 36$$
$$4y^2 = 9x^2 + 36$$
$$y^2 = \frac{9x^2 + 36}{4}$$
$$y = \pm\sqrt{\frac{9x^2 + 36}{4}}$$
$$y = \pm\frac{3}{2}\sqrt{x^2 + 4}$$

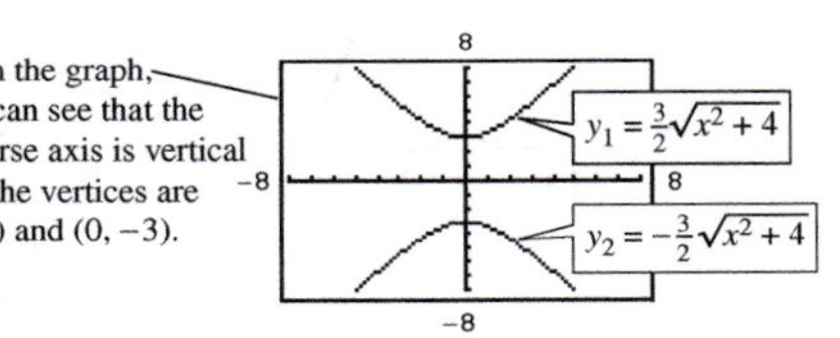

Then use a graphing utility to graph $y_1 = \frac{3}{2}\sqrt{x^2 + 4}$ and $y_2 = -\frac{3}{2}\sqrt{x^2 + 4}$ in the same viewing window.

3.

$9x^2 - 4y^2 + 8y - 40 = 0$	Write original equation.
$9x^2 - 4y^2 + 8y = 40$	Group terms.
$9x^2 - 4(y^2 - 2y + \square) = 40$	Factor out leading coefficients.
$9x^2 - 4(y^2 - 2y + 1) = 40 - 4$	Complete the square.
$9x^2 - 4(y - 1)^2 = 36$	Write in completed square form.
$\frac{x^2}{4} - \frac{(y - 1)^2}{9} = 1$	Divide each side by 36.
$\frac{x^2}{2^2} - \frac{(y - 1)^2}{3^2} = 1$	Write in standard form.

From this equation you can conclude that the hyperbola has a horizontal transverse axis centered at $(0, 1)$. The vertices are at $(-2, 1)$ and $(2, 1)$, and has a conjugate axis with endpoints $(0, 4)$ and $(0, -2)$. To sketch the hyperbola, draw a rectangle through these four points. The asymptotes are the lines passing through the corners of the rectangle.

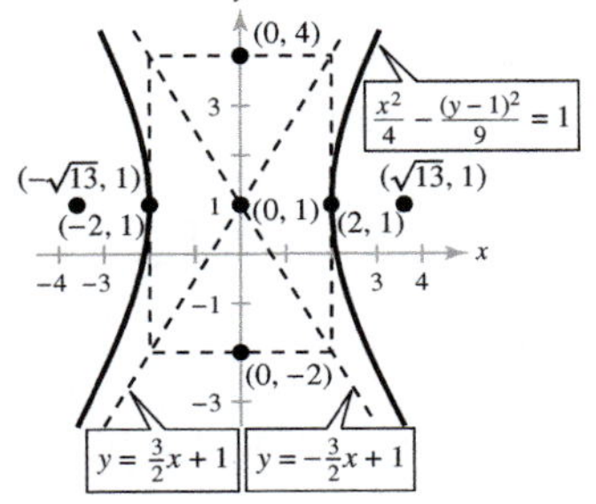

Using $a = 2$ and $b = 3$, you can conclude that the equations of the asymptotes are $y = \frac{3}{2}x + 1$ and $y = -\frac{3}{2}x + 1$.

Finally, you can determine the foci by using the equation $c^2 = a^2 + b^2$. So, you have $c = \sqrt{2^2 + 3^2} = \sqrt{13}$, and the foci are $(\sqrt{13}, 1)$ and $(-\sqrt{13}, 1)$.

4. Using the Midpoint Formula you can determine the center is $(h, k) = \left(\frac{3+9}{2}, \frac{2+2}{2}\right) = (6, 2).$

Furthermore, the hyperbola has a horizontal transverse axis with $a = 3$. From the original equations, you can determine the slopes of the asymptotes to be $m_1 = \frac{b}{a} = \frac{2}{3}$ and $m_2 = -\frac{b}{a} = -\frac{2}{3}$, and because $a = 3$, you can conclude that $b = 2$.

So, the standard form of the equation of the hyperbola is

$$\frac{(x-6)^2}{3^2} - \frac{(y-2)^2}{2^2} = 1.$$

5. Begin by representing the situation in a coordinate plane. The distance between the microphones is 1 mile, or 5280 feet. So, position the point representing microphone A 2640 units to the right of the origin and the point representing microphone B 2640 units to the left of the origin, as shown.

Assuming sound travels at 1100 feet per second, the explosion took place 4400 feet farther from B than from A. The locus of all points that are 4400 feet closer to A than to B is one branch of a hyperbola with foci at A and B. Because the hyperbola is centered at the origin and has a horizontal transverse axis, the standard form of its equation is

$$\frac{x^2}{a^2} - \frac{y^2}{b^2} = 1.$$

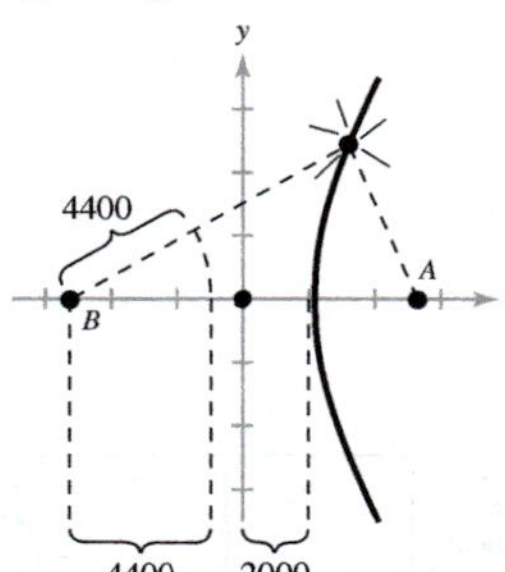

Because the foci are 2640 units from the center, $c = 2640$. Let d_A and d_B be the distances of any point on the hyperbola from the foci at A and B, respectively. From page 713, you have

$|d_B - d_A| = 2a$

$|4400| = 2a$ The points are 4400 feet closer to A than to B

$2200 = a$ Divide each side by 2

So, $b^2 = c^2 - a^2 = 2640^2 - 2200^2 = 2{,}129{,}600$ and you can conclude that the explosion occurred somewhere on the right branch of the hyperbola

$$\frac{x^2}{4{,}840{,}000} - \frac{y^2}{2{,}129{,}600} = 1.$$

6. (a) For the equation $3x^2 + 3y^2 - 6x + 6y + 5 = 0$, you have $A = C = 3$. So, the graph is a circle.

(b) For the equation $2x^2 - 4y^2 + 4x + 8y - 3 = 0$, you have $AC = 2(-4) < 0$. So, the graph is a hyperbola.

(c) For the equation $3x^2 + y^2 + 6x - 2y + 3 = 0$, you have $AC = 3(1) > 0$. So, the graph is an ellipse.

(d) For the equation $2x^2 + 4x + y - 2 = 0$, you have $AC = 2(0) = 0$. So, the graph is a parabola.

Checkpoints for Section 10.5

1. Because $A = 0$, $B = 1$, and $C = 0$, you have $\cot 2\theta = \dfrac{A - C}{B} = 0 \Rightarrow 2\theta = \dfrac{\pi}{2} \Rightarrow \theta = \dfrac{\pi}{4}$ which implies that

$$x = x'\cos\frac{\pi}{4} - y'\sin\frac{\pi}{4} \quad \text{and} \quad y = x'\sin\frac{\pi}{4} + y'\cos\frac{\pi}{4}$$

$$= x'\left(\frac{1}{\sqrt{2}}\right) - y'\left(\frac{1}{\sqrt{2}}\right) \qquad = x'\left(\frac{1}{\sqrt{2}}\right) + y'\left(\frac{1}{\sqrt{2}}\right)$$

$$= \frac{x' - y'}{\sqrt{2}} \qquad = \frac{x' + y'}{\sqrt{2}}.$$

The equation in the $x'y'$-system is obtained by substituting these expressions in the original equation.

$$xy + 6 = 0$$

$$\left(\frac{x' - y'}{\sqrt{2}}\right)\left(\frac{x' + y'}{\sqrt{2}}\right) + 6 = 0$$

$$\frac{(x')^2 - (y')^2}{2} + 6 = 0$$

$$\frac{(x')^2 - (y')^2}{2} = -6$$

$$\frac{(y')^2 - (x')^2}{12} = 1$$

$$\frac{(y')^2}{12} - \frac{(x')^2}{12} = 1$$

$$\frac{(y')^2}{\left(2\sqrt{3}\right)^2} - \frac{(x')^2}{\left(2\sqrt{3}\right)^2} = 1$$

In the $x'y'$-system this is a hyperbola centered at the origin with vertices at $\left(0, \pm 2\sqrt{3}\right)$, as shown. To find the coordinates of the vertices in the xy-system, substitute the coordinates $\left(0, \pm 2\sqrt{3}\right)$ in the equations

$$x = \frac{x' - y'}{\sqrt{2}} \quad \text{and} \quad y = \frac{x' + y'}{\sqrt{2}}.$$

This substitution yields the vertices

$$\left(\frac{0 - 2\sqrt{3}}{\sqrt{2}}, \frac{0 + 2\sqrt{3}}{\sqrt{2}}\right) = \left(-\sqrt{6}, \sqrt{6}\right) \text{ and } \left(\frac{0 - \left(-2\sqrt{3}\right)}{\sqrt{2}}, \frac{0 - 2\sqrt{3}}{\sqrt{2}}\right) = \left(\sqrt{6}, -\sqrt{6}\right)$$

in the xy-system. Note that the asymptotes of the hyperbola have equations $y' = \pm x'$ which correspond to the original x- and y-axes.

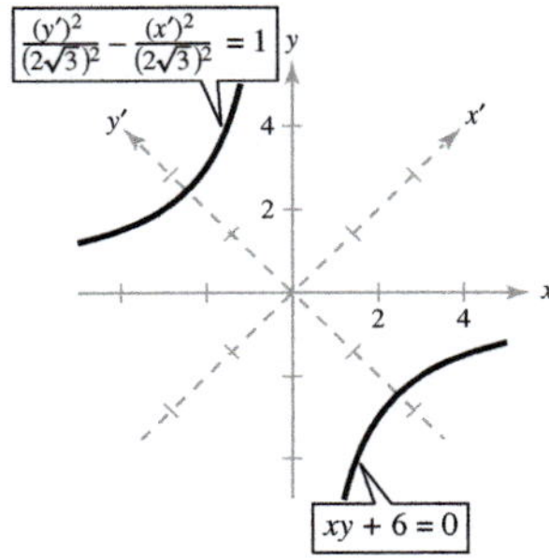

2. Because $A = 12$, $B = 16\sqrt{3}$, and $C = 28$, you have

$$\cot 2\theta = \frac{A - C}{B} = \frac{12 - 28}{16\sqrt{3}} = \frac{-16}{16\sqrt{3}} = \frac{-1}{\sqrt{3}}$$

which implies that $2\theta = \frac{2\pi}{3} \Rightarrow \theta = \frac{\pi}{3}$. The equation in the $x'y'$-system is obtained by making the substitutions

$$\begin{aligned} x &= x'\cos\frac{\pi}{3} - y'\sin\frac{\pi}{3} \\ &= x'\left(\frac{1}{2}\right) - y'\left(\frac{\sqrt{3}}{2}\right) \\ &= \frac{x' - \sqrt{3}y'}{2} \end{aligned}$$

and
$$\begin{aligned} y &= x'\sin\frac{\pi}{3} + y'\cos\frac{\pi}{3} \\ &= x'\left(\frac{\sqrt{3}}{2}\right) + y'\left(\frac{1}{2}\right) \\ &= \frac{\sqrt{3}x' + y'}{2} \end{aligned}$$

in the original equation. So, you have

$$12x^2 + 16\sqrt{3}xy + 28y^2 - 36 = 0$$

$$12\left(\frac{x' - \sqrt{3}y'}{2}\right)^2 + 16\sqrt{3}\left(\frac{x' - \sqrt{3}y'}{2}\right)\left(\frac{\sqrt{3}x' + y'}{2}\right) + 28\left(\frac{\sqrt{3}x' + y'}{2}\right)^2 - 36 = 0$$

which simplifies to

$$\begin{aligned} 36(x')^2 + 4(y')^2 - 36 &= 0 \\ 36(x')^2 + 4(y')^2 &= 36 \\ \frac{(x')^2}{1} + \frac{(y')^2}{9} &= 1 \\ \frac{(x')^2}{1^2} + \frac{(y')^2}{3^2} &= 1. \end{aligned}$$

This is the equation of an ellipse centered at the origin with vertices $(0, \pm 3)$ in the $x'y'$-system as shown.

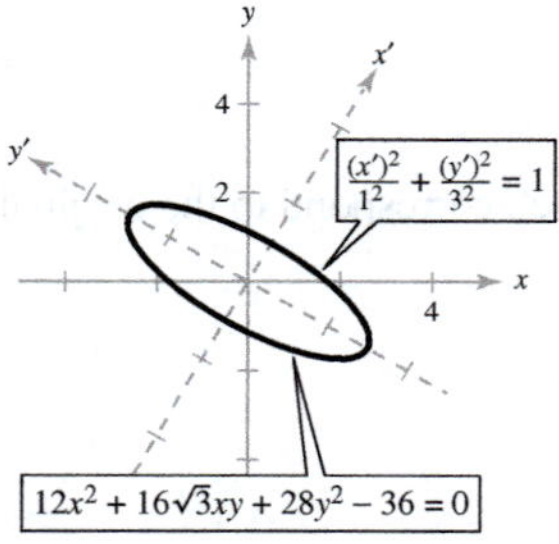

3. 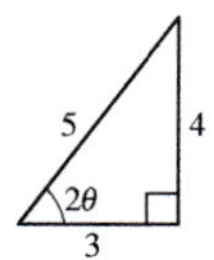

Because $A = 4, B = 4$ and $C = 1$ you have

$$\cot 2\theta = \frac{A - C}{B} = \frac{4 - 1}{4} = \frac{3}{4}.$$

Using this information, draw a right triangle as shown in Figure 10.35. From the figure, you can see that $\cos 2\theta = \frac{3}{5}$. To find the values of $\sin\theta$ and $\cos\theta$, you can use the half-angle formulas in the forms

$$\sin\theta = \sqrt{\frac{1 - \cos 2\theta}{2}} \text{ and } \cos\theta = \sqrt{\frac{1 + \cos 2\theta}{2}}.$$

So,

$$\sin\theta = \sqrt{\frac{1 - \cos 2\theta}{2}} = \sqrt{\frac{1 - \frac{3}{5}}{2}} = \sqrt{\frac{1}{5}} = \frac{1}{\sqrt{5}}$$

$$\cos\theta = \sqrt{\frac{1 + \cos 2\theta}{2}} = \sqrt{\frac{1 + \frac{3}{5}}{2}} = \sqrt{\frac{4}{5}} = \frac{2}{\sqrt{5}}.$$

Consequently, you use the substitutions

$$x = x'\cos\theta - y'\sin\theta = x'\left(\frac{2}{\sqrt{5}}\right) - y'\left(\frac{1}{\sqrt{5}}\right) = \frac{2x' - y'}{\sqrt{5}}$$

and

$$y = x'\sin\theta + y'\cos\theta = x'\left(\frac{1}{\sqrt{5}}\right) + y'\left(\frac{2}{\sqrt{5}}\right) = \frac{x' + 2y'}{\sqrt{5}}.$$

Substituting these expressions in the original equation, you have

$$4x^2 + 4xy + y^2 - 2\sqrt{5}x + 4\sqrt{5}y - 30 = 0$$

$$4\left(\frac{2x' - y'}{\sqrt{5}}\right)^2 + 4\left(\frac{2x' - y'}{\sqrt{5}}\right)\left(\frac{x' + 2y'}{\sqrt{5}}\right) + \left(\frac{x' + 2y'}{\sqrt{5}}\right)^2 - 2\sqrt{5}\left(\frac{2x' - y'}{\sqrt{5}}\right) + 4\sqrt{5}\left(\frac{x' + 2y'}{\sqrt{5}}\right) - 30 = 0$$

which simplifies as follows.

$$5(x')^2 + 10y' - 30 = 0$$

$$5(x')^2 = -10y' + 30$$

$$(x')^2 = -2y' + 6$$

$$(x')^2 = -2(y' - 3)$$

The graph of this equation is a parabola with vertex $(0, 3)$ in the $x'y'$-system. Its axis is parallel to the y'-axis in the $x'y'$-system and because $\sin\theta = \frac{1}{\sqrt{5}}$, $\theta \approx 26.6°$, as shown.

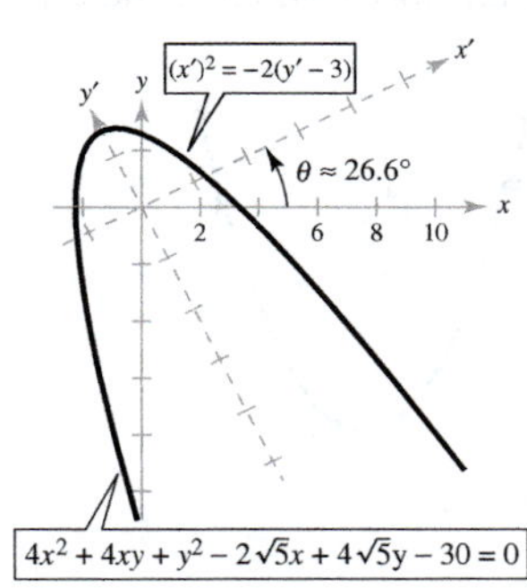

4. Because $B^2 - 4AC = (-8)^2 - 4(2)(8) = 64 - 64 = 0$, the graph is a parabola.

$$2x^2 - 8xy + 8y^2 + 3x + 5 = 0$$
$$8y^2 - 8xy + 2x^2 + 3x + 5 = 0$$
$$8y^2 + (-8x)y + (2x^2 + 3x + 5) = 0$$
$$y = \frac{-(-8x) \pm \sqrt{(-8x)^2 - 4(8)(2x^2 + 3x + 5)}}{2(8)}$$
$$y = \frac{8x \pm \sqrt{64x^2 - 64x^2 - 96x - 160}}{16}$$
$$y = \frac{8x \pm \sqrt{-96x - 160}}{16}$$

So, $y_1 = \dfrac{8x + \sqrt{-96x - 160}}{16}$ and

$y_2 = \dfrac{8x - \sqrt{-96x - 160}}{16}$.

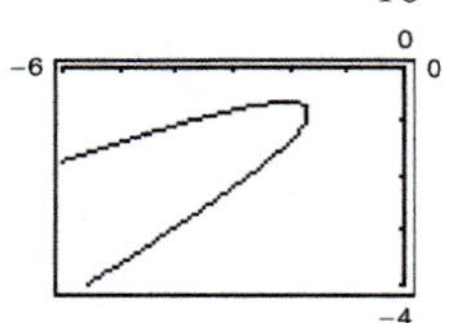

Checkpoints for Section 10.6

1. Using values of t in the specified interval, $-2 \le t \le 2$, the parametric equations yield the points (x, y) shown.

t	x	y
−2	−4	18
−1	−2	6
0	0	2
1	2	6
2	4	18

By plotting these points in the order of increasing values of t, you obtain the curve shown.
The arrows on the curve indicate its orientation as t increases from -2 to 2.

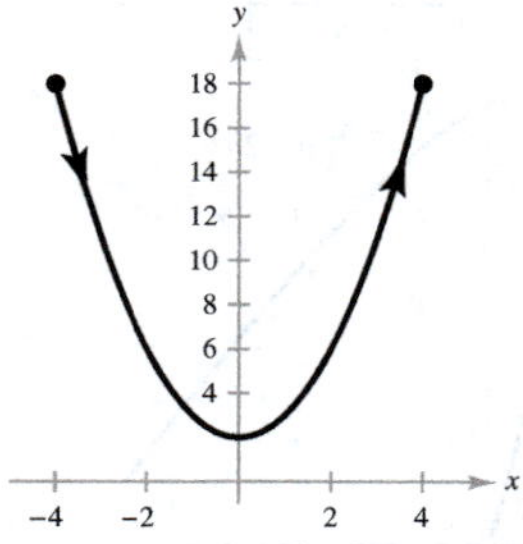

The curve starts at $(-4, 18)$ and ends at $(4, 18)$.

2. Solving for t in the equation for x produces the following.

$$x = \frac{1}{\sqrt{t-1}}$$

$$x^2 = \frac{1}{t-1}$$

$$t - 1 = \frac{1}{x^2}$$

$$t = \frac{1}{x^2} + 1$$

$$t = \frac{1 + x^2}{x^2}$$

Now, substituting in the equation for y, you obtain the following rectangular equation.

$$y = \frac{\dfrac{1+x^2}{x^2} + 1}{\dfrac{1+x^2}{x^2} - 1} = \frac{\dfrac{1 + x^2 + x^2}{x^2}}{\dfrac{1 + x^2 - x^2}{x^2}} = \frac{\dfrac{2x^2+1}{x^2}}{\dfrac{1}{x^2}} = \frac{2x^2+1}{x^2} \cdot \frac{x^2}{1} = 2x^2 + 1$$

From this rectangular equation, you can recognize that the curve is a parabola that opens upward and has its vertex at $(0, 1)$. Also, this rectangular equation is defined for all values of x. The parametric equation for x, however, is defined only when $t > 1$. This implies that you should restrict the domain of x to positive values.

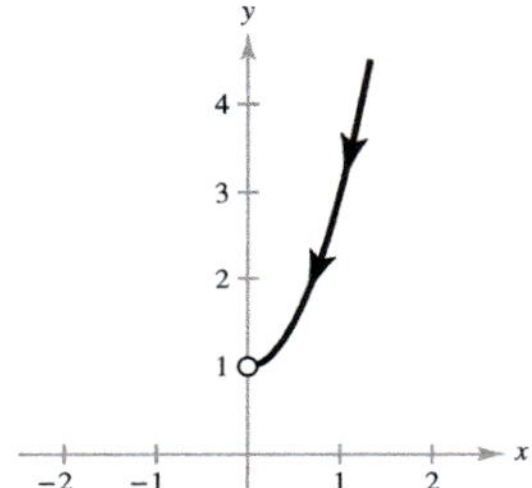

3. (a) Begin by solving for $\cos\theta$ and $\sin\theta$ in the equations

$$x = 5\cos\theta \Rightarrow \cos\theta = \frac{x}{5}$$

$$y = 3\sin\theta \Rightarrow \sin\theta = \frac{y}{3}$$

Use the Pythagorean identity $\sin^2\theta + \cos^2\theta = 1$ to form an equation involving only x and y.

$\sin^2\theta + \cos^2\theta = 1$ Pythagorean identity

$\left(\dfrac{y}{3}\right)^2 + \left(\dfrac{x}{5}\right)^2 = 1$ Substitute $\dfrac{y}{3}$ for $\sin\theta$ and $\dfrac{x}{5}$ for $\cos\theta$.

$\dfrac{y^2}{9} + \dfrac{x^2}{25} = 1$ Simplify.

$\dfrac{x^2}{25} + \dfrac{y^2}{9} = 1$ Rectangular equation

From this rectangular equation, you can see that the graph is an ellipse centered at $(0, 0)$, with horizontal major axis, vertices at $(5, 0)$ and $(-5, 0)$ and minor axis of length $2b = 6$.

Note that the elliptic curve is traced out counterclockwise, starting at $(5, 0)$, as θ increases on the interval $[0, 2\pi)$.

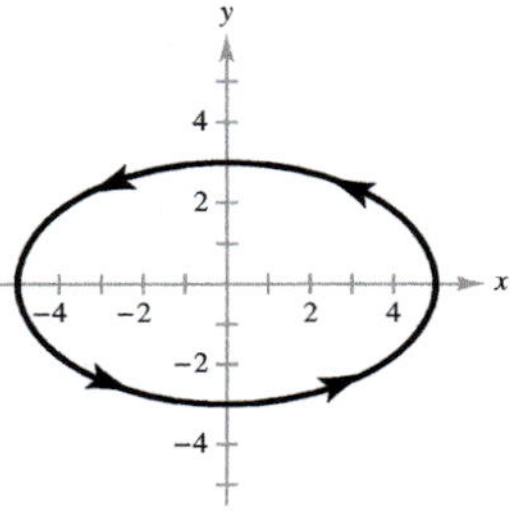

(b) Begin by solving for $\tan \theta$ and $\sec \theta$ in the equations

$x = -1 + \tan \theta \Rightarrow \tan \theta = x + 1$

$y = 2 + 2 \sec \theta \Rightarrow \sec \theta = \dfrac{y - 2}{2}.$

Use the Pythagorean identity $\tan^2 \theta + 1 = \sec^2 \theta$ to form an equation involving only x and y.

$\tan^2 \theta + 1 = \sec^2 \theta$ Pythagorean identity

$(x + 1)^2 + 1 = \left(\dfrac{y - 2}{2}\right)^2$ Substitute $x + 1$ for $\tan \theta$ and $\dfrac{y - 2}{2}$ for $\sec \theta$.

$\left(\dfrac{y - 2}{2}\right)^2 - (x + 1)^2 = 1$ Rewrite.

$\dfrac{(y - 2)^2}{4} - (x + 1)^2 = 1$ Rectangular equation

From this rectangular equation, you can see that the graph is a hyperbola centered at $(-1, 2)$, with a vertical transverse axis. Because $a = 2$, the vertices are $(-1, 4)$ and $(-1, 0)$. However, the restriction on θ, $\dfrac{\pi}{2} \le \theta \le \dfrac{3\pi}{2}$, corresponds to the lower branch of the hyperbola.

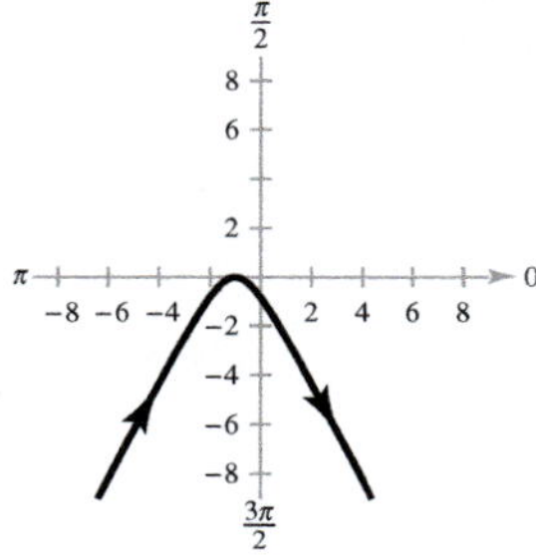

4. (a) Letting $t = x$, you obtain the parametric equations

$x = t$ and $y = x^2 + 2 = t^2 + 2.$

The curve represented by the parametric equations is shown.

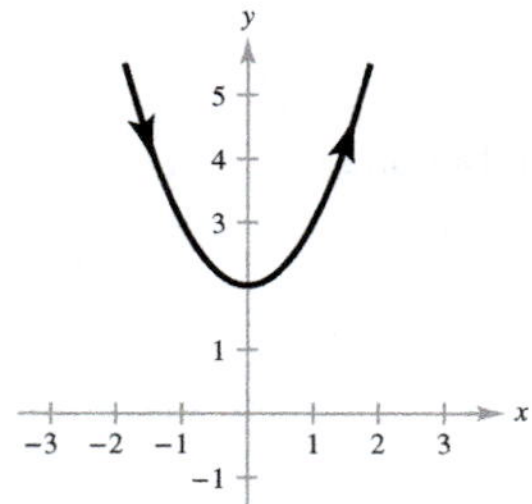

(b) Letting $t = 2 - x$, you obtain the parametric equations

$t = 2 - x \Rightarrow x = 2 - t$ and

$y = x^2 + 2 = t^2 - 4t + 6.$

The curve represented by the parametric equations is shown.

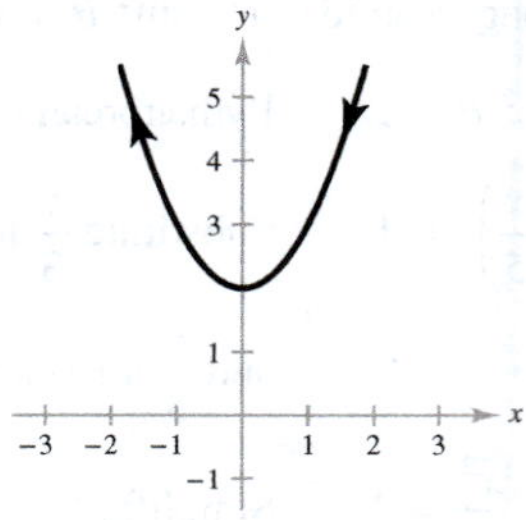

5. As the parameter, let θ be the measure of the circle's rotation, and let the point $P(x, y)$ begin at $(0, 2a)$. When $\theta = 0$, P is at $(0, 2a)$; when $\theta = \pi$, P is at a minimum point $(\pi a, 0)$; and when $\theta = 2\pi$, P is at a maximum point $(2\pi a, 2a)$. From the figure, $\angle APC = \theta$. So,

$$\sin\theta = \sin(\angle APC) = \frac{AC}{a} = \frac{BD}{a}$$

$$\cos\theta = \cos(\angle APC) = \frac{AP}{a}$$

which implies that $BD = a\sin\theta$ and $AP = a\cos\theta$. Because the circle rolls along the x-axis, $OD = \widehat{QD} = a\theta$. Furthermore, because $BA = DC = a$,

$$x = OD + BD = a\theta + a\sin\theta$$

$$y = BA + AP = a + a\cos\theta.$$

So, the parametric equations are $x = a(\theta + \sin\theta)$ and $y = a(1 + \cos\theta)$.

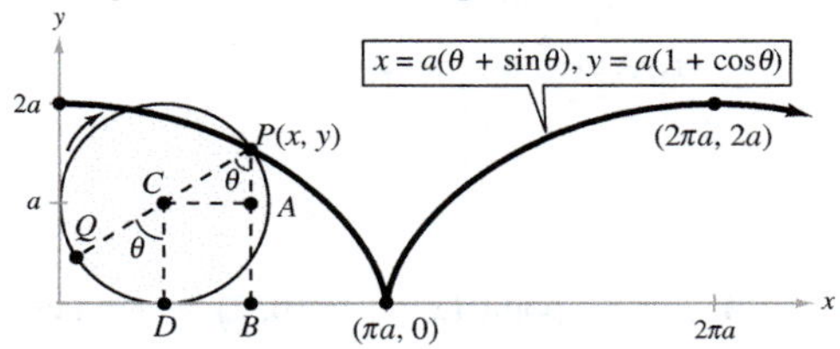

Checkpoints for Section 10.7

1. (a) The point $(r, \theta) = \left(3, \frac{\pi}{4}\right)$ lies three units from the pole on the terminal side of the angle $\theta = \frac{\pi}{4}$.

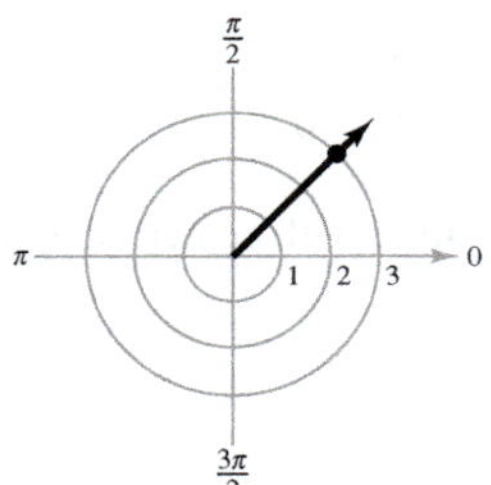

(b) The point $(r, \theta) = \left(2, -\frac{\pi}{3}\right)$ lies two units from the pole on the terminal side of the angle $\theta = -\frac{\pi}{3}$

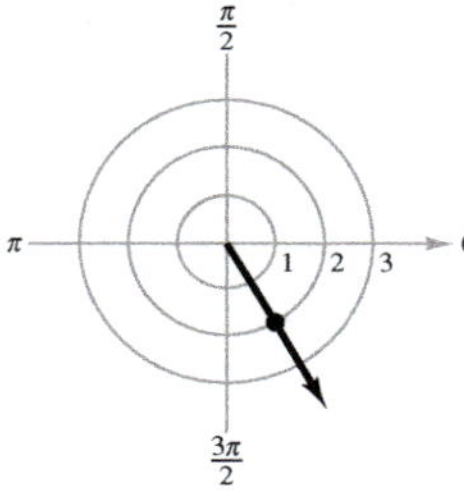

(c) The point $(r, \theta) = \left(2, \frac{5\pi}{3}\right)$ lies two units from the pole on the terminal side of the angle $\theta = \frac{5\pi}{3}$, which coincides with the point $\left(2, -\frac{\pi}{3}\right)$.

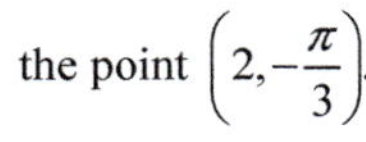

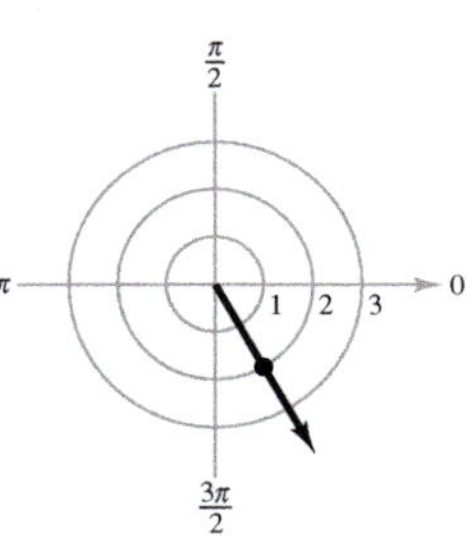

The point is shown. Three other representations are as follows.

$\left(-1, \frac{3\pi}{4} + 2\pi\right) = \left(-1, \frac{11\pi}{4}\right)$ Add 2π to θ.

$\left(1, \frac{3\pi}{4} - \pi\right) = \left(1, -\frac{\pi}{4}\right)$ Replace r with $-r$; subtract π from θ.

$\left(1, \frac{3\pi}{4} + \pi\right) = \left(1, \frac{7\pi}{4}\right)$ Replace r with $-r$; add π to θ.

$$\left(-1, \frac{3\pi}{4}\right) = \left(-1, \frac{11\pi}{4}\right) = \left(1, -\frac{\pi}{4}\right) = \left(1, \frac{7\pi}{4}\right) = \ldots$$

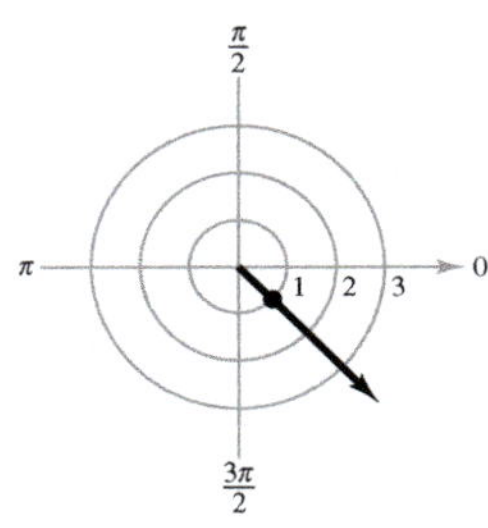

3. For the point $(r, \theta) = (2, \pi)$, you have the following.

$$x = r\cos\theta = 2\cos\pi = (2)(-1) = -2$$

$$y = r\sin\theta = 2\sin\pi = (2)(0) = 0$$

The rectangular coordinates are $(x, y) = (-2, 0)$.

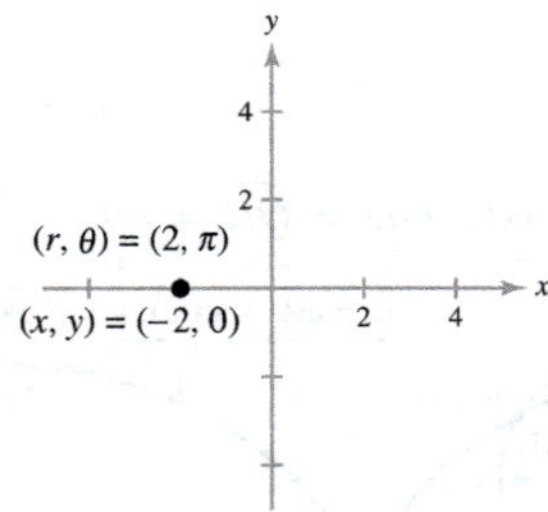

4. For the point $(x, y) = (0, 2)$, which lies on the positive y-axis, you have

$\tan\theta = \dfrac{2}{0}$ is undefined $\Rightarrow \theta = \dfrac{\pi}{2}$.

Choosing a positive value for r,

$$r = \sqrt{x^2 + y^2} = \sqrt{0^2 + 2^2} = 2.$$

So, *one* set of polar coordinates is $(r, \theta) = \left(2, \dfrac{\pi}{2}\right)$ as shown.

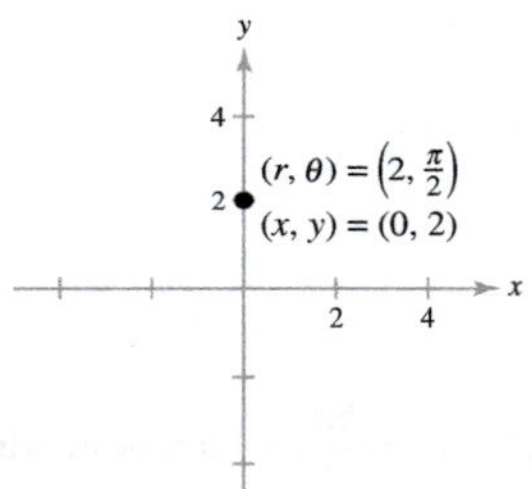

5. (a)

$$r = 7$$

$$r^2 = 49$$

$$x^2 + y^2 = 49$$

The graph consists of all points that are seven units from the pole, which is a circle centered at the origin with a radius of 7.

(b)

$$\theta = \frac{\pi}{4}$$

$$\tan\theta = \tan\frac{\pi}{4}$$

$$\frac{y}{x} = 1$$

$$y = x$$

The graph consists of all points on the line that makes an angle of $\dfrac{\pi}{4}$ with the polar axis and passes through the pole.

(c)

$$r = 6\sin\theta$$

$$r^2 = 6r\sin\theta$$

$$x^2 + y^2 = 6y$$

$$x^2 + y^2 - 6y = 0$$

$$x^2 + y^2 - 6y + 9 = 9$$

$$x^2 + (y - 3)^2 = 9$$

The graph is a circle with center $(0\ 3)$ and radius 3.

Checkpoints for Section 10.8

1. The cosine function is periodic, so you can get a full range of r-values by considering values of θ in the interval $0 \le \theta \le 2\pi$, as shown.

θ	0	$\frac{\pi}{6}$	$\frac{\pi}{4}$	$\frac{\pi}{3}$	$\frac{\pi}{2}$	$\frac{2\pi}{3}$	$\frac{3\pi}{4}$	$\frac{5\pi}{6}$	π
r	6	$3\sqrt{3}$	$3\sqrt{2}$	3	0	-3	$-3\sqrt{2}$	$-3\sqrt{3}$	-6

θ	$\frac{7\pi}{6}$	$\frac{5\pi}{4}$	$\frac{4\pi}{3}$	$\frac{3\pi}{2}$	$\frac{5\pi}{3}$	$\frac{7\pi}{4}$	$\frac{11\pi}{6}$	2π
r	$-3\sqrt{3}$	$-3\sqrt{2}$	-3	0	3	$3\sqrt{2}$	$3\sqrt{3}$	6

By plotting these points, it appears that the graph is a circle of radius 3 whose center is at the point $(x, y) = (3, 0)$.

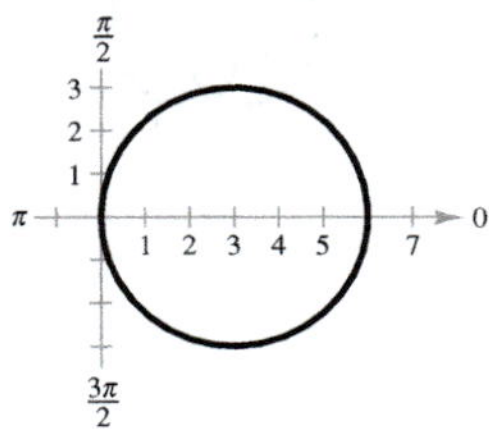

2. Replacing (r, θ) by $(r, \pi - \theta)$ produces the following.

$$\begin{aligned} r &= 3 + 2\sin(\pi - \theta) \\ &= 3 + 2(\sin\pi\cos\theta - \cos\pi\sin\theta) \\ &= 3 + 2[(0)\cos\theta - (-1)\sin\theta] \\ &= 3 + 2(0 + \sin\theta) \\ &= 3 + 2\sin\theta \end{aligned}$$

So, you can conclude that the curve is symmetric with respect to the line $\theta = \pi/2$. Plotting the points in the table and using symmetry with respect to the line $\theta = \pi/2$, you obtain the graph shown, called a dimpled limaçon.

θ	0	$\frac{\pi}{6}$	$\frac{\pi}{4}$	$\frac{\pi}{3}$	$\frac{\pi}{2}$	$\frac{2\pi}{3}$	$\frac{3\pi}{4}$	$\frac{5\pi}{6}$	π
r	3	4	$3 + \sqrt{2}$	$3 + \sqrt{3}$	5	$3 + \sqrt{3}$	$3 + \sqrt{2}$	4	3

θ	$\frac{7\pi}{6}$	$\frac{5\pi}{4}$	$\frac{4\pi}{3}$	$\frac{3\pi}{2}$	$\frac{5\pi}{3}$	$\frac{7\pi}{4}$	$\frac{11\pi}{6}$	2π
r	2	$3 - \sqrt{2}$	$3 - \sqrt{3}$	1	$3 - \sqrt{3}$	$3 - \sqrt{2}$	2	3

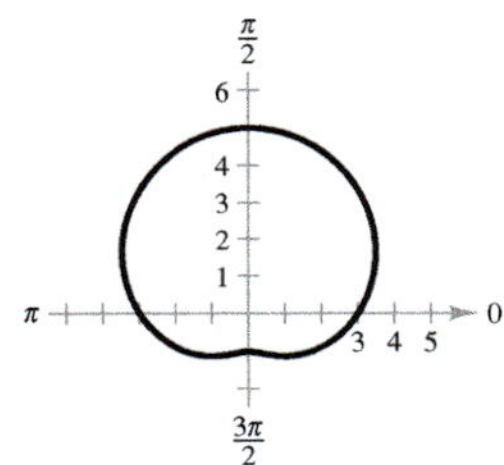

3. From the equation $r = 1 + 2 \sin \theta$, you can obtain the following:

Symmetry: With respect to the line $\theta = \dfrac{\pi}{2}$

Maximum value of $|r|$: $r = 3$ when $\theta = \dfrac{\pi}{2}$

Zero of r: $r = 0$ when $\theta = \dfrac{7\pi}{6}$ and $\dfrac{11\pi}{6}$

The table shows several θ-values in the interval $[0, 2\pi]$

By plotting the corresponding points, you can sketch the graph.

θ	0	$\frac{\pi}{6}$	$\frac{\pi}{4}$	$\frac{\pi}{3}$	$\frac{\pi}{2}$	$\frac{2\pi}{3}$	$\frac{3\pi}{4}$	$\frac{5\pi}{6}$	π
r	1	2	$1 + \sqrt{2}$	$1 + \sqrt{3}$	3	$1 + \sqrt{3}$	$1 + \sqrt{2}$	2	1

θ	$\frac{7\pi}{6}$	$\frac{5\pi}{4}$	$\frac{4\pi}{3}$	$\frac{3\pi}{2}$	$\frac{5\pi}{3}$	$\frac{7\pi}{4}$	$\frac{11\pi}{6}$	2π
r	0	$1 - \sqrt{2}$	$1 - \sqrt{3}$	-1	$1 - \sqrt{3}$	$1 - \sqrt{2}$	0	1

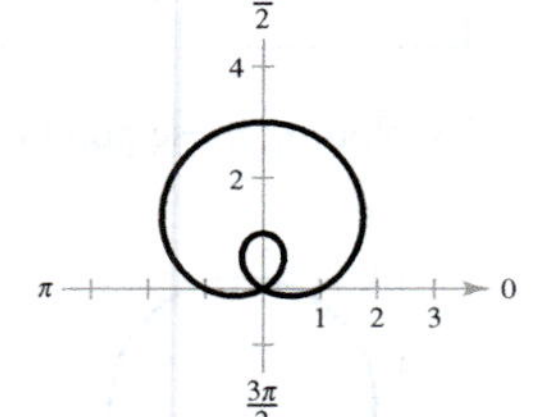

4. From the equation $r = 2 \sin 3\theta$, you can obtain the following.

Symmetry: With respect to the line $\theta = \dfrac{\pi}{2}$

Maximum value of $|r|$: $|r| = 2$ when $3\theta = \dfrac{\pi}{2}, \dfrac{3\pi}{2}, \dfrac{5\pi}{2}$

or $\theta = \dfrac{\pi}{6}, \dfrac{\pi}{2}, \dfrac{5\pi}{6}$

Zeros of r: $r = 0$ when $3\theta = 0, \pi, 2\pi, 3\pi$

or $\theta = 0, \dfrac{\pi}{3}, \dfrac{2\pi}{3}, \pi$

By plotting these points and using symmetry with respect to $\theta = \dfrac{\pi}{2}$, zeros, and maximum values, you obtain the graph.

θ	0	$\frac{\pi}{12}$	$\frac{\pi}{6}$	$\frac{\pi}{4}$	$\frac{\pi}{3}$	$\frac{5\pi}{12}$	$\frac{\pi}{2}$	$\frac{7\pi}{12}$	$\frac{2\pi}{3}$	$\frac{3\pi}{4}$	$\frac{5\pi}{6}$	π
r	0	$\sqrt{2}$	2	$\sqrt{2}$	0	$-\sqrt{2}$	-2	$-\sqrt{2}$	0	$\sqrt{2}$	2	0

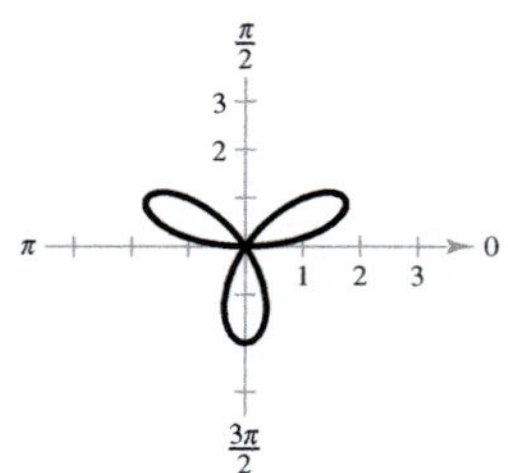

5. *Type of curve*: Rose curve with $n = 3$ petals

Symmetry: With respect to the polar axis

Maximum value of $|r|$: $|r| = 3$ when $\theta = 0, \frac{\pi}{3}, \frac{2\pi}{3}, \pi$

Zeros of r: $r = 0$ when $\theta = \frac{\pi}{6}, \frac{\pi}{2}, \frac{5\pi}{6}$

Using this information together with the points shown, you obtain the graph.

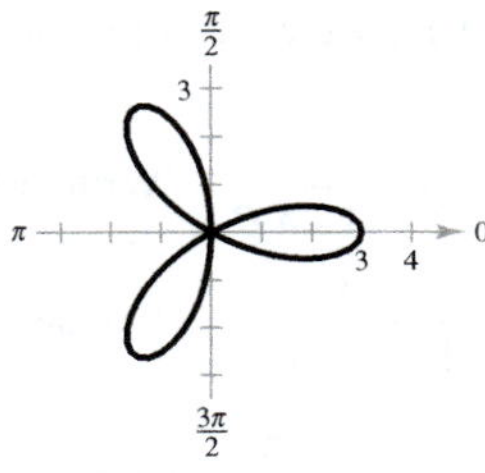

θ	0	$\frac{\pi}{6}$	$\frac{\pi}{4}$	$\frac{\pi}{3}$	$\frac{\pi}{2}$	$\frac{2\pi}{3}$	$\frac{3\pi}{4}$	$\frac{5\pi}{6}$	π
r	3	0	$-\frac{3\sqrt{3}}{2}$	-3	0	3	$\frac{3\sqrt{3}}{4}$	0	-3

6. *Type of curve*: Lemniscate

Symmetry: With respect to the polar axis, the line $\theta = \frac{\pi}{2}$, and the pole

Maximum value of $|r|$: $|r| = 2$ when $\theta = 0, \pi$

Zeros of r: $r = 0$ when $\theta = \frac{\pi}{4}, \frac{3\pi}{4}$

When $\cos 2\theta < 0$, this equation has no solution points.

So, you can restrict the values of θ to three for which $\cos 2\theta \geq 0.$ $0 \leq \theta \leq \frac{\pi}{4}$ and $\frac{3\pi}{4} \leq \theta \leq \pi$

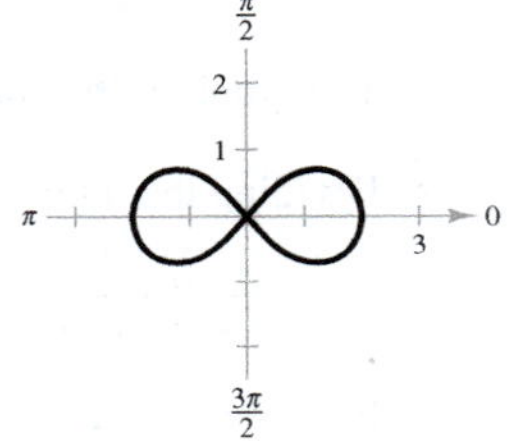

θ	0	$\frac{\pi}{6}$	$\frac{\pi}{4}$	$\frac{3\pi}{4}$	$\frac{5\pi}{6}$	π
$r = \pm 2\sqrt{\cos 2\theta}$	± 2	$\pm\frac{2}{\sqrt{2}}$	0	0	$\pm\frac{2}{\sqrt{2}}$	± 2

Using symmetry and these points, you obtain the graph shown.

Checkpoints for Section 10.9

1. Algebraic Solution

To identify the type of conic, rewrite the equation in the form $r = \dfrac{ep}{1 \pm e \sin\theta}$.

$r = \dfrac{8}{2 - 3\sin\theta}$ Write original equation.

$r = \dfrac{4}{1 - \frac{3}{2}\sin\theta}$ Divide numerator and denominator by 2.

Because $e = \dfrac{3}{2} > 1$, you can conclude the graph is a hyperbola.

Graphical Solution

Use a graphing utility in *polar* mode and be sure to use a square setting as shown.

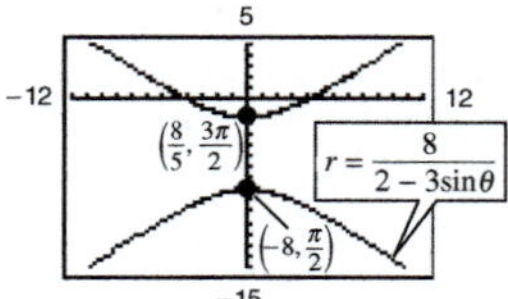

The graph of the conic appears to be a hypberbola.

2. Dividing the numerator and denominator by 2, you have

$$r = \frac{3/2}{1 - 2\sin\theta}.$$

Because $e = 2 > 1$, the graph is a hyperbola.

The transverse axis of the hyperbola lies on the line $\theta = \dfrac{\pi}{2}$, and the vertices occur at $\left(-\dfrac{3}{2}, \dfrac{\pi}{2}\right)$ and $\left(\dfrac{1}{2}, \dfrac{3\pi}{2}\right)$.

Because the length of the transverse axis is 1, you can see that $a = \dfrac{1}{2}$. To find b, write

$$b^2 = a^2\left(e^2 - 1\right) = \left(\frac{1}{2}\right)^2\left[(2)^2 - 1\right] = \frac{3}{4}.$$

So, $b = \dfrac{\sqrt{3}}{2}$. You can use a and b to determine

that the asymptotes of the hyperbola are $y = -1 \pm \dfrac{\sqrt{3}}{3}x$.

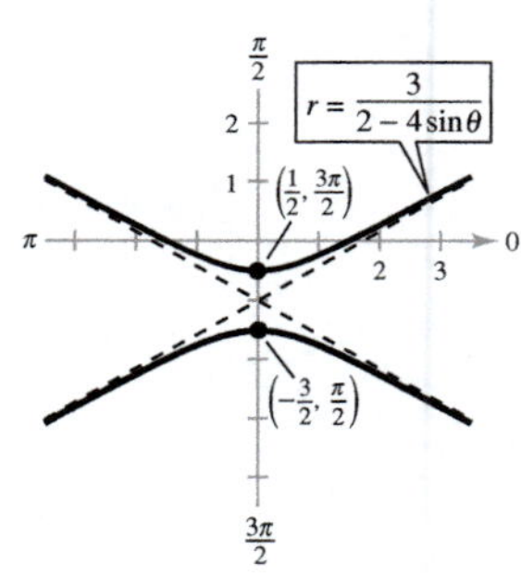

3. Because the directrix is vertical and left of the pole, use an equation of the form

$$r = \frac{ep}{1 - e\cos\theta}.$$

Moreover, because the eccentricity of a parabola is $e = 1$ and the distance between the pole and the directrix is $p = 2$, you have the equation

$$r = \frac{ep}{1 - e\cos\theta} = \frac{2}{1 - \cos\theta}.$$

4. Using a vertical major axis as shown, choose an equation of the form $r = \dfrac{ep}{1 + e \sin \theta}$.

Because the vertices of the ellipse occur when $\theta = \dfrac{\pi}{2}$ and $\theta = \dfrac{3\pi}{2}$, you can determine the length of the major axis to be the sum of the r-values of the vertices.

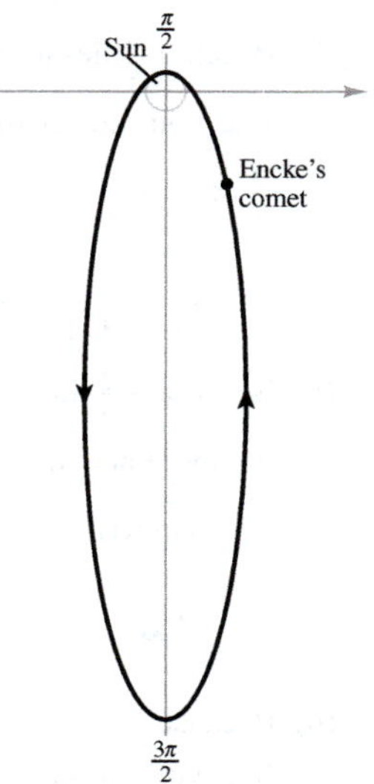

$$2a = \frac{0.847p}{1 + 0.847} + \frac{0.847p}{1 - 0.847} \approx 5.995p \approx 4.420$$

So, $p \approx 0.737$ and $ep \approx (0.847)(0.737) \approx 0.625$.

Using this value of ep in the equation, you have

$$r = \frac{ep}{1 + e \sin \theta} = \frac{0.625}{1 + 0.847 \sin \theta}$$

where r is measured in astronomical units.

To find the closest point to the sun (the focus), substitute $\theta = \dfrac{\pi}{2}$ into this equation.

$$r = \frac{0.625}{1 + 0.847 \sin \pi/2} = \frac{0.625}{1 + 0.847(1)}$$
$$\approx 0.340 \text{ astronomical unit}$$

$x = 2.$

Appendix

Checkpoints for Appendix A.1

1. (a) Natural numbers: $\left\{\frac{6}{3}, 8\right\}$

(b) Whole numbers: $\left\{\frac{6}{3}, 8\right\}$

(c) Integers: $\left\{-22, -1, \frac{6}{3}, 8\right\}$

(d) Rational numbers: $\left\{-22, -7.5, -1 - \frac{1}{4}, \frac{6}{3}, 8\right\}$

(e) Irrational numbers: $\left\{-\pi, \frac{1}{2}\sqrt{2}\right\}$

2. $-1.6 \quad -\frac{3}{4} \quad 0.7 \quad \frac{5}{2}$

$-2 \quad -1 \quad 0 \quad 1 \quad 2 \quad 3 \quad 4 \quad x$

(a) The point representing the real number $\frac{5}{2} = 2.5$ lies halfway between 2 and 3, on the real number line.

(b) The point representing the real number -1.6 lies between -2 and -1 but closer to -2, on the real number line.

(c) The point representing the real number $-\frac{3}{4}$ lies between -1 and 0 but closer to -1, on the real number line.

(d) The point representing the real number 0.7lies between 0 and 1 but closer to 1, on the real number line.

3. (a) Because -5 lies to the left of 1 on the real number line, you can say that -5 is *less than* 1, and write $-5 < 1$.

(b) Because $\frac{3}{2}$ lies to the left of 7 on the real number line, you can say that $\frac{3}{2}$ is *less than* 7, and write $\frac{3}{2} < 7$.

(c) Because $-\frac{2}{3}$ lies to the right of $-\frac{3}{4}$ on the real number line, you can say that $-\frac{2}{3}$ is *greater than* $-\frac{3}{4}$, and write $-\frac{2}{3} > -\frac{3}{4}$.

(d) Because -3.5 lies to the left of 1 on the real number line, you can say that -3.5 is *less than* 1, and write $-3.5 < 1$.

4. (a) The inequality $x > -3$ denotes all real numbers greater than -3.

(b) The inequality $0 < x \leq 4$ means that $x > 0$ and $x \leq 4$. This double inequality denotes all real numbers between 0 and 4, including 4 but not including 0.

5. The interval consists of real numbers greater than or equal to -2 and less than 5.

6. The inequality $-2 \leq x < 4$ can represent the statement "x is less than 4 and at least -2."

7. (a) $|1| = 1$

(b) $-\left|\frac{3}{4}\right| = -\left(\frac{3}{4}\right) = -\frac{3}{4}$

(c) $\frac{2}{|-3|} = \frac{2}{3}$

(d) $-|0.7| = -(0.7) = -0.7$

8. (a) If $x > -3$, then $\frac{|x+3|}{x+3} = \frac{x+3}{x+3} = 1$.

(b) If $x < -3$, then $\frac{|x+3|}{x+3} = \frac{-(x+3)}{x+3} = -1$.

9. (a) $|-3| < |4|$ because $|-3| = 3$ and $|4| = 4$, and 3 is less than 4.

(b) $-|-4| = -|-4|$ because $-|-4| = -4$ and $-|4| = -4$.

(c) $|-3| > -|-3|$ because $|-3| = 3$ and $-|-3| = -3$, and 3 is greater than -3.

10. (a) The distance between 35 and -23 is $|35 - (-23)| = |58| = 58$.

(b) The distance between -35 and -23 is $|-35 - (-23)| = |-12| = 12$.

(c) The distance between 35 and 23 is $|35 - 23| = |12| = 12$.

11. Algebraic Expression: $-2x + 4$

Terms: $-2x$, 4

Coefficients: -2, 4

12. Expression: $4x - 5$

Value of Variable: $x - 0$

Substitute: $4(0) - 5$

Value of Expression: $0 - 5 = -5$

13. (a) $x + 9 = 9 + x$: This statement illustrates the Commutative Property of Addition. In other words, you obtain the same result whether you add x and 9, or 9 and x.

(b) $5(x^3 \cdot 2) = (5x^3)2$: This statement illustrates the Associative Property of Multiplication. In other words, to form the product $5 \cdot x^3 \cdot 2$, it does not matter whether 5 and $(x^3 \cdot 2)$, or $5x^3$ and 2 are multiplied first.

(c) $(2 + 5x^2)y^2 = 2y^2 + 5x^2 \cdot y^2$: This statement illustrates the Distributive Property. In other words, the terms 2 and $5x^2$ are multiplied by y^2.

14. (a) $\dfrac{3}{5} \cdot \dfrac{x}{6} = \dfrac{3x}{30} = \dfrac{3x \div 3}{30 \div 3} = \dfrac{x}{10}$

(b)
$$\begin{aligned}\frac{x}{10} + \frac{2x}{5} &= \frac{x}{10} + \frac{2x}{5} \cdot \frac{2}{2}\\ &= \frac{x}{10} + \frac{2x}{5} = \frac{x}{10} + \frac{2x}{5} \cdot \frac{2}{2}\\ &= \frac{x}{10} + \frac{4x}{10}\\ &= \frac{x + 4x}{10}\\ &= \frac{5x \div 5}{10 \div 5}\\ &= \frac{x}{2}\end{aligned}$$

Checkpoints for Appendix A.2

1. (a) $-3^4 = -(3)(3)(3)(3) = -81$

(b) $(-3)^4 = (-3)(-3)(-3)(-3) = 81$

(c) $3^2 \cdot 3 = 3^{2+1} = 3^3 = (3)(3)(3) = 27$

(d) $\dfrac{3^5}{3^8} = 3^{5-8} = 3^{-3} = \dfrac{1}{3^3} = \dfrac{1}{27}$

2. (a) When $x = 4$, the expression $-x^{-2}$ has a value of

$-x^{-2} = -(4)^{-2} = -\dfrac{1}{4^2} = -\dfrac{1}{16}.$

(b) When $x = 4$, the expression $\dfrac{1}{4}(-x)^4$ has a value of

$\dfrac{1}{4}(-x)^4 = \dfrac{1}{4}(-4)^4 = \dfrac{1}{4}(256) = 64.$

3. (a)
$$\begin{aligned}\left(2x^{-2}y^3\right)\left(-x^4y\right) &= (2)(-1)\left(x^{-2}\right)\left(x^4\right)\left(y^3\right)(y)\\ &= -2x^2y^4\end{aligned}$$

(b) $\left(4a^2b^3\right)^0 = 1,\ a \neq 0,\ b \neq 0$

(c)
$$\begin{aligned}\left(-5z\right)^3\left(z^2\right) &= \left(-5\right)^3\left(z\right)^3 z^2\\ &= -125z^5\end{aligned}$$

(d)
$$\begin{aligned}\left(\frac{3x^4}{x^2y^2}\right)^2 &= \left(\frac{3x^2}{4^2}\right)^2 = \frac{3^2\left(x^2\right)^2}{\left(y^2\right)^2}\\ &= \frac{9x^4}{y^4},\ x \neq 0\end{aligned}$$

4. (a) $2a^{-2} = \dfrac{2}{a^2}$ Property 3

(b)
$$\begin{aligned}\frac{3a^{-3}b^4}{15ab^{-1}} &= \frac{3b^4 \cdot b}{15a \cdot a^3} && \text{Property 3}\\ &= \frac{b^5}{5a^4} && \text{Property 1}\end{aligned}$$

(c)
$$\begin{aligned}\left(\frac{x}{10}\right)^{-1} &= \frac{x^{-1}}{10^{-1}} && \text{Property 7}\\ &= \frac{10}{x} && \text{Property 3}\end{aligned}$$

(d)
$$\begin{aligned}\left(-2x^2\right)^3\left(4x^3\right)^{-1} &= \left(-2\right)^3\left(x^2\right)^3 \cdot 4^{-1} \cdot \left(x^3\right)^{-1} && \text{Property 5}\\ &= \frac{-8x^6}{4x^3} && \text{Properties 3 and 6}\\ &= -2x^3 && \text{Property 2}\end{aligned}$$

5. $45{,}850 = 4.585 \times 10^4$

6. $-2.718 \times 10^{-3} = -0.002718$

7.
$$\begin{aligned}&(24{,}000{,}000{,}000)(0.00000012)(300{,}000)\\ &= \left(2.4 \times 10^{10}\right)\left(1.2 \times 10^{-7}\right)\left(3.0 \times 10^5\right)\\ &= (2.4)(1.2)(3.0)\left(10^8\right)\\ &= 8.64 \times 10^8\\ &= 864{,}000{,}000\end{aligned}$$

8. (a) $-\sqrt{144} = -12$ because $-\left(\sqrt{144}\right) = \left(\sqrt{12^2}\right) = -(12) = -12$.

(b) $\sqrt{-144}$ is not a real number because no real number raised to the second power produces -144.

(c) $\sqrt{\frac{25}{64}} = \frac{5}{8}$ because $\left(\frac{5}{8}\right)^2 = \frac{5^2}{8^2} = \frac{25}{64}$.

(d) $-\sqrt[3]{\frac{8}{27}} = -\frac{2}{3}$ because $-\left(\sqrt[3]{\frac{8}{27}}\right) = -\left(\frac{\sqrt[3]{8}}{\sqrt[3]{27}}\right) = -\left(\frac{2}{3}\right)$.

9. (a) $\frac{\sqrt{125}}{\sqrt{5}} = \sqrt{\frac{125}{5}}$ Property 3

$= \sqrt{25}$ Simplify.

$= 5$ Simplify.

(b) $\sqrt[3]{125^2} = \left(\sqrt[3]{125}\right)^2$ Property 1

$= (5)^2$ Simplify.

$= 25$ Simplify.

(c) $\sqrt[3]{x^2} \cdot \sqrt[3]{x} = \sqrt[3]{x^2 \cdot x}$ Property 2

$= \sqrt[3]{x^3}$ Simplify.

$= x$ Property

(d) $\sqrt{\sqrt{x}} = \sqrt[2\cdot 2]{x}$ Property 4

$= \sqrt[4]{x}$ Simplify.

10. (a) $\sqrt{32} = \sqrt{16 \cdot 2} = \sqrt{4^2 \cdot 2} = 4\sqrt{2}$

(b) $\sqrt[3]{250} = \sqrt[3]{125 \cdot 2} = \sqrt[3]{5^3 \cdot 2} = 5\sqrt[3]{2}$

(c) $\sqrt{24a^5} = \sqrt{4 \cdot 6 \cdot a^4 \cdot a} = \sqrt{4a^4 \cdot 6a}$

$= \sqrt{(2a^2)^2 \cdot 6a}$

$= 2a^2\sqrt{6a}$

(d) $\sqrt[3]{-135x^3} = \sqrt[3]{(-27) \cdot 5 \cdot x^3}$

$= \sqrt[3]{(-3x)^3 \cdot 5}$

$= -3x\sqrt[3]{5}$

11. (a) $3\sqrt{8} + \sqrt{18} = 3\sqrt{4 \cdot 2} + \sqrt{9 \cdot 2}$ Find square factors.

$= 3 \cdot 2\sqrt{2} + 3\sqrt{2}$ Find square roots.

$= 6\sqrt{2} + 3\sqrt{2}$ Multiply.

$= (6 + 3)\sqrt{2}$ Combine like radicals.

$= 9\sqrt{2}$ Simplify.

(b) $\sqrt[3]{81x^5} - \sqrt[3]{24x^2} = \sqrt[3]{27x^3 \cdot 3x^2} - \sqrt[3]{8 \cdot 3x^2}$ Find cube factors.

$= 3x\sqrt[3]{3x^2} - 2\sqrt[3]{3x^2}$ Find cube roots.

$= (3x - 2)\sqrt[3]{3x^2}$ Combine like radicals.

12. (a) $\frac{5}{3\sqrt{2}} = \frac{5}{3\sqrt{2}} \cdot \frac{\sqrt{2}}{\sqrt{2}}$ $\sqrt{2}$ is rationalizing factor.

$= \frac{5\sqrt{2}}{3(2)}$ Multiply.

$= \frac{5\sqrt{2}}{6}$ Simplify.

(b) $\frac{1}{\sqrt[3]{25}} = \frac{1}{\sqrt[3]{25}} \cdot \frac{\sqrt[3]{5}}{\sqrt[3]{5}}$ $\sqrt[3]{5}$ is rationalizing factor.

$= \frac{\sqrt[3]{5}}{\sqrt[3]{125}}$ Multiply.

$= \frac{\sqrt[3]{5}}{5}$ Simplify.

13. $\dfrac{8}{\sqrt{6}-\sqrt{2}} = \dfrac{8}{\sqrt{6}-\sqrt{2}} \cdot \dfrac{\sqrt{6}+\sqrt{2}}{\sqrt{6}+\sqrt{2}}$ Multiply numerator and denominator by conjugate of denominator.

$= \dfrac{8\left(\sqrt{6}+\sqrt{2}\right)}{6+\sqrt{12}-\sqrt{12}-2}$ Use Distributive Property.

$= \dfrac{8\left(\sqrt{6}+\sqrt{2}\right)}{4}$ Simplify.

$= 2\left(\sqrt{6}+\sqrt{2}\right)$ Simplify.

14. $\dfrac{2-\sqrt{2}}{3} = \dfrac{2-\sqrt{2}}{3} \cdot \dfrac{2+\sqrt{2}}{2+\sqrt{2}}$ Multiply numerator and denominator by conjugate of numerator.

$= \dfrac{4+2\sqrt{2}-2\sqrt{2}-2}{3\left(2+\sqrt{2}\right)}$ Multiply.

$= \dfrac{2}{3\left(2+\sqrt{2}\right)}$ Simplify.

15. (a) $\sqrt[3]{27} = 27^{1/3}$

(b) $\sqrt{x^3y^5z} = \left(x^3y^5z\right)^{1/2}$

$= x^{3\cdot 1/2}y^{5\cdot 1/2}z^{1/2}$

$= x^{3/2}y^{5/2}z^{1/2}$

(c) $3x\sqrt[3]{x^2} = 3x\left(x^2\right)^{1/3}$

$= 3x \cdot x^{2/3}$

$= 3x^{1+2/3}$

$= 3x^{5/3}$

16. (a) $\left(x^2-7\right)^{-1/2} = \dfrac{1}{\left(x^2-7\right)^{1/2}} = \dfrac{1}{\sqrt{x^2-7}}$

(b) $-3b^{1/3}c^{2/3} = -3\left(bc^2\right)^{1/3} = -3\sqrt[3]{bc^2}$

(c) $a^{0.75} = a^{3/4} = \sqrt[4]{a^3}$

(d) $\left(x^2\right)^{2/5} = x^{4/5} = \sqrt[5]{x^4}$

17. (a) $(-125)^{-2/3} = \left(\sqrt[3]{-125}\right)^{-2} = (-5)^{-2} = \dfrac{1}{(-5)^2} = \dfrac{1}{25}$

(b) $\left(4x^2y^{3/2}\right)\left(-3x^{-1/3}\right)\left(y^{-3/5}\right) = -12x^{(2)-(1/3)}y^{(3/2)-(3/5)} = -12x^{5/3}y^{9/10}, x \neq 0, y \neq 0$

(c) $\sqrt[3]{\sqrt[4]{27}} = \sqrt[12]{27} = \sqrt[12]{(3)^3} = 3^{3/12} = 3^{1/4} = \sqrt[4]{3}$

(d) $(3x+2)^{5/2}(3x+2)^{-1/2} = (3x+2)^{(5/2)-(1/2)} = (3x+2)^2, x \neq -2/3$

Checkpoints for Appendix A.3

1. Polynomial: $6 - 7x^3 + 2x$

Standard Form: $-7x^3 + 2x + 6$

Degree: 3

Leading Coefficient: -7

2. $\left(2x^3 - x + 3\right) - \left(x^2 - 2x - 3\right)$

$= 2x^3 - x + 3 - x^2 + 2x + 3$

$= 2x^3 - x^2 + (-x + 2x) + (3 + 3)$

$= 2x^3 - x^2 + x + 6$

3. F O I L

$(3x-1)(x-5) = 3x^2 - 15x - x + 5$

$= 3x^2 - 16x + 5$

4. (a) This product has the form

$(u + v)(u - v) = u^2 - v^2.$

$$(3x - 2)(3x + 2) = (3x)^2 - (2)^2$$
$$= 9x^2 - 4$$

(b) This product has the form $(u + v)(u - v) = u^2 - v^2.$

$$(x - 2 + 3y)(x - 2 - 3y) = [(x - 2) + 3y][(x - 2) - 3y]$$
$$= (x - 2)^2 - (3y)^2$$
$$= x^2 - 4x + 4 - 9y^2$$
$$= x^2 - 9y^2 - 4x + 4$$

5. (a) $5x^3 - 15x^2 = 5x^2(x) - 5x^2(3)$ $\quad 5x^2$ is a common factor.

$$= 5x^2(x - 3)$$

(b) $-3 + 6x - 12x^3 = -12x^3 + 6x - 3$

$$= -3(4x^3) + (-3)(-2x) + (-3)(1) \quad -3 \text{ is a common factor.}$$
$$= -3(4x^3 - 2x + 1)$$

(c) $(x + 1)(x^2) - (x + 1)(2) = (x + 1)(x^2 - 2)$ $\quad (x + 1)$ is a common factor.

6. $100 - 4y^2 = 4(25 - y^2)$ $\quad$ 4 is a common factor.

$$= 4[(5)^2 - (y)^2]$$
$$= 4(5 + y)(5 - y) \quad \text{Difference of two squares.}$$

7. $(x - 1)^2 - 9y^4 = (x - 1)^2 - (3y^2)^2$

$$= [(x - 1) + 3y^2][(x - 1) - 3y^2]$$
$$= (x - 1 + 3y^2)(x - 1 - 3y^2)$$

8. $9x^2 - 30x + 25 = (3x)^2 - 2(3x)(5) + 5^2$

$$= (3x - 5)^2$$

9. $64x^3 - 1 = (4x)^3 - (1)^3$

$$= (4x - 1)(16x^2 - 4x + 1)$$

10. (a) $x^3 + 216 = (x)^3 + (6)^3$

$$= (x + 6)(x^2 - 6x + 36)$$

(b) $5y^3 + 135 = 5(y^3 + 27)$

$$= 5[(y)^3 + (3)^3]$$
$$= 5(y + 3)(y^2 - 3y + 9)$$

11. For the trinomial $x^2 + x - 6$, you have $a = 1$, $b = 1$, and $c = -6$. Because b is positive and c is negative, one factor of -6 is positive and one is negative. So, the possible factorizations of $x^2 + x - 6$ are

$(x - 3)(x + 2)$,

$(x + 3)(x - 2)$,

$(x + 6)(x - 1)$, and

$(x - 6)(x + 1)$.

Testing the middle term, you will find the correct factorization to be $(x^2 + x - 6) = (x + 3)(x - 2)$.

12. For the trinomial $2x^2 - 5x + 3$, you have $a = 2$ and $c = 3$, which means that the factors of 3 must have like signs. The possible factorizations are

$(2x + 1)(x + 3)$,

$(2x - 1)(x - 3)$,

$(2x + 3)(x + 1)$, and

$(2x - 3)(x - 1)$.

Testing the middle term, you will find the correct factorization to be $2x^2 - 5x + 3 = (2x - 3)(x - 1)$.

13.
$$\begin{aligned} x^3 + x^2 - 5x - 5 &= \left(x^3 + x^2\right) - (5x + 5) && \text{Group terms.} \\ &= x^2(x + 1) - 5(x + 1) && \text{Factor each group.} \\ &= (x + 1)\left(x^2 - 5\right) && \text{Distributive Property} \end{aligned}$$

14.
$$\begin{aligned} 2x^2 + 5x - 12 &= 2x^2 + 8x - 3x - 12 && \text{Rewrite middle term.} \\ &= \left(2x^2 + 8x\right) - (3x + 12) && \text{Group terms.} \\ &= 2x(x + 4) - 3(x + 4) && \text{Factor groups.} \\ &= (x + 4)(2x - 3) && \text{Distributive Property} \end{aligned}$$

Checkpoints for Appendix A.4

1. (a) The domain of the polynomial $4x^2 + 3, x \geq 0$ is the set of all real numbers that are greater than or equal to 0. The domain is specifically restricted.

(b) The domain of the radical expression $\sqrt{x + 7}$ is the set of all real numbers greater than or equal to -7, because the square root of a negative number is not a real number.

(c) The domain of the rational expression $\dfrac{1 - x}{x}$ is the set of all real numbers except $x = 0$, which would result in division by zero, which is undefined.

2.
$$\begin{aligned} \frac{4x + 12}{x^2 - 3x - 18} &= \frac{4\cancel{(x + 3)}}{(x - 6)\cancel{(x + 3)}} && \text{Factor completely.} \\ &= \frac{4}{x - 6}, \; x \neq -3 && \text{Divide out common factor.} \end{aligned}$$

3.
$$\begin{aligned} \frac{3x^2 - x - 2}{5 - 4x - x^2} &= \frac{3x^2 - x - 2}{-x^2 - 4x + 5} = \frac{(3x + 2)\cancel{(x - 1)}}{-(x + 5)\cancel{(x - 1)}} && \text{Write in standard form.} \\ &= -\frac{3x + 2}{x + 5}, \; x \neq 1 && \text{Divide out common factor.} \end{aligned}$$

4.
$$\begin{aligned} \frac{15x^2 + 5x}{x^3 - 3x^2 - 18x} \cdot \frac{x^2 - 2x - 15}{3x^2 - 8x - 3} &= \frac{5\cancel{x}\cancel{(3x + 1)}}{\cancel{x}(x - 6)\cancel{(x + 3)}} \cdot \frac{(x - 5)\cancel{(x + 3)}}{\cancel{(3x + 1)}(x - 3)} \\ &= \frac{5(x - 5)}{(x - 6)(x - 3)}, \; x \neq -3, x \neq -\frac{1}{3}, x \neq 0 \end{aligned}$$

5. $\dfrac{x^3-1}{x^2-1} \div \dfrac{x^2+x+1}{x^2+2x+1} = \dfrac{x^3-1}{x^2-1} \cdot \dfrac{x^2+2x+1}{x^2+x+1}$ Invert and multiply.

$= \dfrac{\cancel{(x-1)}\cancel{(x^2+x+1)}}{\cancel{(x+1)}\cancel{(x-1)}} \cdot \dfrac{\cancel{(x+1)}(x+1)}{\cancel{x^2+x+1}}$ Factor completely.

$= x+1, x \neq \pm 1$ Divide out common factors.

6. $\dfrac{x}{2x-1} - \dfrac{1}{x+2} = \dfrac{x(x+2)-(2x-1)}{(2x-1)(x+2)}$ Basic definition

$= \dfrac{x^2+2x-2x+1}{(2x-1)(x+2)}$ Distributive Property

$= \dfrac{x^2+1}{(2x-1)(x+2)}$ Combine like terms.

7. The LCD of the ration expression $\dfrac{4}{x} - \dfrac{x+5}{x^2-4} + \dfrac{4}{x+2}$ is $x(x+2)(x-2)$.

$\dfrac{4}{x} - \dfrac{x+5}{(x+2)(x-2)} + \dfrac{4}{x+2} = \dfrac{4(x+2)(x-2)}{x(x+2)(x-2)} - \dfrac{x(x+5)}{x(x+2)(x-2)} + \dfrac{4x(x-2)}{x(x+2)(x-2)}$ Rewrite using the LCD.

$= \dfrac{4(x+2)(x-2)-x(x+5)+4x(x-2)}{x(x+2)(x-2)}$ Distributive Property

$= \dfrac{4x^2-16-x^2-5x+4x^2-8x}{x(x+2)(x-2)}$

$= \dfrac{7x^2-13x-16}{x(x+2)(x-2)}$

8. $\dfrac{\left(\dfrac{1}{x+2}+1\right)}{\left(\dfrac{x}{3}-1\right)} = \dfrac{\left(\dfrac{1+1(x+2)}{x+2}\right)}{\left(\dfrac{x-1(3)}{3}\right)}$ Combine fractions.

$= \dfrac{\left(\dfrac{x+3}{x+2}\right)}{\left(\dfrac{x-3}{3}\right)}$ Simplify.

$= \dfrac{x+3}{x+2} \cdot \dfrac{3}{x-3}$ Invert and multiply.

$= \dfrac{3(x+3)}{(x+2)(x-3)}$

9. $(x-1)^{-1/3} - x(x-1)^{-4/3} = (x-1)^{-4/3}\left[(x-1)^{(-1/3)-(-4/3)} - x\right]$

$= (x-1)^{-4/3}\left[(x-1)^1 - x\right]$

$= -\dfrac{1}{(x-1)^{4/3}}$

10. $$\frac{x^2(x^2-2)^{-1/2}+(x^2-2)^{1/2}}{x^2-2} = \frac{x^2(x^2-2)^{-1/2}+(x^2-2)^{1/2}}{x^2-2}\cdot\frac{(x^2-2)^{1/2}}{(x^2-2)^{1/2}}$$

$$= \frac{x^2(x^2-2)^{0}+(x^2-2)^{1}}{(x^2-2)^{3/2}}$$

$$= \frac{x^2+x^2-2}{(x^2-2)^{3/2}}$$

$$= \frac{2x^2-2}{(x^2-2)^{3/2}}$$

$$= \frac{2(x+1)(x-1)}{(x^2-2)^{3/2}}$$

11. $$\frac{\sqrt{9+h}-3}{h} = \frac{\sqrt{9+h}-3}{h}\cdot\frac{\sqrt{9+h}+3}{\sqrt{9+h}+3}$$

$$= \frac{\left(\sqrt{9+h}\right)^2-(3)^2}{h\left(\sqrt{9+h}+3\right)}$$

$$= \frac{(9+h)-9}{h\left(\sqrt{9+h}+3\right)}$$

$$= \frac{h}{h\left(\sqrt{9+h}+3\right)}$$

$$= \frac{1}{\sqrt{9+h}+3},\ h \neq 0$$

Checkpoints for Appendix A.5

1. (a)

$7 - 2x = 15$	Write original equation.
$-2x = 8$	Subtract 7 from each side.
$x = -4$	Divide each side by -2.

Check:

$$7 - 2x = 15$$
$$7 - 2(-4) \stackrel{?}{=} 15$$
$$7 + 8 \stackrel{?}{=} 15$$
$$15 = 15$$

(b)

$7x - 9 = 5x + 7$	Write original equation.
$2x - 9 = 7$	Subtract $5x$ from each side.
$2x = 16$	Add 9 from each side.
$x = 8$	Divide each side by 2.

Check:

$$7x - 9 = 5x + 7$$
$$7(8) - 9 = 5(8 + 7)$$
$$56 - 9 = 40 + 7$$
$$47 = 47\ \checkmark$$

2.

$\frac{4x}{9} - \frac{1}{3} = x + \frac{5}{3}$	Write original equation.
$(9)\left(\frac{4x}{9}\right) - (9)\left(\frac{1}{3}\right) = (9)x + 9\left(\frac{5}{3}\right)$	Multiply each term by the LCD.
$4x - 3 = 9x + 15$	Simplify.
$-5x = 18$	Combine like terms.
$x = -\frac{18}{5}$	Divide each side by -5.

3.

$$\frac{3x}{x-4} = 5 + \frac{12}{x-4}$$ Write original equation.

$$(x-4)\left(\frac{3x}{x-4}\right) = (x-4)5 + (x-4)\left(\frac{12}{x-4}\right)$$ Multiply each term by LCD.

$$3x = 5x - 20 + 12, x \neq 4$$ Simplify.

$$-2x = -8$$ Divide each side by -2.

$$x = 4$$ Extraneous solution

In the original equation, $x = 4$ yields a denominator of zero. So, $x = 4$ is an extraneous solution, and the original equation has no solution.

4.

$$2x^2 - 3x + 1 = 6$$ Write original equation.

$$2x^2 - 3x - 5 = 0$$ Write in general form.

$$(2x - 5)(x + 1) = 0$$ Factor.

$$2x - 5 = 0 \Rightarrow x = \tfrac{5}{2}$$ Set 1st factor equal to 0.

$$x + 1 = 0 \Rightarrow x = -1$$ Set 2nd factor equal to 0.

The solutions are $x = -1$ and $x = \frac{5}{2}$.

Check: $x = -1$

$$2x^2 - 3x + 1 = 6$$
$$2(-1)^2 - 3(-1) + 1 \stackrel{?}{=} 6$$
$$2(1) + 3 + 1 \stackrel{?}{=} 6$$
$$6 = 6 \checkmark$$

$x = \frac{5}{2}$

$$2x^2 - 3x + 1 = 6$$
$$2\left(\tfrac{5}{2}\right)^2 - 3\left(\tfrac{5}{2}\right) + 1 \stackrel{?}{=} 6$$
$$2\left(\tfrac{25}{4}\right) - \tfrac{15}{2} + 1 \stackrel{?}{=} 6$$
$$6 = 6 \checkmark$$

5. (a)

$$3x^2 = 36$$ Write original equation.

$$x^2 = 12$$ Divide each side by 3.

$$x = \pm\sqrt{12}$$ Extract square roots.

$$x = \pm 2\sqrt{3}$$

The solutions are $x = \pm 2\sqrt{3}$.

Check: $x = -2\sqrt{3}$

$$3x^2 = 36$$
$$3\left(-2\sqrt{3}\right)^2 \stackrel{?}{=} 36$$
$$3(12) \stackrel{?}{=} 36$$
$$36 = 36 \checkmark$$

$x = 2\sqrt{3}$

$$3x^2 = 36$$
$$3\left(2\sqrt{3}\right)^2 \stackrel{?}{=} 36$$
$$3(12) \stackrel{?}{=} 36$$
$$36 = 36 \checkmark$$

(b)

$$(x - 1)^2 = 10$$
$$x - 1 = \pm\sqrt{10}$$
$$x = 1 \pm \sqrt{10}$$

The solutions are $x = 1 \pm \sqrt{10}$.

Check: $x = 1 - \sqrt{10}$

$$(x - 1)^2 = 10$$
$$\left[\left(1 - \sqrt{10}\right) - 1\right]^2 \stackrel{?}{=} 10$$
$$\left(-\sqrt{10}\right)^2 \stackrel{?}{=} 10$$
$$10 = 10 \checkmark$$

$x = 1 + \sqrt{10}$

$$(x - 1)^2 = 10$$
$$\left[\left(1 + \sqrt{10}\right) - 1\right] \stackrel{?}{=} 10$$
$$\left(\sqrt{10}\right)^2 \stackrel{?}{=} 10$$
$$10 = 10 \checkmark$$

6.

$x^2 - 4x - 1 = 0$	Write original equation.
$x^2 - 4x = 1$	Add 1 to each side.
$x^2 - 4x + \underbrace{(2)^2}_{(\text{half of } 4)^2} = 1 + (2)^2$	Add 2^2 to each side.
$(x - 2)^2 = 5$	Simplify.
$x - 2 = \pm\sqrt{5}$	Extract square roots.
$x = 2 \pm \sqrt{5}$	Add 2 to each side.

The solutions are $x = 2 \pm \sqrt{5}$.

Check: $x = 2 - \sqrt{5}$

$$x^2 - 4x - 1 = 0$$
$$\left(2 - \sqrt{5}\right)^2 - 4\left(2 - \sqrt{5}\right) - 1 \stackrel{?}{=} 0$$
$$\left(4 - 4\sqrt{5} + 5\right) - 8 + 4\sqrt{5} - 1 \stackrel{?}{=} 0$$
$$4 + 5 - 8 - 1 \stackrel{?}{=} 0$$
$$0 = 0 \checkmark$$

$x = 2 + \sqrt{5}$ also checks. ✓

7.

$3x^2 - 10x - 2 = 0$	Original equation
$3x^2 - 10x = 2$	Add 2 to each side.
$x^2 - \frac{10}{3}x = \frac{2}{3}$	Divide each side by 3.
$x^2 - \frac{10}{3}x + \left(\frac{5}{3}\right)^2 = \frac{2}{3} + \left(\frac{5}{3}\right)^2$	Add $\left(\frac{5}{3}\right)^2$ to each side.
$\left(x - \frac{5}{3}\right)^2 = \frac{31}{9}$	Simplify.
$x - \frac{5}{3} = \pm\frac{\sqrt{31}}{3}$	Extract square roots.
$x = \frac{5}{3} \pm \frac{\sqrt{31}}{3}$	Add $\frac{5}{3}$ to each side.

The solutions are $\frac{5}{3} \pm \frac{\sqrt{31}}{3}$.

8. $3x^2 + 2x - 10 = 0$ Write original equation.

$$x = \frac{-6 \pm \sqrt{6^2 - 4ac}}{2a}$$ Quadratic Formula

$$x = \frac{-2 \pm \sqrt{(2)^2 - 4(3)(-10)}}{2(3)}$$ Substitute $a = 3$, $b = 2$ and $c = -10$.

$$x = \frac{-2 \pm \sqrt{4 + 120}}{6}$$ Simplify.

$$x = \frac{-2 \pm \sqrt{124}}{6}$$ Simplify.

$$x = \frac{-2 \pm 2\sqrt{31}}{6}$$ Simplify.

$$x = \frac{2\left(-1 \pm \sqrt{31}\right)}{6}$$ Factor our common factor.

$$x = \frac{-1 \pm \sqrt{31}}{3}$$ Simplify.

The solutions are $\frac{-1 \pm \sqrt{31}}{3}$.

Check: $x = \frac{-1 \pm \sqrt{31}}{3}$

$$3x^2 + 2x - 10 = 0$$

$$3\left(\frac{-1 + \sqrt{31}}{3}\right)^2 + 2\left(\frac{-1 + \sqrt{31}}{3}\right) - 10 \stackrel{?}{=} 0$$

$$\frac{1}{3} + \frac{2\sqrt{31}}{3} + \frac{31}{3} + \frac{2\sqrt{31}}{3} - 10 \stackrel{?}{=} 0$$

$$10 - 10 \stackrel{?}{=} 0$$

$$0 = 0 \checkmark$$

The solution $x = \frac{-1 - \sqrt{31}}{3}$ also checks. ✓

9. $18x^2 - 48x + 32 = 0$

$9x^2 - 24x + 16 = 0$

$$x = \frac{-b \pm \sqrt{b^2 - 4ac}}{2a}$$

$$x = \frac{-(-24) \pm \sqrt{(-24)^2 - 4(9)(16)}}{2(9)}$$

$$x = \frac{24 \pm \sqrt{0}}{18}$$

$$x = \frac{4}{3}$$

The quadratic equation has only one solution: $x = \frac{4}{3}$.

10. $9x^4 - 12x^2 = 0$ Write original equation.

$3x^2(3x^2 - 4) = 0$ Factor out common factor.

$3x^2 = 0 \Rightarrow x = 0$ Set 1st factor equal to 0.

$3x^2 - 4 = 0 \Rightarrow 3x^2 = 4$ Set 2nd factor equal to 0.

$$x^2 = \frac{4}{3}$$

$$x = \pm\sqrt{\frac{4}{3}}$$

$$x = \frac{\pm 2\sqrt{3}}{3}$$

Check: $x = 0$

$$9x^4 - 12x^2 = 0$$

$$9(0)^4 - 12(0)^2 \stackrel{?}{=} 0$$

$$0 = 0 \checkmark$$

$$x = \frac{2\sqrt{3}}{3}$$

$$9x^4 - 12x^2 = 0$$

$$9\left(\frac{2\sqrt{3}}{3}\right)^4 - 12\left(\frac{2\sqrt{3}}{3}\right)^2 \stackrel{?}{=} 0$$

$$9\left(\frac{16}{9}\right) - 12\left(\frac{4}{3}\right) \stackrel{?}{=} 0$$

$$16 - 16 \stackrel{?}{=} 0$$

$$0 = 0 \checkmark$$

The solution $x = \dfrac{-2\sqrt{3}}{3}$ also checks. ✓

So, the solutions are $x = 0$ and $x = \pm\dfrac{2\sqrt{3}}{3}$.

11. (a)

$$\begin{aligned} x^3 - 5x^2 - 2x + 10 &= 0 && \text{Write original equation.} \\ x^2(x-5) - 2(x-5) &= 0 && \text{Factor by grouping.} \\ (x-5)(x^2-2) &= 0 && \text{Distributive Property} \\ x - 5 = 0 \Rightarrow x &= 5 && \text{Set 1st factor equal to 0.} \\ x^2 - 2 = 0 \Rightarrow x^2 &= 2 && \text{Set 2nd factor equal to 0.} \\ &= \pm\sqrt{2} \end{aligned}$$

Check: $x = 5$

$$\begin{aligned} x^3 - 5x^2 - 2x + 10 &= 0 \\ (5)^3 - 5(5)^2 - 2(5) + 10 &\stackrel{?}{=} 0 \\ 125 - 125 - 10 + 10 &\stackrel{?}{=} 0 \\ 0 &= 0 \checkmark \end{aligned}$$

$x = \sqrt{2}$

$$\begin{aligned} x^3 - 5x^2 - 2x + 10 &= 0 \\ (\sqrt{2})^3 - 5(\sqrt{2})^2 - 2(\sqrt{2}) + 10 &\stackrel{?}{=} 0 \\ 2\sqrt{2} - 10 - 2\sqrt{2} + 10 &\stackrel{?}{=} 0 \\ 0 &= 0 \checkmark \end{aligned}$$

The solution $x = -\sqrt{2}$ also checks. ✓

So, the solutions are $x = 5$ and $x = \pm\sqrt{2}$.

(b) $6x^3 - 27x^2 - 54x = 0$

$$\begin{aligned} 3x(2x^2 - 9x - 18) &= 0 && \text{Factor out common factor.} \\ 3x(2x+3)(x-6) &= 0 && \text{Factor quadratic factor.} \\ 3x = 0 \Rightarrow x &= 0 && \text{Set 1st factor equal to 0.} \\ 2x + 3 = 0 \Rightarrow x &= -\tfrac{3}{2} && \text{Set 2nd factor equal to 0.} \\ x - 6 = 0 \Rightarrow x &= 6 && \text{Set 3rd factor equal to 0.} \end{aligned}$$

Check: $x = 0$

$$\begin{aligned} 6x^3 - 27x^2 - 54x &= 0 \\ 6(0)^3 - 27(0)^2 - 54(0) &\stackrel{?}{=} 0 \\ 0 &= 0 \checkmark \end{aligned}$$

$x = -\frac{3}{2}$

$$\begin{aligned} 6x^3 - 27x^2 - 54x &= 0 \\ 6\left(-\tfrac{3}{2}\right)^3 - 27\left(-\tfrac{3}{2}\right)^2 - 54\left(-\tfrac{3}{2}\right) &\stackrel{?}{=} 0 \\ 6\left(-\tfrac{27}{8}\right)^3 - 27\left(\tfrac{9}{4}\right)^2 + 27(3) &\stackrel{?}{=} 0 \\ -\tfrac{81}{4} - \tfrac{243}{4} + 81 &\stackrel{?}{=} 0 \\ 0 &= 0 \checkmark \end{aligned}$$

$x = 6$

$$\begin{aligned} 6x^3 - 27x^2 - 54x &= 0 \\ 6(6)^3 - 27(6)^2 - 54(6) &\stackrel{?}{=} 0 \\ 6(216) - 27(36) - 324 &\stackrel{?}{=} 0 \\ 0 &= 0 \checkmark \end{aligned}$$

So, the solutions are $x = 0$, $x = -\frac{3}{2}$, and $x = 6$.

12.

$-\sqrt{40-9x}+2=x$	Write original equation.
$-\sqrt{40-9x}=x-2$	Isolated radical.
$\left(-\sqrt{40-9x}\right)^2=(x-2)^2$	Square each side.
$40-9x=x^2-4x+4$	Simplify.
$0=x^2+5x-36$	Write in general form.
$0=(x-4)(x+9)$	Factor.
$x-4=0 \Rightarrow x=4$	Set 1st factor equal to 0.
$x+9=0 \Rightarrow x=-9$	Set 2nd factor equal to 0.

Check: $x=4$

$$-\sqrt{40-9x}+2=x$$

$$-\sqrt{40-9(4)}+2\stackrel{?}{=}4$$

$$-\sqrt{4}+2\stackrel{?}{=}4$$

$$-2+2\stackrel{?}{=}4$$

$$0\neq 4 \text{ ✗}$$

$x=4$ is an extraneous solution.

$x=-9$

$$-\sqrt{40-9(-9)}+2\stackrel{?}{=}-9$$

$$-\sqrt{121}+2\stackrel{?}{=}-9$$

$$-11+2\stackrel{?}{=}-9$$

$$-9\stackrel{?}{=}-9 \text{ ✓}$$

So, the only solution is $x=-9$.

13. $(x - 5)^{2/3} = 16$ Write original equation.

$\sqrt[3]{(x - 5)^2} = 16$ Rewrite in radical form.

$(x - 5)^2 = 4096$ Cube each side.

$x - 5 = \pm 64$ Extract square roots.

$x = 5 \pm 64$ Add 5 to each side.

$x = -59, x = 69$

Check: $x = -59$

$$(x - 5)^{2/3} = 16$$
$$(-59 - 5)^{2/3} \stackrel{?}{=} 16$$
$$(-64)^{2/3} \stackrel{?}{=} 16$$
$$(-4)^2 \stackrel{?}{=} 16$$
$$16 = 16 \checkmark$$

$x = 69$

$$(x - 5)^{2/3} = 16$$
$$(69 - 5)^{2/3} \stackrel{?}{=} 16$$
$$(64)^{2/3} \stackrel{?}{=} 16$$
$$(4)^2 \stackrel{?}{=} 16$$
$$16 = 16 \checkmark$$

So, the solutions are $x = -59$ and $x = 69$.

14. $|x^2 + 4x| = 7x + 18$

First Equation

$x^2 + 4x = 7x + 18$ Use positive expression.

$x^2 - 3x - 18 = 0$ Write in general form.

$(x + 3)(x - 6) = 0$ Factor.

$x + 3 = 0 \Rightarrow x = -3$ Set 1st factor equal to 0.

$x - 6 = 0 \Rightarrow x = 6$ Set 2nd factor equal to 0.

Second Equation

$-(x^2 + 4x) = 7x + 18$ Use negative expression.

$-x^2 - 4x = 7x + 18$ Distributive Property

$0 = x^2 + 11x + 18$ Write in general form.

Use the Quadratic equation to solve the equation $0 = x^2 + 11x + 18$.

$$x = \frac{-b \pm \sqrt{b^2 - 4ac}}{2a}$$

$$x = \frac{-11 \pm \sqrt{11^2 - 4(1)(18)}}{2(1)}$$

$$x = \frac{-11 \pm \sqrt{49}}{2}$$

$x = -2, -9$

The possible solutions are $x = -9$, $x = -3$, $x = -2$, and $x = 6$.

Because $x = -9$ and $x = -3$ are extraneous, the only solutions are $x = -2$ and $x = 6$.

15. The formula for the volume of a cylindrical container is $V = \pi r^2 h$. To find the height of the container, solve for h.

$$h = \frac{V}{\pi r^2}$$

Then, using $V = 84$ and $r = 3$, find the height.

$$h = \frac{84}{\pi(3)^2}$$

$$h = \frac{84}{9\pi}$$

$$h \approx 2.97$$

So, the height of the container is about 2.97 inches. You can use unit analysis to check that your answer is reasonable.

$$\frac{84 \text{ in.}^3}{9\pi \text{ in.}^2} = \frac{84 \text{ in.} \cdot \cancel{\text{in.}} \cdot \cancel{\text{in.}}}{9\pi \cancel{\text{in.}} \cdot \cancel{\text{in.}}} = \frac{84}{9\pi} \text{ in.} \approx 2.97 \text{ in.}$$

Checkpoints for Appendix A.6

1. (a) $[-1, 3]$ corresponds to $-1 \le x \le 3$. The interval is bounded.

(b) $(-1, 6)$ corresponds to $-1 < x < 6$. The interval is bounded.

(c) $(-\infty, 4)$ corresponds to $x < 4$. The interval is unbounded.

(d) $[0, \infty)$ corresponds to $x \ge 0$. The interval is unbounded.

2.

$7x - 3 \le 2x + 7$	Write original inequality.
$5x \le 10$	Subtract $2x$ and add 3 to each side.
$x \le 2$	Divide each side by 5.

The solution set is all real numbers less than or equal to 2.

3. (a) **Algebraic solution**

$2 - \frac{5}{3}x > x - 6$	Write original inequality.
$6 - 5x > 3x - 18$	Multiply each side by 3.
$-8x > -24$	Subtract $3x$ and subtract 6 from each side.
$x < 3$	Divide each side by -8 reverse the inequality symbol.

The solution set is all real numbers that are less than 3.

(b) **Graphical solution**

Use a graphing utility to graph $y_1 = 2 - \frac{5}{3}x$ and $y_2 = x - 6$ in the same viewing window. Use the *intersect* feature to determine that the graphs intersect at $(3, -3)$. The graph of y_1 lies above the graph of y_2 to the left of their point of intersection, which implies that $y_1 > y_2$ for all $x < 3$.

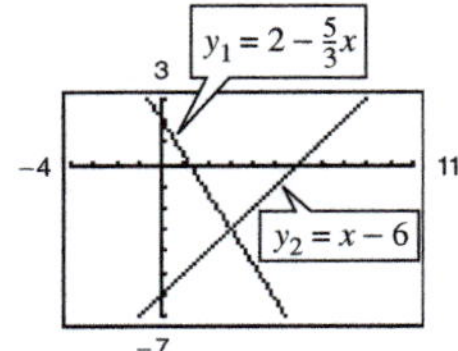

4. $1 < 2x + 7 < 11$ Write original inequality.

$1 - 7 < 2x + 7 - 7 < 11 - 7$ Subtract 7 from each part.

$-6 < 2x < 4$ Simplify.

$-\frac{6}{2} < \frac{2x}{2} < \frac{4}{2}$ Divide each part by 2.

$-3 < x < 2$ Simplify.

The solution set is all real numbers greater than -3 and less than 2, which is denoted by $(-3, 2)$.

−4 −3 −2 −1 0 1 2 3 x

5. $|x - 20| \le 4$ Write original inequality.

$-4 \le x - 20 \le 4$ Write equivalent inequalities.

$-4 + 20 \le x - 20 + 20 \le 4 = 20$ Add 20 to each part.

$16 \le x \le 24$ Simplify.

The solution set is all real numbers that are greater than or equal to 16 and less than or equal to 24, which is denoted by $[16, 24]$.

12 14 16 18 20 22 24 26 28 x

6. Let h represent the number of hours you use the car. Write and solve an inequality.

$10.25h + 8 > 10h + 25$

$0.25h > 17$

$h > 68$

Plan B costs more when you use the car for more than 68 hours in one month.

7. Let x represent the actual weight of your bag. The difference of the actual weight and the weight on the scale is at most $\frac{1}{64}$ pound. That is, $\left|x - \frac{1}{2}\right| \le -\frac{1}{64}$.

You can solve the inequality as follows.

$-\frac{1}{64} \le x - \frac{1}{2} \le \frac{1}{64}$

$\frac{31}{64} \le x \le \frac{33}{64}$

The least your bag can weigh is $\frac{31}{64}$ pound, which would have cost $\left(\frac{31}{64}\text{ pound}\right) \times (\$9.89\text{ per pound}) = \$4.79$.

The most your bag can weigh is $\frac{33}{64}$ pound, which would have cost $\left(\frac{33}{64}\text{ pound}\right) \times (\$9.89\text{ per pound}) = \$5.10$.
So, you might have been under charged by as much as $\$5.10 - \$4.95 = \$0.15$ or over charged as much as $\$4.95 - \$4.79 = \$0.16$.

Checkpoints for Appendix A.7

1. Do not apply radicals term-by-term when adding terms.

Leave as $\sqrt{x^2 + 4}$.

2. $$\begin{aligned} x(x-2)^{-1/2} + 6(x-2)^{1/2} &= (x-2)^{-1/2}\left[x(x-2)^0 + 6(x-2)^1\right] \\ &= (x-2)^{-1/2}[x + 6x - 12] \\ &= (x-2)^{-1/2}(7x - 12) \end{aligned}$$

3. The expression on the left side of the equation is three times the expression on the right side. To make both sides equal, insert a factor of 3.

$$\frac{6x-3}{\left(x^2 - x + 4\right)^2} = (3)\frac{1}{\left(x^2 - x + 4\right)^2}(2x - 1)$$

4. To write the expression on the left side of the equation in the form given on the right side, first multiply the numerator and denominator of the first term by $\frac{1}{9}$. Then multiply the numerator and denominator of the second term by $\frac{1}{25}$.

$$\frac{9x^2}{16} + 25y^2 = \frac{9x^2}{16}\left(\frac{1/9}{1/9}\right) + \frac{25y^2}{1}\left(\frac{1/25}{1/25}\right) = \frac{x^2}{16/25} + \frac{y^2}{1/25}$$

5. $$\frac{-6x}{\left(1 - 3x^2\right)^2} + \frac{1}{3\sqrt{x}} = -6x\left(1 - 3x^2\right)^{-2} + x^{-1/3}$$

6. (a) $$\frac{x^4 - 2x^3 + 5}{x^3} = \frac{x^4}{x^3} - \frac{2x^3}{x^3} + \frac{5}{x^3} = x - 2 + \frac{5}{x^3}$$

(b) $$\begin{aligned} \frac{x^2 - x + 5}{\sqrt{x}} &= \frac{x^2}{x^{1/2}} - \frac{x}{x^{1/2}} + \frac{5}{x^{1/2}} \\ &= x^{3/2} - x^{1/2} + 5x^{-1/2} \end{aligned}$$

Chapter 1 Practice Test Solutions

1. (a) Midpoint: $\left(\dfrac{-3+5}{2}, \dfrac{4+(-6)}{2}\right) = (1, -1)$

(b) Distance: $d = \sqrt{[5-(-3)]^2 + (-6-4)^2}$

$= \sqrt{(8)^2 + (-10)^2}$

$= \sqrt{164}$

$= 2\sqrt{41}$

2. $y = \sqrt{7-x}$

Domain: $x \le 7$

x	7	6	3	-2
y	0	1	2	3

3. $[x-(-3)]^2 + (y-5)^2 = 6^2$

$(x+3)^2 + (y-5)^2 = 36$

4. $m = \dfrac{-1-4}{3-2} = -5$

$y - 4 = -5(x-2)$

$y - 4 = -5x + 10$

$y = -5x + 14$

5. $y = \frac{4}{3}x - 3$

6. $2x + 3y = 0$

$y = -\frac{2}{3}x$

$m_1 = -\frac{2}{3}$

$\perp m_2 = \frac{3}{2}$ through $(4, 1)$

$y - 1 = \frac{3}{2}(x-4)$

$y - 1 = \frac{3}{2}x - 6$

$y = \frac{3}{2}x - 5$

7. $(5, 32)$ and $(9, 44)$

$m = \dfrac{44-32}{9-5} = \dfrac{12}{4} = 3$

$y - 32 = 3(x-5)$

$y - 32 = 3x - 15$

$y = 3x + 17$

When $x = 20$, $y = 3(20) + 17$

$y = \$77.$

8. $f(x-3) = (x-3)^2 - 2(x-3) + 1$

$= x^2 - 6x + 9 - 2x + 6 + 1$

$= x^2 - 8x + 16$

9. $f(3) = 12 - 11 = 1$

$\dfrac{f(x) - f(3)}{x-3} = \dfrac{(4x-11)-1}{x-3}$

$= \dfrac{4x-12}{x-3}$

$= \dfrac{4(x-3)}{x-3}$

$= 4, x \ne 3$

10. $f(x) = \sqrt{36 - x^2} = \sqrt{(6+x)(6-x)}$

Domain: $[-6, 6]$, because $(6+x)(6-x) \ge 0$ on this interval.

Range: $[0, 6]$, because $0 \le (6+x)(6-x) \le 36$ on this interval.

11. (a) $6x - 5y + 4 = 0$

$y = \dfrac{6x+4}{5}$ is a function of x.

(b) $x^2 + y^2 = 9$

$y = \pm\sqrt{9 - x^2}$ is not a function of x.

(c) $y^3 = x^2 + 6$

$y = \sqrt[3]{x^2 + 6}$ is a function of x.

12. Parabola

Vertex: $(0, -5)$

Intercepts: $(0, -5), (\pm\sqrt{5}, 0)$

y-axis symmetry

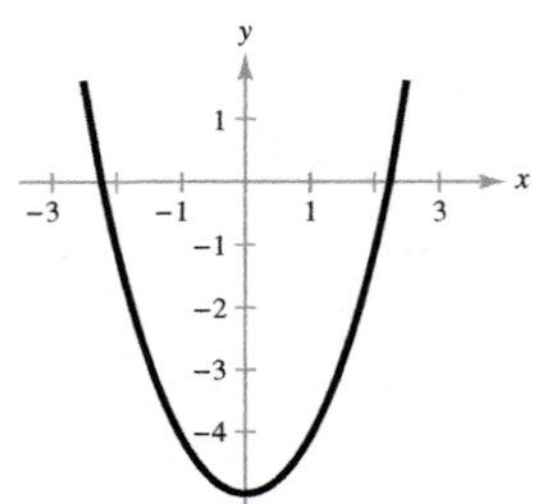

13. Intercepts: $(0, 3), (-3, 0)$

x	-4	-3	-2	-1	0	1	2
y	1	0	1	2	3	4	5

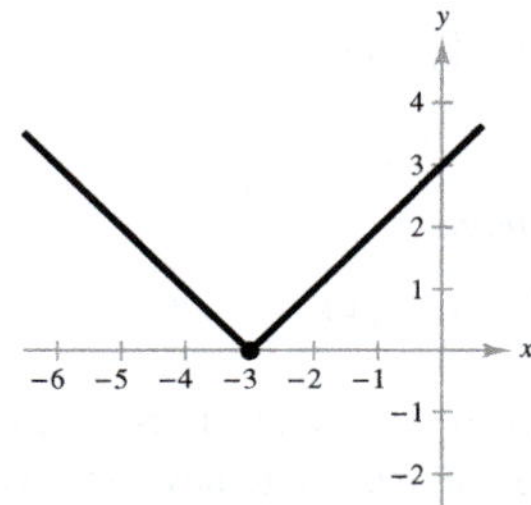

14.

x	-3	-2	-1	0	1	2	3
y	12	6	2	1	3	5	7

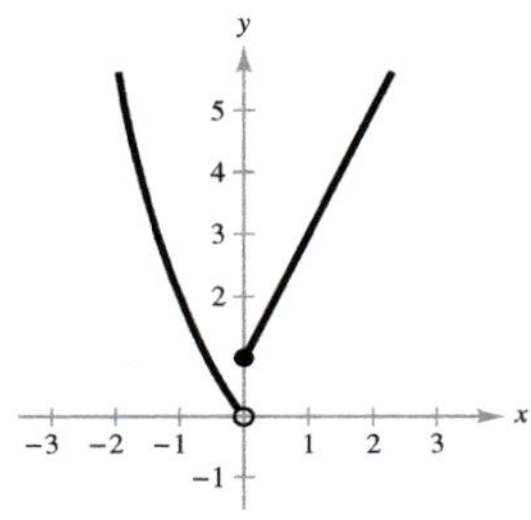

15. (a) $f(x + 2)$

Horizontal shift two units to the left

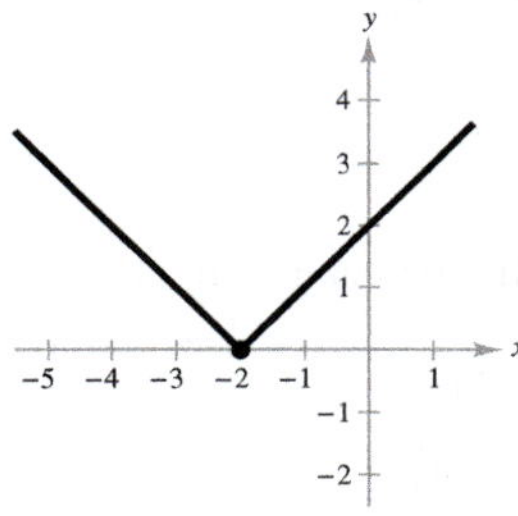

(b) $-f(x) + 2$

Reflection in the x-axis and a vertical shift two units upward

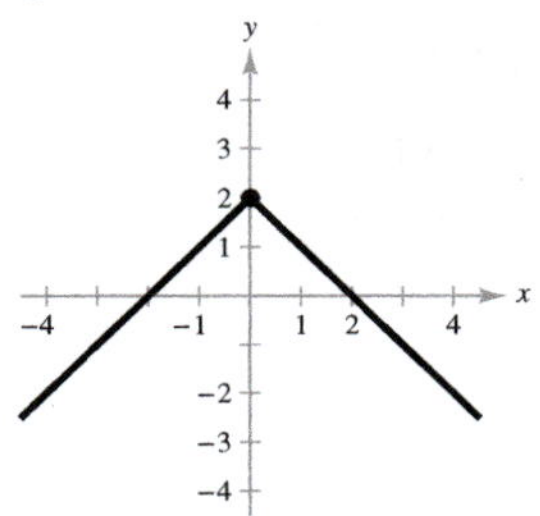

16. (a) $(g - f)(x) = g(x) - f(x)$

$$= (2x^2 - 5) - (3x + 7)$$
$$= 2x^2 - 3x - 12$$

(b) $(fg)(x) = f(x)g(x)$

$$= (3x + 7)(2x^2 - 5)$$
$$= 6x^3 + 14x^2 - 15x - 35$$

17. $f(g(x)) = f(2x + 3)$

$$= (2x + 3)^2 - 2(2x + 3) + 16$$
$$= 4x^2 + 12x + 9 - 4x - 6 + 16$$
$$= 4x^2 + 8x + 19$$

18.

$$f(x) = x^3 + 7$$
$$y = x^3 + 7$$
$$x = y^3 + 7$$
$$x - 7 = y^3$$
$$\sqrt[3]{x - 7} = y$$
$$f^{-1}(x) = \sqrt[3]{x - 7}$$

19. (a) $f(x) = |x - 6|$ does not have an inverse.

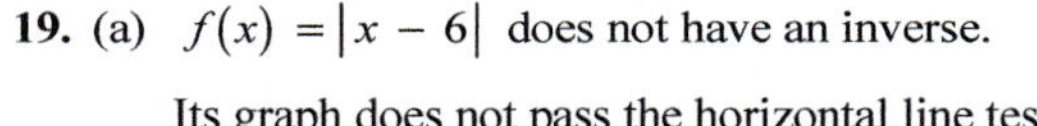

Its graph does not pass the horizontal line test.

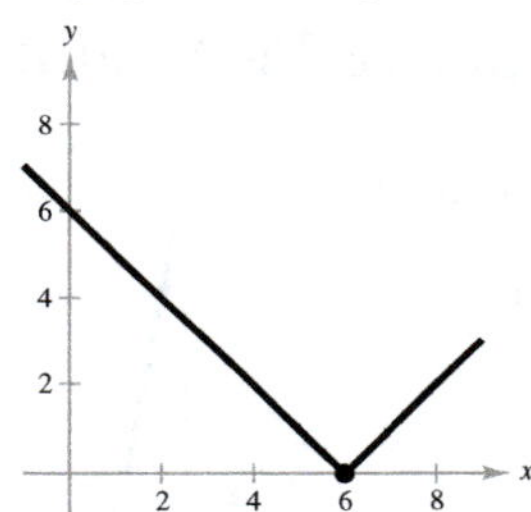

(b) $f(x) = ax + b, a \neq 0$ does have an inverse.

$$y = ax + b$$
$$x = ay + b$$
$$\frac{x - b}{a} = y$$
$$f^{-1}(x) = \frac{x - b}{a}$$

(c) $f(x) = x^3 - 19$ does have an inverse.

$$y = x^3 - 19$$
$$x = y^3 - 19$$
$$x + 19 = y^3$$
$$\sqrt[3]{x + 19} = y$$
$$f^{-1}(x) = \sqrt[3]{x + 19}$$

20. $f(x) = \sqrt{\dfrac{3-x}{x}}, 0 < x \le 3, y \ge 0$

$$y = \sqrt{\frac{3-x}{x}}$$
$$x = \sqrt{\frac{3-y}{y}}$$
$$x^2 = \frac{3-y}{y}$$
$$x^2y = 3 - y$$
$$x^2y + y = 3$$
$$y(x^2+1) = 3$$
$$y = \frac{3}{x^2+1}$$
$$f^{-1}(x) = \frac{3}{x^2+1}, x \ge 0$$

21. False. The slopes of 3 and $\frac{1}{3}$ are not negative reciprocals.

22. True. Let $y = (f \circ g)(x)$. Then $x = (f \circ g)^{-1}(y)$.

Also, $(f \circ g)(x) = y$
$$f(g(x)) = y$$
$$g(x) = f^{-1}(y)$$
$$x = g^{-1}(f^{-1}(y))$$
$$x = (g^{-1} \circ f^{-1})(y)$$

Because $x = x$, we have

$(f \circ g)^{-1}(y) = (g^{-1} \circ f^{-1})(y)$.

23. True. It must pass the vertical line test to be a function and it must pass the horizontal line test to have an inverse.

24. $z = \dfrac{cx^3}{\sqrt{y}}$
$$-1 = \frac{c(-1)^3}{\sqrt{25}}$$
$$-1 = \frac{-c}{5}$$
$$5 = c$$
$$z = \frac{5x^3}{\sqrt{y}}$$

25. $y \approx 0.669x + 2.669$

Chapter 2 Practice Test Solutions

1. x-intercepts: $(1, 0), (5, 0)$

y-intercept: $(0, 5)$

Vertex: $(3, -4)$

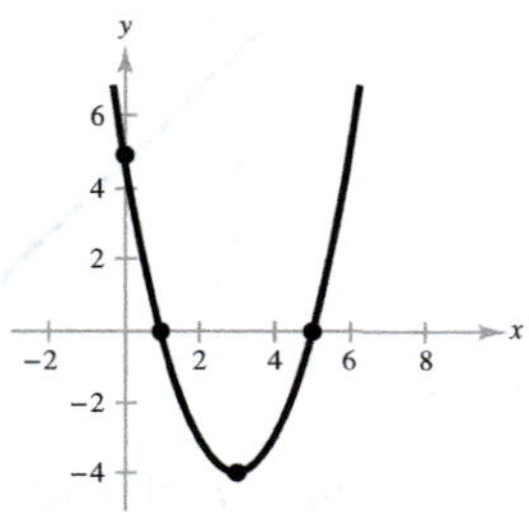

2. $a = 0.01, b = -90$
$$\frac{-b}{2a} = \frac{90}{2(0.01)} = 4500 \text{ units}$$

3. Vertex: $(1, 7)$ opening downward through $(2, 5)$

$y = a(x-1)^2 + 7$ Standard form
$$5 = a(2-1)^2 + 7$$
$$5 = a + 7$$
$$a = -2$$
$$y = -2(x-1)^2 + 7$$
$$= -2(x^2 - 2x + 1) + 7$$
$$= -2x^2 + 4x + 5$$

4. $y = \pm a(x-2)(3x-4)$ where a is any real number
$$y = \pm(3x^2 - 10x + 8)$$

5. Leading coefficient: −3

Degree: 5

Moves down to the right and up to the left

6. $0 = x^5 - 5x^3 + 4x$
$$= x(x^4 - 5x^2 + 4)$$
$$= x(x^2-1)(x^2-4)$$
$$= x(x+1)(x-1)(x+2)(x-2)$$
$$x = 0, x = \pm 1, x = \pm 2$$

7. $f(x) = x(x-3)(x+2)$
$$= x(x^2 - x - 6)$$
$$= x^3 - x^2 - 6x$$

8. Intercepts: $(0, 0), \left(\pm 2\sqrt{3}, 0\right)$

Moves up to the right

Moves down to the left

Origin symmetry

x	-2	-1	0	1	2
y	16	11	0	-11	-16

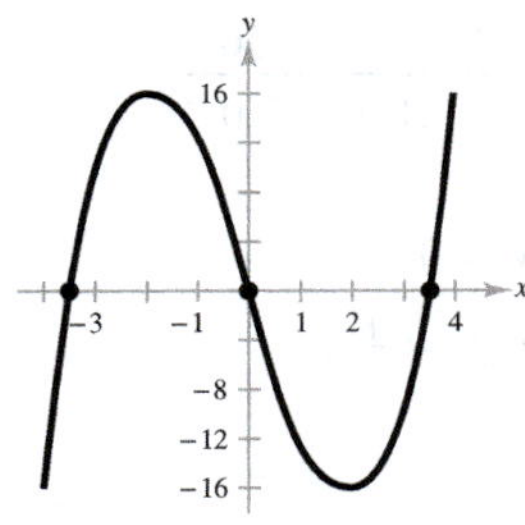

9.
$$\begin{array}{r}
3x^3 + 9x^2 + 20x + 62 + \dfrac{176}{x-3} \\
x - 3\overline{\big)\,3x^4 + 0x^3 - 7x^2 + 2x - 10} \\
\underline{3x^4 - 9x^3} \qquad\qquad\qquad\quad \\
9x^3 - 7x^2 \qquad\qquad\quad \\
\underline{9x^3 - 27x^2} \qquad\qquad\quad \\
20x^2 + 2x \qquad\quad \\
\underline{20x^2 - 60x} \qquad\quad \\
62x - 10 \\
\underline{62x - 186} \\
176
\end{array}$$

10.
$$\begin{array}{r}
x - 2 + \dfrac{5x - 13}{x^2 + 2x - 1} \\
x^2 + 2x - 1\overline{\big)\,x^3 + 0x^2 + 0x - 11} \\
\underline{x^3 + 2x^2 - x} \qquad\quad \\
-2x^2 + x - 11 \\
\underline{-2x^2 - 4x + 2} \\
5x - 13
\end{array}$$

11.
$$\begin{array}{r|rrrrrr}
-5 & 3 & 13 & 0 & 0 & 12 & -1 \\
 & & -15 & 10 & -50 & 250 & -1310 \\
\hline
 & 3 & -2 & 10 & -50 & 262 & -1311
\end{array}$$

$$\frac{3x^5 + 13x^4 + 12x - 1}{x + 5} = 3x^4 - 2x^3 + 10x^2 - 50x + 262 - \frac{1311}{x + 5}$$

12.
$$\begin{array}{r|rrrr}
-6 & 7 & 40 & -12 & 15 \\
 & & -42 & 12 & 0 \\
\hline
 & 7 & -2 & 0 & 15
\end{array}$$

$f(-6) = 15$

13. $0 = x^3 - 19x - 30$

Possible rational roots:
$\pm 1, \pm 2, \pm 3, \pm 5, \pm 6, \pm 10, \pm 15, \pm 30$

$$\begin{array}{r|rrrr}
-2 & 1 & 0 & -19 & -30 \\
 & & -2 & 4 & 30 \\
\hline
 & 1 & -2 & -15 & 0
\end{array}$$

$x = -2$ is a zero.

$0 = (x + 2)(x^2 - 2x - 15)$

$0 = (x + 2)(x + 3)(x - 5)$

Zeros: $x = -2, x = -3, x = 5$

14. $0 = x^4 + x^3 - 8x^2 - 9x - 9$

Possible rational roots: $\pm 1, \pm 3, \pm 9$

$$\begin{array}{r|rrrrr}
3 & 1 & 1 & -8 & -9 & -9 \\
 & & 3 & 12 & 12 & 9 \\
\hline
 & 1 & 4 & 4 & 3 & 0
\end{array}$$

$x = 3$ is a zero.

$0 = (x - 3)(x^3 + 4x^2 + 4x + 3)$

Possible rational roots of $x^3 + 4x^2 + 4x + 3$: $\pm 1, \pm 3$

$$\begin{array}{r|rrrr}
-3 & 1 & 4 & 4 & 3 \\
 & & -3 & -3 & -3 \\
\hline
 & 1 & 1 & 1 & 0
\end{array}$$

$x = -3$ is a zero.

$0 = (x - 3)(x + 3)(x^2 + x + 1)$

The zeros of $x^2 + x + 1$ are $x = \dfrac{-1 \pm \sqrt{3}i}{2}$

(by the Quadratic Formula).

Zeros:

$x = 3, x = -3, x = -\dfrac{1}{2} + \dfrac{\sqrt{3}}{2}i, x = -\dfrac{1}{2} - \dfrac{\sqrt{3}}{2}i$

15. $0 = 6x^3 - 5x^2 + 4x - 15$

Possible rational roots:

$\pm1, \pm3, \pm5, \pm15, \pm\frac{1}{2}, \pm\frac{3}{2}, \pm\frac{5}{2}, \pm\frac{15}{2}, \pm\frac{1}{3}, \pm\frac{5}{3}, \pm\frac{1}{6}, \pm\frac{5}{6}$

16. $0 = x^3 - \frac{20}{3}x^2 + 9x - \frac{10}{3}$

$0 = 3x^3 - 20x^2 + 27x - 10$

Possible rational roots:

$\pm1, \pm2, \pm5, \pm10, \pm\frac{1}{3}, \pm\frac{2}{3}, \pm\frac{5}{3}, \pm\frac{10}{3}$

$$\begin{array}{r|rrrr} 1 & 3 & -20 & 27 & -10 \\ & & 3 & -17 & 10 \\ \hline & 3 & -17 & 10 & 0 \end{array}$$

$0 = (x - 1)(3x^2 - 17x + 10)$

$0 = (x - 1)(3x - 2)(x - 5)$

Zeros: $x = 1, x = \frac{2}{3}, x = 5$

17. Possible rational roots: $\pm1, \pm2, \pm5, \pm10$

$$\begin{array}{r|rrrrr} 1 & 1 & 1 & 3 & 5 & -10 \\ & & 1 & 2 & 5 & 10 \\ \hline & 1 & 2 & 5 & 10 & 0 \end{array}$$

$x = 1$ is a zero.

$$\begin{array}{r|rrrr} -2 & 1 & 2 & 5 & 10 \\ & & -2 & 0 & -10 \\ \hline & 1 & 0 & 5 & 0 \end{array}$$

$x = -2$ is a zero.

$f(x) = (x - 1)(x + 2)(x^2 + 5)$

$= (x - 1)(x + 2)(x + \sqrt{5}i)(x - \sqrt{5}i)$

18. $f(x) = (x - 2)[x - (3 + i)][x - (3 - i)]$

$= (x - 2)[(x - 3) - i][(x - 3) + i]$

$= (x - 2)[(x - 3)^2 - i^2]$

$= (x - 2)[x^2 - 6x + 10]$

$= x^3 - 8x^2 + 22x - 20$

19.

$$\begin{array}{r|rrrr} 3i & 1 & 4 & 9 & 36 \\ & & 3i & 12i - 9 & -36 \\ \hline & 1 & 4 + 3i & 12i & 0 \end{array}$$

20. Vertical asymptote: $x = 0$

Horizontal asymptote: $y = \frac{1}{2}$

x-intercept: $(1, 0)$

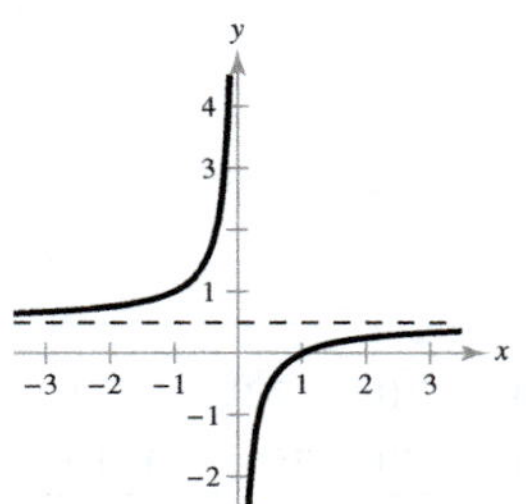

21. $y = 8$ is a horizontal asymptote because the degree of the numerator equals the degree of the denominator. There are no vertical asymptotes.

22. $x = 1$ is a vertical asymptote.

$$\frac{4x^2 - 2x + 7}{x - 1} = 4x + 2 + \frac{9}{x - 1}$$

Thus, $y = 4x + 2$ is a slant asymptote.

23. (a) $(4 - 3i) - (-2 + i) = 4 - 3i + 2 - i = 6 - 4i$

(b) $(4 - 3i)(-2 + i) = -8 + 4i + 6i - 3i^2 = -8 + 10i + 3 = -5 + 10i$

(c) $\dfrac{4 - 3i}{-2 + i} = \dfrac{4 - 3i}{-2 + i} \cdot \dfrac{-2 - i}{-2 - i} = \dfrac{-8 - 4i + 6i + 3i^2}{4 + 1}$

$= \dfrac{-11 + 2i}{5} = -\dfrac{11}{5} + \dfrac{2}{5}i$

24. $x^2 - 49 \le 0$

$(x + 7)(x - 7) \le 0$

Critical numbers: $x = -7$ and $x = 7$

Test intervals: $(-\infty, -7), (-7, 7), (7, \infty)$

Test: Is $x^2 - 49 \le 0$?

Solution set: $[-7, 7]$

25. $\dfrac{x + 3}{x - 7} \ge 0$

Critical numbers: $x = -3$ and $x = 7$

Test intervals: $(-\infty, -3), (-3, 7), (7, \infty)$

Test: Is $\dfrac{x + 3}{x - 7} \ge 0$?

Solution set: $(-\infty, -3] \cup [7, \infty)$

Chapter 3 Practice Test Solutions

1. $x^{3/5} = 8$

$$x = 8^{5/3} = \left(\sqrt[3]{8}\right)^5 = 2^5 = 32$$

2. $3^{x-1} = \frac{1}{81}$

$$3^{x-1} = 3^{-4}$$
$$x - 1 = -4$$
$$x = -3$$

3. $f(x) = 2^{-x} = \left(\frac{1}{2}\right)^x$

x	-2	-1	0	1	2
$f(x)$	4	2	1	$\frac{1}{2}$	$\frac{1}{4}$

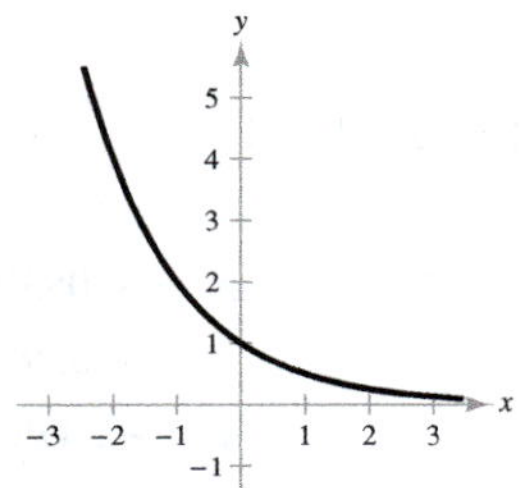

4. $g(x) = e^x + 1$

x	-2	-1	0	1	2
$g(x)$	1.14	1.37	2	3.72	8.39

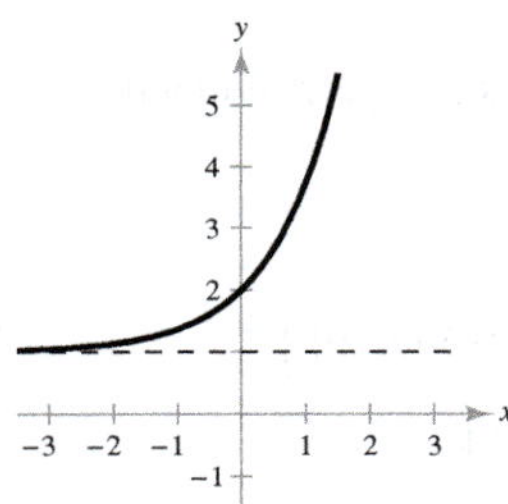

5. (a) $A = P\left(1 + \frac{r}{n}\right)^{nt}$

$$A = 5000\left(1 + \frac{0.09}{12}\right)^{12(3)} \approx \$6543.23$$

(b) $A = P\left(1 + \frac{r}{n}\right)^{nt}$

$$A = 5000\left(1 + \frac{0.09}{4}\right)^{4(3)} \approx \$6530.25$$

(c) $A = Pe^{rt}$

$$A = 5000e^{(0.09)(3)} \approx \$6549.82$$

6. $7^{-2} = \frac{1}{49}$

$$\log_7 \frac{1}{49} = -2$$

7. $x - 4 = \log_2 \frac{1}{64}$

$$2^{x-4} = \frac{1}{64}$$
$$2^{x-4} = 2^{-6}$$
$$x - 4 = -6$$
$$x = -2$$

8. $\log_b \sqrt[4]{\frac{8}{25}} = \frac{1}{4}\log_b \frac{8}{25}$

$$= \frac{1}{4}[\log_b 8 - \log_b 25]$$
$$= \frac{1}{4}\left[\log_b 2^3 - \log_b 5^2\right]$$
$$= \frac{1}{4}[3 \log_b 2 - 2 \log_b 5]$$
$$= \frac{1}{4}\left[3(0.3562) - 2(0.8271)\right]$$
$$= -0.1464$$

9. $5 \ln x - \frac{1}{2}\ln y + 6 \ln z = \ln x^5 - \ln \sqrt{y} + \ln z^6$

$$= \ln\left(\frac{x^5 z^6}{\sqrt{y}}\right), z > 0$$

10. $\log_9 28 = \frac{\log 28}{\log 9} \approx 1.5166$

11. $\log N = 0.6646$

$$N = 10^{0.6646} \approx 4.62$$

12.

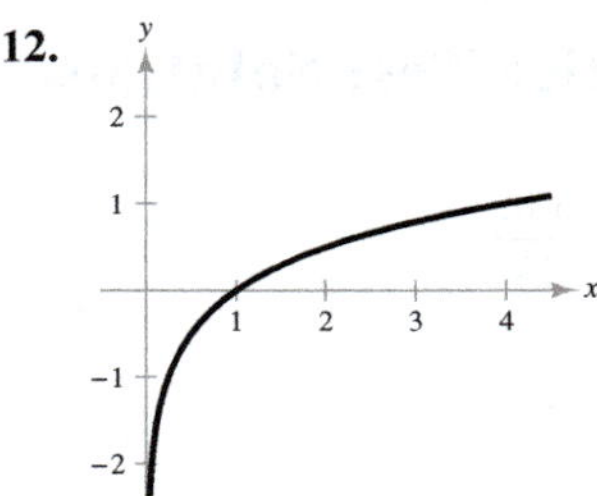

13. Domain:

$$x^2 - 9 > 0$$
$$(x + 3)(x - 3) > 0$$
$$x < -3 \text{ or } x > 3$$

14.

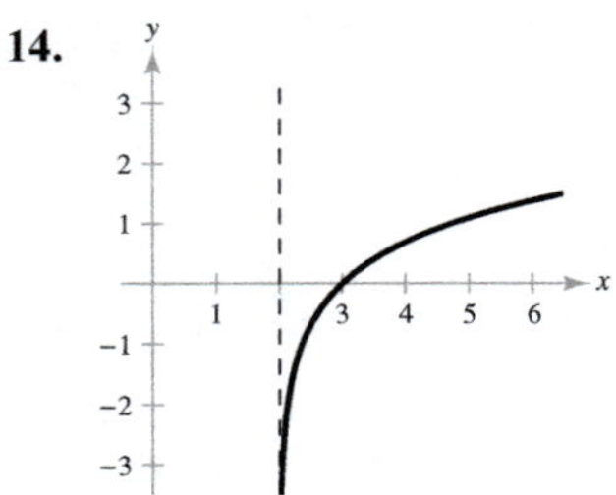

15. False. $\dfrac{\ln x}{\ln y} \neq \ln(x - y)$ because $\dfrac{\ln x}{\ln y} = \log_y x$.

16. $5^3 = 41$

$$x = \log_5 41 = \frac{\ln 41}{\ln 5} \approx 2.3074$$

17.
$$\begin{aligned} x - x^2 &= \log_5 \tfrac{1}{25} \\ 5^{x-x^2} &= \tfrac{1}{25} \\ 5^{x-x^2} &= 5^{-2} \\ x - x^2 &= -2 \\ 0 &= x^2 - x - 2 \\ 0 &= (x + 1)(x - 2) \\ x &= -1 \text{ or } x = 2 \end{aligned}$$

18.
$$\begin{aligned} \log_2 x + \log_2(x - 3) &= 2 \\ \log_2[x(x - 3)] &= 2 \\ x(x - 3) &= 2^2 \\ x^2 - 3x &= 4 \\ x^2 - 3x - 4 &= 0 \\ (x + 1)(x - 4) &= 0 \\ x &= 4 \\ x &= -1 \text{ (extraneous)} \end{aligned}$$

$x = 4$ is the only solution.

19.
$$\begin{aligned} \frac{e^x + e^{-x}}{3} &= 4 \\ e^x(e^x + e^{-x}) &= 12e^x \\ e^{2x} + 1 &= 12e^x \\ e^{2x} - 12e^x + 1 &= 0 \\ e^x &= \frac{12 \pm \sqrt{144 - 4}}{2} \end{aligned}$$

$e^x \approx 11.9161$ or $e^x \approx 0.0839$

$x = \ln 11.9161$ $\qquad x = \ln 0.0839$

$x \approx 2.478$ $\qquad x \approx -2.478$

20.
$$\begin{aligned} A &= Pe^{rt} \\ 12{,}000 &= 6000e^{0.13t} \\ 2 &= e^{0.13t} \\ 0.13t &= \ln 2 \\ t &= \frac{\ln 2}{0.13} \\ t &\approx 5.3319 \text{ years or 5 years 4 months} \end{aligned}$$

Chapter 4 Practice Test Solutions

1. $350° = 350\left(\dfrac{\pi}{180}\right) = \dfrac{35\pi}{18}$

2. $\dfrac{5\pi}{9} = \dfrac{5\pi}{9} \cdot \dfrac{180}{\pi} = 100°$

3.
$$\begin{aligned} 135° \, 14' \, 12'' &= \left(135 + \tfrac{14}{60} + \tfrac{12}{3600}\right)° \\ &\approx 135.2367° \end{aligned}$$

4.
$$\begin{aligned} -22.569° &= -\left(22° + 0.569(60)'\right) \\ &= -22° \, 34.14' \\ &= -\left(22° \, 34' + 0.14(60)''\right) \\ &\approx -22° \, 34' \, 8'' \end{aligned}$$

5. $\cos\theta = \dfrac{2}{3}$

$x = 2, r = 3, y = \pm\sqrt{9 - 4} = \pm\sqrt{5}$

$\tan\theta = \dfrac{y}{x} = \pm\dfrac{\sqrt{5}}{2}$

6. $\sin\theta = 0.9063$

$\theta = \arcsin(0.9063)$

$\theta = 65^\circ = \dfrac{13\pi}{36}$ or $\theta = 180^\circ - 65^\circ = 115^\circ = \dfrac{23\pi}{36}$

7. $\tan 20^\circ = \dfrac{35}{x}$

$x = \dfrac{35}{\tan 20^\circ}$

≈ 96.1617

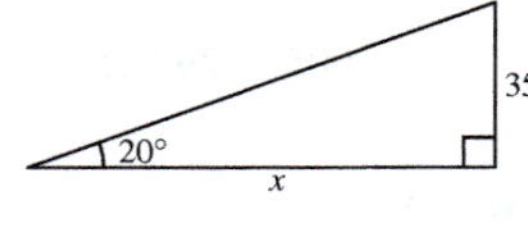

8. $\theta = \dfrac{6\pi}{5}$, θ is in Quadrant III.

Reference angle: $\dfrac{6\pi}{5} - \pi = \dfrac{\pi}{5}$ or 36°

9. $\csc 3.92 = \dfrac{1}{\sin 3.92} \approx -1.4242$

10. $\tan\theta = 6 = \dfrac{6}{1}$, θ lies in Quandrant III.

$y = -6, x = -1, r = \sqrt{36 + 1} = \sqrt{37}$, so

$\sec\theta = \dfrac{\sqrt{37}}{-1} \approx -6.0828.$

11. Period: 4π

Amplitude: 3

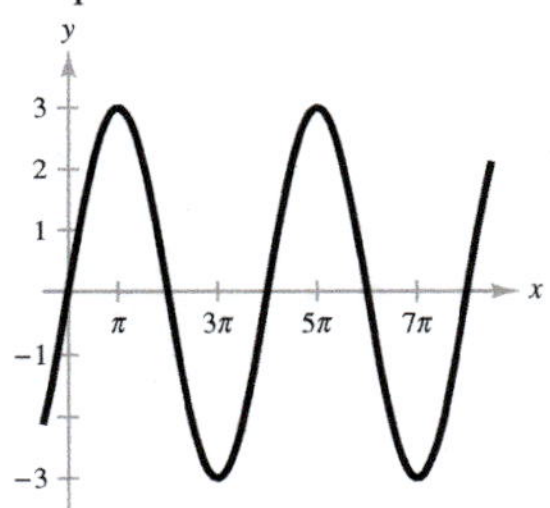

12. Period: 2π

Amplitude: 2

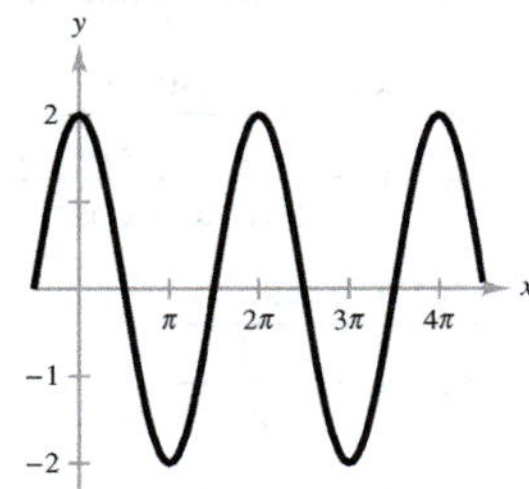

13. Period: $\dfrac{\pi}{2}$

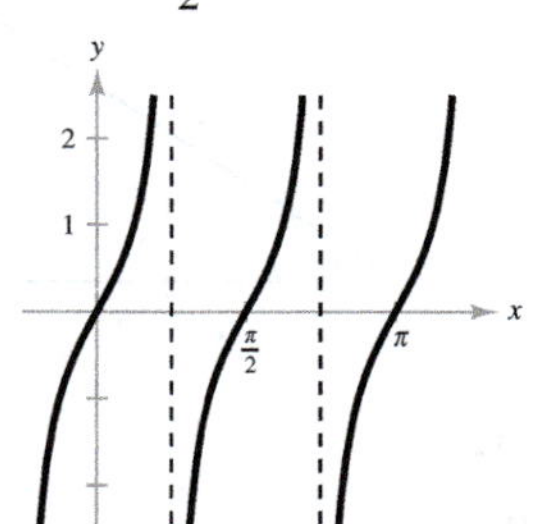

14. Period: 2π

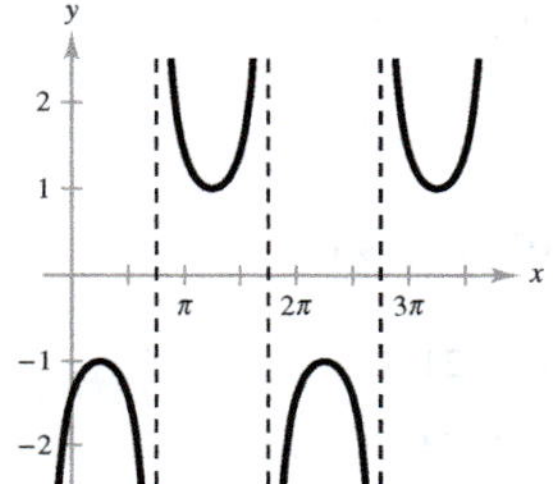

15.

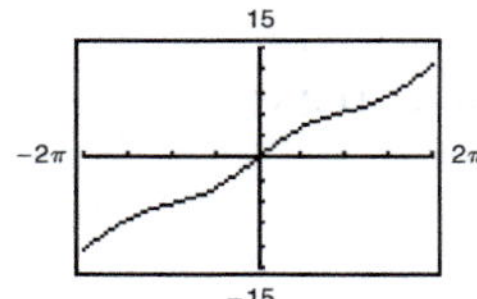

16.

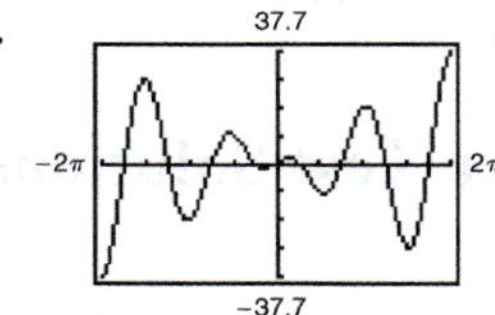

17. $\theta = \arcsin 1$

$\sin\theta = 1$

$\theta = \dfrac{\pi}{2} = 90^\circ$

18. $\theta = \arctan(-3)$

$\tan\theta = -3$

$\theta \approx -1.249 \approx -71.565^\circ$

19. $\sin\left(\arccos \dfrac{4}{\sqrt{35}}\right)$

$\sin\theta = \dfrac{\sqrt{19}}{\sqrt{35}} \approx 0.7368$

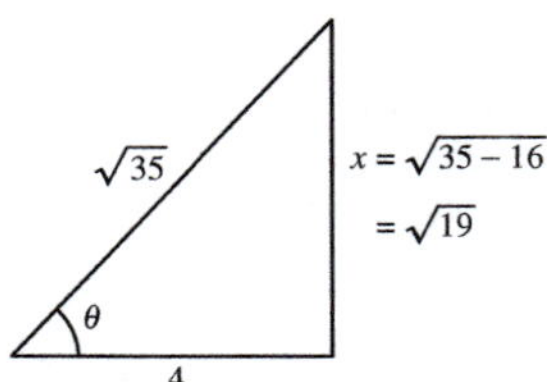

20. $\cos\left(\arcsin \dfrac{x}{4}\right)$

$\cos\theta = \dfrac{\sqrt{16 - x^2}}{4}$

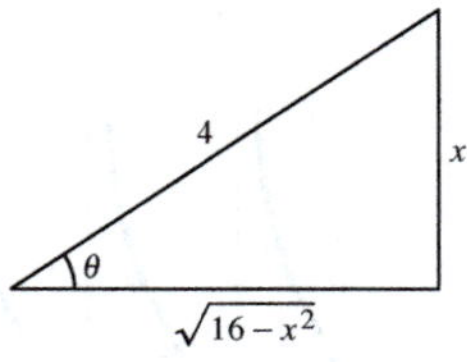

21. Given $A = 40°, c = 12$

$B = 90° - 40° = 50°$

$\sin 40° = \dfrac{a}{12}$

$a = 12 \sin 40° \approx 7.713$

$\cos 40° = \dfrac{b}{12}$

$b = 12 \cos 40° \approx 9.193$

22. Given $B = 6.84°, a = 21.3$

$A = 90° - 6.84° = 83.16°$

$\sin 83.16° = \dfrac{21.3}{c}$

$c = \dfrac{21.3}{\sin 83.16°} \approx 21.453$

$\tan 83.16° = \dfrac{21.3}{b}$

$b = \dfrac{21.3}{\tan 83.16°} \approx 2.555$

23. Given $a = 5, b = 9$

$c = \sqrt{25 + 81} = \sqrt{106} \approx 10.296$

$\tan A = \frac{5}{9}$

$A = \arctan \frac{5}{9} \approx 29.055°$

$B \approx 90° - 29.055° = 60.945°$

24. $\sin 67° = \dfrac{x}{20}$

$x = 20 \sin 67° \approx 18.41$ feet

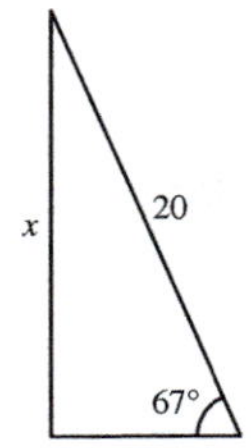

25. $\tan 5° = \dfrac{250}{x}$

$x = \dfrac{250}{\tan 5°}$

≈ 2857.513 feet

≈ 0.541 mi

Chapter 5 Practice Test Solutions

1. $\tan x = \dfrac{4}{11}$, $\sec x < 0 \Rightarrow x$ is in Quadrant III.

$y = -4, x = -11, r = \sqrt{16 + 121} = \sqrt{137}$

$\sin x = -\dfrac{4}{\sqrt{137}} = -\dfrac{4\sqrt{137}}{137}$ $\qquad \csc x = -\dfrac{\sqrt{137}}{4}$

$\cos x = -\dfrac{11}{\sqrt{137}} = -\dfrac{11\sqrt{137}}{137}$ $\qquad \sec x = -\dfrac{\sqrt{137}}{11}$

$\tan x = \dfrac{4}{11}$ $\qquad \cot x = \dfrac{11}{4}$

2. $\dfrac{\sec^2 x + \csc^2 x}{\csc^2 x(1 + \tan^2 x)} = \dfrac{\sec^2 x + \csc^2 x}{\csc^2 x + (\csc^2 x)\tan^2 x}$

$= \dfrac{\sec^2 x + \csc^2 x}{\csc^2 x + \dfrac{1}{\sin^2 x} \cdot \dfrac{\sin^2 x}{\cos^2 x}}$

$= \dfrac{\sec^2 x + \csc^2 x}{\csc^2 x + \dfrac{1}{\cos^2 x}}$

$= \dfrac{\sec^2 x + \csc^2 x}{\csc^2 x + \sec^2 x} = 1$

3. $\ln|\tan\theta| - \ln|\cot\theta| = \ln\dfrac{|\tan\theta|}{|\cot\theta|} = \ln\left|\dfrac{\sin\theta/\cos\theta}{\cos\theta/\sin\theta}\right| = \ln\left|\dfrac{\sin^2\theta}{\cos^2\theta}\right| = \ln|\tan^2\theta| = 2\ln|\tan\theta|$

4. $\cos\left(\frac{\pi}{2} - x\right) = \frac{1}{\csc x}$ is true since $\cos\left(\frac{\pi}{2} - x\right) = \sin x = \frac{1}{\csc x}$.

5. $\sin^4 x + (\sin^2 x)\cos^2 x = \sin^2 x(\sin^2 x + \cos^2 x)$
$= \sin^2 x(1) = \sin^2 x$

6. $(\csc x + 1)(\csc x - 1) = \csc^2 x - 1 = \cot^2 x$

7. $\frac{\cos^2 x}{1 - \sin x} \cdot \frac{1 + \sin x}{1 + \sin x} = \frac{\cos^2 x(1 + \sin x)}{1 - \sin^2 x} = \frac{\cos^2 x(1 + \sin x)}{\cos^2 x} = 1 + \sin x$

8. $\frac{1 + \cos\theta}{\sin\theta} + \frac{\sin\theta}{1 + \cos\theta} = \frac{(1 + \cos\theta)^2 + \sin^2\theta}{\sin\theta(1 + \cos\theta)}$
$= \frac{1 + 2\cos\theta + \cos^2\theta + \sin^2\theta}{\sin\theta(1 + \cos\theta)} = \frac{2 + 2\cos\theta}{\sin\theta(1 + \cos\theta)} = \frac{2}{\sin\theta} = 2\csc\theta$

9. $\tan^4 x + 2\tan^2 x + 1 = (\tan^2 x + 1)^2 = (\sec^2 x)^2 = \sec^4 x$

10. (a) $\sin 105° = \sin(60° + 45°) = \sin 60° \cos 45° + \cos 60° \sin 45°$
$= \frac{\sqrt{3}}{2} \cdot \frac{\sqrt{2}}{2} + \frac{1}{2} \cdot \frac{\sqrt{2}}{2} = \frac{\sqrt{2}}{4}(\sqrt{3} + 1)$

(b) $\tan 15° = \tan(60° - 45°) = \frac{\tan 60° - \tan 45°}{1 + \tan 60° \tan 45°}$
$= \frac{\sqrt{3} - 1}{1 + \sqrt{3}} \cdot \frac{1 - \sqrt{3}}{1 - \sqrt{3}} = \frac{2\sqrt{3} - 1 - 3}{1 - 3} = \frac{2\sqrt{3} - 4}{-2} = 2 - \sqrt{3}$

11. $(\sin 42°)\cos 38° - (\cos 42°)\sin 38° = \sin(42° - 38°) = \sin 4°$

12. $\tan\left(\theta + \frac{\pi}{4}\right) = \frac{\tan\theta + \tan\left(\frac{\pi}{4}\right)}{1 - (\tan\theta)\tan\left(\frac{\pi}{4}\right)} = \frac{\tan\theta + 1}{1 - \tan\theta(1)} = \frac{1 + \tan\theta}{1 - \tan\theta}$

13. $\sin(\arcsin x - \arccos x) = \sin(\arcsin x)\cos(\arccos x) - \cos(\arcsin x)\sin(\arccos x)$
$= (x)(x) - \left(\sqrt{1 - x^2}\right)\left(\sqrt{1 - x^2}\right) = x^2 - (1 - x^2) = 2x^2 - 1$

14. (a) $\cos(120°) = \cos[2(60°)] = 2\cos^2 60° - 1 = 2\left(\frac{1}{2}\right)^2 - 1 = -\frac{1}{2}$

(b) $\tan(300°) = \tan[2(150°)] = \frac{2\tan 150°}{1 - \tan^2 150°} = \frac{-\frac{2\sqrt{3}}{3}}{1 - \left(\frac{1}{3}\right)} = -\sqrt{3}$

15. (a) $\sin 22.5° = \sin\frac{45°}{2} = \sqrt{\frac{1 - \cos 45°}{2}} = \sqrt{\frac{1 - \frac{\sqrt{2}}{2}}{2}} = \frac{\sqrt{2 - \sqrt{2}}}{2}$

(b) $\tan\frac{\pi}{12} = \tan\frac{\frac{\pi}{6}}{2} = \frac{\sin\frac{\pi}{6}}{1 + \cos\left(\frac{\pi}{6}\right)} = \frac{\frac{1}{2}}{1 + \frac{\sqrt{3}}{2}} = \frac{1}{2 + \sqrt{3}} = 2 - \sqrt{3}$

16. $\sin\theta = \frac{4}{5}$, θ lies in Quadrant II $\Rightarrow \cos\theta = -\frac{3}{5}$.

$$\cos\frac{\theta}{2} = \sqrt{\frac{1+\cos\theta}{2}} = \sqrt{\frac{1-\frac{3}{5}}{2}} = \sqrt{\frac{2}{10}} = \frac{1}{\sqrt{5}} = \frac{\sqrt{5}}{5}$$

17. $$\left(\sin^2 x\right)\cos^2 x = \frac{1-\cos 2x}{2}\cdot\frac{1+\cos 2x}{2} = \frac{1}{4}\left[1-\cos^2 2x\right] = \frac{1}{4}\left[1-\frac{1+\cos 4x}{2}\right]$$
$$= \frac{1}{8}\left[2-(1+\cos 4x)\right] = \frac{1}{8}\left[1-\cos 4x\right]$$

18. $$6(\sin 5\theta)\cos 2\theta = 6\left\{\tfrac{1}{2}\left[\sin(5\theta+2\theta)+\sin(5\theta-2\theta)\right]\right\} = 3\left[\sin 7\theta + \sin 3\theta\right]$$

19. $$\sin(x+\pi)+\sin(x-\pi) = 2\left(\sin\frac{\left[(x+\pi)+(x-\pi)\right]}{2}\right)\cos\frac{\left[(x+\pi)-(x-\pi)\right]}{2}$$
$$= 2\sin x\cos\pi = -2\sin x$$

20. $$\frac{\sin 9x+\sin 5x}{\cos 9x-\cos 5x} = \frac{2\sin 7x\cos 2x}{-2\sin 7x\sin 2x} = -\frac{\cos 2x}{\sin 2x} = -\cot 2x$$

21. $$\tfrac{1}{2}\left[\sin(u+v)-\sin(u-v)\right] = \tfrac{1}{2}\left\{(\sin u)\cos v+(\cos u)\sin v-\left[(\sin u)\cos v-(\cos u)\sin v\right]\right\}$$
$$= \tfrac{1}{2}\left[2(\cos u)\sin v\right] = (\cos u)\sin v$$

22. $4\sin^2 x = 1$

$$\sin^2 x = \frac{1}{4}$$
$$\sin x = \pm\frac{1}{2}$$

$\sin x = \frac{1}{2}$ or $\sin x = -\frac{1}{2}$

$x = \frac{\pi}{6}$ or $\frac{5\pi}{6}$ $\quad x = \frac{7\pi}{6}$ or $\frac{11\pi}{6}$

23. $\tan^2\theta + \left(\sqrt{3}-1\right)\tan\theta - \sqrt{3} = 0$

$$(\tan\theta-1)\left(\tan\theta+\sqrt{3}\right) = 0$$

$\tan\theta = 1$ or $\tan\theta = -\sqrt{3}$

$\theta = \frac{\pi}{4}$ or $\frac{5\pi}{4}$ $\quad \theta = \frac{2\pi}{3}$ or $\frac{5\pi}{3}$

24. $\sin 2x = \cos x$

$$2(\sin x)\cos x - \cos x = 0$$
$$\cos x(2\sin x - 1) = 0$$

$\cos x = 0$ or $\sin x = \frac{1}{2}$

$x = \frac{\pi}{2}$ or $\frac{3\pi}{2}$ $\quad x = \frac{\pi}{6}$ or $\frac{5\pi}{6}$

25. $\tan^2 x - 6\tan x + 4 = 0$

$$\tan x = \frac{-(-6)\pm\sqrt{(-6)^2-4(1)(4)}}{2(1)}$$
$$\tan x = \frac{6\pm\sqrt{20}}{2} = 3\pm\sqrt{5}$$

$\tan x = 3+\sqrt{5}$ or $\tan x = 3-\sqrt{5}$

$x \approx 1.3821$ or 4.5237 $\quad x \approx 0.6524$ or 3.7940

Chapter 6 Practice Test Solutions

1. $C = 180^\circ - (40^\circ + 12^\circ) = 128^\circ$

$$a = \sin 40^\circ\left(\frac{100}{\sin 12^\circ}\right) \approx 309.164$$
$$c = \sin 128^\circ\left(\frac{100}{\sin 12^\circ}\right) \approx 379.012$$

2. $\sin A = 5\left(\frac{\sin 150^\circ}{20}\right) = 0.125$

$$A \approx 7.181^\circ$$
$$B \approx 180^\circ - (150^\circ + 7.181^\circ) = 22.819^\circ$$
$$b = \sin 22.819^\circ\left(\frac{20}{\sin 150^\circ}\right) \approx 15.513$$

3. Area $= \frac{1}{2}ab \sin C = \frac{1}{2}(3)(6) \sin 130° \approx 6.894$ square units

4. $h = b \sin A = 35 \sin 22.5° \approx 13.394$

$a = 10$

Since $a < h$ and A is acute, the triangle has no solution.

5. $\cos A = \dfrac{(53)^2 + (38)^2 - (49)^2}{2(53)(38)} \approx 0.4598$

$A \approx 62.627°$

$\cos B = \dfrac{(49)^2 + (38)^2 - (53)^2}{2(49)(38)} \approx 0.2782$

$B \approx 73.847°$

$C \approx 180° - (62.627° + 73.847°)$

$= 43.526°$

6. $c^2 = (100)^2 + (300)^2 - 2(100)(300) \cos 29°$

$\approx 47{,}522.8176$

$c \approx 218$

$\cos A = \dfrac{(300)^2 + (218)^2 - (100)^2}{2(300)(218)} \approx 0.97495$

$A \approx 12.85°$

$B \approx 180° - (12.85° + 29°) = 138.15°$

7. $s = \dfrac{a + b + c}{2} = \dfrac{4.1 + 6.8 + 5.5}{2} = 8.2$

Area $= \sqrt{s(s - a)(s - b)(s - c)}$

$= \sqrt{8.2(8.2 - 4.1)(8.2 - 6.8)(8.2 - 5.5)}$

≈ 11.273 square units

8. $x^2 = (40)^2 + (70)^2 - 2(40)(70)\cos 168°$

$\approx 11{,}977.6266$

$x \approx 190.442$ miles

9. $\mathbf{w} = 4(3\mathbf{i} + \mathbf{j}) - 7(-\mathbf{i} + 2\mathbf{j})$

$= 19\mathbf{i} - 10\mathbf{j}$

10. $\dfrac{\mathbf{v}}{\|\mathbf{v}\|} = \dfrac{5\mathbf{i} - 3\mathbf{j}}{\sqrt{25 + 9}} = \dfrac{5}{\sqrt{34}}\mathbf{i} - \dfrac{3}{\sqrt{34}}\mathbf{j}$

$= \dfrac{5\sqrt{34}}{34}\mathbf{i} - \dfrac{3\sqrt{34}}{34}\mathbf{j}$

11. $\mathbf{u} = 6\mathbf{i} + 5\mathbf{j}, \mathbf{v} = 2\mathbf{i} - 3\mathbf{j}$

$\mathbf{u} \cdot \mathbf{v} = 6(2) + 5(-3) = -3$

$\|\mathbf{u}\| = \sqrt{61}, \qquad \|\mathbf{v}\| = \sqrt{13}$

$\cos \theta = \dfrac{-3}{\sqrt{61}\sqrt{13}}$

$\theta \approx 96.116°$

12. $4(\mathbf{i} \cos 30° + \mathbf{j} \sin 30°) = 4\left(\dfrac{\sqrt{3}}{2}\mathbf{i} + \dfrac{1}{2}\mathbf{j}\right)$

$= \langle 2\sqrt{3}, 2\rangle$

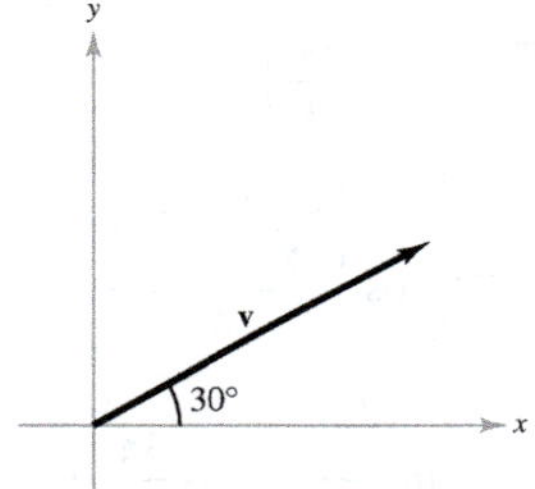

13. $\text{proj}_{\mathbf{v}}\mathbf{u} = \left(\dfrac{\mathbf{u} \cdot \mathbf{v}}{\|\mathbf{v}\|^2}\right)\mathbf{v} = \dfrac{-10}{20}\langle -2, 4\rangle = \langle 1, -2\rangle$

14. $r = \sqrt{25 + 25} = \sqrt{50} = 5\sqrt{2}$

$\tan \theta = \dfrac{-5}{5} = -1$

Because z is in Quadrant IV, $\theta = 315°$.

$z = 5\sqrt{2}(\cos 315° + i \sin 315°)$

15. $\cos 225° = -\dfrac{\sqrt{2}}{2}, \sin 225° = -\dfrac{\sqrt{2}}{2}$

$z = 6\left(-\dfrac{\sqrt{2}}{2} - i\dfrac{\sqrt{2}}{2}\right) = -3\sqrt{2} - 3\sqrt{2}i$

16. $[7(\cos 23° + i \sin 23°)][4(\cos 7° + i \sin 7°)] = 7(4)[\cos(23° + 7°) + i \sin(23° + 7°)]$

$= 28(\cos 30° + i \sin 30°)$

17. $\dfrac{9\left(\cos \dfrac{5\pi}{4} + i \sin \dfrac{5\pi}{4}\right)}{3(\cos \pi + i \sin \pi)} = \dfrac{9}{3}\left[\cos\left(\dfrac{5\pi}{4} - \pi\right) + i \sin\left(\dfrac{5\pi}{4} - \pi\right)\right] = 3\left(\cos \dfrac{\pi}{4} + i \sin \dfrac{\pi}{4}\right)$

18. $(2 + 2i)^8 = \left[2\sqrt{2}(\cos 45° + i \sin 45°)\right]^8 = \left(2\sqrt{2}\right)^8\left[\cos(8)(45°) + i \sin(8)(45°)\right]$

$= 4096[\cos 360° + i \sin 360°] = 4096$

19. $z = 8\left(\cos \frac{\pi}{3} + i \sin \frac{\pi}{3}\right), n = 3$

The cube roots of z are: $\sqrt[3]{8}\left[\cos \frac{\left(\frac{\pi}{3}\right) + 2\pi k}{3} + i \sin \frac{\left(\frac{\pi}{3}\right) + 2\pi k}{3}\right], k = 0, 1, 2$

For $k = 0$: $\sqrt[3]{8}\left[\cos \frac{\frac{\pi}{3}}{3} + i \sin \frac{\frac{\pi}{3}}{3}\right] = 2\left(\cos \frac{\pi}{9} + i \sin \frac{\pi}{9}\right)$

For $k = 1$: $\sqrt[3]{8}\left[\cos \frac{\left(\frac{\pi}{3}\right) + 2\pi}{3} + i \sin \frac{\left(\frac{\pi}{3}\right) + 2\pi}{3}\right] = 2\left(\cos \frac{7\pi}{9} + i \sin \frac{7\pi}{9}\right)$

For $k = 2$: $\sqrt[3]{8}\left[\cos \frac{\left(\frac{\pi}{3}\right) + 4\pi}{3} + i \sin \frac{\left(\frac{\pi}{3}\right) + 4\pi}{3}\right] = 2\left(\cos \frac{13\pi}{9} + i \sin \frac{13\pi}{9}\right)$

20. $x^4 = -i = 1\left(\cos \frac{3\pi}{2} + i \sin \frac{3\pi}{2}\right)$

The fourth roots are: $\sqrt[4]{1}\left[\cos \frac{\left(\frac{3\pi}{2}\right) + 2\pi k}{4} + i \sin \frac{\left(\frac{3\pi}{2}\right) + 2\pi k}{4}\right], k = 0, 1, 2, 3$

For $k = 0$: $\cos \frac{\frac{3\pi}{2}}{4} + i \sin \frac{\frac{3\pi}{2}}{4} = \cos \frac{3\pi}{8} + i \sin \frac{3\pi}{8}$

For $k = 1$: $\cos \frac{\left(\frac{3\pi}{2}\right) + 2\pi}{4} + i \sin \frac{\left(\frac{3\pi}{2}\right) + 2\pi}{4} = \cos \frac{7\pi}{8} + i \sin \frac{7\pi}{8}$

For $k = 2$: $\cos \frac{\left(\frac{3\pi}{2}\right) + 4\pi}{4} + i \sin \frac{\left(\frac{3\pi}{2}\right) + 4\pi}{4} = \cos \frac{11\pi}{8} + i \sin \frac{11\pi}{8}$

For $k = 3$: $\cos \frac{\left(\frac{3\pi}{2}\right) + 6\pi}{4} + i \sin \frac{\left(\frac{3\pi}{2}\right) + 6\pi}{4} = \cos \frac{15\pi}{8} + i \sin \frac{15\pi}{8}$

Chapter 7 Practice Test Solutions

1. $\begin{cases} x + y = 1 \\ 3x - y = 15 \Rightarrow y = 3x - 15 \end{cases}$

$x + (3x - 15) = 1$

$4x = 16$

$x = 4$

$y = -3$

Solution: $(4, -3)$

2. $\begin{cases} x - 3y = -3 \Rightarrow x = 3y - 3 \\ x^2 + 6y = 5 \end{cases}$

$(3y - 3)^2 + 6y = 5$

$9y^2 - 18y + 9 + 6y = 5$

$9y^2 - 12y + 4 = 0$

$(3y - 2)^2 = 0$

$y = \frac{2}{3}$

$x = -1$

Solution: $\left(-1, \frac{2}{3}\right)$

3. $\begin{cases} x + y + z = 6 \Rightarrow z = 6 - x - y \\ 2x - y + 3z = 0 \Rightarrow 2x - y + 3(6 - x - y) = 0 \Rightarrow -x - 4y = -18 \Rightarrow x = 18 - 4y \\ 5x + 2y - z = -3 \Rightarrow 5x + 2y - (6 - x - y) = -3 \Rightarrow 6x + 3y = 3 \end{cases}$

$$\begin{aligned} 6(18 - 4y) + 3y &= 3 \\ -21y &= -105 \\ y &= 5 \\ x &= 18 - 4y = -2 \\ z &= 6 - x - y = 3 \end{aligned}$$

Solution: $(-2, 5, 3)$

4. $x + y = 110 \Rightarrow y = 110 - x$

$xy = 2800$

$$\begin{aligned} x(110 - x) &= 2800 \\ 0 &= x^2 - 110x + 2800 \\ 0 &= (x - 40)(x - 70) \end{aligned}$$

$x = 40$ or $x = 70$

$y = 70 \qquad y = 40$

Solution: The two numbers are 40 and 70.

5. $2x + 2y = 170 \Rightarrow y = \dfrac{170 - 2x}{2} = 85 - x$

$$\begin{aligned} xy &= 1500 \\ x(85 - x) &= 1500 \\ 0 &= x^2 - 85x + 1500 \\ 0 &= (x - 25)(x - 60) \end{aligned}$$

$x = 25$ or $x = 60$

$y = 60 \qquad y = 25$

Dimensions: 60 ft × 25 ft

6. $\begin{cases} 2x + 15y = 4 \Rightarrow 2x + 15y = 4 \\ x - 3y = 23 \Rightarrow 5x - 15y = 115 \end{cases}$

$$\begin{aligned} 7x &= 119 \\ x &= 17 \\ y &= \frac{x - 23}{3} \\ &= -2 \end{aligned}$$

Solution: $(17, -2)$

7. $\begin{cases} x + y = 2 \Rightarrow 19x + 19y = 38 \\ 38x - 19y = 7 \Rightarrow 38x - 19y = 7 \end{cases}$

$$\begin{aligned} 57x &= 45 \\ x &= \tfrac{15}{19} \\ y &= 2 - x \\ &= \tfrac{38}{19} - \tfrac{15}{19} \\ &= \tfrac{23}{19} \end{aligned}$$

Solution: $\left(\frac{15}{19}, \frac{23}{19}\right)$

8. $\begin{cases} 0.4x + 0.5y = 0.112 \Rightarrow 0.28x + 0.35y = 0.0784 \\ 0.3x - 0.7y = -0.131 \Rightarrow 0.15x - 0.35y = -0.0655 \end{cases}$

$0.43x = 0.0129$

$$x = \frac{0.0129}{0.43} = 0.03$$

$$y = \frac{0.112 - 0.4x}{0.5} = 0.20$$

Solution: $(0.03, 0.20)$

9. Let x = amount in 11% fund and
y = amount in 13% fund.

$x + y = 17{,}000 \Rightarrow y = 17{,}000 - x$

$$\begin{aligned} 0.11x + 0.13y &= 2080 \\ 0.11x + 0.13(17{,}000 - x) &= 2080 \\ -0.02x &= -130 \\ x &= \$6500 \text{ at } 11\% \\ y &= \$10{,}500 \text{ at } 13\% \end{aligned}$$

10. $(4, 3), (1, 1), (-1, -2), (-2, -1)$

Use a calculator.

$y = ax + b = \frac{11}{14}x - \frac{1}{7}$

11. $\begin{cases} x + y = -2 \\ 2x - y + z = 11 \\ 4y - 3z = -20 \end{cases}$

$\begin{cases} x + y = -2 \\ -3y + z = 15 \\ 4y - 3z = -20 \end{cases}$ $\quad -2\text{Eq.1} + \text{Eq.2}$

$\begin{cases} x + y = -2 \\ y - 2z = -5 \\ 4y - 3z = -20 \end{cases}$ $\quad \text{Eq.3} + \text{Eq.2}$

$\begin{cases} x + y = -2 \\ y - 2z = -5 \\ 5z = 0 \end{cases}$ $\quad -4\text{Eq.2} + \text{Eq.3}$

$\begin{cases} x + y = -2 \\ y - 2z = -5 \\ z = 0 \end{cases}$

$y - 2(0) = -5 \Rightarrow y = -5$

$x + (-5) = -2 \Rightarrow x = 3$

Solution: $(3, -5, 0)$

12. $\begin{cases} 4x - y + 5z = 4 \\ 2x + y - z = 0 \\ 2x + 4y + 8z = 0 \end{cases}$

$\begin{cases} 2x + 4y + 8z = 0 \\ 2x + y - z = 0 \\ 4x - y + 5z = 4 \end{cases}$ $\quad$ Interchange equations.

$\begin{cases} 2x + 4y + 8z = 0 \\ -3y - 9z = 0 \\ -9y - 11z = 4 \end{cases}$ $\quad \begin{matrix} \\ -\text{Eq.1} + \text{Eq.2} \\ -2\text{Eq.1} + \text{Eq.3} \end{matrix}$

$\begin{cases} 2x + 4y + 8z = 0 \\ -3y - 9z = 0 \\ 16z = 4 \end{cases}$ $\quad -3\text{Eq.2} + \text{Eq.3}$

$\begin{cases} x + 2y + 4z = 0 \\ y + 3z = 0 \\ z = \frac{1}{4} \end{cases}$ $\quad \begin{matrix} \frac{1}{2}\text{Eq.1} \\ -\frac{1}{3}\text{Eq.2} \\ \frac{1}{16}\text{Eq.3} \end{matrix}$

$y + 3\left(\frac{1}{4}\right) = 0 \Rightarrow y = -\frac{3}{4}$

$x + 2\left(-\frac{3}{4}\right) + 4\left(\frac{1}{4}\right) = 0 \Rightarrow x = \frac{1}{2}$

Solution: $\left(-\frac{1}{2}, -\frac{3}{4}, \frac{1}{4}\right)$

13. $\begin{cases} 3x + 2y - z = 5 \\ 6x - y + 5z = 2 \end{cases}$

$\begin{cases} 3x + 2y - z = 5 \\ -5y + 7z = -8 \end{cases}$ $\quad -2\text{Eq.1} + \text{Eq.2}$

$\begin{cases} x + \frac{2}{3}y - \frac{1}{3}z = \frac{5}{3} \\ y - \frac{7}{5}z = \frac{8}{5} \end{cases}$ $\quad \begin{matrix} \frac{1}{3}\text{Eq.1} \\ -\frac{1}{5}\text{Eq.2} \end{matrix}$

Let $a = z$.

Then $y = \frac{7}{5}a + \frac{8}{5}$, and $x + \frac{2}{3}\left(\frac{7}{5}a + \frac{8}{5}\right) - \frac{1}{3}a = \frac{5}{3}$

$$x + \frac{3}{5}a = \frac{3}{5}$$

$$x = -\frac{3}{5}a + \frac{3}{5}.$$

Solution: $\left(-\frac{3}{5}a + \frac{3}{5}, \frac{7}{5}a + \frac{8}{5}, a\right)$ where a is any real number

14. $y = ax^2 + bx + c$ passes through $(0, -1)$, $(1, 4)$, and $(2, 13)$.

At $(0, -1)$: $-1 = a(0)^2 + b(0) + c \Rightarrow c = -1$

At $(1, 4)$: $4 = a(1)^2 + b(1) - 1 \Rightarrow 5 = a + b \Rightarrow 5 = a + b$

At $(2, 13)$: $13 = a(2)^2 + b(2) - 1 \Rightarrow 14 = 4a + 2b \Rightarrow -7 = -2a - b$

$$-2 = -a$$

$$a = 2$$

$$b = 3$$

So, the equation of the parabola is $y = 2x^2 + 3x - 1$.

15. $s = \frac{1}{2}at^2 + v_0t + s_0$ passes through $(1, 12)$, $(2, 5)$, and $(3, 4)$.

At $(1, 12)$: $12 = \frac{1}{2}a + v_0 + s_0$
At $(2, 5)$: $5 = 2a + 2v_0 + s_0$
At $(3, 4)$: $4 = \frac{9}{2}a + 3v_0 + s_0$

$$\begin{cases} a + 2v_0 + 2s_0 = 24 \\ 2a + 2v_0 + s_0 = 5 \\ 9a + 6v_0 + 2s_0 = 8 \end{cases}$$

$$\begin{cases} a + 2v_0 + 2s_0 = 24 & \\ -2v_0 - 3s_0 = -43 & -2\text{Eq.1} + \text{Eq.2} \\ -12v_0 - 16s_0 = -208 & -9\text{Eq.1} + \text{Eq.3} \end{cases}$$

$$\begin{cases} a + 2v_0 + 2s_0 = 24 & \\ -2v_0 - 3s_0 = -43 & \\ 2s_0 = 50 & -6\text{Eq.2} + \text{Eq.3} \end{cases}$$

$$\begin{cases} a + 2v_0 + 2s_0 = 24 & \\ v_0 + \frac{3}{2}s_0 = \frac{43}{2} & -\frac{1}{2}\text{Eq.2} \\ s_0 = 25 & \frac{1}{2}\text{Eq.3} \end{cases}$$

$$s_0 = 25$$
$$v_0 + \frac{3}{2}(25) = \frac{43}{2} \Rightarrow v_0 = 16$$
$$a + 2(-16) + 2(25) = 24 \Rightarrow a = 6$$

So, $s = \frac{1}{2}(6)t^2 - 16t + 25 = 3t^2 - 16t + 25.$

16.  $x^2 + y^2 \geq 9$

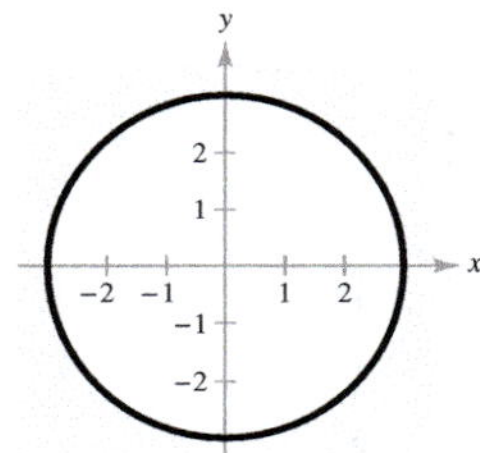

17. $\begin{cases} x + y \leq 6 \\ x \geq 2 \\ y \geq 0 \end{cases}$

y
8
6
4
2
(2, 4)
(2, 0)
(6, 0)
4
6
8
x

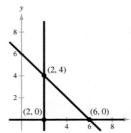

18. Line through $(0, 0)$ and $(0, 7)$: $x = 0$

Line through $(0, 0)$ and $(2, 3)$:

$y = \frac{3}{2}x$ or $3x - 2y = 0$

Line through $(0, 7)$ and $(2, 3)$:

$y = -2x + 7$ or $2x + y = 7$

Inequalities: $\begin{cases} x \geq 0 \\ 3x - 2y \leq 0 \\ 2x + y \leq 7 \end{cases}$

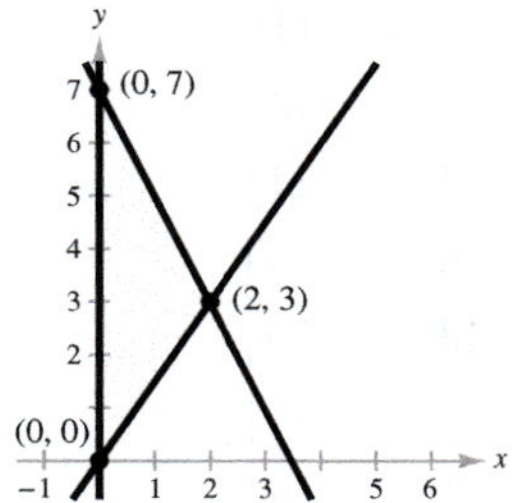

19. Vertices $(0, 0), (0, 7), (6, 0), (3, 5)$

$z = 30x + 26y$

At $(0, 0)$: $z = 0$

At $(0, 7)$: $z = 182$

At $(6, 0)$: $z = 180$

At $(3, 5)$: $z = 220$

The maximum value of z occurs at $(3, 5)$ and is 220.

20. $x^2 + y^2 \le 4$

$(x - 2)^2 + y^2 \ge 4$

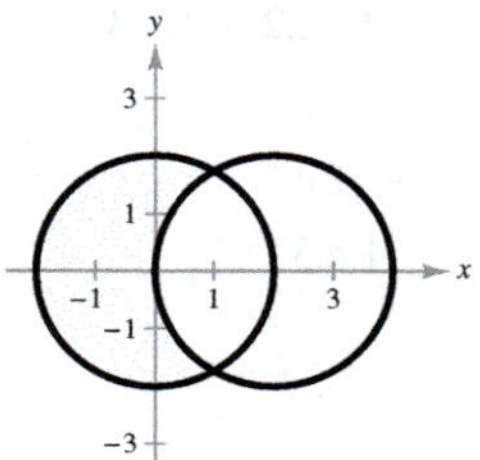

21. $\dfrac{1 - 2x}{x^2 + x} = \dfrac{1 - 2x}{x(x + 1)} = \dfrac{A}{x} + \dfrac{B}{x + 1}$

$1 - 2x = A(x + 1) + Bx$

When $x = 0, 1 = A$.

When $x = -1, 3 = -B \Rightarrow B = -3$.

$\dfrac{1 - 2x}{x^2 + x} = \dfrac{1}{x} - \dfrac{3}{x + 1}$

22. $\dfrac{6x - 17}{(x - 3)^2} = \dfrac{A}{x - 3} + \dfrac{B}{(x - 3)^2}$

$6x - 17 = A(x - 3) + B$

When $x = 3, 1 = B$.

When $x = 0, -17 = -3A + B \Rightarrow A = 6$.

$\dfrac{6x - 17}{(x - 3)^2} = \dfrac{6}{x - 3} + \dfrac{1}{(x - 3)^2}$

Chapter 8 Practice Test Solutions

1. $\begin{bmatrix} 1 & -2 & 4 \\ 3 & -5 & 9 \end{bmatrix}$

$-3R_1 + R_2 \rightarrow \begin{bmatrix} 1 & -2 & 4 \\ 0 & 1 & -3 \end{bmatrix}$

$2R_2 + R_1 \rightarrow \begin{bmatrix} 1 & 0 & -2 \\ 0 & 1 & -3 \end{bmatrix}$

2. $\begin{cases} 3x + 5y = 3 \\ 2x - y = -11 \end{cases}$

$\left[\begin{array}{cc:c} 3 & 5 & 3 \\ 2 & -1 & -11 \end{array}\right]$

$-R_2 + R_1 \rightarrow \left[\begin{array}{cc:c} 1 & 6 & 14 \\ 2 & -1 & -11 \end{array}\right]$

$-2R_1 + R_2 \rightarrow \left[\begin{array}{cc:c} 1 & 6 & 14 \\ 0 & -13 & -39 \end{array}\right]$

$-\frac{1}{13}R_2 \rightarrow \left[\begin{array}{cc:c} 1 & 6 & 14 \\ 0 & 1 & 3 \end{array}\right]$

$-6R_2 + R_1 \rightarrow \left[\begin{array}{cc:c} 1 & 0 & -4 \\ 0 & 1 & 3 \end{array}\right]$

$x = -4, y = 3$

Solution: $(-4, 3)$

3. $\begin{cases} 2x + 3y = -3 \\ 3x - 2y = 8 \\ x + y = 1 \end{cases}$

$\left[\begin{array}{cc:c} 2 & 3 & -3 \\ 3 & 2 & 8 \\ 1 & 1 & 1 \end{array}\right]$

$R_3 \leftrightarrow R_1$: $\left[\begin{array}{cc:c} 1 & 1 & 1 \\ 3 & 2 & 8 \\ 2 & 3 & -3 \end{array}\right]$

$\begin{array}{r} \\ -3R_1 + R_2 \rightarrow \\ -2R_1 + R_3 \rightarrow \end{array} \left[\begin{array}{cc:c} 1 & 1 & 1 \\ 0 & -1 & 5 \\ 0 & 1 & -5 \end{array}\right]$

$-R_2 \rightarrow \left[\begin{array}{cc:c} 1 & 1 & 1 \\ 0 & 1 & -5 \\ 0 & 1 & -5 \end{array}\right]$

$\begin{array}{r} -R_2 + R_1 \rightarrow \\ \\ -R_2 + R_3 \rightarrow \end{array} \left[\begin{array}{cc:c} 1 & 0 & 6 \\ 0 & 1 & -5 \\ 0 & 0 & 0 \end{array}\right]$

$x = 6, y = -5$

Solution: $(6, -5)$

4. $\begin{cases} x + 3z = -5 \\ 2x + y = 0 \\ 3x + y - z = -3 \end{cases}$

$$\begin{bmatrix} 1 & 0 & 3 & \vdots & -5 \\ 2 & 1 & 0 & \vdots & 0 \\ 3 & 1 & -1 & \vdots & 3 \end{bmatrix}$$

$$\begin{matrix} \\ -2R_1 + R_2 \to \\ -3R_1 + R_3 \to \end{matrix} \begin{bmatrix} 1 & 0 & 3 & \vdots & -5 \\ 0 & 1 & -6 & \vdots & 10 \\ 0 & 1 & -10 & \vdots & 18 \end{bmatrix}$$

$$\begin{matrix} \\ \\ -R_2 + R_3 \to \end{matrix} \begin{bmatrix} 1 & 0 & 3 & \vdots & -5 \\ 0 & 1 & -6 & \vdots & 10 \\ 0 & 0 & -4 & \vdots & 8 \end{bmatrix}$$

$$\begin{matrix} \\ \\ -\frac{1}{4}R_3 \to \end{matrix} \begin{bmatrix} 1 & 0 & 3 & \vdots & -5 \\ 0 & 1 & -6 & \vdots & 10 \\ 0 & 0 & 1 & \vdots & -2 \end{bmatrix}$$

$$\begin{matrix} -3R_3 + R_1 \to \\ 6R_3 + R_2 \to \\ \\ \end{matrix} \begin{bmatrix} 1 & 0 & 0 & \vdots & 1 \\ 0 & 1 & 0 & \vdots & -2 \\ 0 & 0 & 1 & \vdots & -2 \end{bmatrix}$$

$x = 1,\ y = -2,\ z = -2$

Solution: $(1, -2, -2)$

5. $$\begin{bmatrix} 1 & 4 & 5 \\ 2 & 0 & -3 \end{bmatrix}\begin{bmatrix} 1 & 6 \\ 0 & -7 \\ -1 & 2 \end{bmatrix} = \begin{bmatrix} (1)(1) + (4)(0) + (5)(-1) & (1)(6) + (4)(-7) + (5)(2) \\ (2)(1) + (0)(0) + (-3)(-1) & (2)(6) + (0)(-7) + (-3)(2) \end{bmatrix} = \begin{bmatrix} -4 & -12 \\ 5 & 6 \end{bmatrix}$$

6. $$\begin{aligned} 3A - 5B &= 3\begin{bmatrix} 9 & 1 \\ -4 & 8 \end{bmatrix} - 5\begin{bmatrix} 6 & -2 \\ 3 & 5 \end{bmatrix} \\ &= \begin{bmatrix} 27 & 3 \\ -12 & 24 \end{bmatrix} - \begin{bmatrix} 30 & -10 \\ 15 & 25 \end{bmatrix} \\ &= \begin{bmatrix} -3 & 13 \\ -27 & -1 \end{bmatrix} \end{aligned}$$

7. $$\begin{aligned} f(A) &= \begin{bmatrix} 3 & 0 \\ 7 & 1 \end{bmatrix}^2 - 7\begin{bmatrix} 3 & 0 \\ 7 & 1 \end{bmatrix} + 8\begin{bmatrix} 1 & 0 \\ 0 & 1 \end{bmatrix} \\ &= \begin{bmatrix} 3 & 0 \\ 7 & 1 \end{bmatrix}\begin{bmatrix} 3 & 0 \\ 7 & 1 \end{bmatrix} - \begin{bmatrix} 21 & 0 \\ 49 & 7 \end{bmatrix} + \begin{bmatrix} 8 & 0 \\ 0 & 8 \end{bmatrix} \\ &= \begin{bmatrix} 9 & 0 \\ 28 & 1 \end{bmatrix} - \begin{bmatrix} 21 & 0 \\ 49 & 7 \end{bmatrix} + \begin{bmatrix} 8 & 0 \\ 0 & 8 \end{bmatrix} \\ &= \begin{bmatrix} -4 & 0 \\ -21 & 2 \end{bmatrix} \end{aligned}$$

8. False.

$$\begin{aligned} (A + B)(A + 3B) &= A(A + 3B) + B(A + 3B) \\ &= A^2 + 3AB + BA + 3B^2 \text{ and, in general, } AB \neq BA. \end{aligned}$$

9.

$$\begin{bmatrix} 1 & 2 & \vdots & 1 & 0 \\ 3 & 5 & \vdots & 0 & 1 \end{bmatrix}$$

$$\begin{array}{r} \\ -3R_1 + R_2 \rightarrow \end{array} \begin{bmatrix} 1 & 2 & \vdots & 1 & 0 \\ 0 & -1 & \vdots & -3 & 1 \end{bmatrix}$$

$$\begin{array}{r} 2R_2 + R_1 \rightarrow \\ \\ \end{array} \begin{bmatrix} 1 & 0 & \vdots & -5 & 2 \\ 0 & -1 & \vdots & -3 & 1 \end{bmatrix}$$

$$\begin{array}{r} \\ -R_2 \rightarrow \end{array} \begin{bmatrix} 1 & 0 & \vdots & -5 & 2 \\ 0 & 1 & \vdots & 3 & -1 \end{bmatrix}$$

$$A^{-1} = \begin{bmatrix} -5 & 2 \\ 3 & -1 \end{bmatrix}$$

10.

$$\begin{bmatrix} 1 & 1 & 1 & \vdots & 1 & 0 & 0 \\ 3 & 6 & 5 & \vdots & 0 & 1 & 0 \\ 6 & 10 & 8 & \vdots & 0 & 0 & 1 \end{bmatrix}$$

$$\begin{array}{r} \\ -3R_1 + R_2 \rightarrow \\ -6R_1 + R_3 \rightarrow \end{array} \begin{bmatrix} 1 & 1 & 1 & \vdots & 1 & 0 & 0 \\ 0 & 3 & 2 & \vdots & -3 & 1 & 0 \\ 0 & 4 & 2 & \vdots & -6 & 0 & 1 \end{bmatrix}$$

$$\begin{array}{r} \\ -R_3 + R_2 \rightarrow \\ \\ \end{array} \begin{bmatrix} 1 & 1 & 1 & \vdots & 1 & 0 & 0 \\ 0 & -1 & 0 & \vdots & 3 & 1 & -1 \\ 0 & 4 & 2 & \vdots & -6 & 0 & 1 \end{bmatrix}$$

$$\begin{array}{r} R_2 + R_1 \rightarrow \\ \\ 4R_2 + R_3 \rightarrow \end{array} \begin{bmatrix} 1 & 0 & 1 & \vdots & 4 & 1 & -1 \\ 0 & -1 & 0 & \vdots & 3 & 1 & -1 \\ 0 & 0 & 2 & \vdots & 6 & 4 & -3 \end{bmatrix}$$

$$\begin{array}{r} \\ -R_2 \rightarrow \\ \frac{1}{2}R_3 \rightarrow \end{array} \begin{bmatrix} 1 & 0 & 1 & \vdots & 4 & 1 & -1 \\ 0 & 1 & 0 & \vdots & -3 & -1 & 1 \\ 0 & 0 & 1 & \vdots & 3 & 2 & -\frac{3}{2} \end{bmatrix}$$

$$\begin{array}{r} -R_3 + R_1 \rightarrow \\ \\ \\ \end{array} \begin{bmatrix} 1 & 0 & 0 & \vdots & 1 & -1 & \frac{1}{2} \\ 0 & 1 & 0 & \vdots & -3 & -1 & 1 \\ 0 & 0 & 1 & \vdots & 3 & 2 & -\frac{3}{2} \end{bmatrix}$$

$$A^{-1} = \begin{bmatrix} 1 & -1 & \frac{1}{2} \\ -3 & -1 & 1 \\ 3 & 2 & -\frac{3}{2} \end{bmatrix}$$

11. (a) $\begin{cases} x + 2y = 4 \\ 3x + 5y = 1 \end{cases}$

$$A = \begin{bmatrix} 1 & 2 \\ 3 & 5 \end{bmatrix}$$

$$A^{-1} = \frac{1}{5 - 6}\begin{bmatrix} 5 & -2 \\ -3 & 1 \end{bmatrix} = \begin{bmatrix} -5 & 2 \\ 3 & -1 \end{bmatrix}$$

$$\begin{bmatrix} x \\ y \end{bmatrix} = A^{-1}B = \begin{bmatrix} -5 & 2 \\ 3 & -1 \end{bmatrix}\begin{bmatrix} 4 \\ 1 \end{bmatrix} = \begin{bmatrix} -18 \\ 11 \end{bmatrix}$$

$x = -18,\ y = 11$

Solution: $(-18, 11)$

(b) $\begin{cases} x + 2y = 3 \\ 3x + 5y = -2 \end{cases}$

Again, $A^{-1} = \begin{bmatrix} -5 & 2 \\ 3 & -1 \end{bmatrix}$.

$$\begin{bmatrix} x \\ y \end{bmatrix} = A^{-1}B = \begin{bmatrix} -5 & 2 \\ 3 & -1 \end{bmatrix}\begin{bmatrix} 3 \\ -2 \end{bmatrix} = \begin{bmatrix} -19 \\ 11 \end{bmatrix}$$

$x = -19,\ y = 11$

Solution: $(-19, 11)$

12. $\begin{vmatrix} 6 & -1 \\ 3 & 4 \end{vmatrix} = 24 - (-3) = 27$

13. $\begin{vmatrix} 1 & 3 & -1 \\ 5 & 9 & 0 \\ 6 & 2 & -5 \end{vmatrix} = -1\begin{vmatrix} 5 & 9 \\ 6 & 2 \end{vmatrix} - 5\begin{vmatrix} 1 & 3 \\ 5 & 9 \end{vmatrix}$

$= -(-44) - 5(-6) = 74$

14. Expand along Row 2.

$$\begin{vmatrix} 1 & 4 & 2 & 3 \\ 0 & 1 & -2 & 0 \\ 3 & 5 & -1 & 1 \\ 2 & 0 & 6 & 1 \end{vmatrix} = \begin{vmatrix} 1 & 2 & 3 \\ 3 & -1 & 1 \\ 2 & 6 & 1 \end{vmatrix} + 2\begin{vmatrix} 1 & 4 & 3 \\ 3 & 5 & 1 \\ 2 & 0 & 1 \end{vmatrix}$$

$= 51 + 2(-29) = -7$

15. $\begin{vmatrix} 6 & 4 & 3 & 0 & 6 \\ 0 & 5 & 1 & 4 & 8 \\ 0 & 0 & 2 & 7 & 3 \\ 0 & 0 & 0 & 9 & 2 \\ 0 & 0 & 0 & 0 & 1 \end{vmatrix} = 6\begin{vmatrix} 5 & 1 & 4 & 8 \\ 0 & 2 & 7 & 3 \\ 0 & 0 & 9 & 2 \\ 0 & 0 & 0 & 1 \end{vmatrix} = 6(5)\begin{vmatrix} 2 & 7 & 3 \\ 0 & 9 & 2 \\ 0 & 0 & 1 \end{vmatrix} = 6(5)(2)\begin{vmatrix} 9 & 2 \\ 0 & 1 \end{vmatrix} = 6(5)(2)(9) = 540$

16. Area $= \dfrac{1}{2}\begin{vmatrix} 0 & 7 & 1 \\ 5 & 0 & 1 \\ 3 & 9 & 1 \end{vmatrix} = \dfrac{1}{2}(31) = \dfrac{31}{2}$

17. $\begin{vmatrix} x & y & 1 \\ 2 & 7 & 1 \\ -1 & 4 & 1 \end{vmatrix} = 3x - 3y + 15 = 0$ or $x - y + 5 = 0$

18. $x = \dfrac{\begin{vmatrix} 4 & -7 \\ 11 & 5 \end{vmatrix}}{\begin{vmatrix} 6 & -7 \\ 2 & 5 \end{vmatrix}} = \dfrac{97}{44}$

19. $z = \dfrac{\begin{vmatrix} 3 & 0 & 1 \\ 0 & 1 & 3 \\ 1 & -1 & 2 \end{vmatrix}}{\begin{vmatrix} 3 & 0 & 1 \\ 0 & 1 & 4 \\ 1 & -1 & 0 \end{vmatrix}} = \dfrac{14}{11}$

20. $y = \dfrac{\begin{vmatrix} 721.4 & 33.77 \\ 45.9 & 19.85 \end{vmatrix}}{\begin{vmatrix} 721.4 & -29.1 \\ 45.9 & 105.6 \end{vmatrix}} = \dfrac{12{,}769.747}{77{,}515.530} \approx 0.1647$

Chapter 9 Practice Test Solutions

1. $a_n = \dfrac{2n}{(n+2)!}$

$a_1 = \dfrac{2(1)}{3!} = \dfrac{2}{6} = \dfrac{1}{3}$

$a_2 = \dfrac{2(2)}{4!} = \dfrac{4}{24} = \dfrac{1}{6}$

$a_3 = \dfrac{2(3)}{5!} = \dfrac{6}{120} = \dfrac{1}{20}$

$a_4 = \dfrac{2(4)}{6!} = \dfrac{8}{720} = \dfrac{1}{90}$

$a_5 = \dfrac{2(5)}{7!} = \dfrac{10}{5040} = \dfrac{1}{504}$

Terms: $\dfrac{1}{3}, \dfrac{1}{6}, \dfrac{1}{20}, \dfrac{1}{90}, \dfrac{1}{504}$

2. $a_n = \dfrac{n+3}{3^n}$

3. $\displaystyle\sum_{i=1}^{6} (2i - 1) = 1 + 3 + 5 + 7 + 9 + 11 = 36$

4. $a_1 = 23, d = -2$

$a_2 = 23 + (-2) = 21$

$a_3 = 21 + (-2) = 19$

$a_4 = 19 + (-2) = 17$

$a_5 = 17 + (-2) = 15$

Terms: 23, 21, 19, 17, 15

5. $a_1 = 12, d = 3, n = 50$

$a_n = a_1 + (n-1)d$

$a_{50} = 12 + (50 - 1)3 = 159$

6. $a_1 = 1$

$a_{200} = 200$

$S_n = \dfrac{n}{2}(a_1 + a_n)$

$S_{200} = \dfrac{200}{2}(1 + 200) = 20{,}100$

7. $a_1 = 7, r = 2$

$a_2 = 7(2) = 14$

$a_3 = 7(2)^2 = 28$

$a_4 = 7(2)^3 = 56$

$a_5 = 7(2)^4 = 112$

Terms: 7, 14, 28, 56, 112

8. $\displaystyle\sum_{n=1}^{10} 6\left(\frac{2}{3}\right)^{n-1}, a_1 = 6, r = \frac{2}{3}, n = 10$

$$S_n = \frac{a_1(1 - r^n)}{1 - r} = \frac{6\left[1 - \left(\frac{2}{3}\right)^{10}\right]}{1 - \frac{2}{3}} = 18\left(1 - \frac{1024}{59{,}049}\right) = \frac{116{,}050}{6561} \approx 17.6879$$

9. $\sum_{n=0}^{\infty} (0.03)^n = \sum_{n=1}^{\infty} (0.03)^{n-1}, a_1 = 1, r = 0.03$

$$S = \frac{a_1}{1-r} = \frac{1}{1-0.03} = \frac{1}{0.97} = \frac{100}{97} \approx 1.0309$$

10. For $n = 1, 1 = \frac{1(1+1)}{2}$.

Assume that $S_k = 1 + 2 + 3 + 4 + \cdots + k = \frac{k(k+1)}{2}$.

$$\text{Then } S_{k+1} = 1 + 2 + 3 + 4 + \cdots + k + (k+1) = \frac{k(k+1)}{2} + k + 1$$
$$= \frac{k(k+1)}{2} + \frac{2(k+1)}{2}$$
$$= \frac{(k+1)(k+2)}{2}.$$

Thus, by the principle of mathematical induction, $1 + 2 + 3 + 4 + \cdots + n = \frac{n(n+1)}{2}$ for all integers $n \geq 1$.

11. For $n = 4$, $4! > 2^4$. Assume that $k! > 2^k$.

Then $(k+1)! = (k+1)(k!) > (k+1)2^k > 2 \cdot 2^k = 2^{k+1}$.

Thus, by the extended principle of mathematical induction, $n! > 2^n$ for all integers $n \geq 4$.

12. ${}_{13}C_4 = \frac{13!}{(13-4)!4!} = 715$

13. $(x+3)^5 = x^5 + 5x^4(3) + 10x^3(3)^2 + 10x^2(3)^3 + 5x(3)^4 + (3)^5$

$= x^5 + 15x^4 + 90x^3 + 270x^2 + 405x + 243$

14. $-{}_{12}C_5 x^7 (2)^5 = -25{,}344x^7$

15. ${}_{30}P_4 = \frac{30!}{(30-4)!} = 657{,}720$

16. $6! = 720$ ways

17. ${}_{12}P_3 = 1320$

18. $P(2) + P(3) + P(4) = \frac{1}{36} + \frac{2}{36} + \frac{3}{36}$

$= \frac{6}{36} = \frac{1}{6}$

19. $P(\text{K, B10}) = \frac{4}{52} \cdot \frac{2}{51} = \frac{2}{663}$

20. Let A = probability of no faulty units.

$P(A) = \left(\frac{997}{1000}\right)^{50} \approx 0.8605$

$P(A') = 1 - P(A) \approx 0.1395$

Chapter 10 Practice Test Solutions

1. $3x + 4y = 12 \Rightarrow y = -\frac{3}{4}x + 3 \Rightarrow m_1 = -\frac{3}{4}$

$4x - 3y = 12 \Rightarrow y = \frac{4}{3}x - 4 \Rightarrow m_2 = \frac{4}{3}$

$$\tan\theta = \left|\frac{(4/3) - (-3/4)}{1 + (4/3)(-3/4)}\right| = \left|\frac{25/12}{0}\right|$$

Since $\tan\theta$ is undefined, the lines are perpendicular (note that $m_2 = -1/m_1$) and $\theta = 90°$.

2. $x_1 = 5, x_2 = -9, A = 3, B = -7, C = -21$

$$d = \frac{|3(5) + (-7)(-9) + (-21)|}{\sqrt{3^2 + (-7)^2}} = \frac{57}{\sqrt{58}} \approx 7.484$$

3. $x^2 - 6x - 4y + 1 = 0$

$$x^2 - 6x + 9 = 4y - 1 + 9$$

$$(x - 3)^2 = 4y + 8$$

$$(x - 3)^2 = 4(1)(y + 2) \Rightarrow p = 1$$

Vertex: $(3, -2)$

Focus: $(3, -1)$

Directrix: $y = -3$

4. Vertex: $(2, -5)$

Focus: $(2, -6)$

Vertical axis; opens downward with $p = -1$

$$(x - h)^2 = 4p(y - k)$$

$$(x - 2)^2 = 4(-1)(y + 5)$$

$$x^2 - 4x + 4 = -4y - 20$$

$$x^2 - 4x + 4y + 24 = 0$$

5. $x^2 + 4y^2 - 2x + 32y + 61 = 0$

$$(x^2 - 2x + 1) + 4(y^2 + 8y + 16) = -61 + 1 + 64$$

$$(x - 1)^2 + 4(y + 4)^2 = 4$$

$$\frac{(x - 1)^2}{4} + \frac{(y + 4)^2}{1} = 1$$

$a = 2, b = 1, c = \sqrt{3}$

Horizontal major axis

Center: $(1, -4)$

Foci: $(1 \pm \sqrt{3}, -4)$

Vertices: $(3, -4), (-1, -4)$

Eccentricity: $e = \dfrac{\sqrt{3}}{2}$

6. Vertices: $(0, \pm 6)$

Eccentricity: $e = \dfrac{1}{2}$

Center: $(0, 0)$

Vertical major axis

$$a = 6, e = \frac{c}{a} = \frac{c}{6} = \frac{1}{2} \Rightarrow c = 3$$

$$b^2 = (6)^2 - (3)^2 = 27$$

$$\frac{x^2}{27} + \frac{y^2}{36} = 1$$

7. $16y^2 - x^2 - 6x - 128y + 231 = 0$

$$16(y^2 - 8y + 16) - (x^2 + 6x + 9) = -231 + 256 - 9$$

$$16(y - 4)^2 - (x + 3)^2 = 16$$

$$\frac{(y - 4)^2}{1} - \frac{(x + 3)^2}{16} = 1$$

$a = 1, b = 4, c = \sqrt{17}$

Center: $(-3, 4)$

Vertical transverse axis

Vertices: $(-3, 5), (-3, 3)$

Foci: $(-3, 4 \pm \sqrt{17})$

Asymptotes: $y = 4 \pm \dfrac{1}{4}(x + 3)$

8. Vertices: $(\pm 3, 2)$

Foci: $(\pm 5, 2)$

Center: $(0, 2)$

Horizontal transverse axis

$a = 3, c = 5, b = 4$

$$\frac{(x - 0)^2}{9} - \frac{(y - 2)^2}{16} = 1$$

$$\frac{x^2}{9} - \frac{(y - 2)^2}{16} = 1$$

9. $5x^2 + 2xy + 5y^2 - 10 = 0$

$A = 5, B = 2, C = 5$

$$\cot 2\theta = \frac{5 - 5}{2} = 0$$

$$2\theta = \frac{\pi}{2} \Rightarrow \theta = \frac{\pi}{4}$$

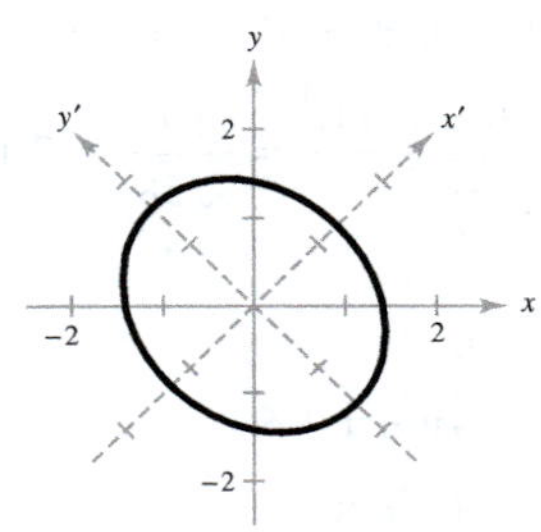

$$x = x' \cos\frac{\pi}{4} - y' \sin\frac{\pi}{4} = \frac{x' - y'}{\sqrt{2}}$$

$$x = x' \cos\frac{\pi}{4} + y' \sin\frac{\pi}{4} = \frac{x' + y'}{\sqrt{2}}$$

$$5\left(\frac{x' - y'}{\sqrt{2}}\right)^2 + 2\left(\frac{x' - y'}{\sqrt{2}}\right)\left(\frac{x' + y'}{\sqrt{2}}\right) + 5\left(\frac{x' + y'}{\sqrt{2}}\right)^2 - 10 = 0$$

$$\frac{5(x')^2}{2} - \frac{10x'y'}{2} + \frac{5(y')^2}{2} + (x')^2 - (y')^2 + \frac{5(x')^2}{2} + \frac{10x'y'}{2} + \frac{5(y')^2}{2} - 10 = 0$$

$$6(x')^2 + 4(y')^2 - 10 = 0$$

$$\frac{3(x')^2}{5} + \frac{2(y')^2}{5} = 1$$

$$\frac{(x')^2}{5/3} + \frac{(y')^2}{5/2} = 1$$

Ellipse centered at the origin

10. (a) $6x^2 - 2xy + y^2 = 0$

$A = 6, B = -2, C = 1$

$B^2 - 4AC = (-2)^2 - 4(6)(1) = -20 < 0$

Ellipse

(b) $x^2 + 4xy + 4y^2 - x - y + 17 = 0$

$A = 1, B = 4, C = 4$

$B^2 - 4AC = (4)^2 - 4(1)(4) = 0$

Parabola

11. Polar: $\left(\sqrt{2}, \frac{3\pi}{4}\right)$

$$x = \sqrt{2}\cos\frac{3\pi}{4} = \sqrt{2}\left(-\frac{1}{\sqrt{2}}\right) = -1$$

$$y = \sqrt{2}\sin\frac{3\pi}{4} = \sqrt{2}\left(\frac{1}{\sqrt{2}}\right) = 1$$

Rectangular: $(-1, 1)$

12. Rectangular: $\left(\sqrt{3}, -1\right)$

$$r = \pm\sqrt{\left(\sqrt{3}\right)^2 + (-1)^2} = \pm 2$$

$$\tan\theta = \frac{\sqrt{3}}{-1} = -\sqrt{3}$$

$$\theta = \frac{2\pi}{3} \text{ or } \theta = \frac{5\pi}{3}$$

Polar: $\left(-2, \frac{2\pi}{3}\right)$ or $\left(2, \frac{5\pi}{3}\right)$

13. Rectangular: $4x - 3y = 12$

Polar: $4r\cos\theta - 3r\sin\theta = 12$

$$r(4\cos\theta - 3\sin\theta) = 12$$

$$r = \frac{12}{4\cos\theta - 3\sin\theta}$$

14. Polar: $r = 5\cos\theta$

$$r^2 = 5r\cos\theta$$

Rectangular: $x^2 + y^2 = 5x$

$$x^2 + y^2 - 5x = 0$$

15. $r = 1 - \cos\theta$

Cardioid

Symmetry: Polar axis

Maximum value of $|r|$: $r = 2$ when $\theta = \pi$

Zero of r: $r = 0$ when $\theta = 0$

θ	0	$\frac{\pi}{2}$	π	$\frac{3\pi}{2}$
r	0	1	2	1

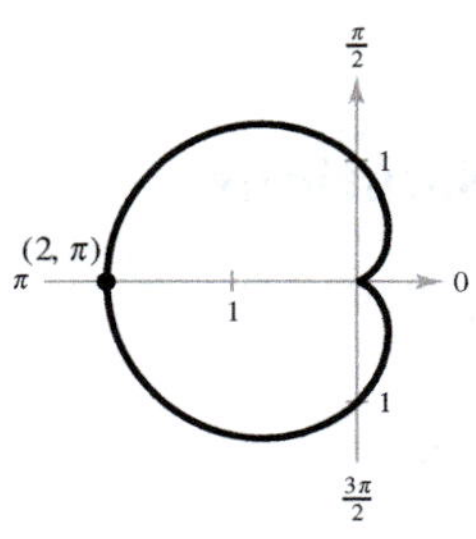

16. $r = 5\sin 2\theta$

Rose curve with four petals

Symmetry: Polar axis, $\theta = \frac{\pi}{2}$, and pole

Maximum value of $|r|$: $|r| = 5$ when

$$\theta = \frac{\pi}{4}, \frac{3\pi}{4}, \frac{5\pi}{4}, \frac{7\pi}{4}$$

Zeros of r: $r = 0$ when $\theta = 0, \frac{\pi}{2}, \pi, \frac{3\pi}{2}$

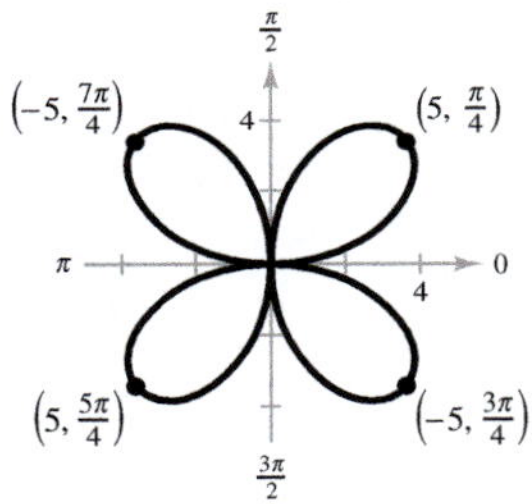

17. $r = \dfrac{3}{6 - \cos\theta}$

$$r = \frac{1/2}{1 - (1/6)\cos\theta}$$

$e = \frac{1}{6} < 1$, so the graph is an ellipse.

θ	0	$\frac{\pi}{2}$	π	$\frac{3\pi}{2}$
r	$\frac{3}{5}$	$\frac{1}{2}$	$\frac{3}{7}$	$\frac{1}{2}$

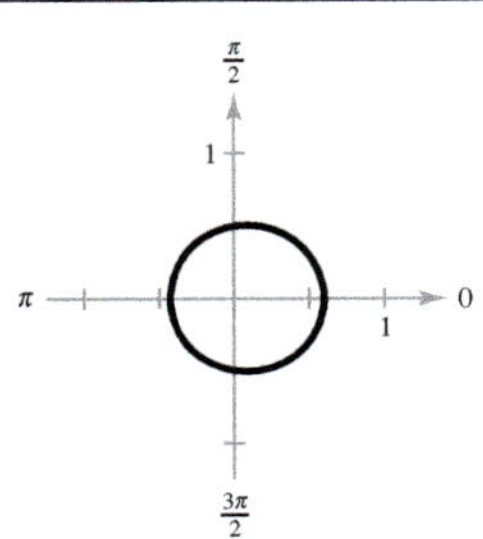

18. Parabola

Vertex: $\left(6, \frac{\pi}{2}\right)$

Focus: $(0, 0)$

$$e = 1$$

$$r = \frac{ep}{1 + e\sin\theta}$$

$$r = \frac{p}{1 + \sin\theta}$$

$$6 = \frac{p}{1 + \sin(\pi/2)}$$

$$6 = \frac{p}{2}$$

$$12 = p$$

$$r = \frac{12}{1 + \sin\theta}$$

19. $x = 3 - 2\sin\theta$, $y = 1 + 5\cos\theta$

$$\frac{x - 3}{-2} = \sin\theta, \frac{y - 1}{5} = \cos\theta$$

$$\left(\frac{x - 3}{-2}\right)^2 + \left(\frac{y - 1}{5}\right)^2 = 1$$

$$\frac{(x - 3)^2}{4} + \frac{(y - 1)^2}{25} = 1$$

20. $x = e^{2t}$, $y = e^{4t}$

$x > 0$, $y > 0$

$$y = \left(e^{2t}\right)^2 = (x)^2 = x^2, x > 0, y > 0$$

PART II

Chapter 1 Chapter Test Solutions

1. Midpoint: $\left(\frac{-2+6}{2}, \frac{5+0}{2}\right) = \left(2, \frac{5}{2}\right)$

Distance: $d = \sqrt{(-2-6)^2 + (5-0)^2}$

$= \sqrt{64 + 25}$

$= \sqrt{89}$

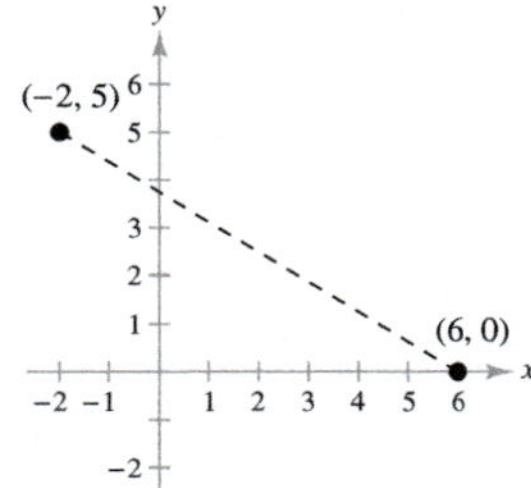

2. $V = \pi r^2 h$

$= \pi(4)^2 h$

$= 16\pi h$

3. $y = 3 - 5x$

x-intercept: $\left(\frac{3}{5}, 0\right)$

y-intercept: $(0, 3)$

No axis or origin symmetry

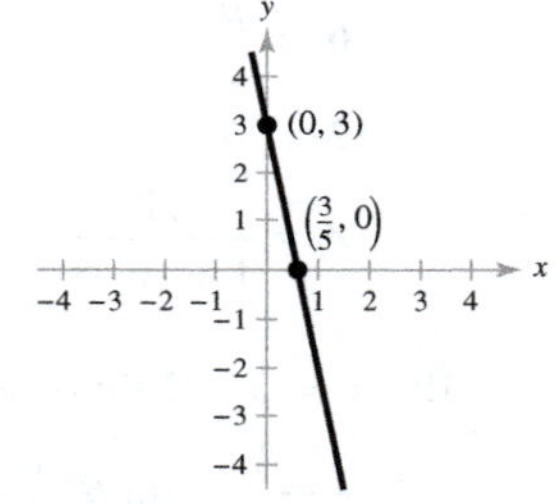

4. $y = 4 - |x|$

x-intercepts: $(\pm 4, 0)$

y-intercept: $(0, 4)$

y-axis symmetry

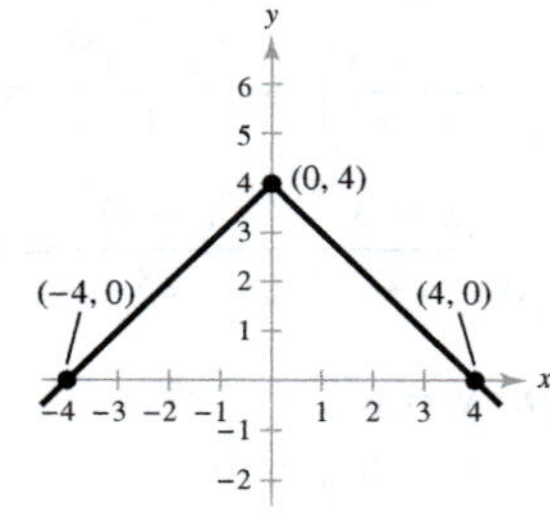

5. $y = x^2 - 1$

x-intercepts: $(\pm 1, 0)$

y-intercept: $(0, -1)$

y-axis symmetry

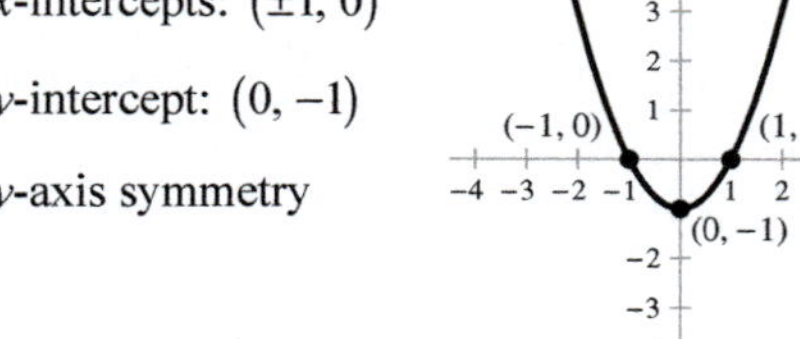

6. Center: $(1, 3)$

Radius: 4

Standard form:

$(x-1)^2 + (y-3)^2 = 16$

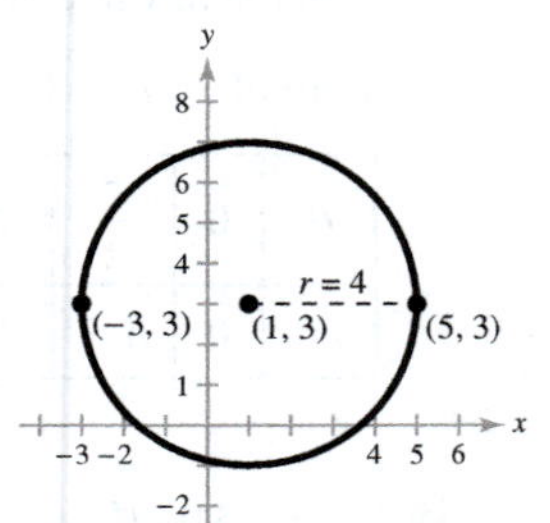

7. $(-2, 5), (1, -7)$

$m = \frac{-7-5}{1-(-2)} = \frac{-12}{3} = -4$

$y - 5 = -4(x - (-2))$

$y - 5 = -4(x + 2)$

$y - 5 = -4x - 8$

$y = -4x - 3$

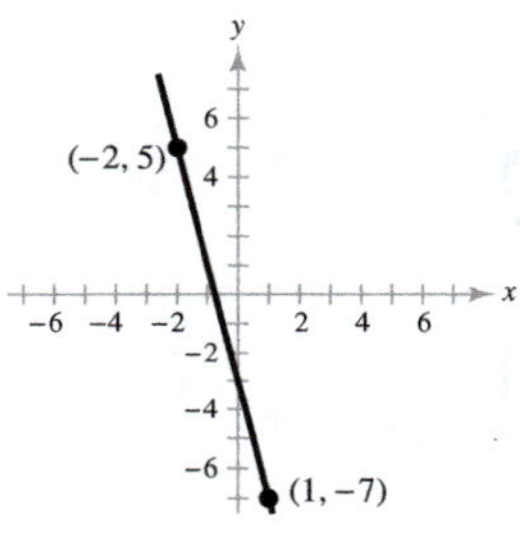

8. $(-4, -7), \left(1, \frac{4}{3}\right)$

$m = \frac{\frac{4}{3} - (-7)}{1 - (-4)} = \frac{\frac{25}{3}}{5} = \frac{5}{3}$

$y - (-7) = \frac{5}{3}(x - (-4))$

$y + 7 = \frac{5}{3}(x + 4)$

$y + 7 = \frac{5}{3}x + \frac{20}{3}$

$y = \frac{5}{3}x - \frac{1}{3}$

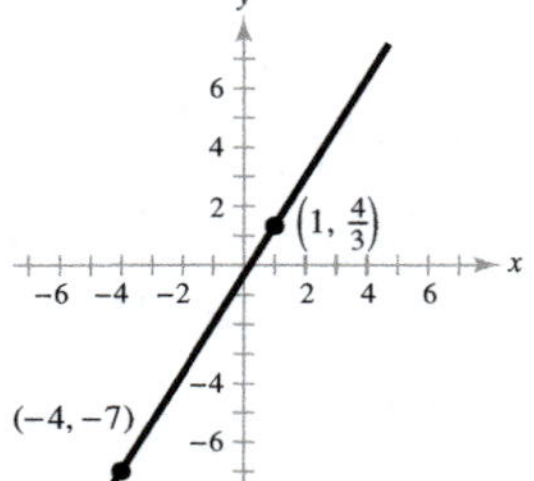

9. $5x + 2y = 3$

$$2y = -5x + 3$$

$$y = -\frac{5}{2}x + \frac{3}{2}$$

(a) Parallel line:

$$m = -\frac{5}{2}$$

$$y - 4 = -\frac{5}{2}(x - 0)$$

$$y - 4 = -\frac{5}{2}x$$

$$y = -\frac{5}{2}x + 4$$

(b) Perpendicular line:

$$m = \frac{2}{5}$$

$$y - 4 = \frac{2}{5}(x - 0)$$

$$y - 4 = \frac{2}{5}x$$

$$y = \frac{2}{5}x + 4$$

10. $f(x) = \dfrac{\sqrt{x + 9}}{x^2 - 81}$

(a) $f(7) = \dfrac{4}{-32} = -\dfrac{1}{8}$

(b) $f(-5) = \dfrac{2}{-56} = -\dfrac{1}{28}$

(c) $f(x - 9) = \dfrac{\sqrt{x}}{(x - 9)^2 - 81} = \dfrac{\sqrt{x}}{x^2 - 18x}$

11. (a) $f(x) = 10 - \sqrt{3 - x}$

$$3 - x \ge 0$$

$$3 \ge x$$

$$x \le 3$$

Domain: All real numbers x such that $x \le 3$

(b) $10 - \sqrt{3 - x} = 0 \Rightarrow 10 = \sqrt{3 - x}$

$$100 = 3 - x$$

$$x = -97$$

12. $f(x) = |x + 5|$

(a)

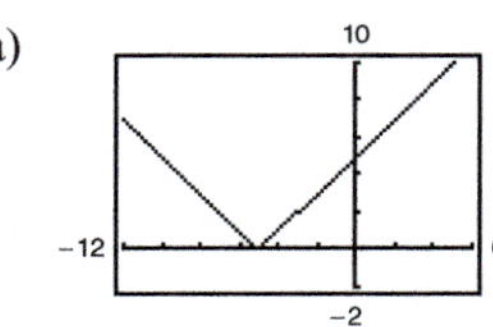

(b) Increasing on $(-5, \infty)$

Decreasing on $(-\infty, -5)$

(c) The function is neither odd nor even.

13. $f(x) = 4x\sqrt{3 - x}$

(a)

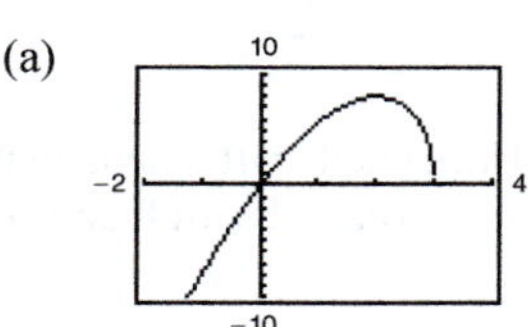

(b) Increasing on $(-\infty, 2)$

Decreasing on $(2, 3)$

(c) The function is neither odd nor even.

14. $f(x) = 2x^6 + 5x^4 - x^2$

(a)

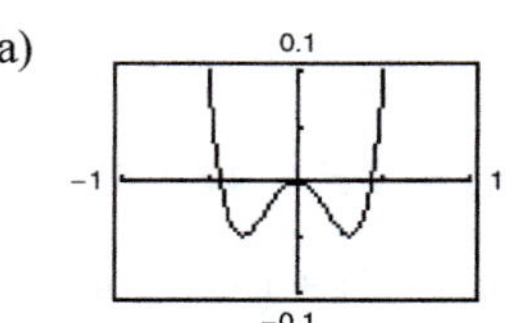

(b) Increasing on $(-0.31, 0), (0.31, \infty)$

Decreasing on $(-\infty, -0.31), (0, 0.31)$

(c) y-axis symmetry $\Rightarrow$ the function is even.

15. $f(x) = \begin{cases} 3x + 7, & x \le -3 \\ 4x^2 - 1, & x > -3 \end{cases}$

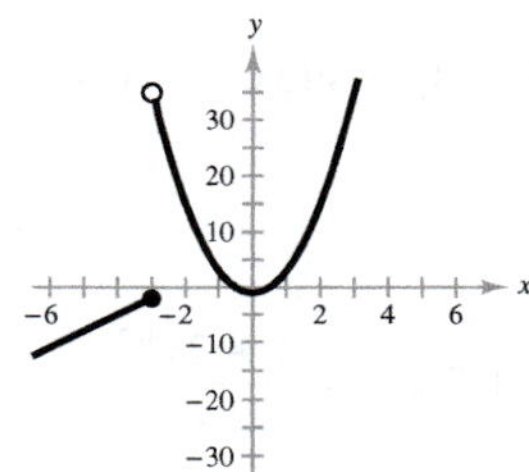

16. $h(x) = 4[\![x]\!]$

(a) The parent function is $f(x) = [\![x]\!]$

(b) The graph of h is a vertical stretch of the graph of f.

(c)

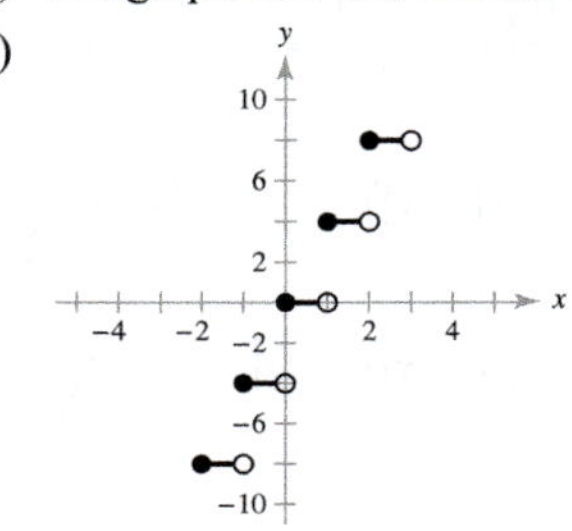

17. $h(x) = \sqrt{x+5} + 8$

(a) Parent function: $f(x) = \sqrt{x}$

(b) Transformation: Horizontal shift 5 units to the left and a vertical shift 8 units upward

(c)

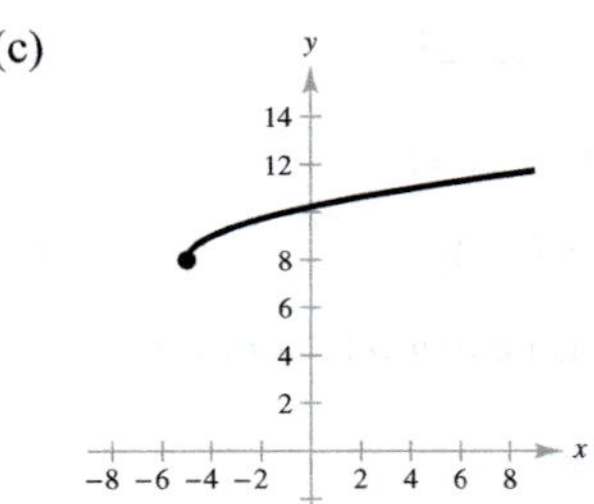

18. $h(x) = -2(x-5)^3 + 3$

(a) Parent function: $f(x) = x^3$

(b) Transformation:

Vertical stretch, reflection in x-axis, horizontal shift 5 units to the right, vertical shift 3 units upward

(c)

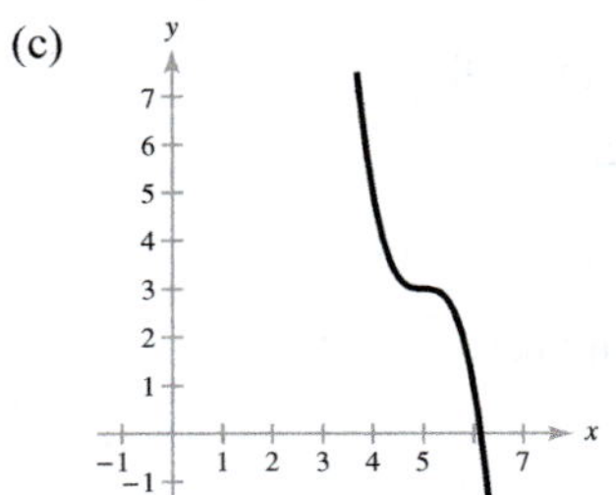

19. $f(x) = 3x^2 - 7,\ g(x) = -x^2 - 4x + 5$

(a) $(f+g)(x) = (3x^2 - 7) + (-x^2 - 4x + 5) = 2x^2 - 4x - 2$

(b) $(f-g)(x) = (3x^2 - 7) - (-x^2 - 4x + 5) = 4x^2 + 4x - 12$

(c) $(fg)(x) = (3x^2 - 7)(-x^2 - 4x + 5) = -3x^4 - 12x^3 + 22x^2 + 28x - 35$

(d) $\left(\dfrac{f}{g}\right)(x) = \dfrac{3x^2 - 7}{-x^2 - 4x + 5},\ x \neq -5, 1$

(e) $(f \circ g)(x) = f(g(x)) = f(-x^2 - 4x + 5) = 3(-x^2 - 4x + 5)^2 - 7 = 3x^4 + 24x^3 + 18x^2 - 120x + 68$

(f) $(g \circ f)(x) = g(f(x)) = g(3x^2 - 7) = -(3x^2 - 7)^2 - 4(3x^2 - 7) + 5 = -9x^4 + 30x^2 - 16$

20. $f(x) = \dfrac{1}{x},\ g(x) = 2\sqrt{x}$

(a) $(f+g)(x) = \dfrac{1}{x} + 2\sqrt{x} = \dfrac{1 + 2x^{3/2}}{x},\ x > 0$

(b) $(f-g)(x) = \dfrac{1}{x} - 2\sqrt{x} = \dfrac{1 - 2x^{3/2}}{x},\ x > 0$

(c) $(fg)(x) = \left(\dfrac{1}{x}\right)(2\sqrt{x}) = \dfrac{2\sqrt{x}}{x},\ x > 0$

(d) $\left(\dfrac{f}{g}\right)(x) = \dfrac{1/x}{2\sqrt{x}} = \dfrac{1}{2x\sqrt{x}} = \dfrac{1}{2x^{3/2}},\ x > 0$

(e) $(f \circ g)(x) = f(g(x)) = f(2\sqrt{x}) = \dfrac{1}{2\sqrt{x}} = \dfrac{\sqrt{x}}{2x},\ x > 0$

(f) $(g \circ f)(x) = g(f(x)) = g\left(\dfrac{1}{x}\right) = 2\sqrt{\dfrac{1}{x}} = \dfrac{2}{\sqrt{x}} = \dfrac{2\sqrt{x}}{x},\ x > 0$

21. $f(x) = x^3 + 8$

Since f is one-to-one, f has an inverse.

$$y = x^3 + 8$$
$$x = y^3 + 8$$
$$x - 8 = y^3$$
$$\sqrt[3]{x - 8} = y$$
$$f^{-1}(x) = \sqrt[3]{x - 8}$$

22. $f(x) = |x^2 - 3| + 6$

Since f is not one-to-one, f does not have an inverse.

23. $f(x) = 3x\sqrt{x} = 3x^{3/2}, x \geq 0$

Because f is one-to-one, f has an inverse.

$$y = 3x^{3/2}$$
$$x = 3y^{3/2}$$
$$\tfrac{1}{3}x = y^{3/2}$$
$$\left(\tfrac{1}{3}x\right)^{2/3} = y, x \geq 0$$
$$f^{-1}(x) = \left(\tfrac{1}{3}x\right)^{2/3}, x \geq 0$$

24.
$$v = k\sqrt{s}$$
$$24 = k\sqrt{16}$$
$$6 = k$$
$$v = 6\sqrt{s}$$

25.
$$A = kxy$$
$$500 = k(15)(8)$$
$$500 = k(120)$$
$$\tfrac{25}{6} = k$$
$$A = \tfrac{25}{6}xy$$

26.
$$b = \frac{k}{a}$$
$$32 = \frac{k}{1.5}$$
$$48 = k$$
$$b = \frac{48}{a}$$

Chapter Test Solutions for Chapter 2

1. (a)

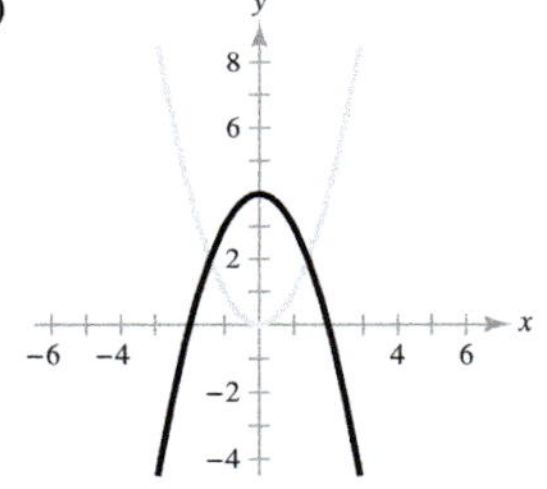

The graph of g is a reflection in the x-axis and a vertical shift up of four units of the graph of $y = x^2$.

(b)

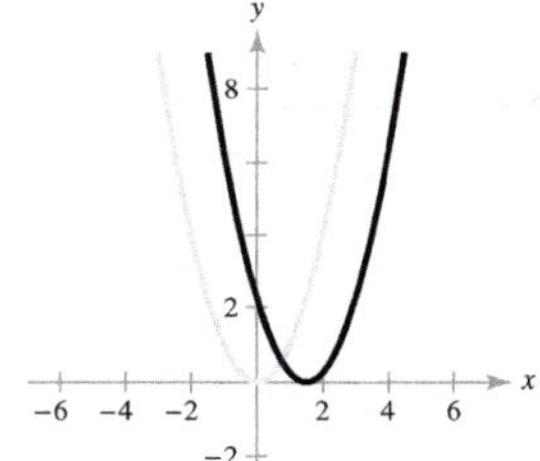

The graph of g is a horizontal shift right $\frac{3}{2}$ units of the graph of $y = x^2$.

2. Vertex: $(3, -6)$

$$y = a(x - 3)^2 - 6$$

Point on the graph: $(0, 3)$

$$3 = a(0 - 3)^2 - 6$$
$$9 = 9a \Rightarrow a = 1$$

So, $y = (x - 3)^2 - 6$.

3. (a) $y = -\frac{1}{20}x^2 + 3x + 5$

$= -\frac{1}{20}(x^2 - 60x + 900 - 900) + 5$

$= -\frac{1}{20}\left[(x - 30)^2 - 900\right] + 5$

$= -\frac{1}{20}(x - 30)^2 + 50$

Vertex: $(30, 50)$

The maximum height is 50 feet.

(b) The constant term, $c = 5$, determines the height at which the ball was thrown. Changing this constant results in a vertical translation of the graph, and therefore, changes the maximum height.

4. $h(t) = -\frac{3}{4}t^5 + 2t^2$

The degree is odd and the leading coefficient is negative.

The graph rises to the left and falls to the right.

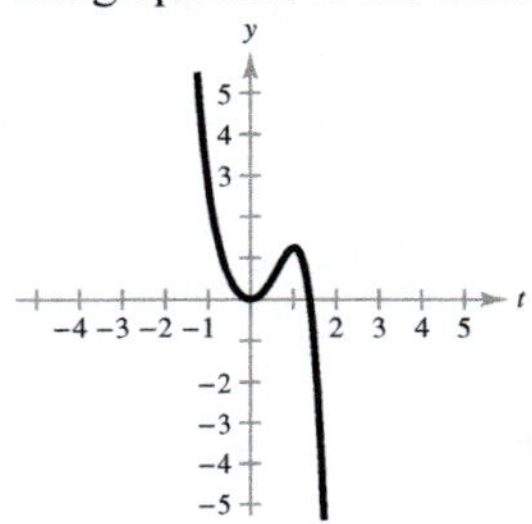

5.
$$\begin{array}{r} 3x + \dfrac{x-1}{x^2+1} \\ x^2 + 0x + 1 \overline{)\,3x^3 + 0x^2 + 4x - 1} \\ \underline{3x^3 + 0x^2 + 3x} \\ x - 1 \end{array}$$

Thus, $\dfrac{3x^3 + 4x - 1}{x^2 + 1} = 3x + \dfrac{x - 1}{x^2 + 1}$.

6.
$$\begin{array}{r|rrrrr} -2 & 2 & 0 & -3 & 4 & -1 \\ & & -4 & 8 & -10 & 12 \\ \hline & 2 & -4 & 5 & -6 & 11 \end{array}$$

So, $\dfrac{2x^4 - 3x^2 + 4x - 1}{x + 2} = 2x^3 - 4x^2 + 5x - 6 + \dfrac{11}{x + 2}$.

7. $f(x) = 2x^3 - 5x^2 - 6x + 15$

$$\begin{array}{r|rrrr} \frac{5}{2} & 2 & -5 & -6 & 15 \\ & & 5 & 0 & -15 \\ \hline & 2 & 0 & -6 & 0 \end{array}$$

$2x^3 - 5x^2 - 6x + 15 = \left(x - \frac{5}{2}\right)(2x^2 - 6)$

$= 2\left(x - \frac{5}{2}\right)(x^2 - 3)$

$= 2\left(x - \frac{5}{2}\right)\left(x + \sqrt{3}\right)\left(x - \sqrt{3}\right)$

Zeros: $x = \frac{5}{2}, x = \pm\sqrt{3}$

8. (a) $\sqrt{-16} - 2(7 + 2i) = 4i - 14 - 4i$

$= -14$

(b) $(5 - i)(3 + 4i) = 15 + 20i - 3i - 4i^2$

$= 15 + 17i - 4(-1)$

$= 19 + 17i$

9. $\dfrac{8}{1 + 2i} = \dfrac{8}{1 + 2i} \cdot \dfrac{1 - 2i}{1 - 2i}$

$= \dfrac{8 - 16i}{1 - 4i^2}$

$= \dfrac{8 - 16i}{1 - 4(-1)}$

$= \dfrac{8 - 16i}{5}$

$= \dfrac{8}{5} - \dfrac{16}{5}i$

10. If $x = 3i$ is a zero, so is $x = -3i$

$$\begin{aligned} f(x) &= x(x - 2)(x - 3i)(x + 3i) \\ &= (x^2 - 2x)(x^2 + 9) \\ &= x^4 - 2x^3 + 9x^2 - 18x \end{aligned}$$

11. If $x = 2 + \sqrt{3}\,i$ is a zero, so is $x = 2 - \sqrt{3}\,i$

$$\begin{aligned} f(x) &= (x - 1)(x - 1)\left[x - \left(2 - \sqrt{3}\,i\right)\right]\left[x - \left(2 + \sqrt{3}\,i\right)\right] \\ &= \left(x^2 - 2x + 1\right)\left[(x - 2) + \sqrt{3}\,i\right]\left[(x - 2) - \sqrt{3}\,i\right] \\ &= \left(x^2 - 2x + 1\right)\left[(x - 2)^2 - \left(\sqrt{3}\,i\right)^2\right] \\ &= \left(x^2 - 2x + 1\right)\left[\left(x^2 - 4x + 4\right) + 3\right] \\ &= \left(x^2 - 2x + 1\right)\left(x^2 - 4x + 7\right) \\ &= x^4 - 6x^3 + 16x^2 - 18x + 7 \end{aligned}$$

12. $f(x) = 3x^3 + 14x^2 - 7x - 10$

Possible rational zeros: $\pm 1, \pm 2, \pm 5, \pm 10, \pm\frac{1}{3}, \pm\frac{2}{3}, \pm\frac{5}{3}, \pm\frac{10}{3}$

$$\begin{array}{r|rrrr} 1 & 3 & 14 & -7 & -10 \\ & & 3 & 17 & 10 \\ \hline & 3 & 17 & 10 & 0 \end{array}$$

$$\begin{aligned} f(x) &= (x - 1)\left(3x^2 + 17x + 10\right) \\ &= (x - 1)(3x + 2)(x + 5) \end{aligned}$$

The zeros of $f(x)$ are $x = -5$, $x = -\frac{2}{3}$, and $x = 1$.

13. $f(x) = x^4 - 9x^2 - 22x - 24$

Possible rational zeros: $\pm 1, \pm 2, \pm 3, \pm 4, \pm 6, \pm 8, \pm 12, \pm 24$

$$\begin{array}{r|rrrrr} -2 & 1 & 0 & -9 & -22 & -24 \\ & & -2 & 4 & 10 & 24 \\ \hline & 1 & -2 & -5 & -12 & 0 \end{array}$$

$$\begin{array}{r|rrrr} 4 & 1 & -2 & -5 & -12 \\ & & 4 & 8 & 12 \\ \hline & 1 & 2 & 3 & 0 \end{array}$$

$$f(x) = (x + 2)(x - 4)\left(x^2 + 2x + 3\right)$$

By the Quadratic Formula the zeros of $x^2 + 2x + 3$ are $x = -1 \pm \sqrt{2}i$. The zeros of f are: $x = -2, 4, -1 \pm \sqrt{2}i$.

14. $h(x) = \dfrac{3}{x^2} - 1 = \dfrac{3 - x^2}{x^2}$

x-intercepts: $(\pm\sqrt{3}, 0)$

Vertical asymptote: $x = 0$

Horizontal asymptote: $y = -1$

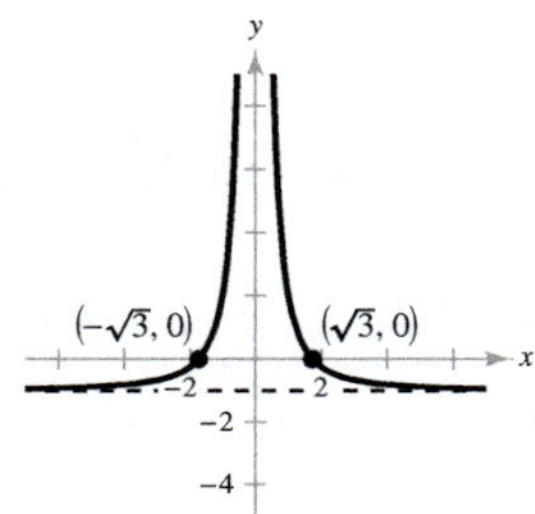

15. $f(x) = \dfrac{2x^2 - 5x - 12}{x^2 - 16}$

$= \dfrac{(2x + 3)(x - 4)}{(x + 4)(x - 4)}$

$= \dfrac{2x + 3}{x + 4}, x \neq 4$

x-intercept: $\left(-\dfrac{3}{2}, 0\right)$

y-intercept: $\left(0, \dfrac{3}{4}\right)$

Vertical asymptote: $x = -4$

Horizontal asymptote: $y = 2$

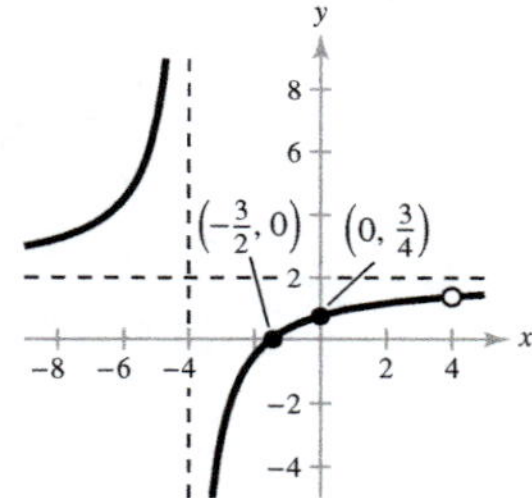

16. $g(x) = \dfrac{x^2 + 2}{x - 1} = x + 1 + \dfrac{3}{x - 1}$

y-intercept: $(0, -2)$

Vertical asymptote: $x = 1$

Slant asymptote: $y = x + 1$

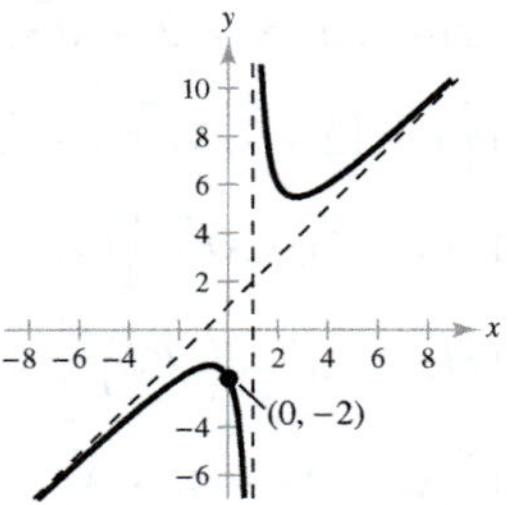

17. $2x^2 + 5x > 12$

$2x^2 + 5x - 12 > 0$

$(2x - 3)(x + 4) > 0$

Critical numbers: $x = \frac{3}{2}, x = -4$

Test intervals: $(-\infty, -4), \left(-4, \frac{3}{2}\right), \left(\frac{3}{2}, \infty\right)$

Test: Is $(2x - 3)(x + 4) > 0$?

Solution set: $(-\infty, -4) \cup \left(\frac{3}{2}, \infty\right)$

In inequality notation: $x < -4$ or $x > \frac{3}{2}$

$\frac{3}{2}$

−5 −4 −3 −2 −1 0 1 2 3 x

18. $\dfrac{2}{x} \leq \dfrac{1}{x + 6}$

$\dfrac{2}{x} - \dfrac{1}{x + 6} \leq 0$

$\dfrac{2(x + 6) - 1(x)}{x(x + 6)} \leq 0$

$\dfrac{x + 12}{x(x + 6)} \leq 0$

Key numbers: $x = 0, -6, -12$

Test intervals: $(-\infty, -12), (-12, -6), (-6, 0), (0, \infty)$

Test: Is $\dfrac{x + 12}{x(x + 6)} \leq 0$?

Solution set: $(-\infty, -12] \cup (-6, 0)$

In inequality notation: $x \leq -12$ or $-6 < x < 0$

−18 −15 −12 −9 −6 −3 0 3 x

Chapter Test Solutions for Chapter 3

1. $0.7^{2.5} \approx 0.410$

2. $3^{-\pi} \approx 0.032$

3. $e^{-7/10} \approx 0.497$

4. $e^{3.1} \approx 22.198$

5. $f(x) = 10^{-x}$

x	-1	$-\frac{1}{2}$	0	$\frac{1}{2}$	1
$f(x)$	10	3.162	1	0.316	0.1

Horizontal asymptote: $y = 0$

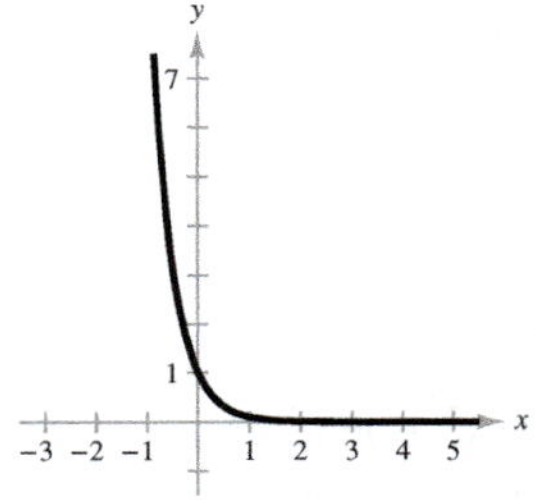

6. $f(x) = -6^{x-2}$

x	-1	0	1	2	3
$f(x)$	-0.005	-0.028	-0.167	-1	-6

Horizontal asymptote: $y = 0$

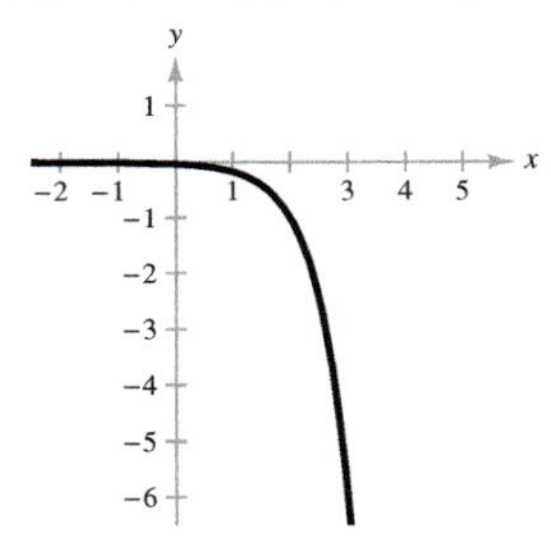

7. $f(x) = 1 - e^{2x}$

x	-1	$-\frac{1}{2}$	0	$\frac{1}{2}$	1
$f(x)$	0.865	0.632	0	-1.718	-6.389

Horizontal asymptote: $y = 1$

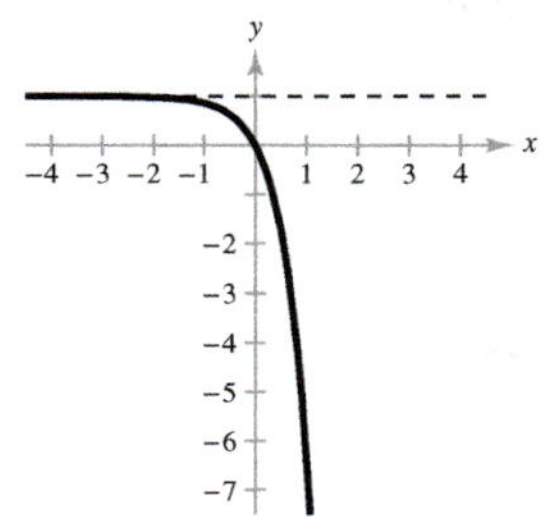

8. (a) $\log_7 7^{-0.89} = -0.89$

(b) $4.6 \ln e^2 = 4.6(2) = 9.2$

9. $f(x) = 4 + \log x$

Domain: $(0, \infty)$

x-intercept:

$$4 + \log x = 0$$
$$\log x = -4$$
$$10^{\log x} = 10^{-4}$$
$$x = 10^{-4}$$
$$(10^{-4}, 0) = (0.0001, 0)$$

Vertical asymptote: $x = 0$

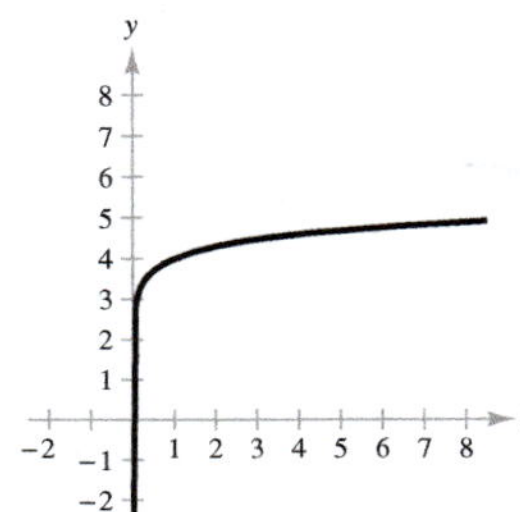

10. $f(x) = \ln(x - 4)$

Domain: $(4, \infty)$

$\ln(x - 4) = 0$

x-intercept: $e^{\ln(x-4)} = e^0$

$x - 4 = 1$

$x = 5$

$(5, 0)$

Vertical asymptote: $x = 4$

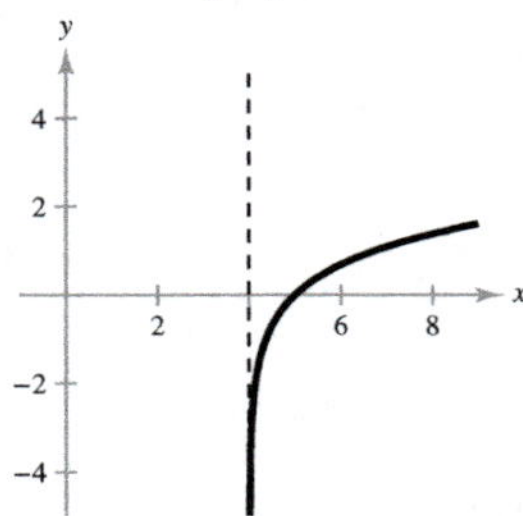

11. $f(x) = 1 + \ln(x + 6)$

Domain: $(-6, \infty)$

x-intercept:

$1 + \ln(x + 6) = 0$

$\ln(x + 6) = -1$

$e^{\ln(x+6)} = e^{-1}$

$x + 6 = e^{-1}$

$x = -6 + e^{-1}$

$(-6 + e^{-1}, 0) \approx (-5.632, 0)$

Vertical asymptote: $x = -6$

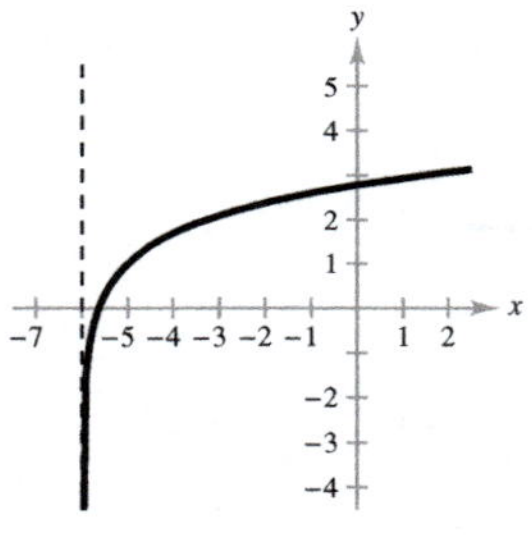

12. $\log_5 35 = \dfrac{\ln 35}{\ln 5} = \dfrac{\log 35}{\log 5} \approx 2.209$

13. $\log_{16} 0.63 = \dfrac{\log 0.63}{\log 16} \approx -0.167$

14. $\log_{3/4} 24 = \dfrac{\log 24}{\log (3/4)} \approx -11.047$

15. $\log_2 3a^4 = \log_2 3 + \log_2 a^4 = \log_2 3 + 4 \log_2|a|$

16. $\ln \dfrac{\sqrt{x}}{7} = \ln(\sqrt{x}) - \ln 7$

$= \ln\sqrt{x} - \ln 7$

$= \ln x^{1/2} - \ln 7$

$= \dfrac{1}{2} \ln x - \ln 7$

17. $\ln \dfrac{10x^2}{y^3} = \ln(10x^2) - \ln y^3$

$= \ln 10 + \ln x^2 - \ln y^3$

$= \ln 10 + 2 \ln x - 3 \ln y$

18. $\log_3 13 + \log_3 y = \log_3 13y$

19. $4 \ln x - 4 \ln y = \ln x^4 - \ln y^4 = \ln \dfrac{x^4}{y^4}$

20. $3 \ln x - \ln(x + 3) + 2 \ln y = \ln x^3 - \ln(x + 3) + \ln y^2 = \ln \dfrac{x^3y^2}{x + 3}$

21. $5^x = \frac{1}{25}$

$5^x = 5^{-2}$

$x = -2$

22. $3e^{-5x} = 132$

$e^{-5x} = 44$

$-5x = \ln 44$

$x = \dfrac{\ln 44}{-5} \approx -0.757$

23. $\dfrac{1025}{8 + e^{4x}} = 5$

$$1025 = 5(8 + e^{4x})$$
$$205 = 8 + e^{4x}$$
$$197 = e^{4x}$$
$$\ln 197 = 4x$$
$$x = \frac{\ln 197}{4} \approx 1.321$$

24. $\ln x = \frac{1}{2}$

$$x = e^{1/2} \approx 1.649$$

25. $18 + 4 \ln x = 7$

$$4 \ln x = -11$$
$$\ln x = -\tfrac{11}{4}$$
$$x = e^{-11/4} \approx 0.0639$$

26. $\log x + \log(x - 15) = 2$

$$\log[x(x - 15)] = 2$$
$$x(x - 15) = 10^2$$
$$x^2 - 15x - 100 = 0$$
$$(x - 20)(x + 5) = 0$$
$$x - 20 = 0 \quad \text{or} \quad x + 5 = 0$$
$$x = 20 \qquad\qquad x = -5$$

The value $x = -5$ is extraneous. The only solution is $x = 20$.

27. $y = ae^{bt}$

$(0, 2745)$: $2745 = ae^{b(0)} \Rightarrow a = 2745$

$$y = 2745e^{bt}$$

$(9, 11{,}277)$: $11{,}277 = 2745e^{b(9)}$

$$\frac{11{,}277}{2745} = e^{9b}$$
$$\ln\left(\frac{11{,}277}{2745}\right) = 9b$$
$$\frac{1}{9}\ln\left(\frac{11{,}277}{2745}\right) = b \Rightarrow b \approx 0.1570$$

So, $y = 2745e^{0.1570t}$.

28. $y = ae^{bt}$

$$\frac{1}{2}a = ae^{b(21.77)}$$
$$\frac{1}{2} = e^{21.77b}$$
$$\ln\left(\frac{1}{2}\right) = 21.77b$$
$$b = \frac{\ln(1/2)}{21.77} \approx -0.0318$$
$$y = ae^{-0.0318t}$$

When $t = 19$: $y = ae^{-0.0318(19)} \approx 0.55a$

So, 55% will remain after 19 years.

29. $H = 70.228 + 5.104x + 9.222 \ln x, \frac{1}{4} \le x \le 6$

(a)

x	H (cm)
$\frac{1}{4}$	58.720
$\frac{1}{2}$	66.388
1	75.332
2	86.828
3	95.671
4	103.43
5	110.59
6	117.38

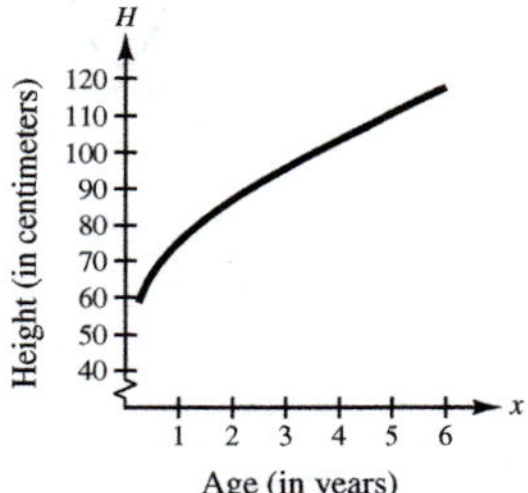

(b) Estimate: 103

When $x = 4$, $H \approx 103.43$ cm.

Chapters 1–3 Cumulative Test Solutions

1.

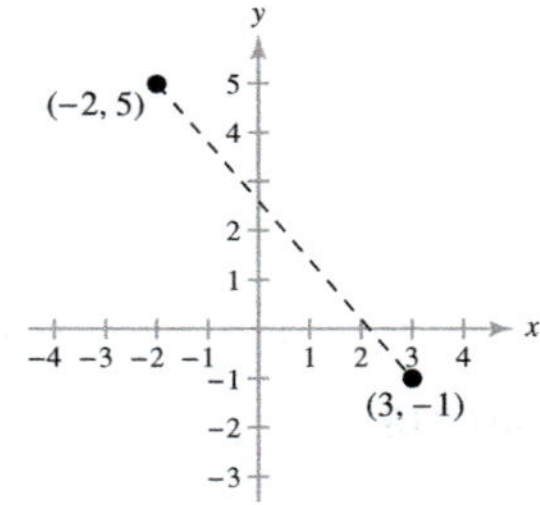

Midpoint: $\left(\dfrac{3+(-2)}{2}, \dfrac{-1+5}{2}\right) = \left(\dfrac{1}{2}, 2\right)$

Distance: $d = \sqrt{(3-(-2))^2 + (-1-5)^2} = \sqrt{(5)^2 + (-6)^2} = \sqrt{25+36} = \sqrt{61}$

2. $x - 3y + 12 = 0$

Line

x-intercept: $(-12, 0)$

y-intercept: $(0, 4)$

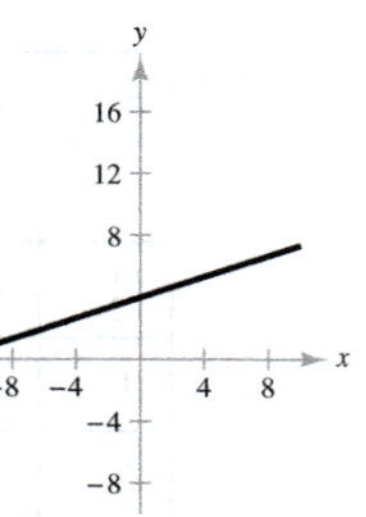

3. $y = x^2 - 9$

Parabola

x-intercepts: $(\pm 3, 0)$

y-intercept: $(0, -9)$

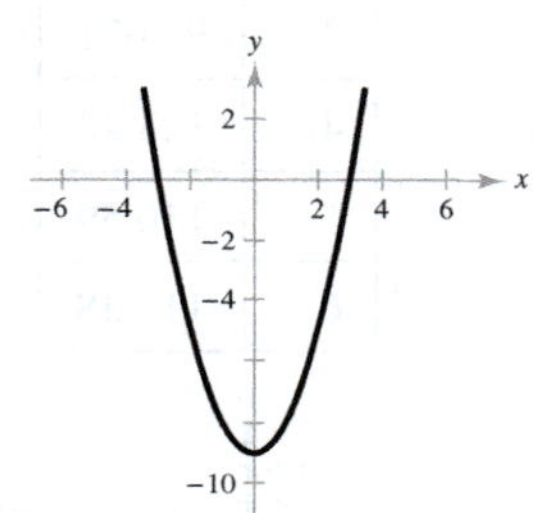

4. $y = \sqrt{4 - x}$

Domain: $x \le 4$

x-intercept: $(4, 0)$

y-intercept: $(0, 2)$

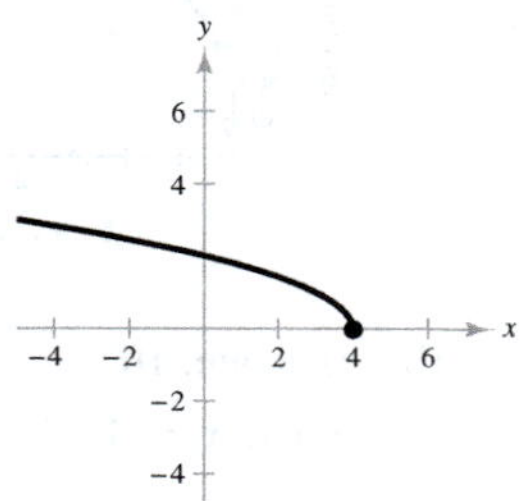

5. $\left(-\dfrac{1}{2}, 1\right)$ and $(3, 8)$

$$m = \frac{8-1}{3-(-1/2)} = \frac{7}{7/2} = 2$$

$$y - 8 = 2(x - 3)$$

$$y - 8 = 2x - 6$$

$$y = 2x + 2$$

6. It fails the Vertical Line Test. For some values of x there correspond two values of y.

7. $f(x) = \dfrac{x}{x-2}$

(a) $f(6) = \dfrac{6}{4} = \dfrac{3}{2}$

(b) $f(2)$ is undefined because division by zero is undefined.

(c) $f(s+2) = \dfrac{s+2}{(s+2)-2} = \dfrac{s+2}{s}$

8. $y = \sqrt[3]{x}$

(a) $r(x) = \frac{1}{2}\sqrt[3]{x}$ is a vertical shrink by a factor of $\frac{1}{2}$.

(b) $h(x) = \sqrt[3]{x} + 2$ is a vertical shift two units upward.

(c) $g(x) = \sqrt[3]{x+2}$ is a horizontal shift two units to the left.

9. $f(x) = x - 4$, $g(x) = 3x + 1$

(a) $(f+g)(x) = f(x) + g(x)$
$= (x-4) + (3x+1)$
$= 4x - 3$

(b) $(f-g)(x) = f(x) - g(x)$
$= (x-4) - (3x+1)$
$= -2x - 5$

(c) $(fg)(x) = f(x)g(x)$
$= (x-4)(3x+1)$
$= 3x^2 - 11x - 4$

(d) $\left(\dfrac{f}{g}\right)(x) = \dfrac{f(x)}{g(x)} = \dfrac{x-4}{3x+1}$

Domain: All real numbers x except $x = -\dfrac{1}{3}$

10. $f(x) = \sqrt{x-1},\ g(x) = x^2 + 1$

(a) $(f+g)(x) = f(x) + g(x)$
$= \sqrt{x-1} + x^2 + 1$

(b) $(f-g)(x) = f(x) - g(x)$
$= \sqrt{x-1} - x^2 - 1$

(c) $(fg)(x) = f(x)g(x)$
$= \sqrt{x-1}\left(x^2+1\right)$
$= x^2\sqrt{x-1} + \sqrt{x-1}$

(d) $\left(\dfrac{f}{g}\right)(x) = \dfrac{f(x)}{g(x)} = \dfrac{\sqrt{x-1}}{x^2+1}$

Domain: all real numbers x such that $x \geq 1$

11. $f(x) = 2x^2,\ g(x) = \sqrt{x+6}$

(a) $(f \circ g)(x) = f(g(x))$
$= f\left(\sqrt{x+6}\right)$
$= 2\left(\sqrt{x+6}\right)^2$
$= 2(x+6)$
$= 2x + 12$

Domain: all real numbers x such that $x \geq -6$

(b) $(g \circ f)(x) = g(f(x))$
$= g\left(2x^2\right)$
$= \sqrt{2x^2 + 6}$

Domain: all real numbers x

12. $f(x) = x - 2,\ g(x) = |x|$

(a) $(f \circ g)(x) = f(g(x))$
$= f(|x|)$
$= |x| - 2$

Domain: all real numbers x

(b) $(g \circ f)(x) = g(f(x))$
$= g(x-2)$
$= |x-2|$

Domain: all real numbers x

13. $h(x) = 3x - 4$

Because h is one-to-one, h has an inverse.

$$y = 3x - 4$$
$$x = 3y - 4$$
$$x + 4 = 3y$$
$$\tfrac{1}{3}(x+4) = y$$
$$h^{-1}(x) = \tfrac{1}{3}(x+4)$$

14. $P = kS^3$

$$750 = k(27)^3 \Rightarrow k = \frac{750}{(27)^3} = \frac{250}{6561}$$

$$P = \frac{250}{6561}S^3$$

When $S = 40$:

$$P = \left(\frac{250}{6561}\right)(40)^3 \approx 2438.65 \text{ kilowatts}$$

15. Vertex: $(-8, 5)$

Point: $(-4, -7)$

$$y - k = a(x-h)^2$$
$$y - 5 = a(x+8)^2$$
$$-7 - 5 = a(-4+8)^2$$
$$-12 = 16a$$
$$-\tfrac{3}{4} = a$$
$$y = -\tfrac{3}{4}(x+8)^2 + 5$$

16. $h(x) = -x^2 + 10x - 21$
$= -\left(x^2 - 10x\right) - 21$
$= -\left(x^2 - 10x + 25 - 25\right) - 21$
$= -\left(x^2 - 10x + 25\right) + 25 - 21$
$= -(x-5)^2 + 4$

Parabola

Vertex: $(5, 4)$

$$-(x-5)^2 + 4 = 0$$
$$4 = (x-5)^2$$
$$\pm 2 = x - 5$$
$$x = 7, 3$$

Intercepts: $(3, 0), (7, 0)$

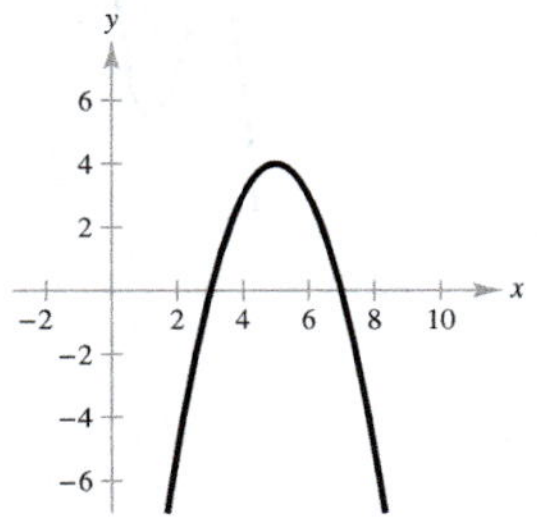

17. $f(t) = -\frac{1}{2}(t-1)^2(t+2)^2$

4th degree polynomial,

Falls to the left

Falls to the right

$$-\frac{1}{2}(t-1)^2(t+2)^2 = 0$$

$$(t-1)^2 = 0 \rightarrow t = 1$$

$$(t+2)^2 = 0 \rightarrow t = -2$$

zeros are of even multiplicity

x-intercepts: $(1, 0), (-2, 0)$

$$f(0) = -\frac{1}{2}(0-1)^2(0+2)^2$$

$$= -\frac{1}{2}(-1)^2(2)^2 = -2$$

y-intercept: $(0, -2)$

18. $g(s) = s^3 - 3s^2$

Cubic

Falls to the left,

Rises to the right

$$s^3 - 3x^2 = 0$$

$$s^2(s-3) = 0$$

$$s = 0 \text{ (even multiplicity)}$$

$$s - 3 = 0 \rightarrow s = 3 \text{ (odd multiplicity)}$$

Intercepts: $(0, 0), (3, 0)$

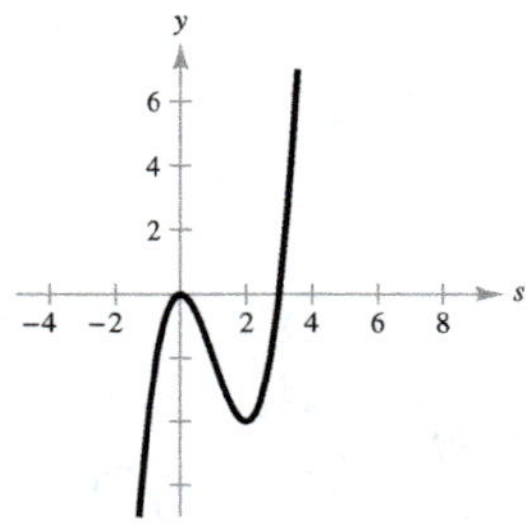

19. $f(x) = x^3 + 2x^2 + 4x + 8$

$$= x^2(x+2) + 4(x+2)$$

$$= (x+2)(x^2+4)$$

$$x + 2 = 0 \Rightarrow x = -2$$

$$x^2 + 4 = 0 \Rightarrow x = \pm 2i$$

The zeros of $f(x)$ are -2 and $\pm 2i$.

20. $f(x) = x^4 + 4x^3 - 21x^2$

$$= x^2(x^2 + 4x - 21)$$

$$= x^2(x+7)(x-3)$$

The zeros of $f(x)$ are $0, -7,$ and 3.

21. $f(x) = 2x^4 - 11x^3 + 30x^2 - 62x - 40$

Possible Rational Zeros:

$\pm 1, \pm 2, \pm 4, \pm 5, \pm 8, \pm 10, \pm 20, \pm 40, \pm \frac{1}{2}, \pm \frac{5}{2}$

By testing (or by looking at the graph of $f(x)$), you see that $x = 4$ and $x = -\frac{1}{2}$ are zeros.

4	2	–11	30	–62	–40
		8	–12	72	40
	2	–3	18	10	0

$-\frac{1}{2}$	2	–3	18	10
		–1	2	–10
	2	–4	20	0

$$f(x) = (x-4)\left(x+\tfrac{1}{2}\right)(2x^2 - 4x + 20)$$

$$= (x-4)\left(x+\tfrac{1}{2}\right)(2)(x^2 - 2x + 10)$$

$$= (x-4)(2x+1)(x^2 - 2x + 10)$$

By Completing the Square (or by the Quadratic Formula), the zeros of $x^2 - 2x + 10$ are $1 \pm 3i$.

$$f(x) = (x-4)(2x+1)(x-1-3i)(x-1+3i)$$

Zeros of $f(x)$: $4, -\frac{1}{2}, 1+3i, 1-3i$

22.

$$\begin{array}{r} 3x - 2 + \dfrac{-3x+2}{2x^2+1} \\ 2x^2 + 0x + 1 \overline{\smash{\big)}\, 6x^3 - 4x^2 + 0x + 0} \\ \underline{6x^3 + 0x^2 + 3x} \\ -4x^2 - 3x + 0 \\ \underline{-4x^2 + 0x - 2} \\ -3x + 2 \end{array}$$

Thus, $\dfrac{6x^3 - 4x^2}{2x^2+1} = 3x - 2 - \dfrac{3x-2}{2x^2+1}$.

23.

2	3	0	2	−5	3
		6	12	28	46
	3	6	14	23	49

Thus,

$$\frac{3x^4 + 2x^2 - 5x + 3}{x - 2} = 3x^3 + 6x^2 + 14x + 23 + \frac{49}{x - 2}.$$

24. $g(x) = x^3 + 3x^2 - 6$

From the graph, you can see that $g(x)$ has one real zero. It is between 1 and 2 because $g(1)$ is negative and $g(2)$ is positive. The zero is $x \approx 1.20$.

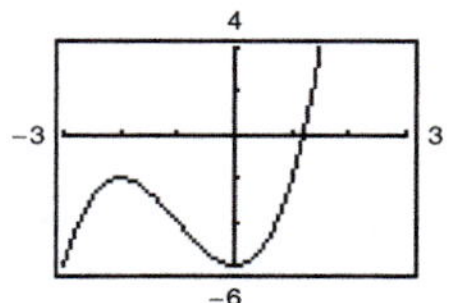

25. $f(x) = \dfrac{2x}{x^2 + 2x - 3}$

$= \dfrac{2x}{(x + 3)(x - 1)}$

Intercept: $(0, 0)$

Vertical asymptotes: $x = -3, x = 1$

Horizontal asymptote: $y = 0$

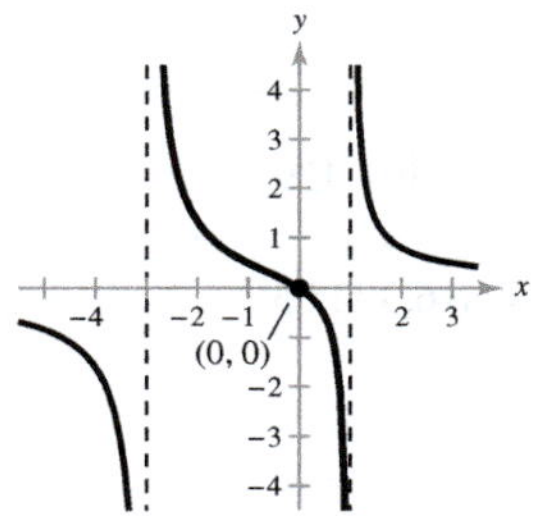

26. $f(x) = \dfrac{x^2 - 4}{x^2 + x - 2}$

$= \dfrac{(x + 2)(x - 2)}{(x + 2)(x - 1)}$

$= \dfrac{x - 2}{x - 1}, x \neq -2$

Vertical asymptote: $x = 1$

Horizontal asymptote: $y = 1$

x-intercept: $(2, 0)$

y-intercept: $(0, 2)$

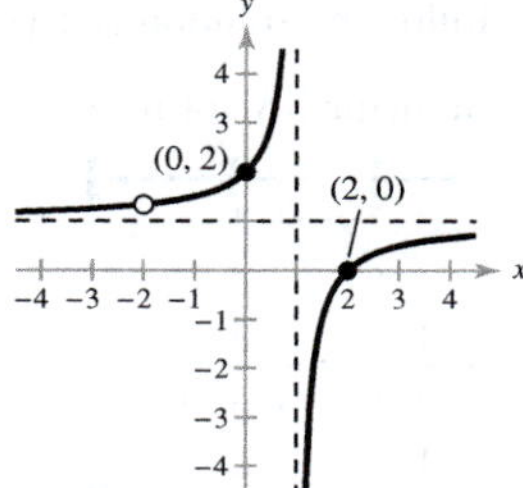

27. $f(x) = \dfrac{x^3 - 2x^2 - 9x + 18}{x^2 + 4x + 3}$

$= \dfrac{x^2(x - 2) - 9(x - 2)}{(x + 1)(x + 3)}$

$= \dfrac{(x - 2)(x^2 - 9)}{(x + 1)(x + 3)}$

$= \dfrac{(x - 2)(x + 3)(x - 3)}{(x + 1)(x + 3)}$

$= \dfrac{(x - 2)(x - 3)}{x + 1}$

$= \dfrac{x^2 - 5x + 6}{x + 1}$

$= x - 6 + \dfrac{12}{x + 1}, x \neq -3$

Vertical asymptote: $x = -1$

Slant asymptote: $y = x - 6$

x-intercepts: $(2, 0), (3, 0)$

y-intercept: $(0, 6)$

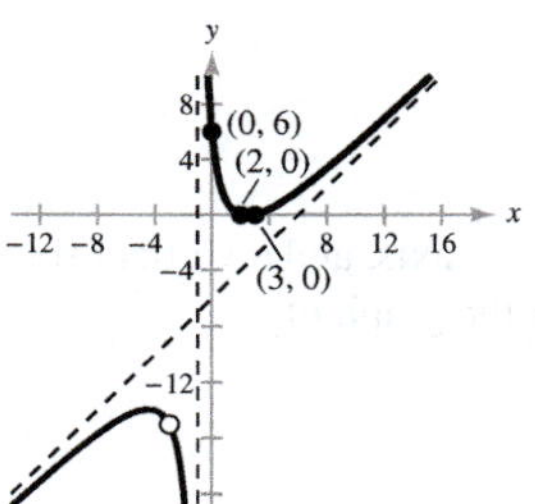

28. $$2x^3 - 18x \le 0$$
$$2x(x^2 - 9) \le 0$$
$$2x(x + 3)(x - 3) \le 0$$

Key numbers: $x = -3, x = 0, x = 3$

Test intervals: $(-\infty, -3), (-3, 0), (0, 3), (3, \infty)$

Test: Is $2x(x + 3)(x - 3) \le 0$?

By testing a value in each interval, you have the following solution set: $(-\infty, -3) \cup (0, 3)$.

In inequality form, $x \le -3$ or $0 \le x \le 3$.

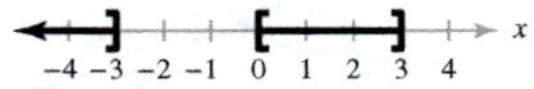

29. $$\frac{1}{x+1} \ge \frac{1}{x+5}$$
$$\frac{1}{x+1} - \frac{1}{x+5} \ge 0$$
$$\frac{4}{(x+1)(x+5)} \ge 0$$

Key numbers: $x = -1, x = -5$

Test intervals: $(-\infty, -5), (-5, -1), (-1, \infty)$

Test: Is $\frac{4}{(x+1)(x+5)} \ge 0$?

By testing a value in each interval, you have the following solution set: $(-\infty, -5) \cup (-1, \infty)$.

In inequality form, $x < -5$ or $x > -1$.

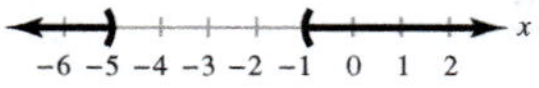

30. $f(x) = \left(\frac{2}{5}\right)^x$

$g(x) = -\left(\frac{2}{5}\right)^{-x+3}$

g is a reflection in the x-axis, a reflection in the y-axis, and a horizontal shift three units to the right of the graph of f.

31. $f(x) = 2.2^x$

$g(x) = -2.2^x + 4$

g is a reflection in the x-axis, and a vertical shift four units upward of the graph of f.

32. $\log 98 \approx 1.991$

33. $\log\left(\frac{6}{7}\right) \approx -0.067$

34. $\ln\sqrt{31} \approx 1.717$

35. $\ln\left(\sqrt{30} - 4\right) \approx 0.390$

36. $$\ln\left(\frac{x^2 - 25}{x^4}\right) = \ln(x^2 - 25) - \ln x^4$$
$$= \ln[(x + 5)(x - 5)] - \ln x^4$$
$$= \ln(x + 5) + \ln(x - 5) - 4\ln x, \ x > 5$$

37. $$2\ln x - \frac{1}{2}\ln(x + 5) = \ln x^2 - \ln\sqrt{x + 5}$$
$$= \ln\frac{x^2}{\sqrt{x + 5}}, \ x > 0$$

38. $$6e^{2x} = 72$$
$$e^{2x} = 12$$
$$2x = \ln 12$$
$$x = \frac{\ln 12}{2} \approx 1.242$$

39. $$e^{2x} - 13e^x + 42 = 0$$
$$(e^x - 6)(e^x - 7) = 0$$
$$e^x - 6 = 0 \Rightarrow e^x = 6 \Rightarrow x = \ln 6 \approx 1.792$$
$$e^x - 7 = 0 \Rightarrow e^x = 7 \Rightarrow x = \ln 7 \approx 1.946$$

40. $$\ln\sqrt{x + 2} = 3$$
$$\tfrac{1}{2}\ln(x + 2) = 3$$
$$\ln(x + 2) = 6$$
$$x + 2 = e^6$$
$$x = e^6 - 2 \approx 401.429$$

41. $A = 2500e^{(0.075)(25)} \approx \$16{,}302.05$

42. $$N = 175e^{kt}$$
$$420 = 175e^{k(8)}$$
$$2.4 = e^{8k}$$
$$\ln 2.4 = 8k$$
$$\frac{\ln 2.4}{8} = k$$
$$k \approx 0.1094$$
$$N = 175e^{0.1094t}$$
$$350 = 175e^{0.1094t}$$
$$2 = e^{0.1094t}$$
$$\ln 2 = 0.1094t$$
$$t = \frac{\ln 2}{0.1094} \approx 6.3 \text{ hours to double}$$

43. Let $P = 32$ and solve for t.

$$32 = 20.913e^{0.0184t}$$
$$\frac{32}{20.913} = e^{0.0184t}$$
$$1.530 \approx e^{0.0184t}$$
$$\ln 1.530 \approx \ln e^{0.0184t}$$
$$0.4253 \approx 0.0184t$$
$$23.1 \approx t$$

According to the model, the population of Texas will reach 32 million during 2023.

Chapter Test Solutions for Chapter 4

1. $\theta = \dfrac{5\pi}{4}$

(a)

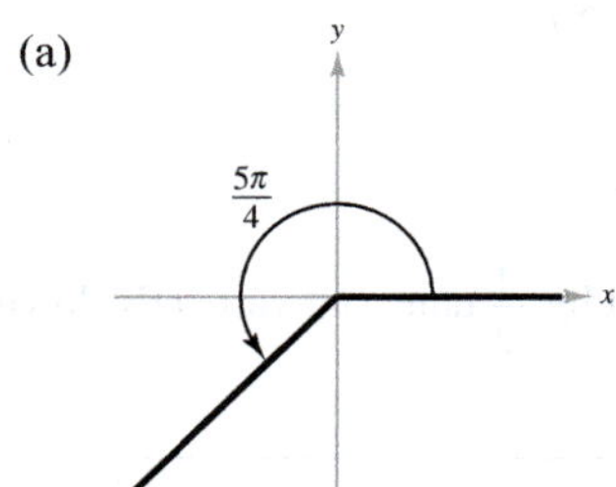

(b) $\dfrac{5\pi}{4} + 2\pi = \dfrac{13\pi}{4}$

$\dfrac{5\pi}{4} - 2\pi = -\dfrac{3\pi}{4}$

(c) $\dfrac{5\pi}{4}\left(\dfrac{180^\circ}{\pi}\right) = 225^\circ$

2. $\dfrac{105 \text{ km}}{\text{hr}} \times \dfrac{1 \text{ hr}}{60 \text{ min}} = 1.75 \text{ km per min}$

diameter $= 1$ meter $= 0.001$ km

radius $= \dfrac{1}{2}$ diameter $= 0.0005$ km

$$\text{Angular speed} = \frac{\theta}{t}$$
$$= \frac{1.75}{2\pi(0.0005)} \cdot 2\pi$$
$$= 3500 \text{ radians per minute}$$

3. $130^\circ = \dfrac{130\pi}{180} = \dfrac{13\pi}{18}$ radians

$$A = \frac{1}{2}r^2\theta = \frac{1}{2}(25)^2\left(\frac{13\pi}{18}\right) \approx 709.04 \text{ square feet}$$

4. Given $\tan\theta = \dfrac{3}{2}$.

$$\text{hyp} = \sqrt{3^2 + 2^2} = \sqrt{13}$$
$$\sin\theta = \frac{\text{opp}}{\text{hyp}} = \frac{3}{\sqrt{13}} = \frac{3\sqrt{13}}{13}$$
$$\cos\theta = \frac{\text{adj}}{\text{hyp}} = \frac{2}{\sqrt{13}} = \frac{2\sqrt{13}}{13}$$
$$\cot\theta = \frac{\text{adj}}{\text{opp}} = \frac{2}{3}$$
$$\sec\theta = \frac{\text{hyp}}{\text{adj}} = \frac{\sqrt{13}}{2}$$
$$\csc\theta = \frac{\text{hyp}}{\text{opp}} = \frac{\sqrt{13}}{3}$$

5. $x = -2, y = 6$

$$r = \sqrt{(-2)^2 + (6)^2} = 2\sqrt{10}$$
$$\sin\theta = \frac{y}{r} = \frac{6}{2\sqrt{10}} = \frac{3}{\sqrt{10}} = \frac{3\sqrt{10}}{10}$$
$$\cos\theta = \frac{x}{r} = \frac{-2}{2\sqrt{10}} = -\frac{1}{\sqrt{10}} = -\frac{\sqrt{10}}{10}$$
$$\tan\theta = \frac{y}{x} = \frac{6}{-2} = -3$$
$$\csc\theta = \frac{r}{y} = \frac{2\sqrt{10}}{6} = \frac{\sqrt{10}}{3}$$
$$\sec\theta = \frac{r}{x} = \frac{2\sqrt{10}}{-2} = -\sqrt{10}$$
$$\cot\theta = \frac{x}{y} = \frac{-2}{6} = -\frac{1}{3}$$

6. $\theta = 205°$

$\theta' = 205° - 180° = 25°$

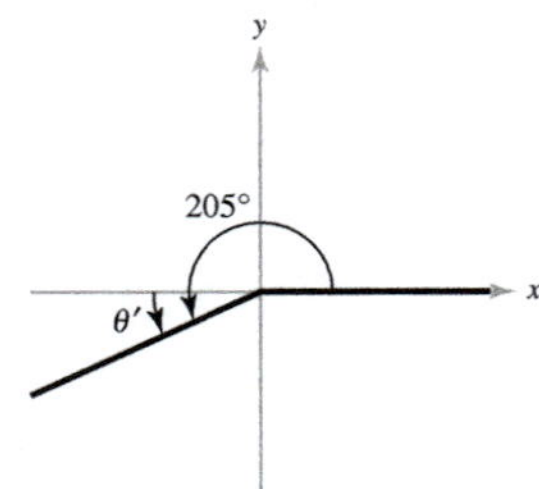

7. $\sec\theta < 0$ and $\tan\theta > 0$

$\dfrac{r}{x} < 0$ and $\dfrac{y}{x} > 0$

Quadrant III

8. $\cos\theta = -\dfrac{\sqrt{3}}{2}$

Reference angle is 30° and θ is in Quadrant II or III.

$\theta = 150°$ or $210°$

9. $\cos\theta = \frac{3}{5}$, $\tan\theta < 0 \Rightarrow \theta$ lies in Quadrant IV.

Let $x = 3, r = 5 \Rightarrow y = -4$.

$\sin\theta = -\frac{4}{5}$

$\cos\theta = \frac{3}{5}$

$\tan\theta = -\frac{4}{3}$

$\csc\theta = -\frac{5}{4}$

$\sec\theta = \frac{5}{3}$

$\cot\theta = -\frac{3}{4}$

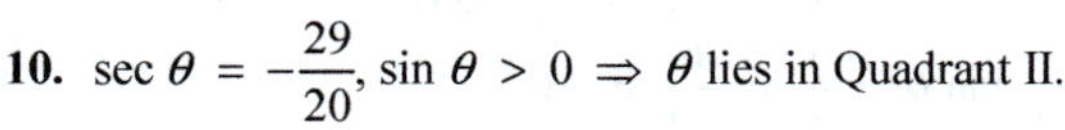

10. $\sec\theta = -\dfrac{29}{20}$, $\sin\theta > 0 \Rightarrow \theta$ lies in Quadrant II.

Let $r = 29, x = -20 \Rightarrow y = 21$.

$\sin\theta = \dfrac{21}{29}$

$\cos\theta = -\dfrac{20}{29}$

$\tan\theta = -\dfrac{21}{20}$

$\csc\theta = \dfrac{29}{21}$

$\cot\theta = -\dfrac{20}{21}$

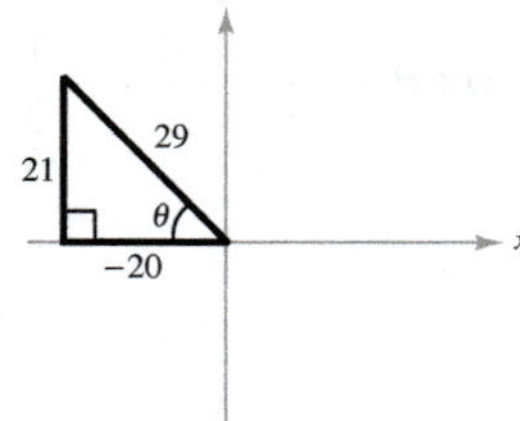

11. $g(x) = -2\sin\left(x - \dfrac{\pi}{4}\right)$

Period: 2π

Amplitude: $|-2| = 2$

Shifted to the right by $\dfrac{\pi}{4}$ units and reflected in the x-axis.

x	0	$\frac{\pi}{4}$	$\frac{3\pi}{4}$	$\frac{5\pi}{4}$	$\frac{7\pi}{4}$
y	$\sqrt{2}$	0	−2	0	2

12. $f(t) = \cos\left(t + \dfrac{\pi}{2}\right) - 1$

Period: 2π

Amplitude: $|1| = 1$

Shifted to the left by $\dfrac{\pi}{2}$ units and vertically down one unit.

t	$-\frac{\pi}{2}$	0	$\frac{\pi}{2}$	π	$\frac{3\pi}{2}$
y	0	−1	−2	−1	0

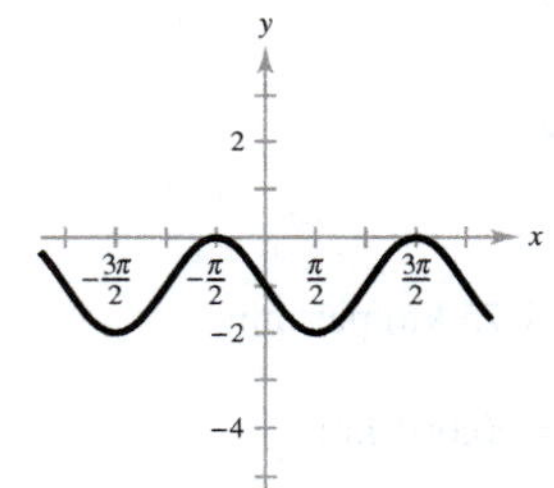

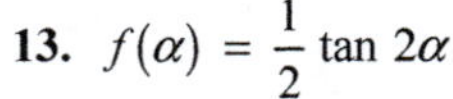

13. $f(\alpha) = \dfrac{1}{2}\tan 2\alpha$

Period: $\dfrac{\pi}{2}$

Asymptotes:

$x = -\dfrac{\pi}{4}, x = \dfrac{\pi}{4}$

α	$-\frac{\pi}{8}$	0	$\frac{\pi}{8}$
$f(\alpha)$	$-\frac{1}{2}$	0	$\frac{1}{2}$

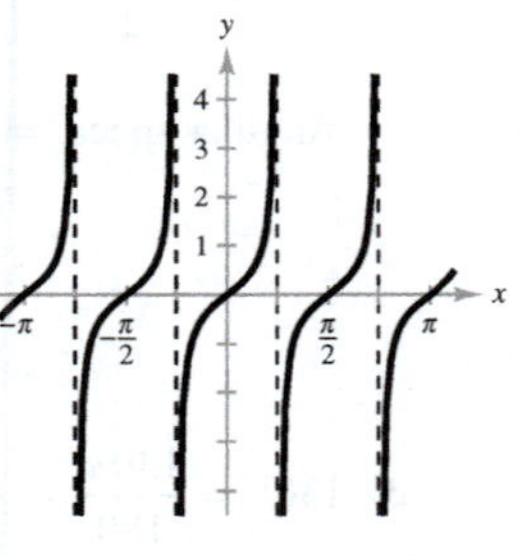

14. $y = \sin 2\pi x + 2\cos \pi x$

Periodic: period = 2

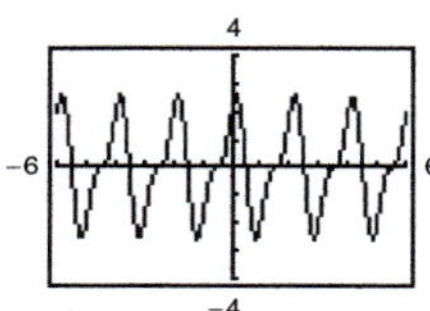

15. $y = 6e^{-0.12x}\cos(0.25x),\ 0 \le x \le 32$

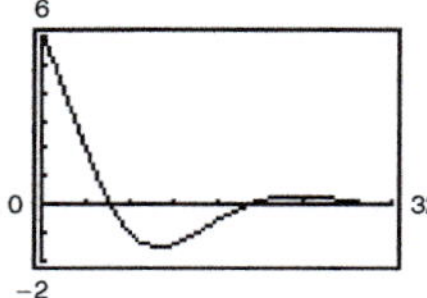

Not periodic

As $x \to \infty$, $y \to 0$.

16. $f(x) = a\sin(bx + c)$

Amplitude: $2 \Rightarrow |a| = 2$

Reflected in the x-axis: $a = -2$

Period: $4\pi = \dfrac{2\pi}{b} \Rightarrow b = \dfrac{1}{2}$

Phase shift: $\dfrac{c}{b} = -\dfrac{\pi}{2} \Rightarrow c = -\dfrac{\pi}{4}$

$$f(x) = -2\sin\left(\frac{x}{2} - \frac{\pi}{4}\right)$$

17. $\cot\left(\arcsin \dfrac{3}{8}\right)$

Let $y = \arcsin \dfrac{3}{8}$. Then $\sin y = \dfrac{3}{8}$ and

$$\cot\left(\arcsin \frac{3}{8}\right) = \cot y = \frac{\sqrt{55}}{3}.$$

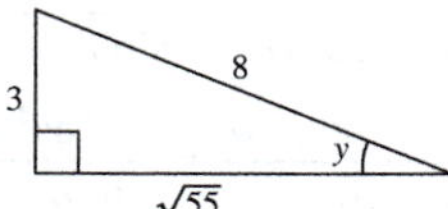

18. $f(x) = 2\arcsin\left(\frac{1}{2}x\right)$

Domain: $[-2, 2]$

Range: $[-\pi, \pi]$

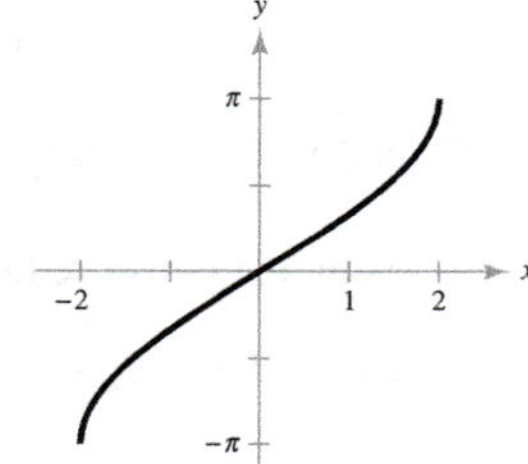

19. $\tan\theta = -\dfrac{110}{90}$

$$\theta = \arctan\left(-\frac{110}{90}\right)$$

$$\theta \approx -50.7$$

$$\theta \approx 309.3°$$

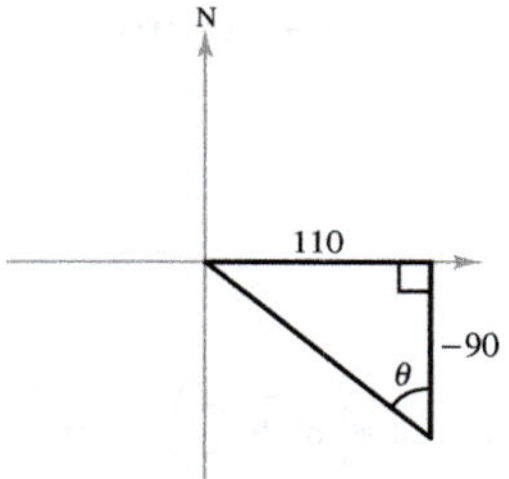

20. $d = a\cos bt$

$a = -6$

$\dfrac{2\pi}{b} = 2 \Rightarrow b = \pi$

$d = -6\cos \pi t$

Chapter Test Solutions for Chapter 5

1. $\csc\theta = \dfrac{5}{2}$, $\tan\theta < 0$

θ is in Quadrant II.

$$\sin\theta = \frac{1}{\csc\theta} = \frac{1}{\frac{5}{2}} = \frac{2}{5}$$

$$\cos\theta = -\sqrt{1 - \sin^2\theta} = -\sqrt{1 - \left(\frac{2}{5}\right)^2} = -\frac{\sqrt{21}}{5}$$

$$\sec\theta = \frac{1}{\cos\theta} = -\frac{5}{\sqrt{21}} = -\frac{5\sqrt{21}}{21}$$

$$\tan\theta = \frac{\sin\theta}{\cos\theta} = \frac{\frac{2}{5}}{-\frac{\sqrt{21}}{5}} = -\frac{2}{\sqrt{21}} = -\frac{2\sqrt{21}}{21}$$

$$\cot\theta = \frac{1}{\tan\theta} = -\frac{\sqrt{21}}{2}$$

2. $\csc^2\beta(1 - \cos^2\beta) = \dfrac{1}{\sin^2\beta}(\sin^2\beta) = 1$

3.
$$\frac{\sec^4 x - \tan^4 x}{\sec^2 x + \tan^2 x} = \frac{(\sec^2 x + \tan^2 x)(\sec^2 x - \tan^2 x)}{\sec^2 x + \tan^2 x}$$
$$= \sec^2 x - \tan^2 x = 1$$

4.
$$\frac{\cos\theta}{\sin\theta} + \frac{\sin\theta}{\cos\theta} = \frac{\cos^2\theta + \sin^2\theta}{\sin\theta\cos\theta} = \frac{1}{\sin\theta\cos\theta}$$
$$= \csc\theta\sec\theta$$

5. $\sin\theta\sec\theta = \sin\theta\dfrac{1}{\cos\theta} = \dfrac{\sin\theta}{\cos\theta} = \tan\theta$

6. $\sec^2 x\tan^2 x + \sec^2 x = \sec^2 x\left(\sec^2 x - 1\right) + \sec^2 x = \sec^4 x - \sec^2 x + \sec^2 x = \sec^4 x$

7.
$$\frac{\csc\alpha + \sec\alpha}{\sin\alpha + \cos\alpha} = \frac{\dfrac{1}{\sin\alpha} + \dfrac{1}{\cos\alpha}}{\sin\alpha + \cos\alpha} = \frac{\dfrac{\cos\alpha + \sin\alpha}{\sin\alpha\cos\alpha}}{\sin\alpha + \cos\alpha} = \frac{1}{\sin\alpha\cos\alpha}$$
$$= \frac{\cos^2\alpha + \sin^2\alpha}{\sin\alpha\cos\alpha} = \frac{\cos^2\alpha}{\sin\alpha\cos\alpha} + \frac{\sin^2\alpha}{\sin\alpha\cos\alpha}$$
$$= \frac{\cos\alpha}{\sin\alpha} + \frac{\sin\alpha}{\cos\alpha} = \cot\alpha + \tan\alpha$$

8. $\tan\left(x + \dfrac{\pi}{2}\right) = \tan\left(\dfrac{\pi}{2} - (-x)\right) = \cot(-x) = -\cot x$

9. Using the power reducing formula for cosine,

$$2\cos^2 5y = 2\left(\frac{1 + \cos(2\cdot 5y)}{2}\right)$$
$$= 1 + \cos 10y.$$

So, $1 + \cos 10y = 2\cos^2 5y$.

10. Using the double angle formula for sines,

$$\frac{1}{2}\sin\left(2\cdot\frac{\alpha}{3}\right) = \frac{1}{2}\cdot 2\sin\frac{\alpha}{3}\cos\frac{\alpha}{3}$$
$$= \sin\frac{\alpha}{3}\cos\frac{\alpha}{3}.$$

So, $\sin\dfrac{\alpha}{3}\cos\dfrac{\alpha}{3} = \dfrac{1}{2}\sin\dfrac{2\alpha}{3}$.

11.
$$4\sin 3\theta\cos 2\theta = 4\cdot\tfrac{1}{2}\left[\sin(3\theta + 2\theta) + \sin(3\theta - 2\theta)\right]$$
$$= 2(\sin 5\theta + \sin\theta)$$

12.
$$\cos\left(\theta + \frac{\pi}{2}\right) - \cos\left(\theta - \frac{\pi}{2}\right) = \cos\theta\cos\frac{\pi}{2} - \sin\theta\sin\frac{\pi}{2} - \left(\cos\theta\cos\frac{\pi}{2} + \sin\theta\sin\frac{\pi}{2}\right)$$
$$= \cos\theta(0) - \sin\theta(1) - \cos\theta(0) - \sin\theta(1)$$
$$= -2\sin\theta$$

13.
$$\tan^2 x + \tan x = 0$$
$$\tan x\left(\tan x + 1\right) = 0$$

$\tan x = 0$ or $\tan x + 1 = 0$

$x = 0, \pi$ $\quad\quad \tan x = -1$

$\quad\quad x = \dfrac{3\pi}{4}, \dfrac{7\pi}{4}$

14.
$$\sin 2\alpha - \cos\alpha = 0$$
$$2\sin\alpha\cos\alpha - \cos\alpha = 0$$
$$\cos\alpha(2\sin\alpha - 1) = 0$$

$\cos\alpha = 0$ or $2\sin\alpha - 1 = 0$

$\alpha = \dfrac{\pi}{2}, \dfrac{3\pi}{2}$ $\quad\quad \sin\alpha = \dfrac{1}{2}$

$\quad\quad \alpha = \dfrac{\pi}{6}, \dfrac{5\pi}{6}$

15.
$$4\cos^2 x - 3 = 0$$
$$\cos^2 x = \frac{3}{4}$$
$$\cos x = \pm\sqrt{\frac{3}{4}} = \pm\frac{\sqrt{3}}{2}$$
$$x = \frac{\pi}{6}, \frac{5\pi}{6}, \frac{7\pi}{6}, \frac{11\pi}{6}$$

16.
$$\csc^2 x - \csc x - 2 = 0$$
$$(\csc x - 2)(\csc x + 1) = 0$$

$\csc x - 2 = 0$ or $\csc x + 1 = 0$

$\csc x = 2$ $\quad\quad \csc x = -1$

$\dfrac{1}{\sin x} = 2$ $\quad\quad \dfrac{1}{\sin x} = -1$

$\sin x = \dfrac{1}{2}$ $\quad\quad \sin x = -1$

$x = \dfrac{\pi}{6}, \dfrac{5\pi}{6}$ $\quad\quad x = \dfrac{3\pi}{2}$

17. $5 \sin x - x = 0$

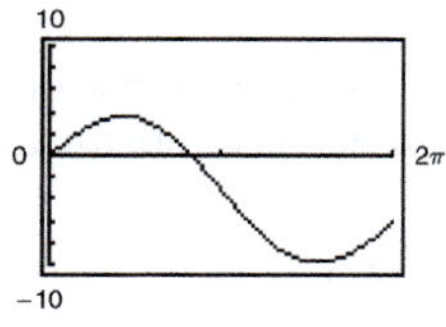

$x \approx 0, 2.596$

18.
$$\begin{aligned} 105° &= 135° - 30° \\ \cos 105° &= \cos(135° - 30°) \\ &= \cos 135° \cos 30° + \sin 135° \sin 30° \\ &= -\cos 45° \cos 30° + \sin 45° \sin 30° \\ &= \left(-\frac{\sqrt{2}}{2}\right)\left(\frac{\sqrt{3}}{2}\right) + \left(\frac{\sqrt{2}}{2}\right)\left(\frac{1}{2}\right) \\ &= \frac{-\sqrt{6} + \sqrt{2}}{4} = \frac{\sqrt{2} - \sqrt{6}}{4} \end{aligned}$$

19. $x = 2, y = -5, r = \sqrt{29}$

$$\sin 2u = 2 \sin u \cos u = 2\left(-\frac{5}{\sqrt{29}}\right)\left(\frac{2}{\sqrt{29}}\right) = -\frac{20}{29}$$

$$\cos 2u = \cos^2 u - \sin^2 u = \left(\frac{2}{\sqrt{29}}\right)^2 - \left(-\frac{5}{\sqrt{29}}\right)^2 = -\frac{21}{29}$$

$$\tan 2u = \frac{2 \tan u}{1 - \tan^2 u} = \frac{2\left(-\frac{5}{2}\right)}{1 - \left(-\frac{5}{2}\right)^2} = \frac{20}{21}$$

20. Let $y_1 = 2.914 \sin(0.017t - 1.321) + 12.134$ and $y_2 = 10$.

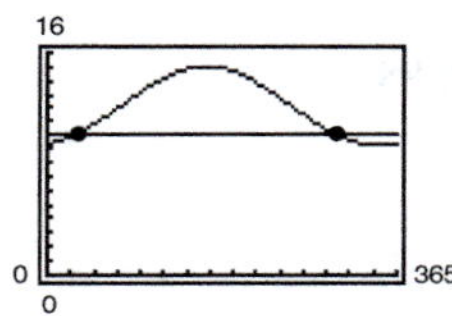

The points of intersection occur when $t \approx 30$ and $t \approx 310$.

The number of days that $D > 10$ hours is 280, from day 30 to day 310.

21.
$$\begin{aligned} 28 \cos 10t + 38 &= 28 \cos\left[10\left(t - \frac{\pi}{6}\right)\right] + 38 \\ \cos 10t &= \cos\left[10\left(t - \frac{\pi}{6}\right)\right] \\ 0 &= \cos\left[10\left(t - \frac{\pi}{6}\right)\right] - \cos 10t \\ &= -2 \sin\left(\frac{10(t - (\pi/6)) + 10t}{2}\right) \sin\left(\frac{10(t - (\pi/6)) - 10t}{2}\right) \\ &= -2 \sin\left(10t - \frac{5\pi}{6}\right) \sin\left(-\frac{5\pi}{6}\right) \\ &= -2 \sin\left(10t - \frac{5\pi}{6}\right)\left(-\frac{1}{2}\right) \\ &= \sin\left(10t - \frac{5\pi}{6}\right) \end{aligned}$$

$$10t - \frac{5\pi}{6} = n\pi \text{ where } n \text{ is any integer.}$$

$$t = \frac{n\pi}{10} + \frac{\pi}{12} \text{ where } n \text{ is any integer.}$$

The first six times the two people are at the same height are: 0.26 minutes, 0.58 minutes, 0.89 minutes, 1.20 minutes, 1.52 minutes, 1.83 minutes.

Chapter 6 Chapter Test Solutions

1. Given two angles and a side opposite one of them, the Law of Cosines cannot be used, therefore use the Law of Sines: AAS.

$A = 24°, B = 68°, a = 12.2$

$C = 180° - A - B = 180° - 24° - 68° = 88°$

$$\frac{a}{\sin A} = \frac{b}{\sin B} \Rightarrow b = \frac{a}{\sin A}(\sin B)$$

$$b = \frac{12.2}{\sin 24°}(\sin 68°) \approx 27.81$$

$$\frac{a}{\sin A} = \frac{c}{\sin C} \Rightarrow c = \frac{a}{\sin A}(\sin C)$$

$$= \frac{12.2}{\sin 24°}(\sin 88°) \approx 29.98$$

2. Given two angles and a side opposite of one of them, the Law of Cosines cannot be used, therefore use the Law of Sines: AAS.

$B = 110°, C = 28°, a = 15.6$

$A = 180° - B - C = 180° - 28° = 42°$

$$\frac{a}{\sin A} = \frac{b}{\sin B} \Rightarrow b = \frac{a}{\sin A}(\sin B)$$

$$= \frac{15.6}{\sin 42°}(\sin 110°) \approx 21.91$$

$$\frac{a}{\sin A} = \frac{c}{\sin C} \Rightarrow c = \frac{a}{\sin A}(\sin C)$$

$$= \frac{15.6}{\sin 42°}(\sin 28°) \approx 10.95$$

3. Given two sides and an angle opposite one of them, the Law of Cosines cannot be used, therefore use the Law of Singes: SSA.

$A = 24°, a = 11.2, b = 13.4$

$$\frac{\sin A}{a} = \frac{\sin B}{b} \Rightarrow \sin B = b\left(\frac{\sin A}{a}\right)$$

$$\sin B = 13.4\left(\frac{\sin 24°}{11.2}\right) \approx 0.4866$$

There are two angles between 0° and 180° where $\sin \theta \approx 0.4866$, $B_1 \approx 29.12°$ and $B_2 \approx 150.88$.

For $B_1 \approx 29.12°$,

$C_1 = 180° - 29.12° - 24° = 126.88°$

$$\frac{c}{\sin C} = \frac{a}{\sin A} \Rightarrow c = \frac{a}{\sin A}(\sin C)$$

$$= \frac{11.2}{\sin 24°}(\sin 126.88°) \approx 22.03$$

For $B_2 \approx 150.88°$,

$C = 180° - 150.88° - 24° = 5.12°.$

$$\frac{c}{\sin C} = \frac{a}{\sin A} \Rightarrow c = \frac{a}{\sin A}(\sin C)$$

$$= \frac{11.2}{\sin 24°}(\sin 5.12°) \approx 2.46$$

4. Given three sides, the Law of Cosines can be used, therefore use the Law of Cosines: SSS.

$a = 6.0, b = 7.3, c = 12.4$

Law of Cosines: SSS

$$\cos C = \frac{a^2 + b^2 - c^2}{2ab}$$
$$= \frac{(6.0)^2 + (7.3)^2 - (12.4)^2}{2(6.0)(7.3)}$$
$$\approx -0.7360$$
$$C \approx 137.39°$$

$$\frac{c}{\sin C} = \frac{a}{\sin A} \Rightarrow \sin A = a\left(\frac{\sin C}{c}\right)$$
$$= 6.0\left(\frac{\sin 137.39°}{12.4}\right) \approx 0.3276$$

$A \approx 19.12°$

$B = 180° - 137.39° - 19.12° = 23.49°$

5. Given two sides and an angle opposite one of them, the Law of Cosines cannot be used, therefore use the Law of Sines: SSA.

$B = 100°, a = 23, b = 15$

$$\frac{\sin A}{a} = \frac{\sin B}{b} \Rightarrow \sin A = a\left(\frac{\sin B}{b}\right)$$
$$= 23\left(\frac{\sin 100°}{15}\right)$$
$$\approx 1.5100$$

Because the range of sine is $[-1, 1]$ there are no values of A such that $\sin A = 1.5100$.

6. Given two angles and the included side, the Law of Cosines can be used, therefore use the Law of Cosines: SAS.

$C = 121°, a = 34, b = 55$

$c^2 = a^2 + b^2 - 2ab \cos C$

$c^2 = (34)^2 + (55)^2 - 2(34)(55) \cos 121°$

$c^2 = 6107.2424$

$c \approx 78.15$

$$\frac{\sin B}{b} = \frac{\sin C}{c} \Rightarrow \sin B = b\left(\frac{\sin C}{c}\right)$$
$$= 55\left(\frac{\sin 121°}{78.15}\right)$$
$$\approx 0.60325$$

So, $B \approx 37.10°$.

$A = 180° - B - C = 180° - 37.10 - 121° = 21.90.$

7. $a = 60, b = 70, c = 82$

$$s = \frac{a + b + c}{2} = \frac{60 + 70 + 82}{2} = 106$$

$$\text{Area} = \sqrt{s(s - a)(s - b)(s - c)} = \sqrt{106(46)(36)(24)} \approx 2052.5 \text{ square meters}$$

8. $b^2 = 370^2 + 240^2 - 2(370)(240)\cos 167°$

$b \approx 606.3$ miles

$\sin A = \dfrac{a \sin B}{b} = \dfrac{240 \sin 167°}{606.3}$

$A \approx 5.1°$

Bearing: $24° + 5.1° = 29.1°$

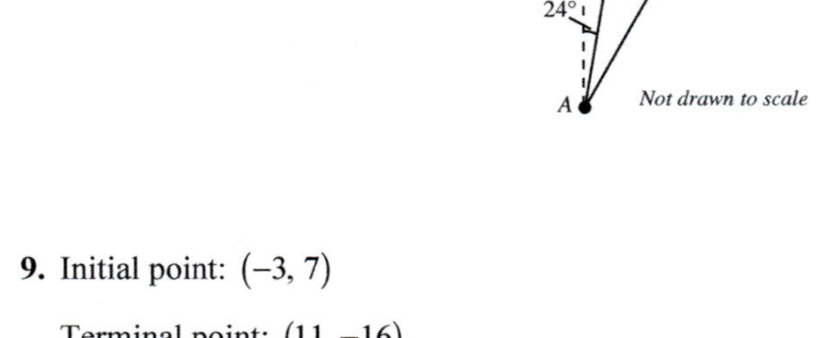

9. Initial point: $(-3, 7)$

Terminal point: $(11, -16)$

$\mathbf{v} = \langle 11 - (-3), -16 - 7 \rangle = \langle 14, -23 \rangle$

10. $\mathbf{v} = 12\left(\dfrac{\mathbf{u}}{\|\mathbf{u}\|}\right) = 12\left(\dfrac{\langle 3, -5 \rangle}{\sqrt{3^2 + (-5)^2}}\right) = \dfrac{12}{\sqrt{34}}\langle 3, -5 \rangle$

$= \dfrac{6\sqrt{34}}{17}\langle 3, -5 \rangle = \left\langle \dfrac{18\sqrt{34}}{17}, -\dfrac{30\sqrt{34}}{17} \right\rangle$

11. $\mathbf{u} = \langle 2, 7 \rangle, \mathbf{v} = \langle -6, 5 \rangle$

$\mathbf{u} + \mathbf{v} = \langle 2, 7 \rangle + \langle -6, 5 \rangle = \langle -4, 12 \rangle$

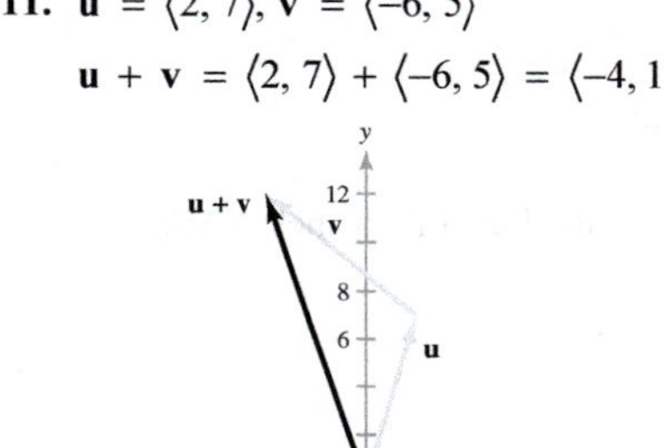

12. $\mathbf{u} = \langle 2, 7 \rangle, \mathbf{v} = \langle -6, 5 \rangle$

$\mathbf{u} - \mathbf{v} = \langle 2, 7 \rangle - \langle -6, 5 \rangle = \langle 8, 2 \rangle$

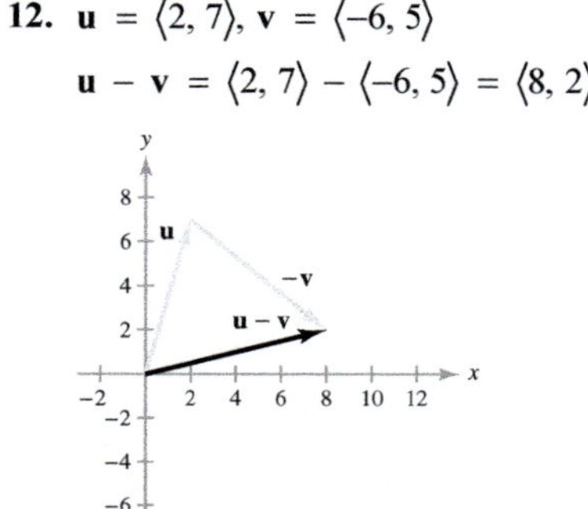

13. $\mathbf{u} = \langle 2, 7 \rangle, \mathbf{v} = \langle -6, 5 \rangle$

$5\mathbf{u} - 3\mathbf{v} = 5\langle 2, 7 \rangle - 3\langle -6, 5 \rangle$

$= \langle 10, 35 \rangle - \langle -18, 15 \rangle$

$= \langle 28, 20 \rangle$

14. $\mathbf{u} = \langle 2, 7 \rangle, \mathbf{v} = \langle -6, 5 \rangle$

$4\mathbf{u} + 2\mathbf{v} = 4\langle 2, 7 \rangle + 2\langle -6, 5 \rangle$

$= \langle 8, 28 \rangle + \langle -12, 10 \rangle$

$= \langle -4, 38 \rangle$

15. The distance between $4 + 3i$ and $1 - i$ is

$d = \sqrt{(1 - 4)^2 + (-1 - 3)^2} = \sqrt{25} = 5$ units.

16. $\mathbf{u} = 250(\cos 45°\,\mathbf{i} + \sin 45°\,\mathbf{j})$

$\mathbf{v} = 130(\cos(-60°)\mathbf{i} + \sin(-60°)\mathbf{j})$

$\mathbf{R} = \mathbf{u} + \mathbf{v} \approx 241.7767\,\mathbf{i} + 64.1934\,\mathbf{j}$

$\|\mathbf{R}\| \approx \sqrt{241.7767^2 + 64.1934^2} \approx 250.15$ pounds

$\tan\theta \approx \dfrac{64.1934}{241.7767} \Rightarrow \theta \approx 14.9°$

17. $\mathbf{u} = \langle -1, 5 \rangle, \mathbf{v} = \langle 3, -2 \rangle$

$\cos\theta = \dfrac{\mathbf{u} \cdot \mathbf{v}}{\|\mathbf{u}\|\|\mathbf{v}\|} = \dfrac{-13}{\sqrt{26}\sqrt{13}} \Rightarrow \theta = 135°$

18. $\mathbf{u} = \langle 6, -10 \rangle, \mathbf{v} = \langle 5, 3 \rangle$

$\mathbf{u} \cdot \mathbf{v} = 6(5) + (-10)(3) = 0$

$\mathbf{u}$ and $\mathbf{v}$ are orthogonal.

19. $\mathbf{u} = \langle 6, 7\rangle, \mathbf{v} = \langle -5, -1\rangle$

$$\mathbf{w}_1 = \text{proj}_{\mathbf{v}}\,\mathbf{u} = \left(\frac{\mathbf{u}\cdot\mathbf{v}}{\|\mathbf{v}\|^2}\right)\mathbf{v} = -\frac{37}{26}\langle -5, -1\rangle = \frac{37}{26}\langle 5, 1\rangle$$

$$\mathbf{w}_2 = \mathbf{u} - \mathbf{w}_1 = \langle 6, 7\rangle - \frac{37}{26}\langle 5, 1\rangle$$

$$= \left\langle -\frac{29}{26}, \frac{145}{26}\right\rangle$$

$$= \frac{29}{26}\langle -1, 5\rangle$$

$$\mathbf{u} = \mathbf{w}_1 + \mathbf{w}_2 = \frac{37}{26}\langle 5, 1\rangle + \frac{29}{26}\langle -1, 5\rangle$$

20. $\mathbf{F} = -500\mathbf{j}, \mathbf{v} = (\cos 12°)\mathbf{i} + (\sin 12°)\mathbf{j}$

$$\mathbf{w}_1 = \text{proj}_{\mathbf{v}}\,\mathbf{F} = \left(\frac{\mathbf{F}\cdot\mathbf{v}}{\|\mathbf{v}\|^2}\right)\mathbf{v} = (\mathbf{F}\cdot\mathbf{v})\mathbf{v}$$

$$= (-500 \sin 12°)\mathbf{v}$$

The magnitude of the force is $500 \sin 12° \approx 104$ pounds.

21. $z = 4 - 4i$

$$r = \sqrt{(4)^2 + (-4)^2} = \sqrt{32} = 4\sqrt{2}$$

$$\tan\theta = -\frac{4}{4} = -1 \Rightarrow \theta = \frac{7\pi}{4}\text{: Quadrant IV}$$

$$4 - 4i = 4\sqrt{2}\left(\cos\frac{7\pi}{4} + i\sin\frac{7\pi}{4}\right)$$

22. $z = 6(\cos 120° + i \sin 120°)$

$$= 6\left(-\frac{1}{2} + \frac{\sqrt{3}}{2}i\right) = -3 + 3\sqrt{3}i$$

23. $$\left[3\left(\cos\frac{7\pi}{6} + i\sin\frac{7\pi}{6}\right)\right]^8 = 3^8\left(\cos\frac{28\pi}{3} + i\sin\frac{28\pi}{3}\right)$$

$$= 6561\left(-\frac{1}{2} - \frac{\sqrt{3}}{2}i\right)$$

$$= -\frac{6561}{2} - \frac{6561\sqrt{3}}{2}i$$

24. $$(3 - 3i)^6 = \left[3\sqrt{2}\left(\cos\frac{7\pi}{4} + i\sin\frac{7\pi}{4}\right)\right]^6$$

$$= \left(3\sqrt{2}\right)^6\left(\cos\frac{21\pi}{2} + i\sin\frac{21\pi}{2}\right)$$

$$= 5832(0 + i)$$

$$= 5832i$$

25. Fourth roots of 256: $256 = 256(\cos 0 + i \sin 0)$

$$\sqrt[4]{256}\left(\cos\frac{0 + 2\pi k}{4} + i\sin\frac{0 + 2\pi k}{4}\right), k = 0, 1, 2, 3$$

$$k = 0\text{: } 4(\cos 0 + i\sin 0) = 4$$

$$k = 1\text{: } 4\left(\cos\frac{\pi}{2} + i\sin\frac{\pi}{2}\right) = 4i$$

$$k = 2\text{: } 4(\cos\pi + i\sin\pi) = -4$$

$$k = 3\text{: } 4\left(\cos\frac{3\pi}{2} + i\sin\frac{3\pi}{2}\right) = -4i$$

26. $x^3 - 27i = 0 \Rightarrow x^3 = 27i$

The solutions to the equation are the cube roots of $27i = 27\left(\cos\frac{\pi}{2} + i\sin\frac{\pi}{2}\right)$.

Cube roots: $\sqrt[3]{27}\left[\cos\frac{\frac{\pi}{2} + 2\pi k}{3} + i\sin\frac{\frac{\pi}{2} + 2\pi k}{3}\right], k = 0, 1, 2$

$k = 0: 3\left(\cos\frac{\pi}{6} + i\sin\frac{\pi}{6}\right) = 3\left(\frac{\sqrt{3}}{2} + \frac{1}{2}i\right) = \frac{3\sqrt{3}}{2} + \frac{3}{2}i$

$k = 1: 3\left(\cos\frac{5\pi}{6} + i\sin\frac{5\pi}{6}\right) = 3\left(-\frac{\sqrt{3}}{2} + \frac{1}{2}i\right) = -\frac{3\sqrt{3}}{2} + \frac{3}{2}i$

$k = 2: 3\left(\cos\frac{3\pi}{2} + i\sin\frac{3\pi}{2}\right) = 3(0 - i) = -3i$

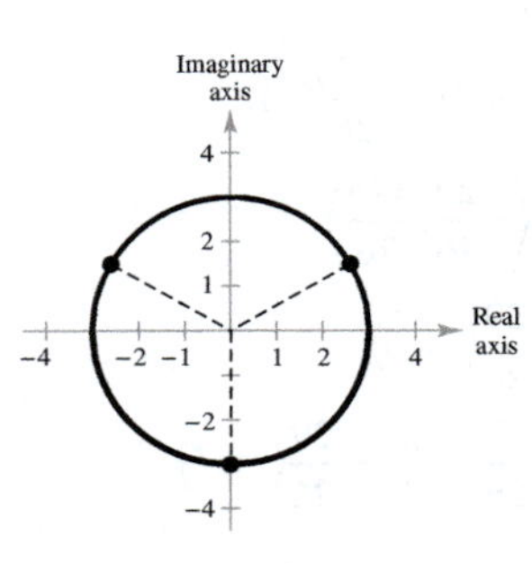

Cumulative Test Solutions for Chapters 4–6

1. (a)

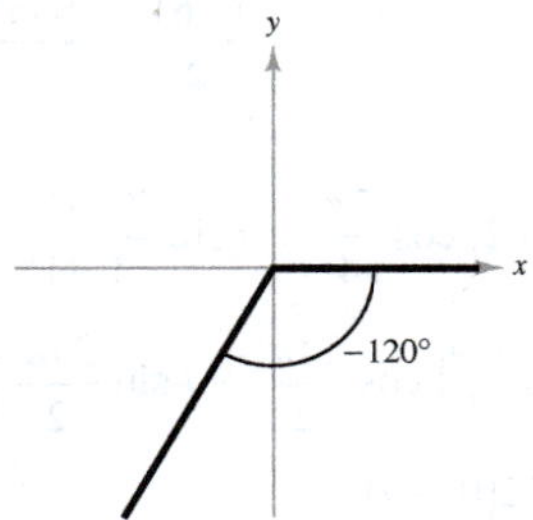

(b) $-120° + 360° = 240°$

(c) $-120\left(\frac{\pi}{180°}\right) = -\frac{2\pi}{3}$

(d) $-120° + 360° = 240°$

$\theta' = 240° - 180° = 60°$

(e) $\sin(-120°) = -\sin 60° = -\frac{\sqrt{3}}{2}$

$\cos(-120°) = -\cos 60° = -\frac{1}{2}$

$\tan(-120°) = \tan 60° = \sqrt{3}$

$\csc(-120°) = \frac{1}{-\sin 60°} = -\frac{2\sqrt{3}}{3}$

$\sec(-120°) = \frac{1}{-\cos 60°} = -2$

$\cot(-120°) = \frac{1}{\tan 60°} = \frac{\sqrt{3}}{3}$

2. $-1.45\left(\frac{180}{\pi}\right) \approx -83.079°$

3. $\tan\theta = \frac{y}{x} = -\frac{21}{20} \Rightarrow r = 29$

Because $\sin\theta < 0$, θ is in Quadrant IV $\Rightarrow x = 20$.

$\cos\theta = \frac{x}{r} = \frac{20}{29}$

4. $f(x) = 3 - 2\sin\pi x$

Period: $\frac{2\pi}{\pi} = 2$

Amplitude: $|a| = |-2| = 2$

Upward shift of 3 units (reflected in x-axis prior to shift)

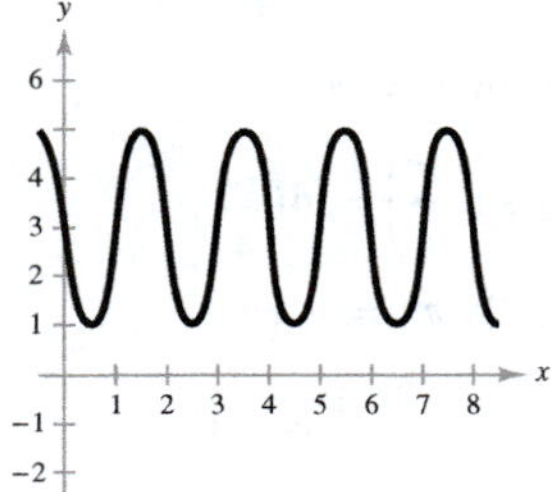

5. $g(x) = \frac{1}{2}\tan\left(x - \frac{\pi}{2}\right)$

Period: π

Asymptotes: $x = 0, x = \pi$

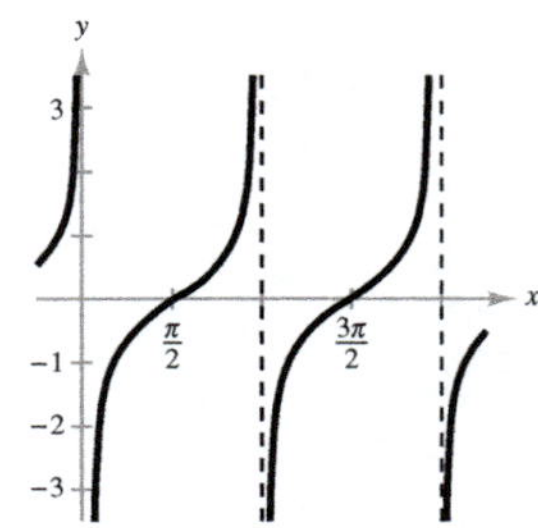

6. $h(x) = -\sec(x + \pi)$

Graph $y = -\cos(x + \pi)$ first.

Period: 2π

Amplitude: 1

Set $x + \pi = 0$ and $x + \pi = 2\pi$ for one cycle.

$x = -\pi \qquad x = \pi$

The asymptotes of $h(x)$ corresponds to the x-intercepts of $y = -\cos(x + \pi)$.

$$x + \pi = \frac{(2n + 1)\pi}{2}$$

$$x = \frac{(2n - 1)\pi}{2} \text{ where } n \text{ is any integer}$$

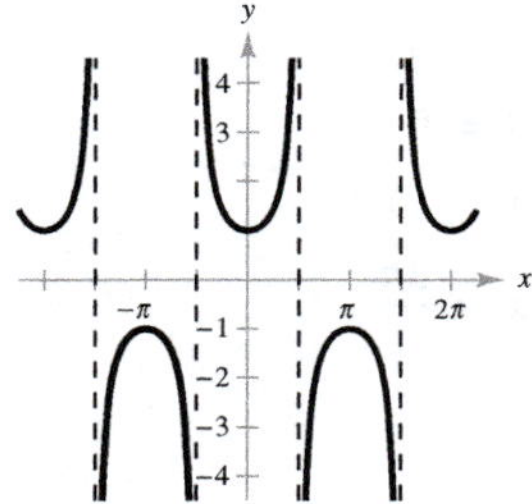

7. $h(x) = a\cos(bx + c)$

Graph is reflected in x-axis.

Amplitude: $a = -3$

Period: $2 = \dfrac{2\pi}{\pi} \Rightarrow b = \pi$

No phase shift: $c = 0$

$h(x) = -3\cos(\pi x)$

8. $f(x) = \dfrac{x}{2}\sin x, -3\pi \le x \le 3\pi$

$$-\frac{x}{2} \le f(x) \le \frac{x}{2}$$

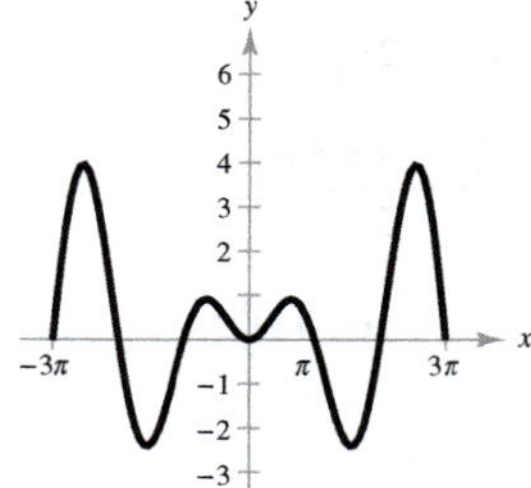

9. $\tan(\arctan 4.9) = 4.9$

10. $\tan\left(\arcsin\dfrac{3}{5}\right) = \dfrac{3}{4}$

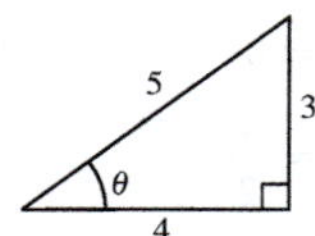

11. $y = \arccos(2x)$

$\sin y = \sin(\arccos(2x)) = \sqrt{1 - 4x^2}$

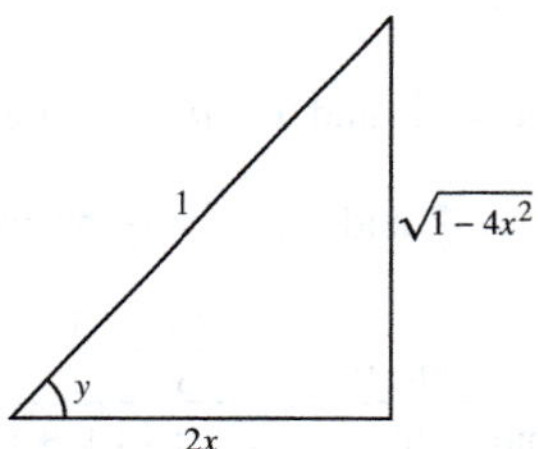

12. $\cos\left(\dfrac{\pi}{2} - x\right)\csc x = \sin x\left(\dfrac{1}{\sin x}\right) = 1$

13.
$$\frac{\sin\theta - 1}{\cos\theta} - \frac{\cos\theta}{\sin\theta - 1} = \frac{\sin\theta - 1}{\cos\theta} - \frac{\cos\theta(\sin\theta + 1)}{\sin^2\theta - 1}$$
$$= \frac{\sin\theta - 1}{\cos\theta} + \frac{\cos\theta(\sin\theta + 1)}{\cos^2\theta} = \frac{\sin\theta - 1}{\cos\theta} + \frac{\sin\theta + 1}{\cos\theta} = \frac{2\sin\theta}{\cos\theta} = 2\tan\theta$$

14. $\cot^2\alpha(\sec^2\alpha - 1) = \cot^2\alpha\tan^2\alpha = 1$

15.
$$\sin(x + y)\sin(x - y) = \tfrac{1}{2}\left[\cos(x + y - (x - y)) - \cos(x + y + x - y)\right]$$
$$= \tfrac{1}{2}[\cos 2y - \cos 2x] = \tfrac{1}{2}\left[1 - 2\sin^2 y - (1 - 2\sin^2 x)\right] = \sin^2 x - \sin^2 y$$

16. $\sin^2 x \cos^2 x = \left(\frac{1 - \cos 2x}{2}\right)\left(\frac{1 + \cos 2x}{2}\right)$

$= \frac{1}{4}(1 - \cos 2x)(1 + \cos 2x)$

$= \frac{1}{4}(1 - \cos^2 2x)$

$= \frac{1}{4}\left(1 - \frac{1 + \cos 4x}{2}\right)$

$= \frac{1}{8}(2 - (1 + \cos 4x))$

$= \frac{1}{8}(1 - \cos 4x)$

17. $2\cos^2 \beta - \cos \beta = 0$

$\cos \beta(2\cos \beta - 1) = 0$

$\cos \beta = 0$ or $2\cos \beta - 1 = 0$

$\beta = \frac{\pi}{2}, \frac{3\pi}{2}$ $\qquad \cos \beta = \frac{1}{2}$

$\beta = \frac{\pi}{3}, \frac{5\pi}{3}$

Answer: $\frac{\pi}{3}, \frac{\pi}{2}, \frac{3\pi}{2}, \frac{5\pi}{3}$

18. $3\tan \theta - \cot \theta = 0$

$3\tan \theta - \frac{1}{\tan \theta} = 0$

$\frac{3\tan^2 \theta - 1}{\tan \theta} = 0$

$3\tan^2 \theta - 1 = 0$

$\tan^2 \theta = \frac{1}{3}$

$\tan \theta = \pm\frac{\sqrt{3}}{3}$

$\theta = \frac{\pi}{6}, \frac{5\pi}{6}, \frac{7\pi}{6}, \frac{11\pi}{6}$

19. $\sin^2 x + 2\sin x + 1 = 0$

$(\sin x + 1)(\sin x + 1) = 0$

$\sin x + 1 = 0$

$\sin x = -1$

$x = \frac{3\pi}{2}$

20. $\sin u = \frac{12}{13} \Rightarrow \cos u = \frac{5}{13}$ and $\tan u = \frac{12}{5}$ because u is in Quadrant I.

$\cos v = \frac{3}{5} \Rightarrow \sin v = \frac{4}{5}$ and $\tan v = \frac{4}{3}$ because v is in Quadrant I.

$$\tan(u - v) = \frac{\tan u - \tan v}{1 + \tan u \tan v} = \frac{\frac{12}{5} - \frac{4}{3}}{1 + \left(\frac{12}{5}\right)\left(\frac{4}{3}\right)} = \frac{16}{63}$$

21. $\tan u = \frac{1}{2}, 0 < u < \frac{\pi}{2}$

$$\tan 2u = \frac{2\tan u}{1 - \tan^2 u} = \frac{2\left(\frac{1}{2}\right)}{1 - \left(\frac{1}{2}\right)^2} = \frac{4}{3}$$

22. $\tan u = \frac{4}{3}, 0 < u < \frac{\pi}{2}$

$\tan u = \frac{4}{3} \Rightarrow \cos u = \frac{3}{5}$

$$\sin\frac{u}{2} = \sqrt{\frac{1 - \cos u}{2}} = \sqrt{\frac{1 - \frac{3}{5}}{2}} = \frac{\sqrt{5}}{5}$$

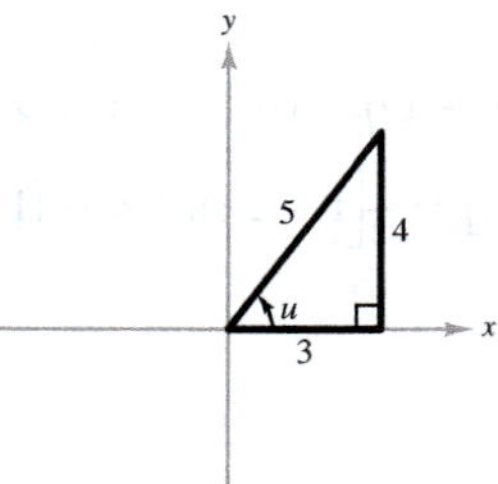

23. $5 \sin\frac{3\pi}{4}\cos\frac{7\pi}{4} = \frac{5}{2}\left[\sin\left(\frac{3\pi}{4} + \frac{7\pi}{4}\right) + \sin\left(\frac{3\pi}{4} - \frac{7\pi}{4}\right)\right]$

$$= \frac{5}{2}\left[\sin\frac{5\pi}{2} + \sin(-\pi)\right]$$

$$= \frac{5}{2}\left(\sin\frac{5\pi}{2} - \sin\pi\right)$$

24. $\cos 9x - \cos 7x = -2\sin\left(\frac{9x + 7x}{2}\right)\sin\left(\frac{9x - 7x}{2}\right)$

$$= -2\sin 8x \sin x$$

25. Given two sides and an angle opposite one of them, the Law of Cosines cannot be used, therefore use the Law of Sines: SSA.

$A = 30°, a = 9, b = 8$

$$\frac{\sin B}{8} = \frac{\sin 30°}{9}$$

$$\sin B = \frac{8}{9}\left(\frac{1}{2}\right)$$

$$B = \arcsin\left(\frac{4}{9}\right)$$

$$B \approx 26.39°$$

$$C = 180° - A - B \approx 123.61°$$

$$\frac{c}{\sin 123.61°} = \frac{9}{\sin 30°}$$

$$c \approx 14.99$$

26. Given two sides and the included angle, the Law of Cosines can be used, therefore use the Law of Cosines: SAS.

$A = 30°, b = 8, c = 10$

$$a^2 = 8^2 + 10^2 - 2(8)(10)\cos 30°$$

$$a^2 \approx 25.4359$$

$$a \approx 5.04$$

$$\cos B = \frac{5.04^2 + 10^2 - 8^2}{2(5.04)(10)}$$

$$\cos B \approx 0.6091$$

$$B \approx 52.48°$$

$$C = 180° - A - B \approx 97.52°$$

27. Because $C = 90°$, use right triangle ratios.

$A = 30°, C = 90°, b = 10$

$B = 180° - 30° - 90° = 60°$

$$\tan 30° = \frac{a}{10} \Rightarrow a = 10\tan 30° \approx 5.77$$

$$\cos 30° = \frac{10}{c} \Rightarrow c = \frac{10}{\cos 30°} \approx 11.55$$

28. Given three sides, the Law of Cosines can be used, therefore use the Law of Cosines: SSS.

$a = 4.7, b = 8.1, c = 10.3$

$$\cos C = \frac{a^2 + b^2 - c^2}{2ab} = \frac{4.7^2 + 8.1^2 - 10.3^2}{2(4.7)(8.1)} = -0.2415 \Rightarrow C \approx 103.98°$$

$$\sin A = \frac{a \sin \mathrm{C}}{c} \approx \frac{4.7\sin 103.98°}{10.3} \approx 0.4428 \Rightarrow A \approx 26.28°$$

$$B \approx 180° - 26.28° - 103.98° = 49.74°$$

29. Given two angles and a side opposite one of them, the Law of Cosines cannot be used, therefore use the Law of Sines: AAS.

$A = 45°, B = 26°, c = 20$

$C = 180° - A - B = 180° - 45° - 26° = 109°$

$$\frac{a}{\sin A} = \frac{c}{\sin C} \Rightarrow a = \frac{c}{\sin C}(\sin A) = \frac{20}{\sin 109°}(\sin 45°) \approx 14.96$$

$$\frac{b}{\sin B} = \frac{c}{\sin C} \Rightarrow b = \frac{c}{\sin C}(\sin B) = \frac{20}{\sin 109°}(\sin 26°) \approx 9.27$$

30. Given two angles and the included side, the Law of Cosines can be used, therefore use the Law of Cosines: SAS.

$a = 1.2, b = 10, C = 80°$

$c^2 = a^2 + b^2 - 2ab \cos C$

$c^2 = (1.2)^2 + (10)^2 - 2(1.2)(10) \cos 80°$

$c^2 \approx 97.2724$

$c \approx 9.86$

$\frac{a}{\sin A} = \frac{c}{\sin C}$

$\sin A = a\left(\frac{\sin C}{c}\right)$

$= 1.2\left(\frac{\sin 80°}{9.86}\right)$

≈ 0.1199

$A \approx 6.88°$

$B \approx 180° - 80° - 6.88° = 93.12°$

31. Area $= \frac{1}{2}(7)(12) \sin 99° = 41.48 \text{ in.}^2$

32. $a = 30, b = 41, c = 45$

$s = \frac{a + b + c}{2} = \frac{30 + 41 + 45}{2} = 58$

Area $= \sqrt{s(s - a)(s - b)(s - c)}$

$= \sqrt{58(28)(17)(13)}$

$\approx 599.09 \text{ m}^2$

33. Terminal point: $(6, 10)$; Initial point: $(-1, 2)$

$\mathbf{u} = \langle 6 - (-1), 10 - 2\rangle = \langle 7, 8\rangle = 7\mathbf{i} + 8\mathbf{j}$

34. $\mathbf{v} = \mathbf{i} + \mathbf{j}$

$\|\mathbf{v}\| = \sqrt{1^2 + 1^2} = \sqrt{2}$

$\mathbf{u} = \frac{\mathbf{v}}{\|\mathbf{v}\|} = \frac{1}{\sqrt{2}}(\mathbf{i} + \mathbf{j}) = \frac{\sqrt{2}}{2}(\mathbf{i} + \mathbf{j})$

35. $\mathbf{u} = 3\mathbf{i} + 4\mathbf{j}, \mathbf{v} = \mathbf{i} - 2\mathbf{j}$

$\mathbf{u} \cdot \mathbf{v} = 3(1) + 4(-2) = -5$

36. $\mathbf{u} = \langle 8, -2\rangle, \mathbf{v} = \langle 1, 5\rangle$

$\mathbf{w}_1 = \text{proj}_{\mathbf{v}}\, \mathbf{u} = \left(\frac{\mathbf{u} \cdot \mathbf{v}}{\|\mathbf{v}\|^2}\right)\mathbf{v} = \frac{-2}{26}\langle 1, 5\rangle = -\frac{1}{13}\langle 1, 5\rangle$

$\mathbf{w}_2 = \mathbf{u} - \mathbf{w}_1 = \langle 8, -2\rangle - \left\langle -\frac{1}{13}, -\frac{5}{13}\right\rangle = \left\langle \frac{105}{13}, -\frac{21}{13}\right\rangle$

$= \frac{21}{13}\langle 5, -1\rangle$

$\mathbf{u} = \mathbf{w}_1 + \mathbf{w}_2 = -\frac{1}{13}\langle 1, 5\rangle + \frac{21}{13}\langle 5, -1\rangle$

37.

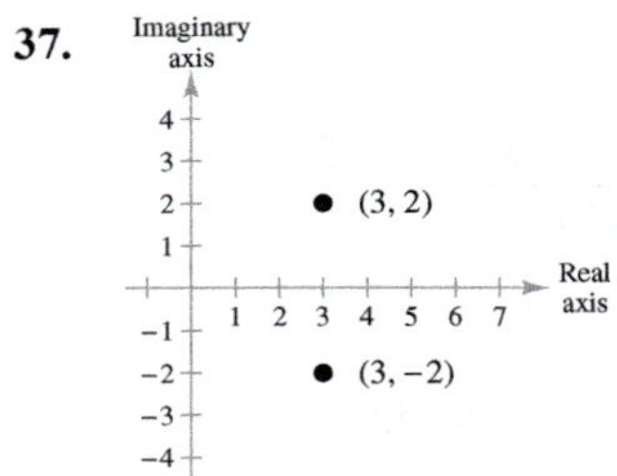

The complex conjugate of $3 - 2i$ is $3 + 2i$.

38. $r = |-2 + 2i| = \sqrt{(-2)^2 + (2)^2} = 2\sqrt{2}$

$\tan \theta = \frac{2}{-2} = -1$

Because $\tan \theta = -1$ and $-2 + 2i$ lies in Quadrant II,

$\theta = \frac{3\pi}{4}$. So, $-2 + 2i = 2\sqrt{2}\left(\cos\frac{3\pi}{4} + i \sin\frac{3\pi}{4}\right)$.

39. $[4(\cos 30° + i \sin 30°)][6(\cos 120° + i \sin 120°)] = (4)(6)[\cos(30° + 120°) + i \sin(30° + 120°) + i \sin(30° + 120°)]$

$= 24(\cos 150° + i \sin 150°)$

40. $1 = 1(\cos 0 + i \sin 0)$

$\sqrt[3]{1} = \sqrt[3]{1}\left[\cos\left(\frac{0 + 2\pi k}{3}\right) + i \sin\left(\frac{0 + 2\pi k}{3}\right)\right], k = 0, 1, 2$

$k = 0: \sqrt[3]{1}\left[\left(\cos\left(\frac{0 + 2\pi(0)}{3}\right) + i \sin\left(\frac{0 + 2\pi(0)}{3}\right)\right)\right] = \cos 0 + i \sin 0 = 1$

$k = 1: \sqrt[3]{1}\left[\left(\cos\left(\frac{0 + 2\pi(1)}{3}\right) + i \sin\left(\frac{0 + 2\pi(1)}{3}\right)\right)\right] = \cos\frac{2\pi}{3} + i \sin\frac{2\pi}{3} = -\frac{1}{2} + \frac{\sqrt{3}}{2}i$

$k = 2: \sqrt[3]{1}\left[\left(\cos\left(\frac{0 + 2\pi(2)}{3}\right) + i \sin\left(\frac{0 + 2\pi(2)}{3}\right)\right)\right] = \cos\frac{4\pi}{3} + i \sin\frac{4\pi}{3} = -\frac{1}{2} - \frac{\sqrt{3}}{2}i$

41. $x^4 + 625 = 0$

$x^4 = -625$

Fourth roots of -625:

$-625 = 625(\cos \pi + i \sin \pi)$

$\sqrt[4]{625}\left(\cos \dfrac{\pi + 2\pi k}{4} + i \sin \dfrac{\pi + 2\pi}{4}\right),\ k = 0, 1, 2, 3$

$k = 0\colon 5\left(\cos \dfrac{\pi}{4} + i \sin \dfrac{\pi}{4}\right) = 5\left(\dfrac{\sqrt{2}}{2} + \dfrac{\sqrt{2}}{2}i\right) = \dfrac{5\sqrt{2}}{2} + \dfrac{5\sqrt{2}}{2}i$

$k = 1\colon 5\left(\cos \dfrac{3\pi}{4} + i \sin \dfrac{3\pi}{4}\right) = 5\left(-\dfrac{\sqrt{2}}{2} + \dfrac{\sqrt{2}}{2}i\right) = -\dfrac{5\sqrt{2}}{2} + \dfrac{5\sqrt{2}}{2}i$

$k = 2\colon 5\left(\cos \dfrac{5\pi}{4} + i \sin \dfrac{5\pi}{4}\right) = 5\left(-\dfrac{\sqrt{2}}{2} - \dfrac{\sqrt{2}}{2}i\right) = -\dfrac{5\sqrt{2}}{2} - \dfrac{5\sqrt{2}}{2}i$

$k = 3\colon 5\left(\cos \dfrac{7\pi}{4} + i \sin \dfrac{7\pi}{4}\right) = 5\left(\dfrac{\sqrt{2}}{2} - \dfrac{\sqrt{2}}{2}i\right) = \dfrac{5\sqrt{2}}{2} - \dfrac{5\sqrt{2}}{2}i$

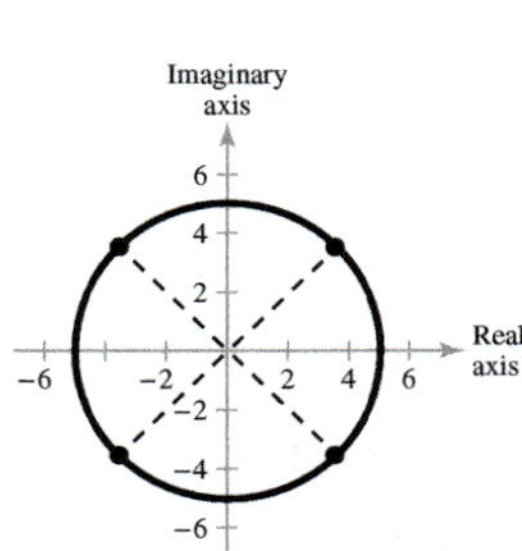

42. Angular speed $= \dfrac{\theta}{t} = \dfrac{2\pi(63)}{1} \approx 395.8$ radians per minute

Linear speed $= \dfrac{s}{t} = \dfrac{42\pi(63)}{1} \approx 8312.7$ inches per minute

43. Area $= \dfrac{\theta r^2}{2} = \dfrac{(105^\circ)\left(\dfrac{\pi}{180^\circ}\right)(12)^2}{2} = 42\pi \approx 131.95\ \text{yd}^2$

44. Height of smaller triangle:

$\tan 16^\circ 45' = \dfrac{h_1}{200}$

$h_1 = 200 \tan 16.75^\circ$

≈ 60.2 feet

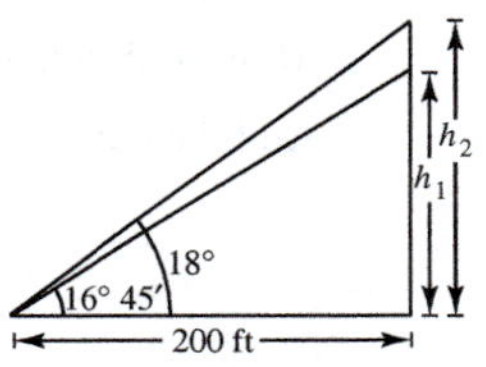

Not drawn to scale

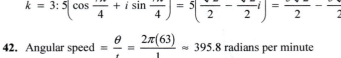

Height of larger triangle:

$\tan 18^\circ = \dfrac{h_2}{200}$

$h_2 = 200 \tan 18^\circ \approx 65.0$ feet

Height of flag: $h_2 - h_1 = 65.0 - 60.2 \approx 5$ feet

45. $\tan \theta = \dfrac{5}{12} \Rightarrow \theta \approx 22.6^\circ$

46. $d = a \cos bt$

$|a| = 4 \Rightarrow a = 4$

$\dfrac{2\pi}{b} = 8 \Rightarrow b = \dfrac{\pi}{4}$

$d = 4 \cos \dfrac{\pi}{4}t$

47. $\mathbf{v}_1 = 500\langle\cos 60^\circ, \sin 60^\circ\rangle = \langle 250, 250\sqrt{3}\rangle$

$\mathbf{v}_2 = 50\langle\cos 30^\circ, \sin 30^\circ\rangle = \langle 25\sqrt{3}, 25\rangle$

$\mathbf{v} = \mathbf{v}_1 + \mathbf{v}_2 = \langle 250 + 25\sqrt{3}, 250\sqrt{3} + 25\rangle \approx \langle 293.3, 458.0\rangle$

$\|\mathbf{v}\| = \sqrt{(293.3)^2 + (458.0)^2} \approx 543.9$

$\tan \theta = \dfrac{458.0}{293.3} \approx 1.56 \Rightarrow \theta \approx 57.4^\circ$

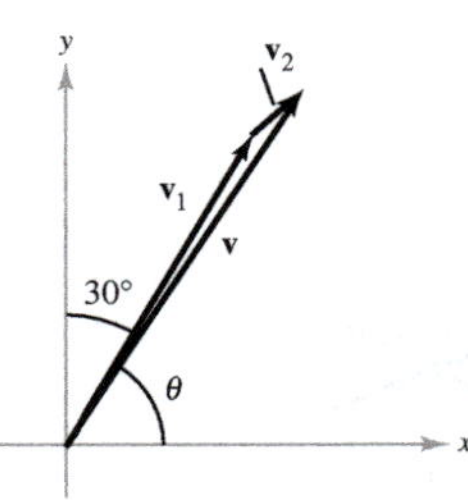

Bearing: $90^\circ - 57.4^\circ = 32.6^\circ$

The plane is traveling on a bearing of 32.6° at 543.9 kilometers per hour.

48. $\mathbf{w} = (85)(10)\cos 60^\circ = 425$ foot-pounds

Chapter Test Solutions for Chapter 7

1. $\begin{cases} x + y = -9 \Rightarrow x = -y - 9 \\ 5x - 8y = 20 \end{cases}$

$$5(-y - 9) - 8y = 20$$
$$-13y = 65$$
$$y = -5$$
$$x - 5 = -9 \Rightarrow x = -4$$

Solution: $(-4, -5)$

2. $\begin{cases} y = x - 1 \\ y = (x - 1)^3 \end{cases}$

$$x - 1 = (x - 1)^3$$
$$x - 1 = x^3 - 3x^2 + 3x - 1$$
$$0 = x^3 - 3x^2 + 2x$$
$$0 = x(x - 1)(x - 2)$$
$$x = 0 \text{ or } x = 1 \text{ or } x = 2$$
$$y = -1 \quad y = 0 \quad y = 1$$

Solutions: $(0, -1), (1, 0), (2, 1)$

3. $\begin{cases} x - y = 4 \Rightarrow x = y + 4 \\ 2x - y^2 = 0 \Rightarrow 2(y + 4) - y^2 = 0 \end{cases}$

$$0 = y^2 - 2y - 8$$
$$0 = (y + 2)(y - 4)$$
$$y = -2 \text{ or } y = 4$$
$$x = 2 \quad x = 8$$

Solutions: $(2, -2), (8, 4)$

4. $\begin{cases} 3x - 6y = 0 \Rightarrow y = \frac{1}{2}x \\ 2x + 5y = 18 \Rightarrow y = -\frac{2}{5}x + \frac{18}{5} \end{cases}$

Solution: $(4, 2)$

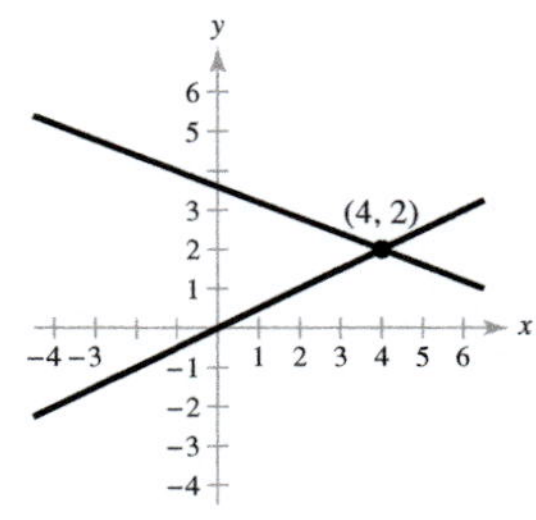

5. $\begin{cases} y = 9 - x^2 \\ y = x + 3 \end{cases}$

Solutions: $(-3, 0), (2, 5)$

6. $\begin{cases} y - \ln x = 4 \Rightarrow y = \ln x + 4 \\ 7x - 2y - 5 = -6 \Rightarrow y = \frac{7}{2}x + \frac{1}{2} \end{cases}$

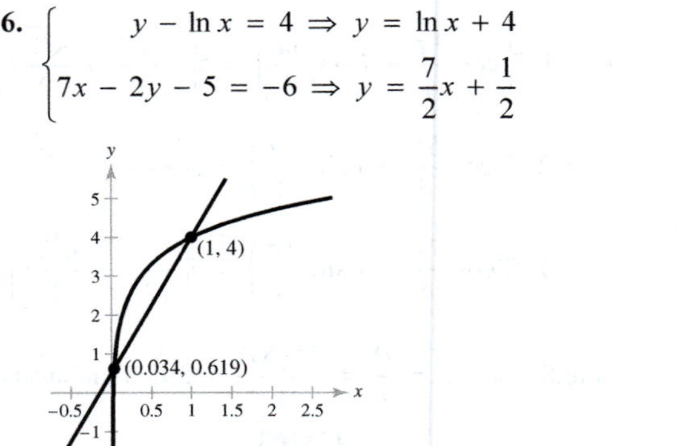

Solutions: $(1, 4)$ and $\approx (0.034, 0.619)$

7. $\begin{cases} 3x + 4y = -26 & \text{Equation 1} \\ 7x - 5y = 11 & \text{Equation 2} \end{cases}$

Multiply Equation 1 by 5: $15x + 20y = -130$

Multiply Equation 2 by 4: $28x - 20y = 44$

Add the equations to eliminate y:
$$\begin{aligned} 15x + 20y &= -130 \\ 28x - 20y &= 44 \\ \hline 43x &= -86 \\ x &= -2 \end{aligned}$$

Back-substitute $x = -2$ into Equation 1:

$$3(-2) + 4y = -26$$
$$y = -5$$

Solution: $(-2, -5)$

8. $\begin{cases} 1.4x - y = 17 & \text{Equation 1} \\ 0.8x + 6y = -10 & \text{Equation 2} \end{cases}$

Multiply Equation 1 by 6: $8.4x - 6y = 102$

Add this to Equation 2 to eliminate y:

$$\begin{aligned} 8.4x - 6y &= 102 \\ 0.8x + 6y &= -10 \\ \hline 9.2x \quad &= 92 \\ x &= 10 \end{aligned}$$

Back-substitute $x = 10$ into Equation 2:

$$\begin{aligned} 0.8(10) + 6y &= -10 \\ 6y &= -18 \\ y &= -3 \end{aligned}$$

Solution: $(10, -3)$

9. $\begin{cases} x - 2y + 3z = 11 \\ 2x \quad\quad - z = 3 \\ \quad 3y + z = -8 \end{cases}$

$\begin{cases} x - 2y + 3z = 11 \\ \quad 4y - 7z = -19 & -2\text{Eq.}1 + \text{Eq.}2 \\ \quad 3y + z = -8 \end{cases}$

$\begin{cases} x - 2y + 3z = 11 \\ \quad y - 8z = -11 & -\text{Eq.}3 + \text{Eq.}2 \\ \quad 3y + z = -8 \end{cases}$

$\begin{cases} x - 2y + 3z = 11 \\ \quad y - 8z = -11 \\ \quad\quad 25z = 25 & -3\text{Eq.}2 + \text{Eq.}3 \end{cases}$

$\begin{cases} x - 2y + 3z = 11 \\ \quad y - 8z = -11 \\ \quad\quad z = 1 & \frac{1}{25}\text{Eq.}3 \end{cases}$

$y - 8(1) = -11 \Rightarrow y = -3$

$x - 2(-3) + 3(1) = 11 \Rightarrow x = 2$

Solution: $(2, -3, 1)$

10. $\begin{cases} 3x + 2y + z = 17 & \text{Equation 1} \\ -x + y + z = 4 & \text{Equation 2} \\ x - y - z = 3 & \text{Equation 3} \end{cases}$

Interchange Equations 1 and 3.

$\begin{cases} x - y - z = 3 \\ -x + y + z = 4 \\ 3x + 2y + z = 17 \end{cases}$

$\begin{cases} x - y - z = 3 \\ \quad\quad 0 \neq 7 & \text{Eq. } 1 + \text{Eq. } 2 \\ 3x + 2y + z = 17 \end{cases}$

Inconsistent

No solution

11. $\dfrac{2x + 5}{x^2 - x - 2} = \dfrac{2x + 5}{(x - 2)(x + 1)} = \dfrac{A}{x - 2} + \dfrac{B}{x + 1}$

$2x + 5 = A(x + 1) + B(x - 2)$

Let $x = 2$: $9 = 3A \Rightarrow A = 3$

Let $x = -1$: $3 = -3B \Rightarrow B = -1$

$\dfrac{2x + 5}{x^2 - x - 2} = \dfrac{3}{x - 2} - \dfrac{1}{x + 1}$

12. $\dfrac{3x^2 - 2x + 4}{x^2(2 - x)} = \dfrac{A}{x} + \dfrac{B}{x^2} + \dfrac{C}{2 - x}$

$3x^2 - 2x + 4 = Ax(2 - x) + B(2 - x) + Cx^2$

Let $x = 0$: $4 = 2B \Rightarrow B = 2$

Let $x = 2$: $12 = 4C \Rightarrow C = 3$

Let $x = 1$: $5 = A + B + C = A + 2 + 3 \Rightarrow A = 0$

$\dfrac{3x^2 - 2x + 4}{x^2(2 - x)} = \dfrac{2}{x^2} + \dfrac{3}{2 - x}$

13. $\dfrac{x^4 + 5}{x^3 - x} = x + \dfrac{x^2 + 5}{x^3 - x}$, use long division first to create a proper fraction.

$\dfrac{x^2 + 5}{x^3 - x} = \dfrac{x^2 + 5}{x(x + 1)(x - 1)} = \dfrac{A}{x} + \dfrac{B}{x - 1} + \dfrac{c}{x - 1}$

$x^2 + 5 = A(x + 1)(x - 1) + Bx(x - 1) + Cx(x + 1)$

Let $x = 0$: $5 = -A \Rightarrow A = -5$

Let $x = -1$: $6 = 2B \Rightarrow B = 3$

Let $x = 1$: $6 = 2C \Rightarrow C = 3$

$\dfrac{x^4 + 5}{x^3 - x} = x - \dfrac{5}{x} + \dfrac{3}{x + 1} + \dfrac{3}{x - 1}$

14. $\dfrac{x^2 - 4}{x^3 + 2x} = \dfrac{x^2 - 4}{x(x^2 + 2)} = \dfrac{A}{x} + \dfrac{Bx + C}{x^2 + 2}$

$$\begin{aligned} x^2 - 4 &= A(x^2 + 2) + (Bx + C)x \\ &= Ax^2 + 2A + Bx^2 + Cx \\ &= (A + B)x^2 + Cx + 2A \end{aligned}$$

Equate the coefficients of like terms:

$1 = A + B, 0 = C, -4 = 2A$

So, $A = -2, B = 3, C = 0$.

$\dfrac{x^2 - 4}{x^3 + 2x} = -\dfrac{2}{x} + \dfrac{3x}{x^2 + 2}$

15. 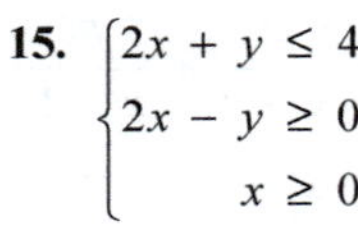

$$\begin{cases} 2x + y \le 4 \\ 2x - y \ge 0 \\ x \ge 0 \end{cases}$$

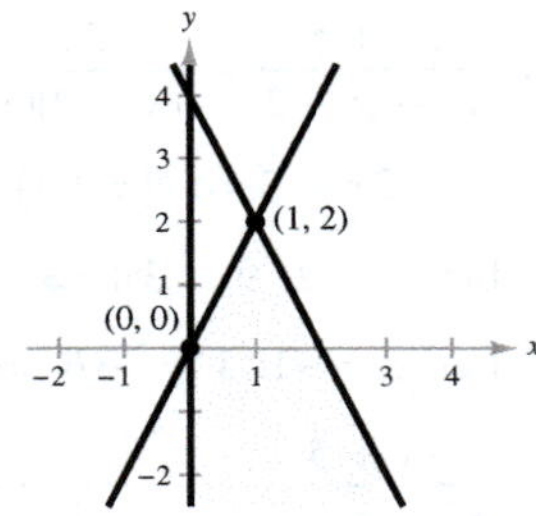

16. $\begin{cases} y < -x^2 + x + 4 \\ y > 4x \end{cases}$

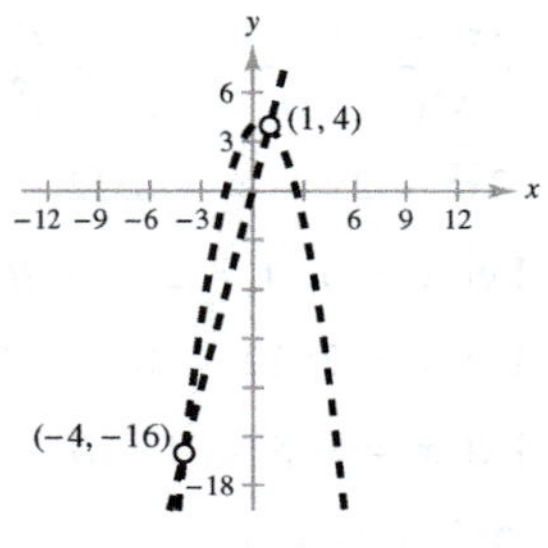

17. $\begin{cases} x^2 + y^2 \le 36 \\ x \ge 2 \\ y \ge -4 \end{cases}$

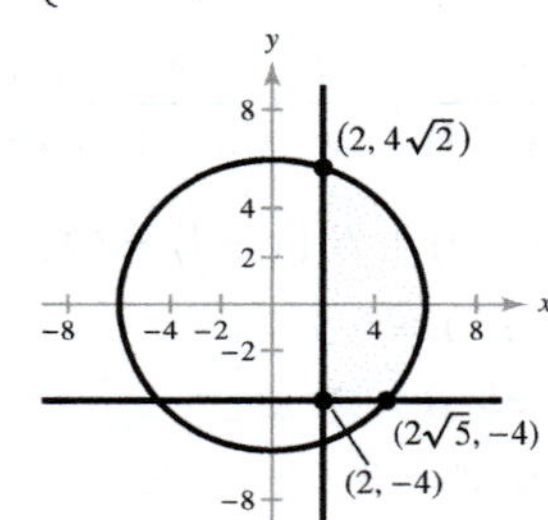

18. Maximize $z = 20x + 12y$ subject to:

$\begin{cases} x \ge 0, y \ge 0 \\ x + 4y \le 32 \\ 3x + 2y \le 36 \end{cases}$

(0, 8)
(8, 6)
(12, 0)
(0, 0)

At $(0, 0)$ we have $z = 0$.

At $(0, 8)$ we have $z = 96$.

At $(8, 6)$ we have $z = 232$.

At $(12, 0)$ we have $z = 240$.

The maximum value, $z = 240$, occurs at $(12, 0)$.

The minimum value, $z = 0$ occurs at $(0, 0)$.

19. Let x = amount of money invested at 4%.
Let y = amount of money invested at 5.5%.

$\begin{cases} x + y = 50{,}000 & \text{Equation 1} \\ 0.04x + 0.055y = 2390 & \text{Equation 2} \end{cases}$

Multiply Equation 1 by −4: $-4x - 4y = -200{,}000$

Multiply Equation 2 by 100: $4x + 5.5y = 239{,}000$

Add these two equations to eliminate x:

$$\begin{aligned} -4x - 4y &= -200{,}000 \\ 4x + 5.5y &= 239{,}000 \\ \hline 1.5y &= 39{,}000 \\ y &= 26{,}000 \end{aligned}$$

Back-substitute $y = 26{,}000$ into Equation 1:

$$\begin{aligned} x + 26{,}000 &= 50{,}000 \\ x &= 24{,}000 \end{aligned}$$

So, \$24,000 should be invested at 4% and \$26,000 should be invested at 5.5%.

20. $y = ax^2 + bx + c$

$(0, 6)$: $6 = c$

$(-2, 2)$: $2 = 4a - 2b + c$

$\left(3, \frac{9}{2}\right)$: $\frac{9}{2} = 9a + 3b + c$

Solving this system yields: $a = -\frac{1}{2}$, $b = 1$, and $c = 6$.

So, $y = -\frac{1}{2}x^2 + x + 6$.

21. Optimize $P = 30x + 40y$ subject to:

$$\begin{cases} x \geq 0, y \geq 0 \\ 0.5x + 0.75y \leq 3750 \\ 2.0x + 1.5y \leq 8950 \\ 0.5x + 0.5y \leq 2650 \end{cases}$$

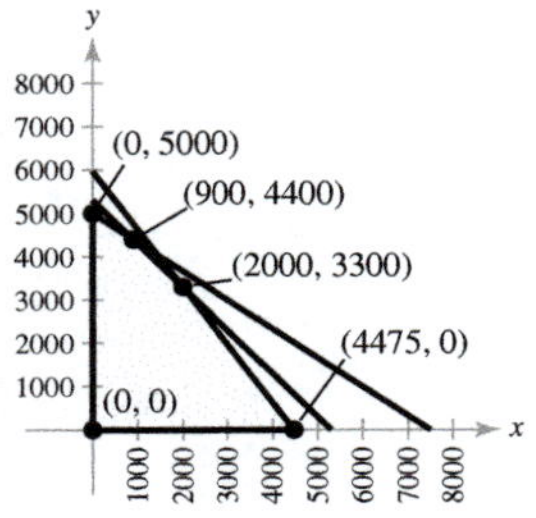

At $(0, 0)$: $P = 0$

At $(0, 5000)$: $P = \$200{,}000$

At $(900, 4400)$: $P = \$203{,}000$

At $(2000, 3300)$: $P = \$192{,}000$

At $(4475, 0)$: $P = \$134{,}250$

The manufacturer should produce 900 units of Model I and produce any of 4400 units of Model II to realize an optimal profit of \$203,000.

Chapter Test Solutions for Chapter 8

1.

$$\begin{bmatrix} 1 & -1 & 5 \\ 6 & 2 & 3 \\ 5 & 3 & -3 \end{bmatrix}$$

$$\begin{matrix} \\ -6R_1 + R_2 \rightarrow \\ -5R_1 + R_3 \rightarrow \end{matrix} \begin{bmatrix} 1 & -1 & 5 \\ 0 & 8 & -27 \\ 0 & 8 & -28 \end{bmatrix}$$

$$\begin{matrix} \\ \\ -R_2 + R_3 \rightarrow \end{matrix} \begin{bmatrix} 1 & -1 & 5 \\ 0 & 8 & -27 \\ 0 & 0 & -1 \end{bmatrix}$$

$$\begin{matrix} \\ \frac{1}{8}R_2 \rightarrow \\ -R_3 \rightarrow \end{matrix} \begin{bmatrix} 1 & -1 & 5 \\ 0 & 1 & -\frac{27}{8} \\ 0 & 0 & 1 \end{bmatrix}$$

$$\begin{matrix} R_2 + R_1 \rightarrow \\ \\ \\ \end{matrix} \begin{bmatrix} 1 & 0 & \frac{13}{8} \\ 0 & 1 & -\frac{27}{8} \\ 0 & 0 & 1 \end{bmatrix}$$

$$\begin{matrix} -\frac{13}{8}R_3 + R_1 \rightarrow \\ \frac{27}{8}R_3 + R_2 \rightarrow \\ \\ \end{matrix} \begin{bmatrix} 1 & 0 & 0 \\ 0 & 1 & 0 \\ 0 & 0 & 1 \end{bmatrix}$$

2.

$$\begin{bmatrix} 1 & 0 & -1 & 2 \\ -1 & 1 & 1 & -3 \\ 1 & 1 & -1 & 1 \\ 3 & 2 & -3 & 4 \end{bmatrix}$$

$$\begin{matrix} \\ R_1 + R_2 \rightarrow \\ -R_1 + R_3 \rightarrow \\ -3R_1 + R_4 \rightarrow \end{matrix} \begin{bmatrix} 1 & 0 & -1 & 2 \\ 0 & 1 & 0 & -1 \\ 0 & 1 & 0 & -1 \\ 0 & 2 & 0 & -2 \end{bmatrix}$$

$$\begin{matrix} \\ \\ -R_2 + R_3 \rightarrow \\ -2R_2 + R_4 \rightarrow \end{matrix} \begin{bmatrix} 1 & 0 & -1 & 2 \\ 0 & 1 & 0 & -1 \\ 0 & 0 & 0 & 0 \\ 0 & 0 & 0 & 0 \end{bmatrix}$$

3.
$$\begin{bmatrix} 4 & 3 & -2 & \vdots & 14 \\ -1 & -1 & 2 & \vdots & -5 \\ 3 & 1 & -4 & \vdots & 8 \end{bmatrix}$$

$$\begin{matrix} 3R_2 + R_1 \rightarrow \\ \\ \\ \end{matrix} \begin{bmatrix} 1 & 0 & 4 & \vdots & -1 \\ -1 & -1 & 2 & \vdots & -5 \\ 3 & 1 & -4 & \vdots & 8 \end{bmatrix}$$

$$\begin{matrix} \\ R_1 + R_2 \rightarrow \\ -3R_1 + R_3 \rightarrow \end{matrix} \begin{bmatrix} 1 & 0 & 4 & \vdots & -1 \\ 0 & -1 & 6 & \vdots & -6 \\ 0 & 1 & -16 & \vdots & 11 \end{bmatrix}$$

$$\begin{matrix} \\ \\ R_2 + R_3 \rightarrow \end{matrix} \begin{bmatrix} 1 & 0 & 4 & \vdots & -1 \\ 0 & -1 & 6 & \vdots & -6 \\ 0 & 0 & -10 & \vdots & 5 \end{bmatrix}$$

$$\begin{matrix} \\ -R_2 \rightarrow \\ -\frac{1}{10}R_3 \rightarrow \end{matrix} \begin{bmatrix} 1 & 0 & 4 & \vdots & -1 \\ 0 & 1 & -6 & \vdots & 6 \\ 0 & 0 & 1 & \vdots & -\frac{1}{2} \end{bmatrix}$$

$$\begin{matrix} -4R_3 + R_1 \rightarrow \\ 6R_3 + R_2 \rightarrow \\ \\ \end{matrix} \begin{bmatrix} 1 & 0 & 0 & \vdots & 1 \\ 0 & 1 & 0 & \vdots & 3 \\ 0 & 0 & 1 & \vdots & -\frac{1}{2} \end{bmatrix}$$

Solution: $\left(1, 3, -\frac{1}{2}\right)$

4. $A = \begin{bmatrix} 6 & 5 \\ -5 & -5 \end{bmatrix}$, $\quad B = \begin{bmatrix} 5 & 0 \\ -5 & -1 \end{bmatrix}$, $\quad C = \begin{bmatrix} 2 & -1 & 4 \\ 0 & 6 & -3 \end{bmatrix}$

(a) $A - B = \begin{bmatrix} 6 & 5 \\ -5 & -5 \end{bmatrix} - \begin{bmatrix} 5 & 0 \\ -5 & -1 \end{bmatrix} = \begin{bmatrix} 6-5 & 5-0 \\ -5-(-5) & -5-(-1) \end{bmatrix} = \begin{bmatrix} 1 & 5 \\ 0 & -4 \end{bmatrix}$

(b) $3C = 3\begin{bmatrix} 2 & -1 & 4 \\ 0 & 6 & -3 \end{bmatrix} = \begin{bmatrix} 3(2) & 3(-1) & 3(4) \\ 3(0) & 3(6) & 3(-3) \end{bmatrix} = \begin{bmatrix} 6 & -3 & 12 \\ 0 & 18 & -9 \end{bmatrix}$

(c) $3A - 2B = 3\begin{bmatrix} 6 & 5 \\ -5 & -5 \end{bmatrix} - 2\begin{bmatrix} 5 & 0 \\ -5 & -1 \end{bmatrix} = \begin{bmatrix} 3(6) - 2(5) & 3(5) - 2(0) \\ 3(-5) - 2(-5) & 3(-5) - 2(-1) \end{bmatrix} = \begin{bmatrix} 8 & 15 \\ -5 & -13 \end{bmatrix}$

(d) $$\begin{aligned} BC &= \begin{bmatrix} 5 & 0 \\ -5 & -1 \end{bmatrix}\begin{bmatrix} 2 & -1 & 4 \\ 0 & 6 & -3 \end{bmatrix} \\ &= \begin{bmatrix} (5)(2) + (0)(0) & (5)(-1) + (0)(6) & (5)(4) + (0)(-3) \\ (-5)(2) + (-1)(0) & (-5)(-1) + (-1)(6) & (-5)(4) + (-1)(-3) \end{bmatrix} \\ &= \begin{bmatrix} 10 & -5 & 20 \\ -10 & -1 & -17 \end{bmatrix} \end{aligned}$$

(e) C^2 is not possible.

5. $A = \begin{bmatrix} 0 & -1 \\ -1 & 0 \end{bmatrix}$

$\mathbf{v} = \langle 2, 3 \rangle = \begin{bmatrix} 2 \\ 3 \end{bmatrix}$

$A\mathbf{v} = \begin{bmatrix} 0 & -1 \\ -1 & 0 \end{bmatrix}\begin{bmatrix} 2 \\ 3 \end{bmatrix} = \begin{bmatrix} -3 \\ -2 \end{bmatrix} = \langle -3, -2 \rangle$ is a reflection of $\mathbf{v}$ in the line $y = -x$.

6. $A = \begin{bmatrix} a & b \\ c & d \end{bmatrix}, \qquad A^{-1} = \dfrac{1}{ad - bc}\begin{bmatrix} d & -b \\ -c & a \end{bmatrix}$

$A = \begin{bmatrix} -4 & 3 \\ 5 & -2 \end{bmatrix}$

$ad - bc = (-4)(-2) - (3)(5) = -7$

$$A^{-1} = -\frac{1}{7}\begin{bmatrix} -2 & -3 \\ -5 & -4 \end{bmatrix} = \begin{bmatrix} \frac{2}{7} & \frac{3}{7} \\ \frac{5}{7} & \frac{4}{7} \end{bmatrix}$$

7.
$$\left[\begin{array}{ccc:ccc} -2 & 4 & -6 & 1 & 0 & 0 \\ 2 & 1 & 0 & 0 & 1 & 0 \\ 4 & -2 & 5 & 0 & 0 & 1 \end{array}\right]$$

$$\begin{array}{r} \\ R_1 + R_2 \rightarrow \\ 2R_1 + R_3 \rightarrow \end{array}\left[\begin{array}{ccc:ccc} -2 & 4 & -6 & 1 & 0 & 0 \\ 0 & 5 & -6 & 1 & 1 & 0 \\ 0 & 6 & -7 & 2 & 0 & 1 \end{array}\right]$$

$$\begin{array}{r} -\frac{1}{2}R_1 \rightarrow \\ -R_3 + R_2 \rightarrow \\ \\ \end{array}\left[\begin{array}{ccc:ccc} 1 & -2 & 3 & -\frac{1}{2} & 0 & 0 \\ 0 & -1 & 1 & -1 & 1 & -1 \\ 0 & 6 & -7 & 2 & 0 & 1 \end{array}\right]$$

$$\begin{array}{r} -2R_2 + R_1 \rightarrow \\ \\ 6R_2 + R_3 \rightarrow \end{array}\left[\begin{array}{ccc:ccc} 1 & 0 & 1 & \frac{3}{2} & -2 & 2 \\ 0 & -1 & 1 & -1 & 1 & -1 \\ 0 & 0 & -1 & -4 & 6 & -5 \end{array}\right]$$

$$\begin{array}{r} \\ -R_2 \rightarrow \\ -R_3 \rightarrow \end{array}\left[\begin{array}{ccc:ccc} 1 & 0 & 1 & \frac{3}{2} & -2 & 2 \\ 0 & 1 & -1 & 1 & -1 & 1 \\ 0 & 0 & 1 & 4 & -6 & 5 \end{array}\right]$$

$$\begin{array}{r} -R_3 + R_1 \rightarrow \\ R_3 + R_2 \rightarrow \\ \\ \end{array}\left[\begin{array}{ccc:ccc} 1 & 0 & 0 & -\frac{5}{2} & 4 & -3 \\ 0 & 1 & 0 & 5 & -7 & 6 \\ 0 & 0 & 1 & 4 & -6 & 5 \end{array}\right]$$

$$A^{-1} = \begin{bmatrix} -\frac{5}{2} & 4 & -3 \\ 5 & -7 & 6 \\ 4 & -6 & 5 \end{bmatrix}$$

8. $\begin{cases} -4x + 3y = 6 \\ 5x - 2y = 24 \end{cases}$

$$\begin{bmatrix} -4 & 3 \\ 5 & -2 \end{bmatrix}\begin{bmatrix} x \\ y \end{bmatrix} = \begin{bmatrix} 6 \\ 24 \end{bmatrix}$$

$$\begin{bmatrix} x \\ y \end{bmatrix} = \begin{bmatrix} -4 & 3 \\ 5 & -2 \end{bmatrix}^{-1}\begin{bmatrix} 6 \\ 24 \end{bmatrix} = \begin{bmatrix} \frac{2}{7} & \frac{3}{7} \\ \frac{5}{7} & \frac{4}{7} \end{bmatrix}\begin{bmatrix} 6 \\ 24 \end{bmatrix} = \begin{bmatrix} 12 \\ 18 \end{bmatrix}$$

Solution: $(12, 18)$

9. $\begin{vmatrix} -6 & 4 \\ 10 & 12 \end{vmatrix} = (-6)(12) - (4)(10) = -112$

10. $\begin{vmatrix} \frac{5}{2} & -\frac{3}{8} \\ -8 & \frac{6}{5} \end{vmatrix} = \left(\frac{5}{2}\right)\left(\frac{6}{5}\right) - \left(-\frac{3}{8}\right)(-8) = 3 - 3 = 0$

11. Expand along Column 3.

$$\begin{vmatrix} 6 & -7 & 2 \\ 3 & -2 & 0 \\ 1 & 5 & 1 \end{vmatrix} = 2\begin{vmatrix} 3 & -2 \\ 1 & 5 \end{vmatrix} + \begin{vmatrix} 6 & -7 \\ 3 & -2 \end{vmatrix} = 2(17) + 9 = 43$$

12. $\begin{cases} 7x + 6y = 9 \\ -2x - 11y = -49 \end{cases} \qquad D = \begin{vmatrix} 7 & 6 \\ -2 & -11 \end{vmatrix} = -65$

$$x = \frac{\begin{vmatrix} 9 & 6 \\ -49 & -11 \end{vmatrix}}{-65} = \frac{195}{-65} = -3$$

$$y = \frac{\begin{vmatrix} 7 & 9 \\ -2 & -49 \end{vmatrix}}{-65} = \frac{-325}{-65} = 5$$

Solution: $(-3, 5)$

13. $\begin{cases} 6x - y + 2z = -4 \\ -2x + 3y - z = 10 \\ 4x - 4y + z = -18 \end{cases} \qquad D = \begin{vmatrix} 6 & -1 & 2 \\ -2 & 3 & -1 \\ 4 & -4 & 1 \end{vmatrix} = -12$

$$x = \frac{\begin{vmatrix} -4 & -1 & 2 \\ 10 & 3 & -1 \\ -18 & -4 & 1 \end{vmatrix}}{-12} = \frac{24}{-12} = -2$$

$$y = \frac{\begin{vmatrix} 6 & -4 & 2 \\ -2 & 10 & -1 \\ 4 & -18 & 1 \end{vmatrix}}{-12} = \frac{-48}{-12} = 4$$

$$z = \frac{\begin{vmatrix} 6 & -1 & -4 \\ -2 & 3 & 10 \\ 4 & -4 & -18 \end{vmatrix}}{-12} = \frac{-72}{-12} = 6$$

Solution: $(-2, 4, 6)$

14. $A = -\frac{1}{2}\begin{vmatrix} -5 & 0 & 1 \\ 4 & 4 & 1 \\ 3 & 2 & 1 \end{vmatrix} = -\frac{1}{2}(-14) = 7$

15. $\begin{matrix} K & N & O \\ C & K & - \\ O & N & - \\ W & O & O \\ D & - & - \end{matrix}\begin{bmatrix} 11 & 14 & 15 \\ 3 & 11 & 0 \\ 15 & 14 & 0 \\ 23 & 15 & 15 \\ 4 & 0 & 0 \end{bmatrix}\begin{bmatrix} 1 & -1 & 0 \\ 1 & 0 & -1 \\ 6 & -2 & -3 \end{bmatrix} = \begin{bmatrix} 115 & -41 & -59 \\ 14 & -3 & -11 \\ 29 & -15 & -14 \\ 128 & -53 & -60 \\ 4 & -4 & 0 \end{bmatrix}$

Message: $[11 \quad 14 \quad 15], [3 \quad 11 \quad 0], [15 \quad 14 \quad 0], [23 \quad 15 \quad 15], [4 \quad 0 \quad 0]$

Encoded Message: 115 −41 −59 14 −3 −11 29 −15 −14 128 −53 −60 4 −4 0

16. Let x = amount of 60% solution and y = amount of 20% solution.

$$\begin{cases} x + y = 100 \Rightarrow y = 100 - x \\ 0.60x + 0.20y = 0.50(100) \Rightarrow 6x + 2y = 500 \end{cases}$$

By substitution,

$$\begin{aligned} 6x + 2(100 - x) &= 500 \\ 6x + 200 - 2x &= 500 \\ 4x &= 300 \\ x &= 75 \\ y &= 100 - x = 25. \end{aligned}$$

75 liters of 60% solution and 25 liters of 20% solution

Chapter Test Solutions for Chapter 9

1. $a_n = \dfrac{(-1)^n}{3n + 2}$

$a_1 = -\dfrac{1}{5}$

$a_2 = \dfrac{1}{8}$

$a_3 = -\dfrac{1}{11}$

$a_4 = \dfrac{1}{14}$

$a_5 = -\dfrac{1}{17}$

2. $\dfrac{3}{1!}, \dfrac{4}{2!}, \dfrac{5}{3!}, \dfrac{6}{4!}, \dfrac{7}{5!}, \ldots$

$a_n = \dfrac{n + 2}{n!}$

3. $8 + 21 + 34 + 47 + \ldots$

$a_5 = 60, a_6 = 73, a_7 = 86$

$S_7 = 8 + 21 + 34 + 47 + 60 + 73 + 86 = 329$

4.
$$\begin{aligned} a_5 &= 45, a_{12} = 24 \\ a_{12} &= a_5 + 7d \\ 24 &= 45 + 7d \\ -21 &= 7d \\ -3 &= d \\ a_1 &= a_5 - 4d \\ a_1 &= 45 - 4(-3) \\ &= 57 \\ a_n &= a_1 + (n - 1)d \\ &= 57 + (n - 1)(-3) \\ &= -3n + 60 \end{aligned}$$

5.
$$\begin{aligned} a_2 &= 14, a_6 = 224 \\ a_6 &= a_2 r^4 \\ 224 &= 14r^4 \\ 16 &= r^4 \\ 2 &= r \\ a_2 &= a_1 r \\ 14 &= a_1(2) \\ 7 &= a_1 \\ a_n &= 7(2)^{n-1} \end{aligned}$$

or

$$\begin{aligned} &= 7 \cdot 2^n 2^{-1} \\ &= \frac{7}{2}(2)^n \end{aligned}$$

6. $\displaystyle\sum_{i=1}^{50}(2i^2+5) = 2\sum_{i=1}^{50} i^2 + \sum_{i=1}^{50} 5$

$= 2\left[\dfrac{50(51)(101)}{6}\right] + 50(5)$

$= 86{,}100$

7. $\displaystyle\sum_{n=1}^{9}(12n-7) = 12\sum_{n=1}^{9} n - \sum_{n=1}^{9} 7$

$= 12\left[\dfrac{9(10)}{2}\right] - 9(7)$

$= 477$

8. $\displaystyle\sum_{i=1}^{\infty} 4\left(\frac{1}{2}\right)^i = \frac{2}{1-\frac{1}{2}} = 4$

9. $\displaystyle\sum_{n=1}^{\infty}\left(-\frac{1}{3}\right)^n = \frac{-\frac{1}{3}}{1-\left(-\frac{1}{3}\right)} = \frac{-\frac{1}{3}}{\frac{4}{3}} = -\frac{1}{4}$

10. $5 + 10 + 15 + \cdots + 5n = \dfrac{5n(n+1)}{2}$

When $n = 1$, $S_1 = 5 = \dfrac{5(1)(2)}{2}$, so the formula is valid.

Assume that

$S_k = 5 + 10 + 15 + \cdots + 5k = \dfrac{5k(k+1)}{2}$, then

$$\begin{aligned} S_{k+1} &= S_k + a_{k+1} \\ &= \frac{5k(k+1)}{2} + 5(k+1) \\ &= \frac{5k(k+1)}{2} + \frac{10(k+1)}{2} \\ &= \frac{5k(k+1) + 10(k+1)}{2} \\ &= \frac{5(k+1)(k+2)}{2} \\ &= \frac{5(k+1)[(k+1)+1]}{2}. \end{aligned}$$

So, the formula is valid for all integers $n \geq 1$.

11. $(x+6y)^4 = x^4 + {}_4C_1x^3(6y) + {}_4C_2x^2(6y)^2 + {}_4C_3x(6y)^3 + {}_4C_4(6y)^4$

$= x^4 + 24x^3y + 216x^2y^2 + 864xy^3 + 1296y^4$

12. 5th Row of Pascal's Triangle: 1 5 10 10 5 1 and 3rd Row of Pascal's Triangle: 1 3 3 1

$$\begin{aligned} 3(x-2)^5 + 4(x-2)^3 &= 3\left[(1)x^5 + (5)x^4(-2) + (10)x^3(-2)^2 + (10)x^2(-2)^3 + (5)x(-2)^4 + (1)(-2)^5\right] \\ &\quad + 4\left[(1)x^3 + (3)x^2(-2) + (3)x(-2)^2 + (1)(-2)^3\right] \\ &= 3(x^5 - 10x^4 + 40x^3 - 80x^2 + 80x - 32) + 4(x^3 - 6x^2 + 12x - 8) \\ &= 3x^5 - 30x^4 + 124x^3 - 264x^2 + 288x - 128 \end{aligned}$$

13. ${}_nC_rx^{n-r}y^r = {}_7C_3(3a)^4(-2b)^3$

$= 35(81a^4)(-8b^3)$

$= -22{,}680a^4b^3$

So, the coefficient of a^4b^3 is $-22{,}680$.

14. (a) ${}_9P_2 = \dfrac{9!}{7!} = 72$

(b) ${}_{70}P_3 = \dfrac{70!}{67!} = 328{,}440$

15. (a) ${}_{11}C_4 = \dfrac{11!}{7!4!} = 330$

(b) ${}_{66}C_4 = \dfrac{66!}{62!4!} = 720{,}720$

16. $(26)(10)(10)(10) = 26{,}000$ distinct license plates

17. $\underbrace{(1)}_{\text{owner}} \cdot \underbrace{(3)(2)}_{\text{bow seats}} \cdot \underbrace{(5)(4)(3)(2)(1)}_{\text{remaining seats}} = 720$ seating arrangements

18. $\frac{20}{300} = \frac{1}{15} \approx 0.0667$

19. $\dfrac{1}{{}_{30}C_4} = \dfrac{1}{27{,}405}$

20. $P(E') = 1 - P(E)$

$= 1 - 0.90$

$= 0.10$ or 10%

Cumulative Test Solutions for Chapters 7–9

1. $\begin{cases} y = 3 - x^2 \\ 2(y - 2) = x - 1 \Rightarrow 2(3 - x^2 - 2) = x - 1 \end{cases}$

$$\begin{aligned} 2(1 - x^2) &= x - 1 \\ 2 - 2x^2 &= x - 1 \\ 0 &= 2x^2 + x - 3 \\ 0 &= (2x + 3)(x - 1) \\ x &= -\tfrac{3}{2} \text{ or } x = 1 \\ y &= \tfrac{3}{4} \qquad y = 2 \end{aligned}$$

Solutions: $\left(-\frac{3}{2}, \frac{3}{4}\right), (1, 2)$

2. $\begin{cases} x + 3y = -6 \Rightarrow 4x + 12y = -24 \\ 2x + 4y = -10 \Rightarrow -6x - 12y = 30 \end{cases}$

$$\begin{aligned} -2x &= 6 \\ x = -3 &\Rightarrow y = -1 \end{aligned}$$

Solution: $(-3, -1)$

3. $\begin{cases} -2x + 4y - z = -16 \\ x - 2y + 2z = 5 \\ x - 3y - z = 13 \end{cases}$

Interchange equations.

$\begin{cases} x - 2y + 2z = 5 & \text{Eq.1} \\ -2x + 4y - z = -16 & \text{Eq.2} \\ x - 3y - z = 13 & \text{Eq.3} \end{cases}$

$\begin{cases} x - 2y + 2z = 5 & \\ 3z = -6 & 2\text{Eq.1} + \text{Eq.2} \\ -y - 3z = 8 & -\text{Eq.1} + \text{Eq.3} \end{cases}$

From Equation 2, $z = -2$. Substituting this into Equation 3 yields $y = -2$. Using these in Equation 1 yields $x = 5$.

Solution: $(5, -2, -2)$

4. $\begin{cases} x + 3y - 2z = -7 \\ -2x + y - z = -5 \\ 4x + y + z = 3 \end{cases}$

$\begin{cases} x + 3y - 2z = -7 & \\ 7y - 5z = -19 & 2\text{Eq. 1} + \text{Eq.2} \\ -11y + 9z = 31 & -4\text{Eq. 1} + \text{Eq.3} \end{cases}$

$\begin{cases} x + 3y - 2z = -7 & \\ y - \frac{5}{7}z = -\frac{19}{7} & \frac{1}{7}\text{Eq.2} \\ -11y + 9z = 31 & \end{cases}$

$\begin{cases} x + \frac{1}{7}z = \frac{8}{7} & -3\text{Eq.2} + \text{Eq.1} \\ y - \frac{5}{7}z = -\frac{19}{7} & \\ \frac{8}{7}z = \frac{8}{7} & 11\text{Eq.2} + \text{Eq.3} \end{cases}$

$\begin{cases} x + \frac{1}{7}z = \frac{8}{7} & \\ y - \frac{5}{7}z = -\frac{19}{7} & \\ z = 1 & \frac{7}{8}\text{Eq.3} \end{cases}$

$\begin{cases} x = 1 & -\frac{1}{7}\text{Eq.3} + \text{Eq.1} \\ y = -2 & \frac{5}{7}\text{Eq.3} + \text{Eq.2} \\ z = 1 & \end{cases}$

Solution: $(1, -2, 1)$

5. $\begin{cases} x + y = 200 \Rightarrow y = 200 - x \\ 0.75x + 1.25y = 0.95(200) \end{cases}$

$$\begin{aligned} 0.75x + 1.25(200 - x) &= 190 \\ 0.75x + 250 - 1.25x &= 190 \\ -0.50x &= -60 \\ x &= 120 \\ y = 200 - x &= 80 \end{aligned}$$

120 pounds of \$0.75 seed and 80 pounds of \$1.25 seed.

6. $y = ax^2 + bx + c$

$(0, 6)$: $6 = a(0)^2 + b(0) + c \Rightarrow c = 6$

$(2, 3)$: $3 = a(2)^2 + b(2) + 6 \Rightarrow 4a + 2b = -3$

$2a + b = -\frac{3}{2}$

$(4, 2)$: $2 = a(4)^2 + b(4) + 6 \Rightarrow 16a + 4b = -4$

$4a + b = -1$

Solving the system:

$\begin{cases} 2a + b = -\frac{3}{2} \\ 4a + b = -1 \end{cases}$ yields $a = \frac{1}{4}$ and $b = -2$.

So, the equation of the parabola is $y = \frac{1}{4}x^2 - 2x + 6$.

7. $\dfrac{2x^2 - x - 6}{x(x^2 + 2)} = \dfrac{A}{x} + \dfrac{Bx + C}{x^2 + 2}$

$$\begin{aligned} 2x^2 - x - 6 &= A(x^2 + 2) + x(Bx + C) \\ &= Ax^2 + 2A + Bx^2 + Cx \\ &= (A + B)x^2 + Cx + 2A \\ 2A &= -6 \rightarrow A = -3 \\ C &= -1 \\ A + B &= 2 \rightarrow B = 5 \end{aligned}$$

$$\frac{2x^2 - x - 6}{x(x^2 + 2)} = -\frac{3}{x} + \frac{5x - 1}{x^2 + 2}$$

8. $\begin{cases} 2x + y \geq -3 \\ x - 3y \leq 2 \end{cases}$

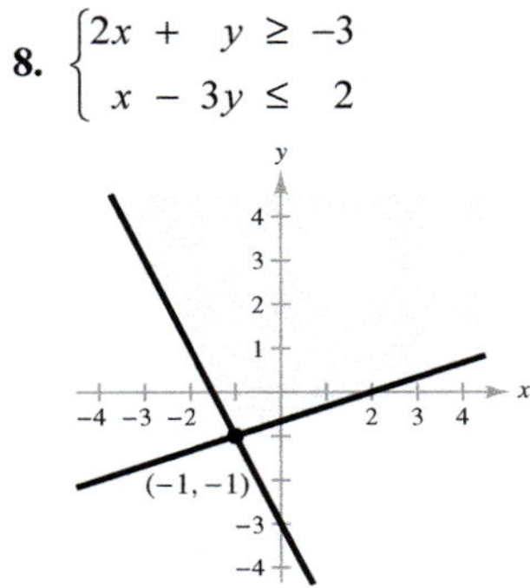

9. $\begin{cases} x - y > 6 \\ 5x + 2y < 10 \end{cases}$

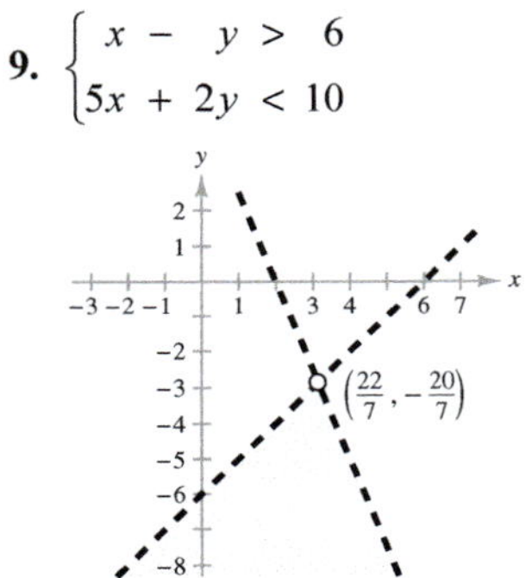

10. Objective function: $z = 3x + 2y$

Subject to: $x + 4y \leq 20$

$2x + y \leq 12$

$x \geq 0, y \geq 0$

At $(0, 0)$: $z = 0$

At $(0, 5)$: $z = 10$

At $(4, 4)$: $z = 30$

At $(6, 0)$: $z = 18$

Minimum of $z = 0$ at $(0, 0)$

Maximum of $z = 20$ at $(4, 4)$

11. $\begin{cases} -x + 2y - z = 9 \\ 2x - y + 2z = -9 \\ 3x + 3y - 4z = 7 \end{cases}$

$$\left[\begin{array}{ccc|c} -1 & 2 & -1 & 9 \\ 2 & -1 & 2 & -9 \\ 3 & 3 & -4 & 7 \end{array}\right]$$

12.

$$\left[\begin{array}{ccc|c} -1 & 2 & -1 & 9 \\ 2 & -1 & 2 & -9 \\ 3 & 3 & -4 & 7 \end{array}\right]$$

$$\begin{array}{r} \\ 2R_1 + R_2 \rightarrow \\ 3R_1 + R_3 \rightarrow \end{array} \left[\begin{array}{ccc|c} -1 & 2 & -1 & 9 \\ 0 & 3 & 0 & 9 \\ 0 & 9 & -7 & 34 \end{array}\right]$$

$$\begin{array}{r} -R_1 \rightarrow \\ \\ -3R_2 + R_3 \rightarrow \end{array} \left[\begin{array}{ccc|c} 1 & -2 & 1 & -9 \\ 0 & 3 & 0 & 3 \\ 0 & 0 & -7 & 7 \end{array}\right]$$

$$\begin{array}{r} \\ \frac{1}{3}R_2 \rightarrow \\ -\frac{1}{7}R_2 \rightarrow \end{array} \left[\begin{array}{ccc|c} 1 & -2 & 1 & -9 \\ 0 & 1 & 0 & 3 \\ 0 & 0 & 1 & -1 \end{array}\right]$$

$$\begin{array}{r} 2R_2 + R_1 \rightarrow \\ \\ \\ \end{array} \left[\begin{array}{ccc|c} 1 & 0 & 1 & -3 \\ 0 & 1 & 0 & 3 \\ 0 & 0 & 1 & -1 \end{array}\right]$$

$$\begin{array}{r} -R_3 + R_1 \rightarrow \\ \\ \\ \end{array} \left[\begin{array}{ccc|c} 1 & 0 & 0 & -2 \\ 0 & 1 & 0 & 3 \\ 0 & 0 & 1 & -1 \end{array}\right]$$

Solution: $(-2, 3, -1)$

In Exercises 13-18,

$$A = \begin{bmatrix} -1 & 3 \\ 6 & 2 \end{bmatrix}, B = \begin{bmatrix} -2 & 5 \\ 0 & -1 \end{bmatrix} \text{ and } C = \begin{bmatrix} 4 & 0 & 1 \\ -3 & 2 & -1 \end{bmatrix}$$

13. $A + B = \begin{bmatrix} -1 & 3 \\ 6 & 2 \end{bmatrix} + \begin{bmatrix} -2 & 5 \\ 0 & -1 \end{bmatrix} = \begin{bmatrix} -3 & 8 \\ 6 & 1 \end{bmatrix}$

14.
$$\begin{aligned} 2A - 5B &= 2A + (-5)B \\ &= 2\begin{bmatrix} -1 & 3 \\ 6 & 2 \end{bmatrix} + (-5)\begin{bmatrix} -2 & 5 \\ 0 & -1 \end{bmatrix} \\ &= \begin{bmatrix} -2 & 6 \\ 12 & 4 \end{bmatrix} + \begin{bmatrix} 10 & -25 \\ 0 & 5 \end{bmatrix} \\ &= \begin{bmatrix} 8 & -19 \\ 12 & 9 \end{bmatrix} \end{aligned}$$

15. $AC = \begin{bmatrix} -1 & 3 \\ 6 & 2 \end{bmatrix}\begin{bmatrix} 4 & 0 & 1 \\ -3 & 2 & -1 \end{bmatrix} = \begin{bmatrix} (-1)(4) + (3)(-3) & (-1)(0) + (3)(2) & (-1)(1) + (3)(-1) \\ (6)(4) + (2)(-3) & (6)(0) + (2)(2) & (6)(1) + (2)(-1) \end{bmatrix} = \begin{bmatrix} -13 & 6 & -4 \\ 18 & 4 & 4 \end{bmatrix}$

16. CB not possible. The number of columns of C is not equal to the number of rows of B.

17. $A^2 = \begin{bmatrix} -1 & 3 \\ 6 & 2 \end{bmatrix}\begin{bmatrix} -1 & 3 \\ 6 & 2 \end{bmatrix} = \begin{bmatrix} (-1)(-1) + (3)(6) & (-1)(3) + (3)(2) \\ (6)(-1) + (2)(6) & (6)(3) + (2)(2) \end{bmatrix} = \begin{bmatrix} 19 & 3 \\ 6 & 22 \end{bmatrix}$

18.
$$\begin{aligned} BA - B^2 &= \begin{bmatrix} -2 & 5 \\ 0 & -1 \end{bmatrix}\begin{bmatrix} -1 & 3 \\ 6 & 2 \end{bmatrix} - \begin{bmatrix} -2 & 5 \\ 0 & -1 \end{bmatrix}\begin{bmatrix} -2 & 5 \\ 0 & -1 \end{bmatrix} \\ &= \begin{bmatrix} -2(-1) + 5(6) & -2(3) + 5(2) \\ 0(-1) + (-1)(6) & 0(3) + (-1)(2) \end{bmatrix} - \begin{bmatrix} -2(-2) + 5(0) & -2(5) + 5(-1) \\ 0(-2) + (-1)(0) & 0(5) + (-1)(-1) \end{bmatrix} \\ &= \begin{bmatrix} 32 & 4 \\ -6 & -2 \end{bmatrix} - \begin{bmatrix} 4 & -15 \\ 0 & 1 \end{bmatrix} \\ &= \begin{bmatrix} 28 & 19 \\ -6 & -3 \end{bmatrix} \end{aligned}$$

19.
$$\left[\begin{array}{ccc:ccc} 1 & 2 & -1 & 1 & 0 & 0 \\ 3 & 7 & -10 & 0 & 1 & 0 \\ -5 & -7 & -15 & 0 & 0 & 1 \end{array}\right]$$

$$\begin{matrix} \\ -3R_1 + R_2 \rightarrow \\ 5R_1 + R_3 \rightarrow \end{matrix} \left[\begin{array}{ccc:ccc} 1 & 2 & -1 & 1 & 0 & 0 \\ 0 & 1 & -7 & -3 & 1 & 0 \\ 0 & 3 & -20 & 5 & 0 & 1 \end{array}\right]$$

$$\begin{matrix} -2R_2 + R_1 \rightarrow \\ \\ -3R_2 + R_3 \rightarrow \end{matrix} \left[\begin{array}{ccc:ccc} 1 & 0 & 13 & 7 & -2 & 0 \\ 0 & 1 & -7 & -3 & 1 & 0 \\ 0 & 0 & 1 & 14 & -3 & 1 \end{array}\right]$$

$$\begin{matrix} -13R_3 + R_1 \rightarrow \\ 7R_3 + R_2 \rightarrow \\ \\ \end{matrix} \left[\begin{array}{ccc:ccc} 1 & 0 & 0 & -175 & 37 & -13 \\ 0 & 1 & 0 & 95 & -20 & 7 \\ 0 & 0 & 1 & 14 & -3 & 1 \end{array}\right]$$

$$\begin{bmatrix} 1 & 2 & -1 \\ 3 & 7 & -10 \\ -5 & -7 & -15 \end{bmatrix}^{-1} = \begin{bmatrix} -175 & 37 & -13 \\ 95 & -20 & 7 \\ 14 & -3 & 1 \end{bmatrix}$$

20. Expand along Row 1.

$$\begin{vmatrix} 7 & 1 & 0 \\ -2 & 4 & -1 \\ 3 & 8 & 5 \end{vmatrix} = 7\begin{vmatrix} 4 & -1 \\ 8 & 5 \end{vmatrix} - 1\begin{vmatrix} -2 & -1 \\ 3 & 5 \end{vmatrix} = 7(28) - 1(-7) = 203$$

21. To produce a reflection in the x-axis, use the matrix $\begin{bmatrix}1 & 0\\0 & -1\end{bmatrix}$ and multiply by each vertex matrix $\begin{bmatrix}0\\2\end{bmatrix}, \begin{bmatrix}0\\5\end{bmatrix}, \begin{bmatrix}3\\2\end{bmatrix}, \begin{bmatrix}3\\5\end{bmatrix}$.

$$\begin{bmatrix}1 & 0\\0 & -1\end{bmatrix}\begin{bmatrix}0\\2\end{bmatrix} = \begin{bmatrix}(1)(0) + (0)(2)\\(0)(0) + (-1)(2)\end{bmatrix} = \begin{bmatrix}0\\-2\end{bmatrix}$$

$$\begin{bmatrix}1 & 0\\0 & -1\end{bmatrix}\begin{bmatrix}0\\5\end{bmatrix} = \begin{bmatrix}(1)(0) + (0)(5)\\(0)(0) + (-1)(5)\end{bmatrix} = \begin{bmatrix}0\\-5\end{bmatrix}$$

$$\begin{bmatrix}1 & 0\\0 & -1\end{bmatrix}\begin{bmatrix}3\\2\end{bmatrix} = \begin{bmatrix}(1)(3) + (0)(2)\\(0)(3) + (-1)(2)\end{bmatrix} = \begin{bmatrix}3\\-2\end{bmatrix}$$

$$\begin{bmatrix}1 & 0\\0 & -1\end{bmatrix}\begin{bmatrix}3\\5\end{bmatrix} = \begin{bmatrix}(1)(3) + (0)(5)\\(0)(3) + (-1)(5)\end{bmatrix} = \begin{bmatrix}3\\-5\end{bmatrix}$$

22. Let x = total sales of gym shoes (in millions),

y = total sales of jogging shoes (in millions),

z = total sales of walking shoes (in millions).

$$\begin{bmatrix}0.079 & 0.064 & 0.029\\0.050 & 0.060 & 0.020\\0.103 & 0.159 & 0.085\end{bmatrix}\begin{bmatrix}x\\y\\z\end{bmatrix} = \begin{bmatrix}479.88\\365.88\\1248.89\end{bmatrix}$$

$$\begin{bmatrix}x\\y\\z\end{bmatrix} = \begin{bmatrix}0.079 & 0.064 & 0.029\\0.050 & 0.060 & 0.020\\0.103 & 0.159 & 0.085\end{bmatrix}^{-1}\begin{bmatrix}479.88\\365.88\\1248.89\end{bmatrix} \approx \begin{bmatrix}2539\\2362\\4418\end{bmatrix}$$

So, sales for each type of shoe amounted to:

Gym shoes: \$2539 million

Jogging shoes: \$2362 million

Walking shoes: \$4418 million

23. $\begin{cases}8x - 3y = -52\\3x + 5y = 5\end{cases}$, $D = \begin{vmatrix}8 & -3\\3 & 5\end{vmatrix} = 49$

$$x = \frac{\begin{vmatrix}-52 & -3\\5 & 5\end{vmatrix}}{49} = \frac{-245}{49} = -5$$

$$y = \frac{\begin{vmatrix}8 & -52\\3 & 5\end{vmatrix}}{49} = \frac{196}{49} = 4$$

Solution: $(-5, 4)$

24. $\begin{cases}5x + 4y + 3z = 7\\-3x - 8y + 7z = -9,\\7x - 5y - 6z = -53\end{cases}$ $D = \begin{vmatrix}5 & 4 & 3\\-3 & -8 & 7\\7 & -5 & -6\end{vmatrix} = 752$

$$x = \frac{\begin{vmatrix}7 & 4 & 3\\-9 & -8 & 7\\-53 & -5 & -6\end{vmatrix}}{752} = \frac{-2256}{752} = -3$$

$$y = \frac{\begin{vmatrix}5 & 7 & 3\\-3 & -9 & 7\\7 & -53 & -6\end{vmatrix}}{752} = \frac{3008}{752} = 4$$

$$z = \frac{\begin{vmatrix}5 & 4 & 7\\-3 & -8 & -9\\7 & -5 & -53\end{vmatrix}}{752} = \frac{1504}{752} = 2$$

Solution: $(-3, 4, 2)$

25. $A = \pm\frac{1}{2}\begin{vmatrix} -2 & 3 & 1 \\ 1 & 5 & 1 \\ 4 & 1 & 1 \end{vmatrix} = -\frac{1}{2}(-18) = 9$

26. $a_n = \dfrac{(-1)^{n+1}}{2n + 3}$

$a_1 = \dfrac{1}{5}$

$a_2 = -\dfrac{1}{7}$

$a_3 = \dfrac{1}{9}$

$a_4 = -\dfrac{1}{11}$

$a_5 = \dfrac{1}{13}$

27. $\dfrac{2!}{4}, \dfrac{3!}{5}, \dfrac{4!}{6}, \dfrac{5!}{7}, \dfrac{6!}{8}, \ldots$

$a_n = \dfrac{(n + 1)!}{n + 3}$

28. $6, 18, 30, 42, \ldots$

$a_n = 12n - 6$

$a_1 = 6, a_{16} = 186$

$S_{16} = \frac{16}{2}(6 + 186) = 1536$

29. (a)

$$\begin{aligned} a_6 &= 20.6 \\ a_9 &= 30.2 \\ a_9 &= a_6 + 3d \\ 30.2 &= 20.6 + 3d \\ 9.6 &= 3d \\ 3.2 &= d \\ a_{20} &= a_9 + 11d = 30.2 + 11(3.2) = 65.4 \end{aligned}$$

(b)

$$\begin{aligned} a_1 &= a_6 - 5d \\ a_1 &= 20.6 - 5(3.2) \\ &= 4.6 \\ a_n &= a_1 + (n - 1)d \\ &= 4.6 + (n - 1)(3.2) \\ &= 3.2n + 1.4 \end{aligned}$$

30. $a_n = 3(2)^{n-1}$

$a_1 = 3$

$a_2 = 6$

$a_3 = 12$

$a_4 = 24$

$a_5 = 48$

31.

$$\begin{aligned} \sum_{i=0}^{\infty} 1.9\left(\frac{1}{10}\right)^{i-1} &= \sum_{i=0}^{\infty} 1.9\left(\frac{1}{10}\right)^{i}\left(\frac{1}{10}\right)^{-1} \\ &= \sum_{i=0}^{\infty} 19\left(\frac{1}{10}\right)^{i} \\ &= \frac{19}{1 - \frac{1}{10}} = 19\left(\frac{10}{9}\right) = \frac{190}{9} \end{aligned}$$

32. 1. When $n = 2$, $3! = 6$ and $2^2 = 4$, thus $3! > 2^2$.

2. Assume

$(k + 1)! > 2^k, k > 2.$

Then, we need to show that $(k + 2)! > 2^{k+1}$.

$(k + 2)! = (k + 1)!(k + 2) > 2^k(2)$ since $k + 2 > 2$.

Thus, $(k + 2)! > 2^{k+1}$.

Therefore, by mathematical induction, the formula is valid for all integers n such that $n \geq 2$.

33.

$$\begin{aligned} (w - 9)^4 &= w^4 + {}_4C_1w^3(-9) + {}_4C_2w^2(-9)^2 + {}_4C_3w(-9)^3 + (-9)^4 \\ &= w^4 - 36w^3 + 486w^2 - 2916w + 6561 \end{aligned}$$

34. ${}_{14}P_3 = \dfrac{14!}{(14 - 3)!} = \dfrac{14!}{11!} = 2184$

35. ${}_{25}P_2 = \dfrac{25!}{(25 - 2)!} = \dfrac{25!}{23!} = 600$

36. $\dbinom{8}{4} = {}_8C_4 = \dfrac{8!}{(8 - 4)!4!} = \dfrac{8!}{4!4!} = 70$

37. ${}_{11}C_6 = \dfrac{11!}{(11 - 6)!6!} = \dfrac{11!}{5!6!} = 462$

38. B A S K E T B A L L

$$\frac{10!}{2!2!2!1!1!1!1!} = 453{,}600 \text{ distinguishable permutations}$$

39. A N T A R C T I C A

$$\frac{10!}{3!2!2!1!1!1!} = 151{,}200 \text{ distinguishable permutations}$$

40. $_{10}P_3 = \dfrac{10!}{(10-3)!} = \dfrac{10!}{7!} = 720$

41. The first digit is 4 or 5, so the probability of picking it correctly is $\frac{1}{2}$. Then there are two numbers left for the second digit so its probability is also $\frac{1}{2}$. If these two are correct, then the third digit must be the remaining number. The probability of winning is

$\left(\frac{1}{2}\right)\left(\frac{1}{2}\right)(1) = \frac{1}{4}.$

Chapter 10 Chapter Test Solutions

1. $4x - 7y + 6 = 0$

$$y = \tfrac{4}{7}x + \tfrac{6}{7}$$

$\tan\theta = \frac{4}{7}$

$\theta \approx 0.5191 \text{ radian} \approx 29.7°$

2. $3x + y = 6 \Rightarrow y = -3x + 6 \Rightarrow m_1 = -3$

$5x - 2y = -4 \Rightarrow y = \frac{5}{2}x + 2 \Rightarrow m_2 = \frac{5}{2}$

$$\tan\theta = \left|\frac{\frac{5}{2} - (-3)}{1 + \frac{5}{2}(-3)}\right| = \frac{\frac{11}{2}}{\frac{13}{2}} = \frac{11}{13}$$

$\theta \approx 0.7023 \text{ radian} \approx 40.2°$

3. $(x_1, y_1) = (2, 9)$

$y = 3x + 4 \Rightarrow 3x - y + 4 = 0 \Rightarrow A = 3, B = -1, C = 4$

$$d = \frac{|(3)(2) + (-1)(9) + 4|}{\sqrt{3^2 + (-1)^2}} = \frac{1}{\sqrt{10}} = \frac{\sqrt{10}}{10}$$

4. $y^2 - 2x + 2 = 0$

$$y^2 = 2(x - 1)$$

Parabola

Vertex: $(1, 0)$

Focus: $\left(\frac{3}{2}, 0\right)$

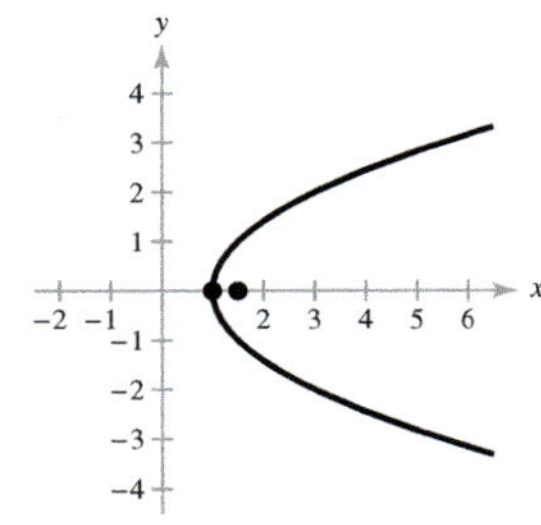

5. $x^2 - 4y^2 - 4x = 0$

$(x-2)^2 - 4y^2 = 4$

$\dfrac{(x-2)^2}{4} - y^2 = 1$

Hyperbola

Center: $(2, 0)$

Horizontal transverse axis

$a = 2, b = 1, c^2 = 1 + 4 = 5 \Rightarrow c = \sqrt{5}$

Vertices: $(0, 0), (4, 0)$

Foci: $\left(2 \pm \sqrt{5}, 0\right)$

Asymptotes: $y = \pm\frac{1}{2}(x - 2)$

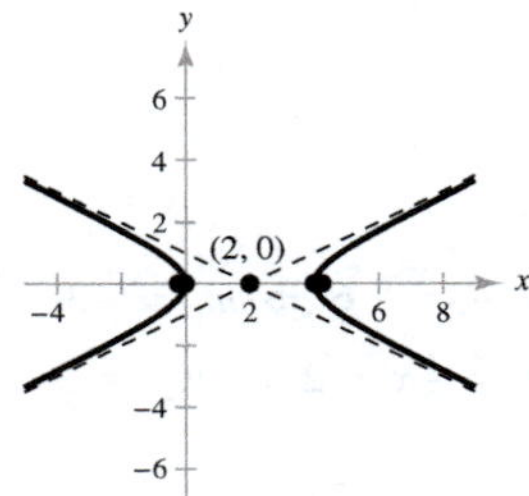

6. $9x^2 + 16y^2 + 54x - 32y - 47 = 0$

$9\left(x^2 + 6x + 9\right) + 16\left(y^2 - 2y + 1\right) = 47 + 81 + 16$

$9(x+3)^2 + 16(y-1)^2 = 144$

$\dfrac{(x+3)^2}{16} + \dfrac{(y-1)^2}{9} = 1$

Ellipse

Center: $(-3, 1)$

$a = 4, b = 3, c = \sqrt{7}$

Foci: $\left(-3 \pm \sqrt{7}, 1\right)$

Vertices: $(1, 1), (-7, 1)$

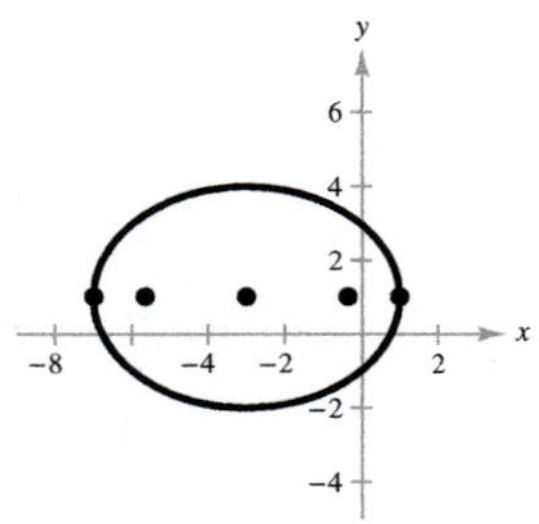

7. $2x^2 + 2y^2 - 8x - 4y + 9 = 0$

$2\left(x^2 - 4x + 4\right) + 2\left(y^2 - 2y + 1\right) = -9 + 8 + 2$

$2(x-2)^2 + 2(y-1)^2 = 1$

$(x-2)^2 + (y-1)^2 = \dfrac{1}{2}$

Circle

Center: $(2, 1)$

Radius: $\sqrt{\dfrac{1}{2}} = \dfrac{\sqrt{2}}{2} \approx 0.707$

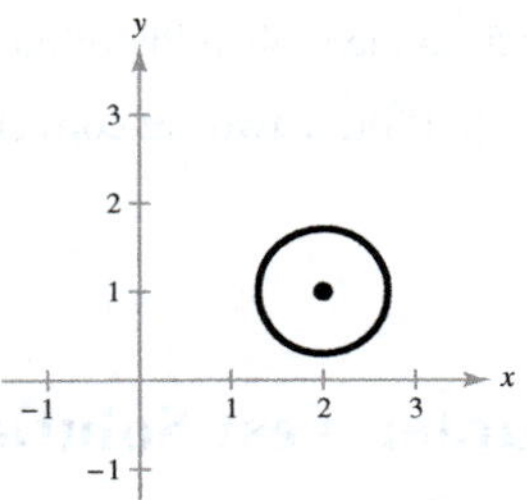

8. Parabola

Vertex: $(3, -4)$

Focus: $(6, -4)$

Horizontal axis

$p = 6 - 3 = 3$

$(y - k)^2 = 4p(x - h)$

$(y - (-4))^2 = 4(3)(x - 3)$

$(y + 4)^2 = 12(x - 3)$

9. Foci: $(0, -2)$ and $(0, 2) \Rightarrow c = 2$

Center: $(0, 0)$

Asymptotes: $y = \pm\frac{1}{9}x$

Vertical transverse axis

$\dfrac{a}{b} = \dfrac{1}{9} \Rightarrow b = 9a$

$c^2 = a^2 + b^2$

$4 = a^2 + (9a)^2$

$4 = 82a^2$

$\dfrac{2}{41} = a^2$

$b^2 = (9a)^2 = 81a^2 = \dfrac{162}{41}$

$\dfrac{y^2}{a^2} - \dfrac{x^2}{b^2} = 1$

$\dfrac{y^2}{2/41} = \dfrac{x^2}{162/41} = 1$

10. $xy + 1 = 0$

$A = 0, B = 1, C = 0$

$$\cot 2\theta = \frac{1 - 1}{1} = 0$$

$$2\theta = 90^\circ$$

$$\theta = 45^\circ$$

$$x = x' \cos 45^\circ - y' \sin 45^\circ = \frac{x' - y'}{\sqrt{2}}$$

$$y = x' \sin 45^\circ + y' \cos 45^\circ = \frac{x' + y'}{\sqrt{\ }}$$

$$\left(\frac{x' - y'}{\sqrt{2}}\right)\left(\frac{x' + y'}{\sqrt{2}}\right) + 1 = 0$$

$$\frac{(x')^2 - (y')^2}{2} + 1 = 0$$

$$\frac{(x')^2}{2} - \frac{(y')^2}{2} = -1$$

$$\frac{(y')^2}{2} - \frac{(x')^2}{2} = 1$$

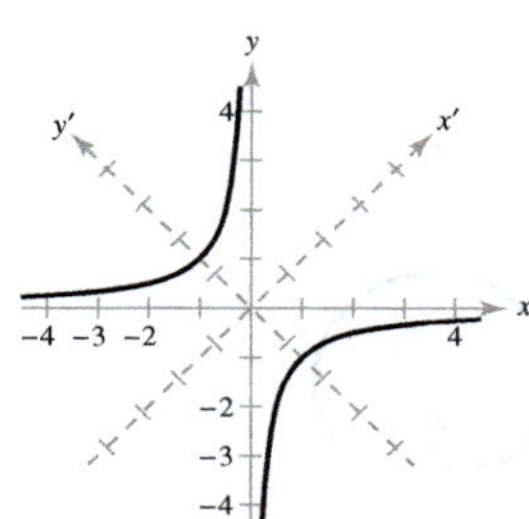

11. $x = 2 + 3\cos\theta$

$y = 2\sin\theta$

$$x = 2 + 3\cos\theta \Rightarrow \frac{x - 2}{3} = \cos\theta$$

$$y = 2\sin\theta \Rightarrow \frac{y}{2} = \sin\theta$$

$$\cos^2\theta + \sin^2\theta = 1$$

$$\frac{(x - 2)^2}{9} + \frac{y^2}{4} = 1$$

θ	0	$\pi/2$	π	$3\pi/2$
x	5	2	-1	2
y	0	2	0	-2

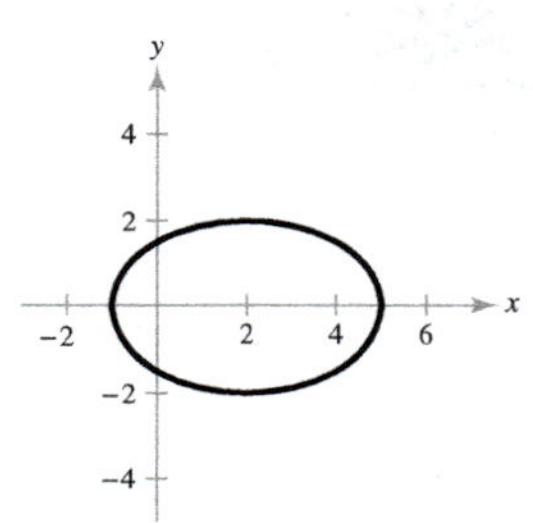

12. $y = 3 - x^2$

(a) $t = x \Rightarrow x = t$ and $y = 3 - t^2$

(b) $t = x + 2 \Rightarrow x = t - 2$ and $y = 3 - (t - 2)^2 = 3 - (t^2 - 4t + 4) = -t^2 + 4t - 1$

13. Polar coordinates: $\left(-2, \frac{5\pi}{6}\right)$

$$x = -2\cos\frac{5\pi}{6} = -2\left(-\frac{\sqrt{3}}{2}\right) = \sqrt{3}$$

$$y = -2\sin\frac{5\pi}{6} = -2\left(\frac{1}{2}\right) = -1$$

Rectangular coordinates: $\left(\sqrt{3}, -1\right)$

14. Rectangular coordinates: $(2, -2)$

$$r = \pm\sqrt{2^2 + (-2)^2} = \pm\sqrt{8} = \pm 2\sqrt{2}$$

$$\tan\theta = -1 \Rightarrow \theta = \frac{3\pi}{4}, \frac{7\pi}{4}$$

Polar coordinates:

$$\left(2\sqrt{2}, \frac{7\pi}{4}\right), \left(-2\sqrt{2}, \frac{3\pi}{4}\right), \left(2\sqrt{2}, -\frac{\pi}{4}\right), \left(-2\sqrt{2}, -\frac{5\pi}{4}\right)$$

15. $x^2 + y^2 = 64$

$r^2 = 64$

$r = 8$

16. $r = \dfrac{4}{1 + \cos\theta}$

$e = 1 \Rightarrow$ Parabola

Vertex: $(2, 0)$

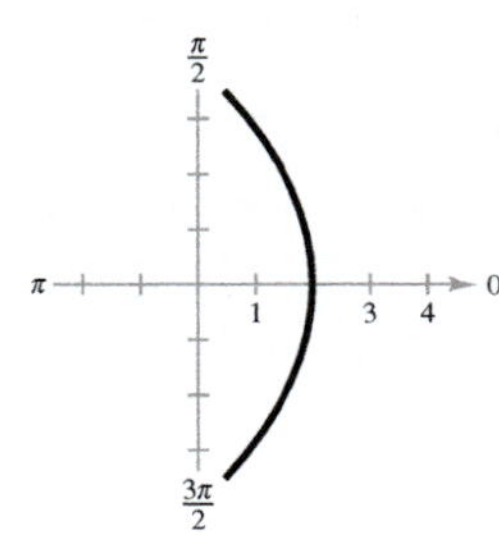

17. $r = \dfrac{4}{2 + \sin\theta}$

$= \dfrac{2}{1 + \frac{1}{2}\sin\theta}$

$e = \dfrac{1}{2} \Rightarrow$ Ellipse

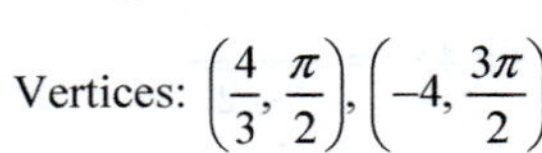

Vertices: $\left(\dfrac{4}{3}, \dfrac{\pi}{2}\right), \left(-4, \dfrac{3\pi}{2}\right)$

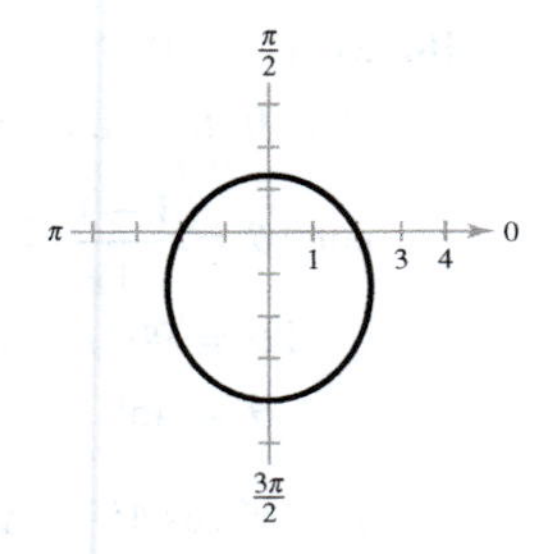

18. $r = 2 + 3\sin\theta$

$\dfrac{a}{b} = \dfrac{2}{3} < 1$

Limaçon with inner loop

θ	0	$\frac{\pi}{2}$	π	$\frac{3\pi}{2}$
r	2	5	2	−1

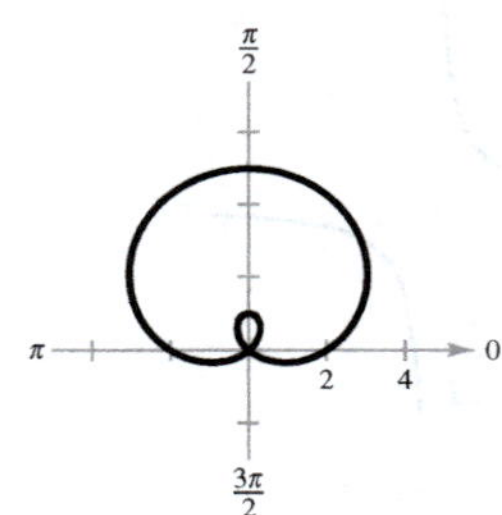

19. $r = 2\sin 4\theta$

Rose curve $(n = 4)$ with eight petals

$|r| = 2$ when

$\theta = \dfrac{\pi}{8}, \dfrac{3\pi}{8}, \dfrac{5\pi}{8}, \dfrac{7\pi}{8}, \dfrac{9\pi}{8}, \dfrac{11\pi}{8}, \dfrac{13\pi}{8}, \dfrac{15\pi}{8}$

$r = 0$ when

$\theta = 0, \dfrac{\pi}{4}, \dfrac{\pi}{2}, \dfrac{3\pi}{4}, \pi, \dfrac{5\pi}{4}, \dfrac{3\pi}{2}, \dfrac{7\pi}{4}, 2\pi$

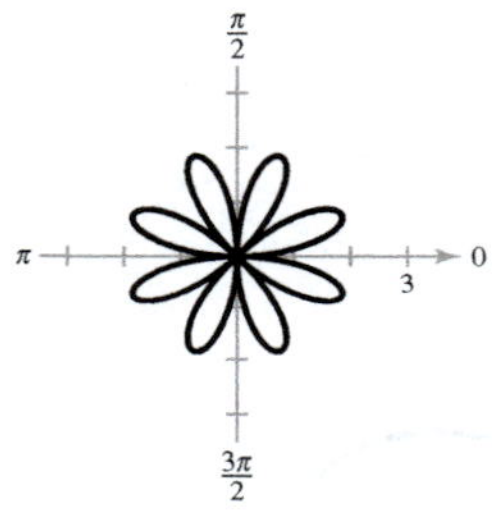

20. Ellipse $e = \dfrac{1}{4}$, focus at the pole, directrix $y = 4$

For a horizontal directrix above the pole:

$r = \dfrac{ep}{1 + e\sin\theta}$

$p =$ distance between the pole and the directrix $\Rightarrow p = 4$

So, $r = \dfrac{(1/4)(4)}{1 + (1/4)\sin\theta} = \dfrac{1}{1 + 0.25\sin\theta}.$

21. Slope: $m = \tan 0.15 \approx 0.1511$

$\sin 0.15 = \dfrac{x}{5280 \text{ feet}}$

$x = 5280 \sin 0.15 \approx 789$ feet

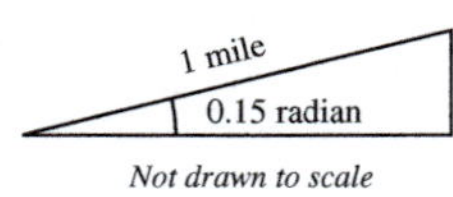

Not drawn to scale

22. $x = (115 \cos \theta)t$ and $y = 3 + (115 \sin \theta)t - 16t^2$

When $\theta = 30° : x = (115 \cos 30°)t$

$y = 3 + (115 \sin 30°)t - 16t^2$

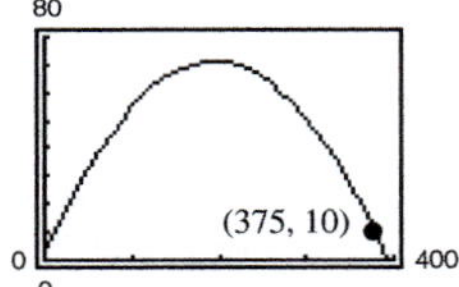

The ball hits the ground inside the ballpark, so it is not a home run.

When $\theta = 35° : x = (115 \cos 35°)t$

$y = 3 + (115 \sin 35°)t - 16t^2$

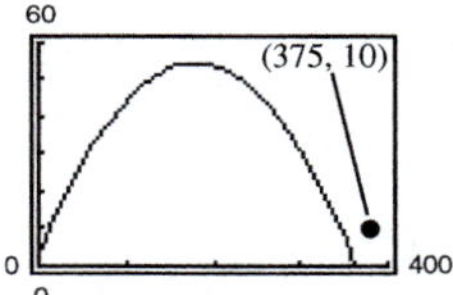

The ball clears the 10 foot fence at 375 feet, so it is a home run.